NUMBERS EXPRESSED IN SCIENTIFIC NOTATION

$$1\ 000\ 000 = 10 \times 10 \times 10 \times 10 \times 10 \times 10 = 10^6$$
$$100\ 000 = 10 \times 10 \times 10 \times 10 \times 10 = 10^5$$
$$10\ 000 = 10 \times 10 \times 10 \times 10 = 10^4$$
$$1000 = 10 \times 10 \times 10 = 10^3$$
$$100 = 10 \times 10 = 10^2$$
$$10 = 10 = 10^1$$
$$1 = 1 = 10^0$$
$$0.1 = 1/10 = 10^{-1}$$
$$0.01 = 1/100 = 1/10^2 = 10^{-2}$$
$$0.001 = 1/1000 = 1/10^3 = 10^{-3}$$
$$0.000\ 1 = 1/10\ 000 = 1/10^4 = 10^{-4}$$
$$0.0\ 000\ 1 = 1/100\ 000 = 1/10^5 = 10^{-5}$$
$$0.00\ 000\ 1 = 1/1\ 000\ 000 = 1/10^6 = 10^{-6}$$

PHYSICAL DATA

Speed of light in a vacuum	$= 2.9979 \times 10^8$ m/s
Speed of sound (20°C, 1 atm)	$= 343$ m/s
Standard atmospheric pressure	$= 1.01 \times 10^5$ Pa
1 light-year	$= 9.461 \times 10^{12}$ km
1 astronomical unit (A.U.), (average Earth–Sun distance)	$= 1.50 \times 10^{11}$ m
Average Earth–Moon distance	$= 3.84 \times 10^8$ m
Equatorial radius of the Sun	$= 6.96 \times 10^8$ m
Equatorial radius of Jupiter	$= 7.14 \times 10^7$ m
Equatorial radius of the Earth	$= 6.37 \times 10^6$ m
Equatorial radius of the Moon	$= 1.74 \times 10^6$ m
Average radius of hydrogen atom	$= 5 \times 10^{-11}$ m
Mass of the Sun	$= 1.99 \times 10^{30}$ kg
Mass of Jupiter	$= 1.90 \times 10^{27}$ kg
Mass of the Earth	$= 5.98 \times 10^{24}$ kg
Mass of the Moon	$= 7.36 \times 10^{22}$ kg
Proton mass	$= 1.6726 \times 10^{-27}$ kg
Neutron mass	$= 1.6749 \times 10^{-27}$ kg
Electron mass	$= 9.1 \times 10^{-31}$ kg
Electron charge	$= 1.602 \times 10^{-19}$ C

STANDARD ABBREVIATIONS

A	ampere	g	gram	M	molarity
amu	atomic mass unit	h	hour	min	minute
atm	atmosphere	hp	horsepower	mph	mile per hour
Btu	British thermal unit	Hz	Hertz	N	newton
C	coulomb	in.	inch	Pa	pascal
°C	degree Celsius	J	joule	psi	pound per square inch
cal	calorie	K	kelvin	s	second
eV	electron volt	kg	kilogram	V	volt
°F	degree Fahrenheit	lb	pound	W	watt
ft	foot	m	meter	Ω	ohm

MasteringPhysics®

This online homework and tutoring system delivers self-paced tutorials that provide individualized coaching, focus on your course objectives, and are responsive to each student's progress. The Mastering system helps instructors maximize class time with customizable, easy-to-assign, and automatically graded assessments that motivate students to learn outside of class and arrive prepared for lecture.

www.masteringphysics.com

PROVEN RESULTS

> The Mastering platform is the only online homework system with research showing that it improves student learning. A wide variety of published papers based on NSF-sponsored research and tests illustrate the benefits of the Mastering program.

Results documented in scientifically valid efficacy papers are available at www.masteringphysics.com/site/results

ENGAGING **EXPERIENCES**

> MasteringPhysics® provides a personalized, dynamic, and engaging experience for each student that strengthens active learning. Survey data show that the immediate feedback and tutorial assistance in MasteringPhysics motivate students to do more homework. The result is that students learn more and improve their test scores.

Interactive Figure Activities

In these activities, science principles come to life as students interact with key figures from the text. Hints and specific wrong-answer feedback help guide students towards mastery of important concepts.

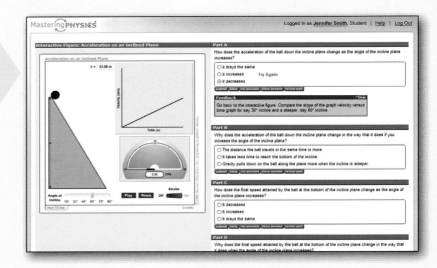

Video Activities

In these activities, students answer multiple-choice questions based on the content of Paul Hewitt's classroom demonstrations.

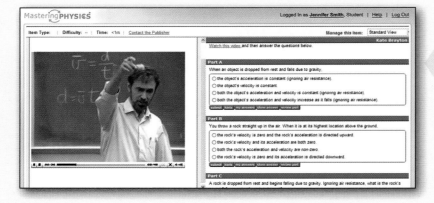

Self-Guided Tutorial Activities

These activities feature extensive, multi-lesson animations that students can work through at their own pace. Students then use what they have learned to answer multiple-choice questions based on the animations.

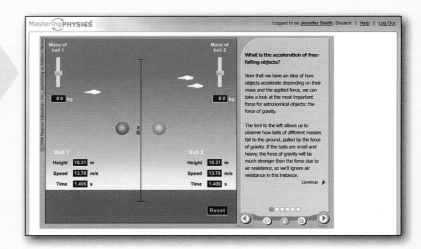

MasteringPhysics Tutorials

MasteringPhysics Tutorials guide students through important topics with self-paced tutorials that provide individualized coaching. Hints are provided, and students are free to choose only the specific help they need. Specific wrong-answer feedback for common wrong answers helps students understand their misconceptions and guides them towards the correct answer. Tutorials for chemistry, astronomy, physics, and earth science are all available in the item library for Conceptual Physical Science.

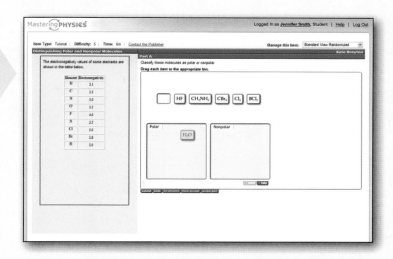

Think and Rank Problems

Think and Rank Problems have been written for every chapter of the Fifth Edition. These problems are assignable within Mastering using an interactive ranking tool, allowing students to drag and drop possible answers into the appropriate order.

End of Chapter Content Available in MasteringPhysics

All end of chapter problems are assignable within MasteringPhysics, including:

- Multiple-choice versions of free response questions that give instructors the option to assign free-response end of chapter questions as they appear in the book, or automatically gradable multiple-choice versions

- Interactive Think and Rank Problems

- Plug and Chug and Think and Solve Problems that are algorithmically coded to provide each student with a unique problem

- All end of chapter problems organized into a format following Bloom's Taxonomy in both the textbook and in MasteringPhysics

❯ The Mastering platform was developed by scientists for science students and instructors, and has a proven history with over 10 years of student use. Mastering currently has more than 1.5 million active registrations with active users in 50 states and in 41 countries.

Learning Outcomes

A learning outcome has been added to each section to help the students focus on the most important concepts in each chapter. Instructors using MasteringPhysics can assign content that is tied to these book-specific learning outcomes.

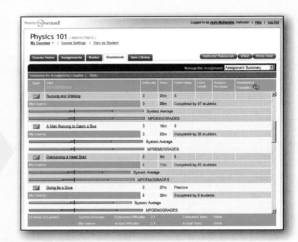

Class Performance on Assignment

Click on a problem to see where your students struggled the most and their most common wrong answers. Compare results at every stage with the national average or with your previous class.

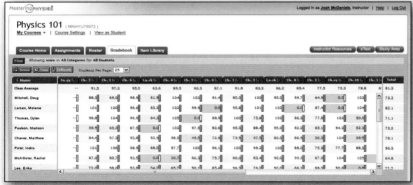

Gradebook

• Every assignment is graded automatically.

• Shades of red highlight vulnerable students and challenging assignments.

Gradebook Diagnostics

This screen provides your favorite weekly diagnostics. With a single click, charts summarize the most difficult problems, vulnerable students, grade distribution, and even improvement in scores over the course.

Conceptual
Physical Science
Fifth Edition

Paul G. Hewitt
City College of San Francisco

John Suchocki
Saint Michael's College

Leslie A. Hewitt

PEARSON

Boston   Columbus   Indianapolis   New York   San Francisco   Upper Saddle River
Amsterdam   Cape Town   Dubai   London   Madrid   Milan   Munich   Paris   Montréal   Toronto
Delhi   Mexico City   São Paulo   Sydney   Hong Kong   Seoul   Singapore   Taipei   Tokyo

Publisher: *James M. Smith*
Project Editor: *Chandrika Madhavan*
Editorial Manager: *Laura Kenney*
Senior Media Producer: *Deb Greco*
Media Producer: *Kate Brayton*
Executive Marketing Manager: *Kerry Chapman*
Associate Director of Production: *Erin Gregg*
Managing Editor: *Corinne Benson*
Production Project Manager: *Mary O'Connell*
Production Service and Composition: *Nesbitt Graphics, Inc.*
Interior Design: *Yin Ling Wong*
Cover Designer: *Mark Ong*
Cover Photo Credit: *Lillian Lee Hewitt*
Photo Research: *Eric Schrader*
Science Image Lead: *Maya Melenchuk*
Illustrations: *Dartmouth Publishing, Inc.*
Manufacturing Buyer: *Jeffrey Sargent*
Manager, Rights and Permissions: *Zina Arabia*
Manager, Cover Visual Research & Permissions: *Karen Sanatar*
Image Permission Coordinator: *Elaine Soares*
Printer and Binder: *RR Donnelley*
Cover Printer: *Lehigh-Phoenix Color*

Library of Congress Cataloging-in-Publication Data

Hewitt, Paul G.
Conceptual physical science / Paul G. Hewitt, John Suchocki, Leslie A. Hewitt. -- 5th ed.
    p. cm.
Includes index.
ISBN 978-0-321-75334-2
    1. Physical sciences--Textbooks. I. Suchocki, John. II. Hewitt, Leslie A. III. Title.
Q158.5.H48 2012
500.2--dc23

2011029644

ISBN-10: 0-321-75334-8; ISBN-13: 978-0-321-75334-2 (Student edition)
ISBN-10: 0-321-77445-0; ISBN-13: 978-0-321-77445-3 (Exam copy)

3 4 5 6 7 8 9 10—RRD—16 15 14 13

*To future elementary school teachers who will inspire students
to value science as a way of knowing about
the world and making sense of it.*

Brief Contents

Detailed Contents

18 Two Classes of Chemical Reactions   454

19 Organic Compounds   489

PART THREE
Earth Science   519

20 Rocks and Minerals   520

21 Plate Tectonics and Earth's Interior   555

22 Shaping Earth's Surface   589

23 Geologic Time—Reading the Rock Record   620

24 The Oceans, Atmosphere, and Climatic Effects   647

The Conceptual Physical Science Photo Album

THIS IS A VERY PERSONAL BOOK, a family undertaking shown in the many photographs throughout. The cover photo was taken in 2010 by physics author Paul's wife Lillian on a Li River cruise in China. Paul is seen with Lillian on page 52, and Lil appears again on pages 165, 193, 243, and 292, and with her pet conure, Sneezlee, on page 279. Lil's mom Siu Bik and dad Wai Tsan Lee are on pages 179 and 220, and Lil's niece Allison Lee Wong and nephew Erik Lee Wong are on page 176. Paul's grown children begin with son Paul on pages 150 and 167 and coauthor Leslie in her student days on page 318. Son Paul's lovely wife Ludmila shows crossed Polaroids on page 286, and their daughter Grace opens the astronomy chapters on page 707. Grace teams up with grandchildren Alexander Hewitt and Megan and Emily Abrams for the series of group photos on page 279. Author Paul's first grandchild Manuel Hewitt swings on page 261.

Paul's sister (and John's mom) Marjorie Hewitt Suchocki (pronounced *Su-hock-ee,* with a silent *c*), a leading process theologian, is shown reflectively on page 270. Paul's brother Steve shows Newton's third law with his daughter Gretchen on page 59. Paul's other brother Dave with his wife Barbara pump water on page 131.

Chemistry author John, who in his "other life" is John Andrew, singer and songwriter, plays his guitar on page 227. He is shown again walking barefoot on red-hot coals on page 164. His wife Tracy is shown with son Ian on page 296 and with son Evan on page 355. Daughter Maitreya is seen eyeing ice cream on page 489 and brushing her teeth with her dear friend Annabelle Creech on page 373. John's nephew Graham Orr on page 397 is seen at ages 7 and 21 demonstrating how water is essential for growth. The Suchocki dog, Sam, pants on page 174. The "just-married" John and Tracy are flanked by John's sisters Cathy Candler and Joan Lucas on page 256. (Tracy's wedding ring is figured prominently on page 348.) Sister Joan is riding her horse on page 25. Nephews and niece Liam, Bo, and Neve Hopwood are seen together in the chemistry part opener on page 293. Cousin George Webster is seen with his scanning electron microscope on page 312. Dear friends from John's years teaching in Hawaii include Rinchen Trashi on page 308 and Kai Dodge and Maile Ventura on page 483. The Suchocki's Vermont friend Nikki Jiraff is seen carbonating water on page 417.

Earth science author Leslie is seen at age 16 illustrating the wonderful idea that we're all made of stardust on page 318. Leslie's husband Bob Abrams is shown on page 613. The late Millie Hewitt, Leslie's mom, illustrates the cooling effect of rapid evaporation on page 167. Leslie's daughters Megan and Emily open the Earth science chapters on page 519. Megan (as a toddler) illustrates magnetic induction on page 216 and does a mineral scratch test on page 528. On page 609 Emily uses a deck of cards to show how ice crystals slip. And, dear to all three authors, our late friend Charlie Spiegel is shown on page 268.

Contributions were made to the physics chapters by renowned physicist Ken Ford, who shows his passion for flying on page 250. Marshall Ellenstein, a contributor, editor, and producer of Paul's DVDs on physics, walks barefoot on broken pieces of glass on page 143. Diane Reindeau, shown on page 240, is another physics contributor.

Physics professor friends include Tsing Bardin illustrating liquid pressure on page 122, while her grandson Francesco Ming Giovannuzzi displays a fireworks sparkler on page 148. Bob Greenler displays a colorful giant bubble on page 264, Ron Hipschman freezes water on page 178, Peter Hopkinson is pictured with his zany mirror on page 290, David Housden shows an impressive circuit display on page 204, John Hubisz demonstrates entropy on page 151, Evan Jones has an LED bulb on page 206, Chelcie Liu shows his novel race tracks in Figure A.3 in Appendix A, Jennie McKelvie makes waves on page 249, Fred Myers shows magnetic force on page 219, Sheron Snyder generates light on page 231, Jim Stith turns his impressive Wimshurst generator on page 195, and Lynda Williams sings her heart out on page 255.

Paul's dear personal friends include Burl Grey on page 21, who stimulated Paul's love of physics a half century ago, and Will Maynez, showing the airtrack he built for CCSF on page 70 and burning a peanut on page 160. Tim Gardner plays with air pressure on page 136 and induction on page 235. Friend from teen years, Paul Ryan sweeps his finger through molten lead on page 180. Friend from college days, Howie Brand illustrates impulse and changes in momentum on page 65. On page 140 another friend from college days, Dan Johnson, crushes a can with atmospheric pressure. Doing the same on a larger scale on page 144 is P. O. Zetterberg with Tomas and Barbara Brage. Tenny Lim, former student and now a design engineer for Jet Propulsion Labs, puts energy into her bow on page 72. Another former student, Helen Yan, now an orbit analyst for Lockheed Martin Corporation and part-time physics instructor at City College of San Francisco, poses with a black-and-white box on page 171. Duane Ackerman's daughter Charlotte opens Part 1 on page 13. Lab manual author Dean Baird's student, Robin Eitelberg, opens Chapter 8 on page 186. The late Jean Curtis demonstrates magnetic levitation on page 227. Science author Suzanne Lyons with children Tristan and Simone illustrate complementary colors on page 291. Ryan Patterson resonates on page 247. Tammy and Larry Tunison demonstrate radiation safety on page 325. Dave Vasquez with his family are barely seen in the solar-powered train on page 81. Young Carlos Vasquez is colorfully shown on page 278. Little cousins Michelle Anna Wong and Miriam Dijamco produce touching music on page 238. Hawaii friend Chiu Man Wu is on page 174 and with daughter Andrea on pages 87 and 267. Former student Cassy Cosme safely breaks bricks with her bare hand on page 65. Anette Zetterberg poses an intriguing thermal expansion puzzle on page 162.

These photographs are of people very dear to the authors and make *Conceptual Physical Science* even more our labor of love.

To the Student

PHYSICAL SCIENCE IS ABOUT THE RULES OF THE PHYSICAL WORLD—physics, chemistry, geology, and astronomy. Just as you can't enjoy a ball game, computer game, or party game until you know its rules, so it is with nature. Nature's rules are beautifully elegant and can be neatly described mathematically. That's why many physical science texts are treated as applied mathematics. But too much emphasis on computation misses something essential—*comprehension*—a gut feeling for the concepts. This book is *conceptual*, focusing on concepts in down-to-earth English rather than in mathematical language. You'll see the mathematical structure in frequent equations, but you'll find them *guides to thinking* rather than recipes for computation.

We enjoy physical science, and you will too—because you'll understand it. Just as a person who knows the rules of botany best appreciates plants, and a person who knows the intricacies of music best appreciates music, you'll better appreciate the physical world about you when you learn its rules.

Enjoy your physical science!

To the Instructor

THIS FIFTH EDITION of *Conceptual Physical Science* with its important ancillaries provides your students an enjoyable and readable introductory coverage of the physical sciences. As in the previous edition, the 28 chapters are divided into four main parts: Physics, Chemistry, Earth Science, and Astronomy. We begin with physics, the basic science that provides a foundation for chemistry, which in turn extends to Earth science and astronomy.

For the nonscience student, this book affords a means of viewing nature more perceptively—seeing that a surprisingly few relationships make up its rules, most of which in Part 1 are the laws of physics unambiguously expressed in equation form. Using equations for problem solving can be kept to a minimum for nonscience students, since this book treats equations as *guides to thinking*. Even students who shy away from mathematics can learn to read equations to see how concepts connect. The symbols in equations are akin to musical notes that guide musicians. A new end-of-chapter feature further boosts student comfort with equations, *Plug and Chug*, which is described below.

For the science student, this same foundation affords a springboard to other sciences such as biology and health-related fields. For more quantitative students, ample end-of-chapter material provides problem-solving activity well beyond the *Plug and Chug* calculations. Many of these *Think and Solve* problems are couched in symbols only—with a Part b that treats numerical values. All problems nevertheless stress the connections in physics.

Physics begins with static equilibrium so that students can start with forces before studying velocity and acceleration. After they achieve success with simple forces, the coverage touches lightly on kinematics, enough preparation for Newton's laws of motion. The pace picks up with the conventional order of mechanics topics followed by heat, thermodynamics, electricity and magnetism, sound, and light. Physics chapters lead to the realm of the atom—a bridge to chemistry.

The chemistry chapters begin with a look at the submicroscopic world of the atom, which is described in terms of subatomic particles and the periodic table. Students are then introduced to the atomic nucleus and its relevance to radioactivity, nuclear power, and astronomy. Subsequent chemistry chapters follow a traditional approach covering chemical changes, bonding, molecular interactions, and the formation of mixtures. With this foundation students are then set to learn the mechanics of chemical reactions and the behavior of organic compounds. As in previous editions, chemistry is related to the student's familiar world—the fluorine in their toothpaste, the Teflon on their frying pans, and the flavors produced by various organic molecules. The environmental aspects of chemistry are also highlighted—from how our drinking water is purified to how atmospheric carbon dioxide influences the pH of rainwater and our oceans.

The Earth science chapters focus on the interconnections among the geosphere, hydrosphere, and atmosphere. Topics for the geosphere chapters begin in a traditional sequence—rocks and minerals, plate tectonics, earthquakes, volcanoes, and the processes of erosion and deposition and their influence on landforms. This foundation material is then revisited in an examination of Earth over geologic time. A study of Earth's oceans leads to a focus on the interactions between the hydrosphere and atmosphere. Heat transfer and the differences in seawater density across the globe set the stage for discussions of atmospheric and oceanic circulation and Earth's overall climate. Concepts from physics are reexamined in the driving forces of weather. We conclude with an exploration of severe weather, which adds depth to the study of the atmosphere.

The applications of physics, chemistry, and Earth science to other massive bodies in the universe culminate in Part 4—Astronomy. Of all the physical sciences, astronomy and cosmology are arguably undergoing the most rapid development. Many recent discoveries are featured in this edition, illustrating how science is more than a growing body of knowledge; it is an arena in which humans actively and systematically reach out to learn more about our place in the universe.

What's New to This Edition

Conceptual Physical Science now comes with a powerful media package including **MasteringPhysics®,** the most widely used, educationally proven, and technologically advanced tutorial and homework system available. MasteringPhysics contains:

Mastering**PHYSICS**

www.masteringphysics.com

- A **library of assignable and automatically graded content,** including tutorials, visual activities, end-of-chapter problems, and test bank questions so instructors can create the most effective homework assignments with just a few clicks. A **color-coded gradebook** and **diagnostic charts** provide unique insight into class performance and summarize the most difficult problems, vulnerable students, grade distribution, and even score improvement over the duration of the course. MasteringPhysics also helps you to identify and report results by **learning outcomes,** or specific measurable goals often used by institutions to assess student progress.

- A **student study area** with Interactive Figures™, award-winning self-guided tutorials, flashcards, and videos.

- The **Pearson eText** is available through MasteringPhysics, either automatically when MasteringPhysics is packaged with new books, or available as a purchased upgrade online. Allowing students access to the text wherever they have access to the Internet, Pearson eText comprises the full text, including figures that can be enlarged for better viewing. With eText, students are also able to pop up definitions and terms to help with vocabulary and the reading of material. Students can also take notes in eText using the annotation feature at the top of each page.

- An **instructor resources section** with PowerPoint lectures, clicker questions, Instructor Manual files, and more.

- Another most significant revision of this Fifth Edition lies with the development of the end-of-chapter review. New questions were added while older ones were either discarded or reworded for improved quality. All questions were then organized following **Bloom's taxonomy** of learning as follows:

Summary of Terms (Knowledge)
    The definitions have been edited to match the definitions given within the chapter. These key terms are now also listed in alphabetical order so that they appear as a mini-glossary for the chapter.

Reading Check Questions (Comprehension)
    These questions frame the important ideas of each section in the chapter. They are meant solely as a review of reading comprehension. They are simple questions and all answers are easily discovered in the chapter.

Activities (Hands-On Application)

The *Activities* are easy-to-perform, hands-on activities designed to help students experience physical science concepts for themselves.

Plug and Chug (Formula Familiarization)

The purpose of these one-step, quick, non-intimidating calculations is familiarization with the equations and formulas in each chapter.

Think and Solve (Mathematical Application)

The *Think and Solve* questions blend simple mathematics with concepts. They allow students to apply the problem-solving techniques featured in the Figuring Physical Science boxed features appearing in many chapters.

Think and Rank (Analysis)

The *Think and Rank* questions ask students to make comparisons of quantities. For example, when asked to rank quantities such as momentum or kinetic energy, students are called to use appreciably more judgment than in providing numerical answers. Some *Think and Rank* questions analyze trends, as in ranking atoms in order of increasing size based upon student understanding of the periodic table. This new feature elicits critical thinking that goes beyond the *Think and Solve* questions.

Exercises (Synthesis)

The *Exercises*, by a notch or two, are the more challenging questions at the end of each chapter. Many require critical thinking, while others are designed to prompt the application of science to everyday situations. All students who want to perform well on exams should be directed to the *Exercises* because these are the questions that directly assess student understanding. Accordingly, many of the *Exercises* have been adapted to a multiple-choice format and integrated into the *Conceptual Physical Science, 5e* test bank. We hope this will allow the instructor to reward those students who put time and effort into the *Exercises.*

Discussion Questions (Evaluation)

The *Discussion Questions* provide students the opportunity to apply the concepts of physical science to real-life situations, such as whether a cup of hot coffee served to you in a restaurant cools faster when cream is added promptly or a few minutes later. Other *Discussion Questions* allow students to present their educated opinions on a number of science-related hot topics, such as the appearance of pharmaceuticals in drinking water or whether it would be a good idea to enhance the ocean's ability to absorb carbon dioxide by adding powdered iron.

- Each chapter review concludes with a set of 10 multiple-choice questions called the **Readiness Assurance Test (RAT)** that students can take for self-assessment. They are advised to study further if they score less than 7 correct answers.

- Also new to this edition, the solutions to the odd-numbered end-of-chapter questions are provided in the back of this book. As before, solutions to all end-of-chapter questions are available to instructors through the *Instructor Manual for Conceptual Physical Science,* which is found in the Instructor Resource Center and in the instructor area of MasteringPhysics.

- This latest edition sports a new and modern-looking page layout design. Integrated into this design are **learning objectives** appearing alongside each chapter section head. Each learning objective begins with an active verb that specifies what the student should be able to do after studying that section, such as "Calculate the energy released by a chemical reaction." These section-specific learning objectives are further integrated into the new MasteringPhysics online tutorial/assessment tool.

- Also in the design, appearing beneath each section head is another new feature, which we call an **Explain This** question. An ET question would be fairly difficult for the student to answer without having read the chapter section. Some require that the student recall earlier material. Others reveal interesting applications of concepts. In all cases the ET question should serve well as a launching point for classroom discussions. The answers to these ET questions appear only in the Instructor Manual.

- The text of all chapters has been edited for accuracy and better readability and also **updated to reflect current events,** such as the nuclear power plant disaster following the 2011 Japanese earthquake and tsunami, the Gulf oil disaster, and the discovery of Fermi clouds arising from the center of our Milky Way galaxy. The structure of the physics and chemistry chapters remains much the same as in the previous edition; however, in chemistry **a new section on nanotechnology** was added in Chapter 14. The order of the Earth science chapters has been reorganized so that Plate Tectonics now follows Rocks and Minerals. In Part 4–Astronomy, the first section of Chapter 28 has been heavily revised.

Ancillary Materials

*C*onceptual Physical Science is now available with **MasteringPhysics**— a homework, tutorial, and assessment system based on years of research into how students work problems and precisely where they need help. Studies show that students who use MasteringPhysics significantly increase their scores compared to doing handwritten homework. MasteringPhysics achieves this improvement by providing students with instantaneous feedback specific to their wrong answers, simpler sub-problems upon request when they get stuck, and partial credit for their method(s). Instructors can also assign end-of-chapter (EOC) problems from every chapter, including multiple-choice questions, section-specific exercises, and general problems. Quantitative problems can be assigned with numerical answers and randomized values or solutions.

Mastering**PHYSICS**

www.masteringphysics.com

The *Instructor Manual for Conceptual Physical Science*, which you'll find to be different from most instructors' manuals, allows for a variety of course designs to fit your taste. It contains many lecture ideas and topics not treated in the textbook as well as teaching tips and suggested step-by-step lectures and demonstrations. It has full-page answers to all the end-of-chapter material in the text.

The *Conceptual Physical Science Practice Book*, our most creative work, guides your students to a sometimes computational way of developing concepts. It spans a wide use of analogies and intriguing situations, all with a user-friendly tone.

The *Computerized Test Bank for Conceptual Physical Science* has more than 2400 multiple-choice questions as well as short-answer and essay questions. The questions are categorized according to level of difficulty. The Test Bank allows you to edit questions, add questions, and create multiple test versions.

The *Laboratory Manual for Conceptual Physical Science* is written by the authors and Dean Baird. In addition to interesting laboratory experiments, it includes a range of activities similar to the activities in the textbook. These guide students to experience phenomena before they quantify the same phenomena in a follow-up laboratory experiment. Answers to the lab manual questions are in the *Instructor Manual*.

Another valuable media resource available to you is the *Instructor Resource DVD for Conceptual Physical Science*. This cross-platform DVD set provides instructors with the largest library available of purpose-built, in-class presentation materials, including all the images from the book in high-resolution JPEG format; Interactive Figures™ and videos; PowerPoint® lecture outlines and clicker questions in PRS-enabled format for each chapter, all of which are written by the authors; and Hewitt's acclaimed Next-Time Questions in PDF format. The *Instructor Resource DVD* provides you with everything you need to prepare for dynamic, engaging lectures in no time.

Lastly, as a supplement for more on algebraic problem solving in physics, consider *Problem Solving in Conceptual Physics*, by Hewitt and Wolf, ISBN 0-321-66258-X.

Go to it! Your conceptual physical science course really can be the most interesting, informative, and worthwhile science course available to your students.

Acknowledgments

We are enormously grateful to Ken Ford for extensive feedback on the first 13 chapters of the previous edition, providing much new and insightful information. We are also grateful to Lillian Lee Hewitt for extensive editorial help. We thank Phil Wolf, who authored many of the *Think and Solve* problems, and David Housden, Evan Jones and John Sperry for contributing their solutions. We are grateful to Marshall Ellenstein and Diane Reindeau, who helped develop the new feature, *Think and Rank*. We appreciate Bruce Novak's insightful edits. For general physics input to previous editions, we remain grateful to Dean Baird, Tsing Bardin, Howie Brand, George Curtis, Paul Doherty, Marshall Ellenstein, Ken Ford, John Hubisz, Dan Johnson, Tenny Lim, Iain McInnes, Fred Myers, Diane Reindeau, Kenn Sherey, Chuck Stone, Larry Weinstein, David Williamson, and Dean Zollman.

For development of chemistry chapters, thanks go to Adedoyin Adeyiga, John Bonte, Emily Borda, Charles Carraher, Natashe Cleveland, Sara Devo, Andy Frazer, Kenneth French, Marcia Gillette, Chu-Ngi Ho, Frank Lambert, Jeremy Mason, Daniel Predecki, Britt Price, Jeremy Ramsey, Kathryn Rust, William Scott, Anne Marie Sokol, Jason Vohs, Bob Widing, and David Yates. Special thanks to Tracy, Ian, Evan, and Maitreya Suchocki for their continued support.

For Earth science feedback we remain thankful to Mary Brown, Ann Bykerk-Kauffman, Oswaldo Garcia, Newell Garfield, Karen Grove, Trayle Kulshan, Jan Null, Katryn Weiss, Lisa White, and Mike Young. For providing several wonderful Earth science photos, we thank Dean Baird (CPS Lab Manual author). A special thank-you to Leslie's husband, Bob Abrams, for his assistance with the Earth science material. Thanks also goes to Leslie's children, Megan and Emily, for their inspiration, their curiosity, and their patience.

For space science we are grateful to Jeffrey Bennett, Megan Donahue, Nicholas Schneider, and Mark Voit for permission to use many of the graphics that appear in their textbook *The Cosmic Perspective*, 6th edition. Also, for reviews of the astronomy chapters we remain grateful to Richard Crowe, Bjorn Davidson, Stacy McGaugh, Michelle Mizuno-Wiedner, John O'Meara, Neil deGrasse Tyson, Joe Wesney, Lynda Williams, and Erick Zackrisson.

For their dedication to this edition, we praise the staff at Pearson in San Francisco. We are especially thankful to Jim Smith, Chandrika Madhavan, and Kate Brayton. We're grateful to Cindy Johnson and the production team at Nesbitt for their patience with our last-minute changes. Thanks to you all!

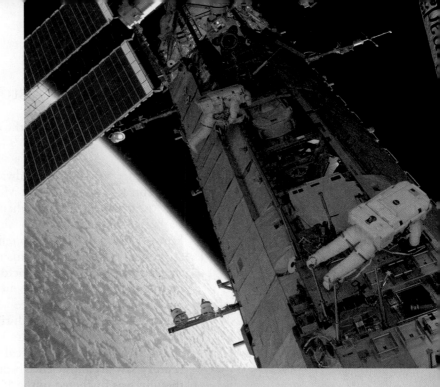

PROLOGUE
The Nature of Science

SCIENCE IS the product of human curiosity about how the world works—an organized body of knowledge that describes the order within nature and the causes of that order. *Science* is an ongoing human activity that represents the collective efforts, findings, and wisdom of the human race, an activity that is dedicated to gathering knowledge about the world and organizing and condensing it into testable laws and theories. In our study of science, we are learning about the rules of nature—how one thing is connected to another and how patterns underlie all we see in our surroundings. Any activity, whether a sports game, computer game, or the game of life, is meaningful only if we understand its rules. Learning about nature's rules is relevant with a capital R!

We will see in this book that science is much more than a body of knowledge. Science is a way of thinking.

LEARNING OBJECTIVE
Acknowledge contributions to science by various cultures.

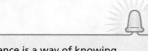

Science is a way of knowing about the world and making sense of it.

FYI In pre-Copernican times the Sun and Moon were viewed as planets. Their planetary status was removed when Copernicus substituted the Sun for Earth's central position. Only then was Earth regarded as a planet among others. More than 200 years later, in 1781, telescope observers added Uranus to the list of planets. Neptune was added in 1846. Pluto was added in 1930—and removed in 2006.

LEARNING OBJECTIVE
Recount how mathematics contributes to success in science.

Scientists have a deep-seated need to know Why? and What if? Mathematics is foremost in their tool kits for tackling these questions.

A Brief History of Advances in Science

EXPLAIN THIS How did the advent of the printing press affect the growth of science?

Science made great headway in Greece in the 4th and 3rd centuries BC and spread throughout the Mediterranean world. Scientific advance came to a near halt in Europe when the Roman Empire fell in the 5th century AD. Barbarian hordes destroyed almost everything in their paths as they overran Europe. Reason gave way to religion, which ushered in what came to be known as the Dark Ages. During this time, the Chinese and Polynesians were charting the stars and the planets. Before the advent of Islam, Arab nations developed mathematics and learned about the production of glass, paper, metals, and various chemicals. Greek science was reintroduced to Europe by Islamic influences that penetrated into Spain during the 10th, 11th, and 12th centuries. Universities emerged in Europe in the 13th century, and the introduction of gunpowder changed the social and political structure of Europe in the 14th century. The 15th century saw art and science beautifully blended by Leonardo da Vinci. Scientific thought was furthered in the 16th century with the advent of the printing press.

The 16th-century Polish astronomer Nicolaus Copernicus caused great controversy when he published a book proposing that the Sun is stationary and that Earth revolves around the Sun. These ideas conflicted with the popular view that Earth was the center of the universe. They also conflicted with Church teachings and were banned for 200 years. The Italian physicist Galileo Galilei was arrested for popularizing the Copernican theory and for his other contributions to scientific thought. Yet a century later, those who advocated Copernican ideas were accepted.

These cycles occur age after age. In the early 1800s, geologists met with violent condemnation because they differed with the account of creation in the book of Genesis. Later in the same century, geology was accepted, but theories of evolution were condemned and the teaching of them was forbidden. Every age has its groups of intellectual rebels who are scoffed at, condemned, and sometimes even persecuted at the time but who later seem beneficial and often essential to the elevation of human conditions. "At every crossway on the road that leads to the future, each progressive spirit is opposed by a thousand men appointed to guard the past."*

Mathematics and Conceptual Physical Science

EXPLAIN THIS What is meant by "Equations are guides to thinking"?

Science and human conditions advanced dramatically after science and mathematics became integrated some four centuries ago. When the ideas of science are expressed in mathematical terms, they are unambiguous. The equations of science provide compact expressions of relationships between concepts. They don't have the multiple meanings that so often confuse the discussion of ideas expressed in common language. When findings in nature are expressed mathematically, they are easier to verify or to disprove by experiment. The mathematical structure of physics is evident in the many equations you will encounter throughout this book. The equations are guides to thinking that

---

* From Count Maurice Maeterlinck's "Our Social Duty."

show the connections between concepts in nature. The methods of mathematics and experimentation led to enormous success in science.*

Scientific Methods

EXPLAIN THIS What else besides the common scientific method advances science?

There is no *one* scientific method. But there are common features in the way scientists do their work. Although no cookbook description of the **scientific method** is really adequate, some or all of the following steps are likely to be found in the way most scientists carry out their work.

1. *Observe.* Closely observe the physical world around you. Recognize a question or a puzzle—such as an unexplained observation.

2. *Question.* Make an educated guess—a **hypothesis**—to answer the question.

3. *Predict.* Predict consequences that can be observed if the hypothesis is correct. The consequences should be *absent* if the hypothesis is not correct.

4. *Test predictions.* Do experiments to see if the consequences you predicted are present.

5. *Draw a conclusion.* Formulate the simplest general rule that organizes the hypothesis, predicted effects, and experimental findings.

Although these steps are appealing, much progress in science has come from trial and error, experimentation without hypotheses, or just plain accidental discovery by a well-prepared mind. The success of science rests more on an attitude common to scientists than on a particular method. This attitude is one of inquiry, experimentation, and humility—that is, a willingness to admit error.

Science is a way to teach how something gets to be known, what is not known, to what extent things are known (for nothing is known absolutely), how to handle doubt and uncertainty, what the rules of evidence are, how to think about things so that judgments can be made, and how to distinguish truth from fraud and from show.
—Richard Feynman

The Scientific Attitude

EXPLAIN THIS Why does falsifying information discredit a scientist but not a lawyer?

It is common to think of a fact as something that is unchanging and absolute. But in science, a **fact** is generally a close agreement by competent observers who make a series of observations about the same phenomenon. For example, although it was once a fact that the universe is unchanging and permanent, today it is a fact that the universe is expanding and evolving. A scientific hypothesis, on the other hand, is an educated guess that is only presumed to be factual until supported by experiment. When a hypothesis has been tested over and over again and has not been contradicted, it may become known as a **law** or *principle*.

If a scientist finds evidence that contradicts a hypothesis, law, or principle, the scientific spirit requires that the hypothesis be changed or abandoned (unless the contradicting evidence, upon testing, turns out to be wrong—which sometimes happens). For example, the greatly respected Greek philosopher

Experiment, not philosophical discussion, decides what is correct in science.

---

* We distinguish between the mathematical structure of science and the practice of mathematical problem solving—the focus of most nonconceptual courses. Note that there are fewer mathematical problems than exercises at the ends of the chapters in this book. The focus is on comprehension before computation.

Aristotle (384–322 BC) claimed that an object falls at a speed proportional to its weight. This idea was held to be true for nearly 2000 years because of Aristotle's compelling authority. Galileo allegedly showed the falseness of Aristotle's claim with one experiment—demonstrating that heavy and light objects dropped from the Leaning Tower of Pisa fell at nearly equal speeds. In the scientific spirit, a single verifiable experiment to the contrary outweighs any authority, regardless of reputation or the number of followers or advocates. In modern science, argument by appeal to authority has little value.*

Scientists must accept their experimental findings even when they would like them to be different. They must strive to distinguish between what they see and what they wish to see, for scientists, like most people, have a vast capacity for fooling themselves.** People have always tended to adopt general rules, beliefs, creeds, ideas, and hypotheses without thoroughly questioning their validity and to retain them long after they have been shown to be meaningless, false, or at least questionable. The most widespread assumptions are often the least questioned. Most often, when an idea is adopted, particular attention is given to cases that seem to support it, while cases that seem to refute it are distorted, belittled, or ignored.

Scientists use the word *theory* in a way that differs from its usage in everyday speech. In everyday speech, a theory is no different from a hypothesis—a supposition that has not been verified. A scientific **theory**, on the other hand, is a synthesis of a large body of information that encompasses well-tested and verified hypotheses about certain aspects of the natural world. Physicists, for example, speak of the quark theory of the atomic nucleus, chemists speak of the theory of metallic bonding in metals, and biologists speak of the cell theory.

Facts are revisable data about the world.

Theories interpret facts.

The theories of science are not fixed; rather, they undergo change. Scientific theories evolve as they go through stages of redefinition and refinement. During the past hundred years, for example, the theory of the atom has been repeatedly refined as new evidence on atomic behavior has been gathered. Similarly, chemists have refined their view of the way molecules bond together, and biologists have refined the cell theory. The refinement of theories is a strength of science, not a weakness. Many people feel that it is a sign of weakness to change their minds. Competent scientists must be experts at changing their minds. They change their minds, however, only when confronted with solid experimental evidence or when a conceptually simpler hypothesis forces them to a new point of view. More important than defending beliefs is improving them. Better hypotheses are made by those who are honest in the face of experimental evidence.

Away from their profession, scientists are inherently no more honest or ethical than most other people. But in their profession, they work in an arena that places a high premium on honesty. The cardinal rule in science is that all hypotheses must be testable—they must be susceptible, at least in principle, to being shown to be *wrong*. Speculations that cannot be tested are regarded as "unscientific." This has the long-run effect of compelling honesty—findings widely publicized among fellow scientists are generally subjected to further testing. Sooner or later, mistakes (and deception) are found out; wishful thinking is exposed. A discredited scientist does not get a second chance in the community of scientists. The penalty for fraud is professional excommunication. Honesty, so important to the progress of science, thus becomes a matter of self-interest to scientists. There is relatively little bluffing in a game in which all bets are called. In fields of study where right and wrong are not so easily established, the pressure to be honest is considerably less.

Before a theory is accepted, it must be tested by experiment and make one or more new predictions—different from those made by previous theories.

---

* But appeal to *beauty* has value in science. More than one experimental result in modern times has contradicted a lovely theory that, upon further investigation, proved to be wrong. This has bolstered scientists' faith that the ultimately correct description of nature involves conciseness of expression and economy of concepts—a combination that deserves to be called beautiful.

** In your education it is not enough to be aware that other people may try to fool you; it is more important to be aware of your own tendency to fool yourself.

In science, it is more important to have a means of proving an idea wrong than to have a means of proving it right. This is a major factor that distinguishes science from nonscience. At first this may seem strange, for when we wonder about most things, we concern ourselves with ways of finding out whether they are true. Scientific hypotheses are different. In fact, if you want to distinguish whether a hypothesis is scientific, look to see if there is a test for proving it wrong. If there is no test for its possible wrongness, then the hypothesis is not scientific. Albert Einstein put it well when he stated, "No number of experiments can prove me right; a single experiment can prove me wrong."

Consider the biologist Charles Darwin's hypothesis that life forms evolve from simpler to more complex forms. This could be proven wrong if paleontologists were to find that more complex forms of life appeared before their simpler counterparts. Einstein hypothesized that light is bent by gravity. This might be proven wrong if starlight that grazed the Sun and could be seen during a solar eclipse were undeflected from its normal path. As it turns out, less complex life forms are found to precede their more complex counterparts and starlight is found to bend as it passes close to the Sun, which support the claims. If and when a hypothesis or scientific claim is confirmed, it is regarded as useful and as a stepping-stone to additional knowledge.

Consider the hypothesis "The alignment of planets in the sky determines the best time for making decisions." Many people believe it, but this hypothesis is not scientific. It cannot be proven wrong, nor can it be proven right. It is *speculation*. Likewise, the hypothesis "Intelligent life exists on other planets somewhere in the universe" is not scientific. Although it can be proven correct by the verification of a single instance of intelligent life existing elsewhere in the universe, there is no way to prove it wrong if no intelligent life is ever found. If we searched the far reaches of the universe for eons and found no life, then that would not prove that it doesn't exist "around the next corner." A hypothesis that is capable of being proven right but not capable of being proven wrong is not a scientific hypothesis. Many such statements are quite reasonable and useful, but they lie outside the domain of science.

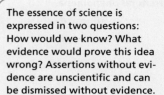

The essence of science is expressed in two questions: How would we know? What evidence would prove this idea wrong? Assertions without evidence are unscientific and can be dismissed without evidence.

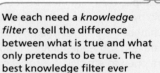

We each need a *knowledge filter* to tell the difference between what is true and what only pretends to be true. The best knowledge filter ever invented for explaining the physical world is science.

CHECKPOINT

Which of these statements is a scientific hypothesis?
(a) Atoms are the smallest particles of matter that exist.
(b) Space is permeated with an essence that is undetectable.
(c) Albert Einstein was the greatest physicist of the 20th century.

Was this your answer?
Only statement (a) is scientific, because there is a test for falseness. The statement not only is *capable* of being proven wrong, but *has* been proven wrong. Statement (b) has no test for possible wrongness and is therefore unscientific. Likewise for any principle or concept for which there is no means, procedure, or test whereby it can be shown to be wrong (if it is wrong). Some pseudoscientists and other pretenders of knowledge will not even consider a test for the possible wrongness of their statements. Statement (c) is an assertion that has no test for possible wrongness. If Einstein was not the greatest physicist, how could we know? Note that because the name Einstein is generally held in high esteem, it is a favorite of pseudoscientists. So we should not be surprised that the name of Einstein, like that of Jesus or of any other highly respected person, is cited often by charlatans who wish to bring respect to themselves and their points of view. In all fields, it is prudent to be skeptical of those who wish to credit themselves by calling upon the authority of others.

Science Has Limitations

EXPLAIN THIS How do the domains of science and the supernatural differ?

Science deals only with hypotheses that are testable. Its domain is therefore restricted to the observable natural world. Although scientific methods can be used to debunk various paranormal claims, they have no way of accounting for testimonies involving the supernatural. The term *supernatural* literally means "above nature." Science works within nature, not above it. Likewise, science is unable to answer philosophical questions, such as "What is the purpose of life?" or religious questions, such as "What is the nature of the human spirit?" Though these questions are valid and may have great importance to us, they rely on subjective personal experience and do not lead to testable hypotheses. They lie outside the realm of science.

SCIENCE AND SOCIETY

Pseudoscience

For a claim to qualify as scientific, it must meet certain standards. For example, the claim must be reproducible by others who have no stake in whether the claim is true or false. The data and subsequent interpretations are open to scrutiny in a social environment where it's okay to have made an honest mistake, but not okay to have been dishonest or deceiving. Claims that are presented as scientific but do not meet these standards are what we call **pseudoscience**, which literally means "fake science." In the realm of pseudoscience, skepticism and tests for possible wrongness are downplayed or flatly ignored.

Examples of pseudoscience abound. Astrology is an ancient belief system that supposes that a person's future is determined by the positions and movements of planets and other celestial bodies. Astrology mimics science in that astrological predictions are based on careful astronomical observations. Yet astrology is not a science because there is no validity to the claim that the positions of celestial objects influence the events of a person's life. After all, the gravitational force exerted by celestial bodies on a person is smaller than the gravitational force exerted by objects making up the earthly

environment: trees, chairs, other people, bars of soap, and so on. Further, the predictions of astrology are not borne out; there just is no evidence that astrology works.

For more examples of pseudoscience, look to television or the Internet. You can find advertisements for a plethora of pseudoscientific products. Watch out for remedies to ailments such as baldness, obesity, and cancer; for air-purifying mechanisms; and for "germ-fighting" cleaning products in particular. Although many such products operate on solid science, others are pure pseudoscience. Buyer beware!

Humans are very good at denial, which may explain why pseudoscience is such a thriving enterprise. Many pseudoscientists do not recognize their efforts as pseudoscience. A practitioner of "absent healing," for example, may truly believe in her ability to cure people she will never meet except through e-mail and credit card exchanges.

She may even find anecdotal evidence to support her contentions. The *placebo effect*, discussed in Section 8.2, can mask the ineffectiveness of various healing modalities. In terms of the human body, what people believe *will* happen often *can* happen because of the physical connection between the mind and body.

That said, consider the enormous downside of pseudoscientific practices. Today more than 20,000 astrologers are practicing in the United States. Do people listen to these astrologers just for the fun of it? Or do they base important decisions on astrology? You might lose money by listening to pseudoscientific entrepreneurs; worse, you could become ill. Delusional thinking, in general, carries risk.

Meanwhile, the results of science literacy tests given to the general public show that most Americans lack a basic understanding of basic concepts of science. Some 63% of American adults are unaware that the mass extinction of the dinosaurs occurred long before the first human evolved; 75% do not know that antibiotics kill bacteria but not viruses; 57% do not know that electrons are smaller than atoms. What we find is a rift—a growing divide—between those who have a realistic sense of the capabilities of science and those who do not understand the nature of science, its core concepts, or, worse, feel that scientific knowledge is too complex for them to understand. Science is a powerful method for understanding the physical world, and a whole lot more reliable than pseudoscience as a means for bettering the human condition.

Science, Art, and Religion

LEARNING OBJECTIVE
Discuss some similarities and differences among science, art, and religion.

EXPLAIN THIS Why is the statement "Never question what this book says" outside the domain of science?

The search for a deeper understanding of the world around us has taken different forms, including science, art, and religion. Science is a system by which we discover and record physical phenomena and think about possible explanations for such phenomena. The arts are concerned with personal interpretation and creative expression. Religion addresses the source, purpose, and meaning of it all. Simply put, science asks *how*, art asks *who*, and religion asks *why*.

Science and the arts have certain things in common. In the art of literature, we find out about what is possible in human experience. We can learn about emotions such as rage and love, even if we haven't yet experienced them. The arts describe these experiences and suggest what may be possible for us. Similarly, a knowledge of science tells us what is possible in nature. Scientific knowledge helps us predict possibilities in nature even before we experience them. It provides us with a way of connecting things, of seeing relationships between and among them, and of making sense of the great variety of natural events around us. While art broadens our understanding of ourselves, science broadens our understanding of our environment.

Science and religion have similarities also. For example, both are motivated by curiosity for the natural. Both have great impact on society. Science, for example, leads to useful technological innovations, while religion provides a foothold for many social services. Science and religion, however, are basically different. Science is concerned with understanding the physical universe, while many religions are concerned with faith in, and the worship of, a supreme being and the creation of human community—not the practice of science. While scientific truth is a matter of public scrutiny, religion is a deeply personal matter. In these respects, science and religion are as different as apples and oranges and do not contradict each other. Science, art, and religion can work very well together, which is why we should never feel forced into choosing one over the other.

When we study the nature of light later in this book, we treat light first as a wave and then as a particle. At first, waves and particles may appear contradictory. You might believe that light can be only one or the other, and that you must choose between them. What scientists have discovered, however, is that light waves and light particles *complement* each other, and that when these two ideas are taken together, they provide a deeper understanding of light. In a similar way, it is mainly people who are either uninformed or misinformed about the deeper natures of both science and religion who feel that they must choose between believing in religion and believing in science. Unless one has a shallow understanding of either or both, there is no contradiction in being religious in one's belief system and being scientific in one's understanding of the natural world.*

Many people are troubled about not knowing the answers to religious and philosophical questions. Some avoid uncertainty by uncritically accepting almost any comforting answer. An important message in science, however, is that uncertainty is acceptable. For example, if you study quantum physics you'll learn that it is not possible to know with certainty both the momentum and position

Art is about cosmic beauty. Science is about cosmic order. Religion is about cosmic purpose.

A truly educated person is knowledgeable in both the arts and the sciences.

* Of course, this does not apply to certain religious extremists who steadfastly assert that one cannot embrace both science and their brand of religion.

of an electron in an atom. The more you know about one, the less you can know about the other. Uncertainty is a part of the scientific process. It's okay not to know the answers to fundamental questions. Why are apples gravitationally attracted to Earth? Why do electrons repel one another? Why do magnets interact with other magnets? Why does energy have mass? At the deepest level, scientists don't know the answers to these questions—at least not yet. We know a lot about where we are, but nothing really about *why* we are. It's okay not to know the answers to such religious questions. Given a choice between a closed mind with comforting answers and an open and exploring mind without answers, most scientists choose the latter. Scientists in general are comfortable with not knowing.

> The belief that there is only one truth and that oneself is in possession of it seems to me the deepest root of all the evil that is in the world.
>
> —Max Born

CHECKPOINT

Which of the following activities involves the utmost human expression of passion, talent, and intelligence: (a) painting and sculpture, (b) literature, (c) music, (d) religion, (e) science?

Was this your answer?
All of them. In this book, we focus on science, which is an enchanting human activity shared by a wide variety of people. With present-day tools and know-how, scientists are reaching further and finding out more about themselves and their environment than people in the past were ever able to do. The more you know about science, the more passionate you feel toward your surroundings. There is science in everything you see, hear, smell, taste, and touch!

LEARNING OBJECTIVE
Relate technology to the furthering of science, and vice versa.

Technology—The Practical Use of Science

EXPLAIN THIS Who thinks of an idea, who develops it, and who uses it?

Science and technology are also different from each other. Science is concerned with gathering knowledge and organizing it. Technology lets humans use that knowledge for practical purposes, and it provides the instruments scientists need to conduct their investigations.

Technology is a double-edged sword. It can be both helpful and harmful. We have the technology, for example, to extract fossil fuels from the ground and then burn the fossil fuels to produce energy. Energy production from fossil fuels has benefited society in countless ways. On the flip side, the burning of fossil fuels damages the environment. It is tempting to blame technology itself for such problems as pollution, resource depletion, and even overpopulation. These problems, however, are not the fault of technology any more than a stabbing is the fault of the knife. It is humans who use the technology, and humans who are responsible for how it is used.

Remarkably, we already possess the technology to solve many environmental problems. The 21st century will likely see a switch from fossil fuels to more sustainable energy sources. We recycle waste products in new and better ways. In some parts of the world, progress is being made toward limiting human population growth, a serious threat that worsens almost every problem faced by humans today. Difficulty in solving today's problems results more from social inertia than from failing technology. Technology is our tool. What we do with this tool is up to us. The promise of technology is a cleaner and healthier world. Wise applications of technology *can* improve conditions on planet Earth.

RISK ASSESSMENT

The numerous benefits of technology are paired with risks. X-rays, for example, continue to be used for medical diagnosis despite their potential for causing cancer. But when the risks of a technology are perceived to outweigh its benefits, it should be used very sparingly or not at all.

Risk can vary for different groups. Aspirin is useful for adults, but for young children it can cause a potentially lethal condition known as *Reye's syndrome*. Dumping raw sewage into the local river may pose little risk for a town located upstream, but for towns downstream the untreated sewage is a health hazard. Similarly, storing radioactive wastes underground may pose little risk for us today, but for future generations the risks of such storage are greater if there is leakage into groundwater. Technologies involving different risks for different people, as well as differing benefits, raise questions that are often hotly debated. Which medications should be sold to the general public over the counter and how should they be labeled? Should food be irradiated in order to put an end to food poisoning, which

kills more than 5000 Americans each year? The risks to all members of society need consideration when public policies are decided.

The risks of technology are not always immediately apparent. No one fully realized the dangers of combustion products when petroleum was selected as the fuel of choice for automobiles early in the last century. From the hindsight of 20/20 vision, alcohols from biomass would have been a superior choice environmentally, but they were banned by the prohibition movements of the day. Because we are now more aware of the environmental costs of fossil-fuel combustion, biomass fuels are making a slow comeback. An awareness of both the short-term risks and the long-term risks of a technology is crucial.

People seem to have a hard time accepting the impossibility of zero risk. Airplanes cannot be made perfectly safe. Processed foods cannot be rendered completely free of toxicity, for all foods are toxic to some degree. You cannot go to the beach without risking skin cancer, no matter how much sunscreen you apply. You cannot

avoid radioactivity, for it's in the air you breathe and the foods you eat, and it has been that way since before humans first walked on Earth. Even the cleanest rain contains radioactive carbon-14, as do our bodies. Between each heartbeat in each human body, there have always been about 10,000 naturally occurring radioactive decays. You might hide yourself in the hills, eat the most natural foods, practice obsessive hygiene, and still die from cancer caused by the radioactivity emanating from your own body. The probability of eventual death is 100%. Nobody is exempt.

Science helps determine the most probable. As the tools of science improve, then assessment of the most probable gets closer to being on target. Acceptance of risk, on the other hand, is a societal issue. Placing zero risk as a societal goal is not only impractical but selfish. Any society striving toward a policy of zero risk would consume its present and future economic resources. Isn't it more noble to accept nonzero risk and to minimize risk as much as possible within the limits of practicality? A society that accepts no risks receives no benefits.

The Physical Sciences: Physics, Chemistry, Earth Science, and Astronomy

LEARNING OBJECTIVE
Compare the fields of physics, chemistry, Earth science, and astronomy.

EXPLAIN THIS Why is physics more fundamental than the other sciences?

Science is the present-day equivalent of what used to be called *natural philosophy*. Natural philosophy was the study of unanswered questions about nature. As the answers were found, they became part of what is now called science. The study of science today branches into the study of living things and nonliving things: the life sciences and the physical sciences. The life sciences branch into such areas as molecular biology, microbiology, and ecology. The *physical sciences* branch into such areas as physics, chemistry, the Earth sciences, and astronomy.

A few words of explanation about each of the major divisions of science: Physics is the study of such concepts as motion, force, energy, matter, heat, sound, light, and the components of atoms. Chemistry builds on physics by telling us how matter is put together, how atoms combine to form molecules, and how the molecules combine to make the materials around us. Physics and

No wars are fought over science.

chemistry, applied to Earth and its processes, make up Earth science—geology, meteorology, and oceanography. When we apply physics, chemistry, and geology to other planets and to the stars, we are speaking about astronomy. Biology is more complex than physical science, for it involves matter that is alive. Underlying biology is chemistry, and underlying chemistry is physics. So physics is basic to both physical science and life science. That is why we begin with physics, then follow with chemistry, then investigate Earth science, and conclude with astronomy. All are treated conceptually, with the twin goals of enjoyment and understanding.

LEARNING OBJECTIVE
Relate learning science to an increased appreciation of nature.

In Perspective

EXPLAIN THIS Who gets the most out of something: one with understanding of it or one without understanding?

Just as you can't enjoy a ball game, computer game, or party game until you know its rules, so it is with nature. Because science helps us learn the rules of nature, it also helps us appreciate nature. You may see beauty in a structure such as the Golden Gate Bridge, but you'll see more beauty in that structure when you understand how all the forces that act on it balance. Similarly, when you look at the stars, your sense of their beauty is enhanced if you know how stars are born from mere clouds of gas and dust—with a little help from the laws of physics, of course. And how much richer it is, when you look at the myriad objects in your environment, to know that they are all composed of atoms—amazing, ancient, invisible systems of particles regulated by an eminently knowable set of laws.

If the complexity of science intimidates you, bear this in mind: All the branches of science rest upon a relatively small number of basic rules. Learn these underlying rules (physical laws), and you have a tool kit to bring to any phenomenon you wish to understand.

Go to it—we live in a time of rapid and fascinating scientific discovery!

For instructor-assigned homework, go to www.masteringphysics.com

SUMMARY OF TERMS (KNOWLEDGE)

Fact A phenomenon about which competent observers who have made a series of observations are in agreement.

Hypothesis An educated guess; a reasonable explanation of an observation or experimental result that is not fully accepted as factual until tested over and over again by experiment.

Law A general hypothesis or statement about the relationship of natural quantities that has been tested over and over again and has not been contradicted; also known as a *principle*.

Pseudoscience Fake science that pretends to be real science.

Science The collective findings of humans about nature, and a process of gathering and organizing knowledge about nature.

Scientific method Principles and procedures for the systematic pursuit of knowledge involving the recognition and formulation of a problem, the collection of data through observation and experiment, and the formulation and testing of hypotheses.

Theory A synthesis of a large body of information that encompasses well-tested and verified hypotheses about certain aspects of the natural world.

READING CHECK QUESTIONS (COMPREHENSION)

1. Briefly, what is science?

A Brief History of Advances in Science

2. Throughout the ages, what has been the general reaction to new ideas about established "truths"?

Mathematics and Conceptual Physical Science

3. What is the role of equations in this course?

Scientific Methods

4. List the steps of the classic scientific method.

The Scientific Attitude

5. In daily life, people are often praised for maintaining some particular point of view, for the "courage of their convictions." A change of mind is seen as a sign of weakness. How is this point of view different in science?

6. What is the test for whether or not a hypothesis is scientific?

7. We see many cases daily of people who are caught misrepresenting things and who soon thereafter are excused and accepted by their contemporaries. How is this different in science?

Science Has Limitations

8. What is meant by the term *supernatural*?

9. What is meant by *pseudoscience*?

Science, Art, and Religion

10. Briefly, how are science and religion similar?

11. Briefly, how are the concerns of science and religion different?

12. Must people choose between science and religion? Explain.

13. Psychological comfort is a benefit of having solid answers to religious questions. What benefit accompanies a position of not knowing answers?

Technology—The Practical Use of Science

14. Briefly distinguish between science and technology.

The Physical Sciences: Physics, Chemistry, Earth Science, and Astronomy

15. Why is physics considered to be the basic science?

In Perspective

16. What is the importance to people of learning nature's rules?

EXERCISES (SYNTHESIS)

17. Which of the following are scientific hypotheses?
    (a) Chlorophyll makes grass green.
    (b) Earth rotates about its axis because living things need an alternation of light and darkness.
    (c) Tides are caused by the Moon.

18. In answer to the question "When a plant grows, where does the material come from?" Aristotle hypothesized by logic that all material came from the soil. Do you consider his hypothesis to be correct, incorrect, or partially correct? What experiments do you propose to support your choice?

DISCUSSION QUESTIONS (EVALUATION)

19. The great philosopher and mathematician Bertrand Russell (1872–1970) wrote about ideas in the early part of his life that he rejected in the latter part of his life. Do you see this as a sign of weakness or as a sign of strength in Bertrand Russell? (Do you speculate that your present ideas about the world around you will change as you learn and experience more, or will further knowledge and experience solidify your present understanding?)

20. Bertrand Russell wrote, "I think we must retain the belief that scientific knowledge is one of the glories of man. I will not maintain that knowledge can never do harm. I think such general propositions can almost always be refuted by well-chosen examples. What I will maintain—and maintain vigorously—is that knowledge is very much more often useful than harmful and that knowledge is very much more often useful than harmful and that fear of knowledge is very much more often harmful than useful." Think of examples to support this statement.

21. Compare life before science and technology "in the good old days" with life in the present time. Be sure to include the fields of medicine, transportation, and communication.

22. Your favorite young relative is wondering about joining a large and growing group in the community, mainly to make new friends. Your advice is sought. Before replying, you learn that the group's charismatic leader tells followers, "Okay, this is how we operate: First, you should NEVER question anything I tell you. Second, you should NEVER question what you read in our literature." What advice do you offer?

PART ONE

Physics

Intriguing! The number of balls released into the array of balls is always the same number emerging from the other side. But why? There's gotta be a reason— mechanical rules of some kind. I'll know why the balls behave so predictably after I learn the rules of mechanics in the following chapters. Best of all, learning these rules will provide a keener intuition for understanding the world around me!

1

CHAPTER 1

Patterns of Motion and Equilibrium

MORE THAN 2000 years ago Greek scientists understood some of the physics we understand today. They had a good grasp of the physics of floating objects and some of the properties of light. But they were confused about motion. One of the first to study motion seriously was Aristotle, the most outstanding philosopher-scientist in ancient Greece. Aristotle attempted to clarify motion by classification.

1.1 Aristotle on Motion

EXPLAIN THIS How did Aristotle classify motion?

LEARNING OBJECTIVE
Establish Aristotle's influence on classifying motion.

Aristotle divided motion into two classes: *natural motion* and *violent motion.* Natural motion had to do with the nature of bodies. Light things like smoke rose, and heavy things like dropped boulders fell. The motions of stars across the night sky were natural. Violent motion, on the other hand, resulted from pushing or pulling forces. Objects whose motions were unnatural were either pushed or pulled. Aristotle believed that natural laws could be understood by logical reasoning.

Two assertions of Aristotle held sway for some 2000 years. One was that heavy objects necessarily fall faster than lighter objects. The other was that moving objects must necessarily have forces exerted on them to keep them moving.

These ideas were completely turned around in the 17th century by Galileo, who held that experiment was superior to logic in uncovering natural laws. Galileo demolished the idea that heavy things fall faster than lighter things in his famous Leaning Tower of Pisa experiment, where he allegedly dropped objects of different weights and showed that—except for the effects of air resistance—they fell to the ground together.

FYI Rather than reading chapters in this book slowly, try reading quickly and more than once. You'll better learn physics by going over the same material several times. With each time, it makes more sense. Don't worry if you don't understand things right away—just keep on reading.

FIGURE 1.1
Galileo's famous demonstration.

FIGURE 1.2
Does a force keep the cannonball moving after it leaves the cannon?

CHECKPOINT

Isn't it common sense to think that Earth is in its proper place and that a force to move it is inconceivable, as Aristotle held, and that the Earth *is* at rest in this universe? (*Think and formulate your own answer. Then check your thinking below.*)

Was this your answer?
Common sense is relative to one's time and place. Aristotle's views were logical and consistent with everyday observations. So unless you become familiar with the physics to follow in this book, Aristotle's views about motion *do* make common sense (and are held by many uneducated people today). But as you acquire new information about nature's rules, you'll likely find your common sense progressing beyond Aristotelian thinking.

ARISTOTLE (384–322 BC)

Aristotle was the foremost philosopher, scientist, and educator of his time. Born in Greece, he was the son of a physician who personally served the king of Macedonia. At age 17, he entered the Academy of Plato, where he worked and studied for 20 years until Plato's death. He then became the tutor of young Alexander the Great.

Eight years later, he formed his own school. Aristotle's aim was to systematize existing knowledge, just as Euclid had systematized geometry. Aristotle made critical observations; collected specimens; and gathered, summarized, and classified almost all of the existing knowledge of the physical world. His systematic approach became the method from which Western science later arose. After his death, his voluminous notebooks were preserved in caves near his home and were later sold to the library at

Alexandria. Scholarly activity ceased in most of Europe through the Dark Ages, and the works of Aristotle were forgotten and lost in the scholarship that continued in the Byzantine and Islamic empires. Several of his texts were reintroduced to Europe during the 11th and 12th centuries and were translated into Latin. The Church, the dominant political and cultural force in Western Europe, at first prohibited the works of Aristotle and then accepted and incorporated them into Christian doctrine.

1.2 Galileo's Concept of Inertia

EXPLAIN THIS Does a hockey puck need a force to keep it sliding?

Galileo tested his revolutionary idea by *experiment*. After studying balls rolling on planes inclined at various angles, he concluded that an object, once moving, continues to move *without* the application of forces. In the simplest sense, a **force** is a push or a pull. Although a force is needed to start an object moving, Galileo showed that once it is moving, no force is needed to keep it moving—except for the force needed to overcome friction (more about friction in Section 1.8). When friction is absent, a moving object needs no force to keep it moving. Galileo reasoned that a ball moving horizontally would move forever if friction were entirely absent. A ball would move of itself—of its own **inertia**, the property by which objects resist changes in motion.

This was the beginning of modern science. Experiment, not philosophical speculation, is the test of truth.

Slope downward—
Speed increases

Slope upward—
Speed decreases

No slope—
Does speed change?

FIGURE 1.3
The motion of balls on various planes.

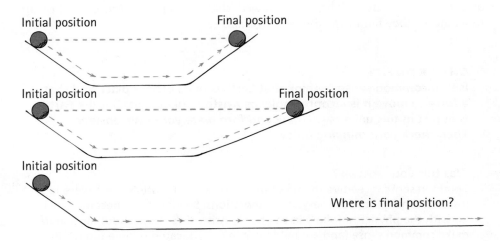

Initial position · · · Final position

Initial position · · · Final position

Initial position · · · Where is final position?

FIGURE 1.4
A ball rolling down an incline tends to roll up to its initial height. The ball must roll a greater distance as the angle of incline on the right is reduced.

GALILEO GALILEI (1564–1642)

Galileo was born in Pisa, Italy, in the same year Shakespeare was born and Michelangelo died. He studied medicine at the University of Pisa and then changed to mathematics. He developed an early interest in motion and was soon at odds with others around him, who held to Aristotelian ideas on falling bodies. He left Pisa to teach at the University of Padua and became an advocate of the new theory of the solar system advanced by the Polish astronomer Copernicus. Galileo was one of the first to build a telescope, and the first to direct it to the nighttime sky and discover mountains on the Moon and the moons of Jupiter. Because he published his findings in Italian instead of in Latin, which was expected of so reputable a scholar, and because of the recent invention of the printing press, his ideas reached many people. He soon ran afoul of the Church and was warned not to teach and not to hold to Copernican views. He restrained himself publicly for nearly 15 years. Then he defiantly published his observations and conclusions, which were counter to Church doctrine. The outcome was a trial in which he was found guilty, and he was forced to renounce his discoveries. By then an old man broken in health and spirit, he was sentenced to perpetual house arrest. Nevertheless, he completed his studies on motion, and his writings were smuggled out of Italy and published in Holland. His eyes had been damaged earlier by viewing the Sun through a telescope, which led to blindness at age 74. He died four years later.

CHECKPOINT

A ball rolling along a level surface slowly comes to a stop. How would Aristotle explain this behavior? How would Galileo explain it? How would you explain it?

Were these your answers?

As mentioned, think about the Checkpoint questions throughout this book *before* reading the answers. When you first formulate your own answers, you'll find yourself learning more—much more!

> Aristotle would probably say that the ball stops because it seeks its natural state of rest. Galileo would probably say that friction overcomes the ball's natural tendency to continue rolling—that friction overcomes the ball's *inertia*, and brings it to a stop. Only you can answer the last question!

FYI Galileo and William Shakespeare were born in the same year, 1564. In 1632 Galileo published his first mathematical treatment of motion—12 years after the Pilgrims landed at Plymouth Rock.

1.3 Mass—A Measure of Inertia

EXPLAIN THIS Why is your mass, but not your weight, the same on Earth as on the Moon?

LEARNING OBJECTIVE
Describe and distinguish between mass and weight.

Every material object possesses inertia; how much depends on its amount of matter—the more matter, the more inertia. In speaking of how much matter something has, we use the term *mass*—the greater the mass of an object, the greater the amount of matter and the greater its inertia. **Mass** is a measure of the inertia of a material object.

Loosely speaking, mass corresponds to our intuitive notion of **weight**. We say something has a lot of matter if it is heavy. That's because we are accustomed to measuring matter by gravitational attraction to Earth. But mass is more fundamental than weight; it is a fundamental quantity that completely escapes the notice of most people. There are times, however, when weight corresponds to our unconscious notion of inertia. For example, if you are trying to determine which of two small objects is heavier, you might shake them back and forth in your hands or move them in some way instead of lifting them. In doing so, you are judging which of the two is more difficult to get moving, seeing which is the more resistant to a *change* in motion. You are really comparing the inertias of the objects.

It is easy to confuse the ideas of mass and weight. We define each as follows:

Mass: The quantity of matter in an object. It is also the measure of the inertia or sluggishness that an object exhibits in response to any effort made to start it, stop it, or change its state of motion in any way.

Weight: The force upon an object due to gravity.

The standard unit of mass is the **kilogram**, abbreviated kg. Weight is measured in units of force (such as pounds). The scientific unit of force is the **newton**, abbreviated N, which we'll use in this book. The abbreviation is written with a capital letter because the unit is named after a person.

Mass and weight are directly proportional to each other.* If the mass of an object is doubled, its weight is also doubled; if the mass is halved, the weight

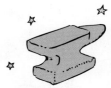

FIGURE 1.5
An anvil in outer space—beyond the Sun for example—may be weightless, but it still has mass.

FIGURE 1.6
The astronaut in space finds it just as difficult to shake the "weightless" anvil as on Earth. If the anvil is more massive than the astronaut, which shakes more—the anvil or the astronaut?

---

* *Directly proportional* means that if you change one thing, the other thing changes proportionally. The constant of proportionality is *g*, the acceleration due to gravity. As we shall soon see, weight = *mg* (or mass × acceleration due to gravity), so 9.8 N = (1 kg)(9.8 m/s^2). In Chapter 4 we'll extend our definition of weight to be the force of a body pressing against a support (for example, against a weighing scale).

FIGURE 1.7
Why will a slow continuous increase in downward force break the string above the massive ball, whereas a sudden increase in downward force breaks the lower string?

FIGURE 1.8
Why does the blow of the hammer not harm her?

Mastering**PHYSICS**

VIDEO: Newton's Law of Inertia

VIDEO: The Old Tablecloth Trick

VIDEO: Toilet Paper Roll

VIDEO: Inertia of a Cylinder

VIDEO: Inertia of an Anvil

VIDEO: Definition of a Newton

is halved. Because of this, mass and weight are often interchanged. Also, mass and weight are sometimes confused because it is customary to measure the quantity of matter in things (their mass) by their gravitational attraction to Earth (their weight). But mass doesn't depend on gravity. Gravity on the Moon, for example, is much less than it is on Earth. Whereas your weight on the surface of the Moon would be much less than it is on Earth, your mass would be the same in both locations.

Don't confuse mass and **volume**, the quantity of space an object occupies. When we think of a massive object, we often think of a big object. An object's size, however, is not necessarily a good way to judge its mass. Which is easier to get moving: a car battery or a king-size pillow? So we find that mass is neither weight nor volume.

A nice demonstration that distinguishes mass from weight is the massive ball suspended on the string shown in Figure 1.7. The top string breaks when the lower string is pulled with a gradual increase in force, but the bottom string breaks when the string is jerked. Which of these cases illustrates the weight of the ball, and which illustrates the mass of the ball? Note that only the top string bears the weight of the ball. So when the lower string is gradually pulled, the tension supplied by the pull is transmitted to the top string. So total tension in the top string is pull plus the weight of the ball. The top string breaks when the breaking point is reached. But when the bottom string is jerked, the mass of the ball—its tendency to remain at rest—is responsible for breakage of the bottom string.

CHECKPOINT

1. Does a 2-kg iron block have twice as much *inertia* as a 1-kg iron block? Twice as much *mass*? Twice as much *volume*? Twice as much *weight* when weighed in the same location?
2. Does a 2-kg iron block have twice as much *inertia* as a 1-kg bunch of bananas? Twice as much *mass*? Twice as much *volume*? Twice as much *weight* when weighed in the same location?
3. How does the mass of a bar of gold vary with location?

Were these your answers?

1. The answer is yes to all questions. A 2-kg block of iron has twice as many iron atoms, and therefore twice the amount of matter, mass, and weight. The blocks consist of the same material, so the 2-kg block also has twice the volume.
2. Two kilograms of *anything* has twice the inertia and twice the mass of 1 kg of anything else. Because mass and weight are proportional in the same location, 2 kg of anything will weigh twice as much as 1 kg of anything. Except for volume, the answer to all the questions is yes. Volume and mass are proportional only when the materials are identical—when they have the same *density*. (Density is mass/volume, as we'll discuss in Chapter 5.) Iron is much more dense than bananas, so 2 kg of iron must occupy less volume than 1 kg of bananas.
3. Not at all! It consists of the same number of atoms no matter what the location. Although its weight may vary with location, it has the same mass everywhere. This is why mass is preferred to weight in scientific studies.

One Kilogram Weighs 10 N

A 1-kg bag of any material at Earth's surface has a weight of about 10 N (more precisely 9.8 N). Away from Earth's surface where the force of gravity is less (on the Moon, for example), the bag would weigh less.

Except in cases where precision is needed, we round off 9.8 and call it 10. So 1 kg of something on Earth's surface weighs about 10 N. If you know the mass in kilograms and want the weight in newtons, multiply the number of kilograms by 10. Or, if you know the weight in newtons, divide by 10 and you'll have the mass in kilograms. As previously mentioned, weight and mass are proportional to each other.

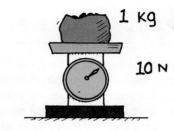

FIGURE 1.9
One kilogram of nails weighs about 10 N, which roughly equals 2.2 lb.

1.4 Net Force

EXPLAIN THIS How can mom and dad push on something to produce a net force of zero?

In simplest terms, a force is a push or a pull. Objects don't speed up, slow down, or change direction unless a force acts. When we say "force," we imply the total force, or *net* force, acting on an object. Often more than one force acts. For example, when you throw a baseball, the forces of gravity, air friction, and the pushing force you apply with your muscles all act on the ball. The **net force** on the ball is the combination of all these forces. It is the net force that changes an object's state of motion.

For example, suppose you pull on a box with a force of 5 N (slightly more than 1 lb). If your friend also pulls with 5 N in the same direction, the net force on the box is 10 N. If your friend pulls on the box with the same magnitude of force as you in the opposite direction, the net force on it is zero. Now if you increase your pull to 10 N and your friend pulls oppositely with 5 N, the net force is 5 N in the direction of your pull. This is shown in Figure 1.10.

The forces in Figure 1.10 are shown by arrows. Forces are vector quantities. A **vector quantity** has both magnitude (how much) and direction (which way). When an arrow represents a vector quantity, the arrow's length represents magnitude and its direction shows the direction of the quantity. Such an arrow is called a **vector**. (You'll find more on vectors in the next chapter, in Appendix B, and in the *Conceptual Physical Science Practice Book*.)

LEARNING OBJECTIVE
Distinguish between force and net force, and give examples.

The relationship between kilograms and pounds is that 1 kg weighs 2.2 lb at Earth's surface. (That means 1 lb is the same as 4.45 N.)

A zero net force on an object doesn't mean that the object must be at rest, but that its state of motion remains unchanged. It can be at rest or moving uniformly in a straight line.

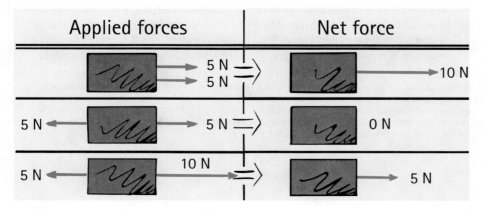

FIGURE 1.10
Net force.

PAUL HEWITT PERSONAL ESSAY

When I was in high school, my counselor advised me not to enroll in science and math classes but instead focus on what seemed to be my gift for art. I took this advice. I was then interested in drawing comic strips and in boxing, neither of which earned me much success. After a stint in the Army, I tried my luck at sign painting, and the cold Boston winters drove me south to Miami, Florida. There, at age 26, I got a job painting billboards and met an intellectual friend, Burl Grey. Like me, Burl had never studied physics in high school. But he was passionate about science in general, and shared his passion with many questions as we painted together.

I remember Burl asking me about the tensions in the ropes that held up the scaffold we were on. The scaffold was simply a heavy horizontal plank suspended by a pair of ropes. Burl twanged the rope nearest his end of the scaffold and asked me to do the same with mine. He was comparing the tensions in both ropes—to determine which was greater. Burl was heavier than I was, and he guessed the tension in his rope was greater. Like a more tightly stretched guitar string, the rope with greater tension twangs at a higher pitch. The finding that Burl's rope had a higher pitch seemed reasonable because his rope supported more of the load.

When I walked toward Burl to borrow one of his brushes, he asked if tensions in the ropes changed. Did tension in his rope increase as I moved closer? We agreed that it should have, because even more of the load was then supported by Burl's rope. How about my rope? Would its tension decrease? We agreed that it would, for it would be supporting less of the total load. I was unaware at the time that I was discussing physics.

Burl and I used exaggeration to bolster our reasoning (just as physicists do). If we both stood at an extreme end of the scaffold and leaned outward, it was easy to imagine the opposite end of the scaffold rising like the end of a seesaw, with the opposite rope going limp. Then there would be no tension in that rope. We then reasoned that the tension in my rope would gradually decrease as I walked toward Burl. It was fun posing such questions and seeing if we could answer them.

A question that we couldn't answer was whether the decrease in tension in my rope when I walked away from it would be *exactly* compensated by a tension increase in Burl's rope. For example, if my rope underwent a decrease of 50 N, would Burl's rope gain 50 N? (We talked pounds back then, but here we use the scientific unit of force, the *newton*—abbreviated N.) Would the gain be *exactly* 50 N? And if so, would this be a grand coincidence? I didn't know the answer until more than a year later, when Burl's stimulation resulted in my leaving full-time painting and going to college to learn more about science.*

There I learned that any object at rest, such as the sign-painting scaffold I worked on with Burl, is said to be in equilibrium. That is, all the forces that act on it balance to zero. So the sum of the upward forces supplied by the supporting ropes indeed do add up to our weights plus the weight of the scaffold. A 50-N loss in one would be accompanied by a 50-N gain in the other.

I tell this true story to make the point that one's thinking is very different when there is a rule to guide it. Now, when I look at any motionless object, I know immediately that all the forces acting on it cancel out. We see nature differently when we know its rules. It makes nature simpler and easier to understand. Without the rules of physics, we tend to be superstitious and see magic where there is none. Quite wonderfully, everything is beautifully connected to everything else by a surprisingly small number of rules. Physics is the study of nature's rules.

* I am indebted to Burl Grey for the stimulation he provided, for when I continued with formal education, it was with enthusiasm. I lost contact with Burl for 40 years. A student in my class at the Exploratorium in San Francisco, Jayson Wechter, who was a private detective, located him in 1998 and put us back in contact. Friendship renewed, we continue in our spirited conversations. It was via Burl that I met my teaching role model, futurist Jacque Fresco, the most talented teacher I've ever met. Now in his 90s, he continues to inspire people toward a positive future through his books, TV documentaries, and most recently by the movie that features his vision, "Future By Design."

1.5 The Equilibrium Rule

EXPLAIN THIS How can the sum of real forces result in no force at all?

I f you tie a string around a 2-lb bag of flour and suspend it on a weighing scale (Figure 1.11), a spring in the scale stretches until the scale reads 2 lb. The stretched spring is under a "stretching force" called *tension*. A scale in a science lab is likely calibrated to read the same force as 9 N. Both pounds and newtons are units of weight, which in turn, are units of *force*. The bag of flour is attracted to Earth with a gravitational force of 2 lb—or equivalently, 9 N. Suspend twice as much flour from the scale and the reading will be 18 N.

Two forces are acting on the bag of flour—tension force acting upward and weight acting downward. The two forces on the bag are equal in magnitude and opposite in direction, and they cancel to zero. Hence the bag remains at rest.

When the net force on something is zero, we say that the object is in *mechanical equilibrium*.* In mathematical notation, the **equilibrium rule** is

$$\Sigma F = 0$$

The symbol Σ stands for "the vector sum of" and F stands for "forces." For a suspended object at rest, like the bag of flour, the rule states that the forces acting upward on the body must be balanced by other forces acting downward to make the vector sum equal zero. (Vector quantities take direction into account, so if upward forces are positive, downward ones are negative; the resulting sum is equal to zero.)

In Figure 1.12 we see the forces of interest to Burl and Paul on their sign-painting scaffold. The sum of the upward tensions is equal to the sum of their weights plus the weight of the scaffold. Note how the magnitudes of the two upward vectors equal the magnitudes of the three downward vectors. Net force on the scaffold is zero, so we say it is in mechanical equilibrium.

LEARNING OBJECTIVE
Describe the rule $\Sigma F = 0$, and give examples.

FIGURE 1.11
Burl Grey, who first introduced the physics author to tension forces, suspends a 2-lb bag of flour from a spring scale, showing its weight and the tension in the string of about 9 N.

FIGURE 1.12
The sum of the upward vectors equals the sum of the downward vectors. $\Sigma F = 0$, and the scaffold is in equilibrium.

CHECKPOINT

If you hang from a trapeze at rest, what is the tension in each of the two supporting vertical ropes?

Was this your answer?
The tension would be half your weight in each rope. In this way, $\Sigma F = 0$.

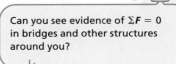

Can you see evidence of $\Sigma F = 0$ in bridges and other structures around you?

---

* We'll see in Appendix A that another condition for mechanical equilibrium is that the net torque equals zero.

1.6   Support Force

EXPLAIN THIS   How does support force relate to your weight?

Consider a book lying at rest on a table. It is in equilibrium. What forces act on the book? One force is that due to gravity—the *weight* of the book. Because the book is in equilibrium, there must be another force acting on it to produce a net force of zero—an upward force opposite to the force of gravity. The table exerts this upward force, called the **support force**. This upward support force, often called the *normal force*, must equal the weight of the book.* If we designate the upward force as positive, then the downward force (weight) is negative, and the sum of the two is zero. The net force on the book is zero. Stating it another way, $\Sigma F = 0$.

To better understand that the table pushes up on the book, compare the case of compressing a spring (Figure 1.13). If you push the spring down, you can feel the spring pushing up on your hand. Similarly, the book lying on the table compresses atoms in the table, which behave like microscopic springs. The weight of the book squeezes downward on the atoms, and they squeeze upward on the book. In this way, the compressed atoms produce the support force.

When you step on a bathroom scale, two forces act on the scale. One is the downward pull of gravity, your weight, and the other is the upward support force of the floor. These forces compress a spring that is calibrated to show your weight (Figure 1.14). In effect, the scale shows the support force. When you weigh yourself on a bathroom scale at rest, the support force and your weight have the same magnitude.

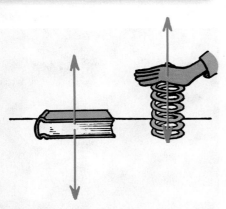

FIGURE 1.13
The table pushes up on the book with as much force as the downward force of gravity on the book. The spring pushes up on your hand with as much force as you exert to push down on the spring.

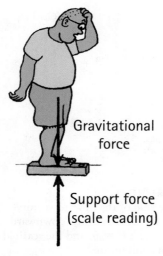

Gravitational force

Support force (scale reading)

FIGURE 1.14
The upward support is as much as the downward gravitational force.

CHECKPOINT

Suppose you stand on two bathroom scales with your weight evenly divided between the two scales. What will each scale read? How about if you stand with more of your weight on one foot than the other?

Were these your answers?
The readings on both scales add up to your weight. This is because the sum of the scale readings, which equals the supporting normal force by the floor, must counteract your weight so the net force on you will be zero. That is, the vector sum $\Sigma F = 0$. If you stand equally on each scale, each will read half your weight. If you lean more on one scale than the other, more than half your weight will be read on that scale but less on the other, so they will still add up to your weight. For example, if one scale reads two-thirds your weight, the other scale will read one-third your weight. In whatever case, $\Sigma F = 0$. Get it?

---

* This force acts at right angles to the surface. When we say "normal to," we are saying "at right angles to," which is why this force is called a normal force.

1.7 Dynamic Equilibrium

LEARNING OBJECTIVE
Distinguish between static and dynamic equilibrium.

EXPLAIN THIS How does a sliding air puck move while no net force acts?

When an object remains stationary, the forces on it add up to zero—it's in equilibrium. More specifically, we say it's in *static equilibrium*. But the state of rest is only one form of equilibrium. An object moving at constant speed in a straight-line path is also in equilibrium. We say it's in *dynamic equilibrium*. Once in motion, if there is no net force to change the object's motion, it moves at an unchanging speed and direction and is in dynamic equilibrium. Whether equilibrium is static or dynamic, $\Sigma F = 0$.

FYI In Chapter 6 we'll discuss thermal equilibrium, and in Appendix A we'll discuss rotational equilibrium.

Interestingly, an object under the influence of only one force cannot be in static or dynamic equilibrium. Net force couldn't be zero. Only when there is no force at all, or when two or more forces combine to zero, can an object be in equilibrium. We can test whether something is in equilibrium by noting whether it undergoes changes in motion.

Consider pushing a desk across a classroom floor. If it moves steadily at constant speed, with no change in its motion, it is in equilibrium. This tells us that another horizontal force acts on the desk—likely the force of friction between the desk and the floor. The fact that the net force on the desk equals zero means that the force of friction must be equal in magnitude and act opposite to our pushing force.

1.8 The Force of Friction

LEARNING OBJECTIVE
Describe friction and its direction when an object slides.

EXPLAIN THIS How much friction acts when you push your desk at constant velocity?

Friction is the resistive force that opposes the motion or attempted motion of an object past another with which it is in contact. It occurs when one object rubs against something else.* Friction occurs for solids, liquids, and gases. An important rule of friction is that it always acts in a direction to oppose motion. If you push a solid block along a floor to the right, the force of friction on the block will be to the left. A boat propelled to the east by its motor experiences water friction to the west. When an object falls downward through the air, the force of friction, **air resistance**, acts upward. Again, for emphasis: friction always acts in a direction to oppose motion.

FIGURE 1.15
When the push on the desk is as great as the force of friction between it and the floor, the net force on the desk is zero and it slides at an unchanging speed.

CHECKPOINT

You push on a piece of furniture and it slides at constant speed across the living room floor. In other words, it is in equilibrium. Two horizontal forces act on it. One is your push and the other is the force of friction that acts in the opposite direction. Which force is greater?

Was this your answer?
Neither, for both forces have the same magnitude. If you call your push positive, then the friction force is negative. Because the pushed furniture is in equilibrium, can you see that the two forces combine to equal zero?

---

* Unlike most concepts in physics, friction is a very complicated phenomenon. The findings are empirical (gained from a wide range of experiments) and the predictions are approximate (also based on experiment).

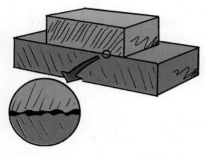

FIGURE 1.16
Friction results from the mutual contact of irregularities in the surfaces of sliding objects. Even surfaces that appear to be smooth have irregular surfaces when viewed at the microscopic level.

The amount of friction between two surfaces depends on the kinds of material and how much they are pressed together. Friction is due to tiny surface bumps and also to the "stickiness" of the atoms on the surfaces of the two materials (Figure 1.16). Friction between a sliding desk you're pushing and a smooth linoleum floor is less than between the desk and a rough floor. And if the surface is inclined, friction is less because it doesn't press as much on the inclined surface (we won't treat inclined surfaces in this chapter).

So we see that when you push horizontally on a piece of furniture and it slides across the floor, both your force and the opposite force of friction affect the motion. When you push hard enough on the sliding furniture to match the friction, the net force on it is zero, and it slides at constant velocity. Notice that we are talking about what we recently learned—that no change in motion occurs when $\Sigma F = 0$.

CHECKPOINT

1. **Suppose you exert a 50-N horizontal force on a heavy desk resting motionless on your classroom floor. The fact that it remains at rest indicates that 50 N isn't great enough to make it slide. How does the force of friction between the desk and floor compare with your push?**
2. **You push harder—say, 55 N—and the desk still doesn't slide. How much friction acts on it?**
3. **You push still harder and the desk moves. Once it is in motion, you push with 60 N, which is just sufficient to keep it sliding at constant velocity. How much friction acts on the desk?**
4. **What net force does a sliding desk experience when you exert a force of 65 N and friction between the desk and the floor is 60 N?**

Were these your answers?

1. The force of friction is 50 N in the opposite direction. Friction opposes the motion that would occur otherwise. The fact that the desk is at rest is evidence that $\Sigma F = 0$.
2. Friction increases to 55 N, and again $\Sigma F = 0$.
3. The force of friction is 60 N, because when the desk is moving at constant velocity, $\Sigma F = 0$.
4. The net force is 5 N, because $\Sigma F = 65 \text{ N} - 60 \text{ N}$. In this case the desk picks up speed. As we will see, it *accelerates*.

1.9 Speed and Velocity

EXPLAIN THIS When can you drive at constant speed while your velocity changes?

Speed

Before the time of Galileo, when measurements of time were vague, people described moving things as simply "slow" or "fast." Galileo measured speed by comparing the distance covered with the *time* it takes to move that distance. He defined **speed** as the distance covered per amount of travel time:

$$\text{Speed} = \frac{\text{distance covered}}{\text{travel time}}$$

For example, if a bicyclist covers 20 kilometers in 1 hour, her speed is 20 km/h. Or, if she runs 6 meters in 1 second, her speed is 6 m/s.

FIGURE 1.17
A common automobile speedometer. Note that speed is shown in units of km/h and mi/h.

TABLE 1.1	APPROXIMATE SPEEDS IN DIFFERENT UNITS

12 mi/h = 20 km/h = 6 m/s (bowling ball)
25 mi/h = 40 km/h = 11 m/s (very good sprinter)
37 mi/h = 60 km/h = 17 m/s (sprinting rabbit)
50 mi/h = 80 km/h = 22 m/s (tsunami)
62 mi/h = 100 km/h = 28 m/s (sprinting cheetah)
75 mi/h = 120 km/h = 33 m/s (batted softball)
100 mi/h = 160 km/h = 44 m/s (batted baseball)

FIGURE 1.18
The greater the distance traveled each second, the faster the horse gallops.

Any combination of units for distance and time can be used for speed—kilometers per hour (km/h), centimeters per day (the speed of a sick snail), or whatever is useful and convenient. The slash symbol (/) is read as "per" and means "divided by." In physics the preferred unit of speed is meters per second (m/s). Table 1.1 compares some speeds in different units.

Instantaneous Speed

Moving things often have variations in speed. A car, for example, may travel along a street at 50 km/h, slow to 0 km/h at a red light, and speed up to only 30 km/h because of traffic. At any instant you can tell the speed of the car by looking at its speedometer. The speed at any instant is the *instantaneous speed*.

Average Speed

In planning a trip by car, the driver often wants to know the travel time. The driver is concerned with the *average speed* for the trip. How is average speed defined?

$$\text{Average speed} = \frac{\text{total distance covered}}{\text{travel time}}$$

Average speed can be calculated rather easily. For example, if you drive a distance of 80 km in 1 h, your average speed is 80 km/h. Likewise, if you travel 320 km in 4 h,

$$\text{Average speed} = \frac{\text{total distance covered}}{\text{travel time}} = \frac{320 \text{ km}}{4 \text{ h}} = 80 \text{ km/h}$$

Note that when a distance in kilometers (km) is divided by a time in hours (h), the answer is in kilometers per hour (km/h).

Because average speed is the entire distance covered divided by the total time of travel, it doesn't indicate the various instantaneous speeds that may have occurred along the way. At any moment during most trips, the instantaneous speed is often different from the average speed.

If we know average speed and travel time, distance traveled is easy to find. A simple rearrangement of the definition above gives

$$\text{Total distance covered} = \text{average speed} \times \text{travel time}$$

For example, if your average speed on a 4-h trip is 80 km/h, then you cover a total distance of 320 km.

If you get a traffic ticket for speeding, is the speed written on your ticket your *instantaneous speed* or your *average speed*?

Mastering**PHYSICS**
VIDEO: Definition of Speed
VIDEO: Average Speed
VIDEO: Velocity
VIDEO: Changing Velocity

CHECKPOINT
1. **What is the average speed of a horse that gallops 100 m in 8 s? How about if it gallops 50 m in 4 s?**
2. **If a car travels with an average speed of 60 km/h for an hour, it will cover a distance of 60 km. (a) How far would the car travel if it moved at this rate for 4 h? (b) For 10 h?**

Were these your answers?
*(Are you reading this before you have reasoned answers in your mind? As mentioned earlier, **think** before you read the answers. You'll not only learn more; you'll enjoy learning more.)*

1. In both cases the answer is 12.5 m/s:

$$\text{Average speed} = \frac{\text{total distance covered}}{\text{travel time}} = \frac{100 \text{ meters}}{8 \text{ seconds}} = \frac{50 \text{ meters}}{4 \text{ seconds}} = 12.5 \text{ m/s}$$

2. The distance traveled is the average speed × time of travel, so
   a. Distance = 60 km/h × 4 h = 240 km
   b. Distance = 60 km/h × 10 h = 600 km

FIGURE 1.19
Although the car can maintain a constant speed along the circular track, it cannot maintain a constant velocity. Why?

Velocity

When we know both the speed and direction of an object, we know its **velocity**. For example, if a vehicle travels at 60 km/h, we know its speed. But, if we say it moves at 60 km/h to the north, we specify its *velocity*. Speed is a description of how fast; velocity is a description of how fast *and* in what direction. As previously mentioned, a quantity such as velocity that specifies direction as well as magnitude is called a *vector quantity*. Velocity is a vector quantity. (Vectors are treated in Appendix B and are nicely developed in the *Conceptual Physical Science Practice Book*.)

Constant speed means steady speed, neither speeding up nor slowing down. Constant velocity, on the other hand, means both constant speed *and* constant direction. Constant direction is a straight line—the object's path doesn't curve. So, constant velocity means motion in a straight line at constant speed—motion, as we will soon see, with no acceleration.

Velocity is "directed" speed.

CHECKPOINT
"She moves at a constant speed in a constant direction." Say the same sentence in fewer words.

Was this your answer?
"She moves at constant velocity."

Motion Is Relative

Everything is always moving. Even when you think you're standing still, you're actually speeding through space. You're moving relative to the Sun and stars—though you are at rest relative to Earth. At this moment your speed relative to the Sun is about 100,000 km/h and even faster relative to the center of our galaxy.

When we discuss the speed or velocity of something, we mean speed or velocity relative to something else. For example, when we say a space shuttle travels at 30,000 km/h, we mean relative to Earth below. Or when we say a racing car reaches a speed of 300 km/h, we mean relative to the track. Unless stated otherwise, all speeds discussed in this book are relative to the surface of Earth. Motion is relative.

FIGURE 1.20
Although you may be at rest relative to Earth's surface, you're moving about 100,000 km/h relative to the Sun.

CHECKPOINT

A hungry mosquito sees you resting in a hammock in a 3-m/s breeze. How fast and in what direction should the mosquito fly in order to hover above you for lunch?

Was this your answer?
The mosquito should fly toward you into the breeze. When just above you, it should fly at 3 m/s in order to hover at rest. Unless its grip on your skin is strong enough after landing, it must continue flying at 3 m/s to keep from being blown off. That's why a breeze is an effective deterrent to mosquito bites.

If you look out an airplane window and view another plane flying at the same speed in the opposite direction, you'll see it flying twice as fast—nicely illustrating relative motion.

1.10 Acceleration

LEARNING OBJECTIVE
Define acceleration, and distinguish it from velocity and speed.

EXPLAIN THIS Why is the word *change* important in describing acceleration?

Most moving things undergo variations in their motion. We say they undergo *acceleration*. The first to clearly formulate the concept of acceleration was Galileo, who developed the concept in his experiments with inclined planes. He found that balls rolling down inclines rolled faster and faster. Their velocity changed as they rolled. Further, the balls gained the same amount of velocity in equal time intervals.

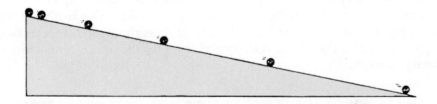

FIGURE 1.21
A ball gains the same amount of speed in equal intervals of time. It undergoes constant acceleration.

Galileo defined the rate of change of velocity as **acceleration:***

$$\text{Acceleration} = \frac{\text{change of velocity}}{\text{time interval}}$$

Acceleration is experienced when you're in a moving car or bus. When the bus driver steps on the gas pedal, the vehicle gains speed. We say that the bus accelerates. Thus, we can see why the gas pedal is called the "accelerator"!

Can you see that a car has three controls that change velocity— the gas pedal (accelerator), the brakes, and the steering wheel?

---

* The Greek letter Δ (delta) is often used as a symbol for "change in" or "difference in." In "delta" notation, $a = \dfrac{\Delta v}{\Delta t}$, where Δv is the change in velocity and Δt is the change in time (the time interval). From this we see that $v = at$. See further development of linear motion in Appendix A.

FIGURE 1.22
We say that a body undergoes acceleration when there is a *change* in its state of motion.

"When you're over the hill, that's when you pick up speed."
—Quincy Jones

Mastering PHYSICS

VIDEO: Definition of Acceleration

VIDEO: Numerical Example of Acceleration

VIDEO: Free Fall: How Fast?

VIDEO: Free Fall: How Far?

VIDEO: Free Fall Acceleration Explained

When the brakes are applied, the vehicle slows. This is also acceleration, because the velocity of the vehicle is changing. When something slows down, we often call this *deceleration*.

Consider driving in a vehicle that steadily increases in speed. Suppose that in 1 s, you steadily increase your velocity from 30 km/h to 35 km/h. In the next second, you go from 35 km/h to 40 km/h, and so on. You change your velocity by 5 km/h each second. We see that

$$\text{Acceleration} = \frac{\text{change of velocity}}{\text{time interval}} = \frac{5 \text{ km/h}}{1 \text{ s}} = 5 \text{ km/h} \cdot \text{s}$$

In this example of straight-line motion, the acceleration is 5 km/h-second (abbreviated as 5 km/h $\cdot$ s).* Note that the unit for time appears twice: once for the unit of velocity and again for the interval of time in which the velocity is changing. Also note that acceleration is not just the change in velocity; it is the change in velocity per second. If either speed or direction changes, or if both change, then velocity changes.

When a vehicle makes a turn, even if its speed does not change, it is accelerating. Can you see why? Acceleration occurs because the vehicle's direction is changing. Acceleration refers to a change in velocity. So acceleration involves a change in speed, a change in direction, or a change in both speed *and* direction. Figure 1.22 illustrates this.

When straight-line motion is being considered, we can use the words *speed* and *velocity* interchangeably in the definition of acceleration. When direction doesn't change, acceleration may be expressed as the rate at which *speed* changes.

Hold a stone above your head (not directly above your head!) and drop it. It accelerates during its fall. When the only force that acts on a falling object is that due to gravity, when air resistance doesn't affect its motion, we say the object is in **free fall**. All freely falling objects in the same vicinity have the same acceleration. Near Earth's surface an object in free fall gains speed at the rate of 10 m/s each second, as shown in Table 1.2.

$$\text{Acceleration} = \frac{\text{change in speed}}{\text{time interval}} = \frac{10 \text{ m/s}}{1 \text{ s}} = 10 \text{ m/s} \cdot \text{s} = 10 \text{ m/s}^2$$

We read the acceleration of free fall as 10 meters per second squared. (More precisely, 9.8 m/s^2.) This is the same as saying that acceleration is 10 meters per second per second. Note again that the unit of time, the second, appears twice. It appears once for the unit of velocity and again for the time during which the velocity changes.

TABLE 1.2	FREE-FALL VELOCITY ACQUIRED AND DISTANCE FALLEN	
Time of Fall (s)	**Velocity Acquired (m/s)**	**Distance Fallen (m)**
0	0	0
1	10	5
2	20	20
3	30	45
4	40	80
5	50	125

* When we divide $\frac{\text{km}}{\text{h}}$ by s $\left(\frac{\text{km}}{\text{h}} \div \text{s}\right)$, we can express the result as $\frac{\text{km}}{\text{h}} \times \frac{1}{\text{s}} = \frac{\text{km}}{\text{h} \cdot \text{s}}$ (some textbooks express this as km/h/s). Or when we divide $\frac{\text{m}}{\text{s}}$ by s $\left(\frac{\text{m}}{\text{s}} \div \text{s}\right)$, we can express this as $\frac{\text{m}}{\text{s}} \times \frac{1}{\text{s}} = \frac{\text{m}}{\text{s} \cdot \text{s}} = \frac{\text{m}}{\text{s}^2}$ (which can also be written as (m/s)/s, or ms^{-2}).

CHECKPOINT

In 2.0 s a car increases its speed from 60 km/h to 65 km/h while a bicycle goes from rest to 5 km/h. Which has the greater acceleration?

Was this your answer?
Both have the same acceleration because both gain the same amount of speed in the same time. Both accelerate at 2.5 km/h·s.

In Figure 1.23, we imagine a freely falling boulder with a speedometer attached to it. As the boulder falls, the speedometer shows that the boulder goes 10 m/s faster each second. This 10 m/s gain each second is the boulder's acceleration. Velocity acquired and distance fallen* are shown in Table 1.2.

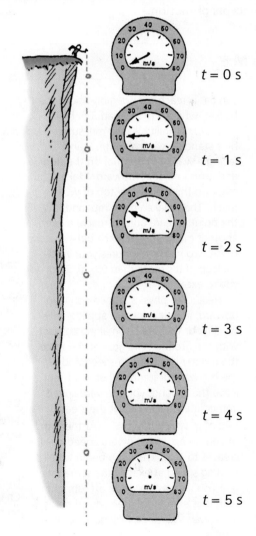

FIGURE 1.23
INTERACTIVE FIGURE

Imagine that a falling boulder is equipped with a speedometer. In each succeeding second of fall, you'd find the boulder's speed increasing by the same amount: 10 m/s. Sketch in the missing speedometer needle at $t = 3$ s, $t = 4$ s, and $t = 5$ s.

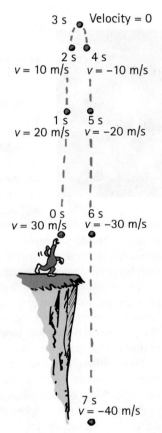

FIGURE 1.24
INTERACTIVE FIGURE

The rate at which velocity changes each second is the same.

* Distance fallen from rest: d = average velocity × time

$$d = \frac{\text{initial velocity} + \text{final velocity}}{2} \times \text{time}$$

$$d = \frac{0 + gt}{2} \times t$$

$$d = \frac{1}{2}gt^2$$

(See Appendix A for further explanation.)

The speed of a vertically thrown ball at the top of its path is zero. Is the acceleration there zero also? (Answer begins with an N.)

How nice—the acceleration due to gravity is 10 m/s each second all the way down. Why this is so, for any mass, awaits you in Chapter 3.

(The acceleration of free fall is further developed in Appendix A and in the *Conceptual Physical Science Practice Book*.) We see that the distance of free fall from rest is directly proportional to the square of the time of fall. In equation form,

$$d = \tfrac{1}{2}gt^2$$

Up-and-down motion is shown in Figure 1.24. The ball leaves the thrower's hand at 30 m/s. Call this the initial velocity. The figure uses the convention of up being + and down being −. The minus sign of downward values of velocity indicates a downward direction. More important, notice that the 1-s interval positions correspond to 10-m/s velocity changes.

Aristotle used logic to establish his ideas of motion, whereas Galileo used experiment. Galileo showed that experiments are superior to logic in testing knowledge. Galileo was concerned with *how* things move rather than *why* they move. The path was paved for Isaac Newton to make further connections of concepts of motion.

HANG TIME

Some athletes and dancers have great jumping ability. Leaping straight up, they seem to "hang in the air," apparently defying gravity. Ask your friends to estimate the **hang time** of the great jumpers—the time a jumper is airborne with his or her feet off the ground. They may estimate 2 or 3 s. But, surprisingly, the hang time of the greatest jumpers is almost always less than 1 s. The perception of a longer time is one of many illusions we have about nature.

People often have a related illusion about the vertical height a human can jump. Most of your classmates probably cannot jump higher than 0.5 m. They can easily step over a 0.5-m fence, but in doing so, their bodies rise only slightly. The height of the barrier is different from the height a jumper's "center of gravity" rises. Many people can leap over a 1-m fence, but only rarely does anybody raise the "center of gravity" of his or her body by 1 m. Even basketball star Kobe Bryant in a standing jump can't raise his body 1.25 m high, although he can easily reach considerably above the basket, which is more than 3 m high.

Jumping ability is best measured by a standing vertical jump. Stand facing a wall with feet flat on the floor and arms extended upward. Make a mark on the wall at the top of your reach. Then make your jump, and, at the point you are able to reach, make another mark. The distance between these two marks measures your vertical leap. If it's more than 0.6 m (2 ft), you're exceptional.

Here's the physics. When you leap upward, jumping force is applied only while your feet are still making contact with the ground. The greater the force, the greater your launch speed and the higher your jump. When your feet leave the ground, your upward speed immediately decreases at the steady rate of *g*, which is 10 m/s². At the top of your jump, your upward speed decreases to zero. Then you begin to fall, gaining speed at exactly the same rate, *g*. If you land as you took off, upright with legs extended, then your time rising equals your time falling; hang time is time up plus time down. While you are airborne, no amount of leg or arm pumping or other bodily motions can change your hang time.

As will be shown in Appendix A, the relationship between time up or down and vertical height is given by

$$d = \tfrac{1}{2}gt^2$$

If the vertical height *d* is known, we can rearrange this expression to read

$$t = \sqrt{\frac{2d}{g}}$$

Quite interestingly, no basketball player on record has exceeded 1.25 m in a vertical standing jump. For the corresponding hang time, let's use 1.25 m for *d*, and the more precise value of 9.8 m/s² for *g*. Solving for *t*, half the hang time (one way), we get

$$t = \sqrt{\frac{2d}{g}} = \sqrt{\frac{2(1.25)\ \text{m}}{9.8\ \text{m/s}^2}} = 0.50\ \text{s}$$

Double this amount (because this is the time for one direction of an up-and-down round trip) and we see that such record-breaking hang time is 1 s.

We're discussing vertical motion here. How about running jumps? We'll see in Chapter 4 that hang time depends only on the jumper's vertical speed at launch. While the jumper is airborne, his or her horizontal speed remains constant while the vertical speed undergoes acceleration. Intriguing physics!

SUMMARY OF TERMS (KNOWLEDGE)

Acceleration The rate at which velocity changes with time; the change in velocity may be in magnitude or direction or both, usually measured in m/s^2.

Air resistance The force of friction acting on an object due to its motion in air.

Equilibrium rule The vector sum of forces acting on a non-accelerating object equals zero: $\Sigma F = 0$.

Force Simply stated, a push or a pull.

Free fall Falling only under the influence of gravity—falling without air resistance.

Friction The resistive force that opposes the motion or attempted motion of an object past another with which it is in contact, or through a fluid.

Hang time The time that one's feet are off the ground during a vertical jump.

Inertia The property of things to resist changes in motion.

Kilogram The unit of mass; one kilogram (symbol kg) is the mass of 1 liter (L) of water at 4°C.

Mass The quantity of matter in an object. More specifically, the measure of the inertia or sluggishness that an object

exhibits in response to any effort made to start it, stop it, deflect it, or change in any way its state of motion.

Net force The combination of all forces that act on an object.

Newton The scientific unit of force.

Speed The distance traveled per time.

Support force The force that supports an object against gravity, often called the *normal force*.

Vector An arrow that represents the magnitude and direction of a quantity.

Vector quantity A quantity whose description requires both magnitude and direction.

Velocity The speed of an object and specification of its direction of motion.

Volume The quantity of space an object occupies.

Weight The force due to gravity on an object. More specifically, the force with which a body presses against a supporting surface.

READING CHECK QUESTIONS (COMPREHENSION)

*Each chapter in this book concludes with a set of questions and exercises, and for some chapters there are problems. The **Reading Check Questions** are designed to help you comprehend ideas and catch the essentials of the chapter material. You'll notice that answers to the questions can be found within the chapters. The **Activities** provide hands-on applications and can be done in or out of class. In Part 1 of this book are **Plug and Chug** problems, very simple one-step "plug-ins" to familiarize you with the formulas of the chapter. Only the most elementary "math" is involved with Plug and Chugs. **Think and Solve** problems, on the other hand, are standard "mathematical problems," some of which are challenging and go much further in applying math applications to chapter material. A new and insightful feature of this edition is the **Think and Rank** tasks, which involve analysis and ranking of the magnitudes of various quantities. The **Exercises** stress thinking rather than mere recall of information and call for a synthesis of the chapter material. In many cases the intention of particular exercises is to help you apply the ideas of physics to familiar situations. Each chapter concludes with **Discussion Questions**, which evaluate the chapter material. Unless you cover only a few chapters in your course, you will likely be expected to tackle only a few Think and Solves, Exercises, and Discussion Questions for each chapter. Every chapter concludes with a **Readiness Assurance Test**, a bank of 10 multiple-choice questions with answers at the bottom of the page.*

1.1 Aristotle on Motion

1. What did Aristotle believe about the relative speeds of fall for heavy and light objects?

2. Did Aristotle believe that forces are necessary to keep objects moving, or did he believe that, once moving, they'd move of themselves?

1.2 Galileo's Concept of Inertia

3. What idea of Aristotle did Galileo discredit with his inclined-plane experiments?

4. Which dominated Galileo's method of extending knowledge: philosophical discussion or experiment?

5. What name is given to the property by which objects resist changes in motion?

1.3 Mass—A Measure of Inertia

6. Which depends on location: weight or mass?

7. Where is your weight greater: on Earth or on the Moon? How about your mass?

8. What are the units of measurement for weight and for mass?

9. A 1-kg object weighs nearly 10 N on Earth. Would it weigh more or less on the Moon?

1.4 Net Force

10. What is the net force on a box pushed to the right with 50 N of force while being pushed to the left with 20 N of force?

11. What two quantities are necessary for a vector quantity?

1.5 The Equilibrium Rule

12. Name the force that occurs in a rope when both ends are pulled in opposite directions.
13. How much tension is there in a vertical rope that holds a 20-N bag of apples at rest?
14. What does $\Sigma F = 0$ mean?

1.6 Support Force

15. Why is the support force on an object often called the normal force?
16. When you weigh yourself, how does the support force of the scale acting on you compare with the gravitational force between you and Earth?

1.7 Dynamic Equilibrium

17. A bowling ball sits at rest and another bowling ball rolls down a lane at constant speed. Which ball, if either, is in equilibrium? Defend your answer.
18. If we push an object at constant velocity, how do we know how much friction acts on the object compared to our pushing force?

1.8 The Force of Friction

19. How does the direction of a friction force compare with the direction of the velocity of a sliding object?

20. If you push to the right on a heavy piece of furniture and it slides, what is the direction of friction on the furniture?
21. Suppose you push to the right on a heavy piece of furniture, but not hard enough to make it slide. Does a friction force act on the furniture?

1.9 Speed and Velocity

22. Distinguish between speed and velocity.
23. Why do we say that velocity is a vector and speed is not?
24. Does the speedometer on a vehicle show average speed or instantaneous speed?
25. How can you be both at rest and moving at 100,000 km/h at the same time?

1.10 Acceleration

26. Distinguish between velocity and acceleration.
27. What is the acceleration of an object that moves at constant velocity? What is the net force on the object in this case?
28. What is the acceleration of an object in free fall at Earth's surface?

ACTIVITIES (HANDS-ON APPLICATION)

29. Your grandparents are likely interested in your educational progress. Perhaps they have little science background and may be mathematically challenged. Write a letter to them, without using equations, and explain the difference between velocity and acceleration. Explain why some of your classmates confuse the two, and give some examples that clear up the confusion.
30. By any method you choose, determine your average walking speed. How do your results compare with those of your classmates?
31. Place a coin on top of a sheet of paper on a desk or table. Pull the paper horizontally with a quick snap. What concept of physics does this illustrate?

32. Place a file card on top of the mouth of a drinking glass. Place a coin over the center of the card. Snap the card horizontally so it flies off the glass. You'll see that the coin drops into the glass. Doesn't this show the same physics concept as in the preceding activity?
33. Stand flatfooted next to a wall and make a mark at the highest point you can reach. Then jump vertically and make another mark at the highest point. The distance between the marks is your vertical jumping distance. Use this distance to calculate your hang time.

PLUG AND CHUG (FORMULA FAMILIARIZATION)

These are "plug-in-the-number" tasks to familiarize you with the main formulas that link the physics concepts of this chapter. They are one-step substitutions, much less challenging than the Think and Solve problems that follow.

$$\text{Average speed} = \frac{\text{total distance covered}}{\text{travel time}}$$

34. Show that the average speed of a rabbit that runs a distance of 30 m in a time of 2 s is 15 m/s.

35. Calculate your average walking speed when you step 1.0 m in 0.5 s.

$$\text{Acceleration} = \frac{\text{change of velocity}}{\text{time interval}} = \frac{\Delta v}{\Delta t}$$

36. Show that the acceleration of a hamster is 5 m/s^2 when it increases its velocity from rest to 10 m/s in 2 s.
37. Show that the acceleration of a car that can go from rest to 100 km/h in 10 s is 10 km/h·s.

Free-fall distance from rest: $d = \frac{1}{2}gt^2$

38. Show that a freely falling rock drops a distance of 45 m when it falls from rest for 3 s.

39. How far will a freely falling object fall from rest in 5 s? In 10 s?

THINK AND SOLVE (MATHEMATICAL APPLICATION)

40. Find the net force produced by a 30-N force and a 20-N force in each of these cases:
    (a) Both forces act in the same direction.
    (b) The two forces act in opposite directions.

41. Lucy Lightfoot stands with one foot on one bathroom scale and her other foot on a second bathroom scale. Each scale reads 350 N. What is Lucy's weight?

42. Henry Heavyweight weighs 1200 N and stands on a pair of bathroom scales so that one scale reads twice as much as the other. What are the scale readings?

43. The sketch shows a painter's scaffold in mechanical equilibrium. The person in the middle weighs 500 N, and the tension in each rope is 400 N. What is the weight of the scaffold?

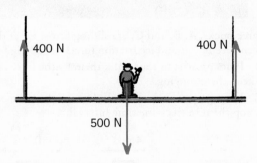

44. A different scaffold that weighs 400 N supports two painters, one 500 N and the other 400 N. The reading in the left scale is 800 N. What is the reading in the right-hand scale?

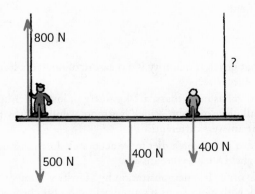

45. A horizontal force of 120 N is required to push a bookcase across a floor at a constant velocity.
    (a) What is the net force acting on the bookcase?
    (b) How much is the friction force that acts on the sliding bookcase?
    (c) How much friction acts on the bookcase when it is at rest on a horizontal surface without being pushed?

46. Reckless Rick driving along the road at 90 km/h bumps into Hapless Harry directly in front of him who is driving at 88 km/h. What is the speed of the collision?

47. An airplane with an airspeed of 60 km/h lands on a runway where the wind speed is 40 km/h.
    (a) What is the landing speed of the plane if the wind is head-on?
    (b) What is the landing speed if the wind is a tailwind, coming from behind the plane?
    (c) What would be the landing speed of the plane in a headwind of 60 km/h?

48. (a) Show that the average speed of a tennis ball is 48 m/s when it travels the full length of the court, 24 m, in 0.5 s.
    (b) How would air resistance affect the travel time?

49. (a) Show that the average speed of Leslie is 10 km/h when she runs to the store 5 km away in 30 min.
    (b) How fast is this in m/s?

50. (a) Show that the acceleration is 7.5 m/s² for a ball that starts from rest and rolls down a ramp and gains a speed of 30 m/s in 4 s.
    (b) Would acceleration be greater or less if the ramp were a bit less steep?

51. Extend Table 1.2 (which gives values from 0 to 5 s) to 6 to 10 s, assuming no air resistance.

52. Lillian rides her bicycle along a straight road at average velocity v.
    (a) Write an equation showing the distance she travels in time t.
    (b) If Lillian's average speed is 7.5 m/s for a time of 5.0 min, show that she travels a distance of 2250 m.

53. A car races on a circular track of radius r.
    (a) Write an equation for the car's average speed when it travels a complete lap in time t.
    (b) The radius of the track is 100 m and the time to complete a lap is 14 s. Show that the average speed around the track is 45 m/s.

54. A ball is thrown straight up with an initial speed of 30 m/s.
    (a) How much time does it take for the ball to reach the top of its trajectory?
    (b) Show that the ball will reach a height of 45 m (neglecting air resistance).

55. A ball is thrown straight up with enough speed so that it is in the air for several seconds.
    (a) What is the velocity of the ball when it reaches its highest point?
    (b) What is its velocity 1 s before it reaches its highest point?

(c)   What is the change in its velocity, Δv, during this 1-s interval?

(d)   What is its velocity 1 s after it reaches its highest point?

(e)   What is the change in velocity, Δv, during this 1-s interval?

(f)   What is the change in velocity, Δv, during the 2-s interval from 1 s before the highest point to 1 s after the highest point? (Caution: We are talking about velocity, not speed.)

(g)   What is the acceleration of the ball during any of these time intervals and at the moment the ball has zero velocity?

56. A school bus slows to a stop with an average acceleration of -2.0 m/s². Show that it takes 5.0 s for the bus to slow from 10.0 m/s to a stop.

57. An airplane starting from rest at one end of a runway accelerates uniformly at 4.0 m/s² for 15 s before takeoff.

(a)   What is its speed of takeoff?

(b)   Show that the plane travels along the runway a distance of 450 m before takeoff.

THINK AND RANK (ANALYSIS)

58. The weights of Burl, Paul, and the scaffold produce tensions in the supporting ropes. Rank the tension in the *left* rope, from greatest to least, in the three situations A, B, and C.

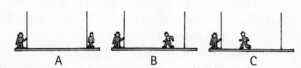

59. Rank the net force on the block from greatest to least in the four situations A, B, C, and D.

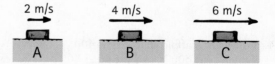

60. Different materials, A, B, C, and D, rest on a table.

(a)   From greatest to least, rank them by how much they resist being set in motion.

(b)   From greatest to least, rank them by the support (normal) force the table exerts on them.

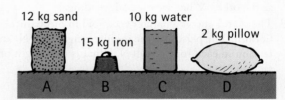

61. Three pucks, A, B, and C, are sliding across ice at the given speeds. Air and ice friction forces are negligible.

(a)   From greatest to least, rank them by the force needed to keep them moving.

(b)   From greatest to least, rank them by the force needed to stop them in the same time interval.

EXERCISES (SYNTHESIS)

62. Knowledge can be gained by philosophical logic and also by experimentation. Which of these did Aristotle favor, and which did Galileo favor?

63. Which of Aristotle's ideas did Galileo discredit in his fabled Leaning Tower of Pisa experiment? With his inclined-plane experiments?

64. A bowling ball rolling along a lane gradually slows as it rolls. How would Aristotle likely interpret this observation? How would Galileo interpret it?

65. A space probe is carried by a rocket into outer space. A friend wonders what keeps the probe moving after the rocket no longer pushes it. What do you say?

66. When a ball rolls down an inclined plane, it gains speed because of gravity. When a ball rolls up an inclined plane, it loses speed because of gravity. Why doesn't gravity play a role when a ball rolls on a horizontal surface?

67. What physical quantity is a measure of how much inertia an object has?

68. Which has more mass: a 2-kg fluffy pillow or a 3-kg small piece of iron? Which has more volume? Why are your answers different?

69. Does a person on a diet more accurately lose mass or lose weight? Defend your answer.

70. A favorite class demonstration by Hewitt is lying on his back with a blacksmith's anvil placed on his chest. When an assistant whacks the anvil with a strong sledge-hammer blow, Hewitt is not injured. How is the physics here similar to that illustrated in Figure 1.8?

71. What is your own mass in kilograms? Your weight in newtons?

72. Gravitational force on the Moon is only $\frac{1}{6}$ that of the gravitational force on Earth. What would be the weight of a 10-kg object on the Moon and on Earth? What would be its mass on the Moon and on Earth?

73. A monkey hangs stationary at the end of a vertical vine. What two forces act on the monkey? Which force, if either, is greater?

74. Can an object be in mechanical equilibrium when only a single force acts on it? Defend your answer.

75. When you push downward on a book at rest on a table, you feel an upward force. Does this force depend on friction? Defend your answer.

76. Nellie Newton hangs at rest from the ends of the rope as shown. How does the reading on the scale compare with her weight?

77. A hockey puck slides across the ice at a constant velocity. Is the sliding puck in equilibrium? Why or why not?

78. If you push horizontally on a carton that contains your new kitchen appliance and it slides across the floor, slightly gaining speed, how does the friction acting on the carton compare with your push?

79. In order to slide a heavy cabinet across the floor at constant speed, you exert a horizontal force of 550 N. Is the force of friction between the cabinet and the floor greater than, less than, or equal to 550 N? Defend your answer.

80. Consider your desk at rest on a your bedroom floor.

    (a) As you and your friend start to lift it, does the support force on the desk provided by the floor increase, decrease, or remain unchanged?

    (b) What happens to the support force on the feet of you and your friend?

81. An empty jug of weight W is at rest on a table. What is the support force exerted on the jug by the table? What is the support force when water of weight w is poured into the jug?

82. What is the acceleration of a mouse that moves across a floor at a constant velocity of 2 m/s?

83. What is the impact speed when a car moving at 100 km/h collides with the rear of another car traveling in the same direction at 98 km/h? When they collide head-on?

84. You're in a car traveling at some specified speed limit. You see another car moving at the same speed toward you. How fast is the car approaching you, compared with the speed limit?

85. Emily Easygo can paddle a canoe in still water at 8 km/h. How successful will she be at canoeing upstream in a river that flows at 8 km/h?

86. Suppose that an object in free fall were somehow equipped with a speedometer. By how much would its speed readings increase with each second of fall?

87. Suppose that the freely falling object in the preceding exercise falls from a rest position and is equipped with an odometer. What equation is most appropriate for determining the distance fallen each second? Do the readings indicate equal or unequal distances of fall for successive seconds? Explain.

88. In the absence of air resistance, a ballplayer tosses a ball straight up.

    (a) By how much does the speed of the ball decrease each second while it is ascending?

    (b) By how much does its speed increase each second while it is descending?

    (c) How does the time of ascent compare with the time of descent?

89. Gracie says acceleration is how fast you go. Alex says acceleration is how fast you get fast. They look to you for confirmation. Who's correct?

90. What is the acceleration of a car that moves at a steady velocity of 100 km/h for 100 s? Why is this question an exercise in careful reading as well as in physics?

91. For a freely falling object dropped from rest, what is its acceleration at the end of 5 s? At the end of 10 s? Defend your answers (and distinguish between velocity and acceleration).

92. Correct your friend who says, "The proposed California Suntrain can easily round a curve at a constant velocity of 160 km/h."

DISCUSSION QUESTIONS (EVALUATION)

93. Asteroids have been moving through space for billions of years. A friend says that initial forces from long ago keep them moving. Do you agree with your friend?

94. In answer to the question "What keeps Earth moving around the Sun?" a friend asserts that inertia keeps it moving. Correct your friend's erroneous answer.

95. Consider a ball at rest in the middle of a toy wagon. When the wagon is pulled forward, the ball rolls against the back of the wagon. A friend asks what force pushes the ball to the back of the wagon. Interpret this observation in terms of inertia.

96. In tearing a paper towel or plastic bag from a roll, discuss why a sharp jerk is more effective than a slow pull.

97. If you're in a car at rest that gets hit from behind, you can suffer a serious neck injury called whiplash. Discuss how whiplash involves the concept of inertia and why cars are equipped with headrests.

98. Why do you lurch forward in a bus that suddenly slows? Why do you lurch backward when it picks up speed? What law applies here?

99. Suppose that you're in a moving car and the engine stops running. You step on the brakes and slow the car to half speed. If you release your foot from the brakes, will the car spontaneously speed up a bit, or will it continue at half speed and slow due to friction? Defend your answer.

100. Harry the painter swings year after year from his bosun's chair. His weight is 500 N, and the rope, unknown to him, has a breaking point of 300 N. Why doesn't the rope break when he is supported as shown on the left? One day, Harry was painting near a flagpole, and, for a change, he tied the free end of the rope to the flagpole instead of to his chair, as shown on the right. Discuss with your friends why Harry took his vacation early.

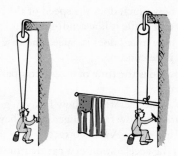

101. Place a heavy book on a table and the table pushes up on the book. A friend reasons that the table can't push upward on the book because if it did, the book would rise above the table. What do you say to your friend? Why doesn't this upward push cause the book to rise from the table?

102. Someone standing at the edge of a cliff (as in Figure 1.24) throws a ball straight up at a certain speed and another ball straight down with the same initial speed. If air resistance is negligible, predict which ball will hit the ground below with greater speed, or whether they will hit at the same speed.

103. When a ball is tossed straight up, it momentarily comes to a stop at the top of its path. Is it in equilibrium during this brief moment? Why or why not?

104. Because Earth rotates once every 24 hours, the west wall in your room moves in a direction toward you at a linear speed that is probably more than 1000 km/h (the exact speed depends on your latitude). When you stand facing the wall, you are carried along at the same speed, so you don't notice it. But when you jump upward, with your feet no longer in contact with the floor, why doesn't the high-speed wall slam into you?

105. If you toss a coin straight upward while riding in a train that travels at uniform and steady motion along a straight-line track, where does the coin land? How about when the train slows while the coin is tossed? When the train rounds a curve?

106. Two balls, A and B, are released simultaneously from rest at the left end of equal-length tracks, as shown. Which ball, A or B, will reach the end of its track first?

107. Refer to the preceding question.

(a) Does ball B roll faster along the lower part of its track than ball A rolls along its straighter track?

(b) Is the speed gained by ball B going down the extra dip the same as the speed it loses going up near the right-hand end—and doesn't this mean that the speeds of balls A and B will be the same at the ends of both tracks?

(c) For ball B, won't the average speed dipping down and up be greater than the average speed of ball A during the same time?

(d) So, overall, does ball A or ball B have the greater average speed? (Do you wish to change your answer to the preceding question?)

> Remember, reading check questions provide you with a self-check of whether or not you grasp the central ideas of the chapter. The exercises, rankings, and problems are extra "pushups" for you to try after you have at least a fair understanding of the chapter and can handle the reading check questions.

READINESS ASSURANCE TEST (RAT)

If you have a good handle on this chapter, if you really do, then you should be able to score 7 out of 10 on this RAT. If you score less than 7, you need to study further before moving on.

Choose the BEST answer to each of the following.

1. Science greatly advanced when Galileo favored
   (a) philosophical discussions over experiment.
   (b) experiment over philosophical discussions.
   (c) nonmathematical thinking.
   (d) imagination.

2. According to Galileo, inertia is a
   (a) force like any other force.
   (b) special kind of force.
   (c) property of all matter.
   (d) concept opposite to force.

3. If gravity between the Sun and Earth suddenly vanished, Earth would continue moving in
   (a) a curved path.
   (b) an outward spiral path.
   (c) an inward spiral path.
   (d) a straight-line path.

4. When a 10-kg block is simultaneously pushed eastward with a force of 20 N and westward with a force of 15 N, the combination of these forces on the block is
   (a) 35 N west.
   (b) 35 N east.
   (c) 5 N east.
   (d) 5 N west.

5. The equilibrium rule, $\Sigma F = 0$, applies to
   (a) objects or systems at rest.
   (b) objects or systems in uniform motion in a straight line.
   (c) both of these
   (d) neither of these

6. When you stand on two bathroom scales, one foot on each scale with weight evenly distributed, each scale will read
   (a) your weight.
   (b) half your weight.
   (c) zero.
   (d) actually more than your weight.

7. Your average speed in skateboarding to your friend's house is 5 m/s. It is possible that your instantaneous speed at some point was
   (a) less than 5 m/s.
   (b) 5 m/s.
   (c) more than 5 m/s.
   (d) any of these

8. During each second of free fall, the speed of an object
   (a) increases by the same amount.
   (b) changes by increasing amounts.
   (c) remains constant.
   (d) doubles.

9. If a falling object gains 10 m/s each second it falls, its acceleration is
   (a) 10 m/s.
   (b) 10 m/s per second.
   (c) both of these
   (d) neither of these

10. A freely falling object has a speed of 30 m/s at one instant. Exactly 1 s later its speed will be
    (a) the same.
    (b) 35 m/s.
    (c) more than 35 m/s.
    (d) 60 m/s.

Answers to RAT

1. c, 2. c, 3. d, 4. c, 5. c, 6. b, 7. d, 8. a, 9. b, 10. c

2 CHAPTER 2
Newton's Laws of Motion

GALILEO'S WORK set the stage for Isaac Newton, who was born shortly after Galileo's death in 1642. By the time Newton was 23, he had developed his famous three laws of motion that completed the overthrow of Aristotelian physics. These three laws first appeared in one of the most famous books of all time, Newton's *Philosophiae Naturalis Principia Mathematica*,* often simply known as the *Principia*. The first law is a restatement of Galileo's concept of inertia; the second law relates acceleration to its cause—force; and the third is the law of action and reaction.

Newton's three laws of motion are the foundation of present-day mechanics. It was Newton's laws that got humans to the Moon.

* The Latin title means "Mathematical Principles of Natural Philosophy." See Newton's biography on page 53.

2.1 Newton's First Law of Motion

EXPLAIN THIS Why isn't inertia a kind of force?

LEARNING OBJECTIVE
State Newton's first law of motion, and relate it to inertia.

Newton's first law of motion, usually called the *law of inertia*, is a restatement of Galileo's idea.

Every object continues in a state of rest or of uniform speed in a straight line unless acted on by a nonzero force.

The key word in this law is *continues*: an object *continues* to do whatever it happens to be doing unless a force is exerted upon it. If the object is at rest, it *continues* in a state of rest. This is nicely demonstrated when a tablecloth is skillfully whipped from beneath dishes sitting on a tabletop, leaving the dishes in their initial state of rest.* On the other hand, if an object is moving, it *continues* to move without changing its speed or direction, as evidenced by space probes that continually move in outer space. This property of objects to resist changes in motion is called **inertia**.

You can think of *inertia* as another word for "laziness" (or resistance to change).

FIGURE 2.1
Inertia in action.

CHECKPOINT

When a space shuttle travels in a nearly circular orbit around Earth, is a force required to maintain its high speed? If suddenly the force of gravity were cut off, what type of path would the shuttle follow?

Was this your answer?
No force in the direction of the shuttle's motion exists. The shuttle "coasts" by its own inertia. The only force acting on it is the force of gravity, which acts at right angles to its motion (toward Earth's center). We'll see later that this right-angled force holds the shuttle in a circular path. If it were cut off, the shuttle would move in a straight-line path at constant speed (constant velocity).

Inertia isn't a kind of force; it's a property of all matter to resist changes in motion.

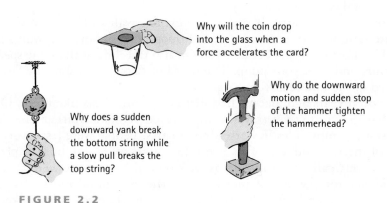

Why will the coin drop into the glass when a force accelerates the card?

Why does a sudden downward yank break the bottom string while a slow pull breaks the top string?

Why do the downward motion and sudden stop of the hammer tighten the hammerhead?

FIGURE 2.2
Examples of inertia.

FIGURE 2.3
Rapid deceleration is sensed by the driver, who lurches forward—inertia in action!

---

* Close inspection shows that brief friction between the dishes and the fast-moving tablecloth starts the dishes moving, but then friction between the dishes and table stops the dishes before they slide very far. If you try this, use unbreakable dishes!

Nicolaus Copernicus (1473–1543)

FIGURE 2.4
Can the bird drop down and catch the worm if Earth moves at 30 km/s?

FIGURE 2.5
When you flip a coin in a high-speed airplane, it behaves as if the airplane were at rest. The coin keeps up with you—inertia in action!

The Moving Earth

As mentioned in the Prologue, the 16th-century Polish astronomer Copernicus caused great controversy when he published a book proposing that Earth revolves around the Sun.* This idea conflicted with the popular view that Earth was the center of the universe. Copernicus's concept of a Sun-centered solar system was the result of years of studying the motion of the planets. He had kept his theory from the public—for two reasons. The first reason was that he feared persecution; a theory so completely different from common opinion would surely be taken as an attack on established order. The second reason was reservations about it himself; he could not reconcile the idea of a moving Earth with the prevailing ideas of motion. The concept of inertia was unknown to him and others of his time. In the final days of his life, at the urging of close friends, he sent his manuscript, *De Revolutionibus Orbium Coelestium,*** to the printer. The first copy of his famous exposition reached him on the day he died—May 24, 1543.

The idea of a moving Earth was much debated. Europeans thought like Aristotle, and the existence of a force big enough to keep Earth moving was beyond their imagination. They had no idea of the concept of inertia. One of the arguments against a moving Earth was the following: Consider a bird sitting at rest on a branch of a tall tree. On the ground below is a fat, juicy worm. The bird sees the worm and drops vertically below and catches it. It was argued that this would be impossible if Earth were moving. A moving Earth would have to travel at an enormous speed to circle the Sun in one year. While the bird would be in the air descending from its branch to the ground below, the worm would be swept far away along with the moving Earth. It seemed that catching a worm on a moving Earth would be an impossible task. The fact that birds do catch worms from tree branches seemed to be clear evidence that Earth must be at rest.

Can you see the error in this argument? The concept of inertia is missing. You see, not only is Earth moving at a great speed, but so are the tree, the branch of the tree, the bird that sits on it, the worm below, and even the air in between. Things in motion remain in motion if no unbalanced forces are acting upon them. So when the bird drops from the branch, its initial sideways motion remains unchanged. It catches the worm quite unaffected by the motion of its total environment.

We live on a moving Earth. If you stand next to a wall and jump up so that your feet are no longer in contact with the floor, does the moving wall slam into you? Why not? It doesn't because you are also traveling at the same speed, before, during, and after your jump. The speed of Earth relative to the Sun is not the speed of the wall relative to you.

Four hundred years ago, people had difficulty with ideas like these. One reason is that they didn't yet travel in high-speed vehicles. Rather, they experienced slow, bumpy rides in horse-drawn carts. People were less aware of the effects of inertia. Today we flip a coin in a high-speed car, bus, or plane and catch the vertically moving coin as we would if the vehicle were at rest. We see evidence for the law of inertia when the horizontal motion of the coin before, during, and after the catch is the same. The coin always keeps up with us.

---

* Copernicus was certainly not the first to think of a Sun-centered solar system. In the fifth century, for example, the Indian astronomer Aryabhata taught that Earth circles the Sun, not the other way around (as the rest of the world believed).
** The Latin title means "On the Revolutions of Heavenly Spheres."

2.2 Newton's Second Law of Motion

EXPLAIN THIS What happens to a car's pickup when you increase your push on it?

Isaac Newton was the first to realize the connection between force and mass in producing acceleration, which is one of the most central rules of nature. He expressed it in his *second law of motion*. **Newton's second law of motion** is:

> **The acceleration produced by a net force on an object is directly proportional to the net force, is in the same direction as the net force, and is inversely proportional to the mass of the object.**

Or, in shorter notation,

$$\text{Acceleration} \sim \frac{\text{net force}}{\text{mass}}$$

By using consistent units such as newtons (N) for force, kilograms (kg) for mass, and meters per second squared (m/s^2) for acceleration, we produce the exact equation:

$$\text{Acceleration} = \frac{\text{net force}}{\text{mass}}$$

In briefest form, where a is acceleration, F is net force, and m is mass:

$$a = \frac{F}{m}$$

Acceleration equals the net force divided by the mass. If the net force acting on an object is doubled, the object's acceleration will be doubled. Suppose instead that the mass is doubled. Then the acceleration will be halved. If both the net force and the mass are doubled, then the acceleration will be unchanged. (These relations are nicely developed in the *Conceptual Physical Science Practice Book*.)

FIGURE 2.6
INTERACTIVE FIGURE MP

Acceleration depends on both the amount of push and the mass being pushed.

Mastering**PHYSICS**

TUTORIAL: Parachuting and Newton's Second Law
VIDEO: Newton's Second Law
VIDEO: Force Causes Acceleration
VIDEO: Friction
VIDEO: Falling and Air Resistance

Force of hand accelerates the brick

Twice as much force produces twice as much acceleration

Twice the force on twice the mass gives the same acceleration

FIGURE 2.7
Acceleration is directly proportional to force.

Force of hand accelerates the brick

The same force accelerates 2 bricks 1/2 as much

3 bricks, 1/3 as much acceleration

FIGURE 2.8
Acceleration is inversely proportional to mass.

When one thing is inversely proportional to another, as one gets bigger, the other gets smaller.

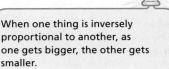

Force *changes* motion, it doesn't *cause* motion.

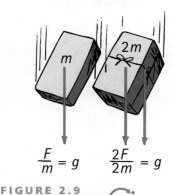

$$\frac{F}{m} = g \qquad \frac{2F}{2m} = g$$

FIGURE 2.9

INTERACTIVE FIGURE (MP)

The ratio of weight (*F*) to mass (*m*) is the same for all objects in the same locality; hence, their accelerations are the same in the absence of air resistance.

When Galileo tried to explain why all objects fall with equal accelerations, wouldn't he have loved to know the rule *a* = *F*/*m*?

CHECKPOINT

1. In the previous chapter we defined acceleration to be the time rate of change of velocity; that is, *a* = (change in *v*)/time. Are we now saying that acceleration is instead the ratio of force to mass—that is, *a* = *F*/*m*? Which is it?
2. A jumbo jet cruises at constant velocity of 1000 km/h when the thrusting force of its engines is a constant 100,000 N. What is the acceleration of the jet? What is the force of air resistance on the jet?
3. Suppose you apply the same amount of force to two separate carts, one cart with a mass of 1 kg and the other with a mass of 2 kg. Which cart will accelerate more, and how much greater will the acceleration be?

Were these your answers?

1. Both are correct. Acceleration is *defined* as the time rate of change of velocity and is *produced by* a force. How much force/mass (usually the cause) determines the rate change in velocity/time (usually the effect). So we must first define acceleration and then define the terms that produce acceleration.
2. The acceleration is zero, as evidenced by the constant velocity. Because the acceleration is zero, it follows from Newton's second law that the net force is zero, which means that the force of air resistance must just equal the thrusting force of 100,000 N and act in the opposite direction. So the air resistance on the jet is 100,000 N. This is in accord with Σ*F* = 0. (Note that we don't need to know the velocity of the jet to answer this question, but only that it is constant—our clue that acceleration, and therefore net force, is zero.)
3. The 1-kg cart will have more acceleration—twice as much, in fact—because it has half as much mass, which means half as much resistance to a change in motion.

When Acceleration Is *g*—Free Fall

Although Galileo founded the concepts of both inertia and acceleration and was the first to measure the acceleration of falling objects, he was unable to explain why objects of various masses fall with equal accelerations. Newton's second law provides the explanation.

We know that a falling object accelerates toward Earth because of the gravitational force of attraction between the object and Earth. As mentioned earlier, when the force of gravity is the only force—that is, when air resistance is negligible—we say that the object is in a state of **free fall**. An object in free fall accelerates toward Earth at 10 m/s² (or, more precisely, at 9.8 m/s²).

The greater the mass of an object, the stronger is the gravitational pull between it and Earth. The double brick in Figure 2.9 for example, has twice the gravitational attraction of the single brick. Why, then, doesn't the double brick fall twice as fast (as Aristotle supposed it would)? The answer is evident in Newton's second law: the acceleration of an object depends not only on the force (weight, in this case), but on the object's resistance to motion—its inertia. Whereas a force produces an acceleration, inertia is a *resistance* to acceleration. So twice the force exerted on twice the inertia produces the same acceleration as half the force exerted on half the inertia. Both accelerate equally. The acceleration due to gravity is symbolized by *g*. We use the symbol *g*, rather than *a*, to denote that acceleration is due to gravity alone.

FIGURING PHYSICAL SCIENCE

Problem Solving

If we know the mass of an object in kilograms (kg) and its acceleration in meters per second per second (m/s²), then the force will be expressed in newtons (N). One newton is the force needed to give a mass of 1 kg an acceleration of 1 m/s². We can arrange Newton's second law to read

Force = mass × acceleration
$$1 \text{ N} = (1 \text{ kg}) \times (1 \text{ m/s}^2)$$

We can see that

$$1 \text{ N} = 1 \text{ kg} \cdot \text{m/s}^2$$

The dot between kg and m/s² means that the units are multiplied.

If we know two of the quantities in Newton's second law, we can calculate the third.

SAMPLE PROBLEM 1

How much force, or thrust, must a 20,000-kg jet plane develop to achieve an acceleration of 1.5 m/s²?

Solution:
Using the equation

Force = mass × acceleration

we can calculate the force:

$$F = ma$$
$$= (20,000 \text{ kg}) \times (1.5 \text{ m/s}^2)$$
$$= 30,000 \text{ kg} \cdot \text{m/s}^2$$
$$= 30,000 \text{ N}$$

Suppose we know the force and the mass, and we want to find the acceleration. For example, what acceleration is produced by a force of 2000 N applied to a 1000-kg automobile? Using Newton's second law, we find that

$$a = \frac{F}{m} = \frac{2000 \text{ N}}{1000 \text{ kg}}$$
$$= \frac{2000 \text{ kg} \cdot \text{m/s}^2}{1000 \text{ kg}} = 2 \text{ m/s}^2$$

If the force is 4000 N, the acceleration is

$$a = \frac{F}{m} = \frac{4000 \text{ N}}{1000 \text{ kg}}$$
$$= \frac{4000 \text{ kg} \cdot \text{m/s}^2}{1000 \text{ kg}} = 4 \text{ m/s}^2$$

Doubling the force on the same mass simply doubles the acceleration.

Physics problems are typically more complicated than these.

SAMPLE PROBLEM 2

Here is a more conceptual problem. It is conceptual because it deals not in numbers, but in concepts directly. The focus is showing symbols for concepts, rather than their numerical values. In the next sample problem, force is *F*, mass is *m*, and acceleration is *a*. This way you build a habit of first thinking in terms of concepts and the symbols that represent them. Part (b) follows up and brings in the numbers after you've done the physics.

A force *F* acts in the forward direction on a carton of chocolates of mass *m*. A friction force *f* opposes this motion. **(a)** Use Newton's second law and show that the acceleration of the carton is

$$\frac{F - f}{m}$$

(b) If the carton's mass is 4.0 kg, the applied force is 12.0 N, and the friction force is 6.0 N, show that the carton's acceleration is 1.5 m/s².

Solution:
(a) We're asked to find the acceleration. From Newton's second law we know that $a = (F_{net})/m$. Here the net force is $F - f$. So the solution is $a = (F - f)/m$ (where all quantities represented are known values). Notice that this answer applies to all situations in which a steady applied force is opposed by a steady frictional force. It covers many possibilities.
(b) Here we simply substitute the numerical values given:

$$a = \frac{F - f}{m} = \frac{12.0 \text{ N} - 6.0 \text{ N}}{4.0 \text{ kg}}$$
$$= 1.5 \frac{\text{N}}{\text{kg}} = 1.5 \text{ m/s}^2$$

(The units N/kg are equivalent to m/s².) Note that the answer, about 15% of *g*, is "reasonable." For more on units of measurement and significant figures, see your Lab Manual.

The ratio of weight to mass for freely falling objects equals the constant *g*. This is similar to the constant ratio of circumference to diameter for circles, which equals the constant *π*. The ratio of weight to mass is identical for both heavy and light objects, just as the ratio of circumference to diameter is the same for both large and small circles (Figure 2.10).

We now understand that the acceleration of free fall is independent of an object's mass. A boulder 100 times as massive as a pebble falls at the same acceleration as the pebble because although the force on the boulder (its weight) is 100 times the force (or weight) on the pebble, its resistance to a change in motion (mass) is 100 times that of the pebble. The greater force offsets the correspondingly greater mass.

Ironically, Galileo couldn't say *why* all bodies fall equally because he never connected the concepts he developed—*acceleration* and *inertia*—with *force*. That connection awaited Newton's second law.

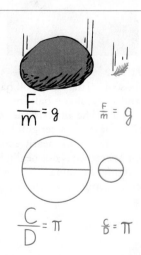

FIGURE 2.10

FIGURE 2.11
Wingsuit fliers nicely mimic the physics that flying squirrels have always enjoyed.

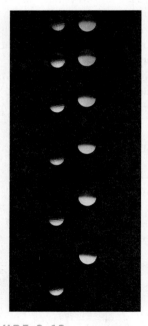

FIGURE 2.12
A stroboscopic study of a golf ball (left) and a Styrofoam ball (right) falling in air. The air resistance is negligible for the heavier golf ball, and its acceleration is nearly equal to *g*. Air resistance is not negligible for the lighter Styrofoam ball, which reaches its terminal velocity sooner.

CHECKPOINT

In a vacuum, a coin and a feather fall equally, side by side. Would it be correct to say that *equal forces of gravity* act on both the coin and the feather in a vacuum?

Was this your answer?
No, no, no—a thousand times no! These objects accelerate equally not because the forces of gravity on them are equal, but because the *ratios* of their weights to masses are equal. Although air resistance is not present in a vacuum, gravity is. (You'd know this if you placed your hand into a vacuum chamber and a cement truck rolled over it!) If you answered yes to this question, let this be a signal to be more careful when you think physics!

When Acceleration of Fall Is Less Than *g*—Non–Free Fall

Most often, air resistance is not negligible for falling objects. Then the acceleration of fall is less. Air resistance depends primarily on two things: speed and frontal area. When a skydiver steps from a high-flying plane, the air resistance on the skydiver's body builds up as the falling speed increases. The result is reduced acceleration. The acceleration can be reduced further by increasing frontal area. A diver does this by orienting his or her body so more air is encountered—by spreading out like a flying squirrel. So air resistance depends on speed and the frontal area encountered by the air.

For free fall, the downward net force is weight—only weight. But when air is present, the downward net force = weight − air resistance. Can you see that the presence of air resistance reduces net force? And that less net force means less acceleration? So as a diver falls faster and faster, the acceleration of fall becomes less and less.* What happens to the net force if air resistance builds up to equal weight? The answer is that net force becomes zero. Here we see $\Sigma F = 0$ again! Then acceleration becomes zero. Does this mean the diver comes to a stop? No! What it means is that the diver no longer gains speed. Acceleration terminates—it no longer occurs. We say the diver has reached **terminal speed**. If we are concerned with direction—down, for falling objects—we say the diver has reached **terminal velocity**.

Terminal speed for a human skydiver varies from about 150 to 200 km/h, depending on weight, size, and orientation of the body. A heavier person has to fall faster for air resistance to balance weight.** The greater weight is more effective in "plowing through" air, resulting in a higher terminal speed for a heavier person. Increasing frontal area reduces terminal speed.

Terminal speeds are reduced when a skydiver wears a wingsuit (Figure 2.11). The wingsuit not only increases a diver's frontal area but also provides a lift similar to that achieved by flying squirrels when they fashion their bodies into "wings." This exhilarating sport, *wingsuit flying*, goes beyond what flying squirrels can accomplish, since a wingsuit flyer can achieve horizontal speeds

---

* In mathematical notation,

$$a = \frac{F_{\text{net}}}{m} = \frac{mg - R}{m}$$

where *mg* is the weight and *R* is the air resistance. Note that when $R = mg$, $a = 0$; then, with no acceleration, the object falls at constant velocity.
** A skydiver's air resistance is proportional to speed squared.

appreciably greater than 170 km/h (100 mph). Looking more like flying bullets than flying squirrels, high-performance wingsuits allow these "bird people" to glide with remarkable precision. To land safely, parachutes are deployed. The large frontal area provided by a parachute produces low speeds (15–25 km/h) for safe landings. Projects for wingsuit flyers to land without a parachute, however, are under way.

CHECKPOINT

A skydiver jumps from a high-flying helicopter. As she falls faster and faster through the air, does her acceleration increase, decrease, or remain the same?

Was this your answer?
Acceleration decreases because the net force on the skydiver decreases. Net force is equal to her weight minus her air resistance, and because air resistance increases with increasing speed, net force and hence acceleration decrease. By Newton's second law,

$$a = \frac{F_{net}}{m} = \frac{mg - R}{m}$$

where mg is her weight and R is the air resistance she encounters. As R increases, both F_{net} and a decrease. Note that if she falls fast enough so that $R = mg$, $a = 0$, so with no acceleration she falls at constant speed.

Consider the interesting demonstration of the falling coin and feather in the glass tube (Figure 2.13). When air is inside, we see that the feather falls more slowly due to air resistance. The feather's weight is very small, so it reaches terminal speed very quickly. Can you see that it doesn't have to fall very far or fast before air resistance builds up to equal its small weight? The coin, on the other hand, doesn't have enough time to fall fast enough for air resistance to build up to equal its weight.

FIGURE 2.13
In a vacuum, a feather and a coin fall at an equal acceleration. When air is present the feather falls much slower, with no acceleration.

CHECKPOINT

Consider two parachutists, a heavy person and a light person, who jump from the same altitude with parachutes of the same size.
1. **Which person reaches terminal speed first?**
2. **Which person has the greater terminal speed?**
3. **Which person reaches the ground first?**
4. **If there were no air resistance, as on the Moon, how would your answers to these questions differ?**

Were these your answers?
To answer these questions, think of a coin and a feather falling in air.
1. Just as a feather reaches terminal speed very quickly, the lighter person reaches terminal speed first.
2. Just as a coin falls faster than a feather through air, the heavier person falls faster and reaches a higher terminal speed.
3. Just like the race between a falling coin and feather, the heavier person falls faster and reaches the ground first.
4. If there were no air resistance there would be no terminal speed at all. Both would be in free fall and hit the ground at the same time.

When Galileo allegedly dropped objects of different weights from the Leaning Tower of Pisa, they didn't actually hit at the same time. They almost did, but because of air resistance, the heavier one hit a split second before the other. But this contradicted the much longer time difference expected by the followers of Aristotle. The behavior of falling objects was never really understood until Newton announced his second law of motion.

2.3 Forces and Interactions

EXPLAIN THIS When you push, what pushes back?

So far, we've treated force in its simplest sense—as a push or pull. In a broader sense, a force is not a thing in itself but is part of an **interaction** between one thing and another. If you push on a wall with your fingers, more is happening than you pushing on the wall. You're interacting with the wall, and the wall is also pushing on you. The fact that your fingers and the wall push on each other is evident in your bent fingers (Figure 2.14). These two forces are equal in magnitude (amount) and opposite in direction. This **force pair** constitutes a single interaction. In fact, you can't push on the wall unless the wall pushes back. A pair of forces is involved: your push on the wall and the wall's push back on you.*

In Figure 2.15, we see a boxer's fist hitting a massive punching bag. The fist hits the bag (and dents it) while the bag hits back on the fist (and stops its motion). This force pair is fairly large. But what if the boxer were hitting a piece of tissue paper? The boxer's fist can exert only as much force on the tissue paper as the tissue paper can exert on the boxer's fist. The fist can't exert any force at all unless what is being hit exerts the same amount of reaction force. An interaction requires a *pair* of forces acting on *two* objects.

When a hammer hits a stake and drives it into the ground, the stake exerts an equal amount of force on the hammer that brings it to an abrupt halt. And when you pull on a cart and it accelerates, the cart pulls back on you, as evidenced perhaps by the tightening of the rope wrapped around your hand. One thing interacts with another; the hammer interacts with the stake, and you interact with the cart.

Which exerts the force and which receives the force? Isaac Newton's answer to this was that neither force has to be identified as "exerter" or "receiver," and he concluded that both objects must be treated equally. For example, when the hammer exerts a force on the stake, the hammer is brought to a halt by the force the stake exerts on it. Both forces are equal and oppositely directed. When you pull the cart, the cart simultaneously pulls on you. This pair of forces, your pull on the cart and the cart's pull on you, make up the single interaction between you and the cart. Such observations led Newton to his third law of motion.

LEARNING OBJECTIVE
Describe how forces always occur in pairs.

FIGURE 2.14
When you lean against a wall, you exert a force on the wall. The wall simultaneously exerts an equal and opposite force on you. Hence you don't topple over.

Mastering**PHYSICS**
VIDEO: Forces and Interaction

FIGURE 2.15
He can hit the massive bag with considerable force. But with the same punch he can exert only a tiny force on the tissue paper in midair.

---

* We tend to think of only living things pushing and pulling. But inanimate things can do likewise. So please don't be troubled about the idea of the inanimate wall pushing on you. It does, just as another person leaning against you would.

2.4 Newton's Third Law of Motion

EXPLAIN THIS How does Newton's third law account for rocket propulsion?

Newton's third law of motion is:

Whenever one object exerts a force on a second object, the second object exerts an equal and opposite force on the first.

We can call one force the *action force*, and the other the *reaction force*. Then we can express Newton's third law in the following form:

To every action there is always an opposed equal reaction.

It doesn't matter which force we call *action* and which we call *reaction*. The important thing is that they are co-parts of a single interaction and that neither force exists without the other. Action and reaction forces are equal in strength and opposite in direction. They occur in pairs and make up one interaction between two things.

When walking, you interact with the floor. Your push against the floor is coupled to the floor's push against you. The pair of forces occurs simultaneously. Likewise, the tires of a car push against the road while the road pushes back on the tires—the tires and the road push against each other. In swimming, you interact with the water that you push backward, while the water pushes you forward—you and the water push against each other. The reaction forces account for our motion in these cases. These forces depend on friction; a person or car on ice, for example, may not be able to exert the action force required to produce the needed reaction force. Neither force exists without the other.

Simple Rule to Identify Action and Reaction

There is a simple rule for identifying action and reaction forces. First, identify the interaction—one thing (object A) interacts with another (object B). Then, action and reaction forces can be stated in the following form:

Action: Object A exerts a force on object B.

Reaction: Object B exerts a force on object A.

The rule is easy to remember. If action is A acting on B, reaction is B acting on A. We see that A and B are simply switched around. Consider the case of your hand pushing on the wall. The interaction is between your hand and the wall. We'll say the action is your hand (object A) exerting a force on the wall (object B). Then the reaction is the wall exerting a force on your hand.

LEARNING OBJECTIVE
Define Newton's third law of motion by giving examples.

Mastering**PHYSICS**

TUTORIAL: Newton's Third Law

When pushing my fingers together I see the same discoloration on each of them. Aha —evidence that each experiences the same amount of force!

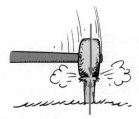

FIGURE 2.16
In the interaction between the hammer and the stake, each exerts the same amount of force on the other.

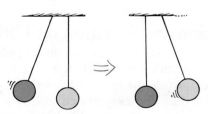

FIGURE 2.17
The impact forces between the blue ball and the yellow ball move the yellow ball and stop the blue ball.

FIGURE 2.18
Earth is pulled up by the boulder with just as much force as the boulder is pulled downward by Earth.

FIGURE 2.19
Action and reaction forces. Note that when the action is "A exerts force on B," the reaction is simply "B exerts force on A."

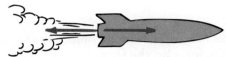

Action: tire pushes on road   Reaction: road pushes on tire

Action: rocket pushes on gas   Reaction: gas pushes on rocket

Action: man pulls on spring   Reaction: spring pulls on man

Action: Earth pulls on ball

Reaction: ball pulls on Earth

Know that an action force and its reaction force always act on *different* objects. Two external forces acting on the same object, even if they are equal and opposite in direction, *cannot* be an action–reaction pair. That's the law!

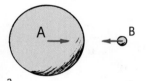

a

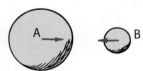

b

c

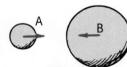

d

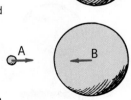

e

FIGURE 2.20
Which falls toward the other: planet A or planet B? Do the accelerations of each relate to their relative masses?

CHECKPOINT
1. A car accelerates along a road. Identify the force that moves the car.
2. Identify the action and reaction forces for the case of an object in free fall (no air resistance).

Were these your answers?
1. It is the road that pushes the car along. Really! Except for air resistance, only the road provides a horizontal force on the car. How does it do this? The rotating tires of the car push back on the road (action). The road simultaneously pushes forward on the tires (reaction). How about that!
2. To identify a pair of action–reaction forces in any situation, first identify the pair of interacting objects. In this case Earth interacts with the falling object via the force of gravity. So Earth pulls the falling object downward (call it *action*). Then *reaction* is the falling object pulling Earth upward. It is only because of Earth's enormous mass that you don't notice its upward acceleration.

Action and Reaction on Different Masses

Quite interestingly, a falling object pulls upward on Earth with as much force as Earth pulls downward on it. The resulting acceleration of a falling object is evident, while the upward acceleration of Earth is too small to detect.

Consider the exaggerated examples of two planetary bodies in parts (a) through (e) in Figure 2.20. The forces between planets A and B are equal in magnitude and oppositely directed in each case. If the acceleration of planet A is unnoticeable in part (a), then it is more noticeable in part (b), where the difference between the masses is less extreme. In part (c), where both bodies have

FIGURE 2.21
INTERACTIVE FIGURE   MP

The force exerted against the recoiling cannon is just as great as the force that drives the cannonball along the barrel. Why, then, does the cannonball undergo more acceleration than the cannon?

equal mass, acceleration of planet A is as evident as it is for planet B. Continuing, we see that the acceleration of planet A becomes even more evident in part (d) and even more so in part (e). So, strictly speaking, when you step off the curb, the street rises ever so slightly to meet you.

When a cannon is fired, an interaction occurs between the cannon and the cannonball. The sudden force that the cannon exerts on the cannonball is exactly equal and opposite to the force the cannonball exerts on the cannon. This is why the cannon recoils (kicks). But the effects of these equal forces are very different. This is because the forces act on different masses. Recall Newton's second law:

$$a = \frac{F}{m}$$

Let F represent both the action and reaction forces, m the mass of the cannon, and m the mass of the cannonball. Different-sized symbols are used to indicate the relative masses and resulting accelerations. Then the acceleration of the cannonball and cannon can be represented in the following way:

$$\text{cannonball: } \frac{F}{m} = a$$

$$\text{cannon: } \frac{F}{m} = a$$

Thus we see why the change in velocity of the cannonball is so large compared with the change in velocity of the cannon. A given force exerted on a small mass produces a large acceleration, while the same force exerted on a large mass produces a small acceleration.

We can extend the idea of a cannon recoiling from the ball it fires to understanding rocket propulsion. Consider an inflated balloon recoiling when air is expelled (Figure 2.22). If the air is expelled downward, the balloon accelerates upward. The same principle applies to a rocket, which continually "recoils" from the ejected exhaust gas. Each molecule of exhaust gas is like a tiny cannonball shot from the rocket (Figure 2.23).

A common misconception is that a rocket is propelled by the impact of exhaust gases against the atmosphere. In fact, before the advent of rockets, it was commonly thought that sending a rocket to the Moon was impossible. Why? Because there is no air above Earth's atmosphere for the rocket to push against. But this is like saying a cannon wouldn't recoil unless the cannonball had air to push against. Not true! Both the rocket and recoiling cannon accelerate because of the reaction forces exerted by the material they fire—not because of any pushes on the air. In fact, a rocket operates better above the atmosphere where there is no air resistance.

FIGURE 2.22
The balloon recoils from the escaping air and climbs upward.

FIGURE 2.23
The rocket recoils from the "molecular cannonballs" it fires and rises.

FYI Gases and fragments shoot out in all directions when a firecracker explodes. When fuel in a rocket burns, a slower explosion, exhaust gases shoot out in one direction.

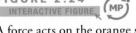

Mastering**PHYSICS**

VIDEO: Action and Reaction on Different Masses

VIDEO: Action and Reaction on Rifle and Bullet

CHECKPOINT

1. **Which pulls harder: the Moon on Earth or Earth on the Moon?**
2. **A high-speed bus and an unfortunate bug have a head-on collision. The force of the bus on the bug splatters it all over the windshield. Is the corresponding force of the bug on the bus greater, less, or the same? Is the resulting deceleration of the bus greater than, less than, or the same as that of the bug?**

Were these your answers?

1. Each pull is the same in magnitude. This is like asking which distance is greater: from New York to San Francisco or from San Francisco to New York. So we see that Earth and the Moon simultaneously pull on each other, each with the *same* amount of force.
2. The magnitudes of the forces are the same, for they constitute an action–reaction force pair that makes up the interaction between the bus and the bug. The accelerations, however, are very different because the masses are different! The bug undergoes an enormous and lethal deceleration, while the bus undergoes a very tiny deceleration—so tiny that the very slight slowing of the bus is unnoticed by its passengers. But if the bug were more massive, as massive as another bus, for example, the slowing down would be quite apparent.

FIGURE 2.24
INTERACTIVE FIGURE MP

A force acts on the orange system and it accelerates to the right.

Defining Your System

An interesting question often arises: if action and reaction forces are equal and opposite, why don't they cancel to zero? To answer this question we must consider the *system* involved. Consider, for example, a system consisting of a single orange (Figure 2.24). The dashed line surrounding the orange encloses and defines the system. The vector that pokes outside the dashed line represents an external force on the system. The system accelerates in accord with Newton's second law. In Figure 2.25 we see that this force is provided by an apple, which doesn't change our analysis. The apple is outside the system. The fact that the orange simultaneously exerts a force on the apple, which is external to the system, may affect the apple (another system), but not the orange. You can't cancel a force on the orange with a force on the apple. So in this case, the action and reaction forces don't cancel.

Now let's consider a larger system, enclosing both the orange and the apple. We see the system bounded by the dashed line in Figure 2.26. Notice that

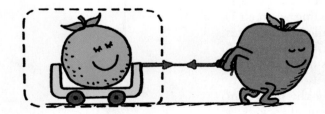

FIGURE 2.25
INTERACTIVE FIGURE MP

The force on the orange, provided by the apple, is not canceled by the reaction force on the apple. The orange still accelerates.

FIGURE 2.26
INTERACTIVE FIGURE MP

In the larger system of orange + apple, action and reaction forces are internal and do cancel. If these are the only horizontal forces, with no external force, no net acceleration of the system occurs.

the force pair is *internal* to the orange–apple system. These forces *do* cancel each other. They play no role in accelerating the system. A force external to the system is needed for acceleration. That's where friction with the floor comes into play (Figure 2.27). When the apple pushes against the floor, the floor simultaneously pushes on the apple—an external force on the system. The system accelerates to the right.

Inside a baseball, trillions of interatomic forces are at play. They hold the ball together, but they play no role in accelerating the ball. Although every one of the interatomic forces is part of an action–reaction pair within the ball, they combine to zero, no matter how many of them there are. A force external to the ball, such as batting it, is needed to accelerate it.

If this is confusing, it may be well to note that Newton had difficulties with the third law himself.

FIGURE 2.27
INTERACTIVE FIGURE

An external horizontal force occurs when the floor pushes on the apple (reaction to the apple's push on the floor). The orange–apple system accelerates.

CHECKPOINT

1. **On a cold, rainy day, your car battery is dead, and you must push the car to move it and get it started. Why can't you move the car by remaining comfortably inside and pushing against the dashboard?**
2. **Does a fast-moving baseball possess force?**

Were these your answers?

1. In this case, the system to be accelerated is the car. If you remain inside and push on the dashboard, the force pair you produce acts and reacts within the system. These forces cancel out, as far as any motion of the car is concerned. To accelerate the car, there must be an interaction between the car and something external—for example, you on the outside pushing against the road.
2. No, a force is not something an object *has*, like mass; it is part of an interaction between one object and another. A speeding baseball may possess the capability of exerting a force on another object when interaction occurs, but it does not possess force as a thing in itself. As we will see in the following chapters, moving things possess momentum and kinetic energy.

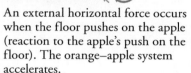

A system may be as tiny as an atom or as large as the universe.

Using Newton's third law, we can understand how a helicopter gets its lifting force. The whirling blades are shaped to force air particles down (action), and the air forces the blades up (reaction). This upward reaction force is called *lift*. When lift equals the weight of the craft, the helicopter hovers in midair. When lift is greater, the helicopter climbs upward.

This is true for birds and airplanes. Birds fly by pushing air downward. The air simultaneously pushes the bird upward. When the bird is soaring, the wing must be shaped so that moving air particles are deflected downward. Slightly tilted wings that deflect oncoming air downward produce lift on an airplane. Air that is pushed downward continuously maintains lift. This supply of air is obtained by the forward motion of the aircraft, which results from propellers or jets that push air backward. When the propellers or jets push air backward, the air simultaneously pushes the propellers or jets forward. We will learn in Chapter 5 that the curved surface of a wing is an airfoil, which enhances the lifting force.

FIGURE 2.28
Ducks fly in a V formation because air pushed downward at the tips of their wings swirls upward, creating an updraft that is strongest off to the side of the bird. A trailing bird gets added lift by positioning itself in this updraft, pushes air downward and creates another updraft for the next bird, and so on. The result is a flock flying in a V formation.

FIGURE 2.29
You cannot touch without being touched—Newton's third law.

We see Newton's third law in action everywhere. A fish propels water backward with its fins, and the water propels the fish forward. The wind caresses the branches of a tree, and the branches caress back on the wind to produce whistling sounds. Forces are interactions between different things. Every contact requires at least a twoness; there is no way that an object can exert a force on nothing. Forces, whether large shoves or slight nudges, always occur in pairs, each opposite to the other. Thus, as author Paul and wife Lillian illustrate in Figure 2.29, we cannot touch without being touched.

LEARNING OBJECTIVE
Summarize and contrast Newton's three laws of motion.

2.5 Summary of Newton's Three Laws

EXPLAIN THIS If the *action* is the force acting on a dropped ball, identify the *reaction*.

In free fall, only a single force acts—the force of gravity. Whenever the force of air resistance also occurs, the falling object is not in free fall.

Newton's first law, the law of inertia: An object at rest tends to remain at rest; an object in motion tends to remain in motion at constant speed along a straight-line path. This property of objects to resist change in motion is called *inertia*. Mass is a measure of inertia. Objects undergo changes in motion only in the presence of a net force.

Newton's second law, the law of acceleration: When a net force acts on an object, the object accelerates. The acceleration is directly proportional to the net force and inversely proportional to the mass. Symbolically, $a \sim F/m$. Acceleration is always in the direction of the net force. When an object falls in a vacuum, the net force is simply the weight, and the acceleration is g (the symbol g denotes that acceleration is due to gravity alone). When an object falls in air, the net force is equal to the weight minus the force of air resistance, and the acceleration is less than g. If and when the force of air resistance equals the weight of a falling object, acceleration terminates, and the object falls at constant speed (called the *terminal speed*).

Newton's third law, the law of action–reaction: Whenever one object exerts a force on a second object, the second object exerts an equal and opposite force on the first. Forces occur in pairs: one is an action and the other is a reaction, which together constitute the interaction between one object and the other. Action and reaction always act on different objects. Neither force exists without the other.

What sports events don't make use of Newton's laws? The answer is simple enough—*none*; they all do.

There has been a lot of new and exciting physics since the time of Isaac Newton. Nevertheless, and quite interestingly, as mentioned at the beginning of the chapter, it was primarily Newton's laws that got us to the Moon. Isaac Newton truly changed our way of viewing the world.

ISAAC NEWTON (1642–1727)

On Christmas Day in the year 1642, the year that Galileo died, Isaac Newton was prematurely born and barely survived. Newton's birthplace was his mother's farmhouse in Woolsthorpe, England. His father died several months before his birth, and he grew up under the care of his mother and grandmother. As a child he showed no particular signs of brightness, and at age $14\frac{1}{2}$ he was taken out of school to work on his mother's farm. As a farmer he was a failure, preferring to read books he borrowed from a neighboring pharmacist. An uncle sensed the scholarly potential in young Isaac and prompted him to study at the University of Cambridge, which he did for five years, graduating without particular distinction.

A plague swept through England, and Newton retreated to his mother's farm—this time to continue his studies. At the farm, at ages 23 and 24, he laid the foundations for the work that was to make him immortal. Seeing an apple fall to the ground led him to consider the force of gravity extending to the Moon and beyond. He formulated the law of universal gravitation. He invented calculus, a very important mathematical tool in science. He extended Galileo's work and developed the three fundamental laws of motion. He also formulated a theory of the nature of light and showed with prisms that white light is composed of all colors of the rainbow. It was his experiments with prisms that first made him famous.

When the plague subsided, Newton returned to Cambridge and soon established a reputation for himself as a first-rate mathematician. His mathematics teacher resigned in his favor and Newton was appointed the Lucasian professor of mathematics. He held this post for 28 years. In 1672 he was elected to the Royal Society, where he exhibited the world's first reflector telescope. It can still be seen, preserved at the library of the Royal Society in London with the inscription: "The first reflecting telescope, invented by Sir Isaac Newton, and made with his own hands."

It wasn't until Newton was 42 that he began to write what is generally acknowledged as the greatest scientific book ever written, the *Philosophiae Naturalis Principia Mathematica*. He wrote the work in Latin and completed it in 18 months. It appeared in print in 1687 and wasn't printed in English until 1729, two years after his death. When asked how he was able to make so many discoveries, Newton replied that he solved his problems by continually thinking very long and hard about them—and not by sudden insight.

At age 46 he was elected a member of Parliament. He attended the sessions in Parliament for two years and never gave a speech. One day he rose and the house fell silent to hear the great man. Newton's "speech" was very brief; he simply requested that a window be closed because of a draft.

A further turn from his work in science was his appointment as warden and then as master of the mint. Newton resigned his professorship and directed his efforts toward greatly improving the workings of the mint, to the dismay of counterfeiters who flourished at that time. He maintained his membership in the Royal Society and was elected president, then re-elected each year for the rest of his life. At age 62, he wrote *Opticks*, which summarized his work on light. Nine years later he wrote a second edition to his *Principia*.

Although Newton's hair turned gray at age 30, it remained full, long, and wavy all his life. Unlike others in his time, he did not wear a wig. He was a modest man, very sensitive to criticism, and he never married. He remained healthy in body and mind into old age. At age 80, he still had all his teeth, his eyesight and hearing were sharp, and his mind was alert. In his lifetime he was regarded by his countrymen as the greatest scientist who ever lived. In 1705 he was knighted by Queen Anne. Newton died at age 85 and was buried in Westminster Abbey along with England's kings and heroes.

Newton "opened up" the universe, showing that the same natural laws that act on Earth govern the larger cosmos as well. For humankind this led to increased humility, but also to hope and inspiration because of the evidence of a rational order. Newton ushered in the Age of Reason. His ideas and insights truly changed the world and elevated the human condition.

HANDS-ON ACTIVITY

If you drop a sheet of paper and a book side by side, the book falls faster than the paper. Why? The book falls faster because of its greater weight compared to the air resistance it encounters. If you place the paper against the lower surface of the raised book and again drop them at the same time, it will be no surprise that they hit the surface below at the same time. The book simply pushes the paper with it as it falls. Now, repeat this, only with the paper on *top* of the book, not sticking over its edge. How will the accelerations of the book and paper compare? Will they separate and fall differently? Will they have the same acceleration? Try it and see! Then explain what happens.

For instructor-assigned homework, go to www.masteringphysics.com

SUMMARY OF TERMS (KNOWLEDGE)

Force pair The action and reaction pair of forces that occur in an interaction.

Free fall Motion under the influence of gravitational pull only.

Inertia The property by which objects resist changes in motion.

Interaction Mutual action between objects during which each object exerts an equal and opposite force on the other.

Newton's first law of motion Every object continues in a state of rest, or in a state of motion in a straight line at constant speed, unless acted on by a net force.

Newton's second law of motion The acceleration produced by a net force on an object is directly proportional to the net force, is in the same direction as the net force, and is inversely proportional to the mass of the object.

Newton's third law of motion Whenever one object exerts a force on a second object, the second object exerts an equal and opposite force on the first object.

Terminal speed The speed at which the acceleration of a falling object terminates when air resistance balances its weight.

Terminal velocity Terminal speed in a given direction (often downward).

READING CHECK QUESTIONS (COMPREHENSION)

2.1 Newton's First Law of Motion

1. State the law of inertia.
2. Is inertia a property of matter or a force of some kind?
3. What concept was missing from people's minds in the 16th century when they couldn't believe Earth was moving?
4. When a bird lets go of a branch and drops to the ground below, why doesn't the moving Earth sweep away from the falling bird?
5. What kind of path would the planets follow if suddenly their attraction to the Sun no longer existed?

2.2 Newton's Second Law of Motion

6. State Newton's second law.
7. Is acceleration directly proportional to force, or is it inversely proportional to force? Give an example.
8. Is acceleration directly proportional to mass, or is it inversely proportional to mass? Give an example.
9. If the mass of a sliding block is tripled at the same time that the net force on it is tripled, how does the resulting acceleration compare to the original acceleration?
10. What is the net force that acts on a 10-N freely falling object?
11. Why doesn't a heavy object accelerate more than a light object when both are freely falling?
12. What is the net force that acts on a 10-N falling object when it encounters 4 N of air resistance? 10 N of air resistance?
13. What two principal factors affect the force of air resistance on a falling object?
14. What is the acceleration of a falling object that has reached its terminal velocity?
15. If two objects of the same size fall through air at different speeds, which encounters the greater air resistance?
16. Why does a heavy parachutist fall faster than a lighter parachutist who wears the same size parachute?

2.3 Forces and Interactions

17. Previously, we stated that a force was a push or pull; now we say it is part of an interaction. Which is it: a push or pull, or part of an interaction?
18. How many forces are required for a single interaction?
19. When you push against a wall with your fingers, they bend because they experience a force. Identify this force.
20. A boxer can hit a heavy bag with great force. Why can't he hit a sheet of tissue paper in midair with the same amount of force?

2.4 Newton's Third Law of Motion

21. State Newton's third law.
22. Consider hitting a baseball with a bat. If we call the force of the bat against the ball the action force, identify the reaction force.

23. If the forces that act on a cannonball and the recoiling cannon from which it is fired are equal in magnitude, why do the cannonball and cannon have very different accelerations?
24. Is it correct to say that action and reaction forces always act on different bodies? Defend your answer.
25. If body A and body B are both within a system, can forces between them affect the acceleration of the system?
26. What is necessary, forcewise, to accelerate a system?
27. Identify the force that propels a rocket into space.
28. How does a helicopter get its lifting force?
29. What law of physics is inferred when we say you cannot touch without being touched?

2.5 Summary of Newton's Three Laws

30. Which of Newton's laws focuses on inertia, which on acceleration, and which on action–reaction?

ACTIVITIES (HANDS-ON APPLICATION)

31. Write a letter to Grandma or Grandpa telling that Galileo introduced the concepts of acceleration and inertia and was familiar with forces, but he didn't see the connection between these three concepts. Tell how Isaac Newton did understand and how the connection explains why heavy and light objects in free fall gain the same speed in the same time. In this letter, you may use an equation or two, as long as you make it clear that an equation is a shorthand notation of ideas you've explained.

32. The net force acting on an object and the resulting acceleration are always in the same direction. You can demonstrate this with a spool. If the spool is pulled horizontally to the right, in which direction will it roll?

33. Hold your hand with the palm down like a flat wing outside the window of a moving automobile. Then slightly tilt the front edge of your hand upward and notice the lifting effect as air is deflected downward from the bottom of your hand. Can you see Newton's laws at work here?

PLUG AND CHUG (FORMULA FAMILIARIZATION)

Do these simple one-step calculations and familiarize yourself with the formulas that link the concepts of force, mass, and acceleration.

$$\textbf{Acceleration: } a = \frac{F_{net}}{m}$$

34. In Chapter 1 acceleration is defined as $a = \Delta v / \Delta t$. Use this formula to show that the acceleration of a cart on an inclined plane that gains 6.0 m/s each 1.2 s is 5.0 m/s².

35. In this chapter we learn that the cause of acceleration is given by Newton's second law: $a = F_{net}/m$. Show that the 5.0-m/s² acceleration of the preceding problem can result from a net force of 15 N exerted on a 3.0-kg cart. (Note: The unit N/kg is equivalent to m/s².)

36. Knowing that a 1-kg object weighs 10 N, confirm that the acceleration of a 1-kg stone in free fall is 10 m/s².

37. A simple rearrangement of Newton's second law gives $F_{net} = ma$. Show that a net force of 84 N is needed to give a 12-kg package an acceleration of 7.0 m/s². (Note: The units kg · m/s² and N are equivalent.)

THINK AND SOLVE (MATHEMATICAL APPLICATION)

38. A Honda Civic Hybrid weighs about 2900 lb. Calculate the weight of the car in newtons and its mass in kilograms. (Note: 0.22 lb = 1 N; 1 kg on Earth's surface has a weight of 10 N.)

39. When two horizontal forces are exerted on the car in the preceding problem, 220 N forward and 180 N backward, the car undergoes acceleration. What additional force is needed to produce non-accelerated motion?

40. An astronaut of mass 120 kg recedes from her spacecraft by activating a small propulsion unit attached to her back. The force generated by a spurt is 30 N. Show that her acceleration is 0.25 m/s^2.

41. Madison pushes with a 160-N horizontal force on a 20-kg crate of coffee resting on a warehouse floor. The force of friction on the crate is 80 N. Show that the acceleration is 4 m/s^2.

42. Sophia exerts a steady 40-N horizontal force on a 8-kg box resting on a lab bench. The box slides against a horizontal friction force of 24 N. Show that the box accelerates at 2 m/s^2.

43. A business jet of mass 30,000 kg takes off when the thrust for each of two engines is 30,000 N. Show that its acceleration is 2 m/s^2.

44. A rocket of mass 100,000 kg undergoes an acceleration of 2 m/s^2. Show that the net force acting on it is 200,000 N.

45. Calculate the horizontal force that must be applied to a 1-kg puck to accelerate on a horizontal friction-free air table with the same acceleration it would have if it were dropped and fell freely.

46. Leroy, who has a mass of 100 kg, is skateboarding at 9.0 m/s when he smacks into a brick wall and comes to a dead stop in 0.2 s.

    (a) Show that his deceleration is 45 m/s^2.

    (b) Show that the force of impact is 4500 N. (Ouch!)

47. Allison exerts a steady net force of 50 N on a 20-kg shopping cart initially at rest for 2.0 s. Find the acceleration of the cart, and show that it moves a distance of 5 m.

48. The heavyweight boxing champion of the world punches a sheet of paper in midair, bringing it from rest up to a speed of 25.0 m/s in 0.050 s. The mass of the paper is 0.003 kg. Show that the force of the punch on the paper is only 1.50 N.

49. Suzie Skydiver with her parachute has a mass of 50 kg.

    (a) Before opening her chute, what force of air resistance will she encounter when she reaches terminal velocity?

    (b) What force of air resistance will she encounter when she reaches a lower terminal velocity after the chute is open?

    (c) Discuss why your answers are the same or different.

50. If you stand next to a wall on a frictionless skateboard and push the wall with a force of 40 N, how hard does the wall push on you? If your mass is 80 kg, show that your acceleration is 0.5 m/s^2.

51. A force F acts in the forward direction on a cart of mass m. A friction force f opposes this motion.

    (a) Use Newton's second law and show that the acceleration of the cart is $(F - f)/m$.

    (b) If the cart's mass is 4.0 kg, the applied force is 12.0 N, and the friction force is 6.0 N, show that the cart's acceleration is 1.5 m/s^2.

52. A firefighter of mass 80 kg slides down a vertical pole with an acceleration of 4 m/s^2. Show that the friction force that acts on the firefighter is 480 N.

53. A rock band's tour bus, mass M, is accelerating away from a stop sign at rate a when a piece of heavy metal, mass $\frac{M}{6}$, falls onto the top of the bus and remains there.

    (a) Show that the bus' acceleration is now $\frac{6}{7}a$.

    (b) If the initial acceleration of the bus is 1.2 m/s^2, show that when the bus carries the heavy metal with it, the acceleration will be 1.0 m/s^2.

THINK AND RANK (ANALYSIS)

54. Boxes of various masses are on a friction-free level table. From greatest to least, rank the (a) net forces on the boxes and (b) accelerations of the boxes.

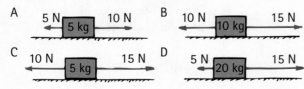

55. In cases A, B, and C, the crate is in equilibrium (no acceleration). From greatest to least, rank the amount of friction between the crate and the floor.

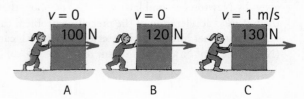

56. Consider a 100-kg box of tools in locations A, B, and C. Rank from greatest to least the (a) masses of the 100-kg box of tools and (b) weights of the 100-kg box of tools.

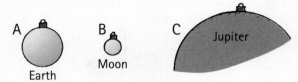

57. Three parachutists, A, B, and C, each have reached terminal velocity at the same altitude.

    (a) From fastest to slowest, rank their terminal velocities.

    (b) From longest to shortest times, rank their order in reaching the ground.

58. The strong man is pulled in the three situations shown. Rank the amount of tension in the rope in his right hand (the one attached to the tree in B and C) from least to greatest.

EXERCISES (SYNTHESIS)

Please do not be intimidated by the large amount of end-of-chapter material in this and some other chapters. If your course covers many chapters, your instructor will likely assign only a few items from each.

59. The auto in the sketch moves forward as the brakes are applied. A bystander says that during the interval of braking, the auto's velocity and acceleration are in opposite directions. Do you agree or disagree?

60. Your empty hand is not hurt when it bangs lightly against a wall. Why does your hand hurt if it is carrying a heavy load? Which of Newton's laws is most applicable here?

61. Why is a massive cleaver more effective for chopping vegetables than a lighter knife that is equally sharp?

62. When you stand on a floor, does the floor exert an upward force against your feet? How much force does it exert? Why aren't you moved upward by this force?

63. A racing car travels along a straight raceway at a constant velocity of 200 km/h. What horizontal forces act, and what is the net force acting on the car?

64. To pull a wagon across a lawn at a constant velocity, you must exert a steady force. Reconcile this fact with Newton's first law, which states that motion with a constant velocity indicates no force.

65. When your car moves along the highway at a constant velocity, the net force on it is zero. Why, then, do you continue running your engine?

66. When you toss a coin upward, what happens to its velocity while ascending? What happens to its acceleration? (Neglect air resistance.)

67. You stand on a weighing scale and read your weight. Then you leap upward from the scale. What happens to your weight reading as you jump?

68. A common saying goes, "It's not the fall that hurts you; it's the sudden stop." Translate this into Newton's laws of motion.

69. If it were not for air resistance, would it be dangerous to go outdoors on rainy days? Defend your answer.

70. What is the net force acting on a 1-kg ball in free fall?

71. What is the net force acting on a falling 1-kg ball if it encounters 2 N of air resistance?

72. For each of the following interactions, identify the action and reaction forces.

    (a) A hammer hits a nail.

    (b) Earth's gravity pulls down on a book.

    (c) A helicopter blade pushes air downward.

73. You hold an apple over your head.

    (a) Identify all the forces acting on the apple and their reaction forces.

    (b) When you drop the apple, identify all the forces acting on it as it falls and the corresponding reaction forces.

74. What is the net force on an apple that weighs 1 N when you hold it at rest above your head? What is the net force on the apple when you release it?

75. Aristotle claimed that the speed of a falling object depends on its weight. We now know that objects in free fall, whatever their weights, undergo the same gain in speed. Why doesn't weight affect acceleration?

76. Why will a cat that falls from the top of a 50-story building hit a safety net below no faster than if it fell from the 20th story?

77. Free fall is motion in which gravity is the only force acting.

    (a) Explain why a skydiver who has reached terminal speed is not in free fall.

    (b) Explain why a satellite circling Earth above the atmosphere is in free fall.

78. How does the weight of a falling body compare with the air resistance it encounters just before it reaches terminal velocity? Just after it reaches terminal velocity?

79. You tell your friend that the acceleration of a skydiver decreases as falling progresses. Your friend then asks if this means that the skydiver is slowing down. What is your answer?

80. Two 100-N weights are attached to a spring scale as shown. Does the scale read 0 N, 100 N, or 200 N, or does it give some other reading? (Hint: Would the reading differ if one of the ropes were tied to the wall instead of to the hanging 100-N weight?)

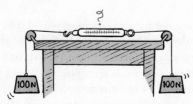

81. When you rub your hands together, can you push harder on one hand than on the other?

82. Can a dog wag its tail without the tail in turn "wagging the dog"? (Consider a dog with a relatively massive tail.)

83. When the athlete holds the barbell overhead, the reaction force is the weight of the barbell on his hand. How does this force vary for the case in which the barbell is accelerated upward? Downward?

84. Consider the two forces acting on the person who stands still—namely, the downward pull of gravity and the upward support of the floor. Are these forces equal and opposite? Do they form an action–reaction pair? Why or why not?

85. Why can you exert greater force on the pedals of a bicycle if you pull up on the handlebars?

86. The strong man will push apart the two initially stationary freight cars of equal mass before he himself drops straight to the ground. Is it possible for him to give either of the cars a greater speed than the other? Why or why not?

87. Suppose two carts, one twice as massive as the other, fly apart when the compressed spring that joins them is released. How fast does the heavier cart roll compared with the lighter cart?

88. If a Mack truck and a motorcycle have a head-on collision, upon which vehicle is the impact force greater? Which vehicle undergoes the greater change in its motion? Defend your answers.

89. Two people of equal mass attempt a tug-of-war with a 12-m rope while standing on frictionless ice. When they pull on the rope, each person slides toward the other. How do their accelerations compare, and how far does each person slide before they meet?

90. Suppose that one person in the preceding exercise has twice the mass of the other. How far does each person slide before they meet?

91. Which team wins in a tug-of-war: the team that pulls harder on the rope or the team that pushes harder against the ground? Explain.

92. The photo shows Steve Hewitt and his daughter Gretchen. Is Gretchen touching her dad, or is he touching her? Explain.

DISCUSSION QUESTIONS (EVALUATION)

93. Discuss whether a stick of dynamite contains force. Similarly, does a boxer's fist contain force? A hammer? Defend your answers.

94. In the orbiting space shuttle, you are handed two identical closed boxes, one filled with sand and the other filled with feathers. Discuss at least a couple of ways that you can tell which is which without opening the boxes?

95. Each of the vertebrae forming your spine is separated from its neighbors by disks of elastic tissue. What happens, then, when you jump heavily on your feet from an elevated position? Can you think of a reason why you are a little shorter in the evening than you are in the morning? (Hint: Think about the hammerhead in Figure 2.2.)

96. Before the time of Galileo and Newton, many scholars thought that a stone dropped from the top of a tall mast on a moving ship would fall vertically and hit the deck behind the mast by a distance equal to how far the ship had moved forward during the time the stone was falling. In light of your understanding of Newton's laws, what do you and your classmates think about this idea?

97. A rocket becomes progressively easier to accelerate as it travels through space. Why is this so? (Hint: About 90% of the mass of a newly launched rocket is fuel.)

98. On which of these hills does the ball roll down with increasing speed and decreasing acceleration along the path? (Use this example if you wish to explain to someone the difference between speed and acceleration.)

99. If you drop an object, its acceleration toward the ground is 10 m/s². If you throw it down instead, will its acceleration after leaving your hand be greater than 10 m/s²? Ignore air resistance. Defend your answer.

100. Can you think of a reason why the acceleration of the object thrown downward through the *air* in the preceding question would actually be less than 10 m/s²?

101. What is the acceleration of a stone at the top of its trajectory when it has been tossed straight upward? (Make sure your answer is consistent with the equation for Newton's second law.)

102. A couple of your friends say that, before a falling body reaches terminal velocity, it *gains* speed while acceleration *decreases*. Do you agree or disagree? Defend your answer.

103. How does the terminal speed of a parachutist before opening a parachute compare to the terminal speed after? Why is there a difference?

104. How does the gravitational force on a falling body compare with the air resistance it encounters before it reaches terminal velocity? After reaching terminal velocity?

105. If and when Galileo dropped two balls from the top of the Leaning Tower of Pisa, air resistance was not really negligible. Assuming that both balls were the same size yet one was much heavier than the other, which ball actually struck the ground first? Discuss your reasoning.

106. A farmer urges his horse to pull a wagon. The horse refuses, saying that to try would be futile, for it would flout Newton's third law. The horse concludes that it can't exert a greater force on the wagon than the wagon exerts on itself and, therefore, the horse wouldn't be able to accelerate the wagon. What explanation can you offer to convince the horse to pull?

107. This is a scenario common with many physics students: you push a heavy car by hand. The car, in turn, pushes back with an opposite but equal force on you. Doesn't this mean that the forces cancel each other, making acceleration impossible? Resolve the misunderstanding underlying this question.

108. If you exert a horizontal force of 200 N to slide a desk across an office floor at a constant velocity, how much friction does the floor exert on the desk? Is the force of friction equal and oppositely directed to your 200-N push? Does the force of friction make up the reaction force to your push? Why or why not?

109. Ken and Joanne are astronauts floating some distance apart in space. They are joined by a safety cord whose ends are tied around their waists. If Ken starts pulling on the cord, will he pull Joanne toward him, or will he pull himself toward Joanne? Explain what happens.

110. If you simultaneously drop a pair of tennis balls from the top of a building, they strike the ground at the same time. If one of the tennis balls is filled with lead pellets, will it fall faster and hit the ground first? Which of the two will encounter more air resistance? Defend your answers.

READINESS ASSURANCE TEST (RAT)

If you have a good handle on this chapter, if you really do, then you should be able to score at least 7 out of 10 on this RAT. If you score less than 7, you need to study further before moving on.

Choose the BEST answer to each of the following.

1. If gravity between Earth and an orbiting communications satellite suddenly vanished, the satellite would move in
   (a) a curved path.
   (b) a straight-line path.
   (c) a path directed toward Earth's surface.
   (d) an outward spiral path.

2. If an object moves along a curved path, then it must be
   (a) accelerating.
   (b) acted on by a force.
   (c) both of these
   (d) none of these

3. A ball rolls down a curved ramp as shown. As its speed increases, its rate of gaining speed
   (a) increases.
   (b) decreases.
   (c) remains unchanged.
   (d) none of these

4. The reason a 10-kg rock falls no faster than a 5-kg rock in free fall is that the
   (a) 10-kg rock has greater acceleration.
   (b) 5-kg rock has greater acceleration.
   (c) force of gravity is the same for both.
   (d) force/mass ratio is the same for both.

5. As mass is added to a pushed object, its acceleration
   (a) increases.
   (b) decreases.
   (c) remains constant.
   (d) quickly reaches zero.

6. The amount of air resistance that acts on a wingsuit flyer (and a flying squirrel) depends on the flyer's
   (a) area.
   (b) speed.
   (c) area and speed.
   (d) acceleration.

7. You drop a soccer ball off the edge of the administration building on your campus. While the soccer ball falls, its speed
   (a) and acceleration both increase.
   (b) increases and its acceleration decreases.
   (c) and acceleration both decrease.
   (d) decreases and its acceleration increases.

8. The amount of force with which a boxer's punch lands depends mostly on the
   (a) physical condition of the boxer.
   (b) mass of what's being hit.
   (c) boxer's attitude.
   (d) none of these

9. When the neck of an air-filled balloon is untied and air escapes, the balloon shoots through the air. The force that propels the balloon is provided by the
   (a) surrounding air.
   (b) ejected air.
   (c) balloon fabric.
   (d) ground beneath the balloon.

10. The force that propels a rocket is provided by
    (a) gravity.
    (b) Newton's laws of motion.
    (c) its exhaust gases.
    (d) the atmosphere against which the rocket pushes.

Answers to RAT

1 b, 2 c, 3 b, 4 d, 5 b, 6 c, 7 b, 8 b, 9 b, 10 c.

CHAPTER 3

Momentum and Energy

WE'VE LEARNED that Galileo's concept of inertia is incorporated into Newton's first law of motion. We discussed inertia in terms of objects at rest and objects in motion. In this chapter, we will consider the inertia of moving objects. When we combine the ideas of inertia and motion, we are dealing with momentum. *Momentum* is a property of moving things. All things have energy, and when moving, they have energy of motion—*kinetic energy*. Things at rest have another kind of energy—*potential energy*. And all objects, whether at rest or moving, have an energy of being—$E = mc^2$. This chapter is about two of the most central concepts in mechanics—momentum and energy.

Mastering**PHYSICS**

VIDEO: Definition of Momentum

FIGURE 3.1
The boulder, unfortunately, has more momentum than the runner.

3.1 Momentum and Impulse

EXPLAIN THIS Why do cannonballs shot from long-barreled cannons experience a greater impulse for the same average force?

We know that it's harder to stop a large truck than a small car when both are moving at the same speed. We say the truck has more momentum than the car. By **momentum**, we mean *inertia in motion* or, more specifically, the mass of an object multiplied by its velocity:

$$\text{Momentum} = \text{mass} \times \text{velocity}$$

Or, in shorthand notation,

$$\text{Momentum} = mv$$

When direction is not an important factor, we can say

$$\text{Momentum} = \text{mass} \times \text{speed}$$

which we still abbreviate *mv*.*

We can see from the definition that a moving object can have a large momentum if it has a large mass, a high speed, or both. A moving truck has more momentum than a car moving at the same speed because the truck has more mass. But a fast car can have more momentum than a slow truck. And a truck at rest has no momentum at all.

If the momentum of an object changes, then either the mass or the velocity or both change. If the mass remains unchanged, as is most often the case, then the velocity changes and acceleration occurs. What produces acceleration? We know the answer is *force*. The greater the net force on an object, the greater its change in velocity and, hence, the greater its change in momentum.

But something else is important in changing momentum: time—how long a time the force acts. If you apply a brief force to a stalled automobile, you produce a change in its momentum. Apply the same force over an extended period of time, and you produce a greater change in the automobile's momentum. A force sustained for a long time produces more change in momentum than does the same force applied briefly. So, both force and time interval are important in changing momentum.

The quantity *force × time interval* is called **impulse**. In shorthand notation,

$$\text{Impulse} = Ft$$

FIGURE 3.2
When you push with the same force for twice the time, you impart twice the impulse and produce twice the change in momentum.

CHECKPOINT

1. **Compare the momentum of a 1-kg cart moving at 10 m/s with that of a 2-kg cart moving at 5 m/s.**
2. **Does a moving object have impulse?**
3. **Does a moving object have momentum?**
4. **For the same force, which cannon imparts a greater impulse to a cannonball: a long cannon or a short one?**

Were these your answers?

1. Both have the same momentum (1 kg × 10 m/s = 2 kg × 5 m/s).
2. No, impulse is not something an object *has*, like momentum. Impulse is what an object can *provide* or what it can *experience* when it interacts

---

* The symbol for momentum is *p*. In most physics textbooks, *p* = *mv*.

with some other object. An object cannot possess impulse, just as it cannot possess force.

3. Yes, but, like velocity, in a relative sense—that is, with respect to a frame of reference, usually Earth's surface. The momentum possessed by a moving object with respect to a stationary point on Earth may be quite different from the momentum it possesses with respect to another moving object.

4. The long cannon imparts a greater impulse because the force acts over a longer time. (A greater impulse produces a greater change in momentum, so a long cannon imparts more speed to a cannonball than a short cannon does.)

3.2   Impulse Changes Momentum

EXPLAIN THIS Why is it a good idea to have your knees bent when you land after a jump?

The greater the impulse exerted on something, the greater the change in momentum. The exact relationship is

$$\text{Impulse} = \text{change in momentum}$$

or in abbreviated notation*

$$Ft = \Delta(mv)$$

where Δ is the symbol for "change in."

The **impulse–momentum relationship** helps us analyze a variety of situations in which momentum changes. Here we will consider some ordinary examples in which impulse is related to increasing and decreasing momentum.

Case 1: Increasing Momentum

To increase the momentum of an object, it makes sense to apply the greatest force possible for as long as possible. A golfer teeing off and a baseball player trying for a home run do both of these things when they swing as hard as possible and follow through with their swings. Following through extends the time of contact.

The forces involved in impulses usually vary from instant to instant. For example, a golf club that strikes a ball exerts zero force on the ball until it comes in contact; then the force increases rapidly as the ball is distorted (Figure 3.3). The force then diminishes as the ball comes up to speed and returns to its original shape. So when we speak of such forces in this chapter, we mean the *average* force.

Case 2: Decreasing Momentum Over a Long Time

If you were in a truck that was out of control and you had to choose between hitting a concrete wall or a haystack, you wouldn't have to call on your knowledge of physics to make up your mind. Common sense tells you to choose the haystack. But knowing the physics helps you understand *why* hitting a soft object is entirely different from hitting a hard one. In the case of hitting either

Mastering**PHYSICS**
VIDEO: Changing Momentum
VIDEO: Decreasing Momentum Over a Short Time

Timing is especially important when changing momentum.

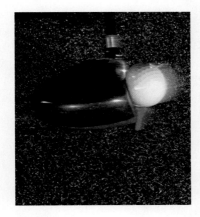

FIGURE 3.3
The force of impact on a golf ball varies throughout the duration of impact.

Mastering**PHYSICS**
TUTORIAL: Momentum and Collisions

---

* This relationship is derived by rearranging Newton's second law to make the time factor more evident. If we equate the formula for acceleration, $a = F/m$, with what acceleration actually is, $a = \Delta v/\Delta t$, we get $F/m = \Delta v/\Delta t$. From this we derive $F\Delta t = \Delta(mv)$. Calling Δt simply t, the time interval, we have $Ft = \Delta(mv)$.

FIGURE 3.4

If the change in momentum occurs over a long time, then the hitting force is small.

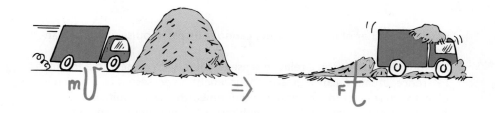

FIGURE 3.5

If the change in momentum occurs over a short time, then the hitting force is large.

Different forces exerted over different time intervals can produce the same impulse:

$$F_t \quad \text{or} \quad _F t$$

the wall or the haystack and coming to a stop, it takes the *same* impulse to decrease your momentum to zero. The same impulse does not mean the same amount of force or the same amount of time; rather it means the same *product* of force and time. By hitting the haystack instead of the wall, you extend the *time during which your momentum is brought to zero*. A longer time interval reduces the force and decreases the resulting deceleration. For example, if the time interval is increased by a factor of 100, the force is reduced to a hundredth. Whenever we wish the force to be small, we extend the time of contact. Hence the reason for padded dashboards and airbags in motor vehicles.

When you jump from an elevated position down to the ground, what happens if you keep your legs straight and stiff? Ouch! Instead, you bend your knees when your feet make contact with the ground. By doing so you extend the time during which your momentum decreases to 10 to 20 times that of a stiff-legged, abrupt landing. The resulting force on your bones is reduced by a factor of 10 to 20. A wrestler thrown to the floor tries to extend his time of impact with the mat by relaxing his muscles and spreading the impact into a series of smaller ones as his foot, knee, hip, ribs, and shoulder successively hit the mat. Of course, falling on a mat is preferable to falling on a solid floor because the mat also increases the time during which the force acts.

The safety net used by circus acrobats is a good example of how to achieve the impulse needed for a safe landing. The safety net reduces the force experienced by a fallen acrobat by substantially increasing the time interval during which the force acts.

If you're about to catch a fast baseball with your bare hand, you extend your hand forward so you'll have plenty of room to let your hand move backward after you make contact with the ball. You extend the time of impact and thereby reduce the force of impact. Similarly, a boxer rides or rolls with the punch to reduce the force of impact (Figure 3.6).

FIGURE 3.6

In both cases, the impulse provided by the boxer's jaw reduces the momentum of the punch. (a) When the boxer moves away (rides with the punch), he extends the time and diminishes the force. (b) If the boxer moves into the glove, the time is reduced and he must withstand a greater force.

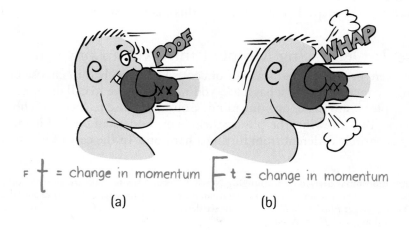

$F\,t$ = change in momentum $F\,t$ = change in momentum

(a)   (b)

Case 3: Decreasing Momentum Over a Short Time

When boxing, if you move into a punch instead of away, you're in trouble. It's the same as if you catch a high-speed baseball while your hand moves toward the ball instead of away upon contact. Or, when your car is out of control, if you drive it into a concrete wall instead of a haystack, you're really in trouble. In these cases of short impact times, the impact forces are large. Remember that for an object brought to rest, the impulse is the same no matter how it is stopped. But if the time is short, the force is large.

The idea of short time of contact explains how a karate expert can split a stack of bricks with the blow of her bare hand (Figure 3.7). She brings her arm and hand swiftly against the bricks with considerable momentum. This momentum is quickly reduced when she delivers an impulse to the bricks. The impulse is the force of her hand against the bricks multiplied by the time during which her hand makes contact with the bricks. By swift execution, she makes the time of contact very brief and correspondingly makes the force of impact huge. If her hand is made to bounce upon impact, as we will soon see, the force is even greater.

FIGURE 3.7
Cassy imparts a large impulse to the bricks in a short time and produces a considerable force.

CHECKPOINT

1. **If the boxer in Figure 3.6 increases the duration of impact to three times as long by riding with the punch, by how much is the force of impact reduced?**
2. **If the boxer instead moves *into* the punch to decrease the duration of impact by half, by how much is the force of impact increased?**
3. **A boxer being hit with a punch contrives to extend time for best results, whereas a karate expert delivers a force in a short time for best results. Isn't there a contradiction here?**

Were these your answers?

1. The force of impact is only a third of what it would have been if he hadn't pulled back.
2. The force of impact is twice what it would have been if he had held his head still. Impacts of this kind account for many knockouts.
3. There is no contradiction because the best results for each are quite different. The best result for the boxer is reduced force, accomplished by maximizing time, and the best result for the karate expert is increased force delivered in minimum time.

Bouncing

If a flowerpot falls from a shelf onto your head, you may be in trouble. If it bounces from your head, you may be in more serious trouble. Why? Because impulses are greater when an object bounces. The impulse required to bring an object to a stop and then to "throw it back again" is greater than the impulse required merely to bring the object to a stop. Suppose, for example, that you catch the falling pot with your hands. You provide an impulse to reduce its momentum to zero. If you throw the pot

FIGURE 3.8
Howie Brand shows that the block topples when the swinging dart bounces from it. When he removes the rubber head of the dart so it doesn't bounce when it hits the block, no tipping occurs.

FIGURING PHYSICAL SCIENCE

Problem Solving

SAMPLE PROBLEM 1

An 8-kg bowling ball rolling at 2 m/s bumps into a padded guardrail and stops.
(a) What is the momentum of the ball just before hitting the guardrail?
(b) How much impulse acts on the ball?
(c) How much impulse acts on the guardrail?

Solution:
(a) The momentum of the ball is $mv = (8 \text{ kg})(2 \text{ m/s}) = 16 \text{ kg} \cdot \text{m/s}$.
(b) In accord with the impulse–momentum relationship, the impulse on the ball is equal to its change in momentum. The momentum changes from 16 kg · m/s to zero. So $Ft = \Delta mv = (16 \text{ kg} \cdot \text{m/s}) - 0 =$

16 kg · m/s = 16 N · s. (Note that the units kg · m/s and N · s are equivalent.)
(c) In accord with Newton's third law, the force of the ball on the padded guardrail is equal and oppositely directed to the force of the guardrail on the ball. Because the time of the interaction is the same for both the ball and the guardrail, the impulses are also equal and opposite. So the amount of impulse on the ball is 16 N · s.

SAMPLE PROBLEM 2

An ostrich egg of mass *m* is thrown at a speed *v* into a sagging bedsheet and is brought to rest in time *t*.
(a) Show that the average force of egg impact is *mv/t*.
(b) If the mass of the egg is 1.0 kg, its speed when it hits the sheet is 2.0 m/s, and it is brought to rest in 0.2 s, show that the average force that acts is 10 N.

(c) Why is breakage less likely with a sagging sheet than with a taut one?

Solution:
(a) From the impulse–momentum equation, $Ft = \Delta mv$, where in this case the egg ends up at rest, $\Delta mv = mv$, and simple algebraic rearrangement gives $F = mv/t$.

(b) $F = \dfrac{mv}{t} = \dfrac{(1.0 \text{ kg})\left(2.0 \frac{m}{s}\right)}{(0.2 \text{ s})}$

$= 10 \text{ kg} \cdot \dfrac{m}{s^2} = 10 \text{ N}$

(c) The time during which the tossed egg's momentum goes to zero is extended when it hits a sagging sheet. Extended time means less force in the impulse that brings the egg to a halt. Less force means less chance of breakage.

upward again, you have to provide additional impulse. This increased amount of impulse is the same that your head supplies if the flowerpot bounces from it.

The fact that impulses are greater when bouncing occurs was used with great success during the California gold rush. The waterwheels used in gold-mining operations were not very effective. A man named Lester A. Pelton recognized a problem with the flat paddles on the waterwheels. He designed a curved paddle that caused the incoming water to make a U-turn upon impact with the paddle. Because the water "bounced," the impulse exerted on the waterwheel was increased. Pelton patented his idea, and he probably made more money from his invention, the Pelton wheel, than any of the gold miners earned. Physics can indeed enrich your life in more ways than one.

FIGURE 3.9
The Pelton wheel. The curved blades cause water to bounce and make a U-turn, which produces a greater impulse to turn the wheel.

Impulse

3.3 Conservation of Momentum

EXPLAIN THIS What stays the same when a pool ball stops after hitting another ball at rest?

Only an impulse external to a system can change the momentum of a system. Internal forces and impulses won't work. For example, consider the cannon being fired in Figure 3.10. The force on the cannonball inside the cannon barrel is equal and opposite to the force causing the cannon to recoil. Because these forces act for the same amount of time, the impulses are also equal and opposite. Recall Newton's third law about action and reaction forces. It applies to impulses, too. These impulses are internal to the system comprising the cannon and the cannonball, so they don't change the momentum of the cannon–cannonball system. Before the firing, the system is at rest and the momentum is zero. After the firing, the net momentum, or total momentum, is *still* zero. Net momentum is neither gained nor lost.

Momentum, like the quantities velocity and force, has both direction and magnitude. It is a *vector quantity*. Like velocity and force, momentum can be canceled. So although the cannonball in the preceding example gains momentum when fired and the recoiling cannon gains momentum in the opposite direction, there is no gain in the cannon–cannonball *system*. The momenta (plural form of *momentum*) of the cannonball and the cannon are equal in magnitude and opposite in direction.* They cancel to zero for the system as a

FIGURE 3.10
INTERACTIVE FIGURE

The net momentum before firing is zero. After firing, the net momentum is still zero, because the momentum of the cannon is equal and opposite to the momentum of the cannonball.

FYI In Figure 3.10, most of the cannonball's momentum is in speed; most of the recoiling cannon's momentum is in mass. So:

$$_mV = m_v$$

---

* Here we neglect the momentum of ejected gases from the exploding gunpowder, which can be considerable. Firing a gun with blanks at close range is a definite no-no because of the considerable momentum of ejecting gases. More than one person has been killed by close-range firing of blanks. In 1998, a minister in Jacksonville, Florida, dramatizing his sermon before several hundred parishioners, including his family, shot himself in the head with a blank round from a .357-caliber Magnum. Although no slug emerged from the gun, exhaust gases did—enough to be lethal. So, strictly speaking, the momentum of the bullet (if any) + the momentum of the exhaust gases is equal to the opposite momentum of the recoiling gun.

FIGURE 3.11
A cue ball hits an eight ball head-on. Consider this event in three systems: (a) An external force acts on the eight-ball system, and its momentum increases. (b) An external force acts on the cue-ball system, and its momentum decreases. (c) No external force acts on the cue-ball + eight-ball system, and momentum is conserved (simply transferred from one part of the system to the other).

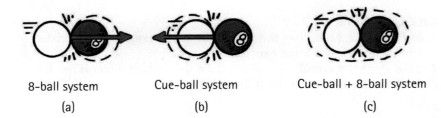

8-ball system   Cue-ball system   Cue-ball + 8-ball system
(a)   (b)   (c)

whole. If no net force or net impulse acts on a system, the momentum of that system cannot change.

When momentum, or any quantity in physics, does not change, we say it is *conserved*. The idea that momentum is conserved when no external force acts is elevated to a central law of mechanics, called the **law of conservation of momentum**, which states:

> **In the absence of an external force, the momentum of a system remains unchanged.**

For any system in which all forces are internal—as, for example, cars colliding, atomic nuclei undergoing radioactive decay, or stars exploding—the net momentum of the system before and after the event is the same.

CHECKPOINT

1. Newton's second law states that if no net force is exerted on a system, no acceleration occurs. Does it follow that no change in momentum occurs?
2. Newton's third law states that the force a cannon exerts on a cannonball is equal and opposite to the force the cannonball exerts on the cannon. Does it follow that the *impulse* the cannon exerts on the cannonball is equal and opposite to the *impulse* the cannonball exerts on the cannon?

Were these your answers?

1. Yes, because no acceleration means that no change occurs in velocity or in momentum (mass × velocity). Another line of reasoning is simply that no net force means there is no net impulse and thus no change in momentum.
2. Yes, because the interaction between both occurs during the same *time* interval. Because time is equal and the forces are equal and opposite, the impulses, *Ft*, are also equal and opposite. Impulse is a vector quantity and can be canceled.

Collisions

The collision of objects clearly illustrates the conservation of momentum. Whenever objects collide in the absence of external forces, the net momentum of both objects before the collision equals the net momentum of both objects after the collision.

$$\text{net momentum}_{\text{before collision}} = \text{net momentum}_{\text{after collision}}$$

This is true no matter how the objects might be moving before they collide.

Momentum is conserved for all collisions, elastic and inelastic (whenever external forces don't interfere).

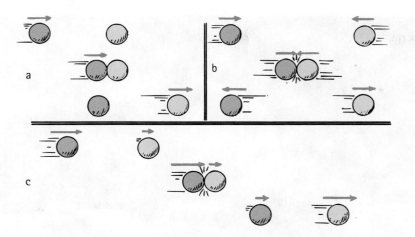

FIGURE 3.12
INTERACTIVE FIGURE   MP

Elastic collisions of equally massive balls. (a) A green ball strikes a yellow ball at rest. (b) A head-on collision. (c) A collision of balls moving in the same direction. In each case, momentum is transferred from one ball to the other.

When a moving billiard ball has a head-on collision with another billiard ball at rest, the moving ball comes to rest and the other ball moves with the speed of the colliding ball. We call this an **elastic collision**; ideally, the colliding objects rebound without lasting deformation or the generation of heat (Figure 3.12). But momentum is conserved even when the colliding objects become entangled during the collision. This is an **inelastic collision**, characterized by deformation, or the generation of heat, or both. In a perfectly inelastic collision, the objects stick together. Consider, for example, the case of a freight car moving along a track and colliding with another freight car at rest (Figure 3.13). If the freight cars are of equal mass and are coupled by the collision, can we predict the velocity of the coupled cars after impact?

Suppose the single car is moving at 10 m/s, and we consider the mass of each car to be m. Then, from the conservation of momentum,

$$(\text{net } mv)_{\text{before}} = (\text{net } mv)_{\text{after}}$$

$$(m \times 10 \text{ m/s})_{\text{before}} = (2m \times V)_{\text{after}}$$

By simple algebra, $V = 5$ m/s. This makes sense: because twice as much mass is moving after the collision, the velocity must be half as much as the velocity before the collision. Both sides of the equation are then equal.

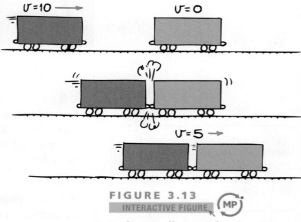

FIGURE 3.13
INTERACTIVE FIGURE   MP

Inelastic collision. The momentum of the freight car on the left is shared with the same-mass freight car on the right after collision.

CONSERVATION LAWS

A conservation law specifies that certain quantities in a system remain precisely constant, regardless of what changes may occur within the system. It is a law of constancy during change. In this chapter, we see that momentum is unchanged during collisions. We say that momentum is conserved. We'll soon learn that energy is conserved as it transforms—the amount of energy in light, for example, transforms completely to thermal energy when the light is absorbed. In Appendix A we'll see that angular momentum is conserved—whatever the rotational motion of a planetary system, its angular momentum remains unchanged so long as it is free of outside influences. In Chapter 8, we'll learn that electric charge is conserved, which means that it can be neither created nor destroyed. When we study nuclear physics, we'll see that these and other conservation laws rule in the submicroscopic world. Conservation laws are a source of deep insights into the simple regularity of nature and are often considered the most fundamental of physical laws. Can you think of things in your own life that remain constant as other things change?

FIGURE 3.14
Will Maynez demonstrates his air track. Blasts of air from tiny holes provide a friction-free surface for the carts to glide on.

Galileo worked hard to produce smooth surfaces to minimize friction. How he would have loved to experiment with today's air tracks!

CHECKPOINT

Consider the air track in Figure 3.14. Suppose a gliding cart with a mass of 0.5 kg bumps into, and sticks to, a stationary cart that has a mass of 1.5 kg. If the speed of the gliding cart before impact is v_{before}, how fast will the coupled carts glide after collision?

Was this your answer?
According to momentum conservation, the momentum of the 0.5-kg cart before the collision = momentum of both carts stuck together afterward.

$$(0.5 \text{ kg}) \, v_{before} = (0.5 \text{ kg} + 1.5 \text{ kg}) \, v_{after}$$

$$v_{after} = \frac{0.5 \text{ kg} \, v_{before}}{(0.5 \text{ kg} + 1.5 \text{ kg})} = \frac{0.5 \text{ kg} \, v_{before}}{2 \text{ kg}} = \frac{v_{before}}{4}$$

This makes sense, because four times as much mass will be moving after the collision, so the coupled carts will glide more slowly. The same momentum means that four times the mass glides $\frac{1}{4}$ as fast.

So we see that changes in an object's motion depend both on force and on how long the force acts. When "how long" means time, we refer to the quantity *force × time* as impulse. But "how long" can mean distance also. When we consider the quantity *force × distance*, we are talking about something entirely different—the concept of *energy*.

LEARNING OBJECTIVE
Describe how the work done on an object relates to its change in energy.

3.4 Energy and Work

EXPLAIN THIS How much faster will you hit the ground if you fall from twice the height?

Perhaps the concept most central to all of science is energy. The combination of energy and matter makes up the universe: matter is substance, and energy is the mover of substance. The idea of matter is easy to grasp.

Matter is stuff that we can see, smell, and feel. Matter has mass and occupies space. Energy, on the other hand, is abstract. We cannot see, smell, or feel most forms of energy. Surprisingly, the idea of energy was unknown to Isaac Newton, and its existence was still being debated in the 1850s. Although energy is familiar to us, it is difficult to define, because it is not only a "thing" but also both a thing and a process—similar to being both a noun and a verb. Persons, places, and things have energy, but we usually observe energy only when it is being transferred or being transformed. It appears in the form of electromagnetic waves from the Sun, and we feel it as thermal energy; it is captured by plants and binds molecules of matter together; it is in the foods we eat, and we receive it by digestion. Even matter itself is condensed, bottled-up energy, as set forth in Einstein's famous formula, $E = mc^2$, which we'll return to in the last part of this book. In general, **energy** is the property of a system that enables it to do *work*.

When you push a crate across a floor you're doing work. By definition, *force* × *distance* equals the concept we call **work**.

When we lift a load against Earth's gravity, work is done. The heavier the load or the higher we lift the load, the more work is being done. Two things enter the picture whenever work is done: (1) application of a force and (2) the movement of something by that force. For the simplest case, in which the force is constant and the motion is in a straight line in the direction of the force,* we define the work done on an object by an applied force as the product of the force and the distance through which the object is moved. In shorter form:

$$\text{Work} = \text{force} \times \text{distance}$$

$$W = Fd$$

If we lift two loads one story up, we do twice as much work as we do in lifting one load the same distance, because the *force* needed to lift twice the weight is twice as much. Similarly, if we lift a load two stories instead of one story, we do twice as much work because the *distance* is twice as great.

We see that the definition of work involves both a force and a distance. A weightlifter who holds a barbell weighing 1000 N overhead does no work on the barbell. She may get really tired holding the barbell, but if it is not moved by the force she exerts, she does no work *on the barbell*. Work may be done on the muscles by stretching and contracting, which is force times distance on a biological scale, but this work is not done on the barbell. Lifting the barbell, however, is a different story. When the weightlifter raises the barbell from the floor, she does work on it.

The unit of measurement for work combines a unit of force (N) with a unit of distance (m); the unit of work is the newton-meter (N·m), also called the *joule* (J), which rhymes with *cool*. One joule of work is done when a force of 1 N is exerted over a distance of 1 m, as in lifting an apple over your head. For larger values, we speak of kilojoules (kJ, thousands of joules), or megajoules (MJ, millions of joules). The weightlifter in Figure 3.16 does work in kilojoules. To stop a loaded truck moving at 100 km/h requires megajoules of work.

The word *work*, in common usage, means physical or mental exertion. Don't confuse the physics definition of work with the everyday notion of work.

FIGURE 3.15
He may expend energy when he pushes on the wall, but if the wall doesn't move, no work is done on the wall. Energy expended becomes thermal energy.

FIGURE 3.16
Work is done in lifting the barbell.

---

* More generally, work is the product of only the component of force that acts in the direction of motion and the distance moved. For example, if a force acts at an angle to the motion, the component of force parallel to the motion is multiplied by the distance moved. When a force acts at right angles to the direction of motion, with no force component in the direction of motion, no work is done. A common example is a satellite in a circular orbit; the force of gravity is at right angles to its circular path and no work is done on the satellite. Hence, it orbits with no change in speed.

FIGURE 3.17
The potential energy of Tenny's drawn bow equals the work (average force × distance) that she did in drawing the bow into position. When the arrow is released, most of the potential energy of the drawn bow will become the kinetic energy of the arrow.

CHECKPOINT

Assuming you have average strength, can you lift a 160-kg object with your bare hands? Can you do 1600 J of work on it?

Were these your answers?
An object with a mass of 160 kg weighs 1600 N, or 352 lb (the weight of a large refrigerator). So no, you cannot lift it without the use of some type of machine. If you can't move it, you can't do work on it. You'd do 1600 J of work on it if you could lift it a vertical distance of 1 m.

Potential Energy

An object may store energy by virtue of its position. The energy that is stored and held in readiness is called **potential energy** (PE) because in the stored state it has the potential for doing work. A stretched or compressed spring, for example, has the potential for doing work. When a bow is drawn, energy is stored in the bow. The bow can do work on the arrow. A stretched rubber band has potential energy because of the relative position of its parts. If the rubber band is part of a slingshot, it is capable of doing work.

The chemical energy in fuels is also potential energy. It is actually energy of position at the submicroscopic level. This energy is available when the positions of electric charges within and between molecules are altered—that is, when a chemical change occurs. Any substance that can do work through chemical action possesses potential energy. Potential energy is found in fossil fuels, electric batteries, and the foods we consume.

Work is required to elevate objects against Earth's gravity. The potential energy due to elevated positions is called *gravitational potential energy*. Water in an elevated reservoir and the raised ram of a pile driver both have gravitational potential energy. Whenever work is done, energy is exchanged.

The amount of gravitational potential energy possessed by an elevated object is equal to the work done against gravity in lifting it. The work done equals the force required to move it upward multiplied by the vertical distance it is moved (remember $W = Fd$). The upward force required while moving at constant velocity is equal to the weight, mg, of the object, so the work done in lifting it through a height h is the product mgh:

$$\text{Gravitational potential energy} = \text{weight} \times \text{height}$$

$$\text{PE} = mgh$$

Note that the height is the distance above some chosen reference level, such as the ground or the floor of a building. The gravitational potential energy, mgh, is relative to that level and depends only on mg and h. We can see, in Figure 3.18,

An average apple weighs 1 N. When it is held 1 m above ground, then relative to the ground it has a PE of 1 J.

FIGURE 3.18
The potential energy of the 10-N ball is the same (30 J) in all three cases because the work done in elevating it 3 m is the same whether it is (a) lifted with 10 N of force, (b) pushed with 6 N of force up the 5-m incline, or (c) lifted with 10 N up each 1-m step. No work is done in moving it horizontally (neglecting friction).

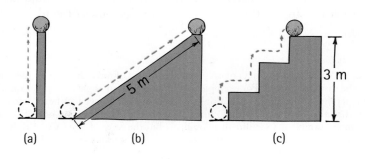

(a)   (b)   (c)

that the potential energy of the elevated ball does not depend on the path taken to get it there.

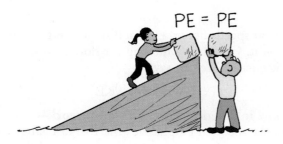

FIGURE 3.19
He raises a block of ice by lifting it vertically. She pushes an identical block of ice up the ramp. Can you see that they do equal amounts of work? And can you see that when both blocks are raised to the same vertical height, they possess the same potential energy?

Kinetic Energy

If you push on an object, you can set it in motion. If an object is moving, then it is capable of doing work. It has energy of motion. We say it has *kinetic energy* (KE). The **kinetic energy** of an object depends on the mass of the object as well as its speed. It is equal to the mass multiplied by the square of the speed, multiplied by the constant $\frac{1}{2}$:

$$\text{Kinetic energy} = \tfrac{1}{2}\,\text{mass} \times \text{speed}^2$$

$$\text{KE} = \tfrac{1}{2}\,mv^2$$

> Gravitational potential energy always involves *two* interacting objects—one relative to the other. The ram of a pile driver, for example, interacts via gravitational force with Earth.

When you throw a ball, you do work on it to give it speed as it leaves your hand. The moving ball can then hit something and push it, doing work on what it hits. The kinetic energy of a moving object is equal to the work required to bring it from rest to that speed, or the work the object can do while being brought to rest:

$$\text{Net force} \times \text{distance} = \text{kinetic energy}$$

or, in equation notation,

$$Fd = \tfrac{1}{2}\,mv^2$$

FIGURE 3.20
The potential energy of the elevated ram of the pile driver is converted to kinetic energy during its fall.

Note that the speed is squared, so if the speed of an object is doubled, its kinetic energy is quadrupled ($2^2 = 4$). Consequently, four times the work is required to double the speed. Likewise, nine times the work is required to triple the speed ($3^2 = 9$). The fact that speed or velocity is squared for kinetic energy clearly distinguishes the concepts of kinetic energy and momentum. What we can say is that in all interactions, whenever work is done, some form of energy increases. Whenever work is done, energy changes.

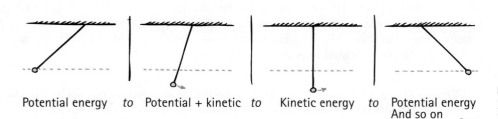

Potential energy   *to*   Potential + kinetic   *to*   Kinetic energy   *to*   Potential energy
And so on

FIGURE 3.21
Energy transitions in a pendulum. PE is relative to the lowest point of the pendulum, when it is vertical.

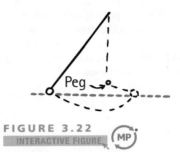

FIGURE 3.22
INTERACTIVE FIGURE (MP)

The pendulum bob will swing to its original height whether or not the peg is present.

FIGURE 3.23
The downhill "fall" of the roller coaster results in its roaring speed in the dip, and this kinetic energy sends it up the steep track to the next summit.

3.5 Work–Energy Theorem

EXPLAIN THIS How much farther will you skid on wet grass if you run twice as fast?

When a car speeds up, its gain in kinetic energy comes from the work done on it. Or, when a moving car slows, work is done to reduce its kinetic energy. We can say*

$$\text{Work} = \Delta \text{KE}$$

Work equals *change* in kinetic energy. This is the **work–energy theorem**.

The work–energy theorem emphasizes the role of change. Some forces can change potential energy. Recall our example of the weightlifter raising the barbell. While he exerts a force through a distance, he does work on the barbell and changes its potential energy. And when the barbell is held stationary, no further work is done and there is no further change in energy. Now if the weightlifter drops the barbell, gravity does work as the barbell is pulled down, increasing its kinetic energy.

If you push against a box on a floor and the box doesn't slide, then no change in its energy tells you that you are not doing work on the box. If you then push harder and the box slides, you are doing work on it. You push in one direction and friction acts in the other direction. The difference is a net force that does work to give the box its kinetic energy.

The work–energy theorem applies to decreasing speed as well. Energy is required to reduce the speed of a moving object or to bring it to a halt. When we apply the brakes to slow a moving car, we do work on it. This work is the friction force supplied by the brakes multiplied by the distance over which the friction force acts. The more kinetic energy something has, the more work is required to stop it.

Interestingly, the friction supplied by the brakes is the same whether the car moves slowly or quickly. Friction between solid surfaces doesn't depend on speed. The variable that makes a difference is the braking distance. A car moving at twice the speed of another takes four times ($2^2 = 4$) as much work to stop. Therefore, it takes four times as much distance to stop. Accident investigators are well aware that an automobile going 100 km/h has four times the kinetic energy it would have at 50 km/h. So a car going 100 km/h skids four times as far when its brakes are locked as it does when going 50 km/h. Kinetic energy depends on speed *squared*.

Automobile brakes convert kinetic energy to heat. Professional drivers are familiar with another way to slow a vehicle—shift to low gear to allow the engine to do the braking. Today's hybrid cars do the same and divert braking energy to electrical storage batteries, where it is used to complement the energy produced by gasoline combustion (Chapter 9 treats how they accomplish this).

Kinetic energy and potential energy are two of the many forms of energy, and they underlie other forms of energy, such as chemical energy, nuclear energy, sound, and light. Kinetic energy of random molecular motion is related to temperature; potential energies of electric charges account for voltage; and kinetic and potential energies of vibrating air define sound intensity. Even light energy originates from the motion of electrons within atoms. Every form of energy can be transformed into every other form.

---

* This can be derived as follows: If we multiply both sides of $F = ma$ (Newton's second law) by d, we get $Fd = mad$. Recall from Chapter 2 that, for constant acceleration, $d = \frac{1}{2}at^2$, so we can say $Fd = ma\left(\frac{1}{2}at^2\right) = \frac{1}{2}maat^2 = \frac{1}{2}m(at)^2$; and substituting $v = at$, we get $Fd = \frac{1}{2}mv^2$. That is, work = KE or, more specifically, $W = \Delta\text{KE}$.

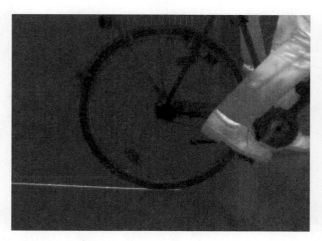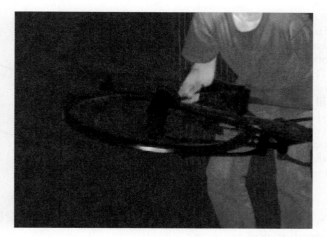

FIGURE 3.24
Because of friction, energy is transferred both into the floor and into the tire when the bicycle skids to a stop. An infrared camera reveals the heated tire track (the red streak on the floor, left) and the warmth of the tire (right). (Courtesy of Michael Vollmer.)

CHECKPOINT
1. **When you are driving at 90 km/h, how much more distance do you need to stop than if you were driving at 30 km/h?**
2. **For the same force, why does a longer cannon impart more speed to a cannonball?**

Were these your answers?
1. Nine times as much distance. The car has nine times as much kinetic energy when it travels three times as fast: $\frac{1}{2}m(3v)^2 = \frac{1}{2}m9v^2 = 9\left(\frac{1}{2}mv^2\right)$. The friction force is ordinarily the same in either case; therefore, nine times as much work requires nine times as much distance.
2. As learned earlier, a longer barrel imparts more impulse because of the longer *time* during which the force acts. The work–energy theorem similarly tells us that the longer the *distance* over which the force acts, the greater the change in kinetic energy. So we see two reasons for cannons with long barrels producing greater cannonball speeds.

Kinetic Energy and Momentum Compared

Momentum and kinetic energy are properties of moving things, but they differ from each other. Like velocity, momentum is a vector quantity and is therefore directional and capable of being canceled entirely. But kinetic energy is a non-vector (scalar) quantity, like mass, and can never be canceled. The momenta of two firecrackers approaching each other may cancel, but when they explode, there is no way their energies can cancel. Energies transform to other forms; momenta do not. Another difference is the velocity dependence of the two. Whereas momentum depends on velocity (mv), kinetic energy depends on the square of velocity $\left(\frac{1}{2}mv^2\right)$. An object that moves with twice the velocity of another object of the same mass has twice the momentum but four times the kinetic energy. So when a car traveling twice as fast crashes, it crashes with four times the energy.

If the distinction between momentum and kinetic energy isn't really clear to you, you're in good company. Failure to make this distinction resulted in disagreements and arguments between the best British and French physicists for almost two centuries.

Energy is nature's way of keeping score. Scams that sell energy-making machines rely on funding from deep pockets and shallow brains!

FYI Scientists have to be open to new ideas. That's how science grows. But a body of established knowledge exists that can't be easily overthrown. That includes energy conservation, which is woven into every branch of science and supported by countless experiments from the atomic to the cosmic scale. Yet no concept has inspired more "junk science" than energy. Wouldn't it be wonderful if we could get energy for nothing, to possess a machine that gives out more energy than is put into it? That's what many practitioners of junk science offer. Gullible investors put their money into some of these schemes. But none of them pass the test of being real science. Perhaps someday a flaw in the law of energy conservation will be discovered. If it ever is, scientists will rejoice at the breakthrough. But so far, energy conservation is as solid as any knowledge we have. Don't bet against it.

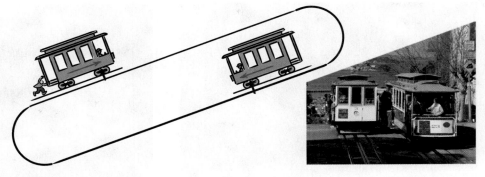

FIGURE 3.25
Cable cars on the steep hills of San Francisco nicely transfer energy to one another via the cable beneath the street. The cable forms a complete loop that connects cars going both downhill and uphill. In this way a car moving downhill does work on a car moving uphill. So the increased gravitational PE of an uphill car is due to the decreased gravitational PE of a car moving downhill.

LEARNING OBJECTIVE
Relate conservation of energy to physics and science in general.

Mastering**PHYSICS**

VIDEO: Bowling Ball and Conservation of Energy

VIDEO: Conservation of Momentum: Numerical Example

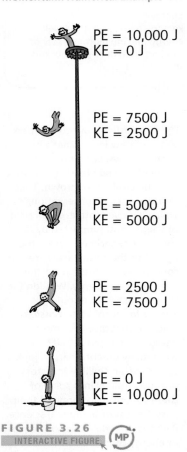

PE = 10,000 J
KE = 0 J

PE = 7500 J
KE = 2500 J

PE = 5000 J
KE = 5000 J

PE = 2500 J
KE = 7500 J

PE = 0 J
KE = 10,000 J

FIGURE 3.26
INTERACTIVE FIGURE

A circus diver at the top of a pole has a potential energy of 10,000 J. As he dives, his potential energy converts to kinetic energy. Note that, at successive positions one-fourth, one-half, three-fourths, and all the way down, the total energy is constant.

3.6 Conservation of Energy

EXPLAIN THIS What is the energy score before and after galaxies collide?

Whenever energy is transformed or transferred, none is lost and none is gained. In the absence of work input or output or other energy exchanges, the total energy of a system before some process or event is equal to the total energy after.

Consider the changes in energy in the operation of the pile driver back in Figure 3.20. Work done to raise the ram, giving it potential energy, becomes kinetic energy when the ram is released. This energy transfers to the piling below. The distance the piling penetrates into the ground multiplied by the average force of impact is almost equal to the initial potential energy of the ram. We say *almost* because some energy goes into heating the ground and ram during penetration. Taking heat energy into account, we find that energy transforms without net loss or net gain. Quite remarkable!

The study of various forms of energy and their transformations has led to one of the greatest generalizations in physics—the **law of conservation of energy**:

Energy cannot be created or destroyed; it may be transformed from one form into another, but the total amount of energy never changes.

When we consider any system in its entirety, whether it be as simple as a swinging pendulum or as complex as an exploding supernova, one quantity isn't created or destroyed: energy. It may change form or it may simply be transferred from one place to another, but the total energy score stays the same. This energy score takes into account the fact that the atoms that make up matter are themselves concentrated bundles of energy. When the nuclei (cores) of atoms rearrange themselves, enormous amounts of energy can be released. The Sun shines because some of this nuclear energy is transformed into radiant energy.

Enormous compression due to gravity and extremely high temperatures in the deep interior of the Sun fuse the nuclei of hydrogen atoms together to form helium nuclei. This is *thermonuclear fusion*, a process that releases radiant energy, a small part of which reaches Earth. Part of the energy reaching Earth falls on plants (and on other photosynthetic organisms), and part of this, in turn, is later stored in the form of coal. Another part supports life in the food chain that begins with plants (and other photosynthesizers), and part of this energy later is stored in oil. Part of the energy from the Sun goes into the evaporation of water from the ocean, and part of this returns to Earth in rain that may be trapped behind a dam. By virtue of its elevated position, the water behind a dam has energy that may be used to power a generating plant below, where it is transformed to electric energy. The energy travels through wires to homes, where it is used for lighting, heating, cooking, and operating electrical gadgets. How wonderful that energy transforms from one form to another!

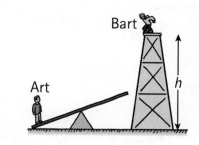

FIGURING PHYSICAL SCIENCE

Problem Solving

SAMPLE PROBLEM

Acrobat Art of mass m stands on the left end of a seesaw. Acrobat Bart of mass M jumps from a height h onto the right end of the seesaw, thus propelling Art into the air.

(a) Neglecting inefficiencies, how does the PE of Art at the top of his trajectory compare with the PE of Bart just before Bart jumps?

(b) Show that ideally Art reaches a height $\frac{M}{m}h$.

(c) If Art's mass is 40 kg, Bart's mass is 70 kg, and the height of the initial jump was 4 m, show that Art rises a vertical distance of 7 m.

Solution:

(a) Neglecting inefficiencies, the entire initial PE of Bart before he drops goes into the PE of Art rising to his peak—that is, at Art's moment of zero KE.

(b) $PE_{Bart} = PE_{Art}$

$Mgh_{Bart} = mgh_{Art}$

$h_{Art} = \frac{M}{m}h.$

(c) $h_{Art} = \frac{M}{m}h = \left(\frac{70 \text{ kg}}{40 \text{ kg}}\right) 4 \text{ m} = 7 \text{ m}.$

3.7 Power

EXPLAIN THIS Why do you run out of breath when running up stairs but not when walking up?

The definition of work says nothing about how long it takes to do the work. The same amount of work is done when carrying a bag of groceries up a flight of stairs, whether we walk up or run up. So why are we more out of breath after running upstairs in a few seconds than after walking upstairs in a few minutes? To understand this difference, we need to talk about a measure of how fast the work is done—*power*. **Power** is equal to the amount of work done per time it takes to do it:

$$\text{Power} = \frac{\text{work done}}{\text{time interval}}$$

The work done in climbing stairs requires more power when the worker is running up rapidly than it does when the worker is climbing slowly. A high-power automobile engine does work rapidly. An engine that delivers twice the power of another, however, does not necessarily move a car twice as fast or twice as far. Twice the power means that the engine can do twice the work in the same amount of time—or it can do the same amount of work in half the time. A powerful engine can produce greater acceleration.

Power is also the rate at which energy is changed from one form to another. The unit of power is the joule per second, called the *watt*. This unit was named in honor of James Watt, the 18th-century developer of the steam engine. One watt (W) of power is used when 1 J of work is done in 1 s. One kilowatt (kW) equals 1000 W. One megawatt (MW) equals 1 million watts.

FYI Your heart uses slightly more than 1 W of power in pumping blood through your body.

FIGURE 3.27
The three main engines of a space shuttle can develop 33,000 MW of power when fuel is burned at the enormous rate of 3400 kg/s. This is like emptying an average-size swimming pool in 20 s.

3.8 Machines

EXPLAIN THIS Why should or shouldn't you invest in a machine that creates energy?

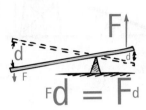

FIGURE 3.28
The lever.

A machine is a device for multiplying forces or simply changing the direction of forces. The principle underlying every machine is conservation of energy. Consider one of the simplest machines, the **lever** (Figure 3.28). At the same time that we do work on one end of the lever, the other end does work on the load. We see that the direction of force is changed: if we push down, the load is lifted up. If the little work done by friction forces is small enough to neglect, the work input equals the work output:

$$\text{Work input} = \text{work output}$$

Because work equals force times distance, **conservation of energy for machines** tells us that *input force × input distance = output force × output distance.*

$$(\text{force} \times \text{distance})_{\text{input}} = (\text{force} \times \text{distance})_{\text{output}}$$

The point of support on which a lever rotates is called the *fulcrum.* When the fulcrum of a lever is relatively close to the load, a small input force produces a large output force. This is because the input force is exerted through a large distance and the load is moved through a correspondingly short distance. So a lever can be a force multiplier. But no machine can multiply work or multiply energy. That's a conservation-of-energy no-no!

FIGURE 3.29
Applied force × applied distance = output force × output distance.

Today, a child can use the principle of the lever to jack up the front end of an automobile. By exerting a small force through a large distance, she can provide a large force that acts through a small distance. Consider the ideal example illustrated in Figure 3.29. Every time she pushes the jack handle down 25 cm, the car rises only a hundredth as far but with 100 times the force.

Another simple machine is a pulley. Can you see that it is a lever "in disguise"? When used as in Figure 3.30, it changes only the direction of the force; but, when used as in Figure 3.31, the output force is doubled. Force is increased and distance is decreased. As with any machine, forces can change while work input and work output are unchanged.

A block and tackle is a system of pulleys that multiplies force more than a single pulley can. With the ideal pulley system shown in Figure 3.32, the man

MasteringPHYSICS®
VIDEO: Machines: Pulleys

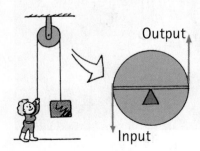

FIGURE 3.30
This pulley acts like a lever with equal arms. It changes only the direction of the input force.

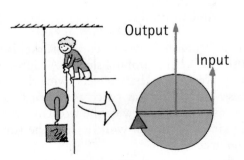

FIGURE 3.31
In this arrangement, a load can be lifted with half the input force. Note that the "fulcrum" is at the left end rather than in the center (as is the case in Figure 3.30).

FIGURE 3.32
Applied force × applied distance = output force × output distance.

pulls 7 m of rope with a force of 50 N and lifts a load of 500 N through a vertical distance of 0.7 m. The energy the man expends in pulling the rope is numerically equal to the increased potential energy of the 500-N block. Energy is transferred from the man to the load.

Any machine that multiplies force does so at the expense of distance. Likewise, any machine that multiplies distance, such as your forearm and elbow, does so at the expense of force. No machine or device can put out more energy than is put into it. No machine can create energy; it can only transfer energy or transform it from one form to another.

3.9   Efficiency

EXPLAIN THIS   What is meant by an *ideal machine*?

The three previous examples were of *ideal machines*; 100% of the work input appeared as work output. An ideal machine would operate at 100% efficiency. In practice, this doesn't happen, and we can never expect it to happen. In any transformation, some energy is dissipated to molecular kinetic energy—thermal energy. This makes the machine and its surroundings warmer.

Efficiency can be expressed by the ratio

$$\text{Efficiency} = \frac{\text{useful energy output}}{\text{total energy input}}$$

Even a lever converts a small fraction of input energy into heat when it rotates about its fulcrum. We may do 100 J of work but get out only 98 J. The lever is then 98% efficient, and we waste 2 J of work input as heat. In a pulley system, a larger fraction of input energy goes into heat. If we do 100 J of work, the forces of friction acting through the distances through which the pulleys turn and rub about their axles may dissipate 60 J of energy as heat. So the work output is only 40 J, and the pulley system has an efficiency of 40%. The lower the efficiency of a machine, the greater the amount of energy wasted as heat.*

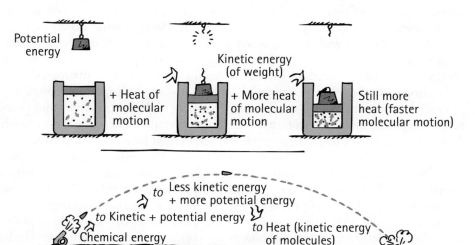

* When you study thermodynamics in Chapter 6, you'll learn that an internal combustion engine *must* transform some of its fuel energy into thermal energy. A fuel cell, on the other hand, doesn't have this limitation. Watch for fuel cell–powered vehicles in the future!

LEARNING OBJECTIVE
Describe efficiency in terms of energy input and output.

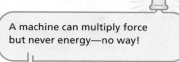

A machine can multiply force but never energy—no way!

FIGURE 3.33
Energy transitions. The graveyard of mechanical energy is thermal energy.

Inventors take heed: When introducing a new idea, first be sure it is in context with what is presently known. For example, it should be consistent with the conservation of energy.

CHECKPOINT

Consider an imaginary miracle car that has a 100% efficient internal combustion engine and burns fuel that has an energy content of 40 megajoules per liter (MJ/L). If the air resistance and overall frictional forces on the car traveling at highway speed are 500 N, show that the distance the car could travel per liter at this speed is 80 km/L.

Was this your answer?
From the definition that *work = force × distance*, simple rearrangement gives *distance = work/force*. If all 40 million J of energy in 1 L were used to do the work of overcoming the air resistance and frictional forces, the distance would be

$$\text{Distance} = \frac{\text{work}}{\text{force}} = \frac{40,000,000 \text{ J/L}}{500 \text{ N}} = 80,000 \text{ m/L} = 80 \text{ km/L}$$

(This is about 190 miles per gallon [mpg].) The important point here is that, even with a hypothetically perfect engine, there is an upper limit of fuel economy dictated by the conservation of energy.

LEARNING OBJECTIVE
Identify and describe the two ultimate sources of energy on Earth.

FYI The power available in sunlight is about 1 kW/m². If all of the solar energy falling on a square meter could be harvested for power production, that energy would generate 1000 W. Some solar cells can convert 40% of the power, or about 400 W/m². Solar power via low-cost thin solar films used in building materials, including roofing and glass, is changing the way we produce and distribute energy.

3.10 Sources of Energy

EXPLAIN THIS How can the Sun be the source of hydroelectric-, wind-, and fossil-fuel power?

Except for nuclear power, the source of practically all our energy is the Sun. Even the energy we obtain from petroleum, coal, natural gas, and wood originally came from the Sun. That's because these fuels are created by photosynthesis—the process by which plants trap solar energy and store it as plant tissue.

Sunlight evaporates water, which later falls as rain; rainwater flows into rivers and into dams where it is directed to generator turbines. Then it returns to the sea, where the cycle continues. Even the wind, caused by unequal warming of Earth's surface, is a form of solar power. The energy of wind can be used to turn generator turbines within specially equipped windmills. Because wind power can't be turned on and off at will, it is presently a supplement to fossil and nuclear fuels for large-scale power production. Harnessing the wind is most practical when the energy it produces is stored for future use, such as in the form of hydrogen.

Hydrogen is the least polluting of all fuels. Most hydrogen in America is produced from natural gas, in a process that uses high temperatures and pressures to separate hydrogen from hydrocarbon molecules. The same is done with other fossil fuels. A downside to separating hydrogen from carbon compounds is the unavoidable production of carbon dioxide, a greenhouse gas. A simpler and cleaner method that doesn't produce greenhouse gases is *electrolysis*—electrically splitting water into its constituent parts. Figure 3.34 shows how you can perform this in the lab or at home: Place two wires that are connected to the terminals of an ordinary battery into a glass of salted water. Be sure the wires don't touch each other. Bubbles of hydrogen form on one wire, and bubbles of oxygen form on the other. A fuel cell is similar, but runs backward. Hydrogen and oxygen gas are combined at electrodes and electric current is produced, along with water. The space shuttle uses fuel cells to meet its electrical needs while producing drinking water for the astronauts. Here on Earth fuel-cell researchers are developing fuel cells for buses, automobiles, and trains.

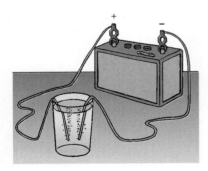

FIGURE 3.34
When electric current passes through conducting water, bubbles of hydrogen form at one wire and bubbles of oxygen form at the other. This is *electrolysis*. A fuel cell does the opposite—hydrogen and oxygen enter the fuel cell and are combined to produce electricity and water.

A hydrogen economy may likely start with railroad trains rather than auto-mobiles being powered by fuel cells. Hydrogen can be obtained via solar cells, many along train tracks and on the rail ties themselves (Figure 3.35). Photo-voltaic cells transform sunlight to electricity. They are familiar in solar-powered calculators, iPods, and flexible solar-powered shingles on rooftops. Solar cells can also supply the energy needed to produce hydrogen. It is important to know that hydrogen is not a *source* of energy. Energy is required to make hydrogen (to extract it from water and carbon compounds). As with electricity, the production of hydrogen needs an energy source; the hydrogen thus produced provides a way of storing and transporting that energy. Again, for emphasis, hydrogen is *not* an energy source.

The most concentrated source of usable energy is that stored in nuclear fuels—uranium and plutonium. For the same weight of fuel, nuclear reactions release about 1 million times more energy than do chemical or food reactions. Watch for renewed interest in this form of power that doesn't pollute the atmosphere. Interestingly, Earth's interior is kept hot because of nuclear power, which has been with us since time zero.

A by-product of nuclear power in Earth's interior is geothermal energy. Geothermal energy is held in underground reservoirs of hot water. Geothermal energy is predominantly limited to areas of volcanic activity, such as Iceland, New Zealand, Japan, and Hawaii. In these locations, heated water near Earth's surface is tapped to provide steam for driving turbogenerators.

In locations where heat from volcanic activity is near the surface and groundwater is absent, another method holds promise for producing electricity: dry-rock geothermal power (Figure 3.36). With this method, water is put into cavities in deep, dry, hot rock. When the water turns to steam, it is piped to a turbine at the surface. After turning the turbine, it is returned to the cavity for reuse. In this way, carbon-free electricity is produced.

As the world population increases, so does our need for energy, especially because per-capita demand is also growing. With the rules of physics to guide them, technologists are presently researching newer and cleaner ways to develop energy sources. But they race to keep ahead of a growing world population and greater demand in the developing world. Unfortunately, as long as controlling population is politically and religiously incorrect, human misery becomes the check to unrestrained population growth. H. G. Wells once wrote (in *The Outline of History*), "Human history becomes more and more a race between education and catastrophe."

FIGURE 3.35
The power harvested by photovoltaic cells can be used to extract hydrogen for fuel-cell transportation. Plans for trains that run on solar power collected on railroad-track ties are presently at the drawing-board stage (see http://www.SuntrainUSA.com).

> **FYI** Another source of energy is tidal power, by which the surging of tides turns turbines to produce power. Interestingly, this form of energy is neither nuclear nor from the Sun. It comes from the rotational energy of our planet.

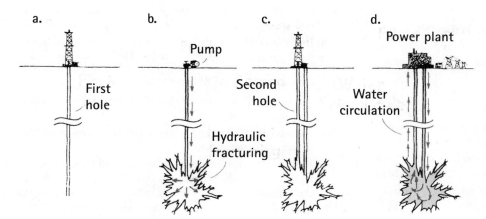

a.   b.   c.   d.

First hole

Pump

Second hole

Hydraulic fracturing

Power plant

Water circulation

FIGURE 3.36
Dry-rock geothermal power. (a) A hole is sunk several kilometers into dry granite. (b) Water is pumped into the hole at high pressure and fractures the surrounding rock to form a cavity with increased surface area. (c) A second hole is sunk to intercept the cavity. (d) Water is circulated down one hole and through the cavity, where it is superheated before rising through the second hole. After driving a turbine, it is recirculated into the hot cavity again, making a closed cycle.

For instructor-assigned homework, go to www.masteringphysics.com MP

SUMMARY OF TERMS (KNOWLEDGE)

Conservation of energy for machines The work output of any machine cannot exceed the work input. In an ideal machine, where no energy is transformed into thermal energy,

$$\text{work}_{input} = \text{work}_{output} \quad \text{and} \quad (Fd)_{input} = (Fd)_{output}$$

Efficiency The percentage of the work put into a machine that is converted into useful work output:

$$\text{Efficiency} = \frac{\text{useful energy output}}{\text{total energy input}}$$

(More generally, efficiency is useful energy output divided by total energy input.)

Elastic collision A collision in which colliding objects rebound without lasting deformation or the generation of heat.

Energy The property of a system that enables it to do work.

Impulse The product of the force acting on an object and the time during which it acts.

Impulse–momentum relationship Impulse is equal to the change in the momentum of an object that the impulse acts upon. In symbol notation:

$$Ft = \Delta(mv)$$

Inelastic collision A collision in which the colliding objects become distorted, generate heat, and possibly stick together.

Kinetic energy Energy of motion, quantified by the relationship:

$$\text{Kinetic energy} = \frac{1}{2}mv^2$$

Law of conservation of energy Energy cannot be created or destroyed; it may be transformed from one form into another, but the total amount of energy never changes.

Law of conservation of momentum In the absence of an external force, the momentum of a system remains unchanged. Hence, the momentum before an event involving only internal forces is equal to the momentum after the event:

$$mv_{\text{before event}} = mv_{\text{after event}}$$

Lever A simple machine consisting of a rigid rod pivoted at a fixed point called the fulcrum.

Machine A device, such as a lever or pulley, that increases (or decreases) a force or simply changes the direction of a force.

Momentum Inertia in motion, given by the product of the mass of an object and its velocity.

Potential energy The energy that matter possesses due to its position:

$$\text{Gravitational PE} = mgh$$

Power The rate of doing work (or the rate at which energy is expended):

$$\text{Power} = \frac{\text{work}}{\text{time}}$$

Work The product of the force and the distance moved by the force:

$$W = Fd$$

(More generally, work is the component of force in the direction of motion multiplied by the distance moved.)

Work–energy theorem The net work done on an object equals the change in kinetic energy of the object:

$$\text{Work} = \Delta\text{KE}$$

READING CHECK QUESTIONS (COMPREHENSION)

3.1 Momentum and Impulse

1. Which has a greater momentum: an automobile at rest or a moving skateboard?
2. When a ball is hit with a given force, why does contact over a long time impart more speed to the ball?

3.2 Impulse Changes Momentum

3. Why is it a good idea to extend your bare hand forward when you are getting ready to catch a fast-moving baseball?
4. Why would it be a bad idea to have the back of your hand up against the outfield wall when you catch a long fly ball?
5. In karate, why is a force that is applied for a short time more effective?
6. In boxing, why is it advantageous to roll with the punch?
7. If a ball has the same speed just before being caught and just after being thrown, in which case does the ball undergo the greatest change in momentum: (a) when it is caught, (b) when it is thrown, or (c) when it is caught and then thrown back?
8. In the preceding question, in which of the cases a, b, or c is the greatest impulse required?

3.3 Conservation of Momentum

9. What does it mean to say that momentum (or any quantity) is *conserved*?
10. Is momentum conserved during an elastic collision? During an inelastic collision?
11. Railroad car A rolls at a certain speed and makes a perfectly elastic collision with car B of the same mass. After the collision, car A is observed to be at rest. How does the speed of car B compare with the initial speed of car A?
12. If the equally massive railroad cars of the preceding question couple together after colliding inelastically, how does their speed after the collision compare with the initial speed of car A?

3.4 Energy and Work

13. When is energy most evident?

14. Cite an example in which a force is exerted on an object without doing work on the object.

15. Which, if either, requires more work: lifting a 50-kg sack a vertical distance of 2 m or lifting a 25-kg sack a vertical distance of 4 m?

16. A car is raised a certain distance on a service-station lift and therefore has potential energy relative to the floor. If it were raised twice as high, how much potential energy would it have?

17. Two cars are raised to the same elevation on service-station lifts. If one car is twice as massive as the other, how do their potential energies compare?

18. If a moving car speeds up until it is going twice as fast, how much kinetic energy does it have compared with its initial kinetic energy?

3.5 Work–Energy Theorem

19. Compared with some original speed, how much work must the brakes of a car supply to stop a car that is moving four times as fast? How does the stopping distance compare?

20. If you push a crate horizontally with a force of 100 N across a 10-m factory floor, and the friction force between the crate and the floor is a steady 70 N, how much kinetic energy does the crate gain?

3.6 Conservation of Energy

21. What will be the kinetic energy of the ram of a pile driver when it suddenly undergoes a 10-kJ decrease in potential energy?

22. An apple hanging from a limb has potential energy because of its height. If the apple falls, what becomes of this energy just before the apple hits the ground? When it hits the ground?

3.7 Power

23. What is the relationship between work and power?

3.8 Machines

24. Can a machine multiply the input force? Input distance? Input energy? (If your three answers are the same, seek help; this question is especially important.)

25. If a machine multiplies force by a factor of 4, what other quantity is diminished, and by how much?

3.9 Efficiency

26. What is the efficiency of a machine that miraculously converts all the input energy into useful output energy?

27. What becomes of energy when efficiency is lowered in a machine?

3.10 Sources of Energy

28. What is the ultimate source of energies for fossil fuels, dams, and windmills?

29. What is the ultimate source of geothermal energy?

30. Can we correctly say that hydrogen is a relatively new source of energy? Why or why not?

ACTIVITIES (HANDS-ON APPLICATION)

31. If your instructor has an air table or air track, play around with carts or air pucks. Most important, predict what will happen before you initiate collisions.

32. When you get a bit ahead in your studies, cut classes some afternoon and visit your local pool or billiards parlor and bone up on momentum conservation. Note that, no matter how complicated the collision of balls, the momentum along the line of action of the cue ball before impact is the same as the combined momenta of all the balls along this direction after impact and that the components of momenta perpendicular to this line of action cancel to zero after impact, the same value as before impact in this direction. You'll see both the vector nature of momentum and its conservation more clearly when rotational skidding, "English," is not imparted to the cue ball. When English is imparted by striking the cue ball off-center, rotational momentum, which is also conserved, somewhat complicates analysis. But, regardless of how the cue ball is struck, in the absence of external forces, both linear and rotational momentum are always conserved. Both pool and billiards offer a first-rate exhibition of momentum conservation in action.

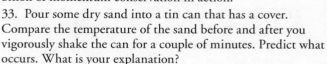

33. Pour some dry sand into a tin can that has a cover. Compare the temperature of the sand before and after you vigorously shake the can for a couple of minutes. Predict what occurs. What is your explanation?

34. Place a small rubber ball on top of a basketball or soccer ball and then drop them together. If vertical alignment nicely remains as they fall to the floor, you'll see that the small ball bounces unusually high. Can you reconcile this with energy conservation?

PLUG AND CHUG (FORMULA FAMILIARIZATION)

Momentum = mv

35. Show that the momentum is 16 kg · m/s for a 2-kg brick parachuting straight downward at a constant speed of 8 m/s.

36. Calculate the momentum of a 10-kg bowling ball rolling at 2 m/s.

Impulse = *Ft*

37. Show that the impulse on a baseball that is hit with 100 N of force in a time of 0.5 s is 50 N · s.

38. Calculate the impulse of a cart when an average force of 10 N is exerted on it for 2.5 s.

Impulse–momentum relationship: *Ft* = Δ(*mv*)

39. Show that an average force of 25 N exerted on a cart for 2 s changes the momentum of the cart by 50 kg · m/s.

40. Rearrange the equation *Ft* = Δ(*mv*) to solve for the force *F*. Show that force *F* is 25 N if it acts for 2 s to cause a 25-kg cart to gain 2 m/s in speed. (1 N = 1 kg · m/s²)

Work = force × distance; *W* = *Fd*

41. Show that 70 J of work is done when a 20-N force moves a cart 3.5 m from its initial position. (1 J = N · m)

42. Calculate the work done in lifting a 100-N block of ice a vertical distance of 5 m.

Gravitational potential energy = mass × acceleration due to gravity × height; PE = *mgh*

43. Show that the gravitational potential energy of a 10-kg boulder raised 5 m above ground level is 500 J. (You can express *g* in units of N/kg because m/s² is equivalent to N/kg.)

44. Calculate the number of joules of potential energy required to elevate a 1.5-kg book 2.0 m.

Kinetic energy = ½*mv*²

45. Show that the kinetic energy of an 84-kg scooter moving at 2 m/s is 168 J. (1 J is equivalent to 1 N · m, which is equivalent to 1 kg · m²/s².)

46. Show that the scooter in the preceding problem will have four times the kinetic energy when its speed is doubled to 4 m/s.

Work–energy theorem: *W* = ΔKE

47. A sustained force of 50 N moves a model airplane 20 m along its runway to provide the required speed for take-off. Show that the kinetic energy at takeoff is 1000 J.

48. Show that 90 J of work is needed to increase the speed of a 20-kg cart by 3 m/s.

Power = $\frac{\text{work done}}{\text{time interval}}$ = $\frac{W}{t}$

49. Show that the power required to give a brick 100 J of potential energy in a time of 2 s is 50 W.

THINK AND SOLVE (MATHEMATICAL APPLICATION)

50. In Chapter 1 we learned the definition of acceleration, $a = \Delta v/\Delta t$, and in Chapter 2 we learned that the cause of acceleration involves net force, where $a = F/m$. Equate these two equations for acceleration and show that, for constant mass, $F\Delta t = \Delta(mv)$.

51. A 10-kg bag of groceries is tossed onto a table at 3 m/s and slides to a stop in 2 s. Modify the equation $F\Delta t = \Delta(mv)$ to show that the force of friction is 15 N.

52. An ostrich egg of mass *m* is tossed at a speed *v* into a sagging bed sheet and is brought to rest in a time *t*.
    (a) Show that the force acting on the egg when it hits the sheet is *mv/t*.
    (b) If the mass of the egg is 1 kg, its initial speed is 2 m/s, and the time to stop is 1 s, show that the average force on the egg is 2 N.

53. A 6-kg ball rolling at 3 m/s bumps into a pillow and stops in 0.5 s.
    (a) Show that the force exerted by the pillow is 36 N.
    (b) How much force does the ball exert on the pillow?

54. At a baseball game a ball of mass *m* = 0.15 kg moving at a speed *v* = 30 m/s is caught by a fan.
    (a) Show that the impulse supplied to bring the ball to rest is 4.5 N · s.
    (b) If the ball is stopped in 0.02 s, show that the average force of the ball on the catcher's hand is 225 N.

55. Judy (mass 40 kg), standing on slippery ice, catches her dog Atti (mass 15 kg) leaping toward her at 3.0 m/s. Use conservation of momentum to show that the speed of Judy and her dog after the catch is 0.8 m/s.

56. A railroad diesel engine weighs four times as much as a freight car. The diesel engine coasts at 5 km/h into a freight car that is initially at rest. Use conservation of momentum to show that after they couple together, the engine + car coast at 4 km/h.

57. A 5-kg fish swimming at 1 m/s swallows an absent-minded 1-kg fish swimming toward it at a velocity that brings both fish to a halt. Show that the speed of the smaller fish before lunch was 5 m/s.

58. Belly-Flop Bernie dives from atop a tall flagpole into a swimming pool below. His potential energy at the top is 10,000 J. Show that when his potential energy reduces to 4000 J, his kinetic energy is 6000 J.

59. A simple lever is used to lift a heavy load. When a 60-N force pushes one end of the lever down 1.2 m, the load rises 0.2 m. Show that the weight of the load is 360 N.

60. In the preceding problem:
    (a) How much work is done by the 60-N force?
    (b) What is the gain in potential energy of the load?
    (c) How does the work done compare with the increased potential energy of the load?

61. In raising a 6000-N piano with a pulley system, the movers note that, for every 2 m of rope pulled down, the piano rises 0.2 m. Ideally, show that the force required to lift the piano is 600 N.

62. The girl steadily pulls her end of the rope upward a distance of 0.4 m with a constant force of 50 N.

    (a) By how much does the potential energy of the block increase?

    (b) Show that the mass of the block is 10 kg.

63. How many watts of power do you expend when you exert a force of 1 N that moves a book 2 m in a time interval of 1 s?

64. Show that 480 W of power is expended by a weightlifter when lifting a 60-kg barbell a vertical distance of 1.2 m in a time interval of 1.5 s.

65. When an average force F is exerted over a certain distance on a shopping cart of mass m, its kinetic energy increases by $\frac{1}{2}mv^2$.

    (a) Use the work–energy theorem to show that the distance over which the force acts is $mv^2/2F$.

    (b) If twice the force is exerted over twice the distance, how does the resulting increase in kinetic energy compare with the original increase in kinetic energy?

66. Emily holds a banana of mass m over the edge of a bridge of height h. She drops the banana and it falls to the river below. Use conservation of energy to show that the speed of the banana just before hitting the water is $v = \sqrt{2gh}$.

67. Starting from rest, Megan zooms down a frictionless slide from an initial height of 4.0 m. Show that her speed at the bottom of the slide is $\sqrt{80}$ m/s, or 8.9 m/s.

THINK AND RANK (ANALYSIS)

68. The balls have different masses and speeds. Rank the following from greatest to least: (a) momentum and (b) the impulses needed to stop the balls.

    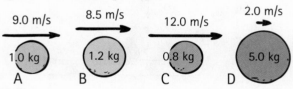

69. Jogging Jake runs along a train flatcar that moves at the velocities shown. In each case, Jake's velocity is given relative to the car. Call direction to the right positive. Rank the following from greatest to least: (a) the magnitude of Jake's momentum relative to the flatcar and (b) Jake's momentum relative to an observer at rest on the ground.

    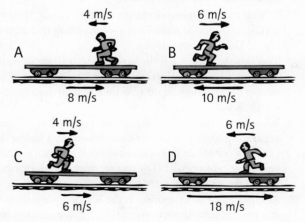

70. Sam pushes crates starting from rest across the floor of his classroom for 3 s with a net force as shown above. For each crate, rank the following from greatest to least: (a) impulse delivered, (b) change in momentum, (c) final speed, and (d) momentum in 3 s.

71. A ball is released from rest at the left of the metal track shown. Assume it has only enough friction to roll, but not to lessen its speed. Rank these quantities from greatest to least at each point: (a) momentum, (b) KE, and (c) PE.

72. The roller coaster ride starts from rest at point A. Rank these quantities from greatest to least at each point A–E: (a) speed, (b) KE, and (c) PE.

    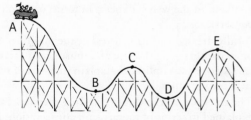

73. Rank the scale readings from greatest to least. (Ignore friction.)

    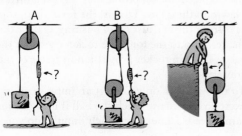

EXERCISES (SYNTHESIS)

74. A lunar vehicle is tested on Earth at a speed of 12 km/h. When it travels at the same speed on the Moon, is its momentum greater, less, or the same?

75. In terms of impulse and momentum, why do airbags in cars reduce the chances of injury in accidents?

76. In grandpa's time automobiles were previously manufactured to be as rigid as possible, whereas autos are now designed to crumple upon impact. Why?

77. If you throw a raw egg against a wall, it will break; but, if you throw it with the same speed into a sagging sheet, the egg won't break. Explain, using concepts from this chapter.

78. In terms of impulse and momentum, when a boxer is being hit, why is it important that he or she move away from the punch? Why is it disadvantageous to move into an oncoming punch?

79. A pair of skaters initially at rest push against each other so that they move in opposite directions. What is the total momentum of the two skaters as they move apart? Is there a different answer if their masses are not the same?

80. Bronco dives from a hovering helicopter and finds his momentum increasing. Does this violate conservation of momentum? Explain.

81. When you are traveling in your car at highway speed, the momentum of a bug is suddenly changed as it splatters onto your windshield. Compared with the change in momentum of the bug, by how much does the momentum of your car change?

82. If you throw a ball horizontally while standing on a skateboard, you roll backward with a momentum that matches that of the ball. Will you roll backward if you hold onto the ball while going through the motions of throwing it? Explain in terms of momentum conservation.

83. You are at the front of a floating canoe near a dock. You leap, expecting to easily land on the dock. Instead you land in the water. Explain in terms of momentum conservation.

84. A fully dressed person is at rest in the middle of a pond on perfectly frictionless ice and must get to shore. How can this be accomplished? Explain in terms of momentum conservation.

85. The examples in the three preceding exercises can be explained in terms of momentum conservation. Now explain them in terms of Newton's third law.

86. In Chapter 2 rocket propulsion was explained in terms of Newton's third law. That is, the force that propels a rocket is from the exhaust gases pushing against the rocket, the reaction to the force the rocket exerts on the exhaust gases. Explain rocket propulsion in terms of momentum conservation.

87. To throw a ball, do you exert an impulse on it? Do you exert an impulse to catch the ball if it's traveling at the same speed? About how much impulse do you exert, in comparison, if you catch the ball and immediately throw it back again? (Imagine yourself on a skateboard.)

88. When vertically falling sand lands in a horizontally moving cart, the cart slows. Ignore any friction between the cart and the tracks. Give two reasons for the slowing of the cart, one in terms of a horizontal force acting on the cart and one in terms of momentum conservation.

89. Freddy Frog drops vertically from a tree onto a horizontally moving skateboard. The skateboard slows. Give two reasons for the slowing, one in terms of a horizontal friction force between Freddy's feet and the skateboard, and one in terms of momentum conservation.

90. In a movie, the hero jumps straight down from a bridge onto a small boat that continues to move with no change in velocity. What physics is being violated here?

91. If your friend pushes a stroller four times as far as you do while exerting only half the force, which one of you does more work? How much more?

92. Which requires more work: stretching a strong spring a certain distance or stretching a weak spring the same distance? Defend your answer.

93. Two people who weigh the same climb a flight of stairs. The first person climbs the stairs in 30 s, and the second person climbs them in 40 s. Which person does more work? Which uses more power?

94. When a cannon with a longer barrel is fired, the force of expanding gases acts on the cannonball for a longer distance. What effect does this have on the velocity of the emerging cannonball? (Do you see why long-range cannons have such long barrels?)

95. At what point in its motion is the KE of a pendulum bob at a maximum? At what point is its PE at a maximum? When its KE is at half its maximum value, how much PE does it possess?

96. A baseball and a golf ball have the same momentum. Which has the greater kinetic energy?

97. A physics instructor demonstrates energy conservation by releasing a heavy pendulum bob, as shown in the sketch, allowing it to swing to and fro. What would happen if, in his exuberance, he gave the bob a slight shove as it left his nose? Explain.

98. On a playground slide, a child has potential energy that decreases by 1000 J while her kinetic energy increases by 900 J. What other form of energy is involved, and how much?

99. Consider the identical balls released from rest on tracks A and B, as shown. When they reach the right ends of the tracks, which will have the greater speed? Why is this question easier to answer than the similar one (Discussion Question 106) in Chapter 1?

100. If a golf ball and a Ping-Pong ball move with the same KE, can you say which has the greater speed? Explain in terms of the definition of KE. Similarly, in a gaseous mixture of heavy molecules and light molecules with the same average KE, can you say which have the greater speed?

101. In the absence of air resistance, a snowball thrown vertically upward with a certain initial KE returns to its original level with the same KE. When air resistance is a factor affecting the snowball, does it return to its original level with the same, less, or more KE? Does your answer contradict the law of energy conservation?

102. You're on a rooftop and you throw one ball downward to the ground below and another upward. The second ball, after rising, falls and also strikes the ground below. If air resistance can be neglected, and if your downward and upward initial speeds are the same, how do the speeds of the balls compare upon striking the ground? (Use the idea of energy conservation to arrive at your answer.)

103. When a driver applies the brakes to keep a car going downhill at a constant speed and constant kinetic energy, the potential energy of the car decreases. Where does this energy go? Where does most of it appear in a hybrid vehicle?

104. When the mass of a moving object is doubled with no change in speed, by what factor is its momentum changed? By what factor is its kinetic energy changed?

105. When the velocity of an object is doubled, by what factor is its momentum changed? By what factor is its kinetic energy changed?

106. Which, if either, has greater momentum: a 1-kg ball moving at 2 m/s or a 2-kg ball moving at 1 m/s? Which has greater kinetic energy?

107. If an object's kinetic energy is zero, what is its momentum?

108. If your momentum is zero, is your kinetic energy necessarily zero also?

109. Two lumps of clay with equal and opposite momenta have a head-on collision and come to rest. Is momentum conserved? Is kinetic energy conserved? Why are your answers the same or different?

110. Consider the swinging-balls apparatus. If two balls are lifted and released, momentum is conserved as two balls pop out the other side with the same speed as the released balls at impact. But momentum would also be conserved if one ball popped out at twice the speed. Explain why this never happens.

DISCUSSION QUESTIONS (EVALUATION)

111. Railroad cars are loosely coupled so that there is a noticeable time delay from the time the first car is moved until the last cars are moved from rest by the locomotive. Discuss the advisability of this loose coupling and slack between cars from the point of view of impulse and momentum.

112. Your friend says that the law of momentum conservation is violated when a ball rolls down a hill and gains momentum. What do you say?

113. An ice sailcraft is stalled on a frozen lake on a windless day. The skipper sets up a fan as shown. If all the wind bounces back from the sail, will the craft be set in motion? If so, in what direction?

114. Will your answer to the preceding question be different if the air is stopped by the sail without bouncing? Discuss.

115. Discuss the advisability of simply removing the sail in the preceding two questions.

116. Suppose that three astronauts outside a spaceship decide to play catch. All three astronauts have the same mass and are equally strong. The first astronaut throws the second astronaut toward the third one and the game begins. Describe the motion of the astronauts as the game proceeds. How long will the game last?

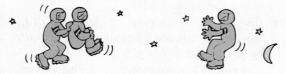

117. Can something have energy without having momentum? Explain. Can something have momentum without having energy? Discuss.

118. Imagine you're in a completely dark room with no windows and you cut a 1-ft^2 round hole in the roof. When the Sun is high in the sky, about 100 W of solar power enters the hole. On the floor where the light hits, you place a beach ball covered with aluminum foil, with the shiny side out. Discuss the illumination in your room compared with that of a 100-W incandescent lightbulb?

119. Discuss the physics that explains how the girl in Figure 3.29 can jack up a car while applying so little force.

120. Why bother using a machine if it cannot multiply work input to achieve greater work output?

121. In the pulley system shown, block A has a mass of 10 kg and is suspended precariously at rest. Assume that the pulleys and string are massless and there is no friction. No friction means that the tension in one part of the supporting string is the same at any other part. Discuss why the mass of block B is 20 kg.

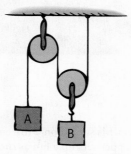

122. Does a car burn more fuel when its lights are turned on? Does the overall consumption of fuel depend on whether the engine is running while the lights are on? Discuss and defend your answers.

123. This may seem like an easy question for a physics type to answer: With what force does a rock that weighs 10 N strike the ground if dropped from a rest position 10 m high? In fact, the question cannot be answered unless you have more information. What information, and why?

124. To combat wasteful habits, we often speak of "conserving energy," by which we mean turning off lights, heating or cooling systems, and hot water when not being used. In this chapter, we also speak of "energy conservation." Distinguish between these two usages.

125. Your friend says that one way to improve air quality in a city is to have traffic lights synchronized so that motorists can travel long distances at constant speed. Discuss the physics that supports this claim.

126. If an automobile had a 100%-efficient engine, transferring all of the fuel's energy to work, would the engine be warm to your touch? Would its exhaust heat the surrounding air? Would it make any noise? Would it vibrate? Would any of its fuel go unused? Discuss.

127. The energy we require to live comes from the chemically stored potential energy in food, which is transformed into other energy forms during the metabolism process. What consequence awaits a person whose combined work and heat output is less than the energy consumed? What happens when the person's work and heat output is greater than the energy consumed? Can an undernourished person perform extra work without extra food? Discuss and defend your answers.

128. A red ball of mass m and a blue ball of mass $2m$ have the same kinetic energy. Explain which of the two has the larger momentum, using equations to guide your discussion.

129. No work is done by gravity on a bowling ball resting or moving on a bowling alley because the force of gravity on the ball acts perpendicular to the surface. But on an incline, the force of gravity has a vector component parallel to the alley, as sketch B shows. Discuss the two ways this component accounts for (a) acceleration of the ball and (b) work done on the ball to change its kinetic energy.

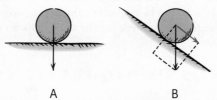

A   B

130. Consider, a bob attached by a string, a simple pendulum, that swings to and fro.

(a) Why doesn't the tension force in the string do work on the pendulum?

(b) Explain, however, why the force due to gravity on the pendulum at nearly every point *does* work on the pendulum?

(c) What is the single position of the pendulum where "no work by gravity" occurs?

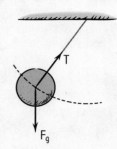

131. Consider a satellite in a circular orbit above Earth's surface. In Chapter 4 we will learn that the force of gravity changes only the direction of motion of a satellite in circular motion (and keeps it in a circle), but does *not* change its speed. Work done on the satellite by the gravitational force is zero. What is your explanation?

READINESS ASSURANCE TEST (RAT)

If you have a good handle on this chapter, if you really do, then you should be able to score at least 7 out of 10 on this RAT. If you score less than 7, you need to study further before moving on.

Choose the BEST answer to each of the following.

1. A 1-kg ball has the same speed as a 10-kg ball. Compared with the 1-kg ball, the 10-kg ball has
   (a) less momentum.
   (b) the same momentum.
   (c) 10 times as much momentum.
   (d) 100 times as much momentum.

2. In the absence of external forces, momentum is conserved in
   (a) an elastic collision.
   (b) an inelastic collision.
   (c) either an elastic or an inelastic collision.
   (d) neither an elastic nor an inelastic collision.

3. If the running speed of Fast Freda doubles, what also doubles is her
   (a) momentum.
   (b) kinetic energy.
   (c) both of these
   (d) neither of these

4. Which of these equations best explains the usefulness of automobile airbags?
   (a) $F = ma$
   (b) $Ft = \Delta(mv)$
   (c) $KE = \frac{1}{2}mv^2$
   (d) $Fd = \Delta \frac{1}{2}mv^2$

5. Which of these equations is most useful for solving a problem that asks for the distance a fast-moving box slides across a post office floor and comes to a stop?
   (a) $F = ma$
   (b) $Ft = \Delta(mv)$
   (c) $KE = \frac{1}{2}mv^2$
   (d) $Fd = \Delta \frac{1}{2}mv^2$

6. How much work is done on a 200-kg crate that is hoisted 2 m in a time of 4 s?
   (a) 400 J
   (b) 1000 J
   (c) 1600 J
   (d) 4000 J

7. A model airplane moves twice as fast as another identical model airplane. Compared to the kinetic energy of the slower airplane, the kinetic energy of the faster airplane is
   (a) the same.
   (b) twice as much.
   (c) four times as much.
   (d) more than four times as much.

8. The ultimate source of energy for wind power, fossil fuels, and biomass is
   (a) nuclear.
   (b) matter itself.
   (c) solar.
   (d) photovoltaic.

9. A billiard ball and a bowling ball have the same speed. Compared with the heavier bowling ball, the lighter billiard ball has
   (a) less momentum and less kinetic energy.
   (b) the same momentum and same kinetic energy.
   (c) more momentum but less kinetic energy.
   (d) less momentum but more kinetic energy.

10. A machine cannot multiply
   (a) forces.
   (b) distances.
   (c) energy.
   (d) but it can multiply all of these.

Answers to RAT

1. c, 2. c, 3. a, 4. b, 5. d, 6. d, 7. c, 8. c, 9. a, 10. c

CHAPTER 4
Gravity, Projectiles, and Satellites

CENTURIES BEFORE Newton's discovery, Aristotle and others believed that all heavenly bodies move in divine circles, requiring no explanation. Newton, however, recognized that a force of some kind must act on the planets; otherwise, their paths would be straight lines. He also recognized that any force on a planet would be directed toward a fixed central point—toward the Sun. This force of gravity was the same force that pulls an apple off a tree. Newton's stroke of intuition, that the force between Earth and an apple is the same as the force that acts between moons and planets and everything else in our universe, was a revolutionary break with the prevailing notion that there were two sets of natural laws: one for earthly events and an altogether different set for motion in the heavens. This union of terrestrial laws and cosmic laws is called the *Newtonian synthesis*.

4.1 The Universal Law of Gravity

EXPLAIN THIS What exactly did Newton discover about gravity?

According to popular legend, Newton was sitting under an apple tree when the idea struck him that gravity extends beyond Earth. Perhaps he looked up through tree branches toward the origin of the falling apple and noticed the Moon. Perhaps the apple hit him on the head, as popular stories tell us. In any event, Newton had the insight to see that the force between Earth and a falling apple is the same force that pulls the Moon in an orbital path around Earth, a path similar to a planet's path around the Sun.

To test this hypothesis, Newton compared the fall of an apple with the "fall" of the Moon. He realized that the Moon falls in the sense that *it falls away from the straight line it would follow if there were no forces acting on it.* Because of its tangential velocity, it "falls around" the round Earth (as we shall investigate later in this chapter). By simple geometry, the Moon's distance of fall per second could be compared with the distance that an apple or anything that far away would fall in one second. Newton's calculations didn't check. Disappointed, but recognizing that brute fact must always win over a beautiful hypothesis, he placed his papers in a drawer, where they remained for nearly 20 years. During this period, he founded and developed the field of geometric optics, for which he first became famous.

Newton's interest in mechanics was rekindled with the advent of a spectacular comet in 1680 and another two years later. He returned to the Moon problem at the prodding of his astronomer friend, Edmund Halley, for whom the second comet was later named. Newton made corrections in the experimental data used in his earlier method and obtained excellent results. Only then did he publish what is one of the most far-reaching generalizations of the human mind: the **law of universal gravitation.**[*]

Everything pulls on everything else in a beautifully simple way that involves only mass and distance. According to Newton, any body attracts any other body with a force that is directly proportional to the product of their masses and inversely proportional to the square of the distance separating them.

This statement can be expressed as

$$\text{Force} \sim \frac{\text{mass}_1 \times \text{mass}_2}{\text{distance}^2}$$

or symbolically as

$$F \sim \frac{m_1 m_2}{d^2}$$

where m_1 and m_2 are the masses of the bodies and d is the distance between their centers. Thus, the greater the masses m_1 and m_2, the greater the force of attraction between them, in direct proportion to the masses.[**] The greater the distance of separation d, the weaker the force of attraction, in inverse proportion to the square of the distance between their centers of mass.

FIGURE 4.1
Could the gravitational pull on the apple reach to the Moon?

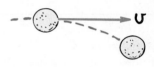

FIGURE 4.2
The tangential velocity of the Moon about Earth allows it to fall around Earth rather than directly into it. If this tangential velocity were reduced to zero, what would be the fate of the Moon?

[*] This is a dramatic example of the painstaking effort and cross-checking that go into the formulation of a scientific theory. Contrast Newton's approach with the failure to "do one's homework," the hasty judgments, and the absence of cross-checking that so often characterize the pronouncements of people advocating less-than-scientific theories.

[**] Note the different role of mass here. Thus far, we have treated mass as a measure of inertia, which is called *inertial mass.* Now we see mass as a property that affects gravitational force, which in this context is called *gravitational mass.* It is experimentally established that the two are equal, and, as a matter of principle, the equivalence of inertial and gravitational mass is the foundation of Einstein's general theory of relativity.

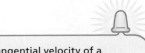

The tangential velocity of a planet or moon moving in a circle is at right angles to the force of gravity.

FIGURE 4.3
As the rocket gets farther from Earth, gravitational strength between the rocket and Earth decreases.

> Just as sheet music guides a musician playing music, equations guide a physical science student to understand how concepts are connected.

Mastering**PHYSICS**

TUTORIAL: Motion and Gravity
TUTORIAL: Orbits and Kepler's Laws

> Just as π relates circumference and diameter for circles, G relates gravitational force with mass and distance.

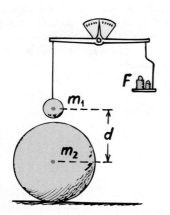

FIGURE 4.4
Von Jolly's method of measuring G. Balls of mass m_1 and m_2 attract each other with a force F equal to the weights needed to restore balance.

CHECKPOINT

1. In Figure 4.2, we see that the Moon falls around Earth rather than straight into it. If the Moon's tangential velocity were zero, how would it move?
2. According to the equation for gravitational force, what happens to the force between two bodies if the mass of one of the bodies is doubled? If both masses are doubled?
3. Gravitational force acts on all bodies in proportion to their masses. Why, then, doesn't a heavy body fall faster than a light body?

Were these your answers?

1. If the Moon's tangential velocity were zero, it would fall straight down and crash into Earth!
2. When one mass is doubled, the force between it and the other one doubles. If both masses double, the force is four times as much.
3. The answer goes back to Chapter 2. Recall Figure 2.9, in which heavy and light bricks fall with the same acceleration because both have the same ratio of weight to mass. Newton's second law ($a = F/m$) reminds us that greater force acting on greater mass does not result in greater acceleration.

The Universal Gravitational Constant, G

The proportionality form of the universal law of gravitation can be expressed as an exact equation when the constant of proportionality G is introduced. G is called the *universal gravitational constant*. Then the equation is

$$F = G\frac{m_1 m_2}{d^2}$$

In words, the force of gravity between two objects is found by multiplying their masses, dividing by the square of the distance between their centers, and then multiplying this result by the constant G. The magnitude of G is identical to the magnitude of the force between a pair of 1-kg masses that are 1 m apart: 0.0000000000667 N. This small magnitude indicates an extremely weak force. In standard units and in scientific notation:*

$$G = 6.67 \times 10^{-11} \text{ N} \cdot \text{m}^2/\text{kg}^2$$

Interestingly, Newton could calculate the product of G and Earth's mass, but not either one alone. Calculating G alone was first done by the English physicist Henry Cavendish in the 18th century, a century after Newton's time.

Cavendish found G by measuring the tiny force between lead masses with an extremely sensitive torsion balance. A simpler method was later developed by Philipp von Jolly, who attached a spherical flask of mercury to one arm of a sensitive balance (Figure 4.4). After the balance was put in equilibrium, a 6-ton lead sphere was rolled beneath the mercury flask. The gravitational force between the two masses was measured by the weight needed on the opposite

*The numerical value of G depends entirely on the units of measurement we choose for mass, distance, and time. The international system of choice uses the following units: for mass, the kilogram; for distance, the meter; and for time, the second. Scientific notation is discussed in the Lab Manual for this book.

end of the balance to restore equilibrium. All the quantities—m_1, m_2, F, and d—were known, from which the constant G was calculated:

$$G = \frac{F}{\left(\dfrac{m_1 m_2}{d^2}\right)} = 6.67 \times 10^{-11} \text{ N/kg}^2/\text{m}^2 = 6.67 \times 10^{-11} \text{ N} \cdot \text{m}^2/\text{kg}^2$$

The force of gravity is the weakest of the four known fundamental forces. (The other three are the electromagnetic force and two kinds of nuclear forces.) We sense gravitation only when masses like that of Earth are involved. If you stand on a large ship, the force of attraction between you and the ship is too weak for ordinary measurement. The force of attraction between you and Earth, however, can be measured. It is your weight.

Your weight depends not only on your mass but also on your distance from the center of Earth. At the top of a mountain, your mass is the same as it is anywhere else, but your weight is slightly less than it is at sea level. That's because your distance from Earth's center is greater.

Once the value of G was known, the mass of Earth was easily calculated. The force that Earth exerts on a mass of 1 kg at its surface is 9.8 N. The distance between the 1-kg mass and the center of Earth is Earth's radius, 6.4×10^6 m. Therefore, from $F = G(m_1 m_2 / d^2)$, where m_1 is the mass of Earth,

$$9.8 \text{ N} = 6.67 \times 10^{-11} \text{ N} \cdot \text{m}^2/\text{kg}^2 \; \frac{1 \text{ kg} \times m_1}{(6.4 \times 10^6 \text{ m})^2}$$

from which the mass of Earth is calculated to be $m_1 = 6 \times 10^{24}$ kg.

In the 18th century, when G was first measured, people all over the world were excited about it. Newspapers everywhere announced the discovery as one that measured the mass of the planet Earth. How exciting that Newton's formula gives the mass of the entire planet, with all its oceans, mountains, and inner parts yet to be discovered. G and the mass of Earth were calculated when a great portion of Earth's surface was still undiscovered.

4.2 Gravity and Distance: The Inverse-Square Law

EXPLAIN THIS How much smaller does your hand look when it is twice as far from your eye?

We can better understand how gravity is diluted with distance by considering how paint from a paint gun spreads with increasing distance (Figure 4.5). Suppose we position a paint gun at the center of a sphere with a radius of 1 m, and a burst of paint spray travels 1 m to produce a square patch of paint that is 1 mm thick. How thick would the patch be if the experiment were done in a sphere with twice the radius? If the same amount of paint travels in straight lines for 2 m, it spreads to a patch twice as tall and twice as wide. The paint is then spread over an area four times as big, and its thickness would be only $\frac{1}{4}$ mm.

Can you see from Figure 4.5 that for a sphere of radius 3 m, the thickness of the paint patch would be only $\frac{1}{9}$ mm? Can you see that the thickness of the paint decreases as the square of the distance increases? This is known as the **inverse-square law**. The inverse-square law holds for gravity and for all phenomena in which the effect from a localized source spreads uniformly throughout the

You can never change only one thing! Every equation reminds us of this—you can't change a term on one side without affecting the other side.

Mastering**PHYSICS**
VIDEO: von Jolly's Method of Measuring the Attraction Between Two Masses

LEARNING OBJECTIVE
Describe the rule by which gravity diminishes with distance.

Mastering**PHYSICS**
VIDEO: Inverse-Square Law

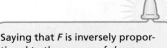

Saying that F is inversely proportional to the *square* of d means, for example, that if d gets bigger by a factor of 3, F gets *smaller* by a factor of 9.

FIGURE 4.5
The inverse-square law. Paint spray travels radially away from the nozzle of the can in straight lines. Like gravity, the "strength" of the spray obeys the inverse-square law.

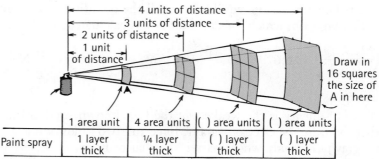

Paint spray	1 area unit	4 area units	( ) area units	( ) area units
	1 layer thick	¼ layer thick	( ) layer thick	( ) layer thick

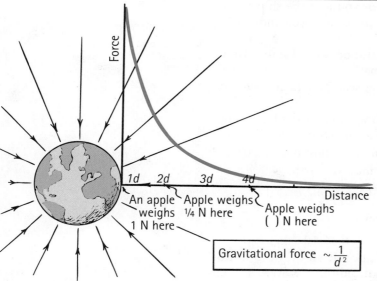

$$\text{Gravitational force} \sim \frac{1}{d^2}$$

FIGURE 4.6
INTERACTIVE FIGURE (MP)

The weight of an apple depends on its distance from Earth's center.

surrounding space: the electric field about an isolated electron, light from a match, radiation from a piece of uranium, and sound from a cricket.

Newton's law of gravity as written applies to particles and spherical bodies, as well as to non-spherical bodies sufficiently far apart. The distance term d in Newton's equation is the distance between the centers of masses of the objects. Note in Figure 4.6 that the apple that normally weighs 1 N at Earth's surface weighs only $\frac{1}{4}$ as much when it is twice the distance from Earth's center. The greater an object's distance from Earth's center, the less the object weighs. A child who weighs 300 N at sea level weighs only 299 N atop Mt. Everest. For greater distances, force is less. For very great distances, Earth's gravitational force approaches zero. The force *approaches* zero, but it never gets there. Even if you were transported to the far reaches of the universe, the gravitational influence of home would still be with you. It may be overwhelmed by the gravitational influences of nearer and/or more massive bodies, but it is there. The gravitational influence of every material object, however small or however far, is exerted through all of space.

FIGURE 4.7
The person's weight (not her mass) decreases as she increases her distance from Earth's center.

CHECKPOINT

1. By how much does the gravitational force between two objects decrease when the distance between their centers is doubled? Tripled? Increased tenfold?
2. Consider an apple at the top of a tree that is pulled by Earth's gravity with a force of 1 N. If the tree were twice as tall, would the force of gravity be only $\frac{1}{4}$ as strong? Defend your answer.

Were these your answers?

1. It decreases to one-fourth, one-ninth, and one-hundredth the original value.
2. No, because an apple at the top of the twice-as-tall apple tree is not twice as far from Earth's center. The taller tree would need a height equal to the radius of Earth (6370 km) for the apple's weight at its top to reduce to $\frac{1}{4}$ N. Before its weight decreases by 1%, an apple or any object must be raised 32 km—nearly four times the height of Mt. Everest. So, as a practical matter, we disregard the effects of everyday changes in elevation.

4.3 Weight and Weightlessness

EXPLAIN THIS   How does your weight change when you're inside an accelerating elevator?

When you step on a bathroom scale, you effectively compress a spring inside. When the pointer stops, the elastic force of the deformed spring balances the gravitational attraction between you and Earth—nothing moves, because you and the scale are in static equilibrium. The pointer is calibrated to show your **weight**. If you stand on a bathroom scale in a moving elevator, you'll find variations in your weight. If the elevator accelerates upward, the springs inside the bathroom scale are more compressed and your weight reading is greater. If the elevator accelerates downward, the springs inside the scale are less compressed and your weight reading is less. If the elevator cable breaks and the elevator falls freely, the reading on the scale goes to zero. According to the scale's reading, you would be **weightless**. Would you really be weightless? We can answer this question only if we agree on what we mean by *weight*.

In Chapter 1 we treated the weight of an object as the force due to gravity upon it. When in equilibrium on a firm surface, weight is evidenced by a support force, or, when in suspension, by a supporting rope tension. In either case, with no acceleration, weight equals *mg*. In future rotating habitats in space, where rotating environments act as giant centrifuges, support force can occur without regard to gravity. So a broader definition of the weight of something is the force it exerts against a supporting floor or a weighing scale. According to this definition, you are as heavy as you feel; in an elevator that accelerates downward, the supporting force of the floor is less and you weigh less. If the elevator is in free fall, your weight is zero (Figure 4.10). Even in this weightless condition, however, a gravitational force is still acting on you, causing your downward acceleration. But gravity now is not felt as weight because there is no support force.

Astronauts in orbit are without a support force and are in a sustained state of weightlessness. They sometimes experience "space sickness" until they become accustomed to a state of sustained weightlessness. Astronauts in orbit are in a state of continual free fall.

The International Space Station (ISS), shown in Figure 4.11, provides a weightless environment. The station facility and astronauts all accelerate equally toward Earth, at somewhat less than 1 *g* because of their altitude. This acceleration is not sensed at all. With respect to the station, the astronauts experience zero *g*. Over extended periods of time, this causes loss of muscle strength and other detrimental changes in the body. Future space travelers, however, need not be subjected to weightlessness. Habitats that lazily rotate as giant wheels or

LEARNING OBJECTIVE
Describe how weight is a support force.

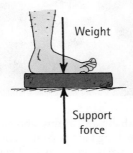

FIGURE 4.8
Two forces act on a weighing scale: a downward force of gravity (your weight, *mg*, if there is no acceleration) and an upward support force. These equal and opposite forces squeeze an inner springlike device that is calibrated to show weight.

FIGURE 4.9
Both are weightless.

Mastering**PHYSICS**
VIDEO: Weight and Weightlessness
VIDEO: Apparent Weightlessness

FIGURE 4.10
Your weight equals the force with which you press against the supporting floor. If the floor accelerates up or down, your weight varies (even though the gravitational force *mg* that acts on you remains the same).

Normal weight

Greater than normal weight

Less than normal weight

Zero weight

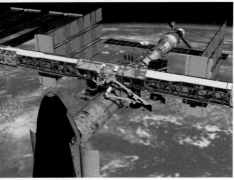

FIGURE 4.11
The inhabitants in this laboratory and docking facility continually experience weightlessness. They are in free fall around Earth. Does a force of gravity act on them?

LEARNING OBJECTIVE
Connect and extend the law of gravity to areas beyond science.

Astronauts inside an orbiting space vehicle have no weight, even though the force of gravity pulling them toward Earth is only slightly less than at ground level.

FYI The space surrounding all objects with mass is energized with a *gravitational field*. Similarly, the space around a magnet is energized with a *magnetic field*, and the space about an electrically charged object is energized with an *electric field*.

Mastering**PHYSICS**
VIDEO: Discovery of Neptune

pods at the end of a tether will likely replace today's nonrotating space habitats. Rotation effectively supplies a support force and nicely provides weight.

CHECKPOINT

In what sense is drifting in space far away from all celestial bodies like stepping off the top of a stepladder?

Was this your answer?
In both cases, you'd experience weightlessness. Drifting in deep space, you would remain weightless because no discernable force acts on you. Stepping off the top of a stepladder, you would be only momentarily weightless because of a momentary lapse of support force.

4.4 Universal Gravitation

EXPLAIN THIS How did Newton's laws affect the U.S. Constitution?

We all know that Earth is round. But why is Earth round? It is round because of gravitation. Everything attracts everything else, and so Earth has attracted itself together as far as it can! Any "corners" of Earth have been pulled in; as a result, every part of the surface is equidistant from the center of gravity. This makes it a sphere. Therefore, we see, from the law of gravitation, that the Sun, the Moon, and Earth are spherical because they have to be (although rotational effects make them slightly ellipsoidal).

If everything pulls on everything else, then the planets must pull on each other. The force that controls Jupiter, for example, is not just the force from the Sun; there are also pulls from the other planets. Their effect is small in comparison with the pull of the much more massive Sun, but it still shows. When Saturn is near Jupiter, its pull disturbs the otherwise smooth path traced by Jupiter. Both planets "wobble" about their expected orbits. The interplanetary forces causing this wobbling are called *perturbations*. By the 1840s, studies of the most recently discovered planet at the time, Uranus, showed that the deviations of its orbit could not be explained by perturbations from all other known planets. Either the law of gravitation was failing at this great distance from the Sun or an unknown eighth planet was perturbing the orbit of Uranus. An Englishman and a Frenchman, J. C. Adams and Urbain Leverrier, respectively, each assumed Newton's law to be valid, and they independently calculated where an eighth planet should be. Adams sent a letter to the Greenwich Observatory in England; at about the same time, Leverrier sent a letter to the Berlin Observatory in Germany. They both suggested that a certain area of the sky be searched for a new planet. Adams' request was delayed by misunderstandings at Greenwich, but Leverrier's request was heeded immediately. The planet Neptune was discovered that very night!

Subsequent tracking of the orbits of both Uranus and Neptune led to the discovery of Pluto in 1930 at the Lowell Observatory in Arizona. Whatever you may have learned in your early schooling, Pluto is no longer a planet. In 2006 Pluto was officially classified as a *dwarf planet*. Other objects of Pluto's size continue to be discovered beyond Neptune.* Pluto takes 248 years to make a single revolution about the Sun, so no one will see it in its discovered position again until 2178.

---

* Quaoar has a moon; Eris is 30% wider than Pluto and also has a moon. Object 2003 EL61 has two moons. Objects nicknamed Sedna and Buffy, discovered in 2005, are nearly the size of Pluto.

Recent evidence suggests that the universe is expanding and accelerating outward, pushed by an antigravity *dark energy* that makes up some 73% of the universe. Another 23% is composed of the yet-to-be-discovered particles of exotic *dark matter*. Ordinary matter, the stuff of stars, cabbages, and kings, makes up only about 4%. The concepts of dark energy and dark matter are late-20th- and 21st-century confirmations. The present view of the universe has progressed appreciably beyond what Newton and those of his time perceived.

Yet few theories have affected science and civilization as much as Newton's theory of gravity. The successes of Newton's ideas ushered in the Enlightenment. Newton had demonstrated that, by observation and reason, people could uncover the workings of the physical universe. How profound that all the moons and planets and stars and galaxies have such a beautifully simple rule to govern them, namely:

$$F = G\frac{m_1 m_2}{d^2}$$

The formulation of this simple rule is one of the major reasons for the success in science that followed, for it provided hope that other phenomena of the world might also be described by equally simple and universal laws.

This hope nurtured the thinking of many scientists, artists, writers, and philosophers of the 1700s. One of these was the English philosopher John Locke, who argued that observation and reason, as demonstrated by Newton, should be our best judge and guide in all things. Locke urged that all of nature and even society should be searched to discover any "natural laws" that might exist. Using Newtonian physics as a model of reason, Locke and his followers modeled a system of government that found adherents in the thirteen British colonies across the Atlantic. These ideas culminated in the Declaration of Independence and the Constitution of the United States of America.

FYI It's widely assumed that when Earth was no longer thought to be the center of the universe, both it and humankind were demoted in importance and were no longer considered special. On the contrary, writings of the time suggest most Europeans viewed humans as filthy and sinful because of Earth's lowly position—farthest from heaven, with hell at its center. Human elevation didn't occur until the Sun, viewed positively, took a center position. We became special by showing we're not so special.

"I can live with doubt and uncertainty and not knowing. I think it is much more interesting to live not knowing than to have answers that might be wrong."
—*Richard Feynman*

4.5 Projectile Motion

EXPLAIN THIS Why do a dropped ball and a ball thrown horizontally hit the ground in the same time?

Without gravity, a rock tossed at an angle skyward would follow a straight-line path. Because of gravity, however, the path curves. A tossed rock, a cannonball, or any object that is projected by some means and continues in motion by its own inertia is called a **projectile**. To the cannoneers of earlier centuries, the curved paths of projectiles seemed very complex. Today these paths are surprisingly simple when we look at the horizontal and vertical components of velocity separately.

The horizontal component of velocity for a projectile is no more complicated than the horizontal velocity of a bowling ball rolling freely on the lane of a bowling alley. If the retarding effect of friction can be ignored, no horizontal force acts on the ball and its velocity is constant. It rolls of its own inertia and covers equal distances in equal intervals of time (Figure 4.12, right). The horizontal component of a projectile's motion is just like the bowling ball's motion along the lane.

The vertical component of motion for a projectile following a curved path is just like the motion described in Chapter 1 for a freely falling object. The vertical component is exactly the same as for an object falling freely straight down, as shown at the left in Figure 4.12. The faster the object falls, the greater the distance covered in each successive second. Or, if the object is projected upward, the vertical distances of travel decrease with time on the way up.

LEARNING OBJECTIVE
Apply the independence of horizontal and vertical motion to projectiles.

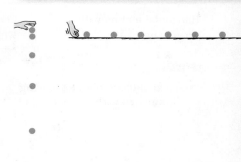

FIGURE 4.12
(*left*) Drop a ball, and it accelerates downward and covers a greater vertical distance each second.
(*right*) Roll it along a level surface, and its velocity is constant because no component of gravitational force acts horizontally.

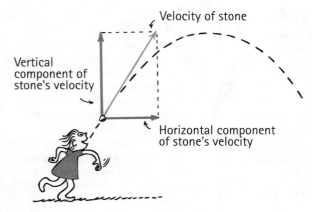

FIGURE 4.13
Vertical and horizontal components of a stone's velocity.

Mastering**PHYSICS**
TUTORIAL: Projectile Motion
VIDEO: Projectile Motion Demo
VIDEO: More Projectile Motion

The curved path of a projectile is a combination of horizontal and vertical motion. Velocity is a vector quantity, and a velocity vector at an angle has horizontal and vertical components, as seen in Figure 4.13. When air resistance is small enough to ignore, the horizontal and vertical components of a projectile's velocity are completely independent of one another. Their combined effect produces the trajectories of projectiles.

Projectiles Launched Horizontally

Projectile motion is nicely analyzed in Figure 4.14, which shows a simulated multiple-flash exposure of a ball rolling off the edge of a table. Investigate it carefully, for there's a lot of good physics there. On the left we notice equally timed sequential positions of the ball without the effect of gravity. Only the effect of the ball's horizontal component of motion is shown. Next we see vertical motion without a horizontal component. The curved path in the third view is best analyzed by considering the horizontal and vertical components of motion separately. There are two important things to notice. The first is that the ball's horizontal component of velocity doesn't change as the falling ball moves forward. The ball travels the same horizontal distance in equal times between each flash. That's because there is no component of gravitational force acting horizontally. Gravity acts only *downward*, so the only acceleration of the ball is *downward*. The second thing to notice is that the vertical positions become farther apart with time. The vertical distances traveled are the same as if the ball were simply dropped. Note that the curvature of the ball's path is the combination of horizontal motion, which remains constant, and vertical motion, which undergoes acceleration due to gravity.

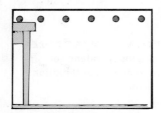

Horizontal motion with *no* gravity

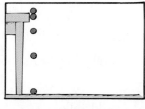

Vertical motion only with gravity

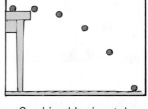

Combined horizontal and vertical motion

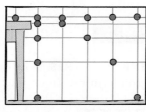

Superposition of the preceding cases

FIGURE 4.14

INTERACTIVE FIGURE MP

Simulated photographs of a moving ball illuminated with a strobe light.

The trajectory of a projectile that accelerates only in the vertical direction while moving at a constant horizontal velocity is a **parabola**. When air resistance is small enough to neglect, as it is for a heavy object without great speed, the trajectory is parabolic.

CHECKPOINT

At the instant a cannon fires a cannonball horizontally over a level range, another cannonball held at the side of the cannon is released and drops to the ground. Which ball, the one fired downrange or the one dropped from rest, strikes the ground first?

Was this your answer?
Both cannonballs hit the ground at the same time, because both fall *the same vertical distance*. Note that the physics is the same as the physics of Figures 4.14 through 4.16. We can reason this another way by asking which one would hit the ground first if the cannon were pointed at an *upward* angle. Then the dropped cannonball would hit first, while the fired ball is still airborne. Now consider the cannon pointing *downward*. In this case, the fired ball hits first. So projected upward, the dropped one hits first; downward, the fired one hits first. Is there some angle at which there is a dead heat, where both hit at the same time? Can you see that this occurs when the cannon is horizontal?

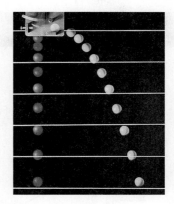

FIGURE 4.15
INTERACTIVE FIGURE

A strobe-light photograph of two golf balls released simultaneously from a mechanism that allows one ball to drop freely while the other is projected horizontally.

Projectiles Launched at an Angle

In Figure 4.17, we see the paths of stones thrown at an angle upward (left) and downward (right). The dashed straight lines at the top show the ideal trajectories of the stones if there were no gravity. Notice that the vertical distance that each stone falls beneath the idealized straight-line path is the same for equal times. This vertical distance is independent of what's happening horizontally.

Figure 4.18 shows specific vertical distances for a cannonball shot at an upward angle. If there were no gravity the cannonball would follow the straight-line path shown by the dashed line. But there is gravity, so this doesn't occur. What happens is that the cannonball continuously falls beneath the imaginary line until it finally strikes the ground. Note that the vertical distance it falls

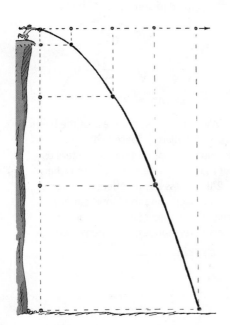

FIGURE 4.16
The vertical dashed line at left is the path of a stone dropped from rest. The horizontal dashed line at the top would be its path if there were no gravity. The curved solid line shows the resulting trajectory that combines horizontal and vertical motion.

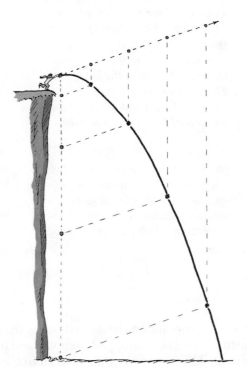

FIGURE 4.17
Whether launched at an angle upward or downward, the vertical distance of fall beneath the idealized straight-line path is the same for equal times.

FIGURING PHYSICAL SCIENCE

Problem Solving

SAMPLE PROBLEM 1
A ball of mass 1.0 kg rolls off of a 1.25-m-high lab table and hits the floor 3.0 m from the base of the table.

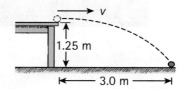

(a) Show that the ball takes 0.5 s to hit the floor.
(b) Show that the ball leaves the table at 6.0 m/s.

Solution:
(a) We want the time of the ball in the air. First, some physics. The time t it takes for any ball to hit the floor would be the same as if it were dropped from rest a vertical distance y. We say from rest because initially it moves horizontally off the desk, with zero velocity in the vertical direction.

From $y = \frac{1}{2}gt^2 \Rightarrow t^2 = \frac{2y}{g}$, we have

$$t = \sqrt{\frac{2y}{g}} = \sqrt{\frac{2(1.25\ m)}{10\ m/s^2}} = 0.5\ s$$

(b) The horizontal speed of the ball as it leaves the table, using time 0.5 s, is

$$v_x = \frac{d}{t} = \frac{x}{t} = \frac{3.0\ m}{0.5\ s} = 6.0\ m/s$$

Notice how the terms of the equations guide the solution. Notice also that the mass of the ball, not showing in the equations, is extraneous information (as would be the color of the ball).

SAMPLE PROBLEM 2
A horizontally moving tennis ball barely clears the net, a distance y above the surface of the court. To land within the tennis court, the ball must not be moving too fast.

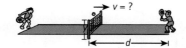

(a) To remain within the court's border, a horizontal distance d from the bottom of the net, ignoring air resistance and any spin effects of the ball, show that the ball's maximum speed over the net is

$$v = \frac{d}{\sqrt{\dfrac{2y}{g}}}$$

(b) Suppose the height of the net is 1.00 m, and the court's border is 12.0 m from the bottom of the net. Use $g = 10\ m/s^2$ and show that the maximum speed of the horizontally moving ball clearing the net is about 27 m/s (about 60 mi/h).
(c) Does the mass of the ball make a difference? Defend your answer.

Solution:
(a) As with Sample Problem 1, the physics concept here involves projectile motion in the absence of air resistance, where horizontal and vertical components of velocity are independent.

We're asked for horizontal speed, so we write

$$v_x = \frac{d}{t}$$

where d is horizontal distance traveled in time t. As with Sample Problem 1, the time t of the ball in flight is the same as if we had just dropped it from rest a vertical distance y from the top of the net. As the ball clears the net, its highest point in its path, its vertical component of velocity is zero.

From $y = \frac{1}{2}gt^2 \Rightarrow t^2 = \frac{2y}{g} \Rightarrow t = \sqrt{\frac{2y}{g}}$,

$$v = \frac{d}{t} = \frac{d}{\sqrt{\dfrac{2y}{g}}}$$

Can you see that solving in terms of symbols better shows that these two problems are one and the same? All the physics occurs in steps (a) and (b) in Sample Problem 1. These steps are combined in step (a) of Sample Problem 2.

(b) $v = \dfrac{d}{\sqrt{\dfrac{2y}{g}}} = \dfrac{12.0\ m}{\sqrt{\dfrac{2(1.00\ m)}{10\ m/s^2}}}$

$$= 26.8\ m/s \approx 27\ m/s$$

(c) We can see that the mass of the ball (in both problems) doesn't show up in the equations for motion, which tells us that mass is irrelevant. Recall from Chapter 2 that mass has no effect on a freely falling object—and the tennis ball is a freely falling object (as is every projectile when air resistance can be neglected).

---

beneath any point on the dashed line is the same vertical distance it would have fallen if it had been dropped from rest and had been falling for the same amount of time. This distance, as introduced in Chapter 1, is given by $d = \frac{1}{2}gt^2$, where t is the elapsed time. For $g = 10\ m/s^2$, this becomes $d = 5t^2$.

We can put it another way: Shoot a projectile skyward at some angle and pretend there is no gravity. After so many seconds t, it should be at a certain point along a straight-line path. But because of gravity, it isn't. Where is it? The answer is that it's directly below this point. How far below? The answer in meters is $5t^2$ (or, more precisely, $4.9t^2$). How about that!

DOING PHYSICAL SCIENCE

Hands-On Dangling Beads

Make your own model of projectile paths. Divide a ruler or a stick into five equal spaces. At position 1, hang a bead from a string that is 1 cm long, as shown. At position 2, hang a bead from

a string that is 4 cm long. At position 3, do the same with a 9-cm length of string. At position 4, use 16 cm of string, and for position 5, use 25 cm of string. If you hold the stick horizontally, you will have a version of Figure 4.16. Hold it at a slight upward angle to show

a version of Figure 4.17 (left). Hold it at a downward angle to show a version of Figure 4.17 (right).

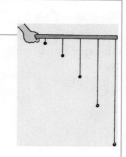

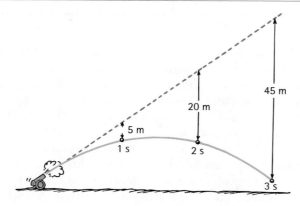

FIGURE 4.18
With no gravity, the projectile would follow a straight-line path (dashed line). But because of gravity, the projectile falls beneath this line the same vertical distance it would fall if it were released from rest. Compare the distances fallen with those given in Table 1.2 in Chapter 1. (With $g = 9.8$ m/s^2, these distances are more precisely 4.9 m, 19.6 m, and 44.1 m.)

CHECKPOINT

1. **Suppose the cannonball in Figure 4.18 were fired faster. How many meters below the dashed line would it be at the end of the 5 s?**
2. **If the horizontal component of the cannonball's velocity is 20 m/s, how far downrange will the cannonball be in 5 s?**

Were these your answers?

1. The vertical distance beneath the dashed line at the end of 5 s is 125 m [looking at magnitudes only: $d = 5t^2 = 5(5)^2 = 5(25) = 125$ m]. Interestingly enough, this distance doesn't depend on the angle of the cannon. If air resistance is neglected, any projectile will fall $5t^2$ meters below where it would have reached if there were no gravity.
2. With no air resistance, the cannonball will travel a horizontal distance of 100 m [$d = v_x t = (20$ m/s$)(5$ s$) = 100$ m]. Note that because gravity acts only vertically and there is no acceleration in the horizontal direction, the cannonball travels equal horizontal distances in equal times. This distance is simply its horizontal component of velocity multiplied by the time (and not $5t^2$, which applies only to vertical motion under the acceleration of gravity).

Figure 4.19 shows the paths of several projectiles, all with the same initial speed but different launching angles. The figure neglects the effects of air resistance, so the trajectories are all parabolas. Notice that these projectiles reach different *altitudes*, or heights above the ground. They also have different *horizontal ranges*, or distances traveled horizontally. The remarkable thing to note from Figure 4.19 is that the same range is obtained from two different launching angles when the angles add up to 90°! An object thrown into the air at an angle of 60°, for example, has the same range as if it were thrown at the same speed at an angle of 30°. For the smaller angle, of course, the object remains in the air for a shorter time. The greatest range occurs when the launching angle is 45°—and when air resistance is negligible.

FIGURE 4.19
INTERACTIVE FIGURE MP

Ranges of a projectile shot at the same speed at different projection angles.

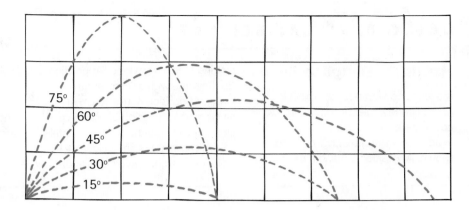

FIGURE 4.20
Maximum range would be attained when a ball is batted at an angle of nearly 45°—but only in the absence of air drag.

Without the effects of air, a baseball would reach the maximum range when it is batted 45° above the horizontal. Without air resistance, the ball rises just like it falls, covering the same amount of ground while rising as while falling. But not so when air resistance slows the ball. Its horizontal speed at the top of its path is lower than its horizontal speed when the ball leaves the bat, *so it covers less ground while falling than while rising.* As a result, for maximum range the ball must leave the bat with more horizontal speed than vertical speed—at about 25° to 34°, considerably less than 45°. Likewise for golf balls. (As Chapter 5 will show, the ball's spin also affects the range.) For heavy projectiles like javelins and the shot, air has less effect on the range. A javelin, being heavy

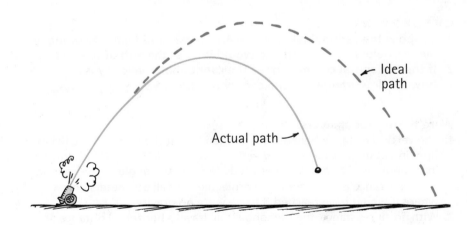

Ideal path

Actual path

FIGURE 4.21
INTERACTIVE FIGURE MP

In the presence of air resistance, the trajectory of a high-speed projectile falls short of the idealized parabolic path.

HANG TIME REVISITED

In Chapter 1, we stated that airborne time during a jump is independent of horizontal speed. Now we see why this is so—horizontal and vertical components of motion are independent of each other. The rules of projectile motion apply to jumping. Once one's feet are off the ground, only the force of gravity acts on the jumper (neglecting air resistance). Hang time depends only on the vertical component of liftoff velocity. However, the action of running can make a difference. When the jumper is running, the liftoff force during jumping can be somewhat increased by the pounding of the feet against the ground (and the ground pounding against the feet in action–reaction fashion), so hang time for a running jump can often exceed hang time for a standing jump. But once the runner's feet are off the ground, only the vertical component of liftoff velocity determines hang time.

and presenting a very small cross-section to the air, follows an almost perfect parabola when thrown. So does a shot. Aha, but *launching speeds* are not equal for heavy projectiles thrown at different angles. When a javelin or a shot is thrown, a significant part of the launching *force* goes into lifting—combating gravity—so launching at 45° means a lower launching speed. You can test this yourself: Throw a heavy boulder horizontally, then at an angle upward—you'll find the horizontal throw to be considerably faster. So the maximum range for heavy projectiles thrown by humans is attained for angles of less than 45°—and not because of air resistance.

CHECKPOINT

1. **A baseball is batted at an angle into the air. Once the ball is air-borne, and neglecting air resistance, what is the ball's acceleration vertically? Horizontally?**
2. **At what part of its trajectory does the baseball have minimum speed?**
3. **Consider a batted baseball following a parabolic path on a day when the Sun is directly overhead. How does the speed of the ball's shadow across the field compare with the ball's horizontal component of velocity?**

Were these your answers?

1. Vertical acceleration is *g* because the force of gravity is vertical. Horizontal acceleration is zero because no horizontal force acts on the ball.
2. A ball's minimum speed occurs at the top of its trajectory. If it is launched vertically, its speed at the top is zero. If launched at an angle, the vertical component of velocity is zero at the top, leaving only the horizontal component. So the speed at the top is equal to the horizontal component of the ball's velocity at any point. Doesn't this make sense?
3. They are the same!

When air resistance is small enough to be negligible, the time that a projectile takes to rise to its maximum height is the same as the time it takes to fall back to its initial level (Figure 4.22). This is because its deceleration by gravity while going up is the same as its acceleration by gravity while coming down. The speed it loses while going up is therefore the same as the speed gained while coming down. So the projectile arrives at its initial level with the same speed it had when it was initially projected.

Baseball games normally take place on level ground. For the short-range projectile motion on the playing field, Earth can be considered flat because the flight of the baseball is not affected by Earth's curvature. For very long-range projectiles, however, the curvature of Earth's surface must be taken into account. We'll now see that, if an object is projected fast enough, it falls all the way around Earth and becomes an Earth satellite.

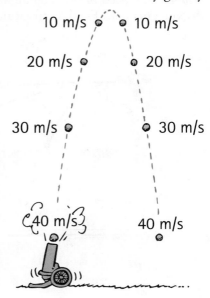

FIGURE 4.22
Without air resistance, speed lost while going up equals speed gained while coming down: Time going up equals time coming down.

FIGURE 4.23
How fast is the ball thrown?

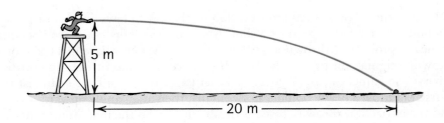

5 m

20 m

CHECKPOINT

The boy on the tower in Figure 4.23 throws a ball 20 m downrange. What is his pitching speed?

Was this your answer?
The ball is thrown horizontally, so the pitching speed is horizontal distance divided by time. A horizontal distance of 20 m is given, but the time is not stated. However, knowing the vertical drop is 5 m, you remember that a 5-m drop takes 1 s! From the equation for constant speed (which applies to horizontal motion), $v = d/t = (20 \text{ m})/(1 \text{ s}) = 20$ m/s. It is interesting to note that the equation for constant speed, $v = d/t$, guides our thinking about the crucial factor in this problem—the *time*.

LEARNING OBJECTIVE
Relate a projectile trajectory that matches Earth's curvature to satellite motion.

Earth's curvature, dropping 5 m for each 8-km tangent, means that if you were floating in a calm ocean, you'd be able to see only the top of a 5-m mast on a ship 8 km away.

4.6 Fast-Moving Projectiles—Satellites

EXPLAIN THIS What does Earth's curvature have to do with Earth satellites?

Consider the girl pitching a ball on the cliff in Figure 4.24. If gravity did not act on the ball, the ball would follow a straight-line path shown by the dashed line. But gravity does act, so the ball falls below this straight-line path. In fact, as just discussed, 1 s after the ball leaves the pitcher's hand it has fallen a vertical distance of 5 m below the dashed line—whatever the pitching speed. It is important to understand this, for it is the crux of satellite motion.

An Earth **satellite** is simply a projectile that falls *around* Earth rather than *into* it. The speed of the satellite must be great enough to ensure that its falling distance matches Earth's curvature. A geometrical fact about the curvature of

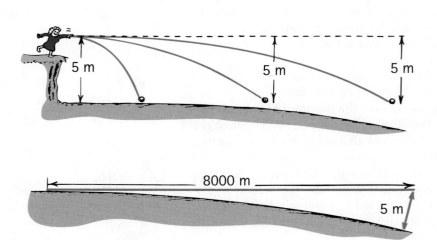

5 m   5 m   5 m

8000 m

5 m

FIGURE 4.24
If you throw a stone at any speed, 1 s later it will have fallen 5 m below where it would have been without gravity.

Earth is that its surface drops a vertical distance of 5 m for every 8000 m tangent to the surface (Figure 4.24). If a baseball could be thrown fast enough to travel a horizontal distance of 8 km during the 1 s it takes to fall 5 m, then it would follow the curvature of Earth. This is a speed of 8 km/s. If this doesn't seem fast, convert it to kilometers per hour and you get an impressive 29,000 km/h (or 18,000 mi/h)!

At this speed, atmospheric friction would burn the baseball—or even a piece of iron—to a crisp. This is the fate of bits of rock and other meteorites that enter Earth's atmosphere and burn up, appearing as "falling stars." That is why satellites, such as the space shuttles, are launched to altitudes of 150 kilometers or more—to be above almost all of the atmosphere and to be nearly free of air resistance. A common misconception is that satellites orbiting at high altitudes are free from gravity. Nothing could be further from the truth. The force of gravity on a satellite 200 kilometers above Earth's surface is nearly as strong as it is at the surface. Otherwise the satellite would go in a straight line and leave Earth. The high altitude positions the satellite not beyond Earth's gravity, but beyond Earth's atmosphere, where air resistance is almost totally absent.

Satellite motion was understood by Isaac Newton, who reasoned that the Moon was simply a projectile circling Earth under the attraction of gravity. This concept is illustrated in a drawing by Newton (Figure 4.27). He compared the motion of the Moon to that of a cannonball fired from the top of a high mountain. He imagined that the mountaintop was above Earth's atmosphere, so that air resistance would not impede the motion of the cannonball. If fired with a low horizontal speed, a cannonball would follow a curved path and soon hit Earth below. If it were fired faster, its path would be less curved and it would hit Earth farther away. If the cannonball were fired fast enough, Newton reasoned, the curved path would become a circle and the cannonball would circle Earth indefinitely. It would be in orbit.

Both the cannonball and the Moon have tangential velocity (parallel to Earth's surface) sufficient to ensure motion *around* Earth rather than *into* it. Without resistance to reduce its speed, the Moon or any Earth satellite "falls" around Earth indefinitely. Similarly, the planets continuously fall around the Sun in closed paths. Why don't the planets crash into the Sun? They don't because of sufficient tangential velocities. What would happen if their tangential velocities were reduced to zero? The answer is simple enough: Their falls would be straight toward the Sun, and they would indeed crash into it. Any objects in the solar system without sufficient tangential velocities have long ago crashed into the Sun. What remains is the harmony we observe.

FIGURE 4.25
Earth's curvature (not to scale).

FIGURE 4.26
If the speed of the stone and the curvature of its trajectory are great enough, the stone may become a satellite.

An Earth sattelite is a projectile in a constant state of free fall. Because of its tangential velocity, it falls around Earth rather than vertically into it.

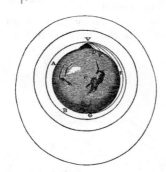

FIGURE 4.27
"The greater the velocity . . . with which (a stone) is projected, the farther it goes before it falls to the Earth. We may therefore suppose the velocity to be so increased, that it would describe an arc of 1, 2, 5, 10, 100, 1000 miles before it arrived at the Earth, till at last, exceeding the limits of the Earth, it should pass into space without touching."
—Isaac Newton, *System of the World*

CHECKPOINT
One of the beauties of physics is that there are usually different ways to view and explain a given phenomenon. Is the following explanation valid? "Satellites remain in orbit instead of falling to Earth because they are beyond the main pull of Earth's gravity."

Was this your answer?
No, no, a thousand times no! If any moving object were beyond the pull of gravity, it would move in a straight line and would not curve around Earth. Satellites remain in orbit because they *are* being pulled by gravity, not because they are beyond it. For the altitudes of most Earth satellites, Earth's gravitational force on a satellite is only a few percent weaker than it is at Earth's surface.

4.7 Circular Satellite Orbits

EXPLAIN THIS Why does kinetic energy and momentum remain constant for a satellite in a circular orbit?

An 8-km/s cannonball fired horizontally from Newton's mountain would follow Earth's curvature and glide in a circular path around Earth again and again (provided the cannoneer and the cannon got out of the way). Fired at a slower speed, the cannonball would strike Earth's surface; fired at a faster speed, it would overshoot a circular orbit, as we will discuss shortly. Newton calculated the speed for circular orbit, and because such a cannon-muzzle velocity was clearly impossible, he did not foresee the possibility of humans launching satellites (and he likely didn't consider multistage rockets).

Note that in circular orbit, the speed of a satellite is not changed by gravity; only the direction changes. We can understand this by comparing a satellite in circular orbit with a bowling ball rolling along a bowling lane. Why doesn't the gravity that acts on the bowling ball change its speed? The answer is that gravity pulls straight downward with no component of force acting forward or backward.

Consider a bowling lane that completely surrounds Earth, elevated high enough to be above the atmosphere and air resistance. The bowling ball rolls at constant speed along the lane. If a part of the lane were cut away, the ball would roll off its edge and would hit the ground below. A faster ball encountering the gap would hit the ground farther along the gap. Is there a speed at which the ball will clear the gap (like a motorcyclist who drives off a ramp and clears a gap to meet a ramp on the other side)? The answer is yes: 8 km/s will be enough to clear that gap—and any gap, even a 360° gap. The ball would be in circular orbit.

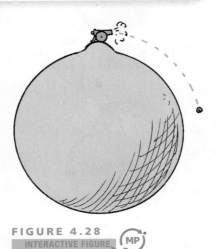

FIGURE 4.28
INTERACTIVE FIGURE (MP)

Fired fast enough, the cannonball goes into orbit.

Mastering**PHYSICS**
VIDEO: Circular Orbits

FIGURE 4.29
(a) The force of gravity on the bowling ball is at 90° to its direction of motion, so it has no component of force to pull it forward or backward, and the ball rolls at constant speed. (b) The same is true even if the bowling alley is larger and remains "level" with the curvature of Earth.

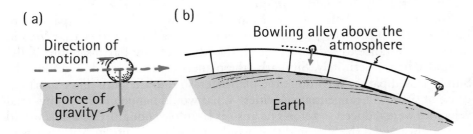

(a) Direction of motion Force of gravity

(b) Bowling alley above the atmosphere Earth

Note that a satellite in circular orbit is always moving in a direction perpendicular to the force of gravity that acts upon it. No component of force is acting in the direction of satellite motion to change its speed. Only a change in direction occurs. So we see why a satellite in circular orbit moves parallel to the surface of Earth at constant speed—a very special form of free fall.

For a satellite close to Earth, the period (the time for a complete orbit about Earth) is about 90 min. For higher altitudes, the orbital speed is less, the distance is more, and the period is longer. For example, communication satellites located in orbit 5.5 Earth radii above the surface of Earth have a period of 24 h. This period matches the period of daily Earth rotation. For an orbit around the equator, these satellites remain above the same point on the ground. The Moon is even farther away and has a period of 27.3 days. The higher the orbit of a satellite, the lower its speed, the longer its path, and the longer its period.*

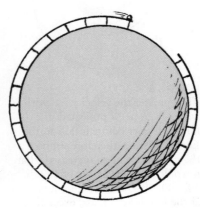

FIGURE 4.30
What speed will allow the ball to clear the gap?

* The speed of a satellite in circular orbit is given by $v = \sqrt{GM/d}$, and the period of satellite motion is given by $T = 2\pi\sqrt{d^3/GM}$, where G is the universal gravitational constant, M is the mass of Earth (or whatever body the satellite orbits), and d is the distance of the satellite from the center of Earth or other parent body.

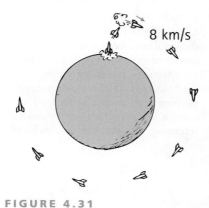

Putting a payload into Earth orbit requires control over the speed and direction of the rocket that carries it above the atmosphere. A rocket initially fired vertically is intentionally tipped from the vertical course. Then, once above the drag of the atmosphere, it is aimed horizontally, whereupon the payload is given a final thrust to orbital speed. We see this in Figure 4.31, where, for the sake of simplicity, the payload is the entire single-stage rocket. With the proper tangential velocity, it falls around Earth, rather than into it, and becomes an Earth satellite.

FIGURE 4.31
The initial thrust of the rocket lifts it vertically. Another thrust tips it from its vertical course. When it is moving horizontally, it is boosted to the required speed for orbit.

CHECKPOINT

1. True or false: Earth satellites normally orbit at altitudes in excess of 150 km to be above both gravity and the atmosphere of Earth.
2. Satellites in close circular orbit fall about 5 m during each second of orbit. Why doesn't this distance accumulate and send satellites crashing into Earth's surface?

Were these your answers?

1. False. Satellites are above the atmosphere and air resistance—*not* gravity! It's important to note that Earth's gravity extends throughout the universe in accord with the inverse-square law.
2. In each second, the satellite falls about 5 m below the straight-line tangent it would have followed if there were no gravity. Earth's surface also curves 5 m beneath a straight-line 8-km tangent. The process of falling with the curvature of Earth continues from tangent line to tangent line, so the curved path of the satellite and the curve of Earth's surface "match" all the way around Earth. Satellites do, in fact, crash to Earth's surface from time to time when they encounter air resistance in the upper atmosphere that decreases their orbital speed.

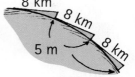

The initial vertical climb gets a rocket quickly through the denser part of the atmosphere. Eventually, the rocket must acquire enough tangential speed to remain in orbit without thrust, so it must tilt until its path is parallel to Earth's surface.

4.8 Elliptical Orbits

EXPLAIN THIS Why does the kinetic energy and momentum of a satellite change in an elliptical orbit?

If a projectile just above the drag of the atmosphere is given a horizontal speed somewhat greater than 8 km/s, it overshoots a circular path and traces an oval path called an **ellipse**.

An ellipse is a specific curve: the closed path taken by a point that moves in such a way that the sum of its distances from two fixed points (called *foci*) is constant. For a satellite orbiting a planet, one focus is at the center of the planet; the other focus could be internal or external to the planet. An ellipse can be easily constructed by using a pair of tacks (one at each focus), a loop of string, and a pen (Figure 4.32). The closer the foci are to each other, the closer the ellipse is to a circle. When both foci are together, the ellipse *is* a circle. So we can see that a circle is a special case of an ellipse.

Whereas the speed of a satellite is constant in a circular orbit, its speed varies in an elliptical orbit. For an initial speed greater than 8 km/s, the satellite

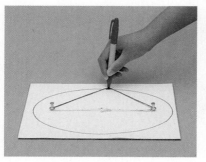

FIGURE 4.32

INTERACTIVE FIGURE   MP

A simple method for constructing an ellipse.

FIGURE 4.33
Elliptical orbit. When the speed of the satellite exceeds 8 km/s, (a) it overshoots a circular path and travels away from Earth against gravity. (b) At its maximum altitude it starts to come back toward Earth. (c) The speed it lost in going away is gained in returning, and the cycle repeats itself.

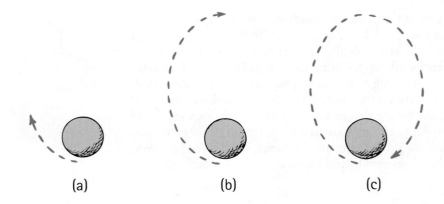

(a)   (b)   (c)

FYI When a spacecraft enters the atmosphere at too steep an angle, more than about 6°, it can burn up. If it comes in too shallow, it could bounce back into space like a pebble skipped across water.

overshoots a circular path and moves away from Earth, against the force of gravity. It therefore loses speed. The speed it loses in receding is regained as it falls back toward Earth, and it finally rejoins its original path with the same speed it had initially (Figure 4.33). The procedure repeats over and over, and an ellipse is traced during each cycle.

Interestingly enough, the parabolic path of a projectile, such as a tossed baseball or a cannonball, is actually a tiny segment of a skinny ellipse that extends within and just beyond the center of Earth (Figure 4.34a). In Figure 4.34b, we see several paths of cannonballs fired from Newton's mountain. All these ellipses have the center of Earth as one focus. As muzzle velocity is increased, the ellipses are less *eccentric* (more nearly circular); and, when muzzle velocity reaches 8 km/s, the ellipse rounds into a circle and does not intercept Earth's surface. The cannonball coasts in circular orbit. At greater muzzle velocities, orbiting cannonballs trace the familiar external ellipses.

FIGURE 4.34
(a) The parabolic path of the cannonball approximates part of an ellipse that extends within Earth. Earth's center is the far focus. (b) All paths of the cannonball are ellipses. For less than orbital speeds, the center of Earth is the far focus; for a circular orbit, both foci are Earth's center; for greater speeds, the near focus is Earth's center.

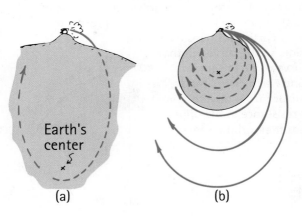

Earth's center

(a)   (b)

CHECKPOINT

The orbital path of a satellite is shown in the sketch. At which of the marked positions A through D does the satellite have the highest speed? The lowest speed?

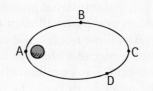

Were these your answers?
The satellite has its highest speed as it whips around A and has its lowest speed at position C. After passing C, it gains speed as it falls back to A to repeat its cycle.

4.9 Escape Speed

EXPLAIN THIS What is your fate if you are launched from Earth at a speed greater than 11.2 km/s?

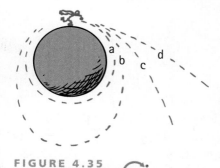

We know that a cannonball fired horizontally at 8 km/s from Newton's mountain would find itself in orbit. But what would happen if the cannonball were instead fired at the same speed *vertically*? It would rise to some maximum height, reverse direction, and then fall back to Earth. Then the old saying "What goes up must come down" would hold true, just as surely as a stone tossed skyward is returned by gravity (unless, as we shall see, its speed is great enough).

In today's spacefaring age, it is more accurate to say, "What goes up *may* come down," for a critical starting speed exists that permits a projectile to escape Earth. This critical speed is called the **escape speed** or, if direction is involved, the *escape velocity*. From the surface of Earth, escape speed is 11.2 km/s. If you launch a projectile at any speed greater than that, it leaves Earth, traveling slower and slower, never stopping due to Earth's gravity.* We can understand the magnitude of this speed from an energy point of view.

How much work would be required to lift a payload against the force of Earth's gravity to a distance extremely far ("infinitely far") away? We might think that the change of potential energy would be infinite because the distance is infinite. But gravity diminishes with distance by the inverse-square law. The force of gravity on the payload would be strong only near Earth. Most of the work done in launching a rocket occurs within 10,000 km or so of Earth. It turns out that the change of potential energy of a 1-km body moved from the surface of Earth to an infinite distance is 62 million J (62 MJ). So to put a payload infinitely far from Earth's surface requires at least 62 million joules of energy per kilogram of load. We won't go through the calculation here, but 62 MJ/kg corresponds to a speed of 11.2 km/s, whatever the total mass involved. This is the escape speed from the surface of Earth.**

If we give a payload any more energy than 62 MJ/kg at the surface of Earth or, equivalently, any more speed than 11.2 km/s, then, neglecting air resistance, the payload will escape from Earth, never to return. As the payload continues outward, its potential energy increases and its kinetic energy decreases. Earth's gravitational pull continuously slows it down but never reduces its speed to zero. The payload escapes.

The escape speeds from various bodies in the solar system are shown in Table 4.1. Note that the escape speed from the surface of the Sun is 620 km/s. Even at 150,000,000 km from the Sun (Earth's distance), the escape speed to break free of the Sun's influence is 42.5 km/s—considerably more than the escape speed from Earth. An object projected from Earth at a speed greater than 11.2 km/s but less than 42.5 km/s will escape Earth but not the Sun. Rather than recede forever, it will take up an orbit around the Sun.

FIGURE 4.35
INTERACTIVE FIGURE **MP**

If Superman tosses a ball 8 km/s horizontally from the top of a mountain high enough to be just above air resistance (a), then about 90 min later he can turn around and catch it (neglecting Earth's rotation). Tossed slightly faster (b), it takes an elliptical orbit and returns in a slightly longer time. Tossed at more than 11.2 km/s (c), it escapes Earth. Tossed at more than 42.5 km/s (d), it escapes the solar system.

* Escape speed from any planet or any body is given by $v = \sqrt{2GM/d}$, where G is the universal gravitational constant, M is the mass of the attracting body, and d is the distance from its center. (At the surface of the body, d would simply be the radius of the body.) For a bit more mathematical insight, compare this formula with the one for orbital speed in the footnote on page 106.
** Interestingly enough, this might well be called the *maximum falling speed*. Any object, however far from Earth, released from rest and allowed to fall to Earth only under the influence of Earth's gravity would not exceed 11.2 km/s. (With air friction, it would be less.)

TABLE 4.1	ESCAPE SPEEDS AT THE SURFACE OF BODIES IN THE SOLAR SYSTEM		
Astronomical Body	Mass (Earth masses)	Radius (Earth radii)	Escape Speed (km/s)
Sun	333,000	109	620
Sun (at a distance of Earth's orbit)		23,500	42.2
Jupiter	318	11	60.2
Saturn	95.2	9.2	36.0
Neptune	17.3	3.47	24.9
Uranus	14.5	3.7	22.3
Earth	1.00	1.00	11.2
Venus	0.82	0.95	10.4
Mars	0.11	0.53	5.0
Mercury	0.055	0.38	4.3
Moon	0.0123	0.27	2.4

FYI You won't fully appreciate the frontiers of physical science unless you're familiar with its foothills.

Just as planets fall around the Sun, stars fall around the centers of galaxies. Those with insufficient tangential speeds are pulled into, and are gobbled up by, the galactic nucleus—usually a black hole.

The first probe to escape the solar system, *Pioneer 10*, was launched from Earth in 1972 with a speed of only 15 km/s. The escape was accomplished by directing the probe into the path of oncoming Jupiter. It was whipped about by Jupiter's great gravitational field, picking up speed in the process—similar to the increase in the speed of a baseball encountering an oncoming bat. Its speed of departure from Jupiter was increased enough to exceed the escape speed from the Sun at the distance of Jupiter. *Pioneer 10* passed the orbit of Pluto in 1984. Unless it collides with another body, it will wander indefinitely through interstellar space. Like a note inside a bottle cast into the sea, *Pioneer 10* contains information about Earth that might be of interest to extraterrestrials, in hopes that it will one day "wash up" and be found on some distant "seashore."

It is important to stress that the escape speed of a body is the initial speed given by a brief thrust, after which there is no force to assist motion. One could escape Earth at *any* sustained speed more than zero, given enough time. For example, suppose a rocket is launched to a destination such as the Moon. If the rocket engines burn out when still close to Earth, the rocket needs a minimum speed of 11.2 km/s. But if the rocket engines can be sustained for long periods of time, the rocket could reach the Moon without ever attaining 11.2 km/s.

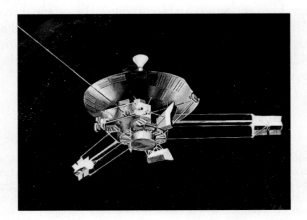

FIGURE 4.36
Pioneer 10, launched from Earth in 1972, passed Pluto in 1984 and is now drifting through the outer reaches of our solar system.

FIGURE 4.37
The European–U.S. spacecraft *Cassini* beams close-up images of Saturn and its giant moon Titan to Earth. It also measures surface temperatures, magnetic fields, and the size, speed, and trajectories of tiny surrounding space particles.

It is interesting to note that the accuracy with which an unoccupied rocket reaches its destination is not accomplished by staying on a planned path or by getting back on that path if the rocket strays off course. No attempt is made to return the rocket to its original path. Instead, the control center in effect asks, "Where is it now and what is its velocity? What is the best way to reach its destination, given its present situation?" With the aid of high-speed computers, the answers to these questions are used to find a new path. Corrective thrusters direct the rocket to this new path. This process is repeated all the way to the goal.*

The mind that encompasses the universe is as marvelous as the universe that encompasses the mind.

* Is there a lesson to be learned here? Suppose you find that you are off course. You may, like the rocket, find it more fruitful to follow a course that leads to your goal as best plotted from your present position and circumstances, rather than try to get back on the course you plotted from a previous position, perhaps under different circumstances.

For instructor-assigned homework, go to www.masteringphysics.com (MP)

SUMMARY OF TERMS (KNOWLEDGE)

Ellipse The oval path followed by a satellite. The sum of the distances from any point on the path to two points called foci is a constant. When the foci are together at one point, the ellipse is a circle. As the foci get farther apart, the ellipse becomes more *eccentric*.

Escape speed The speed that a projectile, space probe, or similar object must reach to escape the gravitational influence of Earth or of another celestial body to which it is attracted.

Inverse-square law The intensity of an effect from a localized source spreads uniformly throughout the surrounding space and weakens with the inverse square of the distance:

$$\text{Intensity} = \frac{1}{\text{distance}^2}$$

Gravity follows an inverse-square law, as do the effects of electric, light, sound, and radiation phenomena.

Law of universal gravitation Every body in the universe attracts every other body with a force that, for two bodies, is directly proportional to the product of their masses and inversely proportional to the square of the distance separating their centers:

$$F = G\frac{m_1 m_2}{d^2}$$

Parabola The curved path followed by a projectile under the influence of constant gravity only.

Projectile Any object that is projected by some means and continues its motion by its own inertia.

Satellite A projectile or small celestial body that orbits a larger celestial body.

Weight The force that an object exerts on a supporting surface (or, if suspended, on a supporting string), which is often, but not always, due to the force of gravity.

Weightless Being without a support force, as in free fall.

READING CHECK QUESTIONS (COMPREHENSION)

1. What did Newton discover about gravity?

4.1 The Universal Law of Gravity

2. In what sense does the Moon "fall"?
3. State Newton's law of universal gravitation in words. Then do the same with one equation.
4. What is the magnitude of the gravitational force between two 1-kg bodies that are 1 m apart?
5. What is the magnitude of the gravitational force between Earth and a 1-kg body at its surface?

4.2 Gravity and Distance: The Inverse-Square Law

6. How does the force of gravity between two bodies change when the distance between them is tripled?
7. Where do you weigh more: at sea level or on top of one of the peaks of the Rocky Mountains? Defend your answer.

4.3 Weight and Weightlessness

8. Would the springs inside a bathroom scale be more compressed or less compressed if you weighed yourself in an elevator that accelerated upward? Downward?
9. Would the springs inside a bathroom scale be more compressed or less compressed if you weighed yourself in an elevator that moved upward at *constant velocity*? Downward at *constant velocity*?
10. Explain why occupants of the International Space Station have no weight, yet are firmly in the grips of Earth's gravity.
11. When is your weight equal to *mg*?

4.4 Universal Gravitation

12. What was the cause of the perturbations discovered in the orbit of the planet Uranus?
13. The perturbations of Uranus led to what greater discovery?
14. What is the status of Pluto in the family of planets?
15. Which is thought to be more prevalent in the universe: dark matter or dark energy?

4.5 Projectile Motion

16. A stone is thrown upward at an angle. Neglecting air resistance, what happens to the horizontal component of its velocity along its trajectory?

17. A stone is thrown upward at an angle. Neglecting air resistance, what happens to the vertical component of its velocity along its trajectory?
18. A projectile is launched upward at an angle of 75° from the horizontal and strikes the ground a certain distance downrange. At what other angle of launch at the same speed would this projectile land just as far away?
19. A projectile is launched vertically at 100 m/s. If air resistance can be neglected, at what speed does it return to its initial level?

4.6 Fast-Moving Projectiles—Satellites

20. What does Earth's curvature have in common with the speed needed for a projectile to orbit Earth?
21. Why is it important that a satellite remain above Earth's atmosphere?
22. When a satellite is above Earth's atmosphere, is it also beyond the pull of Earth's gravity? Defend your answer.
23. If a satellite were beyond Earth's gravity, what path would it follow?

4.7 Circular Satellite Orbits

24. Why doesn't the force of gravity change the speed of a bowling ball as it rolls along a bowling lane?
25. Why doesn't the force of gravity change the speed of a satellite in circular orbit?
26. Is the period longer or shorter for orbits of higher altitude?

4.8 Elliptical Orbits

27. Why does the force of gravity change the speed of a satellite in an elliptical orbit?
28. At what part of an elliptical orbit does an Earth satellite have the highest speed? The lowest speed?

4.9 Escape Speed

29. What happens to a satellite close to Earth's surface if it is given a speed exceeding 11.2 km/s?
30. Although a space vehicle can outrun Earth's gravity, can it get entirely beyond Earth's gravity?

ACTIVITIES (HANDS-ON APPLICATION)

31. With a ballpoint pen write your name on a piece of paper on your desk. No problem. Now try it upside down—for example, with the paper held against a book above your head. Note the pen "doesn't work." Now you see that gravity acts on the ink in the barrel through which the ink flows!

32. Hold your hands outstretched in front of you, one twice as far from your eyes as the other, and make a casual judgment as to which hand looks bigger. Most people see them to be about the same size, while many see the nearer hand as slightly bigger. Almost no one, upon casual inspection,

sees the nearer hand as four times as big. But, by the inverse-square law, the nearer hand should appear to be twice as tall and twice as wide and therefore seem to occupy four times as much of your visual field as the farther hand. Your belief that your hands are the same size is so strong that you likely over-rule this visual information. Now, if you overlap your hands slightly and view them with one eye closed, you'll see the nearer hand as clearly bigger. This raises an interesting question: What other illusions do you have that are not so easily checked?

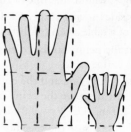

33. Repeat the preceding eyeballing experiment, only this time use two one-dollar bills—one flat or unfolded, and the other folded along its middle lengthwise and again widthwise, so it has $\frac{1}{4}$ the area. Now hold the two bills in front of your eyes. Where do you hold the folded dollar bill so that it looks the same size as the unfolded one? Nice?

34. With stick and strings, make a "trajectory stick" as shown on page 101.

35. With your friends, whirl a bucket of water in a vertical circle fast enough so the water doesn't spill out. As it happens, the water in the bucket *is* falling, but with less speed than you give to the bucket. Tell how your bucket swing is like satellite motion—that satellites in orbit continually fall toward Earth, but not with enough vertical speed to get closer to the curved Earth below. Remind your friends that physics is about finding the connections in nature!

PLUG AND CHUG (FORMULA FAMILIARIZATION)

$$F = G\frac{m_1 m_2}{d^2}$$

36. Using the formula for gravity, show that the force of gravity on a 1-kg mass at Earth's surface is 9.8 N. (The mass of Earth is 6×10^{24} kg, and its radius is 6.4×10^6 m.)

37. Calculate the force of gravity on the same 1-kg mass if it were 6.4×10^6 m above Earth's surface (that is, if it were two Earth radii from Earth's center).

38. Show that the average force of gravity between Earth (mass = 6.0×10^{24} kg) and the Moon (mass = 7.4×10^{22} kg) is 2.1×10^{20} N. The average Earth–Moon distance is 3.8×10^8 m.

39. Show that the force of gravity is 3.6×10^{22} N between Earth and the Sun (Sun's mass = 2.0×10^{30} kg; average Earth–Sun distance = 1.5×10^{11} m).

40. Show that the force of gravity is 4.0×10^{-8} N between a newborn baby (mass = 3.0 kg) and the planet Mars (mass = 6.4×10^{23} kg), when Mars is at its closest to Earth (distance = 5.6×10^{10} m).

41. Calculate the force of gravity between a newborn baby of mass 3.0 kg and the obstetrician of mass 100 kg, who is 0.5 m from the baby. Which exerts more gravitational force on the baby: Mars or the obstetrician? By how much?

THINK AND SOLVE (MATHEMATICAL APPLICATION)

42. Suppose you stood on top of a ladder that was so tall that you were three times as far from Earth's center as you presently are. Show that your weight would be $\frac{1}{9}$ its present value.

43. Show that the gravitational force between two planets is quadrupled if the masses of both planets are doubled but the distance between them stays the same.

44. Show that there is no change in the force of gravity between two objects when their masses are doubled and the distance between them is also doubled.

45. Find the change in the force of gravity between two planets when the distance between them is *decreased* by a factor of 10.

46. Consider a pair of planets for which the distance between them is decreased by a factor of 5. Show that the force between them becomes 25 times as strong.

47. Many people mistakenly believe that the astronauts who orbit Earth are "above gravity." Earth's mass is 6×10^{24} kg, and its radius is 6.38×10^6 m (6380 km). Use the inverse-square law to show that in space-shuttle territory, 200 km above Earth's surface, the force of gravity on a shuttle is about 94% that at Earth's surface.

48. Newton's universal law of gravity tells us that $F = G(m_1 m_2/d^2)$. Newton's second law tells us that $a = F_{net}/m$.
    (a) With a bit of algebraic reasoning show that your gravitational acceleration toward any planet of mass M a distance d from its center is $a = GM/d^2$.
    (b) How does this equation tell you whether or not your gravitational acceleration depends on your mass?

49. A ball is thrown horizontally from a cliff at a speed of 10 m/s. Show that its speed 1 s later is 14.1 m/s.

50. An airplane is flying horizontally with speed 1000 km/h (280 m/s) when an engine falls off. Neglecting air resistance, assume it takes 30 s for the engine to hit the ground.

    (a) Show that the altitude of the airplane is 4.4 km. (Use $g = 9.8$ m/s^2.)

    (b) Show that the horizontal distance the airplane engine falls is 8.4 km.

    (c) If the airplane somehow continues to fly as if nothing had happened, where is the engine relative to the airplane at the moment the engine hits the ground?

51. A satellite at a particular point along an elliptical orbit has a gravitational potential energy of 5000 MJ with respect to Earth's surface and a kinetic energy of 4500 MJ. Later in its orbit the satellite's potential energy is 6000 MJ. Use conservation of energy to find its kinetic energy at that point.

52. A rock thrown horizontally from a bridge hits the water below. The rock travels a smooth parabolic path in time t.

    (a) Show that the height of the bridge is $\frac{1}{2}gt^2$.

    (b) What is the height of the bridge if the time the rock is airborne is 2 s?

    (c) To solve this problem, what information is assumed here that wasn't in Chapter 2?

53. A baseball is tossed at a steep angle into the air and makes a smooth parabolic path. Its time in the air is t, and it reaches a maximum height h. Assume that air resistance is negligible.

    (a) Show that the height reached by the ball is $gt^2/8$.

    (b) If the ball is in the air for 4 s, show that it reaches a height of about 20 m.

    (c) If the ball reached the same height as when tossed at some other angle, would the time of flight be the same?

54. A penny on its side moving at speed v slides off the horizontal surface of a table a vertical distance y from the floor.

    (a) Show that the penny lands a distance $v\sqrt{2y/g}$ from the base of the coffee table.

    (b) If the speed is 3.5 m/s and the coffee table is 0.4 m tall, show that the distance the coin lands from the base of the table is 1.0 m. (Use $g = 9.8$ m/s^2.)

55. Students in a lab measure the speed of a steel ball launched horizontally from a tabletop to be v. The tabletop is distance y above the floor. They place a tall tin coffee can of height $0.1y$ on the floor to catch the ball.

    (a) Show that the can should be placed a horizontal distance from the base of the table of $v\sqrt{2(0.9y)/g}$.

    (b) If the ball leaves the tabletop at a speed of 4.0 m/s, the tabletop is 1.5 m above the floor, and the can is 0.15 m tall, show that the center of the can should be placed a horizontal distance of 2.1 m from the base of the table.

THINK AND RANK (ANALYSIS)

56. The planet and its moon gravitationally attract each other. Rank the forces of attraction between each pair from greatest to least.

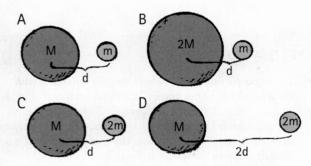

57. Consider the light of multiple candle flames, each of the same brightness. Rank the light that enters your eye from brightest to dimmest for these situations: (a) three candles seen from a distance of 3 m, (b) two candles seen from a distance of 2 m, and (c) one candle seen from a distance of 1 m.

58. Rank the average gravitational forces from greatest to least between (a) the Sun and Mars, (b) the Sun and the Moon, and (c) the Sun and Earth.

59. A ball is tossed off the edge of a cliff with the same speed but at different angles as shown. From greatest to least, rank the (a) initial PEs of the balls relative to the ground below, (b) initial KEs of the balls when tossed, (c) KEs of the balls when they hit the ground below, and (d) times of flight while airborne.

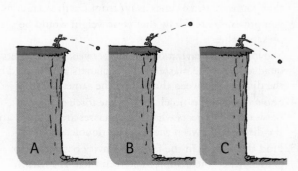

60. The dashed lines show three circular orbits about Earth. Rank from greatest to least (a) their orbital speeds and (b) their times to orbit Earth.

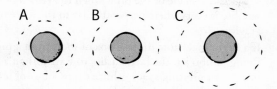

61. The positions of a satellite in elliptical orbit are indicated. Rank these quantities from greatest to least: (a) gravitational force, (b) speed, (c) momentum, (d) KE, (e) PE, (f) total energy (KE + PE), and (g) acceleration.

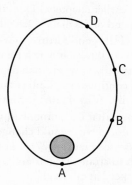

EXERCISES (SYNTHESIS)

62. What would be the path of the Moon if somehow all gravitational forces on it vanished to zero?

63. Upon which is the gravitational force greater: a 1-kg piece of iron or a 1-kg piece of glass? Defend your answer.

64. Consider a space pod somewhere between Earth and the Moon, at just the right distance so that the gravitational attractions to Earth and the Moon are equal. Is this location nearer Earth or the Moon?

65. An astronaut lands on a planet that has the same mass as Earth but half the diameter. How does the astronaut's weight differ from that on Earth?

66. An astronaut lands on a planet that has the same mass as Earth but twice the diameter. How does the astronaut's weight differ from that on Earth?

67. If Earth somehow expanded to have a larger radius, with no change in mass, how would your weight be affected? How would it be affected if Earth instead shrunk? (Hint: Let the equation for gravitational force guide your thinking.)

68. Why do the passengers in high-altitude jet planes feel the sensation of weight while passengers in the International Space Station do not?

69. To begin your wingsuit flight, you step off the edge of a high cliff. Why are you then momentarily weightless? At that point, is gravity acting on you?

70. In synchronized diving, divers remain in the air for the same time. With no air resistance, they would fall together. But air resistance is appreciable, so how do they remain together in fall?

71. What two forces act on you while you are in a moving elevator? When are these forces of equal magnitude, and when are they not?

72. If you were in a freely falling elevator and you dropped a pencil, it would hover in front of you. Is there a force of gravity acting on the pencil? Defend your answer.

73. While you are setting up an experiment, a ball rolls off your lab table. Will the time to hit the floor depend on the speed of the ball as it leaves the table? (Does a faster ball take a longer time to hit the floor?) Defend your answer.

74. A heavy crate accidentally falls from a high-flying airplane just as it flies directly above a shiny red Porsche smartly parked in a car lot. Relative to the Porsche, where does the crate crash?

75. In the absence of air resistance, why doesn't the horizontal component of a projectile's motion change, while the vertical component does change?

76. At what point in its trajectory does a batted baseball have its minimum speed? If air resistance can be neglected, how does this compare with the horizontal component of its velocity at other points?

77. Two golfers each hit a ball at the same speed, but one at 60° with the horizontal and the other at 30° with the horizontal. Which ball goes farther? Which hits the ground first? (Ignore air resistance.)

78. When you jump upward, your hang time is the time your feet are off the ground. Does hang time depend on your vertical component of velocity when you jump, your horizontal component of velocity, or both? Defend your answer.

79. The hang time of a basketball player who jumps a vertical distance of 2 ft (0.6 m) is about 0.6 s. What is the hang time if the player reaches the same height while jumping 4 ft (1.2 m) horizontally?

80. Earth and the Moon are gravitationally attracted to the Sun, but they don't crash into the Sun. A friend says that is because Earth and the Moon are beyond the Sun's main gravitational influence. Other friends look to you for a response. What do you say?

81. Does the speed of a falling object in the absence of air resistance depend on its mass? (Recall the answer to this question in earlier chapters.) Does the speed of a satellite in orbit depend on its mass? Defend your answers.

82. If you've had the good fortune to witness the launching of an Earth satellite, you may have noticed that the rocket starts vertically upward, then departs from a vertical course and continues its climb at an angle. Why does it start vertically? Why doesn't it continue vertically?

83. Newton knew that if a cannonball were fired from a tall mountain, gravity would change its speed all along its trajectory (see Figure 4.27). But if it were fired fast enough to attain circular orbit, gravity would not change its speed at all. Explain.

84. Satellites are normally sent into orbit by firing them in an easterly direction, the direction in which Earth spins. What is the advantage of this?

85. Hawaii, more than any other state in the United States, is the most efficient launching site for nonpolar satellites. Why is this so? (Hint: Look at the spinning Earth from above either pole and compare it to a spinning turntable.)

86. If a space vehicle circled Earth at a distance equal to the Earth–Moon distance, how long would it take for it to make a complete orbit? In other words, what would be its period?

87. Earth is farther away from the Sun in June and closest in December. In which of these two months is Earth moving faster around the Sun?

88. What is the shape of the orbit when the velocity of the satellite is everywhere perpendicular to the force of gravity?

89. If a flight mechanic drops a box of tools from a high-flying jumbo jet, the box crashes to Earth. If an astronaut in an orbiting space vehicle drops a box of tools, does it crash to Earth also? Defend your answer.

90. How could an astronaut in a space vehicle "drop" an object vertically to Earth?

91. If you stopped an Earth satellite dead in its tracks—that is, reduced its tangential velocity to zero—it would simply crash into Earth. Why, then, don't the communication satellites that "hover motionless" above the same spot on Earth crash into Earth?

92. The orbital velocity of Earth about the Sun is 30 km/s. If Earth were suddenly stopped in its tracks, it would simply fall radially into the Sun. Devise a plan whereby a rocket loaded with radioactive wastes could be fired into the Sun for permanent disposal. How fast and in what direction with respect to Earth's orbit should the rocket be fired?

DISCUSSION QUESTIONS (EVALUATION)

93. Comment on whether or not the following label on a consumer product should be cause for concern. *CAUTION: The mass of this product pulls on every other mass in the universe, with an attracting force that is proportional to the product of the masses and inversely proportional to the square of the distance between them.*

94. Newton tells us that gravitational force acts on all bodies in proportion to their masses. Why, then, doesn't a heavy body fall faster than a light body?

95. Okay, a friend says, gravitational force is proportional to mass. Is the force then stronger on a crumpled piece of aluminum foil than on an identical piece of foil that has not been crumpled? Isn't that why, when dropped, the crumpled one falls faster? Defend your answer, and explain why the two fall differently.

96. Two facts: A freely falling object at Earth's surface drops vertically 5 m in 1 s. Earth's curvature "drops" 5 m for each 8-km tangent. Discuss how these two facts relate to the 8-km/s speed necessary to orbit Earth.

97. A new member of your discussion group says that, since Earth's gravity is so much stronger than the Moon's gravity, rocks on the Moon could be dropped to Earth. What is wrong with this assumption?

98. A friend says that astronauts inside the International Space Station are weightless because they're beyond the pull of Earth's gravity. Correct your friend's ignorance.

99. Another new member of your discussion group says the primary reason astronauts in orbit feel weightless is because they are being pulled by other planets and stars. Why do you agree or disagree?

100. Occupants inside future donut-shaped rotating habitats in space will be pressed to their floors by rotational effects. Their sensation of weight feels as real as that due to gravity. Does this indicate that support force need not be related to gravity?

101. An apple falls because of the gravitational attraction to Earth. How does the gravitational attraction of Earth to the apple compare? (Does force change when you interchange m_1 and m_2 in the equation for gravity—$m_2 m_1$ instead of $m_1 m_2$?)

102. A small light source located 1 m in front of a 1-m² opening illuminates a wall behind. If the wall is 1 m behind the opening (2 m from the light source), the illuminated area covers 4 m². How many square meters are illuminated if the wall is 3 m from the light source? 5 m? 10 m?

103. The intensity of light from a central source varies inversely as the square of the distance. If you lived on a planet only half as far from the Sun as our Earth, how would the light intensity compare with that on Earth? How about a planet five times as far away as Earth?

104. Jupiter is more than 300 times as massive as Earth, so it might seem that a body on the surface of Jupiter would weigh 300 times as much as it weighs on Earth. But it so happens that a body weighs scarcely three times as much on the surface of Jupiter as it weighs on the surface of Earth. Discuss why this is so. (Hint: Let the terms in the equation for gravitational force guide your thinking.)

105. When will the gravitational force between you and the Sun be greater: today at noon or tomorrow at midnight? Defend your answer.

106. Explain why the following reasoning is wrong. "The Sun attracts all bodies on Earth. At midnight, when the Sun is directly below, it pulls on you in the same direction as Earth pulls on you; at noon, when the Sun is directly overhead, it pulls on you in a direction opposite to Earth's pull on you. Therefore, you should be somewhat heavier at midnight and somewhat lighter at noon."

107. Which requires more fuel: a rocket going from Earth to the Moon or a rocket returning from the Moon to Earth? Why?

108. Some people dismiss the validity of scientific theories by saying they are "only" theories. The law of universal gravitation is a theory. Does this mean that scientists still doubt its validity? Explain.

109. A friend claims that bullets fired by some high-powered rifles travel for many meters in a straight-line path before they start to fall. Another friend disputes this claim and states that all bullets from any rifle drop beneath a straight-line path a vertical distance given by $\frac{1}{2}gt^2$ as soon as they leave the barrel and that the curved path is apparent for low velocities and less apparent for high velocities. Now it's your turn: Do all bullets drop the same vertical distance in equal times? Explain.

110. A park ranger wants to shoot a monkey hanging from a branch of a tree with a tranquilizing dart. The ranger aims directly at the monkey, not realizing that the dart will follow a parabolic path and thus will fall below the monkey. The monkey, however, sees the dart leave the gun and lets go of the branch to avoid being hit. Will the monkey be hit anyway? Does the velocity of the dart affect your answer, assuming that it is great enough to travel the horizontal distance to the tree before hitting the ground? Defend your answer.

111. A satellite can orbit at 5 km above the Moon's surface but not at 5 km above Earth's surface. Why?

112. As part of their training before going into orbit, astronauts experience weightlessness when riding in an airplane that is flown along the same parabolic trajectory as a freely falling projectile. A classmate says that the gravitational forces on everything inside the plane during this maneuver cancel to zero. Another classmate looks to you for confirmation. What is your response?

113. Would the speed of a satellite in close circular orbit about Jupiter be greater than, equal to, or less than 8 km/s? Defend your answer.

114. A communication satellite with a 24-h period hovers over a fixed point on Earth. Why is it placed in orbit only in the plane of Earth's equator? (Hint: Think of the satellite's orbit as a ring around Earth.)

115. Here's a situation that should elicit good discussion. In an accidental explosion, a satellite breaks in half while in circular orbit about Earth. One half is brought momentarily to rest. What is the fate of the half brought to rest? What happens to the other half? (Hint: Think momentum conservation.)

116. Here's a situation to challenge you and your friends. A rocket coasts in an elliptical orbit around Earth. To attain the greatest amount of KE for escape using a given amount of fuel, should it fire its engines at the apogee (the point at which it is farthest from Earth) or at the perigee (the point at which it is closest to Earth)? (Hint: Let the formula $Fd = \Delta KE$ be your guide to thinking. Suppose the thrust F is brief and of the same duration in either case. Then consider the distance d the rocket would travel during this brief burst at the apogee and at the perigee.)

READINESS ASSURANCE TEST (RAT)

If you have a good handle on this chapter, if you really do, then you should be able to score at least 7 out of 10 on this RAT. If you score less than 7, you need to study further before moving on.

Choose the BEST answer to each of the following.

1. The Moon falls toward Earth in the sense that it falls
   (a) with an acceleration of 10 m/s², as do apples on Earth.
   (b) beneath the straight-line path it would follow without gravity.
   (c) both of these
   (d) neither of these

2. The force of gravity between two planets depends on their
   (a) planetary compositions.
   (b) planetary atmospheres.
   (c) rotational motions.
   (d) none of these

3. Inhabitants of the International Space Station do not have a
   (a) force of gravity on their bodies.
   (b) sufficient mass.
   (c) support force.
   (d) condition of free fall.

4. A spacecraft on its way from Earth to the Moon is pulled equally by Earth and the Moon when it is
   (a) closer to Earth's surface.
   (b) closer to the Moon's surface.
   (c) halfway from Earth to the Moon.
   (d) at no point, since Earth always pulls more strongly.

5. If you tossed a baseball horizontally and with no gravity, it would continue in a straight line. With gravity it falls about
   (a) 1 m below that line.
   (b) 5 m below that line.
   (c) 10 m below that line.
   (d) none of these

6. When no air resistance acts on a projectile, its horizontal acceleration is
   (a) *g.*
   (b) at right angles to *g.*
   (c) centripetal.
   (d) zero.

7. Without air resistance, a ball tossed at an angle of 40° with the horizontal goes as far downrange as one tossed at the same speed at an angle of
   (a) 45°.
   (b) 50°.
   (c) 60°.
   (d) none of these

8. When you toss a projectile sideways, it curves as it falls. It will become an Earth satellite if the curve it makes
   (a) matches the curve of Earth's surface.
   (b) results in a straight line.
   (c) spirals out indefinitely.
   (d) none of these

9. A satellite in elliptical orbit about Earth travels fastest when it moves
   (a) close to Earth.
   (b) far from Earth.
   (c) the same everywhere.
   (d) halfway between the near and far points from Earth.

10. A satellite in Earth orbit is mainly above Earth's
    (a) atmosphere.
    (b) gravitational field.
    (c) both of these
    (d) neither of these

Answers to RAT

1. *b,* 2. *d,* 3. *c,* 4. *b,* 5. *b,* 6. *d,* 7. *b,* 8. *a,* 9. *a,* 10. *a*

5

CHAPTER 5
Fluid Mechanics

LIQUIDS AND gases have the ability to flow; hence, they are called *fluids*. Because they are both fluids we find that they obey similar mechanical laws. How is it that iron boats don't sink in water or that helium balloons don't sink from the sky? What determines whether an object will float or sink in water? How does the Falkirk Wheel above, an alternative to a locks-and-canal system, use very little energy to rotate boats from a lower body of water to a higher one? Why do its two balanced water-filled caissons weigh the same regardless of what the boats weigh? Why is gas compressible while liquid is not? Why is it impossible to breathe through a snorkel when you're under more than a meter of water? Why do your ears pop when riding an elevator? How do hydrofoils and airplanes attain lift? To discuss fluids, it is important to introduce two concepts—*density* and *pressure*.

LEARNING OBJECTIVE
Distinguish among weight, mass, and density.

TABLE 5.1	DENSITIES
Material	(kg/m³)
Solids	
Iridium	22,650
Osmium	22,610
Platinum	21,090
Gold	19,300
Uranium	19,050
Lead	11,340
Silver	10,490
Copper	8,920
Iron	7,870
Aluminum	2,700
Ice	919
Liquids	
Mercury	13,600
Glycerin	1,260
Seawater	1,025
Water at 4°C	1,000
Ethyl alcohol	785
Gasoline	680
Gases (kg/m³ at sea level)	
Dry air:	
at 0°C	1.29
at 10°C	1.25
at 20°C	1.21
Helium	0.178
Hydrogen	0.090
Oxygen	1.43

5.1 Density

EXPLAIN THIS Does squeezing a loaf of bread increase its mass, its density, or both?

An important property of a material, whether in the solid, liquid, or gaseous phase, is the measure of compactness: **density**. We think of density as the "lightness" or "heaviness" of materials of the same size. It is a measure of how much mass occupies a given space; it is the amount of matter per unit volume:

$$\text{Density} = \frac{\text{mass}}{\text{volume}}$$

FIGURE 5.1
When the volume of the bread is reduced, its density increases.

The densities of some materials are listed in Table 5.1. Mass is measured in grams or kilograms, and volume in cubic centimeters (cm^3) or cubic meters (m^3).* A gram of any material has the same mass as 1 cm^3 of water at a temperature of 4°C. So water has a density of 1 g/cm^3. Mercury's density is 13.6 g/cm^3, which means that it has 13.6 times as much mass as an equal volume of water. Iridium, a hard, brittle, silvery-white metal in the platinum family, is the densest substance on Earth.

A quantity known as weight density, commonly used when discussing liquid pressure, is expressed by the amount of weight per unit volume:**

$$\text{Weight density} = \frac{\text{weight}}{\text{volume}}$$

CHECKPOINT
1. **Which has the greater density—1 kg of water or 10 kg of water?**
2. **Which has the greater density—5 kg of lead or 10 kg of aluminum?**
3. **Which has the greater density—an entire candy bar or half of one?**

Were these your answers?
1. The density of any amount of water is the same: 1 g/cm^3 or, equivalently, 1000 kg/m^3, which means that the mass of water that would exactly fill a thimble of volume 1 cm^3 would be 1 g; or the mass of water that would fill a 1-m^3 tank would be 1000 kg. One kilogram of water would fill a tank only a thousandth as large, 1 L, whereas 10 kg would fill a 10-liter tank. Nevertheless, the important concept is that the ratio of mass/volume is the same for *any* amount of water.
2. Density is a *ratio* of weight or mass per volume, and this ratio is greater for any amount of lead than for any amount of aluminum—see Table 5.1.
3. Both the half and the entire candy bar have the same density.

FYI The metals lithium, sodium, and potassium (not in Table 5.1) are all less dense than water and float in water.

* A cubic meter is a sizable volume and contains a million cubic centimeters, so there are a million grams of water in a cubic meter (or, equivalently, a thousand kilograms of water in a cubic meter). Hence, 1 g/cm^3 = 1000 kg/m^3.
** Weight density is common to the United States Customary System (USCS) units, in which 1 ft^3 of fresh water (nearly 7.5 gallons) weighs 62.4 lb. So fresh water has a weight density of 62.4 lb/ft^3. Salt water is slightly denser at 64 lb/ft^3.

5.2 Pressure

EXPLAIN THIS Why does wearing high heels increase the pressure on the floor?

Place a book on a bathroom scale; whether you place it on its back, on its side, or balanced on a corner, it still exerts the same force. The weight reading is the same. Now balance the book on the palm of your hand and you sense a difference—the *pressure* of the book depends on the area over which the force is distributed (Figure 5.2). There is a difference between force and pressure. **Pressure** is defined as the force exerted over a unit of area, such as a square meter or square foot:*

$$\text{Pressure} = \frac{\text{force}}{\text{area}}$$

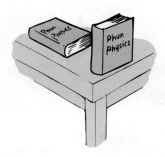

FIGURE 5.2
Although the weight of both books is the same, the upright book exerts greater pressure against the table.

CHECKPOINT

Does a bathroom scale measure weight, pressure, or both?

Was this your answer?
A bathroom scale measures weight, the force that compresses an internal spring or equivalent. The weight reading is the same whether you stand on one or both feet (although the pressure on the scale is twice as much when standing on one foot).

Pressure in a Liquid

When you swim under water, you can feel the water pressure acting against your eardrums. The deeper you swim, the greater the pressure. What causes this pressure? It is simply the weight of the fluids directly above you—water plus air—pushing against you. As you swim deeper, more water is above you. Therefore, there's more pressure. If you swim twice as deep, twice the weight of water is above you, so the water's contribution to the pressure you feel is doubled. Added to the water pressure is the pressure of the atmosphere, which is equivalent to an extra 10.3-m depth of water. Because atmospheric pressure at Earth's surface is nearly constant, the pressure differences you feel under water depend only on changes in depth.

The pressure due to a liquid is precisely equal to the product of weight density and depth:**

$$\text{Liquid pressure} = \text{weight density} \times \text{depth}$$

FIGURE 5.3
This water tower does more than store water. The height of the water above ground level ensures substantial and reliable water pressure to the many homes it serves.

---

* Pressure may be measured in any unit of force divided by any unit of area. The standard international (SI) unit of pressure, the newton per square meter, is called the *pascal* (Pa), after the 17th-century theologian and scientist Blaise Pascal. A pressure of 1 Pa is very small and approximately equals the pressure exerted by a dollar bill resting flat on a table. Science types prefer kilopascals (1 kPa = 1000 Pa).

** This is derived from the definitions of pressure and density. Consider an area at the bottom of a vessel that contains liquid. The weight of the column of liquid directly above this area produces pressure. From the definition *weight density = weight/volume*, we can express this weight of liquid as *weight = weight density × volume*, where the volume of the column is simply the area multiplied by the depth. Then we get

$$\text{Pressure} = \frac{\text{force}}{\text{area}} = \frac{\text{weight}}{\text{area}} = \frac{\text{weight density} \times \text{volume}}{\text{area}} = \frac{\text{weight density} \times (\text{area} \times \text{depth})}{\text{area}}$$

$$= \text{weight density} \times \text{depth}$$

For the total pressure we should add to this equation the pressure due to the atmosphere on the surface of the liquid.

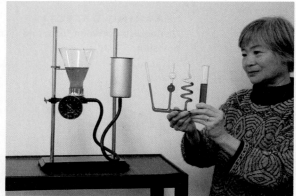

FIGURE 5.4
Liquid pressure is the same for any given depth below the surface, regardless of the shape of the containing vessel.

When blood pressure is measured, notice that it is done in your upper arm—level with your heart.

FIGURE 5.5
The average water pressure acting against the dam depends on the average depth of the water and not on the volume of water held back. The large shallow lake exerts only half the average pressure that the small deep pond exerts.

Mastering**PHYSICS**
VIDEO: Dam and Water

Note that pressure does not depend on the volume of liquid. You feel the same pressure a meter deep in a small pool as you do a meter deep in the middle of the ocean. This is illustrated by the connecting vases shown in Figure 5.4. If the pressure at the bottom of a large vase were greater than the pressure at the bottom of a neighboring narrower vase, the greater pressure would force water sideways and then up the narrower vase to a higher level. We find, however, that this doesn't happen. Pressure depends on depth, not volume.

Water seeks its own level. This can be demonstrated by filling a garden hose with water and holding the two ends upright. The water levels are equal whether the ends are held close together or far apart. Pressure is depth dependent, not volume dependent. So we see there is an explanation for why water seeks its own level.

In addition to being depth dependent, liquid pressure is exerted equally in all directions. For example, if we are submerged in water, it makes no difference which way we tilt our heads—our ears feel the same amount of water pressure. Because a liquid can flow, the pressure isn't only downward. We know pressure acts upward when we try to push a beach ball beneath the water's surface. The bottom of a boat is certainly pushed upward by water pressure. And we know water pressure acts sideways when we see water spurting sideways from a leak in an upright can. Pressure in a liquid at any point is exerted in equal amounts in all directions.

When liquid presses against a surface, a net force is directed perpendicular to the surface (Figure 5.6). If there is a hole in the surface, the liquid spurts at right angles to the surface before curving downward because of gravity (Figure 5.7). At greater depths the pressure is greater and the speed of the exiting liquid is greater.*

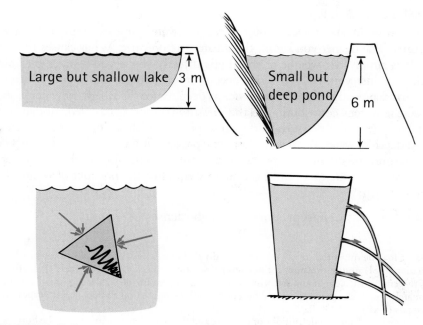

FIGURE 5.6
The forces due to liquid pressure against a surface combine to produce a net force that is perpendicular to the surface.

FIGURE 5.7
The force vectors act in a direction perpendicular to the inner container surface and increase with increasing depth.

FYI Molecules that make up a liquid can flow by sliding over one another. A liquid takes the shape of its container. Its molecules are close together and greatly resist compressive forces, so liquids, like solids, are difficult to compress.

* The speed of liquid exiting the hole is $\sqrt{2gh}$, where h is the depth below the free surface. Interestingly, this is the same speed that water or anything else would have if freely falling the same distance h.

5.3 Buoyancy in a Liquid

LEARNING OBJECTIVE
Relate the buoyant force to pressure differences in a fluid.

EXPLAIN THIS Why is it easier to lift a boulder in water than out of water?

Anyone who has ever lifted a submerged object out of water is familiar with buoyancy, the apparent loss of weight of submerged objects. For example, lifting a large boulder off the bottom of a riverbed is a relatively easy task as long as the boulder is below the surface. When it is lifted above the surface, however, the force required to lift it is considerably more. This is because when the boulder is submerged, the water exerts an upward force on it—opposite in direction to gravity. This upward force is called the **buoyant force** and is a consequence of greater pressure at greater depth. Figure 5.8 shows why the buoyant force acts upward. Pressure is exerted everywhere against the object in a direction perpendicular to its surface. The arrows represent the magnitude and direction of forces at different places. Forces that produce pressures against the sides due to equal depths cancel one another. Pressure is greatest against the bottom of the boulder simply because the bottom of the boulder is deeper. Because the upward forces against the bottom are greater than the downward forces against the top, the forces do not cancel, and there is a net force upward. This net force is the buoyant force.

Mastering**PHYSICS**
VIDEO: Buoyancy

Stick your foot in a swimming pool and your foot is immersed. Jump in and sink and immersion is total—you're submerged.

FIGURE 5.8
The greater pressure against the bottom of a submerged object produces an upward buoyant force.

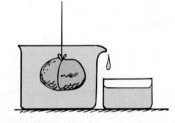

FIGURE 5.9
When a stone is submerged, it displaces a volume of water equal to the volume of the stone.

Water displaced

FIGURE 5.10
The raised level due to placing a stone in the container is the same as if a volume of water equal to the volume of the stone were poured in.

If the weight of the submerged object is greater than the buoyant force, the object sinks. If the weight is equal to the buoyant force acting upward on the submerged object, it remains at any level, like a fish. If the buoyant force is greater than the weight of the completely submerged object, it rises to the surface and floats.

Understanding buoyancy requires understanding the meaning of the expression "volume of water displaced." If a stone is placed in a container that is already up to its brim with water, some water overflows (Figure 5.9). Water is *displaced* by the stone. A little thought tells us that the *volume of the stone*—that is, the amount of space it occupies or its number of cubic centimeters—is equal to the *volume of water displaced*. Place any object in a container partially filled with water, and the level of the surface rises (Figure 5.10). How high? That would be to exactly the level that would be reached by pouring in a volume of water equal to the volume of the submerged object. This is a good method for determining the volume of irregularly shaped objects: *A completely submerged object always displaces a volume of liquid equal to its own volume.*

FIGURE 5.11
A liter of water occupies a volume of 1000 cm^3, has a mass of 1 kg, and weighs 9.8 N. Its density may therefore be expressed as 1 kg/L and its weight density as 9.8 N/L. (Seawater is slightly denser, 1.03 kg/L.)

5.4 Archimedes' Principle

EXPLAIN THIS How can a concrete barge loaded with iron ore float?

The relationship between buoyancy and displaced liquid was first discovered in the third century BC by the Greek scientist Archimedes. It is stated as follows:

An immersed body is buoyed up by a force equal to the weight of the fluid it displaces.

This relationship is called **Archimedes' principle**. It applies to liquids and gases, which are both fluids. If an immersed body displaces 1 kg of fluid, the buoyant force acting on it is equal to the weight of 1 kg.* By *immersed*, we mean either *completely* or *partially submerged*. If we immerse a sealed 1-L container halfway into the water, it displaces half a liter of water and is buoyed up by the weight of half a liter of water. If we immerse it completely (submerge it), it is buoyed up by the weight of a full liter (or 1 kg) of water. Unless the completely submerged container is compressed, the buoyant force equals the weight of 1 kg at *any* depth. This is because, at any depth, it can displace no greater volume of water than its own volume. And the weight of this volume of water (not the weight of the submerged object!) is equal to the buoyant force.

If a 25-kg object displaces 20 kg of fluid upon immersion, its apparent weight equals the weight of 5 kg. Notice in Figure 5.12 that the 3-kg block has an apparent weight equal to the weight of 1 kg when submerged. The apparent weight of a submerged object is its weight out of water minus the buoyant force.

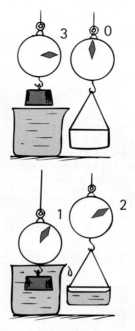

FIGURE 5.12
A 3-kg block weighs more in air than it does in water. When the block is submerged in water, its loss in weight is the buoyant force, which equals the weight of water displaced.

Mastering**PHYSICS**
VIDEO: Archimedes' Principle

CHECKPOINT

1. **Does Archimedes' principle tell us that if an immersed block displaces 10 N of fluid, the buoyant force on the block is 10 N?**
2. **A 1-L container completely filled with lead has a mass of 11.3 kg and is submerged in water. What is the buoyant force acting on it?**
3. **A boulder is thrown into a deep lake. As it sinks deeper and deeper into the water, does the buoyant force on it increase? Decrease?**

Were these your answers?

1. Yes. Looking at it in a Newton's-third-law way, when the immersed block pushes 10 N of fluid aside, the fluid reacts by pushing back on the block with 10 N.
2. The buoyant force is equal to the weight of 1 kg (9.8 N) because the volume of water displaced is 1 L, which has a mass of 1 kg and a weight of 9.8 N. The 11.3 kg of the lead is irrelevant; 1 L of anything submerged in water displaces 1 L and is buoyed upward with a force 9.8 N, the weight of 1 kg. (Get this straight before going further!)
3. Buoyant force remains the same. It doesn't change as the boulder sinks because the boulder displaces the same volume of water at any depth. Because water is practically incompressible, its density is very nearly the same at all depths; hence, the weight of water displaced, or the buoyant force, is practically the same at all depths.

* A kilogram is not a unit of force but a unit of mass. So, strictly speaking, the buoyant force is not 1 kg, but the *weight* of 1 kg, which is 9.8 N. We could as well say that the buoyant force is 1 *kilogram weight*, not simply 1 kg.

Perhaps your instructor will summarize Archimedes' principle by way of a numerical example to show that the difference between the upward-acting and the downward-acting forces on a submerged cube (due to differences of pressure) is numerically identical to the weight of fluid displaced. It makes no difference how deep the cube is placed, because, although the pressures are greater with increasing depths, the *difference* between the pressure up against the bottom of the cube and the pressure exerted downward against the top of the cube is the same at any depth (Figure 5.13). Whatever the shape of the submerged body, the buoyant force is equal to the weight of fluid displaced.

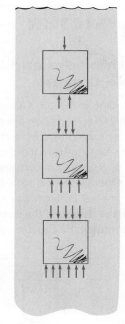

FIGURE 5.13
The difference in the upward and downward forces acting on the submerged block is the same at any depth.

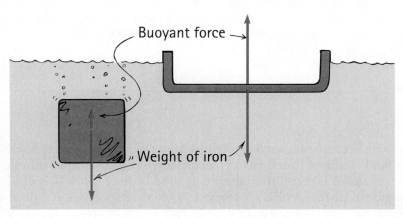

FIGURE 5.14
An iron block sinks, while the same quantity of iron shaped like a bowl floats.

Flotation

Iron is much denser than water and therefore sinks, but an iron ship floats. Why is this so? Consider a solid 1-ton block of iron. Iron is nearly eight times as dense as water, so when it is submerged it displaces only $\frac{1}{8}$ ton of water, which is certainly not enough to prevent it from sinking. Suppose we reshape the same iron block into a bowl, as shown in Figure 5.14. It still weighs 1 ton. When we place it in the water, it settles into the water, displacing a greater volume of water than before. The deeper it is immersed, the more water it displaces and the greater the buoyant force acting on it. When the buoyant force equals 1 ton, the iron sinks no further.

When the iron boat displaces a weight of water equal to its own weight, it floats. This is called the **principle of flotation**:

A floating object displaces a weight of fluid equal to its own weight.

Every ship, submarine, or dirigible airship must be designed to displace a weight of fluid equal to its own weight. Thus, a 10,000-ton ship must be built wide enough to displace 10,000 tons of water before it immerses too deep in the water. The same applies to vessels in air. A dirigible or huge balloon that weighs 100 tons displaces at least 100 tons of air. If it displaces more, it rises; if it displaces less, it descends. If it displaces exactly its weight, it hovers at constant altitude.

Because the buoyant force upon a body equals the weight of the fluid it displaces, denser fluids exert more buoyant force upon a body than less-dense fluids of the same volume. A ship therefore floats higher in salt water than in

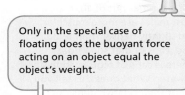

Only in the special case of floating does the buoyant force acting on an object equal the object's weight.

Mastering**PHYSICS**
VIDEO: Flotation

FIGURE 5.15
The weight of a floating object equals the weight of the water displaced by the submerged part.

PHYSICS IN HISTORY

Archimedes and the Gold Crown

According to legend, Archimedes (287–212 BC) had been given the task of determining whether a crown made for King Hiero II of Syracuse was of pure gold or contained some less expensive metals such as silver. Archimedes' problem was to determine the density of the crown without destroying it. He could weigh the crown, but determining its volume was a problem. The story tells us that Archimedes came to the solution when he noted the rise in water level while immersing his body in the public baths of Syracuse. Legend reports that he excitedly rushed naked through the streets shouting "Eureka! Eureka!" ("I have found it! I have found it!").

What Archimedes discovered was a simple and accurate way of finding the volume of an irregular object—the displacement method of determining volumes. Once he knew both the weight and volume, he could calculate the density. Then the density of the crown could be compared with the density of gold. Archimedes' insight preceded Newton's law of motion, from which Archimedes' principle can be derived, by almost 2000 years.

FIGURE 5.16
A floating object displaces a weight of fluid equal to its own weight.

FYI People who can't float are, 9 times out of 10, males. Most males are more muscular and slightly denser than females. Also, cans of diet soda float whereas cans of regular soda sink in water. What does this tell you about their relative densities?

fresh water because salt water is slightly denser than fresh water. In the same way, a solid chunk of iron floats in mercury even though it sinks in water.

The physics of Figure 5.16 is nicely employed by the Falkirk Wheel, a unique rotating boat lift that replaces a series of 11 locks in Scotland. A pair of water-filled caissons are connected on opposite sides of a 35-m-tall wheel. When a boat enters a caisson, the amount of water that overflows weighs exactly as much as the boat. As Figure 5.16 illustrates, each water-filled caisson weighs the same whether or not it carries boats (or multiple boats or even no boats as long as the water in each caisson has the same depth). The wheel always remains balanced as it rotates and lifts boats 18 m from a lower body of water to a higher one (Figure 5.17). So, in spite of its enormous mass, the wheel rotates each half revolution with very little power input.

Notice in our discussion of liquids that Archimedes' principle and the law of flotation were stated in terms of *fluids*, not liquids. That's because although liquids and gases are different phases of matter, they are both fluids, with much the same mechanical principles. Let's turn our attention to the mechanics of gases in particular.

LINK TO EARTH SCIENCE

Floating Mountains

Mountains float on Earth's semiliquid mantle just as icebergs float in water. Both the mountains and icebergs are less dense than the material they float upon. Just as most of an iceberg is below the water surface (90%), most of a mountain (about 85%) extends into the dense semiliquid mantle. If you could shave off the top of an iceberg, the iceberg would be lighter and be buoyed up to nearly its original height before its top was shaved. Similarly, when mountains erode they are lighter, and are pushed up from below to float to nearly their original heights. So when a kilometer of mountain erodes away, some 85% of a kilometer of mountain returns. That's why it takes so long for mountains to weather away. Mountains, like icebergs, are

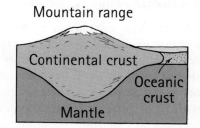

bigger than they appear to be. The concept of floating mountains is *isostacy*—Archimedes' principle for rocks.

FIGURE 5.17
The Falkirk Wheel has two balanced, water-filled caissons, one moving up while the other moves down. The caissons rotate as the wheel turns so the water and boats don't tip out as the wheel makes each half revolution.

5.5 Pressure in a Gas

LEARNING OBJECTIVE
Relate volume and pressure changes for a confined gas.

EXPLAIN THIS Why is holding your breath a no-no for scuba divers ascending to the surface of the water?

The primary difference between a gas and a liquid is the distance between molecules. In a gas, the molecules are far apart and free from the cohesive forces that dominate their motions in the liquid and solid phases. Molecular motions in a gas are less restricted. A gas expands, fills all space available to it, and exerts a pressure against its container. Only when the quantity of gas is very large, such as in Earth's atmosphere or a star, do the gravitational forces limit the size or determine the shape of the mass of gas.

Liquids and gases are both fluids. A gas takes the shape of its container. A liquid does so only below its surface.

Boyle's Law

The air pressure inside the inflated tires of an automobile is considerably greater than the atmospheric pressure outside. The density of air inside is also greater than that of the air outside. To understand the relation between pressure and density, think of the molecules of air (primarily nitrogen and oxygen) inside the tire. The air molecules behave like tiny billiard balls, randomly moving and banging against the inner walls, producing a jittery force that appears to our coarse senses as a steady push. This pushing force, averaged over the wall area, provides the pressure of the enclosed air.

Suppose there are twice as many molecules in the same volume (Figure 5.18). Then the air density is doubled. If the molecules move at the same average speed—or, equivalently, if they have the same temperature—then the number of collisions is doubled. This means that the pressure is doubled. So pressure is proportional to density.

We double the density of air in the tire by doubling the amount of air. We can also double the density of a *fixed* amount of air by compressing it to half its volume. Consider the cylinder with the movable piston in Figure 5.19. If the piston is pushed downward so that the volume is half the original volume, the density of molecules is doubled, and the pressure is correspondingly doubled. Decrease the volume to a third of its original value, and the pressure is increased by a factor of 3, and so forth (provided the temperature remains the same).

Notice in these examples with the piston that the product of pressure and volume remains the same. For example, a doubled pressure multiplied by a halved volume gives the same value as a tripled pressure multiplied by a one-third volume. In general, we can state that the product of pressure and volume for a given mass of gas is a constant *as long as the temperature does not change. Pressure × volume* for a

FIGURE 5.18
When the density of gas in the tire is increased, pressure is increased.

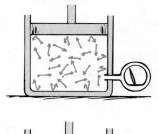

FIGURE 5.19
When the volume of gas is decreased, density and therefore pressure are increased.

sample of gas at some initial time is equal to any *different pressure* × *different volume* of the same sample of gas at some later time. In shorthand notation,

$$P_1V_1 = P_2V_2$$

where P_1 and V_1 represent the original pressure and volume, respectively, and P_2 and V_2 the second pressure and volume. This relationship is called **Boyle's law**, after Robert Boyle, the 17th-century physicist who is credited with its discovery.*

Boyle's law applies to ideal gases. An ideal gas is one in which the disturbing effects of the forces between molecules and the finite size of the individual molecules can be neglected. Air and other gases under normal pressures and temperatures approach ideal gas conditions.

CHECKPOINT

1. **A piston in an airtight pump is withdrawn so that the volume of the air chamber is tripled. What is the change in pressure?**
2. **A scuba diver breathes compressed air beneath the surface of water. If she holds her breath while returning to the surface, what happens to the volume of her lungs?**

Were these your answers?

1. The pressure in the piston chamber is reduced to one-third. This is the principle that underlies a mechanical vacuum pump.
2. When she rises toward the surface, the surrounding water pressure on her body decreases, allowing the volume of air in her lungs to increase—ouch! A first lesson in scuba diving is to not hold your breath when ascending. To do so can be fatal.

Mastering**PHYSICS**
VIDEO: Air Has Weight
VIDEO: Air Is Matter
VIDEO: Air Has Pressure

Interestingly, von Guericke's demonstration preceded knowledge of Newton's third law. The forces on the hemispheres would have been the same if he had used only one team of horses and tied the other end of the rope to a tree!

5.6 Atmospheric Pressure

EXPLAIN THIS How does the weight of air surrounding a planet affect atmospheric pressure at its surface?

We live at the bottom of an ocean of air. The atmosphere, much like the water in a lake, exerts a pressure. One of the most celebrated experiments demonstrating the pressure of the atmosphere was conducted in 1654 by Otto von Guericke, burgermeister of Magdeburg and inventor of the vacuum pump. Von Guericke placed together two copper hemispheres about 0.5 m in diameter to form a sphere, as shown in Figure 5.20. He set a gasket made of a ring of leather soaked in oil and wax between them to make an airtight joint. When he evacuated the sphere with his vacuum pump, two teams of eight horses each were unable to pull the hemispheres apart.

When the air pressure inside a cylinder like that shown in Figure 5.21 is reduced, an upward force is exerted on the piston. This force is large enough to lift a heavy weight. If the inside diameter of the cylinder is 12 cm or greater, a person can be lifted by this force.

---

* A general law that takes temperature changes into account is $P_1V_1/T_1 = P_2V_2/T_2$, where T_1 and T_2 represent the initial and final *absolute* temperatures, measured in SI units called kelvins (see Chapter 6).

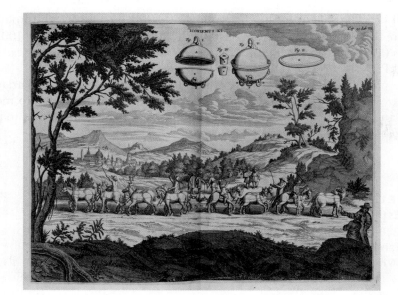

FIGURE 5.20
The famous "Magdeburg hemispheres" experiment of 1654, demonstrating atmospheric pressure. Two teams of horses couldn't pull the evacuated hemispheres apart. Were the hemispheres sucked together or pushed together? By what?

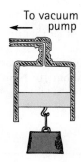

FIGURE 5.21
Is the piston pulled up or pushed up?

FIGURE 5.22
You don't notice the weight of a bag of water while you're submerged in water. Similarly, you don't notice that the air around you has weight.

What do the experiments of Figures 5.20 and 5.21 demonstrate? Do they show that air exerts pressure or that there is a "force of suction"? If we say there is a force of suction, then we assume that a vacuum can exert a force. But what is a vacuum? It is an absence of matter; it is a condition of nothingness. How can nothing exert a force? The hemispheres are not sucked together, nor is the piston holding the weight sucked upward. The pressure of the atmosphere is pushing against the hemispheres and the piston.

Just as water pressure is caused by the weight of water, **atmospheric pressure** is caused by the weight of air. We have adapted so completely to the invisible air that we sometimes forget it has weight. Perhaps a fish "forgets" about the weight of water in the same way. The reason we don't feel this weight crushing against our bodies is that the pressure inside our bodies equals that of the surrounding air. There is no net force for us to sense.

At sea level, 1 m³ of air at 20°C has a mass of about 1.2 kg. To estimate the mass of air in your room, estimate the number of cubic meters in the room, multiply by 1.2 kg/m³, and you'll have the mass. Don't be surprised if it's heavier than your kid sister. If your kid sister doesn't believe air has weight, maybe it's because she's always surrounded by air. Hand her a plastic bag of water and she'll tell you it has weight. But hand her the same bag of water while she's submerged in a swimming pool, and she won't feel the weight. We don't notice that air has weight because we're submerged in air.

Whereas water in a lake has the same density at any level (assuming constant temperature), the density of air in the atmosphere decreases with altitude. Although 1 m³ of air at sea level has a mass of about 1.2 kg, at 10 km, the same volume of air has a mass of about 0.4 kg. To compensate for this, airplanes are pressurized; the additional air needed to fully pressurize a 747 jumbo jet, for example, is more than 1000 kg. Air is heavy, if you have enough of it.

Consider the mass of air in an upright 30-km-tall hollow bamboo pole that has an inside cross-sectional area of 1 cm². If the density of air inside the pole matches the density of air outside, the enclosed mass of air would be about 1 kg. The weight of this much air is about 10 N. So the air pressure at the bottom of the bamboo pole would be about 10 N/cm². Of course, the same is true without the bamboo pole. There are 10,000 cm² in 1 m², so a column of air 1 m² in cross-section that extends up through the atmosphere has a mass of about

FIGURE 5.23
The mass of air that would occupy a bamboo pole that extends to the "top" of the atmosphere is about 1 kg. This air has a weight of about 10 N.

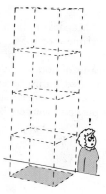

FIGURE 5.24
The weight of air that presses down on a 1-m² surface at sea level is about 100,000 N. So atmospheric pressure is about 10^5 N/m², or about 100 kPa.

Workers in underwater construction work in an environment of compressed air. The air pressure in their underwater chambers is at least as great as the combined pressure of water and atmosphere outside.

FIGURE 5.25
A simple mercury barometer. Mercury is pushed up into the tube by atmospheric pressure.

FIGURE 5.26
Strictly speaking, they do not suck the soda up the straws. They instead reduce pressure in the straws, which allows the weight of the atmosphere to press the liquid up into the straws. Could they drink a soda this way on the Moon?

10,000 kg. The weight of this air is about 100,000 N. This weight produces a pressure of 100,000 N/m²—or equivalently, 100,000 pascals (Pa), or 100 kilopascals (kPa). To be more precise, the average atmospheric pressure at sea level is 101.3 kPa.*

The pressure of the atmosphere is not uniform. Besides altitude variations, there are variations in atmospheric pressure at any one locality due to moving fronts and storms. Measurement of changing air pressure is important to meteorologists in predicting weather.

CHECKPOINT

1. **Estimate the mass of air in kilograms in a classroom that has a 200-m² floor area and a 4-m-high ceiling. (Assume a chilly 10°C temperature.)**
2. **Why doesn't the pressure of the atmosphere break windows?**

Were these your answers?

1. The mass of air is 1000 kg. The volume of air is 200 m² × 4 m = 800 m³; each cubic meter of air has a mass of about 1.25 kg, so 800 m³ × 1.25 kg/m³ = 1000 kg (about a ton).
2. Atmospheric pressure is exerted on *both* sides of a window, so no net force is exerted on the window. If for some reason the pressure is reduced or increased on one side only, as in a strong wind, then watch out!

Barometers

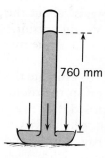

760 mm

An instrument used for measuring the pressure of the atmosphere is called a **barometer**. A simple mercury barometer is illustrated in Figure 5.25. A glass tube, longer than 76 cm and closed at one end, is filled with mercury and tipped upside down in a dish of mercury. The mercury in the tube flows out of the submerged open bottom until the difference in the mercury levels in the tube and the dish is 76 cm. The empty space trapped above, except for some mercury vapor, is a pure vacuum.

The explanation for the operation of such a barometer is similar to that of children balancing on a seesaw. The barometer "balances" when the weight of liquid in the tube exerts the same pressure as the atmosphere outside. Whatever the width of the tube, a 76-cm column of mercury weighs the same as the air that would fill a vertical 30-km tube of the same width. If the atmospheric pressure increases, then the atmosphere pushes down harder on the mercury in the dish and pushes the mercury higher in the tube. Then the increased height of the mercury column exerts an equal balancing pressure.

Water could instead be used to make a barometer, but the glass tube would have to be much longer—13.6 times as long, to be exact. The density of mercury is 13.6 times the density of water. That's why a tube of water 13.6 times

* The pascal is the SI unit of measurement. The average pressure at sea level (101.3 kPa) is often called 1 atmosphere (atm). In British units, the average atmospheric pressure at sea level is 14.7 lb/in² (pounds per square inch, or psi).

longer than one of mercury (of the same cross-section) is needed to provide the same weight as mercury in the tube. A water barometer would have to be 13.6 × 0.76 m, or 10.3 m high—too tall to be practical.

What happens in a barometer is similar to what happens when you drink through a straw. By sucking, you reduce the air pressure in the straw when it is placed in a drink. Atmospheric pressure on the drink then pushes the liquid up into the reduced-pressure region. Strictly speaking, the liquid is not sucked up; it is pushed up the straw by the pressure of the atmosphere. If the atmosphere is prevented from pushing on the surface of the drink, as in the party-trick bottle with the straw through an airtight cork stopper, one can suck and suck and get no drink.

If you understand these ideas, you can understand why there is a 10.3-m limit on the height to which water can be lifted with vacuum pumps. The old-fashioned farm-type pump shown in Figure 5.27 operates by producing a partial vacuum in a pipe that extends down into the water below. Atmospheric pressure on the surface of the water simply pushes the water up into the region of reduced pressure inside the pipe. Can you see that, even with a perfect vacuum, the maximum height to which water can be lifted in this way is 10.3 m?

A small portable instrument that measures atmospheric pressure is the *aneroid barometer* (Figure 5.28). A metal box partially exhausted of air with a slightly flexible lid bends in or out with changes in atmospheric pressure. Motion of the lid is indicated on a scale by a mechanical spring-and-lever system. Atmospheric pressure decreases with increasing altitude, so a barometer can be used to determine elevation. An aneroid barometer calibrated for altitude is called an *altimeter* (altitude meter). Some of these instruments are sensitive enough to indicate a change in elevation as you walk up a flight of stairs.*

Reduced air pressures are produced by pumps, which work by virtue of a gas tending to fill its container. If a space with less pressure is provided, gas flows from the region of higher pressure to the one of lower pressure. A vacuum pump simply provides a region of lower pressure into which the normally fast-moving gas molecules randomly move. The air pressure is repeatedly lowered by piston and valve action (Figure 5.29).

FIGURE 5.27
The atmosphere pushes water from below up into a pipe that is evacuated of air by the pumping action.

When the pump handle is pushed down and the piston is raised, air in the pipe is "thinned" as it expands to fill a larger volume. Atmospheric pressure on the well surface pushes water up into the pipe, causing water to overflow at the spout.

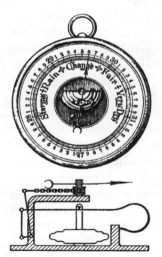

FIGURE 5.28
The aneroid barometer.

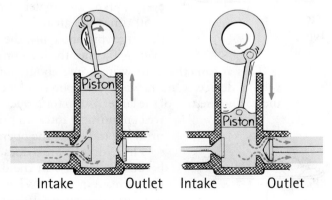

FIGURE 5.29
A mechanical vacuum pump. When the piston is lifted, the intake valve opens and air moves in to fill the empty space. When the piston is moved downward, the outlet valve opens and the air is pushed out. What changes would you make to convert this pump into an air compressor?

---

* Evidence of a noticeable pressure difference over a 1-m or less difference in elevation is any small helium-filled balloon that rises in air. The atmosphere really does push with more force against the lower bottom than against the higher top!

5.7 Pascal's Principle

EXPLAIN THIS How can small pressures in hydraulic machines produce large forces?

One of the most important facts about fluid pressure is that a change in pressure at one part of the fluid is transmitted undiminished to other parts. For example, if the pressure of city water is increased at the pumping station by 10 units of pressure, the pressure everywhere in the pipes of the connected system is increased by 10 units of pressure (providing the water is at rest). This rule is called **Pascal's principle**:

A change in pressure at any point in an enclosed fluid at rest is transmitted undiminished to all points in the fluid.

Pascal's principle was discovered in the 17th century by theologian and scientist Blaise Pascal, for whom the SI unit of pressure, the pascal (1 Pa = 1 N/m^2), is named.

Fill a U-tube with water and place pistons at each end, as shown in Figure 5.30. Pressure exerted against the left piston is transmitted throughout the liquid and against the bottom of the right piston. (The pistons are simply "plugs" that can slide freely but snugly inside the tube.) The pressure that the left piston exerts against the water is exactly equal to the pressure the water exerts against the right piston. This is nothing to write home about. But suppose you make the tube on the right side wider and use a piston of larger area; then the result is impressive. In Figure 5.31 the piston on the right has 50 times the area of the piston on the left (say the left has 100 cm^2 and the right 5000 cm^2). Suppose a 10-kg load is placed on the left piston. Then an additional pressure due to the weight of the load is transmitted throughout the liquid and up against the larger piston. Here is where the difference between force and pressure comes in. The additional pressure is exerted against every square centimeter of the larger piston. Because there is 50 times the area, 50 times as much force is exerted on the larger piston. Thus, the larger piston supports a 500-kg load—50 times the load on the smaller piston!

This *is* something to write home about, for we can multiply forces using such a device. One newton of input produces 50 N of output. By further increasing the area of the larger piston (or reducing the area of the smaller piston), we can multiply force, in principle, by any amount. Pascal's principle underlies the operation of the hydraulic press.

The hydraulic press does not violate energy conservation, because a decrease in the distance moved compensates for the increase in force. When the small piston in Figure 5.31 is moved downward 10 cm, the large piston is raised only one-fiftieth of this, or 0.2 cm. The input force multiplied by the distance moved by the smaller piston is equal to the output force multiplied by the distance moved by the larger piston; this is one more example of a simple machine operating on the same principle as a mechanical lever.

Pascal's principle applies to all fluids, whether gases or liquids. A typical application of Pascal's principle for gases and liquids is the automobile lift seen in many service stations (Figure 5.32). Increased air pressure

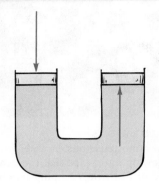

FIGURE 5.30
The force exerted on the left piston increases the pressure in the liquid and is transmitted to the right piston.

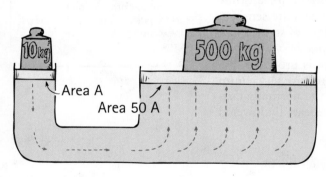

FIGURE 5.31
A 10-kg load on the left piston supports 500 kg on the right piston.

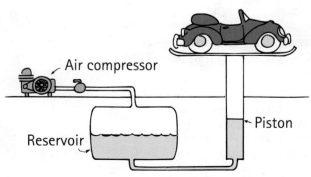

FIGURE 5.32
Pascal's principle in a service station.

produced by an air compressor is transmitted through the air to the surface of oil in an underground reservoir. The oil in turn transmits the pressure to a piston, which lifts the automobile. The relatively low pressure that exerts the lifting force against the piston is about the same as the air pressure in automobile tires.

Hydraulics is employed by modern devices ranging from very small to enormous. Note the hydraulic pistons in almost all construction machines where heavy loads are involved (Figure 5.33).

CHECKPOINT

1. **As the automobile in Figure 5.32 is being lifted, how does the change in oil level in the reservoir compare to the distance the automobile moves?**
2. **If a friend commented that a hydraulic device is a common way of multiplying energy, what would you say?**

Were these your answers?

1. The car moves up a greater distance than the oil level drops, because the area of the piston is smaller than the surface area of the oil in the reservoir.
2. No, no, no! Although a hydraulic device, like a mechanical lever, can multiply *force*, it always does so at the expense of distance. Energy is the product of force and distance. Increase one, decrease the other. *No device has ever been found that can multiply energy!*

FYI Pascal was an invalid at age 18 and remained so until his death at age 39. He is remembered scientifically for hydraulics, which changed the technological landscape more than he imagined. He is remembered theologically for his many assertions, one of which relates to centuries of human landscape: "Men never do evil so cheerfully and completely as when they do so from religious conviction."

5.8 Buoyancy in a Gas

EXPLAIN THIS How high will a helium-filled party balloon rise in air?

LEARNING OBJECTIVE
Describe the application of Archimedes' principle to gases.

A crab lives at the bottom of its ocean floor and looks upward at jellyfish and other lighter-than-water marine life drifting above it. Similarly, we live at the bottom of our ocean of air and look upward at balloons and other lighter-than-air objects drifting above us. A balloon is suspended in air and a jellyfish is suspended in water for the same reason: each is buoyed upward by a displaced weight of fluid equal to its own weight. Objects in water are buoyed upward because the pressure acting up against the bottom of the object exceeds the pressure acting down against the top. Likewise, air pressure acting upward against an object immersed in air is greater than the pressure above pushing down. The buoyancy in both cases is numerically equal to the weight of fluid displaced. Archimedes' principle applies to air just as it does for water:

An object surrounded by air is buoyed up by a force equal to the weight of the air displaced.

We know that a cubic meter of air at ordinary atmospheric pressure and room temperature has a mass of about 1.2 kg, so its weight is about 12 N. Therefore, any 1-m³ object in air is buoyed up with a force of 12 N. If the mass of the 1-m³ object is greater than 1.2 kg (so that its weight is greater than 12 N), it falls to

FIGURE 5.34
All bodies are buoyed up by a force equal to the weight of air they displace. Why, then, don't all objects float like this balloon?

FYI If a balloon is free to expand when rising, it gets larger. But the density of surrounding air decreases. So, interestingly, the greater volume of displaced air doesn't weigh more, and buoyancy remains the same! If a balloon is not free to expand, buoyancy decreases as a balloon rises because of the less dense displaced air. Usually balloons expand when they initially rise, and if they don't eventually rupture, fabric stretching reaches a maximum and balloons settle where buoyancy matches weight.

Mastering**PHYSICS®**
VIDEO: Buoyancy of Air

the ground when released. If an object of this size has a mass of less than 1.2 kg, buoyant force is greater than weight and it rises in the air. Any object that has a mass that is less than the mass of an equal volume of air rises in the air. Stated another way, any object less dense than air rises in air. Gas-filled balloons that rise in air are less dense than air.

No gas at all in a balloon would mean no weight (except for the weight of the balloon's material), but such a balloon would be crushed by atmospheric pressure. The gas used in balloons prevents the atmosphere from collapsing them. Hydrogen is the lightest gas, but it is seldom used because it is highly flammable. In sport balloons, the gas is simply heated air. In balloons intended to reach very high altitudes or to remain aloft for a long time, helium is commonly used. Its density is small enough that the combined weight of the helium, the balloon, and the cargo is less than the weight of air they displace. Low-density gas is used in a balloon for the same reason that cork is used in life preservers. The cork possesses no strange tendency to be drawn toward the water's surface, and the gas possesses no strange tendency to rise. Cork and gases are buoyed upward like anything else. They are simply light enough for the buoyancy to be significant.

Unlike water, the "top" of the atmosphere has no sharply defined surface. Furthermore, unlike water, the atmosphere becomes less dense with altitude. Whereas cork floats to the surface of water, a released helium-filled balloon does not rise to any atmospheric surface. Will a lighter-than-air balloon rise indefinitely? How high will a balloon rise? We can state the answer in several ways. A gas-filled balloon rises only so long as it displaces a weight of air greater than its own weight. Because air becomes less dense with altitude, a lesser weight of air is displaced per given volume as the balloon rises. When the weight of displaced air equals the total weight of the balloon, upward motion of the balloon ceases. We can also say that when the buoyant force on the balloon equals its weight, the balloon ceases rising. Equivalently, when the density of the balloon (including its load) equals the density of the surrounding air, the balloon ceases rising. Helium-filled toy rubber balloons usually break some time after being released into the air when the expansion of the helium they contain stretches the rubber until it ruptures.

CHECKPOINT

Is a buoyant force acting on you? If so, why are you not buoyed up by this force?

Was this your answer?
A buoyant force *is* acting on you, and you *are* buoyed upward by it. You aren't aware of it only because your weight is so much greater.

Large helium-filled dirigible airships are designed so that when they are loaded, they slowly rise in air; that is, their total weight is a little less than the weight of air displaced. When in motion, the ship may be raised or lowered by means of horizontal "elevators."

Thus far we have treated pressure only as it applies to stationary fluids. Motion produces an additional influence.

FIGURE 5.35
Because the flow is continuous, water speeds up when it flows through the narrow and/or shallow part of the brook.

5.9 Bernoulli's Principle

EXPLAIN THIS Why does a spinning baseball curve when thrown?

LEARNING OBJECTIVE
Relate changes in the speed of fluid flow to changes in pressure.

Consider a continuous flow of liquid or gas through a pipe: the volume of fluid that flows past any cross-section of the pipe in a given time is the same as that flowing past any other section of the pipe—even if the pipe widens or narrows. For continuous flow, a fluid speeds up when it goes from a wide to a narrow part of the pipe. This is evident for a broad, slow-moving river that flows more swiftly as it enters a narrow gorge. It is also evident as water flowing from a garden hose speeds up when you squeeze the end of the hose to make the stream narrower.

The motion of a fluid in steady flow follows imaginary *streamlines*, represented by thin lines in Figure 5.36 and in other figures that follow. Streamlines are the smooth paths of bits of fluid. The lines are closer together in narrower regions, where the flow speed is greater. (Streamlines are visible when smoke or other visible fluids are passed through evenly spaced openings, as in a wind tunnel.)

Daniel Bernoulli, an 18th-century Swiss scientist, studied fluid flow in pipes. His discovery, now called **Bernoulli's principle**, can be stated as follows:

Where the speed of a fluid increases, internal pressure in the fluid decreases.

Where streamlines of a fluid are closer together, flow speed is greater and pressure within the fluid is lower. Changes in internal pressure are evident for water containing air bubbles. The volume of an air bubble depends on the surrounding water pressure. Where water gains speed, pressure is lowered and bubbles become bigger. In water that slows, pressure is higher and bubbles are squeezed to a smaller size.

Bernoulli's principle is a consequence of the conservation of energy, although, surprisingly, he developed it long before the concept of energy was formalized.* The full energy picture for a fluid in motion is quite complicated. Simply stated, more speed and kinetic energy mean less pressure, and more pressure means less speed and kinetic energy.

Bernoulli's principle applies to a smooth, steady flow (called *laminar* flow) of constant-density fluid. At speeds above some critical point, however, the flow may become chaotic (called *turbulent* flow) and follow changing, curling paths called *eddies*. This exerts friction on the fluid and dissipates some of its energy. Then Bernoulli's equation doesn't apply well.

The decrease of fluid pressure with increasing speed may at first seem surprising, particularly if you fail to distinguish between the pressure *within* the fluid, internal pressure, and the pressure *by* the fluid on something that interferes with its flow. Internal pressure within flowing water and the external pressure it can exert on whatever it encounters are two different pressures. When the momentum of moving water or anything else is suddenly reduced, the impulse it exerts is relatively huge. A dramatic example is the use of high-speed jets of water to cut steel in modern machine shops. The water has very little internal pressure, but the pressure the stream exerts on the steel interrupting its flow is enormous.

FYI Because the volume of water flowing through a pipe of different cross-sectional areas A remains constant, speed of flow v is high where the area is small and low where the area is large.

This is stated in the equation of continuity:

$$A_1 v_1 = A_2 v_2$$

The product $A_1 v_1$ at point 1 equals the product $A_2 v_2$ at point 2.

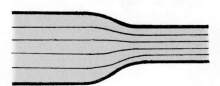

FIGURE 5.36
Water speeds up when it flows into the narrower pipe. The close-together streamlines indicate increased speed and decreased internal pressure.

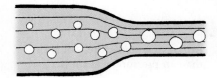

FIGURE 5.37
Internal pressure is greater in slower-moving water in the wide part of the pipe, as evidenced by the more-squeezed air bubbles. The bubbles are bigger in the narrow part because internal pressure there is less.

* In mathematical form: $\frac{1}{2}mv^2 + mgy + pV = $ constant (along a streamline), where m is the mass of some small volume V, v its speed, g the acceleration due to gravity, y its elevation, and p its internal pressure. If mass m is expressed in terms of density ρ, where $\rho = m/V$, and each term is divided by V, Bernoulli's equation reads: $\frac{1}{2}\rho v^2 + \rho gy + p = $ constant. Then all three terms have units of pressure. If y does not change, an increase in v means a decrease in p, and vice versa. Note that when v is zero, Bernoulli's equation reduces to $\Delta p = \rho g \Delta y$ (weight density × depth).

FYI The friction of both liquids and gases sliding over one another is called *viscosity* and is a property of all fluids.

FIGURE 5.38
The paper rises when Tim blows air across its top surface.

> Recall from Chapter 3 that a large change in momentum is associated with a large impulse. So when water from a firefighter's hose hits you, the impulse can knock you off your feet. Interestingly, the pressure *within* that water is relatively small!

FIGURE 5.39
Air pressure above the roof is less than air pressure beneath the roof.

FIGURE 5.40
The vertical vector represents the net upward force (lift) that results from more air pressure below the wing than above the wing. The horizontal vector represents air drag.

Applications of Bernoulli's Principle

Hold a sheet of paper in front of your mouth, as shown in Figure 5.38. When you blow across the top surface, the paper rises. That's because the internal pressure of moving air against the top of the paper is less than the atmospheric pressure beneath it.

Anyone who has ridden in a convertible car with the canvas top up has noticed that the roof puffs upward as the car moves. This is Bernoulli's principle again. The pressure outside—on top of the fabric, where air is moving—is less than the static atmospheric pressure on the inside.

Consider wind blowing across a peaked roof. The wind gains speed as it flows over the roof, as the crowding of streamlines in Figure 5.39 indicates. Pressure along the streamlines is reduced where they are closer together. The greater pressure inside the roof can lift it off the house. During a severe storm, the difference in outside and inside pressure doesn't need to be very much. A small pressure difference over a large area produces a force that can be formidable.

If we think of the blown-off roof as an airplane wing, we can better understand the lifting force that supports a heavy aircraft. In both cases, a greater pressure below pushes the roof or the wing into a region of lesser pressure above. Wings come in a variety of designs. What they all have in common is that air is made to flow faster over the wing's top surface than under its lower surface. This is mainly accomplished by a tilt in the wing, called its *angle of attack*. Then air flows faster over the top surface for much the same reason that air flows faster in a narrowed pipe or in any other constricted region. Most often, but not always, different speeds of airflow over and beneath a wing are enhanced by a difference in the curvature (*camber*) of the upper and lower surfaces of the wing. The result is more-crowded streamlines along the top wing surface than along the bottom. When the average pressure difference over the wing is multiplied by the surface area of the wing, we have a net upward force—lift. Lift is greater when there is a large wing area and when the plane is traveling fast. A glider has a very large wing area relative to its weight, so it does not have to be going very fast for sufficient lift. At the other extreme, a fighter plane designed for high-speed flight has a small wing area relative to its weight. Consequently, it must take off and land at high speeds.

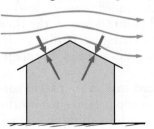

We all know that a baseball pitcher can throw a ball in such a way that it curves to one side as it approaches home plate. This is accomplished by imparting a large spin to the ball. Similarly, a tennis player can hit a ball so it curves. A thin layer of air is dragged around the spinning ball by friction, which is enhanced by the baseball's threads or the tennis ball's fuzz. The moving layer of air produces a crowding of streamlines on one side. Note in Figure 5.41b that the streamlines are more crowded at B than at A for the direction of spin shown. Air pressure is greater at A, and the ball curves as shown.

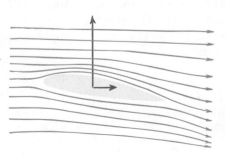

Recent findings show that many insects increase lift by employing motions similar to those of a curving baseball. Interestingly, most insects do not flap their wings up and down. They flap them forward and backward, with a tilt that provides an angle of attack. Between flaps, their wings make semicircular motions to create lift.

a   b

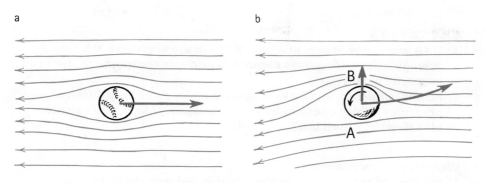

Motion of air relative to ball

FIGURE 5.41
(a) The streamlines are the same on either side of a nonspinning baseball. (b) A spinning ball produces a crowding of streamlines. The resulting "lift" (red arrow) causes the ball to curve (blue arrow).

A familiar sprayer, such as a perfume atomizer, uses Bernoulli's principle. When you squeeze the bulb, air rushes across the open end of a tube inserted into the perfume. This reduces the pressure in the tube, whereupon atmospheric pressure on the liquid below pushes it up into the tube, where it is carried away by the stream of air.

Bernoulli's principle explains why trucks passing closely on the highway are drawn to each other, and why passing ships run the risk of a sideways collision. Water flowing between the ships travels faster than water flowing past the outer sides. Streamlines are closer together between the ships than outside, so water pressure acting against the hulls is reduced between the ships. Unless the ships are steered to compensate for this, the greater pressure against the outer sides of the ships forces them together. Figure 5.43 shows how to demonstrate this in your kitchen sink or bathtub.

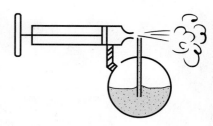

FIGURE 5.42
Why does the liquid in the reservoir go up the tube?

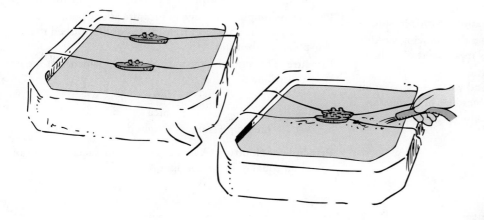

FIGURE 5.43
Try this in your sink. Loosely moor a pair of toy boats side by side. Then direct a stream of water between them. The boats draw together and collide. Why?

Bernoulli's principle plays a small role when your bathroom shower curtain swings toward you in the shower when the water is on full blast. The pressure in the shower stall is reduced with fluid in motion, and the relatively greater pressure outside the curtain pushes it inward. Like so much in the complex real world, this is but one physics principle that applies. More important is the convection of air in the shower. In any case, the next time you're taking a shower and the curtain swings in against your legs, think of Daniel Bernoulli.

FIGURE 5.44
The curved shape of an umbrella can be disadvantageous on a windy day.

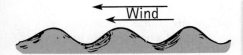

Wind

> ### CHECKPOINT
>
> 1. On a windy day, waves in a lake or the ocean are higher than their average height. How does Bernoulli's principle contribute to the increased height?
> 2. Blimps, airplanes, and rockets operate under three very different principles. Which operates by way of buoyancy? Bernoulli's principle? Newton's third law?

Were these your answers?

1. The troughs of the waves are partially shielded from the wind, so air travels faster over the crests. Pressure there is more reduced than down below in the troughs. The greater pressure in the troughs pushes water into the even higher crests.
2. Blimps operate by way of buoyancy, airplanes by the Bernoulli principle, and rockets by way of Newton's third law. Interesting, Newton's third law also plays a significant role in airplane flight—wing pushes air downward; air pushes wing upward.

For instructor-assigned homework, go to www.masteringphysics.com

SUMMARY OF TERMS (KNOWLEDGE)

Archimedes' principle An immersed body is buoyed up by a force equal to the weight of the fluid it displaces (for both liquids and gases).

Atmospheric pressure The pressure exerted against bodies immersed in the atmosphere resulting from the weight of air pressing down from above. At sea level, atmospheric pressure is about 101 kPa.

Barometer Any device that measures atmospheric pressure.

Bernoulli's principle The pressure in a fluid moving steadily without friction or external energy input decreases when the fluid velocity increases.

Boyle's law The product of pressure and volume is a constant for a given mass of confined gas regardless of changes in either pressure or volume individually, so long as the temperature remains unchanged:

$$P_1 V_2 = P_2 V_2$$

Buoyant force The net upward force that a fluid exerts on an immersed object.

Density The amount of matter per unit volume:

$$\text{Density} = \frac{\text{mass}}{\text{volume}}$$

Weight density is expressed as weight per unit volume.

Pascal's principle A change in pressure at any point in an enclosed fluid at rest is transmitted undiminished to all points in the fluid.

Pressure The ratio of force to the area over which that force is distributed:

$$\text{Pressure} = \frac{\text{force}}{\text{area}}$$

$$\text{Liquid pressure} = \text{weight density} \times \text{depth}$$

Principle of flotation A floating object displaces a weight of fluid equal to its own weight.

READING CHECK QUESTIONS (COMPREHENSION)

1. Give two examples of a fluid.

5.1 Density

2. What happens to the volume of a loaf of bread when it is squeezed? What happens to the mass? What happens to the density?
3. Distinguish between mass density and weight density.

5.2 Pressure

4. Distinguish between force and pressure. Compare their units of measurement.
5. How does the pressure exerted by a liquid change with depth in the liquid? How does the pressure exerted by a liquid change as the density of the liquid changes?

6. Discounting the pressure of the atmosphere, if you swim twice as deep in water, how much more water pressure is exerted on your ears? If you swim in salt water, is the pressure greater than in fresh water at the same depth?

7. How does water pressure 1 m below the surface of a small pond compare to water pressure 1 m below the surface of a huge lake?

8. If you punch a hole in the side of a container filled with water, in what direction does the water initially flow outward from the container?

5.3 Buoyancy in a Liquid

9. Why does buoyant force act upward on an object submerged in water?

10. How does the volume of a completely submerged object compare with the volume of water displaced?

5.4 Archimedes' Principle

11. State Archimedes' principle.

12. What is the difference between being immersed and being submerged?

13. How does the buoyant force on a fully submerged object compare with the weight of the water displaced?

14. What is the mass in kilograms of 1 L of water? What is its weight in newtons?

15. If a 1-L container is immersed halfway in water, what is the volume of the water displaced? What is the buoyant force on the container?

16. Does the buoyant force on a floating object depend on the weight of the object or on the weight of the fluid displaced by the object? Or are these two weights the same for the special case of floating? Defend your answer.

17. What weight of water is displaced by a 100-ton floating ship? What is the buoyant force that acts on this ship?

5.5 Pressure in a Gas

18. By how much does the density of air increase when it is compressed to half its volume?

19. What happens to the air pressure inside a balloon when the balloon is squeezed to half its volume at constant temperature?

5.6 Atmospheric Pressure

20. What is the approximate mass in kilograms of a column of air that has a cross-sectional area of 1 cm^2 and extends from sea level to the upper atmosphere? What is the weight in newtons of this amount of air?

21. How does the downward pressure of the 76-cm column of mercury in a barometer compare with the air pressure at the bottom of the atmosphere?

22. How does the weight of mercury in a barometer tube compare with the weight of an equal cross-section of air from sea level to the top of the atmosphere?

23. Why would a water barometer have to be 13.6 times as tall as a mercury barometer?

24. When you drink liquid through a straw, is it more accurate to say that the liquid is pushed up the straw rather than sucked up? What exactly does the pushing? Defend your answer.

5.7 Pascal's Principle

25. What happens to the pressure in all parts of a confined fluid when the pressure in one part is increased?

26. Does Pascal's principle provide a way to get more energy from a machine than is put into it? Defend your answer.

5.8 Buoyancy in a Gas

27. A balloon that weighs 1 N is suspended in air, drifting neither up nor down. How much buoyant force acts on it? What happens if the buoyant force decreases? Increases?

5.9 Bernoulli's Principle

28. What are streamlines? Is the pressure higher or lower in regions of crowded streamlines?

29. Does Bernoulli's principle refer to internal pressure changes in a fluid, or to pressures that a fluid can exert on objects in the path of the flowing fluid? Or both?

30. What do peaked roofs, convertible tops, and airplane wings have in common when air moves faster across their top surfaces?

ACTIVITIES (HANDS-ON APPLICATION)

31. Try to float an egg in water. Then dissolve salt in the water until the egg floats. How does the density of an egg compare to that of tap water? Salt water?

32. Punch a couple of holes in the bottom of a water-filled container, and water spurts out because of water pressure. Now drop the container, and, as it freely falls, note that the water no longer spurts out. If your friends don't understand this, could you explain it to them?

33. Place a wet Ping-Pong ball in a can of water held high above your head. Then drop the can on a rigid floor. Because of surface tension, the ball is pulled beneath the surface as the can falls. What happens when the can comes to an abrupt stop is worth watching!

34. Try this in the bathtub or when you're washing dishes: Lower a drinking glass, mouth downward, over a small floating object. What do you observe? How deep must the glass be pushed in order to compress the enclosed air to half its volume? (You won't be able to do this in your bathtub unless it's 10.3 m deep!)

35. Compare the pressure exerted by the tires of your car on the road with the air pressure in the tires. For this project, find the weight of your car from the Internet, and then divide it by 4 to get the  approximate weight held up by one tire. You can approximate the area of tire contact with the road by tracing the edges of tire contact on a sheet of paper marked with 1-inch × 1-inch squares beneath the tire. After you get the pressure of the tire on the road, compare it with the air pressure in the tire. Are they nearly equal? Which one is greater?

36. You ordinarily pour water from a full glass into an empty glass simply by placing the full glass above the empty glass and tipping. Have you ever poured air from one glass to another? The procedure is similar. Lower two glasses in water, mouths downward. Let one fill with  water by tilting its mouth upward. Then hold the mouth of the water-filled glass downward above the air-filled glass. Slowly tilt the lower glass and let the air escape, filling the upper glass. You are pouring air from one glass into another!

37. Raise a filled glass of water above the waterline, but with its mouth beneath the surface. Why doesn't the water run out? How tall would a glass have to be before water began to run out? (You won't be able to do this indoors unless you have a ceiling that is at least 10.3 m higher than the waterline.)

38. Place a card over the open top of a glass filled to the brim with water, and then invert the glass. Why does the card stay in place? Try it sideways.

39. Invert a water-filled soft-drink bottle or small-necked jar. Notice that the water doesn't simply fall out but gurgles out of the container instead. Air pressure doesn't allow the water out until some air has pushed its way up inside the bottle to occupy the space above the liquid. How would an inverted, water-filled bottle empty if you tried this on the Moon?

40. Do as Professor Dan Johnson does. Pour about a quarter cup of water into a gallon or 5-liter metal can with a screw top. Place the can *open* on a stove and heat it until the water boils and steam comes out of the opening. Quickly remove the can from the stove and screw the cap on tightly. Allow the can to stand. The steam inside condenses, which can be hastened by cooling the can with a dousing of cold water. What happens to the vapor pressure inside? (Don't do this with a can you expect to use again.)

41. Heat a small amount of water to boiling in an aluminum soft-drink can and invert the can quickly into a dish of cold water. What happens is surprisingly dramatic!

42. Make a small hole near the bottom of an open tin can. Fill the can with water, which then proceeds to spurt from the hole. If you cover the top of the can firmly with the palm of your hand, the flow stops. Explain.

43. Lower a narrow glass tube or drinking straw into water and place your finger over the top of the tube. Lift the tube from the water and then lift your finger from the top of the tube. What happens? (You'll do this often in chemistry experiments.)

44. Blow across the top of a sheet of paper as Tim does in Figure 5.38. Try this with those of your friends who are not taking a physical science course. Then explain it to them!

45. Push a pin through a small card and place it over the hole of a thread spool. Try to blow the card from the spool by blowing through the hole. Try it in all directions.

46. Hold a spoon in a stream of water as shown and feel the effect of the differences in pressure.

PLUG AND CHUG (FORMULA FAMILIARIZATION)

Pressure = weight density × depth
(Neglect the pressure due to the atmosphere in the calculations below.)

47. A 1-m-tall barrel is filled with water (with a weight density of 9800 N/m³). Show that the water pressure on the bottom of the barrel is 9800 N/m² or, equivalently, 9.8 kPa.

48. Show that the water pressure at the bottom of the 50-m-high water tower in Figure 5.3 is 490,000 N/m², or is approximately 500 kPa.

49. The depth of water behind the Hoover Dam is 220 m. Show that the water pressure at the base of this dam is 2160 kPa.

50. The top floor of a building is 20 m above the basement. Show that the water pressure in the basement is nearly 200 kPa greater than the water pressure on the top floor.

THINK AND SOLVE (MATHEMATICAL APPLICATION)

51. Suppose you balance a 2-kg ball on the tip of your finger, which has an area of 1 cm^2. Show that the pressure on your finger is 20 N/cm^2, or is 200 kPa.

52. A 12-kg piece of metal displaces 2 L of water when submerged. Show that its density is 6000 kg/m^3. How does this compare with the density of water?

53. A 1-m-tall barrel is closed on top except for a thin pipe extending 5 m up from the top. When the barrel is filled with water up to the base of the pipe (1 m deep) the water pressure on the bottom of the barrel is 9.8 kPa. What is the pressure on the bottom when water is added to fill the pipe to its top?

54. A rectangular barge, 5 m long and 2 m wide, floats in fresh water. Suppose that a 400-kg crate of auto parts is loaded onto the barge. Show that the barge floats 4 cm deeper.

55. Suppose that the barge in the preceding problem can be pushed only 15 cm deeper into the water before the water overflows to sink it. Show that it could carry three, but not four, 400-kg crates.

56. A merchant in Kathmandu sells you a solid-gold, 1-kg statue for a very reasonable price. When you get home, you wonder whether you got a bargain, so you lower the statue into a container of water and measure the volume of displaced water. Show that for 1 kg of pure gold, the volume of water displaced is 51.8 cm^3.

57. A vacationer floats lazily in the ocean with 90% of her body below the surface. The density of the ocean water is 1025 kg/m^3. Show that the vacationer's average density is 923 kg/m^3.

58. Your friend of mass 100 kg can just barely float in fresh water. Calculate her approximate volume.

59. In the hydraulic pistons shown in the sketch, the small piston has a diameter of 2 cm. The large piston has a diameter of 6 cm. How much more force can the larger piston exert compared with the force applied to the smaller piston?

60. On a perfect fall day, you are hovering at rest at low altitude in a hot-air balloon. The total weight of the balloon, including its load and the hot air in it, is 20,000 N. Show that the volume of the displaced air is about 1700 m^3.

61. What change in pressure occurs in a party balloon that is squeezed to one-third its volume with no change in temperature?

62. A mountain-climber of mass 80 kg ponders the idea of attaching a helium-filled balloon to himself to effectively reduce his weight by 25% when he climbs. He wonders what the approximate size of such a balloon would be. Hearing of your physics skills, he asks you. Share with him your calculations that show the volume of the balloon should be about 17 m^3 (slightly more than 3 m in diameter for a spherical balloon).

63. The weight of the atmosphere above 1 m^2 of Earth's surface is about 100,000 N. Density, of course, becomes less with altitude. But suppose the density of air were a constant 1.2 kg/m^3. Calculate where the top of the atmosphere would be. How does this compare with the nearly 40-km-high upper part of the atmosphere?

64. The wings of a certain airplane have a total bottom surface area of 100 m^2. At a particular speed, the difference in air pressure below and above the wings is 4% of atmospheric pressure. Show that the lift on the airplane is 4 × 10^5 N.

THINK AND RANK (ANALYSIS)

65. Rank the pressures from highest to lowest: (a) bottom of a 20-cm-tall container of salt water, (b) bottom of a 20-cm-tall container of fresh water, and (c) bottom of a 5-cm-tall container of mercury.

66. Rank the following from highest to lowest percentage of its volume above the waterline: (a) basketball floating in fresh water, (b) basketball floating in salt water, and (c) basketball floating in mercury.

67. Think about what happens to the volume of an air-filled balloon on top of water and beneath. Then rank the buoyant force on a weighted balloon in water, from most to least, when the balloon is (a) barely floating with

its top at the surface, (b) pushed 1 m beneath the surface, and (c) 2 m beneath the surface.

68. Rank the volume of air in the glass, from greatest to least, when it is held (a) near the surface as shown, (b) 1 m beneath the surface, and (c) 2 m beneath the surface.

69. Rank the buoyant force supplied by the atmosphere on the following, from greatest to least: (a) an elephant, (b) a helium-filled party balloon, and (c) a skydiver at terminal velocity.

70. Rank from greatest to least the amount of lift on the following airplane wings: (a) area 1000 m² with atmospheric pressure difference of 2.0 N/m², (b) area 800 m² with atmospheric pressure difference of 2.4 N/m², and (c) area 600 m² with atmospheric pressure difference of 3.8 N/m².

EXERCISES (SYNTHESIS)

71. What common liquid covers more than two-thirds of our planet, makes up 60% of our bodies, and sustains our lives and lifestyles in countless ways?

72. You know that a sharp knife cuts better than a dull knife. Do you know why this is so? Defend your answer.

73. Which is more likely to hurt: being stepped on by a 200-lb man wearing loafers or being stepped on by a 100-lb woman wearing high heels?

74. Stand on a bathroom scale and read your weight. When you lift one foot up so you're standing on one foot, does the reading change? Does a scale read force or pressure?

75. Why are people who are confined to bed less likely to develop bedsores on their bodies if they use a waterbed rather than a standard mattress?

76. If water faucets upstairs and downstairs are turned fully on, does more water per second flow out of the downstairs faucet? Or is the volume of water flowing from the faucets the same?

77. How much force is needed to push a nearly weightless but rigid 1-L carton beneath a surface of water?

78. Why is it inaccurate to say that heavy objects sink and light objects float? Give exaggerated examples to support your answer.

79. Why will a block of iron float in mercury but sink in water?

80. The mountains of the Himalayas are slightly less dense than the mantle material upon which they "float." Do you suppose that, like floating icebergs, they are deeper than they are high?

81. Why will a volleyball held beneath the surface of water have more buoyant force than if it is floating?

82. Why does an inflated beach ball pushed beneath the surface of water swiftly shoot above the water surface when released?

83. When the wooden block is placed in the beaker that is brim filled with water, what happens to the scale reading after water has overflowed? Answer the same question for an iron block.

84. Give a reason why canal enthusiasts in Scotland appreciate the physics illustrated in Figure 5.16 (the block of wood floating in a vessel brim-filled with water).

85. The Falkirk Wheel in Scotland (see Figure 5.17) rotates with the same low energy no matter what the weight of the boats it lifts. What would be different in its operation if instead of carrying floating boats it carried scrap metal that doesn't float?

86. A half-filled bucket of water is on a spring scale. Does the reading of the scale increase or remain the same if a fish is placed in the bucket? (Is your answer different if the bucket is initially filled to the brim?)

87. A ship sailing from the ocean into a freshwater harbor sinks slightly deeper into the water. Does the buoyant force on it change? If so, does it increase or decrease?

88. In a sporting goods store you see what appears to be two identical life preservers of the same size. One is filled with Styrofoam and the other one is filled with lead pellets. If you submerge these life preservers in the water, upon which is the buoyant force greater? Upon which is the buoyant force ineffective? Why are your answers different?

89. We can understand how pressure in water depends on depth by considering a stack of bricks. The pressure below the bottom brick is determined by the weight of the entire stack. Halfway up the stack, the pressure is half because the weight of the bricks above is half. To explain atmospheric pressure, we should consider compressible bricks, like foam rubber. Why?

90. How does the density of air in a deep mine compare with the density of air at Earth's surface?

91. The "pump" in a vacuum cleaner is merely a high-speed fan. Would a vacuum cleaner pick up dust from a rug on the Moon? Explain.

92. If you could somehow replace the mercury in a mercury barometer with a denser liquid, would the height of the liquid column be greater or less than with mercury? Why?

93. Would it be slightly more difficult to draw soda through a straw at sea level or on top of a very high mountain? Explain.

94. Your friend says that the buoyant force of the atmosphere on an elephant is significantly greater than the buoyant force of the atmosphere on a small helium-filled balloon. What do you say?

95. Why is it so difficult to breathe when snorkeling at a depth of 1 m, and practically impossible at a depth of 2 m? Why can't a diver simply breathe through a hose that extends to the surface?

96. When you replace helium in a balloon with hydrogen, which is less dense, does the buoyant force on the balloon change if the balloon remains the same size? Explain.

97. A steel tank filled with helium gas doesn't rise in air, but a balloon containing the same helium easily does. Why?

98. Two identical balloons of the same volume are pumped up with air to more than atmospheric pressure and suspended on the ends of a stick that is horizontally balanced. One of the balloons is then punctured. Is there a change in the stick's balance? If so, which way does it tip?

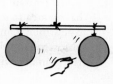

99. It is said that a gas fills all the space available to it. Why, then, doesn't the atmosphere go off into space?

100. Why is there no atmosphere on the Moon?

101. The force of the atmosphere at sea level against the outside of a 10-m^2 store window is about 1 million N. Why doesn't this shatter the window? Why might the window shatter in a strong wind blowing past?

102. Why is the pressure in a car's tires slightly greater after the car has been driven several kilometers?

103. How will two dangling vertical sheets of paper move when you blow between them? Try it and see.

104. When a steadily flowing gas flows from a larger-diameter pipe to a smaller-diameter pipe, what happens to (a) its speed, (b) its pressure, and (c) the spacing between its streamlines?

105. You're having a run of bad luck, and you slip quietly into a small, calm pool as hungry crocodiles lurking at the bottom are relying on Pascal's principle to help them to detect a tender morsel. What does Pascal's principle have to do with their delight at your arrival?

106. What physics principle underlies the following three observations? When passing an oncoming truck on the highway, your car tends to sway toward the truck. The canvas roof of a convertible car bulges upward when the car is traveling at high speeds. The windows of older passenger trains sometimes break when a high-speed train passes by on the next track.

107. How is an airplane able to fly upside down?

DISCUSSION QUESTIONS (EVALUATION)

108. The photo shows physics teacher Marshall Ellenstein walking barefoot on broken glass bottles in his class. What physics concept is Marshall demonstrating, and why is he careful that the broken pieces are small and numerous? (The Band-Aids on his feet are for humor!)

109. Why is blood pressure measured in the upper arm, at the elevation of your heart?

110. Which teapot holds more liquid?

111. A can of diet soda floats in water, whereas a can of regular soda sinks. Discuss this phenomenon first in terms of density and then in terms of weight versus buoyant force.

112. The density of a rock doesn't change when it is submerged in water. Does your density change when you are submerged in water? Discuss and defend your answer.

113. Suppose you wish to lay a level foundation for a home on hilly and bushy terrain. How can you use a garden hose filled with water to determine equal elevations for distant points?

114. If liquid pressure were the same at all depths, would there be a buoyant force on an object submerged in the liquid? Discuss your explanation of this with your friends.

115. Compared to an empty ship, would a ship loaded with a cargo of Styrofoam sink deeper into water or rise in water? Discuss and defend your answer.

116. A barge filled with scrap iron is in a canal lock. If the iron is thrown overboard, does the water level at the side of the lock rise, fall, or remain unchanged? Discuss your explanation with others in your discussion group.

117. A discussion of this raises some eyebrows: Why is the buoyant force on a submerged submarine appreciably greater than the buoyant force on it while it is floating?

118. A balloon is weighted so that it is barely able to float in water. If it is pushed beneath the surface, does it rise back to the surface, stay at the depth to which it is pushed, or sink? Discuss your explanation. (Hint: Does the balloon's density change?)

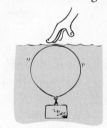

119. When an ice cube in a glass of water melts, does the water level in the glass rise, fall, or remain unchanged? Does your answer change if the ice cube contains many air bubbles? Discuss whether or not your answer changes if the ice cube contains many grains of heavy sand.

120. Count the tires on a large tractor-trailer that is unloading food at your local supermarket, and you may be surprised to count 18 tires. Why so many? (Hint: See Activity 35.)

121. Two teams of eight horses each were unable to pull the Magdeburg hemispheres apart (see Figure 5.20). Why? Suppose two teams of nine horses each could pull them apart. Then would one team of nine horses succeed if the other team were replaced with a strong tree? Discuss and defend your answer.

122. In a classroom demonstration a vacuum pump evacuates air from a large empty oil drum, which slowly and dramatically crumples as shown in the photo. A student friend says that the vacuum sucks in the sides of the drum. What is your explanation?

123. If you bring a bag of potato chips aboard an airplane, you'll note that the unopened bag puffs up as the plane ascends to high altitude. Why? And why is this effect opposite to what happens to the drum in the preceding question?

124. On a sensitive balance, weigh an empty, flat, thin plastic bag. Then weigh the bag filled with air. Will the readings differ? Explain.

125. In the hydraulic arrangement shown, the larger piston has an area that is 50 times that of the smaller piston. The strong man hopes to exert enough force on the large piston to raise the 10 kg that rest on the small piston. Do you think he will be successful? Defend your answer.

126. Invoking ideas from Chapter 2 and this chapter, discuss why is it easier to throw a curve with a tennis ball than a baseball.

127. Your study partner says he doesn't believe in Bernoulli's principle and cites as evidence how a stream of water can knock over a building. The pressure that the water exerts on the building is not reduced, as Bernoulli claims. What distinction is your partner missing?

READINESS ASSURANCE TEST (RAT)

If you have a good handle on this chapter, if you really do, then you should be able to score at least 7 out of 10 on this RAT. If you score less than 7, you need to study further before moving on.

Choose the BEST answer to each of the following.

1. Water pressure at the bottom of a lake depends on the
   (a) weight of water in the lake.
   (b) surface area of the lake.
   (c) depth of the lake.
   (d) all of these

2. The buoyant force that acts on a 20,000-N ship is
   (a) somewhat less than 20,000 N.
   (b) 20,000 N.
   (c) more than 20,000 N.
   (d) dependent on whether it floats in salt or in fresh water.

3. A completely submerged object always displaces its own
   (a) weight of fluid.
   (b) volume of fluid.
   (c) density of fluid.
   (d) all of these
   (e) none of these

4. A rock suspended by a weighing scale weighs 15 N out of water and 10 N when submerged in water. What is the buoyant force on the rock?
   (a) 5 N (b) 10 N (c) 15 N (d) none of these

5. The two caissons of the Falkirk Wheel in Scotland (the device that lifts and lowers ships) remain in balance when
   (a) both are filled to the brim with water.
   (b) ships of different weights float in each.
   (c) water only is in one caisson and a ship is in the other.
   (d) all of these

6. When you squeeze an air-filled party balloon, you reduce its
   (a) volume. (b) mass. (c) weight. (d) all of these

7. Atmospheric pressure is caused by the atmosphere's
   (a) density.
   (b) weight.
   (c) temperature.
   (d) response to solar energy.

8. A hydraulic device multiplies force by 100. This multiplication is done at the expense of
   (a) energy, which is divided by 100.
   (b) the time during which the multiplied force acts.
   (c) the distance through which the multiplied force acts.
   (d) the mechanism providing the force.

9. The flight of a blimp best illustrates
   (a) Archimedes' principle.
   (b) Pascal's principle.
   (c) Bernoulli's principle.
   (d) Boyle's law.

10. As water in a confined pipe speeds up, the pressure it exerts against the inner walls of the pipe
    (a) increases.
    (b) decreases.
    (c) remains constant if flow rate is constant.
    (d) none of these

Answers to RAT

1. c, 2. b, 3. b, 4. a, 5. d, 6. a, 7. b, 8. c, 9. a, 10. b

6

CHAPTER 6

Thermal Energy and Thermodynamics

WHAT'S THE difference between a cup of hot tea and a cup of cool tea? The answer involves molecular motion. In the hot cup the molecules that constitute the tea are moving faster than those in the cooler cup. Matter in all forms is made up of constantly jiggling particles—namely, atoms and/or molecules. When they jiggle at a very slow rate, they form solids. When they jiggle faster, they slide over one another and we have a liquid. When the same particles move so fast that they disconnect and fly loose, we have a gas. When they move still faster, electrons can be torn loose from the atoms, forming a plasma. So whether a substance is a solid, a liquid, a gas, or a plasma depends on the motion of its particles. In this and the following chapter we will investigate the effects of particle motions. We call the energy that a body has by virtue of its energetic jostling of atoms and molecules **thermal energy**.

FIGURE 6.1
Can we trust our sense of hot and cold? Do both fingers feel the same temperature when they are dipped in the warm water? Try this and see (feel) for yourself.

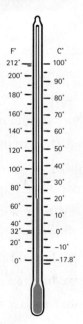

FIGURE 6.2
Fahrenheit and Celsius scales on a thermometer.

FIGURE 6.3
Particles in matter move in different ways. They move from one place to another, they rotate, and they vibrate to and fro. All these modes of motion, plus potential energy, contribute to the overall energy of a substance. Temperature, however, is defined by translational motion.

Mastering**PHYSICS**

VIDEO: Low Temperature with Liquid Nitrogen

6.1 Temperature

EXPLAIN THIS What are two temperatures for ice water?

When you touch a hot stove, thermal energy enters your hand because the stove is warmer than your hand. When you touch a piece of ice, however, thermal energy passes out of your hand and into the colder ice. The quantity that indicates how warm or cold an object is relative to some standard is called **temperature**. We express the temperature of matter by a number that corresponds to the degree of hotness on some chosen scale. A common thermometer measures temperature by means of the expansion and contraction of a liquid, usually mercury or colored alcohol.

The most common temperature scale used worldwide is the Celsius scale, named in honor of the Swedish astronomer Anders Celsius (1701–1744), who first suggested the scale of 100 equal parts (*degrees*) between the freezing point and boiling point of water. The number 0 is assigned to the temperature at which water freezes, and the number 100 to the temperature at which water boils (at standard atmospheric pressure).

The most common temperature scale used in the United States is the Fahrenheit scale, named after its originator, the German physicist D. G. Fahrenheit (1686–1736). On this scale the number 32 is assigned to the temperature at which water freezes, and the number 212 is assigned to the temperature at which water boils. The Fahrenheit scale will become obsolete if and when the United States changes to the metric system.

Arithmetic formulas used for converting from one temperature scale to the other are common in classroom exams. Because such arithmetic exercises are not really physics, we won't be concerned with these conversions (perhaps important in a math class, but not here). Besides, the conversion between Celsius and Fahrenheit temperatures is closely approximated in the side-by-side scales of Figure 6.2.*

Temperature is proportional to the average translational kinetic energy per particle that makes up a substance. By *translational* we mean to-and-fro linear motion. For a gas, we refer to how fast the gas particles are bouncing back and forth; for a liquid, we refer to how fast they slide and jiggle past each other; and for a solid, we refer to how fast the particles move as they vibrate and jiggle in place. Note that temperature does *not* depend on how much of the substance you have. If you have a cup of hot water and then pour half of the water onto the floor, the water remaining in the cup hasn't changed its temperature. The water remaining in the cup contains half the thermal energy that the full cup of water contained, because there are only half as many water molecules in the cup as before. Temperature is a *per-particle property*; *thermal energy* is related to the sum total kinetic energy of all of the particles in your sample.** Twice as much hot water has twice the thermal energy, even though its temperature (the average KE per particle) is the same.

When we measure the temperature of something with a conventional thermometer, thermal energy flows between the thermometer and the object whose temperature we are measuring. When the object and the thermometer have the same average kinetic energy per particle, we say that they are in *thermal*

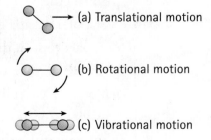

(a) Translational motion

(b) Rotational motion

(c) Vibrational motion

---

* Okay, if you really want to know, the formulas for temperature conversion are $C = \frac{5}{9}(F - 32)$ and $F = \frac{9}{5}C + 32$, where C is the Celsius temperature and F is the Fahrenheit temperature.
** Rather than the term *thermal energy*, physicists prefer the term *internal energy*, to emphasize that the energy is internal to a body.

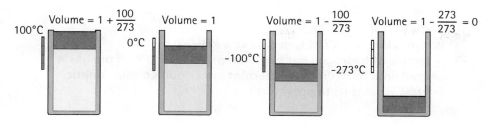

FIGURE 6.4
When pressure is held constant, the volume of a gas changes by $\frac{1}{273}$ of its volume at 0°C with each 1°C change in temperature. At 100°C, the volume is $\frac{100}{273}$ greater than it is at 0°C. When the temperature is reduced to −100°C, the volume is reduced by $\frac{100}{273}$. The rule breaks down near −273°C, where the volume does not really reach zero.

equilibrium. When we measure something's temperature, we are really reading the temperature of the thermometer when it and the object have reached thermal equilibrium.

6.2   Absolute Zero

EXPLAIN THIS   How cold is absolute zero?

As thermal motion increases, a solid object first melts and becomes a liquid. With more thermal motion it then vaporizes. As the temperature further increases, molecules break apart (dissociate) into atoms, and atoms lose some or all of their electrons, thereby forming a cloud of electrically charged particles—a *plasma.* Plasmas exist in stars, where the temperature is millions of degrees Celsius. Temperature has no upper limit.

In contrast, a definite limit exists at the lower end of the temperature scale. Gases expand when heated, and they contract when cooled. Nineteenth-century experiments found something quite amazing. They found that if one starts out with a gas, any gas, at 0°C while pressure is held constant, the volume changes by $\frac{1}{273}$ of the original volume for each degree Celsius change in temperature. When a gas was cooled from 0°C to −10°C, its volume decreased by $\frac{10}{273}$ and it contracted to $\frac{263}{273}$ of its original volume. If a gas at 0°C could be cooled down by 273°C, it would apparently contract $\frac{273}{273}$ volumes and be reduced to zero volume. Clearly, we cannot have a substance with zero volume.

Experimenters got similar results for pressure. Starting at 0°C, the pressure of a gas held in a container of fixed volume decreased by $\frac{1}{273}$ of the original pressure for each Celsius degree its temperature was lowered. If it were cooled to 273°C below zero, it would apparently have no pressure at all. In practice, every gas converts to a liquid before becoming this cold. Nevertheless, these decreases by $\frac{1}{273}$ increments suggested the idea of a lowest temperature: −273°C. That's the lower limit of temperature, **absolute zero**. At this temperature, molecules have lost all available kinetic energy.* No more energy can be removed from a substance at absolute zero. It can't get any colder.

The absolute temperature scale is called the *Kelvin scale*, named after the famous British mathematician and physicist William Thomson, First Baron Kelvin. Absolute zero is 0 K (short for "0 kelvins"; note that the word *degrees* is not used with Kelvin temperatures).** There are no negative numbers on the Kelvin scale. Its temperature divisions are identical to the divisions on the Celsius scale. Thus, the melting point of ice is 273 K, and the boiling point of water is 373 K.

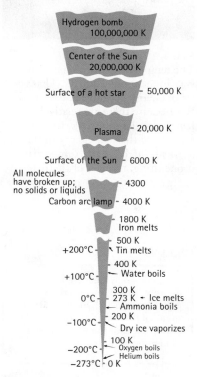

---

* Even at absolute zero, molecules still possess a small amount of kinetic energy, called the *zero-point energy.* Helium, for example, has enough motion at absolute zero to prevent it from freezing. The explanation for this involves quantum theory.
** When Thomson became a baron he took his title from the Kelvin River, which ran through his estate. In 1968 the term *degrees Kelvin* (°K) was officially changed to simply *kelvins* (lowercase k), which is abbreviated K (capital K). The precise value of absolute zero (0 K) is −273.15°C.

FIGURE 6.5
Some absolute temperatures.

CHECKPOINT

1. **Which is larger: a Celsius degree or a kelvin?**
2. **A sample of hydrogen gas has a temperature of 0°C. If the gas is heated until its hydrogen molecules have doubled their kinetic energy, what is its temperature?**

Were these your answers?
1. Neither. They are equal.
2. The 0°C gas has an absolute temperature of 273 K. Twice as much kinetic energy means that it has twice the absolute temperature, or two times 273 K. This would be 546 K, or 273°C.

LEARNING OBJECTIVE
Distinguish between heat and temperature.

FIGURE 6.6
The temperature of the sparks is very high, about 2000°C. That's a lot of energy per molecule of spark. But because there are relatively few molecules per spark, the total amount of thermal energy in the sparks is safely small. Temperature is one thing; transfer of thermal energy is another.

6.3 Heat

EXPLAIN THIS Why do we say that no substances contain heat?

When you place a warm object and a cool object in close proximity, thermal energy transfers in a direction from the warmer object to the cooler object. A physicist defines **heat** as the thermal energy transferred from one thing to another due to a temperature difference.

According to this definition, matter contains *thermal energy*—not heat. Once thermal energy has been transferred to an object or substance, it ceases to be heat. Again, for emphasis: a substance does not contain heat—it contains thermal energy. Heat is thermal energy in transit.

For substances in thermal contact, thermal energy flows from the higher-temperature substance into the lower-temperature substance until thermal equilibrium is reached. This does not mean that thermal energy necessarily flows from a substance with more thermal energy into one with less thermal energy. For example, a bowl of warm water contains more thermal energy than does a red-hot thumbtack. If the tack is placed into the water, thermal energy doesn't flow from the warm water to the tack. Instead, it flows from the hot tack to the cooler water. Thermal energy never flows unassisted from a low-temperature substance into a higher-temperature one.

Temperature is measured in degrees. Heat is measured in joules (or calories). In the U.S. we speak of low-calorie foods and drinks. Most of the world speaks of low-joule foods and drinks.

CHECKPOINT

1. **You apply a flame to 1 L of water for a certain time and its temperature rises by 2°C. If you apply the same flame for the same time to 2 L of water, by how much does its temperature rise?**
2. **If a fast marble hits a random scatter of slow marbles, does the fast marble usually speed up or slow down? Which lose(s) kinetic energy and which gain(s) kinetic energy: the initially fast-moving marble or the initially slow ones? How do these questions relate to the direction of heat flow?**

Were these your answers?
1. Its temperature rises by only 1°C, because 2 L of water contains twice as many molecules, and each molecule receives only half as much energy on the average. So the average kinetic energy, and thus the temperature, increases by half as much.

2. A fast-moving marble slows when it hits slower-moving marbles. It gives up some of its kinetic energy to the slower ones. Likewise with heat. Molecules with more kinetic energy that make contact with slower molecules give some of their excess kinetic energy to the slower ones. The direction of heat flow is from hot to cold. For both the marbles and the molecules, however, the total energy of the system before and after contact is the same.

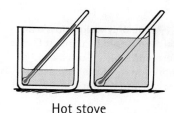

Hot stove

FIGURE 6.7
The pot on the left contains 1 L of water. The pot on the right contains 3 L of water. Although both pots absorb the same quantity of heat, the temperature increases three times as much in the pot with the smaller amount of water.

6.4 Quantity of Heat

EXPLAIN THIS Which is the largest: 1 calorie, 1 Calorie, or 1 joule?

Heat, like work, is energy in transit and is measured in joules. In the U.S. heat has traditionally been measured in calories, another measure of thermal energy. In science courses, the joule is usually preferred. It takes 4.19 J (or equivalently, 1 calorie) of heat to change the temperature of 1 g of water by 1°C.*

The energy ratings of foods and fuels are determined from the energy released when they are burned. (Metabolism is really "burning" at a slow rate.) A common heat unit for labeling food is the kilocalorie (kcal), which is 1000 calories (cal), the heat needed to change the temperature of 1 kg of water by 1°C. To differentiate this unit and the smaller calorie, the food unit is usually called a *Calorie*, with a capital C. So 1 Calorie is really 1000 calories.

What we've learned thus far about heat and thermal energy is summed up in the *laws of thermodynamics*. The word **thermodynamics** stems from Greek words meaning "movement of heat."

LEARNING OBJECTIVE
Distinguish among the units calories, Calories, and joules.

CHECKPOINT
Which raises the temperature of water more: adding 4.19 J or 1 calorie?

Was this your answer?
They have the same effect. This is like asking which is longer: a 1.6-km-long track or a 1-mi-long track. They're the same length, just expressed in different units.

FIGURE 6.8
To the weight watcher, the peanut contains 10 Calories; to the scientist, it releases 10,000 calories (41,900 J) of energy when burned or digested.

6.5 The Laws of Thermodynamics

EXPLAIN THIS How does thermodynamics relate to the conservation of energy?

When thermal energy transfers as heat, the energy lost in one place is gained in another in accord with conservation of energy. When the law of energy conservation is applied to thermal systems, we call it the **first law of thermodynamics**. We state it generally in the following form:

When heat flows to or from a system, the system gains or loses an amount of energy equal to the amount of heat transferred.

LEARNING OBJECTIVE
Describe the three laws of thermodynamics.

* So 1 calorie = 4.19 J. Another common unit of heat is the British thermal unit (Btu). The Btu is defined as the amount of heat required to change the temperature of 1 lb of water by 1°F. One Btu is equal to 1054 J.

FIGURE 6.9
When you push down on the piston, you do work on the air inside. What happens to its temperature?

Absolute zero isn't the coldest you can get. It's the coldest you can hope to approach.

Whatever the system—be it a steam engine, Earth's atmosphere, or the body of a living creature—heat added to it can have two effects. It can increase the system's thermal energy, or it can enable the system to do work on its surroundings (or both). This leads to the following statement of the first law of thermodynamics:

**Heat added = increase in thermal energy
+ external work done by the system**

Suppose that you put an air-filled, rigid, airtight can on a hot plate and add a certain amount of thermal energy to the can. **Warning**: *Do not actually do this*. Because the can has a fixed volume, the walls of the can don't move, so no work is done. All of the heat going into the can increases the thermal energy of the enclosed air, so its temperature rises. Now suppose instead that the can is a flexible container that can expand. The heated air does work as the sides of the can expand, exerting a force for some distance on the surrounding atmosphere. Because some of the added heat goes into doing work, less of the added heat goes into increasing the thermal energy of the enclosed air. Can you see that the temperature of the enclosed air is lower when it does work than when it doesn't do work? The first law of thermodynamics makes good sense.*

The **second law of thermodynamics** restates what we've learned about the direction of heat flow:

Heat never spontaneously flows from a cold substance to a hot substance.

When heat flow is spontaneous—that is, without the assistance of external work—the direction of flow is always from hot to cold. In winter, heat flows from inside a warm home to the cold air outside. In summer, heat flows from the hot air outside into the home's cooler interior. Heat can be made to flow the other way *only* when work is done on the system or by adding energy from another source. This occurs with heat pumps that move heat from cooler outside air into a home's warmer interior, or with air conditioners that remove heat from a home's cool interior to the warmer air outside. Without external effort, the direction of heat flow is always from hot to cold. The second law, like the first, makes logical sense.**

The **third law of thermodynamics** restates what we've learned about the lowest limit of temperature:

No system can reach absolute zero.

As investigators attempt to reach this lowest temperature, it becomes more difficult to get closer to it. In 2010, after nine years of work, a team in Finland recorded a record low of one-billionth of a kelvin (1 picokelvin), tantalizingly close to the unattainable 0 K.

---

* The laws of thermodynamics were the rage back in the 1800s. At that time, horses and buggies were yielding to steam-driven locomotives. There is the story of the engineer who explained the operation of a steam engine to a peasant. The engineer cited in detail the operation of the steam cycle, how expanding steam drives a piston that in turn rotates the wheels. After some thought, the peasant asked, "Yes, I understand all that. But where's the horse?" This story illustrates how difficult it is to abandon our way of thinking about the world when a newer method comes along to replace established ways. Are we different today?
** There is also a *zeroth law of thermodynamics*, which states that if systems A and B are each in thermal equilibrium with system C, then A and B are in thermal equilibrium with each other. The importance of this law was recognized only after the first, second, and third laws had been named, hence the name "zeroth" seemed appropriate.

6.6 Entropy

EXPLAIN THIS   Why does the smell of cookies baking in an oven soon fill the room?

The first law of thermodynamics states that energy can be neither created nor destroyed. It speaks of the *quantity* of energy. The second law speaks of the *quality* of energy, as energy becomes more diffuse and ultimately degenerates into waste.

With this broader perspective, the second law can be stated another way:

In natural processes, high-quality energy tends to transform into lower-quality energy—order tends to disorder.

Processes in which disorder returns to order without external help don't occur in nature. Interestingly, time is given a direction via this thermodynamic rule. Time's arrow always points from order to disorder.*

The idea of ordered energy tending to disordered energy is embodied in the concept of *entropy*.** **Entropy** is the measure of how energy spreads to disorder in a system. When disorder increases, entropy increases. The molecules of an automobile's exhaust, for example, do not spontaneously recombine to form more highly organized gasoline molecules. Warm air that spreads throughout a room when the oven door is open does not spontaneously return to the oven. Whenever a physical system is allowed to spread its energy freely, it always does so in a manner such that entropy increases, while the energy of the system available for doing work decreases.†

However, when work is input to a system, as in living organisms, the entropy of the system can decrease. All living things, from bacteria to trees to human beings, extract energy from their surroundings and use this energy to increase their own organization. The process of extracting energy (for instance, breaking down a highly organized food molecule into smaller molecules) increases entropy elsewhere, so life forms plus their waste products have a net increase in entropy. Energy must be transformed within the living system to support life. When it is not, the organism soon dies and tends toward disorder.

6.7 Specific Heat Capacity

EXPLAIN THIS   Why does a hot frying pan cool faster than equally hot water?

While eating, you've likely noticed that some foods remain hotter much longer than others. Whereas the filling of hot apple pie can burn your tongue, the crust does not, even when the pie has just been removed from the oven. Or a piece of toast may be comfortably eaten a

LEARNING OBJECTIVE
Describe the direction of flow of ordered energy to disordered energy in nature.

FIGURE 6.10
Entropy.

The laws of thermodynamics can be stated this way: You can't win (because you can't get any more energy out of a system than you put into it), you can't break even (because you can't get as much useful energy out as you put in), and you can't get out of the game (entropy in the universe is always increasing).

LEARNING OBJECTIVE
Relate the specific heat capacity of substances to thermal inertia.

---

* In the previous century when movies were new, audiences were amazed to see a train come to a stop inches away from a heroine tied to the tracks. This was filmed by starting with the train at rest, inches away from the heroine, and then moving *backward*, gaining speed. When the film was reversed, the train was seen to move *toward* the heroine. (Next time, watch closely for the telltale smoke that *enters* the smokestack.)

** Entropy can be expressed mathematically. The increase in entropy ΔS of a thermodynamic system is equal to the amount of heat added to the system ΔQ divided by the temperature T at which the heat is added: $\Delta S = \Delta Q/T$.

† Interestingly enough, the American writer Ralph Waldo Emerson, who lived during the time when the second law of thermodynamics was the new science topic of the day, philosophically speculated that not everything becomes more disordered with time and cited the example of human thought. Ideas about the nature of things grow increasingly refined and better organized as they pass through the minds of succeeding generations. Human thought is evolving toward more order.

FIGURING PHYSICAL SCIENCE

Problem Solving

If the specific heat capacity c is known for a substance, then the heat transferred = specific heat capacity × mass × change in temperature. This can be expressed by the formula

$$Q = cm\Delta T$$

where Q is the quantity of heat, c is the specific heat capacity of the substance, m is the mass, and ΔT is the corresponding change in temperature of the substance. When mass m is in grams, using the specific heat capacity of water as 1.0 cal/g·°C gives Q in calories.

SAMPLE PROBLEM 1

What would be the final temperature of a mixture of 50 g of 20°C water and 50 g of 40°C water?

Solution:

The heat gained by the cooler water equals the heat lost by the warmer water. Because the masses of water are the same, the final temperature is midway, 30°C. So we'll end up with 100 g of 30°C water.

SAMPLE PROBLEM 2

Consider mixing 100 g of 25°C water with 75 g of 40°C water. Show that the final temperature of the mixture is 31.4°C.

Solution:

Here we have different masses of water that are mixed together. We equate the heat gained by the cool water to the heat lost by the warm water. We can express this equation formally, then let the expressed terms lead to a solution:

Heat gained by cool water =
  heat lost by warm water

$$cm_1\Delta T_1 = cm_2\Delta T_2$$

ΔT_1 doesn't equal ΔT_2 as in Sample Problem 1 because of different masses of water. Some thinking shows that ΔT_1 is the final temperature T minus 25°C, because T will be greater than 25°C. ΔT_2 is 40°C minus T, because T will be less than 40°C. Then,

$$c(100\ g)(T - 25) = c(75\ g)(40 - T)$$
$$100T - 2500 = 3000 - 75T$$
$$T = 31.4°C$$

SAMPLE PROBLEM 3

Radioactive decay in Earth's interior provides enough energy to keep the interior hot, generate magma, and provide warmth to natural hot springs. This is due to the average release of about 0.03 J/kg each year. Show that the time it takes for a chunk of thermally insulated rock to increase 500°C in temperature (assuming that the specific heat of the rock sample is 800 J/kg·°C) is 13.3 million years.

Solution:

Here we switch to rock, but the same concept applies. And we switch to specific heat capacity expressed in joules per kilogram per degree Celsius. No particular mass is specified, so we'll work with quantity of heat/mass (for our answer should be the same for a small chunk of rock or a huge chunk).

From $Q = cm\Delta T$ we divide by m and get $Q/m = c\Delta T = (800\ J/kg·°C) \times (500°C) = 400,000\ J/kg$. The time required is $(400,000\ J/kg) \div (0.03\ J/kg·yr) = 13.3$ million years. Small wonder it remains hot down there!

FIGURE 6.11
The filling of hot apple pie may be too hot to eat, even though the crust is not.

few seconds after coming from the hot toaster, whereas you must wait several minutes before eating soup that has the same high temperature.

Different substances have different thermal capacities for storing energy. If we heat a pot of water on a stove, we might find that it requires 15 minutes to rise from room temperature to its boiling temperature. But an equal mass of iron on the same stove would rise through the same temperature range in only about 2 minutes. For silver, the time would be less than a minute. Equal masses of different materials require different quantities of heat to change their temperatures by a specified number of degrees.*

As mentioned earlier, a gram of water requires 1 calorie of energy to raise the temperature 1°C. It takes only about one-eighth as much energy to raise the temperature of a gram of iron by the same amount. Water absorbs more heat than iron for the same change in temperature. We say water has a higher **specific heat capacity** (sometimes simply called *specific heat*):

> **The specific heat capacity of any substance is defined as the quantity of heat required to change the temperature of a unit mass of the substance by 1°C.**

---

* In the case of silver and iron, silver atoms are about twice as massive as iron atoms. A given mass of silver contains only about half as many atoms as an equal mass of iron, so only about half the heat is needed to raise the temperature of the silver. Hence, the specific heat of silver is about half that of iron.

We can think of specific heat capacity as thermal inertia. Recall that *inertia* is a term used in mechanics to signify the resistance of an object to a change in its state of motion. Specific heat capacity is like thermal inertia because it signifies the resistance of a substance to a change in temperature.

The High Specific Heat Capacity of Water

Water has a much higher capacity for storing thermal energy than almost any other substance. The reason for water's high specific heat capacity involves the various ways that energy can be absorbed. Energy absorbed by any substance increases the jiggling motion of molecules, which raises the temperature. Or absorbed energy may increase the amount of internal vibration or rotation within the molecules, which adds to the stored energy but does not raise the temperature. Usually absorption of energy involves a combination of both. When we compare water molecules with atoms in a metal, we find many more ways for water molecules to absorb energy without increasing translational kinetic energy. So water has a much higher specific heat capacity than metals—and most other common materials.

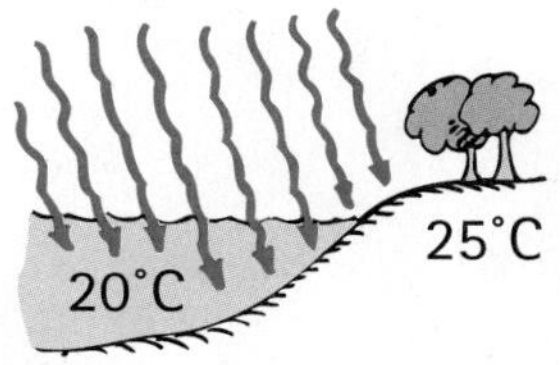

FIGURE 6.12
Because water has a high specific heat capacity and is transparent, it takes more energy to warm the water than to warm the land. Solar energy striking the land is concentrated at the surface, but energy striking the water extends beneath the surface and so is "diluted."

CHECKPOINT

1. **Which has a higher specific heat capacity: water or sand? In other words, which takes longer to warm in sunlight (or longer to cool at night)?**
2. **Why does a piece of watermelon stay cool for a longer time than sandwiches do when both are removed from a picnic cooler on a hot day?**

Were these your answers?

1. Water has the higher specific heat capacity. In the same sunlight, the temperature of water increases more slowly than the temperature of sand. And water cools more slowly at night. (Walking or running barefoot across scorching sand in daytime is a different experience from doing the same in the evening!) The low specific heat capacity of sand and soil, as evidenced by how quickly they warm in the morning Sun and how quickly they cool at night, affects local climates.
2. Water in the melon has more "thermal inertia" than sandwich ingredients, and it resists changes in temperature much more. This thermal inertia is specific heat capacity.

Water's high specific heat capacity affects the world's climate. Look at a globe and notice the high latitude of Europe. Water's high specific heat capacity helps keep Europe's climate appreciably milder than regions of the same latitude in northeastern regions of Canada. Both Europe and Canada receive about the same amount of sunlight per square kilometer. Fortunately for Europeans, the Atlantic Ocean current known as the Gulf Stream carries warm water northeast from the Caribbean Sea, retaining much of its thermal energy long enough to reach the North Atlantic Ocean off the coast of Europe. There the water releases 4.19 J of energy for each gram of water that cools by 1°C. The released energy is carried by westerly winds over the European continent.

Water is useful in the cooling systems of automobiles and other engines because it absorbs a great quantity of heat for small increases in temperature. Water also takes longer to cool.

FIGURE 6.13
Many ocean currents, shown in
blue, distribute heat from the
warmer equatorial regions to the
colder polar regions.

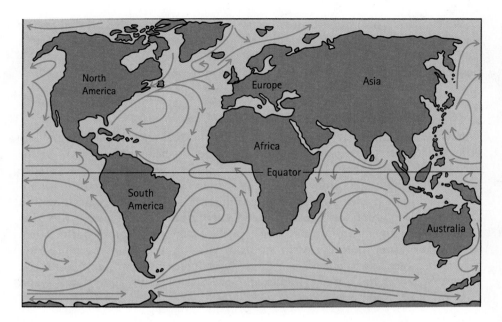

A similar effect occurs in the United States. The winds in North America
are mostly westerly. On the West Coast, air moves from the Pacific Ocean to
the land. In winter months, the ocean water is warmer than the air. Air blows
over the warm water and then moves over the coastal regions. This produces
a warm climate. In summer, the opposite occurs. Air blowing over the water
carries cooler air to the coastal regions. The East Coast benefits less from the
moderating effects of water because the direction of air is from the land to the
Atlantic Ocean. Land, with a lower specific heat capacity, gets hot in the summer but cools rapidly in the winter.

Islands and peninsulas do not have the temperature extremes that are common in interior regions of a continent. The high summer and low winter temperatures common in Manitoba and the Dakotas, for example, are largely due
to the absence of large bodies of water. Europeans, islanders, and people living
near ocean air currents should be glad that water has such a high specific heat
capacity. San Franciscans certainly are!

CHECKPOINT

**Bermuda is close to North Carolina, but, unlike North Carolina, it has a
tropical climate year-round. Why?**

Was this your answer?
Bermuda is an island. The surrounding water warms it when it might otherwise be too cold, and cools it when it might otherwise be too warm.

LEARNING OBJECTIVE
Describe the role of thermal
expansion in common structures.

MasteringPHYSICS

VIDEO: How a Thermostat
Works

6.8 Thermal Expansion

EXPLAIN THIS Why do telephone lines sag in the summer?

As the temperature of a substance increases, its molecules jiggle faster and
move farther apart. The result is *thermal expansion*. Most substances
expand when heated and contract when cooled. Sometimes the changes
aren't noticeable, and sometimes they are. Telephone wires are longer and sag

more on a hot summer day than in winter. Railroad tracks that were laid on cold winter days expand and may even buckle in the hot summer (Figure 6.14). Metal lids on glass fruit jars can often be loosened by heating them under hot water. If one part of a piece of glass is heated or cooled more rapidly than adjacent parts, the resulting expansion or contraction may break the glass. This is especially true of thick glass. Pyrex glass is an exception because it is specially formulated to expand very little with increasing temperature.

Thermal expansion must be taken into account in structures and devices of all kinds. A civil engineer uses reinforcing steel with the same expansion rate as concrete. A long steel bridge usually has one end anchored while the other rests on rockers (Figure 6.15). Notice also that many bridges have tongue-and-groove gaps called *expansion joints* (Figure 6.16). Similarly, concrete roadways and sidewalks are intersected by gaps, which are sometimes filled with tar, so that the concrete can expand freely in summer and contract in winter.

FIGURE 6.14
Thermal expansion. Extreme heat on a July day caused the buckling of these railroad tracks.

FIGURE 6.15
One end of the bridge rides on rockers to allow for thermal expansion. The other end (not shown) is anchored.

FIGURE 6.16
This gap in the roadway of a bridge is called an expansion joint; it allows the bridge to expand and contract.

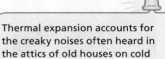

Thermal expansion accounts for the creaky noises often heard in the attics of old houses on cold nights.

The fact that different substances expand at different rates is nicely illustrated with a bimetallic strip (Figure 6.17). This device is made of two strips of different metals welded together, one of brass and the other of iron. When heated, the greater expansion of the brass bends the strip. This bending may be used to turn a pointer, regulate a valve, or close a switch.

   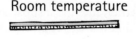   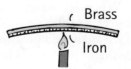

FIGURE 6.17
A bimetallic strip. Brass expands more when heated than iron does, and it contracts more when cooled. Because of this behavior, the strip bends as shown.

A practical application of a bimetallic strip wrapped into a coil is the thermostat (Figure 6.18). When a room becomes too cold, the coil bends toward the brass side and activates an electrical switch that turns on the heater. When the room gets too warm, the coil bends toward the iron side, which breaks the electrical circuit and turns off the heater. Although bimetallic strips nicely illustrate practical physics, electronic sensors now replace them in thermostats and many other thermal devices.

With increases in temperature, liquids expand more than solids. We notice this when gasoline overflows from a car's tank on a hot day. If the tank and its contents expanded at the same rate, no overflow would occur. This is why a gas tank being filled shouldn't be "topped off," especially on a hot day.

To furnace

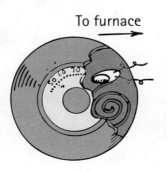

FIGURE 6.18
A pre-electronic thermostat. When the bimetallic coil expands, the drop of liquid mercury rolls away from the electrical contacts and breaks the electrical circuit. When the coil contracts, the mercury rolls against the contacts and completes the circuit.

6.9 Expansion of Water

EXPLAIN THIS Why does ice float?

Water, like most other substances, expands when heated. But interestingly, it *doesn't* expand in the temperature range between 0°C and 4°C. Something quite fascinating happens in this range.

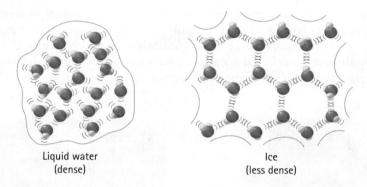

Liquid water
(dense)

Ice
(less dense)

FIGURE 6.19
Water molecules in a liquid are closer together than water molecules frozen in ice, in which they have an open crystalline structure.

FYI Why is ice slippery? Water molecules at the surface of ice have nothing above to cling to. So the hexagonal structure at the surface is weakened and collapses into a thin liquid film, nice for ice skaters.

Ice has a crystalline structure, with open-structured crystals. Water molecules in this open structure have more space between them than they do in the liquid phase (Figure 6.19). This means that ice is less dense than water. When ice melts, not all the open-structured crystals collapse. Some remain in the ice-water mixture, making up a microscopic slush that slightly "bloats" the water—increases its volume slightly (Figure 6.21). This results in ice water being less dense than slightly warmer water. As the temperature of water is increased from 0°C, more of the ice crystals collapse. The melting of these ice crystals further decreases the volume of the water. Two opposite processes occur for the water at the same time—contraction and expansion. Volume decreases as ice crystals collapse, while volume increases due to greater molecular motion. The collapsing effect dominates until the temperature reaches 4°C. After that, expansion overrides contraction because most of the ice crystals have melted (Figure 6.22).

When ice water freezes to become solid ice, its volume increases tremendously. As solid ice cools further, like most substances, it contracts. The density of ice at any temperature is much lower than the density of water, which is why ice floats on water. This behavior of water is very important in nature. If water were most dense at 0°C, it would settle to the bottom of a pond or lake and freeze there instead of at the surface.

FIGURE 6.20
The six-sided structure of a snowflake is a result of the six-sided ice crystals that make it up. The crystals are made when water vapor in a cloud condenses directly to the solid form. Most snowflakes are not as symmetrical as this one.

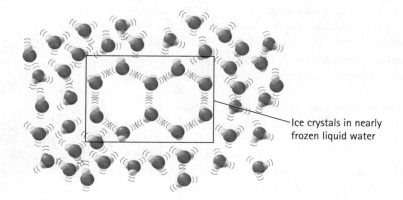

Ice crystals in nearly frozen liquid water

FIGURE 6.21
Close to 0°C, liquid water contains crystals of ice. The open structure of these crystals increases the volume of the water slightly.

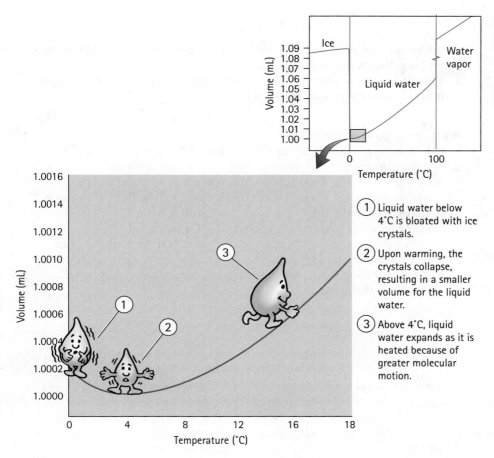

FIGURE 6.22
Between 0°C and 4°C, the volume of liquid water decreases as temperature increases. Above 4°C, thermal expansion exceeds contraction and volume increases as temperature increases.

① Liquid water below 4°C is bloated with ice crystals.

② Upon warming, the crystals collapse, resulting in a smaller volume for the liquid water.

③ Above 4°C, liquid water expands as it is heated because of greater molecular motion.

A pond freezes from the surface downward. In a cold winter the ice is thicker than in a mild winter. Water at the bottom of an ice-covered pond is 4°C, which is relatively warm for organisms that live there. Interestingly, very deep bodies of water are not ice-covered even in the coldest of winters. This is because all the water must be cooled to 4°C before lower temperatures can be reached. For deep water, the winter is not long enough to reduce an entire pond to 4°C. Any 4°C water lies at the bottom. Because of water's high specific heat capacity and poor ability to conduct heat, the bottom of deep bodies of water in cold regions remains at a constant 4°C year round. Fish should be glad that this is so.

Because water is most dense at 4°C, colder water rises and freezes on the surface. This means that fish remain in relative warmth!

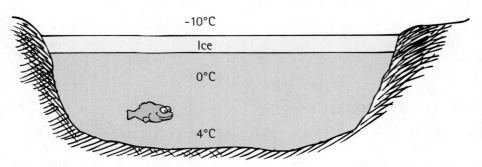

FIGURE 6.23
As water cools, it sinks until the entire pond is at 4°C. Then, as water at the surface cools further, it floats on top and can freeze. Once ice is formed, temperatures lower than 4°C can extend down into the pond.

CHECKPOINT

1. What was the precise temperature at the bottom of Lake Michigan on New Year's Eve in 1901?
2. What's inside the open spaces of the ice crystals shown in Figure 6.19? Is it air, water vapor, or nothing?

Were these your answers?

1. The temperature at the bottom of any body of water that has 4°C water in it is 4°C at the bottom, for the same reason that rocks are at the bottom. Both 4°C water and rocks are more dense than water at any other temperature. Water is a poor heat conductor, so if the body of water is deep and in a region of long winters and short summers, the water at the bottom is likely to remain a constant 4°C year round.
2. There's nothing at all in the open spaces, which can be thought of as empty space—a void. If there were air or vapor in the open spaces, the illustration should show molecules there—oxygen and nitrogen for air and H_2O for water vapor.

LIFE AT THE EXTREMES

Some deserts, such as those on the plains of Spain, the Sahara in Africa, and the Gobi in central Asia, reach surface temperatures of 60°C (140°F). Too hot for life? Not for certain species of ants of the genus *Cataglyphis*, which thrive at this searing temperature. At this extremely high temperature, the desert ants can forage for food without the presence of lizards, which would otherwise prey upon them. Resilient to heat, these ants can withstand higher temperatures than any other creatures in the desert. How they are able to do this involves long legs and heat shock proteins (HSP) in their bodies. They scavenge the desert surface for the corpses of creatures that did not find cover in time, touching the hot sand as little as possible while often sprinting on four legs with two held high in the air. Although their foraging paths zigzag over the desert floor, their return paths are almost straight lines to their nest holes. They attain speeds of 100 body lengths per second. During an average six-day life, most of these ants retrieve 15 to 20 times their weight in food.

From deserts to glaciers, a variety of creatures have invented ways to survive the harshest corners of the world. A species of worm thrives in the glacial ice in the Arctic. Insects in the Antarctic ice pump their bodies full of antifreeze to ward off becoming frozen solid. Some fish that live beneath the ice are able to do the same. Some bacteria thrive in boiling hot springs as a result of having heat-resistant proteins.

An understanding of how creatures survive at the extremes of temperature can provide clues for practical solutions to the physical challenges faced by humans. Astronauts who venture from Earth, for example, will need all the techniques available for coping with unfamiliar environments.

For instructor-assigned homework, go to www.masteringphysics.com

SUMMARY OF TERMS (KNOWLEDGE)

Absolute zero The temperature at which no further energy can be taken from a system.

Entropy The measure of energy dispersal of a system. Whenever energy freely transforms from one form to another, the direction of transformation is toward a state of greater disorder and, therefore, toward one of greater entropy.

First law of thermodynamics A restatement of the law of energy conservation, usually as it applies to systems involving changes in temperature: The heat added to a system is equal to the system's gain in thermal energy plus the work that it does on its surroundings.

Heat The thermal energy that flows from a substance of higher temperature to a substance of lower temperature, commonly measured in calories or joules.

Second law of thermodynamics Heat never spontaneously flows from a cold substance to a hot substance. Also, in natural processes, high-quality energy tends to transform into lower-quality energy—order tends to disorder.

Specific heat capacity The quantity of heat required to raise the temperature of a unit mass of a substance by 1°C.

Temperature A measure of the hotness of substances, related to the average translational kinetic energy per molecule

in a substance, measured in degrees Celsius, degrees Fahrenheit, or kelvins.

Thermal energy The total energy (kinetic plus potential) of the submicroscopic particles that make up a substance.

Thermodynamics The study of thermal energy and its relationship to heat and work.

Third law of thermodynamics No system can reach absolute zero.

READING CHECK QUESTIONS (COMPREHENSION)

6.1 Temperature

1. What are the temperatures for freezing water on the Celsius and Fahrenheit scales? For boiling water?
2. Is the temperature of an object a measure of the total translational kinetic energy of molecules in the object or a measure of the average translational kinetic energy per molecule in the object?
3. Under what conditions can we say that "a thermometer measures its own temperature"?

6.2 Absolute Zero

4. By how much does the pressure of a gas in a rigid vessel decrease when the temperature drops from 0°C to −1°C?
5. What pressure would you expect in a rigid container of 0°C gas if you cooled it to −273°C?
6. What are the temperatures for freezing water and boiling water on the Kelvin temperature scale?
7. How much energy can be removed from a system at 0 K?

6.3 Heat

8. In which direction does thermal energy flow between hot and cold objects?
9. Does a hot object contain thermal energy, or does it contain heat?
10. How does heat differ from thermal energy?
11. What role does temperature play in the direction of thermal energy flow?

6.4 Quantity of Heat

12. Why is heat measured in joules?
13. How many joules are needed to change the temperature of 1 g of water by 1°C?
14. Cite a way that the energy value of foods is determined.
15. Distinguish among a calorie, a Calorie, and a joule.

6.5 The Laws of Thermodynamics

16. Which law of thermodynamics is the conservation of energy applied to thermal systems?

17. What happens to heat added to a system that doesn't increase the temperature of the system?
18. Which law of thermodynamics relates to the direction of heat flow?
19. When can thermal energy in a system move from lower to higher temperatures?
20. Which law of thermodynamics relates to a system reaching 0 K?

6.6 Entropy

21. When disorder in a system increases, does entropy increase or decrease?
22. Under what condition can the entropy of a system be decreased?

6.7 Specific Heat Capacity

23. Which warms faster when heat is applied: iron or silver? Which has the higher specific heat capacity?
24. How does the specific heat capacity of water compare with the specific heat capacities of other common materials?
25. What is the relationship between water's high specific heat capacity and the climate of Europe?

6.8 Thermal Expansion

26. Why does a bimetallic strip bend with changes in temperature?
27. Which generally expands more for an equal increase in temperature: solids or liquids?

6.9 Expansion of Water

28. When the temperature of ice-cold water is increased slightly, does it undergo a net expansion or a net contraction?
29. What is the reason ice is less dense than water?
30. At what temperature do the combined effects of contraction and expansion produce the smallest volume of water?

ACTIVITIES (HANDS-ON APPLICATION)

31. How much energy is in a nut? Burn it and find out. The heat from the flame is energy released when carbon and hydrogen in the nut combine with oxygen in the air (oxidation reactions) to produce CO_2 and H_2O. Pierce a nut (pecan or walnut halves work best) with a bent paper clip that holds the nut above the table surface. Above this, secure a can of water so that you can measure its temperature change when the nut burns. Use about 10^3 cm (10 mL) of water and a Celsius thermometer. As soon as you ignite the nut with a match, place the can of water above it and record the increase in water temperature once the flame burns out. The number of calories released by the burning nut can be calculated by the formula $Q = cm\Delta T$, where c is its specific heat capacity (1 cal/g·°C), m is the mass of water, and ΔT is the change in temperature. The energy in food is expressed in terms of the Calorie, which is 1000 of the calories you'll measure. So to find the number of Calories, divide your result by 1000. (See Think and Solve 36.)

32. Write a letter to your grandparents describing how you're learning to see connections in nature that have eluded you until now, and how you're learning to distinguish related ideas. Use temperature and heat as examples.

PLUG AND CHUG (FORMULA FAMILIARIZATION)

$$Q = cm\Delta T$$

33. Use the formula to show that 300 cal is required to raise the temperature of 30 g of water from 20°C to 30°C. For the specific heat capacity c, use 1 cal/g·°C.

34. Use the same formula to show that 1257 J is required to raise the temperature of the same mass (0.030 kg) of water through the same temperature interval. For the specific heat capacity c, use 4190 J/kg·°C.

35. Show that 300 cal = 1257 J, the same quantity of thermal energy in different units.

THINK AND SOLVE (MATHEMATICAL APPLICATION)

The quantity of heat Q released or absorbed from a substance of specific heat capacity c (which can be expressed in units cal/g·°C or J/kg·°C) and mass m (in g or kg), undergoing a change in temperature ΔT is $Q = cm\Delta T$.

36. Will Maynez burns a 0.6-g peanut beneath 50 g of water, which increases in temperature from 22°C to 50°C. (The specific heat capacity of water is 1.0 cal/g·°C.) (a) Assuming 40% efficiency, show that the peanut's food value is 3500 cal. (b) Then show how the food value in calories per gram is 5.8 kcal/g (or 5.8 Cal/g).

37. Consider a 6.0-g steel nail 8.0 cm long and a hammer that exerts an average force of 600 N on the nail when it is being driven into a piece of wood. The nail becomes warmer. Show that the increase in the nail's temperature is nearly 18°C. (Assume that the specific heat capacity of steel is 450 J/kg·°C.)

38. If you wish to warm 50 kg of water by 20°C for your bath, show that the quantity of heat needed is 1000 kcal (1000 Cal). Then show that this is equivalent to about 4200 kJ.

39. Show that the quantity of heat needed to raise the temperature of a 10-kg piece of steel from 0°C to 100°C is 450,000 J. How does this compare with the heat needed to raise the temperature of the same mass of water through the same temperature difference? The specific heat capacity of steel is 450 J/kg·°C.

40. In lab you submerge 100 g of 40°C nails in 200 g of 20°C water. (The specific heat capacity of iron is 0.12 cal/g·°C.) Equate the heat gained by the water to the heat lost by the nails and show that the final temperature of the water is about 21°C.

To solve the problems below, you will need to know the average coefficient of linear expansion, α, which differs for different materials. We define L to be the length of the object, and α to be the fractional change per unit length for a temperature change of 1°C. For aluminum, $\alpha = 24 \times 10^{-6}$/°C, and for steel, $\alpha = 11 \times 10^{-6}$/°C. The change in length ΔL of a material is given by $\Delta L = L\alpha\Delta T$.

41. Consider a bar 1 m long that expands 0.6 cm when heated. Show that, when similarly heated, a 100-m bar of the same material becomes 100.6 m long.

42. Suppose that the 1.3-km main span of steel for the Golden Gate Bridge had no expansion joints. Show

that for an increase in temperature of 20°C the bridge would be nearly 0.3 m longer.

43. Imagine a 40,000-km steel pipe that forms a ring to fit snugly entirely around the circumference of Earth. Suppose that people along its length breathe on it so as to raise its temperature by 1°C. The pipe gets longer—and is also no longer snug. How high does it stand above ground level? Show that the answer is an astounding 70 m higher! (To simplify, consider only the expansion of its radial distance from the center of Earth, and apply the geometry formula that relates circumference C and radius r, $C = 2\pi r$.)

THINK AND RANK (ANALYSIS)

44. Rank the magnitudes of these units of thermal energy from greatest to least: (a) 1 calorie, (b) 1 Calorie, (c) 1 joule.

45. Three blocks of metal at the same temperature are placed on a hot stove. Their specific heat capacities are listed here. Rank them from fastest to slowest in how quickly each warms up: (a) steel, 450 J/kg·°C; (b) aluminum, 910 J/kg·°C; (c) copper, 390 J/kg·°C.

46. How much the lengths of various substances change with temperature is given by their coefficients of linear expansion, α. The greater the value of α, the greater the change in length for a given change in temperature. Three kinds of metal wires, (a), (b), and (c), are stretched between distant telephone poles. From greatest to least, rank the wires in how much they'll sag on a hot summer day: (a) copper, $\alpha = 17 \times 10^{-6}/°C$; (b) aluminum, $\alpha = 24 \times 10^{-6}/°C$; (c) steel, $\alpha = 11 \times 10^{-6}/°C$.

47. The precise volume of water in a beaker depends on the temperature of the water. Rank from greatest to least the volume of water at these temperatures: (a) 0°C, (b) 4°C, and (c) 10°C.

EXERCISES (SYNTHESIS)

48. Why wouldn't you expect all the molecules in a gas to have the same speed?

49. Consider two glasses, one filled with water and the other half-full, both at the same temperature. In which glass are the water molecules moving faster? In which is there greater thermal energy? In which will more heat be required to increase the temperature by 1°C?

50. Which is greater: an increase in temperature of 1°C or an increase of 1°F?

51. Which has the greater amount of thermal energy: an iceberg or a cup of hot coffee? Defend your answer.

52. On which temperature scale does the average kinetic energy of molecules double when the temperature doubles?

53. When air is rapidly compressed, why does its temperature increase?

54. What happens to the gas pressure within a sealed gallon can when it is heated? When it is cooled? Defend your answers.

55. After a car is driven along a road for some distance, why does the air pressure in the tires increase?

56. When a 1-kg metal pan containing 1 kg of cold water is removed from the refrigerator onto a table, which absorbs more heat from the room: the pan or the water?

57. Does 1 kg of water or 1 kg of iron undergo a greater change in temperature when heat is applied? Defend your answer.

58. Which has the higher specific heat capacity: an object that cools quickly or an object of the same mass that cools more slowly?

59. Desert sand is very hot in the day and very cool at night. What does this tell you about its specific heat capacity?

60. Why does adding the same amount of heat to two different objects not necessarily produce the same increase in temperature?

61. Why does Jello stay cool for a longer time than sandwiches when both are removed from a picnic cooler on a hot day?

62. State an exception to the claim that all substances expand when heated.

63. Would a bimetallic strip of two different metals function if each metal had the same rate of expansion? Is it important that the metals expand at different rates? Defend your answer.

64. In terms of thermal expansion, why is it important that a lock and its key be made of the same or similar materials?

65. Why are incandescent bulbs typically made of very thin glass?

66. For many years a method for breaking boulders was putting them in a hot fire and then dousing them with cold water. Why did this fracture the boulders?

67. An old remedy for separating a pair of nested wedged-together drinking glasses is to run water at different temperatures into the inner glass and over the surface of the outer glass. Which water should be hot and which cold?

68. A metal ball is barely able to pass through a metal ring. When Anette Zetterberg heats the ball, it does not pass through the ring. What happens if she instead heats the ring (as shown): does the size of the hole increase, stay the same, or decrease?

69. Suppose you cut a small gap in a metal ring. If you heat the ring, does the gap become wider or narrower?

70. How does the combined volume of the billions of hexagonal open spaces in the structures of ice crystals in a piece of ice compare with the volume of ice that floats above the water line?

71. A piece of solid iron sinks in a container of molten iron. A piece of solid aluminum sinks in a container of molten aluminum. Why doesn't a piece of solid water (ice) sink in a container of "molten" (liquid) water? Explain, using molecular terms.

DISCUSSION QUESTIONS (EVALUATION)

72. In your room are things such as tables, chairs, and other people. Which of these things has a temperature (a) lower than, (b) greater than, and (c) equal to the temperature of the air?

73. Why can't you tell whether you are running a fever by touching your own forehead?

74. The temperature of the Sun's interior is about 10^7 degrees. Does it matter whether this is degrees Celsius or kelvins? Defend your answer.

75. Which of the laws of thermodynamics tells us what is most probable in nature?

76. If you drop a hot rock into a pail of water, the temperatures of the rock and the water change until both are equal. The rock cools and the water warms. Does this hold true if the hot rock is dropped into the Atlantic Ocean? Defend your answer.

77. On cold winter nights in days past, it was common to bring a hot object to bed with you. Which would keep you warmer through the cold night: a 10-kg iron brick or a 10-kg jug of hot water at the same high temperature? Explain.

78. Why does the presence of large bodies of water tend to moderate the climate of nearby land—making it warmer in cold weather and cooler in hot weather?

79. If the winds at the latitude of San Francisco and Washington, DC, were from the east rather than from the west, why might San Francisco be able to grow only cherry trees and Washington, DC, only palm trees?

80. Compared with conventional water heaters in the United States, why do propane tank-less water heaters, common in other parts of the world, cost up to 60% less to operate?

81. Entropy is a measure of how energy spreads to disorder in a system. Disorder increases and entropy increases. How does this relate to opening a bottle of perfume in the corner of a room?

82. In the preceding question, we see a reason why all the gas molecules in our room don't suddenly rush to one corner, leaving us sitting in a vacuum and gasping for breath. Does the fact that air naturally spreads out mean that entropy increases or decreases?

83. Structural groaning and creaking noises are sometimes heard in the attic of old buildings on cold nights. Give an explanation in terms of thermal expansion.

84. Why is it important that glass mirrors used in astronomical observatories be composed of glass with a low coefficient of expansion?

85. Steel plates are commonly attached to each other with rivets. A rivet is a small metal cylinder, rounded on one end and blunt on the other end. After a hot rivet is inserted into a hole joining the two plates, its blunt end is rounded with a hammer, which is made easier by the hotness of the rivet. How does the hotness of the rivet also help to make a tight fit when it cools?

86. After a machinist quickly slips a hot, snugly fitting iron ring over a very cold brass cylinder, the two cannot be separated intact. Can you explain why this is so?

87. Suppose that water is used in a thermometer instead of mercury. If the temperature is 4°C and then changes, why can't the thermometer indicate whether the temperature is rising or falling?

88. If cooling occurred at the bottom of a pond instead of at the surface, would a lake freeze from the bottom up? Explain.

READINESS ASSURANCE TEST (RAT)

If you have a good handle on this chapter, if you really do, then you should be able to score at least 7 out of 10 on this RAT. If you score less than 7, you need to study further before moving on.

Choose the BEST answer to each of the following.

1. The motion of molecules that most affects temperature is
   (a) translational motion.
   (b) rotational motion.
   (c) internal vibrational motion.
   (d) simple harmonic motion.

2. Whether one object is warmer than another has most to do with
   (a) molecular kinetic energy.
   (b) molecular potential energy.
   (c) heat flow.
   (d) masses of internal particles.

3. Absolute zero corresponds to a temperature of
   (a) 0 K.
   (b) −273°C.
   (c) both of these
   (d) neither of these

4. Thermal energy is normally measured in units of
   (a) calories.
   (b) joules.
   (c) both of these
   (d) neither of these

5. Which two laws of thermodynamics are statements of what *doesn't* happen?
   (a) the first and the second
   (b) the first and the third
   (c) the second and the third
   (d) None; all state what *does* happen.

6. Your garage gets messier by the day. In this case entropy is
   (a) decreasing.
   (b) increasing.
   (c) hanging steady.
   (d) none of these

7. The specific heat capacity of aluminum is more than twice that of copper. If equal quantities of heat are given to equal masses of aluminum and copper, the metal that more rapidly increases in temperature is
   (a) aluminum.
   (b) copper.
   (c) Actually both will increase at the same rate.
   (d) none of these

8. A bimetallic strip used in thermostats relies on the fact that different metals have different
   (a) specific heat capacities.
   (b) thermal energies at different temperatures.
   (c) rates of thermal expansion.
   (d) all of these

9. Water at 4°C will expand when it is slightly
   (a) cooled.
   (b) warmed.
   (c) both
   (d) neither

10. Microscopic slush in water tends to make the water
    (a) more dense.
    (b) less dense.
    (c) more slippery.
    (d) warmer.

Answers to RAT

1. a, 2. a, 3. c, 4. c, 5. c, 6. b, 7. b, 8. c, 9. c, 10. b

CHAPTER 7
Heat Transfer and Change of Phase

WHY DOESN'T coauthor John Suchocki burn his bare feet as he steps (quickly) across red-hot coals? Is it because his feet are wet—as in how no harm occurs when you briefly touch a hot clothes iron with a wetted finger? But he can walk as safely with dry feet. In fact, many fire walkers prefer dry feet because sometimes hot coals stick to wet feet (ouch!). So what is the physics that explains John's feat?

In this chapter we'll learn that when you boil water to make a cup of tea, the process of boiling tends to cool the water rather than heat it. That's right, the water is cooler than it would be if it didn't boil. But if you want to cool your hot hands, you certainly wouldn't put them in boiling water. So in what sense can we say that boiling is a cooling process?

Other intriguing applications of thermal physics make up this chapter. Read on!

7.1 Conduction

EXPLAIN THIS Why is a tile floor cooler to your feet than a rug of the same temperature?

If you hold one end of an iron nail in a flame, the nail quickly becomes too hot to hold. If you hold one end of a short glass rod in a flame, the rod takes much longer before it becomes too hot to hold. In both cases, heat at the hot end travels along the entire length. This method of heat transfer is called **conduction**. Thermal conduction occurs by collisions between particles and their immediate neighbors. Because the heat travels quickly through the nail we say that it is a good *conductor* of heat. Materials that are poor conductors are called *insulators*.

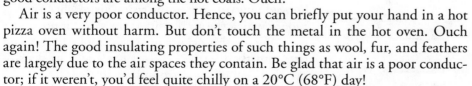

Solids (such as metals) whose atoms or molecules have loosely held electrons are good conductors of heat. These mobile electrons move quickly and transfer energy to other electrons, which migrate quickly throughout the solid. Poor conductors (such as glass, wool, wood, paper, cork, and plastic foam) are made up of molecules that hold tightly to their electrons. In these materials, molecules vibrate in place, and transfer energy only through interactions with their immediate neighbors. Because the electrons are not mobile, energy is transferred much more slowly in insulators.

Wood is a good insulator, and it is often used for cookware handles. Even when a pot is hot, you can briefly grasp the wooden handle with your bare hand without harm. An iron handle of the same temperature would surely burn your hand. Wood is a good insulator even when it's red hot. This explains how fire-walking coauthor John Suchocki can walk barefoot on red-hot wood coals without burning his feet (as shown in the chapter-opener photo). (CAUTION: Don't try this on your own; even experienced fire walkers sometimes receive bad burns when conditions aren't just right.) The main factor here is the poor conductivity of wood—even red-hot wood. Although its temperature is high, very little thermal energy is conducted to the feet. A fire walker must be careful that no iron nails or other good conductors are among the hot coals. Ouch!

Air is a very poor conductor. Hence, you can briefly put your hand in a hot pizza oven without harm. But don't touch the metal in the hot oven. Ouch again! The good insulating properties of such things as wool, fur, and feathers are largely due to the air spaces they contain. Be glad that air is a poor conductor; if it weren't, you'd feel quite chilly on a 20°C (68°F) day!

Snow is a poor conductor because its flakes are formed of crystals that trap air and provide insulation. That's why a blanket of snow keeps the ground warm in winter. Animals in the forest find shelter from the cold in snow banks and in holes in the snow. The snow doesn't provide them with energy—it simply slows down the loss of body heat that the animals generate. The same principle explains why igloos, arctic dwellings built from compacted snow, can shield their inhabitants from the cold.

Interestingly, insulation doesn't prevent the *flow* of thermal energy. Insulation simply slows down the *rate* at which thermal energy flows. Even a warm, well-insulated house gradually cools. Insulation such as rock wool or fiberglass

LEARNING OBJECTIVE
Describe the nature of conduction in solids.

Mastering**PHYSICS**
VIDEO: The Secret to Walking on Hot Coals
VIDEO: Air is a Poor Conductor

FIGURE 7.1
The tile floor feels colder than the wooden floor, even though both are at the same temperature. Tile is a better heat conductor than wood, and it more quickly conducts thermal energy from your feet.

What can be both good and poor at the same time? Answer: Any good insulator is a poor conductor. Or any good conductor is a poor insulator.

FIGURE 7.2
When you stick a nail into ice, does cold flow from the ice to your hand, or does energy flow from your hand to the ice?

FIGURE 7.3
Conduction of heat from Lil's hand to the wine is minimized by the long stem of the wine glass.

FIGURE 7.4
Snow patterns on the roof of a house show areas of conduction and insulation. Bare parts show where heat from the inside has conducted through the roof and melted the snow.

placed in the walls and ceiling of a house slows down the transfer of thermal energy from a warm house to the cooler outside (in winter) and from the warmer outside to the cool house (in summer).

CHECKPOINT

1. **In desert regions that are hot in the day and cold at night, the walls of houses are often made of mud. Why is it important that the mud walls be thick?**
2. **Wood is a better insulator than glass. Yet fiberglass is commonly used to insulate buildings. Why?**

Were these your answers?

1. A wall of appropriate thickness retains the warmth of the house at night by slowing the flow of thermal energy from inside to outside, and it keeps the house cool in the daytime by slowing the flow of thermal energy from outside to inside. Such a wall has *thermal inertia*.
2. Fiberglass is a good insulator, many times better than glass, because of the air that is trapped among its fibers.

LEARNING OBJECTIVE
Describe the nature of convection in fluids.

FYI Convection ovens are simply ovens with a fan inside. Cooking is speeded up by the circulation of heated air.

7.2 Convection

EXPLAIN THIS Why does warm air rise?

On a hot day you can see ripples in the air as hot air rises from an asphalt road. Likewise, if you put an ice cube into a clear glass of hot water, you can see ripples as the cold water from the melting ice cube descends in the glass. Transfer of heat by the motion of fluid as it rises or sinks is called **convection**. Unlike conduction, convection occurs only in fluids (liquids and gases). Convection involves bulk motion of a fluid (currents) rather than interactions at the molecular level.

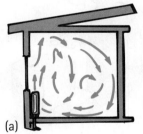

FIGURE 7.5
Convection currents in (a) a gas (air) and (b) a liquid.

We can see why warm air rises. When warmed, air expands, becomes less dense, and is buoyed upward in the cooler surrounding air like a balloon buoyed upward. When the rising air reaches an altitude at which the air density is the same, it no longer rises. We see this occurring when smoke from a fire rises and then settles off as it cools and its density matches that of the surrounding air.

To see for yourself that expanding air cools, do the experiment shown in Figure 7.7. Expanding air really does cool.*

A dramatic example of cooling by expansion occurs with steam expanding through the nozzle of a pressure cooker (Figure 7.8). The combined cooling effects of expansion and rapid mixing with cooler air allow you to hold

---

* Where does the energy go in this case? It goes into work done on the surrounding air as the expanding air pushes outward.

your hand comfortably in the jet of condensed vapor. (CAUTION: If you try this, be sure to place your hand high above the nozzle at first and then lower it slowly to a comfortable distance above the nozzle. If you put your hand directly at the nozzle where no steam is visible, watch out! Steam is invisible and is clear of the nozzle before it expands and cools. The cloud of "steam" you see is actually condensed water vapor, which is much cooler than live steam.)

Cooling by expansion is the opposite of what occurs when air is compressed. If you've ever compressed air with a tire pump, you probably noticed that both the air and the pump became quite hot. Compressing air warms it.

Convection currents stir the atmosphere and produce winds. Some parts of Earth's surface absorb energy from the Sun more readily than others. This results in uneven heating of the air near the ground. We see this effect at the seashore, as Figure 7.9 shows. In the daytime, the ground warms up more than the water. Then warmed air close to the ground rises and is replaced by cooler air that moves in from above the water. The result is a sea breeze. At night, the process reverses because the shore cools off more quickly than the water, and then the warmer air is over the sea. If you build a fire on the beach, you'll see that the smoke sweeps inland during the day and then seaward at night.

FIGURE 7.6
The tip of a heater element submerged in water produces convection currents, which are revealed as shadows (caused by deflections of light in water of different temperatures).

FYI Opening a refrigerator door lets warm air in, which then takes energy to cool. The more empty your fridge, the more cold air is swapped with warm air. So keep your fridge full for lower operating costs—especially if you're an excessive open-and-close-the-door type.

FIGURE 7.7
Blow warm air onto your hand from your wide-open mouth. Now reduce the opening between your lips so the air expands as you blow. Try it now. Do you notice a difference in the temperature of exhaled air? Does air cool as it expands?

FIGURE 7.8
The hot steam expands as it leaves the pressure cooker and is cool to Millie's touch.

CHECKPOINT
Explain why you can hold your fingers beside the candle flame without harm, but not above the flame.

Was this your answer?
Hot air travels upward by air convection. Because air is a poor conductor, very little energy travels sideways to your fingers.

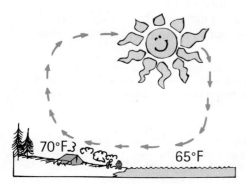

FIGURE 7.9
Convection currents produced by unequal heating of land and water. During the day, warm air above the land rises, and cooler air over the water moves in to replace it. At night, the direction of airflow is reversed, because now the water is warmer than the land.

7.3 Radiation

EXPLAIN THIS   How do we know the temperatures of stars?

Energy travels from the Sun through space and then through Earth's atmosphere and warms Earth's surface. This transfer of energy cannot involve conduction or convection, for there is no medium between the Sun and Earth. Energy must be transmitted some other way—by **radiation.*** The transferred energy is called *radiant energy.*

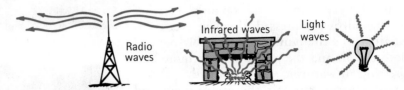

Radiant energy exists in the form of *electromagnetic waves*, ranging from the longest wavelengths to the shortest: radio waves, microwaves, infrared waves (invisible waves below red in the visible spectrum), visible waves, ultraviolet waves, X-rays, and gamma rays. We'll treat waves further in Chapters 11 and 12.

The wavelength of radiation is related to the frequency of vibration. Frequency is the rate of vibration of a wave source. Nellie Newton in Figure 7.11 shakes a rope at a low frequency (top) and at a higher frequency (bottom). Note that shaking at a low frequency produces a long, lazy wave, and shaking at a higher frequency produces a wave of shorter wavelength. We shall see in later chapters that vibrating electrons emit electromagnetic waves. Low-frequency vibrations produce long-wavelength waves, and high-frequency vibrations produce waves with shorter wavelengths.

Emission of Radiant Energy

Every object at any temperature emits radiant energy, spread over a range of frequencies (Figure 7.13). The frequency of the most intense radiation is called the peak frequency $\overline{f}$ and is proportional to the emitter's Kelvin temperature:

$$\overline{f} \sim T$$

If an object is hot enough, some of the radiant energy it emits is in the range of visible light. At a temperature of about 500°C, an object begins to emit the longest waves we can see, red light. Higher temperatures produce a yellowish light. At about 1500°C, all the different waves to which the eye is sensitive are emitted and we see an object as "white hot." A blue-hot star is hotter than a white-hot star, and a red-hot star is less hot. Because a blue-hot star has twice the light frequency of a red-hot star, it has twice the surface temperature of a red-hot star.**

Because the surface of the Sun has a high temperature (by earthly standards), it emits radiant energy at a high frequency—much of it in the visible portion of the

FIGURE 7.10
Types of radiant energy (electromagnetic waves).

FIGURE 7.11
A wave of long wavelength is produced when the rope is shaken gently (at a low frequency). When shaken more vigorously (at a high frequency), a wave of shorter wavelength is produced.

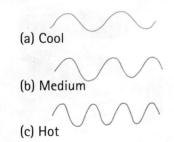

(a) Cool

(b) Medium

(c) Hot

FIGURE 7.12
The frequencies and wavelengths of radiant energy depend on the temperature of the emitters.

---

* The radiation we are talking about here is electromagnetic radiation, including visible light. Don't confuse this with *radioactivity*, a process of the atomic nucleus that we'll discuss in Chapter 13.
**The *amount* of radiant energy Q emitted by an object is proportional to the fourth power of the Kelvin temperature T:

$$Q \sim T^4$$

So, whereas a blue-hot star with twice the peak frequency of a red-hot star has twice the Kelvin temperature, it emits 16 times as much energy as a same-size red-hot star.

The amount of radiation emitted also depends on surface characteristics, which determine the *emissivity* of the object—ranging from close to 0 for very shiny surfaces and close to 1 for very black ones. A perfectly black surface emits what is called *blackbody radiation* and has an emissivity of 1.

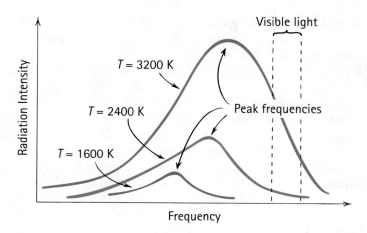

FIGURE 7.13
INTERACTIVE FIGURE MP

Radiation curves for different temperatures. The peak frequency of radiant energy is directly proportional to the absolute temperature of the emitter.

electromagnetic spectrum. The surface of Earth, by comparison, is relatively cool, and so the radiant energy it emits has a frequency lower than that of visible light. The radiation emitted by Earth is in the form of *infrared waves*—below our threshold of sight. Radiant energy emitted by Earth is called **terrestrial radiation**.

The Sun's radiant energy stems from nuclear reactions in its deep interior. Likewise, nuclear reactions in Earth's interior warm Earth (visit the depths of any mine and you'll find that it's warm down there year-round). Much of this thermal energy conducts to the surface and contributes to terrestrial radiation.

All objects—you, your instructor, and everything in your surroundings—continually emit radiant energy over a range of frequencies. Objects with everyday temperatures emit mostly low-frequency infrared waves. When the higher-frequency infrared waves are absorbed by your skin, as when you stand beside a hot stove, you feel the sensation of heat. So it is common to refer to infrared radiation as *heat radiation.* Common infrared sources that give the sensation of heat are the Sun, a lamp filament, and burning embers in a fireplace.

Heat radiation underlies infrared thermometers. You simply point the thermometer at something whose temperature you want, press a button, and a digital temperature reading appears. The radiation emitted by the object in question provides the reading. Typical classroom infrared thermometers operate in the range of about −30°C to 200°C.

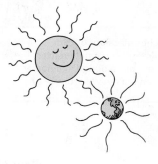

FIGURE 7.14
Both the Sun and Earth emit the same kind of radiant energy. The Sun's glow is visible to the eye; Earth's glow consists of longer waves and isn't visible to the eye.

CHECKPOINT

Which of these do not emit radiant energy: (a) the Sun, (b) lava from a volcano, (c) red-hot coals, (d) this textbook?

Was this your answer?
All the above emit radiant energy—even your textbook, which, like the other substances listed, has a temperature. According to the rule $\bar{f} \sim T$, the book therefore emits radiation whose peak frequency $\bar{f}$ is quite low compared with the radiation frequencies emitted by the other substances. Everything with any temperature above absolute zero emits radiant energy. That's right—*everything*!

FIGURE 7.15
An IR thermometer measures the infrared radiant energy emitted by a body and converts it to temperature.

Absorption of Radiant Energy

If everything is radiating energy, why doesn't everything finally run out of it? The answer is that everything is also *absorbing* energy. Good emitters of radiant energy are also good absorbers; poor emitters are poor absorbers. For example,

FYI Everything around you both radiates and absorbs energy continuously!

FIGURE 7.16
When the black rough-surfaced container and the shiny polished one are filled with hot (or cold) water, the blackened one cools (or warms) faster.

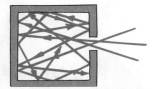

A hot pizza put outside on a winter day is a net emitter. The same pizza placed in a hotter oven is a net absorber.

a radio dish antenna constructed to be a good emitter of radio waves is also, by design, a good receiver (absorber) of them. A poorly designed transmitting antenna is also a poor receiver.

The surface of any material, hot or cold, both absorbs and emits radiant energy. If the surface absorbs more energy than it emits, it is a net absorber and its temperature rises. If it emits more than it absorbs, it is a net emitter and its temperature drops. Whether a surface plays the role of net emitter or net absorber depends on whether its temperature is above or below that of its surroundings. In short, if it's hotter than its surroundings, the surface is a net emitter and cools; if it's colder than its surroundings, it is a net absorber and becomes warmer.

CHECKPOINT

1. **If a good absorber of radiant energy were a poor emitter, how would its temperature compare with the temperature of its surroundings?**
2. **A farmer turns on the propane burner in his barn on a cold morning and heats the air to 20°C (68°F). Why does he still feel cold?**

Were these your answers?

1. If a good absorber were not also a good emitter, there would be a net absorption of radiant energy and the temperature of the absorber would remain higher than the temperature of the surroundings. Things around us approach a common temperature only because good absorbers are, by their nature, also good emitters.
2. The walls of the barn are still cold. The farmer radiates more energy to the walls than the walls radiate back at him, and he feels chilly. (On a winter day, you are comfortable inside your home or classroom only if the walls are warm—not just the air.)

FIGURE 7.17
Radiation that enters the opening has little chance of leaving because most of it is absorbed. For this reason, the opening to any cavity appears black to us.

Reflection of Radiant Energy

Absorption and reflection are opposite processes. A good absorber of radiant energy reflects very little of it, including visible light. Hence, a surface that reflects very little or no radiant energy looks dark. So a good absorber appears dark, and a perfect absorber reflects no radiant energy and appears completely black. The pupil of the eye, for example, allows light to enter with no reflection, which is why it appears black. (An exception occurs in flash photography when pupils appear pink, which occurs when very bright light is reflected off the eye's pink inner surface and back through the pupil.)

Look at the open ends of pipes in a stack; the holes appear black. Look at open doorways or windows of distant houses in the daytime, and they, too, look black. Openings appear black because the light that enters them is reflected back and forth on the inside walls many times and is partly absorbed at each reflection. As a result, very little or none of the light remains to come back out of the opening and travel to your eyes (Figure 7.17).

Good reflectors, on the other hand, are poor absorbers. Clean snow is a good reflector and therefore does not melt rapidly in sunlight. If the snow is dirty, it absorbs radiant energy from the Sun and melts faster. Dropping black soot from an aircraft onto snow-covered mountains is a technique sometimes used in flood control to accomplish controlled melting at favorable times, rather than a sudden runoff of melted snow.

FIGURE 7.18
The hole looks perfectly black and indicates a black interior, when in fact the interior has been painted a bright white.

CHECKPOINT

Which would be more effective in heating the air in a room: a heating radiator painted black or silver?

Was this your answer?
Interestingly, the color of paint is a small factor, so either color can be used. That's because radiators do very little heating by radiation. Their hot surfaces warm surrounding air by conduction, the warmed air rises, and warmed convection currents heat the room. (A better name for this type of heater would be a *convector*.) Now if you're interested in *optimum* efficiency, a silver-painted radiator radiates less, becomes and remains hotter, and does a better job of heating the air.

Emission and absorption in the visible part of the spectrum are affected by color, whereas the infrared part of the spectrum is more affected by surface texture. A dull finish emits/absorbs better in the infrared than a polished one, whatever its color.

7.4 Newton's Law of Cooling

EXPLAIN THIS Why will a hot pizza cool quicker outside on snow than in your room?

LEARNING OBJECTIVE
Relate Newton's law of cooling to everyday thermal occurrences.

Left to themselves, objects hotter than their surroundings eventually cool to match the surrounding temperature. The rate of cooling depends on how much hotter the object is than its surroundings. A hot apple pie cools more each minute if it is put in a cold freezer than if it is left on the kitchen table. That's because in the freezer, the temperature difference between the pie and its surroundings is greater. Similarly, the rate at which a warm house leaks thermal energy to the cold outdoors depends on the difference between the inside and outside temperatures.

The rate of cooling of an object—whether by conduction, convection, or radiation—is approximately proportional to the temperature difference ΔT between the object and its surroundings:

$$\text{Rate of cooling} \sim \Delta T$$

This is known as **Newton's law of cooling**. (Guess who is credited with discovering this?)

The law applies also to warming. If an object is cooler than its surroundings, its rate of warming up is also proportional to ΔT. Frozen food warms up faster in a warm room than in a cold room.

Newton's law of cooling doesn't apply, however, to objects that contain a source of energy, such as a running engine or a radioactive source.

Interestingly, Newton's law of cooling is an empirical relationship, and not a fundamental law like Newton's laws of motion.

THE THERMOS BOTTLE

A common Thermos bottle, a double-walled glass container with a vacuum between its silvered walls, nicely summarizes heat transfer. When a hot or cold liquid is poured into such a bottle, it remains at very nearly the same temperature for many hours. This is because the transfer of thermal energy by conduction, convection, and radiation is severely inhibited.

1. Heat transfer by conduction through the vacuum is impossible. Some thermal energy escapes by conduction through the glass and stopper, but this is a slow process, because glass, plastic, and cork are poor conductors.
2. The vacuum also prevents heat loss through the walls by convection, because there is no air between the walls.

3. Heat loss by radiation is inhibited by the silvered surfaces of the walls, which reflect radiant energy back into the bottle.

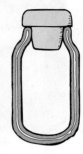

CHECKPOINT

Because a hot cup of tea loses thermal energy more rapidly than a lukewarm cup of tea, would it be correct to say that a hot cup of tea will cool to room temperature before a lukewarm cup of tea will?

Was this your answer?
No! Although the *rate* of cooling is greater for the hotter cup, it has further to cool to reach thermal equilibrium. The extra time is equal to the time it takes to cool to the initial temperature of the lukewarm cup of tea. Cooling rate and cooling time are not the same thing.

LEARNING OBJECTIVE
Describe the similarities between a florist greenhouse and Earth's climate.

7.5 Climate Change and the Greenhouse Effect

EXPLAIN THIS How is solar energy trapped inside an automobile on a sunny day?

An automobile parked in the street in the bright Sun on a hot day with closed windows can get very hot inside—appreciably hotter than the outside air. This is an example of the *greenhouse effect*, so named for the same temperature-raising effect in florists' glass greenhouses. Understanding the greenhouse effect requires knowing about two concepts.

The first concept has been previously stated—that all things radiate, and the wavelength of radiation depends on the temperature of the object emitting the radiation. High-temperature objects radiate short waves; low-temperature objects radiate long waves. The second concept we need to know is that the transparency of things such as air and glass depends on the wavelength of radiation. Air is transparent to both infrared (long) waves and visible (short) waves, unless the air contains excess water vapor and carbon dioxide, in which case it is opaque to infrared. Glass is transparent to visible light waves, but opaque to infrared waves. (The physics of transparency and opacity is discussed in Chapter 11.)

Now to why that car gets so hot in bright sunlight: Compared with the car, the Sun's temperature is very high. This means the waves the Sun radiates are very short. These short waves easily pass through both Earth's atmosphere and the glass windows of the car. So energy from the Sun gets into the car interior, where, except for reflection, it is absorbed. The interior of the car warms up. The

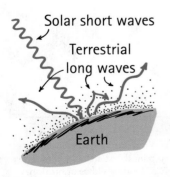

FIGURE 7.19
The hot Sun emits short waves, and the cool Earth emits long waves. Water vapor, carbon dioxide, and other *greenhouse gases* in the atmosphere retain heat that would otherwise be radiated from Earth to space.

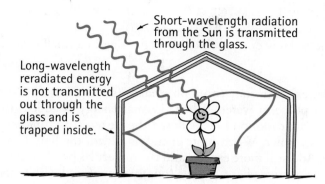

Short-wavelength radiation from the Sun is transmitted through the glass.

Long-wavelength reradiated energy is not transmitted out through the glass and is trapped inside.

FIGURE 7.20
Glass is transparent to short-wavelength radiation but opaque to long-wavelength radiation. Reradiated energy from the plant is of long wavelength because the plant has a relatively low temperature.

car interior radiates its own waves, but because it is not as hot as the Sun, the waves are longer. The reradiated long waves encounter glass that isn't transparent to them. So the reradiated energy remains in the car, which makes the car's interior even warmer (which is why leaving your pet in a car on a hot sunny day is a no-no).

The same effect occurs in Earth's atmosphere, which is transparent to solar radiation. The surface of Earth absorbs this energy, and reradiates part of this as longer-wavelength terrestrial radiation. Atmospheric gases (mainly water vapor and carbon dioxide) absorb and re-emit much of this long-wavelength terrestrial radiation back to Earth. Terrestrial radiation that cannot escape Earth's atmosphere warms Earth. This global warming process is very nice, for Earth would be a frigid $-18°C$ otherwise. Over the last 500,000 years the average temperature of Earth has fluctuated between 19°C and 27°C and is presently at the high point, 27°C—and climbing. Our present environmental concern is that increased levels of carbon dioxide and gases such as methane in the atmosphere may further increase the temperature and produce a new thermal balance unfavorable to the biosphere.

An important credo is "You can never change only one thing." Change one thing, and you change another. A slightly higher Earth temperature means slightly warmer oceans, which means changes in weather and storm patterns. The consensus among scientists is that Earth's climate is warming too fast. We speak of *climate change*. How this will play out is not known. At one extreme, corrections can be made and life will be fine for Earth's inhabitants. At the other extreme, we are reminded of the planet Venus, which in earlier times may have had a climate similar to Earth's. A runaway greenhouse effect is thought to have occurred on Venus, which today has an atmosphere that is 96% carbon dioxide, with an average surface temperature of 460°C. Venus is the hottest planet in the solar system. We certainly don't want to follow its course. Our fate likely lies between these extremes. We don't know.

What we do know is that energy usage is related to population size. We now seriously question the idea of continued growth. (Please read Appendix C, "Exponential Growth and Doubling Time"—very important material.)

CHECKPOINT
What does it mean to say that the greenhouse effect is like a one-way valve?

Was this your answer?
Both the atmosphere of Earth and the glass in a florist greenhouse are transparent to incoming short-wavelength light and block outgoing long waves. Because of the blockage, Earth and the greenhouse get warmer.

A significant role of glass in a florist greenhouse is to prevent convection of cooler outside air with warmer inside air. So the greenhouse effect actually plays a bigger role in global warming than it does in the warming of florist greenhouses.

FIGURE 7.21
When wet, the cloth covering the canteen promotes cooling. As the faster-moving water molecules evaporate from the wet cloth, the temperature of the cloth decreases and cools the metal. The metal, in turn, cools the water within. The water in the canteen can become a lot cooler than the outside air.

FIGURE 7.22
Sam, like other dogs, has no sweat glands (except between his toes). He cools by panting. In this way, evaporation occurs in the mouth and within the bronchial tract.

FYI Water evaporating from your body takes energy with it, which is why you feel cool when emerging from water on a warm and windy day.

7.6 Heat Transfer and Change of Phase

EXPLAIN THIS How does a tub of water freezing in a small room change air temperature?

Matter exists in four common **phases** (states). Ice, for example, is the *solid* phase of water. When thermal energy is added, the increased molecular motion breaks down the frozen structure and it becomes the *liquid* phase, water. When more energy is added, the liquid changes to the *gaseous* phase. Add still more energy, and the molecules break into ions and electrons, giving the *plasma* phase. Plasma (not to be confused with blood plasma) is the illuminating gas found in some TV screens and fluorescent and other vapor lamps. The Sun, stars, and much of the space between them are in the plasma phase. Whenever matter changes phase, a transfer of thermal energy is involved.

Evaporation

Water changes to the gaseous phase by the process of **evaporation**. In a liquid, molecules move randomly at a wide variety of speeds. Think of the water molecules as tiny billiard balls, moving helter-skelter, continually bumping into one another. During their bumping, some molecules gain kinetic energy while others lose kinetic energy. Molecules at the surface that gain kinetic energy by being bumped from below are the ones to break free from the liquid. They leave the surface and escape into the space above the liquid. In this way, they become gas.

When fast-moving molecules leave the water, the molecules left behind are the slower-moving ones. What happens to the overall kinetic energy in a liquid when the high-energy molecules leave? The answer: the average kinetic energy of molecules left in the liquid decreases. The temperature (which measures the average kinetic energy of the molecules) decreases and the water is cooled.

When our bodies begin to overheat, our sweat glands produce perspiration. This is part of nature's thermostat; the evaporation of sweat cools us and helps maintain a stable body temperature. Many animals do not have sweat glands and must cool themselves by other means (Figures 7.22 and 7.23).

CHECKPOINT
Would evaporation be a cooling process if the escaping molecules had the same average energy as those left behind?

Was this your answer?
No. A liquid cools only when more energetic molecules escape. This is similar to billiard balls that gain speed at the expense of others that lose speed. Those that leave (evaporate) are gainers, while losers remain behind and lower the temperature of the water.

In solid carbon dioxide (dry ice), molecules jump directly from the solid to the gaseous phase—that's why it's called *dry ice*. This form of evaporation is called **sublimation**. Mothballs are well known for their sublimation. Even frozen water undergoes sublimation. Because water molecules are so tightly held in a solid, frozen water sublimes much more slowly than liquid water evaporates. Sublimation accounts for the loss of much snow and ice, especially on high, sunny mountain tops. Sublimation also explains why ice cubes left in the freezer for a long time get smaller.

Condensation

The opposite of evaporation is **condensation**—the changing of a gas to a liquid. When gas molecules near the surface of a liquid are attracted to the liquid, they strike the surface with increased kinetic energy and become part of the liquid. This kinetic energy is absorbed by the liquid. The result is increased temperature. So whereas the liquid left behind is cooled with evaporation, with condensation the object upon which the vapor condenses is warmed. Condensation is a warming process.

A dramatic example of warming by condensation is the energy released by steam when it condenses. The steam gives up a lot of energy when it condenses to a liquid and moistens the skin. That's why a burn from 100°C steam is much more damaging than a burn from 100°C boiling water. This energy release by condensation is used in steam-heating systems.

When taking a shower, you may have noticed that you feel warmer in the moist shower region than outside the shower. You can quickly sense this difference when you step outside. Away from the moisture, the rate of evaporation is much higher than the rate of condensation, and you feel chilly. When you remain in the moist shower stall, the rate of condensation is higher and you feel warmer. So now you know why you can dry yourself with a towel much more comfortably if you remain in the shower stall. If you're in a hurry and don't mind the chill, dry yourself off in the hallway.

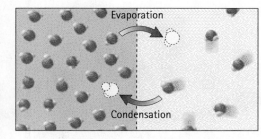

Liquid water   Water vapor

On a July afternoon in dry Phoenix or Santa Fe, you'll feel a lot cooler than in New York City or New Orleans, even when the temperatures are the same. In the drier cities, the rate of evaporation from your skin is much greater than the rate of condensation of water molecules from the air onto your skin. In humid locations, the rate of condensation is higher, perhaps as high as the rate of evaporation. Then, with little or no net evaporation to cool you, you feel uncomfortably warm. (We will explore condensation in the atmosphere when we study weather and climate in Chapter 25.)

a   b   c   d

FIGURE 7.27
The toy drinking bird operates by the evaporation of ether inside its body and by the evaporation of water from the outer surface of its head. The lower body contains liquid ether, which evaporates rapidly at room temperature. As it (a) vaporizes, it (b) creates pressure (inside arrows), which pushes ether up the tube. Ether in the upper part does not vaporize because the head is cooled by the evaporation of water from the outer felt-covered beak and head. When the weight of ether in the head is sufficient, the bird (c) pivots forward, permitting the ether to run back to the body. Each pivot wets the felt surface of the beak and head, and the cycle is repeated.

FIGURE 7.23
Pigs have no sweat glands and therefore cannot cool by the evaporation of perspiration. Instead, they wallow in the mud to cool themselves.

FIGURE 7.24
The exchange of molecules at the interface between liquid and gaseous water.

FIGURE 7.25
If you're chilly outside the shower stall, step back inside and be warmed by the condensation of the excess water vapor there.

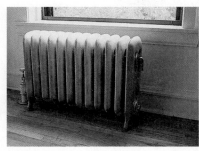

FIGURE 7.26
Thermal energy is released by steam when it condenses inside the "radiator."

CONDENSATION CRUNCH

Put a small amount of water in an aluminum soft-drink can and heat it on a stove until steam issues from the opening. When this occurs, air has been driven out and replaced by steam. Then, with a pair of tongs, quickly invert the can into a pan of water. Crunch! The can is crushed by atmospheric pressure! Why? When the molecules of steam inside the can hit the inner wall, they bounce—the metal certainly doesn't absorb them. But when steam molecules encounter water in the pan, they stick to the water surface. Condensation occurs, leaving a very low pressure in the can, whereupon the surrounding atmospheric pressure crunches the can. Here we see, dramatically, how pressure is reduced by condensation. (This demonstration nicely underlies the condensation cycle of a steam engine—perhaps something for future study.)

MasteringPHYSICS

VIDEO: Condensation is a Warming Process

When we say that we boil water, it is common to mean we are heating it. Actually, the boiling process cools the water.

CHECKPOINT

Place a dish of water anywhere in your room. If the water level in the dish remains unchanged from one day to the next, can you conclude that no evaporation or condensation is occurring?

Was this your answer?
Not at all, for significant evaporation and condensation occur continuously at the molecular level. The fact that the water level remains constant indicates equal rates of evaporation and condensation.

LEARNING OBJECTIVE
Explain the cooling nature of the boiling process.

7.7 Boiling

EXPLAIN THIS Why does heat added to boiling water not increase its temperature?

Evaporation occurs at the surface of a liquid. A change of phase from liquid to gas can also occur beneath the surface under proper conditions. The gas that forms beneath the surface of a liquid produces bubbles. The bubbles are buoyed upward to the surface, where they escape into the surrounding air. This change of phase is called **boiling**.

MasteringPHYSICS

VIDEO: Boiling is a Cooling Process

The pressure of the vapor within the bubbles in a boiling liquid must be great enough to resist the pressure of the surrounding liquid. Unless the vapor pressure is great enough, the surrounding pressures collapse any bubbles that tend to form. At temperatures below the boiling point, the vapor pressure is not great enough. Bubbles do not form until the boiling point is reached.

Boiling, like evaporation, is a cooling process. At first thought, this may seem surprising—perhaps because we usually associate boiling with heating. However, heating water is one thing; boiling it is another. When 100°C water at atmospheric pressure is boiling, it is in thermal equilibrium. The water in the pot is being cooled by boiling as fast as it is being heated by energy from the heat source (Figure 7.29). If cooling did not occur, continued application of heat to a pot of boiling water would raise its temperature.

When pressure on the surface of a liquid increases, boiling is hampered. Then the temperature needed for boiling rises. The boiling point of a liquid depends on the pressure on the liquid—which is most evident with a pressure cooker (Figure 7.30). In such a device, vapor pressure builds up inside and prevents boiling. This results in a water temperature that is higher than the normal boiling point. Note that what cooks the food is the high-temperature water, not the boiling process itself.

Lower atmospheric pressure (as at high altitudes) decreases the boiling temperature. For example, in Denver, Colorado, the "mile-high city," water boils at 95°C instead of at 100°C. If you try to cook food in boiling water that is cooler than 100°C, you must wait a longer time for proper cooking. A three-minute boiled egg in Denver is yucky. If the temperature of the boiling water is very low, food does not cook at all.

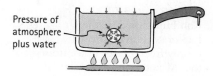

FIGURE 7.28
The motion of vapor molecules in the bubble of steam (much enlarged) creates a gas pressure (called the vapor pressure) that counteracts the atmospheric and water pressure against the bubble.

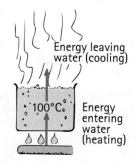

FIGURE 7.29
Heating warms the water from below, and boiling cools it from above.

FIGURE 7.30
The tight lid of a pressure cooker holds pressurized vapor above the water surface, and this inhibits boiling. In this way, the boiling temperature of the water is increased to more than 100°C.

CHECKPOINT

1. **Because boiling is a cooling process, would it be a good idea to cool your hot, sticky hands by dipping them into boiling water?**
2. **Rapidly boiling water has the same temperature as simmering water, both 100°C. Why, then, do the directions for cooking spaghetti often call for rapidly boiling water?**

Were these your answers?

1. No, no, no! When we say boiling is a cooling process, we mean that the water left behind in the pot (and not your hands!) is being cooled relative to the higher temperature it would attain otherwise. Because of the cooling effect of the boiling, the water remains at 100°C instead of getting hotter. A dip in 100°C water would be extremely uncomfortable for your hands!
2. Good cooks know that the reason for the rapidly boiling water is not higher temperature, but simply a way to keep the spaghetti strands from sticking together.

A dramatic demonstration of the cooling effect of evaporation and boiling is shown in Figure 7.31. Here we see a shallow dish of room-temperature water placed on top of an insulating cup in a vacuum jar. When the pressure in the jar is slowly reduced by a vacuum pump, the water begins to boil. As in all evaporation, the highest-energy molecules escape from the water, and the water left behind is cooled. As the pressure is further reduced, more and more of the faster-moving molecules boil away until the remaining liquid water reaches approximately 0°C. Continued cooling by boiling causes ice to form over the surface of the bubbling water. Boiling and freezing occur at the same time! Frozen bubbles of boiling water are a remarkable sight.

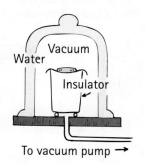

FIGURE 7.31
Apparatus to demonstrate that water freezes and boils at the same time in a vacuum. A gram or two of water is placed in a dish that is insulated from the base by a polystyrene cup.

FIGURE 7.32
Ron Hipschman at the Exploratorium removes a freshly frozen piece of ice from the "Water Freezer" exhibit, a vacuum chamber as depicted in Figure 7.31.

Mastering**PHYSICS**
VIDEO: Pressure Cooker: Boiling and Freezing at the Same Time

FYI Mountaineering pioneers in the 19th century, without altimeters, used the boiling point of water to determine their altitudes.

LEARNING OBJECTIVE
Distinguish between the processes of melting and freezing.

If you spray some drops of coffee into a vacuum chamber, they boil until they freeze. Even after they are frozen, the water molecules continue to evaporate into the vacuum, until little crystals of coffee solids remain. This is how freeze-dried coffee is produced. The low temperature of this process tends to keep the chemical structure of the coffee solids from changing. When hot water is added, much of the original flavor of the coffee is preserved.

7.8 Melting and Freezing

EXPLAIN THIS How can melting and freezing occur at the same time?

Melting occurs when a substance changes phase from a solid to a liquid. To visualize what happens, imagine a group of people holding hands and jumping around. The more violent the jumping, the more difficult it is to keep holding hands. If the jumping is violent enough, continuing to hold hands might become impossible. A similar thing happens to the molecules of a solid when it is heated. As heat is absorbed by the solid, its molecules vibrate more and more violently. If enough heat is absorbed, the attractive forces between the molecules no longer hold them together. The solid melts.

Freezing occurs when a liquid changes to a solid phase—the opposite of melting. As energy is removed from a liquid, molecular motion slows until molecules move so slowly that attractive forces between them bind them together. The liquid freezes when its molecules vibrate about fixed positions and form a solid.

FIGURE 7.33
(a) In a mixture of ice and water at 0°C, ice crystals gain and lose water molecules at the same time. The ice and water are in thermal equilibrium. (b) When salt is added to the water, fewer water molecules enter the ice and it melts along the surface.

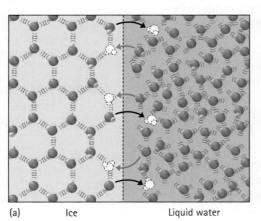

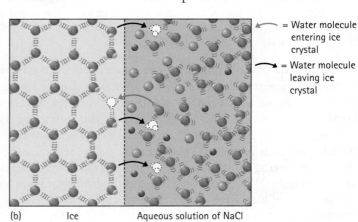

(a)   Ice   Liquid water   (b)   Ice   Aqueous solution of NaCl

= Water molecule entering ice crystal

= Water molecule leaving ice crystal

At atmospheric pressure, ice forms at 0°C. With impurities in the water, the freezing point is lowered. "Foreign" molecules get in the way and interfere with crystal formation. In general, adding anything to water lowers its freezing temperature. Antifreeze is a practical application of this process.

7.9   Energy and Change of Phase

EXPLAIN THIS   How does a refrigerator cool food?

If you heat a solid sufficiently, it melts and becomes a liquid. If you heat the liquid, it vaporizes and becomes a gas. Energy must be put into a substance to change its phase in the direction from solid to liquid to gas. Conversely, energy must be extracted from a substance to change its phase in the direction from gas to liquid to solid (Figure 7.34).

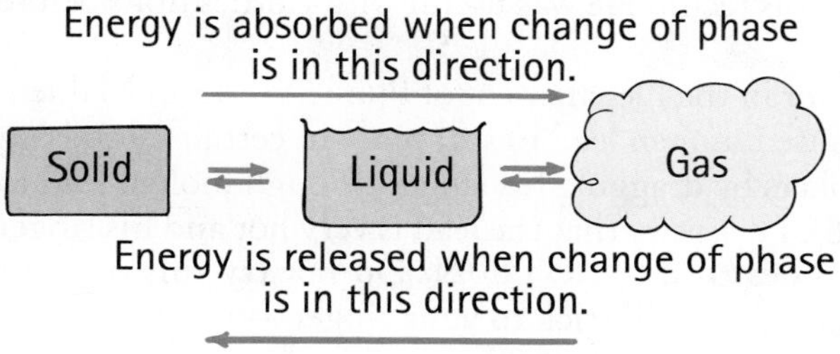

LEARNING OBJECTIVE
Identify the phase changes that require and that expel energy.

Energy is absorbed when change of phase is in this direction.

Solid ⇌ Liquid ⇌ Gas

Energy is released when change of phase is in this direction.

FIGURE 7.34
Energy changes with change of phase.

FYI   Why is rock salt spread on icy roads in winter? A short answer is that salt makes ice melt. Salt in water separates into sodium and chlorine ions. When these ions join water molecules, heat is given off, which melts microscopic parts of an icy surface. The melting process is enhanced by the pressure of automobiles rolling along the salt-covered icy surface, which forces the salt into the ice. The only difference between the rock salt applied to roads in winter and the substance you sprinkle on popcorn is the size of the crystals.

The cooling cycle of a refrigerator nicely illustrates these concepts. A motor pumps a special fluid through the system, where it undergoes the cyclic process of vaporization and condensation. When the fluid vaporizes, thermal energy is drawn from objects stored inside the refrigerator. The gas that forms, with its added energy, condenses to a liquid in outside coils in the back—appropriately called *condensation coils.* The next time you're near a refrigerator, place your hand near the condensation coils in the back and you'll feel the heat that has been extracted from the inside.

The mechanism in a refrigerator is called a "heat pump." It moves heat "uphill" from a cooler to a warmer place. Another example is an air conditioner, used in summertime to extract heat from indoors and move it to a warmer outdoors. In winter, a heat pump can serve as a heater, moving heat from a cooler outdoors to a warmer indoors.

The amount of energy needed to change a unit mass of any substance from solid to liquid (and vice versa) is called the **heat of fusion** for the substance. For water, this is 334 J/g. The amount of energy required to change a unit mass of any substance from liquid to gas (and vice versa) is called the **heat of vaporization** for the substance. For water, this is a whopping 2256 J/g.

In premodern times, farmers in cold climates prevented jars of food from freezing by taking advantage of water's high heat of fusion. They simply kept large tubs of water in their cellars. The outside temperature could drop to well below freezing, but not in the cellars, where water was releasing thermal energy while undergoing freezing. Canned food requires subzero temperatures to freeze because of its salt or sugar content. So farmers had only to replace frozen tubs of water with unfrozen ones, and the cellar temperatures wouldn't fall below 0°C.

Heat of vaporization is either the energy required to separate molecules from the liquid phase or the energy released when gases condense to the liquid phase.

FIGURE 7.35
The energy of sunlight simply and nicely harnessed.

Heat of fusion is either the energy needed to break molecular bonds in a solid and turn it into a liquid or the energy released when molecules in a liquid form bonds to create a solid.

CHECKPOINT

In the process of water vapor condensing in the air, the slower-moving molecules are the ones that condense. Does condensation warm or cool the surrounding air?

Was this your answer?
As slower-moving molecules are removed from the air, there is an increase in the average kinetic energy of molecules that remain in the air. Therefore, the air is warmed. The change of phase is from gas to liquid, which releases energy (Figure 7.34).

Water's high heat of vaporization allows you to briefly touch your wetted finger to a hot skillet on a hot stove without harm. You can even touch it a few times in succession as long as your finger remains wet. Energy that ordinarily would flow into and burn your finger goes instead into changing the phase of the moisture on your finger. This technique was useful with clothes irons before the advent of thermostats.

Paul Ryan, former supervisor in the Department of Public Works in Malden, Massachusetts, has for years used molten lead to seal pipes in certain plumbing operations. He startles onlookers by dragging his finger through molten lead to judge its hotness (Figure 7.36). He is sure that the lead is very hot and his finger is thoroughly wet before he does this. (CAUTION: Do not try this on your own: if the lead is not hot enough, it will stick to your finger—ouch!)

In Chapter 25 we'll discuss the role of thermal energy in climate change.

FIGURE 7.36
Paul Ryan tests the hotness of molten lead by dragging his wetted finger through it.

For instructor-assigned homework, go to www.masteringphysics.com

SUMMARY OF TERMS (KNOWLEDGE)

Boiling A rapid state of evaporation that takes place within the liquid as well as at its surface. As with evaporation, cooling of the liquid results.

Condensation The change of phase from gas to liquid; the opposite of evaporation. Warming of both results.

Conduction The transfer of thermal energy by molecular and electron collisions within a substance.

Convection The transfer of thermal energy in a gas or liquid by means of currents in the heated fluid.

Evaporation The change of phase at the surface of a liquid as it passes to the gaseous phase.

Freezing The process of changing phase from liquid to solid, as from water to ice.

Heat of fusion The amount of energy needed to change a unit mass of any substance from solid to liquid (and vice versa). For water, this is 334 J/g (or 80 cal/g).

Heat of vaporization The amount of energy needed to change a unit mass of any substance from liquid to gas (and vice versa). For water, this is 2256 J/g (or 540 cal/g).

Melting The process of changing phase from solid to liquid, as from ice to water.

Newton's law of cooling The rate of loss of heat from a warm object is proportional to the temperature difference between the object and its surroundings:

$$\text{Rate of cooling} \sim \Delta T$$

Phase The molecular state of a substance: solid, liquid, gas, or plasma.

Radiation The transfer of energy by means of electromagnetic waves.

Sublimation The change of phase directly from solid to gas.

Terrestrial radiation The radiant energy emitted by Earth.

READING CHECK QUESTIONS (COMPREHENSION)

1. What are the three common ways in which heat is transferred?

7.1 Conduction

2. What is the role of "loose" electrons in heat conductors?
3. How is a barefoot fire walker able to walk safely on red-hot wooden coals?
4. Does a good insulator prevent heat from getting through it, or does it simply delay its passage?

7.2 Convection

5. By what means is heat transferred by convection?
6. What happens to the temperature of air when it expands?
7. Why isn't Millie's hand burned when she holds it above the escape valve of the pressure cooker (see Figure 7.8)?
8. Why does the direction of coastal winds change from day to night?

7.3 Radiation

9. How does the peak frequency of radiant energy relate to the absolute temperature of the radiating source?
10. What is terrestrial radiation? How does it differ from solar radiation?
11. Because all objects emit energy to their surroundings, why don't the temperatures of all objects continuously decrease?
12. Why does the pupil of the eye appear black?

7.4 Newton's Law of Cooling

13. Which undergoes a faster rate of cooling: a red-hot poker in a warm oven or a red-hot poker in a cold room? (Or do both cool at the same rate?)
14. Does Newton's law of cooling apply to warming as well as to cooling?

7.5 Climate Change and the Greenhouse Effect

15. What would be the consequence to Earth's climate if the greenhouse effect were completely eliminated?

16. What is meant by the expression "You can never change only one thing"?

7.6 Heat Transfer and Change of Phase

17. What are the four common phases of matter?
18. Do all the molecules in a liquid have about the same speed, or do they have a wide variety of speeds?
19. What is evaporation, and why is it a cooling process?
20. What is sublimation?
21. What is condensation, how does it differ from evaporation, and why is it a warming process?
22. Why is a steam burn more damaging than a burn from boiling water of the same temperature?

7.7 Boiling

23. Distinguish between evaporation and boiling.
24. Why doesn't water boil at 100°C when it is under higher-than-normal atmospheric pressure?
25. Is it the boiling of the water or the higher temperature of the water that cooks food faster in a pressure cooker?

7.8 Melting and Freezing

26. Why does increasing the temperature of a solid make it melt?
27. Why does decreasing the temperature of a liquid make it freeze?
28. Why doesn't water freeze at 0°C when it contains dissolved material?

7.9 Energy and Change of Phase

29. Does a liquid release energy or absorb energy when it changes into a gas? When it changes into a solid?
30. Does a gas release energy or absorb energy when it changes into a liquid? How about a solid changing into a liquid?

ACTIVITIES (HANDS-ON APPLICATION)

31. Write a letter to your grandparents telling them how you're learning about the connections of nature and distinguishing between closely related ideas. Use the concepts of heat and temperature to explain how bringing water to a boil to make tea is actually a process that *cools* the water. Explain how they could convince their tea-time friends of this intriguing concept.

32. If you live where there is snow, do as Benjamin Franklin did more than 200 years ago: Lay samples of light and dark cloth on the snow and note the differences in the rate of melting beneath the samples of cloth.

33. Hold the bottom end of a test tube full of cold water in your hand. Heat the top part in a flame until the water boils. The fact that you can still hold the bottom shows that water

is a poor conductor of heat. This is even more dramatic when you wedge chunks of ice with steel wool at the bottom; then the water above can be brought to a boil without melting the ice. Try it and see.

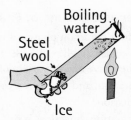

34. Wrap a piece of paper around a thick metal bar and place it in a flame. Note that the paper does not catch fire. Can you figure out why? (Paper generally does not ignite until its temperature reaches 233°C.)

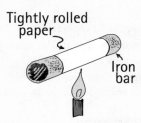

35. Place a Pyrex funnel mouth-down in a saucepan full of water so that the straight tube of the funnel protrudes above the water. Rest a part of the funnel on a nail or coin so that water can get under it. Place the pan on a stove, and observe the water as it begins to boil. Where do the bubbles form first? Why? As the bubbles rise, they expand rapidly and push water ahead of them. The funnel confines the water, which is forced up the tube and driven out at the top. Now do you know how a geyser and a coffee percolator operate?

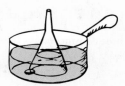

36. Observe the spout of a teakettle full of boiling water. Notice that you cannot see the steam that issues from the spout. The cloud that you see farther away from the spout is not steam, but condensed water droplets. Now hold the flame of a candle in the cloud of condensed steam. Can you explain your observations?

37. You can make rain in your kitchen. Put a cup of water in a Pyrex saucepan and heat it slowly over a low flame. When the water is warm, place a saucer filled with ice cubes on top of the container. As the water below is heated, droplets form at the bottom of the cold saucer and combine until they are large enough to fall, producing a steady "rainfall" as the water below is gently heated. How does this resemble, and differ from, the way that natural rain forms?

38. Measure the temperature of boiling water and the temperature of a boiling solution of salt and water. How do the temperatures compare?

39. Suspend an open-topped container of water in a pan of boiling water, with its top above the surface of the boiling water. You'll note that although water in the inner container can reach 100°C, it can't boil. Can you explain why this is so?

PLUG AND CHUG (FORMULA FAMILIARIZATION)

Quantity of heat: $Q = cm\Delta T$

40. Show that 5000 cal is required to increase the temperature of 50 g water from 0°C to 100°C. The specific heat capacity of water is 1 cal/g · °C.

41. Calculate the quantity of heat absorbed by 20 g of water that warms from 30°C to 90°C.

Heat of fusion: $Q = mL_f$

42. Show that 4000 cal is needed to melt 50 g of 0°C ice. The heat of fusion L_f for water is 80 cal/g.

43. Calculate the quantity of heat needed to melt a 200-g block of ice at 0°C.

Heat of vaporization: $Q = mL_v$

44. Show that 27,000 cal is required to change 50 g of 100°C boiling water into steam. The heat of vaporization for water L_v is 540 cal/g.

45. Calculate the quantity of heat needed to turn 200 g of 100°C water to steam at the same temperature.

46. Show that a total of 36,000 calories is required to change 50 g of 0°C ice to steam at 100°C.

THINK AND SOLVE (MATHEMATICAL APPLICATION)

47. Show that 9300 cal is required to change 15 g of 20°C water to 100°C steam.

48. Show that 100 g of 100°C steam will completely melt 800 g of 0°C ice.

49. The specific heat capacity of ice is about 0.5 cal/g · °C. Suppose it remains at that value all the way to absolute zero.

    (a) Show that the quantity of heat needed to change a 1-g ice cube at absolute zero (−273°C) to 1 g of boiling water is 317 cal.

    (b) How does this number of calories compare with the number of calories required to change the same gram of 100°C boiling water to 100°C steam?

50. A small block of ice at 0°C is subjected to 10 g of 100°C steam and melts completely. Show that the mass of the block of ice can be no more than 80 g.

51. A 10-kg iron ball is dropped onto a pavement from a height of 100 m. Suppose that half of the heat generated goes into warming the ball.

    (a) Show that the temperature increase of the ball is 1.1°C. (The specific heat capacity of iron is 450 J/kg · °C. Use 9.8 N/kg for g.)

    (b) Why is the answer the same for an iron ball of any mass?

52. A block of ice at 0°C is dropped from a height that causes it to completely melt upon impact. Assume that there is no air resistance and that all the energy goes into melting the ice.

    (a) Show that the height necessary for this to occur is at least 34 km. [Hint: Equate the joules of gravitational potential energy to the product of the mass of ice and its heat of fusion ($L_f = 335,000$ J/kg).]

    (b) Explain why the answer doesn't depend on mass?

53. Fifty grams of hot water at 80°C is poured into a cavity in a very large block of ice at 0°C. The final temperature of the water in the cavity becomes 0°C. Show that the mass of ice that melts is 50 g.

54. A 100-g chunk of 80°C iron is dropped into a cavity in a very large block of ice at 0°C. Show that the mass of ice that melts is 11 g. (The specific heat capacity of iron is 0.11 cal/g · °C.)

55. The heat of vaporization of ethyl alcohol L_v is about 200 cal/g. Show that if 4 kg of this refrigerant were allowed to vaporize in a refrigerator, it could freeze 10 kg of 0°C water to ice.

THINK AND RANK (ANALYSIS)

56. From best to worst, rank these materials as heat conductors: (a) copper wire, (b) snow, and a (c) glass rod.

57. From greatest to least, rank the frequency of radiation of these emitters of radiant energy: (a) red-hot star, (b) blue-hot star, and (c) the Sun.

58. Rank the boiling-water temperature from highest to lowest in these locations: (a) Death Valley, (b) Sea level, and (c) Denver, CO (the "mile-high city").

59. From greatest to least, rank the energy needed for these phase changes for equal amounts of H_2O: (a) from ice to ice water, (b) from ice-water to boiling water, and (c) from boiling water to steam.

EXERCISES (SYNTHESIS)

60. What is the purpose of the copper or aluminum layer on the bottom of a stainless-steel pot?

61. In terms of physics, why do restaurants serve baked potatoes wrapped in aluminum foil?

62. Many tongues have been injured by licking a piece of metal on a very cold day. Why would no harm result if a piece of wood were licked on the same day?

63. Wood is a better insulator than glass, yet fiberglass is commonly used as an insulator in wooden buildings. Explain.

64. Visit a snow-covered cemetery and note that the snow does not slope upward against the gravestones but, instead, forms depressions around them, as shown. What is your explanation for this?

65. Wood has a very low conductivity. Does it still have a low conductivity if it is very hot—that is, in the stage of smoldering red-hot coals? Could you safely walk across a bed of red-hot wooden coals with bare feet? Although the coals are hot, does much heat conduct from them to your feet if you step quickly? Could you do the same on pieces of red-hot iron? Explain. (CAUTION: Coals can stick to your feet, so ouch—don't try it!)

66. Double-pane windows have nitrogen gas or very dry air between the panes. Why is ordinary air a poor idea?

67. A friend says that molecules in a mixture of gases in thermal equilibrium have the same average *kinetic energy*. Do you agree or disagree? Defend your answer.

68. A friend says that molecules in a mixture of gases in thermal equilibrium have the same average *speed*. Do you agree or disagree? Defend your answer.

69. What does the high specific heat capacity of water have to do with the convection currents in the air at the seashore?

70. Snow-making machines used at ski areas blow a mixture of compressed air and water through a nozzle. The temperature of the mixture may initially be well above the freezing temperature of water, yet crystals of snow are formed as the mixture is ejected from the nozzle. Explain how this happens.

71. The source of heat of volcanoes and natural hot springs is trace amounts of radioactive minerals in common rock in Earth's interior. Why isn't the same kind of rock at Earth's surface warm to the touch?

72. Is it important to convert temperatures to the Kelvin scale when we use Newton's law of cooling? Why or why not?

73. Why is a water-based white solution, whitewash, sometimes applied to the glass of florists' greenhouses? Would you expect this practice to be more prevalent in winter or summer months?

74. If the composition of the upper atmosphere were changed to permit a greater amount of terrestrial radiation to escape, what effect would this have on Earth's climate?

75. Alcohol evaporates quicker than water at the same temperature. Which produces more cooling: alcohol or the same amount of water on your skin?

76. You can determine the wind direction by wetting your finger and holding it up in the air. Explain.

77. Give two reasons why pouring a cup of hot coffee into a saucer results in faster cooling.

78. Porous canvas bags filled with water are used by travelers in hot weather. When the bags are slung on the outside of a fast-moving car, the water inside is cooled considerably. Explain.

79. If all the molecules in a liquid had the same speed, and some were able to evaporate, would the remaining liquid be cooled? Explain.

80. What is the source of energy that keeps the dunking bird in Figure 7.26 operating?

81. Why does wrapping a bottled beverage in a wet cloth at a picnic often produce a cooler bottle than placing the bottle in a bucket of cold water?

82. Why does the boiling temperature of water decrease when the water is under reduced pressure, such as at a higher altitude?

83. Room-temperature water boils spontaneously in a vacuum—on the Moon, for example. Could you cook an egg in this boiling water? Explain.

84. Your inventor friend proposes a design for cookware that allows boiling to take place at a temperature of less than 100°C so that food can be cooked with the consumption of less energy. Comment on this idea.

85. When boiling spaghetti, is your cooking time reduced if the water is vigorously boiling instead of gently boiling?

86. Why does putting a lid over a pot of water on a stove shorten the time needed for the water to come to a boil, whereas, after the water boils, the use of the lid only slightly shortens the cooking time?

87. When can you add heat to a substance without raising its temperature? Give an example.

88. When can you withdraw heat from a substance without lowering its temperature? Give an example.

89. Air-conditioning units contain no water whatever, yet it is common to see water dripping from operating air conditioners poking outside homes on a hot day. Explain.

90. In the power plant of a nuclear submarine, the temperature of the water in the reactor is above 100°C. How is this possible?

91. Why does a hot dog pant?

DISCUSSION QUESTIONS (EVALUATION)

92. Wrap part of a fur coat around a thermometer. Discuss whether or not the temperature rises.

93. What is the principal reason a feather quilt is so warm on a cold winter night?

94. Friends in your discussion group say that when you touch a piece of ice, the cold flows from the ice to your hand, and that's why your hand is cooled. What is your more enlightened explanation?

95. How do the average kinetic energies of hydrogen and oxygen gases compare when these two gases are mixed at the same temperature? How do their average speeds compare? Discuss how mass is the crux of these questions.

96. Which atoms have the greater average speed in a mixture: U-238 or U-235? How would this affect diffusion through a porous membrane of otherwise identical gases made from these isotopes? Link this to the preceding question.

97. When you are near an incandescent lamp, turn it on and off quickly. You feel its heat, but you find when you touch the bulb that it is not hot. Explain why you felt heat from the lamp.

98. A number of objects at different temperatures placed in a closed room share radiant energy and ultimately come to the same temperature. Would this thermal equilibrium be possible if good absorbers were poor emitters and poor absorbers were good emitters? Defend your answer.

99. You come into a crowded and chilly classroom early in the morning on a cold winter day. Before the end of the hour, the room temperature increases to a comfortable level, even if heat is not provided by the heating system. Why the difference?

100. Why can you drink a cup of boiling-hot tea on top of a high mountain without any danger of burning your mouth? What if you did this in a mineshaft below sea level?

101. Using as a guide the rules that a good absorber of radiation is a good radiator and a good reflector is a poor absorber, state a rule relating the reflecting and radiating properties of a surface.

102. Suppose that, at a restaurant, you are served coffee before you are ready to drink it. In order that it is the hottest when you are ready for it, would you be wiser to add cream to it right away or just before you are ready to drink it? This question should elicit much discussion!

103. What does an air conditioner have in common with a refrigerator?

104. If you wish to save fuel and you're going to leave your warm house for a half-hour or so on a very cold day, should you turn your thermostat down a few degrees, turn it off altogether, or let it remain at the room temperature you desire? This question should elicit much discussion!

105. If you wish to save fuel and you're going to leave your cool house for a half-hour or so on a very hot day, should you turn your air-conditioning thermostat up a bit, turn it off altogether, or let it remain at the room temperature you desire? This question should elicit much discussion!

106. Place a jar of water on a small stand on the bottom of a saucepan full of water. Then the bottom of the jar isn't in contact with the bottom of the pan. When the pan is put on a hot stove, the water in the pan boils but the water in the jar does not. Why?

107. A piece of metal and an equal mass of wood are both removed from a hot oven at equal temperatures and dropped onto blocks of ice. The wood has a higher specific heat capacity than the metal. Which melts more ice before cooling to 0°C?

108. Earth scientists are considering a means of inducing clouds to be a brighter white. What effect would this have on Earth's climate?

109. Why is a tub of water placed in a farmer's canning cellar in cold winters to help prevent canned food from freezing?

110. Why does spraying fruit trees with water before a frost help protect the fruit from freezing?

111. The snow-covered mailboxes raise a question: What physics explains why the light-colored ones are snow covered, while the black ones are free of snow?

READINESS ASSURANCE TEST (RAT)

If you have a good handle on this chapter, if you really do, then you should be able to score at least 7 out of 10 on this RAT. If you score less than 7, you need to study further before moving on.

Choose the BEST answer to each of the following.

1. A fire walker walking barefoot across hot wooden coals depends on wood's
   (a) good conduction.
   (b) poor conduction.
   (c) low specific heat capacity.
   (d) wetness.

2. Thermal convection is linked mostly to
   (a) radiant energy.   (b) fluids.
   (c) insulators.   (d) all of these

3. When air is rapidly compressed, its temperature normally
   (a) increases.
   (b) decreases.
   (c) remains unchanged.
   (d) is unaffected, but not always.

4. An object that absorbs energy well also
   (a) conducts well.   (b) convects well.
   (c) radiates well.   (d) none of these

5. Which of these electromagnetic waves has the lowest frequency?
   (a) infrared   (b) visible
   (c) ultraviolet   (d) gamma rays

6. Compared with terrestrial radiation, the radiation from the Sun has a
   (a) longer wavelength.
   (b) lower frequency.
   (c) both of these
   (d) none of these

7. Glass is transparent to short-wavelength light and is
   (a) opaque to light of longer wavelengths.
   (b) opaque to the same light that is reflected from an interior surface.
   (c) both of these
   (d) none of these

8. When evaporation occurs in a dish of water, the molecules left behind in the water
   (a) are less energetic.
   (b) have decreased average speeds.
   (c) result in lowered temperature.
   (d) all of these

9. When steam changes phase to water, it
   (a) absorbs energy.
   (b) releases energy.
   (c) neither absorbs nor releases energy.
   (d) becomes more conducting.

10. Boiling and freezing can occur at the same time when water is subjected to
    (a) decreased temperatures.
    (b) decreased atmospheric pressure.
    (c) increased temperatures.
    (d) increased atmospheric pressure.

Answers to RAT

1. b, 2. b, 3. a, 4. c, 5. a, 6. d, 7. a, 8. d, 9. b, 10. b

8

CHAPTER 8

Static and Current Electricity

I F YOU are connected to an electrostatic generator, electricity in your hair will be evident when each strand of hair repels other strands. Electricity is everywhere, including the lightning in the sky and the batteries that power your iPad. A study and understanding of electricity require a step-by-step approach, because one concept is the building block for the next. This has been the case in our study of physics thus far, but more so with what now follows. So please give extra care to the study of this material. It can be difficult, confusing, and frustrating if you're hasty, but with careful effort, it can be comprehensible and rewarding. We start with static electricity, electricity at rest, and complete the chapter with current electricity. Let's begin.

8.1 Electric Charge

EXPLAIN THIS What is meant by saying that electric charge is conserved?

Try this: Tie a thread around the middle of a plastic drinking straw and then hang the straw by the thread. Rub half of the hanging straw with a piece of wool. If you rub another straw with wool and bring the rubbed ends of the straws near each other, the two straws *repel*.

If instead you rub a glass test tube with silk and bring the rubbed glass near the hanging straw, the two rubbed ends *attract*. And if you replace the hanging straw with a glass test tube and rub it and another test tube with silk, the two rubbed test tubes *repel*.

The ability of rubbed straws and test tubes to exert forces through space is due to a property we call electric *charge*. Although it may seem like magic, it is no more (or less!) magical than the ability of masses to exert gravitational forces on each other through space.

More than two centuries ago, America's first great scientist, Benjamin Franklin, did similar experiments. He formed the following hypotheses:

1. Every neutral (uncharged) substance has its own appropriate level of electric fluid.

2. Rubbing two materials together transfers "electric fluid" from one material to the other.

3. If an object gains electric fluid, it becomes *positively charged* with electric fluid. Likewise, if an object loses electric fluid, it becomes *negatively charged* with electric fluid.

Franklin couldn't see which fluid was transferred when the glass was rubbed with silk. He decided to call the glass *positively charged*, which means that the silk ended up with a negative charge because it lost electric fluid to the glass. Likewise in our example, the wool ends up positively charged and the plastic straw ends up negatively charged.

Once these charges have been assigned, we can see the most fundamental rule of electrical behavior:

Like charges repel; opposite charges attract.

Electrical forces arise from particles in atoms. In the simple model of the atom proposed in the early 1900s by Ernest Rutherford and Niels Bohr, a positively charged nucleus is surrounded by negatively charged electrons (Figure 8.2). The nucleus attracts the electrons and holds them in orbit, similar to the way the Sun holds the planets in orbit. But with a difference—electrons repel other electrons (whereas gravitational forces only attract).

The following are some important facts about atoms:

1. Every atom has a positively charged nucleus surrounded by negatively charged electrons.

2. All electrons are identical; that is, each has the same mass and the same quantity of negative charge as every other electron.

3. The nucleus is composed of protons and neutrons. (The common form of hydrogen, which has no neutrons, is the only exception.) All protons are positively charged and identical; similarly, all neutrons are identical. A proton has nearly 2000 times the mass of an electron, but its positive charge is equal in magnitude to the negative charge of the electron. A neutron has slightly greater mass than a proton and has no charge.

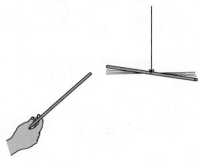

FIGURE 8.1
A plastic straw rubbed with wool is suspended by a thread. When another straw that has also been rubbed with wool is brought nearby, the two straws repel each other.

Mastering**PHYSICS**
TUTORIAL: Electrostatics

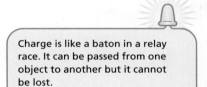

Charge is like a baton in a relay race. It can be passed from one object to another but it cannot be lost.

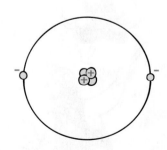

FIGURE 8.2

INTERACTIVE FIGURE

Model of a helium atom. The atomic nucleus is made up of two protons and two neutrons. The positively charged protons attract two negatively charged electrons. What is the net charge of this atom?

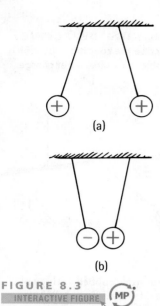

FIGURE 8.3
INTERACTIVE FIGURE MP

(a) Like charges repel.
(b) Unlike charges attract.

4. Atoms usually have as many electrons as protons, so the atom has zero *net* charge.

Just *why* electrons repel electrons and are attracted to protons is beyond the scope of this book. At our level of understanding we simply say that this is nature as we find it—that this electrical behavior is fundamental, or basic.

CHECKPOINT

1. Beneath the complexities of electrical phenomena lies a fundamental rule from which nearly all other electrical effects stem. What is this fundamental rule?
2. How does the charge of an electron differ from the charge of a proton?

Were these your answers?

1. Like charges repel; opposite charges attract.
2. The charges of the two particles are equal in magnitude, but opposite in sign.

Conservation of Charge

Electrons and protons have electric charge. A neutral atom has as many electrons as protons, so it has no net charge. The total positive charge balances the total negative charge exactly. If an electron is removed from an atom, the atom is no longer neutral. The atom then has one more positive charge (proton) than negative charge (electron) and is positively charged.

A charged atom is called an *ion*. A *positive* ion has a net positive charge because it has lost one or more electrons. A *negative* ion has a net negative charge because it has gained one or more extra electrons.

Matter is made of atoms, and atoms are made of electrons and protons (and neutrons as well). An object that has equal numbers of electrons and protons has no net electric charge. But if the numbers do not balance, the object is then electrically charged. An imbalance comes about by adding or removing electrons.

Although the innermost electrons in an atom are held very tightly to the oppositely charged atomic nucleus, the outermost electrons of many atoms are held very loosely and can be easily dislodged. How much energy is required to tear an electron away from an atom varies for different substances. The electrons are held more firmly in rubber or plastic than in wool or fur, for example. Hence, when a plastic straw is rubbed with a piece of wool, electrons transfer from the wool to the plastic straw. The plastic then has an excess of electrons and is negatively charged. The wool, in turn, has a deficiency of electrons and is positively charged. If you rub a glass or plastic rod with silk, you'll find that the rod becomes positively charged. The silk hangs on to electrons more tightly than the glass or plastic rod does. Electrons are rubbed off the rod and onto the silk. In summary:

An object that has unequal numbers of electrons and protons is electrically charged. If it has more electrons than protons, the object is negatively charged. If it has fewer electrons than protons, then it is positively charged.

FIGURE 8.4

When a rubber rod is rubbed with fur, electrons transfer from the fur to the rod. The rod is then negatively charged. Is the fur charged? By how much, compared with the rod? Positively or negatively?

Franklin didn't explain charge transfer in terms of transfer of electrons because electrons were unknown in his day. Later it was found that electrons are neither created nor destroyed but are simply transferred from one material to another. Charge is conserved. In every event, whether large-scale or at the atomic and nuclear level, the principle of *conservation of charge* applies. No case of the creation or destruction of net electric charge has ever been found. The conservation of charge is a cornerstone in physics, ranking with the conservation of energy and momentum.

Any object that is electrically charged has an excess or deficiency of some whole number of electrons—electrons cannot be divided into fractions of electrons. This means that the charge of the object is a whole-number multiple of the charge of an electron. It cannot have a charge equal to the charge of 1.5 or 1000.5 electrons, for example. In all measurements to date, objects have a charge that is a whole-number multiple of the charge of a single electron.

> Conservation of charge is another of the conservation principles. Recall from previous chapters the conservation of momentum and the conservation of energy.

CHECKPOINT

If you scuff electrons onto your shoes while walking across a rug, are you negatively or positively charged?

Was this your answer?
When your rubber- or plastic-soled shoes drag across the rug, they pick up electrons from the rug in the same way you charge a rubber rod by rubbing it with cloth. You have more electrons after you scuff your shoes, so you are negatively charged (and the rug is positively charged).

FIGURE 8.5
Why do you get a slight shock from the doorknob after scuffing across the carpet?

ELECTRONICS TECHNOLOGY AND SPARKS

Electric charge can be dangerous. Two hundred years ago, young boys called *powder monkeys* ran barefooted below the decks of warships to bring sacks of black gunpowder to the cannons above. It was ship law that this task be done barefoot. Why? Because it was important that no static charge build up on the powder on their bodies as they ran to and fro. Bare feet scuffed the decks much less than

shoes and ensured no charge accumulation that might produce an igniting spark and an explosion.

Static charge is a danger in many industries today—not because of explosions, but because delicate electronic circuits may be destroyed by static charges. Some circuit components are sensitive enough to be "fried" by sparks of static electricity. Electronics technicians frequently

wear clothing of special fabrics with ground wires between their sleeves and their socks. Some wear special wristbands that are connected to a grounded surface so that static charges do not build up—when moving a chair, for example. The smaller the electronic circuit, the more hazardous are sparks that may short-circuit the circuit elements.

8.2 Coulomb's Law

EXPLAIN THIS What do the laws of Newton and Coulomb have in common?

LEARNING OBJECTIVE
Relate the inverse-square law to electrical forces.

Electrical force, like gravitational force, decreases inversely as the square of the distance between charges. This relationship, which was discovered by Charles Coulomb in the 18th century, is called **Coulomb's law**. It states that for two charged objects that are much smaller than the distance between them, the force between them varies directly as the product of their charges and inversely as the square of the separation distance. The force acts along a straight line from one charge to the other. Coulomb's law can be expressed as

$$F = k\frac{q_1 q_2}{d^2}$$

> **FYI** Static electricity is a problem at gasoline pumps. Even the tiniest of sparks ignite vapors coming from the gasoline and cause fires—frequently lethal. A good rule is to touch metal to discharge static charge from your body before you fuel. Also, don't use a cell phone when fueling.

IONIZED BRACELETS: SCIENCE OR PSEUDOSCIENCE?

Surveys indicate that most Americans believe that ionized bracelets can reduce joint or muscle pain. Manufacturers claim that ionized bracelets relieve such pain. Are they correct? In 2002, the claim was tested by researchers at the Mayo Clinic in Jacksonville, Florida, who randomly assigned 305 participants to wear an ionized bracelet for 28 days and another 305 participants to wear a placebo bracelet for the same duration. The study volunteers were men and women 18 and older who had self-reported musculoskeletal pain at the beginning of the study.

Neither the researchers nor the participants knew which volunteers wore an ionized bracelet and which wore a placebo bracelet. Both types of bracelets were identical, were supplied by the manufacturer, and were worn according to the manufacturer's recommendations. Interestingly, both groups reported significant relief from pain. No difference was found in the amount of self-reported pain relief between the group wearing the ionized bracelets and the group wearing the placebo bracelets. Apparently, just believing that the bracelet relieves pain does the trick!

Interestingly, the brain initiates the creation of endorphins (which bind to opiate receptor sites) when the person expects to get relief from pain. The placebo effect is very real and measurable via blood titrations. So there's some merit in the old adage that wishing hard for something makes it come true. But this has nothing to do with the physics, chemistry, or biological interaction with the bracelet. Hence, ionized bracelets join the ranks of pseudoscientific devices.

In any society that thrives more on capturing attention than on informing, pseudoscience is big business.

where d is the distance between the charged particles, q_1 represents the quantity of charge of one particle, q_2 represents the quantity of charge of the second particle, and k is the proportionality constant.

The unit of charge is called the **coulomb**, abbreviated C. It turns out that a charge of 1 C is the charge associated with 6.25 billion billion electrons. This might seem like a great number of electrons, but it only represents the amount of charge that flows through a common 100-W lightbulb in a little more than a second.

The proportionality constant k in Coulomb's law is similar to G in Newton's law of gravity. Instead of being a very small number, like G, k is a very large number, approximately

$$k = 9,000,000,000 \text{ N·m}^2/\text{C}^2.$$

In scientific notation, $k = 9.0 \times 10^9 \text{ N·m}^2/\text{C}^2$. The unit N · m^2/C^2 is not central to our interest here; it simply converts the right-hand side of the equation to the unit of force, the newton (N). What is important is the large magnitude of k. If, for example, a pair of like charges of 1 C each were 1 m apart, the force of repulsion between the two would be 9 billion N.* That would be about 10 times the weight of a battleship! Obviously, such quantities of net charge do not usually exist in our everyday environment.

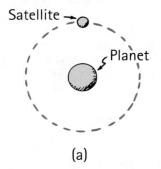
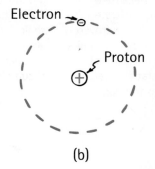

(a)　　　　　　　(b)

FIGURE 8.6
(a) A gravitational force holds the satellite in orbit about the planet, and (b) an electrical force holds the electron in orbit about the proton. In both cases, there is no contact between the bodies. We say that the orbiting bodies interact with the force fields of the planet and proton and are everywhere in contact with these fields. Thus, the force that one electric charge exerts on another can be described as the interaction between one charge and the field set up by the other.

So Newton's law of gravitation for masses is similar to Coulomb's law for electrically charged bodies. The most important difference between gravitational and electrical forces is that electrical forces may be either attractive or repulsive, whereas gravitational forces are only attractive. Coulomb's law underlies the bonding forces between molecules that are essential in the field of chemistry.

---

* Contrast this to the gravitational force of attraction between two 1-kg masses 1 m apart: 6.67×10^{-11} N. This is an extremely small force. For the force to be 1 N, the masses at 1 m apart would have to be nearly 123,000 kg each! Gravitational forces between ordinary objects are exceedingly small, and differences in electrical forces between ordinary objects can be exceedingly huge. We don't sense them because the positives and negatives normally balance out, and, even for objects charged to a high voltage, the imbalance of electrons to protons is typically no more than one part in a trillion trillion.

CHECKPOINT

1. The proton is the nucleus of the hydrogen atom, and it attracts the electron that orbits it. Relative to this force, does the electron attract the proton with less force, more force, or the same amount of force?
2. If a proton at a particular distance from a charged particle is repelled with a given force, by how much does the force decrease when the proton is three times as distant from the particle? Five times as distant?
3. What is the sign of charge of the particle in this case?

Were these your answers?

1. The same amount of force, in accord with Newton's third law—basic mechanics! Recall that a force is an interaction between two things—in this case, between the proton and the electron. They pull on each other equally.
2. In accord with the inverse-square law, at three times the distance, the force decreases to $\frac{1}{9}$ its original value. At five times the distance, the force decreases to $\frac{1}{25}$ of its original value.
3. Positive.

Charge Polarization

If you charge an inflated balloon by rubbing it on your hair and then place the balloon against a wall, it sticks. This is because the charge on the balloon alters the charge distribution in the atoms or molecules in the wall, effectively inducing an opposite charge on the wall. The molecules cannot move from their relatively stationary positions, but their "centers of charge" are moved. The positive part of the atom or molecule is attracted toward the balloon while the negative part is repelled. This has the effect of distorting the atom or molecule (Figure 8.7). The atom or molecule is said to be **electrically polarized**. We will see in Part 2 how polarization plays an important role in chemistry.

CHECKPOINT

You know that a balloon rubbed on your hair sticks to a wall. In a humorous vein, does it follow that your oppositely charged head would also stick to the wall?

Was this your answer?

No, unless you're an airhead (having a head mass about the same as that of an air-filled balloon). The force that holds a balloon to the wall cannot support your heavier head.

8.3 Electric Field

EXPLAIN THIS   What kind of force field surrounds mass? Electric charge?

Electrical forces, like gravitational forces, can act between things that are not in contact with each other. For both electricity and gravity, a force field exists that influences distant charges and masses, respectively. The properties of space surrounding any mass are altered such that another mass introduced to this region experiences a force. This "alteration in space" is called

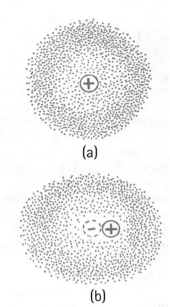

FIGURE 8.7
(a) The center of the negative "cloud" of electrons coincides with the center of the positive nucleus in an atom. (b) When an external negative charge is brought nearby to the right, as on a charged balloon, the electron cloud is distorted so that the centers of negative and positive charge no longer coincide. The atom is electrically polarized.

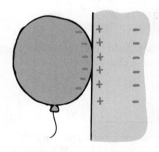

FIGURE 8.8
The negatively charged balloon polarizes molecules in the wooden wall and creates a positively charged surface, so the balloon sticks to the wall.

LEARNING OBJECTIVE
Relate electric field strength with patterns of electrical lines of force.

MICROWAVE OVEN

Imagine an enclosure filled with Ping-Pong balls among a few batons, all at rest. Now imagine that the batons suddenly rotate backward and forward, striking neighboring Ping-Pong balls. Almost immediately, most of the Ping-Pong balls are energized, vibrating in all directions. A microwave oven works similarly. The batons are water molecules made to rotate to and fro in rhythm with microwaves in the enclosure. The Ping-Pong balls are the other molecules that make up the bulk of material being cooked.

H_2O molecules are electrically polarized, with opposite charges on opposite sides. When an electric field is imposed on them, they align with the field as a compass needle aligns with a magnetic field. When the field is made to oscillate, the H_2O molecules oscillate also—and quite energetically when the frequency of the waves matches the natural rotational frequency of the H_2O. So, food is cooked by converting H_2O molecules into flip-flopping energy sources that impart thermal motion to surrounding food molecules. Without polar molecules in the food, a microwave oven wouldn't work. That's why microwaves pass through foam, paper, or ceramic plates and reflect from metals with no effect. They do energize, however, water molecules.

A note of caution is due when boiling water in a microwave oven. Water can sometimes heat faster than bubbles can form, and the water then heats beyond its boiling point—it becomes superheated. If the water is bumped or jarred just enough to cause the bubbles to form rapidly, they'll violently expel the hot water from its container. More than one person has had boiling water blast into his or her face.

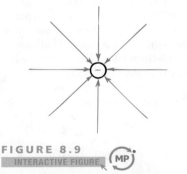

FIGURE 8.9
INTERACTIVE FIGURE (MP)

Electric field representations about a negative charge.

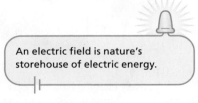

An electric field is nature's storehouse of electric energy.

FIGURE 8.10
INTERACTIVE FIGURE (MP)

Some electric field configurations. (a) Lines of force about a single positive charge. (b) Lines of force for a pair of equal but opposite charges. Note that the lines emanate from the positive charge and terminate on the negative charge. (c) Uniform lines of force between two oppositely charged parallel plates.

its *gravitational field*. We can think of any other mass as interacting with the field and not directly with the mass that produces it. For example, when an apple falls from a tree, we say it is interacting with the mass of Earth, but we can also think of the apple as interacting with the gravitational field of Earth. It is common to think of distant rockets and the like as interacting with gravitational fields rather than bodies responsible for the fields. The field plays an intermediate role in the force between bodies. More important, the field stores energy. So similar to a gravitational field, the space around every electric charge is energized with an **electric field**—an energetic aura that extends through space.*

If you place a charged particle in an electric field, it experiences a force. The direction of the force on a positive charge is the same direction as the field. The electric field about a proton extends radially from the proton. About an electron, the field is in the opposite direction (Figure 8.9). As with electric force, the electric field about a particle obeys the inverse-square law. Some electric field configurations are shown in Figure 8.10, and photographs of field patterns are shown in Figure 8.11. In the next chapter, we'll see how bits of iron similarly align with magnetic fields.

Perhaps your instructor will demonstrate the effects of the electric field that surrounds the charged dome of a Van de Graaff generator (Figure 8.12).

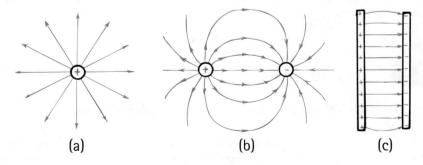

(a)　　　　　(b)　　　　　(c)

---

* An electric field is a vector quantity, having both magnitude and direction. The magnitude of the field at any point is simply the force per unit of charge. If a charge q experiences a force F at some point in space, then the electric field E at that point is $E = F/q$.

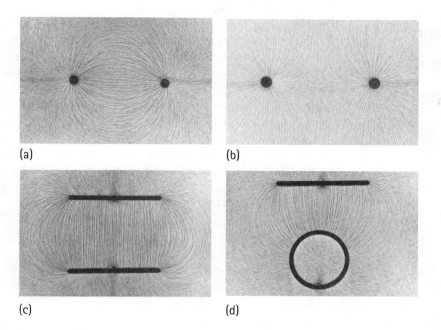

(a)

(b)

(c)

(d)

FIGURE 8.11
Bits of thread suspended in an oil bath line up end-to-end along the direction of the field.
(a) Equal and opposite charges.
(b) Equal like charges.
(c) Oppositely charged plates.
(d) Oppositely charged cylinder and plate.

Static charge on the surface of any electrically conducting surface arranges itself such that the electric field inside the conductor cancels to zero. Note the randomness of threads inside the cylinder of Figure 8.11d, where no field exists.

FYI Whatever the intensity of the electric field about a charged Van de Graaff generator, the electric field inside the dome cancels to zero. This is true for the interiors of all metals that carry static charge.

Charged objects in the field of the dome are either attracted or repelled, depending on their sign of charge.

CHECKPOINT

Both Lillian and the dome of the Van de Graaff generator in Figure 8.12 are charged. Why does Lillian's hair stand out?

Was this your answer?
She and her hair are charged. Each hair is repelled by others around it—evidence that *like charges repel*. Even a small charge produces an electrical force greater than the weight of strands of hair. Fortunately, the electrical force is not great enough to make her arms stand out!

8.4 Electric Potential

EXPLAIN THIS Why aren't you harmed when you touch a 5000-V party balloon?

In our study of energy in Chapter 3, we learned that an object has gravitational potential energy because of its location in a gravitational field. Similarly, a charged object has potential energy by virtue of its location in an electric field. Just as work is required to lift a massive object against the gravitational field of the Earth, work is required to push a charged particle against the electric field of a charged body. This work changes the electric potential energy of the charged particle.* Similarly, work done in compressing a spring increases the potential energy of the spring (Figure 8.14a). Likewise, the work done in pushing a charged particle closer to the charged sphere in Figure 8.14b increases the potential energy of the charged particle. We call the energy possessed by the charged particle that is due to its location **electric potential energy**. If the particle is released,

FIGURE 8.12
Both Lillian and the spherical dome of the Van de Graaff generator are electrically charged.

LEARNING OBJECTIVE
Distinguish between electric potential energy and electric potential.

---

* This work is positive if it increases the electric potential energy of the charged particle and negative if it decreases it.

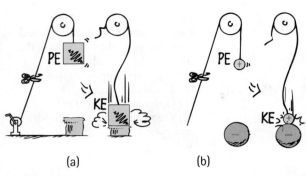

FIGURE 8.13

(a) The PE (gravitational potential energy) of a mass held in a gravitational field. (b) The PE of a charged particle held in an electric field. When the mass and particle are released, how does the KE (kinetic energy) acquired by each compare with the decrease in PE?

In a nutshell: *Electric potential* and *potential* mean the same thing—electric potential energy per unit charge—in units of volts. On the other hand, *potential difference* is the same as *voltage*—the *difference* in electric potential between two points—also in units of volts.

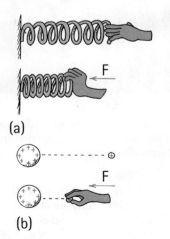

FIGURE 8.14

(a) The spring has more elastic PE when compressed. (b) The small charge similarly has more PE when pushed closer to the charged sphere. In both cases, the increased PE is the result of work input.

it accelerates in a direction away from the sphere, and its electric potential energy changes to kinetic energy.

If we push a particle with twice the charge, we do twice as much work. Twice the charge in the same location has twice the electric potential energy; with three times the charge, there is three times as much potential energy; and so on. When working with electricity, rather than dealing with the total potential energy of a charged body, it is convenient to consider the electric potential energy *per charge*. We simply divide the amount of energy in any case by the amount of charge. The concept of potential energy per charge is called **electric potential**; that is,

$$\text{Electric potential} = \frac{\text{electric potential energy}}{\text{amount of charge}}$$

The unit of measurement for electric potential is the volt, so electric potential is often called *voltage*. A potential of 1 volt (V) equals 1 joule (J) of energy per 1 coulomb (C) of charge:

$$1 \text{ volt} = \frac{1 \text{ joule}}{1 \text{ coulomb}}$$

Thus, a 1.5-V battery gives 1.5 J of energy to every 1 C of charge flowing through the battery. *Electric potential* and *voltage* are the same thing, and they are commonly used interchangeably.

The significance of voltage is that a definite value for it can be assigned to a location. We can speak about the voltages at different locations in an electric field whether or not charges occupy those locations. The same is true of voltages at various locations in an electric circuit. Later in this chapter, we will see that the location of the positive terminal of a 12-V battery is maintained at a voltage 12 V higher than the location of the negative terminal. When a conducting medium connects this voltage difference, any charges in the medium move between these locations.

CHECKPOINT

1. If there were twice as many coulombs in the test charge near the charged sphere in Figure 8.15, would the *electric potential energy* of the test charge relative to the charged sphere be the same, or would it be twice as great? Would the *electric potential* of the test charge be the same, or would it be twice as great?
2. What does it mean to say that the battery in your car is rated at 12 V?

Were these your answers?

1. The result of twice as many coulombs is twice as much *electric potential energy* because it takes twice as much work to put the charge there. But the *electric potential* would be the same. Twice the energy divided by twice the charge gives the same potential as one unit of energy divided by one unit of charge. Electric potential is not the same thing as electric potential energy. Be sure you understand this before you study further.
2. It means that one of the battery terminals is 12 V higher in potential than the other one. We'll soon learn that when a circuit is connected between these terminals, each coulomb of charge in the resulting current is given 12 J of energy as it passes through the battery (and 12 J of energy "spent" in the circuit).

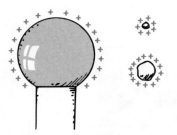

FIGURE 8.15
The larger test charge has more PE in the field of the charged dome, but the electric potential of any amount of charge at the same location is the same.

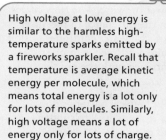

FIGURE 8.16
Although the voltage of the charged balloon is high, the electric potential energy is low because of the small amount of charge.

High voltage at low energy is similar to the harmless high-temperature sparks emitted by a fireworks sparkler. Recall that temperature is average kinetic energy per molecule, which means total energy is a lot only for lots of molecules. Similarly, high voltage means a lot of energy only for lots of charge.

Rub a balloon on your hair, and the balloon becomes negatively charged—perhaps to several thousand volts! That would be several thousand joules of energy, if the charge were 1 C. However, 1 C is a fairly respectable amount of charge. The charge on a balloon rubbed on hair is typically much less than a millionth of a coulomb. Therefore, the amount of energy associated with the charged balloon is very, very small. A high voltage means a lot of energy only if a lot of charge is involved. Electric potential energy differs from electric potential (or voltage).

Mastering**PHYSICS**
VIDEO: Electric Potential
VIDEO: Van deGraff Generator

8.5 Voltage Sources

EXPLAIN THIS Why is an electric battery often called an *electric pump*?

LEARNING OBJECTIVE
Recognize how a potential difference is necessary for electric current.

When the ends of a heat conductor are at different temperatures, heat energy flows from the higher temperature to the lower temperature. The flow ceases when both ends reach the same temperature. Any material having free charged particles that easily flow through it when an electric force acts on them is called an electric **conductor**. By contrast, any material in which charged particles do not easily flow is called an *insulator*. Both heat and electric conductors are characterized by electric charges that are free to move. Similar to heat flow, when the ends of an electric conductor are at different electric potentials—when there is a **potential difference**—charges in the conductor flow from the higher potential to the lower potential. The flow of charge persists until both ends reach the same potential. Without a potential difference, no flow of charge occurs.

To attain a sustained flow of charge in a conductor, some arrangement must be provided to maintain a difference in potential while charge flows from one end to the other. The situation is analogous to the flow of water from a higher reservoir to a lower one

Mastering**PHYSICS**
VIDEO: Caution on Handling Electric Wires
VIDEO: Birds and High-Voltage Wires

FIGURE 8.17
Although the Wimshurst machine can generate thousands of volts, it puts out no more energy than the work that Jim Stith puts into it by cranking the handle.

FIGURE 8.18

(a) Water flows from the reservoir of higher pressure to the reservoir of lower pressure. The flow ceases when the difference in pressure ceases. (b) Water continues to flow because a difference in pressure is maintained with the pump.

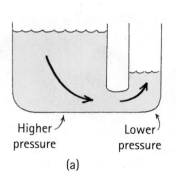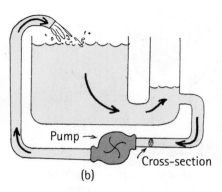

Higher pressure   Lower pressure   Pump   Cross-section

(a)   (b)

> **FYI** Chemical batteries don't respond well to sudden surges of charge. An alternative that does respond well to spurts of energy input is a spinning flywheel. Unlike the ones used by potters for spinning clay, modern flywheels are made of lightweight composite materials that are strong and can be spun at high speeds without coming apart. Rotational kinetic energy is then converted to other forms of energy. Watch for flywheels as energy-storing devices.

FIGURE 8.19

An unusual source of voltage. The electric potential between the head and tail of the electric eel (*Electrophorus electricus*) can be up to 650 V.

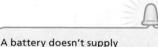

> A battery doesn't supply electrons to a circuit; it instead supplies energy to electrons that already exist in the circuit.

LEARNING OBJECTIVE
Relate the speed of electrons in a circuit to dc and ac.

> **FYI** When a common automobile battery provides an electrical pressure of 12 V to a circuit connected across its terminals, 12 J of energy is supplied to each coulomb of charge that is made to flow in the circuit.

(Figure 8.18a). Water flows in a pipe that connects the reservoirs only as long as a difference in water level exists. The flow of water in the pipe, like the flow of charge in a wire, ceases when the pressures at each end are equal. (We imply this phenomenon when we say that water seeks its own level.) A continuous flow is possible if the difference in water levels—hence the difference in water pressures—is maintained with the use of a suitable pump (Figure 8.18b).

A sustained electric current requires a suitable pumping device to maintain a difference in electric potential—to maintain a voltage. Chemical batteries or generators are "electrical pumps" that can maintain a steady flow of charge. These devices do work to pull negative charges apart from positive ones. In chemical batteries, this work may be done by the chemical disintegration of zinc or lead in acid, and the energy stored in the chemical bonds is converted to electric potential energy.

Generators separate charge by electromagnetic induction, a process we will describe in the next chapter. The work that is done (by whatever means) in separating the opposite charges is available at the terminals of the battery or generator. This energy per charge provides the difference in potential (voltage) that provides the "electrical pressure" to move electrons through a circuit joined to those terminals.

8.6 Electric Current

EXPLAIN THIS What kinds of current are produced by a battery and by a generator?

Just as a water current is a flow of H₂O molecules, **electric current** is a flow of charged particles. In circuits of metal wires, electrons make up the flow of charge. One or more electrons from each metal atom are free to move throughout the atomic lattice. These charge carriers are called *conduction electrons*. Protons, on the other hand, do not move in a solid because they are bound within the nuclei of atoms that are more or less locked in fixed positions. In fluids, however, positive ions as well as electrons may constitute the flow of an electric charge.

An important difference between water flow and electron flow has to do with their conductors. If you purchase a water pipe at a hardware store, the clerk doesn't sell you the water to flow through it. You provide that yourself.

By contrast, when you buy "an electron pipe," an electric wire, you also get the electrons. Every bit of matter, wires included, contains enormous numbers of electrons that swarm about in random directions. When a source of voltage sets them moving, we have an electric current.

The *rate* of electrical flow is measured in *amperes*. An **ampere** is the rate of flow of 1 coulomb of charge per second. (That's a flow of 6.25 billion billion electrons per second.) In a wire that carries 4 amperes to a car headlight bulb, for example, 4 C of charge flows past any cross-section in the wire each second. In a wire that carries 8 amperes, twice as many coulombs flow past any cross-section each second.

The speed of electrons as they drift through a wire is surprisingly slow. This is because electrons continually bump into atoms in the wire. The net speed, or *drift speed,* of electrons in a typical circuit is much less than 1 cm/s. The electric signal, however, travels at nearly the speed of light. That's the speed at which the electric *field* in the wire is established.

Also interesting is that a current-carrying wire has almost no net charge. Under ordinary conditions, there are as many conduction electrons swarming through the atomic lattice as there are positively charged atomic nuclei. The numbers of electrons and protons balance, so whether a wire carries a current or not, the net charge of the wire is normally zero at every moment.

FIGURE 8.20
Each coulomb of charge that is made to flow in a circuit that connects the ends of this 1.5-V flashlight cell is energized with 1.5 J.

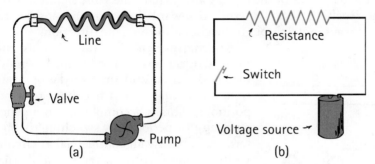

FIGURE 8.21
Analogy between (a) a simple hydraulic circuit and (b) an electric circuit.

There is often some confusion between charge flowing *through* a circuit and voltage placed, or impressed, *across* a circuit. We can distinguish between these ideas by considering a long pipe filled with water. Water flows through the pipe if there is a difference in pressure across (or between) its ends. Water flows from the high-pressure end to the low-pressure end. Only the water flows, not the pressure. Similarly, electric charge flows because of the differences in electrical pressure (voltage). You say that *charges* flow through a circuit because of an applied voltage across the circuit. You don't say that *voltage* flows through a circuit. Voltage doesn't go anywhere, for it is the charges that move. Voltage produces current (if there is a complete circuit).

FIGURE 8.22
The electric field lines between the terminals of a battery are directed through a conductor, which joins the terminals. A thick metal wire is shown here, but the path from one terminal to the other is usually an electric circuit. (If you touch this conducting wire, you won't be shocked, but the wire will heat quickly and may burn your hand!)

HISTORY OF 110 VOLTS

In the early days of electrical lighting, high voltages burned out electric light filaments, so low voltages were more practical. The hundreds of power plants built in the United States prior to 1900 adopted 110 V (or 115 or 120 V) as their standard. The tradition of 110 V was decided upon because it made the bulbs of the day glow as brightly as a gas lamp. By the time electrical lighting became popular in Europe, engineers had figured out how to make lightbulbs that would not burn out so fast at higher voltages. Power transmission is more efficient at higher voltages, so Europe adopted 220 V as its standard. The U.S. remained with 110 V (today, it is officially 120 V) because of the initial huge expense in the installation of 110-V equipment. Interestingly, in ac circuits 120 V is the *root-mean-square* average of the voltage. The actual voltage in a 120-V ac circuit varies between +170 V and −170 volts, delivering the same power to an iron or a toaster as a 120-V dc circuit.

Mastering**PHYSICS**
VIDEO: Alternating Current

FIGURE 8.23
Time graphs of dc and ac.

LEARNING OBJECTIVE
Relate the length and width of wires to electrical resistance.

Direct Current and Alternating Current

Electric current may be direct or alternating. **Direct current (dc)** refers to charges flowing in one direction. A battery produces direct current in a circuit because the terminals of the battery always have the same sign. Electrons move from the repelling negative terminal toward the attracting positive terminal, and they always move through the circuit in the same direction.

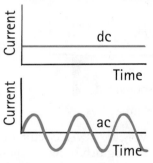

Alternating current (ac) acts as the name implies. Electrons in the circuit are moved first in one direction and then in the opposite direction, alternating to and fro about relatively fixed positions. This is accomplished in a generator or alternator by periodically switching the sign at the terminals. Nearly all commercial ac circuits in the United States involve currents that alternate back and forth at a frequency of 60 cycles per second. This is 60-hertz current [one cycle per second is called a *hertz* (Hz)]. In many countries, 50-Hz current is used. Throughout the world, most residential and commercial circuits are ac because electric energy in the form of ac can easily be stepped up to high voltage to be transmitted great distances with small heat losses, then stepped down to convenient voltages where the energy is consumed. Why this occurs is quite fascinating, and it will be touched on in the next chapter. The rules of electricity in this chapter apply to both dc and ac.

8.7 Electrical Resistance

EXPLAIN THIS What distinguishes a conductor from a *superconductor*?

How much current is in a circuit depends not only on voltage but also on the **electrical resistance** of the circuit. Just as narrow pipes resist water flow more than wide pipes, thin wires resist electric current more than thicker wires. And length contributes to resistance also. Just as long pipes have more resistance than short ones, long wires offer more electrical resistance. And most important is the material from which the wires were made. Copper has a low electrical resistance, while a strip of rubber has an enormous resistance. Temperature also affects electrical resistance. The greater the jostling of atoms within a conductor (the higher the temperature), the greater its resistance.

The resistance of some materials reaches zero at very low temperatures. These materials are referred to as *superconductors*.

Electrical resistance is measured in units called *ohms*. The Greek letter *omega*, Ω, is commonly used as the symbol for the ohm. This unit was named after Georg Simon Ohm, a German physicist who, in 1826, discovered a simple and very important relationship among voltage, current, and resistance.

Superconductors

In common household wiring, flowing electrons collide with atomic nuclei in the wire and convert their kinetic energy to thermal energy in the wire. Early-20th-century investigators discovered that certain metals in a bath of liquid helium at 4 K lost all electrical resistance. The electrons in these conductors traveled pathways that avoided atomic collisions, permitting them to flow indefinitely. These materials are called **superconductors**, having zero electrical resistance to the flow of charge. No current is lost and no heat is generated in superconductivity. For decades, it was generally thought that zero electrical resistance could occur only in certain metals near absolute zero. Then, in 1986, superconductivity was achieved at 30 K, which spurred hopes of finding superconductivity above 77 K, the point at which nitrogen liquefies. Nitrogen is easier to handle than liquid helium, which is needed for creating colder conditions. The historic leap came in the following year with a nonmetallic compound that lost its resistance at 90 K.

Various ceramic oxides have since been found to be superconducting at temperatures above 100 K. These ceramic materials are "high-temperature" superconductors. High-temperature superconductor (HTS) cables, already in use, carry more current at a lower voltage, which means large power transformers can be located farther away from urban centers—allowing the development of green space. Watch for additional growth of HTS cables in delivering electric power.

8.8 Ohm's Law

EXPLAIN THIS What is the source of electrons in a body undergoing electric shock?

The relationship among voltage, current, and resistance is summarized by a statement called **Ohm's law**. Ohm discovered that the amount of current in a circuit is directly proportional to the voltage established across the circuit and is inversely proportional to the resistance of the circuit:

$$\text{Current} = \frac{\text{voltage}}{\text{resistance}}$$

Or, in units form,

$$\text{Amperes} = \frac{\text{volts}}{\text{ohms}}$$

So, for a given circuit of constant resistance, current and voltage are proportional to each other.* This means we'll get twice the current for twice the voltage. The greater the voltage, the greater the current. But if the resistance is

---

* Many texts use *V* as the symbol for voltage, *I* for current, and *R* for resistance, and express Ohm's law as $V = IR$. It then follows that $I = V/R$, or $R = V/I$, so that, if any two variables are known, the third can be found. (The names of the units are often abbreviated: V for volts, A for amperes, and Ω (the capital Greek letter omega) for ohms.)

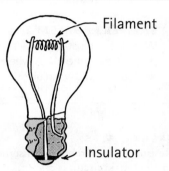

Filament

Insulator

FIGURE 8.24
The conduction electrons that surge to and fro in the filament of this incandescent lamp do not come from the voltage source. They are within the filament to begin with. The voltage source simply provides them with surges of energy. When switched on, the very thin tungsten filament heats up to 3000°C and roughly doubles its resistance.

LEARNING OBJECTIVE
Relate current, voltage, and resistance in electric circuits.

Current is a flow of charge, pressured into motion by voltage and hampered by resistance.

The unit of electrical resistance is the ohm, Ω. Like the song of old, "Ω, Ω on the Range."

FIGURING PHYSICAL SCIENCE

Problem Solving

SAMPLE PROBLEM 1

How much current flows through a lamp with a resistance of 60 Ω when the voltage across the lamp is 12 V?

Solution:
From Ohm's law:

$$\text{Current} = \frac{\text{voltage}}{\text{resistance}} = \frac{12 \text{ V}}{60 \text{ }\Omega} = 0.2 \text{ A}$$

SAMPLE PROBLEM 2

What is the resistance of a toaster that draws a current of 12 A when connected to a 120-V circuit?

Solution:
Rearranging Ohm's law:

$$\text{Resistance} = \frac{\text{voltage}}{\text{current}} = \frac{120 \text{ V}}{12 \text{ A}} = 10 \text{ }\Omega$$

SAMPLE PROBLEM 3

At 100,000 Ω, how much current flows through your body if you touch the terminals of a 12-V battery?

Solution:

$$\text{Current} = \frac{\text{voltage}}{\text{resistance}} = \frac{12 \text{ V}}{100,000 \text{ }\Omega}$$
$$= 0.00012 \text{ A}$$

SAMPLE PROBLEM 4

If your skin is very moist, so that your resistance is only 1000 Ω, and you touch the terminals of a 12-V battery, how much current do you receive?

Solution:

$$\text{Current} = \frac{\text{voltage}}{\text{resistance}} = \frac{12 \text{ V}}{1000 \text{ }\Omega}$$
$$= 0.012 \text{ A}$$

Ouch!

FIGURE 8.25
Resistors. The symbol of resistance in an electric circuit is ‑‑⋀⋀‑‑.

Mastering**PHYSICS**
VIDEO: Ohm's Law

FYI The gas inside an incandescent lightbulb is a mixture of nitrogen and argon. As the tungsten filament is heated, minute particles of tungsten evaporate—much like steam leaving boiling water. Over time, these particles are deposited on the inner surface of the glass, causing the bulb to blacken. Losing its tungsten, the filament eventually breaks and the bulb "burns out." A remedy is to replace the air inside the bulb with a halogen gas, such as iodine or bromine. Then the evaporated tungsten combines with the halogen rather than depositing on the glass, which remains clear. Furthermore, the halogen-tungsten combination splits apart when it touches the hot filament, returning halogen as a gas while restoring the filament by depositing tungsten back onto it. This is why halogen lamps have such long lifetimes.

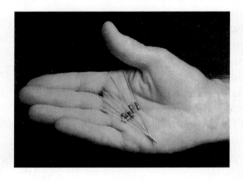

doubled for a circuit, the current is half what it would have been otherwise. The greater the resistance, the smaller the current. Ohm's law makes good sense.

The resistance of a typical lamp cord is much less than 1 Ω, and a typical lightbulb has a resistance of more than 100 Ω. An iron or electric toaster has a resistance of 15 to 20 Ω. The current inside these and all other electrical devices is regulated by circuit elements called resistors (Figure 8.25), whose resistance may be a few ohms or millions of ohms. Resistors heat up when current flows through them, but for small currents the heating is slight.

Electric Shock

The damaging effects of shock are the result of current passing through the human body. What causes electric shock in the body—current or voltage? From Ohm's law, we can see that this current depends on the voltage that is applied and also on the electrical resistance of the human body. The resistance of one's body depends on its condition, and it ranges from about 100 Ω, if it is soaked with salt water, to about 500,000 Ω, if the skin is very dry. If we touch the two electrodes of a battery with dry fingers, completing the circuit from one hand to the other, we offer a resistance of about 100,000 Ω. We usually cannot feel 12 V, and 24 V just barely tingles. If our skin is moist, 24 V can be quite uncomfortable. Table 8.1 describes the effects of different amounts of current on the human body.

In order for you to receive a shock, there must be a *difference* in electric potential between one part of your body and another part. Most of the current passes along the path of least electrical resistance connecting these two points. Suppose you fall from a bridge and manage to grab a high-voltage power line, halting your fall. So long as you touch nothing else of different potential, you

TABLE 8.1	EFFECT OF ELECTRIC CURRENTS ON THE BODY
Current	**Effect**
0.001 A	Can be felt
0.005 A	Is painful
0.010 A	Causes involuntary muscle contractions (spasms)
0.015 A	Causes loss of muscle control
0.070 A	Goes through the heart; causes serious disruption; probably fatal if current lasts for more than 1 s

receive no shock at all. Even if the wire is a few thousand volts above ground potential and you hang by it with two hands, no appreciable charge flows from one hand to the other. This is because there is no appreciable difference in electric potential between your hands. If, however, you reach over with one hand and grab a wire of different potential . . . *zap*! We have all seen birds perched on high-voltage wires. Every part of their bodies is at the same high potential as the wire, so they feel no ill effects.

You can prove something to be unsafe, but you can never prove something to be completely safe.

FIGURE 8.26
The bird can stand harmlessly on one wire of high potential, but it had better not reach over and touch a neighboring wire! Why not?

FIGURE 8.27
The third prong connects the body of the appliance directly to ground. Any charge that builds up on an appliance is therefore conducted to the ground.

Interestingly, the source of electrons in the current that shocks you is your own body. As in all conductors, the electrons are already there. It is the energy given to the electrons that you should be wary of. They are energized when a voltage difference exists across different parts of your body.

Most electric plugs and sockets today are wired with three, instead of two, connections. The principal two flat prongs on an electrical plug are for the current-carrying double wire, one part "live" and the other neutral, while the third round prong is grounded—connected directly to the ground (Figure 8.27). Appliances such as irons, stoves, washing machines, and dryers are connected with these three wires. If the live wire accidentally comes into contact with the metal surface of the appliance, and you touch the appliance, you could receive a dangerous shock. This won't occur when the appliance casing is grounded via the ground wire, which ensures that the appliance casing is at zero ground potential.

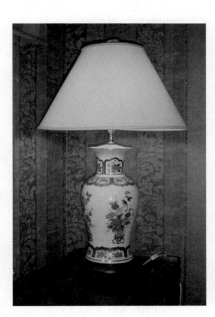

FIGURE 8.28
This table lamp has an insulating body and doesn't need the third (ground) wire.

CHECKPOINT
What causes electric shock: current or voltage?

Was this your answer?
Electric shock *occurs* when current is produced in the body, but the current is *caused* by an impressed voltage.

FYI *Myth:* Lightning never strikes the same place twice. *Fact:* Lightning does favor certain spots, mainly high locations. The Empire State Building is struck by lightning about 25 times every year.

INJURY BY ELECTRIC SHOCK

Many people are killed each year by current from common 120-V electric circuits. If your hand touches a faulty 120-V light fixture while your feet are on the ground, there's likely a 120-V "electrical pressure" between your hand and the ground. Resistance to current is usually greatest between your feet and the ground, and so the current is usually not enough to do serious harm. But if your feet and the ground are wet, there is a low-resistance electrical path between you and the ground. The 120 V across this lowered resistance may produce a harmful current in your body.

Pure water is not a good conductor. But the ions that are normally found in water make it a fair conductor. Dissolved materials in water, especially small quantities of salt, lower the resistance even more. There is usually a layer of salt remaining on your skin from perspiration, which, when wet, lowers your skin resistance to a few hundred ohms or less. Handling electrical devices while taking a bath is a definite no-no.

Injury by electric shock occurs in three forms: (1) burning of tissues by heating, (2) contraction of muscles, and (3) disruption of cardiac rhythm. These conditions are caused by the delivery of

excessive power for too long a time in critical regions of the body.

Electric shock can upset the nerve center that controls breathing. In rescuing shock victims, the first thing to do is remove them from the source of the electricity. Use a dry wooden stick or some other nonconductor so that you don't get electrocuted yourself. Then apply artificial respiration. It is important to continue artificial respiration. There have been cases of victims of lightning who did not breathe without assistance for several hours, but who were eventually revived and who completely regained good health.

LEARNING OBJECTIVE
Distinguish between series and parallel circuits.

8.9 Electric Circuits

EXPLAIN THIS How can a circuit be connected so that the current in each part is the same?

Any path along which electrons can flow is a *circuit*. For a continuous flow of electrons, there must be a complete circuit with no gaps. A gap is usually provided by an electric switch that can be opened or closed to either cut off energy or allow energy to flow. Most circuits have more than one device that receives electric energy. These devices are commonly connected in a circuit in one of two ways: in *series* or in *parallel*. When connected in series, they form a single pathway for electron flow between the terminals of the battery, generator, or wall outlet (which is simply an extension of these terminals). When connected in parallel, they form branches, each of which is a separate path for the flow of electrons. Both series and parallel connections have their own distinctive characteristics. In the following sections, we shall briefly discuss circuits using these two types of connections.

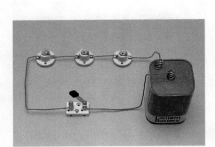

Series Circuits

A simple **series circuit** is shown in Figure 8.29. Three lamps are connected in series with a battery. The same current exists almost immediately in all three lamps when the switch is closed. The current does not "pile up" or accumulate in any lamp but flows *through* each lamp. Electrons that make up this current leave the negative terminal of the battery, pass through each of the resistive filaments in the lamps in turn, and then return to the positive terminal of the battery. (The same amount of current passes through the battery.) This is the only path of the electrons through the circuit. A break anywhere in the path results in an open circuit, and the flow of electrons ceases. Such a break occurs when the switch is opened, when the wire is accidentally cut, or when one of the lamp filaments burns out.

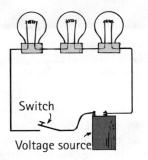

Switch

Voltage source

FIGURE 8.29
INTERACTIVE FIGURE (MP)

A simple series circuit. The 6-V battery provides 2 V across each lamp.

The circuit shown in Figure 8.29 illustrates the following characteristics of series connections:

1. Electric current has a single pathway through the circuit. This means that the current passing through the resistance of each electrical device along the pathway is the same.

2. This current is resisted by the resistance of the first device, the resistance of the second, and that of the third, so the total resistance to current in the circuit is the sum of the individual resistances along the circuit path.

3. The current in the circuit is numerically equal to the voltage supplied by the source divided by the total resistance of the circuit. This is in accord with Ohm's law.

4. The total voltage impressed across a series circuit divides among the individual electrical devices in the circuit so that the sum of the "voltage drops" across the resistance of each individual device is equal to the total voltage supplied by the source. This characteristic follows from the fact that the amount of energy given to the total current is equal to the sum of energies given to each device.

5. The voltage drop across each device is proportional to its resistance. This follows from Ohm's law expressed in the form $V = IR$. For constant current I, the voltage V is directly proportional to the resistance R.

> **CHECKPOINT**
> 1. **What happens to the current in the other lamps if one lamp in a series circuit burns out?**
> 2. **What happens to the brightness of each lamp in a series circuit when more lamps are added to the circuit?**

Were these your answers?
1. If one of the lamp filaments burns out, the path connecting the terminals of the voltage source breaks and current ceases. All lamps go out.
2. Adding more lamps in a series circuit produces a greater circuit resistance. This decreases the current in the circuit and therefore in each lamp, which causes dimming of the lamps. Energy is divided among more lamps, so the voltage drop across each lamp is less.

The rules above hold for ac or dc circuits. It is easy to see the main disadvantage of a series circuit: if one device fails, current in the entire circuit ceases. Some old Christmas tree lights are connected in series. When one bulb burns out, it's fun and games (or frustration) trying to locate which one to replace.

Most circuits are wired so that it is possible to operate several electrical devices at once, each independently of the other. In your home, for example, a lamp can be turned on or off without affecting the operation of other lamps or electrical devices. This is because these devices are connected not in series but in parallel with one another.

Parallel Circuits

A simple **parallel circuit** is shown in Figure 8.30. Three lamps are connected to the same two points, A and B. Electrical devices connected to the same two points of an electrical circuit are said to be *connected in parallel*.

FYI All batteries degrade. The lithium-ion cells popular in notebook computers, cameras, and cell phones erode faster when highly charged and warm. So keep yours at about half charge in a cool or cold environment to extend battery life.

We often think of current flowing through a circuit, but don't say this around somebody who is picky about grammar, for the expression "current flows" is redundant. More properly, charge flows—which is current.

FYI Batteries now deliver power to devices implanted in the human body. A number of approaches have been proposed to tap into the power or fuel sources the body already provides. Watch for their implementation in the near future.

Mastering**PHYSICS**
TUTORIAL: Electricity and Circuits
VIDEO: Electric Circuits

FYI After failing more than 6000 times before perfecting the first electric lightbulb, Thomas Edison stated that his trials were not failures, for he successfully discovered 6000 ways that don't work.

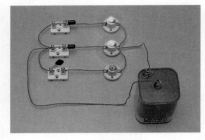

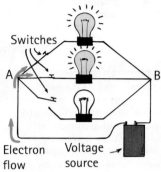

Switches

A — B

Electron flow · Voltage source →

FIGURE 8.30
INTERACTIVE FIGURE

A simple parallel circuit. A 6-V battery provides 6 V across the top two lamps.

FIGURE 8.31
New Zealand physics instructor David Housden constructs a parallel circuit by fastening lamps to extended terminals of a common battery. He asks his class to predict the relative brightnesses of two identical lamps in one wire about to be connected in parallel.

In a parallel circuit, *most* current travels in the path of least resistance—but not all. *Some* current travels in each path.

Electrons leaving the negative terminal of the battery need travel through only one lamp filament before returning to the positive terminal of the battery. In this case, current branches into three separate pathways from A to B. A break in any one path does not interrupt the flow of charge in the other paths. Each device operates independently of the other devices (whether the circuit is ac or dc).

The circuit shown in Figure 8.30 illustrates the following major characteristics of parallel connections:

1. Each device connects the same two points, A and B, of the circuit. The voltage is therefore the same across each device.

2. The total current in the circuit divides among the parallel branches. Because the voltage across each branch is the same, the amount of current in each branch is inversely proportional to the resistance of the branch.

3. The total current in the circuit equals the sum of the currents in its parallel branches.

4. As the number of parallel branches is increased, the overall resistance of the circuit is *decreased*. Overall resistance is lowered with each added path between any two points of the circuit. This means the overall resistance of the circuit is less than the resistance of any one of the branches.

CHECKPOINT

1. **What happens to the current in the other lamps if one of the lamps in a parallel circuit burns out?**

2. **What happens to the brightness of each lamp in a parallel circuit when more lamps are added in parallel to the circuit?**

Were these your answers?

1. If one lamp burns out, the other lamps are unaffected. The current in each branch, according to Ohm's law, is equal to voltage/resistance, and because neither voltage nor resistance is affected in the other branches, the current in those branches is unaffected. The total current in the overall circuit (the current through the battery), however, is decreased by an amount equal to the current drawn by the lamp in question before it burned out. But the current in any other single branch is unchanged.

2. The brightness of each lamp is unchanged as other lamps are introduced (or removed). Only the total resistance and total current in the total circuit changes, which is to say that the current in the battery changes. (There is resistance in a battery also, which we assume is negligible here.) As lamps are introduced, more paths are available between the battery terminals, which effectively decreases total circuit resistance. This decreased resistance is accompanied by an increased current, the same increase that feeds energy to the lamps as they are introduced. Although changes of resistance and current occur for the circuit as a whole, no changes occur in any individual branch in the circuit.

Parallel Circuits and Overloading

Electricity is usually fed into a home by way of two wires called *lines*. These lines are very low in resistance and are connected to wall outlets in each room—sometimes through two or more separate circuits. An electric potential of about

ELECTRIC ENERGY AND TECHNOLOGY

Try to imagine everyday home life before the advent of electric energy. Think of homes without electric lights, refrigerators, heating and cooling systems, telephones, and radio and TV. We may romanticize a better life without these, but only if we overlook the many hours of daily toil devoted to laundry, cooking, and heating homes. We'd also have to overlook how difficult it was to reach a doctor in times of emergency before the advent of the telephone—when all the doctor had in his bag were laxatives, aspirins, and sugar pills—and when infant death rates were staggering.

We have become so accustomed to the benefits of technology that we are only faintly aware of our dependency on dams, power plants, mass transportation, electrification, modern medicine, and modern agricultural science for our very existence. When we enjoy a good meal, we give little thought to the technology that went into growing, harvesting, and delivering the food on our table. When we turn on a light, we give little thought to the centrally controlled power grid that links the widely separated power stations by long-distance transmission lines. These lines serve as the productive life force of industry, transportation, and the electrification of civilization. Anyone who thinks of science and technology as "inhuman" fails to grasp the ways in which they make our lives more human.

110 to 120 V ac is applied across these lines by a transformer in the neighborhood. (A transformer, as we shall see in the next chapter, is a device that steps down the higher voltage supplied by the power utility.) As more devices are connected to a circuit, more pathways for current result. This lowers the combined resistance of the circuit. Therefore, more current exists in the circuit, which is sometimes a problem. Circuits that carry more than a safe amount of current are said to be *overloaded*.

We can see how overloading occurs in Figure 8.32. The supply line is connected to a toaster that draws 8 amperes, a heater that draws 10 amperes, and a lamp that draws 2 amperes. When only the toaster is operating and drawing 8 amperes, the total line current is 8 amperes. When the heater is also operating, the total line current increases to 18 amperes (8 amperes to the toaster plus 10 amperes to the heater). If you turn on the lamp, the line current increases to 20 amperes. Connecting additional devices increases the current still more. Connecting too many devices into the same circuit results in overheating the wires, which can cause a fire.

Safety Fuses

To prevent overloading in circuits, fuses are connected in series along the supply line. In this way the entire line current must pass through the fuse. The fuse shown in Figure 8.33 is constructed with a wire ribbon that heats up and melts at a given current. If the fuse is rated at 20 amperes, it passes 20 amperes, but no more. A current above 20 amperes melts the fuse, which "blows out" and breaks the circuit. Before a blown fuse is replaced, the cause of overloading should be determined and remedied. Sometimes insulation that separates the wires in a circuit wears away and allows the wires to touch. This greatly reduces the resistance in the circuit and is called a *short circuit*.

In modern buildings, fuses have been largely replaced by circuit breakers, which use magnets or bimetallic strips to open a switch when the current is excessive. Utility companies use circuit breakers to protect their lines all the way back to the generators.

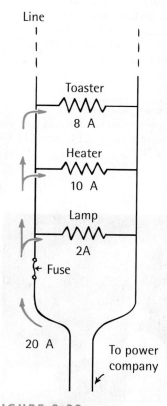

FIGURE 8.32
Circuit diagram for appliances connected to a household circuit.

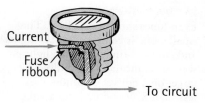

FIGURE 8.33
A safety fuse.

LEARNING OBJECTIVE
Relate current and voltage to power with their units of measurement.

8.10 Electric Power

EXPLAIN THIS Why shouldn't you connect a 120-V hairdryer to 240 V?

The moving charges in an electric current do work. This work, for example, can heat a circuit or turn a motor. The rate at which work is done—that is, the rate at which electric energy is converted into another form, such as mechanical energy, heat, or light—is called **electric power**. Electric power is equal to the product of current and voltage:*

$$\text{Power} = \text{current} \times \text{voltage}$$

If the voltage is expressed in volts and the current in amperes, then the power is expressed in watts. So, in units form,

$$\text{Watts} = \text{amperes} \times \text{volts}$$

The relationship between energy and power is a practical matter. From the definition, power = energy per unit time, it follows that energy = power × time. So an energy unit can be a power unit multiplied by a time unit, such as kilowatt-hours (kWh). One kilowatt-hour is the amount of energy transferred in 1 h at the rate of 1 kW. Therefore in a locality in which electric energy costs 15¢/kWh, a 1000-W clothes iron can operate for 1 h at the cost of 15¢. A refrigerator, typically rated at around 500 W, costs less for an hour, but much more over the course of a month.

FIGURE 8.34
The power and voltage on a compact fluorescent lamp (CFL) read "13 W 120 V."

An incandescent bulb rated at 60 W draws a current of 0.5 A (60 W = 0.5 A × 120 V). A 100-W bulb draws about 0.8 A. Figure 8.34 shows a compact fluorescent lamp (CFL) that fits into a standard lightbulb socket. Interestingly, a 26-W CFL provides about the same amount of light as a 100-W incandescent bulb—only one-quarter of the power for the same light!** In addition to significantly greater efficiencies, CFLs also have increased bulb lifetimes.† Incandescent bulbs are now being replaced by CFLs.

A longer-lasting light source is the light-emitting diode (LED), the most primitive being the little red lights that tell you whether your electronic devices are on or off. Between CFLs and LEDs, watch for common incandescent bulbs to be history.

FIGURE 8.35
Evan Jones shows two LEDs. The smaller one is common in flashlights and emits 15 times as much light per watt as an incandescent bulb. The larger one, not yet common, uses less than 8 W and replaces a same-size incandescent 60-W bulb.

* Recall from Chapter 3 that *power = work/time*; 1 W = 1 J/s. Note that the units for mechanical power and electric power agree (work and energy are both measured in joules):

$$\text{Power} = \frac{\cancel{\text{charge}}}{\text{time}} \times \frac{\text{energy}}{\cancel{\text{charge}}} = \frac{\text{energy}}{\text{time}}$$

** It turns out that the power formula $P = IV$ doesn't apply to CFLs because the alternating voltage and current are out of step with each other (out of phase), and the product of current and voltage is larger than the actual power consumption. How much larger? Check the printed data at the base of a CFL to find out.
† A downside to CFLs is the trace amounts of mercury sealed in their glass tubing, some 4 mg. But the single largest source of mercury emissions in the environment is coal-fired power plants. According to the EPA, when coal power is used to illuminate a single incandescent lamp, more mercury is released into the air than exists in a comparably luminous CFL.

FIGURING PHYSICAL SCIENCE

Problem Solving

SAMPLE PROBLEM 1

If a 120-V line to a socket is limited to 15 A by a safety fuse, will it operate a 1200-W hair dryer?

Solution:

Yes. From the expression *watts = amperes × volts*, we can see that current = 1200 W/120 V = 10 A, so the hair dryer will operate when connected to the circuit. But two hair dryers on the same circuit will blow the fuse.

SAMPLE PROBLEM 2

At 30¢/kWh, what does it cost to operate the 1200-W hair dryer for 1 h?

Solution:

1200 W = 1.2 kW; 1.2 kW × 1 h × 30¢/1 kWh = 36¢.

MAGNETIC THERAPY*

Back in the 18th century, a celebrated "magnetizer" from Vienna, Franz Mesmer, brought his magnets to Paris and established himself as a healer in Parisian society. He healed patients by waving magnetic wands above their heads.

At that time, Benjamin Franklin, the world's leading authority on electricity, was visiting Paris as a U.S. representative. He suspected that Mesmer's patients did benefit from his ritual, but only because it kept them away from the bloodletting practices of other physicians. At the urging of the medical establishment, King Louis XVI appointed a royal commission to investigate Mesmer's claims. The commission included Franklin and Antoine Lavoisier, the founder of modern chemistry. The commissioners designed a series of tests in which some subjects thought they were receiving Mesmer's treatment when they weren't, while others received the treatment but were led to believe they had not. The results of these blind experiments established beyond any doubt that Mesmer's success was due solely to the power of suggestion. To this day, the report is a model of clarity and reason. Mesmer's reputation was destroyed, and he retired to Austria.

Now some two hundred years later, with increased knowledge of magnetism and physiology, hucksters of magnetism are attracting even larger followings. But there is no government commission of Franklins and Lavoisiers to challenge their claims. Instead,

magnetic therapy is another of the untested and unregulated "alternative therapies" given official recognition by Congress in 1992.

Although testimonials about the benefits of magnets are many, there is no scientific evidence whatever for magnets boosting body energy or combating aches and pains. None. Yet millions of therapeutic magnets are sold in stores and catalogs. Consumers are buying magnetic bracelets, insoles, wrist and knee bands, back and neck braces, pillows, mattresses, lipstick, and even water. They are told that magnets have powerful effects on the body, mainly increasing blood flow to injured areas. The idea that blood is attracted by a magnet is bunk. Although blood protein is weakly diamagnetic and is repulsed by magnetic fields, the magnets used in magnetic therapy are much too weak to have any measurable effects on blood flow. Furthermore, most therapeutic magnets are of the refrigerator type, with a very limited range. To get an idea of how quickly the field of these magnets drops off, see how many sheets of paper one of these magnets will hold on a refrigerator or any iron surface. The magnet will fall off after a few sheets of paper separate it from the iron surface. The field doesn't extend much more than one millimeter, and it wouldn't penetrate the skin, let alone into muscles. And even if it did, there is no scientific evidence that magnetism has any beneficial effects on

the body at all. But, again, testimonials are another story.

Sometimes an outrageous claim has some truth to it. For example, the practice of bloodletting in previous centuries was, in fact, beneficial to a small percentage of men. These men suffered the genetic disease *hemochromatosis*, excess iron in the blood (women were less afflicted partly due to menstruation). Although the number of men who benefited from bloodletting was small, testimonials of its success prompted the widespread practice that killed many.

No claim is so outrageous that testimonials can't be found to support it. Claims that the Earth is flat or claims for the existence of flying saucers are quite harmless and may amuse us. Magnetic therapy may likewise be harmless for many ailments, but not when it is used to treat a serious disorder in place of modern medicine. Pseudoscience may be promoted to intentionally deceive or it may be the result of flawed and wishful thinking. In either case, pseudoscience is very big business. The market is enormous for therapeutic magnets and other such fruits of unreason.

Scientists must keep open minds, must be prepared to accept new findings, and must be ready to be challenged by new evidence. But scientists also have a responsibility to inform the public when they are being deceived and, in effect, robbed by pseudoscientists whose claims are without substance.

* *Adapted from Voodoo Science: The Road from Foolishness to Fraud,* by Robert L. Park; Oxford University Press, 2000.

SUMMARY OF TERMS (KNOWLEDGE)

Alternating current (ac) An electric current that repeatedly reverses its direction; the electric charges vibrate about relatively fixed points. In the United States, the vibrational rate is 60 Hz.

Ampere The unit of electric current; the rate of flow of 1 coulomb of charge per second.

Conductor Any material having free charged particles that easily flow through it when an electrical force acts on them.

Coulomb The SI unit of electric charge. One coulomb (symbol C) is equal in magnitude to the total charge of 6.25×10^{18} electrons.

Coulomb's law The relationship among electrical force, charge, and distance: If the charges are alike in sign, the force is repelling; if the charges are unlike, the force is attractive.

Direct current (dc) An electric current flowing in one direction only.

Electric current The flow of electric charge that transports energy from one place to another.

Electric field Defined as force per unit charge, it can be considered an energetic aura surrounding charged objects. About a charged point, the field decreases with distance according to the inverse-square law, like a gravitational field. Between oppositely charged parallel plates, the electric field is uniform.

Electric potential The electric potential energy per amount of charge, measured in volts and often called *voltage*.

Electric potential energy The energy a charge possesses by virtue of its location in an electric field.

Electric power The rate of energy transfer, or the rate of doing work; the amount of energy per unit time, which can be measured by the product of current and voltage:

$$\text{Power} = \text{current} \times \text{voltage}$$

It is measured in watts (or kilowatts), where $1 \text{ A} \times 1 \text{ V} = 1 \text{ W}$.

Electrical resistance The property of a material that resists the flow of an electric current through it, measured in ohms (Ω).

Electrically polarized Term applied to an atom or molecule in which the charges are aligned so that one side has a slight excess of positive charge and the other side a slight excess of negative charge.

Ohm's law The current in a circuit varies in direct proportion to the potential difference or voltage and inversely with the resistance:

$$\text{Current} = \frac{\text{voltage}}{\text{resistance}}$$

A current of 1 A is produced by a potential difference of 1 V across a resistance of $1 \text{ }\Omega$.

Parallel circuit An electric circuit with two or more devices connected in such a way that the same voltage acts across each one, and any single one completes the circuit independently of all the others.

Potential difference The difference in potential between two points, measured in volts and often called *voltage difference*.

Series circuit An electric circuit with devices connected in such a way that the current is the same in each device.

Superconductor Any material with zero electrical resistance, in which electrons flow without losing energy and without generating heat.

READING CHECK QUESTIONS (COMPREHENSION)

8.1 Electric Charge

1. Which part of an atom is positively charged, and which part is negatively charged?
2. How does the charge of one electron compare with the charge of another electron?
3. How do the masses of electrons compare with the masses of protons?
4. How does the number of protons in the atomic nucleus normally compare with the number of electrons that orbit the nucleus?
5. What kind of charge does an object acquire when electrons are stripped from it?
6. What is meant by saying that charge is *conserved*?

8.2 Coulomb's Law

7. How is Coulomb's law similar to Newton's law of gravitation? How is it different?

8. How does a coulomb of charge compare with the charge of a single electron?

9. How does the magnitude of electrical force between a pair of charged particles change when the particles are moved twice as far apart? Three times as far apart?

10. How does an electrically polarized object differ from an electrically charged object?

8.3 Electric Field

11. Give two examples of common force fields.

12. How is the direction of an electric field defined?

8.4 Electric Potential

13. Distinguish between electric potential energy and electric potential in terms of units of measurement.

14. A balloon may easily be charged to several thousand volts. Does that mean it has several thousand joules of energy? Explain.

8.5 Voltage Sources

15. What condition is necessary for a sustained flow of electric charge through a conducting medium?

16. How much energy is given to each coulomb of charge passing through a 6-V battery?

8.6 Electric Current

17. Does electric charge flow *across* a circuit or *through* a circuit? Does voltage flow *across* a circuit or is it *impressed across* a circuit?

18. Distinguish between dc and ac.

19. Does a battery produce dc or ac? Does the generator at a power station produce dc or ac?

8.7 Electrical Resistance

20. Which has greater resistance: a thick wire or a thin wire of the same length?

21. What is the unit of electrical resistance?

8.8 Ohm's Law

22. What is the effect on current through a circuit of steady resistance when the voltage is doubled? When both voltage and resistance are doubled?

23. Which has greater electrical resistance: wet skin or dry skin?

24. What is the function of the third prong on the plug of an electrical appliance?

25. What is the source of electrons that produces a shock when you touch a charged conductor?

8.9 Electric Circuits

26. In a circuit consisting of two lamps connected in series, if the current in one lamp is 1 A, what is the current in the other lamp?

27. If 6 V were impressed across the circuit in Question 26, and the voltage across the first lamp were 2 V, what would be the voltage across the second lamp?

28. How does the total current through the branches of a parallel circuit compare with the current through the voltage source?

29. As more lines are opened at a fast-food restaurant, the resistance to the motion of people trying to get served is reduced. How is this similar to what happens when more branches are added to a parallel circuit?

8.10 Electric Power

30. What is the relationship among electric power, current, and voltage?

ACTIVITIES (HANDS-ON APPLICATION)

31. Write a letter to your favorite uncle and bring him up to speed on your progress with physics. Mention some of the terms in this chapter and tell how learning to distinguish among them contributes to your understanding. Relate the terms to practical examples.

32. Demonstrate charging by friction and discharging from pointed objects with a friend who stands at the far end of a carpeted room. Wearing your leather shoes, scuff your way across the rug until your noses are close together. This can be a delightfully tingling experience, depending on the dryness of the air and how pointed your noses are.

33. Briskly rub a comb against your hair or a woolen garment and then bring it near a small but smooth stream of running water. Is the stream of water charged? (Before you say yes, note the behavior of the stream when an opposite charge is brought nearby.)

34. A car battery is actually a series of cells. A single electric cell can be made by placing two plates of different materials that have different affinities for electrons in a conducting solution. You can make a simple 1.5-V cell by placing a strip of copper and a strip of zinc in a tumbler of salt water. The voltage of a cell depends on the materials used and the conducting solution they are placed in, not the size of the plates.

An easier cell to construct is the citrus cell. Stick a paper clip and a piece of copper wire into a lemon. Hold the ends of the wire close together, but not touching, and place the ends on your tongue. The slight tingle you feel and the metallic taste you experience result from a slight current of electricity pushed by the citrus cell through the wires when your moist tongue closes the circuit.

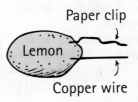

Paper clip

Lemon

Copper wire

PLUG AND CHUG (FORMULA FAMILIARIZATION)

$$\text{Coulomb's law: } F = k\frac{q_1 q_2}{d^2}$$

35. Two point charges each with 0.1 C of charge are 0.1 m apart. Knowing that k is 9×10^9 N·m^2/C^2 (the proportionality constant for Coulomb's law), show that the force between the charges is 9×10^9 N.

$$\text{Ohm's law: } I = \frac{V}{R}$$

36. A toaster has a heating element of 15 Ω and is connected to a 120-V outlet. Show that the current drawn by the toaster is 8 A.

37. When you touch your fingers (resistance 1000 Ω) to the terminals of a 6-V battery, show that the small current moving through your fingers is 0.006 A.

38. Calculate the current in the 240-Ω filament of a bulb connected to a 120-V line.

$$\text{Power} = IV$$

39. An electric toy draws 0.5 A from a 120-V outlet. Show that the toy consumes 60 W of power.

40. Calculate the power of a hair dryer that operates on 120 V and draws a current of 10 A.

THINK AND SOLVE (MATHEMATICAL APPLICATION)

41. Two pellets, each with a charge of 1 microcoulomb (10^{-6} C), are located 3 cm (0.03 m) apart. Show that the electric force between them is 10 N.

42. Two point charges are separated by 4 cm. The attractive force between them is 20 N. Show that when they are separated by 8 cm the force between them is 5 N. (Why can you solve this problem without knowing the magnitudes of the charges?)

43. If the charges attracting each other in the preceding problem have equal magnitudes, show that each charge has a magnitude of 1.9 microcoulombs (1.9×10^{-6} C).

44. A droplet of ink in an industrial ink-jet printer carries a charge of 1.6×10^{-10} C and is deflected onto paper by a force of 3.2×10^{-4} N. Show that the strength of the electric field ($E = F/q$) required to produce this force is 2×10^6 N/C.

45. A 12-V battery moves 4 C of charge from one terminal to the other. Show that the battery does 48 J of work.

46. If you expend 10 J of work to push a 1-C charged particle against an electric field, what will be its change of voltage? When the particle is released, what will be its kinetic energy as it flies past its starting position?

47. The potential difference between a storm cloud and the ground is 100 million volts. If a charge of 2 C flashes in a bolt from cloud to Earth, show that the change of potential energy of the charge is 2×10^8 J.

48. The current driven by voltage V in a circuit of resistance R is given by Ohm's law, $I = V/R$. Show that the resistance of a circuit carrying current I and driven by voltage V is given by the equation $R = V/I$.

49. The same voltage V is impressed on each of the branches of a parallel circuit. The voltage source provides a total current I_{total} to the circuit and "sees" a total equivalent resistance of R_{eq} in the circuit. That is, $V = I_{total}R_{eq}$. The total current is equal to the sum of the currents through each branch of the parallel circuit. In a circuit with n branches, $I_{total} = I_1 + I_2 + I_3 + \cdots + I_n$. Use Ohm's law ($I = V/R$) and show that the equivalent resistance of a parallel circuit with n branches is given by

$$\frac{1}{R_{eq}} = \frac{1}{R_1} + \frac{1}{R_2} + \frac{1}{R_3} + \cdots + \frac{1}{R_n}$$

50. The wattage marked on a lightbulb is not an inherent property of the bulb; rather, it depends on the voltage to which it is connected, usually 110 V or 120 V. Show that the current in a 300-W bulb connected in a 120-V circuit is 2.5 A.

51. Rearrange the formula current = voltage/resistance to express *resistance* in terms of current and voltage. Then consider the following: A certain device in a 120-V circuit has a current rating of 20 A. Show that the resistance of the device is 6 Ω.

52. Using the formula power = current × voltage, show that the current drawn by a 1200-W hair dryer connected to 120 V is 10 A. Then use the same method for the solution to the preceding problem and show that the resistance of the hair dryer is 12 Ω.

53. The power of an electric circuit is given by the formula $P = IV$. Use Ohm's law to express V and show that power can be expressed by the equation $P = I^2R$.

54. A dehumidifier with a resistance of 20 Ω draws 6.0 A when connected to an electrical outlet. Show that the power consumed by the appliance is 720 W.

55. An electric space heater dissipates 1320 W of power via electromagnetic radiation and heat when connected to 120 V. When the current is unknown, the power can be expressed as $P = V^2/R$. Use this formula to show that the resistance of the space heater is about 11 Ω.

56. The total charge that an automobile battery can supply without being recharged is given in ampere-hours. A typical 12-V battery has a rating of 60 ampere-hours (60 A for 1 h, 30 A for 2 h, and so on). Suppose that you forget to turn off the headlights in your parked automobile. If each of the two headlights draws 3 A, show that your battery will go dead in about 10 h.

57. Show that it costs 5¢ to operate a 25-W porch light for 24 h if electric energy costs 8¢/kWh.

58. Suppose you operate a 100-W lamp continuously for 1 week when the power utility rate is 8¢/kWh. Show that the cost is $1.34.

59. An electric dryer connected to a 120-V source draws 8.4 A of current. Show that the amount of heat generated in 1 min is about 60 kJ.

60. For the electric dryer of the preceding problem, show that the number of coulombs that flow through in 1 min is approximately 500 C.

61. An incandescent lightbulb with an operating resistance of 95 Ω is labeled "150 W." Is this bulb designed for use in a 120-V circuit or a 220-V circuit? Defend your answer.

62. In periods of peak demand, power companies lower their voltage in order to save them power (and save you money)! To see the effect, consider a 1200-W toaster that draws 10 A when connected to 120 V. Suppose the voltage is lowered by 10% to 108 V. By how much does the current decrease? By how much does the power decrease? (Caution: The 1200-W label is valid only when 120 V is applied. When the voltage is lowered, the resistance of the toaster, not its power, remains constant.)

THINK AND RANK (ANALYSIS)

63. The three pairs of metal, same-size spheres have different charges on their surfaces as indicated. Each pair is brought together, allowed to touch, and then separated. Rank from greatest to least the total amount of charge on the pairs of spheres after separation.

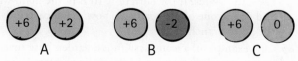

64. Rank the circuits according to the brightness of the identical bulbs, from brightest to dimmest.

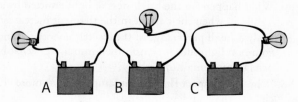

65. All the bulbs in the three circuits are identical. An ammeter is placed in different locations, as shown. Rank the current readings in the ammeter from highest to lowest.

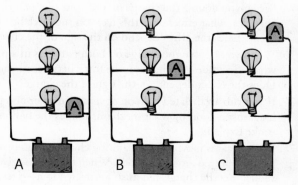

66. All bulbs are identical in the three circuits. An ammeter is connected next to each battery as shown. Rank the current readings in the ammeter from highest to lowest.

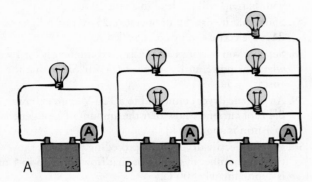

67. All bulbs are identical in the circuits shown. A voltmeter is connected across a single bulb to measure the voltage drop across it. Rank the voltage readings from highest to lowest.

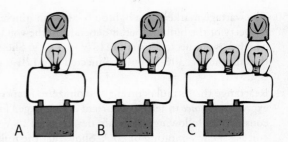

68. Consider the three parts of the circuit: (1) the top branch with two bulbs, (2) the lower branch with one bulb, and (3) the battery.

   (a) Rank the current through each part, from greatest to least.

   (b) Rank the voltage across each, from greatest to least.

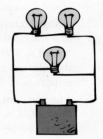

EXERCISES (SYNTHESIS)

69. At the atomic level, what is meant by saying something is electrically charged?

70. Why is charge usually transferred by electrons rather than by protons?

71. Why are objects with vast numbers of electrons normally not electrically charged?

72. Why do clothes often cling together after tumbling in a clothes dryer?

73. If electrons were positive and protons were negative, would Coulomb's law be written the same or differently?

74. When you double the distance between a pair of charged particles, what happens to the force between them? Does it depend on the sign of the charges? What law defends your answer?

75. When you double the charge on only one of a pair of particles, what effect does this have on the force between them? Does the effect depend on the sign of the charge?

76. When you double the charge on both particles in a pair, what effect does this have on the force between them? Does the effect depend on the sign of the charge?

77. If you rub an inflated balloon against your hair and place it against a door, by what mechanism does the balloon stick? Explain.

78. When a car is moved into a painting chamber, a mist of paint is sprayed around its body. When the body is given a sudden electric charge and mist is attracted to it—presto—the car is quickly and uniformly painted. What does the phenomenon of polarization have to do with this?

79. By what specific means do the bits of fine threads align in the electric fields in Figure 8.11?

80. Suppose that the strength of the electric field about an isolated point charge has a certain value at a distance of 1 m. How does the electric field strength compare at a distance of 2 m from the point charge? What law guides your answer?

81. Why is a good conductor of electricity also a good conductor of heat?

82. Why is voltage often referred to as an *electrical pressure*, especially when comparing electric circuits and water flow in pipes?

83. Consider a water pipe that branches into two smaller pipes. If the flow of water is 10 L/min in the main pipe and 4 L/min in one of the branches, how much water per minute flows in the other branch?

84. Consider a circuit with a main wire that branches into two other wires. If the current is 10 A in the main wire and 4 A in one of the branches, how much current is in the other branch?

85. One example of a water system is a garden hose that waters a garden. Another is the cooling system of an automobile. Which of these exhibits behavior more analogous to that of an electric circuit? Explain.

86. What happens to the brightness of light emitted by a lightbulb when the current in the filament increases?

87. Only a small percentage of the electric energy fed into a common lightbulb is transformed into light. What happens to the remaining energy?

88. Why are compact fluorescent lamps (CFLs) more efficient than incandescent lamps?

89. Which is less damaging: plugging a 110-V appliance into a 220-V circuit or plugging a 220-V appliance into a 110-V circuit? Explain.

90. If a current of one- or two-tenths of an ampere were to pass into one of your hands and out the other, you would probably be electrocuted. However, if the same current were to pass into your hand and out the elbow above the same hand, you could survive, even though the current might be large enough to burn your flesh. Explain.

91. Would you expect to find dc or ac in the filament of a lightbulb in your home? In the headlight of an automobile?

92. Are automobile headlights wired in parallel or in series? What is your evidence?

93. As more lanes are added to toll booths, the resistance to vehicles passing through is reduced. How is this similar to what happens when more branches are added to a parallel circuit?

94. Between current and voltage, (a) which remains the same for a 10-Ω and a 20-Ω resistor connected in a series circuit? (b) Which remains the same for a 10-Ω and a 20-Ω resistor connected in a parallel circuit?

95. Comment on the warning sign shown in the sketch.

DANGER!
HIGH RESISTANCE
(1 000 000 000 Ω)

96. What unit of measurement is represented by (a) joule per coulomb, (b) coulomb per second, and (c) watt-second?

97. What is the effect on the current in a wire if both the voltage across it and its resistance are doubled? If both are halved? Let Ohm's law guide your thinking.

98. An electroscope is a simple device consisting of a metal ball that is attached by a conductor to two thin leaves of metal foil protected from air disturbances in a jar, as shown. When the ball is touched by a charged body, the leaves that normally hang straight down spread apart. Why? (Electroscopes are useful not only as charge detectors but also for measuring the quantity of charge: the greater the charge transferred to the ball, the more the leaves diverge.)

99. The leaves of a charged electroscope collapse in time. At higher altitudes, they collapse more rapidly. Can you think of an explanation? (Hint: The existence of cosmic rays was first indicated by this observation.)

100. Suppose an investigator places both a free electron and then a free proton into an electric field between oppositely charged conducting plates.

    (a) How do the forces acting on the electron and proton compare?

    (b) How do their accelerations compare?

    (c) How do their directions of travel compare?

101. Is it correct to say that the energy from a car battery ultimately comes from fuel in the gas tank? Defend your answer.

102. Why are the wingspans of birds a consideration in determining the spacing between parallel wires on power poles?

103. If several bulbs are connected in series to a battery, they may feel warm to the touch even though they are not visibly glowing. What is your explanation?

104. A 1-mi-long copper wire has a resistance of 10 Ω. What is its new resistance when it is shortened by (a) cutting it in half, and (b) doubling it over and using it as if it were one wire of half the length but twice the cross-sectional area?

105. A car's headlight dissipates 40 W on low beam and 50 W on high beam. Is there more or less resistance in the high-beam filament?

DISCUSSION QUESTIONS (EVALUATION)

106. The proportionality constant k in Coulomb's law is huge in ordinary units, whereas the proportionality constant G in Newton's law of gravitation is tiny. What does this indicate about the relative strengths of these two forces?

107. A friend says that the reason one's hair stands out while touching a charged Van de Graaff generator is simply that the hair strands become charged and are light enough so that the repulsion between strands is visible. Do you agree or disagree?

108. Your tutor tells you that an ampere and a volt really measure the same thing, and the different terms only make a simple concept seem confusing. Why should you find another tutor?

109. In which of the circuits does a current exist to light the bulb?

110. Does more current "flow" out of a battery than into it? Does more current "flow" into a lightbulb than out of it? Explain.

111. Sometimes you hear someone say that a particular appliance "uses up" electricity. What is it that the appliance actually consumes, and what becomes of it?

112. Does a lamp with a thick filament draw more current or less current than a lamp with a thin filament? Defend your answer.

113. Is the current in a lightbulb connected to a 220-V source greater or less than that in the same bulb when it is connected to a 110-V source?

114. Is the following label on a household product cause for concern? "Caution: This product contains tiny, electrically charged particles moving at speeds in excess of 100,000,000 kilometers per hour."

115. The equivalent resistance of a pair of resistors depends on how they're connected. Suppose you wish to connect a pair of resistors in a way that their equivalent resistance is less than the resistance of either one. Should you connect them in series or in parallel?

116. A friend says that a battery provides not a source of constant current, but a source of constant voltage. Do you agree or disagree, and why?

117. A friend says that adding bulbs in series to a circuit provides more obstacles to the flow of charge, so there is less current with more bulbs, but adding bulbs in parallel provides more paths so more current can flow. Do you agree or disagree, and why?

118. Consider a pair of flashlight bulbs connected to a battery. Do they glow brighter if they are connected in series or in parallel? Does the battery run down faster if they are connected in series or in parallel?

119. In the circuit shown, how do the brightnesses of the three identical lightbulbs compare? Which lightbulb draws the most current? What happens if bulb A is unscrewed? If bulb C is unscrewed?

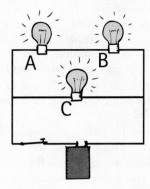

120. As more and more bulbs are connected in series to a flashlight battery, what happens to the brightness of each bulb? Assuming that the heating inside the battery is negligible, what happens to the brightness of each bulb when more and more bulbs are connected in parallel?

121. A battery has internal resistance, so, if the current it supplies goes up, the voltage it supplies goes down. If too many bulbs are connected in parallel across a battery, does their brightness diminish? Explain.

122. Are these three circuits equivalent to one another? Why or why not?

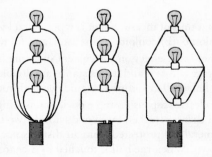

123. Your friend says that electric current takes the path of least resistance. Why is it more accurate in the case of a parallel circuit to say that *most* current travels in the path of least resistance?

124. Consider a pair of incandescent bulbs, a 60-W bulb and a 100-W bulb. If the bulbs are connected in series in a circuit, across which bulb is the greater voltage drop? If the bulbs are connected in parallel?

READINESS ASSURANCE TEST (RAT)

If you have a good handle on this chapter, if you really do, then you should be able to score at least 7 out of 10 on this RAT. If you score less than 7, you need to study further before moving on.

Choose the BEST answer to each of the following.

1. When we say charge is conserved, we mean that charge can
   (a) be saved, like money in a bank.
   (b) not be created or destroyed.
   (c) be created or destroyed, but only in nuclear reactions.
   (d) take equivalent forms.

2. When a pair of charged particles are brought twice as close to each other, the force between them becomes
   (a) twice as strong.
   (b) four times as strong.
   (c) half as strong.
   (d) one quarter as strong.

3. An electric field surrounds all
   (a) electric charge.
   (b) electrons.
   (c) protons.
   (d) all of these

4. Electric potential and electric potential energy are
   (a) one and the same in most cases.
   (b) two terms for the same concept.
   (c) both of these
   (d) neither of these

5. Which statement is correct?
   (a) Voltage flows in a circuit.
   (b) Charge flows in a circuit.
   (c) A battery is the source of electrons in a circuit.
   (d) All are correct.

6. When you double the voltage in a simple electric circuit, you double the
   (a) current.
   (b) resistance.
   (c) both of these
   (d) neither of these

7. If you double both the current and the voltage in a circuit, the power
   (a) remains unchanged if the resistance remains constant.
   (b) halves.
   (c) doubles.
   (d) quadruples.

8. In a simple circuit consisting of a single lamp and a single battery, when the current in the lamp is 2 A, the current in the battery is
   (a) half, 1 A.
   (b) 2 A.
   (c) dependent on the internal battery resistance.
   (d) not enough information to say

9. In a circuit with two lamps in parallel, if the current in one lamp is 2 A, the current in the battery is
   (a) half, 1 A.
   (b) 2 A.
   (c) more than 2 A.
   (d) not enough information to say

10. What is the power rating of a lamp connected to a 12-V source when it carries 2.5 A?
    (a) 4.8 W
    (b) 14.5 W
    (c) 30 W
    (d) none of these

Answers to RAT

1. b, 2. b, 3. d, 4. d, 5. b, 6. a, 7. d, 8. b, 9. c, 10. c.

9

CHAPTER 9
Magnetism and Electromagnetic Induction

EGAN SHOWS not only how a magnet attracts nails but also how nails stuck to her magnet attract nails that dangle below it. In Chapter 8 we discussed a charged balloon that sticks to a wall. Is the physics similar for the nails that stick to the magnet? And what causes magnetism—is it electrical? Does the fact that compass needles point to Earth's poles tell us that Earth is a giant magnet? Is it true that magnetism about planet Earth shields us from harmful cosmic rays? Is Earth's magnetism responsible for the spectacular colors of the aurora borealis? What are magnetic fields, and how are changes in them essential for the operation of electric generators and electric motors? These intriguing questions and more will be answered in this chapter.

9.1 Magnetic Poles

EXPLAIN THIS   In what sense are magnetic poles similar to the sides of a coin?

Anyone who has played around with magnets knows that magnets exert forces on one another. A **magnetic force** is similar to an electrical force in that a magnet can both attract and repel without touching (depending on which end of the magnet is held near another) and the strength of its interaction depends on the distance between magnets. Whereas electric charges produce electrical forces, regions called *magnetic poles* give rise to magnetic forces.

If you suspend a bar magnet at its center by a piece of string, you've got a compass. One end, called the *north-seeking pole*, points northward. The opposite end, called the *south-seeking pole*, points southward. More simply, these are called the *north* and *south* poles. All magnets have both a north and a south pole (some have more than one of each). Refrigerator magnets have narrow strips of alternating north and south poles. These magnets are strong enough to hold sheets of paper against a refrigerator door, but they have a very short range because the north and south poles cancel a short distance from the magnet. In a simple bar magnet, the magnetic poles are located at the two ends. A common horseshoe magnet is a bar magnet bent into a U shape. Its poles are also located at its two ends.

If the north pole of one magnet is brought near the north pole of another magnet, they repel. The same is true of a south pole near a south pole. If opposite poles are brought together, however, attraction occurs:*

Like poles repel; opposite poles attract.

This rule is similar to the rule for the forces between electric charges, in which like charges repel one another and unlike charges attract. But there is a very important difference between magnetic poles and electric charges. Whereas electric charges can be isolated, magnetic poles cannot. Electrons and protons are entities by themselves. A cluster of electrons need not be accompanied by a cluster of protons, and vice versa. But a north magnetic pole never exists without the presence of a south pole, and vice versa. The north and south poles of a magnet are like the head and tail of the same coin.

If you break a bar magnet in half, each half still behaves as a complete magnet. Break the pieces in half again, and you have four complete magnets. You can continue breaking the pieces in half and never isolate a single pole. Even if your pieces were one atom thick, there would still be two poles on each piece, which suggests that the atoms themselves are magnets.

CHECKPOINT
Does every magnet necessarily have a north and a south pole?

Was this your answer?
Yes, just as every coin has two sides, a "head" and a "tail." (Some "trick" magnets have more than two poles, but none has only one.)

---

* The force of interaction between magnetic poles is given by $F \sim (p_1 p_2)/d^2$, where p_1 and p_2 represent magnetic pole strengths and d represents the separation distance between the poles. Note the similarity of this relationship to Coulomb's law and Newton's law of universal gravitation.

Mastering**PHYSICS**
VIDEO: Oersted's Discovery

FIGURE 9.1
A horseshoe magnet.

Any sufficiently advanced technology is indistinguishable from magic.
—Arthur C. Clarke

FYI   Interestingly, the north pole of a magnet points north because it's attracted to Earth's magnetic *south* pole! Earth's magnetic north pole is in Antarctica. Magnetic and geographic poles don't match.

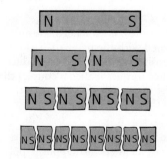

FIGURE 9.2
If you break a magnet in half, you have two magnets. Break these in half, and you have four magnets, each with a north and south pole. Continue breaking the pieces further and further and you find that you always get the same results. Magnetic poles exist in pairs.

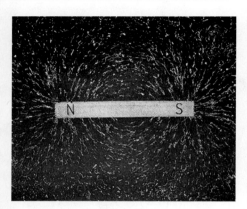

FIGURE 9.3
INTERACTIVE FIGURE

Top view of iron filings sprinkled on
a sheet of paper on top of a magnet.
The filings trace out a pattern of
magnetic field lines in the surrounding space.

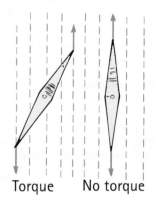

Torque   No torque

FIGURE 9.4
When the compass needle is not
aligned with the magnetic field, the
oppositely directed forces produce a
pair of torques (called a couple) that
twist the needle into alignment.

FIGURE 9.5
The magnetic field patterns for a
pair of magnets. (a) Opposite poles
are nearest to each other. (b) Like
poles are nearest to each other.

Mastering**PHYSICS**

TUTORIAL: Magnetic Fields

9.2 Magnetic Fields

EXPLAIN THIS What is the origin of all magnetic fields?

If you sprinkle some iron filings on a sheet of paper placed on a magnet,
you'll see that the filings trace out an orderly pattern of lines that surround
the magnet. The space around the magnet is energized by a **magnetic field**.
The shape of the field is revealed by magnetic field lines that spread out from
one pole and return to the other pole. It is interesting to compare the field patterns in Figures 9.3 and 9.5 with the electric field patterns in Figures 8.10 and
8.11 in the previous chapter.

The direction of the field outside the magnet is, by convention, from the
north pole to the south pole. Where the lines are closer together, the field is
stronger. We can see that the magnetic field strength is greater at the poles. If
we place another magnet or a small compass anywhere in the field, its poles
tend to align with the magnetic field.

A magnetic field is produced by the motion of electric charge.* Where, then,
is this motion in a common bar magnet? The answer is, in the electrons of the
atoms that make up the magnet. These electrons are in constant motion. Two
kinds of electron motion produce magnetism: electron spin and electron revolution. A common science model views electrons as spinning about their own
axes like tops, while they revolve about the nuclei of their atoms like planets
revolving around the Sun. In most common magnets, electron spin is the main
contributor to magnetism.

Every spinning electron is a tiny magnet. A pair of electrons spinning in
the same direction creates a stronger magnet. A pair of electrons spinning in
opposite directions, however, work against each other. The magnetic fields cancel. This is why most substances are not magnets. In most atoms, the various
fields cancel one another because the electrons spin in opposite directions. In
such materials as iron, nickel, and cobalt, however, the fields do not cancel
each other entirely. Each iron atom has four electrons whose spin magnetism
is uncanceled. Each iron atom, then, is a tiny magnet. The same is true, to a
lesser extent, of nickel and cobalt atoms. Most common magnets are made
from alloys containing iron, nickel, and cobalt, as well as aluminum, in various
proportions.

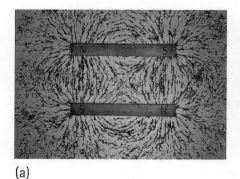

(a)

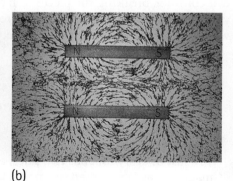

(b)

* Interestingly, because motion is relative, the magnetic field is relative. For example, when an
electron moves by you, a definite magnetic field is associated with the moving electron. But if you
move along with the electron, so that there is no motion relative to you, you find no magnetic
field associated with the electron. Magnetism is relativistic, as first explained by Albert Einstein
when he published his first paper on special relativity, "On the Electrodynamics of Moving
Bodies."

FIGURE 9.6
Fred Myers shows that the magnetic field of a ceramic magnet penetrates flesh and the plastic coating on a paper clip.

FYI Both the spinning motion and the orbital motion of every electron in an atom produce magnetic fields. These fields combine constructively or destructively to produce the magnetic field of the atom. The resulting field is greatest for iron atoms. (Electrons don't actually spin like a rotating planet, but behave as if they were—the concept of spin is a quantum effect.)

Most of the iron objects around you are magnetized to some degree. A filing cabinet, a refrigerator, and even cans of food on your pantry shelf have north and south poles induced by Earth's magnetic field. If you pass a compass from their bottoms to their tops, you can easily identify their poles. (See Activity 33 at the end of this chapter, where you are asked to turn cans upside down and note how many days go by for the poles to reverse themselves.)

9.3 Magnetic Domains

EXPLAIN THIS In what ways can magnets lose their strength over time?

The magnetic field of an individual iron atom is so strong that interactions among adjacent atoms cause large clusters of them to line up with one another. These clusters of aligned iron atoms are called **magnetic domains**. Each domain is perfectly magnetized and is made up of billions of aligned atoms. The domains are microscopic (Figure 9.7), and there are many of them in a crystal of iron.

Not every piece of iron is a magnet, because the domains in ordinary iron are not aligned. In a common iron nail, for example, the domains are randomly oriented. But when you bring a magnet nearby, they can be induced into alignment. (It is interesting to listen, with an amplified stethoscope, to the clickety-clack of domains aligning in a piece of iron when a strong magnet approaches.) The domains align themselves much as electric charges in a piece of paper align themselves (become polarized) in the presence of a charged rod. When you remove the nail from the magnet, ordinary thermal motion causes most or all of the domains in the nail to return to a random arrangement.

Permanent magnets can be made by placing pieces of iron or similar magnetic materials in a strong magnetic field. Alloys of iron differ; soft iron is easier to magnetize than steel. It helps to tap the material to nudge any stubborn domains into alignment. Another way is to stroke the material with a magnet. The stroking motion aligns the domains. If a permanent magnet is dropped or heated outside the strong magnetic field from which it was made, some of the domains are jostled out of alignment and the magnet becomes weaker.

LEARNING OBJECTIVE
Describe magnetic field strength in terms of domain alignment.

FIGURE 9.7
A microscopic view of magnetic domains in a crystal of iron. Each domain consists of billions of aligned iron atoms. In this view, orientation of the domains is random.

FYI A magnetic stripe on a credit card contains millions of tiny magnetic domains held together by a resin binder. Data are encoded in binary code, with zeros and ones distinguished by the frequency of domain reversals.

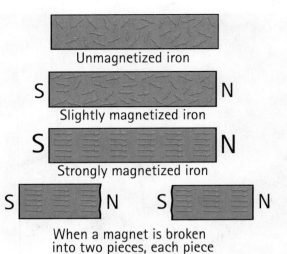

Unmagnetized iron

S ——————————— N
Slightly magnetized iron

S ——————————— N
Strongly magnetized iron

S —— N   S —— N

When a magnet is broken
into two pieces, each piece
is an equally strong magnet.

FIGURE 9.8

INTERACTIVE FIGURE

Pieces of iron in successive stages
of magnetism. The arrows represent domains; the head is a north
pole and the tail is a south pole.
Poles of neighboring domains neutralize each other's effects, except
at the ends.

FIGURE 9.9
Wai Tsan Lee shows iron nails
becoming induced magnets.

CHECKPOINT
1. **Why doesn't a magnet pick up a penny or a piece of wood?**
2. **How can a magnet attract a piece of iron that is not magnetized?**

Were these your answers?
1. A penny and a piece of wood have no magnetic domains that can be
   induced into alignment.
2. Like the compass needle in Figure 9.4, domains in the unmagnetized piece
   of iron are induced into alignment by the magnetic field of the magnet.
   One domain pole is attracted to the magnet and the other domain pole
   is repelled. Does this mean the net force is zero? No, because the force is
   slightly greater on the domain pole closest to the magnet than it is on the
   farther pole. That's why there is a net attraction. In this way, a magnet
   attracts unmagnetized pieces of iron (Figure 9.9).

LEARNING OBJECTIVE
Relate magnetic field strength to
electric wire configurations.

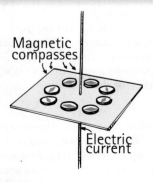

Magnetic
compasses

Electric
current

FIGURE 9.10
The compasses show the circular
shape of the magnetic field surrounding the current-carrying wire.

9.4 Electric Currents and Magnetic Fields

EXPLAIN THIS What increases when a current-carrying wire is bent into a loop?

A moving charge produces a magnetic field. A current of charges, then,
also produces a magnetic field. The magnetic field that surrounds a
current-carrying wire can be demonstrated by arranging an assortment
of compasses around the wire (Figure 9.10). The magnetic field about the
current-carrying wire makes up a pattern of concentric circles. When the current reverses direction, the compass needles turn around, showing that the
direction of the magnetic field changes also.*

---

* Earth's magnetism is generally accepted as being the result of electric currents that accompany
thermal convection in the molten parts of Earth's interior. Earth scientists have found evidence that
Earth's poles periodically reverse places—more than 20 reversals have occurred in the past 5 million years. This is perhaps the result of changes in the direction of electric currents within Earth.

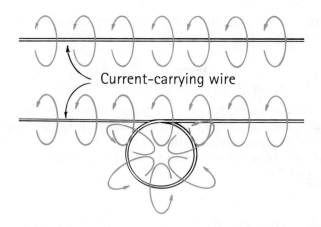

Current-carrying wire

FIGURE 9.11
Magnetic field lines about a current-carrying wire become bunched up when the wire is bent into a loop.

If the wire is bent into a loop, the magnetic field lines become bunched up inside the loop (Figure 9.11). If the wire is bent into another loop that overlaps the first, the concentration of magnetic field lines inside the loops is doubled. It follows that the magnetic field intensity in this region is increased as the number of loops is increased. The magnetic field intensity is appreciable for a current-carrying coil that has many loops.

> **FYI** A long helically wound coil of insulated wire is called a *solenoid*.

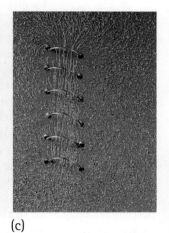

(a) (b) (c)

FIGURE 9.12
Iron filings sprinkled on paper reveal the magnetic field configurations about (a) a current-carrying wire, (b) a current-carrying loop, and (c) a coil of loops.

Electromagnets

If a piece of iron is placed in a current-carrying coil of wire, the alignment of magnetic domains in the iron produces a particularly strong magnet known as an **electromagnet**. The strength of an electromagnet can be increased simply by increasing the current through the coil. Strong electromagnets are used to control charged-particle beams in high-energy accelerators. They also levitate and propel high-speed trains.

Figure 9.13 shows a maglev train, which has no diesel or other conventional engine. Levitation is accomplished by magnetic coils that run along a guideway. The coils repel large magnets on the train's undercarriage. Once the train is levitated a few centimeters, power supplied to the coils propels the train by continuously alternating the electric current fed to the coils, which alternates their magnetic polarity. In this way a magnetic field pulls the vehicle forward, while a magnetic field farther back pushes it forward. The alternating pulls and pushes produce a forward thrust. Maglev trains are already operational. A popular one in China currently carries passengers quickly and quietly at speeds topping

FIGURE 9.13
A magnetically levitated train—a *maglev*. Whereas conventional trains vibrate as they ride on rails at high speeds, maglevs can travel vibration-free at high speeds because they make no physical contact with the guideway they float above.

400 km/h from Shanghai to its distant international airport. Watch for expansion of this growing technology.

Superconducting Electromagnets

Superconductors (see Chapter 8) have the interesting property of expelling magnetic fields. Because magnetic fields cannot penetrate the surface of a superconductor, magnets levitate above them. The reasons for this behavior, which are beyond the scope of this book, involve quantum mechanics.

LEARNING OBJECTIVE
Show how relative directions, fields, and motion affect force.

9.5 Magnetic Forces on Moving Charges

EXPLAIN THIS How does Earth's magnetic field protect us from cosmic radiation?

A charged particle at rest does not interact with a static magnetic field. However, if the charged particle moves in a magnetic field, the magnetic character of a charge in motion becomes evident: The charged particle experiences a deflecting force.* The force is greatest when the particle moves in a direction perpendicular to the magnetic field lines. At other angles, the force is less, and it becomes zero when the particle moves parallel to the field lines. In any case, the direction of the force is always perpendicular to the magnetic field lines and the velocity of the charged particle (Figure 9.16). So a moving charge is deflected when it crosses through a magnetic field, but when it travels parallel to the field, no deflection occurs.

This deflecting force is very different from the forces that occur in other interactions, such as the gravitational forces between masses, the electric forces between charges, and the magnetic forces between magnetic poles. The force that acts on a moving charged particle, such as an electron in an electron beam, does not act along the line that joins the sources of interaction. Instead, it acts perpendicularly both to the magnetic field and to the electron beam.

We are fortunate that charged particles are deflected by magnetic fields. This fact was employed in guiding electrons onto the inner surface of early television tubes to produce pictures. Also, charged particles from outer space are deflected by Earth's magnetic field. Otherwise the harmful cosmic rays bombarding Earth's surface would be much more intense.

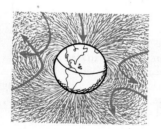

FIGURE 9.15
Earth's magnetic field deflects the many charged particles that make up cosmic radiation.

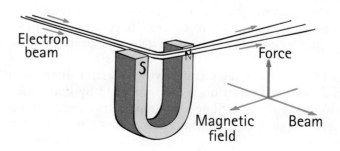

FIGURE 9.16
A beam of electrons is deflected by a magnetic field.

---

* When particles of electric charge q and velocity v move perpendicularly into a magnetic field of strength B, the force F on each particle is simply the product of the three variables: $F = qvB$. For non-perpendicular angles, v in this relationship must be the component of velocity perpendicular to B.

Magnetic Force on Current-Carrying Wires

MasteringPHYSICS

VIDEO: Magnetic Forces on a
Current-Carrying Wire

Simple logic tells you that if a charged particle moving through a magnetic field experiences a deflecting force, then a current of charged particles moving through a magnetic field also experiences a deflecting force. If the particles are deflected while moving inside a wire, the wire is also deflected (Figure 9.17).

If we reverse the direction of current, the deflecting force acts in the opposite direction. The force is strongest when the current is perpendicular to the magnetic field lines. The direction of force is not along the magnetic field lines or along the direction of current. The force is perpendicular to both field lines and current. It is a sideways force—perpendicular to the wire.

We see that, just as a current-carrying wire deflects a magnet such as a compass needle, a magnet deflects a current-carrying wire. When discovered, these complementary links between electricity and magnetism created much excitement. Almost immediately, people began harnessing the electromagnetic force for useful purposes—with great sensitivity in electric meters and with great force in electric motors.

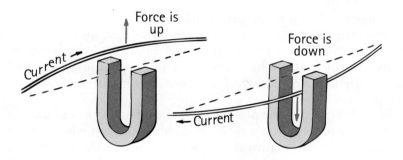

FIGURE 9.17
INTERACTIVE FIGURE   MP

A current-carrying wire experiences a force in a magnetic field. (Can you see that this is a simple extension of Figure 9.16?)

FYI In an advanced course, you'll learn the "simple" right-hand rule.

CHECKPOINT

What law of physics tells you that if a current-carrying wire produces a force on a magnet, a magnet must produce a force on a current-carrying wire?

Was this your answer?
Newton's third law, which applies to all forces in nature.

Electric Meters

The simplest meter to detect electric current is a magnetic compass. The next simplest meter is a compass in a coil of wires (Figure 9.18). When an electric current passes through the coil, each loop produces its own effect on the needle, so even a very small current can be detected. Such a current-indicating instrument is called a *galvanometer*.

A more common design is shown in Figure 9.19. It employs more loops of wire and is therefore more sensitive. The coil is mounted for movement, and the magnet is held stationary. The coil turns against a spring, so the greater

FIGURE 9.18
A very simple galvanometer.

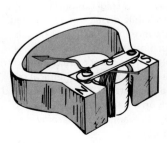

FIGURE 9.19
A common galvanometer design.

FIGURE 9.20
Both the ammeter and the voltmeter are basically galvanometers. (The electrical resistance of the instrument is designed to be very low for the ammeter and very high for the voltmeter.)

the current in its windings, the greater its deflection. A galvanometer may be calibrated to measure current (amperes), in which case it is called an *ammeter*. Or it may be calibrated to measure electric potential (volts), in which case it is called a *voltmeter.**

Electric Motors

If we change the design of the galvanometer slightly so that deflection makes a complete turn rather than a partial rotation, we have an *electric motor*. The principal difference is that the current in a motor is made to change direction each time the coil makes a half rotation. This happens in a cyclic fashion to produce continuous rotation, which has been used to run clocks, operate gadgets, and lift heavy loads.

In Figure 9.21 we see the principle of the electric motor in bare outline. A permanent magnet produces a magnetic field in a region where a rectangular loop of wire is mounted to turn about the axis shown by the dashed line. When a current passes through the loop, it flows in opposite directions in the upper and lower sides of the loop. (It must do this because if charge flows into one end of the loop, it must flow out the other end.) If the upper portion of the loop is forced to the left, then the lower portion is forced to the right, as if it were a galvanometer. But, unlike a galvanometer, the current is reversed during each half revolution by means of stationary contacts on the shaft. The parts of the wire that brush against these contacts are called *brushes*. In this way, the current in the loop alternates so that the forces in the upper and lower regions do not change directions as the loop rotates. The rotation is continuous as long as current is supplied.

We have described here only a very simple dc motor. Larger motors, dc or ac, are usually manufactured by replacing the permanent magnet by an electromagnet that is energized by the power source. Of course, more than a single loop is used. Many loops of wire are wound about an iron cylinder, called an *armature*, which then rotates when the wire carries current.

The advent of electric motors brought to an end much human and animal toil in many parts of the world. Electric motors have greatly changed the way people live.

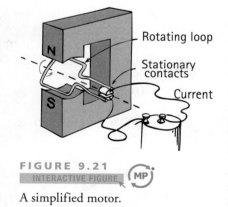

Rotating loop

Stationary contacts

Current

FIGURE 9.21
INTERACTIVE FIGURE MP

A simplified motor.

FYI The galvanometer is named after Luigi Galvani (1737–1798), who, while dissecting a frog's leg, discovered that dissimilar metals touching the leg caused it to twitch. This chance discovery led to the invention of the chemical cell and the battery. The next time you pick up a galvanized pail, think of Luigi Galvani in his anatomy laboratory.

---

* To some degree, measuring instruments change what is being measured—ammeters and voltmeters included. Because an ammeter is connected in series with the circuit it measures, its resistance is made very low. That way, it doesn't appreciably lower the current it measures. Because a voltmeter is connected in parallel, its resistance is made very high, so that it draws very little current for its operation. In the lab part of your course you'll likely learn how to connect these instruments in simple circuits.

MRI: MAGNETIC RESONANCE IMAGING

Magnetic resonance imaging scanners provide high-resolution pictures of the tissues inside a body. Superconducting coils produce a strong magnetic field (up to 60,000 times as strong as the intensity of Earth's magnetic field) that is used to align the protons of hydrogen atoms in the body of the patient.

Like electrons, protons have a "spin" property, so they align with a magnetic field. Unlike a compass needle that aligns with Earth's magnetic field, the proton's axis wobbles about

the applied magnetic field. Wobbling protons are slammed with a burst of radio waves tuned to push the proton's spin axis sideways, perpendicular to the applied magnetic field. When the radio waves pass and the protons quickly return to their wobbling pattern, they emit faint electromagnetic signals whose frequencies depend slightly on the chemical environment in which the proton resides. The signals, which are detected by sensors, are then analyzed by a computer to reveal

varying densities of hydrogen atoms in the body and their interactions with surrounding tissue. The images clearly distinguish between fluid and bone, for example.

MRI was formerly called NMRI (nuclear magnetic resonance imaging), because hydrogen nuclei resonate with the applied fields. Because of public phobia about anything "nuclear," this diagnostic technique is now called MRI. (Tell your friends that every atom in their bodies contains a nucleus!)

CHECKPOINT

What is the major similarity between a galvanometer and a simple electric motor? What is the major difference?

Were these your answers?
A galvanometer and a motor are similar in that they both use coils positioned in a magnetic field. When a current passes through the coils, forces on the wires rotate the coils. The major difference is that the maximum coil rotation in a galvanometer is half a turn, whereas the coil in a motor (which is wrapped on an armature) rotates through many complete turns. This is accomplished by alternating the direction of the current with each half turn of the armature.

9.6 Electromagnetic Induction

EXPLAIN THIS How can a car moving along a paved road activate a traffic signal?

LEARNING OBJECTIVE
Describe how Faraday's law is central to the industrial age.

In the early 1800s, the only current-producing devices were voltaic cells, which produced small currents by dissolving metals in acids. These were the forerunners of modern batteries. The question arose as to whether electricity could be produced from magnetism. The answer was provided in 1831 by two physicists, Michael Faraday in England and Joseph Henry in the United States—each working without knowledge of the other. Their discovery changed the world by making electricity commonplace—powering industries by day and lighting up cities at night.

Faraday and Henry both discovered **electromagnetic induction**—that electric current could be produced in a wire simply by moving a magnet into or out of a coil of wire (Figure 9.22). No battery or

FYI Multiple loops of wire must be insulated, because bare wire loops touching each other make a short circuit. Joseph Henry's wife tearfully sacrificed part of the silk in her wedding gown to cover the wires of Henry's first electromagnets.

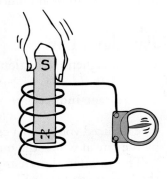

FIGURE 9.22
When the magnet is plunged into the coil, charges in the coil are set in motion, and voltage is induced in the coil.

Voltage is induced in the wire loop whether the magnetic field moves past the wire or the wire moves through the magnetic field.

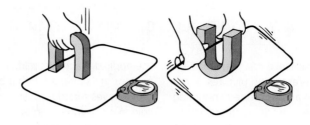

FIGURE 9.24

INTERACTIVE FIGURE MP
When a magnet is plunged into a coil with twice as many loops as another, twice as much voltage is induced. If the magnet is plunged into a coil with three times as many loops, three times as much voltage is induced.

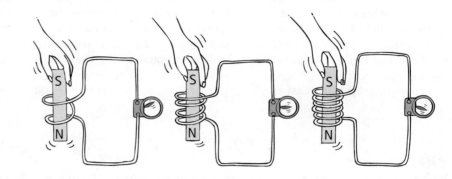

Note that a magnetic field does not induce voltage: a *change* in the field over some *time interval* does. If the field changes in a closed loop, and the loop is an electric conductor, then both voltage and current are induced.

FIGURE 9.25
It is more difficult to push the magnet into a coil with many loops because the magnetic field of each current loop resists the motion of the magnet.

Mastering**PHYSICS**
VIDEO: Applications of Electromagnetic Induction
VIDEO: Faraday's Law

other voltage source was needed—only the motion of a magnet in a wire loop. They discovered that voltage is caused, or induced, by the relative motion between a wire and a magnetic field. Whether the magnetic field moves near a stationary conductor or vice versa, voltage is induced either way (Figure 9.23).

The greater the number of loops of wire moving in a magnetic field, the greater the induced voltage (Figure 9.24). Pushing a magnet into a coil with twice as many loops induces twice as much voltage; pushing into a coil with 10 times as many loops induces 10 times as much voltage; and so on. It may seem that we get something (energy) for nothing simply by increasing the number of loops in a coil of wire, but we don't. We find that it is more difficult to push the magnet into a coil made up of more loops. This is because the induced voltage produces a current, which makes an electromagnet, which repels the magnet in our hand. So we must do more work against this "back force" to induce more voltage (Figure 9.25).

The amount of voltage induced depends on how fast the magnetic field lines are entering or leaving the coil. Very slow motion produces hardly any voltage at all. Rapid motion induces a greater voltage.

Faraday's Law

Electromagnetic induction is summarized by **Faraday's law**:

$$\text{Voltage induced} \ \sim \ \text{number of loops} \times \frac{\text{change in magnetic field}}{\text{time}}$$

The induced voltage in a coil is proportional to the number of loops multiplied by the rate at which the magnetic field changes within those loops.

The amount of *current* produced by electromagnetic induction depends on the resistance of the coil and the circuit that it connects, as well as the induced voltage.* For example, we can plunge a magnet into and out of a closed rubber loop and into and out of a closed loop of copper. The voltage induced in each is the same, providing the loops are the same size and the magnet moves with the same speed. But the current in each is quite different. The electrons in the rubber sense the same voltage as those in the copper, but their bonding to the fixed atoms prevents the movement of charge that so freely occurs in the copper.

FIGURE 9.26
Guitar pickups are tiny coils with magnets inside them. The magnets magnetize the steel strings. When the strings vibrate, voltage is induced in the coils and boosted by an amplifier, and sound is produced by a speaker.

CHECKPOINT

If you push a magnet into a coil, as shown in Figure 9.25, you'll feel a resistance to your push. Why is this resistance greater in a coil with more loops?

Was this your answer?
Simply put, more work is required to provide more energy. You can also look at it this way: When you push a magnet into a coil, you induce electric current and cause the coil to become an electromagnet. The more loops in the coil, the stronger the electromagnet that you produce and the stronger it pushes back against you. (If the electromagnetic coil attracted your magnet instead of repelling it, energy would have been created from nothing and the law of energy conservation would have been violated. So the coil must repel the magnet.)

We have mentioned two ways in which voltage can be induced in a loop of wire: by moving the loop near a magnet and by moving a magnet near the loop. There is a third way—by changing a current in a nearby loop. All three of these cases possess the same essential ingredient—a changing magnetic field in the loop.

We see electromagnetic induction all around us. On the road, we see it operate when a car drives over buried coils of wire to activate a nearby traffic light. When iron parts of a car move over the buried coils, the effect of Earth's magnetic field on the coils is changed, inducing a voltage to trigger the changing of the traffic lights. Similarly, when you walk through the upright coils in the security system at an airport, any metal you carry slightly alters the magnetic field in the coils. This change induces voltage, which sounds an alarm. When the magnetic strip on the back of a credit card is scanned, induced voltage pulses identify the card. Something similar occurs in the recording head of a tape recorder: magnetic domains in the tape are sensed as the tape moves past a current-carrying coil. Electromagnetic induction is at work in computers, coffee makers, kitchen stovetops, cordless electric toothbrushes, and devices galore. As we soon see, it underlies the electromagnetic waves that we call light.

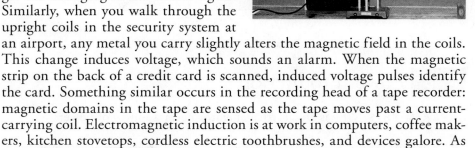

FIGURE 9.27
When Jean Curtis powers the large coil with ac, an alternating magnetic field is established in the iron bar and thence through the metal ring. Current is therefore induced in the ring, which then establishes its own magnetic field, which always acts in a direction to oppose the field producing it. The result is mutual repulsion—levitation.

FYI Shake flashlights need no batteries. Shake the flashlight for 30 seconds or so and generate up to 5 minutes of bright illumination. Electromagnetic induction occurs as built-in magnet slides to and fro between coils that charge a capacitor. When brightness diminishes, shake again. You provide the energy to charge the capacitor.

---

* Current also depends on the *inductance* of the coil. Inductance measures the tendency of a coil to resist a change in current because the magnetism produced by one part of the coil opposes the change of current in other parts of the coil. In ac circuits it is comparable to resistance in dc circuits. To reduce "information overload" we will not treat inductance in this book.

9.7 Generators and Alternating Current

EXPLAIN THIS Why does a generator produce ac rather than dc?

When a magnet is repeatedly plunged into and back out of a coil of wire, the direction of the induced voltage alternates. As the magnetic field strength inside the coil is increased (as the magnet enters), the induced voltage in the coil is directed one way. When the magnetic field strength diminishes (as the magnet leaves), the voltage is induced in the opposite direction. The frequency of the alternating voltage that is induced is equal to the frequency of the changing magnetic field within the loop.

Rather than moving the magnet, it is more practical to move the coil. This is best accomplished by rotating the coil in a stationary magnetic field (Figure 9.28). This arrangement is called a **generator**. It is essentially the opposite of a motor. Whereas a motor converts electric energy into mechanical energy, a generator converts mechanical energy into electric energy.

FIGURE 9.28

INTERACTIVE FIGURE MP

A simple generator. Voltage is induced in the loop when it is rotated in the magnetic field. Brushes convert the ac to dc.

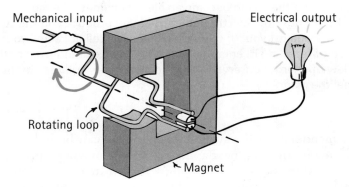

Mechanical input   Electrical output
Rotating loop
Magnet

Because the voltage induced by the generator alternates, the current produced is ac, an alternating current.* The alternating current in our homes is produced by generators standardized so that the current goes through 60 full cycles of change in magnitude and direction each second—60 hertz.

FYI A motor and a generator are actually the same device, with input and output reversed.

FIGURE 9.29
As the loop rotates, the magnitude and direction of the induced voltage (and current) change. One complete rotation of the loop produces one complete cycle in voltage (and current).

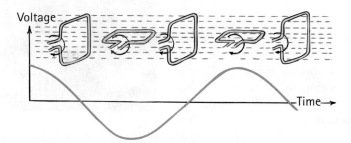

Voltage
Time→

9.8 Power Production

EXPLAIN THIS Why is an electric generator never a source of power?

Fifty years after Faraday and Henry discovered electromagnetic induction, Nikola Tesla and George Westinghouse put those findings to practical use and showed the world that electricity could be generated reliably and in sufficient quantities to light entire cities.

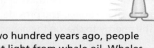
Two hundred years ago, people got light from whale oil. Whales should be glad that humans discovered electricity!

* By means such as appropriately designed *brushes* (contacts that brush against the rotating *armature*, as shown in the figures), the ac in the loop(s) can be converted to dc to make a dc generator.

Tesla built generators that were much like those still in use, but quite a bit more complicated than the simple model we have discussed. Tesla's generators had armatures consisting of bundles of copper wires that were made to spin within strong magnetic fields by means of a turbine, which, in turn, was spun by the energy of steam or falling water. The rotating loops of wire in the armature cut through the magnetic field of the surrounding electromagnets, thereby inducing alternating voltage and current.

We can look at this process from an atomic point of view. When the wires in the spinning armature cut through the magnetic field, oppositely directed electromagnetic forces act on the negative and positive charges. Electrons respond to this force by momentarily swarming relatively freely in one direction throughout the crystalline copper lattice; the copper atoms, which are actually positive ions, are forced in the opposite direction. But the ions are anchored in the lattice, so they barely move at all. Only the electrons move significantly, sloshing back and forth in alternating fashion with each rotation of the armature. The energy produced by this electronic sloshing is tapped at the electrode terminals of the generator.

It's important to know that generators don't produce energy—they simply convert energy from some other form to electric energy. As we discussed in Chapter 3, energy from a source, whether fossil or nuclear fuel or wind or water, is converted to mechanical energy to drive the turbine. The attached generator converts most of this mechanical energy to electric energy. Some people think that electricity is a primary source of energy. It is not. It is a carrier of energy that requires a source.

FIGURE 9.30
Steam drives the turbine, which is connected to the armature of the generator.

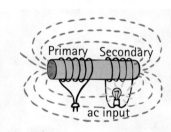

FIGURE 9.31
A simple transformer.

FIGURE 9.32
A practical transformer. Both primary and secondary coils are wrapped on the inner part of the iron core (yellow), which guides alternating magnetic field lines (green) produced by ac in the primary. The alternating field induces ac voltage in the secondary. Thus power at one voltage from the primary is transferred to the secondary at a different voltage.

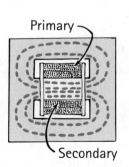

9.9 The Transformer—Boosting or Lowering Voltage

EXPLAIN THIS Which of these can a transformer increase: voltage, current, energy?

When changes in the magnetic field of a current-carrying coil of wire are intercepted by a second coil of wire, voltage is induced in the second coil. This is the principle of the **transformer**—a simple electromagnetic-induction device consisting of an input coil of wire (the primary)

FIGURE 9.33
This common transformer lowers 120 V to 6 V or 9 V. It also converts ac to dc by means of a *diode* inside—a tiny electronic device that acts as a one-way valve.

FIGURE 9.34
A common neighborhood transformer, which typically steps 2400 V down to 240 V for houses and small businesses. Inside the building, the 240 V can divide to a safer 120 V.

LEARNING OBJECTIVE
Describe how the nature of light relates to electromagnetic induction.

and an output coil of wire (the secondary). The coils are wound on an iron core so that the magnetic field of the primary passes through the secondary. The primary is powered by an ac voltage source, and the secondary is connected to some external circuit. Changes in the primary current produce changes in its magnetic field, which extend to the secondary, and, by electromagnetic induction, voltage is induced in the secondary. If the number of turns of wire in both coils is the same, voltage input and voltage output are the same. Nothing is gained. But if the secondary has more turns than the primary, then greater voltage is induced in the secondary. This is a *step-up transformer*. If the secondary has fewer turns than the primary, the ac voltage induced in the secondary is lower than that in the primary. This is a *step-down transformer*.

The relationship between primary and secondary voltages relative to the number of turns is as follows:

$$\frac{\text{Primary voltage}}{\text{Number of primary turns}} = \frac{\text{secondary voltage}}{\text{number of secondary turns}}$$

It might seem that we get something for nothing with a transformer that steps up the voltage, but we don't. When voltage is stepped up, current in the secondary is less than in the primary. The transformer actually transfers energy from one coil to the other. The rate of transferring energy is *power*. The power used in the secondary is supplied by the primary. The primary gives no more than the secondary uses, in accord with the law of energy conservation. If any slight power losses due to heating of the core can be neglected, then

$$\text{Power into primary} = \text{power out of secondary}$$

Electric power is equal to the product of voltage and current, so we can say that

$$(\text{Voltage} \times \text{current})_{\text{primary}} = (\text{voltage} \times \text{current})_{\text{secondary}}$$

The ease with which voltages can be stepped up or down with a transformer is the principal reason that most electric power is ac rather than dc.

9.10 Field Induction

EXPLAIN THIS What is light?

Electromagnetic induction explains the induction of voltages and currents. Actually, the more basic concept of *fields* is at the root of both voltages and currents. The modern view of electromagnetic induction states that electric and magnetic fields are induced. These, in turn, produce the voltages we have considered. So induction occurs whether or not a conducting wire or any material medium is present. In this more general sense, Faraday's law states:

An electric field is induced in any region of space in which a magnetic field is changing with time.

There is a second effect, an extension of Faraday's law. It is the same except that the roles of electric and magnetic fields are interchanged. It is one of nature's many symmetries. This effect, which was advanced by the British physicist James Clerk Maxwell in about 1860, is known as **Maxwell's counterpart to Faraday's law**:

A magnetic field is induced in any region of space in which an electric field is changing with time.

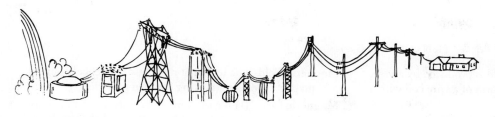

FIGURE 9.35
Voltage generated in power stations is stepped up with transformers before being transferred across country by overhead cables. Then other transformers reduce the voltage before supplying it to homes, offices, and factories.

In each case, the strength of the induced field is proportional to the rates of change of the inducing field. The induced electric and magnetic fields are at right angles to each other (Figure 9.36).

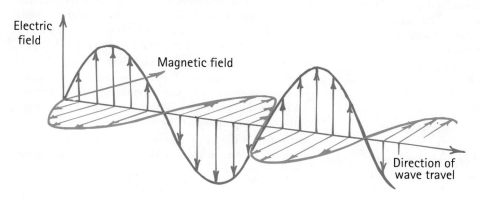

FIGURE 9.36

The electric and magnetic fields of an electromagnetic wave in free space are perpendicular to each other and to the direction of motion of the wave.

Maxwell saw the link between electromagnetic waves and light. If electric charges are set into vibration in the range of frequencies that match those of light, waves are produced that *are* light! Maxwell discovered that light is simply electromagnetic waves in the range of frequencies to which the eye is sensitive.

On the eve of his discovery, Maxwell had a date with the young woman he was later to marry. Story has it that while they were walking in a garden, she remarked about the beauty and wonder of the stars. Maxwell asked her how she would feel if she knew that she was walking with the only person in the world who knew what starlight really was. In fact, at that time, James Clerk Maxwell was the only person in the entire world to know that light of any kind is energy carried in waves of electric and magnetic fields that continually regenerate each other.

The laws of electromagnetic induction were discovered at about the time the American Civil War was being fought. From a long view of human history, there can be little doubt that events such as the American Civil War will pale into provincial insignificance in comparison with the more significant event of the 19th century: the discovery of the electromagnetic laws.

FYI Enormous intergalactic magnetic fields that spread far beyond the galaxies have recently been detected. These giant magnetic fields make up an important part of the cosmic energy store and play a significant role in shaping the evolution of galaxies and large-scale grouping of galaxies.

Each of us needs a knowledge filter to tell us the difference between what is true and what only pretends to be true. The best knowledge filter ever invented is science.

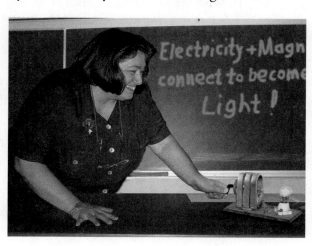

FIGURE 9.37
In turning the crank of the generator, Sheron Snyder does work, which is transformed into voltage and current, which, in turn, are transformed into light.

For instructor-assigned homework, go to www.masteringphysics.com

SUMMARY OF TERMS (KNOWLEDGE)

Electromagnet A magnet whose field is produced by an electric current. It is usually in the form of a wire coil with a piece of iron inside the coil.

Electromagnetic induction The induction of voltage when a magnetic field changes with time.

Faraday's law The law of electromagnetic induction, in which the induced voltage in a coil is proportional to the number of loops multiplied by the rate at which the magnetic field changes within those loops. (The induction of voltage is actually the result of a more fundamental phenomenon: the induction of an electric field.)

$$\text{Voltage induced} \sim \text{number of loops} \times \frac{\text{change in magnetic field}}{\text{time}}$$

Generator An electromagnetic induction device that produces electric current by rotating a coil within a stationary magnetic field.

Magnetic domains Clustered regions of aligned magnetic atoms. When these regions themselves are aligned with one another, the substance containing them is a magnet.

Magnetic field The region of magnetic influence around a magnetic pole or a moving charged particle.

Magnetic force (1) Between magnets, it is the attraction of unlike magnetic poles for each other and the repulsion between like magnetic poles. (2) Between a magnetic field and a moving charge, it is a deflecting force due to the motion of the charge: The deflecting force is perpendicular to the velocity of the charge and perpendicular to the magnetic field lines. This force is greatest when the charge moves perpendicular to the field lines and is smallest (zero) when it moves parallel to the field lines.

Maxwell's counterpart to Faraday's law A magnetic field is induced in any region of space in which an electric field is changing with time. Correspondingly, an electric field is induced in any region of space in which a magnetic field is changing with time.

Transformer A device for transferring electric power from one coil of wire to another by means of electromagnetic induction.

READING CHECK QUESTIONS (COMPREHENSION)

9.1 Magnetic Poles

1. How does the range of refrigerator magnets compare with the range of common bar magnets?
2. In what way is the rule for the interaction between magnetic poles similar to the rule for the interaction between electric charges?
3. In what way are magnetic poles very different from electric charges?

9.2 Magnetic Fields

4. What produces a magnetic field?
5. What two kinds of motion are exhibited by electrons in an atom?

9.3 Magnetic Domains

6. What is a magnetic domain?
7. Why is iron magnetic and wood not magnetic?

9.4 Electric Currents and Magnetic Fields

8. What is the shape of a magnetic field about a current-carrying wire?
9. What happens to the direction of the magnetic field about an electric current when the direction of the current is reversed?
10. Why is the magnetic field strength inside a current-carrying loop of wire greater than the field strength about a straight section of wire?

11. How is the strength of a magnetic field in a coil affected when a piece of iron is placed inside? Defend your answer.

9.5 Magnetic Forces on Moving Charges

12. In what direction relative to a magnetic field does a charged particle move in order to experience maximum deflecting force? Minimum deflecting force?
13. What effect does Earth's magnetic field have on the intensity of cosmic rays striking Earth's surface?
14. What relative direction between a magnetic field and a current-carrying wire results in the greatest force on the wire? In the smallest force?
15. What happens to the direction of the magnetic force on a wire in a magnetic field when the current in the wire is reversed?
16. What is a galvanometer called when it is calibrated to read current? To read voltage?
17. Is it correct to say that an electric motor is a simple extension of the physics that underlies a galvanometer?

9.6 Electromagnetic Induction

18. What important discovery did physicists Michael Faraday and Joseph Henry make?
19. State Faraday's law.
20. What are the three ways in which voltage can be induced in a loop of wire?

9.7 Generators and Alternating Current

21. How does the frequency of induced voltage compare with how frequently a magnet is plunged into and out of a coil of wire?

22. What are the basic differences and similarities between a generator and an electric motor?

23. Is the current produced by a common generator ac or dc?

9.8 Power Production

24. What commonly supplies the energy input to a turbine?

25. Is it correct to say that a generator produces energy? Defend your answer.

9.9 The Transformer—Boosting or Lowering Voltage

26. Is it correct to say that a transformer boosts electric energy? Defend your answer.

27. Which of these does a transformer change: voltage, current, energy, power?

9.10 Field Induction

28. What is induced by the rapid alternation of a magnetic field?

29. What is induced by the rapid alternation of an electric field?

30. What important connection did Maxwell discover about electric and magnetic fields?

ACTIVITIES (HANDS-ON APPLICATION)

31. Write a letter to a relative or friend saying that you have discovered the answer to what has been a mystery for centuries—the nature of light. State how light is related to electricity and magnetism.

32. An iron bar can be magnetized easily by aligning it with the magnetic field lines of Earth and striking it lightly a few times with a hammer. This works best if the bar is tilted down to match the dip of Earth's magnetic field. The hammering jostles the domains so that they can better align with Earth's field. The bar can be demagnetized by striking it when it is in an east–west direction.

33. Earth's magnetic field induces some degree of magnetism in most of the iron objects around you. With a compass you can see that cans of food on your pantry shelf have north and south poles. When you pass the compass from their bottoms to their tops, you can easily identify their poles. Mark the poles, either N or S. Then turn the cans upside down and note how many days it takes for the poles to reverse themselves. Explain to your friends why the poles reverse.

34. Drop a small bar magnet through a vertical plastic pipe, noting its speed of fall. Then do the same with a copper pipe. Whoa! Why the difference?

PLUG AND CHUG (FORMULA FAMILIARIZATION)

$$\text{Transformer relationship:} \quad \frac{\textbf{Primary voltage}}{\textbf{Number of primary turns}} = \frac{\textbf{secondary voltage}}{\textbf{number of secondary turns}}$$

35. The primary of a transformer connected to 120 V has 10 turns. The secondary has 100 turns. Show that the output voltage is 1200 V. This is a step-up transformer.

36. The primary of a transformer connected to 120 V has 100 turns. The secondary has 10 turns. Show that the output voltage is 12 V. This is a step-down transformer.

THINK AND SOLVE (MATHEMATICAL APPLICATION)

37. A video game console requires 6 V to operate correctly. A transformer allows the device to be powered from a 120-V outlet. If the primary has 500 turns, show that the secondary should have 25 turns.

38. A model electric train requires 6 V to operate. When it is connected to a 120-V household circuit, a transformer is needed. If the primary coil of the transformer has 360 turns, show that the secondary coil should have 18 turns.

39. A transformer for a laptop computer converts a 120-V input to a 24-V output. Show that the primary coil has five times as many turns as the secondary coil.

40. If the output current for the transformer in the preceding problem is 1.8 A, show that the input current is 0.36 A.

41. A transformer has an input of 6 V and an output of 36 V. If the input is changed to 12 V, show that the output would be 72 V.

42. An ideal transformer has 50 turns in its primary and 250 turns in its secondary. 12-V ac is connected to the primary. Show that:

    (a) 60 V ac is available at the secondary,

    (b) 6 A of current is in a 10-Ω device connected to the secondary, and

    (c) the power supplied to the primary is 360 W.

43. Neon signs require about 12,000 V for their operation. Consider a neon-sign transformer that operates off 120-V lines. How many more turns should be on the secondary compared with the primary?

44. A power of 100 kW (10^5 W) is delivered to the other side of a city by a pair of power lines, between which the voltage is 12,000 V.

    (a) Use the formula $P = IV$ to show that the current in the lines is 8.3 A.

    (b) If each of the two lines has a resistance of 10 Ω, show that there is a 83-V change of voltage *along* each line. (Think carefully. This voltage change is along each line, *not between* the lines.)

    (c) Show that the power expended as heat in both lines together is 1.38 kW (distinct from power delivered to customers).

    (d) How do your calculations support the importance of stepping voltages up with transformers for long-distance transmission?

THINK AND RANK (ANALYSIS)

45. Bar magnets are moved into the wire coils in identical quick fashion. Voltage induced in each coil causes a current, as indicated on the galvanometer. Neglect the electrical resistance in the loops in the coil, and rank from highest to lowest the readings on the galvanometer.

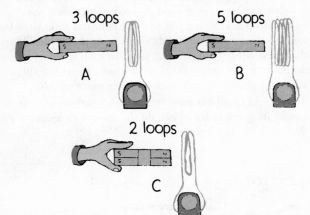

46. The three transformers are each powered with 100 W, and all have 100 turns on the primary. The number of turns on each secondary varies as indicated.

    (a) Rank the voltage output of the secondaries from greatest to least.

    (b) Rank the current in the secondaries from greatest to least.

    (c) Rank the power output in the secondaries from greatest to least.

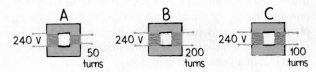

EXERCISES (SYNTHESIS)

47. Many dry cereals are fortified with iron, which is added in the form of small iron particles. How might these particles be separated from the cereal?

48. All atoms have moving electric charges. Why, then, aren't all materials magnetic?

49. To make a compass, point an ordinary iron nail along the direction of Earth's magnetic field (which, in the Northern Hemisphere, is angled downward as well as northward) and repeatedly strike it for a few seconds with a hammer or a rock. Then suspend it at its center of gravity by a string. Why does the act of striking magnetize the nail?

50. If you place a chunk of iron near the north pole of a magnet, attraction will occur. Why will attraction also occur if you place the same iron near the south pole of the magnet?

51. What is different about the magnetic poles of refrigerator magnets compared with those of bar magnets?

52. What kind of force field surrounds a stationary electric charge? What additional field surrounds it when it moves?

53. Will either pole of a magnet attract a paper clip? Explain what is happening inside the attracted paper clip. (Hint: Consider Figure 8.8 back in Chapter 8.)

54. Nails sticking to a magnet is understandable. Why are other nails attracted to the stuck nails, as in Figure 9.9?

55. A friend tells you that aluminum lies beneath the layer of white plastic on a refrigerator door. How could you check to see if this is true (without any scraping)?

56. What is the net magnetic force on a compass needle? By what mechanism does a compass needle align with a magnetic field?

57. We know that a compass points northward because Earth is a giant magnet. Does the northward-pointing needle point northward when the compass is brought to the Southern Hemisphere?

58. Magnet A has twice the magnetic field strength of magnet B, and at a certain distance it pulls on magnet B with a force of 50 N. With how much force does magnet B then pull on magnet A?

59. In Figure 9.17, we see a magnet exerting a force on a current-carrying wire. Does a current-carrying wire exert a force on a magnet? Why or why not?

60. When a current-carrying wire is placed in a strong magnetic field, no force acts on the wire. What orientation of the wire is likely?

61. A strong magnet attracts a paper clip to itself with a certain force. Does the paper clip exert a force on the strong magnet? If not, why not? If so, does it exert as much force on the magnet as the magnet exerts on it? Defend your answers.

62. When steel naval ships are built, the location of the shipyard and the orientation of the ship while in the shipyard are recorded on a brass plaque permanently fixed to the ship. What does this have to do with magnetism?

63. Can an electron at rest in a magnetic field be set into motion by the magnetic field? What if it were at rest in an electric field?

64. Two charged particles are projected into a magnetic field that is perpendicular to their velocities. If the charges are deflected in opposite directions, what does this tell you about the particles?

65. Residents of northern Canada are bombarded by more intense cosmic radiation than are residents of Mexico. Why is this so?

66. When walking in space, why do astronauts keep to altitudes beneath the Van Allen radiation belts?

67. What changes in cosmic-ray intensity at Earth's surface would you expect during periods in which Earth's magnetic field is passing through a zero phase while undergoing pole reversals?

68. When preparing to undergo a magnetic resonance imaging (MRI) scan, why are patients advised to remove metallic objects such as eyeglasses, watches, jewelry, and cell phones?

69. In a mass spectrometer, ions are directed into a magnetic field, where they curve around in the field and strike a detector. If a variety of singly ionized atoms travel at the same speed through the magnetic field, would you expect them all to be deflected by the same amount? Or would you expect different ions to be bent by different amounts?

70. Historically, replacing dirt roads with paved roads reduced rolling friction between vehicles and the surface of the road. Replacing paved roads with steel rails reduced friction further. What will be the next step in reducing friction between vehicles and the surfaces over which they move? What friction will remain after surface friction has been eliminated?

71. A common pickup for an electric guitar consists of a coil of wire around a small permanent magnet, as shown in Figure 9.26. Why will this type of pickup fail with nylon strings?

72. When Tim pushes the wire between the poles of the magnet, the galvanometer registers a pulse. When he lifts the wire, another pulse is registered. How do the pulses differ?

73. Why is a generator armature harder to rotate when it is connected to a circuit and supplying electric current?

74. Does a cyclist coast farther if the headlamp connected to the bike generator is turned off? Explain.

75. If your metal car moves over a wide, closed loop of wire embedded in a road surface, is Earth's magnetic field within the loop altered? Does this produce a current pulse? Can you think of a practical application at a traffic intersection?

76. At the security area of an airport you walk through a metal detector, which consists of a weak ac magnetic field inside a large coil of wire. If you forget to take keys out of your pocket as you pass through the detector, or if you wear a pacemaker, why is an alarm sounded?

77. A piece of plastic tape coated with iron oxide is magnetized more in some parts than in others. When the tape is moved past a small coil of wire, what happens in the coil? What has been a practical application of this?

78. How do the input and output parts of a generator and a motor compare?

79. Your friend says that, if you crank the shaft of a dc motor manually, the motor becomes a dc generator. Do you agree or disagree? Defend your position.

80. If you place a metal ring in a region where a magnetic field is rapidly alternating, the ring may become hot to your touch. Why?

81. How could a lightbulb near, yet not touching, an electromagnet be lit? Is ac or dc required? Defend your answer.

82. Two separate but similar coils of wire are mounted close to each other, as shown. The first coil is connected to a battery and has a direct current flowing through it. The second coil is connected to a galvanometer. How does the galvanometer respond when the switch in the first circuit

is closed? After being closed when the current is steady? When the switch is opened?

Primary   Secondary

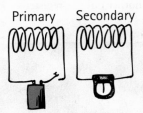

83. Why will more voltage be induced with the apparatus shown in the preceding exercise if an iron core is inserted in the coils?

84. Why won't a transformer work in a dc circuit?

85. What is the principal difference between a step-up transformer and a step-down transformer?

86. In what sense can a transformer be thought of as an electrical lever? What does it multiply? What does it not multiply?

87. Can an efficient transformer step up energy? Defend your answer.

88. A friend says that changing electric and magnetic fields generate each other, and this gives rise to visible light when the frequency of change matches the frequencies of light. Do you agree? Explain.

89. Would electromagnetic waves exist if changing magnetic fields could produce electric fields but changing electric fields could not in turn produce magnetic fields? Explain.

90. Your physics instructor drops a magnet through a long vertical copper pipe and it moves slowly compared with the drop of a nonmagnetized object. Provide an explanation.

91. This exercise is similar to the preceding one. Why will a bar magnet fall slower and reach terminal velocity in a vertical copper or aluminum tube but not in a cardboard tube?

DISCUSSION QUESTIONS (SYNTHESIS)

92. Discuss why a motor also tends to act like a generator.

93. Both English physicist Michael Faraday and American physicist Joseph Henry independently discovered electromagnetic induction at about the same time. In Henry's electrical experiments, his wife donated part of her wedding gown for silk to cover the wires of Henry's electromagnets. What was the purpose of the silk covering?

94. Your lab partner says, "An electron always experiences a force in an electric field, but not always in a magnetic field." If you agree with him, defend his statement.

95. One method for making a compass is to stick a magnetized needle into a piece of cork and float it in a glass bowl full of water, as shown. The needle aligns itself with the horizontal component of Earth's magnetic field. As the north pole of this compass is attracted northward, does the needle float toward the north side of the bowl? Defend your answer.

96. A cyclotron is a device for accelerating charged particles to high speeds as they follow an expanding spiral path. The charged particles are subjected to both an electric field and a magnetic field. One of these fields increases the speed of the charged particles, and the other field causes them to follow a curved path. Discuss which field performs which function.

97. A beam of high-energy protons emerges from a cyclotron. Do you suppose a magnetic field is associated with these particles? Discuss.

98. A magnetic field can deflect a beam of electrons, but it cannot do work on the electrons to change their speed. Discuss why this is so.

99. Why can a hum usually be heard when a transformer is operating?

100. Why doesn't a transformer work with direct current? Why is ac required?

101. Do a pair of parallel current-carrying wires exert forces on each other?

102. A magician places an aluminum ring on a table, underneath which is hidden an electromagnet. When the magician says "abracadabra" (and pushes a switch that starts current flowing through the coil under the table), the ring jumps into the air. Explain his "trick."

103. Discuss what is wrong with this scheme. To generate electricity without fuel, arrange a motor to run a generator that produces electricity that is stepped up with transformers so that the generator can run the motor and simultaneously furnish electricity for other uses.

READINESS ASSURANCE TEST (RAT)

If you have a good handle on this chapter, if you really do, then you should be able to score at least 7 out of 10 on this RAT. If you score less than 7, you need to study further before moving on.

Choose the BEST answer to each of the following.

1. The source of all magnetism is
   (a) tiny bits of iron.
   (b) tiny domains of aligned atoms.
   (c) small lodestones.
   (d) the motion of electrons.

2. Surrounding moving electric charges are
   (a) electric fields.
   (b) magnetic fields.
   (c) both of these
   (d) neither of these

3. A magnetic force acts most strongly on a current-carrying wire when the wire
   (a) carries a very large current.
   (b) is perpendicular to the magnetic field.
   (c) either or both of these
   (d) none of the above

4. A magnetic force acting on a beam of electrons can change
   (a) only the direction of the beam.
   (b) only the energy of the electrons.
   (c) both the direction and the energy.
   (d) neither the direction nor the energy.

5. When you thrust a bar magnet to and fro into a coil of wire, you induce
   (a) direct current.
   (b) alternating current.
   (c) neither dc nor ac.
   (d) alternating voltage only, not current.

6. The underlying physics of an electric motor is that
   (a) electric and magnetic fields repel each other.
   (b) a current-carrying wire experiences force in a magnetic field.
   (c) like magnetic poles both attract and repel each other.
   (d) ac voltage is induced by a changing magnetic field.

7. The essential physics concept in an electric generator is
   (a) Coulomb's law.
   (b) Ohm's law.
   (c) Faraday's law.
   (d) Newton's second law.

8. A transformer works by way of
   (a) Coulomb's law.
   (b) Ohm's law.
   (c) Faraday's law.
   (d) Newton's second law.

9. A step-up transformer in an electric circuit can step up
   (a) voltage.
   (b) energy.
   (c) both of these
   (d) neither of these

10. Electricity and magnetism connect to form
    (a) mass.
    (b) energy.
    (c) ultra high-frequency sound.
    (d) light.

Answers to RAT

1. d, 2. c, 3. 4a, 5. b, 6. b, 7. c, 8. c, 9. a, 10. d

10

CHAPTER 10

Waves and Sound

MANY THINGS in the world about us wiggle and jiggle—the surface of a bell, a string on a violin, the reed in a clarinet, lips on the mouthpiece of a trumpet, and the vocal cords of your larynx when you speak or sing. All these things *vibrate*. When they vibrate in air, they make the air molecules they touch wiggle and jiggle too, in exactly the same way, and these vibrations spread out in all directions, getting weaker, losing energy as heat, until they die out completely. But if these vibrations were to reach your ear instead, they would be transmitted to a part of your brain, and you would hear sound.

10.1 Vibrations and Waves

EXPLAIN THIS How do vibrating electrons produce radio waves?

LEARNING OBJECTIVE
Distinguish among amplitude, wavelength, frequency, and period.

In a general sense, anything that moves back and forth, to and fro, from side to side, in and out, or up and down is vibrating. A **vibration** is a wiggle in time. A wiggle in space and time is a **wave**. A wave extends from one location to another. Light and sound are both vibrations that propagate throughout space as waves, but as waves of two very different kinds. Sound is the propagation of vibrations through a material medium—a solid, a liquid, or a gas. If no medium exists to vibrate, then no sound is possible. Sound cannot travel in a vacuum. But light can, because (as we discuss in Chapter 11) light is a vibration of nonmaterial electric and magnetic fields—a vibration of pure energy. Although light can pass through many materials, it needs none. This is evident when it propagates through the vacuum between the Sun and Earth.

MasteringPHYSICS®

TUTORIAL: Waves and Vibrations

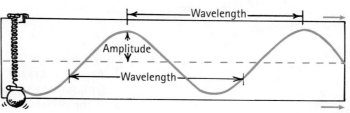

FIGURE 10.1
INTERACTIVE FIGURE

When the bob vibrates up and down, a marking pen traces out a sine curve on the paper, which is moved horizontally at constant speed.

The relationship between a vibration and a wave is shown in Figure 10.1. A marking pen on a bob attached to a vertical spring vibrates up and down and traces a waveform on a sheet of paper that is moved horizontally at constant speed. The waveform is actually a *sine curve*, a pictorial representation of a wave. As for a water wave, the high points are called *crests*, and the low points are the *troughs*. The straight dashed line represents the "home" position, or midpoint, of the vibration. The term **amplitude** refers to the distance from the midpoint to the crest (or to the trough) of the wave. So the amplitude equals the maximum displacement from equilibrium.

The **wavelength** of a wave is the distance from the top of one crest to the top of the next one or, equivalently, the distance between successive identical parts of the wave. The wavelengths of waves at the beach are measured in meters, the wavelengths of ripples in a pond in centimeters, and the wavelengths of light in billionths of a meter (nanometers). All waves have a vibrating source.

How frequently a vibration occurs is described by its *frequency*. The **frequency** of a vibrating pendulum, or of an object on a spring, specifies the number of to-and-fro vibrations it makes in a given time (usually in 1 s). A complete to-and-fro oscillation is one vibration. If it occurs in 1 s, the frequency is one vibration per second. If two vibrations occur in 1 s, the frequency is two vibrations per second.

The unit of frequency is called the **hertz** (Hz), after Heinrich Hertz, who demonstrated the existence of radio waves in 1886. One vibration per second is 1 Hz; two vibrations per second is 2 Hz, and so on. Higher frequencies are measured in kilohertz (kHz), and still higher frequencies in megahertz (MHz). AM radio waves are usually measured in kilohertz, while FM radio waves are measured in megahertz. A station at 960 kHz on the AM radio dial, for example, broadcasts radio waves that have a frequency of 960,000 vibrations per second. A station at 101.7 MHz on the FM dial broadcasts radio waves with a frequency of 101,700,000 hertz. These radio-wave frequencies are the frequencies at which electrons are forced to vibrate in the antenna of a radio station's transmitting tower. Still higher frequencies are measured in gigahertz (GHz), 1 billion vibrations per second. Cell phones operate in the GHz range, which means electrons inside are jiggling in unison billions of times per second! The frequency of the vibrating electrons and the frequency of the wave produced are the same.

The **period** of a wave or vibration is the time it takes for a complete vibration—for a complete cycle. Period can be calculated from frequency, and vice versa.

FIGURE 10.2
The source of any wave is something that vibrates. Electrons in the transmitting antenna vibrate 940,000 times each second and produce 940-kHz radio waves. Radio waves can't be seen or heard, but they send a pattern that tells a radio or a TV set what sounds or pictures to make.

FYI The frequency of a wave matches the frequency of its vibrating source. This is true not only of sound waves, but, as we'll see in the next chapter, of light waves also. The waves we're learning about, strictly speaking, are *periodic waves*—having distinct periods.

Suppose, for example, that a pendulum makes two vibrations in 1 s. Its frequency is 2 Hz. The time needed to complete one vibration—that is, the period of vibration—is $\frac{1}{2}$ s. Or if the vibration frequency is 3 Hz, then the period is $\frac{1}{3}$ s. The frequency and period are the inverse of each other:

$$\text{Frequency} = \frac{1}{\text{period}}$$

Or, vice versa,

$$\text{Period} = \frac{1}{\text{frequency}}$$

CHECKPOINT

1. **An electric razor completes 60 cycles every second. What are (a) its frequency and (b) its period?**
2. **If the difference in height between the crest and trough of a wave is 60 cm, what is the amplitude of the wave?**

Were these your answers?

1. (a) 60 cycles per second or 60 Hz; (b) $\frac{1}{60}$ second.
2. The amplitude is 30 cm, half of the crest-to-trough height distance.

LEARNING OBJECTIVE
Describe how energy is carried in waves.

10.2 Wave Motion

EXPLAIN THIS How does wave speed relate to frequency and wavelength?

If you drop a stone into a calm pond, waves travel outward in expanding circles. Energy is carried by the wave, traveling from one place to another. The water itself goes nowhere. This can be seen by waves encountering a floating leaf. The leaf bobs up and down, but it doesn't travel with the waves. The waves move along, not the water. The same is true for waves of wind over a field of tall grass on a gusty day. Waves travel across the grass, while the individual blades of grass remain in place; they swing to and fro between definite limits, but they go nowhere. When you speak, molecules in air propagate the disturbance through the air at about 340 m/s. The disturbance, not the air itself, travels across the room at this speed. In these examples, when the wave motion ceases, the water, the grass, and the air return to their initial positions. A characteristic of wave motion is that the medium transporting the wave returns to its initial condition after the disturbance has passed.

FIGURE 10.3
Diane Riendeau uses a classroom wave machine to demonstrate how a vibration produces a wave.

Wave Speed

The speed of periodic wave motion is related to the frequency and wavelength of the waves. Consider the simple case of water waves (Figures 10.4 and 10.5). Imagine that we fix our eyes on a stationary point on the water's surface and observe the waves passing by that point. We can measure how much time passes between the arrival of one crest and the arrival of the next one (the period), and we can also observe the distance between crests (the wavelength). We know that speed is defined as distance divided by time. In this case, the distance is one wavelength and the time is one period, so the speed of a wave = wavelength/period.

For example, if the wavelength is 10 m and the time between crests at a point on the surface is 0.5 s, the wave is traveling 10 m in 0.5 s and its speed is 10 m divided by 0.5 s, or 20 m/s.

Because period is the inverse of frequency, the formula **wave speed** = wavelength/period can also be written as

$$\text{Wave speed} = \text{frequency} \times \text{wavelength}$$

This relationship applies to all kinds of waves, whether they are water waves, sound waves, or light waves.

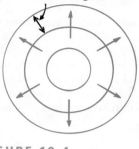

FIGURE 10.4
A top view of water waves.

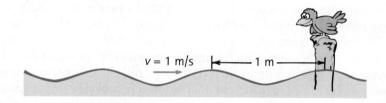

FIGURE 10.5
INTERACTIVE FIGURE   MP

If the wavelength is 1 m, and one wavelength per second passes the pole, then the speed of the wave is 1 m/s.

CHECKPOINT

1. If a train of freight cars, each 10 m long, rolls by you at the rate of three cars each second, what is the speed of the train?
2. If a water wave oscillates up and down three times each second and the distance between wave crests is 2 m, (a) what is its frequency? (b) What is its wavelength? (c) What is its wave speed?

Were these your answers?

1. 30 m/s. We can see this in two ways. According to the definition of speed in Chapter 2, $v = \dfrac{d}{t} = \dfrac{3 \times 10 \text{ m}}{1 \text{ s}} = 30 \text{ m/s}$, because 30 m of train passes you in 1 s. If we compare our train to wave motion, where wavelength corresponds to 10 m and frequency is 3 Hz, then

    Speed = frequency × wavelength = 3 Hz × 10 m = 30 m/s

2. (a) 3 Hz; (b) 2 m; (c) Wave speed = frequency × wavelength = 3/s × 2 m = 6 m/s.

FYI It is customary to express the speed of a wave by the equation $v = f\lambda$, where v is wave speed, f is wave frequency, and λ (the Greek letter lambda) is wavelength.

Be clear about the distinction between *frequency* and *speed*. How frequently a wave vibrates is altogether different from how fast it moves from one location to another.

10.3 Transverse and Longitudinal Waves

EXPLAIN THIS Exactly what is transmitted in all kinds of waves?

Fasten one end of a Slinky to a wall and hold the free end in your hand. If you shake the free end up and down, you produce vibrations that are at right angles to the direction of wave travel. The right-angled, or sideways, motion is called *transverse motion*. This type of wave is called a **transverse wave**. Waves in the stretched strings of musical instruments and on the surfaces of liquids are transverse waves. We will see later that electromagnetic waves, some of which are radio waves and light waves, are also transverse waves.

LEARNING OBJECTIVE
Distinguish between transverse and longitudinal waves.

Mastering**PHYSICS**

VIDEO: Transverse vs. Longitudinal Waves

FIGURE 10.6
INTERACTIVE FIGURE **MP**

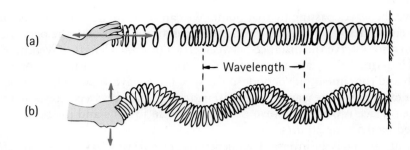

Both waves transfer energy from left to right. (a) When the end of the Slinky is pushed and pulled rapidly along its length, a longitudinal wave is produced. (b) When its end is shaken up and down (or side to side), a transverse wave is produced.

FIGURE 10.7
If you vibrate a Ping-Pong paddle in the midst of a lot of Ping-Pong balls, the balls bounce from one another and also vibrate.

A **longitudinal wave** is one in which the direction of wave travel is along the direction in which the source vibrates. You produce a longitudinal wave with your Slinky when you shake it back and forth along the Slinky's axis (Figure 10.6a). The vibrations are then parallel to the direction of energy transfer. Part of the Slinky is compressed, and a wave of **compression** travels along it. Between successive compressions is a stretched region called a **rarefaction**. Both compressions and rarefactions travel parallel to the Slinky. Together they make up the longitudinal wave. Figure 10.6b shows the generation of a transverse wave.

If you study earthquakes, you'll learn about two types of waves that travel in the ground. One type is longitudinal (P waves), and the other type is transverse (S waves). These travel at different speeds, which provides investigators with a means of determining the source of the waves. Furthermore, the transverse waves cannot travel through liquid matter, while the longitudinal waves can, which provides a means of determining whether matter below ground is molten or solid.

LEARNING OBJECTIVE
Identify compressions and rarefactions in a sound wave.

10.4 Sound Waves

EXPLAIN THIS Why doesn't sound travel in a vacuum?

Think of the air molecules in a room as tiny randomly moving Ping-Pong balls. If you vibrate a Ping-Pong paddle in the midst of the balls, you send a to-and-fro vibration through them. The balls vibrate in rhythm with your vibrating paddle. In some regions they are momentarily bunched up (compressions), and in other regions in between they are momentarily spread out (rarefactions). The vibrating prongs of a tuning fork do the same to air molecules. Vibrations made up of compressions and rarefactions spread from the tuning fork throughout the air, and a *sound wave* is produced.

The wavelength of a sound wave is the distance between successive compressions or, equivalently, the distance between successive rarefactions. Each molecule in the air vibrates to and fro about some equilibrium position as the waves move by.

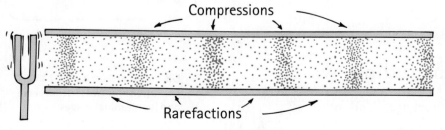

FIGURE 10.8
Compressions and rarefactions travel (both at the same speed and in the same direction) from the tuning fork through the air in the tube. The wavelength is the distance between successive compressions (or rarefactions).

Our subjective impression about the frequency of sound is described as **pitch**. A high-pitched sound, such as that from a tiny bell, has a high vibration frequency. Sound from a large bell has a low pitch because its vibrations are of a low frequency. Pitch is how high or low we perceive a sound to be, depending on the frequency of the sound wave.

The human ear can normally hear pitches from sound ranging from about 20 Hz to about 20,000 Hz. As we age, this range shrinks. So by the time you can afford high-end speakers for your home theater system, you may not be able to tell the difference. Sound waves of frequencies lower than 20 Hz are called *infrasonic waves*, and those of frequencies higher than 20,000 Hz are called *ultrasonic waves*. We cannot hear infrasonic or ultrasonic sound waves.* But dogs and some other animals can.

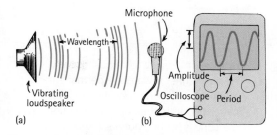

Most sound is transmitted through air, but any elastic substance—solid, liquid, or gas—can transmit sound.** Air is a poor conductor of sound compared with solids and liquids. You can hear the sound of a distant train clearly by placing your ear against the rail. When swimming, have a friend at a distance click two rocks together beneath the surface of water while you are submerged. Observe how well water conducts the sound. Sound cannot travel in a vacuum because there is nothing to compress and expand. The transmission of sound requires a medium.

Pause to reflect on the physics of sound while you are quietly listening to your radio sometime. The radio loudspeaker is a paper cone that vibrates in rhythm with an electrical signal. Air molecules next to the vibrating cone of the speaker are themselves set into vibration. These, in turn, vibrate against neighboring molecules, which, in turn, do the same, and so on. As a result, rhythmic patterns of compressed and rarefied air emanate from the loudspeaker, showering the entire room with undulating motions. The resulting vibrating air sets your eardrum into vibration, which, in turn, sends cascades of rhythmic electrical impulses along nerves in the cochlea of your inner ear and into the brain. And thus you listen to the sound of music.

Speed of Sound

If, from a distance, we watch a person chopping wood or hammering, we can easily see that the blow occurs a noticeable time before its sound reaches our ears. Thunder is often heard seconds after a flash of lightning is seen. These common experiences show that sound requires time to travel from one place to another. The speed of sound depends on wind conditions, temperature, and humidity. It does not depend on the loudness or the frequency of the sound; all sounds travel at the same speed in a given medium. The speed of sound in dry air at 0°C is about 330 m/s, which is nearly 1200 km/s. Water vapor in the air increases this speed slightly. Sound travels faster through warm air than through cold air. This is to be expected, because the faster-moving molecules in warm air bump into each other more frequently and, therefore, can transmit a pulse in less time.† For each 1-degree increase in temperature above 0°C, the speed of sound in air increases by 0.6 m/s. Thus, in air at a normal room temperature of about 20°C, sound travels at about 340 m/s. In water, the speed of sound is about 4 times its speed in air; in steel, about 15 times its speed in air.

FIGURE 10.9
(a) The radio loudspeaker is a paper cone that vibrates in rhythm with an electrical signal. The sound that is produced sets up similar vibrations in the microphone. The vibrations are displayed on an oscilloscope. (b) The waveform on the oscilloscope screen is a graph of pressure against time, showing how air pressure near the microphone rises and falls as sound waves pass. When the loudness increases, the amplitude of the waveform increases.

FYI Elephants communicate with one another with infrasonic waves. Their large ears help them detect these low-frequency sound waves.

FIGURE 10.10
Waves of compressed and rarefied air, generated by the vibrating cone of the loudspeaker, reproduce the sound of music.

FYI A sound wave traveling through the ear canal vibrates the eardrum, which vibrates three tiny bones, which in turn vibrate the fluid-filled cochlea. Inside the cochlea, tiny hair cells convert the pulse into an electrical signal to the brain.

* In hospitals, concentrated beams of ultrasound are used to break up kidney stones and gallstones, eliminating the need for surgery.
** An elastic substance is "springy," has resilience, and can transmit energy with little loss. Steel, for example, is elastic, but lead and putty are not.
† The speed of sound in a gas is about $\frac{3}{4}$ the average speed of its molecules.

LOUDSPEAKERS

The loudspeakers of your radio and other sound-producing systems change electrical signals into sound waves. The electrical signals pass through a coil wound around the neck of a paper cone. This coil, which acts as an electromagnet, is located near a permanent magnet. When current flows one way, magnetic force pushes the electromagnet toward the permanent magnet, pulling the cone inward. When current flows in the opposite direction, the cone is pushed outward. Vibrations in the electrical signal cause the cone to vibrate. Vibrations of the cone then produce sound waves in the air.

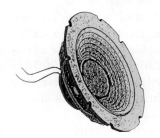

CHECKPOINT

1. Do compressions and rarefactions in a sound wave travel in the same direction or in opposite directions from one another?
2. What is the approximate distance of a thunderstorm when you note a 3-s delay between the flash of lightning and the sound of thunder?

Were these your answers?

1. They travel in the same direction.
2. Assuming the speed of sound in air is about 340 m/s, in 3 s sound travels 340 m/s × 3 s = 1020 m. There is no appreciable time delay for the flash of light, so the storm is slightly more than 1 km away.

LEARNING OBJECTIVE
Distinguish between the reflection and refraction of waves.

10.5 Reflection and Refraction of Sound

EXPLAIN THIS How can differences in air temperature bend sound waves?

Like light, when sound encounters a surface, it can either be returned by the surface or continue through it. When it is returned, the process is **reflection**. We call the reflection of sound an *echo*. The fraction of sound energy reflected from a surface is large if the surface is rigid and smooth, but it is less if the surface is soft and irregular. The sound energy that is not reflected is transmitted or absorbed.

Sound reflects from a smooth surface in the same way that light does—the angle of incidence (the angle between the direction of the sound and the normal to the reflecting surface) is equal to the angle of reflection (Figure 10.11). Sometimes, when sound reflects from the walls, ceiling, and floor of a room, the surfaces are too reflective and the sound becomes garbled. Sound due to multiple reflections is called a **reverberation**. On the other hand, if the reflective surfaces are too absorbent, the sound level is low and the room may sound dull and lifeless. Reflected sound in a room makes it sound lively and full, as you have probably experienced while singing in the shower. The designer of an auditorium or concert hall must find a balance between reverberation and absorption. The study of sound properties is called *acoustics*.

It is often advantageous to position highly reflective surfaces behind the stage to direct sound out to the audience. In some concert halls, reflecting

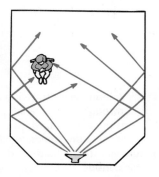

FIGURE 10.11
The angle of incident sound is equal to the angle of reflected sound.

surfaces are suspended above the stage. Ones such as those in Davies Hall in San Francisco are large shiny plastic surfaces that also reflect light. A listener can look up at these reflectors and see the reflected images of the members of the orchestra (the plastic reflectors are somewhat curved, which increases the field of view). Both sound and light obey the same law of reflection. Thus, if a reflector is oriented so that you can see a particular musical instrument, rest assured that you can also hear it. Sound from the instrument follows the line of sight to the reflector and then to you. In some halls, absorbers rather than reflectors are used to improve the acoustics.

Refraction occurs when sound continues through a medium and bends. Sound waves bend when parts of the wave fronts travel at different speeds. This may happen when sound waves are affected by uneven winds, or when sound travels through air of uneven temperatures. On a warm day, the air near the ground may be appreciably warmer than the air above, so the speed of sound near the ground increases. Sound waves therefore tend to bend away from the ground, resulting in sound that does not seem to transmit well (Figure 10.13).

FIGURE 10.12
The plastic plates above the orchestra reflect both light and sound. Adjusting them is quite simple: what you see is what you hear.

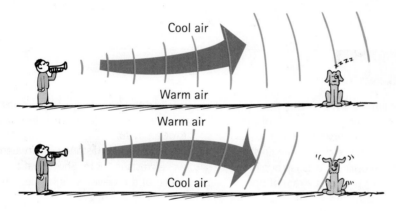

The refraction of sound occurs under water, where the speed of sound varies with temperature. This poses a problem for surface vessels that bounce ultrasonic waves off the bottom of the ocean to chart its features, but it's a blessing to submarines that wish to escape detection. Because the ocean has layers of water that are at different temperatures, the refraction of sound leaves gaps or "blind spots" in the water. This is where submarines hide. If not for refraction, submarines would be much easier to detect.

Physicians use the multiple reflections and refractions of ultrasonic waves to "see" the interior of the body without the use of X-rays. High-frequency sound (ultrasound) that enters the body is reflected more strongly from the organs' exteriors than from their interiors, producing an outline of the organs (Figure 10.15). This ultrasound echo technique is nothing new to bats and dolphins, which can emit ultrasonic squeaks and locate objects by their echoes.

FIGURE 10.13
Sound waves are bent in air of uneven temperatures.

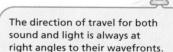
The direction of travel for both sound and light is always at right angles to their wavefronts.

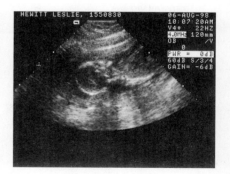

FIGURE 10.15
The 14-week-old fetus that became Megan Hewitt Abrams, who is more recently seen on page 216.

MasteringPHYSICS
VIDEO: Refraction of Sound

FIGURE 10.14
A dolphin emits ultrahigh-frequency sound to locate and identify objects in its environment. It senses distance by the time delay between sending sound and receiving the echo, and it senses direction by differences in time for the echo to reach the dolphin's two ears. A dolphin's main diet is fish. Because fish hear mainly low frequencies, they are not alerted to the fact that they are being hunted.

DOLPHINS AND ACOUSTICAL IMAGING

The dominant sense of the dolphin is hearing, because vision is not a very useful sense in the often murky and dark depths of the ocean. Whereas sound is a passive sense for us, it is an active sense for the dolphin, which sends out sounds and then perceives its surroundings by means of the echoes that return. The ultrasonic waves emitted by a dolphin enable it to "see" through the bodies of other animals and people. Skin, muscle, and fat are almost transparent to dolphins, so they

"see" a thin outline of the body—but the bones, teeth, and gas-filled cavities are clearly apparent. Dolphins can "see" physical evidence of cancers, tumors, and heart attacks—which humans have only recently been able to detect with ultrasound.

What's more fascinating, the dolphin can reproduce the sonic signals that paint the mental image of its surroundings; thus, it is probably able to communicate its experiences to other dolphins by communicating the full

acoustic image of what it has "seen," placing the image directly in the minds of other dolphins. It needs no word or symbol for "fish," for example, but can communicate an image of the real thing—perhaps with emphasis highlighted by selective filtering, as we similarly communicate a musical concert to others via various means of sound reproduction. Small wonder that the language of the dolphin is very unlike our own!

FIGURING PHYSICAL SCIENCE

Problem Solving

SAMPLE PROBLEM 1
An oceanic depth-sounding vessel surveys the ocean floor with ultrasonic sound that travels 1530 m/s in seawater. How deep is the water if the time delay of the echo from the ocean floor is 2 s?

Solution:
The round trip is 2 s, meaning 1 s down and 1 s up. Then,

$d = vt = 1530$ m/s $\times 1$ s $= 1530$ m

(Radar works similarly; microwaves rather than sound waves are transmitted.)

SAMPLE PROBLEM 2
While sitting on the dock of the bay, Otis notices incoming waves with distance d between crests. The incoming crests lap against the pier pilings at a rate of one every 2 s.

(a) Find the frequency of the waves.
(b) Show that the speed of the waves is given by fd.
(c) Suppose the distance d between wave crests is 1.8 m. Show that the speed of the waves is slightly less than 1.0 m/s.

Solution:
(a) The frequency of the waves is given: one per 2 s, or $f = 0.5$ Hz
(b) $v = f\lambda = fd$.
(c) $v = f\lambda = fd = 0.5$ Hz $(1.8$ m$)$
$= 0.5\left(\frac{1}{s}\right)(1.8$ m$) = 0.9$ m/s.

Mastering**PHYSICS**
VIDEO: Resonance
VIDEO: Resonance and Bridges

10.6 Forced Vibrations and Resonance

EXPLAIN THIS Why is the sound different for various items dropped on a floor?

If you strike an unmounted tuning fork, its sound is rather faint. Repeat with the handle of the fork held against a table after striking it, and the sound is louder. This is because the table is forced to vibrate, and its larger surface sets more air in motion. The table is forced into vibration by a fork of any frequency. This is an example of **forced vibration**. The vibration of a factory floor caused by the running of heavy machinery is another example of forced vibration. A more pleasing example is given by the sounding boards of stringed instruments.

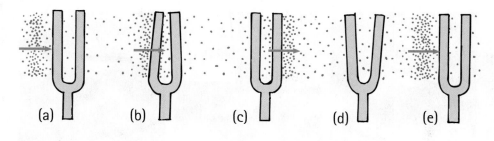

(a)   (b)   (c)   (d)   (e)

FIGURE 10.16
Stages of resonance. (a) The first compression meets the fork and gives it a tiny and momentary push; (b) the fork bends and then (c) returns to its initial position just at the time a rarefaction arrives and (d) overshoots in the opposite direction. Just when it returns to its initial position, (e) the next compression arrives to repeat the cycle. Now it bends farther because it is moving.

If you drop a wrench and a baseball bat on a concrete floor, you easily notice the difference in their sounds. This is because each vibrates differently when striking the floor. They are not forced to vibrate at a particular frequency; instead, each vibrates at its own characteristic frequency. Any object composed of an elastic material, when disturbed, vibrates at its own special set of frequencies, which together form its characteristic sound. We speak of an object's **natural frequency**, which depends on such factors as the elasticity and shape of the object. Bells and tuning forks, of course, vibrate at their own characteristic frequencies. Interestingly, most things, from atoms to planets and almost everything else in between, have springiness to them, and they vibrate at one or more natural frequencies.

When the frequency of forced vibrations on an object matches the object's natural frequency, a dramatic increase in amplitude occurs. This phenomenon is called **resonance**. Literally, *resonance* means "resounding" or "sounding again." Putty doesn't resonate, because it isn't elastic, and a dropped handkerchief is too limp to resonate. In order for something to resonate, it needs both a force to pull it back to its starting position and enough energy to maintain its vibration.

A common experience illustrating resonance occurs when you are on a swing. When pumping a swing, you pump in rhythm with the natural frequency of the swing. More important than the force with which you pump is the timing. Even small pumps, or small pushes from someone else, if delivered in rhythm with the frequency of the swinging motion, produce large amplitudes.

A common classroom demonstration of resonance is illustrated with a pair of tuning forks adjusted to the same frequency and spaced a meter or so apart (Figure 10.17). When one of the forks is struck, it sets the other fork into vibration. This is a small-scale version of pushing a friend on a swing—it's the timing that's important. When a series of sound waves impinge on the fork, each compression gives the prong of the fork a tiny push. Because the frequency of these pushes corresponds to the natural frequency of the fork, the pushes successively increase the amplitude of its vibration. This is because the pushes occur at the right time and repeatedly occur in the same direction as the instantaneous motion of the fork. The motion of the second fork is called a *sympathetic vibration.*

If the forks are not adjusted for matched frequencies, the timing of pushes is off, and resonance doesn't occur. When you tune your radio, you are similarly adjusting the natural frequency of the electronics in the device to match one of the many surrounding signals. The device then resonates to one station at a time, instead of playing all stations at once.

Resonance is not restricted to wave motion. It occurs whenever successive impulses are applied to a vibrating object in rhythm with its natural frequency. Cavalry troops marching across a footbridge near Manchester, England, in 1831 inadvertently caused the bridge to collapse when they marched in rhythm with the bridge's natural frequency. Since then, it is customary to order troops to "break step" when crossing bridges. A more recent bridge disaster was caused by wind-generated resonance (Figure 10.18).

FYI Owls have extremely sensitive ears. Hunting at night, owls tune in to the soft rustles and squeaks of rodents and other small mammals. Like humans, owls locate sound sources by using the fact that sound waves often reach one ear milliseconds before the other. An owl moves its head as it glides toward its prey; when sounds from the target reach both ears at once, the meal is dead ahead. In some owls, one ear is also higher than the other, further sharpening their prey-locating ability.

FIGURE 10.17
Ryan demonstrates resonance with a pair of tuning forks with matched frequencies.

FYI Parrots, like humans, use their tongues to craft and shape sound. Tiny changes in tongue position produce big differences in the sound first produced in the syrinx, a voice box organ nestled between the trachea and lungs.

FIGURE 10.18
In 1940, four months after being completed, the Tacoma Narrows Bridge in the state of Washington was destroyed by wind-generated resonance. The mild gale produced a fluctuating force in resonance with the natural frequency of the bridge, steadily increasing the amplitude until the bridge collapsed.

LEARNING OBJECTIVE
Describe how interference is a property of all wave behavior.

Why does Hollywood persist in playing engine noises whenever a spacecraft in outer space passes by? Wouldn't seeing them float by silently be far more dramatic?

FIGURE 10.19
Constructive and destructive interference in a transverse wave.

10.7 Interference

EXPLAIN THIS In what way can both sound and light be canceled?

An intriguing property of all waves is **interference**. Consider transverse waves. When the crest of one wave overlaps the crest of another, their individual effects add together. The result is a wave of increased amplitude. This is *constructive interference* (Figure 10.19). When the crest of one wave overlaps the trough of another, their individual effects are reduced. The high part of one wave simply fills in the low part of another. This is *destructive interference*.

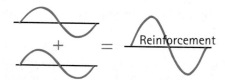
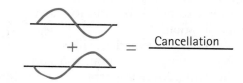

Wave interference is easiest to observe in water. In Figure 10.20, we see the interference pattern produced when two vibrating objects touch the surface of water. We can see the regions in which the crest of one wave overlaps the trough of another to produce a region of zero amplitude. At points along such regions, the waves arrive out of step. We say they are *out of phase* with one another.

FIGURE 10.20
Two sets of overlapping water waves produce an interference pattern.

Interference is a property of all wave motion, whether the waves are water waves, sound waves, or light waves. We see a comparison of interference for transverse waves and for longitudinal waves in Figure 10.22. For the transverse waves of light we see constructive interference where crests and troughs of one wave superimpose on another. Such waves hitting a screen show bright light. Dark light appears where destructive interference occurs—where crests overlap troughs. Similar effects occur for the interference of longitudinal sound waves, shown by the regions of compressions and rarefactions.

Destructive sound interference is at the heart of *antinoise technology*. Some noisy devices such as jackhammers are now equipped with microphones that send the sound of the device to electronic microchips, which create mirror-image wave patterns of the sound signals. This mirror-image sound signal is fed to earphones worn by the operator. In this way, sound compressions (or rarefactions) from the hammer are canceled by mirror-image rarefactions

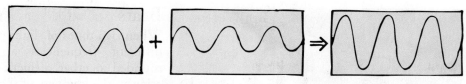

The superposition of two identical transverse waves in phase produces a wave of increased ampitude.

The superposition of two identical longitudinal waves in phase produces a wave of increased intensity.

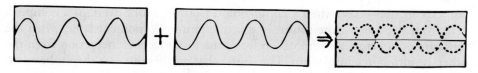

Two identical transverse waves that are out of phase destroy each other when they are superimposed.

Two identical longitudinal waves that are out of phase destroy each other when they are superimposed.

FIGURE 10.22
Constructive (top two panels) and destructive (bottom two panels) wave interference in transverse and longitudinal waves.

FIGURE 10.21
New Zealander Jennie McKelvie showing interference with a classroom ripple tank.

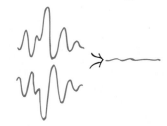

FIGURE 10.23
When a mirror image of a sound signal combines with the sound itself, the sound is canceled.

(or compressions) in the earphones. The combination of signals cancels the jackhammer noise. Antinoise devices are also common in some aircraft, which are much quieter inside than before this technology was introduced. Are automobiles next, perhaps eliminating the need for mufflers?

Sound interference is dramatically illustrated when monaural sound is played by stereo speakers that are out of phase. Speakers are out of phase when the input wires to one speaker are interchanged (positive and negative wire inputs reversed). For a monaural signal, this means that when one speaker is sending a compression of sound, the other is sending a rarefaction. The sound produced is not as full and not as loud as from speakers properly connected in phase. The longer waves are canceled by interference. Shorter waves are canceled as the speakers are brought closer together, and when the two speakers are brought face to face against each other, very little sound is heard! Only the highest frequencies survive cancellation. You must try this experiment to appreciate it.

The interference of light is evident in the bright colors seen in reflections from thin films of gasoline on water. Reflections from the gasoline and water surfaces interfere, canceling colors and producing their complementary colors (discussed in the next chapter).

FIGURE 10.24
When the positive and negative wire inputs to one of the stereo speakers have been interchanged, the speakers are then out of phase. When the speakers are far apart, monaural (not stereo) sound is not as loud as it is from properly phased speakers. When they are brought face to face, very little sound is heard. Interference is nearly complete, as the compressions of one speaker fill in the rarefactions of the other.

FIGURE 10.25
Ken Ford tows gliders in quiet comfort when he wears his noise-canceling earphones. In larger aircraft, sound from the engines is processed and emitted as anti-noise from loudspeakers inside the cabin to provide passengers with a quieter ride.

Mastering**PHYSICS**
VIDEO: Interference and Beats

Beats

When two tones of slightly different frequencies are sounded together, a fluctuation in the loudness of the combined sounds is heard; the sound is loud, then faint, then loud, then faint, and so on. This periodic variation in the loudness of sound is called **beats**, and it is due to interference. If you strike two slightly mismatched tuning forks, one fork vibrates at a different frequency from the other, and the vibrations of the forks are momentarily in step, then out of step, then in again, and so on. When the combined waves reach our ears in step—say, when a compression from one fork overlaps a compression from the other—the sound is at a maximum. A moment later, when the forks are out of step, a compression from one fork meets a rarefaction from the other, resulting in a minimum. The sound that reaches our ears throbs between maximum and minimum loudness and produces a tremolo effect.

Beats can occur with any kind of wave, and they can provide a practical way to compare frequencies. To tune a piano, for example, a piano tuner listens for beats produced between a standard tuning fork and those of a particular string on the piano. When the frequencies are identical, the beats disappear. The members of an orchestra tune up their instruments by listening for beats between their instruments and a standard tone produced by a piano or some other instrument.

FIGURE 10.26
The interference of two sound sources of slightly different frequencies produces beats.

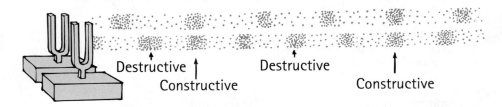

Destructive Destructive

Constructive Constructive

FYI See the production of standing waves at http://www2.biglobe.ne.jp/~norimari/science/JavaEd/e-wave4.html.

Standing Waves

Another fascinating effect of interference is *standing waves*. Tie a rope to a wall and shake the free end up and down. The wall is too rigid to shake, so the waves are reflected back along the rope. By shaking the rope just right, you can cause the incident and reflected waves to interfere and form a **standing wave**, in which parts of the rope, called the *nodes*, are stationary. You can hold your fingers on either side of the rope at a node, and the rope doesn't touch them. Other parts of the rope, however, would make contact with your fingers. The positions on a standing wave with the largest displacements are known as *antinodes*. Antinodes occur halfway between nodes.

Standing waves are produced when two sets of waves of equal amplitude and wavelength pass through each other in opposite directions. Then the waves are steadily in and out of phase with each other and produce stable regions of constructive and destructive interference (Figure 10.27).

Standing waves are set up in the strings of musical instruments when plucked, bowed, or struck. They are produced in the air in an organ pipe, a flute, or a clarinet—and in the air of a soft-drink bottle when air is blown over the top.

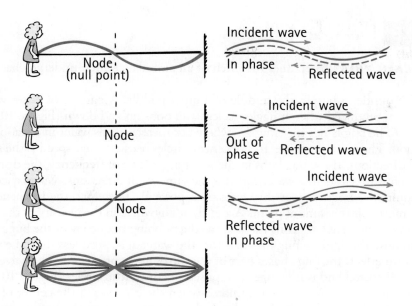

The incident and reflected waves interfere to produce a standing wave.

Standing waves appear in a tub of water or a cup of coffee when sloshed back and forth at the appropriate frequency. Standing waves can be produced with either transverse or longitudinal vibrations.

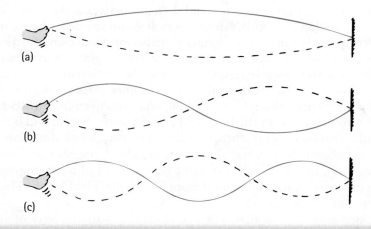

(a) Shake the rope until you set up a standing wave of one loop ($\frac{1}{2}$ wavelength).
(b) Shake with twice the frequency and produce a wave with two loops (1 wavelength).
(c) Shake with three times the frequency and produce three loops ($\frac{3}{2}$ wavelengths).

CHECKPOINT

1. **Is it possible for one wave to cancel another wave so that no amplitude remains?**
2. **Suppose you set up a standing wave of three segments, as shown in Figure 10.28c. If you shake with a frequency twice as great, how many wave segments occur in your new standing wave? How many wavelengths?**

Were these your answers?

1. Yes. This is called destructive interference. When a standing wave is set up in a rope, for example, parts of the rope have no amplitude—the nodes.
2. If you impart twice the frequency to the rope, you produce a standing wave with twice as many segments (six). Because a full wavelength has two segments, you have three complete wavelengths in your standing wave.

The frequency of a "classic" wave—such as a sound wave, water wave, or radio wave—matches the frequency of its vibrating source. (In the quantum world of atoms and photons, the rules are different.)

10.8 Doppler Effect

EXPLAIN THIS When does the pitch of an ambulance siren undergo change?

Consider a bug in the middle of a quiet puddle. A pattern of water waves is produced when it jiggles its legs and bobs up and down (Figure 10.29). The bug is not traveling anywhere but merely treads water in a stationary position. The waves it creates are concentric circles because wave speed is the same in all directions. If the bug bobs in the water at a constant frequency, the distance between wave crests (the wavelength) is the same in all directions. Waves encounter point A as frequently as they encounter point B. Therefore, the frequency of wave motion is the same at points A and B, or anywhere in the vicinity of the bug. This wave frequency remains the same as the bobbing frequency of the bug.

Suppose the jiggling bug moves across the water at a speed less than the wave speed. In effect, the bug chases part of the waves it has produced. The wave pattern is distorted and is no longer composed of concentric circles (Figure 10.30). The center of the outer wave originated when the bug was at the center of that circle. The center of the next smaller wave originated when the bug was at the center of that circle, and so forth. The centers of the circular waves move in the direction of the swimming bug. Although the bug maintains the same bobbing frequency as before, an observer at B would see the waves coming more often. The observer would measure a higher frequency. This is because each successive wave has a shorter distance to travel and therefore arrives at B sooner than if the bug weren't moving toward B. An observer at A, on the other hand, measures a lower frequency because of the longer time between wave-crest arrivals. This occurs because each successive wave travels farther to get to A as a result of the bug's motion. This change in frequency due to the motion of the source (or due to the motion of the receiver) is called the **Doppler effect** (after the Austrian physicist and mathematician Christian Doppler, who lived from 1803 to 1853).

Water waves spread over the flat surface of the water. Sound and light waves, on the other hand, travel in three-dimensional space in all directions like an expanding balloon. Just as circular waves are closer together in front of the swimming bug, spherical sound or light waves ahead of a moving source are closer together and reach an observer more frequently. The Doppler effect holds for all types of waves.

The Doppler effect is evident when you hear the changing pitch of an ambulance or fire-engine siren. When the siren is approaching you, the crests of the sound waves encounter your ear more frequently, and the pitch is higher than normal. And when the siren passes you and moves away, the crests of the waves encounter your ear less frequently, and you hear a drop in pitch.

The Doppler effect also occurs for light. When a light source approaches, its measured frequency increases; when it recedes, its frequency decreases. An increase in light frequency is called a *blueshift*, because the increase is toward a higher frequency, or toward the blue end of the color spectrum. A decrease in frequency is called a *redshift*, referring to a shift toward a lower frequency, or toward the red end of the color spectrum. Galaxies, for example, show a redshift in the light they emit as they move away from us in the expanding universe. Measuring this shift allows us to calculate their speed. A rapidly spinning star shows a redshift on the side turning away from us and a blueshift on the side turning toward us. This enables us to calculate the star's spin rate.

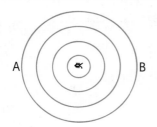

FIGURE 10.29

Top view of water waves made by a stationary bug jiggling in still water.

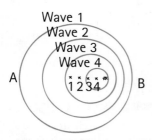

FIGURE 10.30
INTERACTIVE FIGURE MP

Water waves made by a bug swimming in still water toward point B.

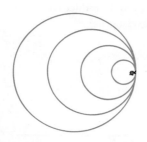

FIGURE 10.31

The wave pattern made by a bug swimming at wave speed.

FIGURE 10.32
INTERACTIVE FIGURE MP

The pitch of sound increases when the source moves toward you, and it decreases when the source moves away.

CHECKPOINT

When a light or sound source moves toward you, is there an increase or a decrease in the wave speed?

Was this your answer?
Neither! The frequency of a wave undergoes a change when the source is moving, not the wave speed.

10.9 Bow Waves and the Sonic Boom

EXPLAIN THIS How can a snapped circus whip produce a sonic boom?

LEARNING OBJECTIVE
Describe the production of bow waves and shock waves.

When a source of waves travels as fast as the waves it produces, a *wave barrier* is produced. Consider the bug in our previous example. If it swims as fast as the waves it makes, the bug keeps up with the waves it produces. Instead of moving ahead of the bug, the waves superimpose on one another directly, forming a hump in front of the bug (Figure 10.31). Thus, the bug encounters a wave barrier. The bug must expend some extra effort to swim over the hump before it can swim faster than wave speed.

The same thing happens when an aircraft travels at the speed of sound. The waves overlap to produce a barrier of compressed air on the leading edges of the wings and on other parts of the aircraft. This produces a "sound barrier," which on older aircraft caused some control problems, and on modern aircraft makes for some interesting visual effects (Figure 10.33). At higher speed the aircraft is *supersonic*. It is like the bug, which, once it has passed its wave barrier, finds the medium ahead relatively smooth and undisturbed.

When the bug swims faster than wave speed, it produces a pattern of overlapping waves, ideally shown in Figure 10.34. The bug overtakes and outruns the waves it produces. The overlapping waves form a V shape, called a **bow wave**, which appears to be dragging behind the bug. Overlapping waves produce the familiar bow wave generated by a speedboat knifing through the water.

Some wave patterns created by sources moving at various speeds are shown in Figure 10.35. Note that after the speed of the source exceeds wave speed, increased speed produces a narrower V shape.*

Whereas a speedboat knifing through the water generates a two-dimensional bow wave at the surface of the water, a supersonic aircraft similarly generates a three-dimensional *shock wave*. Just as a bow wave is produced by overlapping circles that form a V, a **shock wave** is produced by overlapping spheres that form a cone. And just as the bow wave of a speedboat spreads until it reaches

FIGURE 10.33
Condensation of water vapor by rapid expansion of air can be seen in the rarefied region behind the wall of compressed air.

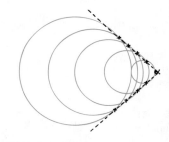

FIGURE 10.34
Idealized wave pattern made by a bug swimming faster than wave speed.

FIGURE 10.35
Idealized patterns made by a bug swimming at successively greater speeds. Overlapping at the edges occurs only when the bug swims faster than wave speed.

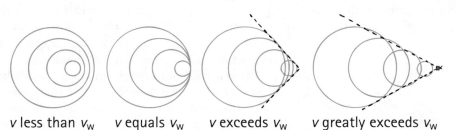

v less than v_W v equals v_W v exceeds v_W v greatly exceeds v_W

---

* Bow waves generated by boats in water are more complex than is indicated here. Our idealized treatment serves as an analogy for the production of the less complex shock waves in air.

FIGURE 10.36
The shock wave of a bullet piercing a sheet of Plexiglas. Light is deflected as the bullet passes through the compressed air that makes up the shock wave, making it visible. Look carefully and see the second shock wave originating at the tail of the bullet.

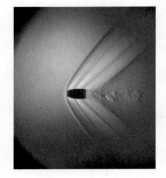

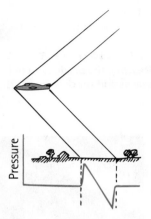

FIGURE 10.37
The shock wave actually consists of two cones—a high-pressure cone with its apex at the bow and a low-pressure cone with its apex at the tail. A graph of the air pressure at ground level between the cones takes the shape of the letter N.

FIGURE 10.38
A shock wave.

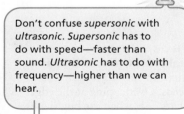

Don't confuse *supersonic* with *ultrasonic*. *Supersonic* has to do with speed—faster than sound. *Ultrasonic* has to do with frequency—higher than we can hear.

FIGURE 10.39
The shock wave has not yet reached listener A, but it is now reaching listener B, and it has already reached listener C.

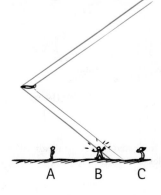

the shore of a lake, the conical wake generated by a supersonic aircraft spreads until it reaches the ground.

The bow wave of a speedboat that passes by can splash and douse you if you are at the water's edge. You could say that, in a sense, you are hit by a "water boom." In the same way, when the conical shell of compressed air that sweeps behind a supersonic aircraft reaches listeners on the ground below, the sharp crack they hear is described as a **sonic boom**.

We don't hear a sonic boom from slower-than-sound (subsonic) aircraft because the sound waves reach our ears gradually and are perceived as a continuous tone. Only when the craft moves faster than sound do the waves overlap to reach the listener in a single burst. The sudden increase in pressure is much the same in effect as the sudden expansion of air produced by an explosion. Both processes direct a burst of high-pressure air to the listener. The ear is hard-pressed to distinguish between the high pressure caused by an explosion and that produced by many overlapping waves.

A water skier is familiar with the fact that next to the high hump of the V-shaped bow wave is a V-shaped depression. The same is true of a shock wave, which consists of two cones: a high-pressure cone generated at the bow of the supersonic aircraft and a low-pressure cone that follows toward (or at) the tail of the aircraft. The edges of these cones are visible in the photograph of the supersonic bullet in Figure 10.36. Between these two cones, the air pressure rises sharply to above atmospheric pressure, then falls below atmospheric pressure before sharply returning to normal beyond the inner tail cone (Figure 10.37). This overpressure, suddenly followed by underpressure, intensifies the sonic boom.

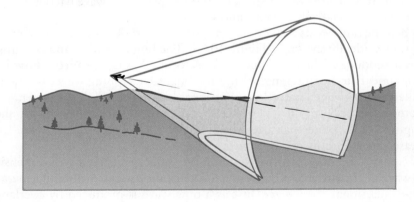

A common misconception is that sonic booms are produced when an aircraft breaks through the sound barrier—that is, just when the aircraft exceeds the speed of sound. This is essentially the same as saying that a boat produces a bow wave when it overtakes its own waves. This is not true. A shock wave and its resulting sonic boom are swept continuously behind an aircraft that is traveling faster than sound, just as a bow wave is swept continuously behind a speedboat. In Figure 10.39, listener B is in the process of hearing a sonic boom. Listener C has already heard it, and listener A will hear it shortly. The aircraft that generated this shock wave may have broken through the sound barrier hours ago!

The moving source need not be "noisy" to produce a shock wave. Once an object is moving faster than the speed of sound, it makes sound. A supersonic bullet passing overhead produces a crack, which is a small sonic boom. If the bullet were larger and disturbed more air in its path, the crack would be more boomlike. When a lion tamer cracks a circus whip, the cracking sound is actually a sonic boom produced because the tip of the whip is traveling faster than the speed of sound. Both the bullet and the whip are not in themselves sound sources, but when they travel at supersonic speeds, they produce their own sound as they generate shock waves.

10.10 Musical Sounds

EXPLAIN THIS   How do musical instruments produce their characteristic sounds?

LEARNING OBJECTIVE
Distinguish noise from musical sounds.

Most of the sounds we hear are noises. The impact of a falling object, the slamming of a door, the roaring of a motorcycle, and most of the sounds from traffic in city streets are noises. Noise corresponds to an irregular vibration of the eardrum produced by an irregularly vibrating source. Graphs that indicate the varying pressure of the air on the eardrum are shown in Figure 10.41. In part (a), we see the erratic pattern of noise. In part (b), the sound of music has shapes that repeat themselves periodically. These are periodic tones, or musical "notes." (But musical instruments can make noise as well!) Such graphs can be displayed on the screen of an oscilloscope when the electrical signal from a microphone is fed into the input terminal of this useful device.

We have no trouble distinguishing between the tone from a piano and the tone from a clarinet of the same musical pitch (frequency). Each of these tones has a characteristic sound that differs in **quality**, or timbre, a mixture of harmonics of different intensities. Most musical sounds are composed of a superposition of many frequencies called **partial tones**, or simply *partials*. The lowest frequency, called the **fundamental frequency**, determines the pitch of the note. Partial tones that are whole multiples of the fundamental frequency are called **harmonics**. A tone that has twice the frequency of the fundamental is the second harmonic, a tone with three times the fundamental frequency is the third harmonic, and so on (Figure 10.42).* The variety of partial tones gives a musical note its characteristic quality.

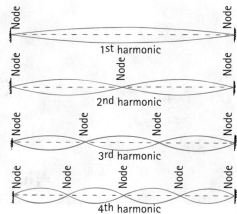

FIGURE 10.40
Physics chanteuse Lynda Williams, physics instructor at Santa Rosa Junior College, puts herself fully into the physics of music.

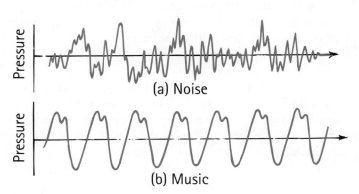

FIGURE 10.41
Graphical representations of noise and music.

FIGURE 10.42
Modes of vibration of a guitar string.

* Not all partial tones present in a complex tone are integer multiples of the fundamental. Unlike the harmonics of woodwinds and brasses, stringed instruments, such as the piano, produce "stretched" partial tones that are nearly, but not quite, harmonics.

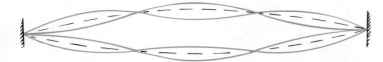

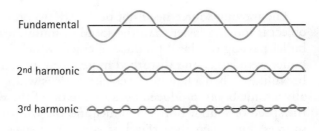

Fundamental

2nd harmonic

3rd harmonic

Composite wave

FIGURE 10.43
A composite vibration of the fundamental mode and the third harmonic.

FIGURE 10.44
Sine waves combine to produce a composite wave.

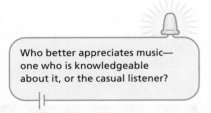

Who better appreciates music—one who is knowledgeable about it, or the casual listener?

Thus, if we strike middle C on the piano, we produce a fundamental tone with a pitch of about 262 Hz and also a blending of partial tones of two, three, four, five, and so on times the frequency of middle C. The number and relative loudness of the partial tones determine the quality of sound associated with the piano. Sound from practically every musical instrument consists of a fundamental and partials. Pure tones, those having only one frequency, can be produced electronically. Electronic synthesizers, for example, produce pure tones and mixtures of these to produce a vast variety of musical sounds.

FIGURE 10.45
Sounds from the piano and clarinet differ in quality.

Piano C

Clarinet C

The quality of a tone is determined by the presence and relative intensity of the various partials. The ear recognizes the different partials and can therefore differentiate the different sounds produced by a piano and a clarinet. A pair of tones of the same pitch with different qualities has either different partials or a difference in the relative intensity of the partials.

Amazingly, when listening to music we can discern what instruments are being played, what notes are playing, and what their relative loudness is. Whether the music is live or electronic, our ears break the overall sound signal into its component parts automatically. How this incredible feat is accomplished has to do with Fourier analysis, which concludes our study of sound.

FIGURE 10.46
Does each listener hear the same music?

FOURIER ANALYSIS

In 1822, the French mathematician and physicist Joseph Fourier made a discovery with application to music. He found that wave motion could be reduced to simple sine waves. A sine wave is the simplest of waves, having a single frequency, as shown in Figures 10.1 and 10.44. All periodic waves, however complicated, can be broken down into constituent sine waves of different amplitudes and frequencies. The mathematical operation for doing this is called Fourier analysis. We will not explain the mathematics here, but we will simply point out that, by such analysis, one can find the pure sine tones that constitute the tone of, say, a violin. When these pure tones are sounded together by selecting the proper keys on an electric organ, they combine to produce the tone of the violin. The lowest-frequency sine wave is the fundamental, and it determines the pitch of the note. The higher-frequency sine waves are the partials, which give the characteristic quality. Thus, the waveform of any musical sound is no more than a sum of simple sine waves.

Because the waveform of music is a multitude of various sine waves, to duplicate sound accurately by radio, tape recorder, or CD player, we should be able to process as large a range of frequencies as possible. The notes of a piano keyboard range from 27 Hz to 4200 Hz, but to duplicate the music of a piano composition accurately, the sound system must have a range of frequencies up to 20,000 Hz. The greater the range of the frequencies of an electrical sound system, the closer the musical output approximates the original sound, hence the wide range of frequencies that can be produced in a high-fidelity sound system.

Our ear performs a sort of Fourier analysis automatically. It sorts out the complex jumble of air pulsations that reach it, and it transforms them into pure tones. And we recombine various groupings of these pure tones when we listen. What combinations of tones we have learned to focus our attention on determines what we hear when we listen to a concert. We can direct our attention to the sounds of the various instruments and discern the faintest tones from the loudest; we can delight in the intricate interplay of instruments and still detect the extraneous noises of others around us. This is a most incredible feat.

For instructor-assigned homework, go to www.masteringphysics.com

SUMMARY OF TERMS (KNOWLEDGE)

Amplitude For a wave or vibration, the maximum displacement on either side of the equilibrium (midpoint) position.

Beats A series of alternate reinforcements and cancellations produced by the interference of two waves of slightly different frequency, heard as a throbbing effect in sound waves.

Bow wave The V-shaped wave made by an object moving across a liquid surface at a speed greater than the wave speed.

Compression A condensed region of the medium through which a longitudinal wave travels.

Doppler effect The change in frequency of wave motion resulting from motion of the sender or the receiver.

Forced vibration The setting up of vibrations in an object by a vibrating force.

Frequency For a vibrating body or medium, the number of vibrations per unit time. For a wave, the number of crests that pass a particular point per unit time.

Fundamental frequency The lowest frequency of vibration, or the first harmonic. In a string, the vibration makes a single segment.

Harmonic A partial tone that is an integer multiple of the fundamental frequency. The vibration that begins with the fundamental vibrating frequency is the first harmonic, twice the fundamental is the second harmonic, and so on in sequence.

Hertz The SI unit of frequency; one hertz (symbol Hz) equals one vibration per second.

Interference A property of all types of waves; a result of superimposing different waves, often of the same wavelength. *Constructive* interference results from crest-to-crest reinforcement; *destructive interference* results from crest-to-trough cancellation.

Longitudinal wave A wave in which the medium vibrates in a direction parallel (longitudinal) to the direction in which the wave travels. Sound consists of longitudinal waves.

Natural frequency The frequency at which an elastic object naturally tends to vibrate, so that minimum energy is required to produce a forced vibration or to continue vibration at that frequency.

Partial tone One of the frequencies present in a complex tone. When a partial tone is an integer multiple of the lowest frequency, it is a harmonic.

Period The time required for a vibration or a wave to make a complete cycle; equal to 1/frequency.

Pitch The subjective impression of the frequency of sound.

Quality The characteristic timbre of a musical sound, which is governed by the number and relative intensities of partial tones.

Rarefaction A rarefied region, or a region of lessened pressure, of the medium through which a longitudinal wave travels.

Reflection The return of a sound wave; an echo.

Refraction The bending of a wave, either through a non-uniform medium or from one medium to another, caused by differences in wave speed.

Resonance The response of a body when a forcing frequency matches its natural frequency.

Reverberation Re-echoed sound.

Shock wave The cone-shaped wave made by an object moving at supersonic speed through a fluid.

Sonic boom The loud sound resulting from a shock wave.

Standing wave A stationary wave pattern formed in a medium when two sets of identical waves pass through the medium in opposite directions.

Transverse wave A wave in which the medium vibrates in a direction perpendicular (transverse) to the direction in which the wave travels. Light consists of transverse waves.

Vibration A wiggle in time.

Wave A wiggle in both space and time.

Wavelength The distance between successive crests, troughs, or identical parts of a wave.

Wave speed The speed with which waves pass a particular point:

$$\text{Wave speed} = \text{frequency} \times \text{wavelength}$$

READING CHECK QUESTIONS (COMPREHENSION)

10.1 Vibrations and Waves

1. What is the source of all waves?
2. Distinguish among these characteristics of a wave: period, amplitude, wavelength, and frequency.
3. How are frequency and period related?

10.2 Wave Motion

4. In one word, what is it that moves from source to receiver in wave motion?
5. Does the medium in which a wave travels move with the wave?
6. What is the relationship among frequency, wavelength, and wave speed?

10.3 Transverse and Longitudinal Waves

7. In what direction are the vibrations in a transverse wave, relative to the direction of wave travel? In a longitudinal wave?
8. In what direction do compressed regions and rarefied regions of a longitudinal wave travel?

10.4 Sound Waves

9. Does sound travel faster in warm air or in cold air? Defend your answer.
10. How does the speed of sound in water compare with the speed of sound in air? How does the speed of sound in steel compare with the speed of sound in air?

10.5 Reflection and Refraction of Sound

11. What is the law of reflection for sound?
12. What is a reverberation?
13. Relate wave speed and bending to the phenomenon of refraction.
14. Does sound tend to bend upward or downward when it travels faster near the ground than higher up?
15. How do dolphins perceive their environment in dark and murky water?

10.6 Forced Vibrations and Resonance

16. Why does a struck tuning fork sound louder when its handle is held against a table?
17. Distinguish between forced vibrations and resonance.
18. When you listen to a radio, why do you hear only one station at a time instead of all stations at once?
19. Why do troops "break step" when crossing a bridge?

10.7 Interference

20. What kinds of waves exhibit interference?
21. Distinguish between constructive interference and destructive interference.
22. What does it mean to say that one wave is out of phase with another?
23. What physical phenomenon underlies beats?
24. What is a node? What is an antinode?

10.8 Doppler Effect

25. In the Doppler effect, does frequency change? Does wavelength change? Does wave speed change?
26. Can the Doppler effect be observed with longitudinal waves, with transverse waves, or with both?

10.9 Bow Waves and the Sonic Boom

27. How do the speed of a wave source and the speed of the waves themselves compare when a wave barrier is being produced? How do they compare when a bow wave is being produced?
28. How does the V shape of a bow wave depend on the speed of the wave source?
29. True or False: A sonic boom occurs only when an aircraft is breaking through the sound barrier. Defend your answer.

10.10 Musical Sounds

30. Distinguish between a musical sound and noise.

ACTIVITIES (HANDS-ON APPLICATION)

31. Tie a rubber tube, a spring, or a rope to a fixed support and shake it to produce standing waves. See how many nodes you can produce.

32. Test to see which of your ears has better hearing by covering one ear and finding how far away your open ear can hear the ticking of a clock; repeat for the other ear. Notice also how the sensitivity of your hearing improves when you cup your hands behind your ears.

33. Do the activity suggested in Figure 10.24 with a stereo sound system. Simply reverse the wire inputs to one of the speakers so that the two are out of phase. When monaural sound is played and the speakers are brought face to face, the lowering of volume is truly amazing! If the speakers are well insulated, you hear almost no sound at all.

34. For this activity, you'll need an isolated loudspeaker (bare of its casing) and a sheet of plywood or cardboard—the bigger the better. Cut a hole in the middle of the sheet that is about the size of the speaker. Listen to music from the isolated speaker, and then hear the difference when the speaker is placed against the hole. The sheet diminishes the amount of sound from the back of the speaker that interferes with sound coming from the front side, producing a much fuller sound. Now you know why speakers are mounted in enclosures.

35. Wet your finger and slowly rub it around the rim of a thin-rimmed, stemmed glass while you hold the base of the glass firmly to a tabletop with your other hand. The friction of your finger excites standing waves in the glass, much like the wave made on the strings of a violin by the friction from a violin bow. Try it with a metal bowl.

36. Swing a buzzer of any kind over your head in a circle. You won't hear the Doppler shift, but your friends off to the side will. The pitch will increase as it approaches them, and decrease when it recedes. Then switch places with a friend so you can hear it too.

37. Make the lowest-pitched vocal sound you are capable of; then keep doubling the pitch to see how many octaves your voice can span. Compare your octaves with those of friends.

38. Blow over the top of two empty bottles and see if the tones produced have the same pitch. Then put one bottle in a freezer and try the procedure again. Sound travels more slowly in the cold, denser air of the cold bottle and the note is lower. Try it and see.

PLUG AND CHUG (FORMULA FAMILIARIZATION)

$$\text{Frequency} = \frac{1}{\text{period}}$$

39. A pendulum swings to and fro every 3 s. Show that its frequency of swing is $\frac{1}{3}$ Hz.

$$\text{Period} = \frac{1}{\text{frequency}}$$

40. Another pendulum swings to and fro at a regular rate of two times per second. Show that its period is 0.5 s.

$$\text{Wave speed} = \text{frequency} \times \text{wavelength} = f\lambda$$

41. A 3-m-long wave oscillates 1.5 times each second. Show that the speed of the wave is 4.5 m/s.

42. Show that a certain 1.2-m-long wave with a frequency of 2.5 Hz has a wave speed of 3.0 m/s.

43. A tuning fork produces a sound with a frequency of 256 Hz and a wavelength in air of 1.33 m. Show that the speed of sound in the vicinity of the fork is 340 m/s.

THINK AND SOLVE (MATHEMATICAL APPLICATION)

44. A nurse counts 72 heartbeats in 1 min. Show that the frequency and period of the heartbeats are 1.2 Hz and 0.83 s, respectively.

45. A weight suspended from a spring is seen to bob up and down over a distance of 20 cm twice each second. What are its (a) frequency, (b) period, and (c) amplitude?

46. We know that speed v = distance/time. Show that when the distance traveled is one wavelength λ and the time of travel is the period T (which equals 1/frequency), you get $v = f\lambda$.

47. A skipper on a boat notices wave crests passing his anchor chain every 5 s. He estimates the distance between wave crests to be 15 m. He also correctly estimates the speed of the waves. Show that his estimation of wave speed is 3 m/s by (a) the classic formula for speed, distance divided by time, and (b) frequency × wavelength.

48. A mosquito flaps its wings at a rate of 600 vibrations per second, which produces the annoying 600-Hz buzz. Given that the speed of sound is 340 m/s, how far does the sound travel between wing beats? In other words, find the wavelength of the mosquito's sound.

49. The highest frequency sound humans can hear is about 20,000 Hz. What is the wavelength of sound in air at this frequency? What is the wavelength of the lowest sounds we can hear, about 20 Hz?

50. Microwave ovens typically cook food using microwaves with a frequency of about 3.00 GHz (gigahertz, 10^9 Hz). Show that the wavelength of these microwaves traveling at the speed of light is 10 cm.

51. For years, marine scientists were mystified by sound waves detected by underwater microphones in the Pacific Ocean. These so-called T waves were among the purest

sounds in nature. Eventually the researchers traced the source to underwater volcanoes whose rising columns of bubbles resonated like organ pipes. A typical T wave has a frequency of 7 Hz. Knowing that the speed of sound in seawater is 1530 m/s, show that the wavelength of a T wave is 219 m.

52. An oceanic depth-sounding vessel surveys the ocean bottom with ultrasonic waves that travel at 1530 m/s in seawater. Show that when the time delay of an echo to the ocean floor below is 4 s, the depth of the water is 3060 m.

53. A bat flying in a cave emits a sound and receives its echo 0.1 s later. Show that the distance to the wall of the cave is 17 m.

54. Susie hammers on a block of wood when she is 85 m from a large brick wall. Each time she hits the block, she hears an echo 0.5 s later. With this information, show that the speed of sound is 340 m/s.

55. Imagine an old hermit who lives in the mountains. Just before going to sleep, he yells "WAKE UP." The sound echoes off the nearest mountain and returns 8 h later. Show that the mountain is almost 5000 km distant.

56. On a keyboard, you strike middle C, whose frequency is 256.0 Hz.

    (a) Show that the period of one vibration of this tone is 0.004 s.

    (b) As the sound leaves the instrument at a speed of 340 m/s, show that its wavelength in air is 1.33 m.

57. (a) If you were so foolish as to play your keyboard instrument underwater, where the speed of sound is 1500 m/s, show that the wavelength of the middle-C tone in water would be 5.86 m.

    (b) Explain why middle C (or any other tone) has a longer wavelength in water than in air.

58. What beat frequencies are possible with tuning forks of frequencies 256, 259, and 261 Hz?

59. As shown in the drawing, the half-angle of the shock-wave cone generated by a supersonic aircraft is 45°. What is the speed of the plane relative to the speed of sound?

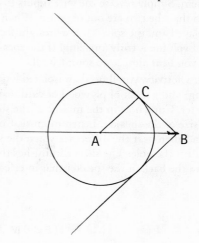

THINK AND RANK (ANALYSIS)

60. All the waves shown have the same speed in the same medium. Use a ruler and rank these waves from greatest to least for (a) amplitude, (b) wavelength, (c) frequency, and (d) period.

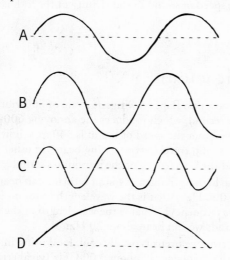

61. Note the four pairs of transverse wave pulses that move toward each other. At some point in time the pulses meet and interact (interfere) with each other. Rank the four pairs, from highest to lowest, on the basis of the height of the peak that results when the centers of the pulses coincide.

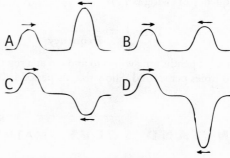

62. The siren of a fire engine is heard in three situations: when the fire engine is traveling (a) toward the listener at 30 km/h, (b) toward the listener at 50 km/h, and (c) away from the listener at 20 km/h. Rank the pitches heard, from highest to lowest.

63. The three shock waves are produced by supersonic aircraft. Rank their speeds from fastest to slowest.

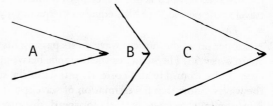

64. Rank the speed of sound through the following media from fastest to slowest: (a) air, (b) steel, (c) water.

65. Rank the beat frequency from highest to lowest for the following pairs of sounds: (a) 132 Hz, 136 Hz; (b) 264 Hz, 258 Hz; (c) 528 Hz, 531 Hz; and (d) 1056 Hz, 1058 Hz.

EXERCISES (SYNTHESIS)

66. A student that you're tutoring says that the two terms *wave speed* and *wave frequency* refer to the same thing. What is your response?

67. You dip your finger at a steady rate into a puddle of water to make waves. What happens to the wavelength if you dip your finger more frequently?

68. How does the frequency of vibration of a small object floating in water compare to the number of waves passing it each second?

69. Red light has a longer wavelength than violet light. Which has the higher frequency?

70. What kind of motion should you impart to the nozzle of a garden hose so that the resulting stream of water approximates a sine curve?

71. What kind of motion should you impart to a stretched coiled spring (or to a Slinky) to produce a transverse wave? A longitudinal wave?

72. A cat can hear sound frequencies up to 70,000 Hz. Bats send and receive ultrahigh-frequency squeaks up to 120,000 Hz. Which animal hears sound of shorter wavelengths: cats or bats?

73. A bat chirps as it flies toward a wall. Is the frequency of the echoed chirps it receives higher, lower, or the same as the emitted ones?

74. A nylon guitar string vibrates in a standing-wave pattern, as shown. What is the wavelength of the wave?

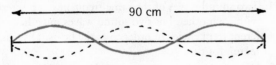

90 cm

75. Why don't you hear the sound of a distant fireworks display until after you see it?

76. If the Moon blew up, why wouldn't we hear the sound?

77. Why would it be futile to attempt to detect sounds from other planets with the use of state-of-the-art audio detectors?

78. A pair of sound waves of different wavelengths reach the listener's ear as shown. Which has the higher pitch: the short-wavelength sound or the long-wavelength sound? Defend your answer.

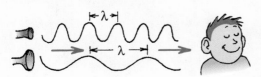

79. The sounds emitted by bats are extremely intense. Why can't humans hear them?

80. In an Olympic competition, a microphone detects the sound of the starter's gun, which is sent electronically to speakers at every runner's starting block. Why?

81. If sound becomes louder, which wave characteristic is likely increasing: frequency, wavelength, amplitude, or speed?

82. What two physics mistakes occur in a science-fiction movie that shows a distant explosion in outer space that you see and hear at the same time?

83. As you pour water into a glass, you repeatedly tap the glass exterior with a spoon. As the tapped glass is being filled, does the pitch of the sound get higher or lower? (What should you do to answer this question?)

84. If the frequency of a sound is doubled, what change occurs in its speed? What change occurs in its wavelength? Defend your answer.

85. Why are marchers following a band at the end of a long parade out of step with marchers near the front?

86. What is the danger posed by people in the balcony of an auditorium stamping their feet in a steady rhythm?

87. Why is the sound of a harp soft in comparison with the sound of a piano?

88. What physics principle does Manuel use when he pumps in rhythm with the natural frequency of the swing?

89. How can a certain note sung by a singer cause a crystal glass to shatter?

90. Walking beside you, your friend takes 50 strides per minute while you take 48 strides per minute. If you begin in step, when will you and your friend again be in step?

91. Suppose a piano tuner hears three beats per second when listening to the combined sound from his tuning fork and the piano note being tuned. After slightly tightening the string, he hears five beats per second. Should the string be loosened or tightened?

92. A railroad locomotive is at rest with its whistle shrieking, and then it starts moving toward you.

    (a) Does the frequency that you hear increase, decrease, or stay the same?

    (b) How about the wavelength reaching your ear?

    (c) How about the speed of sound in the air between you and the locomotive?

93. When you blow your horn while driving toward a stationary listener, he hears an increase in the frequency of the horn. Would the listener hear an increase in the frequency of the horn if he were also in a car traveling at the same speed in the same direction as you? Explain.

94. How does the Doppler effect aid police in detecting speeding motorists?

95. How does the Doppler effect used in radar guns give the speeds of tennis balls and baseballs at sporting events?

96. Astronomers find that light emitted by a particular element at one edge of the Sun has a slightly higher frequency than light from that element at the opposite edge. What do these measurements tell us about the Sun's motion?

97. Would it be correct to say that the Doppler effect is the apparent change in the speed of a wave due to motion of the source? (Why is this question a test of reading comprehension as well as a test of physics knowledge?)

98. Does the conical angle of a shock wave open wider, narrow down, or remain constant as a supersonic aircraft increases its speed?

99. If the sound of an airplane does not originate in the part of the sky where the plane is seen, does this imply that the airplane is traveling faster than the speed of sound? Explain.

100. Why is it that a subsonic aircraft, no matter how loud it may be, cannot produce a sonic boom?

DISCUSSION QUESTIONS (EVALUATION)

101. What does it mean to say that a radio station is "at 101.1 on your FM dial"?

102. At the instant that a high-pressure region is created just outside the prongs of a vibrating tuning fork, what is being created inside between the prongs?

103. If a bell is ringing inside a bell jar, we can no longer hear it when the air is pumped out, but we can still see the bell. Discuss the differences in the properties of sound and light that this indicates.

104. Why is the Moon described as a "silent planet"?

105. If the speed of sound depended on its frequency, discuss why you would not enjoy a concert sitting far from the stage—say, in the second balcony.

106. Discuss why sound travels faster in moist air. Relate to the fact that at the same temperature, water-vapor molecules have the same average kinetic energy as the heavier nitrogen and oxygen molecules in the air. Then discuss how the average speeds of H_2O molecules compare with those of N_2 and O_2 molecules.

107. Why is an echo weaker than the original sound? Discuss the role of distance.

108. A rule of thumb for estimating the distance in kilometers between an observer and a lightning stroke is to divide the number of seconds in the interval between the flash and the sound by 3. Discuss whether or not this rule is correct.

109. If a single disturbance some unknown distance away sends out both transverse and longitudinal waves that travel with distinctly different speeds in the medium, such as in the ground during an earthquake, discuss how the distance to the disturbance is determined.

110. A special device can transmit sound that is out of phase with the sound of a noisy jackhammer to the jackhammer operator by means of earphones. Over the noise of the jackhammer, the operator can easily hear your voice, while you are unable to hear his. Explain.

111. Two sound waves of the same frequency can interfere with each other, but two sound waves must have different frequencies in order to make beats. Discuss the reason for this.

112. Discuss whether or not a sonic boom occurs at the moment when an aircraft exceeds the speed of sound.

READINESS ASSURANCE TEST (RAT)

If you have a good handle on this chapter, if you really do, then you should be able to score at least 7 out of 10 on this RAT. If you score less than 7, you need to study further before moving on.

Choose the BEST answer to each of the following.

1. When we consider the time it takes for a pendulum to swing to and fro, we're talking about the pendulum's
   - (a) frequency.
   - (b) period.
   - (c) wavelength.
   - (d) amplitude.
2. The vibrations along a transverse wave move in a direction
   - (a) parallel to the wave direction.
   - (b) perpendicular to the wave direction.
   - (c) both of these
   - (d) neither of these
3. A common example of a longitudinal wave is
   - (a) sound.
   - (b) light.
   - (c) both of these
   - (d) neither of these
4. The speed of sound varies with
   - (a) amplitude.
   - (b) frequency.
   - (c) temperature.
   - (d) all of these
5. The loudness of a sound is most closely related to its
   - (a) frequency.
   - (b) period.
   - (c) wavelength.
   - (d) amplitude.
6. The vibrations set up in a radio loudspeaker have the same frequencies as the vibrations
   - (a) in the electric signal fed to the loudspeaker.
   - (b) that produce the sound you hear.
   - (c) both of these
   - (d) none of these

7. Sound waves *cannot* be
   - (a) reflected.
   - (b) absorbed.
   - (c) diminished by interference.
   - (d) none of these
8. Noise-canceling devices such as jackhammer earphones make use of sound
   - (a) destruction.
   - (b) interference.
   - (c) resonance.
   - (d) amplification.
9. When a 134-Hz tuning fork and a 144-Hz tuning fork are struck, the beat frequency is
   - (a) 2 Hz.
   - (b) 6 Hz.
   - (c) 8 Hz.
   - (d) more than 8 Hz.
10. A sonic boom *cannot* be produced by
   - (a) an aircraft flying slower than the speed of sound.
   - (b) a whip.
   - (c) a speeding bullet.
   - (d) all of these

Answers to RAT

1. b, 2. b, 3. a, 4. c, 5. d, 6. c, 7. d, 8. b, 9. d, 10. a

CHAPTER 11
Light

LIGHT IS the only thing we can really see. But what is light? We know that during the day, the primary source of light is the Sun, and a secondary source is the brightness of the sky. Other common sources are white-hot filaments in lightbulbs, glowing gases in glass tubes, and flames. We find that light originates from the accelerated motion of electrons. Light is an electromagnetic phenomenon, and it is only a tiny part of a larger whole—a wide range of transverse electromagnetic waves called the *electromagnetic spectrum*. We begin our study of light by investigating its electromagnetic properties, how it interacts with materials, and how it reflects. We'll see its wave nature in how it refracts and how we see its colors, quite spectacularly as rainbows. We'll conclude this exciting chapter with the phenomenon of polarization.

11.1 Electromagnetic Spectrum

EXPLAIN THIS How are electromagnetic waves produced?

LEARNING OBJECTIVE
Describe the nature and range of electromagnetic waves.

If you shake the end of a stick back and forth in still water, you create waves on the water's surface. If you similarly shake an electrically charged rod to and fro in empty space, you create electromagnetic waves in space. We learned in Chapter 9 why this is so: The shaking stick creates an electric current around which is generated a magnetic field, and the changing magnetic field induces an electric field—electromagnetic induction. The changing electric field in turn induces a changing magnetic field. As we learned in Chapter 9, the vibrating electric and magnetic fields regenerate each other to make up an **electromagnetic wave**. This is shown in Figure 11.2 (which is a repeat of Figure 9.36).

Mastering**PHYSICS**
TUTORIAL: Light and Spectroscopy

FIGURE 11.1
If you shake an electrically charged object to and fro, you produce an electromagnetic wave.

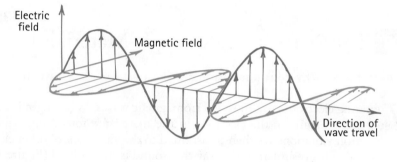

FIGURE 11.2
 INTERACTIVE FIGURE

The electric and magnetic fields of an electromagnetic wave in free space are perpendicular to each other and to the direction of motion of the wave.

In a vacuum, all electromagnetic waves move at the same speed, differing only in frequency. The classification of electromagnetic waves according to frequency, from radio waves to gamma rays, is the **electromagnetic spectrum** (Figure 11.3). The descriptive names of the sections are merely a historical classification, for all waves are the same in their basic nature, differing principally in frequency and wavelength; all of the waves have the same speed. Electromagnetic waves have been measured from 0.01 Hz to radio frequencies up to 108 MHz. Then come ultrahigh frequencies (UHF), followed by microwaves, beyond which are infrared waves, often called *heat waves*.

FIGURE 11.3
 INTERACTIVE FIGURE

The electromagnetic spectrum is a continuous range of waves extending from radio waves to gamma rays.

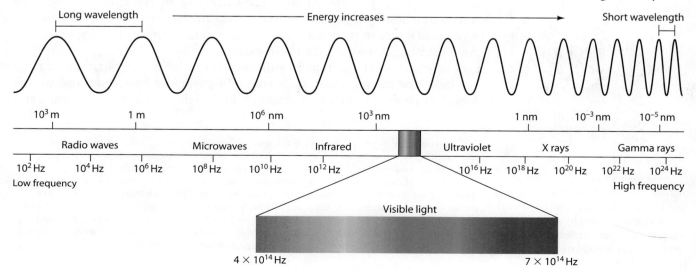

Further still is visible light, which makes up less than a millionth of 1% of the measured electromagnetic spectrum. The higher the frequency of the wave, the shorter its wavelength.*

> Light is energy carried in an electromagnetic wave emitted by vibrating electrons in atoms.

CHECKPOINT
Is it correct to say that a radio wave is a low-frequency light wave? Is a radio wave also a sound wave?

Were these your answers?
Yes and no. Both radio waves and light waves are electromagnetic waves that originate in the vibrations of electrons. Radio waves have lower frequencies than light waves, so a radio wave might be considered a low-frequency light wave (and a light wave might be considered a high-frequency radio wave). But a radio wave is definitely not a sound wave, which we learned in the previous chapter is a mechanical vibration of matter. (Don't confuse a radio wave with the sound that a loudspeaker emits.)

LEARNING OBJECTIVE
Relate the transparency of materials to wave frequencies.

Mastering**PHYSICS®**

VIDEO: Light and Transparent Materials

FIGURE 11.4
Just as a sound wave can force a sound receiver into vibration, a light wave can force the electrons in materials into vibration.

11.2 Transparent and Opaque Materials

EXPLAIN THIS Why don't photons that strike a pane of glass travel through it?

Vibrating electrons emit most electromagnetic waves. When light is incident on matter, some of the electrons in the matter are forced into vibration. Electron vibrations are then transmitted to the vibrations of other electrons in the material. This is similar to the way that sound is transmitted (Figure 11.4).

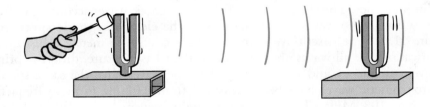

Materials such as glass and water allow light to pass through with almost no absorption, usually in straight lines. These materials are **transparent** to light. To understand how light penetrates a transparent material, visualize the electrons in an atom as if they were connected to the atomic nucleus by springs (Figure 11.5).** An incident light wave sets the electrons into vibration.

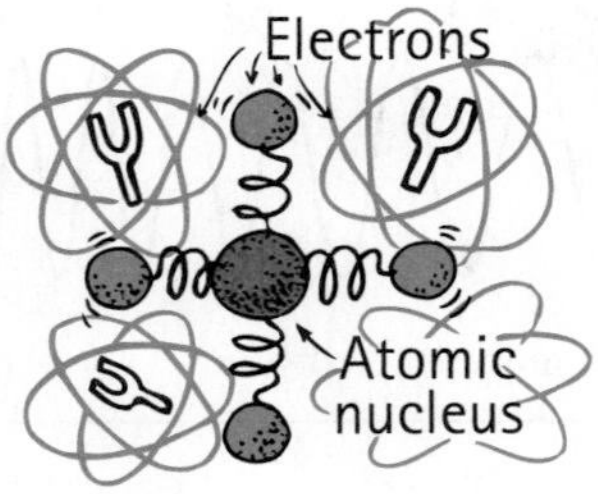

The vibration of electrons in a material is similar to the vibrations of ringing bells and tuning forks. Bells ring at a particular frequency, and tuning forks vibrate at a particular frequency—and so do the electrons of atoms and

FIGURE 11.5
The electrons of atoms have certain natural frequencies of vibration, which can be modeled as particles connected to the atomic nucleus by springs. As a result, atoms and molecules behave somewhat like optical tuning forks.

*The relationship is $c = f\lambda$, where c is the speed of light (constant), f is the frequency, and λ is the wavelength. It is common to describe sound and radio by frequency and light by wavelength. In this book, however, we'll favor the single concept of frequency in describing light.
**Electrons, of course, are not really connected by springs. Here we present a visual "spring model" of the atom to help us understand the interaction of light with matter. The worth of a model lies not in whether it is "true" but in whether it is useful—in explaining observations and predicting new ones. The simplified model that we present here—of an atom whose electrons vibrate as if on springs, with a time interval between absorbing energy and re-emitting it—is quite useful for understanding how light passes through a transparent material.

3 of many atoms

Glass

FIGURE 11.6
A light wave incident on a pane of glass sets up vibrations in the molecules that produce a chain of absorptions and re-emissions, which pass the light energy through the material and out the other side. Because of the time delay between absorptions and re-emissions, the light travels through the glass more slowly than through empty space.

molecules. Different atoms and molecules have different "spring strengths." Electrons in glass have a natural vibration frequency in the ultraviolet range. When ultraviolet rays in sunlight shine on glass, resonance occurs as the wave builds and maintains a large amplitude of electron vibration, just as pushing someone at the resonant frequency on a swing builds a large amplitude. Resonating atoms in the glass can hold on to the energy of the ultraviolet light for quite a long time (about 100 millionths of a second). During this time, the atom undergoes about 1 million vibrations, collides with neighboring atoms, and transfers absorbed energy as thermal energy. Thus, glass is not transparent to ultraviolet. Glass absorbs ultraviolet.

At lower wave frequencies, such as those of visible light, electrons in the glass are forced into vibration at a lower amplitude. The atoms or molecules in the glass hold the energy for less time, with less chance of collision with neighboring atoms and molecules, and less of the energy is transformed to heat. Instead, the energy of vibrating electrons is re-emitted as light. Glass is transparent to all the frequencies of visible light. The frequency of re-emitted light passed from molecule to molecule is identical to the frequency of the original light that produced the vibration. However, there is a slight time delay between absorption and re-emission.

This time delay lowers the average speed of light through the material (Figure 11.6). Light of different frequencies travels at different average speeds through different materials. We say *average speeds*, for the speed of light in a vacuum is a constant 300,000 kilometers per second. We call this speed of light *c*.* The speed of light in the atmosphere is slightly less than it is in a vacuum, but is usually rounded off as *c*. In water, light travels at 75% of its speed in a vacuum, or 0.75*c*. In glass, light travels about 0.67*c*, depending on the type of glass. In a diamond, light travels at less than half its speed in a vacuum, only 0.41*c*. Light travels even slower in a silicon carbide crystal called *carborundum*. When light emerges from these materials into the air, it travels at its original speed.

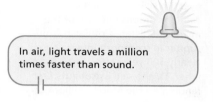
In air, light travels a million times faster than sound.

Light slows when it enters glass?

FIGURE 11.7
When the raised ball is released and hits the others, the ball that emerges from the opposite side is not the same ball that initiated the transfer of energy. Likewise, each photon that emerges from a pane of glass is not the same photon that was incident on the glass. Both the emerging ball and emerging photon are different from, though identical to, the incident ones.

Infrared waves, which have frequencies lower than those of visible light, vibrate not only the electrons but the entire molecules in the structure of the glass and in many other materials. This molecular vibration increases the thermal energy and temperature of the material, which is why infrared waves are often called *heat waves*. Glass is transparent to visible light, but not to ultraviolet and infrared light.

* The more exact value is 299,792 km/s, which is often rounded to 300,000 km/s. (This corresponds to 186,000 mi/s.)

FIGURE 11.8
Clear glass blocks both infrared and ultraviolet, but it is transparent to all the frequencies of visible light.

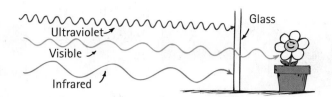

Ultraviolet
Visible
Infrared
Glass

FYI The first person to notice a delay in light travel was the Danish astronomer Ole Roemer, who in 1675 saw the effect of light's finite speed "with his own eyes" in eclipses of one of Jupiter's moons because of the increased distance of Earth from Jupiter in six-month intervals. Nearly 300 years later, in 1969, when TV showed astronauts first landing on the Moon, millions of people in their living rooms noticed the time delay between conversations (at the speed of light) between the astronauts and the earthlings at Mission Control. They noticed the effect of the finite speed of electromagnetic waves "with their own ears."

CHECKPOINT

1. **Why is glass transparent to visible light but opaque to ultraviolet and infrared?**
2. **Pretend that while you are at a social gathering, you make several momentary stops across the room to greet people who are "on your wavelength." How is this analogous to light traveling through glass?**

Were these your answers?

1. The natural frequency of vibration for electrons in glass is the same as the frequency of ultraviolet light, so resonance in glass occurs when ultraviolet waves shine on glass. The absorbed energy is transferred to other atoms as heat, not re-emitted as light, so the glass is opaque at ultraviolet frequencies. In the range of visible light, forced vibration of electrons occurs at smaller amplitudes—vibrations are more subtle. So re-emission of light (rather than the generation of heat) occurs, and the glass is transparent. Lower-frequency infrared light causes whole molecules, rather than electrons, to resonate; again, heat is generated and the glass is opaque.
2. Your average speed across the room would be less because of the time delays associated with your momentary stops. Likewise, the speed of light in glass is less because of the time delays in interactions with atoms along its path.

FIGURE 11.9
Metals are shiny because their free electrons easily vibrate to the oscillations of any incident light, reflecting most of it.

Most things around us are **opaque**—they absorb light without re-emission. Books, desks, chairs, and people are opaque. Energetic vibrations produced by incident light on the atoms of these materials are turned into random kinetic energy—into thermal energy. The materials become slightly warmer.

Metals are opaque to visible light. The outer electrons of atoms in metals are not bound to any particular atom. They are loose and free to wander, with very little restraint, throughout the material (which is why metal conducts electricity and heat so well). When light shines on metal and sets these free electrons into vibration, their energy does not "spring" from atom to atom in the material. It is reflected instead. That's why metals are shiny.

Earth's atmosphere is transparent to some ultraviolet light, to all visible light, and to some infrared light. But the atmosphere is opaque to high-frequency ultraviolet light. The small amount of ultraviolet light that does penetrate causes sunburns. If all ultraviolet light penetrated the atmosphere, we would be fried to a crisp. Clouds are semitransparent to ultraviolet light, which is why you can get a sunburn on a cloudy day. Ultraviolet light is not only harmful to your skin, it is also damaging to tar roofs. Now you know why tarred roofs are often covered with gravel.

Have you noticed that things look darker when they are wet than when they are dry? Light incident on a dry surface, such as sand, bounces directly to your eye. But light incident on a wet surface bounces around inside the transparent

LATERAL INHIBITION

The human eye can do what no camera can do: it can perceive degrees of brightness over a range of about 500 million to 1. The difference in brightness between the Sun and Moon, for example, is about 1 million to 1. But because of an effect called lateral inhibition, we don't perceive the actual differences in brightness. The brightest places in our visual field are prevented from outshining the rest, because whenever a receptor cell on our retina sends a strong brightness signal to our brain, it also signals neighboring cells to dim their responses. In this way, we even out our visual field, which allows us to discern detail in very bright areas and in dark areas as well.

Lateral inhibition exaggerates the difference in brightness at the edges of places in our visual field. Edges, by definition, separate one thing from another. So we accentuate differences rather than similarities. This is illustrated in the pair of shaded rectangles to the right. They appear to be different shades of brightness because of the edge that separates them. But cover the edge with your pencil or your finger, and they look equally bright (try it now)! That's because both rectangles are equally bright; each rectangle is shaded from lighter to darker, moving from left to right. Our eye concentrates on the boundary where the dark edge of the left rectangle joins the light edge of the right rectangle, and our eye–brain system assumes that the rest of the rectangle is the same. We pay attention to the boundary and ignore the rest.

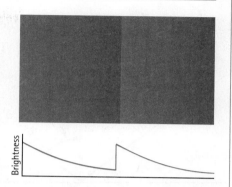

Questions to ponder: Is the way the eye picks out edges and makes assumptions about what lies beyond similar to the way in which we sometimes make judgments about other cultures and other people? Don't we, in the same way, tend to exaggerate the differences on the surface while ignoring the similarities and subtle differences within?

---

wet region before it reaches your eye. What happens with each bounce? Absorption! So sand and other things look darker when wet.

CHECKPOINT
What are two common fates for light shining on a material that isn't absorbed?

Was this your answer?
Transmission and/or reflection. Most light incident on a pane of glass, for example, is transmitted through the pane. But some reflects from its surface. How much transmits and how much reflects varies with the incident angle.

FYI Dark or black skin absorbs ultraviolet radiation before it can penetrate too far. In fair skin, it can travel deeper. Fair skin may develop a tan upon exposure to ultraviolet, which may afford some protection against further exposure. Ultraviolet radiation is also damaging to the eyes.

11.3 Reflection

EXPLAIN THIS Where is your image when you look at yourself in a plane mirror?

LEARNING OBJECTIVE
Describe the law of reflection.

Mastering**PHYSICS**
VIDEO: Image Formation in a Mirror

When this page is illuminated by sunlight or lamplight, electrons in the atoms of the paper are set into vibration. The energized electrons re-emit the light by which we see the page. Light undergoes **reflection** (as we shall soon see, is called *diffuse reflection*). When the page is illuminated by white light, it appears white because the electrons re-emit all the visible frequencies. They reflect all of the light. Very little absorption occurs. The ink on the page is a different story. Except for a bit of reflection, the ink absorbs all the visible frequencies and therefore appears black.

Law of Reflection

Anyone who has played pool or billiards knows that when a ball bounces from a surface, the angle of incidence is equal to the angle of rebound. The same is true of light. This is the **law of reflection**, which holds for all angles:

The angle of reflection equals the angle of incidence.

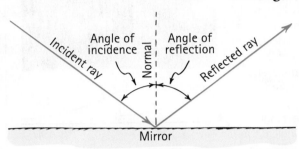

FIGURE 11.10

The law of reflection.

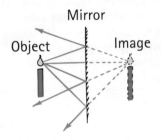

FIGURE 11.11

A virtual image is formed behind the mirror and is located at the position where the extended reflected rays (dashed lines) converge.

Your image behind a plane mirror is as if your twin stood behind a pane of clear glass at a distance as far behind the glass as you are in front of it.

FIGURE 11.12

Marjorie's image is as far behind the mirror as she is in front of it. Note that she and her image have the same color of clothing—evidence that light doesn't change frequency upon reflection. Interestingly, her left-and-right axis is no more reversed than her up-and-down axis. The axis that is reversed, as shown to the right, is her front-and-back axis. That's why it appears that her left hand faces the right hand of her image.

The law of reflection is illustrated with arrows representing light rays in Figure 11.10. Instead of measuring the angles of incident and reflected rays from the reflecting surface, it is customary to measure them from a line perpendicular to the plane of the reflecting surface. This imaginary line is called the *normal*. The incident ray, the normal, and the reflected ray all lie in the same plane.

If you place a candle in front of a mirror, rays of light radiate from the flame in all directions. Figure 11.11 shows only four of the infinite number of rays leaving one of the infinite number of points on the candle. When these rays meet the mirror, they reflect at angles equal to their angles of incidence. The rays diverge from the flame. Note that they also diverge when reflecting from the mirror. These divergent rays appear to emanate from behind the mirror (dashed lines). You see an image of the candle flame at this point. The light rays do not actually come from this point, so the image is called a *virtual image*. The image is as far behind the mirror as the object is in front of the mirror, and image and object have the same size—as long as the mirror is flat. A flat mirror is called a *plane mirror*.

When the mirror is curved, the sizes and distances of object and image are no longer equal. We will not study curved mirrors in this text, except to say that a curved mirror behaves as a succession of flat mirrors, each at a slightly different angular orientation from the one next to it. At each point, the angle of incidence is equal to the angle of reflection (Figure 11.13). Note that in a curved mirror, unlike in a plane mirror, the normals (shown by the dashed black lines) at different points on the surface are not parallel to one another.

Whether the mirror is plane or curved, the eye–brain system cannot ordinarily distinguish between an object and its reflected image. So the illusion that an object exists behind a mirror (or, in some cases, in front of a concave mirror) is merely due to the fact that the light from the object enters the eye in exactly the same manner, physically, as it would have entered if the object really were at the image location.

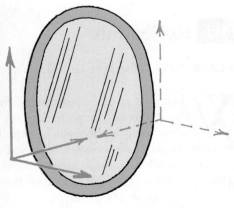

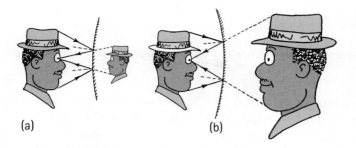

(a) (b)

FIGURE 11.13
(a) The virtual image formed by a convex mirror (a mirror that curves outward) is smaller and closer to the mirror than the object. (b) When the object is close to a concave mirror (a mirror that curves inward like a "cave"), the virtual image is larger and farther away than the object. In either case, the law of reflection holds for each ray.

CHECKPOINT

If you wish to take a picture of your image while standing 5 m in front of a plane mirror, for what distance should you set your camera to provide the sharpest focus?

Was this your answer?
Set the distance for 10 m, the distance between the camera and your image.

FIGURE 11.14
Diffuse reflection. Although reflection of each single ray obeys the law of reflection, the many different surface angles that light rays encounter in striking a rough surface produce reflection in many directions.

Only part of the light that strikes a surface is reflected. For example, on a surface of clear glass and for normal incidence (light perpendicular to the surface), only about 4% is reflected from each surface. On a clean and polished aluminum or silver surface, however, about 90% of the incident light is reflected.

Diffuse Reflection

In contrast to specular reflection is **diffuse reflection**, which occurs when light is incident on a rough surface and reflected in many directions (Figure 11.14). If the surface is so smooth that the distances between successive elevations on the surface are less than about one-eighth the wavelength of the light, there is very little diffuse reflection, and the surface is said to be *polished*. A surface therefore may be polished for radiation of long wavelengths but rough for light of short wavelengths. The wire-mesh "dish" shown in Figure 11.15 is very rough for light waves and is hardly mirrorlike. But for long-wavelength radio waves, it is "polished" and is an excellent reflector.

Light reflecting from this page is diffuse. The page may be smooth to a radio wave, but to a light wave it is rough. Smoothness is relative to the wavelength of the illuminating waves. Rays of light striking this page encounter millions of tiny flat surfaces facing in all directions. The incident light, therefore, is reflected in all directions. This is desirable, for it enables us to see this page and other objects from any direction or position. You can see the road ahead of your car at night, for instance, because of diffuse reflection by the rough road surface. When the road is wet, however, it is smoother with less diffuse reflection, and therefore more difficult to see. Most of our environment is seen by diffuse reflection.

FIGURE 11.15
The open-mesh parabolic dish is a diffuse reflector for short-wavelength light but a polished reflector for long-wavelength radio waves.

FIGURE 11.16
A magnified view of the surface of ordinary paper.

Mastering**PHYSICS**
VIDEO: Model of Refraction

A light ray is always at right
angles to its wavefront.

11.4 Refraction

EXPLAIN THIS   Why does a fish in water seem closer to the surface than it
actually is?

As we learned in Section 11.2, light slows down when it enters glass, and it
travels at different speeds in different materials.* It travels at 300,000 km/s
in a vacuum, at a slightly lower speed in air, and at about three-fourths
that speed in water. Unless the light is perpendicular to the surface of penetration, bending occurs. This is the phenomenon of **refraction**.

To gain a better understanding of the bending of light in refraction, look
at the pair of toy cart wheels in Figure 11.17. The wheels roll from a smooth
sidewalk onto a grass lawn. If the wheels meet the grass at an angle, as the figure shows, they are deflected from their straight-line course. Note that the left
wheel slows first when it interacts with the grass on the lawn. The right wheel
maintains its higher speed while on the sidewalk. It pivots about the slower-
moving left wheel because it travels farther in the same time. So the direction
of the rolling wheels is bent toward the "normal," the black dashed line perpendicular to the grass-sidewalk border in Figure 11.17.

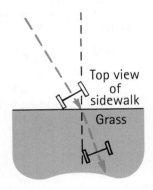

FIGURE 11.17
The direction of the rolling wheels
changes when one wheel slows down
before the other does.

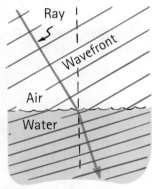

FIGURE 11.18
The direction of the light waves
changes when one part of the wave
slows down before the other part.

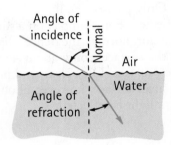

FIGURE 11.19

INTERACTIVE FIGURE

Refraction. The angles of incidence
and refraction are in accord with
Snell's law (see footnote on page 273).

Figure 11.18 shows how a light wave bends in a similar way. Note the
direction of light, indicated by the blue arrow (the light ray). Also note the
wavefronts drawn at right angles to the ray. (If the light source were close,
the wavefronts would appear circular; but if the distant Sun is the source,
the wavefronts are practically straight lines.) The wavefronts are everywhere

---

* Just how much the speed of light differs from its speed in a vacuum is given by the index of
refraction, *n*, of the material:

$$n = \frac{\text{speed of light in vacuum}}{\text{speed of light in material}}$$

For example, the speed of light in a diamond is 124,000 km/s, and so the index of refraction for
diamond is

$$n = \frac{300,000 \text{ km/s}}{124,000 \text{ km/s}} = 2.42$$

For a vacuum, $n = 1$.

at right angles to the light rays. The bending of the wave (sound or light) is caused by a change of speed.*

Figure 11.20 shows a beam of light entering water at the left and exiting at the right. The path would be the same if the light entered from the right and exited at the left. The light paths are reversible for both reflection and refraction. If you see someone's eyes by way of a reflective or refractive device, such as a mirror or a prism, then that person can see you by way of the device also (unless the device is optically coated to produce a one-way effect).

Refraction causes many illusions. One of them is the apparent bending of a stick that is partially submerged in water. The submerged part appears closer to the surface than it actually is. The same is true when you look at a fish in water. The fish appears nearer to the surface and closer than it really is (Figure 11.21). If we look straight down into water, an object submerged 4 meters beneath the surface appears to be only 3 meters deep. Because of refraction, submerged objects appear to be closer.

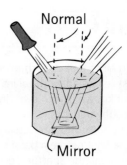

FIGURE 11.20
When light slows down in going from one medium to another, as it does in going from air to water, it bends toward the normal. When it speeds up in traveling from one medium to another, as it does in going from water to air, it bends away from the normal.

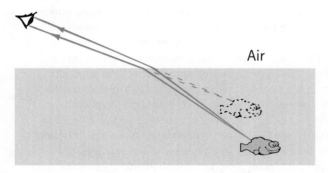

FIGURE 11.21
Because of refraction, a submerged object appears to be nearer to the surface than it actually is.

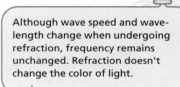
Although wave speed and wavelength change when undergoing refraction, frequency remains unchanged. Refraction doesn't change the color of light.

Refraction occurs in Earth's atmosphere. Whenever we watch a sunset, we see the Sun for several minutes after it has sunk below the horizon (Figure 11.22). Earth's atmosphere is thin at the top and dense at the bottom. Because light travels faster in thin air than in dense air, parts of the wavefronts of

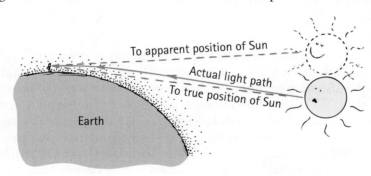

FIGURE 11.22
Because of atmospheric refraction, when the Sun is near the horizon it appears to be higher in the sky.

sunlight at high altitude travel faster than parts closer to the ground. Light rays bend. The density of the atmosphere changes gradually, so light rays bend gradually and follow a curved path. So we gain additional minutes of daylight each day. Furthermore, when the Sun (or Moon) is near the horizon, the rays from the lower edge are bent more than the rays from the upper edge. This shortens the vertical diameter, causing the Sun to appear elliptical (Figure 11.23).

*The quantitative law of refraction, called Snell's law, is credited to Willebrord Snell, a 17th-century Dutch astronomer and mathematician: $n_1 \sin \theta_1 = n_2 \sin \theta_2$, where n_1 and n_2 are the indices of refraction of the media on either side of the surface, and θ_1 and θ_2 are the respective angles of incidence and refraction. If three of these values are known, the fourth can be calculated from this relationship.

For a wave explanation of refraction (and diffraction), read about Huygens' principle, pages 512–515, *Conceptual Physics—11th Edition.*

FIGURE 11.23
The Sun is distorted by differential refraction.

YOUR EYE

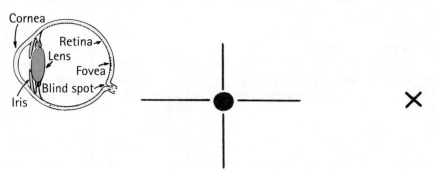

With all of today's technology, the most remarkable optical instrument known is your eye. Light enters through your cornea, which does about 70% of the necessary bending of the light before it passes through your pupil (the aperture, or opening, in the iris). Light then passes through your lens, which provides the extra bending power needed to focus images of nearby objects on your extremely sensitive retina. (Only recently have artificial detectors been made with greater sensitivity to light than the human eye.) An image of the visual field outside your eye is spread over the retina. The retina is not uniform. A spot in the center of the retina, called the fovea, is the region of most acute vision. You see greater detail here than at any other part of your retina. There is also a spot on your retina where the nerves carrying all the information exit the eye on their way to the brain. This is your blind spot.

You can demonstrate that you have a blind spot in each eye. Simply hold this book at arm's length, close your left eye, and look at the round dot and the X to its right with your right eye only. You can see both the dot and the X at this distance. Now move the book slowly toward your face, with your right eye fixed on the dot, and you'll reach a position about 20–25 cm from your eye where the X disappears. When both eyes are open, one eye "fills in" the part to which your other eye is blind. Now repeat with only the left eye open, looking this time at the X, and the dot will disappear. But note that your brain fills in the two intersecting lines. Amazingly, your brain fills in the "expected" view even with one eye closed. Instead of seeing nothing, your brain graciously fills in the appropriate background. Repeat this for small objects on various backgrounds. You not only see what's there—you see what's not there!

The light receptors in your retina do not connect directly to your optic nerve but are instead interconnected with many other cells. Through these interconnections, a certain amount of information is combined and "digested" in your retina. In this way, the light signal is "thought about" before it goes to the optic nerve and then to the main body of your brain. So some brain functioning occurs in your eye. Amazingly, your eye does some of your "thinking."

One of the many beauties of physics is the redness of a fully eclipsed Moon—resulting from the refraction of sunsets and sunrises that completely circle the world. This refracted light shines on an otherwise dark Moon.

A mirage occurs when refracted light appears as if it were reflected light. Mirages are a common sight on a desert when the sky appears to be reflected from water on the distant sand. But when you approach what seems to be water, you find dry sand. Why is this so? The air is very hot close to the sand surface and cooler above the sand. Light travels faster through the thinner hot air near the surface than through the denser cool air above. So wavefronts near the ground travel faster than they do above. The result is upward bending (Figure 11.24). So we see an upside-down view that looks as if reflection were occurring from a water surface. We see a mirage, which is formed by real light and can be photographed (Figure 11.25). A mirage is not, as many people think, a trick of the mind.

FIGURE 11.24
Light from the top of the tree gains speed in the warm and less dense air near the ground. When the light grazes the surface and bends upward, the observer sees a mirage.

FIGURE 11.25
A mirage. The apparent wetness of the road is not a reflection of the sky by water but a refraction of skylight through the warmer and less-dense air near the road surface.

When we look at an object over a hot stove or over a hot pavement, we see a wavy, shimmering effect. This is due to varying densities of air caused by changes in temperature. The twinkling of stars results from similar variations in the sky, where light passes through unstable layers in the atmosphere.

CHECKPOINT

If the speed of light were the same in air of various temperatures and densities, would there still be slightly longer daytimes, twinkling stars at night, mirages, and slightly squashed suns at sunset?

Was this your answer?
No.

FIGURE 11.26
Sunlight passing through a prism separates into a color spectrum. The colors of things depend on the colors of the light that illuminates them.

11.5 Color

EXPLAIN THIS Why do red, green, and blue combine to make white on your TV screen?

Roses are red and violets are blue; colors intrigue artists and physical science types too. To the scientist, the colors of objects are not in the substances of the objects themselves or even in the light they emit or reflect. Color is a physiological experience and is in the eye of the beholder. So when we say that light from a rose petal is red, in a stricter sense we mean that it appears red. Many organisms, including people with defective color vision, do not see the rose as red at all.

Different frequencies of light are perceived as different colors; the lowest frequency we see appears, to most people, as the color red, and the highest appears as violet. Between them range the infinite number of hues that make up the color spectrum of the rainbow. By convention, these hues are grouped into seven colors: red, orange, yellow, green, blue, indigo, and violet. These colors together appear white. The white light from the Sun is a composite of all the visible frequencies.

LEARNING OBJECTIVE
Describe how color depends on the frequency of light.

Mastering**PHYSICS**
TUTORIAL: Color
VIDEO: Colored Shadows
VIDEO: Yellow-Green Peak of Sunlight
VIDEO: Why the Sky is Blue and Sunsets are Red

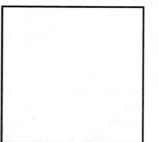

FIGURE 11.27
The square on the top reflects all the colors illuminating it. In sunlight, it is white. When illuminated with blue light, it is blue. The square on the bottom absorbs all the colors illuminating it. In sunlight, it is warmer than the white square.

Except for such light sources as lamps, lasers, and gas discharge tubes, most of the objects around us reflect rather than emit light. They reflect only part of the light that is incident upon them, the part that provides their color.

Selective Reflection

A rose, for example, doesn't emit light; it reflects light. If we pass sunlight through a prism and then place the petal of a deep-red rose in various parts of the spectrum, the petal appears brown or black in all regions of the spectrum except in the red region. In the red part of the spectrum, the petal also appears red, but the green stem and leaves appear black. This shows that the petal has the ability to reflect red light, but it cannot reflect other colors; the green leaves have the ability to reflect green light and, likewise, cannot reflect other colors. When the rose is held in white light, the petals appear red and the leaves appear green, because the petals reflect the red part of the white light and the leaves reflect the green part of the white light. To understand why objects reflect specific colors of light, we turn our attention to the atom.

Light is reflected from objects in a manner similar to the way sound is "reflected" from a tuning fork when another tuning fork nearby sets it into vibration. A tuning fork can be made to vibrate even when the frequencies are not matched, although at significantly reduced amplitudes. The same is true of atoms and molecules. Electrons can be forced into vibration by the vibrating electric fields of electromagnetic waves. Once vibrating, these electrons emit their own electromagnetic waves, just as vibrating acoustical tuning forks emit sound waves.

Interestingly, the petals of most yellow flowers, such as daffodils, reflect red and green as well as yellow. Yellow daffodils reflect a broad band of frequencies. The reflected colors of most objects are not pure single-frequency colors but are a mixture of frequencies.

An object can reflect only frequencies present in the illuminating light. An incandescent lamp emits light of lower average frequencies than sunlight, enhancing any reds viewed in this light. In a fabric having only a little bit of red in it, the red is more apparent under an incandescent lamp than it is under a fluorescent lamp. Fluorescent lamps are richer in the higher frequencies, and so blues are enhanced in their light. How a color appears depends on the light source (Figure 11.30).

Selective Transmission

The color of a transparent object depends on the color of the light it transmits. A red piece of glass appears red because it absorbs colors of white light except red, so red light is transmitted. Similarly, a blue piece of glass appears blue because it transmits primarily blue and absorbs the other colors. These pieces of glass contain dyes or *pigments*—fine particles that selectively absorb light of particular frequencies and selectively transmit others. Light of some of the frequencies is absorbed by the pigments. The rest is re-emitted

FIGURE 11.28
The bunny's dark fur absorbs all the radiant energy in incident sunlight and therefore appears black. Light fur on other parts of the body reflects light of all frequencies and therefore appears white.

FIGURE 11.29
Only energy having the frequency of blue light is transmitted; energy of the other frequencies, or of the complementary color yellow, is absorbed and warms the glass.

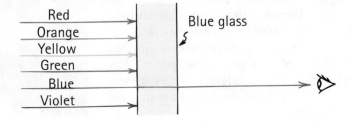

from atom to atom in the glass. The energy of the absorbed light increases the kinetic energy of the atoms, and the glass is warmed. Ordinary window glass doesn't have a color because it transmits light of all visible frequencies equally well.

CHECKPOINT

1. **Why do the leaves of a red rose become warmer than the petals when illuminated with red light?**
2. **When illuminated with green light, why do the petals of a red rose appear black?**

Were these your answers?

1. The leaves absorb rather than reflect red light, so the leaves become warmer.
2. The petals absorb rather than reflect the green light. Because green is the only color illuminating the rose, and green contains no red to be reflected, the rose reflects no color at all and appears black.

FIGURE 11.30
Color depends on the light source.

Mixing Colored Lights

White light is dispersed by a prism into a rainbow-colored spectrum. The distribution of sunlight (Figure 11.31) is uneven, and the light is most intense in the yellow-green part of the spectrum. How fascinating it is that our eyes have evolved to have maximum sensitivity in this range. That's why fire engines and tennis balls are yellow-green for better visibility.

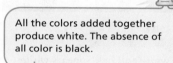

All the colors added together produce white. The absence of all color is black.

All the colors combined produce white. Interestingly, we see white also from the combination of only red, green, and blue light. We can understand this by dividing the solar radiation curve into three regions, as in Figure 11.32. Three types of cone-shaped receptors in our eyes perceive color. Each is stimulated only by certain frequencies of light. Light of lower visible frequencies stimulates the cones that are sensitive to low frequencies and appears red. Light of middle frequencies stimulates the cones that are sensitive to middle frequencies and appears green. Light of higher frequencies stimulates the cones that are sensitive to higher frequencies and appears blue. When all three types of cones are stimulated equally, we see white.

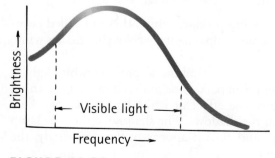

FIGURE 11.31
The radiation curve of sunlight is a graph of brightness versus frequency. Sunlight is brightest in the yellow-green region, which is in the middle of the visible range.

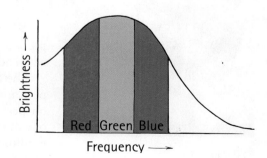

FIGURE 11.32
The radiation curve of sunlight divided into three regions—red, green, and blue. These are the additive primary colors.

It's interesting to note that the "black" you see on the darkest scenes on a TV screen is simply the color of the tube face itself, which is more a light gray than black. Because our eyes are sensitive to the contrast with the illuminated parts of the screen, we see this gray as black.

FIGURE 11.33
INTERACTIVE FIGURE **MP**

Color addition by the mixing of colored lights. When three projectors shine red, green, and blue light on a white screen, the overlapping parts produce different colors. White is produced where all three overlap.

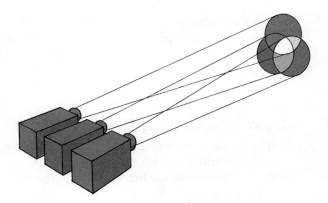

Project red, green, and blue lights on a screen and where they all overlap, white is produced. If two of the three colors overlap, or are added, then another color sensation is produced (Figure 11.33). By adding various amounts of red, green, and blue, the colors to which each of our three types of cones are sensitive, we can produce any color in the spectrum. For this reason, red, green, and blue are called the **additive primary colors**. A close examination of the picture on television screens reveals that the picture is an assemblage of tiny spots, each less than a millimeter across. When the screen is lit, some of the spots are red, some are green, and some are blue; the mixtures of these primary colors at a distance provide a complete range of colors, plus white.

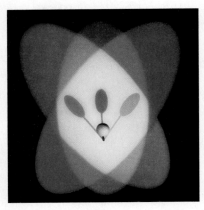

FIGURE 11.34
INTERACTIVE FIGURE **MP**

The white golf ball appears white when it is illuminated with red, green, and blue lights of equal intensities. Why are the shadows cast by the ball cyan, magenta, and yellow?

Complementary Colors

Here's what happens when two of the three additive primary colors are combined:

$$Red + blue = magenta$$
$$Red + green = yellow$$
$$Blue + green = cyan$$

We say that magenta is the opposite of green, cyan is the opposite of red, and yellow is the opposite of blue. The addition of any color to its opposite color results in white.

$$Magenta + green = white \ (= red + blue + green)$$
$$Cyan + red = white \ (= blue + green + red)$$
$$Yellow + blue = white \ (= red + green + blue)$$

When two colors are added together to produce white, they are called **complementary colors**. Every hue has some complementary color that makes white when added to it.

The fact that a color and its complement combine to produce white light is pleasantly used in lighting stage performances. Blue and yellow lights shining on performers, for example, produce the effect of white light—except where one of the two colors is absent, as in the shadows. The shadow of the blue lamp is illuminated by the yellow lamp, and thus it appears yellow. Similarly, the shadow cast by the yellow lamp appears blue. This is a most intriguing effect.

We can see this effect in Figure 11.34, where red, green, and blue lights shine on the golf ball. Note the shadows cast by the ball. The middle shadow is cast by the green spotlight and is not dark because it is illuminated by the red and blue lights, which produces magenta. The shadow cast by the blue light appears

FIGURE 11.35
Carlos Vasquez displays a variety of colors when he is illuminated by only red, green, and blue lamps. Can you account for the other resulting colors that appear?

FIGURING PHYSICAL SCIENCE

SAMPLE PROBLEMS

1. From Figure 11.33 or 11.34, find the complements of cyan, yellow, and red.
2. Red + cyan = _____
3. White − cyan = _____
4. White − red = _____

Solutions:

1. red, blue, cyan
2. white
3. red
4. cyan; interestingly enough, the cyan color of the sea is the result of the removal of red light from white

sunlight. The natural frequency of water molecules coincides with the frequency of infrared light, so infrared is strongly absorbed by water. To a lesser extent, red light is also absorbed by water—enough so that it appears a greenish-blue or cyan color.

yellow because it is illuminated by red and green light. Can you see why the shadow cast by the red light appears cyan?

Mixing Colored Pigments

Every artist knows that if you mix red, green, and blue paint, the result is not white but a muddy dark brown. Mixing red and green paint certainly does not produce yellow, so the rule for adding colored lights doesn't apply here. The mixing of pigments in paints and dyes is entirely different from mixing lights. Pigments are tiny particles that absorb specific colors. For example, pigments that produce the color red absorb the complementary color cyan. So something painted red absorbs cyan, which is why it reflects red. In effect, cyan has been subtracted from white light. Something painted blue absorbs yellow, so it reflects all the colors except yellow. Remove yellow from white and you've got blue. The colors magenta, cyan, and yellow are the **subtractive primary colors**. The variety of colors that you see in the colored photographs in this or any book is the result of magenta, cyan, and yellow dots. Light illuminates the book, and light of some frequencies is subtracted from the light reflected. The rules of color subtraction differ from the rules of light addition.

Ink-jet printers deposit various combinations of cyan, magenta, yellow, and black inks. This is CMYK printing (K indicates black). Interestingly, the three colors can produce black, but that takes more ink and has a color cast—hence the black ink, which does a better job. Examine the color in any of the figures in this or any book with a magnifying glass and see how the overlapping dots of these colors show a wide range of colors. Or look at a billboard up close.

FIGURE 11.36
The color green on a printed page consists of cyan and yellow dots.

FIGURE 11.37
The vivid colors of Sneezlee represent many frequencies of light. The photo, however, is a mixture of only cyan, yellow, magenta, and black (CYMK).

(a)

(b)

(c)

(d)

(e)

(f)

FIGURE 11.38
Only three colors of ink (plus black) are used to print color photographs—(a) magenta, (b) yellow, and (c) cyan, which when combined produce the colors shown in (d). The addition of black (e) produces the finished result (f).

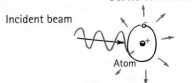

FIGURE 11.39
A beam of light falls on an atom and increases the vibrational motion of electrons in the atom. The vibrating electrons, in turn, re-emit light in various directions. Light is scattered.

FIGURE 11.40
In clean air, the scattering of high-frequency light gives us a blue sky. When the air is full of particles larger than oxygen and nitrogen molecules, light of lower frequencies is also scattered, which adds to the high-frequency scattered light to give us a whitish sky.

Atmospheric soot heats Earth's atmosphere by absorbing light while cooling local regions by blocking sunlight from reaching the ground. Soot particles in the air can trigger severe rains in one region and droughts and dust storms in another.

Why the Sky Is Blue

Not all colors are the result of the addition or subtraction of light. Some colors, like the blue of the sky, are the result of selective scattering.* Consider the analogous case of sound: If a beam of a particular frequency of sound is directed to a tuning fork of a similar frequency, the tuning fork is set into vibration and redirects the beam in multiple directions. The tuning fork *scatters* the sound. A similar process occurs with the scattering of light from atoms and particles that are far apart from one another. This is what happens in the atmosphere.

We know that atoms behave like tiny optical tuning forks and re-emit light waves that shine on them. Very tiny particles act in a similar way. The tinier the particle, the higher the frequency of light it will re-emit. This is similar to the way in which small bells ring with higher notes than larger bells. The nitrogen and oxygen molecules that make up most of the atmosphere are like tiny bells that "ring" with high frequencies when they are energized by sunlight. Like sound from the bells, the re-emitted light is sent in all directions. The re-emitted light is said to be *scattered* in all directions.

Of the visible frequencies of sunlight, violet is scattered the most by nitrogen and oxygen in the atmosphere. Then the other colors are scattered in order: blue, green, yellow, orange, and red. Red is scattered only a tenth as much as violet. Although violet light is scattered more than blue, our eyes are not very sensitive to violet light. Therefore, the blue scattered light is what predominates in our vision, so we see a blue sky!

On clear, dry days, the sky is a much deeper blue than it is on clear, humid days. Places where the upper air is exceptionally dry, such as Italy and Greece, have beautiful blue skies that have inspired painters for centuries. Where the atmosphere contains a lot of particles of dust and other particles larger than oxygen and nitrogen molecules, light of the lower frequencies also undergoes significant scattering. This causes the sky to appear less blue, with a whitish appearance. After a heavy rainstorm, when the airborne particles have been washed away, the sky becomes a deeper blue.

The grayish haze in the skies over large cities is the result of particles emitted by automobile and truck engines and by factories. Even when idling, a typical automobile engine emits more than 100 billion particles per second. Most are invisible, but they act as tiny centers to which other particles adhere. These are the primary scatterers of lower-frequency light. With the largest of these particles, absorption rather than scattering occurs, and a brownish haze is produced. Yuck!

Why Sunsets Are Red

Light that isn't scattered is light that is transmitted. Because red, orange, and yellow light are the least scattered by the atmosphere, light of these low frequencies is better transmitted through the air. Red is scattered the least, and it passes through more atmosphere than any other color. So the thicker the atmosphere through which a beam of sunlight travels, the more time there is to scatter all the higher-frequency parts of the light. As Figure 11.41 shows, sunlight travels through more atmosphere at sunset, which is why sunsets are red.

At noon, sunlight travels through the least amount of atmosphere to reach Earth's surface. Only a small amount of blue is scattered, which makes the Sun appear yellowish. As the day progresses and the Sun descends lower in the sky,

---

* This type of scattering, called *Rayleigh scattering*, occurs whenever the scattering particles are much smaller than the wavelength of incident light and have resonances at frequencies higher than those of the scattered light.

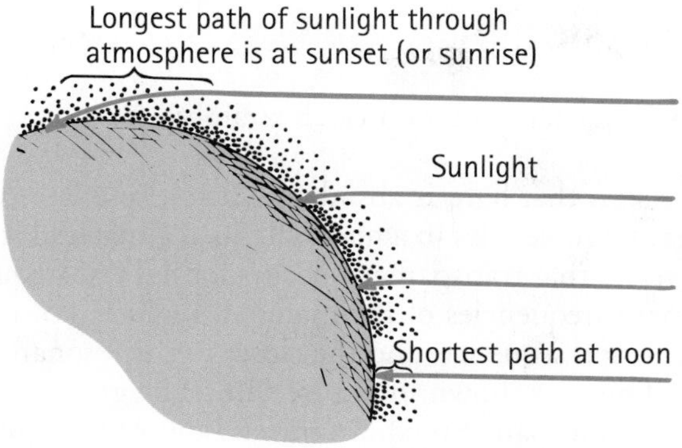

Longest path of sunlight through
atmosphere is at sunset (or sunrise)

Sunlight

Shortest path at noon

FIGURE 11.41
INTERACTIVE FIGURE   MP

Paths of sunlight through the
atmosphere.

as Figure 11.41 indicates, the path through the atmosphere is longer, and more violet and blue are scattered from the sunlight. The Sun becomes progressively redder, going from yellow to orange and finally to a red-orange at sunset. Sunsets and sunrises are unusually colorful following volcanic eruptions because particles larger than atmospheric molecules are more abundant in the air.

Why Clouds Are White

Clouds are made up of clusters of water droplets in a variety of sizes. These clusters of different sizes result in a variety of scattered colors. The tiniest clusters tend to produce blue clouds; slightly larger clusters, green clouds; and still larger clusters, red clouds. The overall result is a white cloud. Electrons close to one another in a cluster vibrate in phase. This results in a greater intensity of scattered light than there would be if the same number of electrons were vibrating separately. Hence, clouds are bright!

Larger clusters of droplets absorb much of the light incident upon them, and so the scattered intensity is less. Therefore, clouds composed of larger clusters darken to a deep gray. Further increase in the size of the clusters causes them to fall as raindrops, and we have rain.

The next time you find yourself admiring a crisp blue sky, or delighting in the shapes of bright clouds, or watching a beautiful sunset, think about all those ultratiny optical tuning forks vibrating away. You'll appreciate these daily wonders of nature even more!

FIGURE 11.42
A cloud is composed of water droplets of various sizes. The tiniest droplets scatter blue light, slightly larger ones scatter green light, and still larger ones scatter red light. The overall result is a white cloud.

CHECKPOINT

1. **If molecules in the sky were to scatter low-frequency light more than high-frequency light, what color would the sky be? What color would sunsets be?**
2. **Distant dark mountains are bluish in color. What is the source of this blueness? (Hint: Exactly what is between us and the mountains we see?)**

Were these your answers?

1. If light of low frequencies were scattered, the noontime sky would appear reddish orange. At sunset, more reds would be scattered by the longer distance traveled by the sunlight, and the sunlight would be predominantly blue and violet. So sunsets would appear blue!
2. If we look at distant dark mountains, very little light from them reaches us, and the blueness of the atmosphere between us and the mountains predominates. The blueness is of the low-altitude "sky" between us and the mountains. That's why distant mountains appear blue.

FIGURE 11.43
The wave appears cyan because seawater absorbs red light. The spray at the crest of the wave appears white because, like clouds, it is composed of a variety of tiny water droplets that scatter all the visible frequencies.

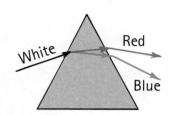

FIGURE 11.44
Dispersion by a prism makes the components of white light visible.

11.6 Dispersion

EXPLAIN THIS Why can't you ever catch a rainbow?

We have seen that light is absorbed when it resonates with electrons of atoms and molecules in a material. Such a material is opaque to light. Also recall that transparency occurs for light of frequencies near (but not at) the resonant frequencies of the material. Light is slowed because of the absorption/re-emission sequence, and the closer to the resonant frequencies, the slower the light. This was shown in Figure 11.6. The grand result is that high-frequency light in a transparent medium travels slower than low-frequency light. Violet light travels about 1% slower in ordinary glass than red light. Lights of colors between red and violet travel at their own respective speeds in glass.

Because light of various frequencies travels at different speeds in transparent materials, different colors of light refract by different amounts. When white light is refracted twice, as in a prism, the separation of light by colors is quite noticeable. This separation of light into colors arranged by frequency is called **dispersion** (Figure 11.44). Because of dispersion, there are rainbows!

Rainbows

For you to see a rainbow, the Sun must shine on drops of water in a cloud or in falling rain. The drops act as prisms that disperse light. When you face a rainbow, the Sun is behind you, in the opposite part of the sky. Seen from an airplane near midday, the bow forms a complete circle. As we will see, all rainbows would be completely round if the ground were not in the way.

You can see how a raindrop disperses light in Figure 11.45. Follow the ray of sunlight as it enters the drop near its top surface. Some of the light here is reflected (not shown), and the remainder is refracted into the water. At this first refraction, the light is dispersed into its spectrum colors, red being deviated the least and violet the most. When the light reaches the opposite side of the drop, each color is partly refracted out into the air (not shown) and partly reflected back into the water. Arriving at the lower surface of the drop, each color is again partly reflected (not shown) and partly refracted back into the air. This refraction at the second surface, like that in a prism, increases the dispersion already produced at the first surface.*

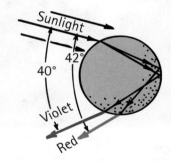

FIGURE 11.45
Dispersion of sunlight by a single raindrop.

FIGURE 11.46
Sunlight incident on two raindrops, as shown, emerges from them as dispersed light. The observer sees the red light from the upper drop and the violet light from the lower drop. Millions of drops produce the entire spectrum of visible light.

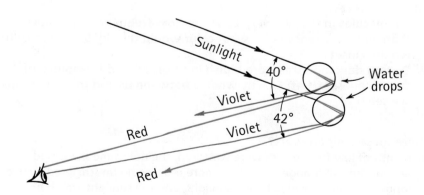

* We're simplifying when we indicate that the red ray disperses at 42°. Actually, the angle between the incoming and outgoing rays can be anywhere between zero and about 42° (zero degrees corresponding to a full 180-degree reversal of the light). The strongest concentration of light intensity for red, however, is near the maximum angle of 42°, as shown in Figures 11.45 and 11.46.

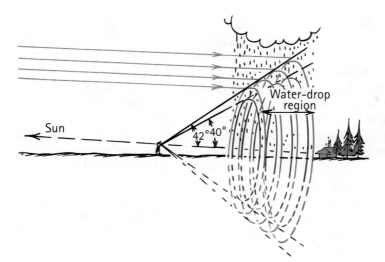

FIGURE 11.47
When your eye is located between the Sun (not shown, off to the left) and the water-drop region, the rainbow you see is the edge of a three-dimensional cone that extends through the water-drop region. Violet is dispersed by drops that form a 40° conical surface; red is seen from drops along a 42° conical surface, with other colors in between. (Innumerable layers of drops form innumerable two-dimensional arcs, like the four sets suggested here.)

Although each drop disperses a full spectrum of colors, an observer is in a position to see only a single color from any one drop (Figure 11.46). If violet light from a single drop reaches an observer's eye, red light from the same drop is incident elsewhere toward the feet. To see red light, one must look to a drop higher in the sky. The color red is seen where the angle between a beam of sunlight and the dispersed light is 42°. The color violet is seen where the angle between the sunbeams and dispersed light is 40°.

Why does the light dispersed by the raindrops form a bow? The answer involves a bit of geometry. First of all, a rainbow is not the flat two-dimensional arc it appears to be. The rainbow you see is actually a three-dimensional cone of dispersed light. The apex of this cone is at your eye. To understand this, consider a glass cone, the shape of those paper cones you sometimes see at drinking fountains. If you held the tip of such a glass cone against your eye, what would you see? You'd see the glass as a circle. Likewise with a rainbow. All the drops that disperse the rainbow's light toward you lie in the shape of a cone—a cone of different layers with drops that deflect red to your eye on the outside, orange beneath the red, yellow beneath the orange, and so on, all the way to violet on the inner conical surface (Figure 11.47). The thicker the region containing water drops, the thicker the conical edge you look through and the more vivid the rainbow.

Mastering**PHYSICS**
VIDEO: The Rainbow

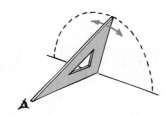

FIGURE 11.48
Only the raindrops along the dashed line disperse red light to the observer at a 42° angle; hence, the light forms a bow.

FIGURE 11.49
Two refractions and a reflection in water droplets produce light at all angles up to about 42°, with the intensity concentrated where we see the rainbow at 40° to 42°. Light doesn't exit the water droplet at angles greater than 42° unless it undergoes two or more reflections inside the drop. Thus the sky is brighter inside the rainbow than outside it. Notice the weak secondary rainbow.

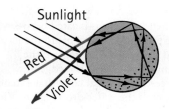

FIGURE 11.50
Double reflection in a drop produces a secondary bow.

Your cone of vision intersects the cloud of drops and creates your rainbow. It is ever so slightly different from the rainbow seen by a person nearby. So, when a friend says, "Look at the pretty rainbow," you can reply, "Okay, move aside so I can see it, too." Everybody sees his or her own personal rainbow.

Another fact about rainbows: A rainbow always faces you squarely. When you move, your rainbow appears to move with you. So you can never approach the side of a rainbow or see it end-on as in the exaggerated view of Figure 11.47. You *can't* reach its end. Thus the saying "looking for the pot of gold at the end of the rainbow" means pursuing something you can never reach.

Often a larger, secondary bow with its colors reversed can be seen arching at a greater angle around the primary bow. We won't treat this secondary bow except to say that it is formed by similar circumstances and is a result of double reflection within the raindrops (Figure 11.50). Because of this extra reflection (and extra refraction loss), the secondary bow is much dimmer and reversed.

> Isn't it true that knowing why rainbows are round and why they're colored *adds* to their beauty? Knowledge doesn't subtract—it adds.

CHECKPOINT

1. **Suppose you point to a wall with your arm extended. Then you sweep your arm around, making an angle of about 42° to the wall. If you rotate your arm in a full circle while keeping the same angle, what shape does your arm describe? What shape does your finger sweep out on the wall?**
2. **If light traveled at the same speed in raindrops as it does in air, would we have rainbows?**

Were these your answers?
1. Your arm describes a cone, and your finger sweeps out a circle. Likewise with rainbows.
2. No.

LEARNING OBJECTIVE
Describe how polarization relates to wave orientation.

11.7 Polarization

EXPLAIN THIS Why do Polaroid glasses reduce glare?

FIGURE 11.51
A vertically plane-polarized plane wave and a horizontally plane-polarized plane wave.

As we learned in Chapter 10, waves can be either longitudinal or transverse. Sound waves are longitudinal, which means the vibratory motion of the medium is along the direction of wave travel. The fact that light waves exhibit **polarization** demonstrates that light waves are transverse.

If you shake a rope either up and down or from side to side as shown in Figure 11.51, you produce a transverse wave along the rope. The plane of vibration is the same as the plane of the wave. If you shake it up and down, the wave vibrates in a vertical plane. If you shake it back and forth, the wave vibrates in a horizontal plane. We say that such a wave is *plane-polarized*—that the waves traveling along the rope are confined to a single plane. Polarization is a property of transverse waves. (Polarization does not occur among longitudinal waves—there is no such thing as polarized sound.)

A single vibrating electron can emit an electromagnetic wave that is plane-polarized. The plane of polarization matches the vibrational direction of the electron. That means that a vertically accelerating electron emits light that is vertically polarized. A horizontally accelerating electron emits light that is horizontally polarized (Figure 11.52).*

---

* Light may also be circularly polarized and elliptically polarized, which are also transverse polarizations. But we will not study these cases.

A common light source, such as an incandescent lamp, a fluorescent lamp, or a candle flame, emits light that is unpolarized. This is because the electrons that emit the light are vibrating in many random directions. There are as many planes of vibration as the vibrating electrons producing them. A few planes are represented in Figure 11.53a. We can represent all these planes by radial lines, shown in Figure 11.53b. (Or, more simply, the planes can be represented by vectors in two mutually perpendicular directions, as shown in Figure 11.53c.) The vertical vector represents all the components of vibration in the vertical direction. The horizontal vector represents all the components of vibration horizontally. The simple model of Figure 11.53c represents unpolarized light. Polarized light would be represented by a single vector.

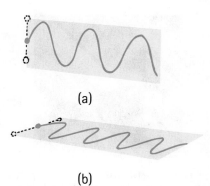

(a)

(b)

FIGURE 11.52
(a) A vertically plane-polarized wave from a charge vibrating vertically. (b) A horizontally plane-polarized wave from a charge vibrating horizontally.

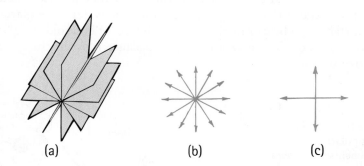

(a)   (b)   (c)

FIGURE 11.53
Representations of plane-polarized waves.

All transparent crystals having a noncubic natural shape have the property of polarizing light. These crystals divide unpolarized light into two internal beams polarized at right angles to each other. Some crystals strongly absorb one beam while transmitting the other (Figure 11.54). This makes them excellent polarizers. Herapathite is such a crystal. Microscopic herapathite crystals are aligned and embedded between cellulose sheets. They make up Polaroid filters, popular in sunglasses. Other Polaroid sheets consist of certain aligned molecules rather than tiny crystals.

If you look at unpolarized light through a Polaroid filter, you can rotate the filter in any direction and the light appears unchanged. But if the light is polarized, rotating the filter allows you to block out more and more of the light until it is completely blocked out. An ideal Polaroid filter transmits 50% of incident unpolarized light. That 50% is polarized. When two Polaroid filters are arranged so that their polarization axes are aligned, light can pass through both, as shown in the rope analogy (Figure 11.55a).

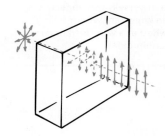

FIGURE 11.54
One component of the incident unpolarized light is absorbed, resulting in emerging polarized light.

Mastering**PHYSICS**

VIDEO: Polarization and 3D Viewing

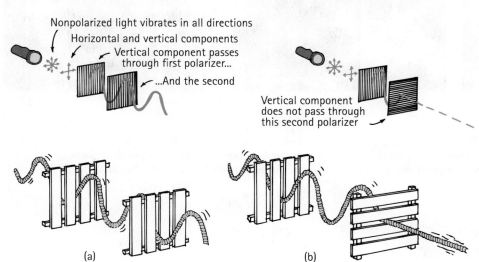

Nonpolarized light vibrates in all directions
Horizontal and vertical components
Vertical component passes through first polarizer...
...And the second

Vertical component does not pass through this second polarizer

(a)   (b)

FIGURE 11.55
A rope analogy illustrates the effect of crossed Polaroids.

FIGURE 11.56
Polaroid sunglasses block out horizontally vibrating light. When the lenses overlap at right angles, light doesn't get through.

If their axes are at right angles to each other (in this case, we say the filters are crossed), almost no light penetrates the pair (Figure 11.55b). (A small amount of shorter wavelengths do get through.) When Polaroid filters are used in pairs like this, the first one is called the *polarizer* and the second one is called the *analyzer*.

Much of the light reflected from nonmetallic surfaces is polarized. The glare from glass or water is a good example. Except for light that hits vertically, the reflected ray has more vibrations parallel to the reflecting surface. The part of the ray that penetrates the surface has more vibrations at right angles to the surface (Figure 11.56). Skipping flat rocks off the surface of a pond provides an appropriate analogy. When the rocks hit parallel to the surface, they are easily reflected by the surface. But when they hit with their faces at right angles to the surface, they "refract" into the water. The glare from reflecting surfaces can be dimmed a lot with the use of Polaroid sunglasses. The polarization axes of the lenses are vertical because most of the glare reflects from horizontal surfaces.

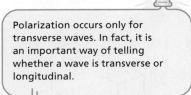

Polarization occurs only for transverse waves. In fact, it is an important way of telling whether a wave is transverse or longitudinal.

FIGURE 11.57
Most glare from nonmetallic surfaces is polarized. Note that the components of incident light parallel to the surface are reflected, and the components perpendicular to the surface pass through the surface into the medium. Because most glare we encounter is from horizontal surfaces, the polarization axes of Polaroid sunglasses are vertical.

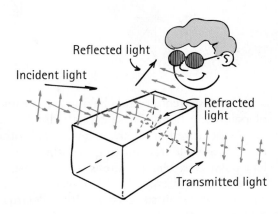

Reflected light

Incident light

Refracted light

Transmitted light

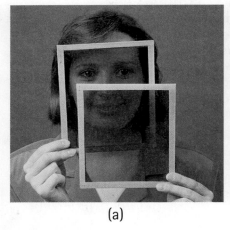

(a)

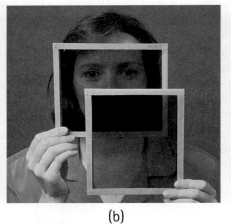

(b)

(c)

FIGURE 11.58
Light is transmitted when the axes of the Polaroids are aligned (a), but absorbed when Ludmila rotates one so that the axes are at right angles to each other (b). When she inserts a third Polaroid at an angle between the crossed Polaroids, light is again transmitted (c). Why? (For the answer, after you have given this some thought see Appendix B, "More About Vectors.")

For instructor-assigned homework, go to www.masteringphysics.com (MP)

SUMMARY OF TERMS (KNOWLEDGE)

Additive primary colors The three colors—red, green, and blue—that, when mixed in certain proportions, can produce any color in the spectrum.

Complementary colors Any two colors that, when mixed, produce white light.

Diffuse reflection Reflection in irregular directions from an irregular surface.

Dispersion The separation of light into colors arranged by frequency.

Electromagnetic spectrum The range of electromagnetic waves that extends in frequency from radio waves to gamma rays.

Electromagnetic wave An energy-carrying wave emitted by vibrating electric charges (often electrons) and composed of oscillating electric and magnetic fields that regenerate each other.

Law of reflection The angle of incidence equals the angle of reflection. The incident and reflected rays lie in a plane that is normal to the reflecting surface.

Opaque The property of absorbing light without re-emission (opposite of transparent).

Polarization The alignment of the transverse electric vectors that make up electromagnetic radiation. Such waves of aligned vibrations are said to be *polarized*.

Reflection The return of light rays from a surface in such a way that the angle at which a given ray is returned is equal to the angle at which it strikes the surface (also called *specular reflection*).

Refraction The bending of an oblique ray of light when it passes from one transparent medium to another. This is caused by a difference in the speed of light in the transparent media. When the change in medium is abrupt (say, from air to water), the bending is abrupt; when the change in medium is gradual (say, from cool air to warm air), the bending is gradual, which accounts for mirages.

Subtractive primary colors The three colors of absorbing pigments—magenta, yellow, and cyan—that, when mixed in certain proportions, can reflect any color in the spectrum.

Transparent The term applied to materials through which light can pass without absorption, usually in straight lines.

READING CHECK QUESTIONS (COMPREHENSION)

11.1 Electromagnetic Spectrum

1. What is the principal difference between a radio wave and light? Between light and an X-ray?
2. How does the frequency of an electromagnetic wave compare with the frequency of the vibrating electrons that produce it?

11.2 Transparent and Opaque Materials

3. In what region of the electromagnetic spectrum is the resonant frequency of electrons in glass?
4. What is the fate of the energy in ultraviolet light incident on glass?
5. What is the fate of the energy in infrared light incident on glass?
6. How does the average speed of light in glass compare with its speed in a vacuum?
7. How does the speed of light that emerges from a pane of glass compare with the speed of light incident on the glass?

11.3 Reflection

8. What is the law of reflection?
9. Compared to the distance of an object in front of a plane mirror, how far behind the mirror is the image?
10. Does the law of reflection hold for curved mirrors? Explain.
11. In what sense does the law of reflection hold for a diffuse reflector?

11.4 Refraction

12. What is the angle between a light ray and its wavefront?
13. What is the relationship between refraction and the speed of light?
14. Does light travel faster in thin air or in dense air? What does this difference in speed have to do with the duration of daylight?
15. What is a mirage?

11.5 Color

16. Which has the higher frequency: red light or blue light?
17. What is the color of the peak frequency of solar radiation? To what color of light are our eyes most sensitive?
18. What are the three primary colors? The three subtractive primary colors?
19. What do we call two colors that add to produce white?
20. Why does the sky normally appear blue?
21. Why does the Sun look reddish at sunrise and sunset?
22. What is the evidence for a cloud being composed of particles having a variety of sizes?
23. What is absorbed by water to give it a cyan color?

11.6 Dispersion

24. Which travels slower in glass: red light or violet light?
25. What prevents rainbows from being seen as complete circles?
26. Why is a secondary rainbow dimmer than a primary bow?

11.7 Polarization

27. Is polarization a property of transverse waves, longitudinal waves, or both?

28. How does the direction of polarization of light compare with the direction of vibration of the electrons that produced it?
29. Will light pass through a pair of Polaroid filters when their air axes are aligned? When their axes are at right angles to each other?
30. How much unpolarized light does an ideal Polaroid filter transmit?

ACTIVITIES (HANDS-ON APPLICATION)

31. Which is your dominant eye? To test which you favor, hold a finger up at arm's length. With both eyes open, look past your finger at a distant object. Now close your right eye. If your finger appears to jump to the right, then your right eye is dominant.

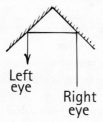

32. Stare at a piece of colored paper for 45 s or so. Then look at a plain white surface. The cones in your retina that are receptive to the color of the paper become fatigued, so you see an afterimage of the complementary color when you look at a white area. This is because the fatigued cones send a weaker signal to the brain. All the colors produce white, but all the colors minus one produce the color that is complementary to the missing color.

33. Simulate your own sunset: Add a few drops of milk to a glass of water and look at a lightbulb through the glass. The bulb appears to be red or pale orange, while light scattered to the side appears blue.

34. Set up two pocket mirrors at right angles and place a coin between them. You'll see four coins. Change the angle of the mirrors and see how many images of the coin you can see. With the mirrors at right angles, look at your face. Then wink. What do you see? You now see yourself as others see you. Hold a printed page up to the double mirrors and compare its appearance with the reflection of a single mirror.

Left eye

Right eye

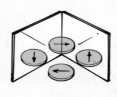

35. Rotate a pair of mirrors, keeping them at right angles to each other. Does your image rotate also? Then place the mirrors 60° apart so that you can see your face. Again rotate the mirrors, and see if your image rotates also. Amazing?

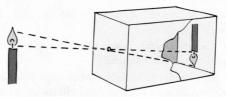

36. Make a pinhole camera, as illustrated. Cut out one end of a small cardboard box, and cover the end with tissue or wax paper. Make a clean-cut pinhole at the other end. (If the cardboard is thick, make it through a piece of aluminum foil placed over an opening in the cardboard.) Aim the camera at a bright object in a darkened room, and you see an upside-down image on the tissue paper. When photographic film was in vogue, students replaced the tissue paper with unexposed photographic film, covering the back so it was light tight, and covering the pinhole with a removable flap, all ready to take a picture. Exposure times differed, depending principally on the kind of film and the amount of light. Lenses on today's commercial cameras are much bigger than pinholes and therefore admit more light in less time—hence the term *snapshot*. For now it will be enough to view images on the tissue or wax paper. Point your camera toward the Sun. And if you do so during a solar eclipse, you'll marvel at the clear crescents on your viewing screen.

37. Write a letter to your grandparents and tell them the reasons for the blueness of the sky, the redness of sunrises and sunsets, and the whiteness of clouds. Explain how knowing the reasons adds to, not subtracts from, your appreciation of nature.

THINK AND SOLVE (MATHEMATICAL APPLICATION)

38. Electrons on a radio broadcasting tower are forced to oscillate up and down an antenna 535,000 times each second. Show that the wavelength of the radio waves that are produced is 561 m.

39. Consider a pulse of laser light aimed at the Moon that bounces back to Earth. The distance between Earth and the Moon is 3.84×10^8 m. Show that the round-trip time for the light is 2.56 s.

40. The nearest star beyond the Sun is Alpha Centauri, which is 4.2×10^{16} m away. If we were to receive a radio message from this star today, show that it would have been sent 4.4 years ago.

41. In 1676, the Danish astronomer Ole Roemer had one of those "Aha!" moments in science. He concluded from accumulated observations of eclipses of Jupiter's moon at different times of the year that light must travel at a finite speed and needed 1300 s to cross the diameter of Earth's orbit around the Sun. Using 300,000,000 km for the diameter of Earth's orbit, calculate the speed of light based on Roemer's 1300-s estimate. How does it differ from a modern value for the speed of light?

42. More than 200 years later, Albert A. Michelson sent a beam of light from a revolving mirror to a stationary mirror 15 km away. Show that the time interval between light leaving and returning to the revolving mirror was 0.0001 s.

43. Blue-green light has a frequency of about 6×10^{14} Hz. Using the relationship $c = f\lambda$, show that its wavelength in air is 5×10^{-7} m. How much larger is this wavelength compared to the size of an atom, which is about 10^{-10} m?

44. A certain radar installation that is used to track airplanes transmits electromagnetic radiation with a wavelength of 3.0 cm.
    (a) Show that the frequency of this radiation is 10 GHz.
    (b) Show that the time required for a pulse of radar waves to reach an airplane 5.0 km away and return would be 3.3×10^{-5} s.

45. A spider hangs by a strand of silk at an eye level 30 cm in front of a plane mirror. You are 65 cm behind the spider. Show that the distance between your eye and the image of the spider in the mirror is 1.25 m.

46. When you walk toward a mirror you see your image approaching you. If you walk at 2 m/s, how fast do you and your image approach each other?

47. When light strikes glass perpendicularly, about 4% of the light is reflected at each surface. Show that the amount of light transmitted through a pane of window glass is about 92%.

48. The average speed of light slows to $0.75c$ when it refracts through a particular piece of plastic.
    (a) What change is there in the light's frequency in the plastic?
    (b) We see in Figure 11.18 that the wavelength of light is shortened when passing through a transparent material. How much does the wavelength change in the piece of plastic?

THINK AND RANK (ANALYSIS)

49. She looks at her face in the handheld mirror. Rank the amount of her face she sees in the three locations, from greatest to least (or is it the same in all positions?).

50. Wheels from a toy cart are rolled from a concrete sidewalk onto the following surfaces: (a) a paved driveway, (b) a grass lawn, and (c) close-cropped grass on a golf-course putting green. Each set of wheels bends at the boundary due to slowing and is deflected from its initial straight-line course. Rank the surfaces according to the amount each set of wheels bends at the boundary, from greatest amount of bending to least.

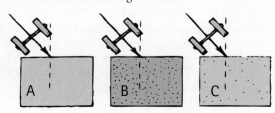

51. Identical rays of light enter three transparent blocks composed of different materials. Light slows down upon entering the blocks. Rank the blocks according to the speed light travels in each, from fastest to slowest.

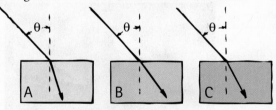

52. Identical rays of light in air are refracted upon entering three transparent materials: (a) water, where the speed of light is $0.75c$; (b) ethyl alcohol; speed $0.7c$; and (c) crown glass; speed $0.6c$. Rank the materials according to how much the light ray bends toward the normal, from most bending to least bending.

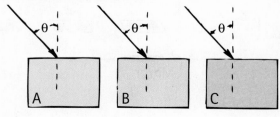

EXERCISES (SYNTHESIS)

53. What is the fundamental source of electromagnetic radiation?

54. What is it, exactly, that waves in a light wave?

55. Which have the longest wavelength: light waves, X-rays, or radio waves?

56. Are the wavelengths of radio and television signals longer or shorter than waves detectable by the human eye?

57. What is the speed of X-rays in a vacuum?

58. How are military people able to see enemy combatants in complete darkness?

59. Do radio waves travel at the speed of sound, at the speed of light, or at some speed in between?

60. What do radio waves and light have in common? What is different about them?

61. If you fire a bullet through a board, it will slow down inside and emerge at a speed that is less than the speed at which it entered. Does light, then, similarly slow down when it passes through glass and also emerge at a lower speed? Defend your answer.

62. Is glass transparent or opaque to light of frequencies that match its own natural frequencies? Explain.

63. Short wavelengths of visible light interact more frequently with the atoms in glass than do longer wavelengths. Does this interaction time tend to speed up or slow down the average speed of light in glass?

64. What determines whether a material is transparent or opaque?

65. You can get a sunburn on a cloudy day, but you can't get a sunburn even on a sunny day if you are behind glass. Also, transition eyeglasses "don't work" inside a car with closed windows. Explain.

66. Peter Hopkinson stands astride a large mirror and boosts class interest with this zany demonstration. How does he accomplish his apparent levitation in midair?

67. The person's eye at point P looks into the mirror. Which of the numbered cards can she see reflected in the mirror?

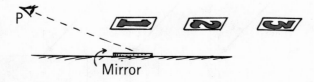

Mirror

68. Trucks often have signs on the back that say, "If you can't see my mirrors, I can't see you." Explain the physics.

69. Why is the lettering on the front of some vehicles—for example, ambulances—"backward"?

70. We see the bird and its reflection. Why don't we see the bird's feet in the reflection?

71. What must be the minimum length of a vertical plane mirror in order for you to see a full view of yourself?

72. What effect does your distance from the plane mirror have on your answer to the preceding exercise? (Try it and see!)

73. Hold a pocket mirror almost at arm's length from your face and note the amount of your face you can see. To see more of your face, should you hold the mirror closer or farther, or would you have to have a larger mirror? (Try it and see!)

74. From a steamy mirror, wipe away just enough moisture to allow you to see your full face. How tall will the wiped area be compared with the vertical dimension of your face? Does distance affect your answer?

75. A pair of toy cart wheels are rolled obliquely from a smooth surface onto two plots of grass, a rectangular plot and a triangular plot, as shown. The ground is on a slight incline, so that, after slowing down in the grass, the wheels will speed up again when emerging onto the smooth surface. Complete the sketches and show some positions of the wheels inside each plot and on the other side of each plot, thereby indicating the direction of travel.

Grass   Grass

76. During a lunar eclipse, the Moon is not completely dark, but is often deep red. Explain this in terms of the refraction of all the sunsets and sunrises around the world.

77. In a dress shop with only fluorescent lighting, a customer insists on taking dresses into the daylight at the doorway to check their color. Is she being reasonable? Explain.

78. The radiation curve of the Sun (see Figure 11.31) shows that the brightest light from the Sun is yellow-green. Why then do we see the Sun as whitish instead of yellow-green? (Hint: Take into account the wideness of the solar radiation curve.)

79. How could you use the spotlights at a play to make the yellow clothes of the performers suddenly change to black?

80. What colors of ink do color ink-jet printers use to produce a full range of colors? Do the colors form by color addition or by color subtraction?

81. The photo shows science author Suzanne Lyons with her son Tristan wearing red and her daughter Simone wearing green. Below that is the negative of the photo, which shows these colors differently. What is your explanation?

82. Check Figure 11.33 to see if the following three statements are accurate. Then provide the missing word in the last statement. (All colors are combined by the addition of light.)

    Red + green + blue = white

    Red + green = yellow = white − blue

    Red + blue = magenta = white − green

    Green + blue = cyan = white − _____

83. What single color of light illuminating a ripe banana will make it appear black?

84. Stare intently for at least a half minute at an American flag. Then turn your gaze to a white wall. What colors do you see in the image of the flag that appears on the wall?

85. Why does smoke from a campfire look bluish against trees near the ground but yellowish against the sky?

86. Tiny particles, like tiny bells, scatter high-frequency waves more than low-frequency waves. Large particles, like large bells, mostly scatter low frequencies. Intermediate-size particles and bells mostly scatter intermediate frequencies. What does this have to do with the whiteness of clouds?

87. Very big particles, like droplets of water, absorb more radiation than they scatter. What does this have to do with the darkness of rain clouds?

88. The atmosphere of Jupiter is more than 1000 km thick. From the surface of this planet, would you expect to see a white Sun?

89. You're explaining to a youngster at the seashore why the water is cyan colored. The youngster points to the whitecaps of overturning waves and asks why they are white. What is your answer?

90. When you stand with your back to the Sun, you see a rainbow as a circular arc. Could you move off to one side and then see the rainbow as the segment of an ellipse rather than the segment of a circle (as Figure 11.47 suggests)? Defend your answer.

91. Two observers standing apart from each other do not see the "same" rainbow. Explain.

92. A rainbow viewed from an airplane may form a complete circle. Where will the shadow of the airplane appear? Explain.

93. What percentage of light is transmitted by two ideal Polaroid filters, one on top of the other with their polarization axes aligned? With their polarization axes at right angles to each other?

94. How can a single Polaroid filter be used to show that the sky is partially polarized? (Interestingly enough, unlike humans, bees and many insects can discern polarized light, and they use this ability for navigation.)

95. Light will not pass through a pair of Polaroid filters when they are aligned perpendicularly. But, if a third Polaroid filter aligned at about 45° to the pair is sandwiched between them, some light does get through. Why?

DISCUSSION QUESTIONS (EVALUATION)

96. In a physics study group, a friend says in a profound tone that light is the only thing we can see. After a few laughs, your friend goes on to say that light is produced by the connection between electricity and magnetism. Is your friend correct?

97. We hear people talk of "ultraviolet light" and "infrared light." Why are these terms misleading? Why are we less likely to hear people talk of "radio light" and "X-ray light"?

98. Light from a camera flash weakens with distance in accord with the inverse-square law. Comment on an airline passenger who takes a flash photo of a city below at nighttime from a high-flying plane.

99. Why doesn't the sharpness of the image in a pinhole camera depend on the position of the viewing screen?

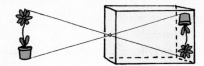

100. If you point the pinhole camera of the preceding exercise at the Sun, you will see a clear, bright solar image on the viewing screen. How does this relate to the circular spots that surround Lillian beneath the sunlit tree shown in the photo?

101. Why will an ideal Polaroid filter transmit 50% of incident nonpolarized light?
102. Why may an ideal Polaroid filter transmit anything from zero to 100% of incident polarized light?

READINESS ASSURANCE TEST (RAT)

If you have a good grasp of this chapter, if you really do, then you should be able to score at least 7 out of 10 on this RAT. If you score less than 7, consider studying further before moving on.

Choose the BEST answer to each of the following.

1. Which of these does *not* belong in the family of electromagnetic waves?
   (a) light
   (b) sound
   (c) radio waves
   (d) X-rays
2. The source of electromagnetic waves is vibrating
   (a) electrons.
   (b) atoms.
   (c) molecules.
   (d) energy fields.
3. The slowing of light in transparent materials has to do with
   (a) the time for absorption and re-emission of light.
   (b) the density of materials.
   (c) different frequency ranges in materials.
   (d) the fundamental difference between light and sound.
4. Whether a particular surface acts as a polished reflector or a diffuse reflector depends on the
   (a) color of reflected light.
   (b) brightness of reflected light.
   (c) wavelength of light.
   (d) angle of incoming light.
5. When a light ray passes at an angle from water into the air, the ray in the air bends
   (a) toward the normal.
   (b) away from the normal.
   (c) either away from or toward the normal.
   (d) parallel to the normal.

6. Refracted light that bends away from the normal is light that has
   (a) slowed down.
   (b) speeded up.
   (c) bounced.
   (d) diffracted.
7. The colors on the cover of your physical science book are due to
   (a) color addition.
   (b) color subtraction.
   (c) color interference.
   (d) scattering.
8. The redness of a sunrise or sunset is due mostly to light that hasn't been
   (a) absorbed.
   (b) transmitted.
   (c) scattered.
   (d) polarized.
9. A rainbow is the result of light in raindrops that undergoes
   (a) internal reflection.
   (b) dispersion.
   (c) refraction.
   (d) all of these
10. Polarization occurs for waves that are
    (a) transverse.
    (b) longitudinal.
    (c) both
    (d) neither

Answers to RAT

1. b, 2. a, 3. a, 4. c, 5. b, 6. b, 7. b, 8. c, 9. d, 10. a

2

Chemistry

Hey Liam, like everyone, I'm made of atoms, which are so small and numerous that I inhale billions of trillions with each breath. Many of these atoms stay and become a part of my body. When I exhale, I'm releasing into the air many of the atoms that were once a part of me.

Gee, Bo, the atoms you exhale are the very ones both our baby sister Neve and I inhale. So your atoms then become a part of us, just as the atoms we exhale eventually become a part of you!

In each breath we inhale, we recycle atoms that once were a part of every person who lived. Hey, in this sense, we're all one!

12

CHAPTER 12

Atoms and the Periodic Table

W E HUMANS have long tinkered with the materials around us and used them to our advantage. Once we learned how to control fire, we were able to create many new substances. Moldable wet clay, for example, was found to harden to ceramic when heated by fire. By 5000 BC, pottery fire pits gave way to furnaces hot enough to convert copper ores to metallic copper. By 1200 BC, even hotter furnaces were converting iron ores to iron. This technology allowed for the mass production of metal tools and weapons and made possible the many achievements of ancient Chinese, Egyptian, and Greek civilizations.

Fast-forward to the 21st century, and we've since learned that all the materials around us are made of remarkably small particles called atoms. We have learned how to manipulate these atoms to produce a vast array of new and useful modern materials. In this chapter, we will explore both the nature of atoms and the amazing chart that tells their story—the periodic table.

12.1 Atoms Are Ancient and Empty

EXPLAIN THIS If atoms are empty, why can't we walk through walls?

Hydrogen, H, the lightest atom, makes up more than 90% of the atoms in the known universe. Most of these hydrogen atoms were formed during the beginning of our universe about 13.7 billion years ago. Heavier atoms are produced in stars, which are massive collections of hydrogen atoms pulled together by gravitational forces. The great pressures deep in a star's interior cause hydrogen atoms to fuse into heavier atoms. With the exception of hydrogen, therefore, all the atoms that occur naturally on Earth—including those in your body—are the products of stars. You are made of stardust, as is everything that surrounds you.

So, atoms are ancient. They have existed through imponderable ages, recycling through the universe in innumerable forms, both nonliving and living. In this sense, you don't "own" the atoms that make up your body—you are simply their present caretaker. Many more caretakers will follow.

Atoms are so small that each breath you exhale contains more than 10 billion trillion of them. This is more than the number of breaths in Earth's atmosphere. Within a few years, the atoms of your breath are uniformly mixed throughout the atmosphere. What this means is that anyone anywhere on Earth inhaling a breath of air takes in numerous atoms that were once part of you. And, of course, the reverse is true: you inhale atoms that were once part of everyone who has ever lived. We are literally breathing one another.

Atoms are so small that they can't be seen with visible light. That's because they are even smaller than the wavelengths of visible light. We could stack microscope on top of microscope and never "see" an atom. Photographs of atoms, such as in Figure 12.1, are obtained with a scanning probe microscope. Discussed further in Section 12.5, this is a nonlight imaging device that bypasses light and optics altogether.

Today we know the atom is made of smaller, subatomic particles—*electrons*, *protons*, and *neutrons*. We also know that atoms differ from one another only in the number of subatomic particles they contain. Protons and neutrons are bound together at the atom's center to form a larger particle—the **atomic nucleus**. The nucleus is a relatively heavy particle that makes up most of an atom's mass. Surrounding the nucleus are the tiny **electrons**, as shown in Figure 12.2.

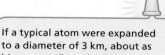

If a typical atom were expanded to a diameter of 3 km, about as big as a medium-sized airport, the nucleus would be about the size of a basketball. Atoms are mostly empty space.

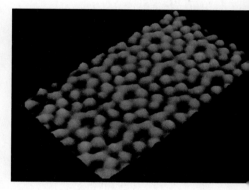

FIGURE 12.1
An image of carbon atoms obtained with a scanning probe microscope.

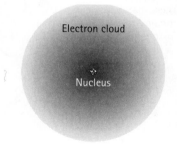

Electron cloud

Nucleus

FIGURE 12.2
Electrons whiz around the atomic nucleus, forming what can be best described as a cloud that is more dense where the electrons tend to spend most of their time. Electrons, however, are invisible to us. Hence, such a cloud can only be imagined. Furthermore, if this illustration were drawn to scale, the atomic nucleus with its protons and neutrons would be too small to be seen. In short, atoms are not well suited to graphical depictions.

CHECKPOINT

A friend claims there are atoms in his brain that were once in the brain of Albert Einstein. Is your friend's claim likely correct or nonsense?

Was this your answer?
Your friend is correct! In addition, there are atoms in your friend's and everyone else's body that were once part of Einstein and everybody else, too! The arrangements of these atoms, however, are now quite different. What's more, the atoms of which you and your friend are composed will be found in the bodies of all the people on Earth who are yet to be.

We and all materials around us are mostly empty space. How can this be? Electrons move about the nucleus in an atom defining the volume of space that the atom occupies. But electrons are very small. If an atom were the size of a baseball stadium, one of its electrons would be smaller than a grain of rice.

FIGURE 12.3
As close as Tracy and Ian are in this photograph, none of their atoms meet. The closeness between us is in our hearts.

Mastering**PHYSICS**

TUTORIAL: Atomic Structure
VIDEO: Evidence for Atoms
VIDEO: Atoms Are Recyclable

LEARNING OBJECTIVE
Recognize the elements of the periodic table as the fundamental building blocks of matter.

Most materials are made from more than one kind of atom. Water, H_2O, for example, is made from the combination of hydrogen and oxygen atoms. These materials are called *compounds*, which we discuss further in Chapter 14.

Furthermore, all the electrons of an atom are widely spaced apart. Atoms are indeed mostly empty space.

So why don't atoms simply pass through one another? How is it that we are supported by the floor despite the empty nature of its atoms? Although subatomic particles are much smaller than the volume of the atom, the range of their electric field is several times larger than that volume. In the outer regions of any atom are electrons, which repel the electrons of neighboring atoms. Two atoms therefore can get only so close to each other before they start repelling (provided they don't join in a chemical bond, as is discussed in Chapter 15).

When the atoms of your hand push against the atoms of a wall, electrical repulsions between electrons in your hand and electrons in the wall prevent your hand from passing through the wall. These same electrical repulsions prevent us from falling through the solid floor. They also allow us the sense of touch. Interestingly, when you touch someone, your atoms and those of the other person do not meet. Instead, atoms from the two of you get close enough so that you sense an electrical repulsion. A tiny, though imperceptible, gap still exists between the two of you (Figure 12.3).

12.2 The Elements

EXPLAIN THIS Why isn't water an element?

You know that atoms make up the matter around you, from stars to steel to chocolate ice cream. Given all these different types of material, you might think that there must be many different kinds of atoms. But the number of different kinds of atoms is surprisingly small. The great variety of substances results from the many ways a few kinds of atoms can be combined. Just as the three colors red, green, and blue can be combined to form any color on a television screen or the 26 letters of the alphabet make up all the words in a dictionary, only a few kinds of atoms combine in different ways to produce all substances. To date, we know of slightly more than 100 distinct atoms. Of these, about 90 are found in nature. The remaining atoms have been created in the laboratory.

Any material made of only one type of atom is classified as an **element**. A few examples are shown in Figure 12.4. Pure gold, for example, is an element—it contains only gold atoms. Nitrogen gas is an element because it contains only nitrogen atoms. Likewise, the graphite in your pencil is an element—carbon. Graphite is made up solely of carbon atoms. All of the elements are listed in a chart called the **periodic table**, shown in Figure 12.5.

As you can see from the periodic table, each element is designated by its **atomic symbol**, which comes from the letters of the element's name. For

FIGURE 12.4
Any element consists of only one kind of atom. Gold consists of only gold atoms, a flask of gaseous nitrogen consists of only nitrogen atoms, and the carbon of a graphite pencil consists of only carbon atoms.

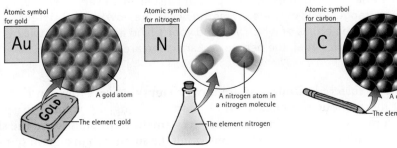

1 H																	2 He
3 Li	4 Be											5 B	6 C	7 N	8 O	9 F	10 Ne
11 Na	12 Mg											13 Al	14 Si	15 P	16 S	17 Cl	18 Ar
19 K	20 Ca	21 Sc	22 Ti	23 V	24 Cr	25 Mn	26 Fe	27 Co	28 Ni	29 Cu	30 Zn	31 Ga	32 Ge	33 As	34 Se	35 Br	36 Kr
37 Rb	38 Sr	39 Y	40 Zr	41 Nb	42 Mo	43 Tc	44 Ru	45 Rh	46 Pd	47 Ag	48 Cd	49 In	50 Sn	51 Sb	52 Te	53 I	54 Xe
55 Cs	56 Ba	57 La	72 Hf	73 Ta	74 W	75 Re	76 Os	77 Ir	78 Pt	79 Au	80 Hg	81 Tl	82 Pb	83 Bi	84 Po	85 At	86 Rn
87 Fr	88 Ra	89 Ac	104 Rf	105 Db	106 Sg	107 Bh	108 Hs	109 Mt	110 Ds	111 Rg	112 Cn	113 Uut	114 Uuq	115 Uup	116 Uuh	117 Uus	118 Uuo

58 Ce	59 Pr	60 Nd	61 Pm	62 Sm	63 Eu	64 Gd	65 Tb	66 Dy	67 Ho	68 Er	69 Tm	70 Yb	71 Lu
90 Th	91 Pa	92 U	93 Np	94 Pu	95 Am	96 Cm	97 Bk	98 Cf	99 Es	100 Fm	101 Md	102 No	103 Lr

FIGURE 12.5
The periodic table lists all the known elements.

example, the atomic symbol for carbon is C, and that for chlorine is Cl. In many cases, the atomic symbol is derived from the element's Latin name. Gold has the atomic symbol Au after its Latin name, *aurum*. Lead has the atomic symbol Pb after its Latin name, *plumbum* (Figure 12.6). Elements with symbols derived from Latin names are usually those that were discovered earliest.

Note that only the first letter of an atomic symbol is capitalized. The symbol for the element cobalt, for instance, is Co, but CO is a combination of two elements: carbon, C, and oxygen, O.

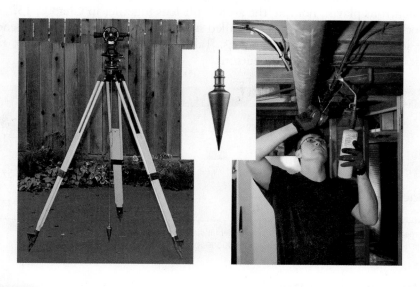

FIGURE 12.6
A plumb bob, a heavy weight attached to a string and used by carpenters and surveyors to establish a straight vertical line, gets it name from the lead (*plumbum*, Pb) that is still sometimes used as the weight. Plumbers got their name because they once worked with lead pipes.

12.3 Protons and Neutrons

EXPLAIN THIS Why aren't we harmed by drinking heavy water, D_2O?

Let us take a closer look at the atom and investigate the particles found in the atomic nucleus. A **proton** carries a positive charge and is relatively heavy—nearly 2000 times as massive as an electron. The proton and electron have the same quantity of charge, but the opposite sign. The number of protons in the nucleus of any atom is equal to the number of electrons whirling about the nucleus. So the opposite charges of protons and electrons balance each other, producing a zero net charge. For example,

LEARNING OBJECTIVE
Describe the structure of the atomic nucleus and how the atomic mass of an element is calculated.

an oxygen atom has a total of eight protons and eight electrons and is thus electrically neutral.

Scientists have agreed to identify elements by **atomic number**, which is the number of protons each atom of a given element contains. The modern periodic table lists the elements in order of increasing atomic number. Hydrogen, with one proton per atom, has atomic number 1; helium, with two protons per atom, has atomic number 2; and so on.

CHECKPOINT
How many protons are there in an iron atom, Fe (atomic number 26)?

Was this your answer?
The atomic number of an atom and its number of protons are the same. Thus, there are 26 protons in an iron atom. Another way to put this is that all atoms that contain 26 protons are, by definition, iron atoms.

If we compare the electric charges and masses of different atoms, we see that the atomic nucleus must be made up of more than just protons. Helium, for example, has twice the electric charge of hydrogen but four times the mass. The added mass is due to another subatomic particle found in the nucleus, the neutron. The **neutron** has about the same mass as the proton, but it has no electric charge. Any object that has no net electric charge is said to be electrically neutral, and that is where the neutron got its name. We discuss the important role that neutrons play in holding the atomic nucleus together in Chapter 13.

Both protons and neutrons are called **nucleons**, a generic term that denotes their location in the atomic nucleus. Table 12.1 summarizes the basic facts about electrons, protons, and neutrons.

TABLE 12.1	SUBATOMIC PARTICLES			
	Particle	Charge	Mass Compared to Electron	Actual Mass* (kg)
	Electron	-1	1	9.11×10^{-31}**
Nucleons $\{$	Proton	$+1$	1836	1.673×10^{-27}
	Neutron	0	1841	1.675×10^{-27}

* Not measured directly but calculated from experimental data.
** 9.11×10^{-31} kg = 0.000000000000000000000000000000911 kg.

Isotopes and Atomic Mass

For any element, no set number of neutrons are in the nucleus. For example, most hydrogen atoms (atomic number 1) have no neutrons. A small percentage, however, have one neutron, and a smaller percentage have two neutrons. Similarly, most iron atoms (atomic number 26) have 30 neutrons, but a small percentage have 29 neutrons. Atoms of the same element that contain different numbers of neutrons are **isotopes** of one another.

We identify isotopes by their mass number, which is the total number of protons and neutrons (in other words, the number of nucleons) in the nucleus. As Figure 12.7 shows, a hydrogen isotope with only one proton is called hydrogen-1, where 1 is the mass number. A hydrogen isotope with one proton and one neutron is therefore hydrogen-2, and a hydrogen isotope with one proton and two

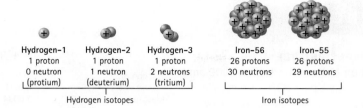

Hydrogen-1
1 proton
0 neutron
(protium)

Hydrogen-2
1 proton
1 neutron
(deuterium)

Hydrogen-3
1 proton
2 neutrons
(tritium)

Hydrogen isotopes

Iron-56
26 protons
30 neutrons

Iron-55
26 protons
29 neutrons

Iron isotopes

FIGURE 12.7
Isotopes of an element have the same number of protons but different numbers of neutrons and hence different mass numbers. The three hydrogen isotopes have special names: protium for hydrogen-1, deuterium for hydrogen-2, and tritium for hydrogen-3. Of these three isotopes, hydrogen-1 is most common. For most elements, such as iron, the isotopes have no special names and are indicated merely by mass number.

neutrons is hydrogen-3. Similarly, an iron isotope with 26 protons and 30 neutrons is called iron-56, and one with only 29 neutrons is iron-55.

An alternative method of indicating isotopes is to write the mass number as a superscript and the atomic number as a subscript to the left of the atomic symbol. For example, an iron isotope with a mass number of 56 and an atomic number of 26 is written:

Mass number $\searrow$ 56

Fe—Atomic symbol

26

Atomic number $\nearrow$

The total number of neutrons in an isotope can be calculated by subtracting its atomic number from its mass number:

$$\begin{array}{r} \text{mass number} \\ -\text{atomic number} \\ \hline \text{number of neutrons} \end{array}$$

For example, uranium-238 has 238 nucleons. The atomic number of uranium is 92, which tells us that 92 of these 238 nucleons are protons. The remaining 146 nucleons must be neutrons:

$$\begin{array}{r} \text{238 protons and neutrons} \\ -\text{92 protons} \\ \hline \text{146 neutrons} \end{array}$$

Atoms interact with one another electrically. Therefore the way any atom behaves in the presence of other atoms is determined largely by the charged particles it contains, especially its outer electrons. Isotopes of an element differ only by mass, not by electric charge. For this reason, isotopes of an element share many characteristics—in fact, as chemicals they cannot be distinguished from one another. For example, a sugar molecule containing seven neutrons per carbon nucleus is digested no differently from a sugar molecule containing six neutrons per carbon nucleus. Interestingly, about 1% of the carbon we eat is the carbon-13 isotope containing seven neutrons per nucleus. The remaining 99% of the carbon in our diet is the more common carbon-12 isotope containing six neutrons per nucleus.

The total mass of an atom is called its **atomic mass**. This is the sum of the masses of all the atom's components (electrons, protons, and neutrons). Because electrons are so much less massive than protons and neutrons, their contribution to atomic mass is negligible. A special unit has been developed for atomic masses. This is the *atomic mass unit*, amu, where 1 atomic mass unit is equal to 1.661×10^{-24} gram, which is slightly less than the mass of a single proton. As shown in Figure 12.8, the atomic masses listed in the periodic table are in atomic mass units. The atomic mass of an element as presented in the periodic table is actually the *average* atomic mass of its various isotopes. Figuring Physical Science on the next page shows how these averages are calculated.

Most water molecules, H_2O, consist of hydrogen atoms with no neutrons. The few that have neutrons, however, are heavier, and because of this difference they can be isolated. Such water is appropriately called "heavy water," which has the formula D_2O.

FIGURING PHYSICAL SCIENCE

Calculating Atomic Mass

About 99% of all carbon atoms are the isotope carbon-12, and most of the remaining 1% are the heavier isotope carbon-13. This small amount of carbon-13 raises the average mass of carbon from 12.000 amu to the slightly greater value of 12.011 amu.

To arrive at the atomic mass presented in the periodic table, first multiply the mass of each naturally occurring isotope of an element by the fraction of its abundance and then add up all the fractions.

SAMPLE PROBLEM 1

Carbon-12 has a mass of 12.0000 amu and makes up 98.89% of naturally occurring carbon. Carbon-13 has a mass of 13.0034 amu and makes up 1.11% of naturally occurring carbon. Use this information to show that atomic mass of carbon shown in the periodic table, 12.011 amu, is correct.

Solution:

Recognize that 98.89% and 1.11% expressed as decimals are 0.9889 and 0.0111, respectively.

	Contributing Mass of ^{12}C	Contributing Mass of ^{13}C	
Fraction of Abundance	0.9889	0.0111	
Mass (amu)	× 12.000	× 13.0034	step 1
	11.867	0.144	

atomic mass = 11.867 + 0.144 = 12.011 step 2

YOUR TURN

Chlorine-35 has a mass of 34.97 amu, and chlorine-37 has a mass of 36.95 amu. Determine the atomic mass of chlorine, Cl (atomic number 17), if 75.53% of all chlorine atoms are the chlorine-35 isotope and 24.47% are the chlorine-37 isotope. (See Figure 12.8 for the answer.)

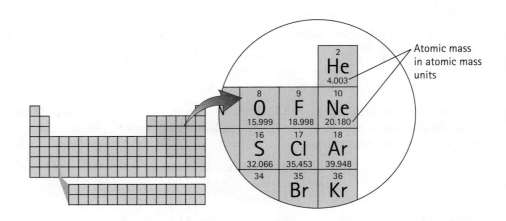

Atomic mass in atomic mass units

FIGURE 12.8
Helium, He, has an atomic mass of 4.003 amu, and neon, Ne, has an atomic mass of 20.180 amu.

CHECKPOINT

Distinguish between mass number and atomic mass.

Was this your answer?
Both terms include the word *mass* and so are easily confused. Focus your attention on the second word of each term, however, and you'll get it right every time. Mass number is a count of the *number* of nucleons in an isotope. An atom's mass number requires no units because it is simply a count. Atomic mass is a measure of the total *mass* of an atom, which is given in atomic mass units.

LEARNING OBJECTIVE
Interpret how elements are organized in the periodic table.

12.4 The Periodic Table

EXPLAIN THIS How is the periodic table like a dictionary?

The periodic table is a listing of all the known elements with their atomic masses, atomic numbers, and atomic symbols. But the periodic table contains much more information. The way the table is organized, for example, tells us much about the elements' properties. Let's look at how the elements are grouped as metals, nonmetals, and metalloids.

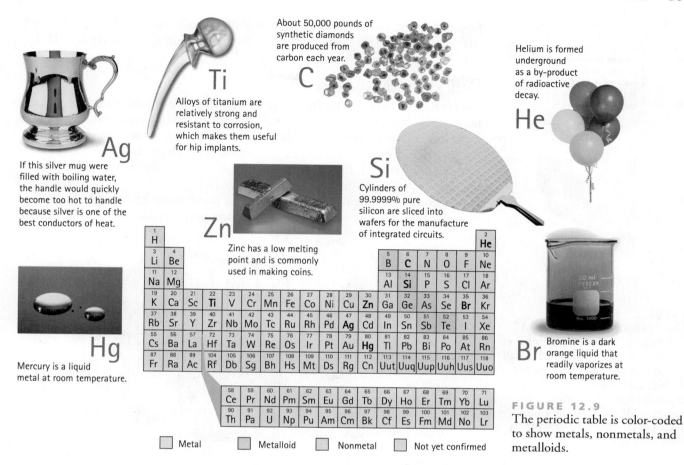

About 50,000 pounds of synthetic diamonds are produced from carbon each year.

Alloys of titanium are relatively strong and resistant to corrosion, which makes them useful for hip implants.

Helium is formed underground as a by-product of radioactive decay.

If this silver mug were filled with boiling water, the handle would quickly become too hot to handle because silver is one of the best conductors of heat.

Cylinders of 99.9999% pure silicon are sliced into wafers for the manufacture of integrated circuits.

Zinc has a low melting point and is commonly used in making coins.

Mercury is a liquid metal at room temperature.

Bromine is a dark orange liquid that readily vaporizes at room temperature.

☐ Metal   ☐ Metalloid   ☐ Nonmetal   ☐ Not yet confirmed

FIGURE 12.9
The periodic table is color-coded to show metals, nonmetals, and metalloids.

As shown in Figure 12.9, most of the known elements are metals, which are defined as elements that are shiny, opaque, and good conductors of electricity and heat. Metals are *malleable*, which means they can be hammered into different shapes or bent without breaking. They are also *ductile*, which means they can be drawn into wires. All but a few metals are solid at room temperature. The exceptions are mercury, Hg; gallium, Ga; cesium, Cs; and francium, Fr, which are all liquids at a warm room temperature of 30°C (86°F). Another interesting exception is hydrogen, H, which takes on the properties of a liquid metal only at very high pressures (Figure 12.10). Under normal conditions, hydrogen behaves like a nonmetallic gas.

The nonmetallic elements, with the exception of hydrogen, are on the right side of the periodic table. Nonmetals are very poor conductors of electricity and heat, and may also be transparent. Solid nonmetals are neither malleable nor ductile. Rather, they are brittle and shatter when hammered. At 30°C (86°F), some nonmetals are solid (carbon, C), others are liquid (bromine, Br), and still others are gaseous (helium, He).

Six elements are classified as metalloids: boron, B; silicon, Si; germanium, Ge; arsenic, As; antimony, Sb; and tellurium, Te. Situated between the metals and the nonmetals in the periodic table, the metalloids have both metallic and nonmetallic characteristics. For example, these elements are weak conductors of electricity, which makes them useful as semiconductors in the integrated circuits of computers. Note from the periodic table how germanium, Ge (atomic number 32), is closer to the metals than to the nonmetals. Because of this positioning, we can deduce that germanium has more metallic properties than silicon, Si (atomic number 14) and is a slightly better conductor of electricity. So we find that integrated circuits fabricated with germanium operate faster than those

FIGURE 12.10
Geoplanetary models suggest that hydrogen exists as a liquid metal deep beneath the surfaces of Jupiter (shown here) and Saturn. These planets are composed mostly of hydrogen. Inside them, the pressure exceeds 3 million times Earth's atmospheric pressure. At this tremendously high pressure, hydrogen is pressed to a liquid-metal phase.

Please put to rest any fear you may have about needing to memorize the periodic table, or even parts of it—better to focus on the many great concepts behind its organization.

fabricated with silicon. Because silicon is much more abundant and less expensive to obtain, however, silicon computer chips remain the industry standard.

Periods and Groups

Two other important ways in which the elements are organized in the periodic table are by horizontal rows and vertical columns. Each horizontal row is called a **period**, and each vertical column is called a **group** (or sometimes a *family*). As shown in Figure 12.11, there are 7 periods and 18 groups.

Across any period, the properties of elements gradually change. This gradual change is called a periodic trend. As is shown in Figure 12.12, one periodic trend is that atomic size tends to decrease as you move from left to right across any period. Note that the trend repeats from one horizontal row to the next. This

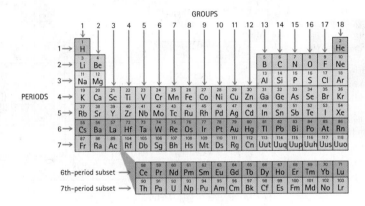

FIGURE 12.11
The 7 periods (horizontal rows) and 18 groups (vertical columns) of the periodic table. Note that not all periods contain the same number of elements. Also note that, for reasons explained later, the sixth and seventh periods each include a subset of elements, which are listed apart from the main body.

phenomenon of repeating trends is called periodicity, a term used to indicate that the trends recur in cycles. Each horizontal row is called a period because it corresponds to one full cycle of a trend.

CHECKPOINT
Which are larger: atoms of cesium, Cs (atomic number 55), or atoms of radon, Rn (atomic number 86)?

Was this your answer?
Perhaps you tried looking at Figure 12.12 to answer this question and quickly became frustrated because the sixth-period elements are not shown. Well, relax. Look at the trends and you'll see that, in any one period, all atoms to the left are larger than those to the right. Accordingly, cesium is positioned at the far left of period 6, and so you can reasonably predict that its atoms are larger than those of radon, which is positioned at the far right of period 6. The periodic table is a road map to understanding the elements.

Down any group (vertical column), the properties of elements tend to be remarkably similar, which is why these elements are said to be "grouped" or "in a family." As Figure 12.13 shows, several groups have traditional names that describe the properties of their elements. Early in human history, people discovered that ashes mixed with water produce a slippery solution useful for removing grease. By the Middle Ages, such mixtures were described as being

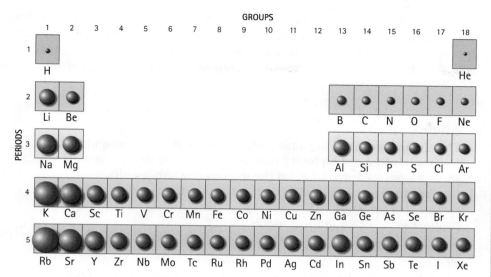

FIGURE 12.12
The size of atoms gradually decreases as we move from left to right across any period. Atomic size is a periodic (repeating) property.

FYI As the tungsten filament inside a lightbulb is heated, minute particles of tungsten evaporate. Over time, these particles are deposited on the inner surface of the bulb, causing the bulb to blacken. As it loses its tungsten, the filament eventually breaks and the bulb "burns out." A remedy is to replace the air inside the bulb with a halogen gas, such as iodine or bromine. In such a *halogen* bulb, the evaporated tungsten combines with the halogen rather than depositing on the bulb, which remains clear. Furthermore, the tungsten becomes unstable and splits from the halogen when it touches the hot filament. The halogen returns as a gas while the tungsten is deposited onto the filament, thereby restoring the filament. This is why halogen lamps have such long lifetimes.

alkaline, a term derived from the Arabic word for ashes, *al-qali*. Alkaline mixtures found many uses, particularly in the preparation of soaps (Figure 12.14). We now know that alkaline ashes contain compounds of group 1 elements, most notably potassium carbonate, also known as potash. Because of this history, group 1 elements, which are metals, are called the *alkali metals*.

Elements of group 2 also form alkaline solutions when mixed with water. Furthermore, medieval alchemists noted that certain minerals (which we now know are made up of group 2 elements) do not melt or change when put in fire. These fire-resistant substances were known to the alchemists as "earth." As a holdover from these ancient times, group 2 elements are known as the *alkaline-earth metals*.

Over toward the right side of the periodic table, elements of group 16 are known as the *chalcogens* ("ore-forming" in Greek) because the top two elements of this group, oxygen and sulfur, are so commonly found in ores. Elements of group 17 are known as the *halogens* ("salt-forming" in Greek) because of their tendency to form various salts. Group 18 elements are all unreactive gases that tend not to combine with other elements. For this reason, they are called the *noble gases*, presumably because the nobility of earlier times were above interacting with common folk.

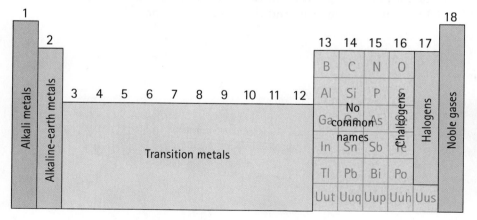

FIGURE 12.13
The common names for various groups of elements.

FIGURE 12.14
Ashes and water make a slippery alkaline solution once used to clean hands.

The elements of groups 3 through 12 are all metals that do not form alkaline solutions with water. These metals tend to be harder than the alkali metals and less reactive with water; hence they are used for structural purposes. Collectively they are known as the *transition metals*, a name that denotes their central position in the periodic table. The transition metals include some of the most familiar and important elements—iron, Fe; copper, Cu; nickel, Ni; chromium, Cr;

silver, Ag; and gold, Au. They also include many lesser-known elements that are nonetheless important in modern technology. People with hip implants appreciate the transition metals titanium, Ti; molybdenum, Mo; and manganese, Mn, because these noncorrosive metals are used in implant devices.

> **CHECKPOINT**
>
> **The elements copper, Cu; silver, Ag; and gold, Au, are three of the few metals that can be found naturally in their elemental state. These three metals have found great use as currency and jewelry for a number of reasons, including their resistance to corrosion and their remarkable colors. How is the fact that these metals have similar properties reflected in the periodic table?**
>
> **Was this your answer?**
> Copper (atomic number 29), silver (atomic number 47), and gold (atomic number 79) are all in the same group in the periodic table (group 11), which suggests they should have similar—though not identical—properties.

A uranium atom is 40 times as heavy as a lithium atom, but only slightly larger in size because its more highly charged nucleus pulls harder on its electrons. But it has more electrons to pull, a balancing act that barely changes the atom's size.

FYI From 1943 to 1986, the Hanford nuclear facility in central Washington state produced 72 tons of plutonium, nearly two-thirds the nation's supply. Creating this much plutonium generated an estimated 450 billion gallons of radioactive and hazardous liquids, which were discharged into the local environment. Today, some 53 million gallons of high-level radioactive and chemical wastes are stored in 177 underground tanks, many of them leaking into the groundwater.

Within the sixth period is a subset of 14 metallic elements (atomic numbers 58 to 71) that are quite unlike any of the other transition metals. A similar subset (atomic numbers 90 to 103) is found within the seventh period. These two subsets are the *inner transition metals*. Inserting the inner transition metals into the main body of the periodic table as in Figure 12.15 results in a long and cumbersome table. So that the table can fit nicely on a standard paper size, these elements are commonly placed below the main body of the table, as shown in Figure 12.16.

The sixth-period inner transition metals are called the *lanthanides* because they fall after lanthanum, La. Because of their similar physical and chemical properties, they tend to occur mixed together in the same locations on Earth. Also because of their similarities, lanthanides are unusually difficult to purify. Recently, the commercial use of lanthanides has increased. Several lanthanide elements, for example, are used in the fabrication of the light-emitting diodes (LEDs) of computer monitors and flat-screen televisions.

FIGURE 12.15
Inserting the inner transition metals between atomic groups 3 and 4 results in a periodic table that is not easy to fit on a standard sheet of paper.

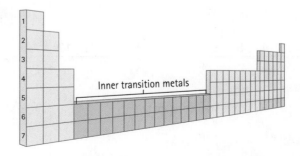

The seventh-period inner transition metals are called the *actinides* because they fall after actinium, Ac. They, too, all have similar properties and hence are not easily purified. The nuclear power industry faces this obstacle because it requires purified samples of two of the most publicized actinides: uranium, U, and plutonium, Pu. Actinides heavier than uranium are not found in nature but are synthesized in the laboratory.

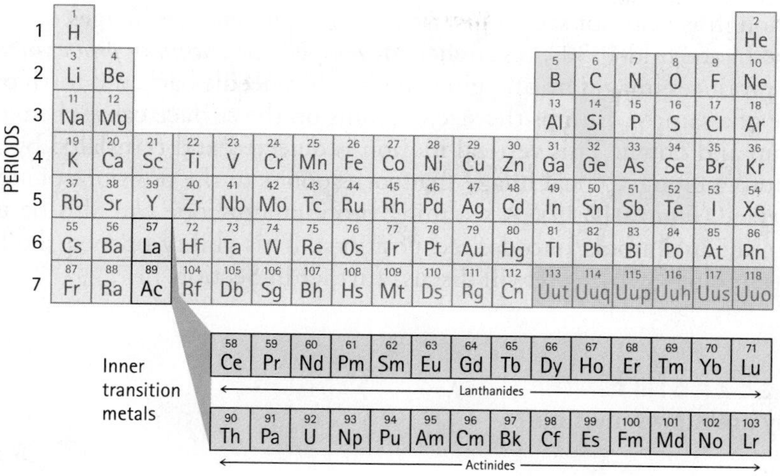

1	1 H																	2 He
2	3 Li	4 Be											5 B	6 C	7 N	8 O	9 F	10 Ne
3	11 Na	12 Mg											13 Al	14 Si	15 P	16 S	17 Cl	18 Ar
4	19 K	20 Ca	21 Sc	22 Ti	23 V	24 Cr	25 Mn	26 Fe	27 Co	28 Ni	29 Cu	30 Zn	31 Ga	32 Ge	33 As	34 Se	35 Br	36 Kr
5	37 Rb	38 Sr	39 Y	40 Zr	41 Nb	42 Mo	43 Tc	44 Ru	45 Rh	46 Pd	47 Ag	48 Cd	49 In	50 Sn	51 Sb	52 Te	53 I	54 Xe
6	55 Cs	56 Ba	57 La	72 Hf	73 Ta	74 W	75 Re	76 Os	77 Ir	78 Pt	79 Au	80 Hg	81 Tl	82 Pb	83 Bi	84 Po	85 At	86 Rn
7	87 Fr	88 Ra	89 Ac	104 Rf	105 Db	106 Sg	107 Bh	108 Hs	109 Mt	110 Ds	111 Rg	112 Cn	113 Uut	114 Uuq	115 Uup	116 Uuh	117 Uus	118 Uuo

PERIODS

Inner transition metals

58 Ce	59 Pr	60 Nd	61 Pm	62 Sm	63 Eu	64 Gd	65 Tb	66 Dy	67 Ho	68 Er	69 Tm	70 Yb	71 Lu

Lanthanides

90 Th	91 Pa	92 U	93 Np	94 Pu	95 Am	96 Cm	97 Bk	98 Cf	99 Es	100 Fm	101 Md	102 No	103 Lr

Actinides

FIGURE 12.16
The typical display of the inner transition metals. The count of elements in the sixth period goes from lanthanum (La, 57) to cerium (Ce, 58) on through to lutetium (Lu, 71) and then back to hafnium (Hf, 72). A similar jump is made in the seventh period.

12.5 Physical and Conceptual Models

EXPLAIN THIS How do we predict the behavior of atoms?

Atoms are so small that the number of them in a baseball is roughly equal to the number of Ping-Pong balls that could fit inside a hollow sphere as big as Earth, as Figure 12.17 illustrates. This number is incredibly large—beyond our intuitive grasp. Atoms are so incredibly small that we can never see them in the usual sense. This is because light travels in waves, and atoms are smaller than the wavelengths of visible light, which is the light that allows the human eye to see things. As illustrated in Figure 12.18, the diameter of an object visible under the highest magnification must be larger than the wavelengths of visible light.

Atoms in a baseball

Ping-Pong balls in Earth

FIGURE 12.17
If Earth were filled with nothing but Ping-Pong balls, the number of balls would be roughly equal to the number of atoms in a baseball. Put differently, if a baseball were the size of Earth, one of its atoms would be the size of a Ping-Pong ball.

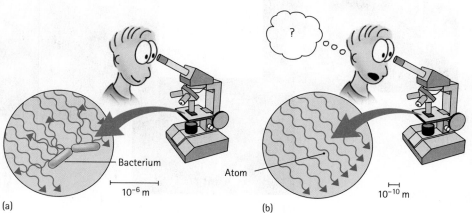

Bacterium

10^{-6} m

Atom

10^{-10} m

(a)　(b)

FIGURE 12.18
Microscopic objects can be seen through a microscope that works with visible light, but submicroscopic particles cannot. (a) A bacterium is visible because it is larger than the wavelengths of visible light. We can see the bacterium through the microscope because the bacterium reflects visible light. (b) An atom is invisible because it is smaller than the wavelengths of visible light and so does not reflect the light toward our eyes.

Although we cannot see atoms *directly*, we can generate images of them *indirectly*. In the mid-1980s, researchers developed the *scanning probe microscope*, which produces images by dragging an ultrathin needle back and forth over the surface of a sample. Bumps the size of atoms on the surface cause the needle to move up and down. This vertical motion is detected and translated by a computer into a topographical image that corresponds to the positions of atoms on the surface (Figure 12.19). A scanning probe microscope can also be used to push individual atoms into desired positions. This ability opened the field of nanotechnology, which we will discuss further in Section 14.7.

CHECKPOINT
Why are atoms invisible?

Was this your answer?
An individual atom is smaller than the wavelengths of visible light and so is unable to reflect that light. Atoms are invisible, therefore, because visible light passes right by them. The atomic images generated by scanning probe microscopes are not photographs taken by a camera. Rather, they are computer renditions generated from the movements of an ultrathin needle.

A very small or very large visible object can be represented with a **physical model**, which is a model that replicates the object at a more convenient scale. Figure 12.20a, for instance, shows a large-scale physical model of a microorganism that a biology student uses to study the microorganism's internal structure. Because atoms are invisible, however, we cannot use a physical model to represent them. In other words, we cannot simply scale up the atom to a larger size, as we might with a microorganism. (A scanning probe microscope merely shows the *positions* of atoms and not actual images of atoms, which do not have the solid surfaces implied in the scanning probe image of Figure 12.19b.) So, rather than describing the atom with a physical model, chemists use what is known as a **conceptual model**, which describes a system. The more accurate a conceptual model, the more accurately it predicts the behavior of the system. The weather is best described using a conceptual model like the one shown

FIGURE 12.19
(a) Scanning probe microscopes are relatively simple devices used to create submicroscopic imagery. (b) An image of gallium and arsenic atoms obtained with a scanning probe microscope. (c) Each dot in the world's tiniest map consists of a few thousand gold atoms, each atom moved into its proper place by a scanning probe microscope.

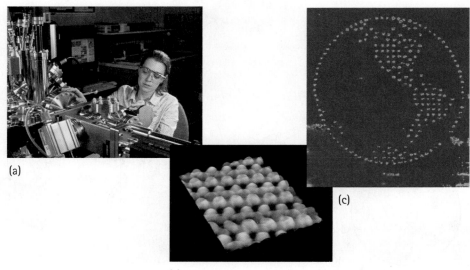

(a)

(b)

(c)

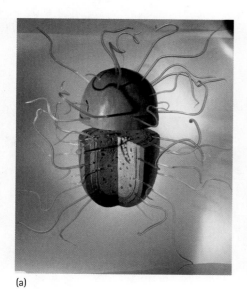

(a)

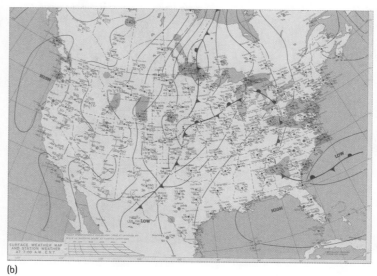

(b)

FIGURE 12.20
(a) This large-scale model of a microorganism is a physical model. (b) Weather forecasters rely on conceptual models such as this one to predict the behavior of weather systems.

in Figure 12.20b. Such a model shows how the various components of the system—humidity, atmospheric pressure, temperature, electric charge, the motion of large masses of air—interact with one another. Other systems that can be described by conceptual models are the economy, population growth, the spread of diseases, and team sports.

CHECKPOINT

A basketball coach describes a playing strategy to her team by way of sketches on a game board. Do the illustrations represent a physical model or a conceptual model?

Was this your answer?
The sketches are a conceptual model the coach uses to describe a system (the players on the court), with the hope of predicting an outcome (winning the game).

Like the weather, the atom is a complex system of interacting components, and it is best described with a conceptual model. You should therefore be careful not to interpret any visual representation of an atomic conceptual model as a re-creation of an actual atom. In Section 12.7, for example, you will be introduced to the planetary model of the atom, wherein electrons are shown orbiting the atomic nucleus much as planets orbit the Sun. This planetary model is limited, however, in that it fails to explain many properties of atoms. Thus newer and more accurate (and more complicated) conceptual models of the atom have since been introduced. In these models, electrons appear as a cloud hovering around the atomic nucleus, but even these models have their limitations. Ultimately, the best models of the atom are purely mathematical.

In this textbook, our focus is on conceptual atomic models that are easily represented by visual images, including the planetary model and a model in which electrons are grouped in units called shells. Despite their limitations, such images are excellent guides to learning about the behavior of atoms, especially for the beginning student. As we discuss in the following sections, scientists developed these models to help explain how atoms emit light.

We can't "see" an atom because it is too small. We can't see the farthest star either. There's much that we can't see. But that doesn't prevent us from thinking about such things or even collecting indirect evidence.

FIGURE 12.21
White light is separated into its color components by (a) a prism and (b) a diffraction grating.

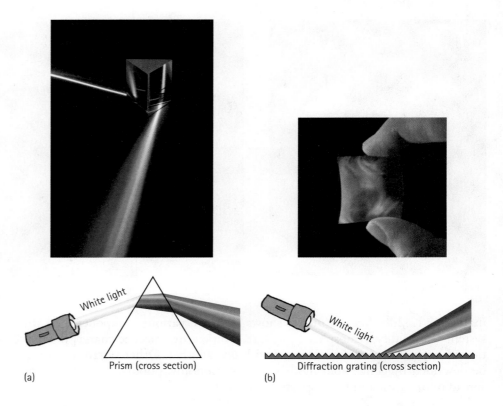

Prism (cross section)

(a)

Diffraction grating (cross section)

(b)

12.6 Identifying Atoms Using the Spectroscope

EXPLAIN THIS How is it possible to tell what stars are made of when they are so very far away?

LEARNING OBJECTIVE
Describe how an atom reveals its identity by the light it emits.

FYI A star's age is revealed by its elemental makeup. The first and oldest stars were composed of hydrogen and helium because those were the only elements available at that time. Heavier elements were produced after many of these early stars exploded in supernovae. Later stars incorporated these heavier elements in their formation. In general, the younger a star, the greater amounts of these heavier elements it contains.

Recall from Chapter 11 that we see white light when all frequencies of visible light reach our eye at the same time. By passing white light through a prism or through a diffraction grating, we can separate the color components of the light, as shown in Figure 12.21. (Remember—each color of visible light corresponds to a different frequency.) A **spectroscope**, shown in Figure 12.22, is an instrument used to observe the color components of any light source. The spectroscope allows us to analyze the light emitted by elements as they are made to glow.

FIGURE 12.22
INTERACTIVE FIGURE **MP**

(a) In a spectroscope, light emitted by atoms passes through a narrow slit before being separated into particular frequencies by a prism or (as shown here) a diffraction grating. (b) This is what the eye sees when the slit of a diffraction-grating spectroscope is pointed toward a white-light source. Spectra of colors appear to the left and right of the slit.

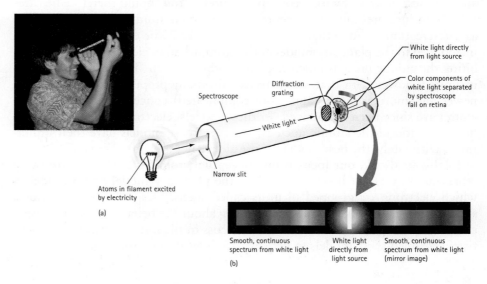

Spectroscope

Diffraction grating

White light directly from light source

Color components of white light separated by spectroscope fall on retina

White light

Narrow slit

Atoms in filament excited by electricity

(a)

Smooth, continuous spectrum from white light

White light directly from light source

Smooth, continuous spectrum from white light (mirror image)

(b)

Light is given off by atoms subjected to various forms of energy, such as heat or electricity. The atoms of a given element emit only certain frequencies of light, however. As a consequence, each element emits a distinctive glow when energized. Sodium atoms emit bright yellow light, which makes them useful as the light source in streetlamps because our eyes are very sensitive to yellow light. To name just one more example, neon atoms emit a brilliant red-orange light, which makes them useful as the light source in neon signs.

When we view the light from a glowing element through a spectroscope, we see that the light consists of a number of discrete (separate from one another) frequencies rather than a continuous spectrum like the one shown in Figure 12.22. The pattern of frequencies formed by a given element—some of which are shown in Figure 12.23—is referred to as that element's **atomic spectrum**. The atomic spectrum is an element's fingerprint. You can identify the elements in a light source by analyzing the light through a spectroscope and looking for characteristic patterns. If you don't have the opportunity to work with a spectroscope in your laboratory, check out the activities at the end of this chapter.

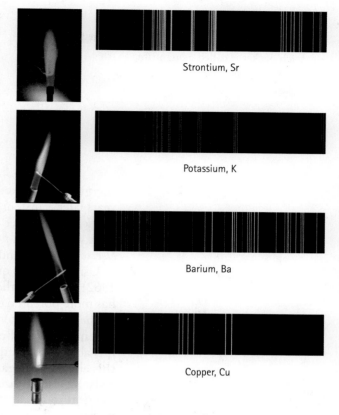

Strontium, Sr

Potassium, K

Barium, Ba

Copper, Cu

FIGURE 12.23
Elements heated by a flame glow their characteristic color. This is commonly called a flame test and is used to test for the presence of an element in a sample. When viewed through a spectroscope, the color of each element is revealed to consist of a pattern of distinct frequencies known as an atomic spectrum.

CHECKPOINT
How might you deduce the elemental composition of a star?

Was this your answer?
Aim a well-built spectroscope at the star and study its spectral patterns. In the late 1800s, this was done with our own star, the Sun. Spectral patterns of hydrogen and some other known elements were observed, in addition to one pattern that could not be identified. Scientists concluded that this unidentified pattern belonged to an element not yet discovered on Earth. They named this element helium after the Greek word for "sun," *helios.*

12.7 The Quantum Hypothesis

EXPLAIN THIS Why do atomic spectra contain only a limited number of light frequencies?

LEARNING OBJECTIVE
Recount how the quantum nature of energy led to Bohr's planetary model of the atom.

An important step toward our present understanding of atoms and their spectra was taken by the German physicist Max Planck (1858–1947). In 1900, Planck hypothesized that light energy is *quantized* in much the same way matter is. The mass of a gold brick, for example, equals some whole-number multiple of the mass of a single gold atom. Similarly, an electric charge is always some whole-number multiple of the charge on a single electron. Mass and electric charge are therefore said to be *quantized* in that they consist of some number of fundamental units.

What Planck did with his **quantum hypothesis** was to recognize that a beam of light energy is not the continuous (nonquantized) stream of energy we think it is. Instead, the beam consists of zillions of small, discrete packets of

FIGURE 12.24
Light is quantized, which means it consists of a stream of energy packets. Each packet is called a quantum, also known as a photon.

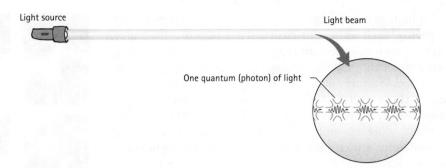

energy, each packet called a **quantum**, as represented in Figure 12.24. A few years later, Einstein recognized that these quanta of light behave much like tiny particles of matter. To emphasize their particulate nature, each quantum of light was called a *photon*, a name coined because of its similarity to the words *electron*, *proton*, and *neutron*.

Using Planck's quantum hypothesis, the Danish scientist Niels Bohr (1885–1962) explained the formation of atomic spectra as follows. First, an electron has more potential energy when it is farther from the nucleus. This is analogous to the greater potential energy an object has when it is held higher above the ground. Second, Bohr recognized that when an atom absorbs a photon of light, it is absorbing energy. This energy is acquired by one of the electrons. Because this electron has gained energy, it must move away from the nucleus.

Bohr also realized that the opposite is true: when a high-potential-energy electron in an atom loses some of its energy, the electron moves closer to the nucleus and the energy lost from the electron is emitted from the atom as a photon of light. Both absorption and emission are illustrated in Figure 12.25.

Just as I can't stand between two adjacent steps, an electron can't exist between two energy levels.

CHECKPOINT
Which has more energy: a photon of red light or a photon of infrared light?

Was this your answer?
Red light has a higher frequency than infrared light, which means a photon of red light has more energy than a photon of infrared light. Recall that a photon is a single discrete packet (a quantum) of radiant energy.

Bohr reasoned that because light energy is quantized, the energy of an electron in an atom must also be quantized. In other words, an electron cannot have just any amount of potential energy. Rather, within the atom there must be a number of distinct energy levels, analogous to steps on a staircase. Where you are on a staircase is restricted to where the steps are—you cannot stand

FIGURE 12.25
An electron is lifted away from the nucleus as the atom it is in absorbs a photon of light and drops closer to the nucleus as the atom releases a photon of light.

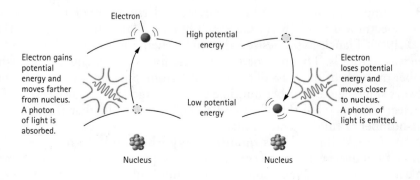

Electron

High potential energy

Electron gains potential energy and moves farther from nucleus. A photon of light is absorbed.

Low potential energy

Electron loses potential energy and moves closer to nucleus. A photon of light is emitted.

Nucleus

Nucleus

at a height that is, say, halfway between any two adjacent steps. Similarly, an atom has only a limited number of permitted energy levels, and an electron can never have an amount of energy between these permitted energy levels. Bohr gave each energy level a **principal quantum number n**, where n is always some integer. The lowest energy level has a principal quantum number $n = 1$. An electron for which $n = 1$ is as close to the nucleus as possible, and an electron for which $n = 2$, $n = 3$, and so forth is farther away, in a stepwise fashion, from the nucleus.

Using these ideas, Bohr developed a conceptual model in which an electron moving around the nucleus is restricted to certain distances from the nucleus, with these distances determined by the amount of energy the electron has. Bohr saw this as similar to how the planets are held in orbit around the Sun at given distances from the Sun. The allowed energy levels for any atom, therefore, could be graphically represented as orbits around the nucleus, as shown in Figure 12.26. Bohr's quantized model of the atom thus became known as the *planetary model*.

Bohr used his planetary model to explain why atomic spectra contain only a limited number of light frequencies. According to the model, photons are emitted by atoms as electrons move from higher-energy outer orbits to lower-energy inner orbits. The energy of an emitted photon is equal to the difference in energy between the two orbits. Because an electron is restricted to discrete orbits, only particular light frequencies are emitted, as atomic spectra show.

Interestingly, any transition between two orbits is always instantaneous. In other words, the electron doesn't "jump" from a higher to a lower orbit the way a squirrel jumps from a higher branch in a tree to a lower one. Rather, an electron takes no time to move between two orbits. Bohr was serious when he stated that electrons could never exist between permitted energy levels!

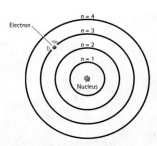

FIGURE 12.26
Bohr's planetary model of the atom, in which electrons orbit the nucleus much as planets orbit the Sun, is a graphical representation that helps us understand how electrons can possess only certain quantities of energy.

CHECKPOINT
Is the Bohr model of the atom a physical model or a conceptual model?

Was this your answer?
The Bohr model is a conceptual model. It is not a scaled-up version of an atom, but instead is a representation that accounts for the atom's behavior.

Bohr's planetary atomic model proved to be a tremendous success. By utilizing Planck's quantum hypothesis, Bohr's model solved the mystery of atomic spectra. Despite its successes, though, Bohr's model was limited because it did not explain why energy levels in an atom are quantized. Bohr was quick to point out that his model was to be interpreted only as a crude beginning, and the picture of electrons whirling about the nucleus like planets about the Sun was not to be taken literally (a warning to which popularizers of science paid no heed).

Recall from Chapter 11 that a photon behaves like a particle when it is being emitted by an atom or being absorbed by photographic film or other detectors, but it behaves like a wave in traveling from a source to the place where it is detected.

12.8 Electron Waves

EXPLAIN THIS How is a plucked guitar string like an electron in an atom?

LEARNING OBJECTIVE
Summarize how electrons, when confined to an atom, behave like self-reinforcing wavelike entities.

If light has both wave properties and particle properties, why can't a material particle, such as an electron, also have both? This question was posed by the French physicist Louis de Broglie (1892–1987) while he was still a graduate student in 1924. His revolutionary answer was that every particle of matter is somehow

Mastering**PHYSICS**
VIDEO: Electron Waves

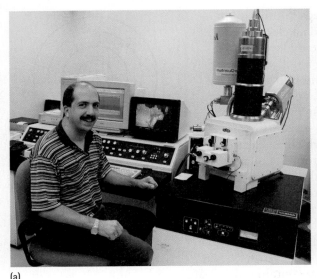

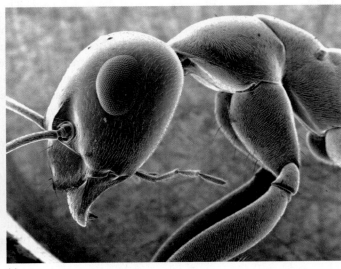

(a)

(b)

FIGURE 12.27

(a) An electron microscope makes practical use of the wave nature of electrons. The wavelengths of electron beams are typically shorter than the wavelengths of visible light by a factor of a thousand, and so the electron microscope can distinguish detail not visible with optical microscopes. (b) Detail of an ant as seen with an electron microscope at a "low" magnification of 200×. Note the remarkable resolution.

endowed with a wave to guide it as it travels. The more slowly an electron moves, the more its behavior is that of a particle with mass. The more quickly it moves, however, the more its behavior is that of a wave of energy. This duality is an extension of Einstein's famous equation $E = mc^2$, which tells us that matter and energy are interconvertible. We talk more about this relationship in the next chapter.

A practical application of the wave properties of fast-moving electrons is the electron microscope, which focuses not visible-light waves, but rather electron waves. Because electron waves are much shorter than visible-light waves electron microscopes can show far greater detail than optical microscopes, as Figure 12.27 shows.

In an atom, an electron moves at very high speeds—on the order of 2 million m/s—and therefore exhibits many of the properties of a wave. An electron's wave nature can be used to explain why electrons in an atom are restricted to particular energy levels. Permitted energy levels are a natural consequence of electron waves closing in on themselves in a synchronized manner.

As an analogy, consider the wire loop shown in Figure 12.28. This loop is affixed to a mechanical vibrator that can be adjusted to create waves of different

FYI Electron waves are three-dimensional, which makes them difficult to visualize, but scientists have come up with ways of visualizing them. This includes *probability clouds* and *atomic orbitals*, which you would learn about in a follow-up course on chemistry.

FIGURE 12.28

For the fixed circumference of a wire loop, only some wavelengths are self-reinforcing. (a) The loop affixed to the post of a mechanical vibrator at rest. Waves are sent through the wire when the post vibrates. (b) Waves created by vibration at particular rates are self- reinforcing. (c) Waves created by vibration at other rates are not self-reinforcing.

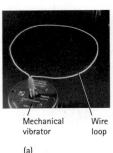

Mechanical vibrator   Wire loop

(a)

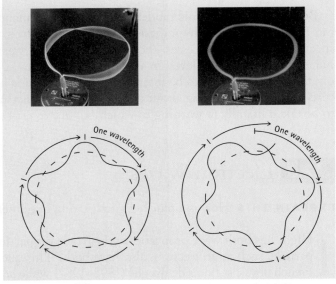

(b) Wavelength is self-reinforcing.   (c) Wavelength produces chaotic motion.

wavelengths in the wire. Waves passing through the wire that meet up with themselves, as shown in Figure 12.28b, form a stationary wave pattern called a standing wave (see Section 10.7). This pattern results because the peaks and valleys of successive waves are perfectly matched, which makes the waves reinforce one another. With other wavelengths, as shown in Figure 12.28c, successive waves are not synchronized. As a result, the waves do not build to great amplitude.

The only waves that an electron exhibits while confined to an atom are those that are self-reinforcing. These resemble a standing wave centered on the atomic nucleus. Each standing wave corresponds to one of the permitted energy levels. Only the frequencies of light that match the difference between any two of these permitted energy levels can be absorbed or emitted by an atom.

The wave nature of electrons also explains why they do not spiral closer and closer to the positive nucleus that attracts them. By viewing each electron orbit as a self-reinforcing wave, we see that the circumference of the smallest orbit can be no smaller than a single wavelength.

CHECKPOINT

What must an electron be doing in order to have wave properties?

Was this your answer?
According to de Broglie, particles of matter behave like waves by virtue of their motion. An electron must therefore be moving in order to have wave properties. In atoms, electrons move at speeds of about 2 million m/s, and so their wave nature is most pronounced.

12.9 The Shell Model

EXPLAIN THIS Why do elements in the same group of the periodic table have similar properties?

LEARNING OBJECTIVE
Show how electrons behave as though they are arranged in a series of shells centered around the atomic nucleus.

For the purposes of a simplified understanding of how atoms behave, we turn to the *shell model*, first made popular by the noted chemist and two-time Nobel laureate Linus Pauling (1901–1994). This model is similar to Bohr's planetary model in that it shows electrons restricted to particular distances from the nucleus. The shell model, however, is a bit more sophisticated because it incorporates the wave nature of electrons. How it does so is beyond the scope of this book, which will present only this model as is. But if you ever wonder why the model has this or that attribute, the answer can be traced back to the wave nature of electrons. The great benefit of learning this particular conceptual model is that it helps us to understand the organization of the periodic table. Furthermore, it builds a strong foundation for understanding how atoms form chemical bonds, which is the main focus of Chapter 15.

According to the shell model, electrons behave as though they are arranged in a series of concentric shells. A **shell** is a region of space around the atomic nucleus within which electrons may reside. An important aspect of this model is that there are at least seven shells and each shell can hold only a limited number of electrons. As shown in Figure 12.29, the innermost shell can hold 2 electrons, the second and third shells 8 each, the fourth and fifth shells 18 each, and the sixth and seventh shells 32 each.

The quality of a song depends on the arrangement of musical notes. In a similar fashion, the properties of an element depend on the arrangements of electrons in its atoms

Mastering**PHYSICS**
TUTORIAL: Bohr's Shell Model

(a) A cutaway view of the seven shells, with the number of electrons each shell can hold indicated. (b) A two-dimensional, cross-sectional view of the shells. (c) An easy-to-draw cross-sectional view that resembles Bohr's planetary model.

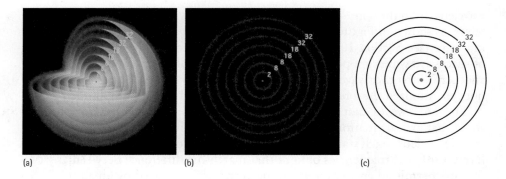

(a)　　　　(b)　　　　(c)

What do poets and scientists have in common? They both use metaphors to help us understand abstract concepts and relationships. The "shell," for example, is a metaphor that helps us visualize an invisible reality. Scientific models are essentially equivalent to the metaphorical language used in poetry.

A series of seven such concentric shells accounts for the seven periods of the periodic table. Furthermore, the number of elements in each period is equal to the shell's capacity for electrons. The first shell, for example, has the capacity for only two electrons. That's why we find only two elements, hydrogen and helium, in the first period (Figure 12.30). Hydrogen is the element whose atoms have only one electron. This one electron resides within the first shell, which is the shell closest to the nucleus. Each helium atom has two electrons, both of which are also within the first shell, which is thus filled to its maximum capacity. Similarly, the second and third shells each have the capacity for eight electrons, so eight elements are found in both the second and the third periods.

The electrons of the outermost occupied shell in any atom are directly exposed to the external environment and are the first to interact with other atoms. Most notably, they are the ones that participate in chemical bonding, as we will discuss in Chapter 15. The electrons in the outermost shell, therefore, are quite important.

The quality of a song depends on the arrangement of musical notes. In a similar fashion, the properties of an element depend on the arrangements of electrons in its atoms, especially the outer-shell electrons. Look carefully at Figure 12.30. Can you see that the outer-shell electrons of atoms above and below one another on the periodic table (within the same group) are similarly organized? For example, atoms of the first group, which include hydrogen, lithium, and sodium,

FIGURE 12.30
The first three periods of the periodic table according to the shell model. Elements in the same period have electrons in the same shells. Elements in the same period differ from one another by the number of electrons in the outermost shell.

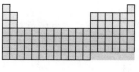

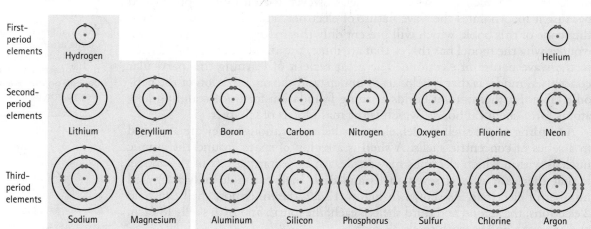

First-period elements	Hydrogen							Helium
Second-period elements	Lithium	Beryllium	Boron	Carbon	Nitrogen	Oxygen	Fluorine	Neon
Third-period elements	Sodium	Magnesium	Aluminum	Silicon	Phosphorus	Sulfur	Chlorine	Argon

each have a single outer-shell electron. The atoms of the second group, including beryllium and magnesium, each have two outer-shell electrons. Similarly, atoms of the last group, including helium, neon, and argon, each have their outermost shell filled to capacity with electrons—two for helium and eight for both neon and argon. In general, the outer-shell electrons of atoms in the same group of the periodic table are similarly organized. This explains why elements of the same group have similar properties—a concept first presented in Section 12.4.

FIGURE 12.31

Two-time Nobel laureate Linus Pauling (1901–1994) was an early proponent of teaching beginning chemistry students a shell model, from which the organization of the periodic table could be described. In 1954, Pauling won the Nobel Prize in Chemistry for his research into the nature of the chemical bond. In 1962, he was awarded the Nobel Peace Prize for his campaign against the testing of nuclear bombs, which introduced massive amounts of radioactivity into the environment.

CHECKPOINT

Do atoms really consist of shells that look like those depicted in Figure 12.29?

Was this your answer?
No. The shell model is *not* a depiction of the "appearance of an atom." Rather, it is a conceptual model that allows us to account for observed behavior. An atom, therefore, does not actually contain a series of concentric shells; it merely behaves as though it does.

Remember that the shell model is not to be interpreted as an actual representation of the atom's physical structure. Rather, it serves as a tool to help us understand and predict how atoms behave. In Chapter 15 we will use a simplified version of this model, called the *electron-dot structure*, to show how atoms join together to form molecules, which are tightly held groups of atoms. In the next chapter, however, we will explore in greater detail the nature of the atomic nucleus, which is a potential source of enormous amounts of energy.

FYI According to Einstein's theory of special relativity, at 60% of the speed of light, gold's innermost electrons experience only 52 seconds for each one of our minutes. A diamond may be forever, but the innermost electrons of gold are 8 s/min slow!

For instructor-assigned homework, go to www.masteringphysics.com

---

SUMMARY OF TERMS (KNOWLEDGE)

Atomic mass The mass of an element's atoms listed in the periodic table as an average value based on the relative abundance of the element's isotopes.

Atomic nucleus The dense, positively charged center of every atom.

Atomic number A count of the number of protons in the atomic nucleus.

Atomic spectrum The pattern of frequencies of electromagnetic radiation emitted by the atoms of an element, considered to be an element's "fingerprint."

Atomic symbol The abbreviation for an element or atom.

Conceptual model A representation of a system that helps us predict how the system behaves.

Electron An extremely small, negatively charged subatomic particle found outside the atomic nucleus.

Element Any material that is made up of only one type of atom.

Group A vertical column in the periodic table, also known as a *family* of elements.

Isotopes Members of a set of atoms of the same element whose nuclei contain the same number of protons but different numbers of neutrons.

Neutron An electrically neutral subatomic particle of the atomic nucleus.

Nucleon Any subatomic particle found in the atomic nucleus; another name for either a proton or a neutron.

Period A horizontal row in the periodic table.

Periodic table A chart in which all the known elements are listed in order of atomic number.

Physical model A representation of an object on some convenient scale.

Principal quantum number _n_ An integer that specifies the quantized energy level of an atomic orbital.

Proton A positively charged subatomic particle of the atomic nucleus.

Quantum A small, discrete packet of light energy.

Quantum hypothesis The idea that light energy is contained in discrete packets called quanta.

Shell A region of space around the atomic nucleus within which electrons may reside.

Spectroscope A device that uses a prism or diffraction grating to separate light into its color components.

READING CHECK QUESTIONS (COMPREHENSION)

12.1 Atoms Are Ancient and Empty

1. Which is the oldest element?
2. Is it possible to see an atom using visible light?
3. What is at the center of every atom?

12.2 The Elements

4. How many types of atoms can you expect to find in a pure sample of any element?
5. Distinguish between an atom and an element.
6. What is the atomic symbol for the element cobalt?

12.3 Protons and Neutrons

7. What role does atomic number play in the periodic table?
8. Distinguish between atomic number and mass number.
9. Distinguish between mass number and atomic mass.

12.4 The Periodic Table

10. Are most elements metallic or nonmetallic?
11. How many periods are there in the periodic table? How many groups?
12. What happens to the properties of elements across any period of the periodic table?

12.5 Physical and Conceptual Models

13. If a baseball were the size of the Earth, about how large would its atoms be?
14. When we use a scanning probe microscope, do we see atoms directly or only indirectly?
15. What is the difference between a physical model and a conceptual model?

12.6 Identifying Atoms Using the Spectroscope

16. What does a spectroscope do to the light coming from an atom?
17. What causes an atom to emit light?
18. Why do we say atomic spectra are like fingerprints of the elements?

12.7 The Quantum Hypothesis

19. What was Planck's quantum hypothesis?
20. Which has more potential energy: an electron close to an atomic nucleus or an electron far from an atomic nucleus?
21. Did Bohr think of his planetary model as an accurate representation of what an atom looks like?

12.8 Electron Waves

22. Who first proposed that electrons exhibit the properties of a wave?
23. About how fast does an electron travel around the atomic nucleus?
24. How does the speed of an electron change its fundamental nature?

12.9 The Shell Model

25. Does the periodic table explain the shell model, or does the shell model explain the periodic table?
26. Which electrons are most responsible for the properties of an atom?
27. What is the relationship between the maximum number of electrons each shell can hold and the number of elements in each period of the periodic table?

ACTIVITIES (HANDS-ON APPLICATION)

28. Fluorescent lights contain spectral lines from the light emission of mercury atoms. Special coatings on the inner surface of the bulb help to accentuate visible frequencies, which can be seen through the diffraction grating reflection of a compact disc. Cut a narrow slit through a piece of thick paper (or thin cardboard) and place it over a bright fluorescent bulb. View this slit at an oblique angle against a CD and look for spectral lines. Place the slit over an incandescent bulb, and you'll see a smooth continuous spectrum (no lines) because the incandescent filament glows at all visible frequencies.

Try looking at different brands of fluorescent bulbs. You'll also be able to see spectral lines in streetlights and fireworks. For those, it is best to use "rainbow" glasses, available from a nature, toy, or hobby store.

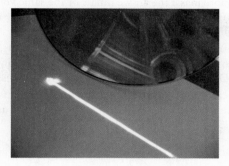

29. You can "quantize" your whistle by whistling down a long tube, such as the tube from a roll of wrapping paper. First, without the tube, whistle from a high pitch to a low pitch. Do it loudly and in a single breath. (If you can't whistle, find someone who can.) Next, try the same thing while holding the tube to your lips. Aha! Note that some frequencies simply cannot be whistled, no matter how hard you try. These frequencies are forbidden because their wavelengths are not a multiple of the length of the tube.

Try experimenting with tubes of different lengths. To hear yourself more clearly, use a flexible plastic tube and twist the outer end toward your ear.

When your whistle is confined to the tube, the consequence is a quantization of its frequencies. When an electron wave is confined to an atom, the consequence is a quantization of the electron's energy.

30. Stretch a rubber band between your thumbs and pluck it with your index finger. Better yet, stretch the rubber band in front of a windy fan to get it vibrating. Note that the area of greatest oscillation is always at the midpoint. This is a self-reinforcing wave that occurs as overlapping waves bounce back and forth from thumb to thumb.

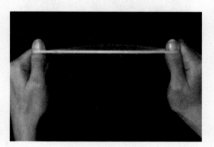

THINK AND SOLVE (MATHEMATICAL APPLICATION)

31. A class of 20 students takes an exam and every student scores 80%. What is the class average? Would the class average be slightly lower, the same, or slightly higher if one of the students instead scored 100%? How is this similar to how we derived the atomic masses of elements?

32. The isotope lithium-7 has a mass of 7.0160 amu, and the isotope lithium-6 has a mass of 6.0151 amu. Given that 92.58% of all lithium atoms found in nature are lithium-7 and 7.42% are lithium-6, show that the atomic mass of lithium, Li (atomic number 3), is 6.941 amu.

33. The element bromine, Br (atomic number 35), has two major isotopes of similar abundance, both around 50%. The atomic mass of bromine is reported in the periodic table as 79.904 amu. Choose the most likely set of mass numbers for these two bromine isotopes: (a) Br-79, Br-81; (b) Br-79, Br-80; (c) Br-80, Br-81.

THINK AND RANK (ANALYSIS)

34. Rank the three subatomic particles in order of increasing mass: (a) neutron, (b) proton, (c) electron.

35. Consider these atoms: helium, He; chlorine, Cl; and argon, Ar. Rank them in terms of their atomic number, from smallest to largest.

36. Consider three 1-g samples of matter: (a) carbon-12, (b) carbon-13, and (c) uranium-238. Rank them in terms of the number of atoms, from most to least.

37. Consider these atoms: helium, He; aluminum, Al; argon, Ar. Rank them, from smallest to largest, in terms of (a) size, (b) number of protons in the nucleus, and (c) number of electrons.

38. Of the atoms sodium, Na; magnesium, Mg; and aluminum, Al, one tends to lose three electrons, another tends to lose two electrons, and another tends to lose one electron. Rank these atoms in order of the number of electrons they tend to lose, from fewest to most.

EXERCISES (SYNTHESIS)

39. A cat strolls across your backyard. An hour later, a dog with its nose to the ground follows the trail of the cat. Explain what is going on from a molecular point of view.

40. If all the molecules of a body remained part of that body, would the body have any odor?

41. Where did the atoms that make up a newborn baby originate?

42. The atoms that make up your body are mostly empty space, and structures such as the chair you're sitting on are composed of atoms that are also mostly empty space. So why don't you fall through the chair?

43. Where did the carbon atoms in Leslie's hair originate? (Shown below is a photo of coauthor Leslie at age 16.)

44. In what sense can you truthfully say that you are a part of every person around you?

45. Is the head of a politician really made of 99.99999999% empty space?

46. If two protons and two neutrons are removed from the nucleus of an oxygen-16 atom, a nucleus of which element remains?

47. If an atom has 43 electrons, 56 neutrons, and 43 protons, what is its approximate atomic mass? What is the name of this element?

48. The nucleus of an electrically neutral iron atom contains 26 protons. How many electrons does this iron atom have?

49. Evidence for the existence of neutrons did not come until many years after the discoveries of the electron and the proton. Give a possible explanation for this.

50. Which has more atoms: a 1-g sample of carbon-12 or a 1-g sample of carbon-13? Explain.

51. Why aren't the atomic masses listed in the periodic table whole numbers?

52. Which contributes more to an atom's mass: electrons or protons? Which contributes more to an atom's size?

53. What is the approximate mass of an oxygen atom in atomic mass units? What is the approximate mass of two oxygen atoms? How about an oxygen molecule?

54. What is the approximate mass of a carbon atom in atomic mass units? How about a carbon dioxide molecule?

55. Which is heavier: a water molecule, H_2O, or a carbon dioxide molecule, CO_2?

56. When we breathe, we inhale oxygen, O_2, and exhale carbon dioxide, CO_2, plus water vapor, H_2O. Which likely has more mass: the air we inhale or the same volume of air we exhale? Does breathing cause you to lose or gain weight?

57. As a tree respires, it takes in carbon dioxide, CO_2, and water vapor, H_2O, from the air while also releasing oxygen, O_2. Does the tree lose or gain weight as it respires? Explain.

58. Which of the following diagrams best represents the size of the atomic nucleus relative to the size of the atom?

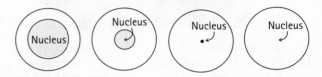

59. A beam of protons and a beam of neutrons of the same energy are both harmful to living tissue. The beam of neutrons, however, is less harmful. Suggest why.

60. Germanium, Ge (atomic number 32), computer chips operate faster than silicon, Si (atomic number 14), computer chips. So how might a gallium, Ga (atomic number 31), chip compare with a germanium chip?

61. Helium, He, is a nonmetallic gas and the second element in the periodic table. Rather than being placed adjacent to hydrogen, H, however, helium is placed on the far right of the table. Why?

62. Name 10 elements you have access to macroscopic samples of as a consumer here on Earth.

63. Strontium, Sr (atomic number 38), is especially dangerous to humans because it tends to accumulate in calcium-dependent bone marrow tissues (calcium, Ca, atomic number 20). How does this fact relate to what you know about the organization of the periodic table?

64. With the periodic table as your guide, describe the element selenium, Se, using as many of this chapter's key terms as you can.

65. As depicted in Figure 12.19, are gallium atoms really red and arsenic atoms green?

66. With scanning probe microscopy technology, we see not actual atoms but rather images of them. Explain.

67. Why isn't it possible for a scanning probe microscope to make images of the inside of an atom?

68. What do the components of a conceptual model have in common?

69. Would you use a physical model or a conceptual model to describe the following: a gold coin, dollar bill, car engine, air pollution, virus, spread of sexually transmitted disease?

70. What is the function of an atomic model?

71. What is the relationship between the light emitted by an atom and the energies of the electrons in the atom?

72. How might you distinguish a sodium-vapor streetlight from a mercury-vapor streetlight?

73. What particle within an atom vibrates to generate electromagnetic radiation? This particle is vibrating back and forth between what?

74. How can a hydrogen atom, which has only one electron, create so many spectral lines?

75. Which color of light comes from a greater energy transition: red or blue?

76. How does the wave model of electrons orbiting the nucleus account for the fact that the electrons can have only discrete energy values?

77. What might the spectrum of an atom look like if the atom's electrons were not restricted to particular energy levels?

78. Some older cars vibrate loudly when moving at particular speeds. For example, at 65 mph the car is quiet, but at 60 mph the car rattles uncomfortably. How is this analogous to the quantized energy levels of an electron in an atom?

79. Does a shell have to contain electrons in order to exist?

80. Place the proper number of electrons in each shell.

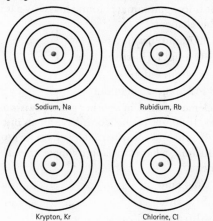

Sodium, Na   Rubidium, Rb

Krypton, Kr   Chlorine, Cl

81. Use the shell model to explain why a potassium atom, K, is larger than a sodium atom, Na.

82. Neon, Ne (atomic number 10), cannot attract any additional electrons. Why?

83. Use the shell model to explain why a lithium atom, Li, is larger than a beryllium atom, Be.

DISCUSSION QUESTIONS (EVALUATION)

84. If matter is made of atoms and atoms are made of subatomic particles, what comes together to create subatomic particles? Where might you find the answer?

85. Astronomical measurements reveal that about 90% of the mass of the universe is invisible to us. This invisible matter, also known as dark matter, is likely to be "exotic" matter—very different from the elements that make up the periodic table (see Section 28.4). We know dark matter is there because of its gravitational effects, but scientists can only guess as to its nature. What do you think dark matter might be made of? How soon might we know the answer?

READINESS ASSURANCE TEST (RAT)

If you have a good handle on this chapter, if you really do, then you should be able to score 7 out of 10 on this RAT. If you score less than 7, you need to study further before moving on.

Choose the BEST answer to each of the following.

1. Which are older: the atoms in the body of an elderly person or the atoms in the body of a baby?
   (a) A baby's are older because this is surely a trick question.
   (b) An elderly person's are older because they have been around much longer.
   (c) They are the same age, which is appreciably older than the solar system.
   (d) It depends on their diet.

2. You could swallow a capsule of germanium, Ge (atomic number 32), without significant ill effects. If a proton were added to each germanium nucleus, however, you would not want to swallow the capsule because the germanium would
   (a) become arsenic.
   (b) become radioactive.
   (c) expand and likely lodge in your throat.
   (d) have a change in flavor.

3. Why aren't the atomic masses given in the periodic table whole numbers?
   (a) Scientists have yet to make the precise measurements.
   (b) That would be too much of a coincidence.
   (c) The atomic masses are average atomic masses.
   (d) Today's instruments are able to measure the atomic masses to many decimal places.

4. If an atom has 43 electrons, 56 neutrons, and 43 protons, what is its approximate atomic mass? What is the name of this element?
   (a) 137 amu; barium
   (b) 99 amu; technetium
   (c) 99 amu; radon
   (d) 142 amu; einsteinium

5. An element found in another galaxy exists as two isotopes. If 80.0% of the atoms have an atomic mass of 80.00 amu and the other 20.0% have an atomic mass of 82.00 amu, what is the approximate *atomic mass* of the element?
   (a) 80.4 amu
   (b) 81.0 amu
   (c) 81.6 amu
   (d) 64.0 amu
   (e) 16.4 amu

6. List the following atoms in order of increasing atomic size: thallium, Tl; germanium, Ge; tin, Sn; phosphorus, P.
   (a) Ge < P < Sn < Tl
   (b) Tl < Sn < P < Ge
   (c) Tl < Sn < Ge < P
   (d) P < Ge < Sn < Tl

7. Which element has chemical properties the most similar to chlorine (Cl, atomic number 17)?
   (a) O
   (b) Na
   (c) S
   (d) Ar
   (e) Br

8. Would you use a physical model or a conceptual model to describe the following: the brain, the mind, the solar system, the beginning of the universe?
   (a) conceptual, physical, conceptual, physical
   (b) conceptual, conceptual, conceptual, conceptual
   (c) physical, conceptual, physical, conceptual
   (d) physical, physical, physical, physical

9. How does the wave model of electrons orbiting the nucleus account for the fact that the electrons can have only discrete energy values?
   (a) Electrons are able to vibrate at only particular frequencies.
   (b) When an electron wave is confined, it is reinforced at only particular frequencies.
   (c) The energy values of an electron occur only where its wave properties have a maximum amplitude.
   (d) The wave model accounts for the shells that an electron may occupy, not its energy levels.

10. How many electrons are in the third shell of sodium, Na (atomic number 11)?
   (a) none
   (b) one
   (c) two
   (d) three

Answers to RAT

1. c, 2. a, 3. c, 4. b, 5. a, 6. d, 7. e, 8. a, 9. b, 10. b

CHAPTER 13

The Atomic Nucleus and Radioactivity

THE ATOMIC nucleus and nuclear processes are one of the most misunderstood and controversial areas of science. Distrust of anything *nuclear*, or anything *radioactive*, is much like the fears of electricity more than a century ago. Indeed, electricity can be dangerous, and even lethal, when improperly handled. But with safeguards and well-informed consumers, society has determined that the benefits of electricity outweigh its risks. Today we are making similar decisions about nuclear technology's risks and benefits. The risks became most evident with the 2011 earthquake and tsunami that destroyed the Japanese Fukushima nuclear power plant. The benefits, however, include the large-scale production of electrical energy with no emission of carbon dioxide, which is a potent greenhouse gas. Should society continue investing in nuclear energy? Now more than ever, it is important that we "know nukes!"

13.1 Radioactivity

EXPLAIN THIS Why is it both impractical and impossible to prevent our exposure to radioactivity?

Elements with unstable nuclei are said to be *radioactive*. They eventually break down and eject energetic particles and emit high-frequency electromagnetic radiation. This process is **radioactivity**, which, because it involves the decay of the atomic nucleus, is often called *radioactive decay*.

A common misconception is that radioactivity is new in the environment, but it has been around far longer than the human race. Interestingly, the deeper you go below Earth's surface, the hotter it gets. At a mere depth of 30 km the temperature is hotter than 500°C. At greater depths it is so hot that rock melts into magma, which can rise to Earth's surface to escape as lava. Superheated subterranean water can escape violently to form geysers or more gently to form a soothing natural hot spring. The main reason it gets hotter down below is that Earth contains an abundance of radioactive isotopes and is heated as it absorbs radiation from these isotopes. So volcanoes, geysers, and hot springs are all powered by radioactivity. Even the drifting of continents (see Chapter 22) is related to Earth's internal radioactivity. Radioactivity is as natural as sunshine and rain.

Alpha, Beta, and Gamma Rays

All isotopes of elements with an atomic number greater than 83 (bismuth) are radioactive. These isotopes, and certain lighter radioactive isotopes, emit three distinct types of radiation, named by the first three letters of the Greek alphabet, α, β, γ— *alpha*, *beta*, and *gamma*. Alpha rays carry a positive electric charge, beta rays carry a negative charge, and gamma rays carry no charge. The three rays can be separated by placing a magnetic field across their paths (Figure 13.2).

An **alpha particle** is the combination of two protons and two neutrons (in other words, it is the nucleus of the helium atom, atomic number 2). Alpha particles are relatively easy to shield because of their relatively large size and their double positive charge (+2). For example, they do not normally penetrate through light materials such as paper or clothing. Because of their great kinetic energies, however, alpha particles can cause significant damage to the surface of a material, especially living tissue. When traveling through only a few centimeters of air, alpha particles pick up electrons and become nothing more

Radioactivity has been around since Earth's beginning.

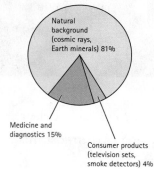

Natural background (cosmic rays, Earth minerals) 81%

Medicine and diagnostics 15%

Consumer products (television sets, smoke detectors) 4%

FIGURE 13.1
Origins of radiation exposure for an average individual in the United States.

FYI Once alpha and beta particles slow down, they combine to form harmless helium. This happens primarily deep underground. As the newly formed helium seeps toward the surface, it becomes concentrated within natural gas deposits. Some natural gas deposits, such as those in Texas, contain as much as 7% helium. This helium is isolated and sold for various applications, such as blimps and helium balloons. Interestingly, natural gas fields within the United States contain about two-thirds of the world's supply of helium.

FIGURE 13.2
INTERACTIVE FIGURE

In a magnetic field, alpha rays bend one way, beta rays bend the other way, and gamma rays don't bend at all. Note that the alpha rays bend less than do the beta rays. This occurs because alpha particles have more inertia (mass) than beta particles.

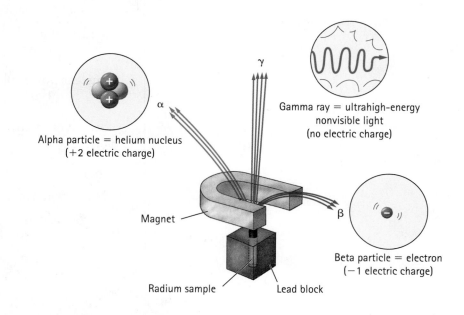

α

Alpha particle = helium nucleus (+2 electric charge)

γ

Gamma ray = ultrahigh-energy nonvisible light (no electric charge)

Magnet

β

Beta particle = electron (−1 electric charge)

Radium sample   Lead block

than harmless helium. As a matter of fact, that's where the helium in a child's balloon comes from—practically all of Earth's helium atoms were at one time energetic alpha particles.

A **beta particle** is an electron ejected from a nucleus. Once ejected, it is indistinguishable from an electron in a cathode ray or electrical circuit, or one orbiting the atomic nucleus. The difference is that a beta particle originates inside the nucleus—from a neutron. As we shall soon see, the neutron becomes a proton once it loses the electron that is a beta particle. A beta particle is normally faster than an alpha particle and carries only a single negative charge (−1). Beta particles are not as easy to stop as alpha particles are, and they can penetrate light materials such as paper or clothing. They can penetrate fairly deeply into skin, where they have the potential for harming or killing living cells. But they are not able to penetrate deeply into denser materials such as aluminum. Beta particles, once stopped, simply become part of the material they are in, like any other electron.

Gamma rays are the high-frequency electromagnetic radiation emitted by radioactive elements. Like visible light, a gamma ray is pure energy. The amount of energy in a gamma ray, however, is much greater than in visible light, ultraviolet light, or even X-rays. Because they have no mass or electric charge and because of their high energies, gamma rays can penetrate most materials. However, they cannot penetrate unusually dense materials such as lead, which absorbs them. Delicate molecules inside cells throughout our bodies that are zapped by gamma rays suffer structural damage. Hence, gamma rays are generally more harmful to us than alpha or beta particles (unless the alphas or betas are ingested).

CHECKPOINT

Pretend you are given three radioactive rocks—one an alpha emitter, one a beta emitter, and one a gamma emitter. You can throw away one, but of the remaining two, you must hold one in your hand and place the other in your pocket. What can you do to minimize your exposure to radiation?

Was this your answer?
Hold the alpha emitter in your hand because the skin on your hand shields you. Put the beta emitter in your pocket because beta particles are likely stopped by the combined thickness of your clothing and skin. Throw away the gamma emitter because gamma rays penetrate your body from any of these locations. Ideally, of course, you should distance yourself as much as possible from all of the rocks.

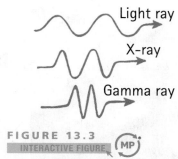

FIGURE 13.3
INTERACTIVE FIGURE MP

A gamma ray is simply electromagnetic radiation, much higher in frequency and energy than light and X-rays.

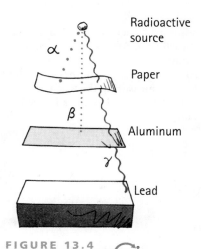

FIGURE 13.4
INTERACTIVE FIGURE MP

Alpha particles are the least penetrating and can be stopped by a few sheets of paper. Beta particles readily pass through paper, but not through a sheet of aluminum. Gamma rays penetrate several centimeters into solid lead.

Mastering**PHYSICS**

TUTORIAL: Nuclear Physics
VIDEO: Radioactive Decay

FIGURE 13.5
The shelf life of fresh strawberries and other perishables is markedly increased when the food is subjected to gamma rays from a radioactive source. The strawberries on the right were treated with gamma radiation, which kills the microorganisms that normally lead to spoilage. The food is only a receiver of radiation and is not transformed into an emitter of radiation, as can be confirmed with a radiation detector.

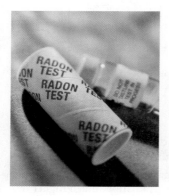

FIGURE 13.6
A commercially available radon test kit for the home. The canister is unsealed in the area to be sampled. Radon seeping into the canister is adsorbed by activated carbon within the canister. After several days, the canister is resealed and sent to a laboratory that determines the radon level by measuring the amount of radiation emitted by the adsorbed radon.

Common rocks and minerals in our environment contain significant quantities of radioactive isotopes because most of them contain trace amounts of uranium. People who live in brick, concrete, or stone buildings are exposed to greater amounts of radiation than people who live in wooden buildings.

The leading source of naturally occurring radiation is radon-222, an inert gas arising from uranium deposits. Radon is a heavy gas that tends to accumulate in basements after it seeps up through cracks in the floor. Levels of radon vary from region to region, depending on local geology. You can check the radon level in your home with a radon detector kit (Figure 13.6). If levels are abnormally high, corrective measures such as sealing the basement floor and walls and maintaining adequate ventilation are recommended.

About one-fifth of our annual exposure to radiation comes from nonnatural sources, primarily medical procedures. Television sets, fallout from nuclear testing, and the coal and nuclear power industries are also contributors. The coal industry far outranks the nuclear power industry as a source of radiation. The global combustion of coal annually releases about 13,000 tons of radioactive thorium and uranium into the atmosphere. Both these minerals are found naturally in coal deposits, so their release is a natural consequence of burning coal. Worldwide, the nuclear power industries generate about 10,000 tons of radioactive waste each year. Most of this waste, however, is contained and *not* released into the environment.

Radiation Dosage

Radiation dosage is commonly measured in *rads* (radiation *a*bsorbed *d*ose), a unit of absorbed energy. One **rad** is equal to 0.01 J of radiant energy absorbed per kilogram of tissue.

The capacity for nuclear radiation to cause damage is not just a function of its level of energy, however. Some forms of radiation are more harmful than others. For example, suppose you have two arrows, one with a pointed tip and one with a suction cup at its tip. Shoot the two of them at an apple at the same speed and both have the same kinetic energy. The one with the pointed tip, however, invariably does more damage to the apple than the one with the suction cup. Similarly, some forms of radiation cause greater harm than other forms, even when we receive the same number of rads from both forms.

The unit of measure for radiation dosage based on potential damage is the **rem** (*r*oentgen *e*quivalent *m*an).* In calculating the dosage in rems, we multiply the number of rads by a factor that corresponds to different health effects of different types of radiation determined by clinical studies. For example, 1 rad of alpha particles has the same biological effect as 10 rads of beta particles.** We call both of these dosages 10 rems:

Particle	Radiation Dosage		Factor		Health Effect
alpha	1 rad	×	10	=	10 rems
beta	10 rad	×	1	=	10 rems

* This unit is named for the discoverer of X-rays, Wilhelm Roentgen.
** This is true even though beta particles have more penetrating power, as discussed earlier.

CHECKPOINT

Would you rather be exposed to 1 rad of alpha particles or 1 rad of beta particles?

Was this your answer?
Multiply these quantities of radiation by the appropriate factor to get the dosages in rems. Alpha: 1 rad × 10 = 10 rems; beta: 1 rad × 1 = 1 rem. The factors show us that, physiologically speaking, alpha particles are 10 times as damaging as beta particles.

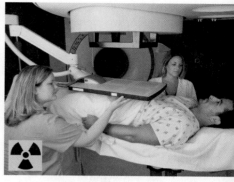

FIGURE 13.7
Nuclear radiation is focused on harmful tissue, such as a cancerous tumor, to selectively kill or shrink the tissue in a technique known as *radiation therapy*. This application of nuclear radiation has saved millions of lives—a clear-cut example of the benefits of nuclear technology. The inset shows the internationally used symbol indicating an area where radioactive material is being handled or produced.

Lethal doses of radiation begin at 500 rems. A person has about a 50% chance of surviving a dose of this magnitude received over a short period of time. During radiation therapy, a patient may receive localized doses in excess of 200 rems each day for a period of weeks (Figure 13.7).

All the radiation we receive from natural sources and from medical procedures is only a fraction of 1 rem. For convenience, the smaller unit *millirem* is used, where 1 millirem (mrem) is 1/1000 of a rem.

The average person in the United States is exposed to about 360 mrem a year, as Table 13.1 indicates. About 80% of this radiation comes from natural sources, such as cosmic rays and Earth itself. A typical chest X-ray exposes a person to 5–30 mrem (0.005–0.030 rem), less than 1/10,000 of the lethal dose. Interestingly, the human body is a significant source of natural radiation, primarily from the potassium we ingest. Our bodies contain about 200 g of potassium. Of this quantity, about 20 mg is the radioactive isotope potassium-40, which is a beta emitter. Radiation is indeed everywhere.

When radiation encounters the intricately structured molecules in the watery, ion-rich brine that makes up our cells, the radiation can create chaos on the atomic scale. Some molecules are broken, and this change alters other molecules, which can be harmful to life processes.

Cells can repair most kinds of molecular damage caused by radiation if the radiation is not too severe. A cell can survive an otherwise lethal dose of radiation if the dose is spread over a long period of time to allow intervals for healing. When radiation is sufficient to kill cells, the dead cells can be replaced by new ones. Sometimes a radiated cell survives with a damaged DNA molecule. New cells arising from the damaged cell retain the altered genetic information, producing a *mutation*. Usually the effects of a mutation are insignificant,

FIGURE 13.8
The film badges worn by Tammy and Larry contain audible alerts for both radiation surge and accumulated exposure. Information from the individualized badges is periodically downloaded to a database for analysis and storage.

TABLE 13.1	ANNUAL RADIATION EXPOSURE
Source	Typical Dose (mrem) Received Annually
Natural Origin	
Cosmic radiation	26
Ground	33
Air (radon-222)	198
Human tissues (K-40; Ra-226)	35
Human Origin	
Medical procedures	
Diagnostic X-rays	40
Nuclear medicine	15
TV tubes, other consumer products	11
Weapons-test fallout	1

FIGURE 13.9
Tracking fertilizer uptake with a radioactive isotope.

FIGURE 13.9
Tracking fertilizer uptake with a radioactive isotope.

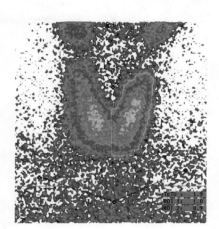

FIGURE 13.10
The thyroid gland, located in the neck, absorbs much of the iodine that enters the body through food and drink. Images of the thyroid gland, such as the one shown here, can be obtained by giving a patient the radioactive isotope iodine-131. These images are useful in diagnosing metabolic disorders.

but occasionally the mutation results in cells that do not function as well as unaffected ones, sometimes leading to a cancer. If the damaged DNA is in an individual's reproductive cells, the genetic code of the individual's offspring may retain the mutation.

Radioactive Tracers

In scientific laboratories radioactive samples of all the elements have been made. This is accomplished by bombardment with neutrons or other particles. Radioactive materials are extremely useful in scientific research and industry. To check the action of a fertilizer, for example, researchers combine a small amount of radioactive material with the fertilizer and then apply the combination to a few plants. The amount of radioactive fertilizer taken up by the plants can be easily measured with radiation detectors. From such measurements, scientists can inform farmers of the proper amount of fertilizer to use. Radioactive isotopes used to trace such pathways are called *tracers*.

In a technique known as medical imaging, tracers are used to diagnose internal disorders. This technique works because the path the tracer takes is influenced only by its physical and chemical properties, not by its radioactivity. The tracer may be introduced alone or along with some other chemical that helps target the tracer to a particular type of tissue in the body.

LEARNING OBJECTIVE
Describe how the strong nuclear force acts to hold nucleons together in the atomic nucleus.

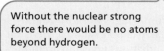

Without the nuclear strong force there would be no atoms beyond hydrogen.

13.2 The Strong Nuclear Force

EXPLAIN THIS Why are larger nuclei less stable than smaller nuclei?

As described in Chapter 12, the atomic nucleus occupies only a tiny fraction of the volume of the atom, leaving most of the atom as empty space. The nucleus is composed of *nucleons*, which, as discussed in Chapter 12, is the collective name for protons and neutrons.

We know that electric charges of like sign repel one another. So how do positively charged protons in the nucleus stay clumped together? This question led to the discovery of an attraction called the **strong nuclear force**, which acts between all nucleons. This force is very strong but over only extremely short distances (about 10^{-15} m, the diameter of a typical atomic nucleus). Repulsive electric interactions, on the other hand, are relatively long-ranged. Figure 13.11 suggests a comparison of the strengths of these two forces over distance. For protons that are close together, as in small nuclei, the attractive strong nuclear force easily overcomes the repulsive electric force. But for protons that are far apart, such as those on opposite edges of a large nucleus, the attractive strong nuclear force may be weaker than the repulsive electric force.

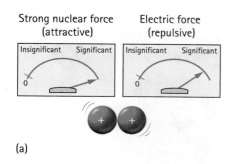

(a)

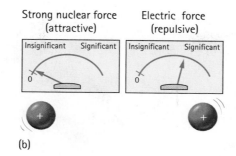

(b)

FIGURE 13.11
INTERACTIVE FIGURE

(a) Two protons near each other experience both an attractive strong nuclear force and a repulsive electric force. At this tiny separation distance, the strong nuclear force overcomes the electric force, and the protons stay together. (b) When the two protons are relatively far from each other, the electric force is more significant and the protons repel each other. This proton–proton repulsion in large atomic nuclei reduces nuclear stability.

A large nucleus is not as stable as a small one. In a helium nucleus, which has two protons, each proton feels the repulsive effect of only one other proton. In a uranium nucleus, however, each of the 92 protons feels the repulsive effects of the other 91 protons! The nucleus is unstable. We see that there is a limit to the size of the atomic nucleus. For this reason, all nuclei with more than 83 protons are radioactive.

CHECKPOINT

Two protons in the atomic nucleus repel each other, but they are also attracted to each other. Why?

Was this your answer?
Although two protons repel each other by the electric force, they also attract each other by the strong nuclear force. Both of these forces act simultaneously. As long as the attractive strong nuclear force is stronger than the repulsive electric force, the protons remain together. When the electric force overcomes the strong nuclear force, however, the protons fly apart from each other.

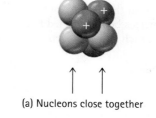

(a) Nucleons close together

(b) Nucleons far apart

FIGURE 13.12
(a) All nucleons in a small atomic nucleus are close to one another; hence, they experience an attractive strong nuclear force. (b) Nucleons on opposite sides of a larger nucleus are not as close to one another, and so the attractive strong nuclear forces holding them together are much weaker. The result is that the large nucleus is less stable.

Neutrons serve as a "nuclear cement" holding the atomic nucleus together. Protons attract both protons and neutrons by the strong nuclear force. Protons also repel other protons by the electric force. Neutrons, on the other hand, have no electric charge and so only attract other protons and neutrons by the strong nuclear force. The presence of neutrons therefore adds to the attraction among nucleons and helps hold the nucleus together (Figure 13.13).

The more protons there are in a nucleus, the more neutrons are needed to help balance the repulsive electric forces. For light elements, it is sufficient to have about as many neutrons as protons. The most common isotope of carbon, C-12, for instance, has equal numbers of each—six protons and six neutrons. For large nuclei, more neutrons than protons are needed. Because the strong

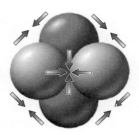

All nucleons, both protons and neutrons, attract one another by the strong nuclear force.

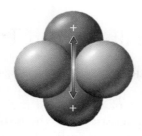

Only protons repel one another by the electric force.

FIGURE 13.13
The presence of neutrons helps hold the nucleus together by increasing the effect of the strong nuclear force, represented by the single-headed arrows.

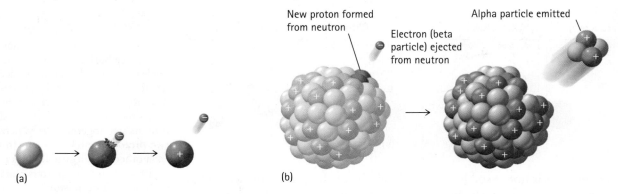

FIGURE 13.14
(a) A neutron near a proton is stable, but a neutron by itself is unstable and decays to a proton by emitting an electron. (b) Destabilized by an increase in the number of protons, the nucleus begins to shed fragments, such as alpha particles.

nuclear force diminishes rapidly over distance, nucleons must be practically touching in order for the strong nuclear force to be effective. Nucleons on opposite sides of a large atomic nucleus are not as attracted to one another. The electric force, however, does not diminish by much across the diameter of a large nucleus and so begins to win out over the strong nuclear force. To compensate for the weakening of the strong nuclear force across the diameter of the nucleus, large nuclei have more neutrons than protons. Lead, for example, has about one and a half times as many neutrons as protons.

So we see that neutrons are stabilizing and large nuclei require an abundance of them. But neutrons are not always successful in keeping a nucleus intact. Interestingly, neutrons are not stable when they are by themselves. A lone neutron is radioactive, and spontaneously transforms to a proton and an electron (Figure 13.14a). A neutron seems to need protons around to keep this from happening. After the size of a nucleus reaches a certain point, the neutrons so outnumber the protons that there are not enough protons in the mix to prevent the neutrons from turning into protons. As neutrons in a nucleus change into protons, the stability of the nucleus decreases because the repulsive electric force becomes increasingly significant. The result is that pieces of the nucleus fragment away in the form of radiation, as indicated in Figure 13.14b.

> **CHECKPOINT**
> **What role do neutrons serve in the atomic nucleus? What is the fate of a neutron when alone or distant from one or more protons?**
>
> **Was this your answer?**
> Neutrons serve as a nuclear cement in nuclei and add to nuclear stability. But when alone or away from protons, a neutron becomes radioactive and spontaneously transforms to a proton and an electron.

LEARNING OBJECTIVE
Recognize how radioactive elements can be identified by the rate at which they decay and how this decay results in the formation of new elements.

Mastering**PHYSICS**
VIDEO: *Half-Life*

13.3 Half-Life and Transmutation

EXPLAIN THIS How is the rate of transmutation related to half-life?

The rate of decay for a radioactive isotope is measured in terms of a characteristic time, the **half-life**. This is the time it takes for half of an original quantity of an element to decay. For example, radium-226 has a half-life of 1620 years, which means that half of a radium-226 sample will be converted to other elements by the end of 1620 years. In the next

1620 years, half of the remaining radium will decay, leaving only one-fourth the original amount of radium. (After 20 half-lives, the initial quantity of radium-226 will be diminished by a factor of about 1 million.)

Half-lives are remarkably constant and not affected by external conditions. Some radioactive isotopes have half-lives that are less than a millionth of a second, while others have half-lives of more than a billion years. Uranium-238 has a half-life of 4.5 billion years. All uranium eventually decays in a series of steps to lead. In 4.5 billion years, half the uranium presently in Earth today will be lead.

It is not necessary to wait through the duration of a half-life in order to measure it. The half-life of an element can be calculated at any given moment by measuring the rate of decay of a known quantity. This is easily done using a radiation detector (Figure 13.16). In general, the shorter the half-life of a substance, the faster it disintegrates, and the more radioactivity per amount is detected.

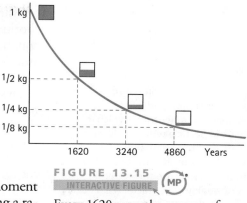

FIGURE 13.15
INTERACTIVE FIGURE

Every 1620 years the amount of radium decreases by half.

CHECKPOINT

1. If a radioactive isotope has a half-life of 1 day, how much of an original sample is left at the end of the second day? The third day?
2. Which gives a higher counting rate on a radiation detector: a radioactive material with a short half-life or a radioactive material with a long half-life?

The radioactive half-life of a material is also the time for its decay rate to reduce to half.

Were these your answers?

1. One-fourth of the original sample is left at the end of the second day— the three-fourths that underwent decay is then a different element altogether. At the end of 3 days, one-eighth of the original sample remains.
2. The material with the shorter half-life is more active and shows a higher counting rate on a radiation detector.

(a)

When a radioactive nucleus emits an alpha or a beta particle, there is a change in atomic number, which means that a different element is formed. (Recall from Chapter 12 that an element is defined by its atomic number, which is the number of protons in the nucleus.) The changing of one chemical element to another is called **transmutation**. Transmutation occurs in natural events and is also initiated artificially in the laboratory.

Natural Transmutation

Consider uranium-238, the nucleus of which contains 92 protons and 146 neutrons. When an alpha particle is ejected, the nucleus loses two protons and two neutrons. Because an element is defined by the number of protons in its nucleus, the 90 protons and 144 neutrons left behind are no longer identified as being uranium. Instead we have the nucleus of a different element—thorium. This transmutation can be written as a nuclear equation:

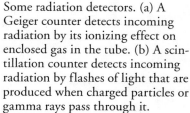

(b)

FIGURE 13.16
Some radiation detectors. (a) A Geiger counter detects incoming radiation by its ionizing effect on enclosed gas in the tube. (b) A scintillation counter detects incoming radiation by flashes of light that are produced when charged particles or gamma rays pass through it.

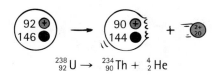

$$^{238}_{92}\text{U} \rightarrow \,^{234}_{90}\text{Th} + \,^{4}_{2}\text{He}$$

We see that $^{238}_{92}\text{U}$ transmutes to the two elements written to the right of the arrow. When this transmutation occurs, energy is released, partly in the form of kinetic energy of the alpha particle ($_2^4$He), partly in the kinetic energy of the thorium atom, and partly in the form of gamma radiation. In this and all such equations, the mass numbers at the top balance ($238 = 234 + 4$) and the atomic numbers at the bottom also balance ($92 = 90 + 2$).

Thorium-234, the product of this reaction, is also radioactive. When it decays, it emits a beta particle. Because a beta particle is an electron, the atomic number of the resulting nucleus is *increased* by 1. So after beta emission by thorium with 90 protons, the resulting element has 91 protons. It is no longer thorium, but the element protactinium. Although the atomic number has increased by 1 in this process, the mass number (protons + neutrons) remains the same. The nuclear equation is

$$_{90}^{234}\text{Th} \rightarrow {}_{91}^{234}\text{Pa} + {}_{-1}^{0}e$$

We write an electron as $_{-1}^{0}e$. The superscript 0 indicates that the electron's mass is insignificant relative to that of protons and neutrons. The subscript -1 is the electric charge of the electron.

So we see that when an element ejects an alpha particle from its nucleus, the mass number of the resulting atom is decreased by 4, and its atomic number is decreased by 2. The resulting atom is an element two spaces back in the periodic table of the elements. When an element ejects a beta particle from its nucleus, the mass of the atom is practically unaffected, meaning there is no change in mass number, but its atomic number increases by 1. The resulting atom belongs to an element one place forward in the periodic table. Gamma radiation results in no change in either the mass number or the atomic number. So we see that radioactive elements can decay backward or forward in the periodic table.

The successions of radioactive decays of $_{92}^{238}$U to $_{82}^{206}$Pb, an isotope of lead, is shown in Figure 13.17. Each gray arrow shows an alpha decay, and each red arrow shows a beta decay. Notice that some of the nuclei in the series can decay in both ways. This is one of several similar radioactive series that occur in nature.

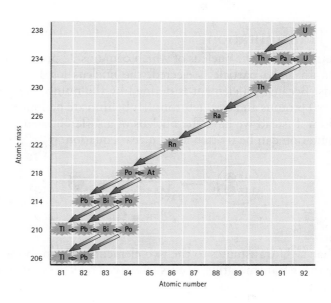

FIGURE 13.17
U-238 decays to Pb-206 through a series of alpha and beta decays.

CHECKPOINT

1. Complete the following nuclear reactions:
   a. $^{226}_{88}\text{Ra} \rightarrow {}^{?}_{?}? + {}^{0}_{-1}e$
   b. $^{209}_{84}\text{Po} \rightarrow {}^{205}_{82}\text{Pb} + {}^{?}_{?}?$
2. What finally becomes of all the uranium that undergoes radioactive decay?

Were these your answers?

1. a. $^{226}_{88}\text{Ra} \rightarrow {}^{226}_{89}\text{Ac} + {}^{0}_{-1}e$
   b. $^{209}_{84}\text{Po} \rightarrow {}^{205}_{82}\text{Pb} + {}^{4}_{2}\text{He}$
2. All uranium ultimately becomes lead. On the way to becoming lead, it exists as a series of elements, as indicated in Figure 13.17.

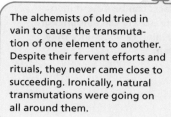

The alchemists of old tried in vain to cause the transmutation of one element to another. Despite their fervent efforts and rituals, they never came close to succeeding. Ironically, natural transmutations were going on all around them.

Artificial Transmutation

Ernest Rutherford, in 1919, was the first of many investigators to succeed in transmuting a chemical element. He bombarded nitrogen gas with alpha particles from a piece of radioactive ore. The impact of an alpha particle on a nitrogen nucleus transmutes nitrogen into oxygen:

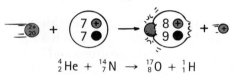

$$^{4}_{2}\text{He} + {}^{14}_{7}\text{N} \rightarrow {}^{17}_{8}\text{O} + {}^{1}_{1}\text{H}$$

Rutherford used a device called a *cloud chamber* to record this event (Figure 13.18). In a cloud chamber, moving charged particles show a trail of ions along their path in a way similar to the ice crystals that show the trail of jet planes high in the sky. From a quarter of a million cloud-chamber tracks photographed on movie film, Rutherford showed seven examples of atomic transmutation. Analysis of tracks bent by a strong external magnetic field showed that when an alpha particle collided with a nitrogen atom, a proton bounced out and the heavy atom recoiled a short distance. The alpha particle disappeared. The alpha particle was absorbed in the process, transforming nitrogen to oxygen.

Since Rutherford's announcement in 1919, experimenters have carried out many other nuclear reactions, first with natural bombarding projectiles from radioactive ores and then with still more energetic projectiles—protons and electrons hurled by huge particle accelerators. Artificial transmutation produces the hitherto unknown synthetic elements at the upper end of the periodic table. All of these artificially made elements have short half-lives. If they ever existed naturally when Earth was formed, they have long since decayed.

FIGURE 13.18
A cloud chamber. Charged particles moving through supersaturated vapor leave trails. When the chamber is in a strong electric or magnetic field, bending of the tracks provides information about the charge, mass, and momentum of the particles.

FIGURE 13.19
Tracks of elementary particles in a bubble chamber, a similar yet more complicated device than a cloud chamber. Two particles have been destroyed at the points where the spirals emanate, and four others created in the collision.

LEARNING OBJECTIVE
Review how the age of ancient artifacts can be determined by measuring the amounts of remaining radioactivity they contain.

13.4 Radiometric Dating

EXPLAIN THIS How does radioactivity allow archeologists to measure the age of ancient artifacts?

Earth's atmosphere is continuously bombarded by cosmic rays, and this bombardment causes many atoms in the upper atmosphere to transmute. These transmutations result in many protons and neutrons being "sprayed out" into the environment. Most of the protons are stopped as they collide with the atoms of the upper atmosphere, stripping electrons from these atoms to become hydrogen atoms. The neutrons, however, keep going for longer distances because they have no electric charge and therefore do not interact electrically with matter. Eventually, many of them collide with the nuclei in the denser lower atmosphere. A nitrogen nucleus that captures a neutron, for instance, becomes an isotope of carbon by emitting a proton:

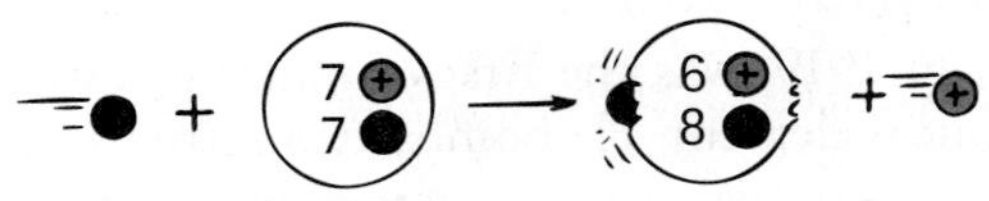

$$_0^1 n + {}_7^{14}N \rightarrow {}_6^{14}C + {}_1^1H$$

This carbon-14 isotope, which makes up less than one-millionth of 1% of the carbon in the atmosphere, is radioactive and has eight neutrons. (The most common isotope, carbon-12, has six neutrons and is not radioactive.) Because both carbon-12 and carbon-14 are forms of carbon, they have the same chemical properties. Both these isotopes can chemically react with oxygen to form carbon dioxide, which is taken in by plants. This means that all plants contain a tiny bit of radioactive carbon-14. All animals eat plants (or at least plant-eating animals) and therefore have a little carbon-14 in them. In short, all living things on Earth contain some carbon-14.

Carbon-14 is a beta emitter and decays back to nitrogen by the following reaction:

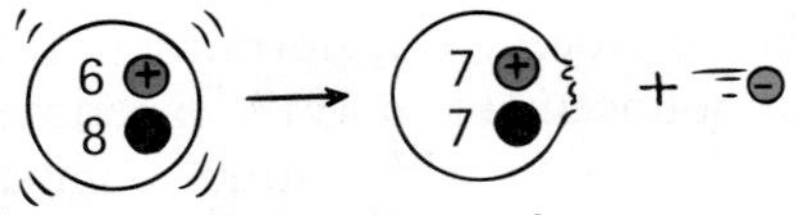

$$_6^{14}C \rightarrow {}_7^{14}N + {}_{-1}^0 e$$

FYI A 1-g sample of carbon from recently living matter contains about 50 trillion billion (5×10^{22}) carbon atoms. Of these carbon atoms, about 65 billion (6.5×10^{10}) are the radioactive C-14 isotope. This gives the carbon a beta disintegration rate of about 13.5 decays per minute.

Because plants continue to take in carbon dioxide as long as they live, any carbon-14 lost by decay is immediately replenished with fresh carbon-14 from the atmosphere. In this way, a radioactive equilibrium is reached at which there is a constant ratio of about one carbon-14 atom to every 800 billion carbon-12 atoms. When a plant dies, replenishment of carbon-14 stops. Then the percentage of carbon-14 decreases at a constant rate given by its half-life. The longer a plant or other organism is dead, therefore, the less carbon-14 it contains relative to the constant amount of carbon-12.

The half-life of carbon-14 is about 5730 years. This means that half of the carbon-14 atoms that are now present in a plant or animal that dies today will decay in the next 5730 years. Half of the remaining carbon-14 atoms will then decay in the following 5730 years, and so forth.

With this knowledge, scientists can calculate the age of carbon-containing artifacts, such as wooden tools or skeletons, by measuring their current level of radioactivity. This process, known as **carbon-14 dating**, enables us to probe as much as 50,000 years into the past. Beyond this time span, too little carbon-14 remains to permit accurate analysis.

22,920 years ago

17,190 years ago

11,460 years ago

5730 years ago

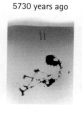

Present

Carbon-14 dating would be an extremely simple and accurate dating method if the amount of radioactive carbon in the atmosphere had been constant over the ages. But it hasn't been. Fluctuations in the Sun's magnetic field as well as changes in the strength of Earth's magnetic field affect cosmic-ray intensities in Earth's atmosphere, which in turn produce fluctuations in the production of C-14. In addition, changes in Earth's climate affect the amount of carbon dioxide in the atmosphere. The oceans are great reservoirs of carbon dioxide. When the oceans are warm, they release more carbon dioxide into the atmosphere than when they are cold. We'll return to the oceans and their important interplay with carbon dioxide in Chapters 18 and 24.

FIGURE 13.20

The amount of radioactive carbon-14 in the skeleton diminishes by half every 5730 years, with the result that today the skeleton contains only a fraction of the carbon-14 it originally had. The red arrows symbolize relative amounts of carbon-14.

CHECKPOINT

Suppose an archaeologist extracts a gram of carbon from an ancient ax handle and finds it one-fourth as radioactive as a gram of carbon extracted from a freshly cut tree branch. About how old is the ax handle?

Was this your answer?
Assuming the ratio of C-14 to C-12 was the same when the ax was made, the ax handle is as old as two half-lives of C-14, or about 11,460 years old.

> One ton of ordinary granite contains about 9 g of uranium and 20 g of thorium. Basalt rocks contain 3.5 g and 7.7 g of the same elements, respectively.

The dating of older, but nonliving, materials is accomplished with radioactive minerals, such as uranium. The naturally occurring isotopes U-238 and U-235 decay very slowly and ultimately become isotopes of lead—but not the common lead isotope Pb-208. For example, U-238 decays through several stages to finally become Pb-206, whereas U-235 finally becomes the isotope Pb-207. Lead isotopes 206 and 207 that now exist were at one time uranium. The older the rock, the higher the percentage of these remnant isotopes.

From the half-lives of uranium isotopes and the percentage of lead isotopes in uranium-bearing rock, it is possible to calculate the date at which the rock was formed. We'll return to isotopic dating when we investigate Earth's dynamic interior in Chapter 21.

13.5 Nuclear Fission

EXPLAIN THIS Why isn't it possible for a nuclear power plant to explode like a nuclear bomb?

In 1938, two German scientists, Otto Hahn and Fritz Strassmann, made an accidental discovery that was to change the world. While bombarding a sample of uranium with neutrons in the hope of creating new, heavier elements, they were astonished to find chemical evidence for the production of

LEARNING OBJECTIVE
Describe the process by which large atomic nuclei can split in half, leading to the production of energy.

Mastering**PHYSICS**
VIDEO: Plutonium

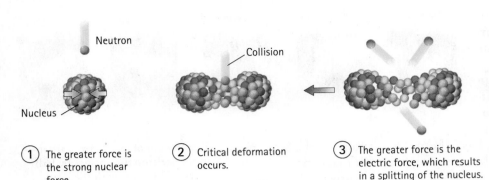

Neutron

Collision

Nucleus

(1) The greater force is the strong nuclear force.

(2) Critical deformation occurs.

(3) The greater force is the electric force, which results in a splitting of the nucleus.

FIGURE 13.21
INTERACTIVE FIGURE

Nuclear deformation may result in repulsive electric forces overcoming attractive nuclear forces, in which case fission occurs.

FYI Otto Hahn, rather than Lise Meitner, received the Nobel Prize for the work on nuclear fission. Notoriously, Hahn didn't even acknowledge Meitner's role. See more about this in the readable book $E = mc^2$, by David Bodanis.

barium, an element with about half the mass of uranium. Hahn wrote of this news to his former colleague Lise Meitner, who had fled from Nazi Germany to Sweden because of her Jewish ancestry. From Hahn's evidence, Meitner concluded that the uranium nucleus, activated by neutron bombardment, had split in half. Soon thereafter, Meitner, working with her nephew, Otto Frisch, also a physicist, published a paper in which the term *nuclear fission* was first coined.

In the nucleus of every atom is a delicate balance between attractive nuclear forces and repulsive electric forces between protons. In all known nuclei, the nuclear forces dominate. In certain isotopes of uranium, however, this domination is tenuous. If a uranium nucleus stretches into an elongated shape (Figure 13.21), the electric forces may push it into an even more elongated shape. If the elongation passes a certain point, the electric forces overwhelm the strong nuclear forces, and the nucleus splits. This is **nuclear fission**.

The energy released by the fission of one U-235 nucleus is relatively enormous—about 7 million times the energy released by the explosion of one TNT molecule. This energy is mainly in the form of kinetic energy of the fission fragments that fly apart from one another, with some energy given to ejected neutrons and the rest to gamma radiation.

A typical uranium fission reaction is

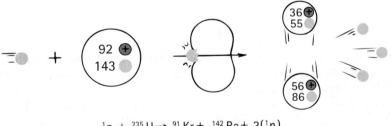

$$_0^1 n + \, _{92}^{235}U \rightarrow \, _{36}^{91}Kr + \, _{56}^{142}Ba + 3(_0^1 n)$$

Note in this reaction that 1 neutron starts the fission of a uranium nucleus and that the fission produces 3 neutrons. (A fission reaction may produce fewer or more than 3 neutrons.) These product neutrons can cause the fissioning of 3 other uranium atoms, releasing 9 more neutrons. If each of these 9 neutrons succeeds in splitting a uranium atom, the next step in the reaction produces 27 neutrons, and so on. Such a sequence, illustrated in Figure 13.22, is called a **chain reaction**—a self-sustaining reaction in which the products of one reaction event stimulate further reaction events.

Why don't chain reactions occur in naturally occurring uranium ore deposits? They would if all uranium atoms fissioned so easily. Fission occurs mainly for the rare isotope U-235, which makes up only 0.7% of the uranium in naturally occurring uranium metal. When the more abundant isotope U-238 absorbs neutrons created by fission of U-235, the U-238 typically does not undergo fission. So any chain reaction is snuffed out by the neutron-absorbing U-238, as well as by the rock in which the ore is imbedded.

If a chain reaction occurred in a baseball-size chunk of pure U-235, an enormous explosion would result. If the chain reaction were started in a smaller

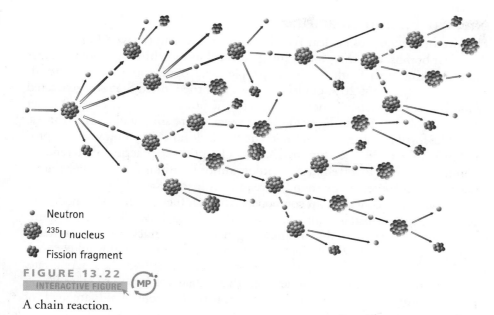

- Neutron
- ^{235}U nucleus
- Fission fragment

FIGURE 13.22

INTERACTIVE FIGURE MP

A chain reaction.

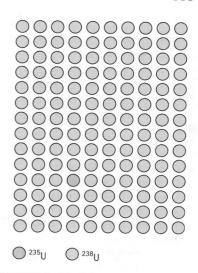

^{235}U ^{238}U

FIGURE 13.23
Only 1 part in 140 of naturally occurring uranium is U-235.

chunk of pure U-235, however, no explosion would occur. This is because of geometry: the ratio of surface area to mass is larger in a small piece than in a large one (just as there is more skin on six small potatoes with a combined mass of 1 kg than there is on a single 1-kg potato). So there is more surface area on a bunch of small pieces of uranium than on a large piece. In a small piece of U-235, neutrons leak through the surface before an explosion can occur. In a bigger piece, the chain reaction builds up to enormous energies before the neutrons get to the surface and escape (Figure 13.24). For masses greater than a certain amount, called the **critical mass**, an explosion of enormous magnitude may take place.

Consider a large quantity of U-235 divided into two pieces, each with a mass less than critical. The units are *subcritical*. Neutrons in either piece readily reach a surface and escape before a sizable chain reaction builds up. But if the pieces are suddenly driven together, the total surface area decreases. If the timing is right and the combined mass is greater than critical, a violent explosion takes place. This is what happens in a nuclear fission bomb (Figure 13.25).

Constructing a fission bomb is a formidable task. The difficulty is separating enough U-235 from the more abundant U-238. Scientists took more than two years to extract enough U-235 from uranium ore to make the bomb that was detonated at Hiroshima in 1945. To this day uranium isotope separation remains a difficult process.

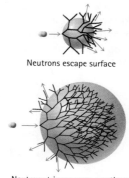

Neutrons escape surface

Neutrons trigger more reactions

FIGURE 13.24
The exaggerated view shows that a chain reaction in a small piece of pure U-235 runs its course before it can cause a large explosion because neutrons leak from the surface too soon. The surface area of the small piece is large relative to the mass. In a larger piece, more uranium and less surface are presented to the neutrons.

CHECKPOINT

A 1-kg ball of U-235 is at critical mass, but the same ball broken up into small chunks is not. Explain.

Was this your answer?
The small chunks have more combined surface area than the ball from which they came (just as the combined surface area of gravel is greater than the surface area of a boulder of the same mass). Neutrons escape via the surface before a sustained chain reaction can build up.

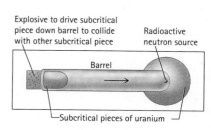

Explosive to drive subcritical piece down barrel to collide with other subcritical piece

Radioactive neutron source

Barrel

Subcritical pieces of uranium

FIGURE 13.25
Simplified diagram of a uranium fission bomb.

FYI With the rise of the German Nazis in the 1930s, many scientists, especially those of Jewish ancestry, fled mainland Europe to America. They included dozens of brilliant theoretical physicists who eventually played key roles in the development of nuclear fission. Of these physicists, Leo Szilard (1898–1964) first envisioned the idea of a chain nuclear reaction. With Albert Einstein's consent, Szilard drafted a letter that was signed by Einstein and delivered to President Roosevelt in 1939. This letter outlined the possibility of the chain reaction and its implications for a nuclear bomb. Within six years the first test nuclear bomb was exploded in the desert in New Mexico. In 1945, Szilard generated a petition in which 68 of the scientists involved in the nuclear program asked President Truman not to drop the atomic bomb on a populous Japanese city, such as Nagasaki.

Nuclear Fission Reactors

The awesome energy of nuclear fission was introduced to the world in the form of nuclear bombs, and this violent image still colors our thinking about nuclear power, making it difficult for many people to recognize its potential usefulness. Currently, about 20% of electric energy in the United States is generated by *nuclear fission reactors* (whereas most electric power is nuclear in some other countries—about 75% in France). These reactors are simply nuclear furnaces. They, like fossil fuel furnaces, do nothing more elegant than boil water to produce steam for a turbine (Figure 13.26). The greatest practical difference is the amount of fuel involved: a mere kilogram of uranium fuel, smaller than a baseball, yields more energy than 30 freight-car loads of coal.

A fission reactor contains four components: nuclear fuel, control rods, moderator (to slow neutrons, which is required for fission), and liquid (usually water) to transfer heat from the reactor to the turbine and generator. The nuclear fuel is primarily U-238 plus about 3% U-235. Because the U-235 isotopes are so highly diluted with U-238, an explosion like that of a nuclear bomb is not possible. The reaction rate, which depends on the number of neutrons that initiate the fission of other U-235 nuclei, is controlled by rods inserted into the reactor. The control rods are made of a neutron-absorbing material, usually the metal cadmium or boron.

Heated water around the nuclear fuel is kept under high pressure to keep it at a high temperature without boiling. It transfers heat to a second lower-pressure water system, which operates the turbine and electric generator in a conventional fashion. In this design, two separate water systems are used so that no radioactivity reaches the turbine or the outside environment.

A significant disadvantage of fission power is the generation of radioactive waste products. Light atomic nuclei are most stable when composed of equal numbers of protons and neutrons, as discussed earlier, and heavy nuclei need more neutrons than protons for stability. For example, U-235 has 143 neutrons but only 92 protons. When uranium fissions into two medium-weight elements, the extra neutrons in their nuclei make them unstable. They are radioactive, most with very short half-lives, but some with half-lives of thousands of years. Safely disposing of these waste products as well as materials made radioactive in the production of nuclear fuels requires special storage casks and procedures. Although fission has been successfully producing electricity for a half century, disposing of radioactive wastes in the United States remains problematic.

The designs for nuclear power plants have progressed over the years. The earliest designs from the 1950's through 1990's are called the Generation I, II, and III reactors. The safety systems of these reactors are "active" in that they rely on active

FIGURE 13.26
Diagram of a nuclear fission power plant. Note that the water in contact with the fuel rods is completely contained, and radioactive materials are not involved directly in the generation of electricity.

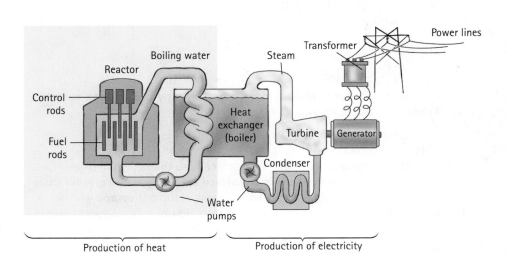

FIGURE 13.27
The nuclear reactor is housed within a dome-shaped containment building that is designed to prevent the release of radioactive isotopes in the event of an accident. The Soviet-built Chernobyl nuclear power plant that reached meltdown in 1986 had no such containment building, so massive amounts of radiation were released into the environment.

measures, such as water pumps, that act to keep the reactor core cool in the event of an accident. Notably, these active measures failed when Japan's Generation II Fukushima Daiichi nuclear plant was hit by a powerful earthquake and tsunami in 2011. While not yet operational, the latest Generation IV nuclear reactors will have fundamentally different designs. For example, they will incorporate passive safety measures that cause the reactor to shut down by itself in the event of an emergency. The fuel source may be the depleted uranium stockpiled from earlier reactors. Furthermore, these reactors can be built as small modular units that generate between 150 and 600 megawatts of power rather than the 1500 megawatts that is the usual output of today's reactors. Smaller reactors are easier to manage and can be used to build a generating capacity suited to the community being served.

The benefits of fission power include plentiful electricity and the conservation of many billions of tons of fossil fuels. Every year these fuels are turned to heat, smoke, and megatons of poisonous gases such as sulfur oxides. Notably, fossil fuels are far more precious as sources of organic molecules, which, as we will discuss in Chapter 19, can be used to create medicines, clothing, automobiles, and much more.

A nuclear power plant "meltdown" occurs when the fissioning nuclear fuels are no longer submerged within a cooling fluid, such as water. The temperature rises to the point that the solid nuclear fuel, and the reaction vessel itself, melt into a liquid phase that has the potential of penetrating through the floor of the containment building.

CHECKPOINT

Coal contains tiny quantities of radioactive materials, enough that more environmental radiation surrounds a typical coal-fired power plant than a fission power plant. What does this indicate about the shielding typically surrounding the two types of power plants?

Was this your answer?
Coal-fired power plants are as American as apple pie, with no required (and expensive) shielding to restrict the emissions of radioactive particles. Nukes, on the other hand, are required to have shielding to ensure strictly low levels of radioactive emissions.

FYI Recent evidence discovered by neutrino research in 2011 indicates that a major source of Earth's internal energy, perhaps half, is due to nuclear fission within Earth's core. This heat-generating process is occurring deep beneath your feet right now! Indeed, power from the atomic nuclei is as old as Earth itself.

The Breeder Reactor

One of the fascinating features of fission power is the breeding of fission fuel from nonfissionable U-238. This breeding occurs when small amounts of fissionable isotopes are mixed with U-238 in a reactor. Fission liberates neutrons that convert the relatively abundant nonfissionable U-238 to U-239, which

An average ton of coal contains 1.3 parts per million (ppm) of uranium and 3.2 ppm of thorium. That's why the average coal-burning power plant is a far greater source of airborne radioactive material than a nuclear power plant.

beta decays to Np-239, which in turn beta decays to fissionable plutonium—Pu-239. So in addition to the abundant energy produced, fission fuel is bred from the relatively abundant U-238 in the process.

Breeding occurs to some extent in all fission reactors, but a reactor specifically designed to breed more fissionable fuel than is put into it is called a *breeder reactor*. Using a breeder reactor is like filling your car's gas tank with water, adding some gasoline, then driving the car and having more gasoline after the trip than at the beginning! The basic principle of the breeder reactor is very attractive, for after a few years of operation a breeder-reactor power plant can produce vast amounts of power while breeding twice as much fuel as its original fuel.

The downside is the enormous complexity of successful and safe operation. The United States gave up on breeders about two decades ago, and only Russia, France, Japan, and India are still investing in them. Officials in these countries point out that the supplies of naturally occurring U-235 are limited. At present rates of consumption, all natural sources of U-235 may be depleted within a century. If countries then decide to turn to breeder reactors, they may well find themselves digging up the radioactive wastes they once buried.

13.6 Mass–Energy Equivalence

EXPLAIN THIS Why does it get easier to pull nucleons away from nuclei heavier than iron?

In the early 1900s, Albert Einstein discovered that mass is actually "congealed" energy. Mass and energy are two sides of the same coin, as stated in his celebrated equation $E = mc^2$. In this equation E stands for the energy that any mass has at rest, m stands for mass, and c is the speed of light. This relationship between energy and mass is the key to understanding why and how energy is released in nuclear reactions.

Is the mass of a nucleon inside a nucleus the same as that of the same nucleon outside a nucleus? This question can be answered by considering the work that would be required to separate nucleons from a nucleus. From physics we know that work, which is expended energy, equals *force × distance*. Think of the amount of force required to pull a nucleon out of the nucleus through a sufficient distance to overcome the attractive strong nuclear force, comically indicated in Figure 13.28. Enormous work would be required. This work is energy added to the nucleon that is pulled out.

According to Einstein's equation, this newly acquired energy reveals itself as an increase in the nucleon's mass. The mass of a nucleon outside a nucleus is greater than the mass of the same nucleon locked inside a nucleus. For example, a carbon-12 atom—the nucleus of which is made up of six protons and six neutrons—has a mass of exactly 12.00000 atomic mass units (amu). Therefore on average, each nucleon contributes a mass of 1 amu. However, outside the nucleus, a proton has a mass of 1.00728 amu and a neutron has a mass of 1.00867 amu. Thus we see that the combined mass of six free protons and six free neutrons—(6 × 1.00728) + (6 × 1.00867) = 12.09570—is greater than the mass of one carbon-12 nucleus. The greater mass reflects the energy required to pull the nucleons apart from one another. Thus, what mass a nucleon has depends on where the nucleon is.

A graph of the nuclear masses for the elements from hydrogen through uranium is shown in Figure 13.29. The graph slopes upward with increasing atomic number as expected: elements are more massive as atomic number increases. The slope curves because there are proportionally more neutrons in the more massive atoms.

LEARNING OBJECTIVE
Show how the mass of a nucleon depends on the identity of the nucleus within which it is contained.

FIGURE 13.28
Work is required to pull a nucleon from an atomic nucleus. This work increases the energy and hence the mass of the nucleon outside the nucleus.

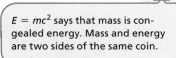

$E = mc^2$ says that mass is congealed energy. Mass and energy are two sides of the same coin.

A more important graph results from the plot of nuclear mass *per nucleon* from hydrogen through uranium (Figure 13.30). This is perhaps the most important graph in this book, for it is the key to understanding the energy associated with nuclear processes.

Note that the masses of the nucleons are different when combined in different nuclei. The greatest mass per nucleon occurs for the proton alone, hydrogen, because it has no binding energy to pull its mass down. Progressing beyond hydrogen, the mass per nucleon is smaller, and is least for one in the nucleus of the iron atom. Beyond iron, the process reverses itself as nucleons have progressively more and more mass in atoms of increasing atomic number. This continues all the way to uranium and elements heavier than uranium.

From Figure 13.30 we can see how energy is released when a uranium nucleus splits into two nuclei of lower atomic number. Uranium, being toward the right-hand side of the graph, is shown to have a relatively large amount of

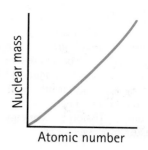

FIGURE 13.29
The plot shows how nuclear mass increases with increasing atomic number.

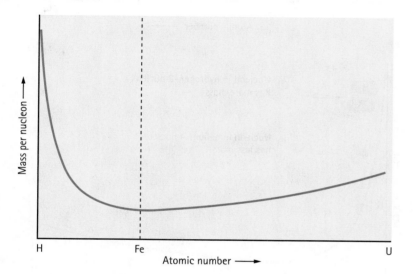

FIGURE 13.30
This graph shows that the average mass of a nucleon depends on which nucleus it is in. Individual nucleons have the greatest mass in the lightest nuclei, the least mass in iron, and intermediate mass in the heaviest nuclei.

mass per nucleon. When the uranium nucleus splits in half, however, smaller nuclei of lower atomic numbers are formed. As shown in Figure 13.31, these nuclei are lower on the graph than uranium, which means that they have a smaller amount of mass per nucleon. Thus, nucleons lose mass in their transition from being in a uranium nucleus to being in one of its fragments. When this decrease in mass is multiplied by the speed of light squared (c^2 in Einstein's equation), the product is equal to the energy yielded by each uranium nucleus as it undergoes fission.

The graph of Figure 13.30 (and Figures 13.31 and 13.32) reveals the energy of the atomic nucleus, a primary source of energy in the universe—which is why it can be considered the most important graph in this book.

CHECKPOINT
Correct the following incorrect statement: When a heavy element such as uranium undergoes fission, there are fewer nucleons after the reaction than before.

Was this your answer?
When a heavy element such as uranium undergoes fission, there aren't fewer nucleons after the reaction. Instead, there's *less mass* in the same number of nucleons.

FIGURE 13.31
The mass of each nucleon in a uranium nucleus is greater than the mass of each nucleon in any one of its nuclear fission fragments. This lost mass is mass that has been transformed into energy, which is why nuclear fission is an energy-releasing process.

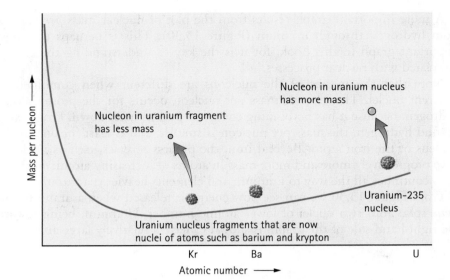

FIGURE 13.32
INTERACTIVE FIGURE

The mass of each nucleon in a hydrogen-2 nucleus is greater than the mass of each nucleon in a helium-4, which results from the fusion of two hydrogen-2 nuclei. This lost mass has been converted to energy, which is why nuclear fusion is an energy-releasing process.

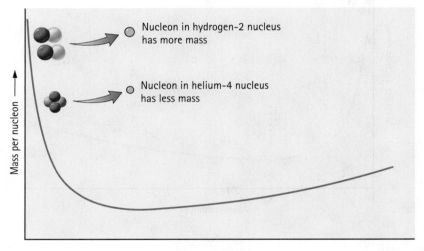

We can think of the mass-per-nucleon graph as an energy valley that starts at hydrogen (the highest point) and slopes steeply to the lowest point (iron), then slopes gradually up to uranium. Iron is at the bottom of the energy valley and is the most stable nucleus. It is also the most tightly bound nucleus; more energy per nucleon is required to separate nucleons from its nucleus than from any other nucleus.

All nuclear power today is by way of nuclear fission. A more promising long-range source of energy is found on the left side of the energy valley.

13.7 Nuclear Fusion

EXPLAIN THIS How does the energy of gasoline come from nuclear fusion?

Notice in the graph of Figure 13.30 that the steepest part of the energy valley goes from hydrogen to iron. Energy is released as light nuclei combine. This combining of nuclei is **nuclear fusion**—the opposite of nuclear fission. We see from Figure 13.32 that, as we move along the list of elements from hydrogen to iron, the average mass per nucleon decreases. Thus when two small nuclei fuse, say two hydrogen isotopes, the mass of the resulting helium-4 nucleus is less than the mass of the two small nuclei before fusion. Energy is released as smaller nuclei fuse.

For a fusion reaction to occur, the nuclei must collide at a very high speed in order to overcome their mutual electric repulsion. The required speeds correspond to the extremely high temperatures found in the core of the Sun and other stars. Fusion brought about by high temperatures is called **thermonuclear fusion**. In the high temperatures of the Sun, approximately 657 million tons of hydrogen are converted into 653 million tons of helium *each second*. The missing 4 million tons of mass are discharged as radiant energy.

Such reactions are, quite literally, nuclear burning. Thermonuclear fusion is analogous to ordinary chemical combustion. In both chemical and nuclear burning, a high temperature starts the reaction; the release of energy by the reaction maintains a high enough temperature to spread the fire. The net result of the chemical reaction is a combination of atoms into more tightly bound molecules. In nuclear fusion reactions, the net result is more tightly bound nuclei.

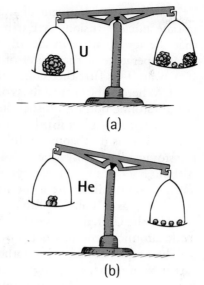

(a)

(b)

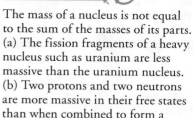

FIGURE 13.33

INTERACTIVE FIGURE

The mass of a nucleus is not equal to the sum of the masses of its parts. (a) The fission fragments of a heavy nucleus such as uranium are less massive than the uranium nucleus. (b) Two protons and two neutrons are more massive in their free states than when combined to form a helium nucleus.

FYI A common reaction is the fusion of H-2 and H-3 nuclei to become He-4 plus a neutron. most of the energy released is in the kinetic energy of the ejected neutron, with the rest of the energy in the kinetic energy of the recoiling He-4 nucleus. Interestingly, without the neutron energy carrier, a fusion reaction won't occur. The intensity of fusion reactions is measured by the accompanying neutron flux.

CHECKPOINT

1. Fission and fusion are opposite processes, yet each releases energy. Isn't this contradictory?
2. To get nuclear energy released from the element iron, should iron be fissioned or fused?
3. Predict whether the temperature of the core of a star increases or decreases when iron and elements of higher atomic number than iron in the core are fused.

Were these your answers?

1. No, no, no! This is contradictory only if the same element is said to release energy by both the processes of fission and fusion. Only the fusion of light elements and the fission of heavy elements result in a decrease in nucleon mass and a release of energy.
2. Neither, because iron is at the very bottom of the "energy valley." Fusing a pair of iron nuclei produces an element to the right of iron on the curve, where mass per nucleon is higher. If you split an iron nucleus, the products lie to the left of iron on the curve—also a higher mass per nucleon. So no energy is released. For energy release, "decrease mass" is the name of the game—any game, chemical or nuclear.
3. In the fusion of iron and any nuclei beyond, energy is absorbed and the star core cools at this late stage of its evolution. This, however, leads to the star's collapse, which then greatly increases its temperature. Interestingly, elements beyond iron are not manufactured in normal fusion cycles in stellar sources, but are manufactured when stars violently explode—supernovae.

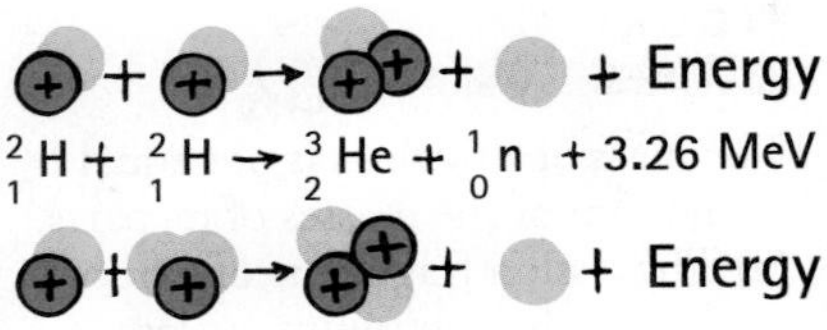

$$\oplus + \oplus \rightarrow \oplus\oplus + \bigcirc + \text{Energy}$$

$$^2_1H + ^2_1H \rightarrow ^3_2He + ^1_0n + 3.26 \text{ MeV}$$

$$\oplus + \oplus \rightarrow \oplus\oplus + \bigcirc + \text{Energy}$$

$$^2_1H + ^3_1H \rightarrow ^4_2He + ^1_0n + 17.6 \text{ MeV}$$

FIGURE 13.34

INTERACTIVE FIGURE (MP)

Fusion reactions of hydrogen isotopes. Most of the energy released is carried by the neutrons, which are ejected at high speeds.

Before the development of the atomic bomb, the temperatures required to initiate nuclear fusion on Earth were unattainable. When researchers found that the temperature inside an exploding atomic bomb is four to five times the temperature at the center of the Sun, the thermonuclear bomb was but a step away. This first thermonuclear bomb, a hydrogen bomb, was detonated in 1952. Whereas the critical mass of fissionable material limits the size of a fission bomb (atomic bomb), no such limit is imposed on a fusion bomb (thermonuclear or hydrogen bomb). Just as there is no limit to the size of an oil-storage depot, there is no theoretical limit to the size of a fusion bomb. Like the oil in the storage depot, any amount of fusion fuel can be stored safely until ignited. Although a mere match can ignite an oil depot, nothing less energetic than an atomic bomb can ignite a thermonuclear bomb. We can see that there is no such thing as a "baby" hydrogen bomb. A typical thermonuclear bomb stockpiled by the United States today, for example, is about 1000 times as destructive as the atomic bomb detonated over Hiroshima at the end of World War II.

The hydrogen bomb is another example of a discovery used for destructive rather than constructive purposes. The potential constructive possibility is the controlled release of vast amounts of clean energy.

Controlling Fusion

Carrying out fusion reactions under controlled conditions requires temperatures of millions of degrees. A variety of techniques exist for attaining high temperatures. No matter how the temperature is produced, a problem is that all materials melt and vaporize at the temperatures required for fusion. One solution is to confine the reaction in a nonmaterial container.

A nonmaterial container is a magnetic field that can exist at any temperature and can exert powerful forces on charged particles in motion. "Magnetic walls" of sufficient strength provide a kind of magnetic straitjacket for hot gases called plasmas. Magnetic compression further heats the plasma to fusion temperatures. At this writing, fusion by magnetic confinement has been only partially successful— a sustained and controlled reaction has so far been out of reach.

Although no nuclear fusion power plants are currently operating, an international project now exists whose goal is to prove the feasibility of nuclear fusion power in the near future. This fusion power project is the International Thermonuclear Experimental Reactor (ITER). After construction at the chosen site in Cadarache, France, the first sustainable fusion reaction may begin as early as 2015 (Figure 13.35). The reactor will house electrically charged hydrogen gas (plasma) heated to more than 100 million °C, which is hotter than the center of the Sun. In addition to producing about 500 MW of power, the reactor could be the energy source for the creation of hydrogen, H_2, which could be used to power fuel cells, such as those incorporated into automobiles.

If people are one day to dart about the universe in the same way we jet about Earth today, their supply of fuel is ensured. The fuel for fusion—hydrogen—is found in every part of the universe, not only in the stars but also in the space between them. About 91% of the atoms in the universe are estimated to be hydrogen. For people of the

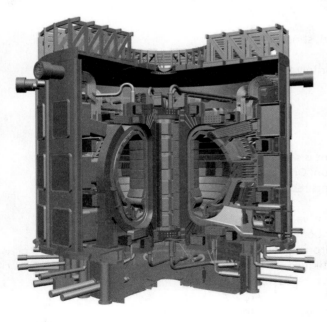

FIGURE 13.35

A cross-sectional view of the ITER (rhymes with "fitter") planned to be built and operating in Cadarache, France, before 2020.

future, the supply of raw materials is also ensured because all the elements known to exist result from the fusing of more and more hydrogen nuclei. Future humans might synthesize their own elements and produce energy in the process, just as the stars have always done.

For instructor-assigned homework, go to www.masteringphysics.com

SUMMARY OF TERMS (KNOWLEDGE)

Alpha particle A subatomic particle consisting of the combination of two protons and two neutrons ejected by a radioactive nucleus. The composition of an alpha particle is the same as that of the nucleus of a helium atom.

Beta particle An electron emitted during the radioactive decay of a radioactive nucleus.

Carbon-14 dating The process of estimating the age of once-living material by measuring the amount of radioactive carbon-14 present in the material.

Chain reaction A self-sustaining reaction in which the products of one reaction event initiate further reaction events.

Critical mass The minimum mass of fissionable material needed for a sustainable chain reaction.

Gamma ray High-frequency electromagnetic radiation emitted by radioactive nuclei.

Half-life The time required for half the atoms in a sample of a radioactive isotope to decay.

Nuclear fission The splitting of the atomic nucleus into two smaller halves.

Nuclear fusion The combining of nuclei of light atoms to form heavier nuclei.

Rad A quantity of radiant energy equal to 0.01 J absorbed per kilogram of tissue.

Radioactivity The high-energy particles and electromagnetic radiation emitted by a radioactive substance.

Rem A unit for measuring the ability of radiation to harm living tissue.

Strong nuclear force The attractive force between all nucleons, effective at only very short distances.

Thermonuclear fusion Nuclear fusion brought about by high temperatures.

Transmutation The changing of an atomic nucleus of one element into an atomic nucleus of another element through a decrease or increase in the number of protons.

READING CHECK QUESTIONS (COMPREHENSION)

13.1 Radioactivity

1. Which type of radiation—alpha, beta, or gamma—results in the greatest change in mass number? The greatest change in atomic number?

2. What is the origin of most of the natural radiation we encounter?

3. Which produces more radioactivity in the atmosphere: coal-fired power plants or nuclear power plants?

4. Is radioactivity on the Earth something relatively new? Defend your answer.

13.2 The Strong Nuclear Force

5. Why doesn't the repulsive electric force of protons in the atomic nucleus cause the protons to fly apart?

6. Which have more neutrons than protons: large nuclei or small nuclei?

7. What role do neutrons play in the atomic nucleus?

13.3 Half-Life and Transmutation

8. In what form is most of the energy released by atomic transmutation?

9. What change in atomic number occurs when a nucleus emits an alpha particle? A beta particle?

10. What is the long-range fate of all the uranium that exists in the world today?

11. What is meant by the half-life of a radioactive sample?

12. What is the half-life of uranium-238?

13.4 Radiometric Dating

13. What happens to a nitrogen atom in the atmosphere that captures a neutron?

14. Why is there more carbon-14 in living bones than in once-living ancient bones of the same mass?

15. Why is lead found in all deposits of uranium ores?

13.5 Nuclear Fission

16. What happens to the uranium-235 nucleus when it is stretched out?

17. Is a chain reaction more likely to occur in two separate pieces of uranium-235 or in the same two pieces stuck together?

18. How is a nuclear reactor similar to a conventional fossil-fuel power plant? How is it different?

13.6 Mass–Energy Equivalence

19. Who discovered that energy and mass are two different forms of the same thing?

20. In which atomic nucleus do nucleons have the least mass?

21. How does the mass per nucleon in uranium compare with the mass per nucleon in the fission fragments of uranium?

13.7 Nuclear Fusion

22. How does the mass of a pair of atoms that have fused compare to the sum of their masses before fusion?

23. What kind of containers are used to contain plasmas at temperatures of millions of degrees?

24. What kind of nuclear power is responsible for sunshine?

ACTIVITIES (HANDS-ON APPLICATION)

25. Throw ten coins onto a flat surface. Move aside all the coins that landed tails-up. Collect the remaining coins. After tossing them once again, remove all coins landing tails-up. Repeat this process until all the coins have been removed. Can you see how this relates to radioactive half-life? In units of "tosses," what is the average half-life of 25 coins? 50 coins? 1 million coins?

26. Repeat the preceding activity, but use 10 dimes and 25 pennies. Let the dimes represent a radioactive isotope, such as carbon-14, while the pennies represent a nonradioactive isotope, such as carbon-12. Remove only the dimes when they land heads-up. Collect all the pennies and add them to the dimes that were heads-up. Does the number of pennies affect the behavior of the dimes?

Someone gives you two sets of coins. The first set contains 10 dimes and 25 pennies. The second set contains 2 dimes and 25 pennies. Which set of coins has gone through a greater number of tosses? Which set provides the most "radioactivity" after a toss? Which set is analogous to a sample of once-living ancient material?

27. Calculate your estimated annual dose of radiation using the EPA's radiation dose calculator available at http://www.epa.gov/radiation/understand/calculate.html.

28. Stand one domino upright so that when it topples it hits two other upright dominos, which also each hit two other upright dominos, and so forth. Arrange as many upright dominos as you can in this fashion so that they fan out as

shown in the photograph. Your challenge is to arrange the dominos so that every one of them falls. Topple the first domino and observe your chain reaction. Focus your attention on the sound.

This dominoes chain reaction occurs on a two-dimensional flat surface. What is the dimensional geometry of a nuclear chain? What would happen if a Ping-Pong ball were tossed into a room in which the floor was covered with thousands of set-to-kill spring-action mouse traps? You can see such an explosive event by using the keywords "mouse trap chain reaction" for an Internet video search.

THINK AND SOLVE (MATHEMATICAL APPLICATION)

29. Radiation from a point source follows an inverse-square law in which the amount of radiation received is proportional to $1/d^2$, where d is distance. If a Geiger counter that is 1 m away from a small source reads 100 counts per minute, what will its reading be 2 m from the source? 3 m from it?

30. Consider a radioactive sample with a half-life of one week. How much of the original sample will be left at the end of the second week? The third week? The fourth week?

31. A radioisotope is placed near a radiation detector, which registers 80 counts per second. Eight hours later, the detector registers 5 counts per second. What is the half-life of the radioactive isotope?

32. Uranium-238 absorbs a neutron and then emits a beta particle. Show that the resulting nucleus is neptunium-239.

THINK AND RANK (ANALYSIS)

33. Rank these three types of radiation by their ability to penetrate this page of your book, from highest to lowest: (a) alpha particle, (b) beta particle, (c) gamma ray.

34. Consider the atoms C-12, C-14, and N-14. From greatest to least, rank them by the number of (a) protons in the nucleus, (b) neutrons in the nucleus, and (c) nucleons in the nucleus.

35. Rank the following isotopes in order of their radioactivity, from the most radioactive to the least radioactive:

(a) nickel-59, half-life 75,000 years; (b) uranium-238, half-life 4.5 billion years; (c) actinium-225, half-life 10 days.

36. Rank the following in order from the most energy released to the least energy released for these hypothetical cases: (a) uranium-235 splitting into two equal fragments, (b) uranium-235 splitting into three equal fragments, (c) uranium-235 splitting into 92 equal fragments.

EXERCISES (SYNTHESIS)

37. Just after an alpha particle leaves the nucleus, would you expect it to speed up? Defend your answer.

38. A pair of protons in an atomic nucleus repel each other, but they are also attracted to each other. Explain.

39. Why do different isotopes of the same element have the same chemical properties?

40. In bombarding atomic nuclei with proton "bullets," why must the protons be given large amounts of kinetic energy in order to make contact with the target nuclei?

41. Why is lead found in all deposits of uranium ores?

42. What does the proportion of lead and uranium in rock tell us about the age of the rock?

43. What are the atomic number and the atomic mass of the element formed when $^{218}_{84}Po$ emits a beta particle? What are they if the polonium emits an alpha particle?

44. Elements heavier than uranium in the periodic table do not exist in any appreciable amounts in nature because they have short half-lives. Yet there are several elements below uranium in the table that have equally short half-lives but do exist in appreciable amounts in nature. How can you account for this?

45. People who work around radioactivity wear film badges to monitor the amount of radiation that reaches their bodies. Each badge consists of a small piece of photographic film enclosed in a lightproof wrapper. What kind of radiation do these devices monitor, and how can they determine the amount of radiation people receive?

46. When food is irradiated with gamma rays from a cobalt-60 source, does the food become radioactive? Defend your answer.

47. Radium-226 is a common isotope on Earth, but it has a half-life of about 1620 years. Given that Earth is about 4.5 billion years old, why is there any radium at all?

48. Is carbon dating advisable for measuring the age of materials a few years old? How about a few thousand years old? A few million years old?

49. Why isn't carbon-14 dating accurate for estimating the age of materials older than 50,000 years?

50. The age of the Dead Sea Scrolls was determined by carbon-14 dating. Could this technique have worked if they had been carved on stone tablets? Explain.

51. If you find that half of 1000 people born in the year 2000 are still living in 2060, does this mean that one-quarter of them will be alive in 2120 and one-eighth of them alive in 2180? What is different about the death rates of people and the "death rates" of radioactive atoms?

52. The uranium ores of the Athabasca Basin deposits of Saskatchewan, Canada, are unusually pure, containing up to 70% uranium oxides. Why doesn't this uranium ore undergo an explosive chain reaction?

53. "Strontium-90 is a pure beta source." How could a physicist test this statement?

54. Why will nuclear fission probably never be used directly for powering automobiles? How could it be used indirectly?

55. Why is carbon better than lead as a moderator in nuclear reactors?

56. How does the mass per nucleon in uranium compare with the mass per nucleon in the fission fragments of uranium?

57. Why doesn't iron yield energy if it undergoes fusion or fission?

58. Uranium-235 releases an average of 2.5 neutrons per fission, while plutonium-239 releases an average of 2.7 neutrons per fission. Which of these elements might you therefore expect to have the smaller critical mass?

59. Which process would release energy from gold: fission or fusion? From carbon? From iron?

60. If a uranium nucleus were to fission into three fragments of approximately equal size instead of two, would more energy or less energy be released? Defend your answer using Figure 13.33.

61. Is the mass of an atomic nucleus greater or less than the sum of the masses of the nucleons that it contains? Why don't the nucleon masses add up to the total nuclear mass?

62. The original reactor built in 1942 was just "barely" critical because the natural uranium that was used contained less than 1% of the fissionable isotope U-235 (half-life 713 million years). What if, in 1942, the Earth had been 9 billion years old instead of 4.5 billion years old? Would this reactor have reached the critical stage with natural uranium?

63. Heavy nuclei can be made to fuse—for instance, by firing one gold nucleus at another one. Does such a process yield energy or cost energy? Explain.

64. Which produces more energy: the fission of a single uranium nucleus or the fusing of a pair of deuterium nuclei? The fission of a gram of uranium or the fusing of a gram of deuterium? (Why do your answers differ?)

65. If a fusion reaction produces no appreciable radioactive isotopes, why does a hydrogen bomb produce significant radioactive fallout?

66. Explain how radioactive decay has always warmed the Earth from the inside and how nuclear fusion has always warmed the Earth from the outside.

67. What percentage of nuclear power plants in operation today are based on nuclear fusion?

68. Sustained nuclear fusion has yet to be achieved and remains a hope for abundant future energy. Yet the energy that has always sustained us has been the energy of nuclear fusion. Explain.

69. Oxygen and two hydrogen atoms combine to form a water molecule. At the nuclear level, if one oxygen and two hydrogen were fused, what element would be produced?

70. If a pair of carbon nuclei were fused and the product emitted a beta particle, what element would be produced?

71. Ordinary hydrogen is sometimes called a perfect fuel because of its almost unlimited supply on Earth, and when it burns, harmless water is the product of the combustion. So why don't we abandon fission energy and fusion energy, not to mention fossil-fuel energy, and just use hydrogen?

DISCUSSION QUESTIONS (EVALUATION)

72. Why might some people consider it a blessing in disguise that fossil fuels are such a limited resource? Centuries from now, what attitudes about the combustion of fossil fuels are our descendants likely to have?

73. The 1986 nuclear power plant accident at Chernobyl, in which dozens of people died and thousands more were exposed to cancer-causing radiation, created fear and outrage worldwide and led some people to call for the closing of all nuclear plants. Yet many people choose to smoke cigarettes in spite of the fact that 2 million people die every year from smoking-related diseases. The risks posed by nuclear power plants are involuntary, risks we must all share like it or not, whereas the risks associated with smoking are voluntary because each smoker chooses to smoke. Why are we so unaccepting of involuntary risk but accepting of voluntary risk?

74. Your friend Paul says that the helium used to inflate balloons is a product of radioactive decay. Your mutual friend Steve says no way. Then there's your friend Alison, fretful about living near a fission power plant. She wishes to get away from radiation by traveling to the high mountains and sleeping out at night on granite outcroppings. Still another friend, Michele, has journeyed to the mountain foothills to escape the effects of radioactivity altogether. While bathing in the warmth of a natural hot spring, she wonders aloud how the spring gets its heat. What do you tell these friends?

75. Speculate about some worldwide changes likely to follow the advent of successful fusion reactors. Compare the advantages and disadvantages of electricity coming from a large central power station versus a network of many smaller solar-based stations owned and operated by individuals.

READINESS ASSURANCE TEST (RAT)

If you have a good handle on this chapter, if you really do, then you should be able to score 7 out of 10 on this RAT. If you score less than 7, you need to study further before moving on.

Choose the BEST answer to each of the following.

1. Which type of radiation from cosmic sources predominates on the inside of a high-flying commercial airplane?
   (a) alpha radiation
   (b) beta radiation
   (c) gamma radiation
   (d) none because all three are abundant

2. Is it possible for a hydrogen nucleus to emit an alpha particle?
   (a) yes, because alpha particles are the simplest form of radiation
   (b) no, because it would require the nuclear fission of hydrogen, which is impossible
   (c) yes, but it does not occur very frequently
   (d) no, because it does not contain enough nucleons

3. A sample of radioactive material is usually a little warmer than its surroundings because
   (a) it efficiently absorbs and releases energy from sunlight.
   (b) its atoms are continuously being struck by alpha and beta particles.
   (c) it is radioactive.
   (d) it emits alpha and beta particles.

4. What evidence supports the contention that the strong nuclear force is stronger than the electric interaction at short internuclear distances?
   (a) Protons are able to exist side by side within an atomic nucleus.
   (b) Neutrons spontaneously decay into protons and electrons.
   (c) Uranium deposits are always slightly warmer than their immediate surroundings.
   (d) Radio interference arises adjacent to any radioactive source.

5. When the isotope bismuth-213 emits an alpha particle, what new element results?
   (a) lead
   (b) platinum
   (c) polonium
   (d) thallium

6. A certain radioactive element has a half-life of 1 h. If you start with a 1-g sample of the element at noon, how much of this same element will be left at 3:00 PM?
   (a) 0.5 g
   (b) 0.25 g
   (c) 0.125 g
   (d) 0.0625 g

7. The isotope cesium-137, which has a half-life of 30 years, is a product of nuclear power plants. How long will it take for this isotope to decay to about one-half of its original amount?
   (a) 0 years
   (b) 15 years
   (c) 30 years
   (d) 60 years
   (e) 90 years

8. If uranium were to split into 90 pieces of equal size instead of 2, would more energy or less energy be released?
   (a) less energy, because of less mass per nucleon
   (b) less energy, because of more mass per nucleon
   (c) more energy, because of less mass per nucleon
   (d) more energy, because of more mass per nucleon

9. Which process would release energy from gold: fission or fusion? From carbon?
   (a) gold: fission; carbon: fusion
   (b) gold: fusion; carbon: fission
   (c) gold: fission; carbon: fission
   (d) gold: fusion; carbon: fusion

10. If an iron nucleus split in two, its fission fragments would have
    (a) more mass per nucleon.
    (b) less mass per nucleon.
    (c) the same mass per nucleon.
    (d) either more or less mass per nucleon.

Answers to RAT

1. c, 2. d, 3. b, 4. a, 5. d, 6. c, 7. a, 8. b, 9. a, 10. a

CHAPTER 14

Elements of Chemistry

A
S YOU progress through this physical science course, you will note an accumulating list of key terms. For example, we say that there are more than 100 kinds of *atoms*, and that any material consisting of a single kind of atom is an *element*. Atoms can link together to form a *molecule*, and a molecule consisting of atoms from different elements is a *compound*. And on and on, one term building on another, as we attempt to describe the nature of matter beyond its casual appearance.

Rather than memorizing key terms with a set of flash cards, you will serve yourself far better by first focusing on the underlying concept each term represents. As you know, it is quite possible to be familiar with a term without truly understanding the underlying concept. So although a vocabulary of science terms is useful for communication, it does not guarantee conceptual understanding. If you focus first on the concepts, the vocabulary will come to you much more naturally.

14.1 Chemistry: The Central Science

EXPLAIN THIS How has chemistry influenced our modern lifestyles?

When you wonder what the land, sky, or ocean is made of, you are thinking about chemistry. When you wonder how a rain puddle dries up, how a car acquires energy from gasoline, or how your body extracts energy from the food you eat, you are again thinking about chemistry. By definition, **chemistry** is the study of matter and the transformations it can undergo. Matter is anything that occupies space. It is the stuff that makes up all material things; anything you can touch, taste, smell, see, or hear is matter. The scope of chemistry, therefore, is very broad.

Chemistry is often described as a central science because it touches all the other sciences. It springs from the principles of physics, and it serves as the foundation for the most complex science of all—biology. Indeed, many of the great advances in the life sciences today, such as genetic engineering, are applications of some very exotic chemistry. Chemistry sets the foundation for the major Earth sciences—geology, oceanography, meteorology. It is also an important component of space science, as described in Figure 14.1. Just as we learned about the origin of the Moon from the chemical analysis of moon rocks in the early 1970s, we are now learning about the history of Mars and other planets from the chemical information gathered by space probes.

Progress in science is made as scientists conduct research. Research is any activity aimed at the systematic discovery and interpretation of new knowledge. Many scientists focus on **basic research**, which leads us to a greater understanding of how the natural world operates. The foundation of knowledge laid down by basic research frequently leads to useful applications. Research that focuses on developing these applications is known as **applied research**. Most chemists choose applied research as their major focus. Applied research in chemistry has provided us with medicine, food, water, shelter, and many of the material goods that characterize modern life. Just a few examples are shown in Figure 14.2.

Over the course of the past century, we excelled at manipulating atoms and molecules to create materials to suit our needs. At the same time, however, we made mistakes in caring for the environment. Waste products were dumped into rivers, buried in the ground, or vented into the air without regard for possible long-term consequences. Many people believed that Earth was so large that its resources were virtually unlimited and that it could absorb wastes without being significantly harmed.

Most nations now recognize this as a dangerous attitude. As a result, government agencies, industries, and concerned citizens are involved in extensive efforts

Mastering**PHYSICS**
TUTORIAL: What is Chemistry?

Practice articulating and paraphrasing the concepts represented by the boldface terms. Do this aloud to yourself (or to a friend), minimizing looking at the book. When you can express these concepts in your own words—in your own "plain English"—you'll have the insight to do well in this course and beyond.

FIGURE 14.1
Special materials of chemistry, such as rocket fuels, metals for spaceships, and fabrics for the space suits, were required to allow astronauts to reach and explore the surface of the Moon.

Industries in the United States employ about 900,000 chemists.

Transparent matrix of processed silicon dioxide

Chemically disinfected drinking water

Caffeine solution

Thermoset polymer

Prescription medicines stored in refrigerator

Chlorofluorocarbon-free refrigerating fluids

Electric energy from a fossil-fuel or nuclear power plant

Metal alloy

Roasting carbohydrates, fats, proteins, and vitamins

Natural gas laced with odoriferous sulfur compounds

Fertilizer-grown vegetables

FIGURE 14.2
Most of the material items in any modern house are shaped by some human-devised chemical process.

FIGURE 14.3
The Responsible Care symbol of the American Chemistry Council; go to responsiblecare.org.

to clean up toxic-waste sites. Such regulations as the international ban on ozone-destroying chlorofluorocarbons have been enacted to protect the environment. Members of the American Chemistry Council, who produce 90% of the chemicals manufactured in the United States, have adopted a program called Responsible Care, in which they have pledged to manufacture without causing environmental damage. The Responsible Care program emblem is shown in Figure 14.3. If we use chemistry wisely, most waste products can be minimized, recycled, engineered into salable commodities, or rendered environmentally benign.

Chemistry has influenced our lives in profound ways, and it will continue to do so in the future. For this reason, it is in everyone's interest to become acquainted with the basic concepts of chemistry.

> **CHECKPOINT**
> **Chemists have learned how to produce aspirin using petroleum as a starting material. Is this an example of basic or applied research?**
>
> **Was this your answer?**
> This is an example of applied research, because the primary goal was to develop a useful commodity. However, the ability to produce aspirin from petroleum depended on an understanding of atoms and molecules developed from many years of basic research.

LEARNING OBJECTIVE
Introduce the molecule as a fundamental unit of matter.

14.2 The Submicroscopic World

EXPLAIN THIS What is found between two adjacent molecules of a gas?

From afar, a sand dune appears to be a smooth, continuous material. Up close, however, the dune reveals itself to be made of tiny particles of sand. In a similar fashion, as discussed in Chapter 12, everything around us—no matter how smooth it may appear—is made of the basic units you know as *atoms*. Atoms are so small, however, that a single grain of sand contains on the order of 125 million trillion of them. There are roughly 250,000 times more atoms in a single grain of sand than there are grains of sand in the dunes shown in Figure 14.4.

As small as atoms are, there is much we have learned about them. We know, for example, that there are more than 100 different types of atoms, and they are listed in the widely recognized periodic table. Some atoms link together to form larger but still incredibly small basic units of matter called **molecules**. As shown in Figure 14.4, for example, two hydrogen atoms and one oxygen atom link together to form a single molecule of water, which you know as H_2O. Water molecules are so small that an 8-oz glass of water contains about a trillion trillion of them.

Our world can be studied at different levels of magnification. At the *macroscopic* level, matter is large enough to be seen, measured, and handled. A handful of sand and a glass of water are macroscopic samples of matter. At the *microscopic* level, physical structure is so fine that it can be seen only with a microscope. A biological cell is microscopic, as is the detail on a dragonfly's wing. Beyond the microscopic level is the **submicroscopic**—the realm of atoms and molecules and an important focus of chemistry.

Recall from Chapter 7 that matter exists in *phases*. At the submicroscopic level, solid, liquid, and gaseous phases are distinguished by how the submicroscopic particles hold together. This is illustrated in Figure 14.5. In solid matter, such as rock, the attractions between particles are strong enough to hold all the particles together in some fixed three-dimensional arrangement. The particles can vibrate about fixed positions, but they cannot move past one another.

Many major advances were made in both the physical sciences and the life sciences over the course of the 20th century. Advances made in the physical sciences, such as our understanding of the chemistry of life, however, will likely propel the life sciences to even more fantastic advances in the 21st century.

FIGURE 14.4
There are far more atoms in a glass of water than there are grains of sand within this towering sand dune.

Oxygen atom
Hydrogen atoms
Water molecule, H₂O

The addition of heat causes these vibrations to increase until, at a certain temperature, the vibrations are rapid enough to disrupt the fixed arrangements. Rock melts into magma (a topic of much discussion in Part 3). Likewise, ice melts into water. The particles can then slip past one another and tumble around much like a bunch of marbles in a bag. This is the liquid phase of matter, and the mobility of the submicroscopic particles gives rise to the liquid's fluid character—its ability to flow and to assume the shape of its container.

Further heating causes the submicroscopic particles in a liquid to move so fast that the attractions they have for one another are unable to hold them together. They then separate from one another, forming a gas. For magma, this doesn't easily happen, because the particles are strongly attracted to one another. Water molecules separate into a gas at 100°C. For a substance like helium, the submicroscopic particles are already in the gaseous phase at room temperature.

Moving at an average speed of 500 m/s (1100 mi/h), the particles of a gas are widely separated from one another. Matter in the gaseous phase therefore occupies much more volume than it does in the solid or liquid phase. Applying pressure to a gas squeezes the gas particles closer together, which decreases the volume. The amount of air an underwater diver needs to breathe for many minutes, for example, can be squeezed (compressed) into a tank small enough to be carried on the diver's back.

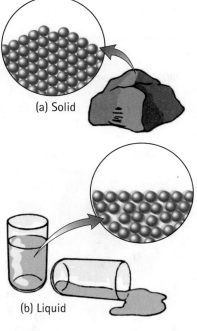
(a) Solid
(b) Liquid

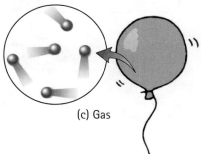
(c) Gas

FYI Coffee and tea are decaffeinated using carbon dioxide in a fourth phase of matter known as a *supercritical fluid*. This phase behaves like a gaseous liquid, which is attained by adding lots of pressure and heat. Supercritical carbon dioxide is relatively easy to produce. To get water to form a supercritical fluid, however, requires pressures in excess of 217 atm and a temperature of 374°C. Supercritical water is very corrosive. Also, so much oxygen can dissolve in supercritical water that flames can burn within this medium, which is ideal for the destruction of toxic wastes.

FIGURE 14.5
The familiar bulk properties of a solid, a liquid, and a gas. (a) The submicroscopic particles of the solid phase vibrate about fixed positions. (b) The submicroscopic particles of the liquid phase slip past one another. (c) The fast-moving submicroscopic particles of the gaseous phase are separated by large average distances.

14.3 Physical and Chemical Properties

EXPLAIN THIS Why are physical changes typically easier to reverse than chemical changes?

Properties that describe the look or feel of a substance, such as color, hardness, density, texture, and phase, are called **physical properties**. Every substance has its own set of characteristic physical properties that we can use to identify that substance (Figure 14.6).

The physical properties of a substance can change when conditions change, but that does not mean that a different substance is created. Cooling liquid water to below 0°C causes the water to transform to solid ice, but the substance is still water, no matter what the phase. The only difference is the relative orientation of the H₂O molecules to one another. In the liquid phase, the water molecules tumble around one another, whereas in the ice phase, they vibrate about fixed positions. The freezing of water is an example of what chemists call a physical change. During a **physical change**, a substance changes its phase or some other physical property, but not its chemical composition, as Figure 14.7 shows.

Gold
Opacity: opaque
Color: yellowish
Phase at 25˚C: solid
Density: 19.3 g/mL

Diamond
Opacity: transparent
Color: colorless
Phase at 25˚C: solid
Density: 3.5 g/mL

Water
Opacity: transparent
Color: colorless
Phase at 25˚C: liquid
Density: 1.0 g/mL

FIGURE 14.6
Gold, diamond, and water can be identified by their physical properties. If a substance has all the physical properties listed under gold, for example, it must be gold.

CHECKPOINT
The melting of gold is a physical change. Why?

Was this your answer?
During a physical change, a substance changes only one or more of its physical properties; its chemical identity does not change. Because melted gold is still gold but in a different form, its melting represents only a physical change.

FIGURE 14.7
Two physical changes. (a) Liquid water and ice may appear to be different substances, but a submicroscopic view shows that both consist of water molecules. (b) At 25°C, the atoms in a sample of mercury are a certain distance apart, yielding a density of 13.53 g/mL. At 100°C, the atoms are farther apart, meaning that each milliliter now contains fewer atoms than at 25°C, and the density is now 13.35 g/mL. The physical property we call density has changed with temperature, but the identity of the substance remains unchanged: mercury is mercury.

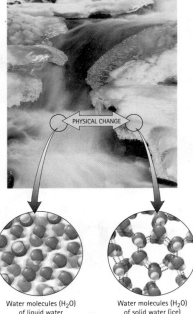

Water molecules (H₂O) of liquid water

Water molecules (H₂O) of solid water (ice)

Atoms of liquid mercury (Hg) at 25°C

Atoms of liquid mercury (Hg) at 100°C (expanded)

(a)   (b)

Methane
Reacts with oxygen to form carbon dioxide and water, giving off lots of heat during the reaction.

Baking soda
Reacts with vinegar to form carbon dioxide and water, absorbing heat during the reaction.

Copper
Reacts with carbon dioxide and water to form the greenish-blue substance called patina.

FIGURE 14.8
The chemical properties of substances allow them to transform to new substances. Natural gas and baking soda transform to carbon dioxide, water, and heat. Copper transforms to patina.

Oxygen

Methane

Water

Carbon dioxide

FIGURE 14.9
INTERACTIVE FIGURE (MP)

The chemical change in which molecules of methane and oxygen transform to molecules of carbon dioxide and water, as atoms break old bonds and form new ones. Although the actual mechanism of this transformation is more complicated than depicted here, the idea that new materials are formed by the rearrangement of atoms is accurate.

Mastering**PHYSICS**
VIDEO: Oxygen Bubble Burst
VIDEO: Fire Water

Chemical properties characterize the ability of a substance to react with other substances or to transform from one substance to another. Figure 14.8 shows three examples. The methane of natural gas has the chemical property of reacting with oxygen to produce carbon dioxide and water, along with appreciable heat energy. Similarly, baking soda has the chemical property of reacting with vinegar to produce carbon dioxide and water while absorbing a small amount of heat energy. Copper has the chemical property of reacting with carbon dioxide and water to form a greenish-blue solid known as *patina*. Copper statues exposed to the carbon dioxide and water in the air become coated with patina. The patina is not copper, it is not carbon dioxide, and it is not water. It is a new substance formed by the reaction of these chemicals with one another.

All three of these transformations involve a change in the way the atoms in the molecules are *chemically bonded* to one another. A **chemical bond** is the force of attraction between two atoms that holds them together. A methane molecule, for example, is made of a single carbon atom bonded to four hydrogen atoms, and an oxygen molecule is made of two oxygen atoms bonded to each other. Figure 14.9 shows the chemical change in which the atoms in a methane molecule and those in two oxygen molecules first pull apart and then form new bonds with different partners, resulting in the formation of molecules of carbon dioxide and water.

Any change in a substance that involves a rearrangement of the way atoms are bonded is called a **chemical change**. Thus the transformation of methane to carbon dioxide and water is a chemical change, as are the other two transformations shown in Figure 14.8.

The chemical change shown in Figure 14.10 occurs when an electric current is passed through water. The energy of the current causes the bonds holding atoms together to break apart. Loose atoms then form new bonds with different atoms, which results in the formation of new molecules. Thus, water molecules are changed to molecules of hydrogen and oxygen, two substances that are very different from water. The hydrogen and oxygen are both gases at room temperature, and they can be seen as bubbles rising to the surface.

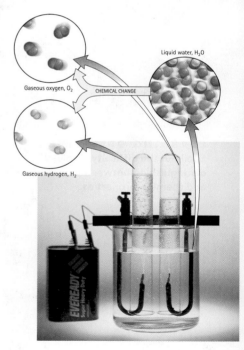

Gaseous oxygen, O_2

Liquid water, H_2O

CHEMICAL CHANGE

Gaseous hydrogen, H_2

FIGURE 14.10
Water can be transformed to hydrogen gas and oxygen gas by applying the energy of an electric current. This is a chemical change, because new materials (the two gases) are formed as the atoms originally found in the water molecules are rearranged.

In the language of chemistry, materials undergoing a chemical change are said to be *reacting*. Methane reacts with oxygen to form carbon dioxide and water. Water reacts when exposed to electricity to form hydrogen gas and oxygen gas. Thus, the term *chemical change* means the same thing as *chemical reaction*. During a **chemical reaction**, new materials are formed by a change in the way atoms are bonded together. We shall explore chemical bonds and the reactions in which they are formed and broken in Chapters 15, 17, and 18.

> A chemical property of a substance is its tendency to change into another substance. For example, it is a chemical property of iron to transform into rust.

CHECKPOINT

Each sphere in the following diagrams represents an atom. Joined spheres represent molecules. One set of diagrams shows a physical change, and the other shows a chemical change. Which is which?

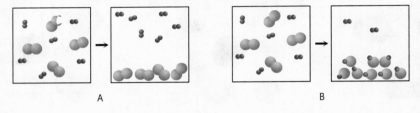

A   B

Was this your answer?

Remember that a chemical change (also known as a chemical reaction) involves molecules breaking apart so that the atoms are free to form new bonds with new partners. Be careful to distinguish this breaking apart from a mere change in the relative positions of a group of molecules. In set A, the molecules before and after the change are the same. They differ only in their positions relative to one another. Set A, therefore, represents only a physical change. In set B, new molecules, consisting of bonded red and blue spheres, appear after the change. These molecules represent a new material, and so set B represents a chemical change.

LEARNING OBJECTIVE
Spell out the difficulty involved in distinguishing between physical and chemical properties.

14.4 Determining Physical and Chemical Changes

EXPLAIN THIS Why is the air over a campfire always moist?

How can you determine whether an observed change is physical or chemical? This can be tricky because in both cases changes in physical appearance occur. Water, for example, looks quite different after it freezes, just as a car looks quite different after it rusts (Figure 14.11). The freezing of water is a physical change because liquid water and frozen water are both forms of water—only the orientation of the water molecules to one another changes. The rusting of a car, by contrast, is the result of the transformation of iron to rust. This is a chemical change because iron and rust are two different materials, each consisting of a different arrangement of atoms. As we shall see in the next two sections, iron is an element, and rust is a compound consisting of iron and oxygen atoms.

Two powerful guidelines can help you assess physical and chemical changes. First, in a physical change, a change in appearance is the result of a new set of conditions imposed on the same material. Restoring the original conditions restores the original appearance: frozen water melts upon warming. Second, in a chemical

FIGURE 14.11
The transformation of water to ice and the transformation of iron to rust both involve changes in physical appearance. The formation of ice is a physical change, whereas the formation of rust is a chemical change.

change, a change in appearance is the result of the formation of a new material that has its own unique set of physical properties. The more evidence you have suggesting that a different material has been formed, the greater the likelihood that the change is a chemical change. Iron is a material that can be used to build cars. Rust is not. This suggests that the rusting of iron is a chemical change.

CHECKPOINT

Evan, shown to the right, has grown an inch in height over the past year. Is this best described as a physical or a chemical change?

Was this your answer?
Are new materials being formed as Evan grows? Absolutely—created out of the food he eats. His body is very different from, say, the peanut butter sandwich he ate yesterday. Yet, through some very advanced chemistry, his body is able to absorb the atoms of that peanut butter sandwich and rearrange them into new materials. Biological growth, therefore, is best described as a chemical change.

Figure 14.12 shows potassium chromate, a material whose color depends on its temperature. At room temperature, potassium chromate is a bright canary yellow. At higher temperatures, it is a deep reddish orange. Upon cooling, the canary color returns, suggesting that the change is physical. With a chemical change, reverting to the original conditions does not restore the original appearance. Ammonium dichromate, shown in Figure 14.13, is an orange material that, when heated, explodes into ammonia, water vapor, and green chromium(III) oxide. When the test tube is returned to the original temperature, there is no trace of orange ammonium dichromate. In its place are new substances having completely different physical properties.

COOL   HOT   COOL

FIGURE 14.12
Potassium chromate changes color as its temperature changes. This change in color is a physical change. A return to the original temperature restores the original bright yellow color.

When heated, orange ammonium dichromate undergoes a chemical change to ammonia, water vapor, and chromium(III) oxide. A return to the original temperature does not restore the orange color, because the ammonium dichromate is no longer there.

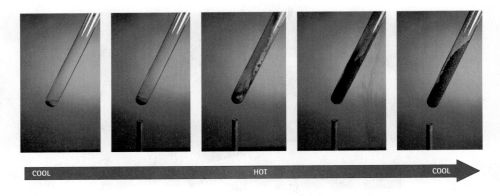

COOL HOT COOL

14.5 Elements to Compounds

EXPLAIN THIS How are compounds different from elements?

As briefly described in Chapter 12, the terms *element* and *atom* are often used in a similar context. You might hear, for example, that gold is an element made of gold atoms. Generally, *element* is used in reference to an entire macroscopic or microscopic sample, and *atom* is used when speaking of the submicroscopic particles in the sample. The important distinction is that elements are made of atoms and not the other way around.

The fundamental unit of an element is indicated by its **elemental formula**. For elements in which the fundamental units are individual atoms, the elemental formula is simply the chemical symbol: Au is the elemental formula for gold, and Li is the elemental formula for lithium, to name just two examples. For elements in which the fundamental units are two or more atoms bonded into molecules, the elemental formula is the chemical symbol followed by a subscript indicating the number of atoms in each molecule. For example, elemental nitrogen, shown in Figure 14.14, commonly consists of molecules containing two nitrogen atoms per molecule. Thus, N_2 is the usual elemental formula given for nitrogen. Similarly, O_2 is the elemental formula for the oxygen we breathe, and S_8 is the elemental formula for sulfur.

> Physical change? Chemical change? It's not always easy to distinguish between the two. Because of many subtleties that are recognized only after years of study and laboratory experience, you'll not soon achieve a firm handle on how to categorize many observed changes. It's okay to learn a little now, and to entrust a lot that remains for some future time or perhaps to others who chose to specialize within this field.

FYI Carbon is the only element that can form bonds with itself indefinitely. Sulfur's practical limit is S_8 and nitrogen's limit is around N_{12}. The elemental formula for a 1-carat diamond, however, is about $C_{10,000,000,000,000,000,000,000}$.

CHECKPOINT

The oxygen we breathe, O_2, is converted to ozone, O_3, in the presence of an electric spark. Is this a physical or chemical change?

Was this your answer?
When atoms regroup, the result is an entirely new substance, and that is what happens here. The oxygen we breathe, O_2, is odorless and life-giving. Ozone, O_3, can be toxic, and it has a pungent smell commonly associated with electric motors. The conversion of O_2 to O_3 is therefore a chemical change. However, both O_2 and O_3 are elemental forms of oxygen.

When atoms of different elements bond to one another, they make a **compound**. Sodium atoms and chlorine atoms, for example, bond to make the compound sodium chloride, commonly known as table salt. Nitrogen atoms and hydrogen atoms join to make the compound ammonia, which is a common household cleaner.

A compound is represented by its **chemical formula**, in which the symbols for the elements are written together. The chemical formula for sodium chloride is NaCl, and the formula for ammonia is NH_3. Numerical subscripts indicate the ratio in which the atoms combine. By convention, the subscript 1 is understood

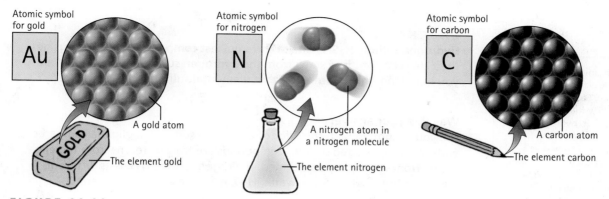

FIGURE 14.14

Any element consists of only one kind of atom. Gold consists of only gold atoms, a flask of gaseous nitrogen consists of only nitrogen atoms, and the carbon of a graphite pencil consists of only carbon atoms.

and omitted. So the chemical formula NaCl tells us that the compound sodium chloride has one sodium atom for every chlorine atom; the chemical formula NH_3 tells us that the compound ammonia has one nitrogen atom for every three hydrogen atoms, as Figure 14.15 shows.

Compounds have physical and chemical properties that are completely different from the properties of their elemental components. The sodium chloride, NaCl, shown in Figure 14.16 is very different from the elemental sodium and the elemental chlorine used in its formation. Elemental sodium, Na, consists of nothing but sodium atoms, which form a soft, silvery metal that can be cut easily with a knife. Its melting point is 97.5°C, and it reacts violently with water. Elemental chlorine, Cl_2, consists of chlorine molecules. This material, a yellow-green gas at room temperature, is very toxic, and it was used as a chemical warfare agent during World War I. Its boiling point is −34°C. The compound sodium chloride, NaCl, is a translucent, brittle, colorless crystal with a melting point of 800°C. Sodium chloride does not react chemically with water the way sodium does; not only is it not toxic to humans, which chlorine is, but the very opposite is true—it is an essential component of all living organisms. Sodium chloride is not sodium, nor is it chlorine; it is uniquely sodium chloride, a tasty chemical when sprinkled lightly over popcorn.

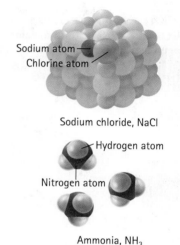

Sodium chloride, NaCl

Ammonia, NH_3

FIGURE 14.15

The compounds sodium chloride and ammonia are represented by their chemical formulas, NaCl and NH_3. A chemical formula shows the ratio of atoms that constitute the compound.

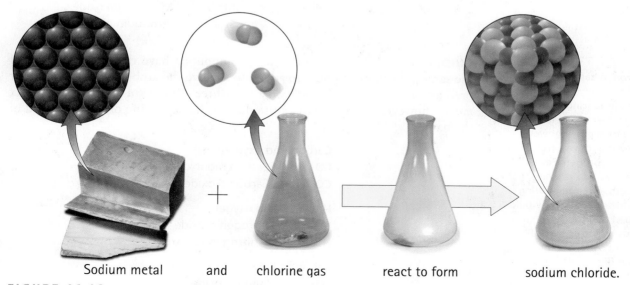

Sodium metal   and   chlorine gas   react to form   sodium chloride.

FIGURE 14.16

Sodium metal and chlorine gas react together to form sodium chloride. Although the compound sodium chloride is composed of sodium and chlorine, the physical and chemical properties of sodium chloride are very different from the physical and chemical properties of either sodium metal or chlorine gas.

A compound is uniquely different from the elements from which it is made. For example, water is a liquid, while the elements that are in it, hydrogen and oxygen, are gases. The harmless compound known as table salt is composed of two very dangerous chemicals: metallic sodium and chlorine gas.

CHECKPOINT

Hydrogen sulfide, H_2S, is one of the smelliest compounds. Rotten eggs get their characteristic bad smell from the hydrogen sulfide they release. Can you infer from this information that elemental sulfur, S_8, is just as smelly?

Was this your answer?

No, you cannot. In fact, the odor of elemental sulfur is negligible compared with that of hydrogen sulfide. Compounds are truly different from the elements from which they are formed. Hydrogen sulfide, H_2S, is as different from elemental sulfur, S_8, as water, H_2O, is from elemental oxygen, O_2.

LEARNING OBJECTIVE
List three guidelines used to name compounds.

14.6 Naming Compounds

EXPLAIN THIS What information is found within the name of a compound?

A system for naming the countless number of possible compounds has been developed by the International Union of Pure and Applied Chemistry (IUPAC). This system is designed so that a compound's name reflects the elements it contains and how those elements are joined. Anyone familiar with the system, therefore, can deduce the chemical identity of a compound from its systematic name.

As you might imagine, this system is very intricate. There is no need for you to learn all its rules. Instead, learning some guidelines will prove most helpful. These guidelines alone will not enable you to name every compound. However, they will acquaint you with how the system works for many simple compounds consisting of only two elements.

GUIDELINE 1 The name of the element farther to the left in the periodic table is followed by the name of the element farther to the right, with the suffix *-ide* added to the name of the latter:

NaCl	Sodium chloride	HCl	Hydrogen chloride
Li_2O	Lithium oxide	MgO	Magnesium oxide
CaF_2	Calcium fluoride	Sr_3P_2	Strontium phosphide

GUIDELINE 2 When two or more compounds have different numbers of the same elements, prefixes are added to remove the ambiguity. The first four prefixes are *mono-* (one), *di-* (two), *tri-* (three), and *tetra-* (four). The prefix *mono-*, however, is commonly omitted from the beginning of the first word of the name:

Carbon and oxygen
CO	Carbon monoxide
CO_2	Carbon dioxide

Nitrogen and oxygen
NO_2	Nitrogen dioxide
N_2O_4	Dinitrogen tetroxide

Sulfur and oxygen
SO_2	Sulfur dioxide
SO_3	Sulfur trioxide

FYI Hydrogen has been touted as the fuel of the future. It burns clean, producing only energy and water vapor. Much would have to happen, however, before we could convert from fossil fuels to hydrogen. For example, we would need an efficient method for generating hydrogen. Ideally, a system would be developed that produces hydrogen using the energy of direct sunlight. Also, the infrastructure for distributing hydrogen would need to be built.

GUIDELINE 3   Many compounds are not usually referred to by their systematic names. Instead, they are assigned common names that are more convenient or have been used traditionally for many years. Some common names are water for H_2O, ammonia for NH_3, and methane for CH_4.

CHECKPOINT
What is the systematic name for NaF?

Was this your answer?
This compound is a cavity-fighting substance added to some toothpastes—sodium fluoride.

14.7   The Advent of Nanotechnology

EXPLAIN THIS   Is nanotechnology the result of basic or applied research?

The age of microtechnology was ushered in some 60 years ago with the invention of the solid-state transistor, a device that serves as a gateway for electronic signals. Engineers were quick to grasp the idea of integrating many transistors together to create logic boards that could perform calculations and run programs. The more transistors they could squeeze into a circuit, the more powerful the logic board. The race thus began to squeeze more and more transistors together into tinier and tinier circuits. The scales achieved were in the realm of the micron (10^{-6} m): thus the term *micro*technology. At the time of the transistor's invention, few people realized the impact microtechnology would have on society—from personal computers to cell phones to the Internet.

Today, we are at the beginning of a similar revolution. Technological advances have recently brought us past the realm of microns to the realm of the nanometer (10^{-9} m), which is the scale of individual atoms and molecules—a realm where we have reached the basic building blocks of matter. Technology that works on this scale where we engineer materials by manipulating individual atoms or molecules is known as **nanotechnology**. No one knows exactly what impact nanotechnology will have on society, but we are quickly coming to realize its vast potential, which is likely to be much greater than that of microtechnology.

Nanotechnology generally concerns the manipulations of objects from 1 to 100 nanometers in size. For perspective, a DNA molecule is about 2.0 nm wide, while a water molecule is only about 0.2 nm wide. Like microtechnology, nanotechnology is interdisciplinary, requiring the cooperative efforts of chemists, engineers, physicists, molecular biologists, and many others. Interestingly, there are already many products on the market that contain components developed through nanotechnology. These include sunscreens, mirrors that don't fog, dental bonding agents, automotive catalytic converters, stain-free clothing, water filtration systems, the heads of computer hard drives, and many more. Nanotechnology, however, is still in its infancy, and it will likely be decades before its potential is fully realized (Figure 14.17). Consider, for example, that personal computers didn't blossom until the 1990s, some 40 years after the first solid-state transistor.

There are two main approaches to building nanoscale materials and devices: top-down and bottom-up. The top-down approach is an extension of

FYI   Before he died in 2005, Rick Smalley, codiscoverer of the buckyball molecule, advocated that carbon nanotubes, if developed into wires, could be an ideal material for efficiently transporting electricity over vast distances. If such an infrastructure were in place, the wind energy of the Great Plains of the United States would be sufficient to supply the electrical needs of the entire country.

FIGURE 14.17
Carbon nanotubes can be nested within each other to provide the strongest fiber known—a thread 1 mm in diameter can support a weight of about 13,000 lb. A network of such strong fibers could be used to build the once science-fictional space elevator.

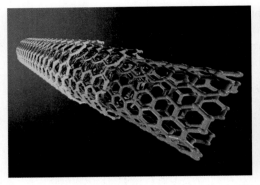

FYI An interesting discovery of nanoscience is that the properties of a material at the level of its atoms can be different from its properties in bulk quantities. A bar of gold, for example, is gold in color. A thin sheet of gold atoms, by contrast, is dark red. There is much research currently being directed toward the discovery of the unique nano properties of materials. Many novel applications of these nano properties are sure to follow.

FIGURE 14.18
A schematic of a scanning probe microscope that detects and characterizes the surface atoms of a material by way of an ultrathin probe tip attached to a miniature cantilever.

microtechnology techniques to smaller and smaller scales. A nanosize circuit board, for example, might be carved from a larger block of material. The bottom-up approach involves building nanosized objects atom by atom. A most important tool for either of these approaches is the **scanning probe microscope**, which detects and characterizes the surface atoms of materials by way of an ultrathin probe tip, as shown in Figures 14.18 and 14.19. The tip is mechanically dragged over the surface. Interactions between the tip and the surface atoms cause movements in a cantilever attached to the tip that are detected by a laser beam and translated by a computer into a topographical image. Scanning probe microscopes can also be used to move individual atoms into desired positions.

Nanotechnology allows the continued miniaturization of integrated circuits needed for ever smaller and more powerful computers. But a computer need not rely on an integrated circuit of nanowires for processing power. A wholly new approach involves designing logic boards in which molecules (not electric circuits) read, process, and write information. One molecule that has proved most promising for such *molecular computation* is DNA, the same molecule that holds our genetic code. An advantage that molecular computing has over conventional computing is that it can run a massive number of calculations in parallel (at the same time). Because of such fundamental differences, molecular computing may one day outshine even the fastest integrated circuits. Molecular computing, in turn, may then be eclipsed by other novel approaches, such as quantum or photon computing, also made possible by nanotechnology.

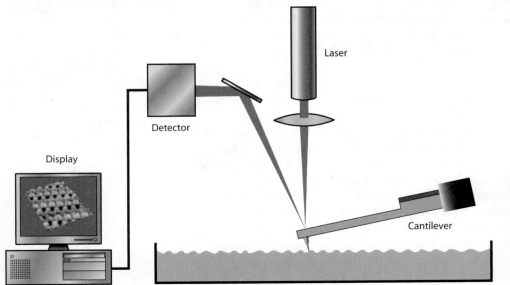
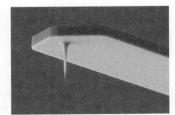

Laser

Detector

Display

Computer

Cantilever

Sample

Probe tip

The ultimate expert on nanotechnology is nature. Living organisms, for example, are complex systems of interacting biomolecules all functioning on the scale of nanometers. In this sense, the living organism is nature's nanomachine. We need look no further than our own bodies to find evidence of the feasibility and power of nanotechnology. With nature as our teacher, we have much to learn. Such knowledge will be particularly applicable to medicine. By becoming nanotechnology experts ourselves, we would be well equipped to understand exact causes of nearly any disease or disorder (aging included) and empowered to develop innovative cures.

What are the limits of nanotechnology? As a society, how will we deal with the impending changes nanotechnology may bring? Consider the possibilities: wall paint that can change color or be used to display video; smart dust that the military could use to seek out and destroy an enemy; solar cells that capture sunlight so efficiently that they render fossil fuels obsolete; robots with so much processing power that we begin to wonder whether they experience consciousness; nanobots that roam our circulatory systems destroying cancerous tumors or arterial plaque; nanomachines that can "photocopy" three-dimensional objects, including living organisms; medicines that more than double the average human life span. Stay tuned for an exciting new revolution in human capabilities.

FIGURE 14.19
An artist's rendition of the interaction between surface carbon atoms and the tip of a scanning probe microscope.

CHECKPOINT

How believable would our present technology be to someone living 200 years ago? How believable might the technology of 200 years in the future be to us right now?

Was this your answer?
Hind sight is 20-20. It's always easy to look back over time and see the progression of events that led to our present state. Much more difficult is it to think forward and project possible scenarios. The technology of 200 years from now may be just as unbelievable to us as our present technology would be unbelievable to someone of 200 years ago.

For instructor-assigned homework, go to www.masteringphysics.com

SUMMARY OF TERMS (KNOWLEDGE)

Applied research Research that focuses on developing applications of knowledge gained through basic research.

Basic research Research that leads us to a greater understanding of how the natural world operates.

Chemical bond The force of attraction between two atoms that holds them together.

Chemical change The formation of new substance(s) by rearranging the atoms of the original material(s).

Chemical formula A notation that indicates the composition of a compound, consisting of the atomic symbols for the different elements of the compound and numerical subscripts indicating the ratio in which the atoms combine.

Chemical property Any property that characterizes the ability of a substance to change into a different substance under specific conditions.

Chemical reaction A term synonymous with *chemical change*.

Chemistry The study of matter and the transformations it can undergo.

Compound A material in which atoms of different elements are bonded to one another.

Elemental formula A notation that uses the atomic symbol and (sometimes) a numerical subscript to denote how many atoms are bonded in one unit of an element.

Molecule An extremely small fundamental structure built of atoms.

Nanotechnology The manipulation of individual atoms or molecules.

Physical change A change in a substance's physical properties but with no change in its chemical identity.

Physical property Any physical attribute of a substance, such as color, density, or hardness.

Scanning probe microscope A tool of nanotechnology that detects and characterizes the surface atoms of materials by way of an ultrathin probe tip, which is detected by laser light as it is mechanically dragged over the surface.

Submicroscopic The realm of atoms and molecules, where objects are too small to be detected by optical microscopes.

READING CHECK QUESTIONS (COMPREHENSION)

14.1 Chemistry: The Central Science

1. Why is chemistry often called the central science?
2. What is the difference between basic research and applied research?
3. What pledge has been made by members of the American Chemistry Council through the Responsible Care program?

14.2 The Submicroscopic World

4. It would take you 31,800 years to count to a trillion. About how many times would you have to do this to count all the atoms in a single grain of sand?
5. Is a biological cell macroscopic, microscopic, or submicroscopic?
6. Are atoms made of molecules or are molecules made of atoms?
7. How are the particles in a solid arranged differently from those in a liquid?
8. How does the arrangement of particles in a gas differ from the arrangements in liquids and solids?
9. Which occupies the greatest volume: 1 g of ice, 1 g of liquid water, or 1 g of water vapor?

14.3 Physical and Chemical Properties

10. What happens to the chemical identity of a substance during a physical change?
11. What is a physical property?
12. What doesn't change during a physical change?
13. What is a chemical property?
14. What is a chemical bond?
15. What changes during a chemical reaction?

14.4 Determining Physical and Chemical Changes

16. Why is the freezing of water considered to be a physical change?
17. Why is it sometimes difficult to decide whether an observed change is physical or chemical?
18. Why is the rusting of iron considered to be a chemical change?
19. What are some of the clues that help us to determine whether an observed change is physical or chemical?

14.5 Elements to Compounds

20. Distinguish between an atom and an element.
21. How many atoms are in a sulfur molecule that has the elemental formula S_8?
22. What is the difference between an element and a compound?
23. How many atoms are there in one molecule of H_3PO_4? How many atoms of each element are there in one molecule of H_3PO_4?
24. What does the chemical formula of a substance tell us about that substance?
25. Are the physical and chemical properties of a compound necessarily similar to those of the elements of which it is composed?

14.6 Naming Compounds

26. Which element within a compound is given first in the compound's name?
27. What is the IUPAC systematic name for the compound KF?
28. What is the chemical formula for the compound titanium dioxide?
29. Why are common names often used for chemical compounds instead of systematic names?

14.7 The Advent of Nanotechnology

30. How soon will nanotechnology give rise to commercial products?
31. What are the two main approaches to building nanoscale materials and devices?
32. Who is the ultimate expert at nanotechnology?

ACTIVITIES (HANDS-ON APPLICATION)

33. A television screen looked at from a distance appears as a smooth continuous flow of images. Up close, however, we see that this is an illusion. What really exists are a series of tiny dots (pixels) that change color in a coordinated way to produce images. Use a magnifying glass to examine closely the screen of a computer monitor or television set.

34. Add a pinch of red Kool-Aid crystals to a still glass of hot water. Add the same amount of crystals to a second still glass of cold water. With no stirring, which would you expect to become uniform in color first: the hot water or the cold water? Why?

35. Air molecules stuck inside an inflated balloon collide continuously with the inner surface of the balloon. Each collision provides a little push outward on the balloon. All the many collisions working together keep the balloon inflated. To get a "feel" for what's happening here, add about a tablespoon of tiny beads to a large balloon (pellets, beans, BBs, or grains of rice also work). Inflate the balloon to its full size and tie it shut. Hold the balloon in the palms of both hands and shake rapidly. Can you feel the collisions? As you shake the balloon wildly, the flying beads represent the gaseous phase. How should you move the balloon so that the beads represent the liquid phase? The solid phase?

36. This activity is for those with access to a gas stove. Place a large pot of cool water on top of the stove, and set the burner on high. What product from the combustion of the natural gas do you see condensing on the outside of the pot? Where did it come from? Would more or less of this product form if the pot contained ice water? Where does this product go as the pot gets warmer? What physical and chemical changes can you identify?

37. When you pour a solution of hydrogen peroxide, H_2O_2, over a cut, an enzyme in your blood decomposes it to produce oxygen gas, O_2, as evidenced by the bubbling that takes place. It is this oxygen at high concentrations at the site of the injury that kills off microorganisms. A similar enzyme is found in baker's yeast.

Wear safety glasses and remove all combustibles, such as paper towels, from a clear countertop area. Pour a small packet of baker's yeast into a tall glass. Add a couple capfuls of 3% hydrogen peroxide and watch oxygen bubbles form. Test for the presense of oxygen by holding a lighted match with a tweezers and putting the flame near the bubbles. Look for the flame to glow more brightly as the escaping oxygen passes over it. Describe oxygen's physical and chemical properties.

THINK AND RANK (ANALYSIS)

38. Rank the following in order of increasing volume: (a) bacterium, (b) virus, (c) water molecule.

39. Rank the following in order of increasing force of attraction between its submicroscopic particles: (a) sugar, (b) water, (c) air.

40. Rank the physical and chemical changes in order of the amount of energy released: (a) the condensation of rain in a thunderstorm, (b) the burning of a gallon of gasoline in a car engine, (c) the explosion of a firecracker.

41. Rank the compounds in order of increasing numbers of atoms: (a) $C_{12}H_{22}O_{12}$, (b) DNA, (c) $Pb(C_2H_3O_2)_2$.

EXERCISES (SYNTHESIS)

42. While visiting a foreign country, a foreign-speaking citizen tries to give you verbal directions to a local museum. After multiple attempts, he is unsuccessful. An onlooker sees your frustration and concludes that you are not smart enough to understand simple directions. Another onlooker sympathizes with you because he knows how difficult it is to navigate through an unfamiliar city. Which onlooker is correct?

43. If someone is able to explain an idea to you using small familiar words, what does this say about how well that person understands the idea?

44. What is the best way to really prove to yourself that you understand an idea?

45. Of physics, chemistry, and biology, which science is the most complex? Please explain.

46. Is chemistry the study of submicroscopic, microscopic, macroscopic, or all three? Defend your answer.

47. You combine 50 mL of small BBs with 50 mL of large BBs and get a total of 90 mL of BBs of mixed sizes. Please explain how this occurs.

48. You combine 50 mL of water with 50 mL of purified alcohol and get a total of 98 mL of the mixture. Please explain how this occurs.

49. Which has stronger attractions among its submicroscopic particles: a solid at 25°C or a gas at 25°C? Explain.

50. The leftmost diagram below shows the moving particles of a gas within a rigid container. Which of the three boxes on the right—(a), (b), or (c)—best represents this material upon the addition of heat?

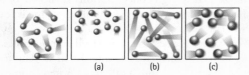
(a)   (b)   (c)

51. The leftmost diagram below shows two phases of a single substance. In the middle box, draw what these particles would look like if heat were taken away. In the box on the right, show what they would look like if heat were added. If each particle represents a water molecule, what is the temperature of the box on the left?

52. A cotton ball is dipped in alcohol and wiped across a tabletop. Explain what happens to the alcohol molecules deposited on the tabletop. Is this a physical or chemical change?

53. A skillet is lined with a thin layer of cooking oil followed by a layer of unpopped popcorn kernels. Upon heating, the kernels all pop, thereby escaping the skillet. Identify any physical or chemical changes.

54. A cotton ball dipped in alcohol is wiped across a tabletop. Would the resulting smell of the alcohol be more or less noticeable if the tabletop were much warmer? Explain.

55. Use Exercise 53 as an analogy to describe what occurs in Exercise 54. Does it make sense to think that the alcohol is made of very tiny particles (molecules) rather than being an infinitely continuous material?

56. Alcohol wiped across a tabletop rapidly disappears. What happens to the temperature of the tabletop? Why?

57. Red Kool-Aid crystals are added to a still glass of water. The crystals sink to the bottom. Twenty-four hours later, the entire solution is red even though no one stirred the water. Explain.

58. With no one looking, you add 5 mL of a cinnamon solution to a blue balloon, which you tie shut. You also add 5 mL of fresh water to a red balloon, which you also tie shut. You heat the two balloons in a microwave until each inflates to about the size of a grapefruit. Your brother then comes along, examines the inflated balloons, and tells you that the blue balloon is the one containing the cinnamon. How did he know?

59. In the winter Vermonters make a tasty treat called "sugar on snow" in which they pour boiled-down maple syrup onto a scoop of clean fresh snow. As the syrup hits the snow, it forms a delicious taffy. Identify the physical changes involved in the making of sugar on snow. Identify any chemical changes.

60. Oxygen, O_2, has a boiling point of 90 K (-183°C), and nitrogen, N_2, has a boiling point of 77 K (-196°C). Which is a liquid and which is a gas at 80 K (-193°C)?

61. Each night you measure your height just before going to bed. When you arise each morning, you measure your height again and consistently find that you are 1 inch taller than you were the night before but only as tall as you were 24 hours ago! Is what happens to your body in this instance best described as a physical change or a chemical change? Be sure to try this activity if you haven't already.

62. State whether each of the following is a physical or chemical property of matter.
    (a) Graphite conducts electricity.
    (b) Bismuth, Bi, loses its iridescence upon melting.
    (c) A copper penny is smushed into an embossed souvenir.

63. State whether each of the following is a physical or chemical property of matter.
    (a) Carbon dioxide escapes when a soda can is opened.
    (b) A bronze statue turns green.
    (c) A silver spoon tarnishes.

64. Classify each change as physical or chemical. Even if you are incorrect in your assessment, you should be able to defend why you chose as you did.
    (a) Grape juice turns to wine.
    (b) Wood burns to ashes.
    (c) Water begins to boil.
    (d) A broken leg mends itself.

65. Classify each change as physical or chemical. Even if you are incorrect in your assessment, you should be able to defend why you chose as you did.
    (a) Grass grows.
    (b) An infant gains 10 lb.
    (c) A rock is crushed to powder.
    (d) A tire is inflated with air.
66. Each sphere in the diagrams below represents an atom. Joined spheres represent molecules. Which box contains a liquid phase? Why can't you assume that box B represents a lower temperature?

A   B

67. Is aging primarily an example of a physical or chemical change?
68. Octane is a component of gasoline. It reacts with oxygen, O_2, to form carbon dioxide and water. Is octane an element or a compound? How can you tell?
69. Oxygen atoms are used to make water molecules. Does this mean that oxygen, O_2, and water, H_2O, have similar properties? Why do we drown when we breathe in water despite all the oxygen atoms present in this material?

70. Oxygen, O_2, is certainly good for you. Does it follow that if small amounts of oxygen are good for you then large amounts of oxygen would be especially good for you?
71. Why isn't water classified as an element?
72. If you eat metallic sodium or inhale chlorine gas, you stand a strong chance of dying. Let these two elements react with each other, however, and you can safely sprinkle the compound on your popcorn for better taste. What is going on?
73. Which of the following boxes contains only an element? Which contains only a compound? How many different types of molecules are shown altogether in the three boxes?

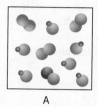 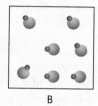 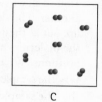

A   B   C

74. What is the chemical formula for the compound dihydrogen sulfide?
75. What is the chemical name for a compound with the formula Ba_3N_2?
76. What is the common name for dioxygen oxide?
77. What is the common name for oxygen oxide?

DISCUSSION QUESTIONS (EVALUATION)

78. Medicines, such as pain relievers and antidepressants, are found in the drinking water supplies of many municipalities. How did these medicines get there? Does it matter that they are there? Should something be done about it? If so, what?
79. Your friend smells cinnamon coming from an inflated rubber balloon containing cinnamon extract. You tell him that the cinnamon molecules are passing through the micropores of the balloon. He accepts the idea that the balloon contains micropores but insists that he is simply smelling cinnamon-flavored air. You explain that scientists have discovered that gases are made of molecules, but that's not good enough for him. He needs to see the evidence for himself. How might you lead him to accept the concept of molecules?
80. The British diplomat, physicist, and environmentalist John Ashton, in speaking to a group of scientists, stated (paraphrased): "There has to be much better communication between the world of science and the world of politics. Consider the different meanings of the word *uncertainty*. To scientists, it means uncertainty over the strength of a signal. To politicians it means 'go away and come back when you're certain.'" Pretend you are a scientist with strong but inconclusive evidence in support of impending climate change. How might you best persuade politicians to take action?

81. The famous 20th-century physicist Richard Feynman (1918–1988) noted: "The laws of science do not limit our ability to manipulate single atoms and molecules." What does?
82. "A calculator is useful but certainly not exciting." Why would someone from 100 years ago vehemently disagree with this statement? We often marvel at a new technology, but how long does this marveling last? How soon before a new technology becomes assumed? Think of other examples. Is technology the source of happiness?
83. How might speculations about potential dangers of nanotechnology threaten public support for it? Consider Michael Crichton's 2002 science-fiction novel *Prey*, in which self-replicating nanobots run amok turning everything they contact into a gray goo.
84. In the past 20 years, the average life expectancy in most nations has risen by a couple of years. So has the "healthy life expectancy," which is a measure of how long people remain in good health. Are the two necessarily related? How so?
85. What effects might a cure for aging have on the problem of overpopulation? How would society be able to support so many people living well into their 100s? What trends do you foresee in company retirement plans? Might there be a greater emphasis on "privatized retirement accounts"? Why?

READINESS ASSURANCE TEST (RAT)

If you have a good handle on this chapter, if you really do, then you should be able to score 7 out of 10 on this RAT. If you score less than 7, you need to study further before moving on.

Choose the BEST answer to each of the following.

1. Chemistry is the study of
   (a) matter.
   (b) transformations of matter.
   (c) only microscopic phenomena.
   (d) only macroscopic phenomena.
   (e) both a and b.

2. Imagine that you can see individual molecules. You watch a small collection of molecules that are moving around slowly while vibrating and bumping against one another. The slower-moving molecules then start to line up, but as they do so, their vibrations increase. Soon all the molecules are aligned and vibrating about fixed positions. What is happening?
   (a) The sample is being cooled and the material is freezing.
   (b) The sample is being heated and the material is melting.
   (c) The sample is being cooled and the material is condensing.
   (d) The sample is being heated and the materal is boiling.
   (e) The sample is unchanged.

3. The phase in which atoms and molecules no longer move is the
   (a) solid phase.
   (b) liquid phase.
   (c) gas phase.
   (d) none of the above.

4. What chemical change occurs when a wax candle burns?
   (a) The wax near the flame melts.
   (b) Carbon soot collects above the flame.
   (c) The molten wax is pulled upward through the wick.
   (d) The heated wax molecules combine with oxygen molecules.
   (e) Two of the above are signs of chemical change.

5. Based on the information given in the following diagrams, which substance has the lower boiling point: one made from molecule A, 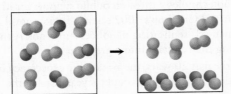, or one made from molecule B, ?

   (a) molecule A, which is the first to transform into a liquid
   (b) molecule B, which is the first to transform into a liquid
   (c) molecule A, which remains in the gaseous phase
   (d) molecule B, which remains in the gaseous phase

6. Does the following transformation represent a physical change or a chemical change?

   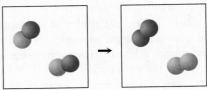

   (a) chemical, because of the formation of elements
   (b) physical, because a new material has been formed
   (c) chemical, because the atoms are connected differently
   (d) physical, because of a change in phase

7. Which is an example of a chemical change?
   (a) water freezing into ice crystals
   (b) aftershave or perfume on your skin generating a smell
   (c) a piece of metal expanding when heated, but returning to its original size when cooled
   (d) breaking a glass window
   (e) gasoline in the engine of a car producing exhaust

8. If you burn 50 g of wood and produce 10 g of ash, what is the total mass of all the products produced from the burning of this wood?
   (a) more than 50 g
   (b) 10 g
   (c) less than 10 g
   (d) 50 g
   (e) none of the above

9. If you have one molecule of TiO_2, how many molecules of O_2 does it contain?
   (a) One, TiO_2 is a mixture of Ti and O_2.
   (b) None, O_2 is a different molecule than TiO_2.
   (c) Two, TiO_2 is a mixture of Ti and 2 O.
   (d) Three, TiO_2 contains three molecules.

10. What is the name of the compound $CaCl_2$?
    (a) carbon chloride
    (b) dichlorocalcium
    (c) calc two
    (d) dicalcium chloride
    (e) calcium chloride

Answers to RAT

1. e, 2. a, 3. d, 4. e, 5. c, 6. c, 7. e, 8. a, 9. b, 10. e

CHAPTER 15

How Atoms Bond and Molecules Attract

WHY DO salt crystals have a distinct cubic shape? As we will see in this chapter, the macroscopic properties of any substance can be traced to how its submicroscopic parts are held together. The sodium and chloride ions in a salt crystal, for example, hold together in a cubic orientation, and, as a result, the macroscopic object we know as a salt crystal is also cubic.

The force of attraction that holds ions or atoms together is the electric force. Chemists refer to this atom-binding force as a chemical bond. In this chapter, we explore three types of chemical bonds: the *ionic bond*, which holds ions together in a crystal; the *metallic bond*, which holds atoms together in a piece of metal, and the *covalent bond*, which holds atoms together in a molecule. Then later in this chapter, we explore how the behavior of ions and molecules gives rise to macroscopic phenomena, such as the mixing of salt and water.

15.1 Electron-Dot Structures

EXPLAIN THIS   Why do the atoms of group 18 resist forming chemical bonds?

An atomic model is needed to help us understand how atoms bond. We begin this chapter with a brief overview of the shell model presented in Section 12.9. Recall how electrons are arranged around an atomic nucleus. Rather than moving in neat orbits like planets around the Sun, electrons are wavelike entities that swarm in various volumes of space called *shells*.

As was shown in Figure 12.29, seven shells are available to the electrons in an atom, and the electrons fill these shells in order, from innermost to outermost. Furthermore, the maximum number of electrons allowed in the first shell is 2, and for the second and third shells it is 8. The fourth and fifth shells can each hold 18 electrons, and the sixth and seventh shells can each hold 32 electrons.* These numbers match the number of elements in each period (horizontal row) of the periodic table. Figure 15.1 shows how this model applies to the first three elements of group 18.

Electrons in the outermost occupied shell of any atom may play a significant role in that atom's chemical properties, including its ability to form chemical bonds. To indicate their importance, we call these electrons **valence electrons** (as described in Section 15.5), and we call the shell they occupy the **valence shell**. Valence electrons can be conveniently represented as a series of dots surrounding an atomic symbol. This notation is called an **electron-dot structure**, or sometimes a *Lewis dot symbol* (in honor of the American chemist G. N. Lewis, who first proposed the concepts of shells and valence electrons).

Figure 15.2 shows the electron-dot structures for the atoms important in our discussions of ionic and covalent bonds. For our discussion of metallic bonds, we'll focus only on the valence electrons of metal atoms and not on their electron-dot structures.

When you look at the electron-dot structure of an atom, you immediately know two important things about that element. You know how many valence electrons it has and how many of these electrons are *paired*. Chlorine, for example, has three sets of paired electrons and one unpaired electron, and carbon has four unpaired electrons:

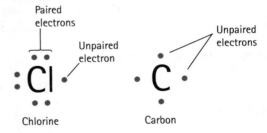

Paired valence electrons are relatively stable. In other words, they usually do not form chemical bonds with other atoms. For this reason, electron pairs in an electron-dot structure are called **nonbonding pairs**. (Do not take this term literally, however, because in Chapter 18 you'll see that under the right conditions, even "nonbonding" pairs can form a chemical bond.)

Valence electrons that are *unpaired*, by contrast, have a strong tendency to participate in chemical bonding. By doing so, they become paired with an electron from another atom. The ionic and covalent bonds discussed in this chapter all result from either a transfer or a sharing of unpaired valence electrons.

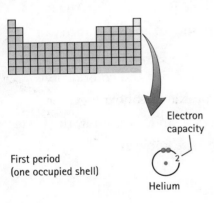

Electron capacity

First period (one occupied shell)

Helium

Second period (two occupied shells)

Neon

Third period (three occupied shells)

Argon

FIGURE 15.1
Occupied shells in the group 18 elements helium through argon. Each of these elements has a filled outermost occupied shell, and the number of electrons in each corresponds to the number of elements in the period to which a particular group 18 element belongs.

* As a point of reference for physicists reading this text, these shells of orbitals are grouped by similar energy levels rather than by principal quantum number. They are the "argonian" shells developed by Linus Pauling to explain chemical bonding and the organization of the periodic table.

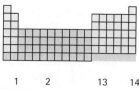

FIGURE 15.2
The valence electrons of an atom are shown in its electron-dot structure. Note that the first three periods here parallel Figure 12.30. Also note that for larger atoms, not all the electrons in the valence shell are valence electrons. Krypton, Kr, for example, has 18 electrons in its valence shell, but only 8 of these are classified as valence electrons.

CHECKPOINT
Where are valence electrons located, and why are they important?

Was this your answer?
Valence electrons are located in the outermost occupied shell of an atom. They are important because they play a leading role in determining the chemical properties of the atom.

Too much detail to learn? What would the scientists of 200 years ago give for the information that today is so readily available to you?

LEARNING OBJECTIVE
Use the periodic table to predict the type of ion an atom tends to form.

15.2 The Formation of Ions

EXPLAIN THIS When does a gain result in a negative?

When the number of protons in the nucleus of an atom equals the number of electrons in the atom, the charges balance and the atom is electrically neutral. If one or more electrons are lost or gained, as illustrated in Figures 15.4 and 15.5, the balance is upset and the atom takes on a net electric charge. Any atom with a net electric charge is an **ion**. When electrons are lost, protons outnumber electrons and the ion has a positive net charge. When electrons are gained, electrons outnumber protons and the ion has a negative net charge.

Chemists use a superscript to the right of the atomic symbol to indicate the magnitude and sign of an ion's charge. Thus, as shown in Figures 15.4 and 15.5, the positive ion formed from the sodium atom is written Na^{1+} and the negative ion formed from the fluorine atom is written F^{1-}. Usually the numeral 1 is omitted when indicating either a 1^+ or 1^- charge. Hence, these two ions are most frequently written Na^+ and F^-.

To give two more examples, a calcium atom that loses two electrons is written Ca^{2+}, and an oxygen atom that gains two electrons is written O^{2-}. (Note that the convention is to write the numeral before the sign, not after it: $2+$, not $+2$.)

We can use the shell model to deduce the type of ion an atom tends to form. According to this model, *atoms tend to lose or gain electrons that result in an outermost occupied shell filled to capacity.* Let's take a moment to consider this point, looking to Figures 15.4 and 15.5 as visual guides.

FIGURE 15.3
Gilbert Newton Lewis (1875–1946) revolutionized chemistry with his theory of chemical bonding, which he published in 1916. He worked most of his life in the chemistry department of the University of California, Berkeley, where he was not only a productive researcher but also an exceptional teacher. Among his teaching innovations was the idea of providing students with problem sets as a follow-up to lectures and readings.

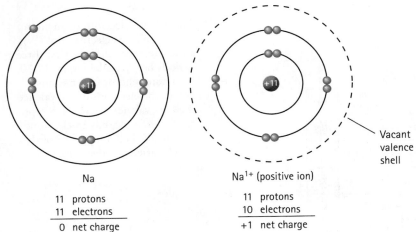

FIGURE 15.4
An electrically neutral sodium atom contains 11 negatively charged electrons surrounding the 11 positively charged protons of the nucleus. When this atom loses an electron, the result is a positive ion.

If an atom has only one or a few electrons in its valence shell, it tends to give up (lose) these electrons so that the next shell inward, which is already filled, becomes the outermost occupied shell. The sodium atom of Figure 15.4, for example, has one electron in its valence shell, which is the third shell. In forming an ion, the sodium atom loses this electron, thereby making the second shell, which is already filled to capacity, the outermost occupied shell. Because the sodium atom has only one valence electron to lose, it tends to form the 1+ ion.

If the valence shell of an atom is almost filled, that atom attracts electrons from another atom and so forms a negative ion. The fluorine atom of Figure 15.5, for example, has one space available in its valence shell for an additional electron. After this additional electron is gained, the fluorine atom achieves a filled valence shell. Fluorine therefore tends to form the 1− ion.

You can use the periodic table as a quick reference when determining the type of ion an atom tends to form. As Figure 15.6 shows, each atom of any group 1 element, for example, has only one valence electron and so tends to form the 1+ ion. Each atom of any group 17 element has room for one additional electron in its valence shell and therefore tends to form the 1− ion. Atoms of the noble-gas elements tend not to form ions of any type because their valence shells are already filled to capacity.

CHECKPOINT
What type of ion does the magnesium atom, Mg, tend to form?

Was this your answer?
The magnesium atom (atomic number 12) is found in group 2 and has two valence electrons to lose (see Figure 15.2). Therefore, it tends to form the 2+ ion.

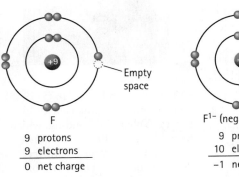

FIGURE 15.5
An electrically neutral fluorine atom contains 9 protons and 9 electrons. When this atom gains an electron, the result is a negative ion.

As is indicated in Figure 15.6, the attraction between an atom's nucleus and its valence electrons is weakest for elements on the left in the periodic table and strongest for elements on the right. From sodium's position in the table, we can see that a sodium atom's single valence electron is not held very strongly, which explains why it is so easily lost. The attraction the sodium nucleus has for its second-shell electrons, however, is much stronger, which is why the sodium atom rarely loses more than one electron.

At the other side of the periodic table, the nucleus of a fluorine atom holds strongly onto its valence electrons, which explains why the fluorine atom tends not to lose any electrons to form a positive ion. Instead, fluorine's nuclear pull on the valence electrons is strong enough to accommodate even an additional electron "imported" from some other atom.

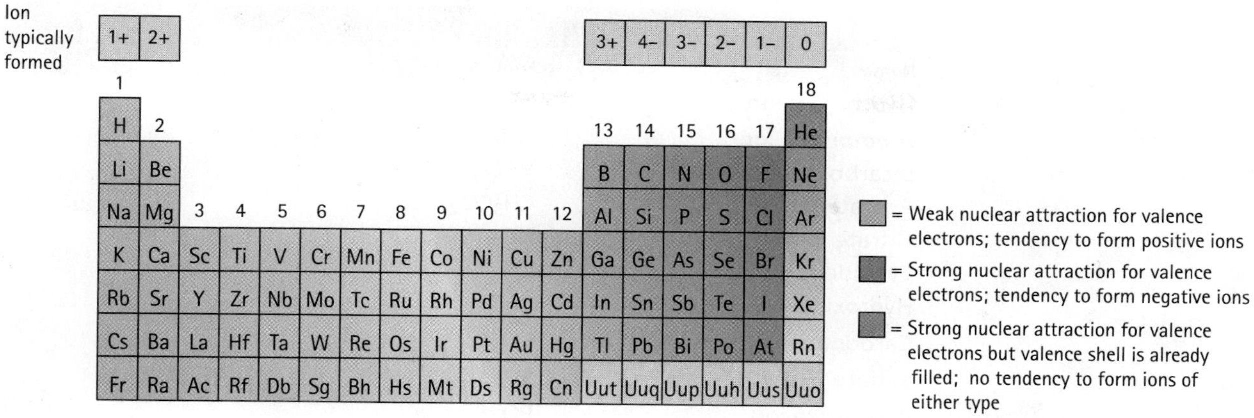

FIGURE 15.6
The periodic table is your guide to the types of ions that atoms tend to form.

The nucleus of a noble-gas atom pulls so strongly on its valence electrons that they are very difficult to remove. Because no space is available in the valence shell of a noble-gas atom, no additional electrons are gained. Thus, a noble-gas atom tends not to form ions of any sort.

CHECKPOINT
Why does the magnesium atom tend to form the 2+ ion?

Was this your answer?
Magnesium is on the left in the periodic table, and so atoms of this element do not hold onto the two valence electrons very strongly. Because these electrons are not held very tightly, they are easily lost, which is why the magnesium atom tends to form the 2+ ion.

Using our shell model to explain the formation of ions works well for groups 1, 2, and 13 through 18. This model is too simplified to work well for the transition metals of groups 3 through 12, however, or for the inner transition metals. In general, these metal atoms tend to form positive ions, but the number of electrons lost varies. For example, depending on conditions, an iron atom may lose two electrons to form the Fe^{2+} ion, or it may lose three electrons to form the Fe^{3+} ion.

Molecules Can Form Ions

So we see that atoms form ions by losing or gaining electrons. Interestingly, molecules can also become ions. In most cases, this occurs whenever a molecule loses or gains a proton—equivalent to the hydrogen ion, H^+. (Recall that a hydrogen atom is a proton together with an electron. The hydrogen ion, H^+, therefore, is simply a proton.) For example, a water molecule, H_2O, can gain a hydrogen ion, H^+ (a proton), to form the hydronium ion, H^3O^+:

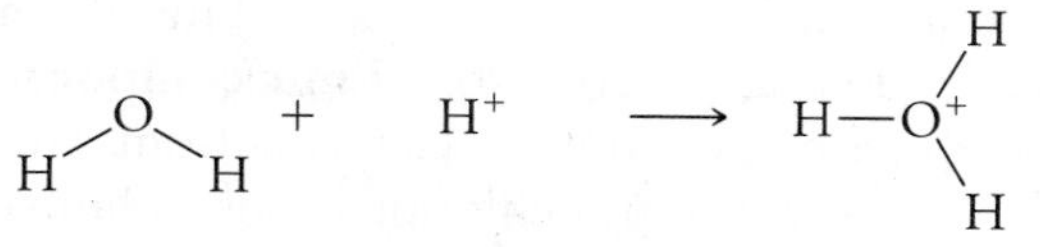

Electrons are negatively charged. So gaining an electron results in a negative ion, and losing an electron results in a positive ion.

FYI What do the ions of the following elements have in common: calcium, Ca; chlorine, Cl; chromium, Cr; cobalt, Co; copper, Cu; fluorine, F; iodine, I; iron, Fe; magnesium, Mg; manganese, Mn; molybdenum, Mo; nickel, Ni; phosphorus, P; potassium, K; selenium, Se; sodium, Na; sulfur, S; zinc, Zn? They are all dietary minerals that are essential for good health but that can be harmful, even lethal, when consumed in excessive amounts.

Water   Hydrogen ion   Hydronium ion
                   (proton)

TABLE 15.1	COMMON POLYATOMIC IONS
Name	Formula
Hydronium ion	H_3O^+
Ammonium ion	NH_4^+
Bicarbonate ion	HCO_3^-
Acetate ion	$CH_3CO_2^-$
Nitrate ion	NO_3^-
Cyanide ion	CN^-
Hydroxide ion	OH^-
Carbonate ion	CO_3^{2-}
Sulfate ion	SO_4^{2-}
Phosphate ion	PO_4^{3-}

Similarly, the carbonic acid molecule, H_2CO_3, can lose two protons to form the carbonate ion, CO_3^{2-}:

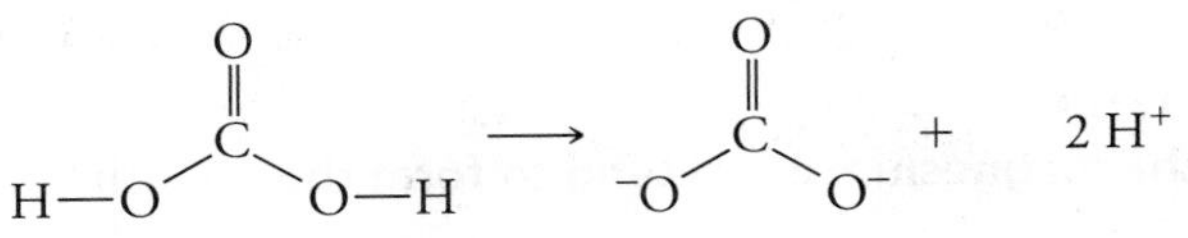

How these reactions occur will be explored in later chapters. For now, you should understand that the hydronium and carbonate ions are examples of **polyatomic ions**, which are molecules that carry a net electric charge. Table 15.1 lists some commonly encountered polyatomic ions.

15.3 Ionic Bonds

LEARNING OBJECTIVE
Describe how ions combine to form ionic compounds.

Mastering**PHYSICS**
VIDEO: Ionic Bonds

EXPLAIN THIS Why do ionic compounds have very high melting points?

When an atom that tends to lose electrons is placed in contact with an atom that tends to gain them, the result is an electron transfer and the formation of two oppositely charged ions. This occurs when sodium and chlorine are combined. As shown in Figure 15.7, the sodium atom loses one of its electrons to the chlorine atom, resulting in the formation of a positive sodium ion and a negative chloride ion. The two oppositely charged ions are attracted to each other by the electric force, which holds them close together. This electric force of attraction between two oppositely charged ions is called an **ionic bond**.

A sodium ion and a chloride ion together make the chemical compound sodium chloride, commonly known as table salt. This and all other chemical compounds containing ions are referred to as **ionic compounds**. All ionic compounds are completely different from the elements from which they are made. As discussed in Section 14.5, sodium chloride is not sodium, nor is it chlorine. Rather, it is a collection of sodium and chloride ions that form a unique material with its own physical and chemical properties.

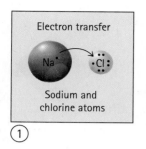

Electron transfer

Na → :Cl:

Sodium and
chlorine atoms

①

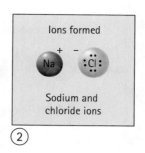

Ions formed

Na⁺ :Cl:⁻

Sodium and
chloride ions

②

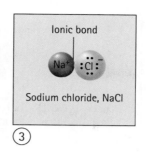

Ionic bond

Na⁺ :Cl:⁻

Sodium chloride, NaCl

③

FIGURE 15.7

(1) An electrically neutral sodium atom loses its valence electron to an electrically neutral chlorine atom. (2) This electron transfer results in two oppositely charged ions. (3) The ions are then held together by an ionic bond. The spheres drawn around these and subsequent illustrations of electron-dot structures indicate the relative sizes of the atoms and ions. Note that the sodium ion is smaller than the sodium atom because the lone electron in the third shell has gone once the ion forms, leaving the ion with only two occupied shells. The chloride ion is larger than the chlorine atom because the addition of that one electron to the third shell makes the shell expand due to the repulsions among the electrons.

CHECKPOINT

Is the transfer of an electron from a sodium atom to a chlorine atom a physical change or a chemical change?

Was this your answer?
Recall from Chapter 14 that only a chemical change involves the formation of new material. Thus, this or any other electron transfer, because it results in the formation of a new substance, is a chemical change.

As Figure 15.8 shows, ionic compounds typically consist of elements that are found on opposite sides of the periodic table. Also, because of how the metals and nonmetals are organized in the periodic table, positive ions are generally derived from metallic elements and negative ions are generally derived from nonmetallic elements.

For all ionic compounds, positive and negative charges must balance. In sodium chloride, for example, there is one sodium $1+$ ion for every chloride $1-$ ion. Charges must also balance in compounds containing ions that carry multiple charges. The calcium ion, for example, carries a charge of $2+$, but the fluoride ion carries a charge of only $1-$. Because two fluoride ions are needed to balance each calcium ion, the formula for calcium fluoride is CaF_2, as Figure 15.9 illustrates. Calcium fluoride occurs naturally in the drinking water of some communities, where it is a good source of the tooth-strengthening fluoride ion, F^-.

An aluminum ion carries a $3+$ charge, and an oxide ion carries a $2-$ charge. Together, these ions make the ionic compound aluminum oxide, Al_2O_3, the

The ionic bond is merely the electrical force of attraction that holds ions of opposite charge together, in accord with Coulomb's law (Chapter 8).

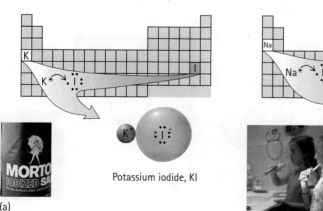

Potassium iodide, KI

(a)

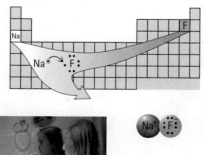

(b)

Sodium fluoride, NaF

FIGURE 15.8

(a) The ionic compound potassium iodide, KI, is added in minute quantities to commercial salt because the iodide ion, I^-, it contains is an essential dietary mineral. (b) The ionic compound sodium fluoride, NaF, is often added to municipal water supplies and toothpastes because it is a good source of the tooth-strengthening fluoride ion, F^-.

FIGURE 15.9
A calcium atom loses two electrons to form a calcium ion, Ca^{2+}. These two electrons may be picked up by two fluorine atoms, transforming the atoms to two fluoride ions. Calcium ions and fluoride ions then join to form the ionic compound calcium fluoride, CaF_2, which occurs naturally as the mineral fluorite.

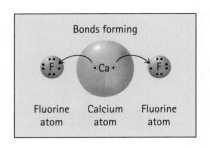

Bonds forming

Fluorine atom / Calcium atom / Fluorine atom

Ionic bonds formed

Calcium fluoride, CaF_2

Fluorite

main component of such gemstones as rubies and sapphires. Figure 15.10 illustrates the formation of aluminum oxide. The three oxide ions in Al_2O_3 carry a total charge of $6-$, which balances the total $6+$ charge of the two aluminum ions. As mentioned earlier, rubies and sapphires differ in color because of the impurities they contain. Rubies are red because of minor amounts of chromium ions, and sapphires are blue because of minor amounts of iron and titanium ions.

CHECKPOINT
What is the chemical formula for the ionic compound magnesium oxide?

Was this your answer?
Because magnesium is a group 2 element, you know a magnesium atom must lose two electrons to form a Mg^{2+} ion. Because oxygen is a group 16 element, an oxygen atom gains two electrons to form an O^{2-} ion. These charges balance in a one-to-one ratio, and so the formula for magnesium oxide is MgO.

FIGURE 15.10
Two aluminum atoms lose a total of six electrons to form two aluminum ions, Al^{3+}. These six electrons may be picked up by three oxygen atoms, transforming the atoms to three oxide ions, O^{2-}. The aluminum and oxide ions then join to form the ionic compound aluminum oxide, Al_2O_3.

An ionic compound typically contains a multitude of ions grouped together in a highly ordered three-dimensional array. In sodium chloride, for example, each sodium ion is surrounded by six chloride ions, and each chloride ion is surrounded by six sodium ions (Figure 15.11). Overall, there is one sodium ion for each chloride ion, but there are no identifiable sodium–chloride pairs. Such an orderly array of ions is known as an ionic crystal. As mentioned at the beginning of this chapter, on the atomic level, the crystalline structure of sodium

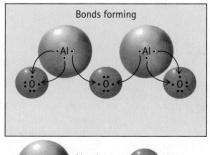

Bonds forming

Aluminum atom / Oxygen atom

Ionic bonds formed

Aluminum oxide, Al_2O_3

Ruby

Sapphire

chloride is cubic, which is why macroscopic crystals of table salt are also cubic. Smash a large cubic sodium chloride crystal with a hammer, and what do you get? Smaller cubic sodium chloride crystals!

Similarly, the crystalline structures of other ionic compounds, such as calcium fluoride and aluminum oxide, are a consequence of how the ions pack together. We will go into more detail about the crystalline structures of minerals in Chapter 20.

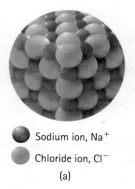

● Sodium ion, Na⁺

● Chloride ion, Cl⁻

(a)   (b)

15.4 Metallic Bonds

EXPLAIN THIS Why aren't alloys described as metallic compounds?

In Section 12.4 you learned about the properties of metals. They conduct electricity and heat, are opaque to light, and deform—rather than fracture—under pressure. Because of these properties, metals are used to build homes, appliances, cars, bridges, airplanes, and skyscrapers. Metal wires across the landscape transmit communication signals and electric power. We wear metal jewelry, exchange metal currency, and drink from metal cans. Yet what gives a metal its metallic properties? We can answer this question by looking at the behavior of its atoms.

The outer electrons of most metal atoms tend to be weakly held to the atomic nucleus. Consequently, these electrons are easily dislodged, leaving behind positively charged metal ions. The many electrons dislodged from a large group of metal atoms flow freely through the resulting metal ions, as is depicted in Figure 15.12. This "fluid" of electrons holds the positively charged metal ions together in the type of chemical bond known as a **metallic bond**.

The mobility of electrons in a metal accounts for the metal's significant ability to conduct electricity and heat. Also, metals are opaque and shiny because the free electrons easily vibrate to the oscillations of any light falling on them, reflecting most of it. Furthermore, the metal ions are not rigidly held to fixed positions, as ions are in an ionic crystal. Rather, because the metal ions are held together by a "fluid" of electrons, these ions can move into various orientations relative to one another, which occurs when a metal is pounded, pulled, or molded into a different shape.

Two or more different metals can be bonded to each other by metallic bonds. This occurs, for example, when molten gold and molten palladium are blended to form a homogeneous solution known as white gold. The quality of the white gold can be modified simply by changing the proportions of gold and palladium. White gold is an example of an **alloy**, which is any mixture composed of two or more metallic elements. By playing around with proportions, metal workers can readily modify the properties of an alloy. For example, in designing the Sacagawea dollar coin, shown in Figure 15.13, the U.S. Mint needed a metal with a gold color—so that it would be easy to recognize—and also have the same electrical characteristics as the Susan B. Anthony dollar coin—so that the new coin could substitute for the Anthony coin in vending machines.

LEARNING OBJECTIVE
Relate the properties of a metal to how the atoms of that metal are chemically bonded.

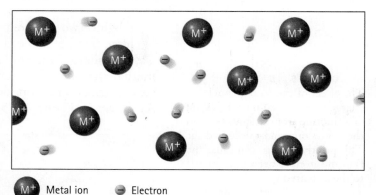

M⁺ Metal ion   ● Electron

FIGURE 15.12

Metal ions are held together by freely flowing electrons. These loose electrons form a kind of "electronic fluid," which flows through the lattice of positively charged ions.

FYI Metal ores contain ionic compounds in which the metal atoms have lost electrons to become positive ions. To convert the ores to metals requires that electrons be given back to the metal ions. This is done by heating the ore with electron-releasing materials, such as carbon, in hot furnaces that reach about 1500°C. The metal emerges in a molten state that can be cast into a variety of useful shapes.

Only a few metals—gold and platinum are two examples—appear in nature in metallic form. Deposits of these natural metals, also known as *native metals*, are quite rare. For the most part, metals found in nature are chemical compounds. Iron, for example, is most frequently found as iron oxide, Fe_2O_3, and copper is found as chalcopyrite, $CuFeS_2$. Geologic deposits containing relatively high concentrations of metal-containing compounds are called *ores*. The metals industry mines these ores from the ground, as shown in Figure 15.14, and then processes them into metals. Although metal-containing compounds occur just about everywhere, only ores are concentrated enough to make the extraction of the metal economical.

FIGURE 15.14
The world's biggest open-pit mine is the copper mine at Bingham Canyon, Utah.

15.5 Covalent Bonds

EXPLAIN THIS A lone proton encounters the lone pair of electrons of an ammonia molecule and forms what?

Imagine two children playing together and sharing their toys. Perhaps a force that keeps the children together is their mutual attraction to the toys they share. In a similar fashion, two atoms can be held together by their mutual attraction for electrons they share. A fluorine atom, for example, has a strong attraction for one additional electron to fill its outermost occupied shell. As shown in Figure 15.15, a fluorine atom can obtain an additional electron

FIGURE 15.15
The effect of the positive nuclear charge (represented by red shading) of a fluorine atom extends beyond the atom's outermost occupied shell. This positive charge can cause the fluorine atom to become attracted to the unpaired valence electron of a neighboring fluorine atom. Then the two atoms are held together in a fluorine molecule by the attraction they both have for the two shared electrons. Each fluorine atom achieves a filled valence shell.

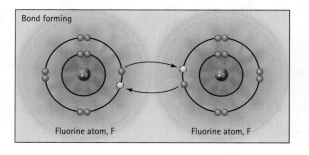

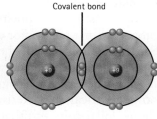

Bond forming

Fluorine atom, F   Fluorine atom, F

Covalent bond

Fluorine molecule, F_2

by holding onto the unpaired valence electron of another fluorine atom. This results in a situation in which the two fluorine atoms are mutually attracted to the same two electrons. This type of electrical attraction in which atoms are held together by their mutual attraction for shared electrons is called a **covalent bond**, in which *co-* signifies sharing and *-valent* indicates that valence electrons are being shared.

A substance composed of atoms held together by covalent bonds is a **covalent compound**. The fundamental unit of most covalent compounds is a **molecule**, which we can now formally define as any group of atoms held together by covalent bonds. Figure 15.16 uses the element fluorine to illustrate this principle.

When writing electron-dot structures for covalent compounds, chemists often use a straight line to represent the two electrons involved in a covalent bond. In some representations, the nonbonding electron pairs are ignored. This occurs in instances where these electrons play no significant role in the process being illustrated. Here are two frequently used ways of showing a fluorine molecule without using spheres to represent the atoms:

$$:\ddot{F}—\ddot{F}:\qquad F—F$$

Remember—the straight line in both versions represents two electrons, one from each atom. Thus, we now have two types of electron pairs to keep track of. The term *nonbonding pair* refers to any pair that exists in the electron-dot structure of an individual atom, and the term *bonding pair* refers to any pair that results from formation of a covalent bond. In a nonbonding pair, both electrons originate in the same atom; in a bonding pair, one electron comes from one of the atoms participating in the covalent bond, and the other electron comes from the other atom participating in the bond.

Recall from Section 15.3 that an ionic bond is formed when an atom that tends to lose electrons makes contact with an atom that tends to gain them. A covalent bond, by contrast, is formed when two atoms that tend to gain electrons are brought into contact with each other. Atoms that tend to form covalent bonds are therefore primarily atoms of the nonmetallic elements in the upper right corner of the periodic table (with the exception of the noble-gas elements, which are very stable and tend not to form bonds).

Hydrogen tends to form covalent bonds because, unlike the other group 1 elements, it has a fairly strong attraction for an additional electron. Two hydrogen atoms, for example, covalently bond to form a hydrogen molecule, H_2, as shown in Figure 15.17.

The number of covalent bonds an atom can form is equal to the number of additional electrons it can attract, which is the number needed to fill its valence shell. Hydrogen attracts only one additional electron, and so it forms only one covalent bond. Oxygen, which attracts two additional electrons, finds them when it encounters two hydrogen atoms and reacts with them to form water, H_2O, as Figure 15.18 shows. In water, not only does the oxygen atom have access to two additional electrons by covalently bonding to two hydrogen atoms, but each hydrogen atom has access to an additional electron by bonding to the oxygen atom. Each atom thus achieves a filled valence shell.

Nitrogen attracts three additional electrons and thus can form three covalent bonds, as occurs in

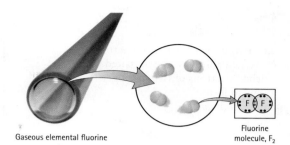

Gaseous elemental fluorine

Fluorine molecule, F_2

FIGURE 15.16

Molecules are the fundamental units of the gaseous covalent compound fluorine, F_2. Notice that in this model of a fluorine molecule, the spheres overlap, whereas the spheres shown earlier for ionic compounds do not. Now you know that this difference in representation is because of the difference in bond types.

Mastering**PHYSICS**

TUTORIAL: Covalent Bonds
VIDEO: Covalent Bonds

FYI Spectroscopic studies of interstellar dust within our galaxy have revealed the presence of more than 120 kinds of molecules, such as hydrogen chloride, HCl; water, H_2O; acetylene, H_2C_2; formic acid, HCO_2H; methanol, CH_3OH; methyl amine, NH_2CH_3; acetic acid, CH_3CO_2H; and even the amino acid glycine, $NH_2CH_2CO_2H$. Notably, about half of these interstellar molecules are carbon-based organic molecules. As discussed in Chapter 13, the atoms originated from the nuclear fusion of ancient stars. How interesting that these atoms then join together to form molecules even in the deep vacuum of outer space. Chemistry is truly everywhere.

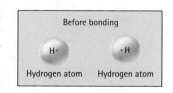

Before bonding

Hydrogen atom   Hydrogen atom

Covalent bond formed

Hydrogen molecule, H_2

FIGURE 15.17

Two hydrogen atoms form a covalent bond as they share their unpaired electrons.

FIGURE 15.18
INTERACTIVE FIGURE MP

The two unpaired valence electrons of oxygen pair with the unpaired valence electrons of two hydrogen atoms to form the covalent compound water.

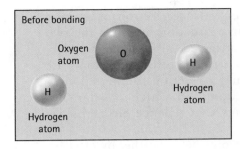

ammonia, NH_3, shown in Figure 15.19. Likewise, a carbon atom can attract four additional electrons and is thus able to form four covalent bonds, as occurs in methane, CH_4. Note that the number of covalent bonds formed by these and other nonmetallic elements parallels the type of negative ions they tend to form (see Figure 15.6). This makes sense because covalent-bond formation and negative-ion formation are both applications of the same concept: nonmetallic atoms tend to gain electrons until their valence shells are filled.

Diamond is a very unusual covalent compound consisting of carbon atoms covalently bonded to one another in four directions. The result is a covalent crystal, which, as shown in Figure 15.20, is a highly ordered, three-dimensional network of covalently bonded atoms. This network of carbon atoms forms a very strong and rigid structure, which is why diamonds are so hard. Also, because a diamond is a group of atoms held together only by covalent bonds, it can be characterized as a single molecule! Unlike most other molecules, a diamond molecule is large enough to be visible to the naked eye, and so it is more appropriately referred to as a macromolecule.

FYI Astronomers have recently discovered an expired star that has a solid core made of diamond. This star-sized diamond is about 4000 km wide, which amounts to about 10 billion trillion trillion carats. It has been named "Lucy" after the Beatles song "Lucy in the Sky with Diamonds." In about 7 billion years, our own star, the Sun, is also likely to crystallize into a huge diamond ball.

CHECKPOINT

How many electrons make up a covalent bond?

Was this your answer?
Two—one from each participating atom.

FIGURE 15.19
(a) A nitrogen atom attracts the three electrons in three hydrogen atoms to form ammonia, NH_3, a gas that can dissolve in water to make an effective cleanser. (b) A carbon atom attracts the four electrons in four hydrogen atoms to form methane, CH_4, the primary component of natural gas. In these and most other cases of covalent-bond formation, the result is a filled valence shell for all the atoms involved.

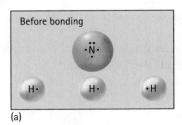
(a)

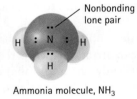
Ammonia molecule, NH_3

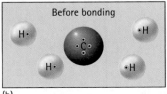
(b)

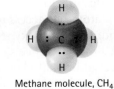
Methane molecule, CH_4

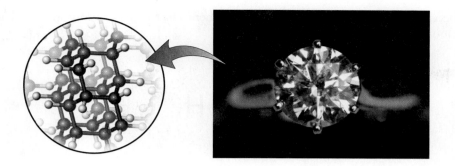

FIGURE 15.20
The crystalline structure of diamond is nicely illustrated with sticks to represent the covalent bonds. The molecular nature of a diamond is responsible for its extreme hardness.

It is possible to have more than two electrons shared between two atoms, and Figure 15.21 shows three examples. Molecular oxygen, O_2, consists of two oxygen atoms connected by four shared electrons. This arrangement is called a *double covalent bond* or, for short, a *double bond*. As another example, the covalent compound carbon dioxide, CO_2, consists of two double bonds connecting two oxygen atoms to a central carbon atom.

Some atoms can form triple covalent bonds, in which six electrons—three from each atom—are shared. One example is molecular nitrogen, N_2. Any double or triple bond is often referred to as a multiple covalent bond. Multiple bonds higher than these, such as the quadruple covalent bond, are not commonly observed.

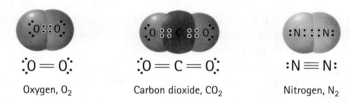

Oxygen, O_2 Carbon dioxide, CO_2 Nitrogen, N_2

FIGURE 15.21
Double covalent bonds in molecules of oxygen, O_2, and carbon dioxide, CO_2, and a triple covalent bond in a molecule of nitrogen, N_2.

15.6 Polar Covalent Bonds

LEARNING OBJECTIVE
Differentiate ionic, polar covalent, and nonpolar covalent chemical bonds.

EXPLAIN THIS   How is the chemical bond between sodium and chlorine mostly ionic but partially covalent?

If the two atoms in a covalent bond are identical, their nuclei have the same positive charge, and therefore the electrons are shared evenly. We can represent these electrons as being centrally located by using an electron-dot structure with the electrons situated exactly halfway between the two atomic symbols. Alternatively, we can draw a cloud in which the positions of the two bonding electrons over time are shown as a series of dots. Where the dots are most concentrated is where the electrons have the greatest probability of being located:

There are always two electrons per covalent bond. A double bond, therefore, consists of four electrons, while a triple bond consists of six electrons.

In a covalent bond between nonidentical atoms, the nuclear charges are different, and consequently the bonding electrons may be shared unevenly. This occurs in a hydrogen–fluorine bond, where electrons are more attracted to fluorine's greater nuclear charge:

> One side of the hydrogen–fluorine bond has a greater density of electrons and is slightly negative, while the opposite side is slightly positive. This makes up a *dipole*, an extension of electric-charge polarization, which was discussed in Chapter 8.

The bonding electrons spend more time around the fluorine atom. For this reason, the fluorine side of the bond is slightly negative and, because the bonding electrons have been drawn away from the hydrogen atom, the hydrogen side of the bond is slightly positive. This separation of charge is called a **dipole** (pronounced *die*-pole) and is represented either by the characters $\delta-$ and $\delta+$ (read "slightly negative" and "slightly positive," respectively) or by a crossed arrow pointing to the negative side of the bond:

$$\overset{\delta+\quad\delta-}{H\!-\!F} \qquad \overset{\longmapsto}{H\!-\!F}$$

So atoms forming a chemical bond engage in a tug-of-war for electrons. How strongly an atom is able to tug on bonding electrons has been measured experimentally and quantified as the atom's **electronegativity**. The range of electronegativities runs from 0.7 to 3.98, as Figure 15.22 shows. The greater an atom's electronegativity, the greater its ability to pull electrons toward itself when bonded. Thus, in hydrogen fluoride, fluorine has a greater electronegativity, or pulling power, than hydrogen.

Electronegativity is greatest for elements at the upper right of the periodic table and lowest for elements at the lower left. Noble gases are not considered in electronegativity discussions because, as previously mentioned, they rarely participate in chemical bonding.

When the two atoms in a covalent bond have the same electronegativity, no dipole is formed (as is the case with H_2) and the bond is classified as a **nonpolar** bond. When the electronegativities of the atoms differ, a dipole may form (as with HF) and the bond is classified as a **polar** bond. Just how polar a bond is depends on the difference between the electronegativity values of the two atoms—the greater the difference, the more polar the bond.

As can be seen in Figure 15.22, the greater the distance between two atoms in the periodic table, the greater the difference in their electronegativities, and hence the greater the polarity of the bond between them. So a chemist can predict which bonds are more polar than others without reading the electronegativities. Bond polarity can be inferred by looking at the relative positions

FIGURE 15.22
The experimentally measured electronegativities of elements.

H 2.2																		He –
Li 0.98	Be 1.57											B 2.04	C 2.55	N 3.04	O 3.44	F 3.98	Ne –
Na 0.93	Mg 1.31											Al 1.61	Si 1.9	P 2.19	S 2.58	Cl 3.16	Ar –
K 0.82	Ca 1.0	Sc 1.36	Ti 1.54	V 1.63	Cr 1.66	Mn 1.55	Fe 1.83	Co 1.88	Ni 1.91	Cu 1.90	Zn 1.65	Ga 1.81	Ge 2.01	As 2.18	Se 2.55	Br 2.96	Kr –
Rb 0.82	Sr 0.95	Y 1.22	Zr 1.33	Nb 1.6	Mo 2.16	Tc 1.9	Ru 2.2	Rh 2.28	Pd 2.20	Ag 1.93	Cd 1.69	In 1.78	Sn 1.96	Sb 2.05	Te 2.1	I 2.66	Xe –
Cs 0.79	Ba 0.89	La 1.10	Hf 1.3	Ta 1.5	W 2.36	Re 1.9	Os 2.2	Ir 2.20	Pt 2.8	Au 2.54	Hg 2.00	Tl 2.04	Pb 2.33	Bi 2.02	Po 2.0	At 2.2	Rn –
Fr 0.7	Ra 0.9	Ac 1.1	Rf –	Db –	Sg –	Bh –	Hs –	Mt –	Ds –	Rg –	Cn –	Uut –	Uuq –	Uup –	Uuh –	Uus –	Uuo –

of the atoms in the periodic table—the farther apart they are, especially when one is at the lower left and one is at the upper right, the greater the polarity of the bond between them.

CHECKPOINT

List these bonds in order of increasing polarity:
P—F, S—F, Ga—F, Ge—F (F, fluorine, atomic number 9; P, phosphorus, atomic number 15; S, sulfur, atomic number 16; Ga, gallium, atomic number 31; Ge, germanium, atomic number 32).
(least polar) _____, _____, _____, _____ (most polar)

Was this your answer?
If you answered the question, or attempted to, before reading this answer, hooray for you! You're doing more than reading the text—you're learning physical science. The greater the difference in electronegativities between two bonded atoms, the greater the polarity of the bond, and so the order of increasing polarity is S—F < P—F < Ge—F < Ga—F.

Note that this answer can be obtained by looking only at the relative positions of these elements in the periodic table rather than by calculating the differences in their electronegativities.

The magnitude of bond polarity is sometimes indicated by the size of the crossed arrow or the $\delta-$ and $\delta+$ symbols used to depict a dipole, as shown in Figure 15.23.

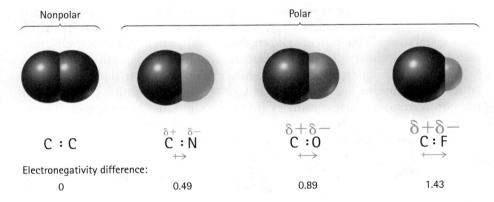

FIGURE 15.23
These bonds are in order of increasing polarity from left to right, a trend indicated by the larger and larger crossed arrows and $\delta-/\delta+$ symbols. Which of these pairs of elements are farthest apart in the periodic table?

Note that the electronegativity difference between atoms in an ionic bond can also be calculated. For example, the bond in NaCl has an electronegativity difference of 2.23, far greater than the difference of 1.43 shown for the C—F bond in Figure 15.23.

What is important to understand here is that there is no black-and-white distinction between ionic and covalent bonds. Rather, there is a gradual change from one to the other as the atoms that bond are farther apart in the periodic table. This continuum is illustrated in Figure 15.24. Atoms on opposite sides of the periodic table have great differences in electronegativity, and hence the bonds between them are highly polar—in other words, ionic. Nonmetallic atoms of the same type have the same electronegativities, and so their bonds are nonpolar covalent. The polar covalent bond with its uneven sharing of electrons and slightly charged atoms is between these two extremes.

FIGURE 15.24
The ionic bond and the nonpolar covalent bond represent the two extremes of chemical bonding. The ionic bond involves a transfer of one or more electrons, and the nonpolar covalent bond involves the equitable sharing of electrons. The character of a polar covalent bond falls between these two extremes.

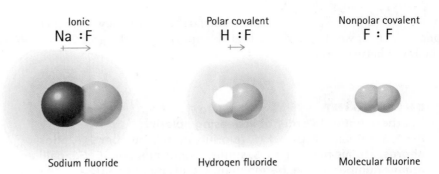

Ionic
Na :F

Polar covalent
H :F

Nonpolar covalent
F : F

Sodium fluoride   Hydrogen fluoride   Molecular fluorine

LEARNING OBJECTIVE
Show how the shape of a molecule affects the molecule's polarity.

Mastering**PHYSICS**®
TUTORIAL: Bonds and Bond Polarity

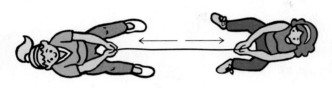

O = C = O

FIGURE 15.25
There is no net dipole in a carbon dioxide molecule, and so the molecule is nonpolar. This is analogous to two people in a tug-of-war. As long as they pull with equal forces but in opposite directions, the rope remains stationary.

15.7 Molecular Polarity

EXPLAIN THIS Which is heavier: carbon dioxide or water?

When all the bonds in a molecule are nonpolar, the molecule as a whole is also nonpolar—as is the case with H_2, O_2, and N_2. When a molecule consists of only two atoms and the bond between them is polar, the polarity of the molecule is the same as the polarity of the bond—as with HF, HCl, and ClF.

Complexities arise when assessing the polarity of a molecule containing more than two atoms. Consider carbon dioxide, CO_2, shown in Figure 15.25. The cause of the dipole in either one of the carbon–oxygen bonds is oxygen's greater pull on the bonding electrons (because oxygen is more electronegative than carbon). At the same time, however, the oxygen atom on the opposite side of the carbon pulls those electrons back to the carbon. The net result is an even distribution of bonding electrons around the entire molecule. So dipoles that are of equal strength but pull in opposite directions in a molecule effectively cancel each other, with the result that the molecule as a whole is nonpolar.

Figure 15.26 illustrates a similar situation for boron trifluoride, BF_3, in which three fluorine atoms are oriented 120° from one another around a central boron atom. Because the angles are all the same, and because each fluorine atom pulls on the electrons of its boron–fluorine bond with the same force, the resulting polarity of this molecule is zero.

FIGURE 15.26
The three dipoles of a boron trifluoride molecule oppose one another at 120° angles, which makes the overall molecule nonpolar. This is analogous to three people pulling with equal force on ropes attached to a central ring. As long as they all pull with equal force and all maintain the 120° angles, the ring remains stationary.

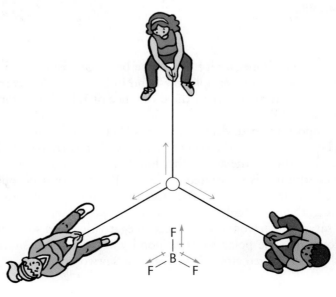

F
|
F—B—F

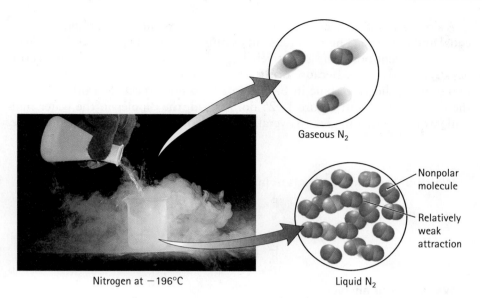

Gaseous N_2

Nonpolar molecule

Relatively weak attraction

Liquid N_2

Nitrogen at −196°C

FIGURE 15.27
Nitrogen is a liquid at temperatures below its chilly boiling point of −196°C. Nitrogen molecules are not very attracted to one another because they are nonpolar. As a result, the small amount of heat energy available at −196°C is enough to separate them and allow them to enter the gaseous phase.

Nonpolar molecules have only relatively weak attractions to other nonpolar molecules. The covalent bonds in a carbon dioxide molecule, for example, are many times stronger than any forces of attraction that might occur between two adjacent carbon dioxide molecules. This lack of attraction between nonpolar molecules explains the low boiling points of many nonpolar substances. Recall from Section 14.2 that boiling is a process wherein the molecules of a liquid separate from one another as they go into the gaseous phase. When only weak attractions exist between the molecules of a liquid, less heat energy is required to liberate the molecules from one another and allow them to enter the gaseous phase. This translates into a relatively low boiling point for the liquid, as, for example, in molecular nitrogen, N_2, shown in Figure 15.27. The boiling points of hydrogen (H_2), oxygen (O_2), carbon dioxide (CO_2), and boron trifluoride, (BF_3), are also quite low for the same reason (see Table 15.2).

There are many instances in which the dipoles of different bonds in a molecule do not cancel each other. Reconsider the rope analogy of Figure 15.26. As long as everyone pulls equally, the ring stays put. Imagine, however, that one person begins to ease off on the rope. Now the pulls are no longer balanced, and the ring begins to move away from the person who is slacking off, as Figure 15.28 shows. Likewise, if one person began to pull harder, the ring would move away from the other two people.

A dipole is a vector quantity possessing both magnitude and direction. When two dipoles are equal and opposite, they effectively cancel each other out.

FIGURE 15.28
If one person eases off in a three-way tug-of-war but the other two continue to pull, the ring moves in the direction of the purple arrow.

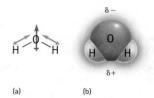

FIGURE 15.29
(a) The individual dipoles in a water molecule add together to give a large overall dipole for the whole molecule, shown in purple. (b) The region around the oxygen atom is therefore slightly negative, and the region around the two hydrogens is slightly positive.

A similar situation occurs in molecules in which polar covalent bonds are not equal and opposite. The most relevant example is water, H_2O. Each hydrogen–oxygen covalent bond has a relatively large dipole because of the great electronegativity difference. Because of the bent shape of the molecule, however, the two dipoles, shown in blue in Figure 15.29, do not cancel each other the way the C—O dipoles in Figure 15.25 do. Instead, the dipoles in the water molecule work together to give an overall dipole, shown in purple, for the molecule.

CHECKPOINT

Which of these molecules is polar and which is nonpolar?

Was this your answer?
Symmetry is often the greatest clue for determining polarity. Because the molecule on the left is symmetrical, the dipoles on the two sides cancel each other. This molecule is therefore nonpolar:

Because the molecule on the right is less symmetrical (more "lopsided"), it is the polar molecule. Because carbon is more electronegative than hydrogen, the dipoles of the two hydrogen–carbon bonds point toward the carbon. Because fluorine is more electronegative than carbon, the dipoles of the carbon–fluorine bonds point toward the fluorines. Because the general direction of all dipole arrows is toward the fluorines, so is the average distribution of the bonding electrons. The fluorine side of the molecule is therefore slightly negative, and the hydrogen side is slightly positive.

A water molecule is a natural dipole—a bit positive on one end and negative on the other. What's the net charge of a dipole?

Figure 15.30 illustrates how polar molecules electrically attract one another and, as a result, are relatively difficult to separate. In other words, polar molecules can be thought of as being "sticky," which is why it takes more energy

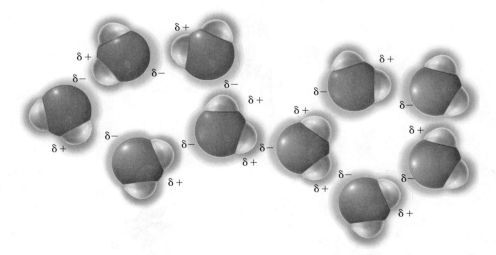

FIGURE 15.30
Water molecules attract one another because each contains a slightly positive side and a slightly negative side. The molecules position themselves such that the positive side of one faces the negative side of a neighbor.

TABLE 15.2	BOILING POINTS OF SOME POLAR AND NONPOLAR SUBSTANCES
Substance	**Boiling point (°C)**
Polar	
Hydrogen fluoride, HF	20
Water, H_2O	100
Ammonia, NH_3	−33
Nonpolar	
Hydrogen, H_2	−253
Oxygen, O_2	−183
Nitrogen, N_3	−196
Boron trifluoride, BF_3	−100
Carbon Dioxide, CO_2	−79

to separate them—to change phase. For this reason, substances composed of polar molecules typically have higher boiling points than substances composed of nonpolar molecules, as Table 15.2 shows.

Water boils at 100°C, whereas carbon dioxide boils at −79°C. This 179°C difference is quite dramatic when you consider that a carbon dioxide molecule is more than twice as massive as a water molecule.

Because molecular "stickiness" can play a lead role in determining a substance's macroscopic properties, molecular polarity is a central concept of chemistry. Figure 15.31 describes an interesting example.

FIGURE 15.31
Oil and water are difficult to mix, as is evident from this oil spill off the coast of Spain in 2002. It's not, however, that oil and water repel each other. Rather, water molecules are so attracted to themselves because of their polarity that they pull themselves together. The non-polar oil molecules are thus excluded and left to themselves. Being less dense than water, oil floats on the surface, where it poses great danger to birds and other wildlife.

15.8 Molecular Attractions

EXPLAIN THIS Is it possible for a fish to die from drowning?

So far you have learned that the atoms of a molecule are held together by covalent bonds. Furthermore, the molecule, behaving as a fundamental unit, may have electrical attractions with neighboring molecules. As discussed in the previous section, the greater the polarity of the molecule, the greater

LEARNING OBJECTIVE
Recognize the important role that molecular interactions play in determining the physical properties of a material.

its attraction to neighboring molecules. This explains how water has such a high boiling point—the water molecules, being quite polar, are so attracted to one another that a lot of energy is required to separate them from one another into the gaseous phase. In this section, we explore further how the physical properties of a material, such as boiling point, can be deduced from the polarity of its molecules. In addition to discussing the attractions among the molecules within a single substance, we'll explore attractions that occur between the fundamental units of different substances, such as water and salt.

As shown in Table 15.3, there are four types of electrical attractions involving molecules. The strength of even the strongest of these attractions is much weaker than any chemical bond. The attraction between two adjacent water molecules, for example, is only about $\frac{1}{20}$ as strong as the chemical bonds holding the hydrogen and oxygen atoms together in the water molecule. Although molecule-to-molecule attractions are relatively weak, their effects on the physical properties of substances are most significant.

> In the first part of this chapter, we talked about how molecules form. Now we see how molecules mix together.

Mastering**PHYSICS**
VIDEO: Polar Attractions

TABLE 15.3	ELECTRICAL ATTRACTIONS BETWEEN A MOLECULE AND ITS NEIGHBOR
Attraction	**Relative Strength**
Ion–dipole	Strongest
Dipole–dipole	
Dipole–induced dipole	
Induced dipole–induced dipole	Weakest

Ions and Dipoles

Recall from Section 15.7 that a *polar* molecule is one in which the bonding electrons are unevenly distributed. One side of the molecule carries a slight negative charge, and the opposite side carries a slight positive charge. This separation of charge makes up a *dipole*.

So what happens to polar molecules, such as water molecules, when they are near an ionic compound, such as sodium chloride? The opposite charges electrically attract one another. A positive sodium ion attracts the negative side of a water molecule, and a negative chloride ion attracts the positive side of a water molecule. This phenomenon is illustrated in Figure 15.32. Such an attraction between an ion and the dipole of a polar molecule is called an *ion–dipole* attraction.

Ion–dipole attractions are much weaker than ionic bonds. However, a large number of ion–dipole attractions can act collectively to disrupt ionic bonds. This is what happens to sodium chloride in water. Attractions exerted by the water molecules break the ionic bonds and pull the ions away from one another. The result, represented in Figure 15.33, is a solution of sodium chloride in water. (A solution in water is called an *aqueous solution*.)

An attraction between two polar molecules is called a *dipole–dipole* attraction. An unusually strong dipole–dipole attraction is the **hydrogen bond**. This attraction occurs between molecules that have a hydrogen atom covalently bonded to a

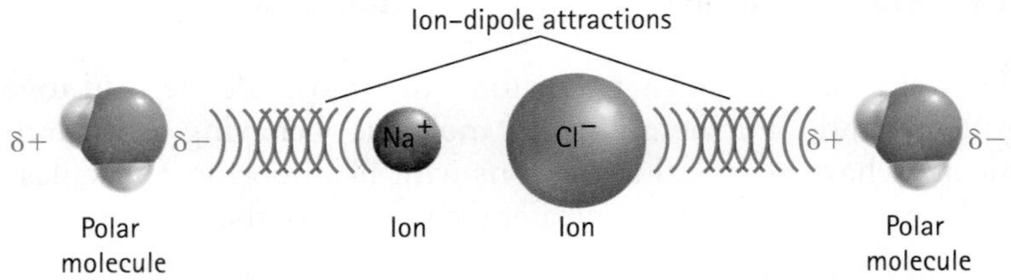

FIGURE 15.32
Electrical attractions are shown as a series of overlapping arcs. The blue arcs indicate negative charge, and the red arcs indicate positive charge.

Ion–dipole attractions

$\delta+$ $\delta-$ Na$^+$ Cl$^-$ $\delta+$ $\delta-$

Polar molecule   Ion   Ion   Polar molecule

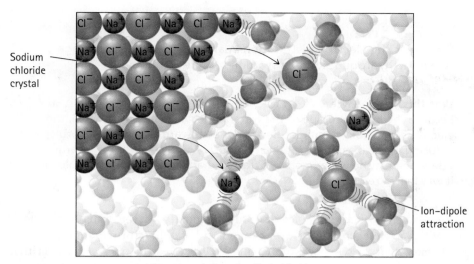

Aqueous solution of sodium chloride

FIGURE 15.33
Sodium and chloride ions tightly bound in a crystal lattice are separated from one another by the collective attraction exerted by many water molecules to form an aqueous solution of sodium chloride.

highly electronegative atom, usually nitrogen, oxygen, or fluorine. Recall from Section 15.6 that the electronegativity of an atom describes how well that atom is able to pull bonding electrons toward itself. The greater the atom's electronegativity, the better it is able to gain electrons and thus the more negative is its charge.

Look at Figure 15.34 to see how hydrogen bonding works. The hydrogen side of a polar molecule (water, in this example) has a positive charge because the more electronegative oxygen atom pulls more strongly on the electrons of the covalent bond. The hydrogen is therefore electrically attracted to a pair of nonbonding electrons on the negatively charged atom of another molecule (in this case, another water molecule). This mutual attraction between hydrogen and the negatively charged atom of another molecule is a *hydrogen bond*.

Even though the hydrogen bond is much weaker than any covalent or ionic bond, the effects of hydrogen bonding can be very pronounced. For example, water owes many of its properties to hydrogen bonds. The hydrogen bond is also of great importance in the chemistry of the large molecules, such as DNA and proteins, that are found in living organisms.

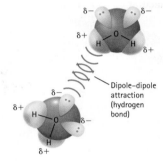

FIGURE 15.34
The dipole–dipole attraction between two water molecules is a hydrogen bond because it involves hydrogen atoms bonded to highly electronegative oxygen atoms.

Induced Dipoles

In many molecules, the electrons are distributed evenly, and so there is no dipole. The oxygen molecule, O_2, is an example. Such a nonpolar molecule can be induced to become a temporary dipole, however, when it is brought close to a water molecule (or to any other polar molecule), as Figure 15.35 illustrates. The slightly negative side of the water molecule pushes the electrons in the oxygen molecule away. Thus, the oxygen molecule's electrons are pushed to the side that is farthest from the water molecule. The result is a temporarily uneven distribution of electrons called an **induced dipole**. The resulting attraction between the permanent dipole (water) and the induced dipole (oxygen) is a *dipole–induced dipole* attraction.

FIGURE 15.35
INTERACTIVE FIGURE   MP

(a) An isolated oxygen molecule has no dipole; its electrons are distributed evenly. (b) An adjacent water molecule induces a redistribution of electrons in the oxygen molecule. (The slightly negative side of the oxygen molecule is shown larger than the slightly positive side because the slightly negative side contains more electrons.)

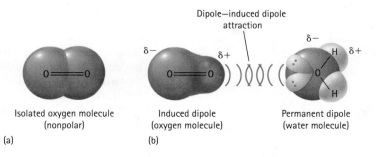

Dipole–induced dipole attraction

Isolated oxygen molecule (nonpolar)   Induced dipole (oxygen molecule)   Permanent dipole (water molecule)

(a)   (b)

FIGURE 15.36
The electrical attraction between water and oxygen molecules is relatively weak, which explains why not much oxygen is able to dissolve in water. For example, water fully aerated at room temperature contains only about 1 oxygen molecule for every 200,000 water molecules. The gills of a fish, therefore, must be highly efficient at extracting molecular oxygen from water.

CHECKPOINT
How does the electron distribution in an oxygen molecule change when the hydrogen side of a water molecule is nearby?

Was this your answer?
Because the hydrogen side of the water molecule is slightly positive, the electrons in the oxygen molecule are pulled toward the water molecule, inducing in the oxygen molecule a temporary dipole in which the larger side is nearest the water molecule (rather than as far away as possible, as it was in Figure 15.35).

Remember—induced dipoles are only temporary. If the water molecule in Figure 15.35b were removed, the oxygen molecule would return to its normal, nonpolar state. As a consequence, dipole–induced dipole attractions are weaker than dipole–dipole attractions. But dipole–induced dipole attractions are strong enough to hold relatively small quantities of oxygen dissolved in water, as depicted in Figure 15.36. This attraction between water and molecular oxygen is vital for fish and other forms of aquatic life that rely on molecular oxygen dissolved in water.

Dipole–induced dipole attractions are also responsible for holding plastic wrap to glass, as shown in Figure 15.37. These wraps are made of very long nonpolar molecules that are induced to have dipoles when placed in contact with glass, which is highly polar. As we discuss next, the molecules of a nonpolar material, such as plastic wrap, can also induce dipoles among themselves. This explains why plastic wrap sticks not only to polar materials such as glass but also to itself.

FIGURE 15.37
Temporary dipoles induced in the normally nonpolar molecules in plastic wrap make it stick to glass.

CHECKPOINT
Distinguish between a dipole–dipole attraction and a dipole–induced dipole attraction.

Was this your answer?
The dipole–dipole attraction is stronger and involves two permanent dipoles. The dipole–induced dipole attraction is weaker and involves a permanent dipole and a temporary one.

Individual atoms and nonpolar molecules, on average, have a fairly even distribution of electrons. Because of the randomness of electron motion, however, at any given moment the electrons in an atom or a nonpolar molecule may be bunched to one side. The result is a temporary dipole, as shown in Figure 15.38.

Just as the permanent dipole of a polar molecule can induce a dipole in a nonpolar molecule, a temporary dipole can do the same thing. This gives rise to the relatively weak *induced dipole–induced dipole* attraction, illustrated in Figure 15.39.

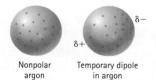

Nonpolar   Temporary dipole
argon   in argon

FIGURE 15.38
The electron distribution in an atom is normally even. At any given moment, however, the electron distribution may be somewhat uneven, resulting in a temporary dipole.

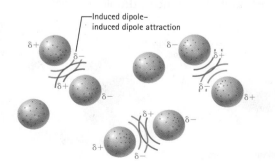

FIGURE 15.39
Because the normally even distribution of electrons in atoms can momentarily become uneven, atoms can be attracted to one another through induced dipole–induced dipole attractions.

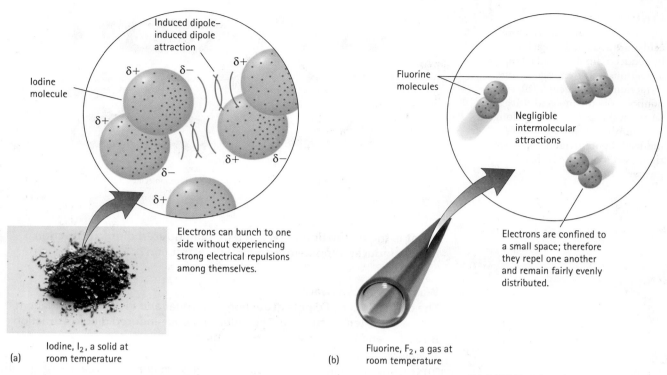

Induced dipole–induced dipole attraction

Iodine molecule

$\delta+$ $\delta-$ $\delta+$

$\delta+$

$\delta-$ $\delta+$

$\delta-$

$\delta+$

Electrons can bunch to one side without experiencing strong electrical repulsions among themselves.

(a)   Iodine, I$_2$, a solid at room temperature

Fluorine molecules

Negligible intermolecular attractions

Electrons are confined to a small space; therefore they repel one another and remain fairly evenly distributed.

(b)   Fluorine, F$_2$, a gas at room temperature

Electrons repel electrons, which means they resist bunching together to one side of the atom. In a large atom, however, the electrons find it fairly easy to do just that. By analogy, consider a cruise ship with only 10 passengers. Because the ship is so large, these 10 passengers could easily congregate to one side. On a much smaller life raft, however, the same 10 passengers would find it necessary to space themselves as evenly apart from each other as possible, lest they tip over. In a similar fashion, larger atoms can form temporary dipoles much more easily than smaller atoms, as is illustrated in Figure 15.40. So larger atoms—and molecules made of larger atoms—have the strongest induced dipole–induced dipole attractions. In other words, they are more "sticky." Iodine, I$_2$, for example, is stickier than fluorine, F$_2$, which explains why iodine is a solid at room temperature while fluorine is a gas, even though they are both nonpolar materials.

Fluorine is one of the smallest atoms, and nonpolar molecules made with fluorine atoms exhibit only very weak induced dipole–induced dipole attractions. This is the principle behind the Teflon nonstick surface. The Teflon molecule, part of which is shown in Figure 15.41, is a long chain of carbon atoms chemically bonded to fluorine atoms, and the fluorine atoms exert essentially no attractions on any material in contact with the Teflon surface—an omelet in a frying pan, for instance.

Induced dipole–induced dipole attractions help explain why natural gas is a gas at room temperature but gasoline is a liquid. The major component of

FIGURE 15.40

(a) Temporary dipoles form more readily in larger atoms, such as those in an iodine molecule, because in larger atoms, electrons bunched to one side are still relatively far apart from one another and not so repelled by the electric force. (b) In smaller atoms, such as those in a fluorine molecule, electrons cannot bunch to one side as well because the repulsive electric force increases as the electrons bunch closer.

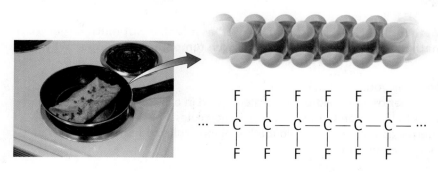

F F F F F F

··· — C — C — C — C — C — C — ···

F F F F F F

FIGURE 15.41

Few things stick to Teflon because of the high proportion of fluorine atoms that it contains. The structure depicted here is only a portion of the full length of the molecule.

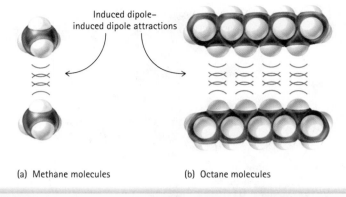

(a) Methane molecules   (b) Octane molecules

FIGURE 15.42
(a) Two nonpolar methane molecules are attracted to each other by induced dipole–induced dipole attractions, but there is only one attraction per molecule. (b) Two nonpolar octane molecules are similar to methane, but they are longer. The number of induced dipole–induced dipole attractions between these two molecules is therefore greater.

CHECKPOINT

What is the distinction between a dipole–induced dipole attraction and an induced dipole–induced dipole attraction?

Was this your answer?
The dipole–induced dipole attraction is stronger and involves a permanent dipole and a temporary one. The induced dipole–induced dipole attraction is weaker and involves two temporary dipoles.

FIGURE 15.43
If the gecko's foot is so sticky, how does the gecko keep its feet clean? Answer: the gecko's foot is extremely nonpolar. Dirt may stick to it briefly, but after a few steps, the dirt sticks better to the surface on which the gecko walks. Of course, there is at least one surface a gecko finds very difficult to climb—Teflon.

natural gas is methane, CH_4, and one of the major components of gasoline is octane, C_8H_{18}. We can see in Figure 15.42 that the number of induced dipole–induced dipole attractions between two methane molecules is appreciably less than the number between two octane molecules. You know that two small pieces of Velcro are easier to pull apart than two long pieces. Like short pieces of Velcro, methane molecules can be pulled apart with little effort. That's why methane has a low boiling point, −161°C, and is a gas at room temperature. Octane molecules, like long strips of Velcro, are relatively difficult to pull apart because of the larger number of induced dipole–induced dipole attractions. The boiling point of octane, 125°C, is therefore much higher than that of methane, and octane is a liquid at room temperature. (The greater mass of octane also plays a role in making its boiling point higher.)

Induced dipole–induced dipole attractions, also known as *dispersion forces*, also explain how the gecko can race up a glass wall and support its entire body weight with only a single toe. A gecko's feet are covered with billions of microscopic hairs called *spatulae*, each of which is about $\frac{1}{300}$ as thick as a human hair. The force of attraction between these hairs and the wall is the weak induced dipole–induced dipole attraction. But because there are so many hairs, the surface area of contact is relatively great, and hence the total force of attraction is enough to prevent the gecko from falling (Figure 15.43). Research is currently under way to develop a synthetic dry glue based on gecko adhesion. Velcro, watch out!

CHECKPOINT

Methanol, CH_3OH, which can be used as a fuel, is not much larger than methane, CH_4, but it is a liquid at room temperature. Suggest why.

Was this your answer?
The polar oxygen–hydrogen covalent bond in each methanol molecule leads to hydrogen bonding between molecules. These relatively strong interparticle attractions hold methanol molecules together as a liquid at room temperature.

For instructor-assigned homework, go to www.masteringphysics.com (MP)

SUMMARY OF TERMS (KNOWLEDGE)

Alloy A mixture of two or more metallic elements.

Covalent bond A chemical bond in which atoms are held together by their mutual attraction for two or more electrons they share.

Covalent compound A substance, such as an element or chemical compound, in which atoms are held together by covalent bonds.

Dipole A separation of charge that occurs in a chemical bond because of differences in the electronegativities of the bonded atoms.

Electron-dot structure A shorthand notation of the shell model of the atom, in which valence electrons are shown around an atomic symbol. The electron-dot structure for an atom or ion is sometimes called a Lewis dot symbol, while the electron-dot structure of a molecule or polyatomic ion is sometimes called a *Lewis structure*.

Electronegativity The ability of an atom to attract a bonding pair of electrons to itself when bonded to another atom.

Hydrogen bond An unusually strong dipole–dipole attraction occurring between molecules that have a hydrogen atom covalently bonded to a small, highly electronegative atom, usually nitrogen, oxygen, or fluorine.

Induced dipole A temporarily uneven distribution of electrons in an otherwise nonpolar atom or molecule.

Ion An atom having a net electric charge because of either a loss or gain of electrons.

Ionic bond A chemical bond in which there is an electric force of attraction between two oppositely charged ions.

Ionic compound A chemical compound containing ions.

Metallic bond A chemical bond in which positively charged metal ions are held together within a "fluid" of loosely held electrons.

Molecule The fundamental unit of a chemical compound, which is a group of atoms held tightly together by covalent bonds.

Nonbonding pairs Two paired valence electrons that are not participating in a chemical bond.

Nonpolar Description of a chemical bond or molecule that has no dipole. In a nonpolar bond or molecule, the electrons are distributed evenly.

Polar Description of a chemical bond or molecule that has a dipole. In a polar bond or molecule, electrons are congregated to one side. This makes that side slightly negative, while the opposite side (lacking electrons) becomes slightly positive.

Polyatomic ion An ionically charged molecule.

Valence electrons The electrons in the outermost occupied shell of an atom.

Valence shell The outermost occupied shell of an atom.

READING CHECK QUESTIONS (COMPREHENSION)

15.1 Electron-Dot Structures

1. How many electrons can occupy the first shell? How many can occupy the second shell?
2. Which electrons are represented by an electron-dot structure?
3. How do the electron-dot structures of elements in the same group in the periodic table compare with one another?

15.2 The Formation of Ions

4. How does an ion differ from an atom?
5. To become a negative ion, does an atom lose or gain electrons?
6. Why does the fluorine atom tend to gain only one electron?

15.3 Ionic Bonds

7. Which elements tend to form ionic bonds?
8. Suppose an oxygen atom gains two electrons to become an oxygen ion. What is its electric charge?
9. What is an ionic crystal?

15.4 Metallic Bonds

10. Do metals more readily gain or lose electrons?
11. What is an alloy?
12. What is a native metal?

15.5 Covalent Bonds

13. Which elements tend to form covalent bonds?
14. How many electrons are shared in a double covalent bond?
15. Within a neutral molecule, how many covalent bonds does an oxygen atom form?
16. Within a polyatomic ion, how many covalent bonds does a negatively charged oxygen form?

15.6 Polar Covalent Bonds

17. What is a dipole?
18. Which element in the periodic table has the greatest electronegativity? Which has the least electronegativity?
19. Which is more polar: a carbon–oxygen bond or a carbon–nitrogen bond?

15.7 Molecular Polarity

20. How can a molecule be nonpolar when it consists of atoms that have different electronegativities?
21. Why do nonpolar substances boil at relatively low temperatures?
22. Which is more symmetrical: a polar molecule or a nonpolar molecule?
23. Why don't oil and water mix?

15.8 Molecular Attractions

24. What is the primary difference between a chemical bond and an attraction between two molecules?

25. Which is stronger: the ion–dipole attraction or the induced dipole–induced dipole attraction?

26. What is a hydrogen bond?

27. Are induced dipoles permanent?

ACTIVITIES (HANDS-ON APPLICATION)

28. View crystals of table salt with a magnifying glass or, better yet, a microscope if one is available. If you have a microscope, crush the crystals with a spoon and examine the resulting powder. Purchase some sodium-free salt, which is potassium chloride, KCl, and examine these ionic crystals, both intact and crushed. Sodium chloride and potassium chloride both form cubic crystals, but there are significant differences. What are they?

29. Use toothpicks and gumdrops or jelly beans of different colors to build models of the molecules shown in Figures 15.17 through 15.19 and Figure 15.21, letting the different colors represent different elements.

   Once you have become proficient at building these models, test your expertise by building models for difluoromethane, CH_2F_2; ethane, C_2H_6; hydrogen peroxide, H_2O_2; and acetylene, C_2H_2. Keep in mind that each carbon atom must have four covalent bonds, each oxygen atom must have two, and each fluorine and hydrogen atom must have only one. Hint: One of these molecules has a triple bond.

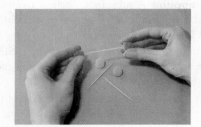

30. To see the action of the ion–dipole attraction, create a static charge on a rubber balloon by rubbing it across your hair. Hold this charged balloon up close to, but not touching, a thin stream of water running from a faucet. Watch the charged balloon divert the path of the falling water. Your balloon is negatively charged because it picks up electrons from your hair. Why would a balloon that was positively charged also attract the stream of water?

THINK AND SOLVE (MATHEMATICAL APPLICATION)

31. Ores of manganese, Mn, sometimes contain the mineral rhodochrosite, $MnCO_3$, which is an ionic compound of manganese ions and carbonate ions. How many electrons has each manganese atom lost to make this compound?

32. What is the electric charge on the calcium ion in calcium chloride, $CaCl_2$?

33. Magnesium ions carry a 2+ charge, and chloride ions carry a 1− charge. What is the chemical formula for the ionic compound magnesium chloride?

34. Barium ions carry a 2+ charge, and nitrogen ions carry a 3− charge. What is the chemical formula for the ionic compound barium nitride?

THINK AND RANK (ANALYSIS)

35. Rank these bonds in order of increasing polarity: (a) C—H, (b) O—H, (c) N—H.

36. Rank these compounds in order of increasing boiling point: (a) fluorine, F_2; (b) hydrogen fluoride, HF; (c) hydrogen chloride, HCl.

37. Rank the compounds in order of increasing symmetry: (a) CH_4, (b) NH_3, (c) H_2O.

38. Rank the following in order of increasing boiling point: (a) CH_4, (b) NH_3, (c) H_2O.

EXERCISES (SYNTHESIS)

39. How do the electron-dot structures of elements in the same group in the periodic table compare with one another?

40. How many more electrons can fit in the valence shell of a fluorine atom?

41. How many more electrons can fit in the valence shell of a hydrogen atom?

42. What happens when hydrogen's electron gets close to the valence shell of a fluorine atom?

43. The valence electron of a sodium atom does not sense the full 11+ of the sodium nucleus. Why not?

44. Why does an atom with few valence electrons tend to lose these electrons rather than gain more?

45. How is the number of unpaired valence electrons in an atom related to the number of bonds that the atom can form?

46. Why is it so easy for a magnesium atom to lose two electrons?

47. Why doesn't the neon atom tend to lose or gain any electrons?

48. Why does an atom with many valence electrons tend to gain electrons rather than lose any?

49. Sulfuric acid, H_2SO_4, loses two protons to form what polyatomic ion? What molecule loses a proton to form the hydroxide ion, OH^-?

50. Which should be larger: the potassium ion, K^+, or the potassium atom, K? Which should be larger: the potassium ion, K^+, or the argon atom, Ar? Explain.

51. Which should be more difficult to pull apart: a sodium ion from a chloride ion or a potassium ion from a chloride ion? Explain.

Shorter distance between positive and negative charges

Longer distance between positive and negative charges

52. Given that the total number of atoms on our planet remains fairly constant, how is it ever possible to deplete a natural resource such as a metal?

53. An artist wants to create a metal sculpture using a mold so that his artwork can be readily mass produced. He wants his sculpture to be exactly 6 inches tall. Should the mold also be 6 inches tall? Why or why not?

54. Which are closer together: the two nuclei within potassium fluoride, KF, or the two nuclei within molecular fluorine, F_2? Explain.

55. Two fluorine atoms join together to form a covalent bond. Why don't two potassium atoms do the same thing?

56. What drives an atom to form a covalent bond: its nuclear charge or the need to have a filled outer shell? Explain.

57. Atoms of nonmetallic elements form covalent bonds, but they can also form ionic bonds. How is this possible?

58. Examine the three-dimensional geometries of PF_5 and SF_4 shown below. Which do you think is the more polar compound?

$$\begin{array}{c} F \\ | \\ F-P{\cdots}F \\ | \\ F \end{array} \qquad \begin{array}{c} F \\ | \\ \cdot\cdot\,S{\cdots}F \\ | \\ F \end{array}$$

PF₅   SF₄

59. In each molecule, which atom carries the greater positive charge: H—Cl, Br—F, C≡O, Br—Br?

60. Which is more polar: a sulfur–bromine (S—Br) bond or a selenium–chlorine (Se—Cl) bond?

61. True or False: The greater the nuclear charge of an atom, the greater is the electronegativity. Explain.

62. True or False: The more shells in an atom, the lower its electronegativity. Explain.

63. Water, H_2O, and methane, CH_4, have about the same mass and differ by only one type of atom. Why is the boiling point of water so much higher than that of methane?

64. Which molecule from each pair should have a higher boiling point (atomic numbers: Cl = 17, O = 8, C = 6, H = 1)?

$$\begin{array}{cc} \text{Cl} \quad\quad \text{Cl} & \text{H} \quad\quad \text{Cl} \\ \text{C==C} & \text{C==C} \\ \text{H} \quad\quad \text{H} & \text{Cl} \quad\quad \text{H} \end{array}$$

$$\begin{array}{cc} \text{Cl} & \text{Cl} \quad\quad \text{H} \\ \text{C==O} & \text{C==C} \\ \text{Cl} & \text{Cl} \quad\quad \text{H} \end{array}$$

65. Three kids sitting equal spaces apart around a table are sharing jelly beans. One of the kids, however, tends only to take jelly beans and rarely gives one away. If each jelly bean represents an electron, who ends up being slightly negative? Who ends up being slightly positive? Is the negative kid just as negative as one of the positive kids is positive? Would you describe this as a polar or nonpolar situation? How about if all three kids were equally greedy?

66. Which is stronger: the covalent bond that holds atoms together within a molecule or the electrical attraction between two neighboring molecules? Explain.

67. The charges with sodium chloride are all balanced—for every positive sodium ion there is a corresponding negative chloride ion. Since its charges are balanced, how can sodium chloride be attracted to water, and vice versa?

68. Why are ion–dipole attractions stronger than dipole–dipole attractions?

69. Chlorine, Cl_2, is a gas at room temperature, but bromine, Br_2, is a liquid. Why?

70. Why is calcium fluoride, CaF_2, a high-melting-point crystalline solid, while stannic chloride, $SnCl_4$, is a volatile liquid?

71. A thin stream of water is pulled to a rubber balloon with a static electric charge. Might a small ice cube also be pulled to a statically charged balloon?

72. Of the two structures at right, one is a typical gasoline molecule and the other is a typical motor oil molecule. Which is which? Base your reasoning not on memorization but rather on what you know about electrical attractions between molecules and the various physical properties of gasoline and motor oil.

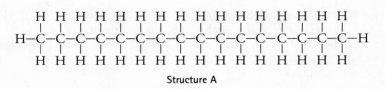

Structure A

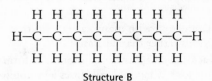

Structure B

DISCUSSION QUESTIONS (EVALUATION)

73. If you've been studying hard, by now you have great insight as to why oil and water don't mix. You understand that the oxygen atom has 8 electrons: 2 in the first shell and 6 in the valence shell. Two of these 6 valence electrons bond to 2 hydrogen atoms, which leaves the 4 remaining valence electrons to form 2 lone pairs (nonbonding electrons). These lone pairs push the hydrogen atoms together, which allows the highly electronegative oxygen atom to pull electrons toward itself. As a result, the oxygen side of the water molecule is slightly negative and the hydrogen side is slightly positive, which makes water a sticky sort of molecule. Because of this stickiness, water has a high boiling point and it effectively excludes less sticky oil molecules. Your understanding of the natural world is now much deeper. What value is there in having these sorts of deeper understandings?

74. What should be done with mining pits after all the ore has been removed? Consider the open-pit copper mine shown in Figure 15.14.

75. What are some of the obstacles people face when trying to recycle materials? How can your community overcome these obstacles? Should the government require that certain materials be recycled? If so, how should this requirement be enforced?

READINESS ASSURANCE TEST (RAT)

If you have a good handle on this chapter, if you really do, then you should be able to score 7 out of 10 on this RAT. If you score less than 7, you need to study further before moving on.

Choose the BEST answer to each of the following.

1. An atom loses an electron to another atom. Is this an example of a physical or chemical change?
   (a) chemical change involving the formation of ions
   (b) physical change involving the formation of ions
   (c) chemical change involving the formation of covalent bonds
   (d) physical change involving the formation of covalent bonds

2. Aluminum ions carry a 3+ charge, and chloride ions carry a 1− charge. What is the chemical formula for the ionic compound aluminum chloride?
   (a) Al_3Cl
   (b) $AlCl_3$
   (c) Al_3Cl_3
   (d) $AlCl$

3. Which would you expect to have a higher melting point: sodium chloride, NaCl, or aluminum oxide, Al_2O_3?
   (a) aluminum oxide, because it is a larger molecule and has a greater number of molecular interactions
   (b) NaCl, because it is a solid at room temperature

   (c) aluminum oxide, because of the greater charges of the ions and hence the greater force of attraction between them
   (d) aluminum oxide, because of the covalent bonds within the molecule

4. Why are ores so valuable?
   (a) They are sources of naturally occurring gold.
   (b) Metals can be efficiently extracted from them.
   (c) They tend to be found in scenic mountainous regions.
   (d) They hold many clues to Earth's natural history.

5. In terms of the periodic table, is there an abrupt or a gradual change between ionic and covalent bonds?
   (a) An abrupt change occurs across the metalloids.
   (b) Actually, any element of the periodic table can form a covalent bond.
   (c) There is a gradual change: the farther apart, the more ionic.
   (d) Whether an element forms one or the other depends on nuclear charge and not on the relative positions in the periodic table.

6. A hydrogen atom does not form more than one covalent bond because it
   (a) has only one shell of electrons.
   (b) has only one electron to share.
   (c) loses its valence electron so readily.
   (d) has such a strong electronegativity.

7. When nitrogen and fluorine combine to form a molecule, the most likely chemical formula is
   (a) N_3F.
   (b) N_2F.
   (c) NF_4.
   (d) NF.
   (e) NF_3.

8. A substance consisting of which molecule shown below should have a higher boiling point?

   $$S{=}C{=}O \qquad O{=}C{=}O$$

   (a) the molecule on the left, SCO, because it comes later in the periodic table
   (b) the molecule on the left, SCO, because it is less symmetrical
   (c) the molecule on the right, OCO, because it is more symmetrical
   (d) the molecule on the right, OCO, because it has more mass

9. The hydrogen bond is a type of
   (a) ionic bond.
   (b) metallic bond.
   (c) covalent bond.
   (d) none of the above

10. Iodine, I_2, has a higher melting point than fluorine, F_2, because its
   (a) atoms are larger.
   (b) molecules are heavier.
   (c) nonpolarity is greater.
   (d) atoms have greater electronegativity.

Answers to RAT

1. *a*, 2. *b*, 3. *c*, 4. *b*, 5. *c*, 6. *b*, 7. *e*, 8. *b*, 9. *d*, 10. *a*

16

CHAPTER 16

Mixtures

HOW CAN fresh water be prepared from seawater? Is it true that a fish can drown in water? When you stir sugar into water, the sugar crystals disappear, but where do they go? What are clouds made of, and what do they have in common with the blood that runs through our veins? When tap water is left boiling on the stove too long, it evaporates completely but leaves a chalky residue in the pot. What is this residue and where did it originate? How is municipal water treated so that it's safe enough to pipe into our homes for consumption? How does a wastewater treatment facility treat wastewater? The answers to these questions involve an understanding of mixtures.

FIGURE 16.1
Earth's atmosphere is a mixture of gaseous elements and compounds. Some of them are shown here.

16.1 Most Materials Are Mixtures

EXPLAIN THIS Does 500 ml of sugar-sweetened water also contain 500 ml of water?

A **mixture** is a combination of two or more substances in which each substance retains its own chemical properties. Most materials we encounter are mixtures: mixtures of elements, mixtures of compounds, or mixtures of elements and compounds. Stainless steel, for example, is a mixture of the elements iron, chromium, nickel, and carbon. Seltzer water is a mixture of a liquid compound, water, and a gaseous compound, carbon dioxide. Our atmosphere, as Figure 16.1 illustrates, is a mixture of the elements nitrogen, oxygen, and argon, plus small amounts of such compounds as carbon dioxide and water vapor.

Tap water is a mixture containing mostly water but also many other compounds. Depending on your location, your water may contain compounds of calcium, magnesium, chlorine, fluorine, iron, and potassium; trace amounts of compounds of lead, mercury, and cadmium; organic compounds; and dissolved oxygen, nitrogen, and carbon dioxide. Although minimizing any toxic components in your drinking water is important, removing all other substances from it is unnecessary, undesirable, and impossible. Some of the dissolved solids and gases give water its characteristic taste, and many of them promote human health: fluoride compounds protect teeth, chlorine destroys harmful bacteria, and as much as 10% of our daily requirement of iron, potassium, calcium, and magnesium is obtained from drinking water (Figures 16.2 and 16.3).

FIGURE 16.2
Tap water provides us with H_2O as well as many other compounds, many of which are flavorful and help us grow, as Graham demonstrates at ages 7 and 21. Bottoms up!

CHECKPOINT

So far, you have learned about three kinds of matter: elements, compounds, and mixtures. Which box below contains only an element? Which contains only a compound? Which contains a mixture?

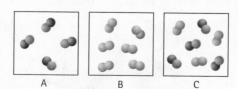

Was this your answer?
The molecules in box A each contain two different types of atoms and so represent a compound. The molecules in box B each consist of the same atoms and so represent an element. Box C is a mixture of the compound and the element.

FIGURE 16.3
Most of the oxygen in the air bubbles produced by an aquarium aerator escapes into the atmosphere. Some of the oxygen, however, mixes with the water. Fish depend on this oxygen to survive. Without this dissolved oxygen, which fish extract from the water with their gills, the fish would promptly drown. So fish don't "breathe" water. They breathe the O_2 that is dissolved in the water.

Mastering**PHYSICS**

VIDEO: Mixtures Can Be Separated by Physical Means

Chemists have devised many ingenious ways of separating the components of a mixture. Most of these techniques employ the simple principle of separating the components by differences in their physical properties

Note how the molecules of the compound and those of the element remain intact in the mixture. That is, when the mixture forms, atoms are not exchanged between the components.

There is a difference between the way substances—either elements or compounds—combine to form mixtures and the way elements combine to form compounds. Each substance in a mixture retains its chemical identity. The sugar molecules in the teaspoon of sugar in Figure 16.4, for example, are identical to the sugar molecules already in the tea. The only difference is that the sugar molecules in the tea are mixed with other substances, mostly water. The formation of a mixture, therefore, is a physical change. In contrast, as discussed in Section 14.5, there is a change in chemical identity when elements join to form compounds. Recall that sodium chloride is not a mixture of sodium and chlorine atoms but is instead a compound, which means it is entirely different from the elements that it contains. The formation of a compound is therefore a chemical change.

Mixtures Can Be Separated by Physical Means

The components of mixtures can be separated from one another by taking advantage of differences in the components' physical properties. A mixture of solids and liquids, for example, can be separated using filter paper through which the liquids pass but the solids do not. This is how coffee is often made: the caffeine and flavor molecules in the hot water pass through the filter and into the coffeepot, while the solid coffee grounds remain behind. This method of separating a solid–liquid mixture is called *filtration*, and it is a common technique used by chemists.

Mixtures can also be separated by taking advantage of a difference in boiling or melting points. Seawater is a mixture of water and a variety of compounds, mostly sodium chloride. Whereas water boils at 100°C, sodium chloride doesn't even melt until 800°C. One way to separate water from the mixture we call seawater, therefore, is to heat the seawater to about 100°C. At this temperature, the liquid water readily transforms to water vapor, but the sodium chloride stays behind dissolved in the remaining water. As the water vapor rises, it can be channeled into a cooler container, where it condenses into a liquid without the dissolved solids. This process of collecting a vaporized substance, called **distillation**, is illustrated in Figure 16.5. Distillation is a very effective, though costly, way of isolating fresh water from seawater. We explore this in further

FIGURE 16.4
Table sugar is a compound consisting only of sucrose molecules. Once these molecules are mixed into hot tea, they become interspersed among the water and tea molecules and form a sugar–tea–water mixture. No new compounds are formed, and so this is an example of a physical change.

Symbol for sugar molecule, which is sucrose, $C_{12}H_{22}O_{11}$ Sugar

Sugar in water

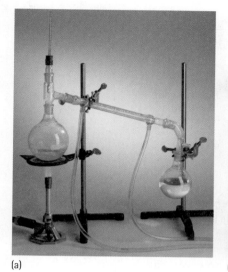

(a)

(b)

FIGURE 16.5
(a) A simple distillation setup used to separate one component from a mixture. The mixture is boiled in the flask on the left. The rising vapors contain only the volatile components of the mixture. The vapors are channeled into a downward-slanting tube kept cool by cold water flowing across its outer surface. The vapors inside the cool tube condense and collect in the flask on the right. (b) A whiskey still functions by the same principle. A mixture containing alcohol is heated to the point where the alcohol, some flavoring molecules, and some water are vaporized. These vapors travel through the copper pipes, where they then condense to a liquid.

detail in Section 16.6. After all the water has been distilled from seawater, dry solids remain. These solids, also a mixture of compounds, contain a variety of commercially valuable compounds, such as sodium chloride and potassium bromide (Figure 16.6).

FIGURE 16.6
At the southern end of San Francisco Bay are areas where the seawater has been partitioned off by earthen dikes. These are evaporation ponds, where the water is allowed to evaporate, leaving behind the solids that were dissolved in the seawater. These solids are further refined for commercial sale. The remarkable colors of the ponds are due to organic pigments made by salt-loving bacteria.

16.2 The Chemist's Classification of Matter

LEARNING OBJECTIVE
Classify the states of matter under the categories of pure and impure.

EXPLAIN THIS Is frozen apple juice an example of a solution, suspension, or heterogeneous mixture?

If a material is **pure**, it consists of only a single element or a single compound. Pure gold, for example, contains nothing but the element gold. Pure table salt contains nothing but the compound sodium chloride. If a material is **impure**, it is a mixture and contains two or more elements or compounds. This classification scheme is shown in Figure 16.7.

Because atoms and molecules are so small, it is impractical to prepare a sample that is truly pure—that is, truly 100% of a single material. For example, if just one atom or molecule out of a trillion trillion were different, then the

FIGURE 16.7
A chemical classification of matter.

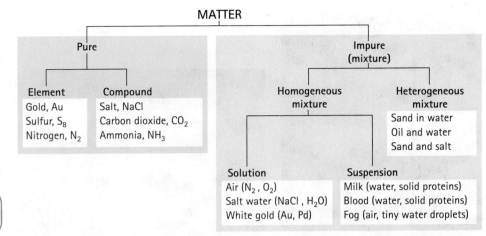

Orange juice may be 100% natural, but it is never 100% pure.

100% pure status would be lost. Samples, however, can be "purified" by various methods such as distillation. *Pure* is understood to be a relative term. When we compare the purity of two samples, the purer one contains fewer impurities. A sample of water that is 99.9% pure has a greater proportion of impurities than does a purer sample of water that is 99.9999% pure.

Sometimes naturally occurring mixtures are labeled as being pure, as in "pure orange juice." Such a statement merely means that nothing artificial has been added. According to a chemist's definition, however, orange juice is anything but pure, as it contains a wide variety of materials, including water, pulp, flavorings, vitamins, and sugars.

Mixtures may be heterogeneous or homogeneous. In a **heterogeneous mixture**, the different components can be seen as individual substances, such as pulp in orange juice, sand in water, or oil globules dispersed in vinegar. The different components are visible. **Homogeneous mixtures** have the same composition throughout. Any one region of the mixture has the same ratio of substances as any other region, and the components cannot be seen as individual identifiable entities. The distinction is shown in Figure 16.8.

A homogeneous mixture may be either a solution or a suspension. In a **solution**, all components are in the same phase. The atmosphere we breathe is a gaseous solution consisting of the gaseous elements nitrogen and oxygen as well as minor amounts of other gaseous materials. Salt water is a liquid solution because both the water and the dissolved sodium chloride are found in a single liquid phase. An example of a solid solution is the alloy white gold, which is a homogeneous mixture of the elements gold and palladium. We shall be discussing solutions in more detail in the next section.

FIGURE 16.8
(a) In heterogeneous mixtures, the different components can be seen with the naked eye. (b) In homogeneous mixtures, the different components are mixed at a much finer level and so are not readily distinguished.

Granite

"Snow" in snow globe   Pizza

(a) Heterogeneous mixtures

Air   Clear seawater   White gold

(b) Homogeneous mixtures

A **suspension** is a homogeneous mixture in which the different components are in different phases, such as solids in liquids or liquids in gases. In a suspension, the mixing is so thorough that the different phases cannot be readily distinguished. Milk is a suspension because it is a homogeneous mixture of proteins and fats finely dispersed in water. Blood is a suspension composed of finely dispersed blood cells in water. Another example of a suspension is clouds, which are homogeneous mixtures of tiny water droplets suspended in air. Shining a light through a suspension, as in Figure 16.9, results in a visible cone as the light is reflected by the suspended components.

The easiest way to distinguish a suspension from a solution in the laboratory is to spin a sample in a centrifuge. This device, spinning at thousands of revolutions per minute, separates the components of suspensions but not those of solutions, as Figure 16.10 shows.

FIGURE 16.9
The path of light becomes visible when the light passes through a suspension.

CHECKPOINT

Impure water can be purified by which of these?
- **(a) removing the impure water molecules**
- **(b) removing everything that is not water**
- **(c) breaking down the water to its simplest components**
- **(d) adding some disinfectant such as chlorine**

Was this your answer?
The answer is (b): impure water can be purified by removing everything that isn't water. H_2O is a compound made of the elements hydrogen and oxygen in a 2-to-1 ratio. Every H_2O molecule is exactly the same as every other, and there's no such thing as an impure H_2O molecule. Just about anything, including you, beach balls, rubber ducks, dust particles, and bacteria, can be found in water. When something other than water is found in water, we say that the water is impure. It is important to see that the impurities are in the water and not part of the water, which means that it is possible to remove them by a variety of physical means, such as filtration or distillation.

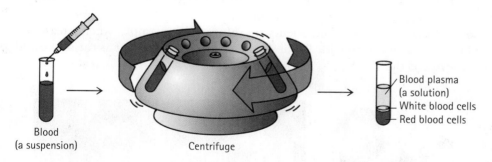

Blood
(a suspension)   Centrifuge

Blood plasma
(a solution)
White blood cells
Red blood cells

FIGURE 16.10
Blood, because it is a suspension, can be centrifuged into its components, which include the blood plasma (a yellowish solution) and white and red blood cells. The components of the plasma, however, cannot be separated from one another here because a centrifuge has no effect on solutions.

16.3 Solutions

EXPLAIN THIS What is the solvent in brown sugar?

What happens when table sugar, known chemically as *sucrose*, is stirred into water? Is the sucrose destroyed? We know it isn't, because it sweetens the water. Does the sucrose disappear because it somehow ceases to occupy space or because it fits within the nooks and crannies of the water? Not so, because adding sucrose changes the volume. This may not be noticeable

LEARNING OBJECTIVE
Describe the components of a solution, and calculate a solution's concentration.

Mastering**PHYSICS**
VIDEO: Solutions

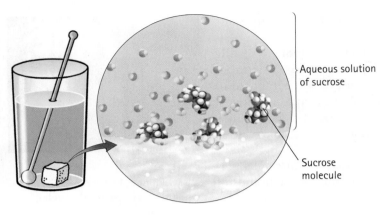

Aqueous solution
of sucrose

Sucrose
molecule

FIGURE 16.11
Water molecules pull the sucrose molecules in a sucrose crystal away from one another. This pulling away does not, however, affect the covalent bonds within each sucrose molecule, which is why each dissolved sucrose molecule remains intact as a single molecule.

To most people, solutions mean finding the answers. To chemists, however, solutions are things that are still all mixed up.

at first, but if you continue adding sucrose to a glass of water, you'll see that the water level rises, just as it would if you were adding sand.

Sucrose stirred into water loses its crystalline form. Each sucrose crystal consists of billions of sucrose molecules packed neatly together. When the crystal is exposed to water (as was first shown in Figure 16.4 and is shown again here in Figure 16.11), an even greater number of water molecules pull on the sucrose molecules via hydrogen bonds formed between the sucrose molecules and the water molecules. With a little stirring, the sucrose molecules soon mix throughout the water. In place of sucrose crystals and water, we have a homogeneous mixture of sucrose molecules in water. As discussed earlier, *homogeneous* means that a sample taken from any part of a mixture is the same as a sample taken from any other part. In our sucrose example, this means that the sweetness of the first sip of the solution is the same as the sweetness of the last sip.

Recall that a homogeneous mixture consisting of a single phase is called a *solution*. Sugar in water is a solution in the liquid phase. Solutions aren't always liquids, however. They can also be solid or gaseous, as Figure 16.12 shows. Gemstones are solid solutions. A ruby, for example, is a solid solution of trace quantities of red chromium compounds in transparent aluminum oxide. A blue sapphire is a solid solution of trace quantities of light green iron compounds and blue titanium compounds in aluminum oxide. Another important example of solid solutions is metal alloys, which are mixtures of different metallic elements. The alloy known as brass is a solid solution of copper and zinc, for instance, and the alloy stainless steel is a solid solution of iron, chromium, nickel, and carbon.

An example of a gaseous solution is the air we breathe. By volume, this solution is 78% nitrogen gas, 21% oxygen gas, and 1% other gaseous materials, including water vapor and carbon dioxide. The air we exhale is a gaseous solution of 75% nitrogen, 14% oxygen, 5% carbon dioxide, and around 6% water vapor. So we see that the air we breathe undergoes a chemical change before being exhaled.

In describing solutions, the component present in the largest amount is the **solvent**, and any other components are **solutes**. For example, when a teaspoon of table sugar is mixed with 1 L of water, we identify the sugar as the solute and the water as the solvent.

(a)

(b)

(c)

FIGURE 16.12
Solutions may occur in (a) the solid phase, (b) the liquid phase, or (c) the gaseous phase.

The process of mixing a solute with a solvent is called **dissolving**. To make a solution, a solute must dissolve in a solvent; that is, the solute and solvent must form a homogeneous mixture. Whether one material dissolves in another is a function of their electrical attractions for each other.

CHECKPOINT

What is the solvent in the gaseous solution we call air?

Was this your answer?
Nitrogen is the solvent, because it is the component that is present in the greatest quantity.

There is a limit to how much of a given solute can be dissolved in a given solvent, as Figure 16.13 illustrates. We know that when you add table sugar to a glass of water, for example, the sugar rapidly dissolves. As you continue to add sugar, however, there comes a point when it no longer dissolves. Instead, it collects at the bottom of the glass, even after stirring. At this point, the water is saturated with sugar, meaning that the water cannot accept any more sugar. When this happens, we have a **saturated solution**, defined as one in which no more solute can be dissolved. A solution that has not reached the limit of solute that will dissolve is called an **unsaturated solution**.

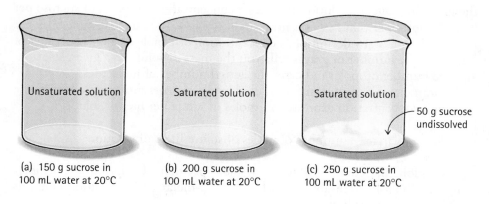

(a) 150 g sucrose in 100 mL water at 20°C
(b) 200 g sucrose in 100 mL water at 20°C
(c) 250 g sucrose in 100 mL water at 20°C

FIGURE 16.13
A maximum of 200 g of sucrose dissolves in 100 mL of water at 20°C. (a) Mixing 150 g of sucrose in 100 mL of water at 20°C produces an unsaturated solution. (b) Mixing 200 g of sucrose in 100 mL of water at 20°C produces a saturated solution. (c) If 250 g of sucrose is mixed with 100 mL of water at 20°C, 50 g of sucrose remains undissolved. (As we discuss later, the concentration of a saturated solution varies with temperature.)

The quantity of solute dissolved in a solution is described in mathematical terms by the solution's **concentration**, which is the amount of solute dissolved per amount of solution:

$$\text{Concentration} = \text{amount of solute/amount of solution}$$

For example, a sucrose–water solution may have a concentration of 1 g of sucrose for every liter of solution. This can be compared with concentrations of other solutions. A sucrose–water solution containing 2 g of sucrose per liter of solution, for example, is more concentrated, and one containing only 0.5 g of sucrose per liter of solution is less concentrated, or more dilute.

Chemists are often more interested in the number of solute particles in a solution than in the number of grams of solute. Submicroscopic particles, however, are so small that the number of them in any observable

FYI How long would it take to count to 1 million? If each count takes 1 s, counting nonstop to a million would take 11.6 days. To count to a billion would take 31.7 years. To count to a trillion would take 31,700 years! Counting to a trillion 602 billion times would take about 2 million times the estimated age of the universe. In short, 602 billion trillion, as discussed on the next page, is an inconceivably large number.

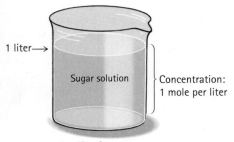

1 liter →

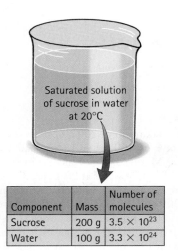

1 mole of sucrose
equals
342 grams of sucrose
equals
6.02×10^{23} molecules of sucrose

FIGURE 16.14
An aqueous solution of sucrose that has a concentration of 1 mole of sucrose per liter of solution contains 6.02×10^{23} sucrose molecules (342 g) in every liter of solution.

Component	Mass	Number of molecules
Sucrose	200 g	3.5×10^{23}
Water	100 g	3.3×10^{24}

FIGURE 16.15
Although 200 g of sucrose is twice as massive as 100 g of water, there are about 10 times as many water molecules in 100 g of water as there are sucrose molecules in 200 g of sucrose. How can this be? Each water molecule is about $\frac{1}{20}$ as massive as (and much smaller than) each sucrose molecule, which means that about 10 times as many water molecules can fit within half the mass.

sample is incredibly large. To avoid awkwardly large numbers, scientists use a unit called the mole. One **mole** of any type of particle is, by definition, 6.02×10^{23} particles. (This superlarge number is about 602 billion trillion, or 602,000,000,000,000,000,000,000 particles. Interestingly, the term *mole* is derived from the Latin word *moles*, meaning heap, mass, or pile.) One mole of gold atoms, for example, is 6.02×10^{23} gold atoms, and 1 mole of sucrose molecules is 6.02×10^{23} sucrose molecules.

Even if you've never heard the term *mole* in your life before now, you are already familiar with the basic idea. Saying "one mole" is just a shorthand way of saying "six point oh two times ten to the twenty-third particles." Just as "a couple of" means 2 of something and "a dozen of" means 12 of something, "a mole of" means 6.02×10^{23} of some elementary unit, such as atoms, molecules, or ions:

- a couple of coconuts = 2 coconuts
- a dozen donuts = 12 donuts
- a mole of molecules = 6.02×10^{23} molecules

One mole of gold atoms, for example, is 6.02×10^{23} gold atoms, and 1 mole of sucrose molecules is 6.02×10^{23} sucrose molecules. A stack containing "1 mole" of pennies would reach a height of about 860 quadrillion km, which is roughly equal to the diameter of our galaxy, the Milky Way. And "1 mole" of marbles would be enough to cover the entire land area of the 50 United States to a depth greater than 1.1 km.

But sucrose molecules are so small that there are 6.02×10^{23} of them in only 342 g of sucrose, which is about a cupful. Thus, because 342 g of sucrose contains 6.02×10^{23} molecules of sucrose, we can use our shorthand wording and say that 342 g of sucrose contains 1 mole of sucrose. As Figure 16.14 shows, therefore, an aqueous solution that has a concentration of 342 g of sucrose per liter of solution also has a concentration of 6.02×10^{23} sucrose molecules per liter of solution or, by definition, a concentration of 1 mole of sucrose per liter of solution. The number of grams tells you the mass of solute in a given solution, and the number of moles indicates the actual number of molecules.

A common unit of concentration used by chemists is **molarity**, which is the solution's concentration expressed in moles of solute per liter of solution:

$$\text{Molarity} = \text{number of moles of solute/liters of solution}$$

A solution that contains 1 mole of solute per liter of solution is a 1-molar solution, which is often abbreviated 1 M. A 2-molar (2 M) solution contains 2 moles of solute per liter of solution.

The difference between referring to the number of molecules of solute and referring to the number of grams of solute can be illustrated by the following question. A saturated aqueous solution of sucrose contains 200 g of sucrose and 100 g of water. Which is the solvent: sucrose or water?

As shown in Figure 16.15, there are 3.5×10^{23} molecules of sucrose in 200 g of sucrose, but there are almost 10 times as many molecules of water in 100 g of water—3.3×10^{24} molecules. As defined earlier, the solvent is the component present in the largest amount, but what do we mean by *amount*? If *amount* means number of molecules, then water is the solvent. If *amount* means mass, then sucrose is the solvent. So the answer depends on how you look at it. From a chemist's point of view, *amount* typically means the number of molecules, and so water is the solvent in this case.

CHECKPOINT

1. **How many moles of sucrose are in 0.5 L of a 2 M solution? How many molecules of sucrose is this?**
2. **Does 1 L of a 1 M solution of sucrose in water contain 1 L of water, less than 1 L of water, or more than 1 L of water?**

Were these your answers?

1. First you need to understand that 2 M means 2 moles of sucrose per liter of solution. To obtain the amount of solute, you should multiply solution concentration by amount of solution:

$$(2 \text{ moles/L})(0.5 \text{ L}) = 1 \text{ mole}$$

which is the same as 6.02×10^{23} molecules.

2. The definition of molarity refers to the number of liters of solution, not to the number of liters of solvent. When sucrose is added to a given volume of water, the volume of the solution increases. So if 1 mole of sucrose is added to 1 L of water, the result is more than 1 L of solution. Therefore, 1 L of a 1 M solution requires less than 1 L of water.

> **FYI** Would a "mole" of stacked pennies stretch across our galaxy? Estimate the answer yourself. First measure the number of stacked pennies in 1 cm. To find the length of one "mole" of stacked pennies, take the number of particles in 1 mole (6.02×10^{23}) and divide it by the number of pennies in 1 cm. Your answer will be in centimeters. To convert to kilometers, divide by the number of centimeters in a kilometer, which is 100,000. Tall stack!

FIGURING PHYSICAL SCIENCE

Calculating for Solutions

From the formula for the concentration of a solution, we can derive equations for the amount of solute and the amount of solution:

Concentration of solution =
    amount of solute/volume of solution
Amount of solute =
            concentration of solution ×
            volume of solution
Volume of solution =
        amount of solute/
        concentration of solution

In solving for any of these values, the units must always match. If concentration is given in grams per liter of solution, for example, the amount of solute must be in grams and the amount of solution must be in liters.

Note that these equations are set up for calculating the volume of solution rather than the volume of solvent. The volume of solution is greater than the volume of solvent because, in addition to containing the solvent, the solution also contains the solute. As discussed at the beginning of Section 16.3, for example, the volume of an aqueous solution of sucrose depends not only on the volume of water but also on the volume of dissolved sucrose.

SAMPLE PROBLEM 1
How many grams of sucrose are in 3 L of an aqueous solution that has a concentration of 2 g of sucrose per liter of solution?

Solution:
This question asks for amount of solute, and so you should use the second of the three formulas given above:

Amount of solute = 2 g/1 L × 3 L = 6 g

SAMPLE PROBLEM 2
A solution you are using in an experiment has a concentration of 10 g of solute per liter of solution. If you pour enough of this solution into an empty laboratory flask to make the flask contain 5 g of the solute, how many liters of the solution have you poured into the flask?

Solution:
This question asks for amount of solution, and you should use the third formula given above:

Volume of solution = 5 g/10 g/L = 0.5 L

SAMPLE PROBLEM 3
At 20°C, a saturated solution of sodium chloride in water has a concentration of about 380 g of sodium chloride per liter of solution. How many grams of sodium chloride are required to make 3 L of a saturated solution?

Solution:
Multiply the solution concentration by the final volume of the solution. This provides the amount of solute required: (380 g/L)(3 L) = 1140 g.

SAMPLE PROBLEM 4
A student is told to use 20 g of sodium chloride to make an aqueous solution that has a concentration of 10 g of sodium chloride per liter of solution. How many liters of solution does she end up with?

Solution:
Divide the amount of solute by the solution concentration to obtain the amount of solution prepared:

20 g/10 g/L = 2 L

16.4 Solubility

EXPLAIN THIS How can oxygen be removed from water?

The **solubility** of a solute is its *ability* to dissolve in a solvent. As can be expected, this ability mainly depends on the submicroscopic attractions between solute particles and solvent particles. If a solute has any appreciable solubility in a solvent, then that solute is said to be **soluble** in that solvent.

Solubility also depends on attractions of solute particles for one another and attractions of solvent particles for one another. As shown in Figure 16.16, for example, a sucrose molecule has many polar hydrogen–oxygen bonds. Sucrose molecules, therefore, can form multiple hydrogen bonds with one another. These hydrogen bonds are strong enough to make sucrose a solid at room temperature and to give it the relatively high melting point of 185°C. In order for sucrose to dissolve in water, the water molecules must first pull sucrose molecules away from one another. This puts a limit on the amount of sucrose that can dissolve in water— eventually, a point is reached at which there are not enough water molecules to separate the sucrose molecules from one another. As we discussed in Section 16.3, this is the point of saturation, and any additional sucrose added to the solution does not dissolve.

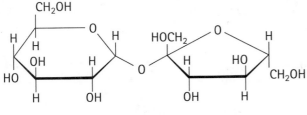

Sucrose

FIGURE 16.16
A sucrose molecule contains many hydrogen–oxygen covalent bonds, in which the hydrogen atoms are slightly positive and the oxygen atoms are slightly negative. These dipoles in any given sucrose molecule result in the formation of hydrogen bonds with neighboring sucrose molecules.

When the molecule-to-molecule attractions among solute molecules are comparable to the molecule-to-molecule attractions among solvent molecules, there is no practical point of saturation. As shown in Figure 16.17, for example, the hydrogen bonds among water molecules are about as strong as those between ethanol molecules. These two liquids therefore mix together quite well and in just about any proportion. We can even add ethanol to water until the ethanol, rather than the water, can be considered the solvent.

A solute that has no practical point of saturation in a given solvent is said to be *infinitely soluble* in that solvent. Ethanol, for example, is infinitely soluble in water. Also, all gases are generally infinitely soluble in other gases because they can be mixed together in just about any proportion.

Let's now look at the other extreme of solubility, in which a solute has very little solubility in a given solvent. An example is oxygen, O_2, in water. In contrast to sucrose, which has a solubility of 200 g per 100 mL of water, only 0.004 g of oxygen can dissolve in 100 mL of water. We can account for oxygen's low

MasteringPHYSICS®
TUTORIAL: Solubility
VIDEO: Solubility Changes with Temperature and Pressure

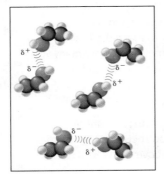

Ethanol

Ethanol and water

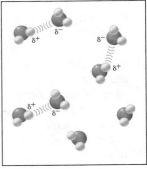

Water

FIGURE 16.17
Ethanol and water molecules are about the same size, and they both form hydrogen bonds. As a result, ethanol and water readily mix with each other.

solubility in water by noting that the only electrical attractions that occur between oxygen molecules and water molecules are relatively weak dipole–induced dipole attractions. More important, however, is the fact that the stronger attraction of water molecules for one another—through the hydrogen bonds that the water molecules form with one another—effectively excludes oxygen molecules from intermingling.

FIGURE 16.18
Glass is frosted by dissolving its outer surface in hydrofluoric acid.

A material that does not dissolve in a solvent to any appreciable extent is said to be **insoluble** in that solvent. We consider many substances insoluble in water, including sand and glass. Just because a material is not soluble in one solvent, however, does not mean it won't dissolve in another. Sand and glass, for example, are soluble in hydrofluoric acid, HF, which is used to give glass the decorative frosted look shown in Figure 16.18. Also, although Styrofoam is insoluble in water, it is soluble in acetone, a solvent used in fingernail polish remover. Pour a little acetone into a Styrofoam cup, and the acetone soon causes the Styrofoam to deform, as demonstrated in Figure 16.19.

CHECKPOINT
Why isn't sucrose infinitely soluble in water?

FIGURE 16.19
Is this cup melting or dissolving?

Was this your answer?
The attraction between two sucrose molecules is much stronger than the attraction between a sucrose molecule and a water molecule. Because of this, sucrose dissolves in water only as long as the number of water molecules far exceeds the number of sucrose molecules. When there are too few water molecules to dissolve any additional sucrose, the solution is saturated.

Solubility Changes with Temperature

You probably know from experience that water-soluble solids usually dissolve better in hot water than in cold water. A highly concentrated solution of sucrose in water, for example, can be made by heating the solution almost to the boiling point. This is how syrups and hard candy are made.

Solubility increases with increasing temperature because hot water molecules have greater kinetic energy and therefore can collide with the solid solute more vigorously. The vigorous collisions help disrupt electrical particle-to-particle attractions in the solid.

Although the solubilities of many solid solutes—sucrose, to name just one example—are greatly increased by temperature increases, the solubilities of other solid solutes, such as sodium chloride, are only mildly affected, as Figure 16.20 shows. This difference involves a number of factors, including the strength of the chemical bonds in the solute molecules and the way those molecules are packed together. Some chemicals, such as calcium carbonate, $CaCO_3$, actually become *less* soluble as the water temperature increases. This explains why the inner surface of tea kettles are often coated with calcium carbonate residues.

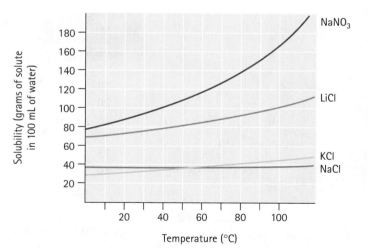

FIGURE 16.20
The solubility of many water-soluble solids increases with temperature, while the solubility of others is only very slightly affected by temperature.

When a sugar solution saturated at a high temperature is allowed to cool, some of the sugar usually comes out of solution and forms a **precipitate**. When this occurs, the solute—sugar in this case—is said to have *precipitated* from the solution.

Let's put on our quantitative thinking caps and consider another example. At 100°C, the solubility of sodium nitrate, $NaNO_3$, in water is 165 g per 100 mL of water. As we cool this solution, the solubility of $NaNO_3$ decreases, as shown in Figure 16.21, and this change in solubility causes some of the dissolved $NaNO_3$ to precipitate (come out of solution). At 20°C, the solubility of $NaNO_3$ is only 87 g per 100 mL of water. So if we cool the 100°C solution to 20°C, 78 g (165 g − 87 g) precipitates, as shown in Figure 16.21.

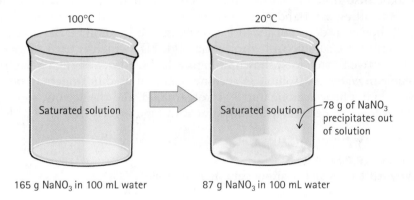

100°C   20°C

Saturated solution   Saturated solution   78 g of $NaNO_3$ precipitates out of solution

165 g $NaNO_3$ in 100 mL water   87 g $NaNO_3$ in 100 mL water

FIGURE 16.21
The solubility of sodium nitrate is 165 g per 100 mL of water at 100°C but only 87 g per 100 mL at 20°C. Cooling a 100°C saturated solution of $NaNO_3$ to 20°C causes 78 g of the solute to precipitate.

> Grease is soluble in paint thinner, which is why paint thinner can be used to clean one's hands of grease. But body oils are also soluble in paint thinner, which is why hands cleaned with paint thinner feel dry and chapped.

Solubility of Gases

In contrast to the solubilities of most solids, the solubilities of gases in liquids *decrease* with increasing temperature, as Table 16.1 shows. This effect occurs because with an increase in temperature, the solvent molecules have more kinetic energy. This makes it more difficult for a gaseous solute to remain in solution because the solute molecules are ejected by the high-energy solvent molecules.

TABLE 16.1	TEMPERATURE-DEPENDENT SOLUBILITY OF OXYGEN GAS IN WATER AT A PRESSURE OF 1 ATM
Temperature (°C)	O_2 Solubility (g O_2/L H_2O)
0	0.067
10	0.052
20	0.044
25	0.039
30	0.037
35	0.033
40	0.031

> Air is a gaseous solution, and one of its minor components is water vapor. The process of this water coming "out of solution" in the form of rain or snow is called "precipitation." The rain or snow is the "precipitate."

Perhaps you have noticed that warm carbonated beverages go flat faster than cold ones. The higher temperature causes the molecules of carbon dioxide gas to leave the liquid solvent at a higher rate.

The solubility of a gas in a liquid also depends on the pressure of the gas immediately above the liquid. In general, a higher gas pressure above the liquid means more of the gas dissolves. A gas at a high pressure has many gas particles crammed into a given volume. The "empty" space in an unopened soft drink bottle, for example, is crammed with carbon dioxide molecules in the gaseous phase. With nowhere else to go, many of these molecules dissolve in the liquid, as shown in Figure 16.22. Alternatively, we might say that the great pressure forces the carbon dioxide molecules into solution. When the bottle is opened, the "head" of highly pressurized carbon dioxide gas escapes. Now the gas pressure above the liquid is lower than before. As a result, the solubility of the carbon dioxide drops, and the carbon dioxide molecules that were once squeezed into the solution begin to escape into the air above the liquid.

The rate at which carbon dioxide molecules leave an opened soft drink is relatively slow. You can increase the rate by pouring in granulated sugar, salt, or sand. The microscopic nooks and crannies on the surfaces of the grains serve as *nucleation sites* where carbon dioxide bubbles can form rapidly and then escape by buoyant forces. Shaking the beverage also increases the surface area of the liquid-to-gas interface, making it easier for the carbon dioxide to escape from the solution. Once the solution is shaken, so much carbon dioxide escapes that the beverage froths over. You also increase the rate at which carbon dioxide escapes when you pour the beverage into your mouth, which abounds in nucleation sites. You can feel the resulting tingly sensation.

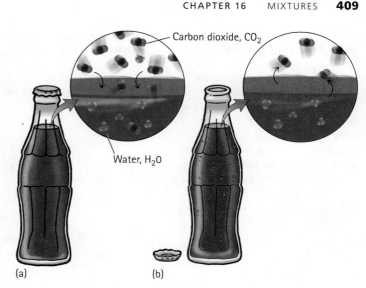

Carbon dioxide, CO_2

Water, H_2O

(a)   (b)

FIGURE 16.22

(a) The carbon dioxide gas above the liquid in an unopened soft drink bottle consists of many tightly packed carbon dioxide molecules that are forced by pressure into solution. (b) When the bottle is opened, the pressure is released, and carbon dioxide molecules originally dissolved in the liquid can escape into the air.

CHECKPOINT

You open two cans of soft drink, one from a warm kitchen shelf and the other from the coldest depths of your refrigerator. Which fizzes more in your mouth?

Was this your answer?
The solubility of carbon dioxide in water decreases with increasing temperature. The warm drink, therefore, fizzes in your mouth more than the cold one does.

16.5 Soaps, Detergents, and Hard Water

EXPLAIN THIS How does washing soda help to clean laundry?

Dirt and grease together make *grime*. Because grime contains many non-polar components, it is difficult to remove from hands or clothing with water alone. To remove most grime, we can use a nonpolar solvent, such as turpentine, which dissolves the grime because of strong induced dipole–induced dipole attractions. Turpentine is good for removing the grime left on hands after such activities as changing a car's motor oil. Rather than washing our dirty hands and clothes with nonpolar solvents, however, we have a more pleasant alternative—soap and water. Soap works because soap molecules have both nonpolar and polar properties. A typical soap molecule has

LEARNING OBJECTIVE
Describe the mechanism by which soaps and detergents clean and how this mechanism is foiled by hard water.

Mastering**PHYSICS**

VIDEO: Soap Works by Being Both Polar and Nonpolar

FYI It is not just dipole–induced dipole attractions that keep carbon dioxide dissolved within water. As we'll discuss in Chapter 18, carbon dioxide reacts with water to form carbonic acid, which is much more soluble in water. When a can of carbonated soda is opened, much of this carbonic acid quickly transforms back into water and carbon dioxide, which quickly bubbles out of solution because of its low solubility.

two parts: a long, nonpolar tail of carbon and hydrogen atoms and a polar head containing at least one ionic bond:

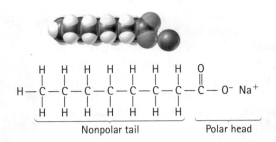

Because most of a soap molecule is nonpolar, it attracts nonpolar grime molecules via induced dipole–induced dipole attractions, as Figure 16.23 illustrates. In fact, grime quickly finds itself surrounded in three dimensions by the nonpolar tails of soap molecules. This attraction is usually sufficient to lift the grime away from the surface being cleaned. With the nonpolar tails facing inward toward the grime, the polar heads are all directed outward, where they are attracted to water molecules by relatively strong ion–dipole attractions. If the water is flowing, the whole conglomeration of grime and soap molecules flows with it, away from your hands or clothes and then down the drain.

For the past several centuries, soaps have been prepared by treating animal fats with sodium hydroxide, NaOH, also known as caustic lye. In this reaction, which is still used today, each fat molecule is broken down into three *fatty acid* soap molecules and one glycerol molecule:

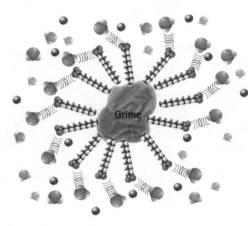

FIGURE 16.23
Nonpolar grime attracts and is surrounded by the nonpolar tails of soap molecules, forming what is called a *micelle*. The polar heads of the soap molecules are attracted by ion–dipole attractions to water molecules, which then carry the soap–grime combination away.

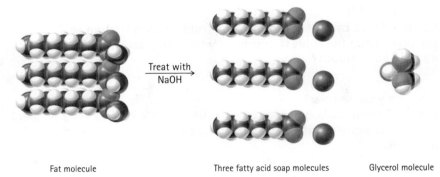

Fat molecule   Three fatty acid soap molecules   Glycerol molecule

In the 1940s, chemists began developing a class of synthetic soaplike compounds, known as *detergents*, that offer several advantages over true soaps, such as stronger grease penetration and lower price.

The chemical structure of detergent molecules is similar to that of soap molecules in that both possess a polar head attached to a nonpolar tail. The polar head in a detergent molecule, however, typically consists of either a sulfate group, $-OSO_3^-$, or a sulfonate group, $-SO_3^-$, and the nonpolar tail can have an assortment of structures.

One of the most common sulfate detergents is sodium lauryl sulfate, a main ingredient of many toothpastes. A common sulfonate detergent is sodium dodecyl benzenesulfonate, also known as a linear alkylsulfonate, or LAS, often found in dishwashing liquids. Both of these detergents are biodegradable, which means that microorganisms can break down the molecules once they are released into the environment.

$$CH_3CH_2CH_2CH_2CH_2CH_2CH_2CH_2CH_2CH_2CH_2—O—\overset{\displaystyle O}{\underset{\displaystyle O}{\overset{\textstyle |}{\underset{\textstyle |}{S}}}}—O^-\ Na^+$$

Sodium lauryl sulfate

$$CH_3CH_2CH_2CH_2CH_2CH_2CH_2CH_2CH_2CH_2CH_2CH_2—C \cdots C—\overset{\displaystyle O}{\underset{\displaystyle O}{\overset{\textstyle |}{\underset{\textstyle |}{S}}}}—O^-\ Na^+$$

Sodium dodecyl benzenesulfonate

FYI "Dry cleaning" is the process of washing clothes without water. The most common dry-cleaning solvent is perchloroethylene, C_2Cl_4, which is gentle on the clothing and can clean a full load in under 10 min. After a washing cycle, the solvent is centrifuged out of the machine, filtered, distilled, and recycled for the next load. Clothes come out of the machine already dry and ready for folding. Perchloroethylene, also known as perc, is relatively safe, but it is mildly carcinogenic and can cause dizziness in those who work with it. An up-and-coming alternative to perc is supercritical carbon dioxide, CO_2.

CHECKPOINT
What type of attractions hold soap or detergent molecules to grime?

Was this your answer?
If you haven't yet formulated an answer, why not back up and reread the question? You've got only four choices: ion–dipole, dipole–dipole, dipole–induced dipole, and induced dipole–induced dipole. The answer is induced dipole–induced dipole attractions, because the interaction is between two nonpolar entities—the grime and the nonpolar tail of a soap or detergent molecule.

Softening Hard Water

Water containing large amounts of calcium and magnesium ions is said to be **hard water**, and it has many undesirable qualities. For example, when hard water is heated, its calcium and magnesium ions tend to bind with negatively charged ions also found in the water to form solid compounds, like those shown in Figure 16.24. These can clog water heaters and boilers. You'll also find coatings of these calcium and magnesium compounds on the inside surfaces of a well-used teakettle (because the solubility of these compounds decreases with increasing temperature, as discussed earlier).

Hard water also inhibits the cleansing actions of soaps and, to a lesser extent, detergents. The sodium ions of soap and detergent molecules carry a 1+ charge, and calcium and magnesium ions carry a 2+ charge (note their positions in the periodic table). The negatively charged portion of the polar head of a soap or detergent molecule is more attracted to the double positive charge of calcium and magnesium ions than to the single positive charge of sodium ions. Soap or detergent molecules, therefore, give up their sodium ions to bind selectively with calcium or magnesium ions:

FIGURE 16.24
Hard water causes calcium and magnesium compounds to build up on the inner surfaces of water pipes, especially those used to carry hot water.

FIGURE 16.25
(a) Sodium carbonate is added to many detergents as a water-softening agent. (b) The doubly positive calcium and magnesium ions of hard water preferentially bind with the doubly negative carbonate ion, freeing the detergent molecules to do their job.

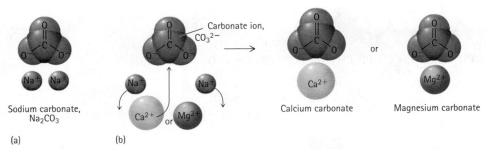

FIGURE 16.26
(1) Negatively charged sites on the unused ion-exchange resin are occupied by sodium ions. (2) As hard water passes over the resin, sodium ions are displaced by calcium and magnesium ions. (3) After the resin becomes saturated with calcium and magnesium ions, it is no longer effective at softening water.

Soap or detergent molecules bound to calcium or magnesium ions tend to be insoluble in water. As they come out of solution, they form a scum, which can appear as a ring around the inside of your bathtub. Because the soap or detergent molecules are tied up with calcium and magnesium ions, more soap or detergent must be added to maintain cleaning effectiveness.

Many detergents today contain sodium carbonate, Na_2CO_3, commonly known as washing soda. The calcium and magnesium ions in hard water are more attracted to the carbonate ion with its two negative charges than they are to a soap or detergent molecule with its single negative charge. With the calcium and magnesium ions bound to the carbonate ion, as shown in Figure 16.25, the soap or detergent is free to do its job. Because it removes the ions that make water hard, sodium carbonate is known as a water-softening agent.

In some homes, the water is so hard that it must be passed through a *water-softening unit*. In a typical unit, illustrated in Figure 16.26, hard water is passed through a large tank filled with tiny beads of a water-insoluble resin known as an *ion-exchange resin*. The surface of the resin contains many negatively charged ions bound to positively charged sodium ions. As calcium and magnesium ions pass over the resin, the ions displace the sodium ions and thereby become bound to the resin. The calcium and magnesium ions can do this because their positive charge (2+) is greater than that of the sodium ions (1+). The

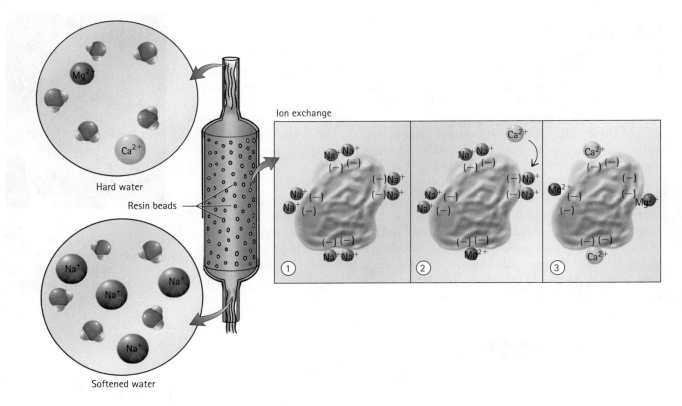

calcium and magnesium ions therefore have a greater attraction for the negative sites on the resin. The net result is that for every calcium or magnesium ion that binds, two sodium ions are set free. In this way, the resin *exchanges* ions. The water that exits the unit is now free of calcium and magnesium ions, but contains sodium ions in their place.

Eventually, all the sites for calcium and magnesium on the resin are filled, and then the resin needs to be either discarded or recharged. It is recharged by flushing it with a concentrated solution of sodium chloride, NaCl. The abundant sodium ions displace the calcium and magnesium ions (ions are *exchanged* once again), freeing up the binding sites on the resin.

> Most modern water softeners are equipped with meters that let you know the rate at which you consume water. This is a great way to keep tabs on your water-conservation efforts.

16.6 Purifying the Water We Drink

LEARNING OBJECTIVE
Identify the industrial means by which water is purified.

EXPLAIN THIS Why is water so difficult to purify?

As was discussed earlier, it is impossible to obtain 100% pure water. However, we can purify water to meet our needs. We do this by taking advantage of the differences in physical properties of water and the solutes or particulates it contains. For the remainder of this chapter, we turn our attention to some of the details involved in the production of drinkable water and the treatment of wastewater.

Water that is safe for drinking is said to be *potable*. In the United States, potable water is currently used for everything from cooking to flushing our toilets. The first step most public utilities take to produce potable water from natural sources is to remove any dirt particles or pathogens, such as bacteria. This is done by mixing the water with certain minerals, such as slaked lime and aluminum sulfate, which coagulates into a gelatinous material, aluminum hydroxide, that intersperses throughout the water (Figure 16.27). A large settling basin is used for this process. Slow stirring causes the gelatinous material to clump together and settle to the bottom of the basin. As these clumps form and settle, they carry with them many of the dirt particles and bacteria. The water is then filtered through sand and gravel.

To improve the odor and flavor of the water, many treatment facilities also *aerate* the water by cascading it through a column of air, as shown in Figure 16.28. Aeration removes many unpleasant-smelling volatile chemicals, such as sulfur compounds. At the same time, air dissolves into the water, giving it a better taste—without dissolved air, the water tastes flat. As a final step, the water is treated with a disinfectant, usually chlorine gas, Cl_2, but sometimes ozone, O_3, and then stored in a holding tank that feeds into the city mains.

> Lack of clean drinking water is one of the world's leading causes of death, especially among children of developing nations.

FIGURE 16.27
Slaked lime, $Ca(OH)_2$, and aluminum sulfate, $Al_2(SO_4)_3$, react to form aluminum hydroxide, $Al(OH)_3$, and calcium sulfate, $CaSO_4$, which together form a gelatinous material.

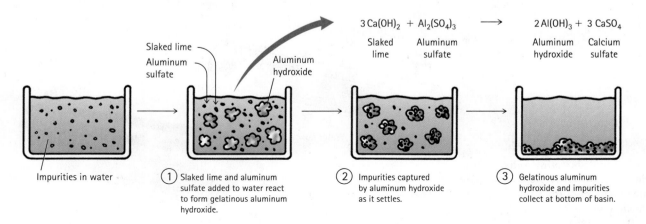

$$3\,Ca(OH)_2 \;+\; Al_2(SO_4)_3 \longrightarrow 2\,Al(OH)_3 \;+\; 3\,CaSO_4$$

Slaked lime   Aluminum sulfate   Aluminum hydroxide   Calcium sulfate

Slaked lime
Aluminum sulfate
Aluminum hydroxide

Impurities in water

① Slaked lime and aluminum sulfate added to water react to form gelatinous aluminum hydroxide.

② Impurities captured by aluminum hydroxide as it settles.

③ Gelatinous aluminum hydroxide and impurities collect at bottom of basin.

FIGURE 16.28
Volatile impurities are removed from drinking water by cascading it through the columns of air within each of these stacks.

Developed countries have the technology and infrastructure to produce vast quantities of water suitable for drinking; as a result, many citizens take their drinking water for granted. The number of public water treatment facilities in developing nations, however, is relatively small. In these locations, many people drink their water in the form of a hot beverage, such as tea, which is disinfected through boiling. Alternatively, disinfecting iodine tablets can be used.

Fuel for boiling and tablets for disinfecting, however, are not always available. As a result, more than 400 people in the world (mostly children) die every hour from preventable diseases or infections such as cholera, typhoid fever, dysentery, and hepatitis, which they contract by drinking contaminated water. In response, several American manufacturers have developed tabletop systems that bathe water with pathogen-killing ultraviolet light. One prototype model, shown in Figure 16.29, disinfects 15 gal/min, weighs about 15 lb, and is powered by photovoltaic solar cells, which permit it to run unsupervised in remote locations.

Aside from pathogens, untreated water from wells or rivers may contain toxic metals that seep into the water supply from natural geologic formations. Many wells in Bangladesh, for example, are made very deep so as to avoid the pathogens that run rampant in the surface waters of this region. The water obtained from these deep wells, however, is highly contaminated with arsenic—a naturally occurring element in Earth's crust. The arsenic is in the underlying rock, which formed from river sediments carried down from the Himalayas. Because this region is so densely populated, as many as 70 million people may be subject to some level of arsenic poisoning, which manifests itself as skin lesions and a higher susceptibility to cancer. Low-cost methods for removing arsenic from well water are greatly needed, as are worldwide recognition of this problem and the political, economic, and social support to overcome it.

> In the early 1990s, an anti-water-chlorination campaign in Peru led the government to stop chlorinating the drinking water. Within months there were 1.3 million new cases of cholera resulting in 13,000 deaths.

CHECKPOINT

At a water treatment facility, how does adding slaked lime and aluminum sulfate to water purify the water?

Was this your answer?
The water entering a water treatment plant is usually a heterogeneous mixture containing suspended solids. The added slaked lime and aluminum sulfate capture these suspended solids, which then sink to the bottom, where they are easily removed.

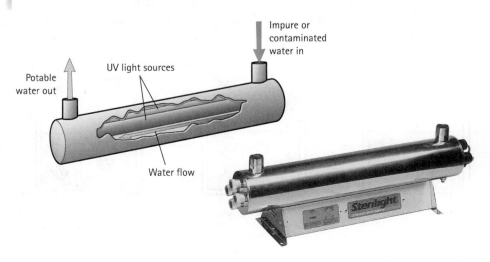

FIGURE 16.29
Small-scale water-disinfecting units, such as the one shown here, hold great value in regions of the world where potable water is scarce.

FIGURE 16.30
Saudi Arabia is the world's leading producer of desalinized water. Its desalination plants, such as the one shown here, have a combined generating capacity of about 4 billion liters per day.

Desalination

With the depletion of sources of natural fresh water in many regions, there has been growing interest in techniques for generating fresh water from Earth's far larger reserves of seawater or from *brackish* (moderately salty) groundwater. Worldwide, *desalination* plants operate in about 120 countries with a combined capacity to produce about 45 billion liters daily. In many areas of the Caribbean, North Africa, and the Middle East, desalinized water is the main source of municipal supply (Figure 16.30).

The two primary methods of removing salts from seawater or brackish water are *distillation* and *reverse osmosis*. These techniques are also highly effective in removing a host of other contaminants, such as pathogens, fertilizers, and pesticides. Distillation and reverse osmosis, therefore, are also used to purify naturally occurring fresh water. Many popular brands of bottled water, for example, contain fresh water that has been treated either by distillation or by reverse osmosis.

Distillation involves vaporizing water with heat and then condensing the vapors into purified liquid water (Section 16.1). More than 60% of Earth's desalinized water is produced using this technique. Because water has such a high boiling point, however, this technique is energy intensive. Today, most distilling plants heat the water by burning large quantities of fossil fuels, which, unfortunately, generates excessive levels of pollution relative to the volume of fresh water produced. Solar distillers avoid the burning of fuels, but they require about 1 m^2 of surface area to produce 4 L of fresh water per day, as shown in Figure 16.31. For a single home or a small village, this surface area requirement may be easily accommodated. For larger urban areas, where open land is scarce,

FYI In 1908, Jersey City, New Jersey, became the first American city to begin chlorinating its drinking water. By 1910, as disinfecting drinking water with chlorine became more widespread, the death rate from typhoid fever dropped to 20 per 100,000. In 1935, the death rate fell to 3 per 100,000. By 1960, fewer than 20 people in the entire United States died from typhoid fever annually.

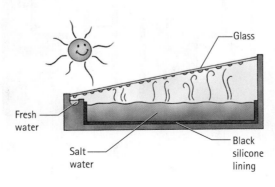

FIGURE 16.31
These solar distillers are popular in the remote communities along the Texas–Mexico border, where the waters from the Rio Grande basin are saline and tainted by the runoff of agricultural chemicals from upstream irrigation.

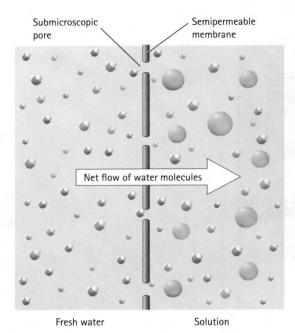

Submicroscopic pore

Semipermeable membrane

Net flow of water molecules

Fresh water   Solution

FIGURE 16.32
INTERACTIVE FIGURE

Osmosis. The submicroscopic pores of a semipermeable membrane allow only water molecules to pass. Because more water molecules exist along the freshwater face of the membrane than along the solution face, more water molecules are available to migrate into the solution than are available to migrate into the fresh water.

solar distillation is less practical, especially when the maintenance costs of vast fields of solar distillers are taken into account.

For many regions, *reverse osmosis* is a preferable method of water desalination. In order to understand reverse osmosis, you must first understand osmosis. Osmosis involves a semipermeable membrane. A **semipermeable membrane** contains submicroscopic pores that allow the passage of water molecules but not of larger solute ions or solute molecules. When a body of fresh water is partitioned from a body of salt water by a semipermeable membrane, water molecules pass from the fresh water into the salt water at a higher rate than they pass from the salt water into the fresh water. The reason for this is the presence of more water molecules along the freshwater face of the membrane than along the saltwater face. The result is a net movement of fresh water into the body of salt water, as illustrated in Figure 16.32. This net flow of water across a semipermeable membrane into a more concentrated solution is called **osmosis**.

The result of osmosis is a buildup in volume of the salt water and a decrease in volume of the fresh water. These changes in volume, in turn, allow for a buildup in pressure, called *osmotic pressure*. For the system in Figure 16.33a, osmotic pressure is the consequence of the salt water's greater height. As osmotic pressure builds, the rate at which water molecules are able to pass from the salt water into the fresh water increases. The water molecules in the salt water are being squeezed back across the membrane by the osmotic pressure. Eventually, the rates of water molecules passing in both directions across the membrane are the same and the system reaches equilibrium, as shown in Figure 16.33b. If external pressure is applied to the salt water, even more water molecules are squeezed across the membrane from the salt water into the fresh water, as shown in Figure 16.33c. Water forced across a semipermeable membrane into a less concentrated solution is **reverse osmosis**. So we see that reverse osmosis is a mechanism for generating fresh water from salt water.

The osmotic pressure for seawater, however, is an astounding 24.8 atm (365 psi). Generating pressures greater than this has its share of technical difficulties and is an energy-intensive process. Nonetheless, engineers have

FIGURE 16.33
(a) Osmosis results in a greater volume of salt water, which causes the pressure to increase on the salt side of the membrane. (b) When the pressure on the salt side becomes high enough, equal numbers of water molecules pass in both directions. (c) The application of external pressure forces water molecules to pass from the salt water to the fresh water, so that now the salt-to-fresh rate exceeds the fresh-to-salt rate.

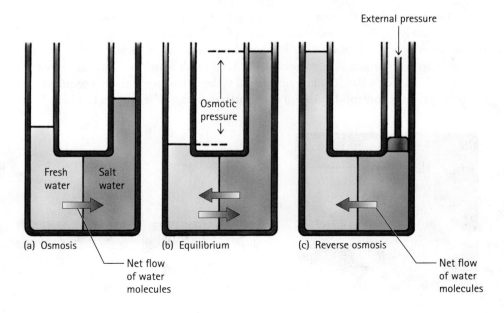

External pressure

Osmotic pressure

Fresh water   Salt water

(a) Osmosis   (b) Equilibrium   (c) Reverse osmosis

Net flow of water molecules   Net flow of water molecules

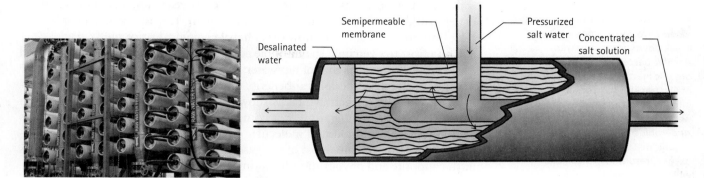

Semipermeable
membrane

Pressurized
salt water

Concentrated
salt solution

Desalinated
water

FIGURE 16.34

An industrial reverse osmosis unit consists of many semipermeable membranes packed around highly pressurized salt water. As desalinated water is pushed out one side, the remaining salt water, which is now even more concentrated, exits on the other side. A network of reverse osmosis units operating parallel to one another can produce enormous volumes of fresh water from salt water.

succeeded in building durable reverse osmosis units, shown in Figure 16.34, that can be networked together to generate fresh water from seawater at rates of millions of gallons per day. Reverse osmosis desalination facilities treating brackish waters, which require much lower external pressures, are proportionately more economical.

CHECKPOINT

Biological membranes, including cucumber membranes, are semipermeable. A cucumber shrivels to a smaller size when it is left in a solution of salt water. Is this an example of osmosis or of reverse osmosis?

Was this your answer?
No external pressure is involved, which rules out reverse osmosis. Instead, the shriveling of the cucumber tells us that the cucumber's cells are losing water to the more concentrated salt water. This is osmosis, whereby water molecules migrate across a semipermeable membrane into regions of higher salt concentrations. If you added a few other ingredients to the solution, such as spices and the right kinds of microorganisms, you would have a pickle.

Bottled Water

Desalinated seawater and brackish water are important new sources of fresh water. Although this fresh water is more costly than fresh water from natural sources, one could argue that the higher cost reflects fresh water's true value. In the United States, natural sources of fresh water are relatively plentiful, allowing companies to sell fresh water at rates of a fraction of a penny per liter. Nonetheless, consumers are still willing to purchase bottled water at up to $2 per liter! Each year, Americans spend about $400 million on bottled water, and the market continues to grow rapidly. It is easy to project a growing reliance on distillation and reverse osmosis.

In 2011, an estimated 170 billion liters of bottled water were consumed worldwide, but mostly in developed nations where tap water is potable. It is difficult to estimate the amount of energy consumed in shipping this bottled water from its source to the customer, but because water is so dense, the amount is likely huge. Assuming it takes about 100 mL of gasoline to ship each liter of bottled water, worldwide this would translate to about 25 million tons of carbon dioxide, not to mention other pollutants. Studies show that only about 20% of drinking water bottles are recycled. In California alone, more than

FIGURE 16.35
Carbonating your own water is not only fun, it's cheaper and more ecologically sound than purchasing your soda from a store.

FYI Naturally occurring water is alive with bacteria, which break down organic matter. *Aerobic bacteria* decompose organic matter only in the presence of O_2, transforming organic matter into odorless carbon dioxide, water, nitrates, and sulfates. *Anaerobic bacteria* decompose organic matter in the absence of oxygen, resulting in methane, which is flammable, and foul-smelling nitrogen- and sulfur-containing compounds. Cesspools owe their wretched stink to a lack of dissolved oxygen and the resulting anaerobic decomposition.

1 billion water bottles wind up in the trash each year. The plastic from these bottles could be used to make 74 million square feet of carpet. People who drink bottled water should know of its hefty ecological price tag.

In an effort to overcome the multiple ecological negatives of bottled water and a price up to 1000 times that of tap water, many bottled water marketers are now focusing on supposed peripheral benefits of their product. This includes the addition of dissolved oxygen, which as shown in Table 16.1, can be no more than 0.039 g/L at room temperature. For comparison, a single breath of air contains about 100 times the amount of molecular oxygen found in a half-liter of "oxygenated" water. Furthermore, most of the gases accumulated in your gut simply pass out the opposite end of your mouth, assuming you haven't burped. Worse still are claims of bottled water that contains "functional" water in which the structure of water has been modified using "subtle energy" to make it more nutritious. Subtle energy, of course, cannot be detected by current science, but the effects on your health, they claim, can be most dramatic. Buyer beware!

Most bottled water sold today is simply municipal water that has been purified via reverse osmosis. Many homeowners have discovered that it is less expensive and more ecologically sound to install a small reverse osmosis unit within their own home. For fun, carbonators can also be installed so that you can have your very own soda fountain, as shown in Figure 16.35.

LEARNING OBJECTIVE
Identify the four stages of wastewater treatment

16.7 Wastewater Treatment

EXPLAIN THIS What chemical holds the record for saving the most lives?

The contents of the sewer systems that underlie most municipalities must be treated before being released into a body of water. The level of treatment depends in great part on whether the treated water is to be released into a river or into the ocean. Wastewater destined for a river requires the highest level of treatment for the benefit of communities downstream. However, in a facility located in a region surrounded by very deep ocean water, as is the facility shown in Figure 16.36, treatment requirements are less stringent.

Human waste loses its form by the time it reaches the wastewater facility, and the wastewater appears as a murky stream. In this stream, however, are many insoluble products—including small plastic items, such as tampon applicators, and gritty material, such as coffee grounds and sand. Hardened balls of grease from discarded cooking fats are also found. The initial step in all wastewater treatments, therefore, is to screen out these insolubles. (You should know that wastewater treatment experts point out that these insolubles—even cooking grease—should be disposed of as solid waste and not washed down the drain or flushed down the toilet.)

After screening, the next level of municipal wastewater treatment is *primary* treatment. In primary treatment, screened wastewater enters a large settling basin, where suspended solids settle out as sludge (Figure 16.37). After a period of time, the sludge is removed from the bottom of the settling basin and is often sent directly to a landfill as solid wastes. Some facilities, however, are equipped with large furnaces in which dried sludge is burned, sometimes along with other municipal wastes, such as paper products. The resulting ash is more compact and takes up less space in a landfill.

Wastewater effluent from primary treatment, as well as higher levels of treatment, is commonly disinfected with either chlorine gas or ozone before its release into the environment. A great advantage of using chlorine gas is that it

FIGURE 16.36
In Honolulu, about 280 million liters of wastewater pass through the largest of several wastewater facilities each day. This water can be piped to depths of hundreds of meters below sea level, whence it continues to flow toward the bottom of the ocean. Water treatment requirements are therefore much less stringent than those at mainland facilities, where the effluent is not so easily discarded.

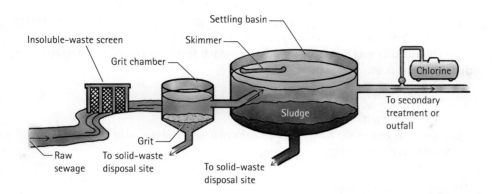

FIGURE 16.37
A schematic for primary-level waste-water treatment. The rotating skimmer on the settling basin removes buoyant materials and artifacts not captured by the screening process.

remains in the water for an extended time after leaving the facility. This provides residual protection against diseases. The chlorine, however, reacts with organic compounds within the effluent to form chlorinated hydrocarbons, many of which are known carcinogens (cancer-causing agents). Also, chlorine kills only bacteria, leaving viruses unharmed. Ozone is more advantageous in that it kills both bacteria and viruses. Also, no carcinogenic by-products result from treating wastewater effluent with ozone. A disadvantage of ozone, however, is that it provides no residual protection for the effluent once it is released. Most facilities in the United States use chlorine for disinfecting, whereas European facilities tend to favor ozone. In a few locations, chlorine and ozone gases have been replaced by strong ultraviolet lamps, which, like ozone, kill both bacteria and viruses but provide no long-term residual protection.

The potential for pathogens to grow in primary effluent is extremely high and, by virtue of the Clean Water Act of 1972, the release of primary effluent is not permitted in most places. A frequently used *secondary* level of treatment, shown in Figure 16.38, involves passing the primary effluent first through an aeration tank. This supplies the oxygen necessary for continued decomposition of organic matter by oxygen-dependent bacteria, known as *aerobic bacteria*. The effluent is then sent into a tank where any fine particles not removed in primary treatment can settle. Because sludge from this settling step is high in aerobic bacteria, some of it is recycled back to the aeration tank to increase efficiency. The remainder of the sludge is hauled off to the landfill or an incinerator.

Many municipalities also require a third level, a *tertiary* level, of wastewater treatment. A number of tertiary processes are used, and most involve filtrations of some sort. A common method is to pass secondary-level effluent through a bed of finely powdered carbon, which captures most particulate matter and many of the organic molecules not removed in earlier stages.

FYI In Hong Kong, about 80% of all toilets flush using seawater. Developed since the 1960s, this system now saves the equivalent of about 25% of freshwater consumption. Also, effluent from freshwater activities, such as personal hygiene and dishwashing, is treated and reused for watering city trees and other nonpotable uses.

For dwellings in remote locations, such as summer cabins, many people choose composting toilets, which use no water. Rather, they allow human waste to decompose aerobically (with oxygen) as air is vented over the waste, which is buried in peat moss. Dried, odor-free compost, which is removed every few months, is useful as a garden fertilizer.

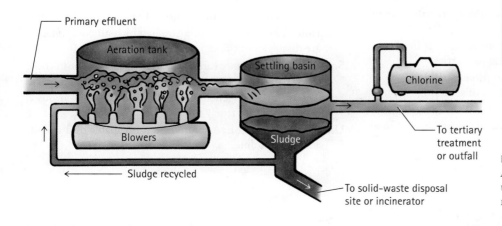

FIGURE 16.38
A schematic for secondary-level treatment of wastewater from a municipal system.

The advantage of tertiary-level treatment is greater protection of our water resources. Unfortunately, tertiary treatment is costly and is normally used only when the need is vital. Primary and secondary levels of treatment are also not without great cost.

CHECKPOINT

Distinguish the main functions of primary, secondary, and tertiary wastewater treatment.

Was this your answer?
Primary wastewater treatment removes the bulk of solid waste and sludge from the sewage effluent using screening devices and large settling basins. Secondary treatment decreases the biochemical oxygen demand of the effluent by aeration. Tertiary treatment removes pathogens and wastes not removed by earlier treatments by filtering the effluent through beds of powdered carbon or other fine particles.

For instructor-assigned homework, go to www.masteringphysics.com

SUMMARY OF TERMS (KNOWLEDGE)

Concentration A quantitative measure of the amount of solute dissolved in a solution.

Dissolving The process of mixing a solute in a solvent to produce a homogeneous mixture.

Distillation A purifying process in which a vaporized substance is collected by exposing it to cooler temperatures over a receiving flask, which collects the condensed purified liquid.

Hard water Water containing large amounts of calcium and magnesium ions.

Heterogeneous mixture A mixture in which the different components can be seen as individual substances.

Homogeneous mixture A mixture in which the components are so finely mixed that any one region of the mixture contains the same ratio of substances as any other region.

Impure The state of a material that is a mixture of more than one element or compound.

Insoluble Not capable of dissolving to any appreciable extent in a given solvent.

Mixture A combination of two or more substances in which each substance retains its chemical properties.

Molarity A common unit of concentration equal to the number of moles of a solute per liter of solution.

Mole The very large number 6.02×10^{23}; usually used in reference to the number of atoms, ions, or molecules in a macroscopic amount of a material.

Osmosis The net flow (diffusion) of water across a semipermeable membrane from a region of low solute concentration to a region of high solute concentration.

Precipitate A solute that has come out of solution.

Pure The state of a material that consists solely of a single element or compound.

Reverse osmosis A technique for purifying water by forcing it through a semipermeable membrane into a region of lower solute concentration.

Saturated solution A solution containing the maximum amount of solute that will dissolve in its solvent.

Semipermeable membrane A membrane containing submicroscopic pores that allow the passage of water molecules but not of larger solute ions or solute molecules.

Solubility The ability of a solute to dissolve in a given solvent.

Soluble Capable of dissolving to an appreciable extent in a given solvent.

Solute Any component in a solution that is not the solvent.

Solution A homogeneous mixture in which all components are dissolved in the same phase.

Solvent The component in a solution that is present in the largest amount.

Suspension A homogeneous mixture in which the various components are finely mixed, but not dissolved.

Unsaturated solution A solution that is capable of dissolving additional solute.

READING CHECK QUESTIONS (COMPREHENSION)

16.1 Most Materials Are Mixtures

1. What defines a material as being a mixture?
2. How can the components of a mixture be separated from one another?
3. How does distillation separate the components of a mixture?

16.2 The Chemist's Classification of Matter

4. Why isn't it practical to have a macroscopic sample that is 100% pure?
5. How is a solution different from a suspension?
6. How can a solution be separated from a suspension?

16.3 Solutions

7. What happens to the volume of a sugar solution as more sugar is dissolved in it?
8. Distinguish between a solute and a solvent.
9. What does it mean to say that a solution is concentrated?
10. Is concentration typically given with the volume of solvent or the volume of solution?

16.4 Solubility

11. Why does the solubility of a gas solute in a liquid solvent decrease with increasing temperature?
12. Why do sugar crystals dissolve faster when crushed?

13. Is sugar a polar or nonpolar substance?

16.5 Soaps, Detergents, and Hard Water

14. Which portion of a soap molecule is nonpolar?
15. What is the difference between a soap and a detergent?
16. What component of hard water makes it hard?
17. Why are soap molecules so attracted to calcium and magnesium ions?

16.6 Purifying the Water We Drink

18. Why is treated water sprayed into the air before it is piped to users?
19. What are two ways in which people disinfect water in areas where municipal treatment facilities are not available?
20. What naturally occurring element has been contaminating the water supply of Bangladesh?

16.7 Wastewater Treatment

21. Why can wastewater treatment requirements in Hawaii be less stringent than those in most locations on the U.S. mainland?
22. What is the first step in treating raw sewage?
23. Why don't all municipalities require third-level treatment of wastewater?

ACTIVITIES (HANDS-ON APPLICATION)

24. To see the gases dissolved in your water, fill a clean cooking pot with water and let it stand at room temperature for several hours. Note the tiny bubbles that adhere to the inner sides of the pot. Where did these tiny bubbles come from? What do you suppose they contain? For further experimentation, repeat this activity in two pots set side by side. In one pot, use warm water from the kitchen faucet. In the second pot, use boiled water that has cooled down to the same temperature.

25. Put on your safety glasses and add several cups of tap water to a cooking pot. Boil the water to dryness. (Turn off the burner before the water is all gone. The heat from the pot will finish the evaporation. Watch out for splattering!) Examine the resulting residue by scraping it off with a knife. These are the solids you ingest with every glass of water you drink.

26. Black ink contains pigments of many different colors. Acting together, these pigments absorb all the frequencies of visible light. Because no light is reflected, the ink appears black. We can use molecular attractions to separate the components of black ink through a technique that is called paper chromatography.

What you need: black felt-tip pen or black water-soluble marker; piece of porous paper, such as paper towel, table napkin, or coffee filter; solvent, such as water, acetone (fingernail-polish remover), rubbing alcohol, or white vinegar

Procedure:

1. Place a concentrated dot of ink at the center of the piece of porous paper.
2. Carefully place one drop of solvent on top of the dot, and watch the ink spread radially with the solvent. Because the different components of the ink have different affinities for the solvent (based on the attractions between component molecules and solvent molecules), they travel with the solvent at different rates.
3. Just after the drop of solvent is completely absorbed, add a second drop at the same location as the first one, then a third, and so on until the ink components have separated to your satisfaction.

Paper chromatography was originally developed to separate plant pigments from one another. The separated pigments had different colors, which is how this technique got its name—*chroma* is Latin for "color." Mixtures need not be colored, however, to be separable by chromatography. All that's required is that the components have distinguishable affinities for the moving solvent and the stationary medium, such as paper, through which the solvent will pass.

27. Just because a solid dissolves in a liquid doesn't mean the solid no longer occupies space. Fill a glass to its brim with the warm water, and then carefully pour all the water into the larger container. Add a couple tablespoons of sugar to the empty glass. Return half of the warm water to the glass and stir to dissolve all the sugar. Return the remaining water, and as you get close to the top, ask a friend to predict whether the water level will be lower than, about the same as, or higher than before. If your friend doesn't understand the result, ask him or her what would happen if you had added the sugar to the glass when the glass was full of water.

28. You can build a relatively inefficient but fun-to-watch distiller at home. Avoid the steam produced in this activity—steam burns can be particularly harmful.

Step 1: Fill a cooking pot to a depth of about 1 cm. Add food coloring or some salt or both.

Step 2: Place a heavy ceramic coffee mug in the center of the pot. The height of the mug should be at least 1 inch below the height of the pot.

Step 3: Lay plastic wrap loosely across the top of the pot and secure it with a rubber band. The seal should not be airtight—leave two of the edges open to prevent a build-up of pressure. Trim away excess plastic wrap. Place an ice cube at the center.

Step 4: Put on safety glasses and turn the burner on low to bring the water to a low boil. Look for signs of cloud formation below the ice cube. Once the water is boiling, the mug may jostle.

Step 5: Turn off the heat. Remove the melted ice with a sponge. Why isn't the solute carried over into the mug? How much distilled water are you able to collect per ice cube? How might you modify your distiller so that it works well using only sunlight?

29. Here's a quick recipe for rock candy. In a cooking pot make a hot saturated solution of sugar in water. Start by mixing sugar and water in a 2:1 ratio by volume. Add more sugar or water as necessary to obtain a clear runny syrup. Let cool for 10 minutes. Roll a wet skewer stick or weight (such as a metal nut) attached to a string in some granulated sugar. Pour the warm sugar syrup into a jar. Submerge the skewer or weight in the sugar syrup. For support, consider setting a clothespin or a pencil across the rim. Cover the top and store in a cool place. The longer you wait, the larger the crystals will be.

THINK AND SOLVE (MATHEMATICAL APPLICATION)

30. Assume the total number of molecules in a sample of liquid is about 3 million trillion. One million trillion of these are molecules of some poison, while 2 million trillion are water molecules. What percentage of all the molecules in the glass are water?

31. Assume the total number of molecules in a glass of liquid is about 1,000,000 million trillion. One million trillion of these are molecules of some poison, while 999,999 million trillion of these are water molecules. What percentage of all the molecules in the glass are water?

32. You drink a small glass of water that is 99.9999% pure water and 0.0001% some poison. Assume the glass contains about a 1,000,000 million trillion molecules, which is about 30 mL. How many poison molecules did you just drink? Should you be concerned?

33. How much sodium chloride, in grams, is needed to make 15 L of a solution that has a concentration of 3.0 g of sodium chloride per liter of solution?

34. If water is added to 1 mole of sodium chloride in a flask until the volume of the solution is 1 L, what is the molarity of the solution? What is the molarity when water is added to 2 moles of sodium chloride to make 0.5 L of solution?

35. A student is told to use 20.0 g of sodium chloride to make an aqueous solution that has a concentration of 10.0 g/L (grams of sodium chloride per liter of solution). Assuming that 20.0 g of sodium chloride has a volume of 7.50 mL, show that she will need about 1.99 L of water to make this solution. In making this solution, should she add the solute to the solvent or the solvent to the solute?

THINK AND RANK (ANALYSIS)

36. Rank the following solutions in order of increasing concentration. solution A, 0.5 mole of sucrose in 2.0 L of solution; solution B, 1.0 mole of sucrose in 3.0 L of solution; solution C, 1.5 moles of sucrose in 4.0 L of solution.

37. Rank the compounds in order of increasing solubility in water:

CH_3CH_2—OH
  Ethanol

$CH_3CH_2CH_2CH_2$—OH
  Butanol

$CH_3CH_2CH_2CH_2CH_2CH_2$—OH
  Hexanol

38. List these compounds in order of increasing boiling point: CI_4, CBr_4, CCl_4, CF_4.

EXERCISES (SYNTHESIS)

39. How might you separate a mixture of sand and salt? How about a mixture of iron and sand?

40. Mixtures can be separated into their components by taking advantage of differences in the chemical properties of the components. Why might this separation method be less convenient than taking advantage of differences in the physical properties of the components?

41. Why can't the elements of a compound be separated from one another by physical means?

42. Many dry cereals are fortified with iron, which is added to the cereal in the form of small iron particles. How might these particles be separated from the cereal?

43. Classify each of the following as a homogeneous mixture, heterogeneous mixture, element, or compound:
(a) table salt, (b) blood, (c) steel, (d) planet Earth.

44. Some bottled water is now advertised as containing extra quantities of "Vitamin O," which is a marketing gimmick for selling oxygen, O_2. Might this bottle water actually contain extra quantities of oxygen, O_2? How much more O_2 than one might find in regular bottled water? How might the amount of oxygen we absorb through our lungs compare to the amount we might absorb through our stomach—after burping?

45. Classify each of the following as an element, compound, or mixture, and justify your classifications: salt, stainless steel, tap water, sugar, vanilla extract, butter, maple syrup, aluminum, ice, milk, cherry-flavored cough drops.

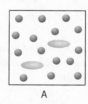

A   B   C

46. Which of the boxes above best represents a suspension?

47. Which of the boxes above best represents a solution?

48. Which of the boxes above best represents a compound?

49. Explain why, for these three substances, the solubility in 20°C water goes down as the molecules get larger but the boiling point goes up.

Substance	Boiling point/ Solubility
CH_3—O^H	65°C infinite
$CH_3CH_2CH_2CH_2$—O^H	117°C 8 g/100 mL
$CH_3CH_2CH_2CH_2CH_2$—O^H	138°C 2.3 g/100 mL

50. The boiling point of 1,4-butanediol is 230°C. Would you expect this compound to be soluble or insoluble in room-temperature water? Explain.

$$H-O-CH_2CH_2CH_2CH_2-O-H$$
1,4-Butanediol

51. Based on atomic size, which would you expect to be more soluble in water: helium, He, or nitrogen, N_2?

52. If nitrogen, N_2, were pumped into your lungs at high pressure, what would happen to its solubility in your blood?

53. The air a scuba diver breathes is pressurized to counteract the pressure exerted by the water surrounding the diver's body. Breathing the high-pressure air causes excessive amounts of nitrogen to dissolve in body fluids, especially the blood. If a diver ascends to the surface too rapidly, the nitrogen bubbles out of the body fluids (much as carbon dioxide bubbles out of a soda immediately after the container is opened). This results in a painful and potentially lethal medical condition known as the *bends*. Why does breathing a mixture of helium and oxygen rather than air help divers avoid getting the bends?

54. Account for the observation that ethanol, C_2H_5OH, dissolves readily in water but dimethyl ether, CH_3OCH_3, which has the same number and kinds of atoms, does not.

Ethanol

Dimethyl ether

55. At 10°C, which is more concentrated: a saturated solution of sodium nitrate, $NaNO_3$, or a saturated solution of sodium chloride, $NaCl$? (See Figure 16.20.)

56. Why is rain or snow called precipitation?

57. The volume of many liquid solvents expands with increasing temperature. What happens to the concentration of a solution made with such a solvent as the temperature of the solution is increased?

58. Hydrogen chloride, HCl, is a gas at room temperature. Would you expect this material to be very soluble or not very soluble in water?

59. Two plastic bottles of fresh seltzer water are opened. Three-fourths of the first bottle are poured out for drinking, while only one-fourth of the second bottle is poured. Both bottles are then tightly resealed. The next day they are both re-opened, and one is less fizzy. Which one? Why?

60. Which should weigh more: 100 mL of fresh water or 100 mL of fresh sparkling seltzer water? Why? Which should weigh more: 100 mL of flat seltzer water at 20°C or the same 100 mL of flat seltzer water brought to 80°C? Why?

61. Why can 500 mL of fresh water absorb more gaseous carbon dioxide than 500 mL of sugar water at the same temperature?

62. Would you expect to find more dissolved oxygen in ocean water around the northern latitudes or in ocean water close to the equator? Why?

Fatty acid molecules can align to form a barrier called a bilipid layer, shown below. In this schematic, the ionic end of the fatty acid is shown as a circle, and the nonpolar chain is shown as a squiggly line. Use this diagram for Exercises 63 and 64.

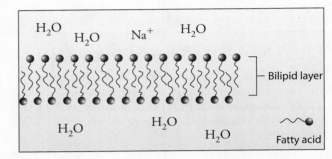

63. Why do nonpolar molecules have a difficult time passing through the bilipid layer? How about polar molecules?

64. Fatty acid molecules can also align to form a bilipid layer that extends in three-dimensions. Shown below is a cross-section of this structure. This is similar to the micelle shown in Figure 16.23, though notably different because it contains an inner compartment of water. What is this structure called? (Hint: It forms the basis of all life.)

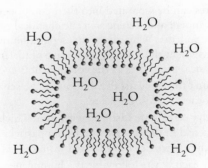

65. How is it that water and soap are attracted to each other not only by ion–dipole attraction but also by dipole–induced dipole attraction?

66. A scum forms on the surface of boiling hard water. What is this scum? Why does it form? And why do hot water heaters lose their efficiency quicker in households with hard water?

67. Calcium and magnesium ions are more attracted to sodium carbonate than to soap. Why?

68. Phosphate ions, PO_4^{3-}, were once added to detergents to assist in cleaning. What function did they serve? These ions are no longer added to detergents because they cause excessive growth of algae in aquatic habitats that receive the wastewater. What chemical has replaced them?

69. Cells at the top of a tree have a higher concentration of sugars than cells at the bottom. How might this fact assist a tree in moving water upward from its roots?

70. Why is flushing a toilet with clean water from a municipal supply about as wasteful as flushing it with bottled water?

71. What reverses in reverse osmosis?

72. Why is it significantly less costly to purify fresh water through reverse osmosis than to purify salt water through reverse osmosis?

73. Some people fear drinking distilled water because they have heard it leaches minerals from the body. Using your knowledge of chemistry, explain why these fears have no basis and how distilled water is in fact very good for drinking.

74. Many homeowners get their drinking water piped up from wells dug into their property. Sometime this well water smells bad because of trace quantities of the gaseous compound hydrogen sulfide, H_2S. How might this odor be removed from water already taken from the tap?

75. Is the decomposition of food by bacteria in our digestive systems aerobic or anaerobic? What evidence supports your answer? How do composting toilets remove the bad smells of human waste?

DISCUSSION QUESTIONS (EVALUATION)

76. Oxygen, O_2, dissolves quite well in a class of compounds known as liquid perfluorocarbons—so well that oxygenated perfluorocarbons can be inhaled in a liquid phase, as is demonstrated by the rodent shown below the water-bound goldfish. Do you suppose perfluorocarbon molecules are polar or nonpolar? Why would the rodent drown if it were brought up to the water layer and the goldfish die if they swam down into the perfluorocarbon layer? How might perfluorocarbons be used to clean our lungs or serve as artificial blood? When is it okay to sacrifice the lives of animals for scientific research?

77. Why are people so willing to buy bottled water when it is so expensive, both financially and environmentally?

78. It is possible to tow icebergs to coastal cities as a source of fresh water. What obstacles—technological, social, environmental, and political—do you foresee for such an endeavor?

79. In reference to human nature, Jerome Delli Priscoli, a social scientist with the U.S. Army Corps of Engineers, stated, "The thirst for water may be more persuasive than the impulse toward conflict." Do you agree or disagree with his statement? Might our universal need for water be our salvation or our demise? Consider the current water disputes between countries such as Sudan and Egypt or Turkey and Iraq.

80. Search the Internet to find what major rivers in the world no longer reach the oceans. What impact does beef consumption have on water use? How are these two issues related?

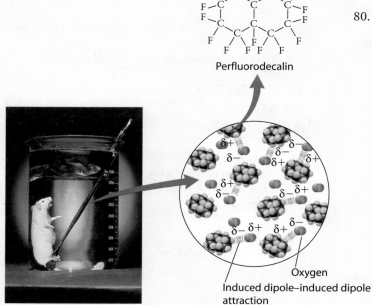

Perfluorodecalin

Oxygen

Induced dipole–induced dipole attraction

READINESS ASSURANCE TEST (RAT)

If you have a good handle on this chapter, if you really do, then you should be able to score 7 out of 10 on this RAT. If you score less than 7, you need to study further before moving on.

Choose the BEST answer to each of the following.

1. Someone argues that he or she doesn't drink tap water because it contains thousands of molecules of some impurity in each glass. How would you respond in defense of the water's purity, if it indeed does contain thousands of molecules of some impurity per glass?
   - (a) Impurities aren't necessarily bad; in fact, they may be good for you.
   - (b) The water contains water molecules, and each water molecule is pure.
   - (c) There's no defense. If the water contains impurities, it should not be drunk.
   - (d) Compared to the billions and billions of water molecules, a thousand molecules of something else is practically nothing.

2. What is the difference between a compound and a mixture?
   - (a) Both consist of atoms of different elements.
   - (b) Their atoms are bonded together in different ways.
   - (c) The components of a mixture are not chemically bonded together.
   - (d) One is a solid and the other is a liquid.

3. The air in your house is an example of a
   - (a) homogeneous mixture because it is mixed very well.
   - (b) heterogeneous mixture because of the dust particles it contains.
   - (c) homogeneous mixture because it is all at the same temperature.
   - (d) heterogeneous mixture because it consists of different types of molecules.

4. Half-frozen fruit punch is always sweeter than the same fruit punch completely melted because
   - (a) the sugar sinks to the bottom.
   - (b) only the water freezes while the sugar remains in solution.
   - (c) the half-frozen fruit punch is warmer.
   - (d) sugar molecules solidify as crystals.

5. Why is sodium chloride, NaCl, insoluble in gasoline? Consider the electrical attractions.
   - (a) Because this molecule is so small, there is not much opportunity for the gasoline to interact with it through any electrical attractions.
   - (b) Because gasoline is a very polar molecule, the salt can form only dipole–induced dipole bonds, which are very weak, giving it a low solubility in gasoline.
   - (c) Because gasoline is so strongly attracted to itself, the salt, NaCl, is excluded.
   - (d) Salt is composed of ions that are too attracted to themselves. Gasoline is nonpolar, so salt and gasoline will not interact very well.

6. Fish don't live very long in water that has just been boiled and brought back to room temperature. Why?
   - (a) There is now a higher concentration of dissolved CO_2 in the water.
   - (b) The nutrients in the water have been destroyed.
   - (c) Since some of the water evaporated while boiling, the salts in the water are now more concentrated. This has a negative effect on the fish.
   - (d) The boiling process removes the air that was dissolved in the water. Upon cooling, the water does not have its usual air content—hence, the fish drown.

7. How many moles of sugar (sucrose) are there in 5 L of sugar water that has a concentration of 0.5 M?
   - (a) 5.5 moles
   - (b) 5.0 moles
   - (c) 2.5 moles
   - (d) 1.5 moles

8. What is an advantage of using chlorine gas to disinfect drinking water supplies?
   - (a) It provides residual protection against pathogens.
   - (b) It gives the water a fresh taste.
   - (c) Residual chlorine in water helps to whiten teeth.
   - (d) Excess chlorine is absorbed in our bodies as a mineral supplement.
   - (e) All of the above.

9. Why do red blood cells, which contain an aqueous solution of dissolved ions and minerals, burst when placed in fresh water?
   - (a) The dissolved ions provide a pressure that eventually bursts open the cell.
   - (b) More water molecules enter the cell than leave the cell.
   - (c) The fresh water acts to dissolve the blood cell wall.
   - (d) All of the above.

10. A stagnant pond smells worse than a babbling brook because
   - (a) odors are not transported downstream.
   - (b) of the type of aquatic life it attracts.
   - (c) it lacks sufficient dissolved oxygen.
   - (d) All of the above.

1. d, 2. c, 3. b, 4. b, 5. d, 6. d, 7. c, 8. a, 9. b, 10. c.

Answers to RAT

17

CHAPTER 17

How Chemicals React

THE HEAT of a lightning bolt causes multiple chemical reactions in the atmosphere, including one in which nitrogen and oxygen react, leading to the formation of nitric acid, HNO_3, and nitrous acid, HNO_2. These acids are carried by rain into the ground, where they transform into nitrate ions that plants use for growing—a process that involves further chemical reactions. We, in turn, eat the plants, or plant-eating animals, to support life-sustaining chemical reactions within ourselves.

Scientists have learned how to control chemical reactions to produce many useful materials—including fertilizers from air, metals from rocks, and pharmaceuticals from petroleum. These materials and the thousands of others produced by human-controlled chemical reactions have dramatically improved our living conditions.

The goal of this chapter is to provide the basics of chemical reactions, which were introduced in Chapter 14.

17.1 Chemical Equations

EXPLAIN THIS How can 50 g of wood burn to produce more than 50 g of products?

As was discussed in Chapter 14, during a chemical reaction, atoms rearrange to create one or more new compounds. This activity is neatly summed up in written form as a **chemical equation**. A chemical equation shows the reacting substances, called **reactants**, to the left of an arrow that points to the newly formed substances, called **products**:

$$\text{reactants} \longrightarrow \text{products}$$

Typically, reactants and products are represented by their elemental or chemical formulas. Sometimes molecular models or, simply, names may be used instead. Phases are also often shown: (*s*) for solid, (*l*) for liquid, and (*g*) for gas. Compounds dissolved in water are designated (*aq*) for aqueous solution. Lastly, numbers are placed in front of the reactants or products to show the ratio in which they either combine or form. These numbers are called *coefficients*, and they represent numbers of individual atoms and molecules. For instance, to represent the chemical reaction in which carbon combines with oxygen to produce carbon dioxide, we write the chemical equation using coefficients of 1:

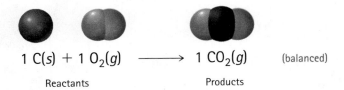

$$1\ C(s) + 1\ O_2(g) \longrightarrow 1\ CO_2(g) \quad \text{(balanced)}$$

Reactants   Products

One of the most important principles of chemistry is the **law of mass conservation**. The law of mass conservation states that matter is neither created nor destroyed during a chemical reaction.* The atoms present at the beginning of a reaction merely rearrange to form new molecules. This means that no atoms are lost or gained during any reaction. The chemical equation must therefore be *balanced*. In a balanced equation, each atom must appear on both sides of the arrow the same number of times. The equation for the formation of carbon dioxide is balanced because each side shows one carbon atom and two oxygen atoms. You can count the number of atoms in the models to see this for yourself.

In another chemical reaction, two hydrogen gas molecules, H_2, react with one oxygen gas molecule, O_2, to produce two molecules of water, H_2O, in the gaseous phase:

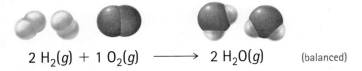

$$2\ H_2(g) + 1\ O_2(g) \longrightarrow 2\ H_2O(g) \quad \text{(balanced)}$$

---

* For all practical purposes this law holds true. Technically, however, any energy released or absorbed by a chemical reaction arises from the transformation of matter into energy, or vice versa. The amount of matter lost or gained in a chemical reaction, however, is so small that, for all practical purposes, we can ignore this detail. Not so for the nuclear reactions discussed in Chapter 13. For these nuclear reactions, matter/energy conversions are much more pronounced.

This equation for the formation of water is also balanced—there are four hydrogen and two oxygen atoms before and after the arrow.

A coefficient in front of a chemical formula tells us the number of times that element or compound must be counted. For example, 2 H_2O indicates two water molecules, which contain a total of four hydrogen atoms and two oxygen atoms.

By convention, the coefficient 1 is omitted so that the above chemical equations are typically written

$$C(s) + O_2(g) \longrightarrow CO_2(g) \quad \text{(balanced)}$$

$$2\,H_2(g) + O_2(g) \longrightarrow 2\,H_2O(g) \quad \text{(balanced)}$$

CHECKPOINT

How many oxygen atoms are indicated by the following balanced equation?

$$3\,O_2(g) \longrightarrow 2\,O_3(g)$$

Was this your answer?
Before the reaction, these six oxygen atoms are found in three O_2 molecules. After the reaction, these same six atoms are found in two O_3 molecules.

An unbalanced chemical equation shows the reactants and products without the correct coefficients. For example, the equation

$$NO(g) \longrightarrow N_2O(g) + NO_2(g) \quad \text{(not balanced)}$$

is not balanced because there are one nitrogen atom and one oxygen atom before the arrow, but three nitrogen atoms and three oxygen atoms after the arrow.

You can balance unbalanced equations by adding or changing coefficients to produce correct ratios. (It's important not to change subscripts, however, because to do so changes the compound's identity—H_2O is water, but H_2O_2 is hydrogen peroxide!) For example, to balance the equation above, add a 3 before the NO:

$$3\,NO(g) \longrightarrow N_2O(g) + NO_2(g) \quad \text{(balanced)}$$

Now there are three nitrogen atoms and three oxygen atoms on each side of the arrow, and the law of mass conservation is not violated.

Practicing chemists develop a skill for balancing equations. This skill involves creative energy and, like other skills, improves with experience. More important than being an expert at balancing equations is knowing why they need to be balanced. And the reason is the law of mass conservation, which tells us that atoms are neither created nor destroyed in a chemical reaction—they are simply rearranged. So every atom present before the reaction must be present after the reaction, even though the groupings of atoms are different.

FYI Chemical explosions typically involve the transformation of an unstable solid or liquid chemical into more stable gases that occupy much more volume. Upon detonation, 1 mole of nitroglycerin, $C_3H_5N_3O_9$, produces 7.25 moles of gases including carbon dioxide, CO_2; nitrogen, N_2; oxygen, O_2; and water vapor, H_2O. The volume change is dramatic—from less than 0.3 L to about 170 L, which is an increase of about 600%. For nitroglycerin and similar high explosives, these gases expand at supersonic speeds, creating a powerful and destructive shock wave.

CHECKPOINT

Write a balanced equation for the reaction showing hydrogen gas, H$_2$, and nitrogen gas, N$_2$, forming ammonia gas, NH$_3$:

$$\_\_\_H_2(g) + \_\_\_N_2(g) \longrightarrow \_\_\_NH_3(g)$$

Was this your answer?

Initially, we see two hydrogen atoms before the reaction arrow and three on the right. This can be remedied by placing a coefficient of 3 by the hydrogen, H$_2$, and a coefficient of 2 by the ammonia, NH$_3$. This makes for six hydrogen atoms both before and after the reaction arrow. Meanwhile, the coefficient of 2 by the ammonia also makes for two nitrogen atoms after the arrow, which balances out the two nitrogen atoms appearing before the arrow. The full balanced equation, therefore, is:

$$3\ H_2(g) \quad + \quad N_2(g) \quad \longrightarrow \quad 2\ NH_3(g)$$

Chemists use many methods to balance equations. Look to the *Conceptual Physical Science—5th Edition Practice Book* supplement for an example. Your instructor may also share with you his or her favorite methods. For more practice balancing equations, see the questions at the end of this chapter.

17.2 Counting Atoms and Molecules by Mass

EXPLAIN THIS How many molecules of carbon dioxide are there in 44 g of carbon dioxide?

LEARNING OBJECTIVE
Calculate the mass of reactants needed to produce a given mass of products.

In any chemical reaction, a specific number of reactants react to form a specific number of products. For example, when carbon and oxygen combine to form carbon dioxide, they always combine in the ratio of one carbon atom to one oxygen molecule. A chemist who wants to carry out this reaction in the laboratory would be wasting chemicals and money if she were to combine, say, four carbon atoms for every one oxygen molecule. The excess carbon atoms would have no oxygen molecules to react with and would remain unchanged.

How is it possible to measure out a specific number of atoms or molecules? Rather than counting these particles individually, chemists can use a scale that measures the mass of bulk quantities. Because different atoms and molecules have different masses, however, a chemist can't simply measure out equal masses of each. Say, for example, he needs the same number of carbon atoms as oxygen molecules. Measuring equal masses of the two materials would not provide equal numbers.

You know that 1 kg of Ping-Pong balls contains more balls than 1 kg of golf balls, as Figure 17.1 illustrates. Likewise, because different atoms and molecules have different masses, there are different numbers of them

FIGURE 17.1
The number of balls in a given mass of Ping-Pong balls is very different from the number of balls in the same mass of golf balls.

Equal masses

in a 1-g sample of each. Because carbon atoms are less massive than oxygen molecules, there are more carbon atoms in 1 g of carbon than there are oxygen molecules in 1 g of oxygen. So, clearly, equal masses of these two particles do not yield equal numbers of carbon atoms and oxygen molecules.

If we know the *relative masses* of different materials, we can measure equal numbers. Golf balls, for example, are about 20 times as massive as Ping-Pong balls, which is to say the relative mass of golf balls to Ping-Pong balls is 20 to 1. Measuring out 20 times as much mass of golf balls as Ping-Pong balls, therefore, gives equal numbers of each, as is shown in Figure 17.2.

The mass of one Ping-Pong ball is 2 g.

The mass of one golf ball is 40 g.

A Ping-Pong ball is 2/40, or 1/20, as massive as a golf ball.

Number of Ping-Pong balls = Number of golf balls

FIGURE 17.2
The number of golf balls in 200 g of golf balls equals the number of Ping-Pong balls in 10 g of Ping-Pong balls.

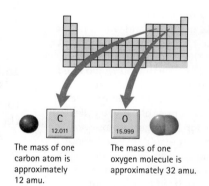

The mass of one carbon atom is approximately 12 amu.

The mass of one oxygen molecule is approximately 32 amu.

A carbon atom is 12/32, or 3/8, as massive as an oxygen molecule.

Number of carbon atoms = Number of oxygen molecules

CHECKPOINT

A customer wants to buy a 1 : 1 mixture of blue and red jelly beans. Each blue bean is twice as massive as each red bean. If the clerk measures out 5 lb of red beans, how many pounds of blue beans must she measure out?

Was this your answer?
Because each blue jelly bean has twice the mass of each red one, the clerk needs to measure out twice as much mass of blues in order to have the same count, which means 10 lb of blues. If the clerk did not know that the blue beans were twice as massive as the red ones, she would not know what mass of blues was needed for the 1 : 1 ratio. Likewise, a chemist would be at a loss in setting up a chemical reaction if she did not know the relative masses of the reactants.

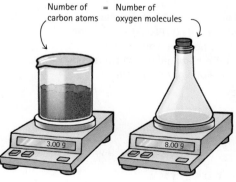

FIGURE 17.3
To have equal numbers of carbon atoms and oxygen molecules requires measuring out $\frac{3}{8}$ as much carbon as oxygen.

The masses of elements shown in the periodic table are relative masses. Using these masses we can measure out equal numbers of atoms or molecules. For example, as illustrated in Figure 17.3, the mass of carbon is 12.011 amu. (As discussed in Section 12.3, one *atomic mass unit* [amu] equals 1.661×10^{-24} gram.) The **formula mass** of a substance is the sum of the atomic masses of the elements in its chemical formula. Therefore, the formula mass of an oxygen molecule, O_2, is about 32 amu (15.999 amu + 15.999 amu).

A carbon atom, therefore, is about $\frac{12}{32}$ as massive as an oxygen molecule. To measure out equal numbers of carbon atoms and oxygen molecules we could measure out 12 g of carbon and 32 g of molecular oxygen. Any proportion equal to $\frac{12}{32}$, such as $\frac{6}{16}$ or $\frac{3}{8}$, would do. For example, 3 g of carbon would have the same number of particles as 8 g of molecular oxygen.

CHECKPOINT

1. Reacting 3 g of carbon, C, with 8 g of molecular oxygen, O_2, results in 11 g of carbon dioxide, CO_2. Does it follow that 1.5 g of carbon will react with 4 g of oxygen to form 5.5 g of carbon dioxide?
2. Will reacting 5 g of carbon with 8 g of oxygen also result in 11 g of carbon dioxide?

Were these your answers?

1. Yes. The quantities are only half as much, but their ratio is the same as when 11 g of carbon dioxide is formed: $1.5 : 4 : 5.5 = 3 : 8 : 11$.
2. Yes. Many students make the common error of thinking that no reaction will occur if the proper ratios of reactants are not provided. You should understand, however, that in a 5-g sample of carbon, 3 g of carbon is available for reacting. This 3 g will react with the 8 g of oxygen to form 11 g of carbon dioxide. There will be 2 g of carbon unreacted after the reaction. Reacting this remaining 2 g of carbon would require more oxygen.

Converting between Grams and Moles

Atoms and molecules react in specific ratios. In the laboratory, however, chemists work with bulk quantities of materials, which are measured by mass. Chemists therefore need to know the relationship between the mass of a given sample and the number of atoms or molecules contained in that mass. The key to this relationship is the *mole*. Recall from Section 16.3 that the mole is a unit equal to 6.02×10^{23}. This number is known as **Avogadro's number**, in honor of the 18th-century scientist Amedeo Avogadro.

As Figure 17.4 illustrates, if you express the numeric value of the atomic mass of any element in *grams*, the number of atoms in a sample of the element having this mass is always 6.02×10^{23}, which is 1 mole. For example,

FYI An Avogadro's number of grains of sand would fill the United States to a depth of about 2 m. There are about 7 billion people on Earth. You would need about 94 trillion Earth-size populations to have an Avogadro's number of people. If you collected 1 million hydrogen atoms every second, it would take you about 19 billion years to come up with a whole gram of hydrogen—the universe itself is only about 13 billion years old. A stack of an Avogadro's number of pennies would be about 800,000 trillion km, which is about the diameter of our galaxy. Placed side by side, these pennies would reach to the Andromeda galaxy, which is about a million light-years away.

FIGURE 17.4
Express the numeric value of the atomic mass of any element in grams, and that many grams contains 6.02×10^{23} atoms.

a 22.990-g sample of sodium metal, Na (atomic mass = 22.990 amu), contains 6.02×10^{23} sodium atoms, and a 207.2-g sample of lead, Pb (atomic mass = 207.2 amu), contains 6.02×10^{23} lead atoms.

The same concept holds for compounds. Express the numeric value of the formula mass of any compound in grams, and a sample having that mass contains 6.02×10^{23} molecules of that compound. For example, there are 6.02×10^{23} O_2 molecules in 31.998 g of molecular oxygen, O_2 (formula mass = 31.998 amu), and 6.02×10^{23} CO_2 molecules in 44.009 g of carbon dioxide, CO_2 (formula mass = 44.009 amu).

CHECKPOINT

1. **How many atoms are there in a 6.941-g sample of lithium, Li (atomic mass = 6.941 amu)?**
2. **How many molecules are there in an 18.015-g sample of water, H_2O (formula mass = 18.015 amu)?**

Were these your answers?

1. Because this number of grams of lithium is numerically equal to the atomic mass, the sample contains 6.02×10^{23} lithium atoms, which is 1 mole.
2. Because this number of grams of water is numerically equal to the formula mass, the sample contains 6.02×10^{23} water molecules, which is 1 mole.

The **molar mass** of any substance, be it element or compound, is defined as the mass of 1 mole of the substance. Thus the units of molar mass are grams per mole. For instance, the atomic mass of carbon is 12.011 amu, which means that 1 mole of carbon has a mass of 12.011 g, and we say that the molar mass of carbon is 12.011 g/mole. The molar mass of molecular oxygen, O_2 (formula mass = 31.998 amu), is 31.998 g/mole. For convenience, values such as these are often rounded off to the nearest whole number. The molar mass of carbon, therefore, might also be presented as 12 g/mole, and that of molecular oxygen as 32 g/mole.

CHECKPOINT

What is the molar mass of water (formula mass = 18 amu)?

Was this your answer?

From the formula mass, you know that 1 mole of water has a mass of 18 g. Therefore the molar mass is 18 g/mole.

Because 1 mole of any substance always contains 6.02×10^{23} particles, the mole is an ideal unit for chemical reactions. For example, 1 mole of carbon (12 g) reacts with 1 mole of molecular oxygen (32 g) to give 1 mole of carbon dioxide (44 g).

In many instances, the ratio in which chemicals react is not 1 : 1. As shown in Figure 17.5, for example, 2 moles (4 g) of molecular hydrogen react with 1 mole (32 g) of molecular oxygen to give 2 moles (36 g) of water. Note how the coefficients of the balanced chemical equation can be conveniently interpreted

FIGURING PHYSICAL SCIENCE

Masses of Reactants and Products

The coefficients of a chemical equation tell us the ratio by which reactants react and products form. The following equation, for example, tells us that every 1 mole of methane, CH_4, reacts to produce 2 moles of water, H_2O.

$$CH_4 + 2\,O_2 \longrightarrow CO_2 + 2\,H_2O$$

So if you were given 16 g of methane, CH_4, how many grams of water, H_2O, would form? From the text you should know that 16 g of methane, CH_4, is 1 mole (formula mass 16 amu). So 16 g of methane would yield 2 moles of water. But how many grams is 2 moles of water? Well, if 1 mole of water, H_2O, equals 18 g (formula mass 18 amu), then 2 moles equals 36 g. Thus, 16 g of methane, CH_4, reacting with oxygen, O_2, would yield 36 g of water, H_2O. Let's look at this process from a step-by-step mathematical point of view.

SAMPLE PROBLEM 1

What mass of water is produced when 16 g of methane, CH_4 (formula mass 16 amu), reacts with oxygen, O_2, in the reaction below?

$$CH_4 + 2\,O_2 \longrightarrow CO_2 + 2\,H_2O$$

Solution:

Step 1. Convert the given mass to moles:

Conversion factor

$$(16\ \text{g}\ CH_4)\left(\frac{1\ \text{mole}\ CH_4}{16\ \text{g}\ CH_4}\right) = 1\ \text{mole}\ CH_4$$

Step 2. Use the coefficients of the balanced equation to find out how many moles of H_2O are produced from this many moles of CH_4:

Conversion factor

$$(1\ \text{mole}\ CH_4)\left(\frac{2\ \text{moles}\ H_2O}{1\ \text{mole}\ CH_4}\right) = 2\ \text{moles}\ H_2O$$

Step 3. Now that you know how many moles of H_2O are produced, convert this value to grams of H_2O:

Conversion factor

$$(2\ \text{moles}\ H_2O)\left(\frac{18\ \text{g}\ H_2O}{1\ \text{mole}\ H_2O}\right) = 36\ \text{g}\ H_2O$$

The method of converting grams of a substance to moles (step 1), then from moles of this substance to moles of that substance (step 2), followed by moles of that substance to grams (step 3) is called *stoichiometry*. Using stoichiometry, a scientist can calculate the amounts of reactants or products in any chemical reaction. The methods of stoichiometry are developed much further in general chemistry courses. For this course, all you need to do is be familiar with what stoichiometry is all about, which is keeping tabs on atoms and molecules as they react to form products. Nonetheless, for a special assignment, you might try your analytical-thinking skills on the following problems. First try to deduce the answers based on what you know about the law of mass conservation, and then follow the steps given here to check your answers.

SAMPLE PROBLEM 2

Show that 44 g of carbon dioxide, CO_2, is produced when 16 g of methane, CH_4, reacts with oxygen, O_2. How many grams of oxygen, O_2, are needed for this reaction?

Solution:

The 16 g of methane, CH_4, is 1 mole, which reacts with oxygen to produce

as the number of moles of reactants or products. A chemist therefore need only convert these numbers of moles to grams in order to know how much mass of each reactant he or she should measure out to have the proper proportions.

Cooking and chemistry are similar in that both require measuring ingredients. Just as a cook looks to a recipe to find the necessary quantities measured by the cup or the tablespoon, a chemist looks to the periodic table to find the necessary quantities measured by the number of grams per mole for each element or compound.

FIGURE 17.5
Two moles of H_2 react with 1 mole of O_2 to give 2 moles of H_2O. This is the same as saying 4 g of H_2 reacts with 32 g of O_2 to give 36 g of H_2O or, equivalently, that 12.04×10^{23} H_2 molecules react with 6.02×10^{23} O_2 molecules to give 12.04×10^{23} H_2O molecules.

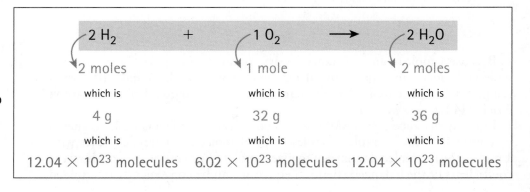

1 mole of carbon dioxide, CO_2. One mole of carbon dioxide (formula mass 44 amu) is 44 g. So the 16 g of methane reacts with oxygen to produce 44 g of carbon dioxide plus 36 g of water. The mass of the products (44 g + 36 g = 80 g) must be equal to the mass of the reactants (16 g + ? = 80 g). So we can calculate that the methane reacted with 64 g of oxygen, which, interestingly enough, is 2 moles, as shown in the equation.

SAMPLE PROBLEM 3

How many grams of ozone (O_3, 48 amu) can be produced from 64 g of oxygen (O_2, 32 amu) in the reaction below?

$$3\,O_2 \longrightarrow 2\,O_3$$

Solution:

According to the law of mass conservation, the amount of mass in the products must equal the amount of mass in the reactants. Given that this reaction involves only one reactant and one product, you should not be surprised to learn that 64 g of reactant produces 64 g of product. Here are the stepwise calculations:

Step 1. Convert grams of O_2 to moles of O_2:

$$(64 \text{ g } O_2)\left(\frac{1 \text{ mole } O_2}{32 \text{ g } O_2}\right) = 2 \text{ moles } O_2$$

Step 2. Convert moles of O_2 to moles of O_3:

$$(2 \text{ moles } O_2)\left(\frac{2 \text{ moles } O_3}{3 \text{ moles } O_2}\right)$$
$$= 1.33 \text{ moles } O_3$$

Step 3. Convert moles of O_3 to grams of O_3:

$$(1.33 \text{ moles } O_3)\left(\frac{48 \text{ g } O_3}{1 \text{ mole } O_3}\right) = 64 \text{ g } O_3$$

SAMPLE PROBLEM 4

What mass of nitrogen monoxide (NO, 30 amu) is formed when 28 g of nitrogen (N_2, 28 amu) reacts with 32 g of oxygen (O_2, 32 amu) in the reaction below?

$$N_2 + O_2 \longrightarrow 2 \text{ NO}$$

How about when 28 g of nitrogen, N_2, is combined with 40 g of oxygen, O_2?

Solution:

There are several ways to answer this problem. One way would be to recognize that 28 g of N_2 is 1 mole of N_2 and 32 g of O_2 is 1 mole of O_2. According to the balanced equation, combining 1 mole of N_2 with 1 mole of O_2 yields 2 moles of NO. The mass of 2 moles of NO is

$$(2 \text{ moles NO})\left(\frac{30 \text{ g NO}}{1 \text{ mole NO}}\right) = 60 \text{ g NO}$$

which is the sum of the masses of the reactants, as it must be because of the law of mass conservation.

Combining 40 g of oxygen with 28 g of nitrogen would be 8 g too much oxygen. Only 32 g of this oxygen would react with the nitrogen, producing 60 g of NO, leaving 8 g of oxygen left over unreacted. Stoichiometry is an area of chemistry rich in opportunities for analytical thinking.

17.3 Reaction Rates

EXPLAIN THIS Why does blowing into a campfire make the fire burn brighter?

LEARNING OBJECTIVE
Describe the requirements that must be met in order for a chemical reaction to occur.

A balanced chemical equation helps determine the amount of products that can be formed from given amounts of reactants. But the equation tells us little about what occurs on the submicroscopic level during the reaction. In this and the following section, we explore that submicroscopic level to show how the *rate* of a reaction can be changed, either by changing the concentration or the temperature of the reactants or by adding what is known as a *catalyst*.

Some chemical reactions, such as the rusting of iron, are slow, while others, such as the burning of gasoline, are fast. The speed of any reaction is indicated by its reaction rate, which is an indicator of how quickly the reactants transform to products. As shown in Figure 17.6, initially a flask may contain only reactant molecules. Over time, these reactants form product molecules, and, as a result, the concentration of product molecules increases. The **reaction rate**, therefore, can be defined either as how quickly the concentration of products increases or as how quickly the concentration of reactants decreases.

FIGURE 17.6

Over time, the reactants in this reaction flask may transform to products. If this happens quickly, the reaction rate is high. If this happens slowly, the reaction rate is low.

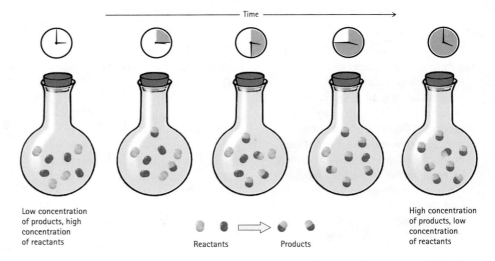

Low concentration of products, high concentration of reactants

Reactants → Products

High concentration of products, low concentration of reactants

FIGURE 17.7

INTERACTIVE FIGURE MP

During a reaction, reactant molecules collide with one another.

What determines the rate of a chemical reaction? The answer is complex, but one important factor is that reactant molecules must physically come together. Because molecules move rapidly, this physical contact is appropriately described as a collision. We can illustrate the relationship between molecular collisions and reaction rate by considering the reaction of gaseous nitrogen and gaseous oxygen to form gaseous nitrogen monoxide, as shown in Figure 17.7.

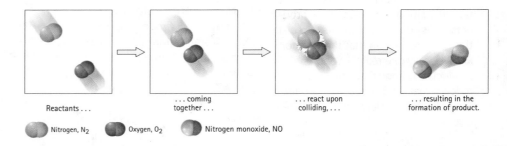

Reactants . . .   . . . coming together . . .   . . . react upon colliding, ... resulting in the formation of product.

Nitrogen, N_2 Oxygen, O_2 Nitrogen monoxide, NO

Because reactant molecules must collide in order for a reaction to occur, the rate of a reaction can be increased by increasing the number of collisions. An effective way to increase the number of collisions is to increase the concentration of the reactants. Figure 17.8 shows that with higher concentrations, more molecules are in a given volume, which makes collisions between molecules more probable. As an analogy, consider a group of people on a dance floor—as the number of people increases, so does the rate at which they bump into one another. An increase in the concentration of nitrogen and oxygen molecules, therefore, leads to a greater number of collisions between these molecules; hence, a greater number of nitrogen monoxide molecules form in a given period of time.

Not all collisions between reactant molecules lead to products, however, because the molecules must collide in a certain orientation in order to react. Nitrogen and oxygen, for example, are much more likely to form nitrogen monoxide when the molecules collide in the parallel orientation shown in Figure 17.7. When they collide in the perpendicular orientation shown in Figure 17.9, nitrogen monoxide does not form. For larger molecules, which can have numerous orientations, this orientation requirement is even more restrictive.

A second reason that not all collisions lead to product formation is that the reactant molecules must also collide with enough kinetic energy to break their bonds. Only then is it possible for the atoms in the reactant molecules to change

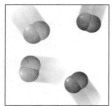

Less concentrated

More concentrated

FIGURE 17.8

INTERACTIVE FIGURE MP

The more concentrated a sample of nitrogen and oxygen, the greater the probability that N_2 and O_2 molecules will collide and form nitrogen monoxide.

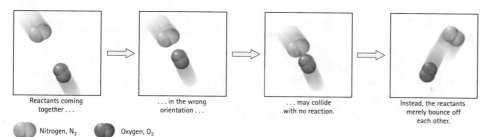

FIGURE 17.9
INTERACTIVE FIGURE   MP

The orientation of reactant molecules in a collision can determine whether a reaction occurs. A perpendicular collision between N_2 and O_2 tends not to result in formation of a product molecule.

Reactants coming together . . .

. . . in the wrong orientation . . .

. . . may collide with no reaction.

Instead, the reactants merely bounce off each other.

Nitrogen, N_2 Oxygen, O_2

bonding partners and form product molecules. The bonds in N_2 and O_2 molecules, for example, are quite strong. In order for these bonds to be broken, collisions between the molecules must contain enough energy to break the bonds. As a result, collisions between slow-moving N_2 and O_2 molecules, even those that collide in the proper orientation, may not form NO, as is shown in Figure 17.10.

The higher the temperature of a material, the faster its molecules move and the more forceful the collisions between them. Higher temperatures, therefore, increase reaction rates. The nitrogen and oxygen molecules that make up our atmosphere, for example, are continually colliding with one another. At the ambient temperatures of our atmosphere, however, these molecules do not generally have sufficient kinetic energy for the formation of nitrogen monoxide. The heat of a lightning bolt, however, dramatically increases the kinetic energy of these molecules to the point that a large portion of the collisions in the vicinity of the bolt result in the formation of nitrogen monoxide. The nitrogen monoxide formed in this manner undergoes further reactions to form nitrate ions that plants depend on for survival, as was discussed in the opening of this chapter. This is an example of *nitrogen fixation*, which you may have explored already in a course on the life sciences.

The life sciences involve fantastic applications of chemistry, nitrogen fixation being just one example. Others include photosynthesis, cellular respiration, and molecular genetics. So there are distinct advantages to learning about chemistry and other physical sciences *before* advancing to the life sciences.

CHECKPOINT

An internal-combustion engine works by drawing a mixture of air and gasoline vapors into a chamber. The action of a piston then compresses these gases into a smaller volume before ignition by the spark of a spark plug. What is the advantage of squeezing the vapors to a smaller volume?

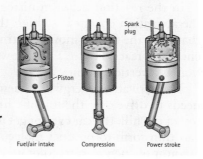

Spark plug

Piston

Fuel/air intake   Compression   Power stroke

Was this your answer?

Squeezing the vapors to a smaller volume effectively increases their concentration and, hence, the number of collisions between molecules. This, in turn, promotes the chemical reaction. As discussed in Section 7.2, compression also increases the temperature, which further favors the chemical reaction.

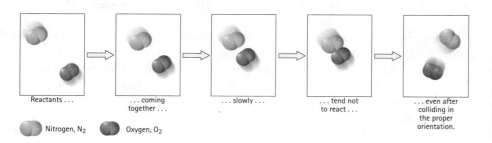

Reactants . . .

. . . coming together . . .

. . . slowly . . .

. . . tend not to react . . .

. . . even after colliding in the proper orientation.

Nitrogen, N_2 Oxygen, O_2

FIGURE 17.10
INTERACTIVE FIGURE   MP

Slow-moving molecules may collide with insufficient force to break their bonds. As a result, they cannot react to form product molecules.

FIGURE 17.11
Reactant molecules must gain a minimum amount of energy, called the activation energy, E_a, before they can transform to product molecules.

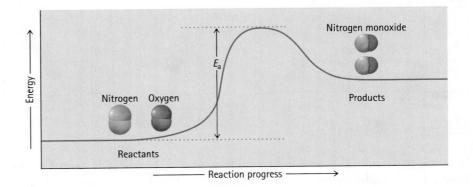

Nitrogen monoxide

E_a

Nitrogen Oxygen

Products

Reactants

Reaction progress

Mastering**PHYSICS**
TUTORIAL: Equilibrium

The energy required to break bonds can also come from the absorption of electromagnetic radiation. As the radiation is absorbed by reactant molecules, the atoms in the molecules may start to vibrate so rapidly that the bonds between them are easily broken. In many instances, direct absorption of electromagnetic radiation is sufficient to break chemical bonds and to initiate a chemical reaction. The common atmospheric pollutant nitrogen dioxide, NO_2, for example, may transform to nitrogen monoxide and atomic oxygen merely on exposure to sunlight:

$$NO_2 + \text{sunlight} \longrightarrow NO + O$$

Whether they result from collisions, absorption of electromagnetic radiation, or both, broken bonds are a necessary first step in most chemical reactions. The energy required for this initial breaking of bonds can be viewed as an *energy barrier*. The minimum energy required to overcome this energy barrier is known as the **activation energy** (E_a).

In the reaction between nitrogen and oxygen to form nitrogen monoxide, the activation energy is so high (because the bonds in N_2 and O_2 are strong) that only the fastest-moving nitrogen and oxygen molecules possess sufficient energy to react. Figure 17.11 shows the activation energy in this chemical reaction as a vertical hump.

The activation energy of a chemical reaction is analogous to the energy a car needs to drive over the top of a hill. Without sufficient energy to climb to the top of the hill, the car cannot get to the other side. Likewise, reactant molecules can transform to product molecules only if the reactant molecules possess an amount of energy equal to or greater than the activation energy.

At any given temperature, there is a wide distribution of kinetic energies in reactant molecules. Some are moving slowly, and others are moving quickly. As we discussed in Section 6.1, the temperature of a material is related to the average of all these kinetic energies. The few fast-moving reactant molecules in Figure 17.12 have enough energy to pass over the energy barrier and are the first to transform to product molecules.

Kinetic energies sufficient to overcome energy barrier

Kinetic energies not sufficient to overcome energy barrier

FIGURE 17.12
Because fast-moving reactant molecules possess sufficient energy to pass over the energy barrier, they are the first ones to transform to product molecules.

When the temperature of the reactants is increased, the number of reactant molecules possessing sufficient energy to pass over the barrier also increases, which is why reactions are generally faster at higher temperatures. Conversely, at lower temperatures, fewer molecules have sufficient energy to pass over the barrier. Hence, reactions are generally slower at lower temperatures.

CHECKPOINT
What kitchen device is used to lower the rate at which microorganisms grow on food?

Was this your answer?
The refrigerator! Microorganisms, such as bread mold, are everywhere and difficult to avoid. By lowering the temperature of microorganism-contaminated food, the refrigerator decreases the rate of the chemical reactions that these microorganisms depend on for growth, thereby increasing the food's shelf life.

> In order for two chemicals to be able to react, they must first collide in the proper orientation. Second, they must have sufficient kinetic energy to initiate the breaking of chemical bonds so that new bonds can form. These are all aspects of a broad theory known as the *molecular-kinetic theory.*

Most chemical reactions are influenced by temperature in this manner, including reactions that occur in living bodies. The body temperature of animals that regulate their internal temperature, such as humans, is fairly constant. However, the body temperature of some animals, such as the alligator shown in Figure 17.13, rises and falls with the temperature of the environment. On a warm day, the chemical reactions occurring in an alligator are "up to speed," and the animal is more active. On a chilly day, however, the chemical reactions proceed at a lower rate, and, as a consequence, the alligator's movements are unavoidably sluggish.

FIGURE 17.13
This alligator became immobilized on the pavement after being caught in the cold night air. By midmorning, shown here, the temperature had warmed sufficiently to allow the alligator to get up and walk away.

17.4 Catalysts

LEARNING OBJECTIVE
Discuss how a catalyst can speed up a chemical reaction, using the destruction of stratospheric ozone as an example.

EXPLAIN THIS Chew a salt-free soda cracker for a few minutes and the cracker begins to taste sweet. Why?

As discussed in the previous section, increasing the concentration or the temperature of the reactants can cause a chemical reaction to go faster. A third way to increase the rate of a reaction is to add a **catalyst**, which is any substance that increases the rate of a chemical reaction by lowering its activation energy. The catalyst may participate as a reactant, but it is then regenerated as a product and is thus available to catalyze subsequent reactions.

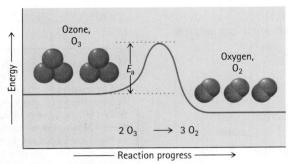

(a) Without catalyst

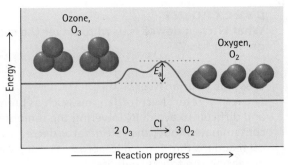

(b) With chlorine catalyst

FIGURE 17.14

(a) The relatively high activation energy (energy barrier) indicates that only the most energetic ozone molecules can react to form oxygen molecules. (b) The presence of chlorine atoms lowers the activation energy, which means more reactant molecules have sufficient energy to form product. The chlorine allows the reaction to proceed in two steps, and the two smaller activation energies correspond to these steps. (Note that the convention is to write the catalyst above the reaction arrow.)

FYI Before the fall of the Soviet Union, numerous oil-drilling sites in Siberia were allowed to vent natural gas freely into the atmosphere, presumably because the natural gas had no commercial value. After the fall of the Soviet Union, the wells were capped to prevent this venting. Within weeks, instruments at the Mauna Loa weather observatory on the other side of the planet noted a significant drop in atmospheric levels of methane and its by-product, carbon dioxide. The effect that we humans have on global atmospheric conditions is measurable.

The conversion of ozone, O_3, to oxygen, O_2, is normally sluggish because the reaction has a relatively high activation energy, as shown in Figure 17.14a. However, when chlorine atoms act as a catalyst, the energy barrier is lowered, as shown in Figure 17.14b, and the reaction can proceed faster.

Chlorine atoms lower the energy barrier of this reaction by providing an alternate pathway involving intermediate reactions, each having a lower activation energy than the uncatalyzed reaction. This alternate pathway involves two steps. Initially, the chlorine reacts with the ozone to form chlorine monoxide and oxygen:

$$Cl \ + \ O_3 \longrightarrow ClO \ + \ O_2$$

Chlorine   Ozone   Chlorine   Oxygen
monoxide

The chlorine monoxide then reacts with another ozone molecule to re-form the chlorine atom as well as to produce two additional oxygen molecules:

$$ClO \ + \ O_3 \longrightarrow Cl \ + \ 2\,O_2$$

Chlorine   Ozone   Chlorine   Oxygen
monoxide

Although chlorine is depleted in the first reaction, it is regenerated in the second reaction. As a result, there is no net consumption of chlorine. At the same time, however, because of the lower energy of activation for these reactions, ozone molecules are rapidly converted to oxygen molecules.

Chlorine atoms in the stratosphere catalyze the destruction of Earth's ozone layer. Evidence indicates that chlorine atoms are generated in the stratosphere as a by-product of human-made chlorofluorocarbons (CFCs), once widely produced as the cooling fluid of refrigerators and air conditioners. Destruction of the ozone layer is a serious concern because of its role in protecting us from the Sun's harmful ultraviolet rays. One chlorine atom in the ozone layer is estimated to catalyze the transformation of 100,000 ozone molecules to oxygen molecules in the one or two years before the chlorine atom is removed by natural processes.

Chemists have been able to harness the power of catalysts for numerous beneficial purposes. The exhaust that comes from an automobile engine, for example, contains a wide assortment of pollutants, such as nitrogen monoxide, carbon monoxide, and uncombusted fuel vapors (hydrocarbons). To reduce the amount of these pollutants entering the atmosphere, most automobiles are equipped with *catalytic converters*, as shown in Figure 17.15. Metal catalysts in a converter speed up reactions that convert exhaust pollutants to less toxic substances. Nitrogen monoxide is transformed to nitrogen and oxygen, carbon monoxide is transformed to carbon dioxide, and unburned fuel is converted to carbon dioxide and water vapor. Because catalysts are not consumed by the reactions they facilitate, a single catalytic converter may continue to operate effectively for the lifetime of the car.

Catalytic converters, along with microchip-controlled fuel–air ratios, have led to a significant drop in the per-vehicle emission of pollutants. A typical car in 1960 emitted about 11 g of uncombusted fuel, 4 g of nitrogen oxide, and 84 g of carbon monoxide per mile traveled. An improved vehicle in 2000 emitted less than 0.5 g of uncombusted fuel, less than 0.5 g of nitrogen oxide, and only about 3 g of carbon monoxide per mile traveled. This improvement, however, has been offset by an increase in the number of cars being driven, as exemplified by the traffic jam shown in Figure 17.16. It is also offset by the growing popularity of SUVs (sport-utility vehicles), which bypass pollution requirements.

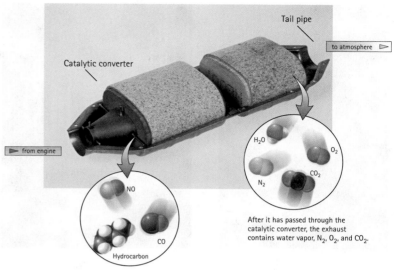

Before it reaches the catalytic converter, the exhaust contains such pollutants as NO, CO, and hydrocarbons.

After it has passed through the catalytic converter, the exhaust contains water vapor, N_2, O_2, and CO_2.

The chemical industry depends on catalysts because they lower manufacturing costs by lowering required temperatures and by providing greater product yields without being consumed. Indeed, more than 90% of all manufactured goods are produced with the assistance of catalysts. Without catalysts, the price of gasoline would be much higher, as would be the prices of such consumer goods as rubber, plastics, pharmaceuticals, automobile parts, clothing, and food grown with chemical fertilizers. Living organisms rely on special types of catalysts known as *enzymes*, which allow exceedingly complex biochemical reactions to occur with ease. You may learn more about the nature and behavior of enzymes in a life science course.

FIGURE 17.15

A catalytic converter reduces the pollution caused by automobile exhaust by converting such harmful combustion products as NO, CO, and hydrocarbons to harmless N_2, O_2, and CO_2. The catalyst is typically platinum, Pt; palladium, Pd; or rhodium, Rd.

CHECKPOINT

How does a catalyst lower the activation energy of a chemical reaction?

Was this your answer?
The catalyst provides an alternate and easier-to-achieve pathway along which the chemical reaction can proceed.

17.5 Energy and Chemical Reactions

EXPLAIN THIS What changes during a chemical reaction?

As we have discussed in the preceding two sections, reactants must have a certain amount of energy in order to overcome the activation energy so that a chemical reaction can proceed. Once a reaction is complete, however, there may be either a net release or a net absorption of energy. Reactions in which there is a net release of energy are called **exothermic**. Rocket ships lift off into space and campfires glow red hot as a result of exothermic reactions. Reactions in which there is a net absorption of energy are called **endothermic**. Photosynthesis, for example, involves a series of endothermic reactions that are driven by the energy of sunlight. Both exothermic and endothermic reactions, illustrated in Figure 17.17, can be understood through the concept of bond energy.

FIGURE 17.16

The exhaust from automobiles today is much cleaner than before the advent of the catalytic converter, but many more cars are on the road. In 1960, there were about 74 million registered motor vehicles in the United States. By 2008, there were about 250 million.

LEARNING OBJECTIVE

Calculate the amount of energy released or absorbed by a chemical reaction using the bond energies of reactants and products.

During a chemical reaction, chemical bonds are broken and atoms rearrange to form new chemical bonds. Such breaking and forming of chemical bonds involve changes in energy. As an analogy, consider a pair of magnets. To separate them requires an input of "muscle energy." Conversely, when the two separated magnets collide, they become slightly warmer than they were, and this warmth is evidence of energy released. The magnets must absorb energy if they are to break apart, and release energy as they come together. The same principle applies to atoms. To pull bonded atoms apart requires an energy input. When atoms combine, there is an energy output, usually in the form of faster-moving atoms and molecules, electromagnetic radiation, or both.

FIGURE 17.17

Chemical reactions that occur when wood is burning have a net release of energy. Chemical reactions that occur in a photosynthetic plant have a net absorption of energy.

Remember, in a chemical reaction, the bonds being formed are different from the bonds that were broken. The bond energies of the bonds being formed, therefore, are also different from those of the bonds that were broken.

The amount of energy required to pull two bonded atoms apart is the same as the amount released when they are brought together. This energy, whether it is the energy that is absorbed as a bond breaks or the energy that is released as a bond forms, is called **bond energy**. Each chemical bond has its own characteristic bond energy. The hydrogen–hydrogen bond energy, for example, is 436 kJ/mole. This means that 436 kJ of energy is absorbed as 1 mole of hydrogen–hydrogen bonds break apart, and 436 kJ of energy is released upon the formation of 1 mole of hydrogen–hydrogen bonds. Different bonds involving different elements have different bond energies, as Table 17.1 shows. You can refer to the table as you study this section, but please do not memorize these bond energies. Instead, focus on understanding what they mean.

TABLE 17.1	SELECTED BOND ENERGIES		
Bond	**Bond Energy (kJ/mole)**	**Bond**	**Bond Energy (kJ/mole)**
H—H	436	N—N	159
H—C	414	O—O	138
H—N	389	Cl—Cl	243
H—O	464	C=O	803
H—F	569	N=O	631
C—O	351	O=O	498
H—Cl	431	C≡C	837
C—C	347	N≡N	946

By convention, a positive bond energy represents the amount of energy absorbed as a bond breaks, and a negative bond energy represents the amount of energy released as a bond forms. Thus, when you are calculating the net energy released or absorbed during a reaction, you'll need to be careful about plus and minus signs. It is standard practice when doing such calculations to assign a plus sign to energy absorbed and a minus sign to energy released. For instance, when dealing with a reaction in which 1 mole of H—H bonds are broken, you'll write +436 kJ to indicate energy absorbed, and when dealing with the formation of 1 mole of H—H bonds, you'll write −436 kJ to indicate energy released. We'll do some sample calculations in a moment.

CHECKPOINT
Do all covalent single bonds have the same bond energy?

Was this your answer?
No. Bond energy depends on the types of atoms bonding. The H–H single bond, for example, has a bond energy of 436 kJ/mole, but the H–O single bond has a bond energy of 464 kJ/mole.

Exothermic Reaction: Net Release of Energy

For any chemical reaction, the total amount of energy absorbed in breaking bonds in reactants is always different from the total amount of energy released as bonds form in the products. Consider the reaction in which hydrogen and oxygen react to form water:

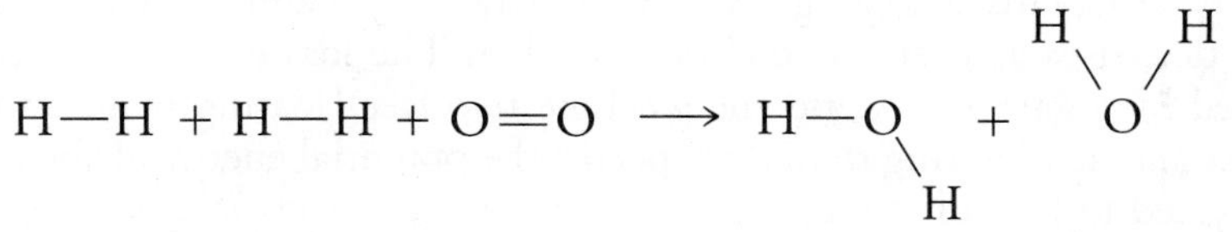

$$\text{H}-\text{H} + \text{H}-\text{H} + \text{O}=\text{O} \longrightarrow \text{H}-\underset{\text{H}}{\text{O}} + \underset{\text{O}}{\overset{\text{H} \quad \text{H}}{}}$$

In the reactants, hydrogen atoms are bonded to hydrogen atoms, and oxygen atoms are double-bonded to oxygen atoms. The total amount of energy absorbed as these bonds break is $+1370$ kJ.

Type of Bond	Number of Moles	Bond Energy	Total Energy
H—H	2	+436 kJ/mole	+872 kJ
O=O	1	+498 kJ/mole	+498 kJ
		Total energy absorbed:	+1370 kJ

In the products there are four hydrogen–oxygen bonds. The total amount of energy released as these bonds form is -1856 kJ.

Type of Bond	Number of Moles	Bond Energy	Total Energy
H—O	4	−464 kJ/mole	−1856 kJ
		Total energy released:	−1856 kJ

The amount of energy released in this reaction exceeds the amount of energy absorbed. The net energy of the reaction is found by adding the two quantities:

$$\text{Net energy of reaction} = \text{energy absorbed} + \text{energy released}$$
$$= +1370 \text{ kJ} + (-1856 \text{ kJ})$$
$$= -486 \text{ kJ}$$

The negative sign on the net energy indicates that there is a net release of energy, and so the reaction is exothermic. For any exothermic reaction, energy can be considered a product and is thus sometimes included after the arrow of the chemical equation:

$$2 \text{ H}_2 + \text{O}_2 \longrightarrow 2 \text{ H}_2\text{O} + \text{energy}$$

In an exothermic reaction, the potential energy of atoms in the product molecules is lower than their potential energy in the reactant molecules. This is illustrated in the reaction profile shown in Figure 17.18. The lowered potential

FYI NASA scientists routinely test various materials for their durability against atomic oxygen, O, which is abundant in the low orbit of the space shuttle. They discovered that atomic oxygen effectively transforms surface organic materials into gaseous carbon dioxide. The scientists realized atomic oxygen's usefulness for restoring paintings damaged by smoke or other organic contaminants. Together with art conservationists they used atomic oxygen to restore certain damaged paintings, and it worked spectacularly.

FIGURE 17.18
In an exothermic reaction, the product molecules are at a lower potential energy than the reactant molecules. The net amount of energy released by the reaction is equal to the difference in potential energies of the reactants and products.

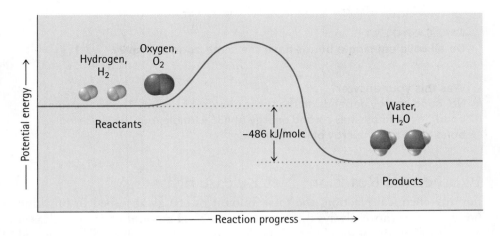

FIGURE 17.19
A space shuttle uses exothermic chemical reactions to lift off from Earth's surface.

energy of the atoms in the product molecules is due to their being more tightly held together. This is analogous to two attracting magnets, whose potential energy decreases as they come closer together. The loss of potential energy is balanced by a gain in kinetic energy. Like two free-floating magnets coming together and accelerating to higher speeds, the potential energy of the reactants is converted to faster-moving atoms and molecules, electromagnetic radiation, or both. This kinetic energy released by the reaction is equal to the difference between the potential energy of the reactants and the potential energy of the products, as is indicated in Figure 17.18.

It is important to understand that the energy released by an exothermic reaction is not created by the reaction. This is in accord with the *law of conservation of energy*, which tells us that energy is neither created nor destroyed in a chemical reaction (or any process). Instead, energy is merely converted from one form to another. During an exothermic reaction, energy that was once in the form of the potential energy of chemical bonds is released as the kinetic energy of fast-moving molecules and/or as electromagnetic radiation.

The amount of energy released in an exothermic reaction depends on the amounts of the reactants. The reaction of large amounts of hydrogen and oxygen, for example, provides the energy to lift the space shuttle shown in Figure 17.19 into orbit. There are two compartments in the large central tank to which the orbiter is attached—one filled with liquid hydrogen and the other filled with liquid oxygen. Upon ignition, these two liquids mix and react chemically to form water vapor, which produces the needed thrust as it is expelled out the rocket cones. Additional thrust is obtained by a pair of solid-fuel rocket boosters containing a mixture of ammonium perchlorate, NH_4ClO_4, and powdered aluminum. On ignition, these chemicals react to form products that are expelled at the rear of the rocket. The balanced equation representing this reaction is

$$3\,NH_4ClO_4 + 3\,Al \longrightarrow Al_2O_3 + AlCl_3 + 3\,NO + 6\,H_2O + energy$$

Recall from Chapter 2 that for every action there is an opposite and equal reaction. A rocket is thrust upward, for example, only as its exhaust chemicals are thrust downward.

CHECKPOINT
Where does the net energy released in an exothermic reaction go?

Was this your answer?
This energy goes into increasing the speeds of reactant atoms and molecules and often into electromagnetic radiation.

Endothermic Reaction: Net Absorption of Energy

When the amount of energy released in product formation is *less* than the amount of energy absorbed when reactant bonds break, the reaction is endothermic. An example is the reaction of atmospheric nitrogen and oxygen to form nitrogen monoxide, which is the same reaction used for many of the discussions earlier in this chapter:

$$N \equiv N + O = O \longrightarrow N = O + N = O$$

The amount of energy absorbed as the chemical bonds in the reactants break is

Type of Bond	Number of Moles	Bond Energy	Total Energy
N≡N	+1	+946 KJ/mole	+946 KJ
O=O	+1	+498 kJ/mole	+498 kJ
		Total energy absorbed:	+1444 kJ

The amount of energy released upon the formation of bonds in the products is

Type of Bond	Number of Moles	Bond Energy	Total Energy
N=O	2	−631 kJ/mole	−1262 KJ
		Total energy released:	−1262 KJ

As before, the net energy of the reaction is found by adding the two quantities:

$$\text{Net energy of reaction} = \text{energy absorbed} + \text{energy released}$$
$$= +1444 \text{ kJ} + (-1262 \text{ kJ})$$
$$= +182 \text{ kJ}$$

The positive sign indicates a net *absorption* of energy, meaning the reaction is endothermic. For any endothermic reaction, energy can be considered a reactant and is thus sometimes included before the arrow of the chemical equation:

$$\text{Energy} + N_2 + O_2 \longrightarrow 2\,NO$$

In an endothermic reaction, the potential energy of atoms in the product molecules is higher than their potential energy in the reactant molecules. This is illustrated in the reaction profile shown in Figure 17.20. Raising the potential energy of the atoms in the product molecules requires a net input of

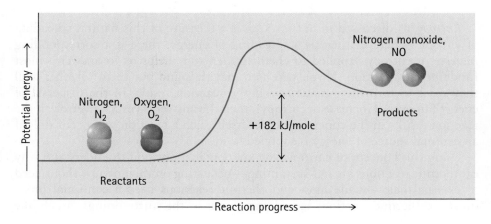

FIGURE 17.20
In an endothermic reaction, the product molecules are at a higher potential energy than the reactant molecules. The net amount of energy absorbed by the reaction is equal to the difference in potential energies of the reactants and products.

energy, which must come from some external source, such as electromagnetic radiation, electricity, or heat. Thus, nitrogen and oxygen react to form nitrogen monoxide only with the application of much heat, as occurs adjacent to a lightning bolt or in an internal-combustion engine.

17.6 Chemical Reactions Are Driven by Entropy

EXPLAIN THIS Why are exothermic reactions self-sustaining?

Energy tends to disperse. It flows from where it is concentrated to where it is spread out. The energy of a hot pan, for example, does not stay concentrated in the pan once the pan is taken off the stove. Instead, the energy spreads away from the pan and into the cooler surroundings. Similarly, the concentrated chemical energy in gasoline disperses in the formation of many very hot, smaller molecules that scatter explosively. Some of this released thermal energy is used by the engine to get the car moving. The rest spreads into the engine block and radiator fluid and then out the exhaust pipe.

A quick way to determine whether a reaction might be favorable is to assess whether the reaction leads to an overall dispersal of energy, which is the same thing as an increase in entropy.

Because energy naturally tends to disperse, a reaction that leads to an increase in entropy will likely occur, while a reaction that leads to a decrease in entropy will <u>not</u> likely occur.

Scientists consider this tendency of energy to disperse to be one of the central reasons for both physical and chemical changes. In other words, changes that result in energy spreading out tend to occur on their own—they are favored. This includes the cooling down of a hot pan and the burning of ignited gasoline. In both cases, there is a dispersal of energy to the environment.

The opposite holds true, too. Changes that result in the concentration of energy do *not* tend to occur—they are not favored. Heat from the room, for example, will never spontaneously move into a pan to heat it up. Likewise, low-energy exhaust molecules coming out of a car's tailpipe will not spontaneously come back together to form higher-energy gasoline molecules. The natural flow of energy is always a one-way trip from where it is concentrated to where it is less concentrated, or "spread out."

That energy tends to disperse is spelled out by the *second law of thermodynamics*, which can be paraphrased as follows:

Any process that happens by itself results in the net dispersal of energy. For example, heat naturally flows from a higher-temperature object to a lower-temperature object because, in doing so, energy is dispersed from where it is concentrated (a hot pan) to where it is spread out (the cooler kitchen).

Entropy, as discussed in Section 6.6, is a measure of this natural spreading of energy. Wherever there is a spreading of energy, there is a corresponding *increase* in entropy. Applied to chemistry, entropy helps us to answer a most fundamental question: If you take two materials and put them together, will they react to form new materials? If the reaction results in the dispersal of energy (an overall increase in entropy), then the answer is yes. Conversely, if the reaction results in the concentration of itself, such a reaction will occur only if an external source of energy is supplied to it.

Using this concept of entropy, you are now in a position to understand why exothermic reactions are self-sustaining—occurring on their own without need of external help—while most endothermic reactions need a continual prodding. Exothermic reactions spread energy out to the surroundings, much like a cooling hot pan. This is an increase in entropy; hence, exothermic reactions are favored to occur. An endothermic reaction, by contrast, requires that energy from the surroundings be absorbed by the reactants. This is a concentration of energy, which is counter to energy's natural tendency to disperse. Endothermic reactions, therefore, can progress from reactants to products only with the continual input of energy. But from where might this energy come? The answer is: from some self-sustaining exothermic reaction occurring elsewhere.

The classic example is photosynthesis, which is an endothermic reaction in which plants use solar energy to create carbohydrates and oxygen from carbon dioxide and water, as represented by this equation:

$$\text{Sunlight} + 6\,CO_2(g) + 6\,H_2O(g) \longrightarrow C_6H_{12}O_6(s) + 6\,O_2(g)$$

| Carbon | Water | Carbohydrate | Oxygen |

In photosynthesis, energy dispersed from the sun becomes contained within the carbohydrate and oxygen products, which, of course, are the primary fuels of living organisms (Figure 17.21). Likewise, most modern materials, such as plastics, synthetic fibers, pharmaceuticals, fertilizers, and metals such as iron and aluminum, are made or purified using endothermic reactions. Our ability to produce these new and useful materials has been the hallmark of modern chemistry. Creating these products, however, necessarily requires the input of energy, which we must obtain from some external source, such as electricity from a power plant that burns fossil or nuclear fuels.

FIGURE 17.21
The Sun is truly a "hothouse"—dispersing enormous amounts of energy from exothermic nuclear reactions. A tiny fraction of the Sun's energy is used to drive photosynthesis, which is vital for plants and plant-eating creatures like us.

CHECKPOINT
Sugar crystals form naturally within a supersaturated solution of sugar water. Does the formation of these crystals result in an increase or a decrease in entropy?

Was This Your Answer
The formation of these sugar crystals results in an increase in entropy. Your clue to an increase in entropy here is that the crystals form "on their own," a spontaneous process and thus one that must result in an entropy increase. Recall from Section 8.6 that energy is released when molecules come together to form a solid (the heat of freezing). This release of heat involves the spreading out of energy, which is, by definition, an increase in entropy.

While on the subject of the second law of thermodynamics, we would be remiss not to think about the close relationship between entropy and our psychological sense of time. That energy tends to spread out is part of our human experience. We *expect* a hot pan to cool, just as we *expect* hot gases to come out of an exhaust pipe. But what if we saw the reverse? For example, what would we think if we saw smoke moving *into* a smokestack? Or what if we saw a diver fly out of the water and rise upward to the diving board? If we were watching these energy-concentrating events on video, we would quickly conclude that the video was running backward. However, if in real life we actually saw such things—if we could survive the shock—we would immediately sense that time itself was running backward. Thus, when we watch smoke shoot up out of a smokestack and a diver diving into a pool, as shown in Figure 17.22, we have the sense that time is moving forward. The second law of thermodynamics, therefore, gives us our psychological sense of time; it is the "arrow of time."

FIGURE 17.22
As the diver dives, his potential energy is converted into kinetic energy. As he splashes into the water, this kinetic energy is spread to make the water molecules move faster and heat up just a little. Will this dissipated energy reconcentrate itself to push him back through the air to the diving platform? The second law of thermodynamics says no.

FYI There are examples of endothermic reactions that proceed spontaneously, absorbing heat from the environment. A classic example is the mixing of ions (salt) in water. In such cases, the environment disperses energy to the dissolving ions, which then spread this energy more widely into the volume of the solvent.

CHECKPOINT

The energy of a diver diving into a pool is dispersed as lots of moving water and a little heat after the diver hits the water. How, then, can the diver get back up to the platform?

Was this your answer?
As a living organism, the diver has a supply of biochemical energy, obtained ultimately from photosynthesis, that he can tap to climb upward against gravity to get back to the diving platform.

For instructor-assigned homework, go to www.masteringphysics.com

SUMMARY OF TERMS (KNOWLEDGE)

Activation energy The minimum energy required in order for a chemical reaction to proceed.

Avogadro's number The number of particles—6.02×10^{23}—contained in 1 mole of anything.

Bond energy The amount of energy required to pull two bonded atoms apart, which is the same as the amount of energy released when the two atoms are brought together into a bond.

Catalyst Any substance that increases the rate of a chemical reaction without itself being consumed by the reaction.

Chemical equation A representation of a chemical reaction in which reactants are shown to the left of an arrow that points to the products.

Endothermic Description of a chemical reaction in which there is a net absorption of energy.

Exothermic Description of a chemical reaction in which there is a net release of energy.

Formula mass The sum of the atomic masses of the elements in a chemical formula.

Law of mass conservation Matter is neither created nor destroyed during a chemical reaction; atoms merely rearrange, without any apparent loss or gain of mass, to form new molecules.

Molar mass The mass of 1 mole of a substance.

Products The new materials formed in a chemical reaction.

Reactants The reacting substances in a chemical reaction.

Reaction rate A measure of how quickly the concentration of products in a chemical reaction increases or the concentration of reactants decreases.

READING CHECK QUESTIONS (COMPREHENSION)

17.1 Chemical Equations

1. What is the purpose of coefficients in a chemical equation?

2. How many oxygen atoms are indicated on the right side of this balanced chemical equation?

$$4 \, Cr(s) + 3 \, O_2(g) \longrightarrow 2 \, Cr_2O_3(g)$$

3. Why is it important that a chemical equation be balanced?

4. Why is it important never to change a subscript in a chemical formula when balancing a chemical equation?

17.2 Counting Atoms and Molecules by Mass

5. Why don't equal masses of golf balls and Ping-Pong balls contain the same number of balls?

6. Why don't equal masses of carbon atoms and oxygen molecules contain the same number of particles?

7. What is the formula mass of nitrogen monoxide, NO, in atomic mass units?

8. If you had 1 mole of marbles, how many marbles would you have? How about 2 moles?

9. How many moles of water are there in 18 g of water?

10. How many molecules of water are there in 18 g of water?

11. Why is saying you have 1 mole of water molecules the same as saying you have 6.02×10^{23} water molecules?

12. What is the mass of an oxygen atom in atomic mass units?

17.3 Reaction Rates

13. Why don't all collisions between reactant molecules lead to product formation?

14. What generally happens to the rate of a chemical reaction with increasing temperature?

15. Which reactant molecules are the first to pass over the energy barrier?

16. What term is used to describe the minimum amount of energy required in order for a reaction to proceed?

17.4 Catalysts

17. What catalyst is effective in the destruction of atmospheric ozone, O_3?

18. What does a catalyst do to the energy of activation for a reaction?

19. What net effect does a chemical reaction have on a catalyst?

20. Why are catalysts so important to our economy?

17.5 Energy and Chemical Reactions

21. If it takes 436 kJ to break a bond, how many kilojoules are released when the same bond is formed?

22. What is released by an exothermic reaction?

23. What is absorbed by an endothermic reaction?

17.6 Chemical Reactions Are Driven by Entropy

24. As energy disperses, where does it go?

25. What is always increasing?

26. Why are exothermic reactions self-sustaining?

ACTIVITIES (HANDS-ON APPLICATION)

27. If you ever have the opportunity to play with an electric train set, be sure to smell the engine car after it has been operating. You will note a slight "electric smell." This is the smell of ozone gas, which is created as the oxygen in the air is zapped with electrical sparks. Why is this smell sometimes apparent during a lightning storm or when you pull your fuzzy sweater off in the dry winter? Is the formation of ozone from oxygen an endothermic or exothermic reaction?

28. An Alka-Seltzer antacid tablet reacts vigorously with water. But how does this tablet react to a solution of half water and half corn syrup? Propose an explanation involving the relationship between reaction speed and the frequency of molecular collisions.

29. Baker's yeast contains a biological catalyst known as *catalase*, which catalyzes the transformation of hydrogen peroxide, H_2O_2, into oxygen, O_2, and water, H_2O. Write a balanced equation for this reaction. Add a couple milliliters of 3% hydrogen peroxide to a glass containing a small amount of baker's yeast. What happens? Why?

Recall, from Section 15.8, that chemical bonds and intermolecular attractions are both consequences of the electric force, the difference being that chemical bonds are generally many times

stronger than molecule-to-molecule attractions. So, just as the formation and breaking of chemical bonds involve energy, so do the formation and breaking of molecular attractions. For molecule-to-molecule attractions, the amount of energy absorbed or released per gram of material is relatively small. Physical changes involving the formation or breaking of molecule-to-molecule attractions, therefore, are much safer to perform, which makes them more suitable for an out-of-laboratory activity. Experience the exothermic and endothermic nature of physical changes for yourself by performing the following two activities.

30. Hold some room-temperature water in the cupped palm of your hand over a sink. Pour an equal amount of room-temperature rubbing alcohol into the water. Is this mixing an exothermic or endothermic process? What's going on at the molecular level?

31. Add lukewarm water to two plastic cups. (Do *not* use insulating Styrofoam cups.) Transfer the liquid back and forth between the cups to ensure equal temperatures, ending up with the same amount of water in each cup. Add several tablespoons of table salt to one cup and stir. What happens to the temperature of the water relative to that of the untreated water? (Hold the cups up to your cheeks to tell.) Is this an exothermic or endothermic process? What's going on at the molecular level?

THINK AND SOLVE (MATHEMATICAL APPLICATION)

32. Show that there are 1.0×10^{22} carbon atoms in a 1-carat pure diamond that has a mass of 0.20 g.

33. How many gold atoms are there in a 5.00-g sample of pure gold, Au (197 amu)?

34. Show that 1 mole of $KClO_3$ contains 122.55 g.

35. Small samples of oxygen gas needed in the laboratory can be generated by a number of simple chemical reactions, such as

$$2\ KClO_3(s) \longrightarrow 2\ KCl(s) + 3\ O_2(g)$$

According to this balanced chemical equation, how many moles of oxygen gas are produced from the reaction of 2 moles of $KClO_3$ solid?

36. Small samples of oxygen gas needed in the laboratory can be generated by a number of simple chemical reactions, such as

$$2\ KClO_3(s) \longrightarrow 2\ KCl(s) + 3\ O_2(g)$$

What mass of oxygen (in grams) is produced when 122.55 g of $KClO_3$ (formula mass = 122.55 amu) takes part in this reaction?

37. Show that the formula mass of 2-propanol, C_3H_8O, is 60 amu, that the formula mass of propene, C_3H_6, is 42 amu, and that the formula mass of water, H_2O, is 18 amu.

38. How many grams of water, H_2O, and propene, C_3H_6, can be formed from the reaction of 6.0 g of 2-propanol, C_3H_8O?

$$C_3H_8O \longrightarrow C_3H_6 + H_2O$$

39. A 16-g sample of methane, CH_4, is combined with a 16-g sample of molecular oxygen, O_2, in a sealed container. Upon ignition, what is the maximum amount of carbon dioxide, CO_2, that can be formed?

$$CH_4 + 2\ O_2 \longrightarrow CO_2 + 2\ H_2O$$

40. Use the bond energies in Table 17.1 and the accounting format shown in Section 17.5 to determine whether these reactions are exothermic or endothermic:

$$H_2 + Cl_2 \longrightarrow 2\ HCl$$
$$2\ HC\equiv CH + 5\ O_2 \longrightarrow 4\ CO_2 + 2\ H_2O$$

41. Use the bond energies in Table 17.1 and the accounting format shown in Section 17.5 to determine whether these reactions are exothermic or endothermic:

(a)

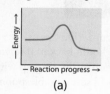

(b)

THINK AND RANK (ANALYSIS)

42. Rank the following in order of increasing number of atoms: (a) 52 g of vanadium, V; (b) 52 g of chromium, Cr; (c) 52 g of manganese, Mn.

43. Rank the following reaction profiles in order of increasing reaction speed.

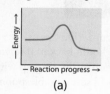 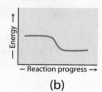

     (a)   (b)   (c)

44. Rank these covalent bonds in order of increasing bond strength: (a) $C\equiv C$, (b) $C=C$, (c) $C-C$.

45. Rank the following in order of increasing entropy: a deck of playing cards (a) at 45°C, new and unshuffled sitting in a room at 25°C; (b) at 233°C, new and unshuffled sitting in a room at 25°C; (c) at 25°C, used and shuffled sitting in a room at 25°C.

EXERCISES (SYNTHESIS)

46. Balance these equations:
    (a) ____$Fe(s)$ + ____$O_2(g)$ $\longrightarrow$ ____ $Fe_2O_3(s)$
    (b) ____$H_2(g)$ + ____$N_2(g)$ $\longrightarrow$ ____$NH_3(g)$
    (c) ____$Cl_2(g)$ + ____$KBr(aq)$ $\longrightarrow$ ____$Br_2(l)$ + ____$KCl\ (aq)$
    (d) ____$CH_4(g)$ + ____$O_2(g)$ $\longrightarrow$ ____$CO_2(g)$ + ____$H_2O(l)$

47. Balance these equations:
    (a) ____$Fe(s)$ + ____$S(s)$ $\longrightarrow$ ____$Fe_2S_3(s)$
    (b) ____$P_4(s)$ + ____$H_2(g)$ $\longrightarrow$ ____$PH_3(g)$
    (c) ____$NO(g)$ + ____$Cl_2(g)$ $\longrightarrow$ ____$NOCl(g)$
    (d) ____$SiCl_4(l)$ + ____$Mg(s)$ $\longrightarrow$ ____$Si(s)$ + ____$MgCl_2(s)$

48. Which of these two chemical equations is balanced?

$$2\,C_4H_{10}(g) + 13\,O_2(g) \longrightarrow 8\,CO_2(g) + 10\,H_2O(l)$$

$$4\,C_6H_7N_5O_{16}(s) + 19\,O_2(g) \longrightarrow 24\,CO_2(g)$$
$$+ 20\,NO_2(g) + 14\,H_2O(g)$$

Use the following illustration to answer Exercises 49–51.

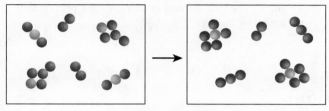

49. Assume the illustrations above are two frames of a movie—one from before the reaction and the other from after the reaction. How many diatomic molecules are represented in this movie?

50. There is an excess of at least one of the reactant molecules. Which one?

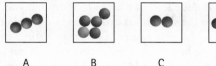

A   B   C   D   E

51. Which equation best describes each reaction shown above?
    (a) $2\,AB_2 + 2\,DCB_3 + B_2 \longrightarrow 2\,DBA_4 + 2\,CA_2$
    (b) $2\,AB_2 + 2\,CDA_3 + B_2 \longrightarrow 2\,C_2A_4 + 2\,DBA$
    (c) $2\,AB_2 + 2\,CDA_3 + A_2 \longrightarrow 2\,DBA_4 + 2\,CA_2$
    (d) $2\,BA_2 + 2\,DCA_3 + A_2 \longrightarrow 2\,DBA_4 + 2\,CA_2$

52. The reactants shown schematically on the left represent methane, CH_4, and water, H_2O. Write the full balanced chemical equation that is depicted.

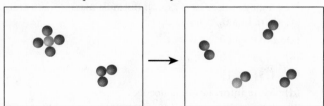

53. The reactants shown schematically on the left represent iron oxide, Fe_2O_3, and carbon monoxide, CO. Write the full balanced chemical equation that is depicted.

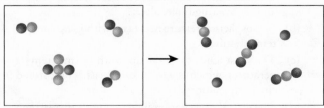

54. Which has more atoms: 17.031 g of ammonia, NH_3, or 72.922 g of hydrogen chloride, HCl?

55. How many moles of molecules are there in: (a) 28 g of nitrogen, N_2; (b) 32 g of oxygen, O_2; (c) 32 g of methane, CH_4; (d) 38 g of fluorine, F_2?

56. How many moles of atoms are there in: (a) 28 g of nitrogen, N_2; (b) 32 g of oxygen, O_2; (c) 16 g of methane, CH_4; (d) 38 g of fluorine, F_2?

57. What is the mass of a water molecule in atomic mass units?

58. What is the mass of a water molecule in grams?

59. Is it possible to have a sample of oxygen that has a mass of 14 amu? Explain.

60. Which has the greater mass: 1.204×10^{24} molecules of molecular hydrogen or 1.204×10^{24} molecules of water?

61. How many grams of gallium are there in a 145-g sample of gallium arsenide, GaAs?

62. How many atoms of arsenic are there in a 145-g sample of gallium arsenide, GaAs?

63. How is it possible for a jet airplane carrying 110 tons of jet fuel to emit 340 tons of carbon dioxide?

64. Does a refrigerator prevent or delay the spoilage of food? Explain.

65. Why does a glowing splint of wood burn only slowly in air but burst into flames when placed in pure oxygen?

66. Give two reasons heat is often added to chemical reactions performed in the laboratory.

67. Explain the connection between photosynthetic life on Earth and the ozone layer.

68. Does the ozone pollution from automobiles help alleviate the ozone hole over the South Pole? Defend your answer.

69. Chlorine is put into the atmosphere by volcanoes in the form of hydrogen chloride, HCl, but this form of chlorine does not remain in the atmosphere for very long. Why?

70. In the following reaction sequence for the catalytic formation of ozone from molecular oxygen, which chemical compound is the catalyst: nitrogen monoxide or nitrogen dioxide?

$$O_2 + 2\,NO \longrightarrow 2\,NO_2$$
$$2\,NO_2 \longrightarrow 2\,NO + 2\,O$$
$$2\,O + 2\,O_2 \longrightarrow 2\,O_3$$

71. Are the chemical reactions that take place in a disposable battery exothermic or endothermic? What evidence supports your answer? Is the reaction going on in a rechargeable battery while it is recharging exothermic or endothermic?

72. What role does entropy play in chemical reactions?

73. Why do exothermic reactions typically favor the formation of products?

74. Under what conditions will a hot pie not lose heat to its surroundings?

75. As the Sun shines on a snow-capped mountain, much of the snow sublimes instead of melts. How is this favored by entropy?

76. Exothermic reactions are favored because they release heat to the environment. Would an exothermic reaction be more favored or less favored if it were carried out within a superheated chamber?

77. Estimate whether entropy increases or deceases with the following reaction. Use data from Table 17.1 to confirm your estimation.

$$2\ C(s)\ +\ 3\ H_2(g)\ \longrightarrow\ C_2H_6(g)$$

78. According to the second law of thermodynamics, exothermic reactions, such as the burning of wood, are favored because they result in the dispersal of energy. Wood, however, will not spontaneously burn even when exposed to pure oxygen, O_2. Why not?

79. In the laboratory, endothermic reactions are usually preformed at elevated temperatures, while exothermic reactions are usually performed at lower temperatures. What are some possible reasons for this?

80. Wild plants readily grow "all by themselves," yet the molecules of the growing plant have *less* entropy than the materials used to make the plant. How is it possible for there to be this *decrease* in entropy for a process that occurs all by itself?

DISCUSSION QUESTIONS (EVALUATION)

81. Many people hear about atmospheric ozone depletion and wonder why we don't simply replace the ozone that has been destroyed. Knowing about chlorofluorocarbons and knowing how catalysts work, explain how this would not be a lasting solution.

82. Throughout the history of life on Earth, there have been at least six major mass extinctions. The one that killed off the dinosaurs occurred about 65 million years ago, and it is thought to have been the result of the impact of a large asteroid. The largest mass extinction of them all, however, was the Ordovician mass extinction, which occurred about 450 million years ago. The cause of this mass extinction is uncertain, but scientists have recently demonstrated how this extinction may have been initiated by an intense burst of gamma rays produced by the explosion of a nearby star. A burst as short as 10 s could have led to the loss of Earth's protective ozone layer, thereby exposing life on Earth to dangerous ultraviolet rays from our Sun. The probability of another nearby star exploding soon is quite low. But take a look around you. Scientists point to the sixth mass extinction as occurring right now. Discuss possible causes of this mass extinction. How long might this mass extinction take? What creatures might survive? Should humans do anything to minimize this mass extinction, or should we just accept it as a natural course of Earth's history?

READINESS ASSURANCE TEST (RAT)

If you have a good handle on this chapter, if you really do, then you should be able to score 7 out of 10 on this RAT. If you score less than 7, you need to study further before moving on.

Choose the BEST answer to each of the following.

1. What coefficients balance this equation?

   ____$P_4(s)$ + ____$H_2(g)$ $\longrightarrow$ ____$PH_3(g)$

   (a) 4, 2, 3
   (b) 1, 6, 4
   (c) 1, 4, 4
   (d) 2, 10, 8

2. What is the formula mass of sulfur dioxide, SO_2?

   (a) about 16 amu
   (b) about 32 amu
   (c) about 60 amu
   (d) about 64 amu

3. Which has the greatest number of atoms?

   (a) 28 g of nitrogen, N_2
   (b) 32 g of oxygen, O_2
   (c) 16 g of methane, CH_4
   (d) 38 g of fluorine, F_2

4. How many molecules of aspirin (formula mass = 180 amu) are there in a 0.250-g sample?

   (a) 6.02×10^{23}
   (b) 8.38×10^{20}
   (c) 1.51×10^{23}
   (d) More information is needed.

5. The yeast in bread dough feeds on sugar to produce carbon dioxide. Why does the dough rise faster in a warmer area?

   (a) There is a greater number of effective collisions among reacting molecules.
   (b) Atmospheric pressure decreases with increasing temperature.
   (c) The yeast tends to "wake up" with warmer temperatures, which is why baker's yeast is best stored in the refrigerator.
   (d) The rate of evaporation increases with increasing temperature.

6. What can you deduce about the activation energy of a reaction that takes billions of years to go to completion? How about a reaction that takes only fractions of a second?

   (a) The activation energy of both these reactions must be very low.

   (b) The activation energy of both these reactions must be very high.

   (c) The slow reaction must have a high activation energy, while the fast reaction must have a low activation energy.

   (d) The slow reaction must have a low activation energy, while the fast reaction must have a high activation energy.

7. What role do CFCs play in the catalytic destruction of ozone?

   (a) Ozone is destroyed upon binding to a CFC molecule that has been energized by ultraviolet light.

   (b) There is no strong scientific evidence that CFCs play a significant role in the catalytic destruction of ozone.

   (c) CFC molecules activate chlorine atoms into their catalytic action.

   (d) CFC molecules migrate to the upper stratosphere where they generate chlorine atoms upon being destroyed by ultraviolet light.

8. Is the synthesis of ozone, O_3, from oxygen, O_2, an example of an exothermic or endothermic reaction? Why?

   (a) exothermic, because ultraviolet light is emitted during its formation

   (b) endothermic, because ultraviolet light is emitted during its formation

   (c) exothermic, because ultraviolet light is absorbed during its formation

   (d) endothermic, because ultraviolet light is absorbed during its formation

9. How much energy, in kilojoules, is released or absorbed from the reaction of 1 mole of nitrogen, N_2, with 3 moles of molecular hydrogen, H_2, to form 2 moles of ammonia, NH_3? Consult Table 17.1 for bond energies.

   (a) $+899$ kJ/mol

   (b) -993 kJ/mol

   (c) $+80$ kJ/mol

   (d) -80 kJ/mol

10. How is it possible to cause an endothermic reaction to proceed when the reaction causes energy to become less dispersed?

    (a) The reaction should be placed in a vacuum.

    (b) The reaction should be cooled down.

    (c) The concentration of the reactants should be increased.

    (d) The reaction should be heated.

Answers to RAT

1. b, 2. d, 3. c, 4. b, 5. a, 6. c, 7. d, 8. d, 9. c, 10. d

18

CHAPTER 18

Two Classes of Chemical Reactions

D URING A chemical reaction, the atoms of the reactants change partners to form new materials we call products. As wood burns, for example, the atoms of cellulose molecules break away from each other in order to combine with the atoms of oxygen molecules to form carbon dioxide, water vapor, plus lots of heat.

In this chapter, we explore two main classes of chemical reactions: acid–base reactions and oxidation–reduction reactions. Acid–base reactions involve the transfer of *protons* from one reactant to another. These sorts of reactions within your stomach help you digest your food. They play a key role in global climate. Most consumer goods can trace their origins to acid–base chemical reactions. Oxidation–reduction reactions involve the transfer of one or more *electrons* from one reactant to another. The burning of wood is an oxidation–reduction reaction, as are the reactions your body uses to transform the food you eat into biochemical energy. Oxidation–reduction reactions are responsible for the rusting of a car. They are also the source of a battery's electric energy.

18.1 Acids Donate Protons; Bases Accept Them

LEARNING OBJECTIVE
Identify when a chemical behaves like an acid or a base.

EXPLAIN THIS Why are many pharmaceuticals treated with hydrogen chloride?

Mastering**PHYSICS**

TUTORIAL: The Nature of Acids and Bases

The term *acid* comes from the Latin *acidus*, which means "sour." The sour taste of vinegar and citrus fruits is due to the presence of acids. Acids are essential in the chemical industry. For example, more than 85 billion pounds of sulfuric acid are produced annually in the United States, making this the number-one manufactured chemical. Sulfuric acid is used to make fertilizers, detergents, paint dyes, plastics, pharmaceuticals, and storage batteries, as well as to produce iron and steel. It is so important in the manufacturing of goods that its production is considered a standard measure of a nation's industrial strength. Figure 18.1 shows only a few of the acids we commonly encounter.

(a)
(b)
(c)
(d)

FIGURE 18.1
Examples of acids. (a) Citrus fruits contain many types of acids, including ascorbic acid, $C_6H_8O_6$, which is vitamin C. (b) Vinegar contains acetic acid, $C_2H_4O_2$, and can be used to preserve foods. (c) Many toilet bowl cleaners are formulated with hydrochloric acid, HCl. (d) All carbonated beverages contain carbonic acid, H_2CO_3; many also contain phosphoric acid, H_3PO_4.

Bases are characterized by their bitter taste and slippery feel. Interestingly, bases themselves are not slippery. Rather, they cause skin oils to transform into slippery solutions of soap. Most commercial preparations for unclogging drains contain sodium hydroxide, NaOH (also known as lye), which is extremely basic and hazardous when concentrated. Bases are also heavily used in industry. Each year in the United States, about 25 billion pounds of sodium hydroxide are manufactured for use in the production of various chemicals and in the pulp and paper industry. Solutions containing bases are often called *alkaline*, a term derived from the Arabic *al-qali* ("the ashes"). Ashes are slippery when wet because of the presence of the base potassium carbonate, K_2CO_3. Figure 18.2 shows some familiar bases.

Acids and bases may be defined in several ways. For our purposes, an appropriate definition is the one suggested in 1923 by the Danish chemist Johannes Brønsted (1879–1947) and the English chemist Thomas Lowry (1874–1936).

(a)
(b)
(c)
(d)

FIGURE 18.2
Examples of bases. (a) Reactions involving sodium bicarbonate, $NaHCO_3$, cause baked goods to rise. (b) Ashes contain potassium carbonate, K_2CO_3. (c) Soap is made by reacting bases with animal or vegetable oils. The soap itself, then, is slightly alkaline. (d) Powerful bases, such as sodium hydroxide, NaOH, are used in drain cleaners.

FIGURE 18.3
The hydronium ion's positive charge is a consequence of the extra proton this molecule has acquired. Hydronium ions, which play a role in many acid–base reactions, are polyatomic ions, which, as mentioned in Section 15.2, are molecules that carry a net electric charge.

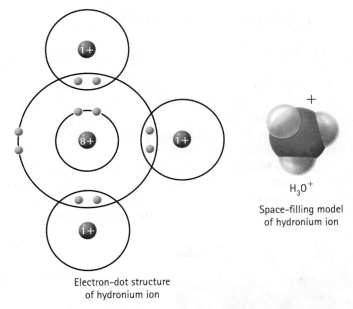

Space-filling model
of hydronium ion

Electron-dot structure
of hydronium ion

Total protons	11+
Total electrons	10−
Net charge	1+

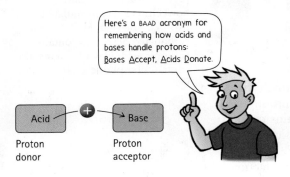

Here's a BAAD acronym for remembering how acids and bases handle protons: Bases Accept, Acids Donate.

Acid → + → Base

Proton
donor

Proton
acceptor

The hydrogen ion, H⁺, does not readily exist in water because any hydrogen ion formed is quickly picked up by a water molecule and transformed to the hydronium ion, H₃O⁺.

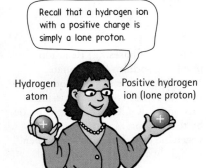

Recall that a hydrogen ion with a positive charge is simply a lone proton.

Hydrogen
atom

Positive hydrogen
ion (lone proton)

In the Brønsted–Lowry definition, an **acid** is any chemical that donates a hydrogen ion, H^+, and a **base** is any chemical that accepts a hydrogen ion. Recall that a hydrogen atom consists of one electron surrounding a one-proton nucleus. A hydrogen ion, H^+, formed from the loss of an electron, therefore, is nothing more than a lone proton. Thus, it is also sometimes said that an acid is a chemical that donates a proton and a base is a chemical that accepts a proton.

Consider what happens when hydrogen chloride is mixed into water:

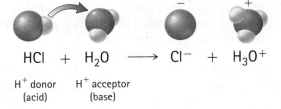

$$HCl + H_2O \longrightarrow Cl^- + H_3O^+$$

H⁺ donor   H⁺ acceptor
(acid)   (base)

Hydrogen chloride donates a hydrogen ion to one of the nonbonding electron pairs on a water molecule, resulting in a third hydrogen bonded to the oxygen. In this case, hydrogen chloride behaves as an acid (proton donor) and water behaves as a base (proton acceptor). The products of this reaction are a chloride ion and a **hydronium ion**, H_3O^+, which, as Figure 18.3 shows, is a water molecule with an extra proton.

When added to water, ammonia behaves as a base as its nonbonding electrons (Section 15.1) accept a hydrogen ion from water, which, in this case, behaves as an acid:

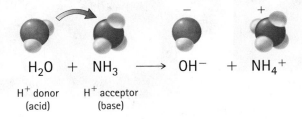

$$H_2O + NH_3 \longrightarrow OH^- + NH_4^+$$

H⁺ donor   H⁺ acceptor
(acid)   (base)

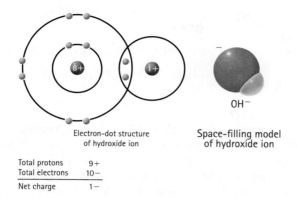

Electron-dot structure
of hydroxide ion

Space-filling model
of hydroxide ion

Total protons	9+
Total electrons	10−
Net charge	1−

FIGURE 18.4
Hydroxide ions have a net negative
charge, which is a consequence of
having lost a proton. Like hydronium ions, they play a part in many
acid–base reactions.

This reaction results in the formation of an ammonium ion and a **hydroxide
ion**, which, as shown in Figure 18.4, is a water molecule without the nucleus of
one of the hydrogen atoms.

An important aspect of the Brønsted–Lowry definition is that it uses a *behavior* to define a substance as an acid or a base. We say, for example, that hydrogen
chloride *behaves* as an acid when mixed with water, which *behaves* as a base.
Similarly, ammonia *behaves* as a base when mixed with water, which under this
circumstance *behaves* as an acid. Because acid–base is seen as a behavior, there
is really no contradiction when a chemical like water behaves as a base in one
instance but as an acid in another instance. By analogy, consider yourself. You
are who you are, but your behavior changes depending on whom you are with.
Likewise, it is a chemical property of water to behave as a base (to accept H^+)
when mixed with hydrogen chloride and as an acid (to donate H^+) when mixed
with ammonia.

The products of an acid–base reaction can also behave as acids or as bases.
An ammonium ion, for example, may donate a hydrogen ion back to a hydroxide ion to re-form ammonia and water:

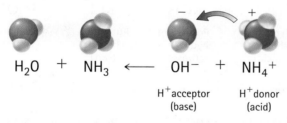

$$H_2O \ + \ NH_3 \ \longleftarrow \ OH^- \ + \ NH_4^+$$

Forward and reverse acid–base reactions proceed simultaneously and can therefore be represented as occurring at the same time by using two oppositely facing
arrows:

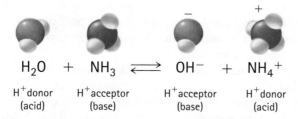

$$H_2O \ + \ NH_3 \ \rightleftharpoons \ OH^- \ + \ NH_4^+$$

When the equation is viewed from left to right, the ammonia behaves as a base
because it accepts a hydrogen ion from the water, which therefore acts as an
acid. Viewed in the reverse direction, the equation shows that the ammonium
ion behaves as an acid because it donates a hydrogen ion to the hydroxide ion,
which therefore behaves as a base.

How we behave depends on who
we're with. Likewise for chemicals.

CHECKPOINT

Identify the acid or base behavior of each participant in the reaction:

$$H_2PO_4^- + H_3O^+ \rightleftharpoons H_3PO_4 + H_2O$$

Was this your answer?

In the forward reaction (left to right), $H_2PO_4^-$ gains a hydrogen ion to become H_3PO_4. In accepting the hydrogen ion, $H_2PO_4^-$ is behaving as a base. It gets the hydrogen ion from the H_3O^+, which is behaving as an acid. In the reverse direction, H_3PO_4 loses a hydrogen ion to become $H_2PO_4^-$ and is thus behaving as an acid. The recipient of the hydrogen ion is H_2O, which is behaving as a base as it transforms to H_3O^+.

A Salt Is the Ionic Product of an Acid–Base Reaction

In everyday language, the word *salt* implies sodium chloride, NaCl, table salt. In the language of chemistry, however, **salt** is a general term meaning any ionic compound formed from the reaction between an acid and a base. Hydrogen chloride and sodium hydroxide, for example, react to produce the salt sodium chloride and water:

$$HCl \quad + \quad NaOH \quad \longrightarrow \quad NaCl \quad + \quad H_2O$$

Hydrogen   Sodium   Sodium   Water
chloride   hydroxide   chloride
(acid)   (base)   (salt)

Similarly, the reaction between hydrogen chloride and potassium hydroxide yields the salt potassium chloride and water:

$$HCl \quad + \quad KOH \quad \longrightarrow \quad KCl \quad + \quad H_2O$$

Hydrogen   Potassium   Potassium   Water
chloride   hydroxide   chloride
(acid)   (base)   (salt)

Potassium chloride is the main ingredient in "salt-free" table salt, as noted in Figure 18.5.

Salts are generally far less corrosive than the acids and bases from which they are formed. A corrosive chemical has the power to disintegrate a material or wear away its surface. Hydrogen chloride is a remarkably corrosive acid, which makes it useful for cleaning toilet bowls and etching metal surfaces. Sodium hydroxide is a very corrosive base used for unclogging drains. Mixing hydrogen chloride and sodium hydroxide together in equal portions, however, produces an aqueous solution of sodium chloride—salt water, which is not nearly as destructive as either starting material.

There are as many salts as there are acids and bases. Sodium cyanide, NaCN, is a deadly poison. "Saltpeter," which is potassium nitrate, KNO_3, is useful as a fertilizer and in the formulation of gunpowder. Calcium chloride, $CaCl_2$, is commonly used to de-ice walkways, and sodium fluoride, NaF, helps prevent tooth decay. The acid–base reactions forming these salts are shown in Table 18.1.

The reaction between an acid and a base is called a **neutralization** reaction. As can be seen in the color-coding of the neutralization reactions in Table 18.1, the positive ion of a salt comes from the base and the negative ion comes from the acid. The remaining hydrogen and hydroxide ions join to form water.

FIGURE 18.5

"Salt-free" table salt substitutes contain potassium chloride in place of sodium chloride. Caution is advised in using these products, however, because excessive quantities of potassium salts can lead to serious illness. Furthermore, sodium ions are a vital component of our diet and should never be totally excluded. For a good balance of these two important ions, you might inquire about commercially available half-and-half mixtures of sodium chloride and potassium chloride, such as the one shown here.

TABLE 18.1   ACID–BASE REACTIONS AND THE SALTS FORMED

Acid		Base		Salt		Water
HCN Hydrogen cyanide	+	NaOH Sodium hydroxide	$\longrightarrow$	NaCN Sodium cyanide	+	H_2O
HNO_3 Nitric acid	+	KOH Potassium hydroxide	$\longrightarrow$	KNO_3 Potassium nitrate	+	H_2O
2 HCl Hydrogen chloride	+	$Ca(OH)_2$ Calcium hydroxide	$\longrightarrow$	$CaCl_2$ Calcium chloride	+	$2H_2O$
HF Hydrogen fluoride	+	NaOH Sodium hydroxide	$\longrightarrow$	NaF Sodium fluoride	+	H_2O

FYI What makes one acid strong and another weak? Briefly, it involves the stability of the negative ion that remains after the proton has been donated. Hydrogen chloride is a strong acid because the chloride ion can accommodate the negative charge rather well. Acetic acid, however, is a weaker acid because the resulting oxygen ion is less able to accommodate the negative charge.

Not all neutralization reactions result in the formation of water. In the presence of hydrogen chloride, for example, the drug pseudoephedrine behaves as a base by accepting H^+ from a hydrogen chloride. The negative Cl^- then joins the pseudoephedrine–H^+ ion to form the salt pseudoephedrine hydrochloride, which is a nasal decongestant, shown in Figure 18.6. This salt is soluble in water and can be absorbed through the digestive system.

CHECKPOINT

Is a neutralization reaction best described as a physical change or a chemical change?

Was this your answer?
New chemicals are formed during a neutralization reaction, meaning the reaction is a chemical change.

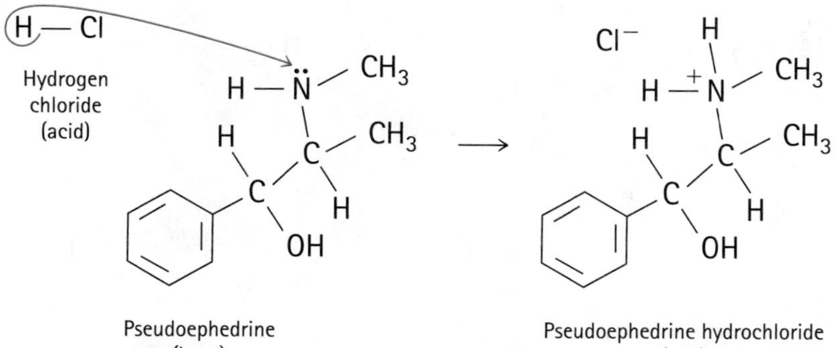

FIGURE 18.6
Hydrogen chloride and pseudoephedrine react to form the salt *pseudoephedrine hydrochloride*, which, because of its solubility in water, is readily absorbed into the body.

18.2 Relative Strengths of Acids and Bases

EXPLAIN THIS Why is hydrogen chloride, HCl, such a strong acid?

In general, the stronger an acid, the more readily it donates hydrogen ions. Likewise, the stronger a base, the more readily it accepts hydrogen ions. An example of a strong acid is hydrogen chloride, HCl, and an example of a strong base is sodium hydroxide, NaOH. The corrosiveness of these materials is a result of their strength.

LEARNING OBJECTIVE
Describe how the strength of an acid or base affects the number of ions in solution.

Mastering**PHYSICS**
TUTORIAL: Strong and Weak Acids and Bases

MasteringPHYSICS

VIDEO: Some Acids and Bases
Are Stronger Than Others

One way to assess the strength of an acid or base is to measure how much of it remains after it has been added to water. If little remains, the acid or base is strong. If a lot remains, the acid or base is weak. To illustrate this concept, consider what happens when the strong acid hydrogen chloride is added to water and what happens when the weak acid acetic acid, $C_2H_4O_2$ (the active ingredient of vinegar), is added to water.

Being an acid, hydrogen chloride donates hydrogen ions to water, forming chloride ions and hydronium ions. Because HCl is such a strong acid, nearly all of it is converted to these ions, as is shown in Figure 18.7.

Because acetic acid is a weak acid, it has much less tendency to donate hydrogen ions to water. When this acid is dissolved in water, only a small portion of the acetic acid molecules are converted to ions, a process that occurs as the polar O—H bonds are broken (the C—H bonds of acetic acid are unaffected by the water because of their nonpolarity). The majority of acetic acid molecules remain intact in their original non-ionized form, as shown in Figure 18.8.

Figures 18.7 and 18.8 show the submicroscopic behavior of strong and weak acids in water. As molecules and ions are too small to see, how then does a chemist measure the strength of an acid? One way is by measuring a solution's

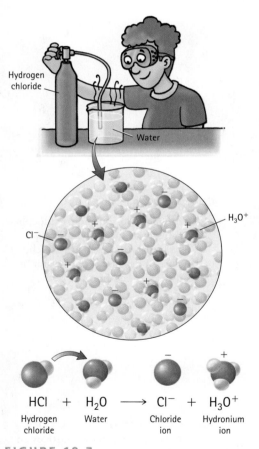

$$HCl \quad + \quad H_2O \quad \longrightarrow \quad Cl^- \quad + \quad H_3O^+$$
Hydrogen   Water   Chloride   Hydronium
chloride   ion   ion

FIGURE 18.7
Immediately after gaseous hydrogen chloride is added to water, it reacts with the water to form hydronium ions and chloride ions. That very little HCl remains (none shown here) lets us know that HCl acts as a strong acid.

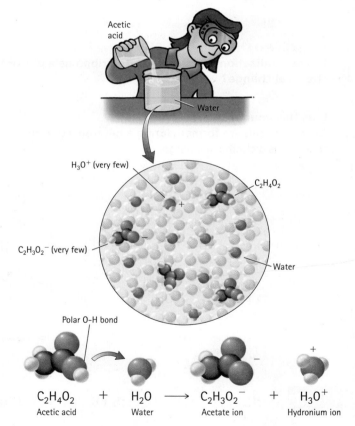

$$C_2H_4O_2 \quad + \quad H_2O \quad \longrightarrow \quad C_2H_3O_2^- \quad + \quad H_3O^+$$
Acetic acid   Water   Acetate ion   Hydronium ion

FIGURE 18.8
When liquid acetic acid is added to water, only a few acetic acid molecules react with water to form ions. The majority of the acetic acid molecules remain in their non-ionized form, which implies that acetic acid is a weak acid.

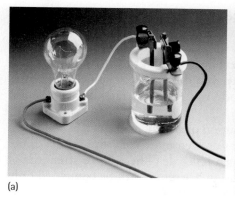
(a)

(b)

(c)

ability to conduct an electric current, as Figure 18.9 illustrates. Pure water contains practically no ions to conduct electricity. When a strong acid is dissolved in water, many ions are generated, as indicated in Figure 18.7. The presence of these ions allows for the flow of a large electric current. A weak acid dissolved in water generates only a few ions, as indicated in Figure 18.8. The presence of fewer ions means there can be only a small electric current.

This same trend is seen with strong and weak bases. Strong bases, for example, tend to accept hydrogen ions more readily than weak bases do. In solution, a strong base allows the flow of a large electric current, and a weak base allows the flow of a small electric current.

FIGURE 18.9
(a) The pure water in this circuit cannot conduct electricity because it contains practically no ions. The lightbulb in the circuit therefore remains unlit. (b) Because HCl is a strong acid, nearly all of its molecules break apart in water, giving a high concentration of ions, which can conduct an electric current that lights the bulb. (c) Acetic acid, $C_2H_4O_2$, is a weak acid; in water, only a small portion of its molecules break up into ions. Because fewer ions are generated, only a weak current exists, and the bulb is therefore dimmer.

CHECKPOINT
According to the aqueous solutions illustrated here, which is the stronger base: NH$_3$ or NaOH?

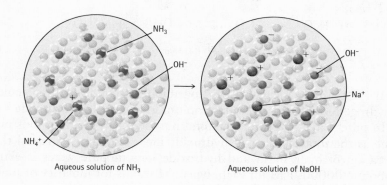

Aqueous solution of NH_3 Aqueous solution of NaOH

Was this your answer?
The solution on the right contains the greater number of ions, meaning that sodium hydroxide, NaOH, is the stronger base. Ammonia, NH_3, is the weaker base, indicated by the relatively few ions in the solution on the left.

Just because an acid or base is strong doesn't mean a solution of that acid or base is corrosive. The corrosive action of an acidic solution is caused by the hydronium ions rather than by the acid that generated those hydronium ions. Similarly, the corrosive action of a basic solution results from the hydroxide ions it contains, regardless of the base that generated those hydroxide ions. A *very* dilute solution of a strong acid or a strong base may have little corrosive

LEARNING OBJECTIVE
Calculate the pH of a solution given the hydronium ion concentration.

action because in such solutions there are only a few hydronium or hydroxide ions. (Almost all the molecules of the strong acid or base break up into ions, but because the solution is dilute, only a few acid or base molecules are present to begin with. As a result, there are only a few hydronium or hydroxide ions.) You shouldn't be too alarmed, therefore, when you discover that some toothpastes are formulated with small amounts of sodium hydroxide, one of the strongest bases known.

On the other hand, a concentrated solution of a weak acid, such as the acetic acid in vinegar, may be just as corrosive as or even more corrosive than a dilute solution of a strong acid, such as hydrogen chloride. The relative strengths of two acids in solution or two bases in solution, therefore, can be compared only when the two solutions have the same concentration.

18.3 Acidic, Basic, and Neutral Solutions

EXPLAIN THIS Why can't water be absolutely pure?

A substance whose ability to behave as an acid is about the same as its ability to behave as a base is said to be **amphoteric**. Water is a good example. Because it is amphoteric, water can react with itself. In behaving as an acid, a water molecule donates a hydrogen ion to a neighboring water molecule, which, in accepting the hydrogen ion, is behaving as a base. This reaction produces a hydroxide ion and a hydronium ion, which react together to re-form the water molecule:

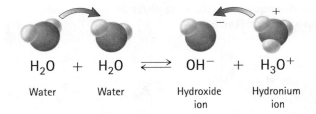

$$H_2O \quad + \quad H_2O \quad \rightleftharpoons \quad OH^- \quad + \quad H_3O^+$$

Water   Water   Hydroxide   Hydronium
                                  ion   ion

When a water molecule gains a hydrogen ion, a second water molecule must lose a hydrogen ion. So for every hydronium ion formed, a hydroxide ion also forms. In pure water, therefore, the total number of hydronium ions must be the same as the total number of hydroxide ions. Experiments reveal that the concentration of hydronium and hydroxide ions in pure water is extremely low—about 0.0000001 M for each, where M stands for molarity or moles per liter (Section 16.3). Water by itself, therefore, is a very weak acid as well as a very weak base, as evidenced by the unlit lightbulb in Figure 18.9a.

CHECKPOINT
Do water molecules react with one another?

Was this your answer?
Yes, but not to any large extent. When they do react, they form hydronium and hydroxide ions. (Note: Make sure you understand this point because it serves as a basis for the rest of this section.)

Further experiments reveal an interesting rule pertaining to the concentrations of hydronium and hydroxide ions in any solution that contains water. The concentration of hydronium ions in any aqueous solution multiplied by the concentration of the hydroxide ions in the solution always equals the constant K_w, which is a very, very small number:

Concentration H_3O^+ × concentration OH^- = K_w = 0.00000000000001

Concentration is usually given as molarity, which is indicated by abbreviating this equation using brackets:

$$[H_3O^+] \times [OH^-] = K_w = 0.00000000000001$$

The brackets mean this equation is read "the molarity of H_3O^+ times the molarity of OH^- equals K_w." Writing in scientific notation, we have

$$[H_3O^+][OH^-] = K_w = 1.0 \times 10^{-14}$$

For pure water, the value of K_w is the concentration of hydronium ions, 0.0000001 M, multiplied by the concentration of hydroxide ions, 0.0000001 M, which can be written in scientific notation as

$$[1.0 \times 10^{-7}][1.0 \times 10^{-7}] = K_w = 1.0 \times 10^{-14}$$

The constant value of K_w is quite important because it means that *no matter what is dissolved in the water*, the product of the hydronium-ion and hydroxide-ion concentrations always equals 1.0×10^{-14}. So if the concentration of H_3O^+ goes up, the concentration of OH^- must go down, and the product of the two remains 1.0×10^{-14}.

CHECKPOINT

1. In pure water, the hydroxide-ion concentration is 1.0×10^{-7} M. What is the hydronium-ion concentration?
2. What is the concentration of hydronium ions in a solution if the concentration of hydroxide ions is 1.0×10^{-3} M?

Were these your answers?

1. 1.0×10^{-7} M, because in pure water $[H_3O^+] = [OH^-]$
2. 1.0×10^{-11} M, because $[H_3O^+][OH^-]$ must equal $1.0 \times 10^{-14} = K_w$

Any solution containing an equal number of hydronium and hydroxide ions is said to be **neutral**. Pure water is an example of a neutral solution—not because it contains so few hydronium or hydroxide ions, but because it contains equal numbers of these ions. A neutral solution is also formed when equal quantities of acid and base are combined, which explains why acids and bases are said to *neutralize* each other.

The balance of hydronium and hydroxide ions in a neutral solution is upset by adding either an acid or a base. Add an acid, and the water will react with that acid to produce more hydronium ions. Many of these additional hydronium ions neutralize hydroxide ions, which then become fewer. The final result is that the hydronium-ion concentration is higher than the hydroxide-ion concentration. Such a solution is said to be **acidic**.

Add a base to water, and the reverse happens. The water will react with that base to produce more hydroxide ions. Many of these additional hydroxide ions neutralize hydronium ions, which then become fewer. The final result is that the hydronium-ion concentration is lower than the hydroxide-ion concentration. Such a solution is said to be **basic**, or sometimes *alkaline*. This behavior is summarized in Figure 18.10.

FIGURE 18.10
The relative concentrations of hydronium and hydroxide ions determine whether a solution is acidic, basic, or neutral.

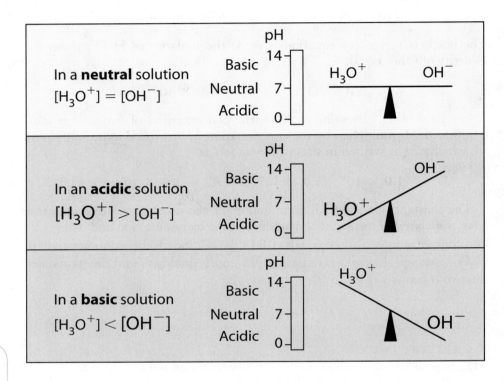

FYI The outer surface of hair is made of microscopic scale-like structures called cuticles that, like window shutters, can open and close. Alkaline solutions cause the cuticles to open up, which makes the hair "porous." Acidic solutions cause the cuticles to close down, which makes the hair "resistant." A beautician can control how long hair retains artificial coloring by modifying the pH of the hair-coloring solution. With an acidic solution, the cuticles close shut so that the dye binds only to the outside of each shaft of hair. This results in a temporary hair coloring, which may come off with the next hair washing. With an alkaline solution, the dye can penetrate through the cuticles into the hair for a more permanent effect.

CHECKPOINT

How does adding ammonia, NH$_3$, to water make a basic solution when there are no hydroxide ions in the formula for ammonia?

Was this your answer?

Ammonia indirectly increases the hydroxide-ion concentration by reacting with water:

$$NH_3 + H_2O \longrightarrow NH_4^+ + OH^-$$

This reaction raises the hydroxide-ion concentration, which has the effect of lowering the hydronium-ion concentration. With the hydroxide-ion concentration now higher than the hydronium-ion concentration, the solution is basic.

The pH Scale Is Used to Describe Acidity

Mastering**PHYSICS**
TUTORIAL: The pH Scale

The *pH scale* is a numeric scale used to express the acidity of a solution. Mathematically, **pH** is equal to the negative logarithm of the hydronium-ion concentration:

$$pH = -\log[H_3O^+]$$

Note again that brackets are used to represent molar concentrations, meaning $[H_3O^+]$ is read "the molar concentration of hydronium ions." For understanding the logarithm function, see Figuring Physical Science on page 466.

Consider a neutral solution that has a hydronium-ion concentration of 1.0×10^{-7} M. To find the pH of this solution, we first take the logarithm of this value, which is -7 (see Figuring Physical Science). The pH is by definition the negative of this value, which means $-(-7) = 7$. Hence, in a neutral solution, in which the hydronium-ion concentration equals 1.0×10^{-7}M, the pH is 7.

Acidic solutions have pH values less than 7. For an acidic solution in which the hydronium-ion concentration is 1.0×10^{-4} M, for example, pH = $-\log(1.0 \times 10^{-4}) = 4$. The more acidic a solution is, the higher its hydronium-ion concentration and the lower its pH.

Basic solutions have pH values greater than 7. For a basic solution in which the hydronium-ion concentration is 1.0×10^{-8} M, for example, pH = $-\log(1.0 \times 10^{-8}) = 8$. The more basic a solution is, the lower its hydronium-ion concentration and the higher its pH.

Figure 18.11 shows typical pH values of some familiar solutions, and Figure 18.12 shows two common ways of determining pH values.

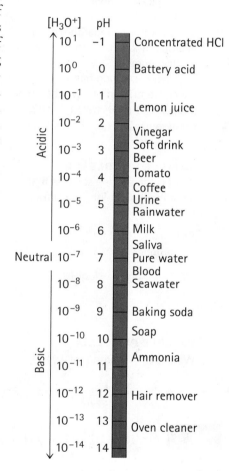

$[H_3O^+]$	pH	
10^1	-1	Concentrated HCl
10^0	0	Battery acid
10^{-1}	1	
		Lemon juice
10^{-2}	2	Vinegar
10^{-3}	3	Soft drink
		Beer
10^{-4}	4	Tomato
		Coffee
10^{-5}	5	Urine
		Rainwater
10^{-6}	6	Milk
		Saliva
10^{-7}	7	Pure water
		Blood
10^{-8}	8	Seawater
10^{-9}	9	Baking soda
10^{-10}	10	Soap
10^{-11}	11	Ammonia
10^{-12}	12	Hair remover
10^{-13}	13	Oven cleaner
10^{-14}	14	

Acidic / Neutral / Basic

FIGURE 18.11
The pH values of some common solutions.

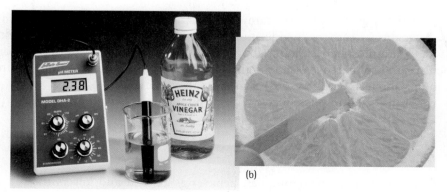

(a)

(b)

FIGURE 18.12
(a) The pH of a solution can be measured electronically using a pH meter. (b) A rough estimate of the pH of a solution can be obtained with litmus paper, which is coated with a dye that changes color with pH.

FIGURING PHYSICAL SCIENCE

Logarithms and pH

The logarithm of a number can be found on any scientific calculator by keying in the number and pressing the [log] button. The calculator finds the power to which 10 is raised to give the number. The logarithm of 10^2, for example, is 2 because that is the power to which 10 is raised to give the number 10^2. If you know that 10^2 is equal to 100, then you'll understand that the logarithm of 100 also is 2. Check this out on your calculator. Similarly, the logarithm of 1000 is 3 because 10 raised to the third power, 10^3, equals 1000. (Note: We speak here of the base-10 logarithm, not the natural logarithm of base e.)

Any positive number, including a very small one, has a logarithm. The logarithm of 0.0001, which equals 10^{-4}, for example, is −4 (the power to which 10 is raised to equal this number).

SAMPLE PROBLEM 1
What is the logarithm of 0.01?

Solution:
The number 0.01 is 10^{-2}, the logarithm of which is −2 (the power to which 10 is raised).

The concentration of hydronium ions in most solutions is typically much less than 1 M. Recall, for example, that in neutral water the hydronium-ion concentration is 0.0000001 M (10^{-7} M). The logarithm of any number smaller than 1 (but greater than zero) is a negative number. The definition of pH includes the minus sign so as to transform the logarithm of the hydronium-ion concentration to a positive number.

When a solution has a hydronium-ion concentration of 1 M, the pH is 0 because 1 M $= 10^0$ M. A 10 M solution has a pH of −1 because 10 M $= 10^1$ M.

SAMPLE PROBLEM 2
What is the pH of a solution that has a hydronium-ion concentration of 0.001 M?

Solution:
The number 0.001 is 10^{-3}, so

$$pH = -\log[H_3O^+]$$
$$= -\log 10^{-3}$$
$$= -(-3) = 3$$

SAMPLE PROBLEM 3
What is the logarithm of 10^5?

Solution:
"What is the logarithm of 10^5?" can be rephrased as "To what power is 10 raised to give the number 10^5?" The answer is 5.

SAMPLE PROBLEM 4
What is the logarithm of 100,000?

Solution:
You should know that 100,000 is the same as 10^5. Thus the logarithm of 100,000 is 5.

SAMPLE PROBLEM 5
What is the pH of a solution with a hydronium-ion concentration of 10^{-9} M? Is this solution acidic, basic, or neutral?

Solution:
The pH is 9, which means this is a basic solution:

$$pH = -\log[H_3O^+]$$
$$= -\log 10^{-9}$$
$$= -(-9)$$
$$= 9$$

LEARNING OBJECTIVE
Discuss how the pH of rain and the oceans is affected by atmospheric carbon dioxide.

FYI Above temperatures of 374°C and pressures of 218 atm, water transforms into a state of matter known as a supercritical fluid, which resembles both a liquid and a gas. For a neutral solution of supercritical water, the pH equals about 2, which means that it is highly corrosive. Research is under way to learn how supercritical water might be used to destroy toxic chemicals, such as chemical warfare agents.

18.4 Acidic Rain and Basic Oceans

EXPLAIN THIS How does burning fossil fuels lower the pH of the ocean?

As previously mentioned, rainwater is naturally acidic. One source of this acidity is carbon dioxide, the same gas that gives fizz to soda drinks. The atmosphere contains about 829 billion tons of CO_2, most of it from such natural sources as volcanoes and decaying organic matter but a growing amount (about 230 billion tons) from human activities.

Water in the atmosphere reacts with carbon dioxide to form *carbonic acid*:

$$CO_2(g) + H_2O(\ell) \longrightarrow H_2CO_3(aq)$$

| Carbon dioxide | Water | Carbonic acid |

Carbonic acid, as its name implies, behaves as an acid and lowers the pH of water. The CO_2 in the atmosphere brings the pH of rainwater to about 5.6—noticeably below the neutral pH value of 7. Because of local fluctuations, the normal pH of rainwater varies between 5 and 7. This natural acidity of rainwater may accelerate the erosion of land, and under certain circumstances it can lead to the formation of underground caves.

By convention, *acid rain* is a term used for rain with a pH lower than 5. Acid rain is created when airborne pollutants, such as sulfur dioxide, are absorbed by atmospheric moisture. Sulfur dioxide is readily converted to sulfur trioxide, which reacts with water to form *sulfuric acid*:

$$2\ SO_2(g)\ +\ O_2(g)\ \longrightarrow\ SO_3(g)$$
Sulfur   Oxygen   Sulfur
dioxide   trioxide

$$SO_3(g)\ +\ H_2O(\ell)\ \longrightarrow\ H_2SO_4(aq)$$
Sulfur   Water   Sulfuric
trioxide   acid

Each year about 20 million tons of SO_2 are released into the atmosphere by the combustion of sulfur-containing coal and oil. Sulfuric acid is much stronger than carbonic acid, and, as a result, rain laced with sulfuric acid eventually corrodes metal, paint, and other exposed substances. Each year, the damage costs billions of dollars. The cost to the environment is also high (Figure 18.13). Many rivers and lakes receiving acid rain become less capable of sustaining life. Much vegetation that receives acid rain doesn't survive. This is particularly evident in heavily industrialized regions.

FYI Acid rain remains a serious problem in many regions of the world. Significant progress, however, has been made toward fixing the problem. In the United States, for example, sulfur dioxide and nitrogen oxide emissions have been reduced by nearly half since 1980. Also, in 2005 the EPA implemented the Clean Air Interstate Rule (CAIR), which is designed to reduce levels of these pollutants even further, especially for areas downwind of heavily industrialized regions.

Mastering**PHYSICS**

VIDEO: Rainwater is Acidic and Ocean Water is Basic

VIDEO: Beach Sand Composition

CHECKPOINT

When sulfuric acid, H_2SO_4, is added to water, what makes the resulting aqueous solution corrosive?

Was this your answer?
Because H_2SO_4 is a strong acid, it readily forms hydronium ions when dissolved in water. Hydronium ions are responsible for the corrosive action.

(b)

FIGURE 18.13
(a) These two photographs show the same obelisk in New York City's Central Park before and after the effects of acid rain. (b) Many forests downwind from heavily industrialized areas, such as in the northeastern United States and in Europe, have been noticeably hard-hit by acid rain.

(a)

FIGURE 18.14
(a) The damaging effects of acid rain do not appear in bodies of fresh water lined with calcium carbonate, which neutralizes any acidity.
(b) Lakes and rivers lined with inert materials are not protected.

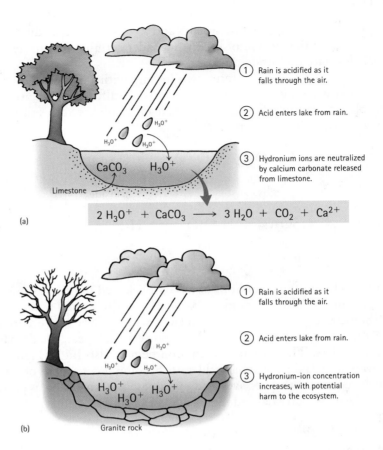

(a)
① Rain is acidified as it falls through the air.

② Acid enters lake from rain.

③ Hydronium ions are neutralized by calcium carbonate released from limestone.

$$2\,H_3O^+ + CaCO_3 \longrightarrow 3\,H_2O + CO_2 + Ca^{2+}$$

CaCO₃   H₃O⁺

Limestone

(b)
① Rain is acidified as it falls through the air.

② Acid enters lake from rain.

③ Hydronium-ion concentration increases, with potential harm to the ecosystem.

Granite rock

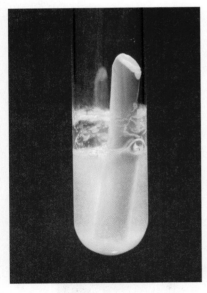

FIGURE 18.15
Most chalks are made from calcium carbonate, which is the same chemical found in limestone. The addition of even a weak acid, such as the acetic acid of vinegar, produces hydronium ions that react with the calcium carbonate to form several products, the most notable being carbon dioxide, which rapidly bubbles out of solution. Try this for yourself! If the bubbling is not as vigorous as shown here, then the chalk is made of other mineral components.

The environmental impact of acid rain depends on local geology, as Figure 18.14 illustrates. In certain regions, such as the midwestern United States, the ground contains significant quantities of the alkaline compound calcium carbonate (limestone), deposited when these lands were submerged under oceans, as has occurred several times over the past 500 million years. Acid rain pouring into these regions is often neutralized by the calcium carbonate before any damage is done. (Figure 18.15 shows calcium carbonate neutralizing an acid.) In the northeastern United States and many other regions, however, the ground contains very little calcium carbonate and is composed primarily of chemically less reactive materials, such as granite. In these regions, the effect of acid rain on lakes and rivers accumulates.

One demonstrated solution to this problem is to raise the pH of acidified lakes and rivers by adding calcium carbonate—a process known as *liming*. The cost of transporting the calcium carbonate, coupled with the need to monitor treated water systems closely, limits liming to only a small fraction of the vast number of water systems already affected. Furthermore, as acid rain continues to pour into these regions, the need to lime also continues.

A longer-term solution to acid rain is to prevent most of the generated sulfur dioxide and other pollutants from entering the atmosphere in the first place. Toward this end, smokestacks have been designed or retrofitted to minimize the quantities of pollutants released. Though this process is costly, the positive effects of these adjustments have been demonstrated. An ultimate long-term solution, however, would be a shift from fossil fuels to cleaner energy sources, such as nuclear and solar energy.

CHECKPOINT
What kind of lakes are protected against the negative effects of acid rain?

Was this your answer?
Lakes that have a floor consisting of basic minerals, such as limestone, are more resistant to acid rain because the chemicals of the limestone (mostly calcium carbonate, $CaCO_3$) neutralize any incoming acid.

It should come as no surprise that the amount of carbon dioxide put into the atmosphere by human activities is growing. What is surprising, however, is that studies indicate that the atmospheric concentration of CO_2 is not increasing proportionately. A likely explanation has to do with the oceans (Figure 18.16). When atmospheric CO_2 dissolves in any body of water—a raindrop, a lake, or the ocean—it forms carbonic acid. In fresh water, this carbonic acid transforms back to water and carbon dioxide, which are released back into the atmosphere. Carbonic acid in the ocean, however, is quickly neutralized by dissolved alkaline substances such as calcium carbonate (the ocean is alkaline, pH $\approx$ 8.2). The products of this neutralization eventually end up on the ocean floor as insoluble solids. Thus, carbonic acid neutralization in the ocean prevents CO_2 from being released back into the atmosphere. The ocean, therefore, is a carbon dioxide *sink*—most of the CO_2 that goes in doesn't come out. So pushing more CO_2 into our atmosphere means pushing more of it into our vast oceans. This is another of the many ways in which the oceans regulate our global environment.

Nevertheless, as Figure 18.17 shows, the concentration of atmospheric CO_2 is increasing. Carbon dioxide is being produced faster than the ocean can absorb it, and this may alter Earth's environment. Carbon dioxide is a *greenhouse gas*, which means it helps keep Earth's surface warm by preventing infrared radiation from escaping into outer space. Without greenhouse gases in the atmosphere, Earth's surface would average a frigid $-18°C$. However, with increasing concentration of CO_2 in the atmosphere, we may experience higher average temperatures. Higher temperatures may significantly alter global weather patterns as well as raise the average sea level, as the polar ice caps melt and the volume of seawater increases because of thermal expansion. Global climate change will be explored in greater detail in Chapter 25.

> The pollution humans release knows no political boundaries. Iron smelters operating in China, for example, release pollutants that are readily detected in Seattle, Washington.

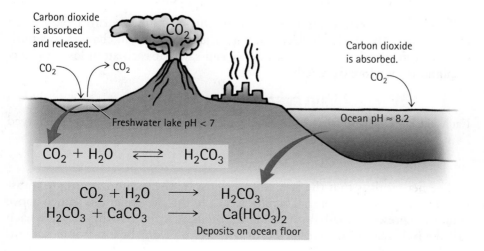

FIGURE 18.16
Carbon dioxide forms carbonic acid on entering any body of water. In fresh water, this reaction is reversible, and the carbon dioxide is released back into the atmosphere. In the alkaline ocean, the carbonic acid is neutralized to such compounds as calcium bicarbonate, $Ca(HCO_3)_2$, which precipitate to the ocean floor. As a result, most of the atmospheric carbon dioxide that enters our oceans remains there.

FIGURE 18.17
Researchers at the Mauna Loa weather observatory in Hawaii have recorded increasing concentrations of atmospheric carbon dioxide since they began collecting data in the 1950s. This famous graph is known as the Keeling curve, after the scientist Charles Keeling, who initiated this project and first noted the trends. Interestingly, the oscillations within the Keeling curve reflect seasonal changes in CO_2 levels.

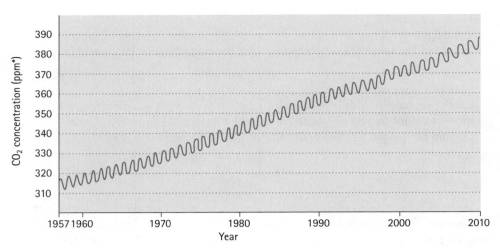

*ppm = parts per million, which tells us the number of carbon dioxide molecules for every million molecules of air

So the pH of rain depends in great part on the concentration of atmospheric CO_2, which depends on the pH of the oceans. These systems are interconnected with global temperatures, which naturally connect to the countless living systems on Earth. How true it is: all the parts are intricately connected, down to the level of atoms and molecules!

18.5 Losing and Gaining Electrons

EXPLAIN THIS   Why is the chlorine atom such a strong oxidizing agent?

Oxidation is the process whereby a reactant loses one or more electrons. **Reduction** is the opposite process, whereby a reactant gains one or more electrons. Oxidation and reduction are complementary processes that occur at the same time. They always occur together; you cannot have one without the other. The electrons lost by one chemical in an oxidation reaction don't simply disappear; they are gained by another chemical in a reduction reaction.

An oxidation–reduction reaction occurs when sodium and chlorine react to form sodium chloride, as shown in Figure 18.18. The equation for this reaction is

$$2\,Na(s) \;+\; Cl_2(g) \longrightarrow 2\,NaCl(s)$$

To see how electrons are transferred in this reaction, we can look at each reactant individually. Each electrically neutral sodium atom changes to a positively charged ion. At the same time, we can say that each atom loses an electron and is therefore oxidized:

$$2\,Na(s) \longrightarrow 2\,Na^+ + 2e^- \quad \text{Oxidation}$$

Each electrically neutral chlorine molecule changes to two negatively charged ions. Each of these atoms gains an electron and is therefore reduced:

$$Cl_2 + 2e^- \longrightarrow 2Cl^- \quad \text{Reduction}$$

The net result is that the two electrons lost by the sodium atoms are transferred to the chlorine atoms. Therefore, each of the two equations shown above actually represents one-half of an entire process, which is why they are each called a **half reaction**. In other words, an electron won't be lost from a sodium atom without the presence of a chlorine atom available to pick up that electron. Both half reactions are required to represent the *whole* oxidation–reduction process. Half reactions are useful for showing which reactant loses electrons and which reactant gains them, which is why half reactions are used throughout this chapter.

Because the sodium causes reduction of the chlorine, the sodium is acting as a *reducing agent*. A reducing agent is any reactant that causes another reactant to be reduced. Note that sodium is oxidized when it behaves as a reducing agent—it loses electrons. Conversely, the chlorine causes oxidation of the sodium and so is acting as an *oxidizing agent*. Because it gains electrons in the process, an oxidizing agent is reduced. Just remember that **l**oss of **e**lectrons is **o**xidation, and **g**ain of **e**lectrons is **r**eduction. Here is a helpful mnemonic adapted from a once-popular children's story: **Leo** the lion went "**ger.**"

Different elements have different oxidation and reduction tendencies—some lose electrons more readily, while others gain electrons more readily, as Figure 18.19 illustrates.

FIGURE 18.18
In the exothermic formation of sodium chloride, sodium metal is oxidized by chlorine gas, and chlorine gas is reduced by sodium metal.

> **CHECKPOINT**
> **True or false?**
> 1. **Reducing agents are oxidized in oxidation–reduction reactions.**
> 2. **Oxidizing agents are reduced in oxidation–reduction reactions.**
>
> **Were these your answers?**
> Both statements are true.

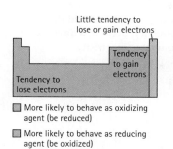

FIGURE 18.19
The ability of an atom to gain or lose electrons is indicated by its position in the periodic table. Those at the upper right tend to gain electrons, and those at the lower left tend to lose them.

Whether a reaction classifies as an oxidation–reduction reaction is not always immediately apparent. The chemical equation, however, can provide some important clues. First, look for changes in the ionic states of elements. Sodium metal, for example, consists of neutral sodium atoms. In the formation of sodium chloride, these atoms transform into positively charged sodium ions, which occurs as sodium atoms lose electrons (oxidation). A second way to identify a reaction as an oxidation–reduction reaction is to look to see whether an element is gaining or losing *oxygen* atoms. As the element gains the oxygen, it is losing electrons to that oxygen because of the oxygen's high electronegativity. The gain of oxygen, therefore, is oxidation (loss of electrons), while the loss of oxygen is reduction (gain of electrons). For example, hydrogen, H_2, reacts with oxygen, O_2, to form water, H_2O, as follows:

$$H{-}H + H{-}H + O{=}O \longrightarrow H{-}O{-}H + H{-}O{-}H$$

Note that the element hydrogen becomes attached to an oxygen atom through this reaction. The hydrogen, therefore, is oxidized.

A third way to identify a reaction as an oxidation–reduction reaction is to see whether an element is gaining or losing hydrogen atoms. The gain of hydrogen is reduction, while the loss of hydrogen is oxidation. For the formation of water shown above, we see that the element oxygen is gaining hydrogen atoms, which means that the oxygen is being reduced—that is, the oxygen is gaining electrons from the hydrogen, which is why the oxygen atom within water is slightly negative as discussed in Section 15.7. The three ways of identifying a reaction as an oxidation–reduction type of reaction are summarized in Figure 18.20.

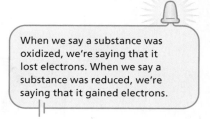

When we say a substance was oxidized, we're saying that it lost electrons. When we say a substance was reduced, we're saying that it gained electrons.

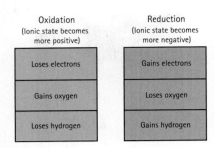

FIGURE 18.20
Oxidation results in a greater positive charge, which can be achieved by losing electrons, gaining oxygen atoms, or losing hydrogen atoms. Reduction results in a greater negative charge, which can be achieved by gaining electrons, losing oxygen atoms, or gaining hydrogen atoms.

LEARNING OBJECTIVE
Recognize where oxidation and reduction occur in a device that generates electricity.

FIGURE 18.21
A nail made of iron placed in a solution of Cu^{2+} ions oxidizes to Fe^{2+} ions, which dissolve in the water. At the same time, copper ions are reduced to metallic copper, which coats the nail. (Negatively charged ions, such as chloride ions, Cl^-, must also be present to balance these positively charged ions in solution.)

CHECKPOINT
In the following equation, is carbon oxidized or reduced?

$$CH_4 + 2\,O_2 \longrightarrow CO_2 + 2\,H_2O$$

Was this your answer?
As the carbon of methane, CH_4, forms carbon dioxide, CO_2, it is losing hydrogen and gaining oxygen, which tells us that the carbon is being oxidized.

18.6 Harnessing the Energy of Flowing Electrons

EXPLAIN THIS Why is lithium a preferred metal for the making of batteries?

Electrochemistry is the study of the relationship between electric energy and chemical change. It involves either the use of an oxidation–reduction reaction to produce an electric current or the use of an electric current to produce an oxidation–reduction reaction.

To understand how an oxidation–reduction reaction can generate an electric current, consider what happens when a reducing agent is placed in direct contact with an oxidizing agent: electrons flow from the reducing agent to the oxidizing agent. This flow of electrons is an electric current, which is a form of kinetic energy that can be harnessed for useful purposes.

Iron atoms, Fe, for example, are better reducing agents than copper ions, Cu^{2+}. So when a piece of iron metal and a solution containing copper ions are placed in contact with each other, electrons flow from the iron to the copper ions, as Figure 18.21 illustrates. The result is the oxidation of iron atoms and the reduction of copper ions.

The elemental iron and copper ions need not be in physical contact for electrons to flow between them. If they are in separate containers but bridged by a conducting wire, the electrons can flow from the iron through the wire to the copper ions. The resulting electric current in the wire can be attached to some

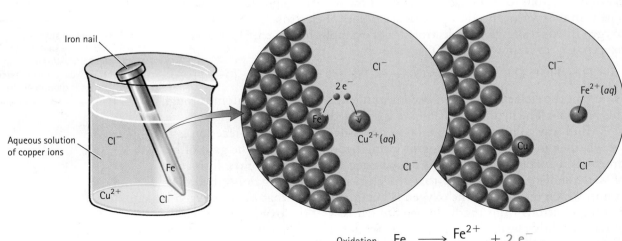

Oxidation $Fe \longrightarrow Fe^{2+} + 2\,e^-$
Reduction $Cu^{2+} + 2\,e^- \longrightarrow Cu$

useful device, such as a lightbulb. But alas, an electric current is not sustained by this arrangement.

The reason the electric current is not sustained is shown in Figure 18.22. An initial flow of electrons through the wire immediately results in a buildup of electric charge in both containers. The container on the left builds up positive charge as it accumulates Fe^{2+} ions from the nail. The container on the right builds up negative charge as electrons accumulate on this side. This situation prevents any further migration of electrons through the wire. Recall that electrons are negative, and so they are repelled by the negative charge in the right container and attracted to the positive charge in the left container. The net result is that the electrons do not flow through the wire, and the bulb remains unlit.

The solution to this problem is to allow ions to migrate into either container so that neither builds up any positive or negative charge. This is accomplished with a *salt bridge*, which may be a U-shaped tube filled with a salt, such as sodium nitrate, $NaNO_3$, and closed with semiporous plugs. Figure 18.23 shows how a salt bridge allows the ions it holds to enter either container, permitting the flow of electrons through the conducting wire and creating a complete electric circuit.

Batteries

So we can see that, with the proper setup, it is possible to harness electric energy from an oxidation–reduction reaction. The apparatus shown in Figure 18.23 is one example. Such devices are called *voltaic cells*. Instead of two containers, a voltaic cell can be an all-in-one, self-contained unit, in which case it is called a *battery*. Batteries are either disposable or rechargeable, and here we explore some examples of each. Although the two types differ in design and composition, they function by the same principle: two materials that oxidize and reduce each other are connected by a medium through which ions travel to balance an external flow of electrons.

Let's look at disposable batteries first. The common *dry-cell battery*, which was invented in the 1860s, is still used today, and it is probably the cheapest disposable energy source for flashlights, toys, and the like. The basic design consists of a zinc cup filled with a thick paste of ammonium chloride, NH_4Cl; zinc chloride, $ZnCl_2$; and manganese dioxide, MnO_2. Immersed in this paste is a porous stick of graphite that projects to the top of the battery, as shown in Figure 18.24.

This side immediately builds up a positive charge that attracts electrons, preventing them from migrating.

This side immediately builds up a negative charge that repels electrons, preventing them from entering.

FIGURE 18.22

An iron nail is placed in water and connected by a conducting wire to a solution of copper ions. Nothing happens, because this arrangement results in a buildup of charge that prevents the further flow of electrons.

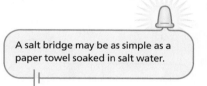

A salt bridge may be as simple as a paper towel soaked in salt water.

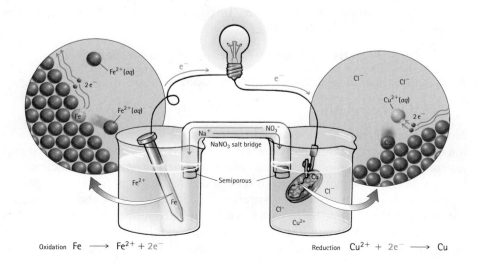

Oxidation   Fe $\longrightarrow$ Fe^{2+} + 2e$^-$ Reduction Cu^{2+} + 2e$^-$ $\longrightarrow$ Cu

FIGURE 18.23

The salt bridge completes the electric circuit. Electrons freed as the iron is oxidized pass through the wire to the container on the right. Nitrate ions, NO_3^-, from the salt bridge flow into the left container to balance the positive charges of the Fe^{2+} ions that form, thereby preventing any buildup of positive charge. Meanwhile, Na^+ ions from the salt bridge enter the right container to balance the Cl^- ions "abandoned" by the Cu^{2+} ions as the Cu^{2+} ions pick up electrons to become metallic copper.

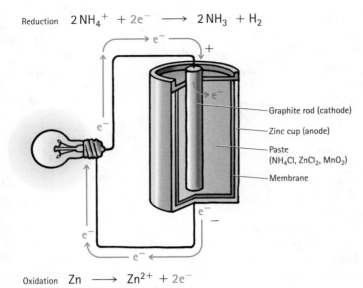

Reduction $2 NH_4^+ + 2e^- \longrightarrow 2 NH_3 + H_2$

- Graphite rod (cathode)
- Zinc cup (anode)
- Paste (NH_4Cl, $ZnCl_2$, MnO_2)
- Membrane

Oxidation $Zn \longrightarrow Zn^{2+} + 2e^-$

FIGURE 18.24
A common dry-cell battery with a graphite rod immersed in a paste of ammonium chloride, manganese dioxide, and zinc chloride.

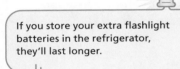

If you store your extra flashlight batteries in the refrigerator, they'll last longer.

Graphite is a good conductor of electricity. Chemicals in the paste receive electrons at the graphite stick and so are reduced. The reaction for the ammonium ions is

$$2NH_4^+(aq) + 2e^- \longrightarrow 2NH_3(g) + H_2(g) \quad \text{Reduction}$$

An **electrode** is any material that conducts electrons into or out of a medium in which electrochemical reactions are occurring. The electrode where chemicals are reduced is called a **cathode**. For any battery, such as the one shown in Figure 18.24, the cathode is always positive ($+$), which indicates that electrons are naturally attracted to this location. The electrons gained by chemicals at the cathode originate at the **anode**, which is the electrode where chemicals are oxidized. For any battery, the anode is always negative ($-$), which indicates that electrons are streaming away from this location. The anode in Figure 18.24 is the zinc cup, where zinc atoms lose electrons to form zinc ions:

$$Zn(s) \longrightarrow Zn^{2+}(aq) + 2e^- \quad \text{Oxidation}$$

The reduction of ammonium ions in a dry-cell battery produces two gases—ammonia, NH_3, and hydrogen, H_2—that need to be removed to avoid a pressure buildup and a potential explosion. Removal is accomplished by having the ammonia and hydrogen react with the zinc chloride and manganese dioxide:

$$ZnCl_2(aq) + 2 NH_3(g) \longrightarrow Zn(NH_3)_2Cl_2(s)$$

$$2 MnO_2(s) + H_2(g) \longrightarrow Mn_2O_3(s) + H_2O(l)$$

The life of a dry-cell battery is relatively short. Oxidation causes the zinc cup to deteriorate, and eventually the contents leak out. Even while the battery is not operating, the zinc corrodes as it reacts with ammonium ions. This zinc corrosion can be inhibited by storing the battery in a refrigerator. As discussed in Section 17.3, chemical reactions slow down with decreasing temperature. Chilling a battery, therefore, slows down the rate at which the zinc corrodes, which increases the life of the battery.

Another type of disposable battery, the more expensive *alkaline battery*, shown in Figure 18.25, avoids many of the problems of dry-cell batteries by operating in a strongly alkaline paste. In the presence of hydroxide ions, the zinc oxidizes to insoluble zinc oxide:

$$Zn(s) + 2 OH^-(aq) \longrightarrow ZnO(s) + H_2O(l) + 2e^- \quad \text{Oxidation}$$

At the same time, manganese dioxide is reduced:

$$2 MnO_2(s) + H_2O(l) + 2e^- \longrightarrow Mn_2O_3(s) + 2 H_2O(aq) \quad \text{Reduction}$$

Note how these two reactions avoid the use of the zinc-corroding ammonium ion (which means alkaline batteries last a lot longer than dry-cell batteries) and also prevent formation of any gaseous products. Furthermore, these reactions are better suited to maintaining a given voltage during longer periods of operation.

The small lithium disposable batteries used for calculators and cameras are variations of the alkaline battery. In the lithium battery, lithium metal is used as the source of electrons rather than zinc. Not only can lithium maintain a higher voltage than zinc, but it also is about $\frac{1}{13}$ as dense, which allows for a lighter battery.

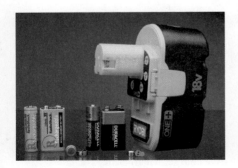

FIGURE 18.25
Alkaline batteries last a lot longer than dry-cell batteries and give a steadier voltage, but they are more expensive.

Disposable batteries have relatively short lives because electron-producing chemicals are consumed. The main feature of *rechargeable* batteries is the reversibility of the oxidation and reduction reactions. A noteworthy example is the nickel metal hydride, NiMH, battery. Charging this battery causes the nickel metal to extract hydrogen from water to form the negatively charged hydride ion, shown below as H: where the two dots represent two electrons.*

$$H_2O + \quad Ni \quad + \quad 2e^- \quad \longrightarrow \quad H\!:\!Ni \quad + \quad HO^-$$

Water   Nickel metal   Nickel hydride   Hydroxide ion

The role of the nickel is to stabilize the two electrons on the hydrogen, which, because it contains an added electron, is called the *hydride* ion, much as chlorine with an extra electron is called the *chloride* ion, using the *-ide* suffix. A fully charged battery thus contains an abundance of nickel hydride. As the battery provides electricity, the hydride ion releases electrons, which allows it to join with the hydroxide ion to reform water:

$$H\!:\!Ni \quad + \quad HO^- \quad \longrightarrow \quad H_2O \quad + \quad Ni \quad + \quad 2e^-$$

Nickel hydride   Hydroxide ion   Water   Nickel metal

So recharging a rechargeable battery simply means regenerating the chemicals, such as nickel hydride, that can release electrons on demand. For the NiMH battery, this chemical is nickel hydride, $H\!:\!Ni$. For a traditional car battery, this chemical is simply lead, Pb, which transforms into lead sulfate, $PbSO_4$, as it releases electrons. As the car battery is recharged, the $PbSO_4$ is transformed back into lead, Pb.

Rechargeable lithium-ion batteries have found a wide range of applications, such as powering laptop computers and cell phones. Safer lithium phosphate iron batteries are used for hybrid cars, such as the popular Toyota Prius shown in Figure 18.26. Hybrids have improved gas mileage because as the car slows down, its kinetic energy is transformed into the electric potential energy of the battery rather than being wasted as heat from the car's brake pads. The captured electric energy of the battery is subsequently used to assist the gas-powered engine to get the car moving. Also, the hybrid's battery system allows the engine to shut off when the car is merely idling or moving slowly, as in heavy traffic.

Continued improvements in battery technology are permitting next-generation hybrids, known as *plug-in hybrids*, which have much larger batteries and smaller fuel tanks. These hybrids can be plugged into electrical outlets, charged overnight, and driven the next day for up to 60 mi without using any gasoline. This is significant because the typical driver in the United States drives less than 40 mi a day. Furthermore, utility companies can sell electricity at cheaper rates at night because their massive generators are underutilized at that time. Alternatively, the plug-in hybrid can be charged via residential photovoltaic panels or a small wind turbine. Cars that remain plugged in during the day would be available to contribute energy back into the grid during peak demand—and owners of such cars could receive rebates for this energy. Furthermore, during an electricity blackout, a family's plug-in hybrid would store enough energy to serve as emergency backup for household needs. Plug-in hybrids with their large and highly efficient batteries offer much in the way of moving individuals and the nation as a whole toward energy conservation and independence.

Aside from the initial charge of a brand-new battery, the energy in a car battery ultimately comes from fuel in the gas tank through the process of recharging.

FIGURE 18.26

As of 2010, more than 1.2 million Prius hybrids had been sold worldwide, about 266,000 of them in the United States. Look now for hybrid vehicles that can be plugged into your home electrical outlet, charged at night, and driven the next day using no gasoline for up to 60 mi.

---

* The nickel in this reaction is actually an intermetallic alloy of nickel and various rare-earth elements, such as lanthanum, La.

Because fuel cells can do so much, there is little doubt that they will be spread throughout the marketplace in the not-too-distant future.

FIGURE 18.27
INTERACTIVE FIGURE **MP**

The hydrogen–oxygen fuel cell.

CHECKPOINT
What chemicals are produced as a nickel metal hydride battery recharges?

Was this your answer?
Nickel hydride, H : Ni, and hydroxide ions, HO$^-$

Fuel Cells

A *fuel cell* is a device that converts the chemical energy of a fuel to electric energy. Fuel cells are by far the most efficient means of generating electricity. A hydrogen–oxygen fuel cell is shown in Figure 18.27. It has two compartments, one for entering hydrogen fuel and the other for entering oxygen fuel, separated by a set of porous electrodes. Hydrogen is oxidized on contact with hydroxide ions at the hydrogen-facing electrode (the anode). The electrons from this oxidation flow through an external circuit and provide electric power before meeting up with oxygen at the oxygen-facing electrode (the cathode). The oxygen readily picks up the electrons (in other words, the oxygen is reduced) and reacts with water to form hydroxide ions. To complete the circuit, these hydroxide ions migrate across the porous electrodes and through an ionic paste of potassium hydroxide, KOH, to join with hydrogen at the hydrogen-facing electrode.

As the oxidation equation shown at the top of Figure 18.27 demonstrates, the hydrogen and hydroxide ions react to produce energetic water molecules that arise in the form of steam. This steam may be used for heating or for generating electricity in a steam turbine. Furthermore, the water that condenses from the steam is pure water, suitable for drinking!

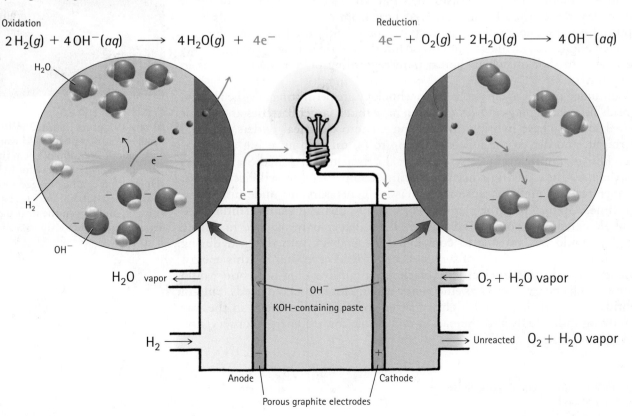

Oxidation
$$2\,H_2(g) + 4\,OH^-(aq) \longrightarrow 4\,H_2O(g) + 4e^-$$

Reduction
$$4e^- + O_2(g) + 2\,H_2O(g) \longrightarrow 4\,OH^-(aq)$$

H_2O

H_2

OH^-

H_2O vapor ←

OH^-
KOH-containing paste

H_2 →

Anode

Cathode

Porous graphite electrodes

← $O_2 + H_2O$ vapor

→ Unreacted $O_2 + H_2O$ vapor

Although fuel cells are similar to dry-cell batteries, they don't run down as long as fuel is supplied. The space shuttle uses hydrogen–oxygen fuel cells to meet its electrical needs. The cells also produce more than 100 gal of drinking water for the astronauts during a typical week-long mission. Back on Earth, researchers are developing fuel cells for buses and automobiles. As shown in Figure 18.28, experimental fuel-cell buses are already operating in several cities, such as Vancouver, British Columbia, and Chicago, Illinois. These vehicles produce very few pollutants and can run much more efficiently than vehicles that burn fossil fuels.

In the future, commercial buildings as well as individual homes may be outfitted with fuel cells as an alternative to receiving electricity (and heat) from regional power stations. Researchers are also working on miniature fuel cells that could replace the batteries used for portable electronic devices, such as cell phones and laptop computers. Such devices could operate for extended periods of time on a single "ampoule" of fuel available at your local supermarket.

Amazingly, a car powered by a hydrogen–oxygen fuel cell requires only about 3 kg of hydrogen to travel 500 km. However, this quantity of hydrogen gas at room temperature and atmospheric pressure would occupy a volume of about 36,000 L, the volume of about four midsize cars! Thus, the major hurdle to the development of fuel-cell technology lies not with the cell but with the fuel. This volume of gas could be compressed to a much smaller volume, as it is in the experimental buses in Vancouver.

Compressing a gas requires energy, however—and, as a consequence, the inherent efficiency of the fuel cell is lost. Chilling hydrogen to its liquid phase, which occupies much less volume, poses similar problems. Instead, researchers are looking for novel ways of providing fuel cells with hydrogen. In one design, hydrogen is generated within the fuel cell from chemical reactions involving liquid fuels, such as methanol, CH_3OH. Alternatively, certain porous materials, including the recently developed carbon nanofibers shown in Figure 18.29, can hold large volumes of hydrogen on their surfaces, behaving in effect like hydrogen "sponges." The hydrogen is "squeezed" out of these materials on demand by controlling the temperature—the warmer the material, the more hydrogen that is released.

FIGURE 18.28
Because this bus is powered by a fuel cell, its tailpipe emits mostly water vapor.

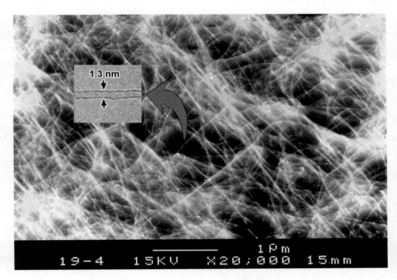

FIGURE 18.29
Carbon nanofibers consist of near-submicroscopic tubes of carbon atoms. They outclass almost all other known materials in their ability to absorb hydrogen molecules. With carbon nanofibers, a volume of 36,000 L of hydrogen can be reduced to a mere 35 L. Carbon nanofibers are a recent discovery, however, and much research is still required to confirm their applicability to hydrogen storage and to develop the technology.

CHECKPOINT

As long as fuel is available to it, a given fuel cell can supply electric energy indefinitely. Why can't batteries do the same?

Was this your answer?
Batteries generate electricity as the chemical reactants they contain are reduced and oxidized. Once these reactants are consumed, the battery can no longer generate electricity. A rechargeable battery can be made to operate again, but only after the energy flow is interrupted so that the reactants can be replenished.

FIGURE 18.30
The electrolysis of water produces hydrogen gas and oxygen gas in a 2:1 ratio by volume, in accord with the chemical formula for water: H_2O. In order for this process to work, ions must be dissolved in the water so that electric charge can be conducted between the electrodes.

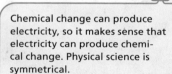

Chemical change can produce electricity, so it makes sense that electricity can produce chemical change. Physical science is symmetrical.

18.7 Electrolysis

EXPLAIN THIS How might electrolysis be used to raise a sunken ship?

Electrolysis is the use of electric energy to produce chemical change. The recharging of a car battery is an example of electrolysis. Another, shown in Figure 18.30, is passing an electric current through water, a process that breaks the water down into its elemental components:

$$\text{Electric energy} + 2\,H_2O(l) \longrightarrow 2\,H_2(g) + O_2(g)$$

Electrolysis is used to purify metals from metal ores. An example is aluminum, the third most abundant element in Earth's crust. Aluminum occurs naturally bonded to oxygen in an ore called bauxite. Aluminum metal wasn't known until about 1827, when it was prepared by reacting bauxite with hydrochloric acid. This reaction gave the aluminum ion, Al^{3+}, which was reduced to aluminum metal, with sodium metal acting as the reducing agent:

$$Al^{3+} + 3\,Na \longrightarrow Al + 3\,Na^+$$

This chemical process was expensive. The price of aluminum at that time was about $100,000 per pound, and it was considered a rare and precious metal. In 1855, aluminum dinnerware and other items were exhibited in Paris with the crown jewels of France. Then, in 1886, two men working independently, Charles Hall (1863–1914) in the United States and Paul Heroult (1863–1914) in France, almost simultaneously discovered a process whereby aluminum could be produced from aluminum oxide, Al_2O_3, a main component of bauxite. In what is now known as the Hall–Heroult process, shown in Figure 18.31, a strong electric current is passed through a molten mixture of aluminum oxide and cryolite, Na_3AlF_6, a naturally occurring mineral. The fluoride ions of the cryolite react with the aluminum oxide to form various aluminum fluoride ions, such as $AlOF_3^{2-}$, which are then oxidized to the aluminum hexafluoride ion, AlF_6^{3-}. The Al^{3+} in this ion is then reduced to elemental aluminum, which collects at the bottom of the reaction chamber. This process, which is still in use by manufacturers today, greatly facilitated mass production of

FIGURE 18.31
The melting point of aluminum oxide (2030°C) is too high for efficiently electrolyzing to aluminum metal. When the oxide is mixed with the mineral cryolite, the melting point of the oxide drops to a more reasonable 980°C. A strong electric current passed through the molten aluminum oxide–cryolite mixture generates aluminum metal at the cathode, where aluminum ions pick up electrons and are thus reduced to elemental aluminum.

$$\text{Oxidation}\quad 2\,AlOF_3^{2-} + 6\,F^- + C \longrightarrow 2\,AlF_6^{3-} + CO_2 + 4e^-$$

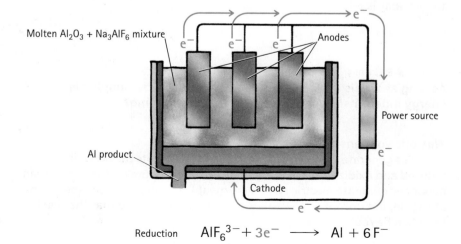

$$\text{Reduction}\quad AlF_6^{3-} + 3e^- \longrightarrow Al + 6\,F^-$$

aluminum metal, and, by 1890, the price of aluminum had dropped to about $2 per pound.

Today, worldwide production of aluminum is about 16 million tons annually. For each ton produced from ore, about 16,000 kWh of electric energy is required, as much as a typical American household consumes in 18 months. Processing recycled aluminum, on the other hand, consumes only about 700 kWh for every ton. Thus, recycling aluminum not only reduces litter but also helps reduce the load on power companies, which in turn reduces air pollution. Furthermore, reserves of high-quality aluminum oxide ores are already depleted in the United States. Recycling aluminum, therefore, also helps minimize the need for developing new bauxite mines in foreign countries.

CHECKPOINT

Is the exothermic reaction in a hydrogen–oxygen fuel cell an example of electrolysis?

Was this your answer?
No. During electrolysis, electric energy is used to produce chemical change. In the hydrogen–oxygen fuel cell, chemical change is used to produce electric energy.

18.8 Corrosion and Combustion

EXPLAIN THIS Do our bodies gradually oxidize or reduce the food molecules we eat?

If you look to the upper right of the periodic table, you will find one of the most common oxidizing agents—oxygen. In fact, if you haven't guessed already, the term *oxidation* is derived from the name of this element. Oxygen can pluck electrons from many other elements, especially those that lie at the lower left of the periodic table. Two common oxidation–reduction reactions involving oxygen as the oxidizing agent are *corrosion* and *combustion*.

CHECKPOINT

Oxygen is a good oxidizing agent, but so is chlorine. What does this indicate about their relative positions in the periodic table?

Was this your answer?
Chlorine and oxygen must lie in the same area of the periodic table. Both have strong effective nuclear charges and are strong oxidizing agents.

Corrosion is the process whereby a metal deteriorates. Corrosion caused by atmospheric oxygen is a widespread and costly problem. About one-quarter of the steel produced in the United States, for example, goes into replacing corroded iron at a cost of billions of dollars annually. Iron corrodes when it reacts with atmospheric oxygen and water to form iron oxide trihydrate, which is

FYI Our bodies require lots of energy for living. We get this energy from special high-energy molecules, such as ATP, which the body produces by oxidizing food molecules with oxygen. If you were to stop breathing, say by choking, your cells would be deprived of oxygen and no longer able to produce these high-energy molecules. The result is a prompt death. But instead of dying, why doesn't the body simply turn off until oxygen becomes available again? Lethal damage occurs because many cellular mechanisms continue to operate even at very low oxygen levels. With some parts working and others not working, the cell is thrown so far off balance that it dies. The trick is to make sure that all cellular processes shut down together. This explains how people who fall into frozen waters can sometimes be resuscitated even though they haven't been breathing for more than an hour—their cells were shut down uniformly because of the rapid onslaught of the extreme cold.

LEARNING OBJECTIVE
Compare and contrast the processes of corrosion and combustion.

Researchers have discovered that mice and other animals breathing certain concentrations of hydrogen sulfide gas, H_2S, enter a state of suspended animation where the body temperature fluctuates only a few degrees above the surrounding temperature. In effect, the animal becomes cold-blooded, which is what happens to bears and ground squirrels when they hibernate. The hydrogen sulfide apparently mimics molecular oxygen, O_2. Cells absorb and try to use the H_2S as though it were O_2, but without the oxidative powers of O_2, the cell's machinery simply shuts down. That the cell shuts down uniformly is key to the subsequent revival of the organism, as discussed in the FYI on page 479. If applicable to humans, hydrogen sulfide–induced suspended animation holds many possibilities, including protection against lethal cellular damage caused by strokes, heart attacks, or other critical injuries in which either blood flow or blood supply is severely limited. This technology may also help donor organs remain viable for longer periods before transplantation.

FIGURE 18.32
Rust itself does not harm the iron structures on which it forms. The loss of metallic iron ruins the structural integrity of these objects.

FIGURE 18.33
The galvanized nail (*bottom*) is protected from rusting by the sacrificial oxidation of zinc.

the naturally occurring reddish-brown substance you know as rust, shown in Figure 18.32:

$$4\,\text{Fe} + 3\,\text{O}_2 + 3\,\text{H}_2\text{O} \longrightarrow 2\,\text{Fe}_2\text{O}_3 \cdot 3\,\text{H}_2\text{O}$$

$\quad\quad$ Iron $\quad\quad$ Oxygen $\quad\quad$ Water $\quad\quad\quad\quad$ Rust

Another common metal oxidized by oxygen is aluminum. The product of aluminum oxidation is aluminum oxide, Al_2O_3, which is not water soluble. Because of its insolubility, aluminum oxide forms a protective coat that shields the metal from further oxidation. This coat is so thin that it's transparent, which is why aluminum maintains its metallic shine.

A protective, water-insoluble oxidized coat is the principle underlying a process called *galvanization*. Zinc has a slightly greater tendency to oxidize than does iron. For this reason, many iron objects, such as the nail pictured in Figure 18.33, are *galvanized* by coating them with a thin layer of zinc. The zinc oxidizes to zinc oxide, an inert, insoluble substance that protects the iron underneath it from rusting.

In a technique called *cathodic protection*, iron structures can be protected from oxidation by placing them in contact with certain metals, such as zinc or magnesium, that have a greater tendency to oxidize. This forces the iron to accept electrons, which means that it is behaving as a cathode. (Rusting occurs only where iron behaves as an anode.) Ocean tankers, for example, are protected from corrosion by strips of zinc affixed to their hulls, as shown in Figure 18.34. Similarly, outdoor steel pipes are protected by being connected to magnesium rods inserted into the ground.

The metals used for cathodic protection are "sacrificing" themselves to be anodes (to lose electrons) so that the desired metal, such as the copper pipe, is spared from oxidation. These sacrificing metals, therefore, are sometimes called *sacrificial anodes*.

FIGURE 18.34
Zinc strips help protect the iron hull of an oil tanker from oxidizing. The zinc strips shown here are attached to the hull's exterior surface.

Yet another way to protect iron and other metals from oxidation is to coat them with a corrosion-resistant metal, such as chromium, platinum, or gold. *Electroplating* is the operation of coating one metal with another by electrolysis, and it is illustrated in Figure 18.35. The object to be electroplated is connected to a negative battery terminal and then submerged in a solution containing ions of the metal to be used as the coating. The positive terminal of the battery is connected to an electrode made of the coating metal. The circuit is completed when this electrode is submerged in the solution. Dissolved metal ions are attracted to the negatively charged object, where they pick up electrons and are deposited as metal atoms. The ions in solution are replenished by the forced oxidation of the coating metal at the positive electrode.

Combustion is an oxidation–reduction reaction between a nonmetallic material and molecular oxygen. Combustion reactions are characteristically exothermic (energy-releasing). A violent combustion reaction is the formation of water from hydrogen and oxygen. As discussed in Section 17.5, the energy from this reaction is used to power rockets into space. More common examples of combustion include the burning of wood and fossil fuels. The combustion of these and other carbon-based chemicals forms carbon dioxide and water. Consider, for example, the combustion of methane, the major component of natural gas:

$$CH_4 \quad + \quad 2\,O_2 \quad \longrightarrow \quad CO_2 \quad + \quad 2\,H_2O \quad + \quad energy$$

Methane   Oxygen   Carbon   Water
             dioxide

In combustion, electrons are transferred when polar covalent bonds are formed in place of nonpolar covalent bonds, or vice versa. (This is in contrast with the other examples of oxidation–reduction reactions presented in this chapter, which involve the formation of ions from atoms or, conversely, atoms from ions.) This concept is illustrated in Figure 18.36, on the next page, which compares the electronic structures of the combustion starting material, molecular oxygen, and the combustion product, water. Molecular oxygen is a nonpolar covalent compound. Although each oxygen atom in the molecule has a fairly strong electronegativity, the four bonding electrons are pulled equally by both atoms and thus cannot congregate on one side or the other. After combustion, however, the electrons are shared between the oxygen and hydrogen atoms in a water molecule and are pulled to the oxygen. This gives the oxygen a slight negative charge, which is another way of saying it has gained electrons and has thus been reduced. At the same time, the hydrogen atoms in the water molecule develop a slight positive charge, which is another way of saying they have lost electrons and have thus been oxidized. This gain of electrons by oxygen and loss of electrons by hydrogen is an energy-releasing process. Typically, the energy is released either as molecular kinetic energy (heat) or as light (the flame).

Interestingly, combustion oxidation–reduction reactions occur throughout your body. You can visualize a simplified model of your metabolism by reviewing Figure 18.36 and substituting a food molecule for the methane. Food molecules relinquish their electrons to the oxygen molecules you inhale. The products are carbon dioxide, water vapor, and energy. You exhale the carbon dioxide and water vapor, but much of the energy from the reaction is used to keep your body warm and to drive the many other biochemical reactions necessary for life.

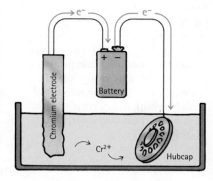

FIGURE 18.35

As electrons flow into the hubcap and give it a negative charge, positively charged chromium ions move from the solution to the hubcap and are reduced to chromium metal, which deposits as a coating on the hubcap. The solution is supplied with ions as chromium atoms in the cathode are oxidized to Cr^{2+} ions.

FYI There are two kinds of matches: the "strike anywhere" type usually having a "bull's-eye" looking tip, and the "safety match," which requires you to strike the match on a strip on the packaging. Both involve the burning of sulfur within the tip of the match. Getting the sulfur to burn using only the oxygen in the air, however, is difficult, which is why the sulfur is blended with an oxidizing agent, such as potassium chlorate, $KClO_3$. For the "strike anywhere" match, a third ingredient, red phosphorus, P_4, is included. The heat of friction causes the red phosphorus to convert into white phosphorus—an alternate form of phosphorus that burns rapidly in air. This initiates the reduction–oxidation reaction between the sulfur and the potassium chlorate, which in turn ignites the burning of the matchstick. Safety matches work the same way, except that the red phosphorus is embedded within the striking strip, which is the only place where the match can be lit.

FIGURE 18.36

(a) Neither atom in an oxygen molecule can preferentially attract the bonding electrons. (b) The oxygen atom of a water molecule pulls the bonding electrons away from the hydrogen atoms on the water molecule, making the oxygen slightly negative and the two hydrogens slightly positive.

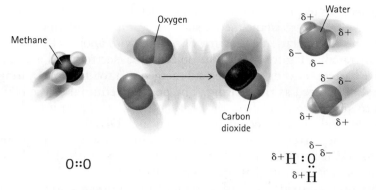

(a) Reactant oxygen atoms share electrons equally in O_2 molecules.

(b) Product oxygen atoms pull electrons away from H atoms in H_2O molecules and are reduced.

For instructor-assigned homework, go to www.masteringphysics.com

SUMMARY OF TERMS (KNOWLEDGE)

Acid A substance that donates hydrogen ions.

Acidic Description of a solution in which the hydronium-ion concentration is higher than the hydroxide-ion concentration.

Amphoteric Description of a substance that can behave as either an acid or a base.

Anode The electrode where chemicals are oxidized.

Base A substance that accepts hydrogen ions.

Basic Description of a solution in which the hydroxide-ion concentration is higher than the hydronium-ion concentration; also sometimes called *alkaline*.

Cathode The electrode where chemicals are reduced.

Combustion An exothermic oxidation–reduction reaction between a nonmetallic material and molecular oxygen.

Corrosion The deterioration of a metal, typically caused by atmospheric oxygen.

Electrochemistry The study of the relationship between electric energy and chemical change.

Electrode Any material that conducts electrons into or out of a medium in which electrochemical reactions are occurring.

Electrolysis The use of electric energy to produce chemical change.

Half reaction Half of an oxidation–reduction reaction, represented by an equation showing electrons as either reactants or products.

Hydronium ion A polyatomic ion made by adding a proton (hydrogen ion) to a water molecule.

Hydroxide ion A polyatomic ion made by removing a proton (hydrogen ion) from a water molecule.

Neutral Description of a solution in which the hydronium-ion concentration is equal to the hydroxide-ion concentration.

Neutralization A reaction between an acid and a base.

Oxidation The process whereby a reactant loses one or more electrons.

pH A measure of the acidity of a solution, equal to the negative logarithm of the hydronium-ion concentration.

Reduction The process whereby a reactant gains one or more electrons.

Salt An ionic compound commonly formed from the reaction between an acid and a base.

READING CHECK QUESTIONS (COMPREHENSION)

18.1 Acids Donate Protons; Bases Accept Them

1. What are the Brønsted–Lowry definitions of acid and base?
2. When an acid is dissolved in water, what ion does the water form?
3. When a chemical loses a hydrogen ion, is it behaving as an acid or a base?

18.2 Relative Strengths of Acids and Bases

4. What does it mean to say that an acid is strong in aqueous solution?
5. Why does a solution of a strong acid conduct electricity better than a solution of a weak acid having the same concentration?
6. When can a solution of a weak base be more corrosive than a solution of a strong base?

18.3 Acidic, Basic, and Neutral Solutions

7. Is water a strong acid or a weak acid?

8. What is true about the relative concentrations of hydronium and hydroxide ions in an acidic solution? How about a neutral solution? A basic solution?

9. What does the pH of a solution indicate?

10. As the hydronium-ion concentration of a solution increases, does the pH of the solution increase or decrease?

18.4 Acidic Rain and Basic Oceans

11. What is the product of the reaction between carbon dioxide and water?

12. What does sulfur dioxide have to do with acid rain?

13. How do humans generate the air pollutant sulfur dioxide?

14. Why aren't atmospheric levels of carbon dioxide rising as rapidly as might be expected based on the increased output of carbon dioxide resulting from human activities?

18.5 Losing and Gaining Electrons

15. Which elements have the greatest tendency to behave as oxidizing agents?

16. What elements have the greatest tendency to behave as reducing agents?

17. Write an equation for the half reaction in which a potassium atom, K, is oxidized.

18. What happens to a reducing agent as it reduces?

18.6 Harnessing the Energy of Flowing Electrons

19. What is electrochemistry?

20. What is the purpose of the salt bridge in a voltaic cell?

21. What type of reaction occurs at the cathode? At the anode?

22. What is the prime difference between a battery and a fuel cell?

23. What else do fuel cells produce besides electricity?

18.7 Electrolysis

24. What is electrolysis, and how does it differ from what goes on inside a battery?

25. What is an example of a metal produced primarily by electrolysis?

26. In what year was the efficient electrolysis of aluminum discovered?

18.8 Corrosion and Combustion

27. What metal coats a galvanized nail?

28. What are some differences between corrosion and combustion?

29. What is iron forced to accept during cathodic protection?

30. What happens to the polarity of oxygen atoms as they transform from molecular oxygen, O_2, into water molecules, H_2O?

ACTIVITIES (HANDS-ON APPLICATION)

31. The pH of a solution can be approximated with a *pH indicator*, which is any chemical whose color changes with pH. Many pH indicators are found in plants; the pigment of red cabbage is a good example. This pigment is red at low pH values (acidic), light purple at slightly acidic pH values, blue at neutral pH values, light green at moderately alkaline pH values, and dark green at very alkaline pH values.

Boil shredded red cabbage in water for about 5 minutes. Strain the broth from the cabbage and allow it to cool. Add the cooled blue broth to at least three cups of clear water so that each cup is less than half-filled or to three white porcelain bowls. Add a teaspoon of white vinegar to one cup and a teaspoon of baking soda to the second cup. Watch for color changes. Add nothing to the third cup so that it remains blue.

What color is the red cabbage before being boiled? Are the juices in red cabbage more or less acidic than vinegar? What would happen to the baking soda solution if you were to slowly add vinegar to it? What color would you get if a teaspoon of concentrated broth were added to a glass of water?

32. Make a concentrated solution of red cabbage extract by boiling a cup of shredded red cabbage in a cup of water for about 5 minutes. Pour this extract into a large transparent container, such as a vase or a 2-L plastic bottle with the top cut off. Add a tablespoon of white vinegar to acidify the solution. Ask yourself or a classmate what would happen to the pH of this solution if you were to fill the container with plain water. Would the pH go up or down, or stay the same? Test your prediction. As you dilute the solution, the color grows lighter, but what happens to its hue? How can adding plain water change the pH of a solution? Might plain water be used to make this solution alkaline?

33. Add about an inch of water to a large test tube followed by a couple drops of phenolphthalein pH indicator, which you will likely need to obtain from your classroom. Add a small pinch of washing soda, which contains sodium carbonate, Na_2CO_3. Upon mixing, the washing soda turns the solution basic, as evidenced by the pink color that forms. Neutralize this base by adding an acid, but not just any acid—use the

acid of your breath. Bubble your breath into the solution through a straw until the pink color disappears. What acid are you adding? How does this activity relate to the acidity of rain? Why do you want to add only a small pinch of washing soda and not a tablespoon?

34. Copper metal reacts slowly with the oxygen in air to form reddish copper (I) oxide, Cu_2O, which is a compound that coats the surface of older pennies, making them look tarnished. When such a penny is placed in a solution of salt in vinegar, the copper (I) oxide acts as a base and reacts with the vinegar to form copper salts. This effectively cleans the penny. The copper salts can then be transformed back into copper metal when exposed to an iron nail.

Stir about half a teaspoon of salt into about half a cup of white distilled vinegar. Use a nonmetal container, such as a ceramic or plastic bowl. Dip a tarnished penny halfway into the solution and notice the rapid cleaning effect. Add at least a dozen tarnished pennies to the solution. As they get cleaned, this will increase the concentration of copper ions in solution. Sandpaper an iron nail to give it a clean surface, and then rest the nail in the vinegar solution for about 10 minutes. Watch for the formation of copper metal on the nail.

Are copper ions positively or negatively charged? What is the charge on the copper atoms in Cu_2O? What is the charge on the oxygen? What must copper ions gain in order to transform back into a metallic form? What was the iron of the nail able to do for the copper ions?

35. Silver tarnishes because it reacts with the small amounts of smelly hydrogen sulfide, H_2S, we put into the air as we digest our food. In this reaction, the silver loses electrons to the sulfur. You can reverse this reaction by allowing the silver to get its electrons back from aluminum.

Flatten some aluminum foil on the bottom of a cooking pot. Fill the pot halfway with water and bring the water to a boil. When the water boils, remove the pot from its heat source. Add a couple tablespoons of baking soda to the hot water. Slowly immerse a tarnished piece of silver in the water and allow the silver to touch the aluminum foil. You should see an immediate effect once the silver and aluminum make contact. (Add more baking soda if you don't.) If your silver piece is very tarnished, you may notice the unpleasant odor of hydrogen sulfide as it is released back into the air.

As silver tarnishes, is it oxidized or reduced? Is baking soda, $NaHCO_3$, an ionic compound? Why is it needed in this activity? Does aluminum behave as an oxidizing agent or a reducing agent as it restores the silver to its untarnished state?

36. A battery is made by connecting a metal that tends to lose electrons with another metal that tends to gain electrons. Two metals that make for a good battery are copper and zinc, both of which are found in any post-1982 penny. Zinc has a low melting point, which makes it easy to remove from the inside of the penny. Get a ceramic bowl and then, wearing safety glasses, grab a post-1982 penny with metal tongs or pliers and place that penny over a blue flame, such as from a Bunsen burner or a kitchen gas stove. After a minute or so, the penny will appear to blister. At this point quickly bring the penny over the ceramic bowl. Tap the tongs or pliers on the bowl's edge to cause the molten zinc to drip into the bowl. After it cools, clean it off in running water. Touch the smooth side of the zinc to the very tip of your tongue alongside a new clean penny. As the two metals make contact on the wet surface of your tongue, you will be able to sense a small tingling, warming sensation. That's the electric current running between the two metals. Notice this happens only when the zinc and penny are together. You have made a battery out of a penny! Why is the effect more pronounced when your tongue is wet with salt water?

37. You can see the electrolysis of water by immersing the top of a disposable 9-V battery in salt water. The bubbles that form contain hydrogen gas produced as the water decomposes. Why does this activity work better with salt water than with tap water? Sharpen both ends of two pencils. Hold a pencil tip to each electrode while submerging the opposite ends of the pencils into salt water. Why does one electrode put out twice as much gas as the other?

THINK AND SOLVE (MATHEMATICAL APPLICATION)

38. Show that the hydroxide-ion concentration in an aqueous solution is 1×10^{-4} M when the hydronium-ion concentration is 1×10^{-10} M. Recall that $10^a \times 10^b = 10^{(a+b)}$.

39. When the hydronium-ion concentration of a solution is 1×10^{-10} M, what is the pH of the solution? Is the solution acidic or basic? When the hydronium-ion concentration of a solution is 1×10^{-4} M, what is the pH of the solution? Is the solution acidic or basic?

40. Show that an aqueous solution having a pH of 5 has a hydroxide-ion concentration of 1×10^{-9} M.

41. When the pH of a solution is 1, the concentration of hydronium ions is 10^{-1} M $= 0.1$ M. Assume that the volume of this solution is 500 mL. What is the pH after 500 mL of water is added? You will need a calculator with a logarithm function to answer this question.

42. Show that the pH of a solution is -0.301 when its hydronium-ion concentration equals 2 moles/L. Is the solution acidic or basic?

43. Each year about 1.6×10^7 (16 million) metric tons (mt) of aluminum are produced. How many grams is this? (Recall that 1 mt is 1000 kg.)

44. Use the following balanced chemical equation to show that the production of 1.6×10^7 metric tons of aluminum, Al, through electrolysis each year produces about 2.0×10^7 metric tons of carbon dioxide, CO_2. Interestingly, the total mass of our atmosphere is only about 5×10^{15} metric tons.

$$4 Al_2O_3 + 3 C \longrightarrow 4 Al + 3 CO_2$$

THINK AND RANK (ANALYSIS)

45. Rank the solutions in order of increasing concentration of hydronium ions, H_3O^+: (a) hydrogen chloride, HCl (concentration $= 2$ M); (b) acetic acid, CH_3COOH (concentration $= 2$ M); (c) ammonia, NH_3 (concentration $= 2$ M).

46. The three chemicals listed below are all very weak acids because they all have a difficult time losing a hydrogen ion, H^+. Upon losing this hydrogen ion, the central atom of each of these molecules takes on a negative charge. Holding onto this negative charge isn't easy, especially when there are positively charged hydrogen ions, H^+, floating about ready to combine with this negative charge. Review the concept of electronegativity in Section 15.6, and rank the acidity of these molecules in order from strongest to weakest: (a) ammonia, NH_3; (b) water, H_2O; (c) methane, CH_4.

47. Rank in order of decreasing pH the rain that fell on the Hawaiian island of Kauai on the mornings of (a) January 18, 1778; (b) December 7, 1941; (c) May 8, 2010. Assume there were no active volcanoes on or around Kauai during these times.

48. Review the concept of electronegativity in Section 15.6, and rank these elements from the weakest to strongest oxidizing agent: (a) chlorine, Cl; (b) sulfur, S; (c) sodium, Na.

49. Review the concept of electronegativity in Section 15.6, and rank these elements from the weakest to strongest reducing agent: (a) chlorine, Cl; (b) sulfur, S; (c) sodium, Na.

50. Rank the following molecules from least oxidized to most oxidized:

Ethane   Ethanol

Acetaldehyde   Acetic aid

EXERCISES (SYNTHESIS)

51. An acid and a base react to form a salt, which consists of positive and negative ions. Which forms the positive ions: the acid or the base? Which forms the negative ions?

52. Identify each substance in these reactions as an acid or a base:

(a) $H_3O^+ + Cl^- \rightleftharpoons H_2O + HCl$

(b) $H_2PO_4^- + H_2O \rightleftharpoons H_3O^+ + HPO_4^{2-}$

(c) $HSO_4^- + H_2O \rightleftharpoons OH^- + H_2SO_4$

(d) $O^{2-} + H_2O \rightleftharpoons OH^- + OH^-$

53. Many of the smelly molecules of cooked fish are alkaline compounds. How might these smelly molecules be conveniently transformed into less smelly salts just before you eat the fish?

54. Does the phosphate ion, a common additive to automatic dishwasher detergent, tend to behave as an acid or a base? Explain.

$$
\begin{array}{c}
\quad O \\
\quad \| \\
{}^{-}O - P - O^{-} \\
\quad | \\
\quad O^{-}
\end{array}
$$

Phosphate ion

55. The main component of bleach is sodium hypochlorite, $NaOCl$, which consists of sodium ions, Na^{+}, and hypochlorite ions, ^{-}OCl. What products are formed when this compound reacts with the hydrochloric acid, HCl, of toilet bowl cleaner?

56. How readily an acid donates a hydrogen ion is a function of how well the acid is able to accommodate the resulting negative charge it gains after donating. Which should be the stronger acid: water or hypochlorous acid? Explain.

$$
\begin{array}{c}
H \diagdown O \\
\quad \diagdown \\
\quad H
\end{array}
\qquad
\begin{array}{c}
H \diagdown O \\
\quad \diagdown \\
\quad Cl
\end{array}
$$

Water   Hypochlorous acid

57. Which should be a stronger base: ammonia, NH_3, or nitrogen trifluoride, NF_3?

$$
\begin{array}{c}
\quad \cdot\cdot \\
H - N - H \\
\quad | \\
\quad H
\end{array}
\qquad
\begin{array}{c}
\quad \cdot\cdot \\
F - N - F \\
\quad | \\
\quad F
\end{array}
$$

Ammonia   Nitrogen Trifluoride

58. Some molecules are able to stabilize a negative charge by passing it from one atom to the next by a flip-flopping of double bonds. This occurs when the negative charge is one atom away from an oxygen double bond as shown. Note that the curved arrows indicate the movement of electrons.

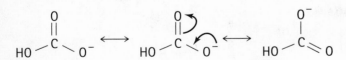

Why then is sulfuric acid so much stronger an acid than carbonic acid?

$$
\begin{array}{c}
\quad O \\
\quad \| \\
HO \diagup C \diagdown OH
\end{array}
\qquad
\begin{array}{c}
\quad O \\
\quad \| \\
HO - S - OH \\
\quad \| \\
\quad O
\end{array}
$$

Carbonic acid

Sulfuric acid

59. The amphoteric reaction between two water molecules is endothermic, which means the reaction requires the input of heat energy in order to proceed:

$$Energy + H_2O + H_2O \rightleftharpoons H_3O^{+} + OH^{-}$$

The warmer the water, the more heat energy is available for this reaction, and the more hydronium and hydroxide ions are formed.

(a) Which has a lower pH: pure water that is hot or pure water that is cold?

(b) Is it possible for water to be neutral but have a pH less than or greater than 7.0?

60. In a neutral solution of supercritical water (374°C, 218 atm) the pH equals about 2. What is the concentration of hydronium ions in this neutral solution? What is the concentration of hydroxide ions? Why is supercritical water so corrosive?

61. What is the concentration of hydronium ions in a solution that has a pH of -3? Why is such a solution impossible to prepare?

62. Can an acidic solution be made less acidic by adding an acidic solution?

63. Bubbling carbon dioxide into water causes the pH of the water to go down (become more acidic) because of the formation of carbonic acid. Will the pH also drop when carbon dioxide is bubbled into a solution of 1 M hydrochloric acid, HCl?

64. Pour vinegar onto beach sand from the Caribbean and the result is a lot of froth and bubbles. Pour vinegar onto beach sand from California, however, and nothing happens. Why?

65. What happens to the pH of soda water as it loses its carbonation?

66. Lakes lying in silicon dioxide, SiO_2, containing granite basins tend to become acidified by acid rain more readily than lakes lying in calcium carbonate, $CaCO_3$, limestone basins. Explain why.

67. How might warmer oceans accelerate global warming?

68. What element is oxidized in the equation and what element is reduced?

$$I_2 + 2\,Br^{-} \longrightarrow 2\,I^{-} + Br_2$$

69. What element behaves as the oxidizing agent in the equation and what element behaves as the reducing agent?

$$Sn^{2+} + 2\,Ag \longrightarrow Sn + 2\,Ag^{+}$$

70. Hydrogen sulfide, H_2S, burns in the presence of oxygen, O_2, to produce water, H_2O, and sulfur dioxide, SO_2. Through this reaction, is sulfur oxidized or reduced?

$$2\,H_2S + 3\,O_2 \longrightarrow 2\,H_2O + 2\,SO_2$$

71. Unsaturated fatty acids, such as $C_{12}H_{22}O_2$, react with hydrogen gas, H_2, to form saturated fatty acids, such as $C_{12}H_{24}O_2$. Are the unsaturated fatty acids being oxidized or reduced in this process?

72. A major source of chlorine gas, Cl_2, is the electrolysis of brine, which is concentrated salt water, $NaCl(aq)$. What other two products result from this electrolysis reaction? Write the balanced chemical equation.

73. Pennies manufactured after 1982 are made of zinc metal, Zn, in a coat of copper metal, Cu. Zinc is more easily oxidized than copper. Why, then, don't these pennies quickly corrode?

74. The general chemical equation for photosynthesis is shown below. Through this reaction are the oxygens of the water molecules, H_2O, oxidized or reduced?

$$6\ CO_2 + 6\ H_2O \longrightarrow C_6H_{12}O_6 + 6\ O_2$$

75. A chemical equation for the combustion of propane, C_3H_8, is shown. Through this reaction is the carbon oxidized or reduced?

$$C_3H_8 + 5\ O_2 \longrightarrow 3\ CO_2 + 4\ H_2O$$

76. Chemical equations need to be balanced not only in terms of the number of atoms but also by the charge. In other words, just as there should be the same number of atoms before and after the arrow of an equation, there should be the same charge. Take this into account to balance this chemical equation:

$$Sn^{2+} + Ag \longrightarrow Sn + Ag^+$$

77. Study Exercise 76 before attempting to balance both the atoms and charges of this chemical equation:

$$Fe^{3+} + I^- \longrightarrow Fe^{2+} + I_2$$

78. Water is 88.88% oxygen by mass. Oxygen is exactly what a fire needs to grow brighter and stronger. So why doesn't a fire grow brighter and stronger when water is added to it?

79. Why is the air over an open flame always moist?

80. Iron atoms have a greater tendency to oxidize than do copper atoms. Is this good news or bad news for a home in which much of the plumbing consists of iron and copper pipes connected together? Explain.

81. Copper atoms have a greater tendency to be reduced than iron atoms do. Was this good news or bad news for the Statue of Liberty, whose copper exterior was originally held together by steel rivets?

82. The type of iron that the human body needs for good health is the Fe^{2+} ion. Cereals fortified with iron, however, usually contain small grains of elemental iron, Fe. What must the body do to this elemental iron to make use of it: oxidation or reduction?

83. When ingested, grain alcohol, C_2H_6O, is metabolized into acetaldehyde, C_2H_4O, which is a toxic substance that causes headaches as well as joint pains typical of a "hangover." Is the grain alcohol oxidized or reduced as it transforms into acetaldehyde?

84. Your body creates chemical energy from the food you eat, which contains carbohydrates, lipids, and proteins. These are large molecules that first need to be broken down into simpler molecules, such as monosaccharides, $C_6H_{12}O_6$, fatty acids, $C_{14}H_{28}O_2$; and amino acids, $C_2H_5NO_2$. Of these molecules, which has the greatest supply of hydrogen atoms per molecule? Which is the most reduced in terms of the fewest number of oxygen atoms per carbon atom? Which can react with the most oxygen molecules to produce the most energy? Which provides the most dietary Calories per gram?

85. Do the digestion and subsequent metabolism of foods and drugs tend to make the molecules of the foods and drugs more or less polar?

86. Why is it easier for the body to excrete a polar molecule than it is to excrete a nonpolar molecule? What chemistry does the body use to get rid of molecules it no longer needs?

DISCUSSION QUESTIONS (EVALUATION)

87. Scientists have experimented with ways of enhancing the ocean's ability to absorb atmospheric carbon dioxide. Adding powdered iron to a small plot of the ocean, they found, has the effect of fostering the growth of microorganisms that increase the rate at which carbon dioxide is absorbed. Might this be a solution to the problem of global warming? Why or why not?

88. Why are atmospheric CO_2 levels routinely up to 15 ppm higher in the spring than in the fall? Why are seasonal fluctuations in atmospheric CO_2 levels much more pronounced in the Northern Hemisphere than in the Southern Hemisphere?

89. Can industries be trusted to self-regulate the amount of pollution they produce? Is government really necessary to enforce these regulations? Shouldn't the conscience of the consumer and the economic advantages of sustainable practices be sufficient to motivate industries to protect the environment?

90. In the centralized model for generating electricity, a relatively small number of power plants produce the massive amounts of electricity that everyone needs. In the decentralized model, electricity is generated by numerous smaller substations, which may include personal wind turbines or photovoltaics. What are the advantages and disadvantages of each model? When should one be favored over the other?

READINESS ASSURANCE TEST (RAT)

If you have a good handle on this chapter, if you really do, then you should be able to score 7 out of 10 on this RAT. If you score less than 7, you need to study further before moving on.

Choose the BEST answer to each of the following.

1. What is the relationship between the hydroxide ion and a water molecule?
   (a) A hydroxide ion is a water molecule plus a proton.
   (b) A hydroxide ion and a water molecule are the same things.
   (c) A hydroxide ion is a water molecule minus a hydrogen nucleus.
   (d) A hydroxide ion is a water molecule plus two extra electrons.

2. What happens to the corrosive properties of an acid and a base after they neutralize each other? Why?
   (a) The corrosive properties are neutralized because the acid and base no longer exist.
   (b) The corrosive properties are embedded in the noncorrosive salt.
   (c) The corrosive properties are doubled because the acid and base are combined in the salt.
   (d) The corrosive properties are neutralized by the salt.

3. Sodium hydroxide, NaOH, is a strong base, which means that it readily accepts hydrogen ions. What products are formed when sodium hydroxide accepts a hydrogen ion from a water molecule?
   (a) sodium ions, hydroxide ions, and water
   (b) sodium ions, hydroxide ions, and hydronium ions
   (c) sodium ions and hydronium ions
   (d) sodium ions and water

4. A weak acid is added to a concentrated solution of hydrochloric acid. Does the solution become more or less acidic? Why?
   (a) more acidic, because there are more hydronium ions being added to the solution
   (b) less acidic, because the solution becomes more dilute with a less concentrated solution of hydronium ions being added to it
   (c) no change in acidity, because the concentration of the hydrochloric acid is too high to be changed by the weak solution
   (d) less acidic, because the concentration of hydroxide ions will increase

5. Why do we use the pH scale to indicate the acidity of a solution rather than simply stating the concentration of hydronium ions?
   (a) The pH scale includes the concentrations of hydronium and hydroxide ions.
   (b) The general public understands the pH scale.
   (c) It is more accurate to use the pH scale.
   (d) The pH scale is more convenient, since the concentration of hydronium ions is usually so low.

6. When the hydronium-ion concentration equals 1 mole/L, what is the pH of the solution? Is the solution acidic or basic?
   (a) pH = 0; acidic
   (b) pH = 1; acidic
   (c) pH = 10; basic
   (d) pH = 7; neutral

7. When lightning strikes, nitrogen molecules, N_2, and oxygen molecules, O_2, in the air react to form nitrates, NO_3^{1-}, which come down in the rain to help fertilize the soil. Is this an example of oxidation or reduction?
   (a) The formation of nitrates is an example of reduction.
   (b) The formation of nitrates is an example of oxidation.
   (c) Both. The nitrogen is oxidized as it reacts with the oxygen while the oxygen is reduced.
   (d) Neither. Although the bonds of both the N_2 and O_2 molecules are broken to form NO_3^{1-}, neither oxidation nor reduction occurs.

8. What element is oxidized in the equation and what element is reduced?

$$Sn^{2+} + 2\,Ag \longrightarrow Sn + 2\,Ag^+$$

   (a) The tin ion, Sn^{2+}, is oxidized, while the silver, Ag, is reduced.
   (b) The tin ion, Sn^{2+}, is reduced, while the silver, Ag, is oxidized.
   (c) Both the tin ion, Sn^{2+}, and the silver, Ag, are reduced.
   (d) Both the tin ion, Sn^{2+}, and the silver, Ag, are oxidized.

9. How does an atom's electronegativity relate to its ability to become oxidized?
   (a) The greater the electronegativity of an atom, the greater its ability to become oxidized.
   (b) The lower the electronegativity of an atom, the lower its ability to become oxidized.
   (c) The greater the electronegativity of an atom, the lower its ability to become oxidized.
   (d) Electronegativity does not affect the atom's ability to become oxidized.

10. Why does a battery that has thick zinc walls last longer than one that has thin zinc walls?
   (a) Thick zinc walls prevent the battery from overheating.
   (b) Thick zinc walls prevent electrons from being lost into the surrounding environment.
   (c) Thick zinc walls hold in the battery acid longer.
   (d) The zinc walls are transformed into zinc ions as the battery provides electricity.

Answers to RAT

1. c, 2. a, 3. a, 4. b, 5. d, 6. a, 7. c, 8. b, 9. c, 10. d

Vanillin

Tetramethylpyrazine

CHAPTER 19
Organic Compounds

CARBON ATOMS are perhaps the most versatile of all atoms. Add to this the fact that carbon atoms can also bond with atoms of other elements, and you see the possibility of an endless number of different carbon-based molecules. Each molecule has its own unique set of physical, chemical, and biological properties. The flavor of vanilla, for example, is perceived when the compound *vanillin* is absorbed by the sensory organs in the nose. The flavor of chocolate is generated when a selection of compounds, such as *tetramethylpyrazine*, are absorbed in the nose.

The study of carbon-containing compounds has come to be known as **organic chemistry**.

Because organic compounds are so closely tied to living organisms and because they have many applications—flavorings, fuels, polymers, medicines, agriculture, and more—it is important to have a basic understanding of them. We begin with the simplest organic compounds—those consisting of only carbon and hydrogen.

LEARNING OBJECTIVE
Identify the structures of hydrocarbons.

Mastering**PHYSICS**

TUTORIAL: Introduction to Organic Molecules

TUTORIAL: Organic Molecules and Isomers

19.1 Hydrocarbons

EXPLAIN THIS How is a road like an oil spill?

Organic compounds that contain only carbon and hydrogen atoms are called **hydrocarbons**, which differ from one another by the number of carbon and hydrogen atoms they contain. The simplest hydrocarbon is methane, CH_4, with only one carbon per molecule. Methane is the main component of natural gas. The hydrocarbon octane, C_8H_{18}, has eight carbons per molecule and is a component of gasoline. The hydrocarbon polyethylene contains hundreds of carbon and hydrogen atoms per molecule. Polyethylene is a plastic used to make many familiar items, such as milk containers and plastic bags.

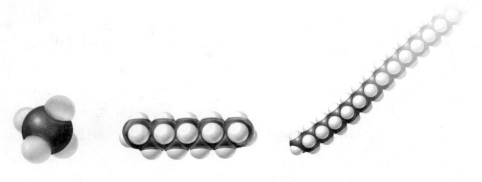

Methane, CH_4 Octane, C_8H_{18} Polyethylene

In looking at the stick structures, remember that each corner or end represents a carbon atom and that each carbon atom must be bonded four times. Because hydrogen atoms are assumed, they're not usually depicted.

Hydrocarbons also differ in the way the carbon atoms connect to one another. Figure 19.1 shows the three hydrocarbons pentane, isopentane, and neopentane. These hydrocarbons have the same molecular formula, C_5H_{12}, but they are structurally different from one another.

We can see the different structural features of pentane, isopentane, and neopentane more clearly by drawing the molecules in two dimensions, as shown in the middle row of Figure 19.1. Alternatively, we can represent them by the

FIGURE 19.1
These three hydrocarbons all have the same molecular formula. We can see their different structural features by highlighting the carbon framework in two dimensions. Easy-to-draw stick structures that use lines for all carbon–carbon covalent bonds are also useful.

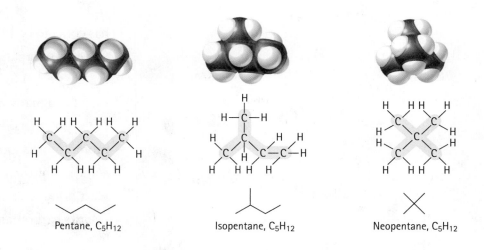

Pentane, C_5H_{12} Isopentane, C_5H_{12} Neopentane, C_5H_{12}

stick structures shown in the bottom row. A stick structure is a commonly used shorthand notation for representing an organic molecule. Each line (stick) represents a covalent bond, and carbon atoms are understood to exist at the end of any line or wherever two or more straight lines meet (unless another type of atom is drawn at the end of the line). Any hydrogen atoms bonded to the carbons are also typically not shown. Instead, their presence is only implied, so that the focus can remain on the skeletal structure that is formed by the carbon atoms.

Molecules such as pentane, isopentane, and neopentane have the same molecular formula, which means they have the same number of the same kinds of atoms. The way these atoms are put together, however, is different. We say that they each have their own **configuration**, which refers to how the atoms are connected. Different configurations result in different chemical structures. Molecules with the same molecular formula but different configurations (and hence different structures) are known as **structural isomers**. Structural isomers differ from each other and have different physical and chemical properties. For example, pentane has a boiling point of 36°C, isopentane's boiling point is 30°C, and neopentane's boiling point is 10°C.

The number of possible structural isomers for a chemical formula increases rapidly as the number of carbon atoms increases. There are three structural isomers for compounds that have the formula C_5H_{12}, 18 for C_8H_{18}, 75 for $C_{10}H_{22}$, and a whopping 366,319 for $C_{20}H_{42}$!

A carbon-based molecule can have different spatial orientations called **conformations**. Flex your wrist, elbow, and shoulder joints, and you'll find your arm passing through a range of conformations. Likewise, organic molecules can twist and turn about their carbon–carbon single bonds and thus have a range of conformations. The structures in Figure 19.2, for example, are different conformations of pentane. In the language of organic chemistry, we say that the *configuration* of a molecule, such as pentane, has a broad range of *conformations*. Change the configuration of pentane, however, and you no longer have pentane. Rather, you have a different structural isomer, such as isopentane, which has its own range of different conformations.

Which of the two is easier to change: the conformation or the configuration of your arm?

CHECKPOINT

Which carbon–carbon bond was rotated to go from the "before" conformation of isopentane to the "after" conformation?

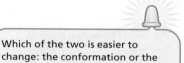

Before   After

Was this your answer?
The best way to answer any question about the conformation of a molecule is to play around with molecular models that you can hold in your hand. In this case, bond (c) rotates in such a way that the carbon at the right end of bond (d) comes up out of the plane of the page, momentarily points straight at you, and then plops back into the plane of the page below bond (c). This rotation is similar to that of the arm of an arm wrestler who, with the arm just above the table while on the brink of losing, suddenly gets a surge of strength and swings the opponent's arm (and his or her own) through a half-circle arc and wins.

Before   After

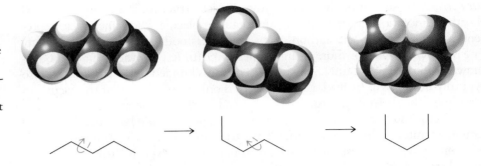

FIGURE 19.2
Three conformations for a molecule of pentane. The molecule looks different in each conformation, but the five-carbon framework is the same in all three conformations. In a sample of liquid pentane, the molecules are found in all conformations—not unlike a bucket of worms.

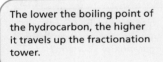

The lower the boiling point of the hydrocarbon, the higher it travels up the fractionation tower.

Hydrocarbons are obtained primarily from coal and petroleum. Most of the coal and petroleum that exists today was formed between 280 million and 395 million years ago when plant and animal matter decayed in the absence of oxygen. At that time, Earth was covered with extensive swamps that, because they were close to sea level, periodically became submerged. The organic matter of the swamps was buried beneath layers of marine sediments and was eventually transformed into either coal or petroleum.

Coal is a solid material that contains many large, complex hydrocarbon molecules. Most of the coal mined today is used to produce steel and to generate electricity at coal-burning power plants.

Petroleum, also called crude oil, is a liquid readily separated into its hydrocarbon components through a process known as *fractional distillation*, shown in Figure 19.3. The crude oil is heated in a pipe still to a temperature high enough to vaporize most of the components. The hot vapor flows into the bottom of a fractionating tower, which is warmer at the bottom than at the top. As the vapor rises in the tower and cools, the various components begin to condense. Hydrocarbons that have high boiling points, such as tar and lubricating stocks, condense first at warmer temperatures. Hydrocarbons that have low boiling points, such as gasoline, travel to the cooler regions at the top of the tower before condensing. Pipes drain the various liquid hydrocarbon fractions from the tower. Natural gas, which is primarily methane, does not condense. It remains a gas and is collected at the top of the tower.

Differences in the strength of molecular attractions explain why different hydrocarbons condense at different temperatures. As discussed in Section 15.8, in our comparison of induced dipole–induced dipole attractions in methane and octane, larger hydrocarbons experience many more of these attractions than smaller hydrocarbons do. For this reason, the larger hydrocarbons condense readily at high temperatures and so are found at the bottom of the tower. Smaller molecules, because they experience fewer attractions to their neighbors, condense only at the cooler temperatures found at the top of the tower.

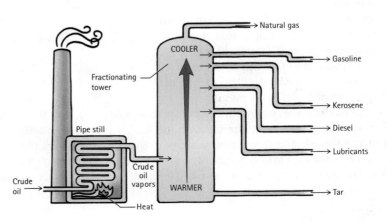

FIGURE 19.3
A schematic for the fractional distillation of petroleum into its useful hydrocarbon components.

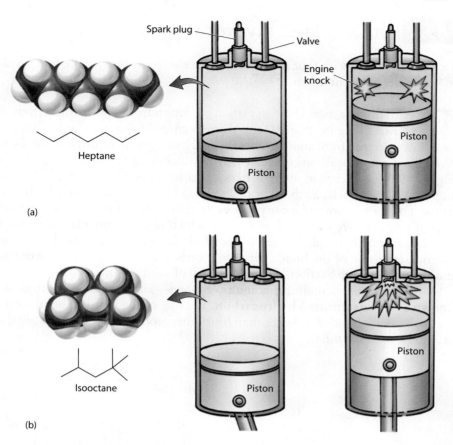

(a)

Heptane

Spark plug

Valve

Engine knock

Piston

Piston

Piston

(b)

Isooctane

Piston

The gasoline obtained from the fractional distillation of petroleum consists of a wide variety of hydrocarbons having similar boiling points. Some of these components burn more efficiently than others in a car engine. The straight-chain hydrocarbons, such as heptane, burn too quickly, causing what is called *engine knock*, as illustrated in Figure 19.4. Gasoline hydrocarbons that have more branching, such as isooctane, burn slowly and result in the engine running more smoothly. These two compounds, heptane and isooctane, are used as standards in assigning *octane ratings* to gasoline. An octane rating of 100 is arbitrarily assigned to isooctane, and heptane is assigned an octane rating of 0. The antiknock performance of a particular gasoline is compared with that of various mixtures of isooctane and heptane, and an octane rating is assigned. Figure 19.5 shows the octane information that appears on a typical gasoline pump.

FIGURE 19.5
Octane ratings are posted on gasoline pumps. Interestingly, the engines of modern cars are designed to run best on 87 octane grade fuel. With higher octane fuels, not only do you lose performance, but you lose your money as well.

CHECKPOINT

Which structural isomer shown in Figure 19.1 should have the highest octane rating?

Was this your answer?

The structural isomer with the greatest amount of branching in the carbon framework likely has the highest octane rating, making neopentane the clear winner. For your information, the ratings are as follows:

Compound	Octane Rating
Pentane	61.7
Isopentane	92.3
Neopentane	116

19.2 Unsaturated Hydrocarbons

EXPLAIN THIS With four unpaired valence electrons, how can carbon bond to only three adjacent atoms?

Recall from Section 15.1 that carbon has four unpaired valence electrons. As shown in Figure 19.6, each of these electrons is available for pairing with an electron from another atom, such as hydrogen, to form a covalent bond. In all the hydrocarbons discussed so far, including the methane shown in Figure 19.6, each carbon atom is bonded to four neighboring atoms by four single covalent bonds. Such hydrocarbons are known as **saturated hydrocarbons**. The term *saturated* means that each carbon has as many atoms bonded to it as possible. We now explore cases in which one or more carbon atoms in a hydrocarbon are bonded to fewer than four neighboring atoms. This occurs when at least one of the bonds between a carbon and a neighboring atom is a multiple bond. (See Section 15.5 for a review of multiple bonds.)

A hydrocarbon containing a multiple bond—either double or triple—is known as an **unsaturated hydrocarbon**. Because of the multiple bond, two of the carbons are bonded to fewer than four other atoms. These carbons are thus said to be *unsaturated*.

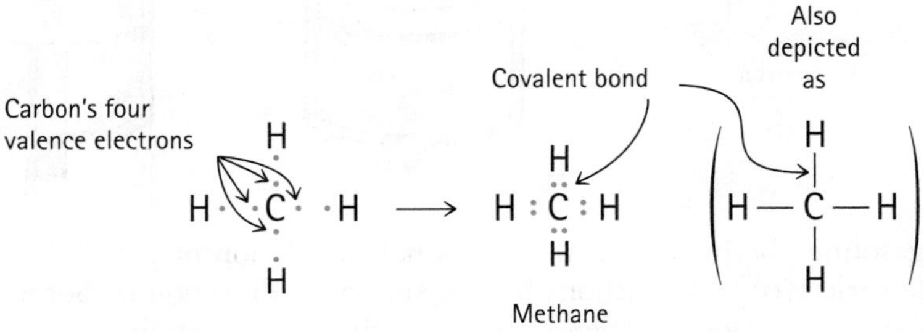

Figure 19.7 compares the saturated hydrocarbon butane with the unsaturated hydrocarbon 2-butene. The number of atoms that are bonded to each of the two middle carbons of butane is four, whereas each of the two middle carbons of 2-butene is bonded to only three other atoms—a hydrogen and two carbons.

An important unsaturated hydrocarbon is benzene, C_6H_6, which may be drawn as three double bonds contained within a flat hexagonal ring, as shown in Figure 19.8a. Unlike the double-bond electrons in most other unsaturated hydrocarbons, the electrons of the double bonds in benzene are not fixed between any two carbon atoms. Instead, these electrons can move freely around the ring. This is commonly represented by drawing a circle within the ring, as shown in Figure 19.8b, rather than by individual double bonds.

Many organic compounds contain one or more benzene rings in their structure. Because many of these compounds are fragrant, any organic molecule that contains a benzene ring is classified as an **aromatic compound** (even if it

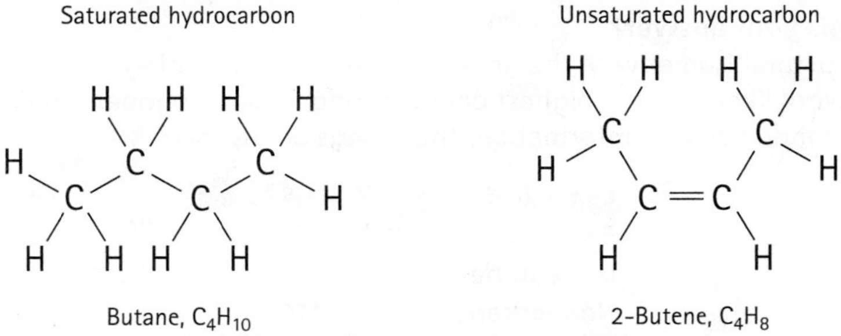

FIGURE 19.6
Carbon has four valence electrons. Each electron pairs with an electron from a hydrogen atom in the four covalent bonds of methane.

FIGURE 19.7
The carbons of the hydrocarbon butane are *saturated*, each being bonded to four other atoms. Because of the double bond, two of the carbons of the unsaturated hydrocarbon 2-butene are bonded to only three other atoms, which makes the molecule an unsaturated hydrocarbon.

Saturated hydrocarbon

Butane, C_4H_{10}

Unsaturated hydrocarbon

2-Butene, C_4H_8

is not particularly fragrant). Figure 19.9 shows three examples. Toluene, a common solvent used as a paint thinner, is toxic and gives airplane glue its distinctive odor. Some aromatic compounds, such as naphthalene, contain two or more benzene rings fused together. At one time, mothballs were made of naphthalene. Most mothballs sold today, however, are made of the less toxic 1,4-dichlorobenzene.

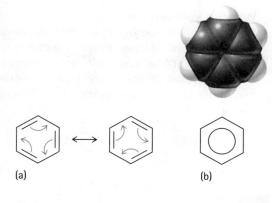

FIGURE 19.8
(a) The double bonds of benzene, C_6H_6, can migrate around the ring. (b) For this reason, they are often represented by a circle within the ring.

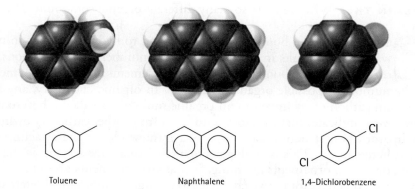

Toluene   Naphthalene   1,4-Dichlorobenzene

FIGURE 19.9
The structures for three odoriferous organic compounds that contain one or more benzene rings: toluene, naphthalene, and 1,4-dichlorobenzene.

An example of an unsaturated hydrocarbon containing a triple bond is acetylene, C_2H_2. A confined flame of acetylene burning in oxygen is hot enough to melt iron, which makes acetylene a choice fuel for welding (Figure 19.10).

Acetylene

FIGURE 19.10
The unsaturated hydrocarbon acetylene, C_2H_2, when burned in this torch, produces a flame that is hot enough to melt iron.

CHECKPOINT
Prolonged exposure to benzene increases the risk of developing certain cancers. The structure of aspirin contains a benzene ring. Does this indicate that prolonged exposure to aspirin increases a person's risk of developing cancer?

Benzene ring

Aspirin

Was this your answer?
No. Although benzene and aspirin both contain a benzene ring, these two molecules have different overall structures and quite different chemical properties. Each carbon-containing organic compound has its own set of unique physical, chemical, and biological properties. Although benzene may cause cancer, aspirin works as a safe remedy for headaches.

LEARNING OBJECTIVE
Discuss the significance of hetero-atoms in organic compounds.

Mastering**PHYSICS**

TUTORIAL: Functional Groups

The chemistry of hydrocarbons is surely interesting, but start adding heteroatoms to these organic molecules and the chemistry becomes extraordinarily interesting. The organic chemicals of living organisms, for example, all contain heteroatoms.

19.3 Functional Groups

EXPLAIN THIS Why are there so many different organic compounds?

Carbon atoms can bond to one another and to hydrogen atoms in many ways, which results in an incredibly large number of hydrocarbons. But carbon atoms can bond to atoms of other elements as well, further increasing the number of possible organic molecules. In organic chemistry, any atom other than carbon or hydrogen in an organic molecule is called a **heteroatom**, where *hetero-* indicates that the atom is different from either carbon or hydrogen.

A hydrocarbon structure can serve as a framework for the attachment of various heteroatoms. This is analogous to a Christmas tree serving as the scaffolding on which ornaments are hung. Just as the ornaments give character to the tree, so do heteroatoms give character to an organic molecule. Heteroatoms have profound effects on the properties of an organic molecule.

Consider ethane, C_2H_6, and ethanol, C_2H_6O, which differ from each other by only a single oxygen atom. Ethane has a boiling point of $-88°C$, making it a gas at room temperature, and it does not dissolve in water very well. Ethanol, by contrast, has a boiling point of $+78°C$, making it a liquid at room temperature. It is infinitely soluble in water, and it is the active ingredient of alcoholic beverages. Consider further ethylamine, C_2H_7N, which has a nitrogen atom on the same basic two-carbon framework. This compound is a corrosive, pungent, highly toxic gas—very different from either ethane or ethanol.

Organic molecules are classified according to the functional groups they contain. A **functional group** is defined as a combination of atoms that behave as a unit. Most functional groups are distinguished by the heteroatoms they contain; some common groups are listed in Table 19.1.

The remainder of this chapter introduces the classes of organic molecules shown in Table 19.1. The role heteroatoms play in determining the properties of each class is the underlying theme. As you study this material, focus on understanding the chemical and physical properties of the various classes of compounds; doing so will give you a greater appreciation of the remarkable diversity of organic molecules and their many applications.

CHECKPOINT

What is the significance of heteroatoms in an organic molecule?

Was this your answer?
Heteroatoms largely determine an organic molecule's "personality."

TABLE 19.1   FUNCTIONAL GROUPS IN ORGANIC MOLECULES

General Structure	Class	General Structure	Class
—C—OH *Hydroxyl group*	Alcohols	O ‖ C—H *Aldehyde group*	Aldehydes
C=C / C—OH *Phenolic group*	Phenols	O ‖ C—N *Amide group*	Amides
—C—O—C— *Ether group*	Ethers	O ‖ C—OH *Carboxyl group*	Carboxylic acids
—C—N *Amine group*	Amines	O ‖ C—O—C— *Ester group*	Esters
O ‖ —C—C—C— *Ketone group*	Ketones		

19.4 Alcohols, Phenols, and Ethers

EXPLAIN THIS What do alcohols, phenols, and ethers have in common?

Alcohols are organic molecules in which a *hydroxyl group* is bonded to a saturated carbon. The hydroxyl group consists of an oxygen bonded to a hydrogen. Because of the polarity of the oxygen–hydrogen bond, low-mass alcohols are often soluble in water, which is itself very polar. Three common alcohols and their melting and boiling points are listed in Table 19.2.

More than 11 billion pounds of methanol, CH_3OH, are produced annually in the United States. Most of it is used for making formaldehyde and acetic acid, important starting materials in the production of plastics. In addition, methanol is used as a solvent, an octane booster, and an anti-icing agent in gasoline. Sometimes called *wood alcohol* because it can be obtained from wood, methanol should never be ingested because, in the body, it is metabolized to formaldehyde and formic acid. Formaldehyde is harmful to the eyes, can lead to blindness, and was once used to preserve dead biological specimens. Formic acid, the active ingredient in an ant bite, can lower the pH of the blood to dangerous levels. Methanol has its own inherent toxicities. Ingesting only about 15 mL (about 3 Tbsp) of methanol may lead to blindness, and about 30 mL can cause death.

Ethanol, C_2H_5OH, on the other hand, is the "alcohol" of alcoholic beverages, and it is one of the oldest chemicals manufactured by humans. Ethanol is

LEARNING OBJECTIVE
Review the general properties of alcohols, phenols, and ethers.

—C—OH
Hydroxyl group

TABLE 19.2   SIMPLE ALCOHOLS

Structure	Scientific Name	Common Name	Melting Point (°C)	Boiling Point (°C)
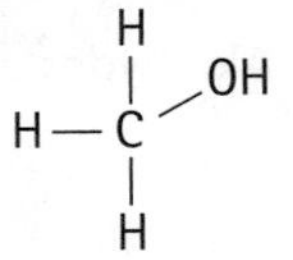 (methanol structure)	Methanol	Methyl alcohol	−97	65
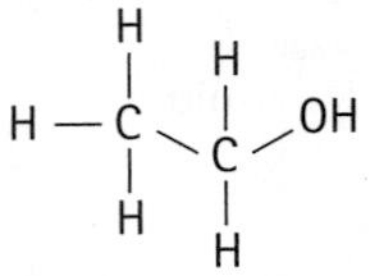 (ethanol structure)	Ethanol	Ethyl alcohol	−115	78
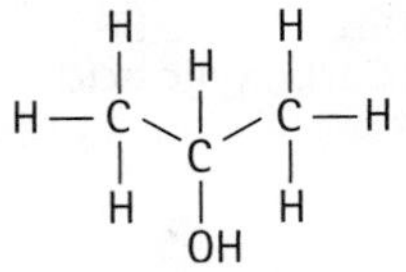 (2-propanol structure)	2-Propanol	Isopropyl alcohol	−126	97

prepared by feeding the sugars of various plants to certain yeasts, which produce ethanol through a biological process known as *fermentation*. Ethanol is also widely used as an industrial solvent. For many years, ethanol intended for this purpose was made by fermentation, but today industrial-grade ethanol is more cheaply manufactured from petroleum by-products, such as ethene, as Figure 19.11 illustrates.

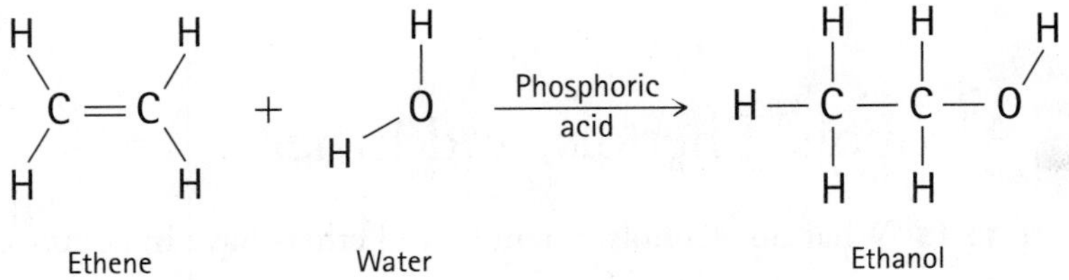

FIGURE 19.11
Ethanol can be synthesized from the unsaturated hydrocarbon ethene, with phosphoric acid as a catalyst.

Ethene   Water   Ethanol

The liquid produced by fermentation has an ethanol concentration no greater than about 12% because at this concentration the yeast cells begin to die. This is why most wines have an alcohol content of about 12%—they are produced solely by fermentation. To attain the higher ethanol concentrations found in "hard" alcoholic beverages such as gin and vodka, the fermented liquid must be distilled. In the United States, the ethanol content of distilled alcoholic beverages is measured as *proof*, which is twice the percentage of ethanol. An 86-proof whiskey, for example, is 43% ethanol by volume. The term *proof* evolved from a crude method once employed to test alcohol content. Gunpowder was wetted with a beverage of suspect alcohol content. If the beverage was primarily water, the powder did not ignite. If the beverage contained a significant amount of ethanol, the powder burned, thus providing "proof" of the beverage's worth.

A third well-known alcohol is isopropyl alcohol, also called 2-propanol. This is the rubbing alcohol you buy at the drugstore. Although 2-propanol has a relatively high boiling point, it evaporates readily, leading to a pronounced cooling effect when it is applied to skin—an effect once used to reduce fevers. (Isopropyl alcohol is very toxic if ingested. See the activities at the end of this chapter to understand why. In place of isopropyl alcohol, washcloths wetted with cold water are nearly as effective in reducing fever, and they are far safer.)

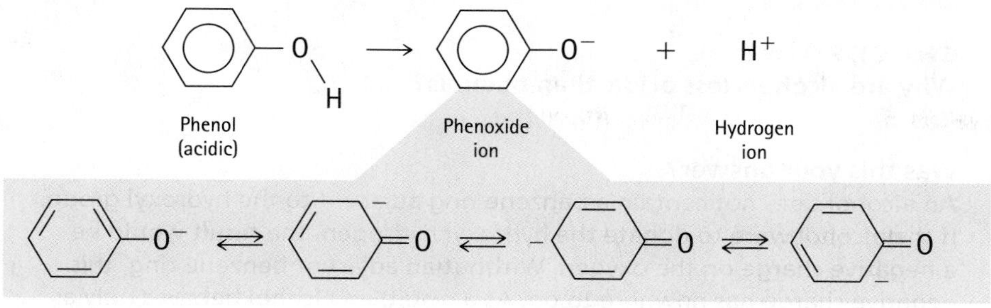

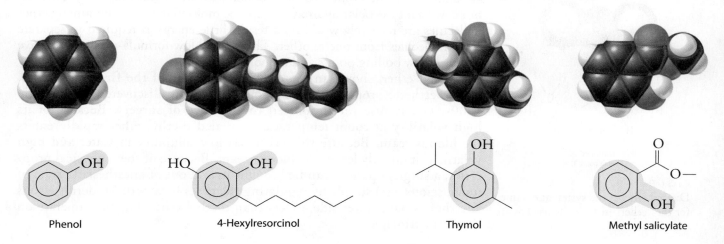

Phenol (acidic) → **Phenoxide ion** + **Hydrogen ion**

FIGURE 19.12
The negative charge of the phenoxide ion can migrate to select positions on the benzene ring. This mobility helps accommodate the negative charge, which is why the phenolic group readily donates a hydrogen ion.

You are probably most familiar with the use of isopropyl alcohol as a topical disinfectant.

Phenols contain a phenolic group, which consists of a hydroxyl group attached to a benzene ring. Because of the presence of the benzene ring, the hydrogen of the hydroxyl group is readily lost in an acid–base reaction, which makes the phenolic group mildly acidic.

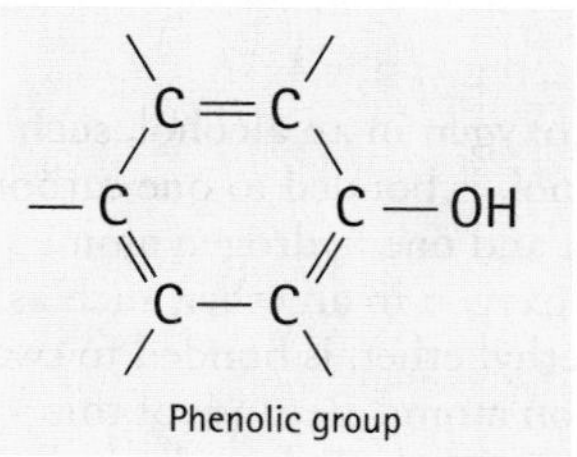

The reason for this acidity is illustrated in Figure 19.12. How readily an acid donates a hydrogen ion is a function of how well the acid can accommodate the resulting negative charge it gains after donating the hydrogen ion. After phenol donates the hydrogen ion, it becomes a negatively charged phenoxide ion. The negative charge of the phenoxide ion, however, is not restricted to the oxygen atom. Recall that the electrons of the benzene ring can migrate around the ring. In a similar manner, the electrons responsible for the negative charge of the phenoxide ion can migrate around the ring, as shown in Figure 19.12. Just as several people can easily hold a hot potato by quickly passing it around, the phenoxide ion can easily hold the negative charge because the charge gets passed around. Because the negative charge of the ion is so nicely accommodated, the phenolic group is more acidic than it would be otherwise.

The simplest phenol, shown in Figure 19.13, is called phenol. In 1867, Joseph Lister (1827–1912) discovered the antiseptic value of phenol, which, when applied to surgical instruments and incisions, greatly increased surgery survival rates. Phenol was the first purposefully used antibacterial solution, or *antiseptic*. Phenol damages healthy tissue, however, and so a number of milder phenols have since been introduced. The phenol 4-hexylresorcinol, for example, is commonly used in throat lozenges and mouthwashes. This compound has even greater antiseptic properties than phenol, and yet it does not damage tissue. Listerine brand mouthwash (named after Joseph Lister) contains the antiseptic phenols thymol and methyl salicylate.

Phenolic group

We're classifying organic molecules based on the functional groups they contain. As you will see shortly, however, organic molecules may contain more than one type of functional group. A single organic molecule, therefore, might be classified as both a phenol and an ether.

FIGURE 19.13
Every phenol contains a phenolic group (highlighted in blue).

Phenol   4-Hexylresorcinol   Thymol   Methyl salicylate

CHECKPOINT

Why are alcohols less acidic than phenols?

Was this your answer?
An alcohol does not contain a benzene ring adjacent to the hydroxyl group. If the alcohol were to donate the hydroxyl hydrogen, the result would be a negative charge on the oxygen. Without an adjacent benzene ring, this negative charge has nowhere to go. As a result, an alcohol behaves only as a very weak acid, much the way water does.

FIGURE 19.14
The oxygen in an alcohol, such as ethanol, is bonded to one carbon atom and one hydrogen atom. The oxygen in an ether, such as dimethyl ether, is bonded to two carbon atoms. Because of this difference, alcohols and ethers of similar molecular mass have vastly different physical properties.

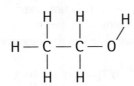

Ethanol: Soluble in water, boiling point 78°C

Dimethyl ether: Insoluble in water, boiling point −25°C

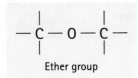

Ether group

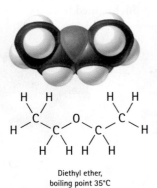

Diethyl ether, boiling point 35°C

FIGURE 19.15
Diethyl ether is the systematic name for the "ether" historically used as an anesthetic.

Ethers are organic compounds structurally related to alcohols. The oxygen atom in an ether group, however, is bonded not to a carbon and a hydrogen but rather to two carbons. As we see in Figure 19.14, ethanol and dimethyl ether have the same chemical formula, C_2H_6O, but their physical properties are vastly different. Whereas ethanol is a liquid at room temperature (boiling point 78°C) and mixes quite well with water, dimethyl ether is a gas at room temperature (boiling point −25°C) and is much less soluble in water.

Ethers are not very soluble in water because without the hydroxyl group they are unable to form strong hydrogen bonds with water (Section 15.8). Furthermore, without the polar hydroxyl group, the molecular attractions among ether molecules are relatively weak. As a result, little energy is required to separate ether molecules from one another. This is why low-formula-mass ethers have relatively low boiling points and evaporate so readily.

Diethyl ether, shown in Figure 19.15, was one of the first anesthetics. The anesthetic properties of this compound were discovered in the early 1800s, and its use revolutionized the practice of surgery. Because of its high volatility at room temperature, inhaled diethyl ether rapidly enters the bloodstream. Because this ether has low solubility in water and high volatility, it quickly leaves the bloodstream. Because of these physical properties, a surgical patient can be brought in and out of anesthesia (a state of unconsciousness) simply by regulating the gases breathed. Modern gaseous anesthetics have fewer side effects than diethyl ether, but they operate on the same principle.

19.5 Amines and Alkaloids

EXPLAIN THIS Why are rainforests of great interest to pharmaceutical companies?

Amines are organic compounds that contain the amine group—a nitrogen atom bonded to one, two, or three saturated carbons. Amines are typically less soluble in water than are alcohols because the nitrogen–hydrogen bond is not quite as polar as the oxygen–hydrogen bond. The lower polarity of amines also means their boiling points are typically somewhat lower than those of alcohols of similar formula mass. Table 19.3 lists three simple amines.

Amine group

Structure	Name	Melting Point (°C)	Boiling Point (°C)
	Ethylamine	−18	17
	Diethylamine	−50	55
	Triethylamine	−7	89

TABLE 19.3   THREE SIMPLE AMINES

One of the most notable physical properties of many low-formula-mass amines is their offensive odor. Figure 19.16 shows two appropriately named amines, putrescine and cadaverine, which are partly responsible for the odor of decaying flesh.

Amines are typically alkaline because the nitrogen atom readily accepts a hydrogen ion from water, as Figure 19.17 illustrates.

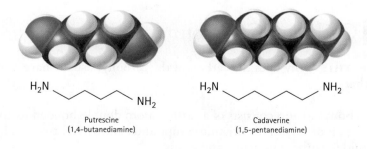

Putrescine
(1,4–butanediamine)

Cadaverine
(1,5-pentanediamine)

FIGURE 19.16
Low-formula-mass amines such as these tend to have offensive odors.

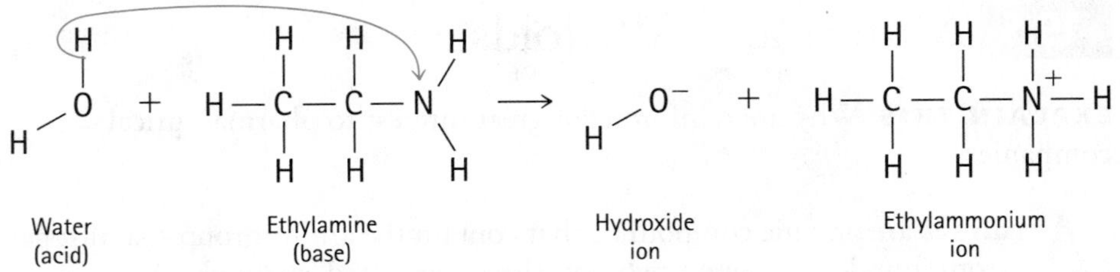

FIGURE 19.17
Ethylamine acts as a base and accepts a hydrogen ion from water to become the ethylammonium ion. This reaction generates a hydroxide ion, which increases the pH of the solution.

Water (acid) + Ethylamine (base) → Hydroxide ion + Ethylammonium ion

A group of naturally occurring complex molecules that are alkaline because they contain nitrogen atoms are often called *alkaloids*. Because many alkaloids have medicinal value, there is great interest in isolating these compounds from plants or marine organisms that contain them. As shown in Figure 19.18, an alkaloid reacts with an acid to form a salt that is usually quite soluble in water. This is in contrast to the non-ionized form of the alkaloid, known as a *free base*, which is typically insoluble in water.

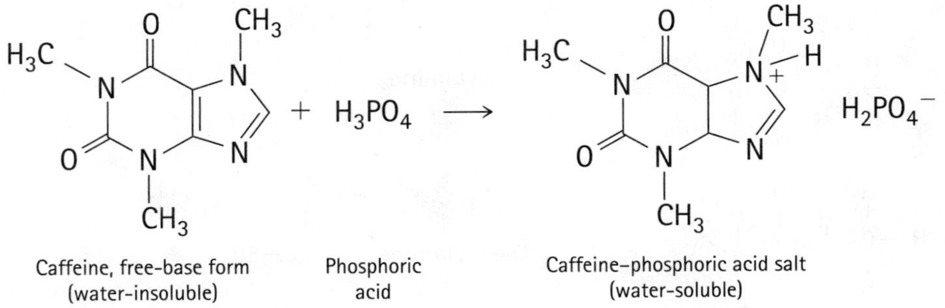

FIGURE 19.18
All alkaloids are bases that react with acids to form salts. An example is the alkaloid caffeine, shown here reacting with phosphoric acid.

Caffeine, free-base form (water-insoluble) + Phosphoric acid → Caffeine–phosphoric acid salt (water-soluble)

Most alkaloids exist in nature not in their free-base form but rather as the salts of naturally occurring acids known as *tannins*, a group of phenol-based organic acids that have complex structures. The alkaloid salts of these acids are usually much more soluble in hot water than in cold water. The caffeine in coffee and tea exists in the form of the tannin salt, which is why coffee and tea are more effectively brewed in hot water. As Figure 19.19 relates, tannins are also responsible for the stains caused by these beverages.

> Most pharmaceuticals that can be administered orally contain nitrogen heteroatoms in the water-soluble salt form.

FIGURE 19.19
Tannins are responsible for the brown stains in coffee mugs or on a coffee drinker's teeth. Because tannins are acidic, they can be readily removed with an alkaline cleanser. Use a little laundry bleach on the mug, and brush your teeth with baking soda.

CHECKPOINT
Why do most caffeinated soft drinks also contain phosphoric acid?

Was this your answer?
Phosphoric acid, as shown in Figure 19.18, reacts with caffeine to form the caffeine–phosphoric acid salt, which is much more soluble in cold water than the naturally occurring tannin salt.

LEARNING OBJECTIVE
Review the general properties of carbonyl compounds.

19.6 Carbonyl Compounds

EXPLAIN THIS Why does the carbon of the carbonyl usually have a slightly positive charge?

A **carbonyl group** consists of a carbon atom double-bonded to an oxygen atom. It occurs in the organic compounds known as ketones, aldehydes, amides, carboxylic acids, and esters.

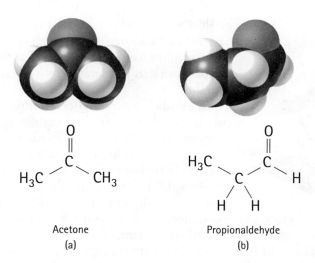

O
‖
C
H₃C   CH₃

Acetone
(a)

O
‖
H₃C   C   H
   C
   H   H

Propionaldehyde
(b)

FIGURE 19.20
(a) When the carbon of a carbonyl group is bonded to two carbon atoms, the result is a ketone. An example is acetone. (b) When the carbon of a carbonyl group is bonded to at least one hydrogen atom, the result is an aldehyde. An example is propionaldehyde.

A **ketone** is a carbonyl-containing organic molecule in which the carbonyl carbon is bonded to two carbon atoms. A familiar example of a ketone is *acetone*, which is often used in fingernail-polish remover and is shown in Figure 19.20a. In an **aldehyde**, the carbonyl carbon is bonded either to one carbon atom and one hydrogen atom, as in Figure 19.20b, or, in the special case of formaldehyde, to two hydrogen atoms.

Many aldehydes are particularly fragrant. A number of flowers, for example, owe their pleasant odor to the presence of simple aldehydes. The smells of lemons, cinnamon, and almonds are due to the aldehydes citral, cinnamaldehyde, and benzaldehyde, respectively. The structures of these three aldehydes are shown in Figure 19.21. Another aldehyde, vanillin, which was introduced at the beginning of this chapter, is the key flavoring molecule derived from seed pods of the vanilla orchid. You may have noticed that vanilla seed pods and vanilla extract are fairly expensive. Imitation vanilla flavoring is less expensive because it is merely a solution of the compound vanillin, which is economically synthesized from waste chemicals from the wood-pulp industry. Imitation vanilla does not taste the same as natural vanilla extract, however, because, in addition to vanillin, many other flavorful molecules contribute to the complex taste of

O
‖
C
—C   C—

Ketone group

O
‖
C
      H

Aldehyde group

FIGURE 19.21
Aldehydes are responsible for many familiar fragrances.

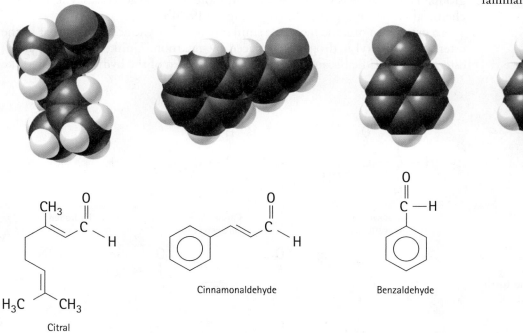

Citral

Cinnamonaldehyde

Benzaldehyde

Vanillin

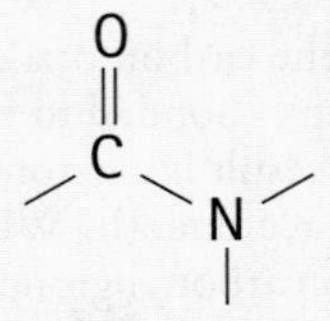

O
‖
C
N

Amide group

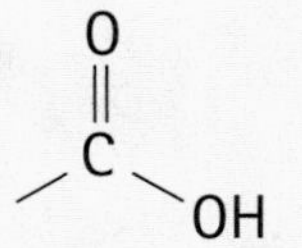

O
‖
C
OH

Carboxyl group

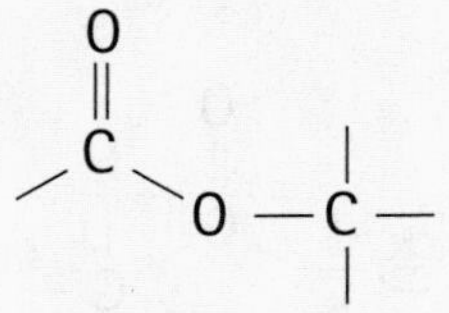

O
‖
C
O — C —

Ester group

CH₂CH₃
O   N
‖   CH₂CH₃
C

CH₃

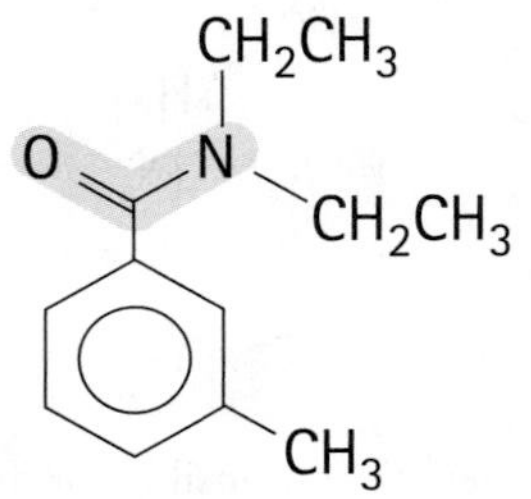

N,N-Diethyl-*m*-toluamide
(DEET)

FIGURE 19.22
N,N-diethyl-*m*-toluamide is an example of an amide. Amides contain the amide group, shown highlighted in blue.

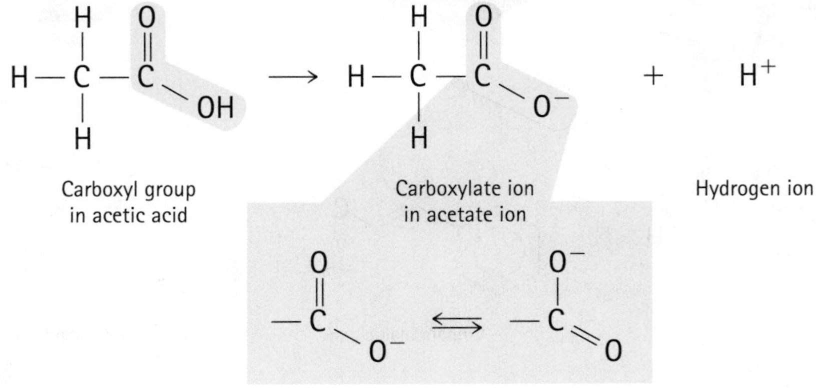

natural vanilla. Many books manufactured in the days before "acid-free" paper smell of vanilla because of the vanillin formed and released as the paper ages, a process that is accelerated by the acids the paper contains.

An **amide** is a carbonyl-containing organic molecule in which the carbonyl carbon is bonded to a nitrogen atom. The active ingredient of most mosquito repellents is an amide whose chemical name is *N,N*-diethyl-*m*-toluamide but is commercially known as DEET, shown in Figure 19.22. This compound is actually not an insecticide. Rather, it causes certain insects, especially mosquitoes, to lose their sense of direction, which effectively protects DEET wearers from being bitten.

A **carboxylic acid** is a carbonyl-containing organic molecule in which the carbonyl carbon is bonded to a hydroxyl group. As its name implies, this functional group can donate hydrogen ions. Organic molecules that contain it are therefore acidic. An example is acetic acid, $C_2H_4O_2$, which, after water, is the main ingredient of vinegar. You may recall that this organic compound was used as an example of a weak acid back in Chapter 18.

As with phenols, the acidity of a carboxylic acid results in part from the ability of the functional group to accommodate the negative charge of the ion that forms after the hydrogen ion has been donated. As shown in Figure 19.23, a carboxylic acid transforms to a carboxylate ion as it loses the hydrogen ion. The negative charge of the carboxylate ion can then pass back and forth between the two oxygens. This spreading out helps accommodate the negative charge.

An interesting example of an organic compound that contains both a carboxylic acid and a phenol is salicylic acid, found in the bark of willow trees and illustrated in Figure 19.24a. At one time brewed for its antipyretic (fever-reducing) effect, salicylic acid is an important analgesic (painkiller), but it causes nausea and stomach upset because of its relatively high acidity, a result of the presence of two acidic functional groups. In 1899, Friedrich Bayer and Company, in Germany, introduced a chemically modified version of this compound in which the acidic phenolic group was transformed into an ester functional group. The result was the less acidic and more tolerable acetylsalicylic acid, the chemical name for aspirin, shown in Figure 19.24b.

An **ester** is an organic molecule similar to a carboxylic acid except that in the ester, the hydroxyl hydrogen is replaced by a carbon. Unlike carboxylic acids, esters are not acidic because they lack the hydrogen of the hydroxyl group. Like

FIGURE 19.23
The negative charge of the carboxylate ion can pass back and forth between the two oxygen atoms of the carboxyl group.

H   O
│   ‖
H — C — C   H   O   H   O
│   OH   →   │   ‖   │   ‖
H   H — C — C   +   H⁺
Carboxyl group   │   O⁻
in acetic acid   H
Carboxylate ion   Hydrogen ion
in acetate ion

O   O⁻
‖   │
— C   ⇌   — C
O⁻   O

Carboxyl
group

Phenolic
group

(a)   Salicylic acid

FIGURE 19.24
(a) Salicylic acid, which is found in the bark of willow trees, is an example of a molecule that contains both a carboxyl group and a phenolic group. (b) Aspirin, acetylsalicylic acid, is less acidic than salicylic acid because it no longer contains the acidic phenolic group, which has been converted to an ester.

Carboxyl
group

Ester

(b)   Aspirin
(acetylsalicylic acid)

aldehydes, many simple esters have notable fragrances and are often used as flavorings. Some familiar ones are listed in Table 19.4.

Esters are fairly easy to synthesize by dissolving a carboxylic acid in an alcohol and then bringing the mixture to a boil in the presence of a strong acid, such as sulfuric acid, H_2SO_4. The synthesis of methyl salicylate from salicylic

FYI In the 1800s most salicylic acid used by people was produced not from willow bark but from coal tar. Tar residues within the salicylic acid had a nasty taste. This, combined with salicylic acid's stomach irritation, led many to view the salicylic acid cure to be worse than the disease. Felix Hoffman, a chemist working at Bayer, added the acetyl group to the phenol group of salicylic acid in 1897. According to Bayer, Hoffman was inspired by his father, who had been complaining about salicylic acid's side effects. To market the new drug, Bayer invented the name *aspirin*, in which "a" is for acetyl; "spir" is for the spirea flower, another natural source of salicylic acid; and "in" is used as a common suffix for medications. After World War I, Bayer, a German company, lost the rights to use the name *aspirin*. Bayer didn't regain these rights until 1994 for a steep price of $1 billion.

TABLE 19.4	SOME ESTERS AND THEIR FLAVORS AND ODORS	
Structure	Name	Flavor/Odor
	Ethyl formate	Rum
	Isopentyl acetate	Banana
	Octyl acetate	Orange
	Ethyl butyrate	Pineapple

acid and methanol is one example. Methyl salicylate is responsible for the smell of wintergreen and is a common ingredient of hard candies.

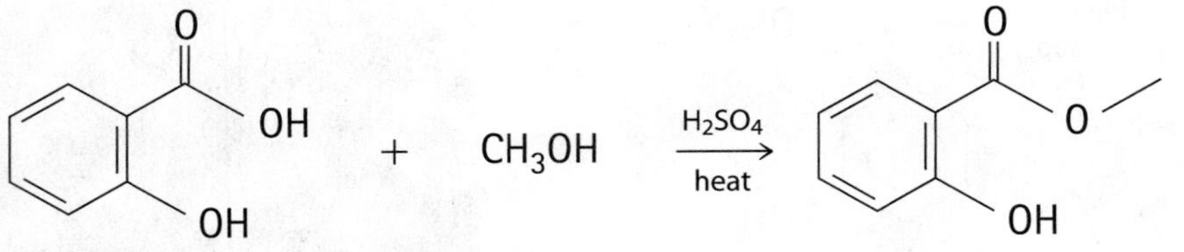

salicylic acid   methanol   methyl salicylate (wintergreen)

There is much more to organic chemistry than just learning functional groups and their general properties. Many, if not most, practicing organic chemists dedicate much of their time to the synthesis of organic molecules that have practical applications, such as for agriculture or pharmaceuticals. Often these target molecules are organic compounds that have been isolated from nature, where they can be found in only small quantities. To create large amounts of these chemicals, the organic chemist devises a pathway through which the compound can be synthesized in the laboratory from readily available smaller compounds. Once synthesized, the compound produced in the laboratory is chemically identical to that found in nature. In other words, it will have the same physical and chemical properties and will also have the same biological effects, if any.

> We eat organic chemicals daily. In fact, organic chemicals are the only things we eat, except for some important minerals, such as the ions of sodium and calcium.

CHECKPOINT

Identify all the functional groups in these four molecules (ignore the sulfur group in penicillin G):

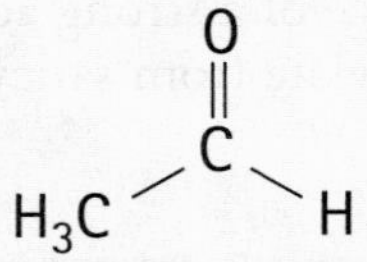

Acetaldehyde

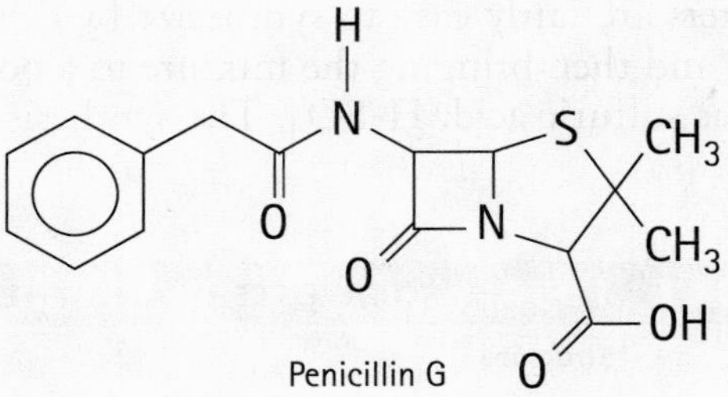

Penicillin G

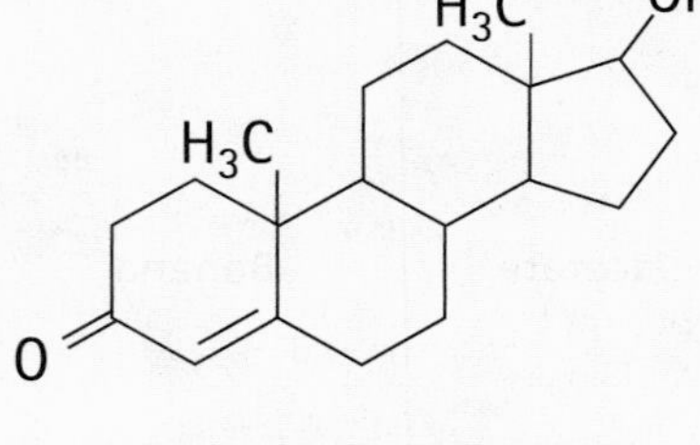

Testosterone

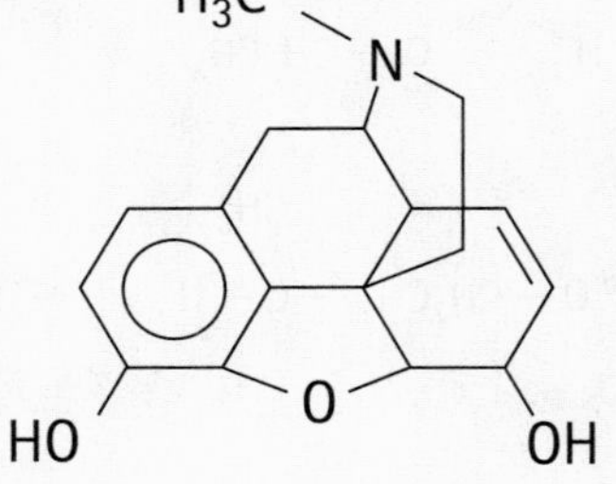

Morphine

Were these your answers?
Acetaldehyde: aldehyde. Penicillin G: amide (two amide groups) and carboxylic acid. Testosterone: alcohol and ketone. Morphine: alcohol, phenol, ether, and amine.

19.7 Polymers

EXPLAIN THIS Why are plastics generally so inexpensive?

Polymers are exceedingly long molecules that consist of repeating molecular units called **monomers**, as Figure 19.25 illustrates. Monomers have relatively simple structures consisting of anywhere from 4 to 100 atoms per molecule. When monomers are chained together, they can form polymers consisting of hundreds of thousands of atoms per molecule. These large molecules are still too small to be seen with the unaided eye. They are, however, giants in the submicroscopic world—if a typical polymer molecule were as thick as a kite string, it would be 1 km long.

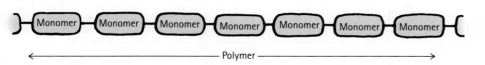

FIGURE 19.25
A polymer is a long molecule consisting of many smaller monomer molecules linked together.

Many of the molecules that constitute living organisms are polymers, including DNA, proteins, the cellulose of plants, and the complex carbohydrates of starchy foods. For now, we focus on the human-made polymers, also known as synthetic polymers, that make up the class of materials that are commonly known as plastics.

We will begin by exploring the two major types of synthetic polymers used today—*addition polymers* and *condensation polymers*.

As shown in Table 19.5, addition and condensation polymers have a wide variety of uses. Solely the product of human design, these polymers pervade modern living. In the United States, for example, synthetic polymers have surpassed steel as the most widely used material.

TABLE 19.5	ADDITION AND CONDENSATION POLYMERS		
Polymers	**Repeating Unit**	**Common Uses**	**Recycling Code**
Addition			
Polyethylene (PE)	$\cdots\overset{\overset{\displaystyle H}{\vert}}{\underset{\underset{\displaystyle H}{\vert}}{C}}-\overset{\overset{\displaystyle H}{\vert}}{\underset{\underset{\displaystyle H}{\vert}}{C}}\cdots$	Plastic bags, bottles	♻ 2 HDPE ♻ 4 LDPE
Polypropylene (PP)	$\cdots\overset{\overset{\displaystyle H}{\vert}}{\underset{\underset{\displaystyle H}{\vert}}{C}}-\overset{\overset{\displaystyle H}{\vert}}{\underset{\underset{\displaystyle CH_3}{\vert}}{C}}\cdots$	Indoor–outdoor carpets	♻ 5 PP
Polystyrene (PS)	$\cdots\overset{\overset{\displaystyle H}{\vert}}{\underset{\underset{\displaystyle H}{\vert}}{C}}-\overset{\overset{\displaystyle H}{\vert}}{\underset{\underset{\displaystyle \bigcirc}{\vert}}{C}}\cdots$	Plastic utensils, insulation	♻ 6 PS
Polyvinyl chloride (PVC)	$\cdots\overset{\overset{\displaystyle H}{\vert}}{\underset{\underset{\displaystyle H}{\vert}}{C}}-\overset{\overset{\displaystyle H}{\vert}}{\underset{\underset{\displaystyle Cl}{\vert}}{C}}\cdots$	Shower curtains, tubing	♻ 3 V

(continued)

TABLE 19.5	CONTINUED		
Polymers	Repeating Unit	Common Uses	Recycling Code

Condensation

| Polyethylene terephthalate | 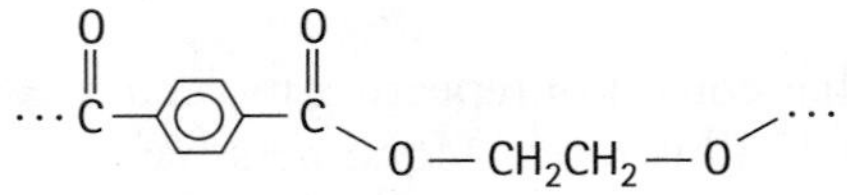 | Clothing, plastic bottles | ♲ PET |
| Melamine–formaldehyde resin (Melmac, Formica) | 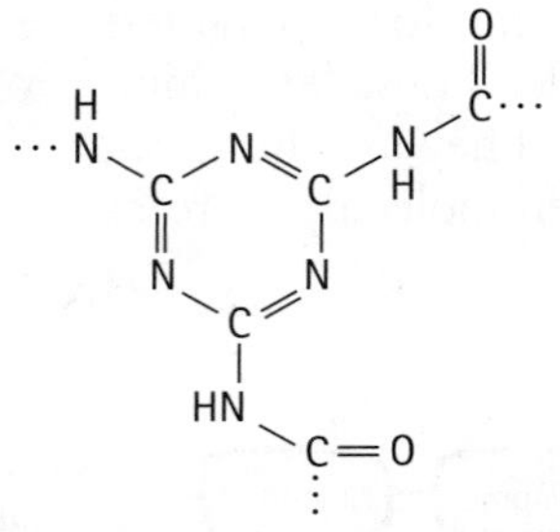 | Dishes, countertops | Not recycled |

Addition Polymers

Addition polymers form simply by the joining of monomer units. For this to happen, each monomer must contain at least one double bond. As shown in Figure 19.26, polymerization occurs when two of the electrons from each double bond split away from each other to form new covalent bonds with neighboring monomer molecules. During this process, no atoms are lost, so the total mass of the polymer is equal to the sum of the masses of all the monomers.

Nearly 12 million tons of polyethylene is produced annually in the United States; that's about 90 lb per U.S. citizen. The monomer from which it is synthesized, ethylene, is an unsaturated hydrocarbon produced in large quantities from petroleum.

Two principal forms of polyethylene are produced by using different catalysts and reaction conditions. High-density polyethylene (HDPE), shown schematically in Figure 19.27a, consists of long strands of straight-chain molecules packed closely together. The tight alignment of neighboring strands makes

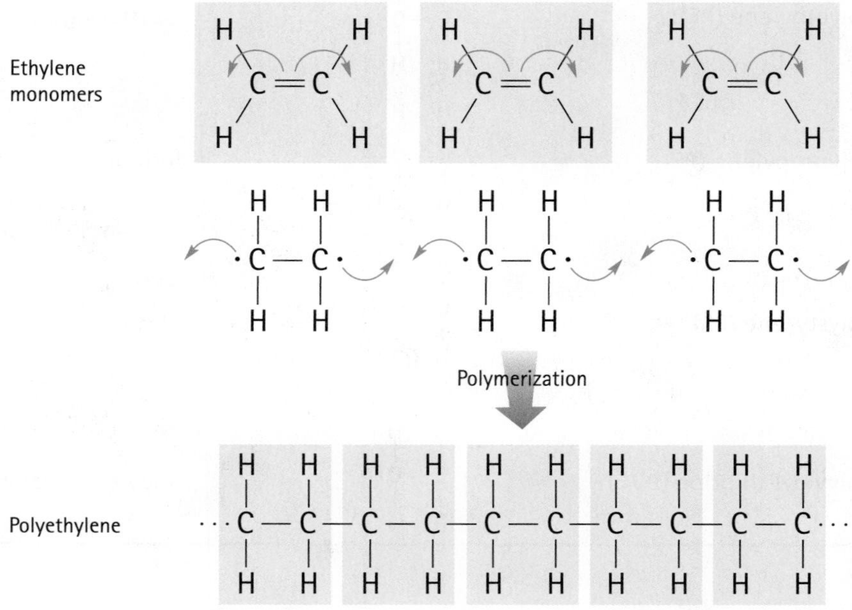

FIGURE 19.26

The addition polymer polyethylene is formed as electrons from the double bonds of ethylene monomer molecules split away and become unpaired valence electrons. Each unpaired electron then joins with an unpaired electron of a neighboring carbon atom to form a new covalent bond that links two monomer units.

HDPE a relatively rigid, tough plastic useful for such things as bottles and milk jugs. Low-density polyethylene (LDPE), shown in Figure 19.27b, is made of strands of highly branched chains, an architecture that prevents the strands from packing closely together. This makes LDPE more bendable than HDPE and gives it a lower melting point. HDPE holds its shape in boiling water; LDPE deforms. LDPE is most useful for such items as plastic bags, photographic film, and electrical-wire insulation.

Other addition polymers are created by using different monomers. The only requirement is that the monomer must contain a double bond. The monomer propylene, for example, yields polypropylene, as shown in Figure 19.28. Polypropylene is a tough plastic material useful for pipes, hard-shell suitcases, and appliance parts. Fibers of polypropylene are used for upholstery, indoor–outdoor carpets, and even thermal underwear.

(a) Molecular strands of HDPE

(b) Molecular strands of LDPE

FIGURE 19.27
(a) The polyethylene strands of HDPE can pack closely together, much like strands of uncooked spaghetti. (b) The polyethylene strands of LDPE are branched, which prevents the strands from packing well.

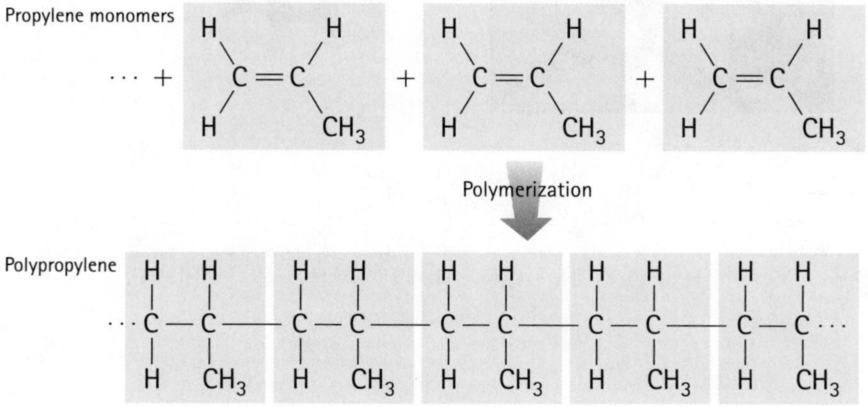

Propylene monomers

Polymerization

Polypropylene

FIGURE 19.28
Propylene monomers polymerize to form polypropylene.

Figure 19.29 shows that using styrene as the monomer yields polystyrene. Transparent plastic cups are made of polystyrene, as are thousands of other household items. Blowing gas into liquid polystyrene generates Styrofoam, which is widely used for coffee cups, packing material, and insulation.

The addition polymer polytetrafluoroethylene, shown in Figure 19.31, is what you know as Teflon. In contrast to the chlorine-containing Saran, fluorine-containing Teflon has a nonstick surface because the fluorine atoms tend not to experience any molecular attractions. In addition, because carbon–fluorine bonds are unusually strong, Teflon can be heated to high temperatures before decomposing. These properties make Teflon an ideal coating for cooking surfaces. It is also relatively inert, which is why many corrosive chemicals are shipped or stored in Teflon containers.

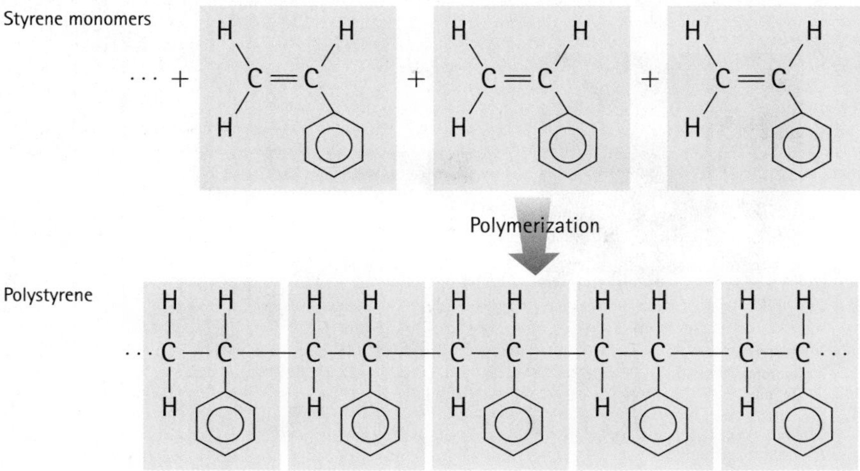

Styrene monomers

Polymerization

Polystyrene

FIGURE 19.29
Styrene monomers polymerize to form polystyrene. Another important addition polymer is polyvinyl chloride (PVC), which is tough and easily molded. Floor tiles, shower curtains, and pipes are most often made of PVC, shown in Figure 19.30.

FYI So if nothing sticks to Teflon, how is Teflon made to adhere to a pan as a coating? That's a trade secret, but rumor has it that there are microscopic pits in the metal pan that help the Teflon adhere physically. Of course, we all know that the Teflon is fairly easy to scrape out of the pan, which is why manufacturers recommend that you stir-fry with a wooden utensil.

CHECKPOINT

What do all monomers that are used to make addition polymers have in common?

Was this your answer?
A double covalent bond between two carbon atoms

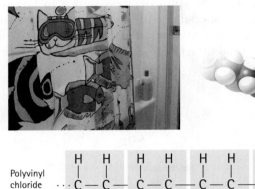

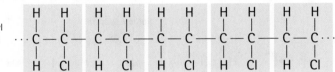

Polyvinyl chloride (PVC)

FIGURE 19.30
Another important addition polymer is polyvinyl chloride (PVC), which is used to fabricate many household items.

Condensation Polymers

A **condensation polymer** is formed when the joining of monomer units is accompanied by the loss of a small molecule, such as water or hydrochloric acid. Any monomer capable of becoming part of a condensation polymer must have a functional group on each end. When two such monomers come together to form a condensation polymer, one functional group of the first monomer links with one functional group of the other monomer. The result is a two-monomer unit that has two terminal functional groups, one from each of the two original monomers. Each of these terminal functional groups in the two-monomer unit is now free to link with one of the functional groups of a third monomer, and then a fourth, and so on. In this way a polymer chain is built.

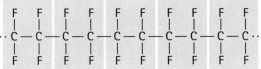

Polytetra-fluoroethylene (Teflon)

FIGURE 19.31
The fluorine atoms in polytetrafluoroethylene tend not to experience molecular attractions, which is why this addition polymer is used as a nonstick coating and lubricant.

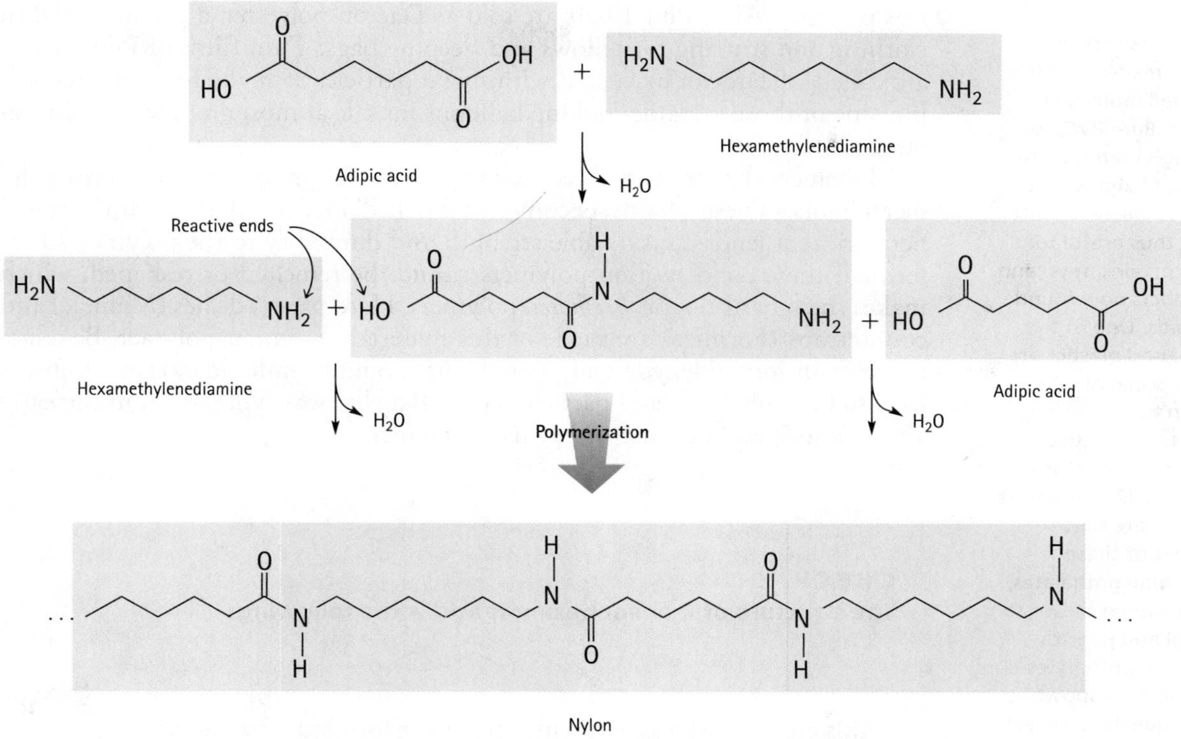

FIGURE 19.32
Adipic acid and hexamethylenediamine polymerize to form the
condensation copolymer nylon.

Figure 19.32 shows this process for the condensation polymer called nylon, which was created in 1937 by DuPont chemist Wallace Carothers (1896–
1937). This polymer is composed of two different monomers, which classifies
it as a *copolymer*. One monomer is adipic acid, which contains two reactive end
groups, both carboxyl groups. The second monomer is hexamethylenediamine,
in which two amine groups are the reactive end groups. One end of an adipic
acid molecule and one end of a hexamethylenediamine molecule can be made
to react with each other, splitting off a water molecule in the process. After two
monomers have joined, reactive ends still remain for further reactions, which
leads to a growing polymer chain. Aside from its use in hosiery, nylon also finds
important uses in the manufacture of ropes, parachutes, clothing, and carpets.

Another widely used condensation polymer is polyethylene terephthalate
(PET), which is formed from the copolymerization of ethylene glycol and
terephthalic acid, as shown in Figure 19.33. Plastic soda bottles are made from

FIGURE 19.33
Terephthalic acid and ethylene
glycol polymerize to form the condensation copolymer polyethylene
terephthalate.

Polyethylene terephthalate (PET)

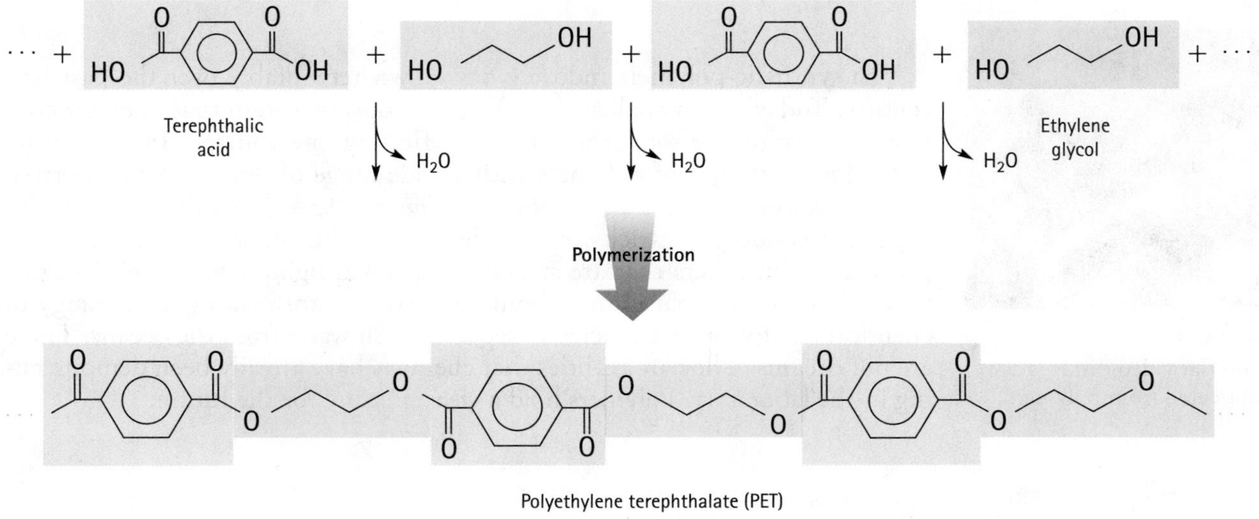

this polymer. Also, PET fibers are sold as Dacron polyester, a product used in clothing and stuffing for pillows and sleeping bags. Thin films of PET, which are called Mylar, can be coated with metal particles to make magnetic recording tape or those metallic-looking balloons for sale at most grocery store checkout counters.

Monomers that contain three reactive functional groups can also form polymer chains. These chains become interlocked in a rigid three-dimensional network that lends considerable strength and durability to the polymer. Once formed, these condensation polymers cannot be remelted or reshaped, which makes them hard-set, or *thermoset*, polymers. Hard plastic dishes (Melmac) and countertops (Formica) are made of this material. A similar polymer, Bakelite, made from formaldehyde and phenols that contain multiple oxygen atoms, is used to bind plywood and particle board. Bakelite was synthesized in the early 1900s, and it was the first widely used polymer.

CHECKPOINT

The structure of 6-aminohexanoic acid is the following:

$$H_2N \underset{O}{\diagup\!\!\!\diagdown} OH$$

Is this compound a suitable monomer for forming a condensation polymer? If so, what is the structure of the polymer formed, and what small molecule is split off during the condensation?

Were these your answers?
Yes, because the molecule has two reactive ends. You know both ends are reactive because they are the ends shown in Figure 19.32. The only difference here is that both types of reactive ends are on the same molecule. Monomers of 6-aminohexanoic acid combine by splitting off water molecules to form the polymer known as nylon-6:

The synthetic-polymers industry has grown remarkably over the past half century. Today, it is a challenge to find any consumer item that does *not* contain a plastic of one sort or another. Try finding one yourself. In the future, watch for new kinds of polymers with a wide range of remarkable properties. One interesting application is shown in Figure 19.34. We already have polymers that conduct electricity, others that emit light, others that replace body parts, and still others that are stronger but much lighter than steel. Imagine synthetic polymers that mimic photosynthesis by transforming solar energy to chemical energy, or that efficiently separate fresh water from the oceans. These are not dreams. They are realities that chemists have already been demonstrating in the laboratory. Polymers hold a clear promise for the future.

FIGURE 19.34
Flexible and flat video displays can now be fabricated from polymers.

For instructor-assigned homework, go to www.masteringphysics.com

SUMMARY OF TERMS (KNOWLEDGE)

Addition polymer A polymer formed by the joining together of monomer units with no atoms being lost as the polymer forms.

Alcohol An organic molecule that contains a hydroxyl group bonded to a saturated carbon.

Aldehyde An organic molecule containing a carbonyl group, the carbon of which is bonded either to one carbon atom and one hydrogen atom or to two hydrogen atoms.

Amide An organic molecule containing a carbonyl group, the carbon of which is bonded to a nitrogen atom.

Amine An organic molecule containing a nitrogen atom bonded to one or more saturated carbon atoms.

Aromatic compound Any organic molecule containing a benzene ring.

Carbonyl group A carbon atom double-bonded to an oxygen atom; found in ketones, aldehydes, amides, carboxylic acids, and esters.

Carboxylic acid An organic molecule containing a carbonyl group, the carbon of which is bonded to a hydroxyl group.

Condensation polymer A polymer formed by the joining together of monomer units accompanied by the loss of small molecules, such as water.

Configuration A description of how the atoms within a molecule are connected. For example, two structural isomers consist of the same number and same kinds of atoms, but in different configurations.

Conformation One of a wide range of possible spatial orientations of a particular configuration.

Ester An organic molecule containing a carbonyl group, the carbon of which is bonded to one carbon atom and one oxygen atom bonded to another carbon atom.

Ether An organic molecule containing an oxygen atom bonded to two carbon atoms.

Functional group A specific combination of atoms that behaves as a unit in an organic molecule.

Heteroatom Any atom other than carbon or hydrogen in an organic molecule.

Hydrocarbon A chemical compound containing only carbon and hydrogen atoms.

Ketone An organic molecule containing a carbonyl group, the carbon of which is bonded to two carbon atoms.

Monomers The small molecular units from which a polymer is formed.

Organic chemistry The study of carbon-containing compounds.

Phenol An organic molecule in which a hydroxyl group is bonded to a benzene ring.

Polymer A long organic molecule made of many repeating units.

Saturated hydrocarbon A hydrocarbon containing no multiple covalent bonds, with each carbon atom bonded to four other atoms.

Structural isomers Molecules that have the same molecular formula but different chemical structures.

Unsaturated hydrocarbon A hydrocarbon containing at least one multiple covalent bond.

READING CHECK QUESTIONS (COMPREHENSION)

19.1 Hydrocarbons

1. How do two structural isomers differ from each other?
2. How are two structural isomers similar to each other?
3. What physical property of hydrocarbons is used in fractional distillation?
4. What types of hydrocarbons are more abundant in higher-octane gasoline?

19.2 Unsaturated Hydrocarbons

5. To how many atoms is a saturated carbon atom bonded?
6. What is the difference between a saturated hydrocarbon and an unsaturated hydrocarbon?
7. How many multiple bonds must a hydrocarbon have in order to be classified as unsaturated?
8. What kind of ring do aromatic compounds contain?

19.3 Functional Groups

9. What is a heteroatom?
10. Why do heteroatoms make such a difference in the physical and chemical properties of an organic molecule?

19.4 Alcohols, Phenols, and Ethers

11. Why are low-formula-mass alcohols soluble in water?
12. What distinguishes an alcohol from a phenol?
13. What distinguishes an alcohol from an ether?

19.5 Amines and Alkaloids

14. Which heteroatom is characteristic of an amine?
15. Do amines tend to be acidic, neutral, or basic?
16. Are alkaloids found in nature?
17. What are some examples of alkaloids?

19.6 Carbonyl Compounds

18. Which elements make up the carbonyl group?
19. How are ketones and aldehydes related to each other? How are they different from each other?
20. How are amides and carboxylic acids related to each other? How are they different from each other?
21. From what naturally occurring compound is aspirin prepared?

19.7 Polymers

22. What happens to the double bond of a monomer participating in the formation of an addition polymer?

23. What is released in the formation of a condensation polymer?

24. Why is plastic wrap made of polyvinylidene chloride stickier than plastic wrap made of polyethylene?

25. What is a copolymer?

ACTIVITIES (HANDS-ON APPLICATION)

26. Two carbon atoms connected by a single bond can rotate relative to each other. This ability to rotate can give rise to numerous conformations (spatial orientations) of an organic molecule. Is it also possible for two carbon atoms connected by a double bond to rotate relative to each other?

Hold two toothpicks side by side and attach one jellybean to each end such that each jellybean has both toothpicks poked into it. Hold one jellybean while rotating the other. What kind of rotations are possible? Relate what you observe to the carbon–carbon double bond. Which structure of Figure 19.7 do you suppose has more possible conformations: butane or 2-butene? What do you suppose is generally true about the ability of atoms connected by a carbon–carbon triple bond to twist relative to each other?

27. A property of polymers is their glass transition temperature, T_g, which is the approximate temperature below which

the polymer is hard and rigid, but above which the polymer is soft and flexible. The T_g of polyethylene is a chilly $-125°C$, which is why polyethylene food wrap is flexible at ambient temperatures. Consider the two polymers polyethylene terephthalate (PETE) and polystyrene (PS). Which do you suppose has the higher T_g? Dip some plastics of these two polymers in boiling water to find out. A common polymer used to make chewing gum is polyvinyl acetate, with a T_g of about 28°C, which is below body temperature but above room temperature. That's why most chewing gums are hard until they soften up in your warm mouth. Drink ice water while chewing gum and note how it quickly hardens.

28. Isopropyl alcohol, also known as rubbing alcohol, is very toxic if ingested. This is because it acts to destroy the digestive proteins and other important biomolecules in your stomach. Do this activity to see firsthand the destructive action of isopropyl alcohol on proteins. Crack open an egg and place the egg white and the yolk into two separate bowls. Pour a capful of isopropyl alcohol into the egg white and observe what happens. In the second bowl, stir the yolk with a fork. Add another capful of isopropyl alcohol to the stirred yolk and observe what happens. The same sort of destruction would occur to your own stomach proteins, as well as various tissues, upon ingesting the isopropyl alcohol. Not good! Our skin, however, is more impervious to the destructive powers of isopropyl alcohol, which therefore serves as a good topical antiseptic.

THINK AND RANK (ANALYSIS)

29. Rank the following molecules in order of the phase they form at room temperature: solid, liquid, gas.

(a) H_3C — (benzene ring) — C with CH_3, H, and C(=O)OH group

(b) $CH_3CH_2CH_2CH_3$

(c) $CH_3CH_2CH_2CH_2 — OH$

30. Rank the following hydrocarbons in order of increasing number of hydrogen atoms:

Cyclobutane (a)   Butane (b)   2-Butene (c)

31. Rank the following hydrocarbons in order of increasing number of hydrogen atoms:

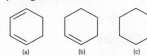

(a)   (b)   (c)

32. Rank the following organic molecules in order of increasing solubility in water:

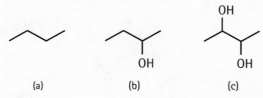

(a) (b) (c)

33. Rank the following organic molecules in order of increasing solubility in water:

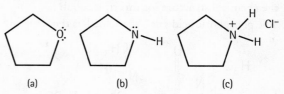

(a) (b) (c)

EXERCISES (SYNTHESIS)

34. What property of carbon allows for the formation of so many different organic molecules?

35. Why does the melting point of hydrocarbons increase as the number of carbon atoms per molecule increases?

36. Draw all the structural isomers for hydrocarbons that have the molecular formula C_6H_{14}.

37. How many structural isomers are shown here?

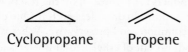

38. What do the compounds cyclopropane and propene have in common?

Cyclopropane   Propene

39. According to Figure 19.3, which has a higher boiling point: gasoline or kerosene?

40. The temperatures in a fractionating tower at an oil refinery are important, but so are the pressures. Where might the pressure in a fractionating tower be highest: at the bottom or at the top? Defend your answer.

41. There are five atoms in the methane molecule, CH_4. One of these five is a carbon atom, which is $\frac{1}{5} \times 100 = 20\%$ carbon. What is the percentage carbon in ethane, C_2H_6? Propane, C_3H_8? Butane, C_4H_{10}?

42. Do heavier hydrocarbons tend to produce more or less carbon dioxide upon combustion compared to lighter hydrocarbons? Why?

43. What are the chemical formulas for the following structures?

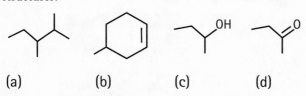

(a) (b) (c) (d)

44. What do these two structures have in common?

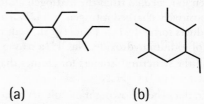

(a)   (b)

45. Remember that carbon–carbon single bonds can rotate but carbon–carbon double bonds cannot rotate. How many different structures are shown below?

46. Why do ethers typically have lower boiling points than alcohols?

47. What is the percent volume of water in 80-proof vodka?

48. One of the skin-irritating components of poison oak is tetrahydrourushiol:

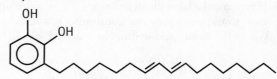

The long, nonpolar hydrocarbon tail embeds itself in a person's oily skin, where the molecule initiates an allergic response. Scratching the itch spreads tetrahydrourushiol molecules over a greater surface area, causing the zone of irritation to grow. Is this compound an alcohol or a phenol? Defend your answer.

49. Cetyl alcohol, $C_{16}H_{34}O$, is a common ingredient of soaps and shampoos. It was once commonly obtained from whale oil, which is where it gets its name ("cetyl" is derived from *cetacean*). Review the discussion of soaps in Section 16.5, and then draw the likely chemical structure for this alcohol.

50. A common inactive ingredient in products such as sunscreen lotions and shampoo is triethylamine, also known as TEA. What is the chemical structure for this compound?

51. A common inactive ingredient in products such as sunscreen lotions and shampoo is triethanolamine. What is the chemical structure for this tri-alcohol?

52. The phosphoric acid salt of caffeine has the structure

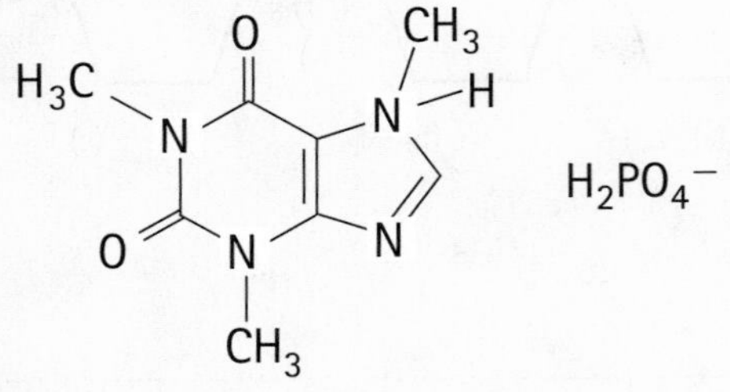

Caffeine–phosphoric acid salt

This molecule behaves as an acid in that it can donate a hydrogen ion, created from the hydrogen atom bonded to the positively charged nitrogen atom. What are all the products formed when 1 mole of this salt reacts with 1 mole of sodium hydroxide, NaOH, a strong base?

53. Draw all the structural isomers for amines that have the molecular formula C_3H_9N.

54. In water, does the following molecule act as an acid, a base, neither, or both?

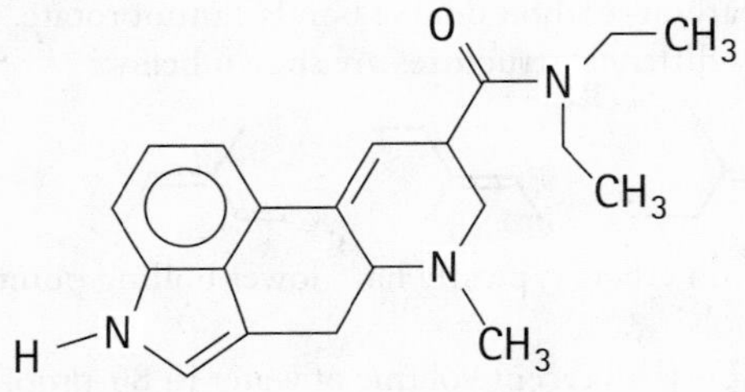

Lysergic acid diethylamide

55. If you saw the label phenylephrine HCl on a decongestant, would you worry that consuming it would expose you to the strong acid hydrochloric acid? Explain.

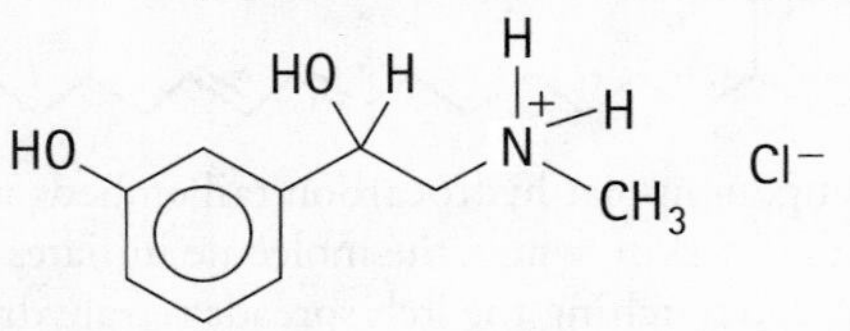

Phenylephrine–hydrochloric acid salt

56. An amino acid is an organic molecule that contains both an amine group and a carboxyl group. At an acidic pH, which structure is more likely? Explain your answer.

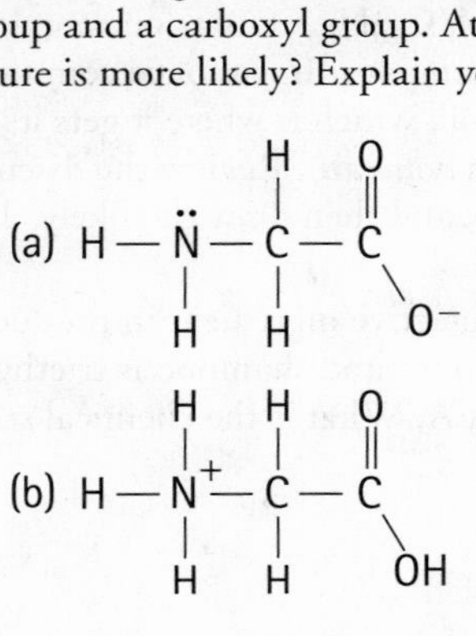

57. Identify the following functional groups in this organic molecule: amide, ester, ketone, ether, alcohol, aldehyde, amine.

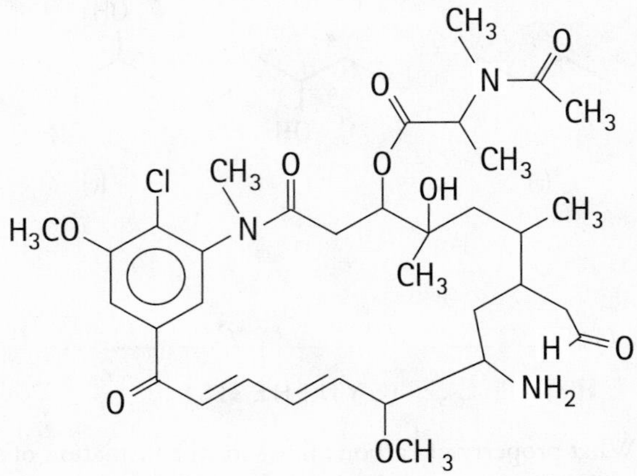

58. Suggest an explanation for why aspirin has a sour taste.

59. Benzaldehyde is a fragrant oil. If stored in an uncapped bottle, this compound will slowly tranform into benzoic acid along the surface. Is this an oxidation or a reduction?

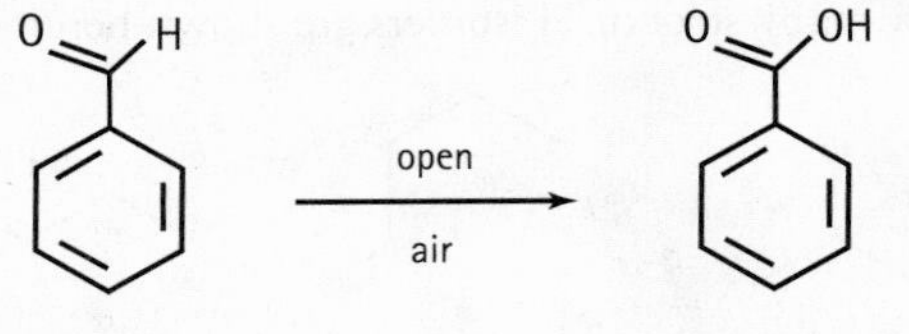

Benzaldehyde   Benzoic acid

60. What products are formed upon the reaction of benzoic acid with sodium hydroxide, NaOH? One of these products is a common food preservative. What is its name?

61. The disodium salt of ethylenediaminetetraacetic acid, also known as EDTA, has a great affinity for lead ions, Pb^{2+}. Why? Can you think of any useful applications of this chemistry?

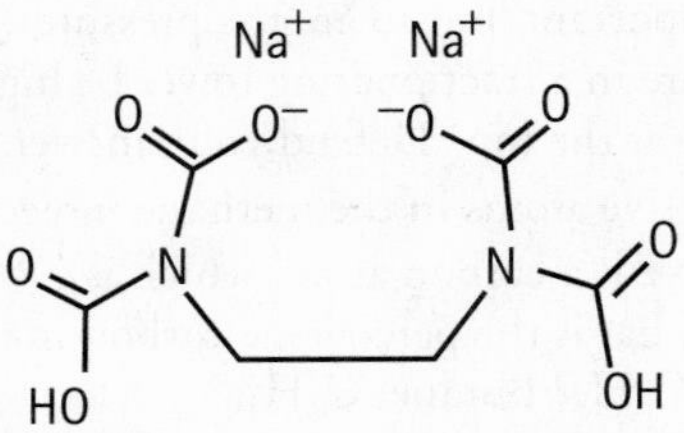

EDTA

62. The amino acid lysine is shown here. What functional group must be removed in order to produce cadaverine, shown in Figure 19.16?

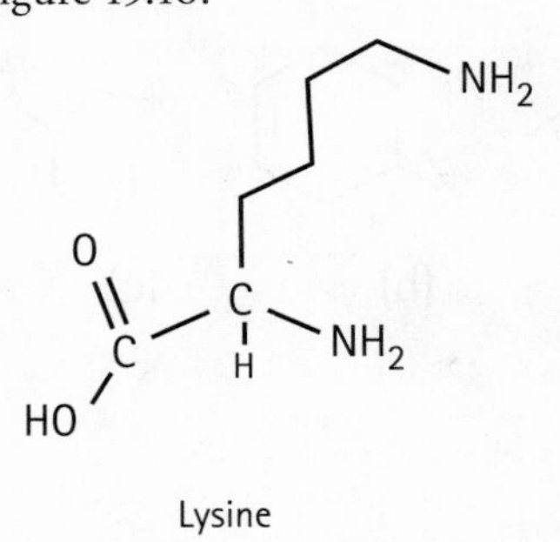

Lysine

63. Would you expect polypropylene to be denser or less dense than low-density polyethylene? Why?

64. Hydrocarbons release a lot of energy when ignited. Where does this energy come from?

65. The polymer styrene–butadiene rubber (SBR), shown here, is used for making tires as well as bubble gum. Is it an addition polymer or a condensation polymer?

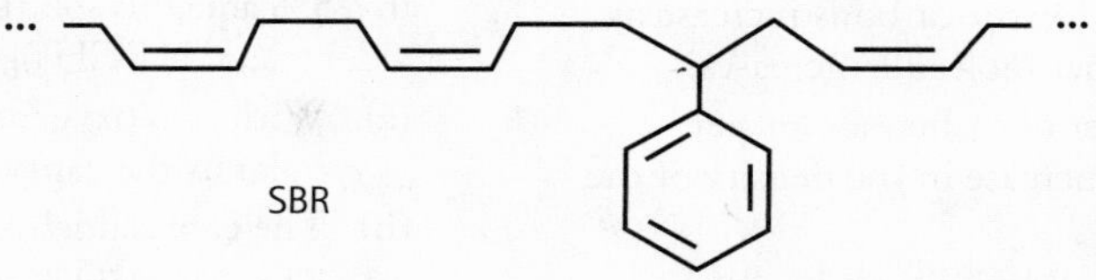

SBR

66. Citral and camphor are both 10-carbon odoriferous natural products made from the joining of two isoprene units plus the addition of a carbonyl functional group. Their chemical structures are shown. Find and circle the two isoprene units in each of these molecules.

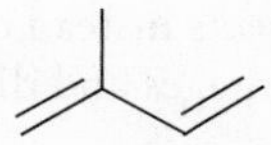

Isoprene
(2-methyl-1,3-butadiene)

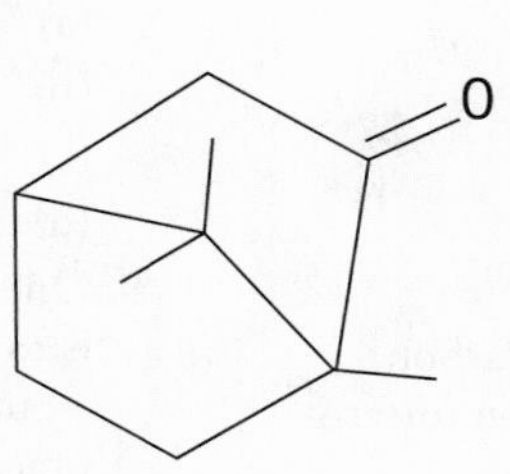

Camphor

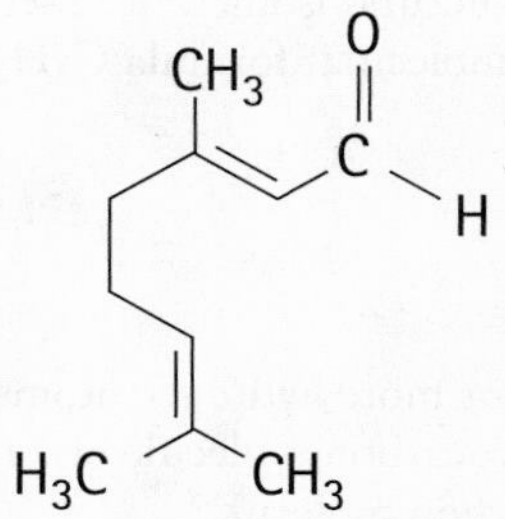

Citral

67. Many of the natural product molecules synthesized by photosynthetic plants are formed by the joining together of isoprene monomers via an addition polymerization. A good example is the nutrient beta-carotene. How many isoprene units are needed to make one beta-carotene molecule? Find and circle these units in the beta-carotene structure shown.

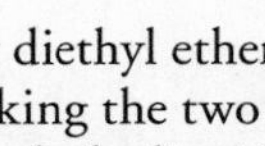

Isoprene
(2-methyl-1,3-butadiene)

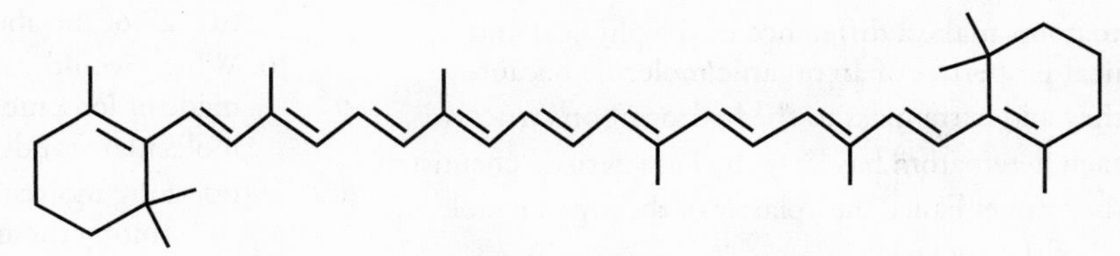

beta-Carotene

DISCUSSION QUESTIONS (EVALUATION)

68. The solvent diethyl ether can be mixed with water but only by shaking the two liquids together. After the shaking is stopped, the liquids separate into two layers, much like oil and vinegar. The free-base form of the alkaloid caffeine is readily soluble in diethyl ether but not in water. Suggest what might happen to the caffeine of a caffeinated beverage if the beverage was first made alkaline with sodium hydroxide and then shaken with some diethyl ether.

69. Alkaloid salts are not very soluble in the organic solvent diethyl ether. What might happen to the free-base form

of caffeine dissolved in diethyl ether if gaseous hydrogen chloride, HCl, were bubbled into the solution?

70. Go online and look up the total synthesis of the anti-cancer drug Taxol. With this major accomplishment in mind, discuss the relative merits of specializing in a single area versus becoming an expert in many different areas. When in life do we have the opportunity of simultaneously narrowing our focus while expanding our horizons?

READINESS ASSURANCE TEST (RAT)

If you have a good handle on this chapter, if you really do, then you should be able to score 7 out of 10 on this RAT. If you score less than 7, you need to study further before moving on.

Choose the BEST answer to each of the following.

1. Why does the melting point of hydrocarbons increase as the number of carbon atoms per molecule increases?
   (a) An increase in the number of carbon atoms per molecule also means an increase in the density of the hydrocarbon.
   (b) The induced dipole–induced dipole molecular attractions are stronger.
   (c) Larger hydrocarbon chains tend to be branched.
   (d) The molecular mass also increases.

2. How many structural isomers are there for hydrocarbons that have the molecular formula C_4H_{10}?
   (a) none
   (b) one
   (c) two
   (d) three

3. Which contains more hydrogen atoms: a five-carbon saturated hydrocarbon molecule or a five-carbon unsaturated hydrocarbon molecule?
   (a) The unsaturated hydrocarbon has more hydrogen atoms.
   (b) The saturated hydrocarbon has more hydrogen atoms.
   (c) They both have the same number of hydrogen atoms.
   (d) It depends whether the unsaturation is due to a double or triple bond.

4. Heteroatoms make a difference in the physical and chemical properties of an organic molecule because
   (a) they add extra mass to the hydrocarbon structure.
   (b) each heteroatom has its own characteristic chemistry.
   (c) they can enhance the polarity of the organic molecule.
   (d) all of the above

5. Why might a high-formula-mass alcohol be insoluble in water?
   (a) A high-formula-mass alcohol is too attracted to itself to be soluble in water.
   (b) The bulk of a high-formula-mass alcohol likely consists of nonpolar hydrocarbons.
   (c) Such an alcohol is likely in a solid phase.
   (d) In order for two substances to be soluble in each other, their molecules need to be of comparable mass.

6. Alkaloid salts are not very soluble in the organic solvent diethyl ether. What might happen to the free-base form of caffeine (an alkaloid) dissolved in diethyl ether if gaseous hydrogen chloride, HCl, were bubbled into the solution?
   (a) A second layer of water would form.
   (b) Nothing; the HCl gas would merely bubble out of solution.
   (c) The diethyl ether–insoluble caffeine salt would form as a white precipitate.
   (d) The acid–base reaction would release heat, which would cause the diethyl ether to start evaporating.

7. Explain why caprylic acid, $CH_3(CH_2)_6COOH$, dissolves in a 5% aqueous solution of sodium hydroxide but caprylaldehyde, $CH_3(CH_2)_6CHO$, does not.
   (a) With two oxygens, the caprylic acid is about twice as polar as the caprylaldehyde.
   (b) The caprylaldehyde is a gas at room temperature.
   (c) The caprylaldehyde behaves as a reducing agent, which neutralizes the sodium hydroxide.
   (d) The caprylic acid reacts to form the water-soluble salt.

8. How many oxygen atoms are bonded to the carbon of the carbonyl of an ester functional group?
   (a) none
   (b) one
   (c) two
   (d) three

9. One solution to the problem of our overflowing landfills is to burn plastic objects instead of burying them. What would be some advantages and disadvantages of this practice?
   (a) disadvantage: toxic air pollutants; advantage: reduced landfill volume
   (b) disadvantage: loss of a vital petroleum-based resource; advantage: generation of electricity
   (c) disadvantage: discourages recycling; advantage: provides new jobs
   (d) all of the above

10. Which would you expect to be more viscous: a polymer made of long molecular strands or one made of short molecular stands? Why?
    (a) long molecular strands because they tend to tangle among themselves
    (b) short molecular strands because of a higher density
    (c) long molecular strands because of a greater molecular mass
    (d) short molecular strands because their ends are typically polar

Answers to RAT

1. b, 2. c, 3. b, 4. d, 5. b, 6. c, 7. d, 8. c, 9. d, 10. a

PART THREE

Earth Science

Hey Megan, the seashore is the perfect place to see interactions of the geosphere, hydrosphere, and atmosphere!

You're right, Emily, it is the perfect place. We are standing on the solid geosphere, but all the while, the hydrosphere and atmosphere are at work weathering the rock we stand on. The hydrosphere is where life on Earth began, and the atmosphere provides the oxygen animals need and the carbon dioxide plants need. Plus the atmosphere shields us from harmful UV rays. Our planet is unique in our solar system. It is our home and we need to learn more about it to be able to preserve it.

20

CHAPTER 20

Rocks and Minerals

OUR EARTH is an interconnected system that can be organized into "spheres"—the geosphere, hydrosphere, and atmosphere. Each sphere is separate, but each sphere touches and is interconnected to the other spheres. The *geosphere* is the rocks and minerals that make up our planet, and their relationship to Earth's internal and external processes. The *hydrosphere* includes Earth's fresh water—rivers, glaciers, and groundwater—and Earth's saline water—the oceans. The *atmosphere* envelops our planet and gives Earth its weather and climate.

The elements that make up Earth are found in its rocks and minerals. These rocks and minerals form Earth's *crust, mantle,* and *core.* The crust rides atop **tectonic plates** that move in response to heat flow and convection in Earth's interior. As these plates move, Earth's surface changes—rock is created and rock is deformed. And as rock at Earth's surface is touched by water and air, it begins to erode and weather and is broken up into smaller rocks. The process continues as the three spheres interact to influence landforms at Earth's surface.

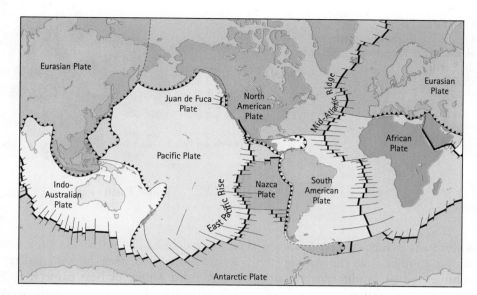

FIGURE 20.1
Earth's crust is a mosaic of tectonic plates that move in response to heat flow and convection in Earth's interior. As the plates move, Earth's surface changes.

20.1 The Geosphere Is Made Up of Rocks and Minerals

EXPLAIN THIS Why are Earth's elements unevenly distributed?

We begin our study of Earth science by examining the ground beneath our feet—Earth's geosphere. The geosphere is made up of rocks, and rocks are made up of minerals. Knowing about rocks and minerals helps us understand the structure and makeup of our geosphere, much as knowing about concrete, steel, and glass helps us understand the design and architecture of a building. For example, the kinds of minerals found in volcanic rocks provide evidence that molten rock erupted from Earth's interior to the surface. And the size and type of minerals in a metamorphic rock can reveal the rate of crystallization and the conditions of formation (the temperature and pressure) that occurred deep below Earth's surface.

Minerals are the building blocks of rocks, and elements, in turn, are the building blocks of minerals. You may recall from our discussion of the periodic table in Chapter 12 that there are 112 known elements, and that many of these elements are rare. But out of this very large number, it may surprise you to learn that just eight elements make up 98% of Earth's entire mass (Figure 20.2)! All of the other elements combined make up the remaining 2%.

Earth's elements are not distributed evenly. For example, most of Earth's iron is concealed deep in the planet's interior, where it forms the central core. Lighter elements, such as silicon and oxygen, are mostly distributed in the mid-to-outer portions of the planet. To explain this lopsided distribution we need to examine the very beginnings of Earth.

Our solar system formed about 4.5 billion years ago, when dust, gases, and rocky and metallic debris orbiting the newly forming Sun collided and coalesced into the planets, asteroids, and comets we know today. One such rocky mass became our Earth, which formed as chunks of all sizes accumulated. When first formed, the elements were distributed evenly throughout because that is the way they accumulated. But all that was about to change.

With each collision, heat was released because of the conversion of kinetic energy to heat energy—impact heating. As Earth grew, gravitational attraction

LEARNING OBJECTIVE
Compare the composition and density of elements at Earth's surface and in Earth's interior.

FYI Earth processes do not occur in isolation—events in one sphere affect one or both of the other spheres.

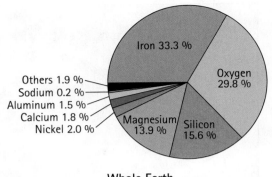

FIGURE 20.2
Only eight of the chemical elements are found in abundance on Earth.

toward the planet's center attracted even more debris. The attraction became strong enough that the young Earth actually squeezed itself into a smaller volume, which produced even more heat. A third source of heat came from the decay of naturally occurring, widely distributed radioactive elements. Although the amount of heat generated in any cubic meter of rock is small, when we think of the trillions of cubic meters of rock within Earth, the amount of heat produced can be quite large. These three sources of heat—*impact heating, gravitational contraction heating,* and *radioactive decay heating*—acted together to bring young Earth to its melting point.

So, in a molten or nearly molten state and under the influence of gravity, dense, heavy iron-rich material sank to Earth's center and less-dense, silicon- and oxygen-rich material rose toward the surface (see Chapter 5). The same type of density segregation occurs in a mixture of oil and water. The heavier water sinks to form a layer at the bottom and the less-dense oil rises to form a layer at the top. In Earth, density segregation led to the formation of a dense, iron-rich *core,* a less-dense, rocky *mantle,* and an even less-dense, rocky *crust* (Figure 20.3).

Figure 20.4 shows the current composition of Earth's crust. When you compare the composition of the crust to that of Earth as a whole, you see that the same few elements appear in both. The percentages, however, are quite different. As expected, the crust is composed of mostly lighter elements. In fact, almost half the mass of Earth's crust is the element oxygen (O) and about a fourth is the element silicon (Si)!

FIGURE 20.3
Earth has a layered internal structure. The layers—the crust, mantle, and core—differ in composition and density. As a whole, Earth's average density is 5.5 g/cm³. Because samples of Earth's crust average 2.7 g/cm³, material below the crust must have much higher densities. Although not directly measured, the mantle averages 4.5 g/cm³, and the core (solid inner and liquid outer) has an estimated average density of 13.5 g/cm³.

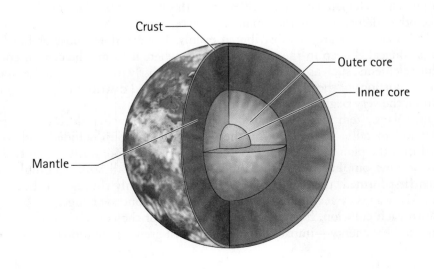

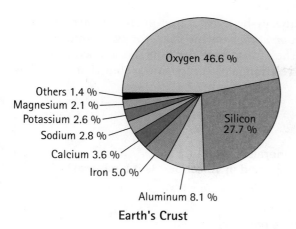

FIGURE 20.4
Percentages of elements in Earth's crust, by mass. Oxygen and silicon make up more than 75% of Earth's crust.

CHECKPOINT

1. Earth's interior is separated into layers, much as oil and vinegar in salad dressing separate into their respective layers. What caused this to happen?
2. How is the composition of the "whole Earth" different from that of "Earth's crust"?

Were these your answers?

1. Gravity. Heat from collision impacts, gravitational contraction, and radioactive decay softened our planet to a point at which its components could easily move around. In this state, dense elements sank toward Earth's center and lighter elements migrated upward to the surface.
2. Iron makes up one-third of Earth as a whole, but most of this iron is in Earth's interior. Oxygen and silicon dominate Earth's outer layers. And as we will soon see, this is why the most common group of rock-forming minerals—the silicate group—is dominated by these two elements.

FYI The dominance of oxygen in Earth's crust becomes even more apparent when you consider its abundance, not by mass, but in terms of numbers of atoms: 63 out of every 100 atoms making up Earth's crust are oxygen atoms. Oxygen isn't just important as a constituent of air—in the solid state, it makes up most of Earth's crust.

20.2 Minerals

EXPLAIN THIS How is a mineral different from a rock?

In everyday usage, we consider minerals part of our diet ("vitamins and minerals"), and minerals provide the raw materials needed for industry (aluminum for cans, iron for steel, etc.). From these two simple examples, it is easy to see the importance of minerals in the geosphere and in our lives. But what exactly is a mineral?

A mineral is a naturally formed, inorganic, crystalline solid, composed of an ordered arrangement of atoms with a specific chemical composition.

The definition is fairly straightforward. To be *naturally formed* means that it is not manufactured in a laboratory. So cubic zirconia and other synthetic gems are not minerals. *Inorganic* means that it is not made from materials that were once part of a living thing. So coal, made from decayed plant material, is not a mineral. To be a *crystalline solid* means that the atoms that make up a mineral are always arranged in an orderly geometric pattern. For example, glass, though solid, has no crystal structure. It is *amorphous* and so is not a mineral.

LEARNING OBJECTIVE
Define a mineral.

FYI Minerals found in rocks and in dietary supplements are similar yet different. The minerals in rocks are naturally occurring, inorganic, crystalline solids, with a specific chemical composition. Minerals in dietary supplements are human-made inorganic compounds that contain elements necessary for life functions. However, the elements used to make dietary supplements ultimately come from the naturally occurring minerals of Earth's crust.

FYI   The term *organic* is derived from *organism*. Some organisms produce minerals. For example, many marine organisms make calcium carbonate minerals (mostly aragonite) for their shells, and oysters can make pearls. The minerals opal and fluorite and some phosphates can also be precipitated by organisms. For example, teeth and bone material contain the mineral apatite. These minerals may be produced by organisms, but they are not made of the living organism. As a rule, organic compounds—those containing carbon, oxygen, and hydrogen atoms—all together—are not minerals. So although the hydrocarbons of petroleum and coal are referred to as mineral fuels, they are not minerals. And despite being a naturally occurring crystalline solid with a very ordered internal arrangement of atoms, table sugar ($C_{12}H_{22}O_{11}$) is not a mineral.

The same types of minerals always have the same geometric arrangement of atoms. A *specific chemical composition* means that, for two samples to be considered the same mineral, they must have the same basic chemical composition. Minerals typically have a range of compositions, but that range has fixed limits.

CHECKPOINT
1. **Are synthetic diamonds minerals?**
2. **Obsidian is a kind of glass formed in volcanoes. Is it a mineral?**

Were these your answers?
1. No. To be a mineral, it must be naturally formed.
2. No. Obsidian, though it is naturally formed and has a specific range of chemical compositions, is a type of glass—it is amorphous and so does not have a crystalline structure. It does not meet *all* the criteria needed to be a mineral.

20.3 Mineral Properties

EXPLAIN THIS   Why are some minerals soft and others hard?

LEARNING OBJECTIVE
Describe the physical properties used to identify minerals.

> Minerals differ from one another in their combination of elements and/or in the internal arrangement of their constituent atoms.

Minerals are classified by chemical composition (which elements are present) and crystal structure (how the elements are arranged). A mineral's observable physical properties depend on its inner microscopic properties. Microscopic properties such as composition, crystal structure, and the strength of chemical bonds determine a mineral's crystal form, hardness (resistance to scratching), fracture or cleavage (how a mineral breaks), color, and density. Most minerals can be identified by these easily observable physical properties. Other physical properties that can help identify minerals are luster (the way a mineral reflects light) and streak (the color of a mineral in its powdered form). In this section, we discuss the physical properties of minerals as expressions of their inner structures.

(a) Crystalline structure form of halite

Chlorine ion
Sodium ion
Basic structural form

(b) Grains of the mineral halite (table salt)

FIGURE 20.5
The structural form of the mineral halite (table salt) is cubic. This form is repeated over and over in three dimensions. The internal order of halite crystals is reflected in its macroscopic mineral grains.

Crystal Form

Have you ever seen table salt (halite) under a magnifying glass? If so, you may have marveled at its perfect geometric form (Figure 20.5). Crystals are well known for the striking geometric shapes that they can exhibit. A crystal's shape, or its **crystal form**, is an expression of the orderly arrangement of its atoms. When you look at a fully formed crystal, what you see is the actual arrangement

(a)   (b)   (c)   (d)   (e)   (f)

of atoms in its structure. Each type of mineral has a unique composition and crystal form (Figure 20.6). Unfortunately, well-shaped crystals are rare in nature because of space constraints—most crystals grow in cramped spaces.

Just as buildings are made of different materials—some stone, some brick, some wood—minerals are made of different elements. Some minerals have the same combination of elements, but their atoms are arranged differently, which makes them different minerals. So, carrying our analogy with buildings further, different architecture using the same materials can result in very different minerals. Sometimes, two or more minerals contain the same elements in the same proportions, but their atoms are arranged differently. As a result, their crystalline structure and the properties they display are different. Such minerals are called **polymorphs** (*poly* = many, *morph* = form) of each other. Graphite and diamond are polymorphs because they both consist entirely of the same element, carbon, but the carbon atoms are arranged differently. As a result, graphite and diamond show vastly different properties (Figure 20.7). Because the formation of these similar-yet-different minerals depends on temperature and pressure, a polymorph is a good indicator of the geological conditions at the time and place of its formation.

FIGURE 20.6
Many minerals are easily recognized by their crystal form. (a) Amethyst, the purple variety of quartz, has a hexagonal crystal form with pointed ends. (b) Pyrite, or "fool's gold," typically forms cubic crystals marked with parallel lines called striations. (c) Rosasite has radiating bluish-green crystals that group into balls. (d) Rhodochrosite (whose name means "rose-colored") has a rhombohedral crystal form. Some minerals have distinctive growth patterns. (e) The mineral hematite often grows in a grape-clustered form. (f) Asbestos minerals have a fibrous form.

CHECKPOINT

1. **Many minerals can be identified by their physical properties—crystal form, hardness, fracture, cleavage, luster, color, streak, and density. Why is identifying a mineral by its crystal form usually difficult?**
2. **What is a crystal?**

Were these your answers?
1. Well-shaped crystals are rare in nature because minerals typically grow in cramped spaces.
2. A crystal is a solid that has a crystalline structure—the atoms, ions, or molecules within it are arranged in a definite repeating pattern.

FIGURE 20.7
Both graphite and diamond are pure carbon. (a) Diamond, the hardest substance known, has a tightly packed symmetrical structure. (b) Graphite has an open, layered structure and is a very soft mineral. When you rub graphite between your fingers, individual graphite molecules glide over one another like cards in a deck, giving it a slippery feel. This slippery effect is why graphite is used as a dry lubricant. Graphite also glides easily when it is stroked onto paper, leaving a mark—hence its use in pencils. (Graphite is also preferable to lead in pencils because it is less toxic.)

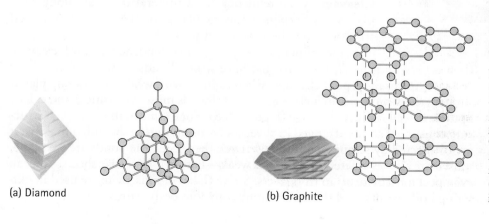

(a) Diamond   (b) Graphite

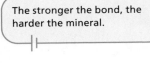

The physical properties of a mineral all relate back to the mineral's chemistry.

TABLE 20.1	MOHS SCALE OF HARDNESS	
Mineral	**Hardness**	**Object of Similar Hardness**
Talc	1	
Gypsum	2	Fingernail (2.5)
Calcite	3	Copper wire or coin (3.5)
Fluorite	4	
Apatite	5	Steel knife blade, glass (5.5)
Feldspar	6	Unglazed porcelain tile (6.5)
Quartz	7	
Topaz	8	
Corundum	9	
Diamond	10	

Hardness

Hardness does not refer to how easily a mineral breaks, but rather to its resistance to scratching. For example, a quartz crystal can scratch a feldspar crystal because quartz is harder than feldspar. The ability of one mineral to scratch another and the resistance of a mineral to being scratched are measures of hardness. We use the **Mohs scale of hardness** (Table 20.1) to compare the hardnesses of different minerals.

Why are some minerals harder than others? Hardness depends on the strength of a mineral's chemical bonds—the stronger its bonds, the harder the mineral. The factors that influence bond strength are ionic charge, atom or ion size, and packing (Chapter 15). Strong bonds are generally found between highly charged ions—the greater the attraction, the stronger the bond. Size affects bond strength as well, because small atoms and ions can generally pack closer together than large atoms and ions. Closely packed atoms and ions have a smaller distance between one another, and thus they form stronger bonds because they attract one another with more force. Gold, with its large atoms, is soft. Its atoms are rather loosely packed and loosely bonded. Diamond, with its small carbon atoms and tightly packed structure, is very hard—the hardest mineral known (see Figure 20.7a).

The stronger the bond, the harder the mineral.

Cleavage and Fracture

If you shatter the mineral calcite with a hammer, the surfaces where it broke are smooth and flat. This "clean" type of breakage occurs parallel to a mineral's *planes of weakness*—planes along which chemical bonds are weak or few in number. **Cleavage** is the tendency for a mineral to break along such planes of weakness. Cleavage planes are determined by crystal structure and chemical bond strength.

Some minerals show a greater tendency toward cleavage than others. In general, minerals that have strong bonds between planar (flat) crystal surfaces show poor cleavage, whereas those with weak bonds along planar surfaces show more distinct cleavage. The minerals muscovite (mica) and calcite both have well-defined cleavage (Figure 20.8). For example, mica's crystal structure consists of atoms arranged in sheets. The atoms *within* the individual sheets are connected by strong bonds, but *between* the sheets, the bonds are weak. So, muscovite cleaves where its bonding is weak—between its planar sheets. You can even peel muscovite off in thin layers. Shiny flakes of muscovite are used in glittering body paints, and they add shimmer to auto body paints as well.

(a)   (b)

FIGURE 20.8
A mineral's cleavage is very useful in its identification. (a) Muscovite, a mineral of the mica group, has perfect cleavage in one direction. It breaks apart into sheets. (b) Calcite (calcium carbonate) has perfect cleavage in three directions (though *not* at right angles, like a cube). It breaks apart into smaller rhombohedral shapes.

Minerals that have no planar alignment of bonds, like quartz, cannot display cleavage and always **fracture**. A fracture that is smooth and curved, so that it resembles broken glass, is called *conchoidal*. The minerals quartz and olivine display smooth conchoidal fractures (Figure 20.9). But most minerals fracture irregularly. The degree and type of cleavage or fracture are useful guides for identifying minerals.

CHECKPOINT

1. **When pieces of calcite and fluorite are scraped together, which scratches which?**
2. **The mineral muscovite displays very distinct cleavage, yet the mineral quartz fractures. How does this relate to each mineral's crystal structure?**

Were these your answers?

1. Looking at Table 20.1, we see that fluorite is harder than calcite. So fluorite scratches calcite.
2. Muscovite forms as a layered sheetlike structure. The bonds between the different layers are weaker than the bonds within the individual layers. Micaceous minerals cleave between the layers. Quartz has a more complicated structure with no layering and no planes of weakness. Therefore, quartz fractures.

FIGURE 20.9
When quartz breaks, it develops a curved, smooth surface that resembles broken glass—a conchoidal fracture. This quartz specimen shows its crystal form and conchoidal fracture.

Color

Although color is an obvious feature of a mineral, it is not a reliable means of identification. Some minerals—copper and turquoise are two examples—have a distinctive color. But most minerals either occur in a variety of colors or can be colorless.

Chemical impurities in a mineral affect color. For example, the common mineral quartz, SiO_2, can be found in many colors, depending on slight impurities. It can be clear and colorless if it has no impurities, or it can be milky white from tiny fluid inclusions. Rose-colored quartz results from small amounts of titanium; purple quartz (amethyst) results from small amounts of iron. The color of the mineral corundum, Al_2O_3, is commonly white or grayish. But impurities in corundum give us rubies and sapphires (Figure 20.10).

Streak is the color of a mineral in its powdered form. When a mineral is rubbed across an unglazed porcelain plate, it leaves behind a thin layer of powder—a streak. Although different samples of the same mineral can have different colors, the color of the mineral's streak is always the same (Figure 20.11). For example, the mineral hematite varies in color (red, brown, or black), but it always makes a reddish-brown streak. Magnetite can be gray or brown to black, but it always makes a black streak. Streak can be used for identifying minerals that have a metallic or semi-metallic luster. Minerals that do not have a metallic luster generally leave behind a white streak, which is not useful for identification.

Density

Density is a property of all matter, minerals included. In practical terms, the density of a mineral tells us how heavy a mineral feels for its size. More specifically,

---

FYI The luster of a mineral is the way its surface appears when it reflects light. Minerals that contain metals tend to have a shiny luster. Minerals such as quartz have a glassy luster, and minerals such as talc have a pearly luster.

Ruby

Sapphire

FIGURE 20.10
The mineral corundum (Al_2O_3) comes in a variety of colors as a result of chemical impurities. The addition of small amounts of chromium in place of aluminum produces the red gemstone *ruby*. With the addition of small amounts of iron and titanium, the result is the blue gemstone *sapphire*.

FIGURE 20.11
The streak test can be used to identify minerals that have a metallic or semi-metallic luster.

TABLE 20.2	DENSITY OF VARIOUS MINERALS (g/cm³)		
Borax	1.7	Pyrite	5.0
Quartz	2.65	Hematite	5.26
Talc	2.8	Copper	8.9
Mica	3.0	Silver	10.5
Chromite	4.6	Gold	19.3

a mineral's density is the ratio of its mass to its volume. The densities of some minerals are shown in Table 20.2.

Gold's particularly high density of 19.3 g/cm³ is nicely taken advantage of by miners panning for gold. Fine gold pieces hidden in a mixture of mud and sand settle to the bottom of the pan when the mixture is swirled in water. Water and less-dense materials spill out when the mixture is swirled. After a succession of douses and swirls, only the substance with the highest density remains in the pan—gold!

LEARNING OBJECTIVE
Distinguish between silicate and nonsilicate minerals.

20.4 Classification of Rock-Forming Minerals

EXPLAIN THIS Why are most rocks made of silicate minerals?

More than 4000 known minerals exist on Earth, and new ones are discovered every year. With so many minerals, how can they all be classified in a simple, systematic fashion? First of all, most minerals are actually quite rare. In fact, only about a dozen minerals make up most of the rocks exposed at Earth's surface. These are the *rock-forming minerals.*

Minerals are classified by their chemical composition. Doing so produces two main categories: the **silicates** and the **nonsilicates** (Figure 20.12). Look back at Figure 20.4 and you will understand the reason for this simple division. Oxygen is the most abundant element in Earth's crust, and silicon is the second most abundant. Minerals that contain both silicon (Si) and oxygen (O) as part of their chemical composition are called silicates. Minerals that do not contain these two elements are called nonsilicates.

Silicon has a great affinity for oxygen. In fact, silicon has such a strong tendency to bond with oxygen that silicon is never found in nature as a pure element; it is *always* chemically combined with oxygen. The combination of silicon and oxygen is called simply silica (SiO_2). The silicates are the most common mineral group, making up more than 90% of Earth's crust. Most silicates also contain the rest of the eight most common elements, which include Fe, Mg, Ca, and Al, but the basic building block of *all* silicates is Si and O.

The silicates are subdivided into two groups: those that contain iron and/or magnesium *(ferromagnesian)* and those that do not *(nonferromagnesian)*. Because of the presence of iron and/or magnesium, ferromagnesian silicates tend to be dense and dark in color. Nonferromagnesian silicates do not contain significant amounts of iron or magnesium; therefore, they generally have relatively low densities and are light in color. The most abundant mineral in the crust is feldspar, a nonferromagnesian silicate that contains aluminum, sodium, potassium, and/or calcium, plus silicon and oxygen. Feldspar makes up more than 50% of the Earth's crust. Quartz (SiO_2), the second most common mineral in Earth's crust, is composed of only silicon and oxygen. If you have ever collected rocks and minerals, you probably have some quartz and feldspar specimens in your collection.

The density of a mineral depends on a number of factors—the masses of the mineral's constituent atoms and the packing of these atoms, which, in turn, is a function of the atoms' sizes.

FYI All silicate minerals have the same fundamental structure of atoms, the silicon-oxygen tetrahedron—four oxygen atoms joined to one silicon atom $(SiO_4)^{4-}$. The powerful bond that unites the oxygen and silicon ions is akin to the cement that holds Earth's crust together.

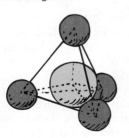

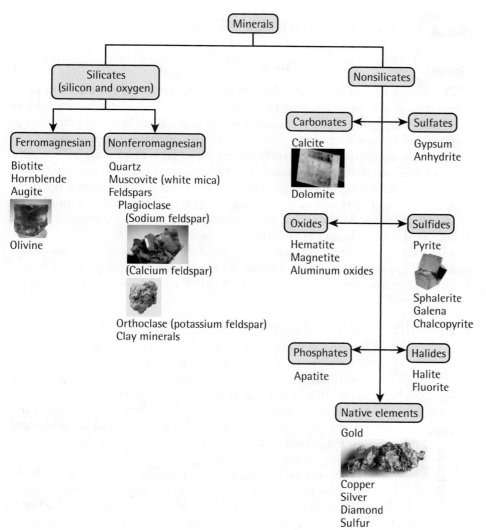

FIGURE 20.12
Classification of common
rock-forming minerals.

Nonsilicate minerals make up just 8% of Earth's crust by mass. The nonsilicates include the carbonates, sulfates, halides, oxides, sulfides, phosphates, and native elements such as gold and silver. The carbonates are the most abundant nonsilicate minerals. Two common carbonate minerals are calcite and dolomite—the main minerals found in the group of rocks called *limestone*.

Carbonates have many uses—buildings are often made from quarried limestone, and carbonates are used to make cement. Sulfates and halides are nonmetallic mineral resources. For example, the sulfate mineral gypsum is used for making plaster, and the halide mineral halite is common table salt. Oxides and sulfides are important metallic mineral resources. Mineral deposits rich in valuable metals are economically important. Such deposits are called *ores*. For example, hematite and magnetite are valuable iron oxide ores used for steelmaking and construction. Galena, a lead sulfide ore, is used in wireless communication systems, and the sulfide chalcopyrite is an important copper ore. Phosphates, though not as common as the other mineral groups, are an important agricultural resource. For example, the mineral apatite is used to make fertilizer.

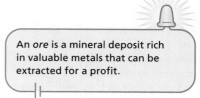

An *ore* is a mineral deposit rich in valuable metals that can be extracted for a profit.

FYI The rate of cooling influences crystal size. Rapid cooling forms a large number of small crystals. Very slow cooling, on the other hand, allows elements to migrate over great distances and allows smaller crystals to merge, so relatively large crystals form. When molten material is cooled so quickly that atoms do not have time to arrange into their respective crystal pattern, the solid formed is *glass*. Its atoms are as unordered as those in ordinary window glass.

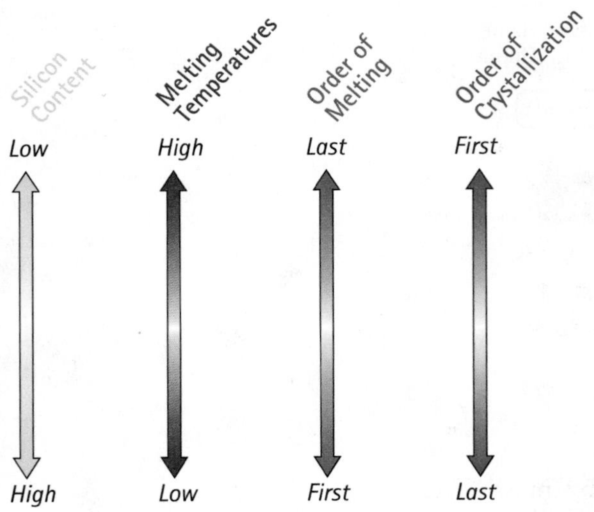

FIGURE 20.13
High-silica-content minerals have low melting points. They are the first minerals to melt and the last minerals to crystallize. Low-silica-content minerals have high melting points. They are the last minerals to melt and the first minerals to crystallize.

Just as ice melts at the same temperature at which water freezes, a mineral's melting point is the same temperature at which it begins to crystallize from hot magma.

20.5 The Formation of Minerals

EXPLAIN THIS Why does the composition of magma change as it cools?

So far we have explored the definition of a mineral and the different properties and classification of minerals. Now we turn our attention to how minerals form. Understanding how minerals form is a steppingstone to understanding how rocks form. Rocks, after all, are made of minerals.

Minerals are formed by **crystallization**—the growth of a crystalline solid from a liquid or gas. Crystallization starts when atoms begin to bond with each other in a particular geometric pattern. As the numbers of atoms and bonds increase, a single crystal forms, with edges reflecting the shape of the underlying geometric pattern. As more and more atoms bond to the microscopic crystal, repeating the underlying pattern, the crystal grows.

Minerals commonly crystallize from two different sources: from **magma**—molten rock—and from water solutions. As we shall see, *igneous* rocks are formed from magma, and certain *sedimentary* rocks are formed from water solutions.

Crystallization in Magma

Magma is composed primarily of the elements found in the silicate group of minerals—namely, silicon and oxygen, plus aluminum, potassium, sodium, calcium, iron, and magnesium. When magma starts to cool, atoms in the hot liquid lose kinetic energy. Attractive forces then pull the atoms into orderly crystalline structures. Minerals crystallize from cooling magma in a systematic fashion, based on their respective melting points.

The crystallization points of silicate minerals strongly depend on the amount of silica they contain. As magma cools, the first minerals to crystallize have the highest (hottest) melting points and the lowest amount of *silica*; the last minerals to crystallize have the lowest (coldest) melting points but the highest silica content (Figure 20.13). Conversely, high-silica minerals melt at lower temperatures than do low-silica minerals. Consider the minerals quartz and feldspar. When a rock with both quartz and feldspar melts, the quartz melts before the feldspar because quartz contains more silica than feldspar (in fact, quartz is pure silica!). In a cooling magma, on the other hand, quartz crystallizes after, and at a lower temperature than, feldspar.

Crystallization occurs step by step in a temperature-dependent sequence. With each step, the composition of the remaining liquid magma changes. The magma becomes depleted in the constituents of minerals that have already crystallized and enriched in the constituents of minerals that have yet to crystallize. So, later-forming minerals contain more silica and different suites of elements—in varying proportions—than early-forming minerals.

To simplify understanding the process of crystallization in a cooling magma, consider the following analogy. Suppose you have all the pieces for a game of checkers—12 red pieces and 12 black ones. All 24 pieces are mixed together to form a single group, which represents the magma before any minerals have formed. So the "liquid" consists of 50% red pieces and 50% black pieces. Now suppose 3 red and 2 black pieces are removed to form the first "mineral." Now 9 red and 10 black pieces remain in the "liquid," which is now represented by only 19 checkers. The "liquid" is now roughly 47% red and 53% black. Thus, the "liquid" has become depleted in red and enriched in black—its composition has changed. If the black checkers represent silica molecules, can you now see that the crystallization process enriches magma in silica? This crystallization process allows a single magma to generate a range of igneous minerals and rocks.

A familiar example of crystallization is ice crystals that form in water when the temperature drops below 0°C. So just as there is water and ice, there is magma and rock.

CHECKPOINT

1. **Olivine and pyroxene are both ferromagnesian silicate minerals that crystallize from cooling magma. In the sequence of crystallization, olivine crystallizes first, pyroxene second. Which mineral contains more silica?**

2. **Having a high melting point can be thought of as requiring a higher (hotter) temperature to melt. A low melting point means that the temperature does not need to be as high for the mineral to melt. How does having a high melting point translate to crystallizing first?**

Were these your answers?

1. The mineral pyroxene contains more silica than olivine. As magma cools, minerals with less silica crystallize before minerals with more silica.

2. When the temperature is high, the constituents of possible minerals are in the liquid state. Remember that water freezes at the same temperature at which ice melts. So think of the melting point as being equivalent to the freezing—that is, crystallization—point. Ice forms when the water temperature drops below the freezing/melting point. When magma begins to cool, minerals with high melting temperatures crystallize first, because the magma temperature drops below their respective freezing/melting points. The minerals with lower melting points do not form yet—they stay in the liquid state. They cannot begin to crystallize until lower temperatures are reached. So minerals with low melting points crystallize last. When all is said and done, the liquid magma freezes to become rock.

FIGURING PHYSICAL SCIENCE

Silica Enrichment in Magma

Magma becomes enriched in silica, SiO_2, as crystals such as olivine grow within it. We can express this in exact quantitative terms with a few calculations. Consider this example: Suppose we begin with 1000 kg of magma, of which 500 kg is silica. Its temperature cools, and 325 kg of olivine crystallize from it.

Before crystallization, the mass percentage of silica in the magma is

$$\frac{500 \text{ kg silica}}{1000 \text{ kg magma}} \times 100\% = 50\%$$

The chemical formula of olivine is $MgFeSiO_4$. The elements needed to make these crystals inside the magma, therefore, come from one unit of

MgO, plus one unit of FeO, plus one unit of SiO_2.

To find the mass percentage of silica in olivine, we divide the formula mass of silica (SiO_2), which is 60.0 amu, by the formula mass of olivine, $MgFeSiO_4$, which is 172 amu. We get

$$\frac{60 \text{ amu}}{172 \text{ amu}} \times 100\% = 34.8\%$$

which rounds off to 35%

Since we now know the mass percentage of silica in olivine, we can figure out how much silica was removed from the magma when 325 kg of olivine crystallized:

325 kg olivine × 0.35 = 114 kg silica

Now that the olivine has crystallized, the mass percentage of silica in the

remaining magma is less than 50%, but how much less? We have

500 kg silica in original magma
− 114 kg silica in olivine
386 kg silica in remaining magma

The total mass of magma is different now, too:

1000 kg magma
− 325 kg olivine
675 kg magma

So, the mass percentage of silica in the remaining magma is

$$\frac{386 \text{ kg silica}}{675 \text{ kg magma}} \times 100\% = 57\%$$

The magma has been enriched in silica. These concepts are revisited in Exercise 37 at the end of the chapter.

FIGURE 20.14

Calcium carbonate precipitating from dripping water in a cave forms icicle-shaped stalactites hanging down from the ceiling and cone-shaped stalagmites protruding upward from the ground.

Crystallization in Water Solutions

Minerals crystallize from water solutions in two main settings. The first is associated with the final stages of magma crystallization—magma generally contains from 1% to 6% water. When a body of magma is nearly solidified, this very hot water circulates through fractures in the new rock and often into the surrounding rock—*hydrothermal activity*. These water solutions usually contain many dissolved mineral constituents. The solutions become chemically saturated as the temperature decreases, causing various minerals to *precipitate*. These minerals are often deposited in cracks, and sometimes within the rock matrix itself. Many of the important ore deposits we find today were formed in this manner.*

Similar to hydrothermal minerals, **chemical sediments** are also formed by the precipitation of mineral constituents from water solutions. But in this second setting, chemical sediments form where temperatures are much cooler than Earth's interior, such as in a body of water on Earth's surface. Chemical sediments fall into two categories: carbonates and evaporites.

Carbonates are minerals and rocks composed mostly of calcium carbonate, $CaCO_3$, which has the mineral name *calcite*. Dolomite, $CaMg(CO_3)_2$, is also a common carbonate mineral. Carbonates can form in two ways—by inorganic precipitation and, as we will see later in this chapter, as a result of biologic activity. Many seashells are composed of calcium carbonate secreted by organisms. Cave dripstones, such as stalactites and stalagmites, provide a great example of calcium carbonate precipitating inorganically from dripping water (Figure 20.14). They form because groundwater (Chapter 22) picks up calcium and carbonate ions as it moves through limestone formations in a cave. The dripping water is saturated with dissolved calcite, so the dripstones form as evaporation removes very small amounts of water during the dripping process.

Evaporites are minerals and rocks precipitated when a restricted body of seawater, or the water of a salty lake, evaporates. Examples are gypsum, anhydrite, and halite. These names apply both to individual minerals and to rocks made of a single type of evaporite mineral. Evaporites precipitate out of water solutions in a way that is very similar to the crystallization of minerals from magma. The difference, though, is that **solubility** rather than melting point determines which minerals crystallize first. As evaporation proceeds, the minerals with the lowest solubility—the most difficult to dissolve—such as gypsum precipitate first, followed by the minerals that dissolve more easily (those with higher solubilities), such as anhydrite and then halite. Although carbonates make up the bulk of chemical sediments, evaporites are a small but important group.

CHECKPOINT

1. When water evaporates from a body of water, what type of mineral is formed?
2. A mineral with low solubility does not dissolve easily. A mineral with high solubility dissolves easily. How is this factor of "dissolvability" related to crystallization?

---

* Hydrothermal activity is rich in rare metals. Such metals are chemically happier "hiding out" with the water until they are so concentrated that they can form their own minerals—such as gold ore.

Were these your answers?

1. Evaporite minerals precipitate out of solution as water evaporates.
2. Minerals that dissolve easily stay dissolved longer in a solution than those that do not dissolve so easily. So as a restricted body of water dries up, the first minerals to crystallize are those that *do not* dissolve so easily—those with lower solubilities. Minerals that are easily dissolved (high solubility) remain in solution longer—they are the last to crystallize.

Now that we have seen the different ways that minerals form, we can begin to learn about the combinations of minerals called rock. We know that minerals formed from the crystallization of magma make up igneous rock, and minerals formed from the precipitation or evaporation of water make up some types of sedimentary rock. Now we will also see that igneous rock breaks down to form sedimentary rock, and that a third rock type—metamorphic rock—forms from rocks that already exist. Minerals, in their many forms, are the building blocks of the many different rocks on Earth.

20.6 Rock Types

EXPLAIN THIS What is the difference between a rock and a mineral?

LEARNING OBJECTIVE
Define the three categories of rocks.

A **rock** is defined as an aggregate of minerals (Figure 20.15). Some rocks are aggregates of fossil shell fragments, solid organic matter, or any combination of two or three of these components. Just as we think of minerals as chemical mixtures or compounds, rocks can be thought of as physical mixtures. In some rocks, the grains are "cemented" together; in others, the grains are tightly interlocked. In many rocks you can see mineral crystals. Granite, one of the most common rocks in Earth's continental crust, contains visible crystals of the minerals feldspar, quartz, hornblende, and others (Figure 20.16). On the other hand, in rocks such as basalt, shale, or slate, individual grains are difficult to distinguish—they are too small to be seen with the unaided eye.

Rocks are divided into the following three categories based on how they were formed:

Igneous rocks are formed by the cooling and crystallization of hot, molten rock. The word *igneous* means "formed by fire." *Plutonic* igneous rocks are formed when molten rock below Earth's surface—magma—cools. Granite is a common plutonic rock. *Volcanic* igneous rocks are formed when molten rock at Earth's surface—lava—cools. Basalt is a common volcanic rock.

Sedimentary rocks are formed at or near Earth's surface from the cementation or compaction of *sediment*—rock, mineral, shell, or solid organic fragments carried by water, wind, or ice and deposited in low-lying areas. Sedimentary rocks also form when minerals precipitate out of water solutions at or near Earth's surface. Sandstone, shale, and limestone are common sedimentary rocks.

(a)   Basalt   Granite

(b)   Sandstone   Limestone

(c)   Marble   Slate

FIGURE 20.15
Rocks are made of minerals. There are three main types of rock. (a) Basalt and granite are igneous rocks. (b) Sandstone and limestone are sedimentary rocks. (c) Marble and slate are metamorphic rocks.

FIGURE 20.16
A rock is an aggregate of one or more minerals, and sometimes shell fragments and/or solid organic matter. This granite is an aggregate of minerals—the minerals feldspar, quartz, and hornblende.

Feldspar
(Mineral)
+

Quartz
(Mineral)
+

Hornblende
(Mineral)

Granite
(Rock)

Metamorphic rocks are formed from older, preexisting rocks (igneous, sedimentary, or metamorphic) that were transformed in Earth's interior by high temperature, high pressure, or both—without melting. The word *metamorphic* means "changed in form." For example, marble is metamorphosed limestone, and slate is metamorphosed shale.

LEARNING OBJECTIVE
Describe the physical and chemical conditions that give rise to different igneous rocks.

20.7 Igneous Rocks

EXPLAIN THIS Why does Earth have a great variety of igneous rocks?

Earth's crust consists primarily of various rocks of igneous origin. On the continents, the most common igneous rocks are granite and andesite. On the ocean floor, basalt is the most common igneous rock. All igneous rock originated as magma.

Generation of Magma

We've learned that many minerals form from cooling magma. But where does magma come from? Do we have magma because Earth's interior is molten? The answer is no. Earth's interior is mostly solid, not molten. And the one layer of Earth that is liquid is composed of molten iron, not magma, and is not encountered until a depth of almost 3000 km.

Simply stated, magma is derived from rocks that have melted. Just as water cools to form ice, magma cools and solidifies to form the minerals that eventually become rock. But how does rock melt to become magma? Temperatures recorded in mines and drill holes indicate that Earth's temperature increases throughout most of the continental crust at an average of 30°C for each kilometer of depth (Figure 20.17). But increased temperature is not enough to cause rocks to melt, even though the temperature at sufficient depth is actually much hotter than that of magma.

If the temperature at depth is hotter than magma, why are the rocks at depth solid? The answer can be found by using water as an analogy. Recall (from Section 7.7) that a phase change occurs when water is heated to the boiling point. We know that water boils at 100°C at sea level. If we increase the pressure, the temperature needs to be higher for boiling to occur. So phase changes depend on pressure as well as temperature. In the case of hot rocks at depth, they are under enormous pressure from the weight of the rock above—enough pressure to prevent melting, even at hotter-than-magma temperatures.

But rocks sometimes melt to form magma, and there are three reasons. The dominant reason, in terms of the amount of magma produced, is that hot rock rises upward from depth to levels where pressure is reduced enough to induce melting.*

Another mechanism for generating magma is the addition of water to rock, which lowers the rock's melting point. To understand fluid-induced melting, we can use the behavior of water as an analogy once again. A "foreign" substance in water lowers its freezing point (it must be colder to freeze). By the same token, the foreign substance also lowers ice's melting point. For example, salt placed on 0°C ice causes the ice to melt even though the temperature has not changed. The salt lowers the freezing point of water, so 0°C is no longer cold enough to keep the ice frozen. In the case of rock, water is the foreign substance that lowers the melting point of the rock—enough to cause melting.

These two magma-generating mechanisms produce magma that rises upward through preexisting rock. And this gives us the third mechanism—namely, that rock can melt when its temperature rises, often because of the presence of hotter materials that have risen from deeper areas. We will explore the first two mechanisms in more detail in Chapter 21, where we can link them to the plate tectonic model. For now, we will restrict our discussion to the general formation of magma from melting rock.

Keep in mind that rock is an aggregate of solids, so the melting of rock into magma occurs over a broad temperature range. As rock is heated, the first minerals to melt are those with the lowest melting points. This is similar to the process of crystallization from magma, but the changes of phase occur in reverse order. If all the minerals in a rock could melt simultaneously, the composition of the resulting magma would be the same as the composition of the original rock. But melting does not occur this way—**partial melting** is the rule.

Magma produced by the partial melting of rock is made from only the constituents of minerals that have melted—the ones with the lowest melting points. And because high-silica minerals have low melting points and are the first to melt, the resulting magma contains more silica. This "new" magma has a different composition from the partially melted rock that produced it. So partial melting results in magmas of many different compositions and—because these magmas cool to form igneous rocks—a variety of igneous rocks.

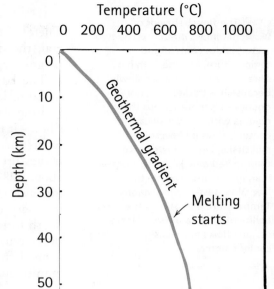

FIGURE 20.17
The temperature inside Earth increases about 30°C for each kilometer of depth from the surface deep into the continental crust (the gradient is much lower in the deep crust and mantle). This increase of temperature with depth is known as the *geothermal gradient.*

---

* The layer below Earth's crust, the mantle, behaves like an elastic solid. Although movement is extremely slow, rock actually flows in convection currents that allow hot materials to rise and cool materials to sink.

FYI The flow behavior of magma is influenced by two factors: silica content and temperature. Magma with high silica content flows more slowly because it is thicker and more gooey—it is more *viscous* than magma with a lower silica content. (This is like a spilled milkshake, which flows slower than spilled milk.) Basaltic magma is an important example of a low-silica, fast-flowing magma. Temperature also affects the ability of magma to flow. Hotter magma flows more easily than cooler magma.

Three Types of Magma, Three Major Igneous Rocks

There are three major types of magma—*basaltic, andesitic,* and *granitic*. In general, the different magmas occur in different geologic settings. These three magma types will be tied to the plate tectonic model in Chapter 21.

The largest region of Earth's interior is the solid mantle, which is composed of low-silica-content igneous rocks. When these rocks undergo partial melting, basaltic magma is formed. Basaltic magma is still relatively low in silica—it is about 50% silica. When solidified at Earth's surface, basaltic lava forms the dark igneous rock known as *basalt*, which is the kind of rock that makes up the Hawaiian Islands and the oceanic crust. Andesitic magma is about 60% silica. The rock known as *andesite*, which is produced from andesitic lava, gets its name from the Andes Mountains in South America, where it is very common. Granitic magma, at about 70% silica, cooling slowly at depth forms *granite* and other similar granitic rocks. Of all the igneous rocks in the crust, oceanic and continental crust combined, approximately 80% was formed from basaltic magma, 10% from andesitic magma, and 10% from granitic magma.

CHECKPOINT

1. If 80% of all igneous rocks are formed from basaltic magma, why do we see so much granite?
2. If basaltic magma is produced from the partial melting of mantle rock, how is most granitic magma produced?

Were these your answers?

1. Basalt is the most common igneous rock on the ocean floor. Look at a globe to see that the oceans cover about 71% of Earth's surface. We see so much granite because it is the most common igneous rock on continental land.
2. When magma rises toward Earth's surface, it makes contact with the surrounding rock. Rock in the continental crust contains more silica than rock in the mantle. These silica-rich rocks are partially melted by the rising magma and incorporated into the melt, which increases the magma's silica content. Additionally, the process of crystallization enriches magma in silica. As magma crystallizes and the liquid becomes separated from the crystals, the liquid portion of the magma changes, becoming more silica enriched. Eventually granitic magma (70% silica) is produced, and when this cools, granitic-type rocks form.

Igneous Rocks at Earth's Surface

Igneous rocks are divided into two general categories based on where they formed. Igneous rocks formed by the eruption of molten rock at Earth's surface are called **volcanic** (or *extrusive*) **rocks**. Molten magma that moves upward from inside Earth and flows onto the surface is called **lava**. The term *lava* refers both to the molten rock itself and to the solid rock that forms from it.

Lava may be extruded through fractures and fissures (long, planar cracks) in Earth's surface or through a central vent—a volcano. Eruptions from a volcano are more familiar to us because they are very dramatic to see, but the outpourings of lava from fissures are much more common. Most fissure eruptions occur underwater as basaltic lava erupts where the ocean floor is spreading apart. Fissure eruptions also occur on land. Episodic and unusually voluminous

FIGURE 20.18
The flood basalts that produced the Columbia Plateau covered more than 200,000 km^2 of the preexisting land surface.

lava outpourings known as *flood basalts* have flooded large areas of Earth throughout its history, creating extensive lava plains. The Columbia Plateau in the Pacific Northwest is the result of extensive flood basalts (Figure 20.18), as is the Deccan Plateau in India.

Volcanoes are vents where magma rises to Earth's surface and erupts as lava. Three main types of volcanoes exist—shield, cinder cone, and composite (Figure 20.19).

Shield volcanoes are built by a steady supply of easily flowing basaltic lava that flows out in all directions to make a broad, gently sloping cone. Mauna Loa in Hawaii, the largest volcano on Earth, is a shield volcano standing 4145 m above sea level and more than 9750 m above the deep ocean floor (Figure 20.20).

Cinder cone volcanoes are very steep but small in comparison to shield volcanoes. Cinder cones are not restricted to a particular type of lava. They are formed from the piling up of ash, cinders, and rocks that have been explosively erupted from a single vent to form a symmetrical, steep-sided cone. Two well-known examples of cinder cones are Sunset Crater in Arizona and Parícutin in Mexico.

A *composite cone* (also known as a stratovolcano) is a volcano built up of alternating layers of lava, ash, and mud. Composite cones have a steep-sided summit and gently sloping lower flanks. Mt. St. Helens in Washington is an example of an active composite cone. In 1980, it erupted violently, disrupting the lives of thousands of people and transforming more than 200 mi^2 of lush forest into a burned, gray landscape.

FYI Volcanoes not only influence the geosphere and the atmosphere, but they can even affect air travel! In April 2010, a volcanic eruption in Iceland created a huge cloud of ash that drifted all over Europe. Because the volcano was covered by ice, heat from the eruption instantly boiled to steam, which enhanced the production of very fine ash. The volume of ash was so great that it posed a serious hazard to aircraft engines. Some of the busiest airports in Europe were forced to close down for several days.

FIGURE 20.19
The three types of volcanoes. (a) Shield volcanoes, such as Mauna Loa, have broad, gentle slopes that average between 1° and 10° (from the horizontal). (b) Cinder cones, such as Sunset Crater in Arizona, generally have smooth steep slopes of 25° to 40° and bowl-shaped summit craters. (c) Composite cones, such as picturesque Mt. Fuji, are also very steep. On average, the slope of a composite cone starts out at 30° at the summit and gradually flattens to 10° at the base.

(a)

(b)

(c)

FIGURE 20.20
(a) Mauna Loa, a shield volcano on the island of Hawaii, is the largest volcano on Earth. (b) When compared with other large volcanoes, its immense size and volume are dramatic.

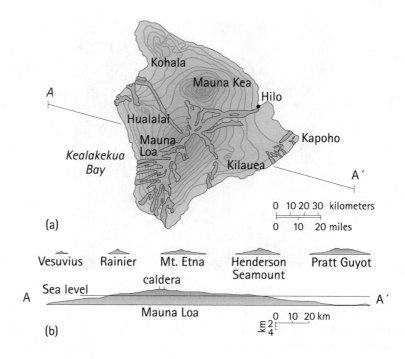

(a)

(b)

Composite volcanoes tend to erupt explosively because their magmas and lavas usually do not flow easily. This thick magma traps volcanic gases, which increases the pressure inside the volcano. We can compare the gases in a magma to gases in a bottle of carbonated soda. If we cover the top of the bottle and shake vigorously, the gases separate from the soda and form bubbles. When we remove the cover, pressure is released and gases and liquid explode from the bottle. The gases in magma behave in much the same way. In a volcanic blast, the pressure and temperature increase, and the whole mass of viscous magma and overlying rock explodes into dust and rubble. When combined with abundant volcanic ash, this mixture can expand and destroy everything in its path. Examples of this kind of volcanic activity occurred at Mt. Vesuvius in AD 79, at Mt. Pelee in 1902, and at Mt. St. Helens in 1980.

Three principal types of volcanic rocks exist—basalt, andesite, and *rhyolite*. Rhyolite forms when granitic lava erupts at Earth's surface. Rhyolite comes in several different forms (Figure 20.21). It can be a typical fine-grained volcanic rock, or the lava can cool quickly to form volcanic glass—pumice if the rock is riddled with tiny holes from former gas bubbles, obsidian if it lacks such holes, or volcanic ash if it is erupted explosively.

FYI Volcanic rocks (such as basalt) form at Earth's surface where they cool quickly; they tend to have microscopic crystals. Plutonic rocks (such as granite) form below Earth's surface where they cool slowly; they tend to have much larger crystals that are easily seen without magnification.

FIGURE 20.21
The volcanic rock rhyolite comes in several forms. (a) Fine-grained rhyolite. (b) Pumice. (c) Obsidian.

(a)   (b)   (c)

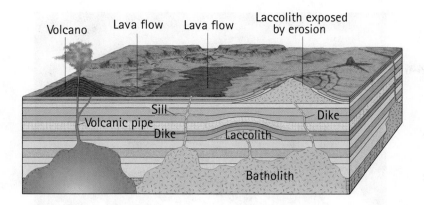

FIGURE 20.22
Intrusive igneous features in cross-sectional view.

Igneous Rocks Beneath Earth's Surface

When magma cools beneath Earth's surface, the igneous rock that forms is called **plutonic** (or *intrusive*) rock. The word *plutonic* is derived from Pluto, the mythological god of the underworld. All intrusive igneous rock bodies are called *plutons*. Being intrusive, plutonic rocks can be studied only after they are exposed by uplift and erosion at Earth's surface. The most common plutonic rock is granite.

Plutons occur in a great variety of shapes and sizes, ranging from small pipe-like *dikes* to large, expansive *batholiths* (Figure 20.22). Batholiths, the largest plutons, are defined as having more than 100 km^2 of surface exposure. Batholiths are created by numerous intrusive events over millions of years—they form the cores of many major mountain systems around the world. Generally speaking, they are the crystallized magma chambers that fed long-since-eroded volcanoes. Many modern mountains are actually the exposed batholith cores of larger mountains that eroded away long ago. Two of the largest batholiths in North America are the Coast Range batholith and the Sierra Nevada batholith (Figure 20.23). It is interesting to note that the Sierra Nevada is gaining height with time—long after all the magma has crystallized—because its rate of uplift is greater than its rate of erosion.

20.8 Sedimentary Rocks

EXPLAIN THIS How does beach sand become rock?

Sedimentary rocks are the most common rocks in the uppermost part of the crust. They cover two-thirds of Earth's surface, forming a thin, extensive blanket over older igneous and metamorphic rocks. Because sedimentary rocks contain the remains of organisms and older rocks, they provide information about geological events that have occurred over time at Earth's surface.

The Formation of Sedimentary Rock

Sedimentary rock forms in a long process with four stages: *weathering, erosion, deposition,* and *sedimentation*. **Weathering** is the disintegration or decomposition of rock—in place—at or near Earth's surface. Agents such as water, wind, ice, and reactive chemicals weather the rock—breaking it into smaller pieces, cracking its surface, rounding and smoothing its edges and corners, and sometimes transforming its chemical composition. Two types of weathering

FIGURE 20.24
The rocks on this mountain peak have been split apart by mechanical weathering. As water freezes and expands in cracks, the rock splits and breaks apart.

FYI When water freezes, the water molecules arrange themselves into six-sided crystalline structures that have much open space. So water expands upon freezing. Out in nature, the freezing process mechanically breaks rock apart. Ice wedging occurs as water seeps into small cracks, freezes, and expands, thereby widening the cracks. Eventually, with repeated freezing, cracks can widen to split the rock!

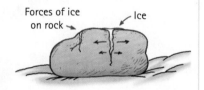

Forces of ice on rock →   ← Ice

occur—*mechanical* and *chemical*. Both produce sediment. *Mechanical weathering* physically breaks rocks into smaller and smaller pieces (Figure 20.24). For example, the freezing and thawing of ice can widen preexisting cracks in rock. In *chemical weathering*, reactions with water decompose rock, analogous to the way bacteria decomposes organic matter but with different chemical reactions (Figure 20.25). Because liquid water and water vapor are everywhere (excepting liquid water in polar regions), chemical weathering produces more sediment than mechanical weathering.

As rock weathers, it also erodes. **Erosion** is the process by which weathered rock particles are removed and transported away by water, wind, or ice. The principal difference between weathering and erosion is that erosion does not occur in place—it involves movement. Running water is the dominant agent of erosion. Pieces of mechanically weathered rock are normally quite angular and jagged when they are first produced. During transportation, especially by water, the various particles collide with one another and break. This decreases their size and rounds off their sharp edges. When transportation stops, deposition and sedimentation begin.

Deposition is the stage in which eroded particles come to rest. Sediments are deposited in horizontal layers, with each successive layer younger than the one beneath it. The larger a sediment particle, the stronger a current must be to carry it. As flowing water slows down, larger particles are the first to be

FIGURE 20.25
These granite rocks have been worn into rounded shapes by chemical weathering. Rainwater chemically decays the rock's outer layers, making the rock easier to erode. Rainwater physically erodes the rock by washing away its weakened outer layers, leaving rounded boulders behind. The rounding occurs because chemical weathering is faster where surface area is greater, and corners and edges have the most surface area.

deposited, while smaller particles remain with the flow. In this way, sediments are sorted according to size as they are deposited (Figure 20.26).

In the process of **sedimentation**, sediments are deposited horizontally one layer at a time. As the deposited sediment accumulates, it begins to change into sedimentary rock. We say that the sediments *lithify*—they undergo *lithification*, a term that means "conversion into rock" (*lith* = rock). Lithification occurs through the processes of *compaction* and, usually, *cementation*. Compaction is the first step. As the weight of overlying sediments presses down on deeper layers, sediment particles are squeezed and compacted together. This compaction squeezes much of the water out of the pores between the sediment particles.

The remaining "pore water" often contains dissolved compounds, such as silica, calcium carbonate, and iron oxide. These compounds can precipitate from solution and partially fill the pore spaces with mineral matter. The mineral matter glues the particles together and acts as a cementing agent. This is the process of cementation. Silica cement, the most durable, produces some of the hardest and most resistant sedimentary rocks. When iron oxide acts as a cementing agent, it produces the red or orange stain often seen in sedimentary rocks. The rock colors of Bryce Canyon National Park in Utah provide a beautiful example of iron oxide stain (Figure 20.27).

(a) Well-sorted sediments

(b) Poorly sorted sediments

FIGURE 20.26
A deposit that contains particles of similar shapes and sizes is called *well sorted*. A *poorly sorted* deposit contains particles of many different shapes and sizes. In general, poorly sorted sediments traveled a short distance before being deposited, and well-sorted, well-rounded sediments traveled a long distance before being deposited.

FYI Sediment shape and size give us clues to the method of sediment transport. Glacial deposits are poorly sorted and very angular because they are trapped in ice during transport. Wind-blown deposits tend to be very well sorted and small. Wind can generally move only small particles.

FIGURE 20.27
The red and orange colors in the sedimentary rocks at Bryce Canyon in Utah are caused by the presence of iron oxide.

CHECKPOINT
Which type of weathering produces the most sediment: mechanical or chemical?

Was this your answer?
Look to the agents of weathering to guide your thinking. Water is the primary agent of weathering. Liquid water and water vapor interact chemically with rock to break it down. Less commonly, freezing and thawing physically assault and disintegrate rock. In this way, water performs a double duty. Because liquid water and water vapor are present nearly everywhere, chemical weathering is the main producer of sediment.

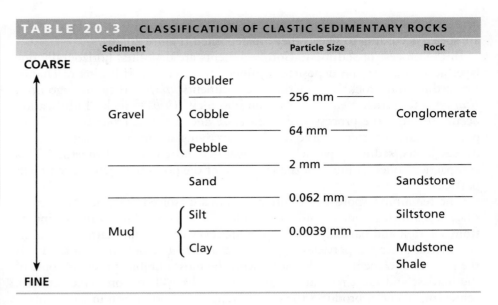

TABLE 20.3   CLASSIFICATION OF CLASTIC SEDIMENTARY ROCKS

Sediment		Particle Size	Rock
COARSE ↑			
Gravel	Boulder		
		256 mm	Conglomerate
	Cobble		
		64 mm	
	Pebble		
		2 mm	
	Sand		Sandstone
		0.062 mm	
Mud	Silt		Siltstone
		0.0039 mm	
	Clay		Mudstone Shale
FINE ↓			

(a)

(b)

(c)

FIGURE 20.28

Clastic sedimentary rocks. (a) Shale, the most abundant sedimentary rock, is composed of very fine mud-sized particles; (b) sandstone is composed of sand-sized particles; and (c) conglomerate is made up of a poorly sorted variety of rounded particles, mostly gravel sized, but also with sands and muds.

Classifying Sedimentary Rocks

Sedimentary rocks are divided into two groups, based on the kinds of sediments they contain: *clastic* and *chemical*.

Clastic Sedimentary Rocks Sedimentary rocks composed of small pieces of weathered rock, shell debris, and/or solid organic matter are called clastic sedimentary rocks. Clastic rocks are classified by particle size (Table 20.3). The classification scheme is meaningful because it allows us to visualize the depositional environment in which the rock was formed. For example, large particle sizes are deposited in energetic environments, such as a raging river, and small particle sizes settle out of the water in tranquil environments, such as the middle of a lake or the center of the ocean. Going from very small grain size to larger grain size, the three most abundant clastic sedimentary rocks are *shale* (which for our purposes includes mudstones), *sandstone*, and *conglomerate* (Figure 20.28).

Shale (or *mudstone*) is formed by the compaction of superfine silt and clay-sized particles. It is finely layered and has the ability to split into thin flakes parallel to the depositional layers. The extremely fine particle size indicates deposition in a low-energy environment characterized by quiet waters, such as deep ocean basins, flood plains, deltas, lakes, or lagoons. The color of shale provides clues to the environment of formation. Gray to black shale indicates the presence of organic matter, which can be preserved only in environments with little or no oxygen, such as swampy areas. If sufficient oxygen had been present in the depositional environment, bacteria would have decomposed the organic matter very quickly. Black shale is commercially important because it is the main source rock from which crude oil formed. Red to brown shale contains ferric oxide (red) or ferric hydroxide (brown). These oxides indicate an oxygenated depositional environment, where deep water was well mixed with water from near the surface. Green shale simply does not contain organic matter, ferric oxide, or ferric hydroxide.

Sandstones are classified into three types, based on their mineral makeup. When quartz is the primary mineral, the rock is simply called *quartz sandstone*. Quartz sandstone is composed of well-sorted, well-rounded quartz particles. Sandstone that contains considerable amounts of the mineral feldspar is called *arkose*. The particles in arkose tend to be angular and not as well sorted as those in quartz sandstone. Sandstone made of a mixture of minerals and angular rock fragments is called *graywacke* (pronounced "gray-wack-ee"). Sandstones form in a variety of environments, including dunes, beaches, marine sand bars, river channels, canyons, and underwater canyons—all places where moderate-energy waters deposit similarly sized grains.

Conglomerate is composed of gravel-sized and smaller rounded rock and mineral fragments. Large rock fragments must have been transported by water currents strong enough to carry them, which indicates dynamic, high-energy environments, such as rapids and fast-moving streams. Because these strong currents round out the rock fragments, the roundness of their edges and corners is a good indication of the distances they have traveled. Conglomerates are often found in river channels and along rapidly eroding coastlines.

Chemical Sedimentary Rocks Recall that chemical sediments, and in turn chemical rocks, form by the precipitation of minerals from water solutions. Some carbonates, such as travertine, are formed by the inorganic precipitation of calcium carbonate (Figure 20.29). Evaporites, such as halite, are formed by chemical precipitation caused by the evaporation of salty waters. In general, chemical sedimentary rocks form where there are no clastic sediments.

Precipitation also occurs indirectly when water-dwelling organisms take up dissolved constituents and use them to build shells or hard body parts or to perform other life processes in which solids are created and then discharged into the surrounding water. Chemical sediments created by organisms are called *biochemical sediments.*

Limestone, made up of the mineral calcite ($CaCO_3$), is the most abundant carbonate rock. About 90% of all limestone rocks form as a result of biologic activity. Many marine organisms make their shells out of calcite. When such organisms die, their shells accumulate on the sea floor. The shells begin to dissolve, forming a noncrystalline *ooze* of calcium carbonate. This ooze eventually crystallizes into calcite, which then forms limestone. Because of compaction and how easily calcium carbonate dissolves, the original textures and structures of the seashells are often obliterated. Sometimes, however, the shells and shell fragments can still be seen as in the fossiliferous limestone, coquina (Figure 20.30).

Warm climates favor carbonate deposition—carbonates are actually more soluble in cold water than in warm water. Evaporite deposits require a dry climate that causes the evaporation of lakes or seawater. As the water dries out, evaporite minerals precipitate and are left behind. Vast carbonate and evaporite deposits on the continents are evidence that expansive, shallow seas have periodically covered the land surfaces in the past.

Fossils: Clues to Life in the Past

Because sedimentary rocks (clastic and chemical) form at Earth's surface, they often contain the remains of life forms—fossils. Fossils not only tell the story of life on Earth, they also give us important clues about Earth's geologic past. As we shall see in Chapter 23, fossils can indicate where and when sediments were deposited. They also help us match rocks from different places that are of similar geologic age. Some fossils are made of whole organisms, but most fossils are just parts of an organism. Other fossils are simply an impression, or print, made in the rock before it hardened. Plants commonly leave their impression as a thin film of carbon. Organisms have become fossilized in many ways (Figure 20.31).

FIGURE 20.29
These "jaws" of travertine at Pamukkale in Turkey were created by the precipitation of calcium carbonate, $CaCO_3$, from a hot mineral spring.

FIGURE 20.30
Coquina is just one example of a biochemical limestone. This coquina, composed of shells and shell fragments, is a fossiliferous limestone—full of fossils!

(a)

(b)

(c)

(d)

FIGURE 20.31
Examples of fossilization. (a) Permineralization occurs when mineral-rich waters fill the porous remains of an organism—like petrified wood. (b) An impression or cast is made by an organism (or part of an organism) that was buried quickly before it could decompose. Its shape is preserved as an impression or cast. (c) Replacement occurs when mineral matter replaces the remains of an organism. Pyrite has replaced the original shell in this specimen. (d) Carbonization occurs when an organism is preserved as a thin film of carbon.

LINK TO BIOLOGY

Fossil Fuels

When ancient plants and animals died, most of the organic matter they were made of was quickly decomposed by bacteria and converted to nutrients consumed by other organisms. Material that escaped bacterial decay was either preserved as sparsely distributed organic matter or converted to biochemical sediments to become coal, oil, or gas.

Coal, oil, and gas are all fossils, in the sense that they are the remains of past organisms. However, these remains have been so changed over time that the forms and even the composition of the accumulated organisms are beyond recognition.

Most coal deposits were formed about 300 million years ago, when steamy swamps covered much of Earth's surface. As plants and trees died, their remains sank to the bottoms of stagnant swamps. These oxygen-poor environments hindered decay, and plant matter accumulated layer upon layer to form a dense and soggy organic material called *peat*. Over time, sediments of sand and clay buried the peat. With more burial, the increase in heat and pressure caused the peat to lithify to form lignite and bituminous coals. Although coal is composed of organic matter rather than minerals, it is considered a chemical sedimentary rock. When it is subjected to even more pressure and heat, coal can transform into the metamorphic rock called *anthracite*, which is the highest grade of coal.

The source of oil and gas is fossilized microscopic organic matter found in buried marine sediments. When buried sediment with considerable amounts of this organic matter is subjected to low heat over a long enough period of time, chemical changes take place that create oil. Under the pressure of the overlying sediments, tiny oil droplets are squeezed out of the source rocks and into overlying porous rocks. The porous rocks—commonly sandstones—become oil reservoirs. Just as in the metamorphism of rocks, deeper burial results in higher temperatures. If the temperature gets high enough, natural gas is generated rather than oil.

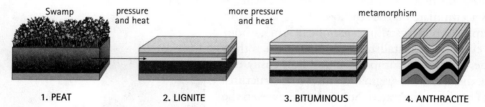

The formation of coal. (1) Peat forms from plant remains in a stagnant swamp. (2) Peat is buried by sediments. Over time, and subjected to increased heat and pressure, peat transforms into lignite—a soft brown coal. (3) More time, burial, heat, and pressure lead to the formation of bituminous coal—a soft, black, high-energy coal used for power production. (4) Metamorphism converts bituminous coal to anthracite, a hard, black, high-energy coal.

CHECKPOINT
1. **What makes coal exceptional among rocks?**
2. **Is coal a chemical sedimentary rock or a clastic sedimentary rock?**

Were these your answers?
1. Coal is composed of organic matter rather than minerals.
2. Coal is a chemical sedimentary rock or, more specifically, a biochemical rock.

20.9 Metamorphic Rocks

EXPLAIN THIS Why are gems usually found in metamorphic rocks?

LEARNING OBJECTIVE
Describe the physical and chemical conditions that give rise to different metamorphic rocks.

What happens when a mass of rock is brought to a location that has much higher temperature and pressure than the environment in which it formed? The changes in physical and chemical conditions to which the rock is exposed can transform the rock. New rock is made from old. The new rock is stable under the new conditions, although the preexisting rock was not.

The changes in rocks that happen as physical and chemical conditions change are called **metamorphism**. All rocks, whether igneous, sedimentary, or metamorphic, can undergo metamorphism. An everyday example of metamorphism is potter's clay. Potter's clay is soft at room temperature. But when heated, it becomes a hard ceramic. Similarly, limestone subjected to enough heat and pressure becomes marble. And shale is metamorphosed to slate. Rocks may also be drastically stretched or compressed. It is important to note that during metamorphism, minerals do not melt. Once minerals melt, metamorphism has ended and igneous activity has begun. In metamorphism, change occurs instead by *recrystallization* of preexisting minerals or by *mechanical deformation* of rock.

Recrystallization occurs when the minerals in a rock change because the rock was subjected to higher temperatures and pressures than the conditions under which it formed. The constituents of the metamorphosing minerals actually migrate and recombine to form new minerals. Recrystallization may occur with or without the exchange of fluid. For example, consider sedimentary rocks that contain such fluids as water or carbon dioxide. The fluids in the rock, which are enclosed in pore spaces, can act as catalysts to initiate or speed up metamorphic reactions. If temperature and pressure are high enough, the rock loses pore space as the fluid in the rock is squeezed out. The released fluid can then chemically react with the surrounding rock, contributing constituents to new minerals that are forming. These fluids can also initiate magma generation far away from the site of metamorphism as they migrate upward and drive fluid-induced melting. Metamorphic reactions can also occur without the involvement of fluids. This happens in the case of low-temperature metamorphism.

Mechanical deformation occurs when a rock is subjected to physical stress. It may or may not involve elevated temperatures. For example, surface rocks that become deeply buried are subjected to increased pressure. Such stress may cause the rocks to flow like a plastic, bending them into intricate folds. Or the increased pressure may deform and flatten the rock, shear it, or break it and grind it into fragments. Such physical stress occurs deep in Earth's crust.

Plants and animals were energized by the Sun before they became fossil fuels. So the energy from fossil fuels is delayed solar power.

CHECKPOINT
Is recrystallization, due to elevated temperature and pressure, the opposite of partial melting?

Was this your answer?
No. The process of crystallization from magma, not recrystallization, is the opposite of partial melting. Recrystallization occurs within a rock because of exposure to high temperatures and/or pressures, without melting. Rock may not even undergo any chemical changes during recrystallization.

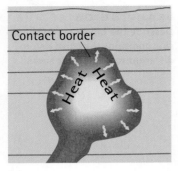

Contact border

Heat Heat

(a)

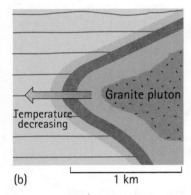

Granite pluton

Temperature decreasing

(b) |—— 1 km ——|

FIGURE 20.32
(a) Contact metamorphism is the result of rising molten magma that intrudes a rock body. (b) Surrounding the solidified intrusive rock is a zone of alteration. Alteration is greatest at the contact area, and it decreases farther away from the contact area.

Types of Metamorphism: Contact and Regional

The most common types of metamorphism are *contact metamorphism* and *regional metamorphism*. Each type of metamorphism is characterized by differences in mechanical deformation and recrystallization.

Contact metamorphism occurs when a body of rock is intruded by magma (Figure 20.32). The high temperature of the magma produces a zone of alteration that surrounds the intrusion. The alteration is greatest at the contact, which is the interface between the intrusive rock and the surrounding rock. The width of the altered zone may range from a few centimeters to several hundred meters. Around a small intrusive body, such as a dike, the altered zone is very narrow and may resemble "baked" rock, with a texture and appearance like ordinary brick. But with a larger intrusive body, such as a batholith, the altered zone may be 100 m thick or more. One of the most common changes is an increase in crystal size due to recrystallization. Crystal size is greatest at the contact and decreases with increasing distance from that point. The water content of the rock also changes with distance from the contact. At the contact, where temperature is high, water content is low because it has boiled away. So we find dry, high-temperature minerals, such as garnet and pyroxene, at the contact. Farther away, we find water-rich, low-temperature minerals, such as muscovite and chlorite (another sheet silicate). Contact metamorphism is typically associated with high temperatures and high water content—lots of chemical activity and little or no mechanical deformation.

Regional metamorphism is the alteration of rock by both heat and pressure over an entire region rather than just near a contact between rock bodies. During the process of mountain building, Earth's crust is severely compressed into a mass of highly deformed rock. This deformation can be seen in the folded and fractured rock layers in many mountain ranges (Figure 20.33). Regionally metamorphosed rocks are found in all the major mountain belts of the world. Regional metamorphism combines recrystallization with mechanical deformation. Large sections of rock can be heated if the rock is buried deeply enough, simply because Earth is hotter at greater depths.

The effects of regional metamorphism are most pronounced in the cores of deformed mountains. Rocks develop a distinct "foliated" texture, which forms because of the great pressures generated by converging tectonic plates (Chapter 21). Zoned sequences of minerals are also characteristic of regional metamorphism. For example, one geographic area may have rocks with one set of minerals, while the adjacent area has rocks with a different set of minerals. Because of

FIGURE 20.33
This satellite photo reveals regional-scale folding of metamorphic rocks in the Appalachian Mountains of central Pennsylvania.

the large-scale nature of regional metamorphism, these zones tend to be broad and extensive. Areas of regional metamorphism are the hunting grounds of gem prospectors, because the heat and pressure that accompany these changes can produce beautiful minerals.

Classifying Metamorphic Rocks

Metamorphic rocks are defined by their appearance and the minerals they contain. For classification and identification, metamorphic rocks can be divided into two groups: *foliated* and *nonfoliated*.

Foliated Metamorphic Rocks When rock is subjected to increased pressure, some of its minerals realign into parallel planes as they recrystallize. The face of each of these parallel planes is perpendicular to the main direction of the compressive force. This leads to a layered appearance called *foliation*. Foliation is a prominent visual feature of regionally metamorphosed rocks, and it is very different from the layering seen in sedimentary rock. Deposition does not cause the foliated texture in metamorphic rocks. Rather, sheet-structured minerals, such as the micas, grow and orient themselves with their sheets perpendicular to the direction of maximum pressure (Figure 20.34). The new rock, which now has parallel flakes, or plates, of mica, is said to be foliated. The most common foliated metamorphic rocks—slate, schist, and gneiss—are derived from sedimentary rocks that have the appropriate chemical composition to favor mica formation (Figure 20.35).

Slate is the "lowest-grade" foliated metamorphic rock, which means that it was formed under relatively low temperature and pressure. Slate, which is metamorphosed shale, is a foliated rock composed of very small particles and tiny mica flakes. The most obvious characteristic of slate is its excellent rock cleavage, which allows it to be split into thin slabs. The best pool tables and chalkboards are made from slate quarried in metamorphic areas where slaty cleavage is well developed. Slate is also commonly used as roofing tile and floor tile.

Schist is one of the most easily recognizable metamorphic rocks because it is scaly looking with large micas that reflect light like tinted windows. Schist forms under higher temperature and pressure conditions than slate, which causes the mineral grains to grow large enough to be identified with the naked eye. Schists usually contain about 50% platy minerals—most commonly muscovite and biotite. The larger mica flakes give the rock a highly reflective surface that is quite striking. Schists are named according to the major minerals in the rock (biotite schist, staurolite-garnet schist, and so on).

Gneiss (pronounced "nice") is a foliated metamorphic rock that contains alternating layers of dark platy minerals and lighter granular minerals. The layers give this metamorphic rock its characteristic banded appearance.

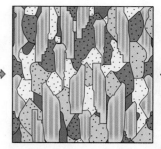

FIGURE 20.34
As compressive forces squeeze platy and sheet-structured minerals, the grains align themselves perpendicular to the main direction of force. Arrows indicate the direction of compressive force.

FIGURE 20.35
Common foliated metamorphic rocks: (a) slate, (b) schist, and (c) gneiss.

(a)   (b)   (c)

(a)

(b)

FIGURE 20.36
Nonfoliated metamorphic rocks:
(a) marble and (b) quartzite.

This appearance results from even greater temperature and pressure conditions than those that create schist. The most common granular minerals found in gneiss are quartz and feldspar. These are also the most common granular minerals in granite. In fact, some gneisses are actually metamorphosed granites.

Nonfoliated Metamorphic Rocks Nonfoliated metamorphic rocks can form because of increased temperature and pressure or because of increased temperature alone. Even under high pressure, foliation cannot develop if the rock lacks the chemical composition needed for micas (or other elongated crystals) to form. Similarly, even if the chemical composition contains the necessary constituents but the pressure is not high enough, such as in contact metamorphism, foliation cannot develop. Two common nonfoliated rocks are marble and quartzite.

Marble (Figure 20.36a) is a crystalline, metamorphosed limestone. Pure marble is white and is virtually 100% calcite, which is neither platy nor elongated. Because of its color and its relative softness (hardness 3), marble is a popular building stone. Often the limestone from which marble formed contained impurities that produce various colors in the marble. Thus, marble can vary in color from pink to gray, green, or even black.

Quartzite (Figure 20.36b) is metamorphosed quartz sandstone, and it is therefore very hard (hardness 7). Quartz is another mineral that is not platy or elongated. The recrystallization of quartzite can be so complete that when struck the rock splits across the original quartz particles, rather than between them. Although pure quartzite is white, it commonly contains impurities that can cause it to be a variety of colors, such as pink, green, or light gray.

CHECKPOINT
1. **Under conditions of extreme temperature, when can a rock no longer undergo metamorphism?**
2. **Why does recrystallization occur in metamorphic rock?**

Were these your answers?
1. When it melts. Once a rock melts, it becomes magma. And when magma cools to form rock, the new rock, by definition, is igneous rock.
2. Recrystallization occurs because the rock is exposed to high temperatures or pressures.

LEARNING OBJECTIVE
Describe how rocks are never truly destroyed but are recycled.

Mastering**PHYSICS**
TUTORIAL: The Rock Cycle Activity
VIDEO: The Rock Cycle

20.10 The Rock Cycle

EXPLAIN THIS How can a metamorphic rock change into an igneous rock?

Earth is a dynamic, ever-changing, and active planet. Made up of atoms and molecules, Earth's elements combine to make minerals, which are formed by the process of crystallization from either magma or water solutions. The minerals formed are determined by the elements present and the conditions that lead to their formation. These factors in turn determine the arrangement of atoms in each mineral and the strength of the bonds that hold the atoms together. More than 90% of Earth's minerals are silicates—composed predominantly of silicon and oxygen plus other elements such as aluminum, iron, calcium, sodium, potassium, and magnesium. Minerals combine to make rocks—the igneous, sedimentary, and metamorphic rocks that we see all around us.

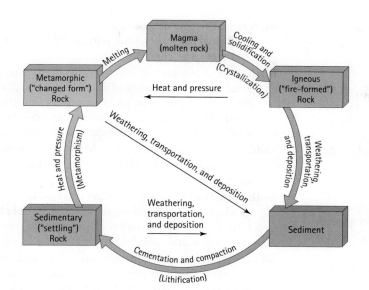

FIGURE 20.37

The rock cycle: Igneous rock, subjected to heat and pressure far below Earth's surface, may become metamorphic rock; metamorphic or sedimentary rocks at Earth's surface may decompose to become sediment that in turn becomes new sedimentary rock. Whatever the route, molten rock rises from the depths of Earth, cools, and solidifies to form a crust that, over eons, is reworked by shifting and erosion, only to return eventually to become magma in Earth's interior.

Most minerals (and hence, rocks) are formed by the crystallization of magma. And magma forms when rock melts. The type of magma formed depends on the type of rocks and minerals that melt. A single magma can transform into different magmas (basaltic, andesitic, and granitic) during migration and crystallization, and then into different types of igneous rocks.

Although most of Earth's crust is composed of igneous and metamorphic rock, the rock we see at the surface is mainly sedimentary. Sedimentary rock forms from the remains of rock that has been weathered and eroded. Sedimentary rock provides a record of environmental and biological changes on Earth's surface. And when sedimentary rock is buried deep within Earth or involved in mountain building, great temperatures and pressures can transform it into metamorphic rock. Under the proper conditions, metamorphic rock can melt and become magma, which eventually solidifies as igneous rock to complete the **rock cycle** (Figure 20.37).

The rock cycle varies in its paths. Igneous rock, for example, may be subjected to heat and pressure far below Earth's surface to become metamorphic rock. Or metamorphic or sedimentary rocks at Earth's surface may decompose to form sediment that becomes new sedimentary rock. There are many possible variations in the cycle.

But what about Earth's interior—what is going on inside our planet? We will now turn our attention to the exploration of Earth's interior.

For instructor-assigned homework, go to www.masteringphysics.com

SUMMARY OF TERMS (KNOWLEDGE)

Chemical sediments Sediments that form by the precipitation of minerals from water on Earth's surface.

Cleavage The tendency of a mineral to break along planes of weakness.

Crystal form The outward expression of the orderly internal arrangement of atoms in a crystal.

Crystallization The growth of a solid from a liquid or gas whose atoms come together in specific chemical proportions and crystalline arrangements.

Density The ratio of the mass of a substance to its volume.

Deposition The stage of sedimentary rock formation in which eroded particles come to rest.

Erosion The wearing away of rocks, and the processes by which rock particles are transported by water, wind, or ice.

Fracture A break that does not occur along a plane of weakness.

Igneous rocks Rocks formed by the cooling and crystallization of hot, molten rock material called magma (or lava).

Lava Molten magma that moves upward from inside Earth and flows onto the surface. The term *lava* refers both to the molten rock itself and to the solid rocks that form from it.

Magma Molten rock in Earth's interior.

Mechanical deformation Metamorphism caused by stress, such as increased pressure.

Metamorphic rocks Rocks formed from preexisting rocks that have been changed or transformed by high temperature, high pressure, or both.

Metamorphism The changes in rock that happen as physical and chemical conditions change.

Mineral A naturally formed, inorganic, crystalline solid composed of an ordered arrangement of atoms with a specific chemical composition.

Mohs scale of hardness A ranking of a mineral's hardness, which is its resistance to scratching.

Nonsilicate A mineral that does not contain silica (silicon + oxygen).

Partial melting The incomplete melting of rocks, resulting in magmas of various compositions.

Plutonic rock Intrusive igneous rock formed from magma that cools beneath Earth's surface. Granite is a plutonic rock.

Polymorphs Two or more minerals that contain the same elements in the same proportions but have different crystal structures.

Recrystallization A process that occurs when rocks are subjected to high temperatures and pressures and go through a change in minerals; often accompanied by the loss of H_2O or CO_2.

Rock An aggregate of minerals. Some rocks are aggregates of fossil shell fragments, solid organic matter, or any combination of these components.

Rock cycle A sequence of events involving the formation, destruction, alteration, and reformation of rocks as a result of the generation and movement of magma; the weathering, erosion, transportation, and deposition of sediment; and the metamorphism of preexisting rocks.

Sedimentary rocks Rocks formed from the accumulation of weathered material (sediments) that has been eroded by water, wind, or ice.

Sedimentation The stage of sedimentary rock formation in which deposited sediments accumulate and change (lithify) into sedimentary rock through the processes of compaction and, usually, cementation.

Silicate A mineral that contains both silicon and oxygen and (usually) other elements in its chemical composition; silicates are the largest and most common rock-forming mineral group.

Solubility A measure of the ease with which a mineral can be dissolved. Low-solubility minerals are difficult to dissolve; high-solubility minerals are easier to dissolve.

Streak The name given to the color of a mineral in its powdered form.

Tectonic plates Sections into which Earth's crust is broken up; they move in response to heat flow and convection in Earth's interior.

Volcanic rocks Extrusive igneous rocks formed by the eruption of molten rock at Earth's surface. Basalt is a volcanic rock.

Weathering Disintegration or decomposition of rock at or near Earth's surface.

READING CHECK QUESTIONS (COMPREHENSION)

20.1 The Geosphere Is Made Up of Rocks and Minerals

1. How did density segregation contribute to Earth's internal layers?
2. What three sources of heat contributed to the melting and density segregation of early Earth?
3. What is the most abundant element for Earth as a whole?
4. What is the most abundant element in Earth's crust? What is the second most abundant element?

20.2 Minerals

5. What is a mineral?
6. What does *inorganic* mean in the definition of a mineral?

20.3 Mineral Properties

7. What physical properties are used to identify minerals?
8. Most mineral samples do not display their crystal forms. Why not?
9. What is a polymorph?

20.4 Classification of Rock-Forming Minerals

10. What is the difference between a silicate mineral and a nonsilicate mineral?
11. Silicate minerals are subdivided into ferromagnesian silicates and nonferromagnesian silicates. What two factors contribute to this subdivision?
12. What is the most abundant mineral in Earth's crust? What is the second most abundant mineral?

20.5 The Formation of Minerals

13. Describe the process of crystallization.

14. What are two sources from which minerals crystallize?

15. As minerals crystallize in cooling magma, which minerals are the first to crystallize: the minerals with lower amounts of silica or the minerals with higher amounts of silica?

16. When water evaporates from a body of water, what type of sediment is left behind?

20.6 Rock Types

17. Name the three major types of rocks, and describe the conditions of their origin.

20.7 Igneous Rocks

18. What are the most common igneous rocks, and where do they generally occur?

19. What is meant by partial melting?

20. With respect to the silica content of the parent rock, what type of magma does partial melting produce?

21. What determines a rock's melting point?

22. In Earth's interior, does temperature increase or decrease with depth?

20.8 Sedimentary Rocks

23. How does weathering produce sediment? Distinguish between weathering and erosion.

24. What is a clastic sedimentary rock?

25. What are the three most common clastic sedimentary rocks?

26. What is the most abundant carbonate rock?

27. How are most carbonate rocks formed?

20.9 Metamorphic Rocks

28. What is metamorphism? What causes it?

29. Distinguish between foliated and nonfoliated metamorphic rocks.

30. In contact metamorphism, water-rich, low-temperature minerals are found far away from the contact zone. Give two examples of such minerals.

ACTIVITIES (HANDS-ON APPLICATION)

31. To see how various elements can separate from one another, fill an empty large-mouthed jar (for example, a peanut butter jar) with sand, pebbles, coarse gravel, and small Styrofoam pellets (perlite). Cap the jar and shake. After a vigorous shake, observe how the different materials settle. Is there a pattern? Shake the mixture a few times to allow the materials to settle. Which materials settle toward the bottom of the jar? Which materials migrate upward? How does the settling of various materials in the peanut butter jar mimic the distribution of Earth's elements?

32. Rock candy is a great example of crystallization in a supersaturated solution. Dissolve as much sugar as possible in some boiling water, then allow the solution to cool. To form crystals you need a place for them to nucleate. Tie some string to a weight and lower the string into the solution *before* it cools. The longer you leave the string undisturbed in the solution, the larger the crystals of sugar. Although rock candy is a crystalline solid, it is not classified as a mineral. Why?

33. Look at some crystals of table salt under a microscope or a magnifying glass and observe their generally cubic shapes. There's no machine at the salt factory specifically designed to give salt crystals these cubic shapes, as opposed to round or triangular ones. The cubic shape occurs naturally and is a reflection of how the atoms of salt are organized—cubically. Smash a few of these salt cubes, and then look at them again carefully. What you'll see are smaller salt cubes! Use the cleavage properties of crystals to explain these results.

34. A physical property of any material is its density—its mass per volume. Pennies made after 1982 contain both copper and zinc. Pennies made before 1982 are pure copper. Zinc is less dense than copper, so post-1982 pennies are less dense and have less mass than pre-1982 pennies. Dig into your penny collection and find 20 pre-1982 and 20 post-1982 pennies. Measure their masses on a sensitive scale, such as a home postage scale. Alternatively, hold the pennies in opposite hands to see if you can feel the difference in their masses. How few pennies can you hold and still feel the difference? Try holding single pennies on your left and right index fingers. Can you tell the difference with your eyes closed? Try this with a friend.

35. The freezing, then thawing, of three common refrigerator items—ice, butter, and cheese—will help solidify your understanding of partial melting.

Step 1. Mix cubes of ice, butter, and cheese together with a little water, and put this mixture in the freezer. Once this mixture has frozen, it will be your "rock."

Step 2. Put this "rock" in a bowl in the refrigerator overnight, and then examine the contents of the bowl the next day.

(a) In this mixture: What melts? What doesn't melt? How is the melt different from the original "rock"?

(b) Now let's think about the partial melting of real rocks: What melts? What doesn't melt? How is the melt different from the original rock?

(c) Now consider the cooling and solidification of the melt: How are the first-formed igneous rocks different from the rock that originally melted?

THINK AND SOLVE (MATHEMATICAL APPLICATION)

36. Gold has a density of 19.3 g/cm³. A 5-gal pail of water (density of water = 1.0 g/cm³) has a mass of about 18 kg. What is the mass of a 5-gal pail of gold?

Refer to "Figuring Physical Science: Silica Enrichment in Magma" on page 531 to answer the following problem.

37. What are the mass percentages of the oxides MgO, FeO, and SiO₂ in pyroxene, MgFeSi₂O₆? Give your answers in whole numbers. Hint: Use the Periodic Table to find atomic mass units, then express the formula for pyroxene as a sum of its constituent atoms.
    (a) How many kilograms of silica are in 225 kg of pyroxene?
    (b) If 325 kg of olivine and 225 kg of pyroxene have crystallized out of the 1000 kg of magma, what is the mass percentage of silica in the remaining liquid?

THINK AND RANK (ANALYSIS)

38. When Earth first formed, its elements were distributed evenly. Rank, from first to last, the episodes that brought about Earth's uneven distribution of elements: (a) gravitational attraction, (b) radioactive decay heating, (c) density segregation, (d) impact heating.

39. Mineral hardness depends on chemical bond strength. From hardest to softest, rank the following minerals: (a) quartz, (b) halite, (c) diamond, (d) gold.

40. Rank the magma types in order of increasing (low to high) silica content: (a) basaltic, (b) granitic, (c) andesitic.

41. Each of the following statements describes one or more characteristics of a particular metamorphic rock. Name the rock for each statement, then rank them from low to high grade:
    (a) foliated rock, sometimes derived from granite
    (b) foliated rock, possessing excellent rock cleavage; generally used in making blackboards
    (c) foliated rock containing about 50% platy minerals; named according to the major minerals in the rock

42. Chemical bond strength greatly influences certain physical properties of a mineral. Rank these properties as to how much they are affected by chemical bonding: (a) color, (b) density, (c) hardness, (d) cleavage.

43. Most minerals can be identified by their physical properties. Rank the following properties used for mineral identification from most useful to least useful: (a) color, (b) hardness, (c) streak, (d) cleavage.

44. Rank these rock-forming minerals from most abundant to least abundant: (a) silicates, (b) carbonates, (c) sulfates, (d) oxides.

45. Different minerals crystallize at different times. Rank the minerals in their order of crystallization: (a) quartz, (b) feldspar, (c) olivine, (d) pyroxene.

46. In partial melting, rocks and minerals with a low melting temperature melt more easily. Rank these minerals in order of partial melting: (a) quartz, (b) feldspar, (c) olivine, (d) pyroxene.

47. Rank the following stages of sedimentary rock formation from first to last: (a) erosion, (b) lithification, (c) weathering, (d) deposition.

EXERCISES (SYNTHESIS)

48. What do we call minerals that have the same combination of elements but a different arrangement of elements?

49. Silicon is essential for the computer industry in making microchips. Can silicon be mined directly from Earth? Defend your answer.

50. What two minerals make up most of the sand in the world?

51. What two mineral groups provide most of the ore that society needs?

52. Can metamorphic rocks exist on an island of purely volcanic origin? Defend your answer.

53. If a rock contains mineral A (30% silica) and mineral B (25% silica), which would melt first as temperature rises?

54. If a magma contains molten forms of mineral A (30% silica) and mineral B (25% silica), which would crystallize first as the magma cools?

55. The factors that influence bond strength influence mineral hardness. What are these factors?

56. If a magma contains molten forms of mineral A (30% silica) and mineral B (25% silica), which would crystallize last as the magma cools?

57. If a rock contains mineral A (30% silica) and mineral B (25% silica), which would melt last as temperature increases?

58. Is it possible for crystallization to enrich magma in more than just silica? Defend your answer.

59. If high-silica minerals are the last to crystallize, why aren't high-silica minerals the last to melt?

60. Why is halite commonly the last mineral to precipitate from evaporating seawater?

61. Would you expect to find any fossils in limestone? Why or why not?

62. How do chemical sediments produce rock? Name two rock types that form by chemical sedimentation.

63. Why is color not always the best way to identify a mineral?

64. While you are hiking in the wilderness, you find a shiny, glassy-looking mineral. What physical test could you use to determine whether this mineral is a diamond?

65. What makes gold so soft (easily scratched) while diamond is so much harder?

66. Imagine that we have a liquid with a density of 3.5 g/cm^3. Knowing that objects of higher density will sink in the liquid, will a piece of quartz sink or float in the liquid? How about a piece of chromite?

67. Is cleavage the same thing as crystal form? Why or why not?

68. Is Earth's interior mostly magma? Explain.

69. Relate the shape and sorting of sand particles to the way in which they were most likely transported.

70. In which parts of Earth's crust (oceanic and/or continental crust) do we find the two common igneous rocks, basalt and granite?

71. Which of these is a true statement about silicate minerals? (a) Melting point decreases as silica percentage increases. (b) Melting point increases as silica percentage increases.

72. If a magma contains molten forms of quartz and olivine (a silicate mineral), which crystallizes first as the magma cools?

73. Are the Hawaiian Islands made up primarily of igneous, sedimentary, or metamorphic rock?

74. Where does most magma originate?

75. What patterns of alteration are characteristic of contact metamorphism?

76. What are the two processes by which rock is changed during metamorphism?

77. Are high-silica minerals "easier" to melt than those with low silica content?

78. What mainly determines a rock's initial melting temperature?

79. What general rock feature does a geologist look for in a sedimentary rock to determine the distance the rock has traveled from its place of origin?

80. What feature of clastic sedimentary rock enables the flow of oil after it has been formed?

81. Which of these rocks—granite, sandstone, limestone, or halite—is the first to weather in a wet (humid) climate? Why?

82. In what two ways does sediment turn into sedimentary rock?

83. In a conglomerate rock, why are pebbles of granite very common and pebbles of marble relatively uncommon?

84. Cite two examples of sedimentary rocks that provide information about past geologic events at Earth's surface.

85. What is a fossil? How are fossils used in the study of geology?

86. What kind of weathering is imposed on a rock when it is smashed into small pieces? When it is dissolved in acid?

87. Which type(s) of rock is (are) made from previously existing rock? Which type does not require high temperature and pressure for its formation?

88. What properties of slate make it good roofing material?

89. Name two mica minerals that can give a metamorphic rock its foliation.

90. How is foliation different from sedimentary layering?

91. Why do we find folded and fractured rock layers in zones of regional metamorphism?

92. Where on Earth's surface are lava flows most common?

93. What feature helps distinguish schist and gneiss from quartzite and marble?

94. How does gneiss differ from granite?

95. Why is schist so easily recognized?

96. If a rock contains both quartz and pyroxene (a silicate mineral), which melts first as the rock is heated?

97. Why does magma composition change as it cools?

DISCUSSION QUESTIONS (EVALUATION)

98. What is the difference between the minerals that make up a rock and the minerals we find in common dietary supplements? Is it possible to supplement our diet by simply eating rocks?

99. If the volcanic glass obsidian is not considered a mineral, why is it considered a rock?

100. We have learned that silica content is a key factor in a mineral's melting point. What other two factors can change a rock's melting point?

101. Which type of rock is most sought by petroleum prospectors: igneous, sedimentary, or metamorphic? Why?

102. Our dependency on fossils fuels has many repercussions—both economic and environmental. For example, the 2010 explosion of BP's Gulf of Mexico oil rig is, to date, the largest accidental oil spill in history. What measures can be taken to prevent further oil spill disasters?

READINESS ASSURANCE TEST (RAT)

If you have a good handle on this chapter, if you really do, then you should be able to score at least 7 out of 10 on this RAT. If you score less than 7, you need to study further before moving on.

Choose the BEST answer to each of the following.

1. The silicates are the largest mineral group because silicon and oxygen are
   (a) the hardest elements on Earth's surface.
   (b) the two most abundant elements in Earth's crust.
   (c) found in the common mineral quartz.
   (d) stable at Earth's surface.

2. Compaction and cementation of sediments lead to
   (a) magma generation.
   (b) lithification.
   (c) formation of pore water.
   (d) metamorphism.

3. Why are silicon and oxygen concentrated near Earth's surface while iron is concentrated at the core?
   (a) Earth's materials separated early in its history through the process of density segregation.
   (b) Silicon and oxygen are less dense than iron.
   (c) both of these
   (d) neither of these

4. Which minerals crystallize first from cooling magma?
   (a) minerals with the lowest melting point
   (b) minerals with the highest melting point
   (c) minerals with the highest solubility
   (d) minerals with high silica content

5. In Earth's interior, what two factors increase with depth?
   (a) pressure and water content
   (b) temperature and pressure
   (c) temperature and water content
   (d) silica content and water content

6. In a sedimentary rock, the degree of particle roundness can indicate
   (a) the duration and/or length of travel.
   (b) where the sediment particles originated.
   (c) where the particles were deposited.
   (d) how the particles were cemented and compacted.

7. The characteristics of regional metamorphism include
   (a) folded and faulted rock layers.
   (b) distinctly foliated rocks.
   (c) zoned sequences of minerals.
   (d) all of these

8. Coarse-grained plutonic igneous rocks are created because
   (a) lava intrudes deep into Earth's interior.
   (b) minerals cooled and grew quickly.
   (c) minerals cooled and grew over long periods of time.
   (d) larger minerals are more stable than smaller ones.

9. What most strongly influences a mineral's hardness?
   (a) the geometry of a mineral's atomic structure
   (b) the strength of a mineral's chemical bonds
   (c) the silica content
   (d) the number of planes of weakness

10. Sedimentary rocks often contain the remains of ancient life forms—fossils. And fossils in these rocks help us understand
    (a) Earth's geologic and biologic history.
    (b) Earth's early formation.
    (c) the zoned sequences of minerals.
    (d) the differentiation of life forms.

Answers to RAT

1. b, 2. b, 3. c, 4. b, 5. b, 6. a, 7. d, 8. c, 9. b, 10. a

CHAPTER 21

21 Plate
Tectonics
and Earth's
Interior

I F IT were possible to dig a hole straight through Earth, what would we find in its interior? Because digging such a hole is impossible, what tools and techniques can we use to explore Earth's insides? Investigation begins with the rocks on the surface, which tell us a great deal about Earth's interior. Earthquakes and volcanic eruptions are also a link to the inner workings of our planet. Observations and careful measurements of their behavior can offer additional clues. All these features and processes are external expressions of Earth's internal processes.

LEARNING OBJECTIVE
Describe how seismic waves travel
through Earth's interior.

21.1 Seismic Waves

EXPLAIN THIS How do earthquakes reveal Earth's internal composition?

Earthquakes, besides being fearsome and destructive events, provide a key to understanding Earth's internal structure. An **earthquake** is the shaking or trembling of the ground that happens when rock under Earth's surface moves or breaks. These internal movements generate waves that travel through Earth's interior and across Earth's surface—*seismic waves*. The speed at which seismic waves travel and the paths that they take provide scientists a view into Earth's interior. What has been discovered is a layered planet. The major layers of Earth are the *crust, mantle, outer core,* and *inner core* (Figure 21.1).

Recall from Chapter 10 that a wave's speed depends on the medium through which it travels. We learned that sound waves generated by clicking two submerged rocks together travel faster through water than through air. And sound waves travel even faster through a solid. Just like sound waves, the speed of seismic waves depends on the elasticity and density of the material through which they travel. The greater the elasticity and density, the greater the wave speed. So measuring the speeds of seismic waves provides clues about Earth's composition.

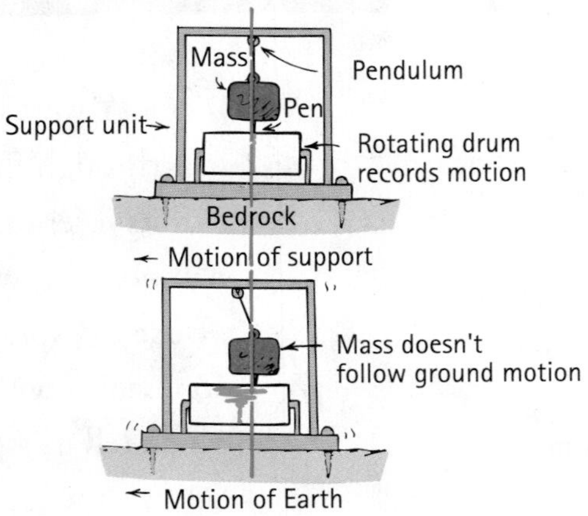

Energy released during an earthquake travels in the form of seismic waves and radiates in all directions within Earth's interior. As the energy travels to Earth's surface, the ground shakes and moves. This ground movement is recorded on a *seismograph* (Figure 21.2). From the combination of seismograph records—*seismograms* from many different earthquakes—Earth's interior is revealed. So just as an X-ray or CAT scan reveals the interior of your body, a seismogram reveals Earth's interior.

There are two types of seismic waves: **body waves**, which travel through Earth's interior, and **surface waves**, which travel along Earth's surface (Figure 21.3). Body waves are further classified as either **primary waves** (P-waves) or **secondary waves** (S-waves). Primary waves, like sound waves, are longitudinal—they compress and expand the rock as they move through it. Like vibrations in a bell, primary waves move out in all directions from their source. Primary waves are the fastest of all seismic waves and so are the first to register on a seismograph. Because both solids and fluids can compress and expand, P-waves can travel through any type of material—solid rock, magma, water, or air. Secondary waves, like the waves produced on a vibrating violin string, are transverse—they vibrate the particles of their medium up and down and from side to side, perpendicular to the direction of wave travel. Because S-waves travel more slowly than P-waves, they are the

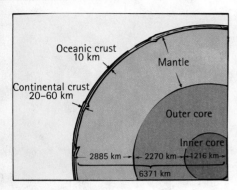

FIGURE 21.1
Cross-section of Earth's interior showing the four major layers and their approximate thicknesses.

FIGURE 21.2
Diagram of a seismograph. When Earth moves, the support unit attached to the ground also moves, but because of inertia, the mass at the end of the pendulum tends to stay in place. A pen attached to the mass marks the relative displacement on the slowly rotating drum beneath. In this way, the seismograph records ground movement.

FIGURE 21.3
INTERACTIVE FIGURE **MP**

Block diagrams show the effects of seismic waves. The yellow portion on the left side of each diagram represents the undisturbed area. (a) Primary body waves alternately compress and expand the material through which they travel—as shown by the different spacings between the vertical lines—similar to the action of a spring. (b) Secondary body waves cause the material through which they travel to oscillate up and down and from side to side. (c) Love surface waves whip back and forth like secondary body waves, but only in the horizontal direction. (d) Rayleigh surface waves have a rolling, up-and-down motion, similar to ocean waves.

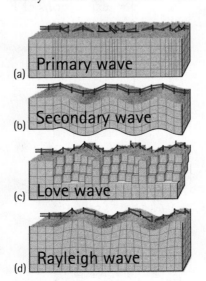

second waves to register on a seismograph. S-waves cannot move through fluids—they travel only through solids.

There are also two types of surface waves: *Rayleigh waves* and *Love waves*. Rayleigh waves have an up-and-down rolling motion, and Love waves have a side-to-side, whiplike motion. Both types of surface waves travel more slowly than P-waves and S-waves, and therefore they are the last to register on a seismograph.

In Earth's interior, seismic waves are reflected by the "surfaces" between differing materials. And when seismic waves pass into a different material, their wave speed changes, causing the wave to refract (Chapter 10). Geoscientists study the reflection, refraction, and speeds of the various types of seismic waves to piece together a story about Earth's interior. Seismic-wave research has revealed the architecture of Earth's internal layers (Figure 21.4).

21.2 Earth's Internal Layers

EXPLAIN THIS What is a possible reason for the absence of S-waves on the opposite side of an earthquake's epicenter?

Near the beginning of the 20th century, Irish geologist Richard Oldham was examining records of a massive earthquake in India when he discovered that its S-waves traveled some distance through Earth and then stopped. He also observed that the P-waves traveled as far as the S-waves into Earth but then refracted at an angle and lost speed. Because S-waves cannot travel through liquid, but P-waves can (at a reduced speed), Oldham deduced that the earthquake waves had encountered an internal boundary—he had discovered Earth's core. The year was 1906.

Three years later, Croatian seismologist Andrija Mohorovičić (pronounced "moho-rovu-chick") analyzed seismic readings from a recent earthquake. He recognized a sharp increase in the speed of seismic waves at another boundary, one that lay at a shallower depth below Earth's surface. Knowing that wave speed depends on the properties of the material through which the wave passes, Mohorovičić concluded that the increase in speed was due to a density change within Earth—wave speed increased as it passed from a lower-density solid to a higher-density solid. Mohorovičić's seismographic data had literally drawn a map of the upper boundary of Earth's *mantle*, a layer of denser rock underlying the less-dense crust. This boundary, known as the **Mohorovičić discontinuity** (called the "Moho" for short), separates Earth's crust from rocks of different composition in the mantle below.

In 1913, Beno Gutenberg reinforced Oldham's earlier findings by observing that both P-waves and S-waves are strongly influenced by a pronounced boundary approximately 2900 km deep—the core–mantle boundary. Seismic observations revealed that when P-waves reach this depth, they are reflected and refracted so strongly that the boundary actually casts a P-wave shadow over part of Earth (Figure 21.5). The shadow is a region where no waves are detected. The wave shadow develops between 105° and 140° from the surface location of the earthquake (the *epicenter*) and has no direct penetration of seismic waves. Because the boundary is so distinct, it marks an important change in the density of the materials present in Earth's interior. Both the overall density of Earth and the speed with which seismic waves travel through the core suggest that the core is composed of iron, a material that is much denser than the silicate rocks that make up the mantle. Furthermore, the sharp boundary between

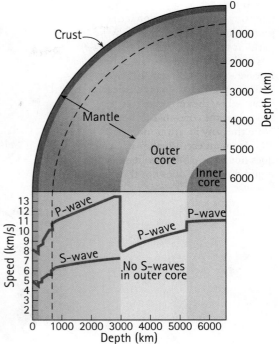

FIGURE 21.4
Cross-section of Earth's internal layers, showing the increases and decreases of P-wave and S-wave velocity in the different layers.

LEARNING OBJECTIVE
Explain how reflection and refraction of seismic waves reveal boundaries in Earth's interior.

FYI How does density relate to elasticity? Elasticity is related to how rigid and springy a material is. It is a measure of a solid's ability to recover its shape once a deforming force is removed from it. For example, steel has a high elasticity, while fresh bread has a low elasticity. (A rubber band, interestingly, is called elastic because it returns to its original shape when released.) The density of rock beneath Earth's surface increases because it is compressed by the weight of material above it. The more the rock is compressed, the more rigid and elastic it becomes. So there is a strong connection between density and elasticity within Earth's interior.

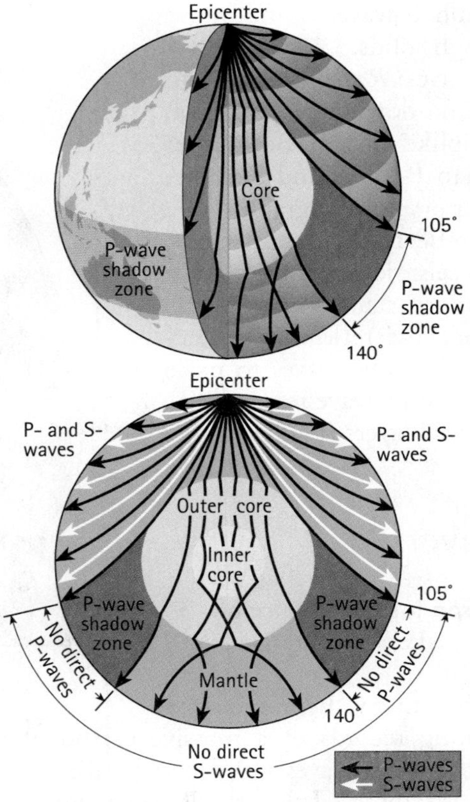

FIGURE 21.5

Cutaway and cross-sectional diagrams showing the change in wave paths at the major internal boundaries and the P-wave shadow. The P-wave shadow between 105° and 140° from an earthquake's epicenter is caused by the refraction of the P-waves at the core–mantle boundary. Note that any location more than 105° from an earthquake's epicenter does not receive S-waves because the liquid outer core does not transmit S-waves.

FIGURE 21.5

Masteringᴘʜʏꜱɪᴄꜱ

VIDEO: The Mantle and Crust

the mantle and core casts an S-wave shadow that is even more extensive than the P-wave shadow—S-waves are unable to pass through the core. Knowing that S-waves travel only through solids, English seismologist Sir Harold Jeffreys confirmed in 1926 that the core, or part of it, must be liquid.

Taken together, the discoveries of Oldham, Mohorovičić, Gutenberg, and Jeffreys indicate that Earth consists of three layers of materials of different composition: the *crust*, *mantle*, and *core*. Each layer is a concentric sphere, so that Earth's overall structure resembles that of a boiled egg.

This simple picture of Earth's layers was refined in 1936 by Inge Lehmann, a Danish seismologist. Her research showed that P-waves refract not only at the core–mantle boundary, but also at a certain depth within the core, where they gain speed. This change in wave speed indicates that the inner region of the core must be solid. So the core was found to have two parts—a liquid outer core of molten iron and a solid-iron inner core. Adding Lehmann's work to earlier findings shows the complete and current picture of Earth's internal layered structure.

Do you suppose these layers in Earth's interior influence the geologic changes our planet experiences? The answer is yes, as you will now see.

CHECKPOINT

What evidence supports the theory that Earth's inner core is solid and its outer core is liquid?

Was this your answer?
The differences between how P-waves and S-waves move through Earth's interior. As both waves encounter the boundary at 2900 km below the surface, a very pronounced wave shadow develops. P-waves are both reflected and refracted at that boundary, but S-waves are only reflected. S-waves cannot travel through liquids, implying that the outer core is liquid. As P-waves move through the outer core, there is a depth at which their speed suddenly increases. Knowing that waves travel faster in solids, we infer the existence of a solid inner core.

FYI Much that we know about Earth's interior was learned as a result of the Cold War between the United States and the former Soviet Union. In the 1960s, when testing of nuclear weapons was very common, underground nuclear explosions were found to produce seismic waves. Both countries installed sensitive seismographic stations to monitor their opponent's activities. It was the seismograms of this network of stations that revealed details of the unseen structure of our planet.

The Core

Earth's **core** is composed mainly of iron and smaller amounts of nickel. In the inner core, the iron and nickel are solid. Although the inner core is indeed very hot, intense pressure from the weight of the rest of Earth prevents the inner core from melting (just as a pressure cooker prevents high-temperature water from boiling, as discussed in Chapter 7).

Because less weight is exerted on the outer core, the pressure is lower there, resulting in a liquid phase of iron and nickel. The molten outer core flows at the rate of several kilometers per year. This flow is evident far outside Earth's surface. The flowing molten outer core produces a flowing electric charge—an electric current. This electric current powers Earth's magnetic field. The magnetic field is not stable but has changed throughout geologic time. Recall from Chapter 9 that there have been times when Earth's magnetic field has diminished to zero, only to build up again with the poles reversed. These magnetic-pole reversals probably result from changes in the direction of fluid flow in the molten outer core of Earth.

> **FYI** The density of rocks at Earth's surface is 2.7–3.0 g/cm^3, whereas the average density of Earth as a whole is 5.5 g/cm^3. Thus, surface rocks are not representative of the planet's interior. To account for Earth's high average density, the density of the core must be at least 10 g/cm^3. This and other reasons suggest that the core is composed of iron and smaller amounts of nickel, the most abundant of the heavier elements.

CHECKPOINT

Iron's normal melting point is 1535°C, yet Earth's inner core temperature is at least 5000°C. Why doesn't the solid inner core melt?

Was this your answer?
The intense pressure from the weight of Earth above crushes atoms together so tightly that even high temperature cannot budge them. Because of the pressure, melting cannot occur and the inner core stays solid.

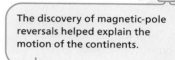

The discovery of magnetic-pole reversals helped explain the motion of the continents.

The Mantle

Surrounding the core of the planet is the **mantle**, a rocky layer some 2900 km thick. The mantle is Earth's thickest layer and makes up about 84% of its volume. From top to bottom, the mantle's composition is relatively uniform—composed of hot, iron-rich silicate rocks. In general, these mantle rocks behave like an elastic solid. And in most parts of the mantle, the rocks can actually flow, even though they are solid. This behavior—the ability of rocks to flow without breaking—is called *plasticity*. So, even though it is fairly uniform in composition, the mantle varies in its physical properties. How do we know that it varies? The answer is seismic studies!

Information from seismic studies divides the mantle into two portions: the lower mantle and the upper mantle. The lower mantle extends from the outer core to a depth of about 700 km (the dashed line in Figure 21.4). The lower mantle is completely solid because pressure in this region is too great for melting to occur.

The upper mantle, which extends from the 700-km depth upward to the crust–mantle boundary, has two zones (Figure 21.6). The lower zone of the upper mantle is called the **asthenosphere**. This zone is solid but contains small amounts of liquid derived from the partial melting of mantle rocks. The asthenosphere is especially plastic, and it flows more easily than the lower mantle. Plastic flow in the mantle takes the form of slowly moving *convection currents*—hot material rises, cools, and then sinks (Figure 21.7). As we will soon see, the constant flowing movements in the asthenosphere greatly affect the surface features of our planet.

Above the asthenosphere is the **lithosphere**. The lithosphere is about 100 km thick and includes the entire crust and the

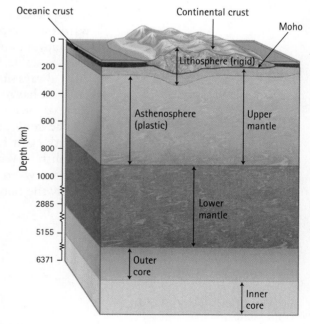

FIGURE 21.6

The core, mantle, and crust each have a different composition. The mantle extends from the core–mantle boundary to the base of the crust. The lower mantle is essentially a single unit from a depth of 2900 km up to about 700 km. Although Earth's upper mantle is of a similar composition, it is divided into two distinct units. The lower zone of the upper mantle is the plastic asthenosphere. The top zone of the upper mantle, and the entire crust, together form the rigid lithosphere.

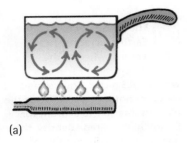

(a)

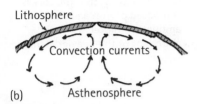

(b)

FIGURE 21.7

(a) A familiar example of convection is seen when water is heated in a pan. (b) A simple model showing convection currents in the asthenosphere.

uppermost part of the mantle. Unlike the asthenosphere, the lithosphere is rigid and brittle and does not flow. The lithosphere is, in a sense, riding on top of the asthenosphere like a raft on a pond. The lithosphere moves along with the motions of the material beneath it in the asthenosphere. The motions in the mantle, however, are not uniform. Because of this, the brittle lithosphere is broken into many individual pieces called *plates*.

Mantle convection currents move at a leisurely pace, taking hundreds of millions of years to complete one loop. Even so, heat-driven motion in Earth's interior shapes and reshapes many of our surface features. The lithospheric plates are always in motion. Their movement causes earthquakes, volcanic activity, and the deformation of large masses of rock that create mountains.

The Crustal Surface

The top part of the lithosphere is the **crust**. The crust is subdivided into *continental crust* and *oceanic crust*. These two types of crust differ in density, composition, and thickness. The crust of the ocean basins is compact—it's about 10 km thick and composed of dense basaltic rocks. Continental crust is thicker (20–60 km) and composed of granitic rocks. Because granitic rocks are less dense than basaltic rocks, most of the continental crust is above sea level.

If continental crust is so much thicker than oceanic crust, why are the ocean basins underwater and the continents high and dry? The answer is found in their density differences and buoyancy (Chapter 5). Remember, the entire lithosphere—the uppermost mantle and crust— "floats" on the asthenosphere. Objects float because a *buoyant force* acts on them. In the case of the "floating lithosphere," the buoyant force is produced by the underlying mantle. The upward push of the asthenosphere opposes the downward pull of gravity. When the buoyant and gravitational forces are in balance, the vertical position of the crust is stable. This is the principle of **isostasy**. So the less-dense continental crust always sits higher than the more-dense oceanic crust, even if the continental crust has more mass. This is isostatic balance—the upward-acting buoyant force of the asthenosphere equals the weight of the entire lithosphere.

The concept of isostasy can be made clear with an analogy. Imagine that Earth's crust is a cargo ship and that the mantle is the ocean. The ship will establish its vertical position in the water when the net force on it is zero. This happens when the gravitational force pulling the ship downward (its weight) equals the buoyant force pushing it upward (Figure 21.8). When the ship is

FIGURE 21.8

Isostasy: The vertical position of the crust is stable when the gravitational and buoyant forces balance. Denser oceanic crust therefore has a lower vertical position than the less-dense continental crust, just as the loaded ship sits lower in the water than the unloaded ship.

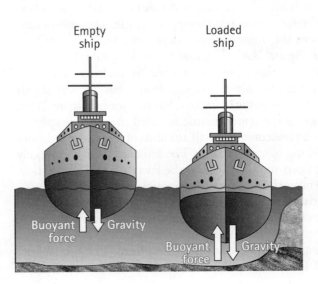

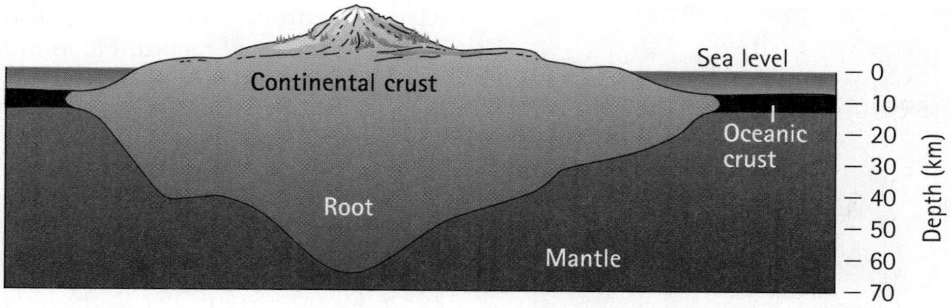

FIGURE 21.9
Continental crust is thicker and less dense than oceanic crust. As such, continental crust floats higher on the mantle than oceanic crust. To achieve isostatic balance, the higher the crust, the deeper the roots.

loaded, it is denser and floats lower in the water—more of it is submerged than when it is empty. Likewise, the crust's vertical position in the mantle rises and falls according to variations in density (Figure 21.9). So thin, dense oceanic crust sits lower in the mantle than thicker, less-dense continental crust.

CHECKPOINT

1. How does Earth's crust behave like a ship floating in water?
2. Why is Earth's crust thicker beneath a mountain?

Were these your answers?

1. The crust's vertical position, as well as the ship's, is determined by the balance of the buoyant and gravitational forces acting on it.
2. Just as most of an iceberg is below sea level, likewise for mountains. Mountains sink until the upward buoyant force balances the downward gravitational force.

21.3 Continental Drift—An Idea Before Its Time

EXPLAIN THIS Why do we find coal deposits in frigid Antarctica?

Have you ever noticed on a map that Africa and South America fit together like pieces of a jigsaw puzzle (Figure 21.10)? One person who took this observation seriously was German naturalist Alfred Wegener (Figure 21.11). Wegener prepared a detailed hypothesis to explain this observation. His hypothesis, known as **continental drift**, stated that the world's continents were once joined together as a single supercontinent that he called *Pangaea*.

Wegener supported his hypothesis with impressive geologic, biologic, and climatologic evidence. He proposed that the boundary of each continent was not at its shoreline but at the edge of its *continental shelf* (the gently sloping platform between the shoreline and the steep slope that leads to the deep ocean floor). When Wegener put South America and Africa together along their continental shelves, the fit was nearly perfect. He then investigated rocks that are now separated by the Atlantic Ocean. The rocks were similar in both age and type. In addition, many mountain chains were found to be very similar to mountain chains across the ocean. Mountains in North America matched up with mountains in Europe, and mountains in South America matched up with those in

LEARNING OBJECTIVE
Describe, and give examples of, the evidence Alfred Wegener used to support his hypothesis of continental drift.

How old were you when you first noticed that the shoreline margins of South America and Africa fit together like a jigsaw puzzle?

FIGURE 21.10
The jigsaw-puzzle fit between continents is even better at the continental shelves than at the shorelines of the continents.

FYI Scientists of the early 20th century believed that oceans and continents were geographically fixed. They regarded the surface of the planet as a static skin spread over a molten, gradually cooling interior. They also believed that the cooling of the planet resulted in its contraction, which caused the outer skin to contort and wrinkle into mountains and valleys.

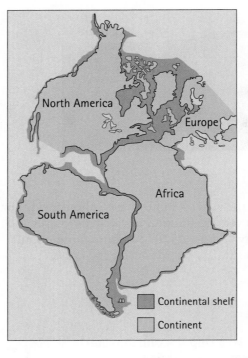

North America

Europe

Africa

South America

☐ Continental shelf
☐ Continent

FIGURE 21.11
Alfred Wegener (1880–1930) was a brilliant interdisciplinary scientist. In 1915 he published his hypothesis of continental drift, which eventually led to the discipline of plate tectonics. His interests included not only meteorology and climatology, but also astronomy, geology, geophysics, oceanography, and paleontology. Throughout his life, Wegener had a fascination with exploring the Arctic, and he was fortunate to survive several Arctic adventures. In 1930 at age 50, however, his luck ran out. Wegener died while crossing an ice sheet on an expedition of Greenland. His body still remains as part of the Greenland glacier. The life of Alfred Wegener—a productive life indeed!

Africa (Figure 21.12). Wegener also looked at the fossil record. He found fossils of identical land-dwelling animals in South America and Africa but nowhere else. This was a strange finding because, today, animals and plants of these regions are notable for their striking differences. And fossils of nearly identical trees are found in South America, India, Australia, and Antarctica.*

Even stronger evidence for a supercontinent was found by studying paleoclimatic (ancient climate) data. More than 300 million years ago, a huge continental ice sheet covered parts of South America, southern Africa, India, and southern Australia (Figure 21.13). Evidence of this ice sheet is found in thousands of well-preserved glacial striations. As a glacier, or ice sheet, moves over the land, it gouges the surface. These gouges, called striations, reveal the direction of ice flow. If these continents were in their present positions, the ice sheet would have had to cover the entire Southern Hemisphere and, in some places, would have had to cross the equator! An ice sheet that extensive would have made the world climate very cold. But there is no evidence of glaciation in the Northern Hemisphere at that time. In fact, the time of glaciation in the Southern Hemisphere was a time of subtropical climate in the Northern Hemisphere. To account for this inconsistency, Wegener proposed that Pangaea existed 300 million years ago, with South Africa located over the South Pole. This reconstruction would bring all the glaciated regions into close proximity near the South Pole and place the modern northern continents nearer to the tropics.

Wegener described continental drift in his book *The Origin of Continents and Oceans*, published in 1915. Although he used evidence from many different scientific disciplines, his well-founded hypothesis was ridiculed by the scientific community. Opponents complained that Wegener failed to provide a suitable driving force to account for the continental movements. (Wegener wrongly proposed that the tidal influence of the Moon could produce the needed force. He also proposed that the continents broke through Earth's crust like

* One fossil plant assemblage that offers strong support for Wegener's idea is the *Glossopteris* flora, which was named after the dominant gymnosperm tree found in the prehistoric southern temperate forests of South America, India, Australia, and Antarctica. Because the seeds from these trees were too large to be distributed by wind, the wide distribution of this flora supports Wegener's hypothesis that the continents were once joined together.

icebreakers cutting through ice.) Without a convincing explanation for his hypothesis, it was dismissed. Only later, in the light of newfound discoveries, did the scientific community accept Wegener's concept.

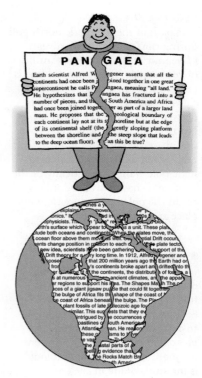

FIGURE 21.12
Wegener likened the fossil and rock matches to finding two pieces of torn newspaper with matching contours and lines of type. If the edges and the lines of type fit together, the two pieces of newspaper must have originally been one.

CHECKPOINT

1. **What evidence might lead someone with no understanding of science to suspect that the continents were once connected?**
2. **Scientists of Wegener's day rejected continental drift because they couldn't imagine how massive, rocky continents could grind through the solid rock of the ocean floor. What did Wegener's contemporaries evidently not know about the mantle?**

Were these your answers?

1. The most obvious evidence is the matching of the edges of the African and South American continents, which can be seen on any world map or globe.
2. Wegener's contemporaries assumed that if the continents were to move, they would have to push through solid rock. They did not know that the mantle has a "plastic-like" layer, the asthenosphere, over which "floating" continents can readily slide.

21.4 Acceptance of Continental Drift

EXPLAIN THIS How do magnetic pole reversals support seafloor spreading?

One of the first key discoveries in support of continental drift came about through studies of Earth's magnetic field. We know from Chapter 9 that Earth is a like a huge magnet, with its magnetic north and south poles near the geographic poles. Because certain minerals align themselves with the magnetic field when a rock is formed, many rocks have a preserved imprint of changes in Earth's magnetism over geologic time. These changes include times when the magnetic north and south poles were reversed. This magnetism from Earth's geologic past is known as **paleomagnetism**.

In the 1950s, a plot of the positions of the magnetic north pole through time revealed that over the past 500 million years, the position of the pole had apparently wandered extensively throughout the world (Figure 21.14). It seemed that either the magnetic poles migrated through time or the continents had drifted. Because the apparent path of polar movement varied from continent to continent, it was more plausible that the continents had moved. Thus, the hypothesis of continental drift was revived, but a mechanism to explain how the movement occurred was still lacking.

Just as Wegener's hypothesis was supported by evidence in different disciplines, the mechanism to explain continental drift was assembled from several different disciplines. Piece by piece, it all came together in the 1960s, as a result of seafloor exploration.

A key player who helped solve the puzzle of continental drift was Harry Hess (Figure 21.15), a geology professor who also served as a naval captain during World War II. To help his attack ship maneuver near shore during beach landings, Hess used a *fathometer*, an innovative depth sounder, to map the underwater topography. But Hess, a scientist as well as a sailor, continued using the fathometer in the open sea to collect data about the deep ocean bottom. Through his wartime scientific surveying, Hess constructed a detailed profile

LEARNING OBJECTIVE
Summarize the evidence for seafloor spreading, and describe the process of subduction.

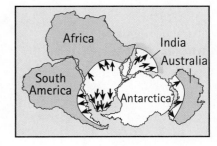

FIGURE 21.13
Glacial striations in rock outcrops in South America, Africa, Antarctica, India, and Australia provide paleoclimate evidence that these continents were once positioned together. The arrows depict the direction of glacial movement.

The path of the magnetic north pole during the last 500 million years. (The unit *m.y.a.* stands for "millions of years ago.") The lower red line is derived from evidence collected in Europe, and the upper red line is derived from evidence collected in North America. One would expect that these two lines would overlie each other. Thus, either the magnetic pole wanders erratically or the continents have moved. But how could the pole be in more than one place at the same time?

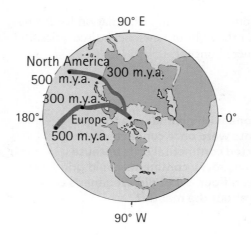

> **FYI** *Paleo-* means "old" or "ancient." As a combining prefix, *paleo-* is used to describe things that occurred in the past. For example, *paleoclimate* is a way to describe ancient climates, *paleomagnetism* describes ancient magnetic data, and *paleontology* is the study of life in ancient geologic time.

> **FYI** Measurements of the ocean floor began during World Wars I and II as echo-sounding devices—primitive sonar systems—began to measure ocean depth by recording the time it took for a sound signal (commonly called a "ping") from the ship to bounce off the ocean floor and return. The returned signals revealed that the ocean floor was much more rugged than previously thought.

Mastering**PHYSICS**
TUTORIAL: Mantle Convection and Seafloor Spreading

FIGURE 21.15
Harry Hess (1906–1969). Hess's hypothesis of seafloor spreading helped establish a mechanism for Wegener's hypothesis of continental drift.

of the ocean floor across the North Pacific Ocean. His findings expanded upon and substantiated other discoveries and emerging ideas.

With the improved technology of the 1950s, oceanographers could map the ocean floor in greater detail. Huge mountain ranges running down the middle of the Atlantic, Pacific, and Indian Oceans were discovered (Figure 21.16). The Mid-Atlantic Ridge, for example, was found to wind down the center of the Atlantic Ocean basin parallel to the American, European, and African coastlines. The ridge stretches 19,312 km, and its highest peaks emerge above sea level to form oceanic islands, such as Iceland and the Azores (Figure 21.17). In the center of the ridge and all along its length is a deep fissure—a volcanic *rift zone*. Another ocean-floor feature discovered was the deep *ocean trenches* (long, deep troughs in the seafloor) near some continental landmasses, particularly around the edges of the Pacific. So, it was revealed that some of the deepest parts of the ocean are actually near some of the continents, and some of the shallowest waters are in the middle of the oceans, at the mid-ocean ridges.

As the topography of the ocean floor was revealed in the Atlantic Ocean basin, a different kind of mapping was being done in the Pacific basin—the intensity of Earth's magnetic field. The magnetic surveys of the ocean floor revealed a curious pattern of stripes—alternating strong and weak magnetic fields. The zebralike pattern was found to run parallel to the coastlines and in other parts of the ocean floor.

With the discovery of the midocean rifts, Hess was inspired to look back at his data from years before. In 1960, he proposed that the seafloor is not permanent but is constantly being renewed. Hess hypothesized that the ocean ridges are located above upwelling convection cells in the mantle. As material from the mantle oozes upward, new lithosphere is formed. The old lithosphere is simultaneously destroyed in the deep ocean trenches near the edges of some continents. Thus, in a conveyor-belt fashion, new lithosphere forms at a spreading center and older lithosphere is pushed out from the ridge crest, eventually to be recycled back into the mantle at a deep ocean trench (Figure 21.18). Hess called his hypothesis **seafloor spreading**.

Support for Hess's theory came from paleomagnetic studies of the ocean floor. In 1962, magnetic surveys of the Atlantic basin began, with results quite similar to those found in the Pacific basin. Two English geologists, Fred Vine and Drummond Matthews, puzzled over Hess's hypothesis of seafloor spreading and their new magnetic data. They proposed that as new basalt is extruded at an oceanic ridge, it is magnetized according to the existing magnetic field. So, what the magnetic surveys of the ocean's floor actually revealed was not magnetic intensity but magnetic direction. The alternating stripes paralleling

FIGURE 21.16
The first detailed map of the ocean floor was created by Marie Tharp and Bruce Heezen. Based on sonar readings, the map reveals enormous mountain ranges in the middle of the oceans and deep ocean trenches near some continental landmasses.

FIGURE 21.17
The Mid-Atlantic Ridge runs down the center of the Atlantic Ocean. Its highest peaks emerge above the water in several places, creating oceanic islands such as Iceland. This photo shows the exposed rift valley on Iceland.

either side of the rift areas indicate periods of normal and reversed polarity (Figure 21.19). As in a very slow magnetic tape recording, the magnetic history of Earth is recorded in the spreading ocean floors. Since the dates of pole reversal can be determined, the magnetic pattern of the spreading seafloor documents both the age of the seafloor and the rate at which it spreads.

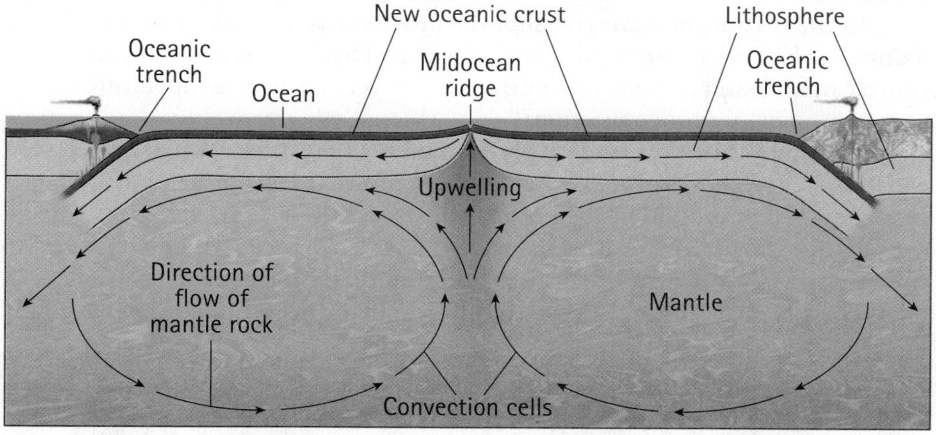

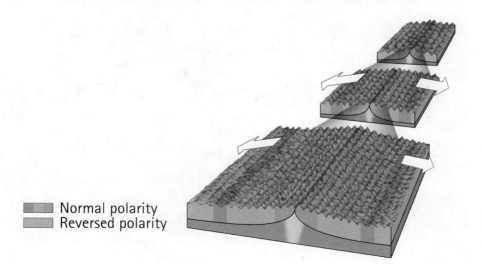

FIGURE 21.18
In conveyor-belt fashion, new lithosphere is formed at the midocean ridges ("spreading centers") as old lithosphere is recycled back into the asthenosphere at a deep ocean trench.

Normal polarity
Reversed polarity

FIGURE 21.19
As new material is extruded at an oceanic ridge (spreading center), it is magnetized according to the existing magnetic field. Magnetic surveys show alternating stripes of normal and reversed polarity paralleling both sides of the rift area. Like a very slow magnetic tape recording, the Earth's magnetic history is recorded in the spreading ocean floors.

The collaborative efforts of Tharp, Heezen, Hess, Vine, and Matthews (and others) convinced the scientific community, once and for all, that seafloor spreading and continental drift must in fact occur. Hess's hypothesis of seafloor spreading provided the mechanism to explain continental drift. The time was right for the revolutionary concepts that follow. The tide of scientific opinion had indeed switched in favor of a mobile Earth.

21.5 The Theory of Plate Tectonics

EXPLAIN THIS How is magma generation related to movement of the plates?

Plate tectonics describes the motions of Earth's lithosphere that create ocean basins, mountain ranges, earthquake belts, and other large-scale features of Earth's surface. The theory of plate tectonics states that Earth's outer shell, the lithosphere, is divided into eight relatively large plates and a number of smaller ones (Figure 21.20). These lithospheric plates ride atop the relatively plastic asthenosphere below. So Wegener's hypothesis of continental drift was on the right track. The continents really do move—they move because they are embedded within the drifting tectonic plates.

Ultimately, lithospheric plates move in response to convection in Earth's interior. Recall from Chapter 7 that heat naturally moves from warmer regions to cooler regions. Inside Earth, heat moves from the hot core and mantle to the cooler crust. Heat flow from the core to the mantle is mostly due to conduction. In the mantle, however, most of the heat flow is due to convection. As hot mantle rock rises, it expands. Then, closer to Earth's surface, the rock cools, contracts, and sinks. From this example, can you see how temperature differences are a key part of convection?

Interestingly, however, gravity plays an even greater part. When heated rock expands it becomes less dense, and when cooled rock contracts it becomes more dense. Convection in the hot mantle occurs because gravity pulls cooled, denser rock downward relative to heated, less-dense rock, which continues to rise upward. As less-dense rock rises and eventually forms new lithosphere, it takes the place of sinking, dense, old lithosphere being pulled downward by

FYI Three essential bits of information are contained in the preserved magnetic record: (1) the polarity of Earth's magnetic field at the time the rock was formed, (2) the direction to the magnetic pole from the rock's location at the time the rock was formed, and (3) the magnetic latitude of the rock's location at the time the rock was formed. Once the magnetic latitude of a rock and the direction of the magnetic poles are known, the position of the magnetic pole at the time of formation can be determined.

If the seafloors spread, continents must move.

MasteringPHYSICS
VIDEO: Plate Tectonics

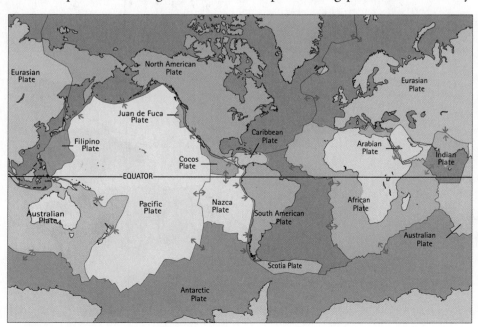

FIGURE 21.20
The lithosphere is divided into eight large plates and a number of smaller ones.

gravity—convection! So gravity and heat are what cause mantle convection. And the lithospheric plates move because they are the upper part of the mantle convection cells.

Earth's lithospheric plates move in a conveyor-belt manner in response to mantle convection. The plates move in different directions and at different speeds. Plates carrying continents, such as the North American Plate, generally move slower, while oceanic plates, such as the Pacific Plate, tend to move much faster. Over geologic time, the various plates have pulled apart, crashed, merged, and separated from one another. Because of these interactions, the edges of plates—the *plate boundaries*—are regions of intense geologic activity (Figure 21.21). While interiors of plates are relatively quiet, most earthquakes, volcanic eruptions, and mountain building events occur where plates meet. There are three types of tectonic plate boundaries:

1. Divergent boundaries—*where plates move away from each other*

2. Convergent boundaries—*where plates move toward each other*

3. Transform boundaries—*where plates slide past each other*

Divergent Plate Boundaries

Heat-driven convection cells in the mantle operate in symmetrical loops. Where adjacent upward-moving convection cells diverge at the surface, lithospheric plates spread apart. Tension is the dominant force where plates move away from each other. These spreading centers are **divergent plate boundaries** (see Figure 21.21a). At the divergent boundary, the asthenosphere is very near the surface and the lithosphere is very thin.*

Mid-ocean ridges mark the locations of most divergent plate boundaries. At the ridge crests, gravity gives plates a small push away from divergent boundaries. Think of the ocean floor's topography—the highest elevations are at the mid-ocean ridges. The ridge area is high compared to the surrounding seafloor because of the lift it gets from rising mantle rock—rock that is convecting upward from the asthenosphere below. Because of the elevation difference between the ridge and the adjacent seafloor, gravity causes the plates to slide down and outward from the mid-ocean ridge like cookies sliding off a tilted cookie sheet.

Magma generated at a divergent boundary is from the partial melting of mantle rock brought upward with rising convection currents. Melting occurs because pressure is reduced on this rock as it nears the surface. Basaltic lava erupts where the plates diverge, partially filling the rift between the diverging plates. When cooled, the basalt becomes new oceanic crust. As the two plates continue moving apart, mantle rock beneath the new crust in the uppermost asthenosphere cools and hardens. Melded to the fresh crust above, new lithosphere is formed, which slowly moves away from the spreading center, cools, contracts, and becomes denser. Oceanic lithosphere is thin and young near the spreading ridge. At increasing distances, lithosphere is progressively thicker and older—equally on both sides of the ridge.

The Mid-Atlantic Ridge is the divergent boundary between the North American and Eurasian Plates in the North Atlantic, and the South American and African Plates in the South Atlantic. The rate of spreading at the Mid-Atlantic Ridge ranges between 1 cm and 6 cm per year. Although this spreading may seem slow, through geologic time the effect has been tremendous. Over the past

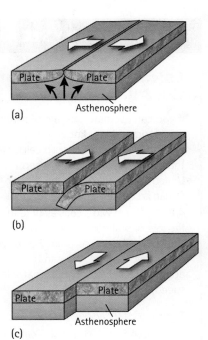

FIGURE 21.21
Plate boundaries are regions of intense geologic activity. They are also the sites of lithospheric formation and destruction. Named for the movement they accommodate, the three types of plate boundaries are (a) divergent, (b) convergent, and (c) transform boundaries.

Mastering**PHYSICS**

TUTORIAL: Plate Boundaries and Plate Tectonics

* Theoretically, the lithosphere has zero thickness at the exact location of the divergent boundary and the asthenosphere is at the surface.

FIGURING PHYSICAL SCIENCE

Calculating the Age of the Atlantic Ocean

If you can estimate the rate of seafloor spreading and you know the present width of an ocean, you can calculate the ocean basin's age. Between the United States and Africa, the Atlantic Ocean is currently about 4830 km or 4.8×10^8 cm wide. Let's assume that the rate of seafloor spreading in the Atlantic has been a constant 2.5 cm/yr over geologic time. We can then apply the familiar equation that relates speed, time, and distance:

Time = distance/speed

= $(4.8 \times 10^8 \text{ cm})/(2.5 \text{ cm/yr})$

= 1.92×10^8 yr

~ 190 million years

Based on these estimates, the age of the Atlantic Ocean is about 190 million years.

SAMPLE PROBLEM

The Red Sea is presently a narrow body of water located over a divergent plate boundary. The plates began to diverge apart from one another 30 million years ago. Knowing the age and the width, what is the average rate of spreading?

Solution:

If we take the current width of the Red Sea to be 300 km or 3.0×10^7 cm,

and the time of spreading to be 30 million years, the speed at which the seafloor spreads is then:

Speed = distance/time

= $(3.0 \times 10^7 \text{ cm})/(3.0 \times 10^7 \text{ yr})$

= 1 cm/yr

At this rate, it will take about 400 million years for the Red Sea to be as wide as the Atlantic Ocean. But Earth is dynamic and ever changing—and spreading rates can change. For example, the Ethiopian rift zone spread by about 8 m in 2006! At the same time, 2.5 km^3 of magma intruded into the crust (enough to fill 2000 football stadiums)! Earth in action!

190 million years, seafloor spreading has transformed a tiny waterway through Africa, Europe, and the Americas into the vast Atlantic Ocean of today!

Spreading centers are not restricted to the ocean floors but also develop on land. Hot rock in Earth's interior rising beneath continental landmasses generates tension in Earth's crust, causing it to stretch and bend upward (upwarping). Gaps in the crust are produced, and large slabs of rock slide and sink down into these gaps. The large down-dropped valleys generated by this process are called either **rifts** or *rift valleys* (Figure 21.22). The Great Rift Valley of East Africa is an excellent example of such a feature; if the spreading continues, it may be the beginning of a new ocean basin.

FIGURE 21.22
Formation of a rift valley and its transformation into an ocean basin. (a) Rising magma uplifts continental crust, causing the surface to crack. (b) Rift valley forms as crust is pulled apart. Africa's Great Rift Valley is in this stage today. (The two sides of the valley move away from each other because they are located above mantle convection cells that have the same circulation pattern as the cells in Figure 21.18.) (c) Water from the ocean drains in as the rift drops below sea level, forming a linear sea, so called because it is usually long and narrow. (d) Over millions of years, the rift continues to widen and becomes an ocean basin.

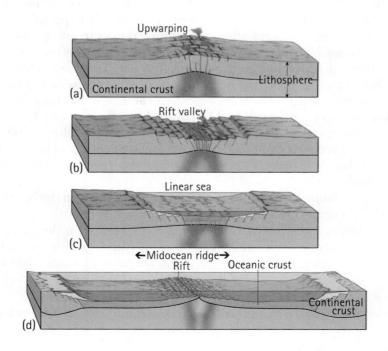

Convergent Plate Boundaries

Convergent plate boundaries, as the name implies, are where plates come together, or converge. Convergent boundaries are areas of compressive stress and, depending on the nature of the plate interactions, the recycling or destruction of lithosphere. These regions of plate collisions are also regions of great mountain building. The type of convergence—or "slow collision"—that takes place depends on the type of lithosphere that is involved. The three kinds of convergent plate boundaries are:

1. Oceanic–oceanic convergence (Figure 21.23a)

2. Oceanic–continental convergence (Figure 21.23b)

3. Continental–continental convergence (Figure 21.23c)

Oceanic–oceanic convergence occurs when two oceanic plates meet and the older (and therefore cooler and denser) plate slides beneath the younger, less-dense plate. The process in which one plate bends and descends beneath the other is called **subduction**, and the area where this occurs is called a *subduction zone*. At Earth's surface, subduction zones are marked by deep ocean trenches that run parallel to the edges of convergent boundaries. The Marianas Trench,

FYI East Africa may be the site of Earth's next major ocean! Spreading processes have already torn Saudi Arabia away from Africa to form the Red Sea. With continued inland spreading, the eastern edge of the present-day African continent will also separate, the Indian Ocean will flood into the area, and the easternmost corner of Africa (the Horn of Africa) will become a large island!

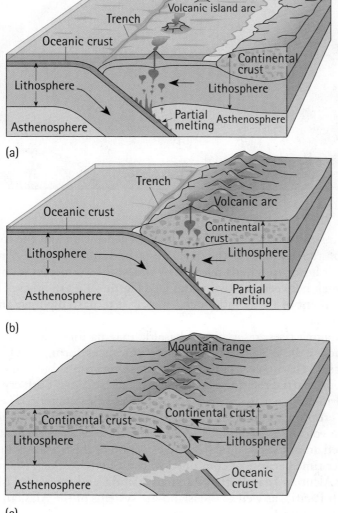

(a)

(b)

(c)

FIGURE 21.23
The three types of convergent margins: (a) oceanic–oceanic, (b) oceanic–continental, and (c) continental–continental.

for example, is where the Pacific Plate and the slower-moving Philippine Plate collide. Dropping 11,000 m (7 mi) below sea level, the Marianas Trench is the deepest location on Earth's crust. If the world's highest mountain, Mt. Everest, were sunk to the bottom of the Marianas Trench, there would still be more than a mile of ocean above it!

Subduction is an important part of mantle convection. Each downgoing plate, or slab, controls the downward part of a convection cell. So, once again, gravity plays a large role. As gravity pulls the oldest edge of the subducting slab into Earth's interior, the rest of the subducting plate is also pulled trench-ward—this process is called *slab-pull*. Slab-pull is the dominant force driving plate tectonics. The longer the subduction zone, the more weight there is to pull on the rest of the downgoing plate. So, longer subduction zones mean faster plate movement. You can visualize a plate being pulled into the asthenosphere as being like a tablecloth being pulled slowly off a table (Figure 21.24).

FIGURE 21.24
Simplified view of convection cells within the mantle, showing slab-pull. The descending part of the plate—the "slab"—heats up, but generally does not melt, as temperature and pressure rise deeper in the mantle. The descending slab is a location of intense metamorphism.

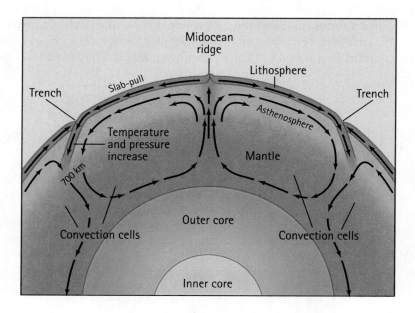

The process of subduction induces magma generation and the formation of volcanoes on the seafloor. As the descending, seawater-saturated plate is pulled downward, it heats up. Fluids released by the wet rock interact with the wedge of mantle rock between the two plates (see Figure 21.23a). These fluids lower the melting point of the mantle rock, causing it to partially melt and generate basaltic magma.

As the magma rises, reservoirs of basaltic magma form in the crust. When magma reaches the ocean floor, undersea volcanoes form, which often start out with basaltic lava. Most of the basaltic magma erupts, but some undergoes crystallization. Partial melting of oceanic crust also occurs. Both crystallization and partial melting act to increase the magma's silica content. The volcanoes then begin erupting andesitic magma, which allows them to grow significantly higher. The volcanoes eventually break the surface of the ocean as a series of islands called an *island arc*. The size and elevation of the islands in an island arc increase over time because of continued volcanic activity. Such island arcs have formed the Aleutian Islands, the Marianas Islands, and the Tonga island group in the South Pacific, as well as the island-arc systems of the Alaskan Peninsula, the Philippines, and Japan.

Oceanic–continental convergence occurs when an oceanic plate collides with a continental plate (see Figure 21.23b). In this case, the denser oceanic plate subducts beneath the less-dense continental plate. A deep ocean trench forms offshore where the converging plates meet. As with oceanic–oceanic convergence, magma is generated. But rising through the thicker continental crust takes longer, allowing more time for crystallization and silica enrichment. And as magma rises through the overlying continental plate, varying amounts of continental crust melt and are mixed into the magma, which also increases the silica content. With a higher silica content, the magma takes on a more andesitic or granitic composition. Most of the erupted lava is andesitic, and such eruptions can be quite intense and violent. The majority of the granitic magma does not erupt but solidifies underground to form intrusive plutonic rock—granite. The Andes Mountains of western South America formed in this way. The Andes continue to grow higher due to the ongoing subduction of the oceanic Nazca Plate beneath the South American Plate. As the Nazca Plate is pulled downward, marine sediments are scraped off onto the granitic roots of the Andes. This material adds thickness and buoyancy to the mountains, which allows the Andes to rise upward more rapidly than wind and rain can erode them.

In the western United States, examples of such volcanic activity are found in the Sierra Nevada, an ancient volcanic range, and the Cascade Range, which is currently active. The Sierra Nevada were produced by subduction of the ancient Farallon Plate beneath the North American Plate. The Sierra Nevada batholith is a

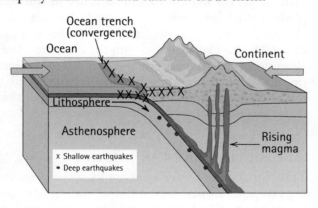

FIGURE 21.25
Earthquakes at a subduction zone get deeper and deeper in the direction of subduction.

remnant of the original volcanic range, while the California Coast Range has remnants of the sediments that accumulated in the trench. The Cascade Range, produced from the subduction of the Juan de Fuca Plate (a piece of the old Farallon Plate) beneath the North American Plate, includes the volcanoes Mt. Rainier, Mt. Shasta, and Mt. St. Helens. The 1980 eruption of Mt. St. Helens gives testimony that the Cascade Range is still quite active.

As you may expect, active subduction zones are areas of intense earthquakes. Earthquakes occur along the subduction zones as the subducted plate grinds against the overriding plate and the subducted plate is compressed and fractured. Earthquakes become steadily deeper and deeper in the direction of subduction (Figure 21.25).

Continental–continental convergence occurs when two continental plates collide. In this case, both plates are composed of buoyant granitic-type rocks. Because the colliding plates have similar densities, neither sinks below the other—so in continental–continental convergence there is no subduction. Instead, the convergence is more like a head-on collision (see Figure 21.23c). Compression causes the plates to break and fold up on each other, making the crust very thick. Intensely compressed and metamorphosed rock defines the zone where continental plates meet. In contrast with convergence involving two oceanic plates or one continental and one oceanic plate, volcanic activity is not a characteristic feature of continental–continental collisions—but earthquakes are.

The collision between continental plates has produced some of the most famous mountain ranges, one majestic example being the snow-capped

FYI The main difference between magma generation at convergent and divergent boundaries is the distance between the site of magma generation and Earth's surface. Magma generation at divergent boundaries is very near the surface, so basaltic magma moves upward unimpeded. At convergent boundaries, magma generation occurs deeper in the mantle, and the rising magma is impeded when it encounters the overlying lithosphere. These two factors increase the travel time to the surface for the mantle-derived basalt. More time means more significant crystallization can occur, resulting in the development of a variety of igneous rocks.

Past

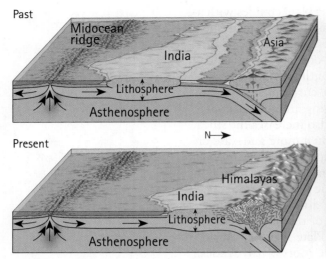

Present

N →

The continental–continental collision of India with Asia produced—and is still producing—the Himalayas.

Himalayas, the highest mountain range in the world. This chain of towering peaks is still being thrust upward as India continues crunching up against Asia (Figure 21.26). The European Alps were formed in a similar fashion when part of the African Plate collided with the Eurasian Plate some 40 million years ago. Relentless pressure between the two plates continues, and it is slowly closing up the Mediterranean Sea. In America, the Appalachian Mountains were produced from a continental–continental collision that ultimately resulted in the formation of the supercontinent Pangaea.

Transform Plate Boundaries

Transform plate boundaries are locations where two plates are neither colliding nor pulling apart but are rather sliding horizontally past one another. The fracture zone that forms a transform plate boundary is called a *transform fault*. Most transform faults are found in the ocean basin and connect offsets in the mid-ocean ridges. Look at the Mid-Atlantic Ridge in Figure 21.27 and notice

> **FYI** The San Francisco earthquake of 1906 did quite a bit of damage. Interestingly, most of the damage was due to the fires that burned the city afterward. The reason—all the water mains had burst and broken during the earthquake, so there was no water to put out the fires!

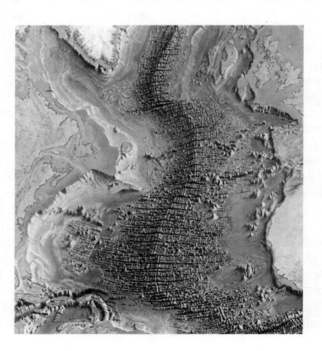

Most transform boundaries occur in ocean basins where they offset oceanic ridges—for example, the Mid-Atlantic Ridge.

how it is broken up into segments. The offset ridge segments are connected by transform faults that "transform" the motion from one ridge segment to another. Between ridge segments, lithosphere coming from one ridge moves in the opposite direction of lithosphere coming from the other ridge (Figure 21.28). Can you see that the two slabs of lithosphere are on different plates? Now look at Figure 21.28 in the area of the inactive fracture zone. The sections of lithosphere on opposite sides of the fracture zone are part of the same plate—both sides are moving in the same direction. But along the transform fault, lithosphere is moving in opposite directions. Can you now see that the transform fault is indeed a plate boundary? Can you also see that fracture zones are former transform faults?

Because there is no tension or compression between the plates, there is no creation or destruction of the lithosphere. A transform plate boundary simply accommodates horizontal plate movement.

Although most transform plate boundaries are short and located within the ocean basins, a few are quite long, such as the San Andreas Fault in California (Figure 21.29). The San Andreas Fault stretches for 1500 km from Cape Mendocino in northern California to the East Pacific Rise in the Gulf of California. The Pacific Plate is moving northwest at a rate of about 5.0 cm/yr relative to the North American Plate. The San Andreas Fault accommodates about 70% of this motion, or about 3.5 cm/yr. The rest of the motion occurs along other faults (such as the Hayward Fault). Grinding and crushing take place as the two plates move past each other. When sections of the plates become locked together, stress builds up until it is relieved in the form of an earthquake. On April 18, 1906, the Pacific Plate lurched about 6 m northward over a 434-km stretch of the fault, releasing the built-up stress and resulting in the catastrophic San Francisco earthquake.

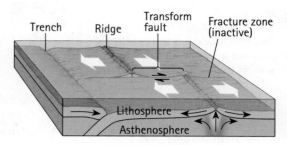

FIGURE 21.28
Transform faults allow two plates to slide past each other at places where two ridge segments are offset.

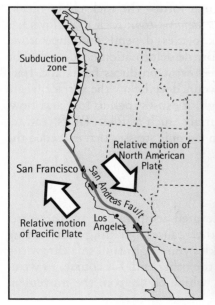

(a)

(b)

FIGURE 21.29
(a) The San Andreas Fault is a transform plate boundary famous for its earthquakes. The slice of California moving northwesterly lies on the Pacific Plate, while the rest of California sits on the North American Plate. (b) In this photo of the San Andreas Fault, notice the long valley created by many years of rock grinding along the fault.

21.6 Continental Evidence for Plate Tectonics

EXPLAIN THIS How can hard and rigid rock layers become crumpled and folded?

Convection in Earth's mantle and slab-pull cause the lithospheric plates to move slowly, but constantly. The interaction between plate boundaries creates stress, which produces strain in the rock. There are three types of stress and three resulting types of strain:

Compressional stress occurs when slabs of rock are pushed together. The resulting stain is shortening of the rock bodies.

Tensional stress occurs when slabs of rock are pulled apart. The resulting strain is extension of the rock bodies.

Shear stress occurs when slabs of rock are both pulled and pushed. The resulting strain is the sliding of one slab of rock past the other in opposite directions, without any noticeable shortening or extension.

Rocks respond to these stresses in three different ways. Rock can undergo *elastic deformation*, *brittle deformation*, or *plastic deformation*. Elastic doesn't necessarily mean "stretchy." A stressed elastic material simply returns to its original shape when the stress is removed. In a stretched rubber band, elastic deformation is observed after it is released—it returns to its original shape. If the rubber band breaks when stretched, it means that the *elastic limit* was exceeded. But the entire rubber band doesn't fall to pieces, does it? When the elastic limit is exceeded, brittle deformation happens in one or two places. Then the rubber band snaps back (and usually stings your fingers!). The remnants of the rubber underwent *elastic rebound*. Because rock is an elastic material, rocks also return to their original shape after stress is removed—unless the elastic limit is exceeded.

When stress exceeds the elastic limit of the rock, the rock, or at least the part that failed, permanently loses its original form—the rock either breaks or flows. Rock that breaks undergoes *brittle deformation*; rock that flows has *plastic deformation*. The response of the rock to stress depends on temperature, pressure, and the composition of the rock. Brittle deformation occurs near the surface where temperature and pressure are low; it produces faults and fractured rocks. Plastic deformation typically occurs deep below the surface where temperature and pressure conditions are high; it causes rock to fold and flow. The geologic structures we see at Earth's surface, such as folds, faults, and related mountains, are examples of strain from tectonic stresses that exceeded the strength of the rock.

FYI Rocks respond differently to stress. Some rocks are strong and some are weak; some rocks are prone to break and others are more likely to flow plastically. Increased pressure, temperature, and water content greatly affect a rock's response to stress. High pressure compresses mineral grains against one another, making the rock stronger, more difficult to break, and more prone to plastic flow. High temperature causes increased vibration of molecules in the rock, weakening rock strength and enhancing plastic flow. Wet rocks are generally weaker than dry rocks—water lubricates mineral grains and enhances microscopic slips that contribute to plastic flow. So rock deep below the surface is prone to plastic flow, but rock at the surface is more apt to fracture and break.

Folds

Deep in the crust under conditions of elevated temperature and pressure, compressive stress that pushes rock together causes buckling and folding (Figure 21.30). This is similar to the wrinkles you might find in a throw rug when you push one end of the rug toward the other end. Of course, to wrinkle sections of rock requires strong forces, which come from the movement of lithospheric plates.

Recall that sediments are deposited horizontally layer by layer, with the bottom layer deposited first. Thus the bottom layer is the oldest in the sequence of deposited layers, and the top layer is the youngest. As these originally flat-lying sedimentary rock layers are subjected to compressive stress, they tilt and

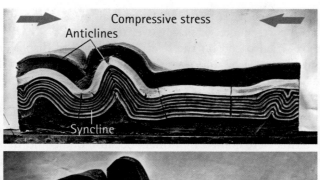

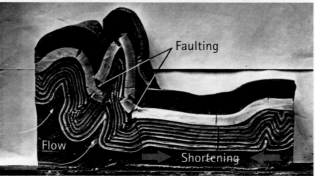

FIGURE 21.30
These photos show a cross-section of wax layers subjected to compressive stress. The wax layers represent sedimentary layers. When compressed from opposite sides, the wax layers become shortened and deform into folds. With more stress, the wax layers continue to deform until they break, forming faults.

become folded. Rock can be up-folded or down-folded. Each fold has an axis, with the rock layers on one side of the axis a mirror image of the rock layers on the other side. You can imagine the axis as a plane extending downward into Earth, as Figure 21.31 shows. When the layers tilt in toward the fold axis, the fold is called a **syncline**. The rocks at the center, or core, of a syncline are the youngest, and as you move horizontally away from the axis, the rocks get older and older. If the fold layers tilt away from the axis, the fold is called an **anticline**. The rocks in the core of an anticline are oldest, and as you move horizontally away from the axis, the rocks get younger. Another way to think about this concept is that anticlines are pushed upward into an arch, and synclines are pushed downward into a sag.

FIGURE 21.31
Anticline and syncline folds. Layer 1 is the oldest rock, and layer 6 is the youngest. The limbs of an anticline (an up-fold) tilt away from the axis of the fold (a marble would roll away from the axis), and the rock layers are oldest at the core of the fold. The limbs of a syncline (a down-fold) tilt toward the axis of the fold (a marble would roll toward the axis), and the rocks are youngest at the core.

CHECKPOINT

Why are rocks at the core of a syncline younger than those farther out from the core, while the opposite is true for an anticline?

Was this your answer?
Think of the rug example. Assume the top surface of the rug is younger than the lower surface. When you push the rug, it can (1) fold upward or (2) fold downward. In the first case, the bottom surface makes up the core—an anticline. In the second case, the top surface makes up the core—a syncline. Makes sense!

Faults

When stress is stronger than rock and conditions are not hot enough or pressure not high enough for plastic deformation, brittle deformation occurs, and rock can fracture into separate blocks. If one block moves relative to the other

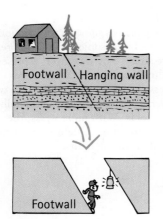

FIGURE 21.32
The terms *footwall* and *hanging wall* were commonly used by miners because one could hang a lamp on a hanging wall and stand on a footwall.

block, the fracture is called a **fault**. Movement along a fault can occur rapidly in the form of an earthquake, or slowly over time.

Faults are classified based on the relative direction of displacement (movement). Look at Figure 21.32 and note the oblique line in the top drawing; this line represents a fault. Imagine that you could pull the block diagram apart at the fault, as shown in the lower drawing. The half containing the fault surface where someone could stand is the *footwall* block. The fault surface of the other half is inclined and would make standing impossible; this is the *hanging wall* block. These terms were coined by miners because one could hang a lamp on a hanging wall, and one could stand on a footwall. Movement on this type of fault is mostly up-and-down motion—the hanging wall and footwall move vertically along the fault plane.

For a fault resulting from compressional forces, the hanging wall is pushed upward along the fault plane relative to the footwall, as Figure 21.33 shows. This type of fault is called a *reverse fault*. The Rocky Mountain foreland (east of the highest peaks), the Canadian Rockies, and the Appalachian Mountains, to name a few, were formed in part by reverse faulting.

FIGURE 21.33
A reverse fault. In a zone of compressional faulting, the hanging wall is pushed up relative to the footwall. (a) A reverse fault before erosion; (b) the same reverse fault after erosion.

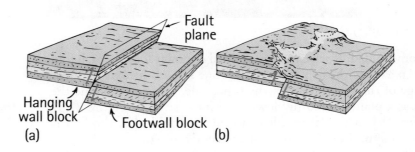

Stress in rock also occurs because of tension. Tensional forces, which pull at rocks, are the opposite of compressional forces, which push. Tension causes the hanging wall to drop downward along the fault plane relative to the adjacent footwall, producing a *normal fault* (Figure 21.34). Virtually the entire state of Nevada, eastern California, southern Oregon, southern Idaho, and western Utah are greatly affected by normal faulting.

FIGURE 21.34
A normal fault. In a zone of tensional faulting, the hanging wall drops down relative to the footwall, forming a normal fault. (a) A normal fault before erosion; (b) the same normal fault after erosion.

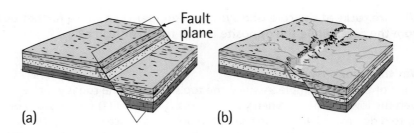

Faults that exhibit horizontal movement, in which blocks of rock slip past one another with very little vertical displacement, are called *strike-slip faults* (Figure 21.35). Some of the world's most famous faults, such as the San Andreas Fault in California, are strike-slip faults. Sticking and slipping of rock blocks along the San Andreas Fault (really a zone of faults) generate many of the earthquakes California is noted for, including the great San Francisco quake of 1906.

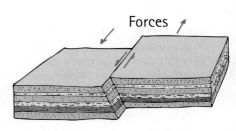

CHECKPOINT

1. **Reverse faults are the result of compressional forces. What happens to Earth's crust in a zone of reverse faulting?**
2. **Normal faults result from tensional stress. What surface feature may we expect to find in a zone of normal faulting?**

Were these your answers?
1. Compressional stress pushes rocks together. So in a zone of reverse faulting we would find shortening of the crust. Look at Figure 21.33; can you see how compression shortens the crust?
2. Tensional stress pulls the crust apart. So in a zone of normal faulting we would find extension of the crust. Look at Figure 21.34; can you see how tension extends and elongates the crust?

FIGURE 21.35
The relative movement of a strike-slip fault is horizontal. The rock bodies are neither shortened nor extended.

Earthquakes

Most earthquakes are related to Earth's tectonic movement. Interactions at plate boundaries create stress, which produces strain in the rock. Strain begins at depth as elastic deformation. With continued stress, rocks at depth store up elastic energy. When the buildup of stress exceeds the rock's elastic limit, the rock breaks and slips into a new position—a fault forms. Faults are not smooth planes—they have irregular surfaces that can interlock and resist movement. When rocks on opposite sides of a fault become stuck and locked into position, the locked-up rock stores elastic energy. Then when stress builds to the point where it exceeds the strength of the fault, the rock suddenly snaps—an earthquake!

The sudden release of stored elastic energy is very similar to the recoil of a spring. The released energy radiates away from the initial break site as seismic waves. This break site, the point of origin where the rock initially slips, is called the *focus* (Figure 21.36). The point on Earth's surface directly above the focus is called the *epicenter*.

FYI 2010 was a devastating year for earthquakes along strike-slip faults! Earthquakes in Haiti and New Zealand were both of similar size and a result of strike-slip fault movement. So why, then, was the damage from these two earthquakes so different? Two reasons: preparedness and economics. New Zealand has frequent earthquakes (140,000 minor quakes a year!) and a developed economy. As such, it has strict building codes, public earthquake education, and earthquake-ready disaster plans. In essence, New Zealand was better prepared! For Haiti, earthquakes are rare (the last earthquake was in 1770) and it is also the poorest country in the Western Hemisphere. Haiti was unprepared—no building codes, no earthquake education, no earthquake disaster plans, and no money. Thousands of people died, thousands were injured and left homeless, and Haiti's infrastructure was severely crippled. As an impoverished country, Haiti cannot rebuild itself—relief efforts for Haiti will need to continue for many years.

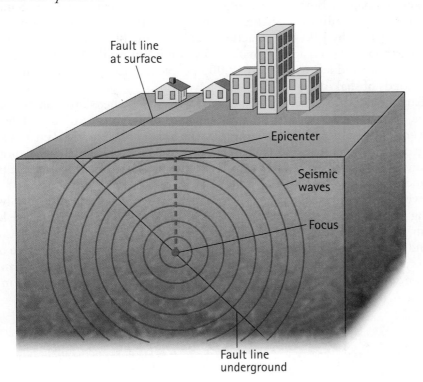

Fault line at surface

Epicenter

Seismic waves

Focus

Fault line underground

FIGURE 21.36
The actual underground location where fault rupture occurs is the *focus*. Directly overhead at the ground surface is the *epicenter*. Seismic waves radiate out from the focus in all directions.

FYI The 2010 Chilean earthquake was so powerful that scientists claim it shortened the length of the day by 1.26 microseconds! You may ask, how could this happen? The mid-latitude location of Chile and the gentle dip of the fault (thrust fault*) shifted Earth's overall mass distribution. Earth's figure axis (the axis about which Earth's mass is balanced) moved by 2.7 milliarcseconds (about 8 cm or 3 inches)! Think of how an ice skater in a spin varies that spin by the extension of her arms. A change in the skater's distribution of mass changes the spin rate. In a similar way, the shift of Earth's mass distribution resulted in a change of speed in Earth's spin.

FYI The Mercalli scale measures earthquake intensity. This 12-degree scale uses observation of damage to estimate an earthquake's strength. For example, intensity 1 is barely felt, intensity 5 is broken chimneys, and, intensity 12 is total damage. Because the Mercalli scale is subject to an observer's interpretation, it is a useful yardstick but not very precise.

FYI Earthquakes cause the ground to shake and rupture. As the ground shakes, so do buildings on top of the land. It is often said that earthquakes don't kill people, but falling buildings do. Earthquakes can cause general property damage, lack of basic necessities, collapse of buildings, loss of life, higher insurance premiums, disease, landslides and avalanches, road and bridge damage, and fires generated by broken gas and electric lines.

The stored energy released in an earthquake does not come as a single large quake. Remember—fault surfaces are irregular and not all movement is at the same time. Although most of the energy is released during the main quake, some of the energy comes before—*foreshocks*—and some comes after—*aftershocks*. You have probably heard these terms in news reports after an earthquake. You have probably also heard the term *magnitude*. Magnitude refers to the earthquake's size and the amount of energy it releases. Large earthquakes have a greater magnitude and release more energy than smaller earthquakes.

Because seismic waves weaken with increasing distance from an earthquake, the strongest ground shaking is generally at the epicenter. But ground shaking is not the only hazard people face during an earthquake. Earthquakes can also trigger landslides. Another hazard is liquefaction, in which wet sediment during an earthquake behaves like a fluid instead of a solid.

The soil and underlying rock can actually increase the size, or amplitude, of seismic waves away from the epicenter. Waves carry energy, and larger-amplitude waves carry more energy than smaller-amplitude waves. Seismic waves can be amplified when they are forced to slow down by certain soil and rock types. The seismic energy that would have otherwise moved quickly through an area "piles up" instead. This increases the amplitude of the waves, which causes stronger ground shaking.

Seismic waves travel faster in hard, rigid rock and slower in softer, less-rigid rock—and they travel even slower in unconsolidated sediment or artificial fill. Put another way, it takes less time to move seismic waves through a kilometer of hard material than through a kilometer of soft material. Seismic waves linger and grow larger in less-rigid rock material.

For a comparison of rigid rock versus less-rigid rock, the 1989 Loma Prieta earthquake in California is a great example. Parts of San Francisco on rigid bedrock had small wave amplitudes, so very little damage occurred to structures built on bedrock. On the other hand, structures built on soft mud close to the bay suffered severe damage or collapse. For example, the Cypress Freeway structure collapsed, and many homes in San Francisco's Marina District were severely damaged. Both of these damage sites contained structures built on unconsolidated sediment. Shaken by the quake, the loose, soft sediment behaved like a slushy jello—liquifaction—and was unable to support structures. The earthquake lasted approximately 15 seconds but resulted in 67 deaths and 4000 injuries and left more than 12,000 people homeless. A lot of damage for 15 seconds of shaking!

Earthquakes are generally associated with areas where lithospheric plates meet. So, not surprisingly, the majority of earthquakes occur in just a few narrow zones (Figure 21.37). For example, the area of the Pacific Rim is a zone of great seismic activity, as is the area of the Mid-Atlantic Ridge. Compare Figure 21.37 to Figure 21.20; can you see that most earthquakes occur at plate boundaries?

Even so, earthquakes can happen anywhere in the world, and not always along plate boundaries. Remember—our Earth is very old, and there have been many changes to its surface. So, faults and related earthquakes can also be found intraplate—far from present plate boundaries— in areas that were once closer to plate boundaries. Because of their geologic history, these intraplate areas are zones of weakness. One such area is the New Madrid seismic zone in the

* A thrust fault has the same type of motion as a reverse fault but differs in the dip angle of the fault plane. A reverse fault has a steep dip, <45°; a thrust fault has a gentle dip, >45°.

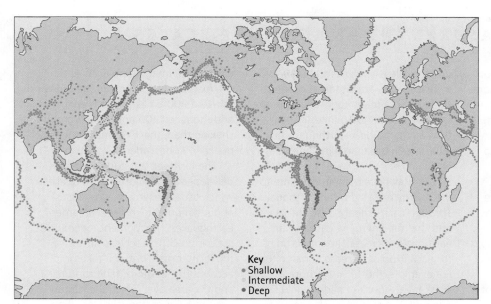

FIGURE 21.37
Most earthquakes occur in just a few narrow zones.

FYI The area of the Pacific Rim that encircles much of the Pacific Ocean is often referred to as the "Ring of Fire." About 80% of the world's big earthquakes occur in the Ring of Fire, and because subduction is related to volcanic activity, about 75% of the world's volcanoes are located here as well.

FYI The 2011 Japan earthquake literally changed Earth's surface. GPS data shows that land near the epicenter is 13 feet (4 m) wider than before, and some areas of northeastern Japan are closer to North America by about 2.4 m (7.9 ft). Earth's axis shifted by 25 cm (9.8 in) increasing Earth's rotation, thereby shortening daylength by 1.8 microseconds! Tectonics in action!

central Mississippi Valley. In the winter of 1811–1812, a series of strong earthquakes along this zone permanently changed the course of the Mississippi River! With an estimated magnitude of 8.0 on the Richter scale, the earthquakes were so strong they caused church bells to ring in Boston, Massachusetts, 1600 km (1000 mi) away! Fortunately, because the region was sparsely settled at the time, the loss of human life and property was minimal.

Earthquakes vary in their size and destructiveness. At divergent plate boundaries, earthquakes are generally mild and shallow. At transform boundaries, earthquakes range from mild to moderate. And at convergent boundaries, they tend to be moderate to very strong. Out of all the plate boundaries, the strongest earthquakes occur along convergent plate boundaries where subduction occurs.

Devastating earthquakes can occur with all three types of faults: reverse, normal, or strike-slip. The 1906 San Francisco earthquake, approximately magnitude 7.8 on the Richter scale, resulted in 700 deaths and extensive fire damage. The 1989 Loma Prieta earthquake near Santa Cruz, California, magnitude 7.1, caused 62 deaths and more than $6 billion in damage. The catastrophic 2010 Haiti earthquake, magnitude 7.0, resulted in an estimated 300,000 deaths and a million people left homeless! Strike-slip faulting was involved in all three of these earthquakes.

Although earthquakes associated with normal faults are quite common, the most catastrophic earthquakes occur along reverse faults, usually those at subduction zones. For example, the 1964 Anchorage, Alaska, earthquake (magnitude 9.2) was the largest earthquake in North America and the second largest ever recorded. The quake and its aftermath caused 131 deaths and $300 million in damage.* Some of the greatest tragedies of recent times include the August 1999 earthquake in Turkey (magnitude 7.6; 17,000 deaths), the January 2001 earthquake in India (magnitude 7.9; 20,000 deaths), the December 2004 earthquake in Sumatra (magnitude 9.2; 280,000 deaths in 14 countries*), the February 2010 earthquake in Chile (magnitude 8.8, 486 deaths), and the March 2011 earthquake and tsunami in Japan (magnitude 9.0; 14,500 deaths and approx. 11,000 people missing)! Look at Figure 21.37 and you will see how these devastating earthquakes align with plate boundaries.

---

* The death toll was largely due to great seismic sea waves, or tsunami. A tsunami is generated from the displacement of water as a result of an earthquake, a submarine landslide, or an underwater volcanic eruption.

EARTHQUAKE MEASUREMENT—MAGNITUDE SCALES

Every year, hundreds of thousands of earthquakes occur. Although most are small and go undetected, the danger of large earthquakes certainly exists. Earthquake-prone regions experience large earthquakes about every 50 to 100 years.

The energy of earthquakes is described by a *magnitude scale*. The Richter magnitude scale, which measures the energy released by an earthquake in terms of ground shaking, was the first widely used magnitude scale. Richter magnitude is based on the maximum seismic-wave amplitude of an earthquake, as recorded by a seismograph. The scale is logarithmic. This means that each 1-point increase

on the scale is equivalent to a 10-fold increase in the amplitude of ground shaking. So a magnitude-6 earthquake shakes the ground 10 times more than a magnitude-5 earthquake but 100 times more than a magnitude-4 earthquake.

But how does magnitude relate to the energy released by an earthquake? Through careful analysis, scientists find that the energy released by an earthquake increases about 30 times for each 1-point increase in magnitude. For example, the 1964 Anchorage, Alaska, earthquake (magnitude 9.2), released 30 times as much energy as, and produced 10 times more ground shaking than, the 1923 Tokyo

earthquake (magnitude 8.2). And the magnitude-9.2 Anchorage quake released 900 times as much energy as, and produced 100 times more ground shaking than, the 1995 magnitude-7.2 quake in Kobe, Japan.

The Richter magnitude scale was developed for California earthquakes, which usually have a shallow focus and are moderate in size. For larger earthquakes, the moment magnitude scale is used. This scale is similar to the Richter scale but also takes into account the length of fault rupture and the area over which the rupture occurred. For earthquakes less than magnitude-6.5, both scales yield equivalent magnitudes.

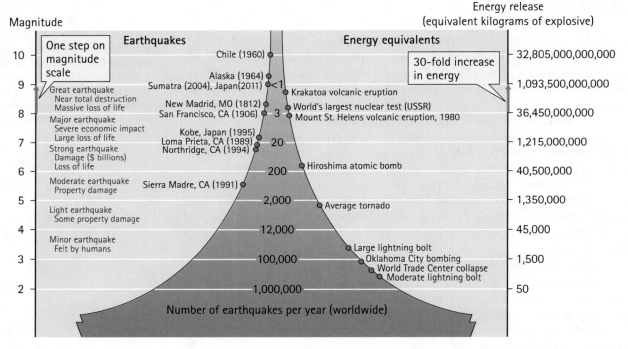

This diagram shows earthquake magnitude compared to the energy released by an earthquake and the energy released by natural and human-caused events. It also shows how many earthquakes of a given magnitude occur each year. For example, the 1906 San Francisco earthquake had a magnitude of 7.8, and about three earthquakes of this size occur on a yearly basis.

Understanding earthquakes is obviously of major importance to society. Unfortunately, earthquakes and fault movement occur with little or no warning and thus are very difficult to predict. The best that can be done is to calculate the probability that earthquakes will occur within a given timeframe. Such probability estimates are based on the past occurrences of earthquakes in

TABLE 21.1	NOTABLE EARTHQUAKES			
Year	Location	Magnitude	Estimated Deaths	Comments
1811	New Madrid, MO	8.0	few	
1906	San Francisco, CA	7.8	700	Fire caused extensive damage
1923	Tokyo, Japan	8.2	150,000	Fire caused extensive destruction
1960	Southern Chile	9.5	5,700	The largest earthquake ever recorded
1964	Anchorage, AK	9.2	131	More than $300 million in damage
1970	Peru	7.9	66,000	Great rockslide
1971	San Fernando, CA	6.5	65	More than $5 billion in damage
1975	Liaoning, China	7.5	few	First major earthquake to be predicted
1976	Tangshan, China	7.6	500,000	
1985	Mexico City, Mexico	8.0	9,000	
1989	Loma Prieta, CA	7.1	62	More than $6 billion in damage
1994	Northridge, CA	6.7	57	More than $25 billion in damage
1995	Kobe, Japan	7.2	5,500	Between $95 and $147 billion in damage
1999	Izmit, Turkey	7.6	17,000	
2001	India	7.9	20,000	
2001	El Salvador	7.7	1,000	
2003	Bam, Iran	6.6	>30,000	
2004	Sumatra, Indonesia	9.2	>280,000	Damage in 14 countries, mostly due to tsunami
2007	Sumatra, Indonesia	6.4	70	
2008	Sichuan, China	7.9	50,000	
2010	Port-Au-Prince, Haiti	7.0	>300,000	Complete devastation, $13.2 billion in damage
2010	Southern Chile	8.8	486	Shifted the entire South American Plate westward
2011	Japan	9.0	>14,500	Major destruction due to quake and tsunami. Radiation leak from nuclear power plant.

the target area. Table 21.1 lists some of the world's most notable earthquakes according to their impact on society.

21.7 The Theory That Explains the Geosphere

EXPLAIN THIS Why are earthquakes and volcanoes found near plate boundaries?

LEARNING OBJECTIVE
Identify the different types of plate boundaries on a globe.

Before the theory of plate tectonics was proposed, the underlying causes of processes such as mountain building, folding, and faulting were poorly understood. Plate tectonics explains the where and why of many geologic phenomena. Indeed, it can be thought of as a unifying theory because it links causes and effects.

Why are the Appalachian Mountains located where they are? What about the Sierra Nevada? The Rocky Mountains? The Alps? The plate tectonics model gives an answer: all large mountain-building events take place near convergent plate boundaries.

We can also relate the formation of the three different rock types to the plate tectonics theory. Although we are simplifying the following discussion of rock formation for the sake of clarity, all rocks are tied to plate interaction in one manner or another. First, consider the most common igneous volcanic rock. The formation of huge volumes of basalt is linked to divergent plate boundaries, where mantle rock partially melts to form new basaltic oceanic crust.

LINK TO PHYSICS

Wave Motion—Tsunami

A tsunami is a wave, or series of waves, generated in a large body of water by any type of powerful disturbance that vertically displaces the water column. Earthquakes, landslides, explosions, and even meteorites can generate a tsunami. When an earthquake causes the disturbance, a tsunami is some-times referred to as a seismic sea wave.

Most tsunami* arise when an earthquake occurs on an underwater reverse fault. Rapid movement of the seafloor—generally upward—quickly thrusts the water above the uplifted

area upward. The huge displaced mass of water then drops back down to sea level and a large wave is generated—a tsunami (Figure 21.38).

For all water waves, the depth over which wave energy travels is equal to one-half of the wave's wavelength. And this makes the behavior of tsunami different from that of most other ocean waves. For example, wind-driven ocean waves usually have wavelengths shorter than 150 m. So, most ocean waves affect only the uppermost part of the water column. Tsunami, however, have very long wavelengths—100 km or more.

One-half of a tsunami's wavelength—about 50 km—is far deeper than the average ocean depth—about 4 km. So, the energy within a tsunami is spread over the entire water column, not just the uppermost layer. This is an enormous amount of energy! In the open ocean, tsunami are barely visible because of their huge wavelength. The ocean rises perhaps 1 m higher than usual, but that 1-m rise is spread over 100 km! And tsunami are fast—they travel through the open ocean at 800 km/h!

The true power of a tsunami becomes apparent as the wave

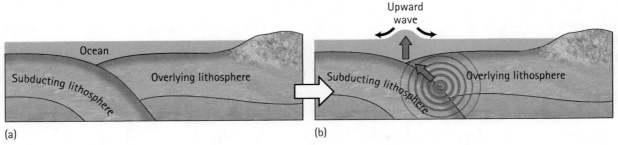

(a)   (b)

FIGURE 21.38
Most tsunami are generated by earthquakes in subduction zones. The edge of the overriding plate is pulled down-ward by the downgoing slab (a). The buildup of elastic energy is released suddenly, causing the earthquake. When the edge of the bent plate snaps back into place, the entire water column is flung upward (b) and then drops back down, causing the tsunami.

Magma is also generated at subduction zones. Plates pulled downward be-come heated and release fluids. These fluids interact with the overlying mantle rock and lithosphere, causing the material to partially melt. Some of this magma crystallizes. The processes of partial melting and crystallization generate a range of magma compositions with varying silica content—basaltic, andesitic, and granitic magmas. The formation of andesite and granite is linked to sub-duction processes at convergent boundaries. And, wherever plates collide there is intense heat and pressure. What kind of rock is associated with heat and pres-sure? Metamorphic rock. Notably, continental–continental convergence zones are where extensive metamorphism occurs.

What about sedimentary rocks? Many sedimentary rocks are also linked to plate motions. As mountains grow by virtue of plate collisions, they also begin to weather and erode. The sediments produced are transported downslope, where they accumulate, layer upon layer, and eventually become sedimentary rock.

(a)   (b)

FIGURE 21.39

The 2011 Japan tsunami pummeled the coast of Japan killing thousands of people and swallowing entire towns. The wave energy traveled throughout the Pacific Ocean basin affecting coastal areas as far away as Hawaii, Washington, and California.

approaches shore. Wave speed depends on water depth, so as the tsunami enters shallow water, the wave slows down. The energy that was spread over 4 km is now squeezed into the space of a few tens of meters, and then even less. Where does this energy go? Up! Wave height can grow from a mere meter to up to 30 m!

And tsunami don't "break" like ordinary ocean waves. As viewed from shore, a tsunami can appear as a very rapidly rising tide—a fast-moving wall of water. But if a wave trough gets to shore first instead of a crest, it appears as a rapidly receding, very low tide. Unfortunately, this can cause people to

investigate the odd sudden appearance of exposed seafloor, complete with flopping fish that have been temporarily stranded. The wave crest quickly follows, drowning unsuspecting onlookers. In many ways, a tsunami on land is like an enormous sledgehammer delivering an extreme blow.

The 2004 Sumatra tsunami generated by subduction of the India Plate beneath the Burma Plate devastated coastal areas throughout the Indian Ocean basin, killing approximately 280,000 people in 14 countries. The 2011 Japan tsunami initiated by subduction of the Pacific Plate beneath the North American Plate was also

catastrophic—16,000 lives were lost, and thousands of people were left homeless with towns literally obliterated. The huge waves that toppled the seawalls designed to protect the Fukushima Nuclear Power Plants helped trigger Japan's nuclear crisis—a level 7 nuclear accident comparable to that of Chernobyl. The energy of the Sumatra and Japan earthquakes (magnitudes 9.2 and 9.0, respectively) is comparable to the energy released by approximately 9,400 gigatons of TNT (approx. 600 times the energy of the bomb at Hiroshima).

* *Tsunami* is a Japanese word meaning "harbor wave." The plural and singular forms of the word are the same.

Lastly, virtually all earthquake and volcanic activity can be tied directly to plate tectonics. These energetic responses to plate interactions are almost always found where plates interact; earthquakes are found at all types of plate boundaries, and volcanoes are concentrated where plates either collide or pull apart.

So, the tectonic interaction between lithospheric plates, which occurs mostly at their boundaries, explains the origin of mountain chains, the formation and destruction of the ocean floors, the three types of rocks found on Earth, and the global distribution of earthquakes and volcanoes. The internal motions that change Earth's surface do so in a cycle. The study of geology uses observable processes that occur today to understand what may have occurred in the past. This concept is commonly stated as: The present is the key to the past. So, what has happened in the past provides clues to what may happen in the future. The Earth is indeed a dynamic planet.

For instructor-assigned homework, go to www.masteringphysics.com

SUMMARY OF TERMS (KNOWLEDGE)

Anticline An up-fold in rock with relatively old rocks at the fold core; rock age decreases with horizontal distance from the fold core.

Asthenosphere A subdivision of the upper mantle situated below the lithosphere, a zone of plastic, easily deformed rock.

Body wave A type of seismic wave that travels through Earth's interior.

Continental drift A hypothesis by Alfred Wegener that the world's continents are mobile and have moved to their present positions as the ancient supercontinent Pangaea broke apart.

Convergent plate boundary A plate boundary where tectonic plates move toward one another; an area of compressive stress where lithosphere is recycled into the mantle or shortened by folding and faulting.

Core The central layer of Earth's interior, divided into an outer liquid core and an inner solid core.

Crust Earth's outermost layer.

Divergent plate boundary A plate boundary where lithospheric plates move away from one another—a spreading center; an area of tensional stress where new lithospheric crust is formed.

Earthquake The shaking or trembling of the ground that happens when rock under Earth's surface moves or breaks.

Fault A fracture along which movement of rock on one side relative to rock on the other side has occurred.

Isostasy The process by which oceanic crust and continental crust come into vertical equilibrium, with respect to the mantle; the dense oceanic crust sits lower in the mantle than the less-dense continental crust.

Lithosphere The entire crust plus the rigid portion of the mantle that is above the asthenosphere.

Mantle The middle layer in Earth's interior, between the crust and the core.

Mohorovičić discontinuity (Moho) The crust–mantle boundary; marks one of the depths where the speed of P-waves traveling through Earth increases.

Paleomagnetism The natural, ancient magnetization in a rock that can be used to determine the polarity of Earth's magnetic field and the rock's location of formation.

Plate tectonics The theory that Earth's lithosphere is broken into pieces (plates) that move over the asthenosphere; boundaries between plates are where most earthquakes and volcanoes occur and where lithosphere is created and recycled.

Primary wave (P-wave) A longitudinal body wave that compresses and expands the material through which it moves; it travels through solids, liquids, and gases and is the fastest seismic wave.

Rift (rift valley) A long, narrow gap that forms as a result of two plates diverging.

Seafloor spreading The moving apart of two oceanic plates at a rift in the seafloor.

Secondary wave (S-wave) A transverse body wave that vibrates the material through which it moves side to side or up and down; it cannot travel through liquids and so does not travel through Earth's outer core.

Subduction The process in which one tectonic plate bends and descends beneath another plate at a convergent boundary.

Surface wave A type of seismic wave that travels along Earth's surface.

Syncline A down-fold in rock with relatively young rocks at the fold core; rock age increases with horizontal distance from the fold core.

Transform plate boundary A plate boundary where two plates are sliding horizontally past each other, without appreciable vertical movement.

READING CHECK QUESTIONS (COMPREHENSION)

21.1 Seismic Waves

1. How do P-waves travel through Earth's interior? How do S-waves travel through Earth's interior?
2. Can S-waves travel through liquids?.
3. Name the two types of surface waves, and describe the motion of each.

21.2 Earth's Internal Layers

4. What was Andrija Mohorovičić's major contribution to Earth science?
5. How did seismic waves contribute to the discovery of Earth's core?
6. What is the evidence that Earth's inner core is solid?

7. What is the evidence that Earth's outer core is liquid?
8. In what ways are the asthenosphere and the lithosphere different from each other?
9. How does continental crust differ from oceanic crust?
10. Why does continental crust stand higher on the mantle than oceanic crust?

21.3 Continental Drift—An Idea Before Its Time

11. What key evidence did Alfred Wegener use to support his hypothesis of continental drift?
12. Wegener proposed that the world's continents had at one time all been joined together into one supercontinent. What was the name of this supercontinent?

21.4 Acceptance of Continental Drift

13. What role did paleomagnetism play in supporting the theory of continental drift?

14. Where are the deepest parts of the ocean?

15. What was the major discovery of Harry Hess?

16. How is the ocean floor similar to a gigantic, slow-moving tape recorder?

17. In what way does seafloor spreading support continental drift?

21.5 The Theory of Plate Tectonics

18. Name and describe the three types of plate boundaries.

19. The lithosphere moves because of convection currents in the mantle. What causes the convection currents?

20. What is a rift? Give an example.

21. What kind of boundary separates the South American Plate from the African Plate?

22. What are the three types of plate collisions that occur at convergent boundaries?

23. What is a transform boundary?

21.6 Continental Evidence for Plate Tectonics

24. Are folded rocks primarily the result of compressional or tensional forces?

25. Distinguish between anticlines and synclines.

26. What is the difference between reverse faults and normal faults?

27. Which kind of fault forms primarily from tension in Earth's crust? Primarily from compression?

28. What happens to rock when stress exceeds a rock's elastic limit?

29. Where are most of the world's volcanoes formed?

30. What is the source of a tsunami's huge amount of energy?

ACTIVITIES (HANDS-ON APPLICATION)

31. A Slinky is a great example for seeing how seismic waves travel through Earth's interior. Place the Slinky on the floor. With two people firmly holding the ends of the Slinky, stretch it out to about 5 ft in length (do not overstretch). On one end, pull the Slinky backward and then push forward rapidly. Stop, and observe. How do the coils move? Which type of seismic wave does this motion mimic? Once the Slinky has stopped moving, jerk the toy to the side so that it moves like a slithering snake. How do the coils move? Which type of seismic wave does this motion mimic?

32. Look for a very old window, and note the lens effect in the bottom part of the glass. Glass has both solid and liquid properties; in fact, it is often thought of as a very viscous liquid. Over many years, its downward flow due to gravity is evidenced by the increased thickness near the bottom of the pane. Tie this observation of "plastic" behavior into our discussion of plate tectonics. Which parts of Earth behave plastically? Which parts are rigid? What do these ideas have to do with plate tectonics?

33. As the ground shakes, so do buildings on top of the land. It is often said that earthquakes don't kill people, but falling buildings do. Older, unreinforced brick buildings and wood-frame buildings can collapse during an earthquake. Reinforcing these older buildings helps to strengthen their structure. We can explore the strength of buildings by using five straws and some tape.

Step 1: Make a square frame by taping four straws together. Hold the frame upright on a flat surface.

Step 2: Hold the bottom straw in place with one hand. Push the top straw to the right with your other hand. Push as far as you can without breaking the frame.

Step 3: Tape the fifth straw horizontally across the middle of the frame. Repeat Step 2.

Step 4: Untape the fifth straw and retape it diagonally across the square from corner to corner. Repeat Step 2.

Which frame is stronger: the four-straw or the five-straw frame? Why? How does the diagonal placement of the fifth straw affect the frame's strength?

THINK AND SOLVE (MATHEMATICAL APPLICATION)

34. The weight of the ocean floor bearing down on the lithosphere is increased by the weight of ocean water. Relative to the weight of the 10-km-thick basaltic ocean crust (density 3 g/cm^3), how much weight does a 3-km-deep ocean (density 1 g/cm^3) contribute? Express your answer as a percentage of the crust's weight.

35. If the mid-Atlantic Ocean is spreading at 2.5 cm per year, how many years has it taken for it to reach its present width of about 5000 km?

36. The Richter magnitude scale is logarithmic, meaning that each increase of 1 point on the scale corresponds to an increase by a factor of 10 in the amplitude of the seismic waves recorded on a seismograph. An earthquake that measures magnitude 8 on the scale has how many times more ground-shaking effect than a quake that measures magnitude 6?

37. If you know the rate of movement along a fault, the amount of offset over a period of time can be calculated. The basic relationship is

$$\text{Rate} = \text{distance/time}$$

Movement along the San Andreas Fault is about 3.5 cm/yr. If a fence were built across the fault in 2005, how far apart will the two sections of the now-broken fence be in 2025?

38. The San Andreas Fault separates the northwest-moving Pacific Plate, on which Los Angeles sits, from the North American Plate, on which San Francisco sits. If the plates slide past each other at a rate of 3.5 cm/yr, how long will it take the two cities to form one large city? (The present distance between Los Angeles and San Francisco is 600 km.)

THINK AND RANK (ANALYSIS)

39. List the parts of Earth's crust in order of generally increasing density: (a) oceanic crust, (b) continental crust, (c) asthenosphere, (d) lithosphere.

40. Going from slowest to fastest, rank the following seismic waves: (a) Rayleigh waves, (b) P-waves, (c) S-waves, (d) Love waves.

41. Earth's crust and its core differ greatly in density. After a quick review of Sections 20.1 and 21.2, rank the following elements in order of increasing density: (a) iron, (b) silicon, (c) nickel, (d) aluminum.

42. Consult Figure 21.37 and rank these countries in terms of earthquake frequency (from greatest to least): (a) Greenland, (b) North America, (c) Indonesia, (d) Australia.

43. Earthquakes at plate boundaries vary in their size and destructiveness. Rank the types of plate boundaries according to increasing earthquake destructiveness: (a) divergent boundaries, (b) transform boundaries, (c) convergent boundaries.

EXERCISES (SYNTHESIS)

44. Compare the relative speeds of primary and secondary seismic waves. Which type of material can each travel through?

45. How can seismic waves indicate whether regions inside Earth are solid or liquid?

46. How do seismic waves indicate layering of materials in Earth's interior?

47. What is the evidence that Earth's inner core is solid?

48. Speculate on why the lower part of the lithosphere is rigid and the asthenosphere is plastic, even though they are both part of the mantle.

49. Even though the inner and outer cores are both composed of predominantly iron and nickel, the inner core is solid and the outer core is liquid. Why?

50. What does the P-wave shadow tell us about Earth's composition?

51. If Earth's mantle is composed of rock, how can we say that the crust floats on the mantle?

52. Why is Earth's crust thicker beneath a mountain range?

53. Which extends farther into the mantle: the continental crust or the oceanic crust? Why?

54. How do erosion and wearing away of a mountain affect the depth to which the crust extends into the lithosphere?

55. Describe how the different paths of polar wandering helped establish that continents move over geologic time.

56. Why are most earthquakes generated near plate boundaries?

57. What is the driving force for mountain building in the Andes?

58. At what type of plate boundary were the Appalachian Mountains produced?

59. Why do mountains tend to form in long narrow ranges?

60. What kind of plate boundary separates the North American Plate from the Pacific Plate?

61. Relate the formation of metamorphic rocks to plate tectonics. Would you expect to find metamorphic rocks at all three types of plate boundaries? Why or why not?

62. Why does granite form frequently at oceanic–continental convergent boundaries but infrequently at oceanic–oceanic convergent boundaries?

63. Cite one line of evidence that suggests that subduction once occurred off the coast of California.

64. Distinguish between continental drift and plate tectonics.

65. Are the present-day ocean basins a permanent feature on our planet? Discuss why or why not.

66. Are the present-day continents a permanent feature on our planet? Discuss why or why not.

67. Why are the most ancient rocks found on the continents, not on the ocean floor?

68. Upon crystallization, certain minerals (the most important being magnetite) align themselves in the direction of the surrounding magnetic field, providing a magnetic fossil imprint. How does the seafloor's magnetic record support the theory of continental drift?

69. How are the theories of seafloor spreading and continental drift supported by paleomagnetic data?

70. What kind of boundaries are associated with centers of seafloor spreading?

71. What is meant by magnetic pole reversals? What useful information do they tell us about Earth's history?

72. How is Earth's crust like a conveyor belt?

73. Where does most of an earthquake's damage generally occur?

74. What type of fault is associated with the 1964 earthquake in Alaska?

75. The Mercalli scale measures earthquake intensity. The Richter scale measures earthquake magnitude. Which scale is the more precise measurement? Why?

76. What is a very likely cause of the existence of Earth's magnetic field?

77. Lithospheric rock is continuously created and destroyed. Where do the creation and destruction take place? Do the rates of the two processes balance each other?

78. Subduction is the process of one lithospheric plate descending beneath another. Why does the oceanic portion of the lithosphere undergo subduction while the continental portion does not?

79. How much more does the ground shake during a magnitude-6.6 earthquake than it does during a magnitude-5.6 earthquake?

80. In 1960, a large tsunami struck the Hawaiian Islands without warning, devastating the coastal town of Hilo. Since that time, a tsunami warning station has been established for the coastal areas of the Pacific. Why do you think these stations are located around the Pacific Rim?

81. How did the Himalaya Mountains originate? How did the Andes Mountains originate?

82. What is the direct source of energy responsible for earthquakes in southern California?

83. How do faults and folds support the idea that lithospheric plates move?

84. Reverse faults are created by compressional forces. Where in the United States do we find evidence of reverse faults?

85. Normal faults are created by tensional forces. Where in the United States do we find evidence of normal faults?

86. Strike-slip faults show horizontal motion. Where in the United States do we find strike-slip faulting?

87. If you found folded beds of sedimentary rock in the field, what detail would you need to know in order to tell whether the fold was an anticline or a syncline?

88. Does the fact that the mantle is beneath the crust necessarily mean that the mantle is denser than the crust? Explain.

89. Magma is generated at divergent and convergent plate boundaries. What type of magma is dominant at each boundary? Why are they different?

90. Where is Earth's longest mountain range located?

91. How old is the Atlantic Ocean thought to be? For how many years has lava been extruding at the Mid-Atlantic Ridge?

92. What type of lava erupts at divergent boundaries? What type erupts at convergent boundaries?

93. In an earthquake, does the release of energy usually happen all at once? Defend your answer.

DISCUSSION QUESTIONS (EVALUATION)

94. During an earthquake, what type of land surface is safer: rigid bedrock or sandy soil? Explain your thinking.

95. Where do most of the world's earthquakes occur? Discuss features of various earthquakes in terms of: geology—type of plate boundary, fault, and rock type; and geography—economic development, preparedness, politics, and recovery efforts.

96. As global temperatures increase, the polar ice caps melt. What influence will this have on the isostatic balance of the lithosphere?

97. The FYI about the 2010 Chilean earthquake suggests that the quake caused a shift in Earth's distribution of mass, which changed Earth's rotation and resulted in a change in the day length. If Earth's rotation increases, what is the effect on day length? If Earth's rotation decreases, what is the effect on day length? In what other way can day length be affected?

98. What clues can we use to recognize the boundaries between ancient plates no longer in existence?

99. What geologic features are explained by plate tectonics?

READINESS ASSURANCE TEST (RAT)

If you have a good handle on this chapter, if you really do, then you should be able to score at least 7 out of 10 on this RAT. If you score less than 7, you need to study further before moving on.

Choose the BEST answer to each of the following.

1. At divergent boundaries, basaltic magma is generated by the
   (a) crystallization of mantle magma.
   (b) partial melting of continental crust.
   (c) partial melting of mantle rock.
   (d) addition of water to mantle rock.

2. The hypothesis of continental drift is *not* supported by
   (a) seafloor spreading.
   (b) paleomagnetism.
   (c) isostasy.
   (d) glacial striations.

3. Which of the following statements is false?
   (a) The mantle includes part of the crust.
   (b) The lithosphere includes the entire crust.
   (c) The mantle includes part of the lithosphere.
   (d) The mantle includes the entire asthenosphere.

4. Seismic waves increase in speed when
   (a) they pass through a liquid.
   (b) rocks become denser and less rigid.
   (c) they form a wave shadow.
   (d) the elasticity of the rock increases.

5. Convection in the mantle is caused primarily by
   (a) heat moving from the core to the crust.
   (b) conduction.
   (c) gravity and temperature differences.
   (d) friction of the overlying lithosphere.

6. Earthquakes are caused by the
   (a) friction between diverging plates.
   (b) sudden release of energy that is stored elastically in deformed rocks.
   (c) expansion of Earth's crust.
   (d) combined motion of tectonic plates.

7. Seafloor spreading provided a driving force for continental drift because
   (a) the youngest seafloor is found near the continents.
   (b) seafloor spreading pushes continents apart.
   (c) mantle convection causes slippage.
   (d) subduction creates the youngest seafloor.

8. Lithosphere is created at _____ boundaries and destroyed at _____ boundaries.
   (a) convergent; divergent
   (b) divergent; convergent
   (c) divergent; transform
   (d) convergent; transform

9. Subduction occurs as a result of
   (a) gravity pulling the older and denser lithosphere downward.
   (b) horizontal plate accommodation.
   (c) upwelling of hot mantle material along the trench.
   (d) lubrication from the generation of andesitic magma.

10. Rocks buckle and fold when subjected to
    (a) tensional force.
    (b) the release of stored elastic energy.
    (c) stretching of Earth's crust.
    (d) compressional force.

Answers to RAT

1. c, 2. c, 3. a, 4. d, 5. c, 6. b, 7. b, 8. b, 9. a, 10. d.

CHAPTER 22

Shaping Earth's Surface

IMAGINE VIEWING Earth from space and surveying its entire surface. What would you see? The first thing you would notice is that most of Earth's surface—about 71%—is covered by water. The remaining 29% is taken up by continental landmasses. Erosion, transportation, and deposition of sediment shape and sculpt the surface of these landmasses.

The hydrosphere and the atmosphere both play roles in this reworking of the surface—water, wind, and ice are the agents that do most of the work. We approach our investigation of Earth's surface by first exploring how these agents flow. In turn, we then look at their effects on Earth's surface.

LEARNING OBJECTIVE
Describe how Earth's water
moves through the water cycle.

22.1 The Hydrologic Cycle

EXPLAIN THIS What are the two dominant processes in the hydrologic cycle?

Slightly more than 97% of Earth's water is in the oceans, and a little more than 2% is frozen in polar ice caps and glaciers. The remainder, less than 1%, consists of water vapor in the atmosphere, water in the ground, and water in rivers and lakes (Figure 22.1).

Water on Earth is constantly circulating, driven by the heat of the Sun and the force of gravity. As the Sun's energy evaporates ocean water, a cycle begins (Figure 22.2). Evaporation moves water molecules from Earth's surface to become part of the atmosphere. The resulting moist air may be transported great distances by wind. Some of the water molecules condense to form clouds and then precipitate as rain or snow. The total amount of water vapor in the atmosphere remains relatively constant: evaporation and precipitation are in balance.

If precipitation falls on the ocean, the cycle is complete—from ocean back to ocean. The cycle is longer and more complex when precipitation falls on land. In this case, water may drain to small streams, then to rivers, and then journey back into the ocean. Or it may percolate into the ground, or evaporate back into the atmosphere before reaching the ocean. Also, water falling on land may become part of a snow pack or glacier. Although snow or ice may lock water up for many years, such water eventually melts or evaporates and returns to the cycle. This natural circulation of water—from the oceans to the air, to the ground, to the oceans, and then back to the atmosphere—is called the **hydrologic cycle**.*

The rain or snow that falls on the continents is Earth's only natural supply of fresh water. More than three-quarters of Earth's fresh water is locked up in the polar ice caps and glaciers. Surprisingly, most of the freely flowing fresh water is not in lakes and rivers, but rather beneath Earth's surface. As rain falls and sinks into the ground, it percolates downward. Some of the percolating water fills the open pore spaces between sediment grains. This water is now called **groundwater**.

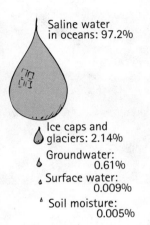

Saline water
in oceans: 97.2%

Ice caps and
glaciers: 2.14%

Groundwater:
0.61%

Surface water:
0.009%

Soil moisture:
0.005%

FIGURE 22.1
Distribution of Earth's water supply.

FYI Most of Earth's surface area is ocean, so it makes sense that evaporation and precipitation are greatest over the oceans. It also makes sense that over the continents, precipitation exceeds evaporation. Balance is maintained between the amount of water taken up into—evaporation—the atmosphere (85% from oceans and 15% from continents) and the amount taken out of—precipitation—the atmosphere (75% to the oceans and 25% to the continents).

FIGURE 22.2
The hydrologic cycle. Water evaporated at Earth's surface enters the atmosphere as water vapor, condenses into clouds, precipitates as rain or snow, and falls back to the surface, only to evaporate again and go through the cycle yet another time.

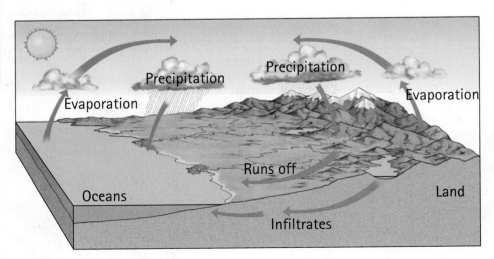

Precipitation

Precipitation

Evaporation

Evaporation

Runs off

Oceans

Land

Infiltrates

---

* This key concept is another conservation principle. Recall that in Chapter 3 we learned about conservation of momentum and energy; in Chapter 9, we learned about the conservation of electric charge; and in Chapter 13, we learned about the conservation of nucleons in nuclear reactions. So now we learn that the amount of water on Earth is conserved. A lack of it in one place means an abundance someplace else, which in most instances is the ocean.

TABLE 22.1	**WATER RESOURCE RESIDENCE TIMES**
Location	Average Residence Time
Atmosphere	1–2 weeks
Ocean	
Shallow depths	100–150 years
Deep depths	30,000–40,000 years
Continents	
Rivers	2–3 weeks
Lakes	10–100 years
Shallow groundwater	up to 100s of years
Deep groundwater	up to 1000s of years
Glaciers	10,000–20,000 years

FYI The time for a certain parcel of water to complete the hydrologic cycle is its *residence time*—the average length of time a water molecule spends in a particular region. Water in polar ice and glaciers has a long residence time. For all practical purposes, the thousands-of-years residence time of deep groundwater means that if it is withdrawn, it will not be replenished. Water is truly a precious natural resource.

CHECKPOINT

1. **What percentage of Earth's water supply is fresh water?**
2. **The volume of water evaporated from all of Earth's land surface is 60,000 km³/yr, but the volume precipitated over the land surface is 96,000 km³/yr. The volume that precipitates each year is 36,000 km³ more than the volume that evaporates—so why isn't all the land flooded?**

Were these your answers?

1. Less than 3%, as you can see by adding the freshwater values in Figure 22.1: 2.14% (glaciers) + 0.61% (groundwater) + 0.009% (surface water) + 0.005% (soil moisture) = 2.764%.
2. The excess water works its way back to the oceans. Excess water to the oceans does not cause sea level to rise because evaporation (85%) exceeds precipitation (75%) by 10%. Balance is maintained between the amount of water evaporated and precipitated over the oceans (85% − 75% = 10%) and the amount precipitated and evaporated over land (25% − 15% = 10%).

Mastering**PHYSICS**

TUTORIAL: The Hydrologic Cycle Activity
VIDEO: The Hydrologic Cycle

22.2 Groundwater

EXPLAIN THIS Why does a water table mimic the surface topography?

LEARNING OBJECTIVE
Describe how water moves underground.

The liquid water in lakes, ponds, rivers, streams, springs, and puddles is the only fresh water that meets our eye, but all these water sources together hold only about 1.5% of Earth's non-ice fresh water. The other 98.5% resides in porous regions beneath Earth's surface.

Have you ever noticed how during a rainstorm sandy ground soaks up rain like a sponge? The water literally disappears into the ground. The type of surface material influences the ease with which water goes into the ground. Some soils, such as sand, soak up water easily. Other soils, such as clay, do not. Rocky surfaces with little or no soil are the poorest absorbers of water, with water penetration occurring mostly through cracks in the rock. Rain that does not soak into the ground becomes *runoff*, which then finds its way to bodies of water such as lakes and rivers or evaporates.

FIGURE 22.3
The unsaturated zone is above the saturated zone. Water in the unsaturated zone does not completely fill the open pore spaces. This is soil moisture. Water in the saturated zone completely fills all open pore spaces. This is groundwater.

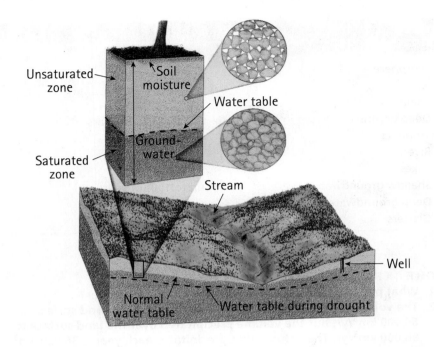

FYI As you will find out, water, wind, and ice are important agents that sculpt Earth's surface. Gravity, however, also plays a major role in altering the landscape. *Mass wasting* refers to the downslope movement of rock and soil under the direct influence of gravity. The combined effects of mass wasting and running water produce stream valleys—the running water cuts downward and mass wasting widens the valley. A prime example is the stream valley we call the Grand Canyon! Water helps set the stage for mass wasting to occur, but gravity is *the* controlling force.

Water beneath the ground exists as groundwater and *soil moisture*. The region where water has completely filled all open pore spaces is called the *saturated zone* (Figure 22.3). Above the saturated zone is the *unsaturated zone*, where soil moisture resides. Pore spaces in the unsaturated zone are not completely filled with water—they contain a significant amount of air. Like water in a swimming pool, the pressure in groundwater increases with depth. Just as we can pump water from a swimming pool, we can pump groundwater from the ground. However, the presence of air in pore spaces prevents us from withdrawing water from the unsaturated zone.

The amount of water that the ground can hold depends on the porosity of the soil or rock. **Porosity** is the volume of open pore space in a soil, sediment, or rock sample compared to the total volume of solids plus voids. Porosity depends on the size and shape of the soil or sediment particles and on how tightly these particles are packed. For example, a soil composed of rounded particles of similar size has a higher porosity than a soil composed of rounded particles of various sizes. This is because the smaller particles fill up the spaces between the larger particles, thereby reducing the overall porosity of the soil.

Porosity represents the maximum *amount* of underground water at a given location. But porosity does not tell us how groundwater *moves*. The **hydraulic conductivity**—a measure of permeability—tells us the degree to which geologic material can transmit water. If the pore spaces are extremely small and poorly connected (as is the case with flattened clay particles), water may barely move at all. Think of it this way: it's a lot easier to sip soda through a large straw than through one of those very small straws intended for stirring coffee. Likewise, it is difficult for water to move through the pores of clay. The hydraulic conductivity—or permeability—of clay is almost zero, even though the porosity of most clays is very high. In contrast, sand and gravel have large, open, well-connected pore spaces, and water moves freely from one pore space to the next. Thus, sand and gravel are highly porous and highly permeable (Figure 22.4).

FIGURING PHYSICAL SCIENCE

Porosity

Porosity tells us the ratio of open space to the total volume of soil, sediment, or rock sample:

$$\text{Porosity} = \frac{\text{volume of open space}}{\text{volume of open space } + \text{ volume of solids}}$$

SAMPLE PROBLEM

The volume of solids in a sediment sample is 975 cm^3 and the volume of open space is 325 cm^3. What is the porosity?

Solution:

$$\text{Porosity} = \frac{325 \text{ cm}^3}{325 \text{ cm}^3 + 975 \text{ cm}^3} = 0.25$$

So the volume of open space is only one-fourth of the total volume.

CHECKPOINT
Why is a sandy soil better for water flow than a clay soil?

Was this your answer?
Water flows easily through sandy soil because it is generally composed of rounded particles with pore spaces that are large and well connected; thus sandy soil is very permeable. We mean the same thing when we say that sandy soil has a high hydraulic conductivity. Water doesn't flow very easily through clay soils because clay is composed of flattened particles with small, poorly connected pore spaces between them; thus clay soils have a low hydraulic conductivity. Water flow is easier in soils with higher hydraulic conductivities.

(a)

(b)

FIGURE 22.4
Porosity and permeability. (a) The sediment particles in clay are small, flat, and tightly packed, with small, poorly connected pore spaces. Thus, clays have high porosity but low permeability (and low hydraulic conductivity). (b) Sediment particles in sand or gravel are relatively uniform in size and shape, with large and well-connected pore spaces. This allows water to flow freely. So sands and gravels can have both high porosity and high permeability (and high hydraulic conductivity).

The Water Table

When digging a hole in the ground, we find that wetness of the soil varies with depth. Just below the surface we encounter the unsaturated zone, where pore spaces are partially filled with water (Figure 22.3). As we descend farther, we enter the saturated zone, where pore spaces are completely filled with water. If our hole is entirely within the unsaturated zone, it does not fill with water. If we dig our hole deeper into the saturated zone, it partially fills with water. The upper boundary of the saturated zone is called the **water table** (Figure 22.3). The level of the water in our hole is the same level as the water table. In fact, the level of water in our hole *is* the water table at that location.

The depth of the water table beneath Earth's surface varies with precipitation and climate. It ranges from zero in marshes and swamps to hundreds of meters in some parts of the deserts. The water table also tends to rise and fall with the surface topography (Figure 22.5). At lakes and perennial streams (streams that flow all year), the water table is above the land surface.

Aquifers and Springs

Any water-bearing underground region through which groundwater can flow is called an *aquifer.* These reservoirs of groundwater underlie the land surface in many places and contain an enormous amount of water. More than half the land area in the United States is underlain by aquifers. One such aquifer is the

FIGURE 22.5

The water table roughly parallels the ground surface. In times of drought, the water table falls, reducing stream flow and drying up wells. The water table also falls if the rate at which water is pumped out of a well exceeds the rate at which the groundwater is replaced.

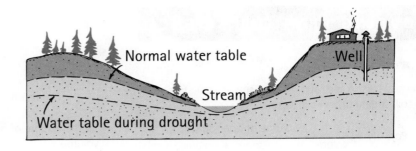

Want to see the water table? Most ponds and lakes are simply a place where the land surface dips below the water table.

Ogallala aquifer, which stretches from South Dakota to Texas and from Colorado to Arkansas!

Aquifers can be either confined or unconfined. In an *unconfined* aquifer, the soil or sediment above the water table is permeable. So water that soaks into the ground flows directly into the aquifer. Water added to an aquifer is called *recharge*. All aquifers are at least partially unconfined, as illustrated in Figure 22.6.

In a *confined* aquifer, the soil or sediment above the aquifer has a low permeability. So water that soaks into the ground cannot flow directly into the aquifer. An aquifer is considered to be confined if it is sandwiched between continuous low-permeability layers (Figure 22.6). Confining layers* occur in sedimentary deposits that have alternating sand and clay layers, or alternating sandstone and shale layers. So water recharge to a confined aquifer comes from water infiltration at the unconfined portions of the aquifer at higher elevation.

As we learned in Chapter 5, pressure in water depends on the height of water above. Water anywhere in the confined portion of an aquifer is below the level of the water table in the recharge area. So groundwater in the confined aquifer—under pressure from water above—flows out through openings at lower elevations. This is an **artesian system**. If the opening is natural and water flows out of the ground, it is an *artesian spring*. If the opening has been drilled, it is an *artesian well*. When first tapped, some artesian wells blast water tens of meters in the air!

Because rock layers are not always continuous, sometimes a low-permeability layer can stop and hold the downward-percolating water above the water table. When this happens, a *perched* water table is created. Wherever the water

FIGURE 22.6

An artesian system is formed when groundwater in an aquifer confined between layers of low-permeability rock rises to the surface through any opening that taps the aquifer. Water flows freely if the water table height in the recharge area is greater than the height of the opening (flowing artesian well and artesian spring). If the height of the opening is greater than the height of the water table in the recharge area, the water does not flow (nonflowing artesian well).

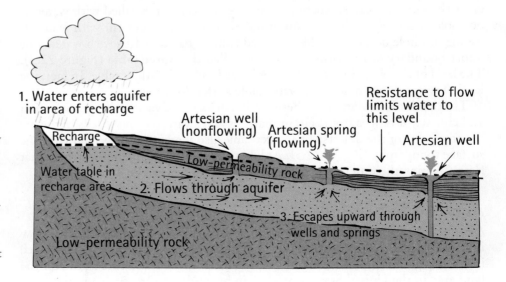

* Geologists call these layers *aquitards*.

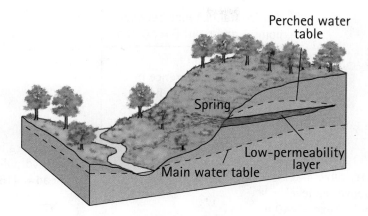

FIGURE 22.7
When the water table intersects the land surface, groundwater is released. From the perched water table, water is released via a spring; from the main water table, water is released by or into a stream.

table meets the land surface, groundwater emerges from an aquifer as a spring, stream, or lake (Figure 22.7). Springs can generally be found where the water table (or a perched water table) intersects the surface along a slope, such as on a hillside or coastal cliff. Because water tends to leak out of the ground through cracks and breaks in a rock, springs are often associated with faults. In fact, field geologists can often locate faults by looking for springs.

CHECKPOINT
1. **What is an aquifer?**
2. **What principal condition is required for an artesian system to occur?**

Were these your answers?
1. An aquifer is a body of rock or sediment through which groundwater moves easily.
2. The principal condition required for an artesian system to occur is the presence of confining layers. Confining layers allow groundwater in confined aquifers to be under higher pressure than groundwater in unconfined aquifers. This allows water to rise above the top of the confined aquifer at natural or human-made openings.

Groundwater Movement

How does water move through the ground? We have learned that a material's permeability affects groundwater flow. The higher the permeability, the easier the flow. But there is another factor that affects the flow of groundwater—gravity. All water flows downhill because of gravity. For example, in a stream at Earth's surface, water flows from areas of high elevation to areas of low elevation. Likewise for groundwater, but beneath the ground it is the elevation of the water table that drives water flow. The elevation of a water table above a particular location, usually sea level, is called the *hydraulic head*. This is the same elevation to which water in an unconfined aquifer rises in a well (Figure 22.8). Recall from Chapter 5 that liquid pressure is directly proportional to the depth of the liquid. Hence, the higher the hydraulic head above a particular location, the greater the water pressure at that location. The downward slope of a water table is called the *hydraulic gradient*. It can be expressed like any slope: "rise over run," or in this case, the difference in hydraulic head between two points divided by the horizontal distance between those points.

FIGURE 22.8
The hydraulic gradient is the difference in hydraulic head between any two locations divided by the horizontal distance between the locations. In this example, we have (440 m − 415 m)/1000 m.

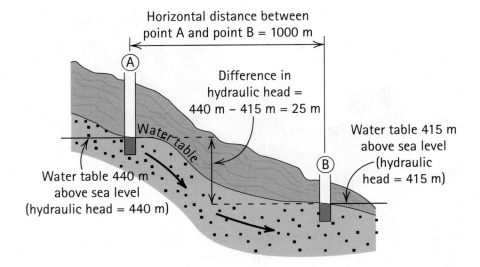

Horizontal distance between point A and point B = 1000 m

Difference in hydraulic head = 440 m − 415 m = 25 m

Water table 415 m above sea level (hydraulic head = 415 m)

Water table 440 m above sea level (hydraulic head = 440 m)

Water table

FYI Groundwater flow, heat, and electric current all move in response to pressure differences. It's nice when different concepts connect in much the same way.

Groundwater flows in response to pressure differences—it flows from high head to low head. Hence, groundwater flow through an aquifer is directly proportional to the hydraulic gradient. This can be expressed as

Groundwater flow rate ~ hydraulic gradient

Flow rate also depends on the hydraulic conductivity of the soil and the cross-sectional area of the aquifer. The cross-sectional area of an aquifer by definition is always perpendicular to the flow direction (similarly for a water pipe, where the cross-sectional area is a circle defined by the diameter of the pipe). Hydraulic conductivity values are high for gravel and lower for fine sand or silt. When the hydraulic conductivity and cross-sectional area are introduced, the proportion can be expressed as the exact equation

Groundwater flow rate = hydraulic conductivity × cross-sectional area × hydraulic gradient

This relationship was first recognized by the French engineer Henri Darcy in 1856 and is aptly termed *Darcy's law.*

Topography plays an equally important role in groundwater flow because it creates the hydraulic gradient. Groundwater moves from regions where the water table is high to regions where the water table is low, essentially flowing "downhill" underground (Figure 22.9).

So, the movement of groundwater is a result of two factors: hydraulic conductivity—the more permeable the material, the faster the flow—and gravity—water moves down the hydraulic gradient.

The speed of groundwater movement is generally very slow compared to the flow speed in rivers and streams. The more permeable the aquifer, the faster

FIGURE 22.9
Groundwater flows from a high-hydraulic-head area, such as beneath a hill, to a low-hydraulic-head area, such as beneath a stream valley. The curved arrows indicate flow, which show that the stream is fed from below.

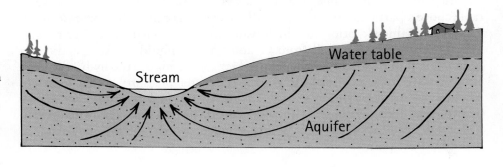

Water table

Stream

Aquifer

the flow; the greater the hydraulic gradient, the faster the flow. The speed and route of groundwater flow can be measured by introducing dye into a well and noting the time it takes the dye to travel to the next well. In most aquifers, groundwater speed is only a few centimeters per day, sufficient to keep underground reservoirs full.

22.3 The Work of Groundwater

EXPLAIN THIS   Why are carbonate rocks so susceptible to groundwater erosion?

LEARNING OBJECTIVE
Identify three ways groundwater can alter the land surface.

Flowing groundwater—no matter how slow—can cause large changes in landscapes. These changes can occur because of humans, but more often than not they occur without any human interference.

Land Subsidence

Most wells are drilled so that groundwater can be pumped from the ground. In areas where groundwater withdrawal has been extreme, the land surface is lowered—it *subsides*. The problem of land subsidence is most noticeable where the subsurface is composed of thick layers of poorly consolidated sediments rather than rock. These thick layers generally have many smaller layers of easily compressed, water-bearing clays sandwiched between a series of sandy aquifers. Recall that clay has a very low hydraulic conductivity. As water is pumped from the aquifers, water slowly leaks out of the clay layers to replenish the aquifers, which usually continue to be pumped. As the clays lose water, they compact, causing the land surface to subside.

FIGURING PHYSICAL SCIENCE

Darcy's Law

Knowing the dimensions and hydraulic conductivity of an aquifer, and the hydraulic head at the beginning and end of the aquifer, we can use Darcy's law to calculate the ground water flow rate. Again, Darcy's law is stated as follows:

Groundwater flow rate = hydraulic conductivity
    $\times$ cross-sectional area $\times$ hydraulic gradient

Consider an aquifer with a length of 10 km that is between two lakes. Measured perpendicular to the length, a vertical slice of the aquifer has a cross-sectional area of 200,000 m^2. The elevation of the water surface at Lake A is 215 m above sea level. The elevation of the water surface at Lake B is 210 m above sea level. The hydraulic conductivity of the aquifer is 10 m/day.

SAMPLE PROBLEM 1
What is the hydraulic gradient in the aquifer?

SAMPLE PROBLEM 2
What is the groundwater flow rate in the aquifer?

Solution to Problem 1:
The lake surfaces give us two locations where we know hydraulic head. First, we need to get the distances and hydraulic heads in the same units:

$$10 \text{ km} \times \frac{1000 \text{ m}}{1 \text{ km}} = 10{,}000 \text{ m}$$

Then

$$\text{Hydraulic gradient} = \frac{215 \text{ m} - 210 \text{ m}}{10{,}000 \text{ m}}$$
$$= 0.0005 \text{ (units cancel out)}$$

Solution to Problem 2:

Groundwater flow rate = 10 m/day $\times$ 200,000 m^2 $\times$ 0.0005
    = 1000 m^3/day

FIGURE 22.10
The Leaning Tower of Pisa. Construction began in about 1173 and was suspended when builders realized the foundation was inadequate. Work was later resumed, however, and the 60-m tower was completed 200 years later. Deviation from the vertical is about 4.6 m. The tower's foundation has been stabilized by groundwater withdrawal management, so now the tower should remain stable for years to come.

FIGURE 22.11
The land surface in California's San Joaquin Valley lowered more than 9 m (30 ft) over a 50-year period because of the withdrawal of groundwater and the resulting compaction of sediments.

FYI Impressive caves and caverns are found at Carlsbad Caverns in New Mexico, Blanchard Springs in Arkansas, Mammoth Cave in Kentucky, Adelsberg Cave in Austria, and Good Luck Cave in Borneo.

Probably the most well-known example of land subsidence is the Leaning Tower of Pisa in Italy, built on the unconsolidated sediments from the Arno River. Over the years, the withdrawal of groundwater has resulted in subsidence and the increased tilt of the tower (Figure 22.10). In the United States, large amounts of groundwater have been pumped for irrigation in the San Joaquin Valley of California. This process has caused the water table to drop 75 m in 20 years, lowering the land surface by as much as 9 m (Figure 22.11). Because water for irrigation is now provided by canals, the sandy aquifers are slowly recharging (refilling with water), but most of the land subsidence caused by the compaction of the clay layers cannot be reversed.

CHECKPOINT
Why is land subsidence most evident in regions where the underlying geology is a series of clay layers sandwiched between sandy aquifers?

Was this your answer?
Clay layers lose water and compact as water is pumped from the adjacent aquifers. Compaction causes the land to subside.

Carbonate Dissolution

Groundwater is stored in vast limestone deposits that underlie millions of square kilometers of Earth's surface. But the groundwater slowly "eats away" at the limestone in which it is housed. Recall from Chapter 20 that limestone is made of the mineral calcite ($CaCO_3$). Rainwater chemically reacts with carbon dioxide in the air and soil to produce carbonic acid. When it seeps downward into limestone, the slightly acidic groundwater partially dissolves the rock. As groundwater steadily dissolves the limestone, it creates unusual erosional features, such as sinkholes and caverns. It is in limestone that we find the only true underground rivers—in other rocks and soils, underground water is found only in pore spaces, not in large, open channels.

Caverns and Caves The dissolving action of underground water has carved out magnificent caverns and caves (a cavern is simply a large cave). Groundwater flow in limestone aquifers occurs mostly through fractures in the rock, rather than through pores. Rainwater (enriched in carbonic acid) soaking into the limestone flows downward through the fractures toward the water table, dissolving rock as it goes. As groundwater flows toward its outlet, say a stream, the slightly acidic water dissolves the surrounding limestone, expanding the fractures and forming underground channels and caves (Figure 22.12a). The stream water is also acidic, and it dissolves and deepens the stream channel while groundwater continues to expand the caves. As the water level in the stream and the level of the water table drop, water drains from the first set of channels and caves. Groundwater continues to dissolve the limestone, forming a new, lower level of channels and caves (Figures 22.12b and 22.12c).

Water dripping from the cave ceiling, now rich in dissolved calcium carbonate, creates icicle-shaped stalactites as water evaporates and the calcium carbonate

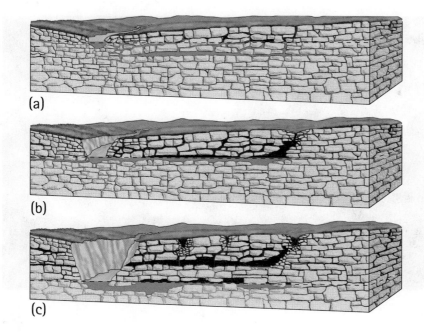

(a)

(b)

(c)

FIGURE 22.12
The formation of a cave begins with a layer of carbonate rock, mildly acidic groundwater, and an enormous span of time. (a) Groundwater makes its way toward a stream. (b) As the stream valley deepens because of erosion, the water table is lowered. The carbonate rock is eaten away as acidified water erodes and enlarges the existing fractures into small caves. (c) Further deepening of the stream valley causes the water table to drop even lower; water in the cave seeps downward, leaving an empty cave above a lowered groundwater level.

FIGURE 22.13
Cave dripstone formations.

FIGURE 22.14
Karst topography covered by vegetation makes up the rolling hills in south central Kentucky.

precipitates. Some of the water solution drips off the end of the stalactites to build corresponding cone-shaped stalagmites on the floor (Figure 22.13).

Sinkholes   Sinkholes are funnel-shaped cavities in the ground that are open to the sky. They are formed in much the same way as caves. Groundwater dissolves limestone and eventually the surface collapses in on itself. Some sinkholes are caves whose roofs have collapsed. Some sinkholes are formed by drought conditions or excessive groundwater pumping.

Karst Regions   When sinkholes, caves, and caverns define the land surface, the terrain is called *karst topography*, named after the Karst region of Yugoslavia, where weathering and erosion of limestone characterize the landscape. The pattern of streams in this type of landscape is very irregular; streams and rivers disappear into the ground and reappear as springs. Some karst areas appear as soft, rolling hills with large depressions that dot the landscape; the depressions are old sinkholes now covered with vegetation (Figure 22.14). In general, karst areas have sharp, rugged surfaces and thin to nonexistent soils as a result of high runoff and dissolution of surface material.

22.4 Surface Water and Drainage Systems

EXPLAIN THIS Why does stream discharge generally increase downstream?

Streams—by which we mean all flowing surface water, from the Mississippi River to the shallowest woodland creek—are dynamic systems that affect both the surface of the land and the people who live on that land. Streams have many benefits to offer: they provide energy, irrigation, and a means of transportation.

Streams also carve out and alter the landscape. The Grand Canyon is testimony to the mighty erosive power of the Colorado River. For millions of years, the river has been carving out the canyon walls, cutting deeper and deeper into the rock as it makes its way to the ocean. Surface water plays another important yet contrasting role as it shapes the landscape—it deposits sediments. In this

LEARNING OBJECTIVE
Identify the variables of stream flow and their effect on stream speed.

FIGURE 22.15
The karst landscape of China has been an inspiration to classical Chinese brush artists for centuries.

FIGURE 22.15
The karst landscape of China has been an inspiration to classical Chinese brush artists for centuries.

FYI Karst regions can be found throughout the world: in the Mediterranean basin; in sections of the Alps and the Pyrenees; in southern China; and in the United States in Kentucky, Missouri, Florida, and Tennessee.

way, surface water is both a destroyer and a creator of sediments and sedimentary rocks.

Stream Flow Geometry

Streams come in a variety of forms—straight or curved, fast or slow. At their headwaters (the stream origin), stream channels are narrow and water flows quickly through deeply incised, V-shaped mountain valleys. Farther downstream, channels widen so that water flows into and along broad, low valleys.

Three variables influence the speed of water in a stream—*stream gradient, stream discharge,* and *channel geometry.* The **gradient** is the vertical drop in the elevation of the stream channel divided by the horizontal distance for that drop. If we look at a long profile of a stream (Figure 22.16), we see that the gradient is steep near the stream's headwaters and gentler, almost horizontal, near its mouth. Because of gravity, stream speed tends to be greater where the stream gradient is steep. Downstream, discharge and channel geometry also influence stream speed.

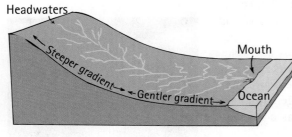

FIGURE 22.16
The long profile of a stream. At a stream's headwaters, the gradient is steep, the channels are narrow and shallow, and the stream flow is rapid. As the stream progresses downslope, the gradient decreases, the channel widens, and discharge increases.

Discharge is the volume of water that passes a given location in a channel in a certain amount of time. It is directly proportional to the cross-sectional area of the channel—the width times the depth—and to the *average* stream speed:

$$\text{Discharge} = \text{cross-sectional area} \times \text{average stream speed}$$

Or put another way,

$$\text{Average stream speed} = \frac{\text{discharge}}{\text{cross-sectional area}}$$

Channel geometry—the shape of the channel—greatly influences stream speed. Consider two streams that have the same cross-sectional area but different channel shapes. Water flowing in a stream touches the channel bottom and sides. Friction between the water and the channel slows the stream speed. So, the shape of a channel determines the amount of water in contact with the channel. The greater the contact area, the greater the friction (Figure 22.17).

FYI Squeeze the end of a garden hose and you'll see that water speeds up when the passage of flow becomes narrower. Likewise for the flow of streams.

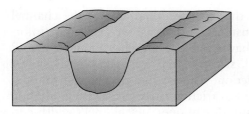
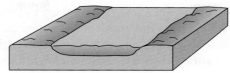

(a) Rounded, deep channel   (b) Wide, shallow channel

FIGURE 22.17
(a) In a rounded, deep channel, the speed of water flow is relatively high because less water is in contact with the channel (there is less friction). (b) Wide, shallow channels tend to have slower flows because more water is in contact with the channel (there is more friction).

Mastering**PHYSICS**

TUTORIAL: Shaping Planetary Surfaces

If the stream channel is rounded and deep, as opposed to flat-bottomed and relatively shallow, the stream speed will be faster because there is less water in contact with the channel.

Stream speed also varies within the channel. Flow speed is slower along the streambed (due to friction), and flow speed is faster near the water's surface. In a large stream flowing in a straight channel, the maximum flow speed is found midchannel (Figure 22.18b). In a stream running through a bending, looping channel, the maximum flow speed is found toward the *outside* of each bend (Figures 22.18a and 2.18c).

It is understandable to think that in the headwaters of a stream, where the gradient is steep, stream speed is also high. In fact, these upland sections of a stream are often called "rapids." But at the headwaters, water moves erratically through narrow, boulder-strewn channels. With greater channel contact, there is more friction; and with erratic flow movement, water moves in all directions. So even though water at the headwaters may be furiously churning, the average downstream speed of the water may not be as fast as it appears—at least when compared to the average stream speed farther downslope.

As a stream progresses downslope, the gradient gradually decreases, the channel typically widens, and, because tributary streams feed into it, discharge increases. With these changes, what happens to stream speed? Let's look back at the equations for stream speed and discharge—a change in one variable changes the other factors. For example, as discharge increases, the width or depth of the channel must increase, the stream speed must increase or a combination of these two factors must increase. If discharge increases but channel dimensions remain the same, stream speed will, of course, increase.

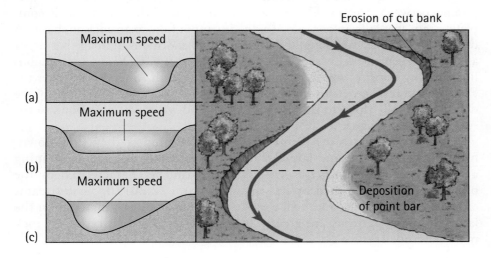

FIGURE 22.18
In a stream that bends (a and c), maximum flow speed is toward the outside of each bend and slightly below the surface. In a straight-channel stream (b), maximum speed is mid-channel and near the water's surface. Erosion of the stream channel occurs where stream speed is greatest (cut bank); deposition occurs where stream flow slows (point bar).

FYI Channel dimension, gradient, discharge, sediment load, and velocity all work together as they influence stream flow. As such, a river is a system of interdependence in which any change in one variable creates change in the entire system.

But what if discharge doubles and the cross-sectional area of the channel doubles? Will stream speed remain constant? In this case we need more information. The cross-sectional area may have doubled, but what about its shape? In a wider and more shallow channel, stream speed may get slightly slower due to greater channel contact and friction; whereas in a deeper and rounder channel, with less channel contact and less friction, stream speed may slightly increase. As a rule, as a stream progresses downslope, the channel's width and depth increase, discharge increases, and stream speed increases. And, as we will see in Section 22.5, sediment load also increases. So the average downstream speed of water in a wide and gently flowing river may be much faster than in the "rapids" where it began.

CHECKPOINT

1. Consider a stream in which discharge doubles downslope, but the channel stays the same size and shape. Now look at the equation for stream speed. What happens to stream speed?
2. Now consider a stream in which discharge doubles and the cross-sectional area of the channel also doubles. Assume that no water enters from side tributaries and friction is insignificant. What happens to stream speed?
3. If the gradient of a stream decreases and the cross-sectional area stays the same, does the discharge stay the same? Assume that no water enters from side tributaries and friction is insignificant.

Were these your answers?
1. Stream speed doubles.
2. Cross-sectional area and discharge of the stream channel increase by the same percentage, so stream speed does not increase.
3. No, because the decreased gradient slows the stream speed. The discharge decreases.

Drainage Basins and Networks

A stream is one small segment of a much larger system called a *drainage basin*. A drainage basin is defined as the total area that contributes water to a given stream. A drainage basin can cover a vast area or be as small as 1 km^2. Drainage basins are separated from one another by *divides*, lines tracing out the highest ground between streams. Under most circumstances, the separation is complete—rain that falls on one side of a divide cannot flow to an adjacent basin. A divide can be either very long, if it separates two enormous drainage basins, or a mere ridge separating two small gullies. The *Continental Divide*, a continuous line running north to south down the length of North America, separates the Pacific basin on the west from the Atlantic basin on the east. Water west of the divide eventually flows to the Pacific Ocean, and water east of it flows to the Atlantic Ocean (Figure 22.19).

As mentioned, streams merge with other streams as they flow downhill, becoming larger and larger. The entire assembly of streams draining a region is called a *drainage network*, which can be characterized by the branching pattern formed by its streams (Figure 22.20). Because streams erode the land surface and hence erode the rocks and rock material on the land, drainage patterns are greatly influenced by the rock type and rock material eroded.

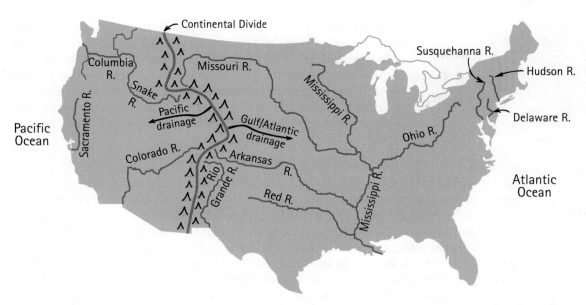

FIGURE 22.19
The Continental Divide in North America separates the Pacific basin on the west from the Atlantic basin on the east.

CHECKPOINT
Distinguish between a drainage basin and a drainage network.

Was this your answer?
A drainage basin is the total area that contributes water to a stream. It includes all the streams as well. A drainage network involves only the streams that drain water from the basins. So a drainage network is part of a drainage basin.

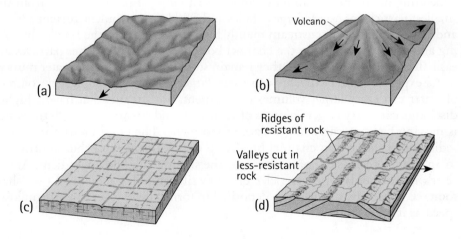

(a) (b) Volcano

(c) (d) Ridges of resistant rock / Valleys cut in less-resistant rock

FIGURE 22.20
Different drainage patterns develop according to surface material and surface structure: (a) dendritic, (b) radial, (c) rectangular, (d) trellis.

22.5 The Work of Surface Water

LEARNING OBJECTIVE
Describe how a river changes the surrounding land.

EXPLAIN THIS Where is a stream's erosive power the greatest: at the headwaters or downstream?

Landscape evolution—progressive changes to Earth's surface—is driven by surface water flow. The manner of flow has a large effect on how water alters the landscape. The flow characteristics of moving water are of two types:

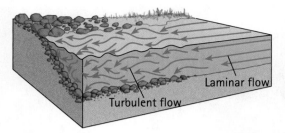

FIGURE 22.21
Laminar flow is slow and steady, with no mixing of sediment in the channel. Turbulent flow is fast and jumbled, stirring up everything in the flow.

turbulent and laminar (Figure 22.21). When water moves erratically downstream, stirring everything with which it comes in contact, the flow is **turbulent**. When water flows steadily downstream with no mixing of sediment, the flow is **laminar**. In general, slow, shallow flows tend to be laminar and faster moving flows tend to be turbulent. Whether a flow is laminar or turbulent depends on the nature and geometry of the stream channel and the speed of the flow.

Erosion and Transport of Sediment

We learned in Chapter 20 that weathering and erosion create and move sediment. Erosion by water is the most common way clastic sediments are carried away from the places in which they formed. Surface water erodes sediment and rocks, transporting them downstream and eventually depositing them in another place. In this way, surface water reshapes our landscape.

Flowing water erodes stream channels in several different ways. First, stream water contains many dissolved substances that *chemically weather* and erode the rocks they encounter. Another powerful mechanism for erosion is *hydraulic action*—the sheer force of running water. Swiftly flowing streams and streams at flood stage have strong erosive power as they break up and loosen great quantities of sediment and rock. The most powerful type of erosion, however, is *abrasion*. Abrasion occurs when sediments and particles actually scour a channel, much like sandpaper scraping on wood. When powered by turbulently spiraling water, rock particles rotate like drill bits and carve out deep potholes (Figure 22.22). The faster the current, the greater the turbulence and the greater the erosion.

Erosion is only the beginning of the story of how surface water alters Earth's surface. Streams carry more than just water—they transport great amounts of sediment from one location to another. In general, laminar flows can lift and carry only the very smallest and lightest particles. A turbulent flow, however, depending on its speed, can move and carry a range of particle sizes—from the smallest particles of clay to large pebbles and cobbles. A turbulent current gathers and moves particles downstream mainly by lifting them into the flow or by rolling and sliding them along the channel bottom. The smaller, finer particles are easily lifted into the flow, and they remain suspended to make the water murky.

As expected, faster currents can carry larger particles. Also, larger volumes of water can carry larger volumes of sediment. So, streams that have a higher discharge can carry larger volumes of sediment, and streams in which the water is moving fast can carry larger sizes of sediment. The continuous abrasion of sediment in the stream channel breaks up the sediments and thus contributes to an overall decrease in particle size as the stream flows downgradient. At the river's mouth, only finer particles of sand, silt, and clay remain. As we shall soon see, these tiny particles are deposited to form a delta when the stream loses speed as it enters the sea.

FIGURE 22.22
The pothole in this metamorphic gneiss was carved by the abrasive action of rotating rock particles carried by a turbulent water current.

> ### CHECKPOINT
> **Which is more effective in transporting sediment: laminar or turbulent flow? Defend your answer.**
>
> **Was this your answer?**
> Turbulent flow, because the water's motion is irregular and sediments have a greater tendency to remain in suspension. Turbulent flow carries these sediments because of the energy of its churning water. In laminar flow, water moves steadily in a straight-line path with no mixing of sediment in the channel.

Erosional and Depositional Environments

Eventually, particles that are being transported by surface water drop out of suspension—they are deposited. This happens when the water loses energy and slows. As a river gradually loses energy, larger particles are deposited first and then smaller ones, causing most surface water deposits to be well sorted.

The most dominant feature of deposited sediments is the way the particles of sediment are laid down, layer upon horizontal layer. These layers are referred to as *beds*. Varying in both thickness and area, each bed represents one episode of deposition. For example, flooding in a particular year might produce a layer of sediment next to a river. A flood any time after that produces an overlying layer. Some beds represent deposition over millennia.

The deposition and erosion of sediments occur in many different environments, including oceans and shorelines, rivers and streams, deserts, and deltas. Each environment in which erosion, transportation, and deposition occur has its own specific characteristics.

Stream Valleys and Floodplains

As rainfall hits the ground, it loosens soil and washes it away. As more and more rain falls and the ground continues to lose soil, gullies form. Once water and soil particles funnel into such a gully, a stream channel is created. This erosive action may be extremely rapid, as in the erosion of unconsolidated sediments, or very slow, as in the erosion of solid rock. Water's erosive power enables a stream to widen and deepen its channel, to transport sediment away, and, in time, to create a valley. In high mountain areas, the erosive action of a stream cuts down into the underlying rock to form a narrow, V-shaped valley. Because the valley is narrow, the stream channel dominates the whole valley bottom. Fast-moving rapids and beautiful waterfalls are characteristic of V-shaped mountain stream valleys (Figure 22.23).

When a fast-flowing mountain stream leaves its narrow valley, it abruptly emerges onto a broad, relatively flat plain. The speed of the flow drops suddenly and the stream dumps its load of sediment. Streams of this type often flow no farther. The sediment deposits are generally fan-shaped and grow outward as additional sediment is deposited (Figure 22.24). The steep upper slopes of such deposits are dominated by boulders, cobbles, and gravels, while the base area and the adjacent plain are composed of sand, silt, and mud.

FIGURE 22.23
At a stream's headwaters, steep gradients contribute to fast-moving rapids. When there is an abrupt change in gradient, we see a beautiful cascading waterfall.

FIGURE 22.24
A fan-shaped clastic sedimentary deposit in Death Valley, California.

FIGURE 22.25
The evolution of a stream valley and development of a floodplain. (a) At the headwaters, the V-shaped stream valley is characterized by steep gradients and fast-moving water that cuts down into the stream channel. Features in this area include cascading rapids and waterfalls. (b) Downstream, with a gentler gradient, the stream focuses its erosive action in a side-to-side sinuous manner, thereby widening the stream valley. (c) Farther downstream, meandering increases and further widens the stream valley to form a large floodplain.

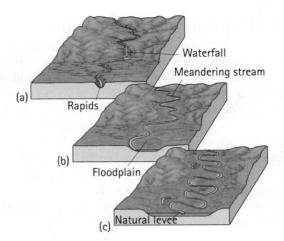

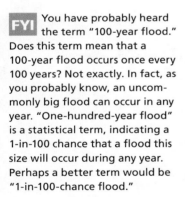

FYI You have probably heard the term "100-year flood." Does this term mean that a 100-year flood occurs once every 100 years? Not exactly. In fact, as you probably know, an uncommonly big flood can occur in any year. "One-hundred-year flood" is a statistical term, indicating a 1-in-100 chance that a flood this size will occur during any year. Perhaps a better term would be "1-in-100-chance flood."

Stream speed plays an important role in both erosion and deposition. As a stream flows downhill, its gradient becomes gentler and its speed slows. The focus of its energy changes from eroding downward (deepening the channel) to eroding laterally in a side-to-side motion. As a result, the stream develops a more sinuous, *meandering* form (Figure 22.25). Recall that as a stream bends and curves (meanders), the flow speed in the stream channel shifts so that the maximum speed is always toward the outside of each bend (look back at Figure 22.18). Rapidly moving water is very effective in eroding material from the outside of the bend, creating a steep bank called a **cut bank**. Material eroded from the cut bank is transported downstream, where it may eventually be deposited in areas where the stream speed decreases. Sandy **point bars** form on the insides of bends by this process. So, as a stream meanders back and forth across a river valley, sediment is eroded from one area and deposited in another area.

The meandering movement creates a wide belt of almost flat land—a **floodplain** (Figure 22.26). As the name implies, this section of the river valley becomes flooded with water and sediments when a river overflows its banks. In a flood, as discharge and flow speed increase, so does the stream's ability to carry sediment. Thus, when a stream overflows its banks, sediment-rich water spills out onto the floodplain. The speed of the water quickly decreases as it spreads out over the wide, flat, floodplain, and a progression of large to small particles is deposited. As expected, larger, coarse-grained sediments are deposited along the edges of the channel and smaller, fine-grained sediments are deposited farther away from the stream channel on the floodplain. The larger particles deposited close to the stream channel form *natural levees* that help confine future floodwaters (Figure 22.27). The widening of the valley, as shown in Figure 22.26, occurs because sediment deposited by the stream, especially during floods, progressively fills the valley.

FIGURE 22.26
Cross section of a river valley. A floodplain is created when a river overflows its banks. Sands and gravels settle out first and act as natural levees to confine the river. Because the finer silt and clay particles can flow as a suspended load, they move beyond the levees and settle on the floodplain.

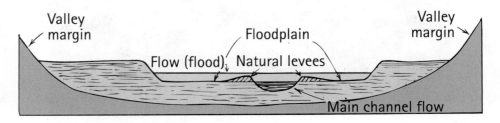

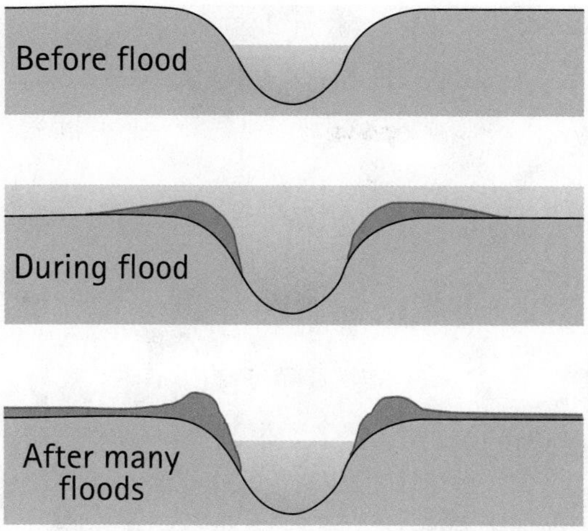

Before flood

During flood

After many floods

FIGURE 22.27
In a flood, increased discharge and flow speed help a stream carry not only a large sediment load but also larger particles. Larger, coarse-grained sediment deposited close to the stream channel forms natural levees that confine the stream between flood stages. Successive floods increase the height of the levees and may even raise the overall elevation of the channel bed so that it is higher than the surrounding floodplain.

> **FYI** The Mississippi River flood of 2011 is one of the largest and most damaging floods of the past century. From Illinois to Louisiana, thousands of square miles of farmland and residential areas were submerged by water. Flooding along the Mississippi is common—humans have been trying to control the mighty river since the great flood of 1927. The taming of this great river is a classic struggle of man versus nature. In the end, the river will win—the Mississippi River will make its own path to the Gulf of Mexico.

CHECKPOINT

Floodplains are often prime agricultural areas. Why would people want to work and live in areas so prone to flooding?

Was this your answer?

People live and work in floodplain areas because such plains are next to rivers that provide easy access to water, food, and a means of transportation. Also, because of periodic flooding, floodplain soils are often extremely fertile and thus serve as prime farmland. As for the factor of danger—don't most people associate danger with anyone but themselves?

> **FYI** Some of the world's greatest rivers have huge deltas at their mouths. Millions of years ago, the mouth of the Mississippi River was where Cairo, Illinois, is today. Since that time, the delta has extended 1600 km south to the city of New Orleans. Less than 5000 years ago, the site of New Orleans was underwater in the Gulf of Mexico!

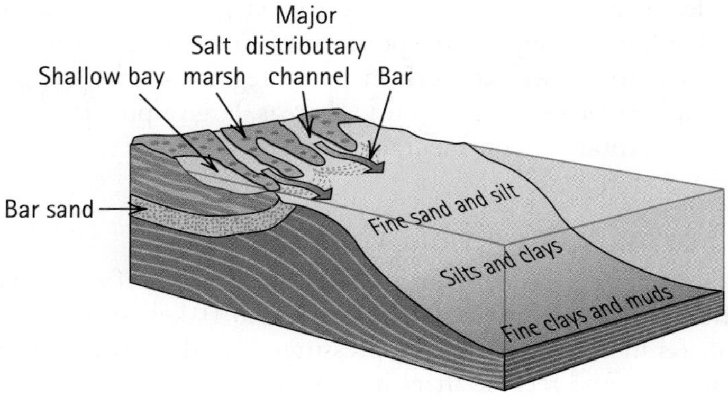

Deltas: The End of the Line for a River

As a stream flows into a standing body of water, such as a sea, bay, or lake, the moving water gradually loses its forward momentum. With reduced energy, stream speed slows and the stream loses its ability to carry sediment. These changes cause the stream to dump its sediment load. In this way, the mouth of the stream and the area immediately offshore become filled with sediment. The dumped sediment forms a fan-shaped deposit called a **delta** (Figure 22.28). Sediment is deposited in order of decreasing weight, with heavy, coarse particles deposited first, at and near the shoreline. Light, fine particles are deposited farther offshore. With the continual addition of incoming sediment, the delta progressively builds itself outward as an extension of land into the body of water.

Shallow bay   Salt marsh   Major distributary channel   Bar

Bar sand

Fine sand and silt

Silts and clays

Fine clays and muds

FIGURE 22.28
Deltas are areas of land generation. As streams flow to the sea, they carry sediment. These sediments are deposited in order of decreasing weight, with heavy, coarse particles settling at or near the shoreline and light, fine particles settling farther offshore. Layer upon layer, the depositional platform called a delta takes form.

FIGURE 22.29
Satellite image of the Mississippi Delta. Note how smaller streams are formed branching off from the main river.

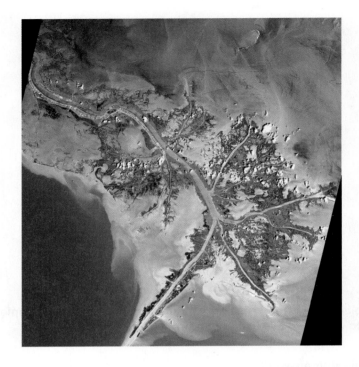

> What is the ultimate destination of all water flow and, hence, the eventual site of deposition of most sediments? Water flows eventually to the ocean, and sediments settle to the ocean floor.

Deltas begin to form underwater, but the addition of incoming sediment eventually causes the delta to emerge as new land. As the main stream channel becomes choked with sediment, more energy is required for water to push through the accumulated sediment than to go around it. So new, smaller channels form off the main channel like branches on a tree. These *distributaries* allow water to flow unimpeded to the standing body of water. As the delta continues to extend outward, the distributaries also become clogged, as sediment continues to arrive from upstream. The first set of distributaries thus form distributaries of their own, taking on the appearance of branching fingers. When the fingers become clogged, the branching process repeats. As streams continue to flow to the standing water and as successive sediment layers are deposited one on top of the other, the delta continues to build and expand outward (Figure 22.29). Thus, delta environments are areas where new land is continuously created.

LEARNING OBJECTIVE
Describe the conditions of glacier formation and movement.

22.6 Glaciers and Glaciation

EXPLAIN THIS How are icebergs related to glaciers?

The mightiest rivers on Earth are frozen solid and normally flow a sluggish few centimeters per day. These great icy rivers are called **glaciers**. Glaciers covered significant portions of Earth several times in the distant past. Glaciation is still at work in many regions of the world, its agents being small alpine glaciers in mountainous areas, large alpine ice fields, and the huge Arctic and Antarctic continental ice sheets.

> **FYI** Glaciers surge when the muddy till below the glacier is warmed by Earth's internal heat. Warmth is retained because insulation against the cold atmosphere is provided by the thick overlying ice. Melting occurs with meltwater seeping into the till, which softens and moves easily under the weight of the overlying ice.

Glacier Formation and Movement

The ice of a glacier is formed from recrystallized snow. After snowflakes fall, their accumulation slowly changes the individual flakes to rounded lumps of icy material. As more snow falls, the pressure exerted on the bottom layers of icy snow compacts and recrystallizes it into glacial ice.

This ice does not become a true glacier, however, until it moves under its own weight. This happens when the thickness of the ice is about 50 m. The pressure exerted by the overlying material causes ice crystals at the base of the glacier to deform *plastically* and flow downslope. This plastic deformation can be likened to what happens to a deck of playing cards. When the deck is pushed from one end, as in Figure 22.30, individual cards slide past one another, shifting the entire deck. Plastic deformation in a glacier is greatest at the base of the ice, where pressure is greatest.

Plastic flow from slipping ice crystals is not the only way that glaciers move. The melting point of ice decreases as pressure increases. When melted ice—*meltwater*—forms at the base of the glacier, a process called *basal sliding* comes into play.* This second mechanism of glacial movement results in the entire glacier sliding downslope, with the meltwater acting as a lubricant. The net speed of the glacial ice increases from the base up. So the fastest movement is at the glacier's surface (Figure 22.31). Carried along by basal sliding and by internal plastic flow, the surface of the glacier behaves like a rigid, brittle mass that may fracture. Huge, gaping cracks called *crevasses* may develop in this surface ice (Figure 22.32). These cracks can extend to great depths and can therefore be quite dangerous for people attempting to cross a glacier.

FIGURE 22.30
When a deck of cards is pushed from one side, the individual playing cards slide past one another, thus shifting the whole deck.

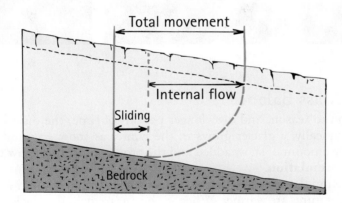

FIGURE 22.31
Cross section of a glacier. Glacial movement has two components: internal flow and sliding resulting from lubrication by meltwater. Movement is slowest at the base because of frictional drag and fastest at the surface. The upper parts of the glacier are carried along in piggyback fashion by plastic flow within the ice.

Average glacier speed varies from glacier to glacier and can range from only a few centimeters to a few hundred centimeters per day. Such slow speeds are measured by placing a line of markers across the ice and recording their changes in position over a period of time, ranging from days to years. Ice moves fastest in the center and more slowly at the edges because of frictional drag (Figure 22.33). Some glaciers experience surges, or periods of much more rapid movement. These surges are probably caused by periodic melting of the base and sudden redistribution of mass. The flow rate in these relatively brief surges can be 100 times the normal rate. Viewed from above, flow bands of rock debris and ice normally have a parallel pattern, but during a surge the flow bands become intricately folded (Figure 22.34).

FIGURE 22.32
Crevasses can extend to very great depths and can be treacherous to cross.

* Meltwater may result from the pressure of the overlying ice (the melting point of ice decreases as pressure increases), from the internal heat of Earth, or from the generation of heat from frictional drag as the glacier moves. Whatever the causes of its formation, meltwater contributes to the movement of a glacier.

FIGURE 22.33
Top view of a glacier. Movement is fastest at the center and gradually decreases along the edges because of friction.

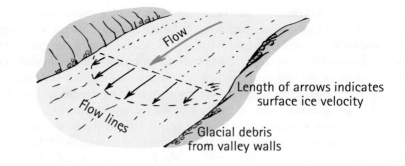

FIGURE 22.34
Glacial flows: (a) normal flow; (b) surge flow.

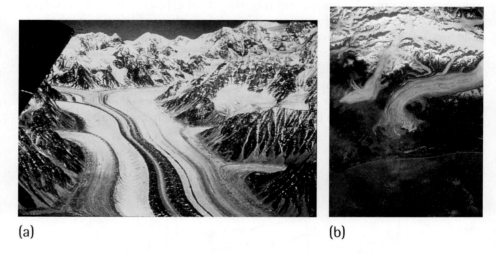

(a)   (b)

Glacial Mass Balance

From season to season, and over longer periods of time, the mass of a glacier changes. Typically, a glacier grows in the winter as snow accumulates on its surface. The amount of snow added and the process of adding snow to a glacier is called **accumulation**.

As ice accumulates and begins to flow downhill, it may move to an altitude where temperatures are warmer. When ice begins to melt, the glacier loses some of its mass. A glacier may also lose mass as it moves downslope to a shoreline, where ice may break off, or *calve*, to form icebergs that float away to sea. Melting and calving are the two primary ways that a glacier loses mass. Although less noticeable, glaciers may also lose mass as the ice *sublimates* to water vapor. By whatever means, the total amount of ice lost and the process of losing ice is called **ablation** (Figure 22.35).

FIGURE 22.35
Accumulation on a glacier takes place at high elevations as snow falls on the glacier and turns to ice. Ablation takes place at lower elevations as ice melts or calves into icebergs or is lost through sublimation.

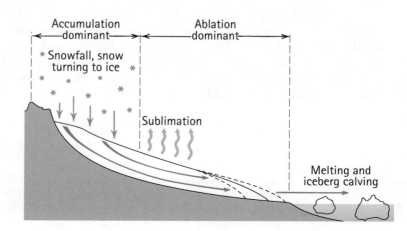

When accumulation equals ablation, the size of the glacier remains constant. For example, in a mountain glacier, accumulation occurs with winter snowfall in the farther-back, higher-elevation parts of the glacier, and ablation occurs in the lower portions, where spring and summer melting is greatest. When the accumulation rate and ablation rate are equal, the melting of the lower portions is offset by the downslope flow of ice from higher elevations. As a result the location of the front edge of the glacier does not change. When accumulation exceeds ablation, the glacier advances—it grows. When ablation exceeds accumulation, the glacier retreats—it shrinks. Naturally, in all these cases, the ice of the glacier always flows downslope.

> **FYI** With global warming, ablation is exceeding accumulation in more and more places. As glaciers calve into the sea, or as ice sheets on land melt, sea level will rise.

CHECKPOINT

Under what conditions does the front of a glacier remain at the same location from year to year?

Was this your answer?
The front of a glacier remains at the same location when the rate of growth (accumulation) equals the rate of shrinking (ablation). In the spring, as ice at the glacier's front melts away, the glacier retreats upslope. At the same time, the increased mass from the prior winter's accumulation causes the glacier to move forward. When the rate at which this forward movement matches the rate of melting, the location of the front edge doesn't change.

22.7 The Work of Glaciers

EXPLAIN THIS What do striations, roches moutonnées, and drumlins have in common?

> **LEARNING OBJECTIVE**
> Identify the glacial landforms created by erosion and deposition.

Like flowing water in streams, glaciers can erode as well as deposit sediment. Both processes produce characteristic landforms, which yield clues about long-gone glaciers.

Glacial Erosion and Erosional Landforms

Glaciers are powerful agents of erosion. Glaciation has created the beautiful landscapes of Tibet, Nepal, and Bhutan in Asia; the Alps of Switzerland; the fjords of Norway; and Yosemite Valley and the Great Lakes in North America. In many ways, a glacier is like a plow, as it scrapes and plucks up rock and sediment. It is also like a sled, as it carries its heavy load to distant places. As it moves across Earth's surface, a glacier loosens and lifts up blocks of rock, incorporating them into the ice. The large rock fragments carried at the bottom of a glacier scrape the underlying bedrock and leave long, parallel scratches (like sled tracks) aligned in the direction of ice flow (Figure 22.36). These scratches are called *striations*.

The two main types of glaciers, *alpine* and *continental*, have different erosional effects and produce dissimilar landforms. Alpine glaciers develop in mountainous areas and are often confined to individual valleys, while continental glaciers cover much larger areas. Alpine glaciers occur in most high mountain chains in the world, such as the Cascades, the Rockies, the Andes, and the Himalayas. The erosional features of alpine glaciation are depicted in Figure 22.37.

FIGURE 22.36
Striations mark the presence of a former glacier.

FIGURE 22.37
(a) The many erosional features of alpine glaciation. (b) The Matterhorn—named for its characteristic "horn" feature. (c) Hanging valleys are a spectacular feature found in areas that have been shaped by alpine glacial erosion. Bridalveil Falls in Yosemite National Park spills out of a hanging valley into the larger valley that was once occupied by the main glacier.

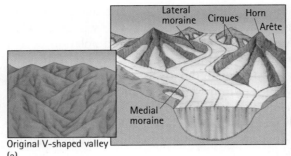

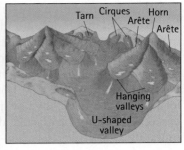

Original V-shaped valley
(a)

(b)

(c)

FIGURE 22.38
Small asymmetrical hills called *roches moutonnées* show the direction of continental glacial movement. On the side of the hill facing the approaching glacier, the slope is smooth and gentle. On the side facing away from the approaching glacier, the slope is rough and steep with a plucked appearance.

Continental glaciers spread over the land surface, smoothing and rounding the underlying topography. Although striations are produced by both alpine and continental glaciers, they have played a larger role in the study of ancient continental glaciers. Because a continental glacier scours very large tracts of land, it tends to leave behind few obvious valleys (making it difficult to determine the glacier's direction of flow). By mapping striations on land once covered by continental glaciers, geologists can decipher the flow direction of the ice. Flow direction is also indicated by small, asymmetrical hills (Figure 22.38) that are known by the French term *roches moutonnées* (singular, *roche moutonnée*). On the "upstream" side of the ice flow, the hill's slope is smooth and striated from the abrasion of ice on bedrock. On the "downstream" side, the slope is rough and steep because the moving ice plucked rock fragments away from cracks in the bedrock.

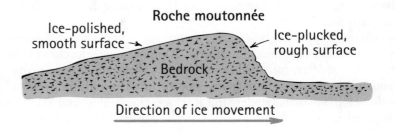

Roche moutonnée

Ice-polished, smooth surface →

Ice-plucked, rough surface

Bedrock

Direction of ice movement

Glacial Sedimentation and Depositional Landforms

As a glacier advances across the land, it acquires and transports great quantities of debris. When the glacier retreats, this debris is left behind as it is melted out of the ice. Because a glacier abrades and picks up everything in its path, glacial deposits are characteristically composed of unsorted, angular rock fragments in a variety of shapes and sizes.

This wide range of particle sizes is a hallmark that differentiates glacial sediment from the much-better-sorted material deposited by streams and winds. Glacial deposits are collectively called **drift**, a term that dates back to the 19th century, when it was conjectured that all such debris had been "drifted in" by the great Biblical Flood.

Drift is deposited in two main ways. When glacial sediment is released into meltwater, it is carried and deposited like any other waterborne sediment; thus, it is well sorted. This type of drift is called *outwash*. Material deposited directly by melting ice forms an unsorted mixture of clay and bouldery rock debris—this is called *till*. Many of the old stone walls and fences of New England are found in areas where the surface material is glacial till. Settlers who tried to farm this land had to remove all the larger boulders before they could plow, and they piled them along the edges of their fields. Often, large boulders that drastically differ from the local bedrock are found in glacial deposits. The large boulders provide proof of a glacier's ability to transport heavy loads for long distances. If a bedrock outcrop that matches the rock type of the out-of-place boulder can be found, then the distance and direction of glacial transport can be estimated.

The most common landform created by glaciers is the *moraine*, a ridge-shaped landform that marks the boundaries of ice flow. Of all the different types of moraines, probably the most important is the *terminal moraine*, as it marks the farthest point of a glacier's advance (Figure 22.39).

Another distinctive landform consisting of glacial sediments is the *drumlin*, an elongated hill shaped like the back of a whale. Formed by continental glaciation and lined up in the direction of ice flow, drumlins have a steep, blunt end in the direction from which the ice came and a tapered gentle slope on the downstream side (Figure 22.40). Perhaps the most famous drumlin in the United States is Bunker Hill in Massachusetts.

Many of the world's lakes, small and large, are the products of glacial action. Glaciers deepened valleys and deposited sediments that acted as dams, blocking stream drainage within some valleys and creating lakes. The Finger Lakes in upstate New York, the "10,000 Lakes" of Minnesota, and the Great Lakes of North America are all products of glacial action.

Geologist Bob Abrams observes the grandeur of the Juneau Ice Field in Alaska.

Geologic features are best viewed from an airplane. Next time you fly in an airplane, request a window seat and enjoy the geology below.

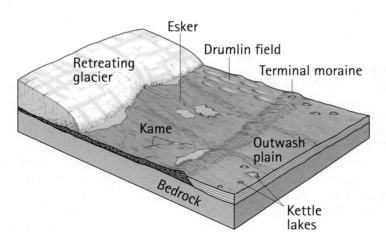

Esker
Drumlin field
Retreating glacier
Terminal moraine
Kame
Outwash plain
Bedrock
Kettle lakes

FIGURE 22.39
Glacial depositional landforms. Of special importance is the terminal moraine, which marks the farthest point of a glacier's advance.

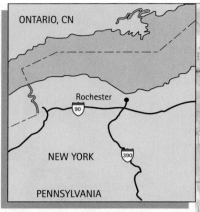

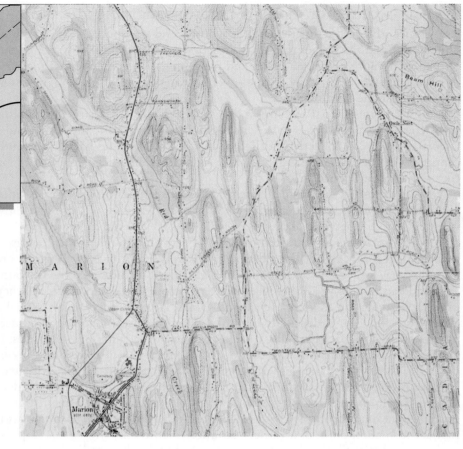

FIGURE 22.40
Topographic map showing numerous oval-shaped drumlins in upstate New York. Drumlins are steep and blunt on the side that faced the approaching glacier but tapered and gently sloping on the downflow side. Looking at the map, can you tell the direction of continental ice flow?

FYI A unique feature of sand dune formation is cross-bedding, which is found on the leeward side of a dune. The direction of cross-bedding indicates the direction of the wind (or water current) that deposited the sediments. A particularly great place to see ancient cross-bedding is in Zion National Park in Utah. Cross-bedding is also a common feature in river deltas and certain stream-channel deposits.

CHECKPOINT
What land surface forms can be used to determine the direction of ice flow?

Was this your answer?
The direction of ice flow can be determined from striations (long, parallel scratches aligned in the direction of ice flow), *roches moutonnées* (small, asymmetrical hills), and *drumlins* (elongated hills shaped like the back of a whale).

LEARNING OBJECTIVE
Describe the process of wind erosion.

22.8 The Work of Air

EXPLAIN THIS What is a wind shadow?

Water is the dominant agent of change altering our natural landscape, but air plays a role too. If you've ever been in a windstorm or at the beach on a windy day, you may have felt the sandblasting effect of the wind. Once in the air, particles of sediment can be carried great distances by the wind. Red dust from the Sahara is found on glaciers in the Swiss Alps and on islands in the Caribbean Ocean. Fine grains of quartz from central Asia blow onto the Hawaiian Islands.

FIGURE 22.41
Generated by blowing winds, ripple marks are narrow ridges of sand separated by wider troughs. They are small, elongated sand dunes. Large sand dunes can be seen in the background of the photograph.

In the desert, winds move over surfaces of dry sand, picking up the small, more easily transported particles but leaving the large, harder-to-move particles behind. The small particles bounce across the desert floor, knocking more particles into the air, to form *ripple marks*, which are actually tiny sand dunes (Figure 22.41). Ripple marks can also be formed by the movement of sand grains in water currents, as seen in shallow streams or under the waves at beaches.

Sand dunes begin to form where airflow is blocked by an obstacle, such as a rock or a clump of vegetation. As wind sweeps over and around the obstacle, the wind speed slows, causing sand grains to fall out of the air in the wind shadow (Figure 22.42). The falling sand forms a mound and blocks the flow of air even more. With more sand and more wind, the mound grows into a dune, which, with continued growth, begins to "move" downwind. The dune moves because grains on the windward slope are blown up and over the crest of the dune, falling on the leeward slope. In this way, wind removes sand from the back of the dune and redeposits it on the front of the dune. Over time, this continuous process moves the entire dune.

FYI Even though desert environments lack moisture, water is still the main cause of erosion and transportation of sediments. Scarce as water is in the desert, when a heavy rain falls, rainwater does not have time to soak into the ground and instead causes powerful flash floods. These flash floods transport and then deposit great quantities of debris and sediment at the bases of mountain slopes and on the floors of wide valleys and basins.

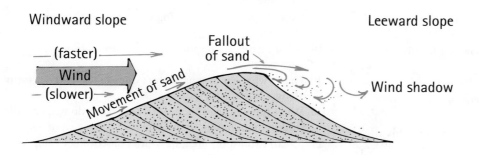

FIGURE 22.42
Formation of a sand dune. When airflow is obstructed, wind speed drops and, as a result, sand grains settle in the wind shadow. With more wind, more sand settles, and a dune is formed. As the dune grows, sand grains on the windward slope move up and over the crest to fall on the leeward slope, which slowly causes movement of the whole dune downwind.

For instructor-assigned homework, go to www.masteringphysics.com

SUMMARY OF TERMS (KNOWLEDGE)

Ablation The amount of ice lost, and the process of losing ice, from a glacier.

Accumulation The amount of snow added, and the process of adding snow, to a glacier.

Artesian system A system in which confined groundwater under pressure can rise above the upper boundary of an aquifer.

Channel geometry The shape of a stream channel; the cross sectional area.

Cut bank A steep bank on the outside bend of a river's channel; an area of erosion.

Delta An accumulation of sediments, commonly forming a triangular or fan-shaped plain, deposited where a stream flows into a standing body of water.

Discharge The volume of water that passes a given location in a stream channel in a certain amount of time.

Drift A general term for all glacial deposits.

Floodplain A wide plain of almost flat land on either side of a stream channel. Submerged during flood stage, the plain is built up by sediments discharged during floods.

Glacier A large mass of ice formed by the compaction and recrystallization of snow, moving downslope under its own weight.

Gradient The vertical drop in the elevation of a stream channel divided by the horizontal distance for that drop; the steepness of the slope.

Groundwater Underground water in the saturated zone.

Hydraulic conductivity A measure of the ability of a porous rock or sediment to transmit fluid.

Hydrologic cycle The natural circulation of all states of water from ocean to atmosphere to land and then back to ocean.

Laminar flow Water flowing smoothly and fairly slowly in straight lines with no mixing of sediment.

Point bar A sandy, gentle bank on the inside bend of a river's channel; an area of deposition.

Porosity The volume of open space in rock or sediment compared to the total volume of solids plus open space.

Sand dune A landform created when airflow is blocked by an obstacle, slowing air speed and therefore promoting the deposition of airborne sand.

Turbulent flow Water flowing rapidly and erratically in a jumbled manner, stirring up everything it touches.

Water table The upper boundary of the saturated zone, below which every pore space is completely filled with water.

READING CHECK QUESTIONS (COMPREHENSION)

22.1 The Hydrologic Cycle

1. Where does most of Earth's precipitation occur?
2. As water is precipitated onto the land, where does it go? Where does most of the water on land end up?

22.2 Groundwater

3. Distinguish between *porosity* and *hydraulic conductivity*.
4. If a hole is dug in the unsaturated zone, does it fill with water? Why or why not?
5. Compare and contrast the unsaturated zone with the saturated zone.
6. What types of soil allow the greatest amount of rainfall to soak in?
7. What is an artesian system, and how is it formed?

22.3 The Work of Groundwater

8. Describe at least one consequence of overpumping groundwater.
9. How does rainwater become acidic? How does this affect limestone?
10. How do stalactites and stalagmites form?
11. Name three erosional features caused by groundwater in carbonate rocks.
12. What is the difference between a cave and a cavern?

22.4 Surface Water and Drainage Systems

13. What is meant by *stream gradient*, and how does it affect stream velocity?
14. What happens to stream speed when the discharge of a stream increases? What happens to discharge when the speed of a stream increases?
15. How does the shape of a stream channel affect flow?

22.5 The Work of Surface Water

16. Which transports more sediment: laminar flow or turbulent flow? Why?
17. Describe three ways flowing water erodes a stream channel.
18. What factors are responsible for the formation of a stream valley?
19. Under what conditions do curvy, meandering rivers form along a floodplain?
20. What types of streams and stream valleys do we generally find in high, mountainous regions?
21. Streams transport great quantities of sediment from one place to another. What is the size range of the particles that can be carried by a fast-moving stream?
22. What is a delta?

22.6 Glaciers and Glaciation

23. How is a glacier formed?

24. In what two main ways do glaciers flow?

25. Under what conditions does a glacier front advance? How about retreat?

22.7 The Work of Glaciers

26. What is the most common depositional landform created by glaciers?

27. What erosional features are likely found in an area of alpine glaciation? See Figure 22.37.

28. What land features are formed from glacial deposits?

22.8 The Work of Air

29. How do sand dunes migrate?

30. How are ripple marks formed?

ACTIVITIES (HANDS-ON APPLICATION)

31. As a glacier moves across the land, it loosens and picks up rocks in its pathway. These rocks, embedded at the glacier's base, scrape the underlying bedrock to form striations. To see how this happens, make your own mini-glacier. Cover the bottom of a small plastic container with sand, fill the container with water, then freeze until solid. Remove your glacier from its container and hold it between gloved fingers. With the sand-side down, rub the ice over a piece of soap. What happens to the surface of the soap?

32. As wind blows across the land surface, it picks up small particles of sand. You can demonstrate how wind moves sediment. Cover the bottom of a cake pan with a flat layer of cornmeal (about 2 cm deep). With a straw, gently blow across the surface of the cornmeal. What happens to the surface? Place a small obstruction (small pebble, shell, screw) in the center of the pan and continue to blow. What happens behind the obstruction?

THINK AND SOLVE (MATHEMATICAL APPLICATION)

33. We know that most of Earth's water is in the oceans. The remaining 2.8% is Earth's freshwater supply. Of this freshwater supply, what percentage is found in the polar ice caps? In groundwater? In streams, lakes, and rivers? Hint: Calculate the freshwater supplies to be equal to 100%.

34. A particular stream widens as it progresses downstream. Using your answers for parts (a) and (b), briefly describe the changes in discharge.

    (a) If the cross-sectional area of the stream is 1 m^2 and the stream speed is 0.5 m/s, what is the stream's discharge?

    (b) If the cross-sectional area of the stream increases to 2 m^2 and the stream speed remains 0.5 m/s, what is the stream's discharge?

Refer to the Figuring Physical Science box on page 597 for Problems 35, 36, and 37.

35. A pumping well was drilled and completed in a sand aquifer. Before the pump was turned on, the hydraulic gradient and the flow rate were measured to be 0.0001 and 1 m^3/day, respectively. With the pump turned on, the gradient becomes 10 times larger. By how much does the flow rate increase?

36. Darcy's law gives us the *volume flow rate*—volume per time (for example, cubic meters per day, m^3/day). Another way to express volume is on a *per-unit-area* basis. For example, if we have a 1-m^3 cube, we know

its base is 1 m^2—the cross-sectional area of the cube—and its height is 1 m. If we fill the cube with water, we can say we have 1 m of water per unit area. Darcy's law can be rearranged to calculate the volume flow rate per unit area. This is called the *specific discharge*, which has units of length per time (for example, meters per day, m/day). Completing parts (a) through (c) will illustrate how specific discharge relates to the volume flow rate. (Hint: Assume that the units for hydraulic conductivity are m/day and the units for cross-sectional area are m^3.)

    (a) Suppose 1 m^3 of water is pumped from a well into an empty cylindrical tank. If the water level is 2 m above the base, what is the cross-sectional area of the tank? Hint: Volume of a cylinder $= \pi r^2 h =$ area of base $\times$ height.

    (b) If it takes half a day to pump the 1 m^3 of water into the tank, what is the flow rate in terms of both volume per time and specific discharge?

    (c) Write Darcy's law so that it calculates the specific discharge.

Please make sure that you understand Problem 36 before doing Problem 37.

37. The hydraulic head at point A is 209 m. At point B, which is 300 m from point A, the hydraulic head is 210 m. The aquifer is composed of sand with a hydraulic conductivity of 150 m/day. Groundwater flows directly from point B to point A. What is the specific discharge?

THINK AND RANK (ANALYSIS)

38. Rank, from greatest to least amounts, the distribution of Earth's freshwater: (a) groundwater; (b) polar ice caps and glaciers; (c) streams, rivers, and lakes; (d) soil moisture.

39. Put in order, from beginning to end, the steps in stream formation: (a) runoff forms, (b) raindrops strike the ground, (c) water begins to run downslope, (d) gullies form.

40. In descending order, rank the rocks in terms of their susceptibility to chemical erosion: (a) marble, (b) limestone, (c) sandstone, (d) quartzite.

41. A stream becomes a river as it moves down gradient. Give the sequence in the evolution of a stream: (a) meandering pattern, (b) rapids, (c) delta.

42. Rank, from greatest to least, these forces of stream channel erosion: (a) chemical erosion, (b) abrasion, (c) hydraulic action.

43. Streams carry sediment as well as water. List, the following types of sediment in order of deposition: (a) boulders and cobbles, (b) pebbles and gravel, (c) sand, (d) clays and mud.

44. A delta is the end of a river. Going from offshore toward the shoreline, rank these according to sediment size: (a) cobbles, (b) pebbles, (c) sand, (d) clay.

45. Glaciers are like icy rivers that flow downgradient. Rank, from first to last, the stages in glacial flow: (a) basal sliding, (b) crevasses, (c) accumulation and thickening of glacial ice, (d) calving.

EXERCISES (SYNTHESIS)

46. What percentage of Earth's supply of water is fresh water, and where is most of it located?

47. Where does most rainfall on Earth finally end up before becoming rain again?

48. In a confined aquifer, water in a well can rise above the top of the aquifer. What is this system called?

49. In an unconfined aquifer, how high can water rise in a well that is not pumped?

50. How is the local hydrologic cycle affected by the practice of drawing drinking water from a river and then returning sewage to the same river?

51. Is water in the unsaturated zone called groundwater? Why or why not?

52. In an aquifer, if the water table next to a stream is lower than the water level in the stream, does groundwater flow into the stream or does stream water flow into the ground? Explain.

53. As runoff into streams increases, which variables of stream flow increase?

54. What is meant by channel geometry?

55. What happens to stream speed if the discharge in a stream doubles while the channel remains the same size and shape?

56. If discharge is held constant and the width of the stream channel decreases, what happens to stream speed? What effect will this have on the stream channel's depth?

57. What three variables influence the speed of stream flow?

58. How does "frictional drag" play a role in the external movement of a glacier? How does this drag affect the internal movement?

59. What are some possible sources of stream flow?

60. In the formation of a river delta, why are larger particles deposited first, followed by smaller particles farther out? Defend your answer.

61. What causes the formation of distributaries off the main channel of a river delta?

62. What is a sinkhole? What factors contribute to its formation?

63. Which of the three agents of transportation—wind, water, and ice—transports the largest boulders? Why?

64. Which of the three agents of transportation—water, ice, and wind—transports only small rocks?

65. Carbonate rocks form mainly in marine environments. Why do we find abundant carbonate deposits on continental land?

66. Are underground rivers ever found in nature? Defend your answer.

67. Do you think a stream with laminar flow can become turbulent without the volume of water in the stream increasing? Defend your answer.

68. Describe the formation of stalactites.

69. Why is surface water both a creator and a destroyer of sediments and sedimentary rocks?

70. Why do point bars form on the inside bends of meandering streams?

71. How is a roche moutonnée different from a drumlin?

72. What well-known landscapes have been carved by glaciers?

73. How do deposits from glacial ice differ from rocks deposited by rivers?

74. Removal of groundwater can cause subsidence. If removal of groundwater is stopped, will the land likely rise again to its original level? Defend your answer.

75. Must a stream's speed increase in order for it to carry more sediment? Explain.

76. What distinguishes a huge block of ice from a glacier?

77. Why do crevasses form on the surfaces of glaciers?

78. Does all the ice in a glacier move at the same speed? Explain.

79. What is the significance of the large, out-of-place boulders that are sometimes found in glacial deposits?

80. How are sand dunes formed?

DISCUSSION QUESTIONS (EVALUATION)

81. Which agents of erosion are assisted by the force of gravity? Give examples.

82. How does an increase in stream gradient and discharge affect a stream's sediment load?

83. Fresh drinking water is a precious resource. As water is depleted from the ground and from streams and rivers, what other resources can be tapped to get fresh water?

84. With global warming and melting of the polar ice caps, how will sea level be affected?

85. With increased development and the growth of cities, the natural landscape is replaced by roads, buildings, housing developments, and parking lots. What overall effect can this have on a river and the watershed?

READINESS ASSURANCE TEST (RAT)

If you have a good handle on this chapter, if you really do, then you should be able to score at least 7 out of 10 on this RAT. If you score less than 7, you need to study further before moving on.

Choose the BEST answer to each of the following.

1. Downstream, the amount of discharge in a stream usually
   (a) becomes turbulent.
   (b) meanders.
   (c) decreases.
   (d) increases.

2. Most of Earth's fresh water is found
   (a) in lakes.
   (b) in ice caps and glaciers.
   (c) in rivers.
   (d) underground.

3. The work of surface water does all of the following except
   (a) erosion.
   (b) deposition.
   (c) land subsidence.
   (d) delta formation.

4. The maximum amount of water a particular soil can hold is determined by its
   (a) porosity.
   (b) permeability.
   (c) degree of saturation.
   (d) amount of recharge.

5. Precipitation that does not soak into the ground or evaporate becomes
   (a) groundwater.
   (b) the water table.
   (c) soil moisture.
   (d) runoff.

6. Sand dunes form as wind
   (a) disperses sand.
   (b) blows sand from the back to the front of the dune.
   (c) blows sand from the front to the back of the dune.
   (d) interrupts the normal sequence of deposition.

7. Deltas form as
   (a) periodic flooding clogs stream channels.
   (b) erosion clogs stream channels.
   (c) the stream gradient decreases.
   (d) streams enter a standing body of water.

8. What factors affect stream speed?
   (a) discharge
   (b) channel length
   (c) stream gradient
   (d) both (b) and (c)
   (e) both (a) and (c)

9. Snow converts to glacial ice when subjected to
   (a) decreasing temperature.
   (b) pressure.
   (c) rain.
   (d) basal sliding.

10. Underground water in the saturated zone is called
    (a) groundwater.
    (b) soil moisture.
    (c) the water table.
    (d) an artesian system.

Answers to RAT

1. d, 2. b, 3. c, 4. a, 5. d, 6. c, 7. d, 8. e, 9. b, 10. a

CHAPTER 23

Geologic Time— Reading the Rock Record

E ARTH IS some 4.5 billion years old. This vast span of time, called *geologic time*, is difficult to comprehend. Imagine that we can compress 4.5 billion years into a single year so that our planet would have begun forming in the primordial solar system on January 1. Earth's oldest rocks would appear at the end of February, simple bacterial life would appear at the end of March, complex plants and animals would emerge in late October, and dinosaurs would rule Earth in mid-December but disappear by December 26. Humans would appear at 11:50 PM on the evening of December 31, and all of recorded human history would take place in the last minute of New Year's Eve!

Earth's history is recorded in the rocks of its crust. This rock record is like a long and detailed diary of Earth-shaping events. The book, however, is incomplete. Many pages are missing, or torn and difficult to read. Fortunately, enough pages have been preserved to provide an account of the remarkable events of Earth's 4.5 billion years of history.

23.1 The Rock Record—Relative Dating

EXPLAIN THIS How can rock layers be used to tell time?

Sedimentary rock layers are deposited layer on top of layer. This sequence of layering provides evidence of relative rock ages. The lower layers were formed before the upper layers, so they are older than the upper layers. Perhaps the world's most spectacular display of the rock record is the Grand Canyon of the Colorado River in Arizona (Figure 23.1). The many layers of rock exposed in the canyon walls and the thickness of these layers are testimony to great geologic activity over millions of years. The conditions under which the sedimentary layers were deposited varied—changing from season to season, year to year, and millennia to millennia. Some layers reveal climatic cycles that span centuries, other layers indicate times when the land surface became submerged beneath shallow seas, while still other layers show periods of increased rainfall accompanied by gradual uplift of the entire area. Millions of years after the top layer was deposited, the Colorado River cut through the accumulated layers like a knife cutting into a layer cake. The result of the river's erosive power is the canyon we see today.

In the Grand Canyon, and elsewhere, geologists use several key principles to determine the relative ages of rocks. **Relative dating** is the ordering of rocks in sequence by their comparative ages. Relative dating doesn't tell the actual date when a rock layer formed, but rather its timing relative to other episodes in Earth's past. The relative dating principles are listed here.

1. **Original horizontality:** Layers of sediment are deposited horizontally. Layers that are tilted or folded must have been moved into that position by disturbances, such as earthquakes and mountain building, after deposition.

2. **Superposition:** In an undeformed (*horizontal*) sequence of sedimentary rocks, each layer is older than the one above and younger than the one below. Like the newspapers in a recycling bin, older papers are found below newer, more recent papers.

3. **Cross-cutting relationships:** An igneous intrusion or fault that cuts through preexisting rock is younger than the rock through which it cuts (Figure 23.2).

4. **Inclusions:** Inclusions are pieces of one rock contained within another. Any inclusion is older than the rock containing it, just as small pieces of rock incorporated in a slab of concrete were formed before the concrete was formed (Figure 23.3).

These four principles are fairly straightforward and can be used to determine the ages of rock formations relative to one another—which formation was formed

FIGURE 23.1
The lowermost layers of the Grand Canyon are older than the uppermost layers, which illustrates the principle of superposition.

FYI Geologists use observations of present-day geologic processes to understand what happened in the geologic past. This idea, referred to as *uniformitarianism*, states that the natural laws (like the laws of physics) we know about today have been constant over time. This concept is simply expressed as "The present is the key to the past".

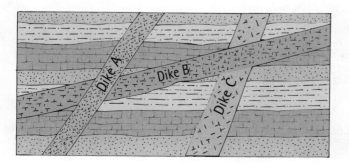

FIGURE 23.2
Dikes cutting into a rock body are younger than the rock into which they cut. In the diagram, dike A cuts into dike B, and dike B cuts into dike C. From the principle of cross-cutting relationships, A is the youngest dike, B the next youngest, and C the oldest of the three. The horizontal layers, which are cut by all three dikes, are all older than C.

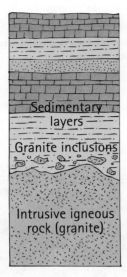

FIGURE 23.3
The rocks locked in the sedimentary layer existed before the sedimentary layer formed, illustrating the principle of inclusions.

first, second, and so on—for a particular area or rock outcrop. But to reconstruct the relative age of rocks for a larger area, additional information is needed. Once again, the Grand Canyon provides the ideal example. The rocks of the Grand Canyon not only chronicle a great span of time, but because the rock layers stretch continuously for hundreds of miles, they also show the range and extent of the depositional area. But not all rock layers stretch on like those found in the Grand Canyon. Over spans of time, rock layers can be broken by faulting and/or folding, or covered by younger sediments, so that all we see are geographically separated rock outcrops. In cases like this, we use the principle of lateral continuity.

5. **Lateral continuity:** Sedimentary layers are deposited in all directions over a large area unless some sort of obstruction or barrier limits their deposition. Faulting, folding, and erosion can separate originally continuous layers into isolated outcrops.

Lateral continuity can be used to match isolated rock outcrops to other rock outcrops over a large area (Figure 23.4). If the various rock layers in isolated outcrops have similar characteristics (such as color, mineralogy, grain size, and fossils) and the vertical sequence of layers is consistent, we can assume that the layers were at one time continuous. So, when combined with other relative dating principles—namely, superposition and original horizontality—lateral continuity allows relative age relationships to be applied over a much larger area.

As discussed in Chapter 20, fossils are the remains or impressions of ancient animals and/or plants preserved in rock. Today it is common knowledge that fossils record the evolution of life on Earth, but in the 1700s fossils were thought of as mere curiosities. Then along came William "Strata" Smith (1769–1839). Working as a surveyor for a canal project, Smith observed that certain rock layers contained different kinds of fossils, and that these fossil-bearing layers followed a consistent, predictable sequence. Smith noted the succession of rock types and the fossils within each layer and then used the fossils to correlate (match up) rock layers at various locations. He discovered that fossils could be used to chronologically order the vertical sequence of rock layers. With this knowledge, Smith established the principle of faunal succession.

6. **Faunal succession:** Fossil organisms follow one another in a definite, irreversible time sequence. Fossil communities change through time as some species become extinct and new ones appear. Such changes are reflected in the rock record. In this way, fossils provide a key tool for recognizing the relative age of rocks.

FIGURE 23.4
Lateral continuity can be used to determine the relative age of rocks in widely separated areas. Outcrops A and B share similar characteristics, but are separated by 500 km. Is it possible that these outcrops were, at one time, continuous? By looking at the sequence of layers, can you tell which rocks are older and which are younger?

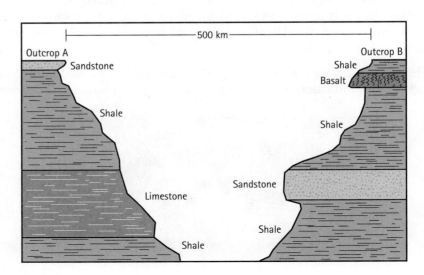

Smith's observation that fossils are found in rocks in a definite order not only helped refine relative dating, but also helped with the correlation of rock layers on a worldwide scale. Fossils—identified and categorized—could be correlated to specific phases in Earth's geologic history. Once scientists established a time period, the fossils in the rock could be used to identify other rocks of the same age in other regions of Earth.

In many ways, studying the rock record is like a detective studying a crime scene. Both studies involve looking for clues. Just as detectives have their methods for solving crimes, Earth scientists rely on the principles of relative dating—age relationships and the correlation of rock layers. They correlate the rocks, the sequences of rock layers, and the fossils within the rock layers. As with a crime scene, the case cannot be solved until sufficient evidence is found. The Earth science detective often has little evidence on which to operate, and sometimes the evidence is completely lost. Earth's internal processes can fold and distort rock layers and external processes can weather and erode rock layers. Whatever the cause, the result is lost evidence.

FYI Fossils tell us about past environments. For example, certain present-day corals are found in warm tropical waters. When a similar fossilized coral is found, we can safely assume that the area where it was found was once covered by a warm and tropical shallow sea. Fossils help unravel Earth's history.

Gaps in the Rock Record

The deposition and formation of sedimentary rock layers are always happening. Go to any riverbed or beach, and you will see sediment being transported and deposited. Yet a continuous sequence of rock layers from Earth's formation to the present time is not found anywhere on Earth. Although deposition is ongoing, so are the processes of weathering and erosion, folding and faulting, and crustal uplift. These processes remove rock layers or interrupt deposition, which creates time gaps in the rock record (Figure 23.5). These gaps, called **unconformities**, are found by observing the relationships of layers and fossils.

The most easily recognized of all unconformities is an **angular unconformity**. In an angular unconformity, tilted or folded sedimentary rocks are covered by younger, relatively horizontal rock layers. They are easy to recognize because the rock layers below the unconformity are at an angle relative to the rock layers above the unconformity. An angular unconformity forms when older, previously horizontal rock layers are uplifted and tilted by movements within Earth (Figure 23.6). During and after the uplift, erosion wears down the tilted layers so that rocks at the surface are eroded to a more or less even plane—a flat land surface. When the period of erosion is over, new sediment layers are deposited over the older, eroded, tilted layers. The angular unconformity is the "erosional surface" that separates the older, tilted layers from the

Mastering**PHYSICS**
TUTORIAL: Formation of an Angular Unconformity

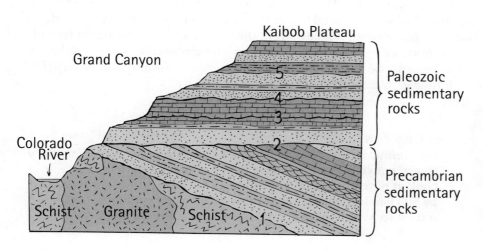

FIGURE 23.5
The age of the Grand Canyon can be deciphered by its sequence of rock layers. As in other places, the sequence is not continuous, and there are time gaps. (1) A nonconformity separating older metamorphic rocks from sedimentary layers. (2) An angular unconformity separating older tilted layers from horizontal layers. Time gaps are also represented between horizontal sedimentary layers. The disconformities (3)–(5) are difficult to identify, and they often require both a good eye and a knowledge of fossils.

FIGURE 23.6
INTERACTIVE FIGURE

The sequence of events that create an angular unconformity.

(a) Sediments are deposited layer upon layer beneath the sea.

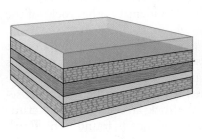

(b) During mountain building solidified sediment layers become folded and deformed. Erosion begins.

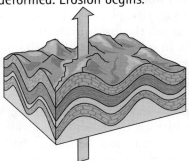

(c) As mountain building wanes, the exposed surface is eroded to a more or less even plain.

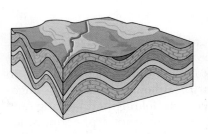

(d) As the land subsides below sea level, younger sediments are deposited on the former erosional surface.

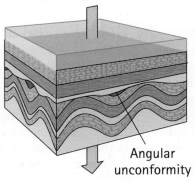

Angular unconformity

younger, horizontal layers. It represents the long interval of time during which uplift and erosion took place. Can you see that angular unconformities usually represent ancient mountain-building events? The part of the rock record representing this long interval is now missing because of erosion, and the unconformity is the evidence that remains.

When overlying sedimentary rocks are found on an eroded surface of metamorphic or igneous plutonic rocks, the unconformity is called a *nonconformity*. The igneous or metamorphic rocks formed deep beneath Earth's surface but were present at the surface when the overlying sedimentary rocks were deposited on top of them. Therefore, a nonconformity shows that a great deal of uplift and erosion occurred before the sedimentary layers were deposited, with a large stretch of time "missing" from the rock record. A more subtle type of unconformity, a *disconformity*, is a time gap between parallel layers of sedimentary rock. Because the rocks above and below the disconformity can be very similar, such time gaps are often quite difficult to identify.

FYI The oldest rocks found are the Acasta gneisses in northwestern Canada. Zircon crystals in the gneiss are dated at 4.03 billion years—the original crystallization age before metamorphism. The oldest mineral found is a zircon crystal from a sandstone in Australia. The zircon is dated at 4.4 billion years! The survival of this zircon crystal may be evidence that our planet cooled much faster than previously thought. Discovery breeds questions. Science is not static; it is ongoing with new things discovered and learned every day.

CHECKPOINT
1. If a granitic intrusion, such as a dike, cuts into or across sedimentary layers, which is older: the granite or the sedimentary layers?
2. Look at outcrops A and B in Figure 23.4. The outcrops are separated by 500 km. Is it possible that these rock layers were once continuous? If so, which rocks are older and which are younger?

Were these your answers?

1. The intrusion is new rock in the making. Therefore, the sedimentary layers are older than the intrusions that cut into them.

2. Yes, it is possible. If continuous, the sandstone capping outcrop A matches the sandstone sandwiched between the two shales in outcrop B. So all rock layers above sandstone B are younger. For outcrop A, original horizontality and superposition tell us that the bottom shale is oldest, followed by limestone, shale, then sandstone. We assume the same sequence in outcrop B even though we cannot see the limestone layer. If fossils were present, we could further support our conclusion.

23.2 Radiometric Dating

EXPLAIN THIS   How do scientists know the age of the Earth?

elative dating tells us which parts of Earth's crust are older or younger, but it doesn't tell us the actual age of a rock—the amount of time that has passed since the rock was formed. The actual age of a rock can be determined by **radiometric dating**, a process that measures the ratio of radioactive isotopes to their decay products. The half-life of a radioactive isotope is the amount of time it takes for half of the material to decay to its daughter product (Figure 23.7). Some common radioactive isotopes used for dating and estimates of geologic time are listed in Table 23.1.

To date geologically young objects, and especially for dating organic matter, carbon-14 is the isotope of choice. Carbon-14 has a relatively short half-life (5760 years), so it is useful for dating geologically recent events, within the last 50,000 years or so (see Section 13.4). To date older materials, radioactive elements such as uranium are used. Many common rocks contain trace amounts of uranium, and only a small amount is needed to perform the laboratory analysis. Uranium-238 decays to its stable daughter isotope, lead-206, and uranium-235 decays to the stable isotope lead-207. No other natural sources exist for these two isotopes of lead; therefore, any lead-206 and lead-207 found in a rock today were at one time uranium. If, for example, a sample contains equal numbers of uranium-235 and lead-207 atoms, the age of the sample is one uranium-235 half-life—704 million years. If, on the other hand, a sample of uranium contains only a small amount of lead-207, the sample is younger than one half-life (a single half-life has not yet passed).

LEARNING OBJECTIVE
Explain radiometric dating and how it is used to determine the age of a rock.

FYI Radiometric dating is based on the assumption that once a mineral has crystallized, any daughter product found within it originated only from the decay of the unstable parent—that is, no daughter product was initially present. Another important assumption is that there is no "leakage" of parent or daughter products into or out of the mineral. For example, if a mineral is reheated by metamorphism, its "time clock" is reset, complicating the age estimation. Radiometric dating and other dating techniques are subject to some uncertainty. In fact, for relative and radiometric dating, cross-checking increases accuracy.

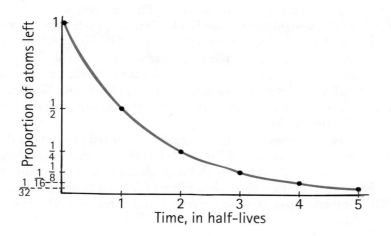

FIGURE 23.7
The amount of parent material remaining versus the amount of time elapsed—measured in half-lives—since decay began. When the parent material decays to half its original amount, the time is one half-life. When one-quarter of the parent material remains, the time is two half-lives; when one-eighth remains, the time is three half-lives, and so on.

Radioactivity is everywhere! In fact, all elements that have an atomic number greater than 82 (lead) are radioactive, but not necessarily dangerously so.

TABLE 23.1	ISOTOPES MOST COMMONLY USED FOR RADIOMETRIC DATING	
Radioactive Parent	**Stable Daughter Product**	**Half-Life Value**
Uranium-238	Lead-206	4.5 billion years
Uranium-235	Lead-207	704 million years
Potassium-40	Argon-40	1.3 billion years
Carbon-14	Nitrogen-14	5760 years

CHECKPOINT

1. **Could carbon-14 be used for dating 100-million-year-old rocks?**
2. **How can we determine the age of sedimentary rock layers?**

Were these your answers?

1. No. Carbon-14 has a half-life of 5760 years and can be used to date only relatively younger rocks. Any carbon-14 (from calcite, for example) in rocks this old would have long since been reduced to undetectable amounts.
2. If we know the maximum age (meaning the rock can be no older than the age of the datable minerals within it) of an overlying and an underlying rock layer, we can bracket the age of the sedimentary layer in between by using the principle of superposition.

FYI Radiometric dating can give us the age of minerals and/ or organic matter, but it cannot directly give us the age of sedimentary rocks. Remember, sedimentary rocks are made from the remains of preexisting rocks. Therefore, we can date only the minerals in the rock, not the sedimentary rock itself. The rock can be no older than the age of the datable minerals within it (principle of inclusions). So how do we date sedimentary rock layers? We use relative dating combined with absolute dates from radiometric dating to provide limits on the age of the rock. The more techniques, the better the date.

23.3 Geologic Time

EXPLAIN THIS Why are periods such an important subdivision?

The timeline for the history of Earth is called the *geologic time scale* (Figure 23.8). The scale is based on the relative ages of rock layers and their fossils. Recall the principle of faunal succession: fossil plants and animals are found in the rock record in a chronological sequence. Because blocks of time are represented by different types of fossils, the geologic time scale divides Earth's 4.5-billion-year history into time units of different sizes.

Eons are the largest unit of geologic time. The eon we are living in began about 543 million years ago. It is called the Phanerozoic, which means "visible life." The Phanerozoic eon is subdivided into three eras: the Paleozoic era (time of ancient life), the Mesozoic era (time of middle life), and the Cenozoic era (time of recent life). Each of the three eras is further divided into periods, which are further divided into epochs. *Periods* are the fundamental time interval because each period represents a major change in life forms. With radiometric dating, scientists can assign an actual age to the various periods. This gives accuracy to the time scale.

Note that the vast majority of Earth's history occurred prior to the Paleozoic era. This vast span of time, called **Precambrian time**, accounts for about 4 billion years of Earth's history! Although Precambrian time is sometimes referred to as a "super-eon," it does not have a formal rank—it is simply Precambrian time. This immense span of time is divided into three eons: the Hadean, the Archean, and the Proterozoic.

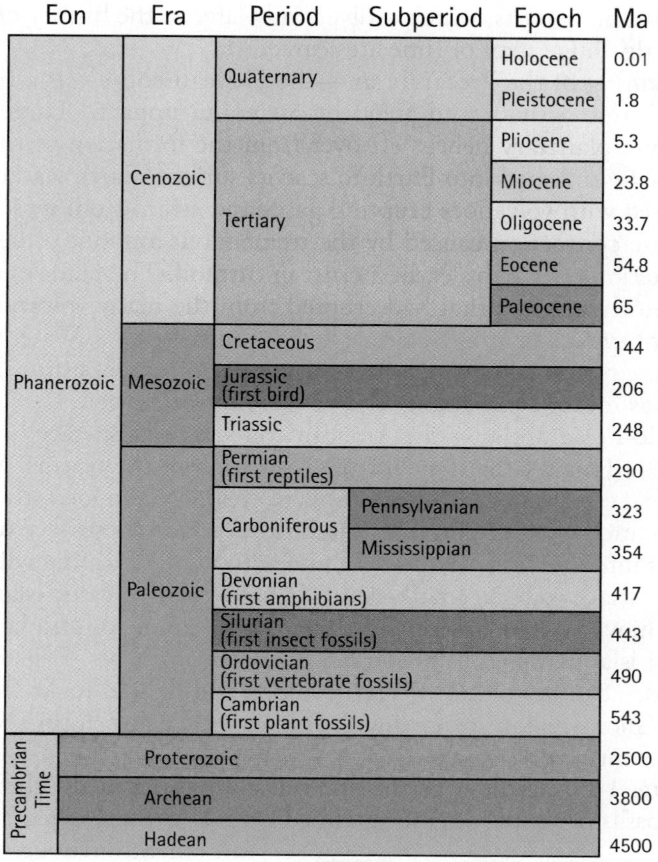

Eon	Era	Period	Subperiod	Epoch	Ma
Phanerozoic	Cenozoic	Quaternary		Holocene	0.01
				Pleistocene	1.8
		Tertiary		Pliocene	5.3
				Miocene	23.8
				Oligocene	33.7
				Eocene	54.8
				Paleocene	65
	Mesozoic	Cretaceous			144
		Jurassic (first bird)			206
		Triassic			248
	Paleozoic	Permian (first reptiles)			290
		Carboniferous	Pennsylvanian		323
			Mississippian		354
		Devonian (first amphibians)			417
		Silurian (first insect fossils)			443
		Ordovician (first vertebrate fossils)			490
		Cambrian (first plant fossils)			543
Precambrian Time		Proterozoic			2500
		Archean			3800
		Hadean			4500

FIGURE 23.8
INTERACTIVE FIGURE

The geologic time scale divides Earth's history into time units of different size. Units of time on this scale are Ma, which stands for *Mega-annum*, or "one million years (ago)." For example, the Paleozoic era began about 543 Ma, or 543 million years ago.

CHECKPOINT

1. Describe the present time in Earth's history in terms of all units of the geologic time scale from eons to epochs.
2. The time units on the geologic scale are Ma. What does *Ma* stand for?

Were these your answers?

1. We are living in the Phanerozoic eon, in the Cenozoic era, in the Quaternary period, and in the Holocene epoch.
2. *Ma* stands for *Mega-annum*, which means "one million years (ago)."

23.4 Precambrian Time (4500 to 543 Million Years Ago)

EXPLAIN THIS What do banded iron formations have to do with photosynthesis?

LEARNING OBJECTIVE
Describe the evolution of Earth's atmosphere during Precambrian time.

Precambrian time ranges from about 4.5 billion years ago, when Earth formed, to about 543 million years ago, when abundant macroscopic life appeared. The Precambrian—the time about which we know the least—comprises almost 90% of Earth's history! Most of the rocks that formed in this early part of Earth's history have been eroded away, metamorphosed, or recycled into Earth's interior. Organisms of that time did not have easily

FIGURE 23.9
This artwork of the Archean eon of the Precambrian shows characteristic features of the time, including space debris, such as meteorites and comets, crashing into Earth's surface. The bright green material on the edge of the water is primitive bacteria (*archaea*), while cyanobacteria thrive in the darker green, round structures called stromatolites in the water.

fossilized hard body parts, which evolved only later in the history of life. Thus, fossils from this huge span of time are scarce.

The beginning of the Precambrian—the Hadean eon—was a time of considerable volcanic activity and frequent meteorite impact.* Large and small chunks of interplanetary debris left over from the formation of the solar system continually smashed into Earth to scar its surface. Earth was an oceanless planet covered with volcanoes erupting gases and steam from its scorching interior. Intense convection caused by the tremendous amount of heat escaping from the interior left Earth's earliest crust in turmoil. The Hadean atmosphere consisted mostly of gases that had erupted from the many volcanoes. Carbon dioxide may have made up 80% or more of this atmosphere. Water vapor made up most of the balance, with molecular nitrogen, ammonia, sulfur dioxide, and nitric oxide as minor constituents. There was no free oxygen.

In the mid-Precambrian—the Archean eon—Earth's surface began to cool (Figure 23.9). This set the stage for the formation of the oceans. With cooler temperatures, water vapor in the atmosphere condensed to form clouds. Out of the clouds poured torrents of rain, enough to cover Earth's surface with shallow seas. Earth's interior was still quite hot and active, and erupting volcanoes dotted Earth's surface to form small islands. These small volcanic islands, carried by tectonic movement, collided with other small islands to form larger islands. These larger islands were the precursors of early continents. Evidence of these early collisions can be found in certain folded and faulted rocks (for example, the Acasta Gneiss region in northwest Canada) that now form the "cores" of present-day continents.

Life emerged during the Archean, and this event touched all other events that followed. Fossils of simple organisms called *stromatolites* are dated at 3.5 billion

years old. Stromatolites** are algae-like colonies of cyanobacteria interlayered with carbonate sediments (see Figure 23.9). The success of these early organisms depended on their ability to survive in the primitive, oxygen-poor environment. These microorganisms evolved a simple version of photosynthesis. Photosynthetic organisms combine CO_2 and the Sun's energy to make simple sugars—a biologically usable energy source.

Oxygen is a waste product of the photosynthesis reaction. The expelled oxygen, however, did not yet accumulate in the atmosphere; instead, it dissolved in seawater and went into the production of certain rocks. The dissolved oxygen reacted with dissolved iron, which was also abundant in the early oceans. The reaction produced

* A meteorite is any solid rock object from interplanetary space that has fallen to Earth's surface without being vaporized during its passage through the atmosphere. We will learn more about these objects in Chapter 26, when we study the formation of the solar system.
** Stromatolites still exist today—they are valuable "living fossils." They can be found in hypersaline lakes and protected marine lagoons, such as Shark's Bay in Western Australia.

LINK TO CHEMISTRY

Banded Iron Formations

Earth's rocks reveal our planet's history. Even Earth's early atmosphere can be determined by studying minerals in the rock record. Four billion years ago, the atmosphere was about 80% carbon dioxide, 10% nitrogen, and 10% water vapor. Today the atmosphere is about 78% nitrogen and 21% oxygen. Quite a difference! What happened to all the carbon dioxide? Where did all the oxygen we have today come from? One process sums it up: photosynthesis.

Much of the early carbon dioxide was dissolved in seawater, where it reacted with dissolved calcium—and to some degree, magnesium—to form limestone. So the first limestones had their origins during the earliest stages of Earth's atmosphere. As photosynthesizing organisms died, their remains sank to the bottom of the ocean to be incorporated into the sediment—more limestone.

Oxygen also dissolves in water. When oxygen production began because of photosynthesis, much of it dissolved in seawater. But before the increased concentration of dissolved oxygen, the oxygen-poor environment allowed dissolved iron to accumulate in the oceans. So dissolved iron was abundant in the early oceans. Then when dissolved oxygen became plentiful, it reacted with the dissolved iron, causing iron in the water to precipitate. So when the air contained little or no oxygen, there was little oxygen in the seawater, but large amounts of dissolved iron. When oxygen concentrations rose to a certain threshold level, iron precipitated out of the seawater to form a layer of solid iron oxide minerals on the seafloor. When oxygen levels decreased, so did iron precipitation. This cyclical pattern of varying oxygen levels is depicted in the alternating layers of iron oxide minerals and iron-free sediment—the banded iron formation.

This process continued from about 2.6 to 1.9 billion years ago, a period of around 700 million years. That's how the iron-rich rocks formed, rocks that now support our industrial society. The world's iron and steel industry is based almost exclusively on iron ores associated with Precambrian banded iron formations.

solid iron oxides, which chemically precipitated to form *banded iron formations*. These ancient rock formations are a testimony to the great amount of oxygen that was expelled during this time.

The Proterozoic marks the end of Earth's longest time span—and this eon lasted another 2 billion years! Let us highlight a few changes that occurred during this expanse of time, starting with the configuration of land. All continental landmasses merged to form a single supercontinent—*Rodinia* (Russian for "homeland"). Rodinia stretched between the North and South Poles and then split to form other continental configurations. Climatically, Earth cooled to the point at which glaciers covered much of its surface. The reasons for such drastic cooling are unclear, but the fact that many of the continents were near either the North Pole or South Pole is certainly a factor.

Life in the Proterozoic flourished in new directions. Evolution produced single-celled organisms with a nucleus (eukaryotes). Fossils of such are found in rocks dated at approximately 1.5 billion years ago. Another important change was the debut of multicellular plants and animals (700 Ma). Proterozoic rocks in southern Australia contain diverse fossils of soft-bodied animals and provide us with the first evidence of an animal community that lived in shallow marine water.

Probably the most important highlight is the accumulation of free oxygen in the atmosphere. The early Proterozoic atmosphere was still mostly nitrogen, with a little water vapor and carbon dioxide. But free oxygen released by photosynthesizing plants in the oceans began to collect in the air. Up until this time, expelled oxygen had been sequestered in the great iron deposits around the world. But as oxygen production outpaced its uptake by iron, free oxygen began to enrich the atmosphere. Then, with sufficient free oxygen (O_2) in the atmosphere, a primitive ozone layer (O_3) began to develop above Earth's surface. This ozone layer was very significant—it reduced the amount of harmful ultraviolet (UV) radiation reaching Earth's surface. The accumulation of free oxygen in Earth's atmosphere and the added protection from the newly formed ozone layer led to the emergence of new life.

CHECKPOINT

1. **In what two ways was the development of free oxygen essential to the development of new life forms?**
2. **Where did the carbon dioxide that characterized the early atmosphere go?**

Were these your answers?

1. Free oxygen, in the form of O_3, provided protection from harmful UV rays; in the form of O_2, it provided oxygen for respiration.
2. Most of the carbon dioxide dissolved in the oceans, combined with dissolved calcium, and formed limestone ($CaCO_3$). Some was consumed by one-celled photosynthesizing organisms, which when dead, were also incorporated into rocks.

LEARNING OBJECTIVE
Describe the significant life changes during each period in the Paleozoic.

23.5 The Paleozoic Era (543 to 248 Million Years Ago)

EXPLAIN THIS What can happen to sea level and shallow marine life when all landmasses are joined together?

FYI Many Paleozoic rocks are economically important. For example, much of the limestone quarried for building and industrial purposes, as well as the coal deposits of western Europe and eastern United States, were formed during the Paleozoic.

More is known about the **Paleozoic era** than the Precambrian, but the Paleozoic is very short in comparison. The Paleozoic era began about 543 million years ago, and it lasted about 295 million years. During this time, sea levels rose and fell worldwide several times. This allowed shallow seas to partially cover the continents and marine life to flourish. Varying sea levels greatly influenced the progression and diversification of life forms—from marine invertebrates to fishes, amphibians, and reptiles. An important event in the Paleozoic era was the evolution of shelled organisms. In fact, it is because of shelled organisms that we know so much more about the Paleozoic than Precambrian time. Shelled organisms have hard parts that are more likely to be preserved as fossils. The Paleozoic era is divided into six periods, each characterized by changes in life forms and changes in land configurations.

The Cambrian Period (543 to 490 Million Years Ago)

The Cambrian period marks the beginning of the Paleozoic era. Global temperatures were significantly warmer than during the preceding icy Proterozoic. In fact, fossil evidence suggests that temperatures were much warmer than they are today. As for land configuration, there were no continental lands located at the poles and no significant ice formation. The ice sheets and glaciers from the

Proterozoic melted, which resulted in rising sea levels. Low-elevation areas of the continents were flooded, and much of the land became covered by shallow seas. This flooding expanded the habitats for early Paleozoic marine organisms. These organisms thrived and evolved to produce a great diversity of life. Hence, this part of Earth's history has become known as the *Cambrian explosion*.

Almost all major groups of marine organisms came into existence during the Cambrian, as shown by abundant fossil evidence. The most important event of this time was that organisms evolved the ability to secrete calcium carbonate and calcium phosphate for the formation of outer skeletons, or shells. This helped organisms become less vulnerable to predators, and it provided protection against harmful UV rays. With outer skeletons, many organisms moved to shallower marine habitats.

So because of hard body parts, the fossil record of Cambrian life is well preserved and dominated by the skeletons of shallow marine invertebrates. A variety of organisms flourished, including the trilobite, the armored "cockroaches" of the Cambrian sea (Figure 23.11).

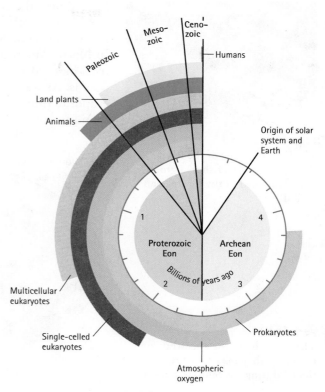

FIGURE 23.10
This "clock" shows the time when major groups of organisms appeared on Earth. The Paleozoic marked the development of a variety of animal and plant life.

The Ordovician Period (490 to 443 Million Years Ago)

During the Ordovician, the ancestral continental landmasses of South America, Africa, Australia, Antarctica, and India were merging together to form a new supercontinent called *Gondwanaland*. Throughout the Ordovician, Gondwanaland shifted southward, finally settling on top of the South Pole. With so much landmass over the South Pole, the latter part of the Ordovician was one of the coldest times in Earth history. Ice and massive glaciers covered much of Gondwanaland, draining the shallow seas. As sea level dropped, many shallow-water invertebrates were deprived of their habitat.

Fossil records show that the early to mid-Ordovician period was a time of great diversity and abundant marine life. The Ordovician also marks the earliest unquestionable arrival of vertebrates, including the jawless fishes (Figure 23.12). The end of the Ordovician period is marked by a surge in the rate of extinctions, probably the result of the widespread cooling and glaciation described above. The extinctions mainly affected shallow-water marine groups because of habitat loss, while deep-water organisms were relatively untouched.

The Silurian Period (443 to 417 Million Years Ago)

During the Silurian period, Gondwanaland remained close to the South Pole, and the ancestral continents of North America, Europe, and Siberia were

FIGURE 23.11
Trilobites are the dominant fossils of the Cambrian period.

> **FYI** How do warming temperatures raise sea levels?
> Several ways: When ice caps and glaciers on land melt, the meltwater flows into the ocean, raising its level (the North Pole ice cap floats on open ocean; its melting would not raise sea level). Also, ocean water expands in volume as its temperature rises—this is an example of thermal expansion, which was discussed in Chapter 6.

FIGURE 23.12

The hagfish is one of the few remaining jawless fishes, a group that flourished in the Ordovician period.

located near the equator (Figure 23.13). The world's climate began to stabilize and warm, which resulted in the melting of many large glaciers and a general rise in sea level. Shallow seas moved in to cover many continental interiors. In North America, inland shallow seas were surrounded by coral reefs. These reefs blocked the circulation of water between the inland seas and the open oceans. As water in the inland seas evaporated, deposits of gypsum and other evaporite minerals were left behind. Evaporite beds formed in the Silurian are found today in Ohio, Michigan, and New York.

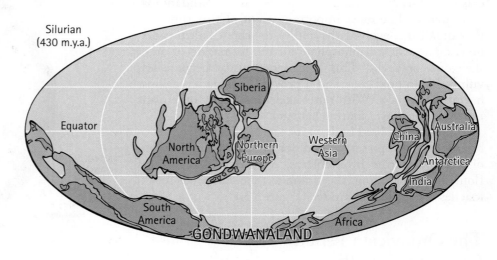

FIGURE 23.13

This map shows the ancestral continents as they may have been positioned during the Silurian period.

The Silurian period brought the emergence of terrestrial life—plants. The earliest known land plants with a well-developed "circulatory" system* appeared during the Silurian. These plants were closely connected to their water origins and inhabited only low wetlands. As plants moved ashore, so did other terrestrial organisms. Air-breathing scorpions and millipedes were common land animals during this period in Earth's history.

The Devonian Period (417 to 354 Million Years Ago)

During the Devonian period, the ancestral continents of Europe and North America merged to form another supercontinent known as *Laurasia* (ancestral Europe, North America, Siberia, and North China), which was positioned near the equator. The supercontinent of Gondwanaland remained in the Southern Hemisphere.

The Devonian climate was generally warm and moist. Plants spread over the land surface, and lowland forests of trees and ferns flourished. In the seas, fishes diversified into many new groups, which is why the Devonian is known as the "age of the fishes." Some groups, such as sharks and bony fishes, are still present today. Among the bony fishes, lobe-finned fishes are of particular interest because they gave rise to land-living, terrestrial vertebrates. Some lobe-finned fishes evolved internal nostrils, which enabled them to breathe air. In addition, the fins of these fishes were lobed and muscular, enabling the animals to support their bodies and "walk." Today, lungfishes and the coelacanth (pronounced "SEE-la-kanth") are the only lobe-finned fishes still in existence. The first truly terrestrial vertebrates, descendants of lobe-finned fishes, appeared during the late Devonian. These vertebrates share many features with the

FYI The coelacanth was thought to have become extinct after the Mesozoic era. However, in 1938, the first living specimen was caught off the coast of East Africa. Since then, other specimens have been discovered in the Madagascar area. The coelacanth is now considered a "living fossil."

---

* The "circulatory" system that distributes water and other resources for a plant is called a *vascular system*. We use the common term *circulatory* because it is easy to visualize.

amphibians of today. For example, like today's amphibians, they laid unshelled eggs and could live only in wet environments (Figure 23.14).

The Carboniferous Period (354 to 290 Million Years Ago)

The Carboniferous period includes both the Mississippian and Pennsylvanian subperiods. During this time, the Paleozoic ocean between Laurasia and Gondwanaland began to close. Throughout the Carboniferous and into the Permian, these two large landmasses began merging to become the supercontinent of **Pangaea** (Greek for "all lands"). The collision of these landmasses contributed to great mountain building—the Appalachian Mountains in North America, the Hercynian and Caledonian Mountains in Europe, and the Ural Mountains in Russia.

Pangaea stretched more or less continuously from pole to pole, with a large part of Gondwanaland at the South Pole and the northernmost tip of Laurasia at the North Pole. Such a stretch of land contributed to alternating periods of ice ages and warming throughout the Carboniferous. By the late Carboniferous, southern Gondwanaland was completely buried by large sheets of ice. The majority of the Laurasian landmass was still located along the equator and was characterized by warm, moist, tropical conditions that contributed to lush vegetation, forests, and swamps.

The name *Carboniferous*, or "carbon-bearing," refers to the swampy conditions that produced widespread coal deposits and characterized this part of Earth's history. In fact, dense swamps covered large portions of what are now North America, Europe, China, and Siberia (Figure 23.15). As plants and trees died, their remains settled to the bottoms of the stagnant swamps and decayed to produce coal (see "Link to Biology: Fossil Fuels" in Chapter 20). Most of the coal used today is derived from these Carboniferous coal swamps.

The Carboniferous period witnessed many life changes. Insects underwent rapid changes that led to diverse forms, including giant cockroaches and dragonflies with wingspans of 80 cm. The first amniotes, the vertebrate group that includes today's reptiles and mammals, also appeared at this time. Amniotes are characterized by a shelled, or *amniote*, egg. The amniote egg provides a completely self-contained environment for an embryo. The shell protects the embryo from drying out, allowing animals to complete the transition, begun by amphibians in the Devonian period, from aquatic environments to land. Thanks to the amniote egg, reptiles do not need to lay their eggs in water the way amphibians do.

The Permian Period (290 to 248 Million Years Ago)

The Permian period marks the end of the Paleozoic era. During this period, all the world's landmasses were still joined together as Pangaea (Figure 23.16). Like a chain reaction, the formation of Pangaea had many consequences. Continental collisions that started in the Carboniferous continued to bring about crustal disturbances that affected not only the edges of the continents but also their inner regions. The ancestral Rocky Mountains,* for example, owe their formation to the dramatic collision that produced Pangaea. In turn, mountain uplift blocked moisture-laden winds from the continental interiors, causing

FIGURE 23.14
This artwork of a late Devonian forest shows an *Acanthostega tetrapod*, an amphibian, climbing over a rock while a dragonfly flies above. Tetrapods flourished in the Devonian, having evolved from fishes.

FIGURE 23.15
Warm, moist climatic conditions contributed to the lush vegetation and swampy forests of the Carboniferous period. These forests produced most of the coal deposits around the world.

---

* The present-day Rocky Mountains formed much later, around 70 million years ago.

FIGURE 23.16
With the collision of continental landmasses, the supercontinent Pangaea was formed.

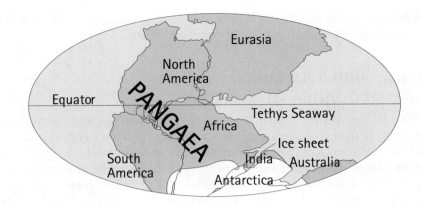

Think of continental collisions like the crumpling of a flat cardboard cereal box. Initially the box is a "flat" surface that takes up only horizontal space. If you push the ends of the box together, the length decreases and the height increases. The box becomes "folded." That's what happens when continents collide; stretches of long, flat crust are shortened and pushed upward into mountains, becoming folded in the process. And mountains, like the crumpled box, occupy both horizontal and vertical space.

much of Pangaea's interior to become dry and arid. So as land configurations changed, so did the Permian climate.

At the beginning of the Permian period, glaciers continued to cover much of the Southern Hemisphere, and tropical equatorial regions were covered by swampy forests. By the mid-Permian, however (after Pangaea was fully formed), the climate became warmer and milder—glaciers decreased and continental interiors became drier. All landmasses together also changed the level of the oceans—sea level was lowered. When sea level is lowered, more of the continental shelf (the area between land and the open ocean) is exposed. Because so much life lives in these shallow areas, these organisms began to lose their habitat. This is but one of the factors that may have led to the extinction of many life forms at the end of the period.

The late Permian is noted as being one of the greatest extinctions of animals in Earth's history. Marine invertebrates were affected more than terrestrial life. About 95% of all marine species and 70% of all land species became extinct. The cause of the extinction is not well understood. One likely explanation, of course, is the domino effect caused by the formation of Pangaea. The redistribution of land and water, the changes in landmass elevations, the change in the world's climate (temperature and precipitation), and the lowering of sea level all come into play.

The new land configuration radically decreased the amount of shallow marine environments. Overall, a large, single landmass has less coastline than that of the same landmass broken into several pieces. Each former piece had been encircled by coast. Put together, most of the former coasts of individual pieces became inland parts of Pangaea. All landmasses joined together also altered the oceanic and atmospheric circulation patterns that existed before the joining, which in turn altered the climate. And the long duration of lowered sea level, about 20–25 million years, undoubtedly placed more stress on already-stressed marine habitats. Whatever happened took a less drastic toll on terrestrial life. Terrestrial life, although affected, continued to evolve, and it expanded rapidly as new land habitats emerged, perhaps due in part to the lowered sea level.

The evolution of reptiles continued throughout the Permian. The reptiles must have been well suited to their environment, for they ruled Earth for 200 million years! (By comparison, modern humans have inhabited Earth for about 100,000 years.) Two major groups of reptiles appeared during the Permian: the diapsids and the synapsids. The synapsids, which include ancestors of the earliest mammals, dominated the Permian. The diapsids were less noticeable than the synapsids in the Permian, but it was the diapsids that eventually gave rise to the dinosaurs early in the Mesozoic era.

CHECKPOINT

1. As continental landmasses merged together, coastlines and continental shelf areas were reduced by two different processes. Look at Figure 23.16 and explain the two possible processes.
2. Why would the consolidation of landmasses into the supercontinent Pangaea increase the seasonal variation they experienced?

Were these your answers?

1. Some of the former coastlines and continental shelves were no longer next to the ocean—they were now in the middle of Pangaea! With all landmasses crunched together, less water was displaced so sea level went down, meaning some former continental shelf areas became dry land.
2. The high specific heat capacity of water has a moderating effect on temperature (as explained in Chapter 6). A consolidated landmass has a large interior area far away from the moderating effect of water, thus increasing the seasonal temperature variation.

23.6 The Mesozoic Era (248 to 65 Million Years Ago)

LEARNING OBJECTIVE
Describe how the breakup of Pangaea influenced climate, sea level, and certain life forms.

EXPLAIN THIS What can happen when members of the same species become separated and begin to thrive in a new environment?

The **Mesozoic era** is separated from the Paleozoic era because of profound changes in fossils. Reptiles that survived the Permian extinction at the end of the Paleozoic evolved to become the rulers of the Mesozoic world. The Mesozoic era consists of three periods—the Triassic, Jurassic, and Cretaceous—and is informally known as the "age of reptiles." The most significant evolutionary event of this era was the rise of the dinosaurs. Mammals evolved from reptiles early in the Mesozoic, but they were relatively small and insignificant compared to the great dinosaurs.

Land plants greatly diversified during the Mesozoic. True pines and redwoods appeared, and they rapidly spread throughout the land. Flowering plants arose in the Cretaceous period, and they diversified so quickly that by the end of the period, they were the dominant plants. The emergence of the flowering plants also accelerated the evolution and specialization of insects.

The major geologic event of the Mesozoic was the breakup of Pangaea. Just as Pangaea was formed in different phases throughout the Paleozoic, Pangaea broke apart phase by phase throughout the Mesozoic and into the Cenozoic (Figure 23.17). The first phase, during the Triassic/Jurassic (~200 m.y.a.), began with the formation of a *rift zone* between what are today's North American and African continents. Out of the rift zone, basaltic lava erupted to form new crust that further separated the diverging continental landmasses. As the rift zone spread, Europe split from North Africa. In this manner, Laurasia became completely separated from Gondwanaland. Also during this time, rifting occurred within Gondwanaland. For example, Africa split from Antarctica and Madagascar as a volcanic rift zone formed in the western Indian Ocean basin.

The second phase of the breakup occurred in the early Cretaceous, about 140 million years ago, as Gondwanaland continued to fragment. The rifting and separation of South America from Africa opened up the South Atlantic

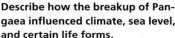

200 million years ago
Mesozoic era

65 million years ago
Cenozoic era

Present

FIGURE 23.17
Stages during the breakup of Pangaea.

FYI The Andes Mountains and the Sierra Nevada have their beginnings in the great volcanic activity that rimmed the eastern Pacific basin during the Mesozoic.

FYI New oceanic crust forms in rift zones as the seafloor spreads and basaltic lava rises from Earth's interior. When spreading is slow, ocean floors are cold and dense. They "stand" lower, so ocean basins are deep. When spreading is fast, ocean floors are warmer and less dense. They stand "higher," so ocean basins are shallow. When ocean basins stand high, seawater can flood the lower elevations of continents. In this way, shallow seas form on top of continental crust.

Ocean basin. Similarly, ancestral India/Madagascar rifted away from Antarctica/Australia to open up the Indian Ocean basin. Of all the former continental unions that existed in Paleozoic time, only that of Europe and Asia survive to the present time.

Recall the mountain building that resulted from Pangaea's formation. As rifting occurred *within* Pangaea, subduction ensued along Pangaea's margins. Of special interest is the subduction of the Pacific oceanic crust beneath the North American and South American continental crust. This produced widespread deformation, volcanism, and mountain building along the entire western coast—from "present-day" Alaska to Chile.

The separation of landmasses had several worldwide consequences. Coastlines were once again more extensive, which increased shallow marine habitat. Land-based life—plants and animals—also changed. As formerly connected habitats became separated, members of the same species also became separated. Now in new, unconnected habitats, organisms that shared a common ancestor or were closely related began to diverge from one another—they became less and less alike. In this way, new species evolved according to their new habitats.

Climate changed as well. Because most landmasses were no longer centered at the equator, the circulation pattern of the oceans and the atmosphere changed. Warm water from equatorial regions circulated northward to warm northern landmasses. Also, the rifting activity that initiated Pangaea's breakup resulted in a worldwide rise of sea level. So with shallow seas covering the continents, climate conditions became very mild. Overall, the Mesozoic climate was much warmer than conditions today.

The Cretaceous Extinction

The end of the Cretaceous period, 65 million years ago, brought another mass extinction that killed more than 60% of Earth's species. Many dinosaurs, flying reptiles, and marine reptiles were wiped out, as were other organisms, both on land and in the seas.

The cause of this great extinction is still a source of some debate among scientists. The most popular explanation is a hypothesis by Luis and Walter Alvarez (1980). They propose that the extinction was caused by the impact of a very large meteorite. The meteorite hit Earth with such force that it wreaked havoc in many ways. The impact caused large, widespread earthquakes and huge shock waves. It also released enormous amounts of heat, which resulted in extreme firestorms. In the meteorite impact's aftermath, a gigantic light-blocking cloud of dust enveloped Earth. The dust cloud spread throughout the atmosphere and lasted long enough to stop the process of photosynthesis. So, the cloud not only chilled Earth but also devastated the food supply. Evidence for this theory is found in the rock record. As the dust settled, a layer of iridium-enriched sediment was deposited. The element iridium is rare at Earth's surface, but it is abundant in meteorites. So the iridium layer that marks the boundary between the Cretaceous and Tertiary periods (called the K-T boundary) is hypothesized to have been spread worldwide by the extraterrestrial impact. The Chicxulub crater (dated at 65 million years), located off the Yucatan coast in Mexico, is the hypothesized site of impact (Figure 23.18). Both the K-T boundary layer and the Chicxulub crater coincide with the time of the great Cretaceous extinction.

An alternative to the Alvarez hypothesis suggests that the iridium layer may have been generated from massive volcanic eruptions. The ash and debris from these eruptions also could have blocked out the Sun. A third possibility is that

FIGURE 23.18

Artwork of the Chicxulub crater on the Yucatan Peninsula of Mexico, soon after its formation. The crater is about 180 km in diameter. The impact that produced the crater may have caused the mass extinction that ended the reign of the dinosaurs, 65 million years ago at the end of the Cretaceous.

large-scale volcanic eruptions could have been caused by the impact of an extra-terrestrial object. Whatever the cause, the Cretaceous extinction dramatically marked the close of the Mesozoic era.

> The shifting of Earth's tectonic plates changes not only land configurations but also climate and even life forms! When you change one factor, many other factors change. Hmmm . . . Like Newton's third law: You cannot touch without being touched.

CHECKPOINT

1. **Could the breakup of Pangaea account for the extinction at the end of the Cretaceous?**
2. **In terms of the number of species destroyed, which was the bigger extinction: the Permian or the Cretaceous?**

Were these your answers?

1. There is no doubt that the reconfiguring of landmasses when Pangaea broke up put added stress on many areas of life. Climate changed, shallow marine habitats changed, species became separated and diverged from one another, moving landmasses triggered widespread volcanic activity, and so on. But the breakup of Pangaea cannot explain the high concentration of iridium at the K-T boundary.
2. The Permian was the bigger mass extinction, having killed about 95% of all marine species and 70% of all land species. The Cretaceous extinction resulted in the demise of about 60% of Earth's species.

23.7 The Cenozoic Era (65 Million Years Ago to the Present)

> **LEARNING OBJECTIVE**
> List the tectonic and life form highlights of the Cenozoic.

EXPLAIN THIS Why is there a bend in the Hawaiian Island/Emperor Seamount chain?

The **Cenozoic era**, known as the "age of mammals," is made up of two periods—the Tertiary and the Quaternary. From oldest to youngest, these two periods are broken into the Paleocene, Eocene, Oligocene, Miocene, and Pliocene epochs (for the Tertiary period) and the Pleistocene and Holocene epochs (for the Quaternary period). We are currently living in the Holocene epoch.

The third and final phase of Pangaea's breakup took place during the early Cenozoic. North America and Greenland split away from Europe, and Antarctica split from Australia. Earth's surface was very active, and landmass collisions were numerous. Considerable mountain-building activity occurred when Africa-Arabia collided with Europe to produce the Alps and India collided with Asia to produce the Himalayas. As shown in Figure 23.19, India moved toward Asia at a rate of 15–20 cm per year—a tectonic speed record! As India partially wedged itself below Asia, it generated an unusually thick accumulation of continental crust. Due to isostasy, this continental crust atop continental crust provided additional height to the Himalayas.

The late-Mesozoic collision that began the formation of the Sierra Nevada and Andes Mountains continued (and continues) into the Cenozoic. Off the western coastline of North America, a midocean ridge formed in the Pacific Ocean—the Pacific Ridge system. This ridge separated two different oceanic plates—the Pacific Plate on the west and the Farallon Plate on the east

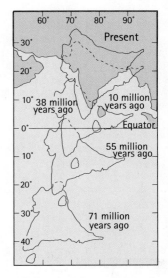

FIGURE 23.19
The formation of the Himalayas was a result of the collision of India with Asia. Because this was a continent-to-continent collision, the Himalayas have an unusually large thickness of continental crust.

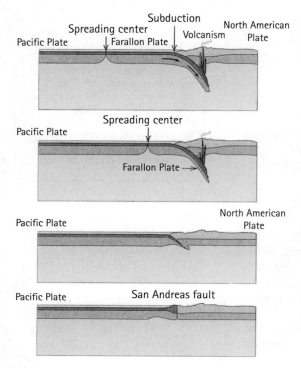

FIGURE 23.20

Subduction sequence of the Farallon Plate beneath the North American Plate. As the spreading center moved toward the North American Plate, the San Andreas fault began to form.

(Figure 23.20). As the Farallon Plate subducted below the North American Plate, the Pacific Ridge began to approach the North American continental margin. This not only reactivated mountain building throughout the west but also contributed to widespread deformation. The collision of the Pacific Ridge with the westward moving North American Plate occurred about 30 million years ago and gave birth to the San Andreas fault (Figure 23.21). In time, as the fault grew, Baja California was torn away from the Mexican mainland, and the Gulf of California was created. Because these plates are still moving, western California and Baja California will eventually become completely detached from the mainland or will find themselves joined to western Canada.

The Hawaiian Island/Emperor Seamount chain (Figure 23.22) provides additional evidence of Cenozoic deformation: the change in direction of the Pacific Plate. The bend in the chain of islands occurred between 30 and 40 million years ago (mid-Tertiary) when the direction of plate motion changed from nearly due north to northwesterly. The change in direction occurred at about the same time as the collision of northern Mexico (the North American Plate) with the Pacific Ridge system.

Climates cooled during much of the Cenozoic era, resulting in the widespread glaciation that characterized the Pleistocene.

FIGURE 23.21

INTERACTIVE FIGURE MP

The San Andreas fault is the result of an encounter between the North American Plate and the Pacific Ridge system. As the fault grew longer, the area of Baja California was torn from the Mexican mainland.

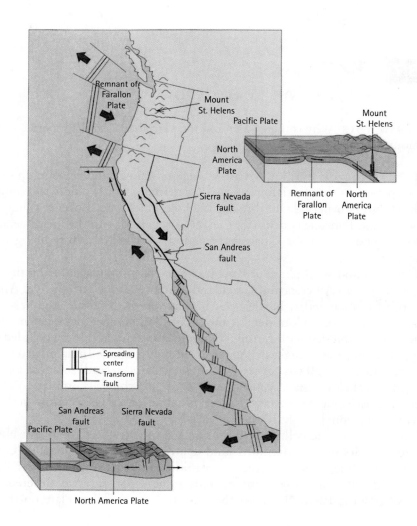

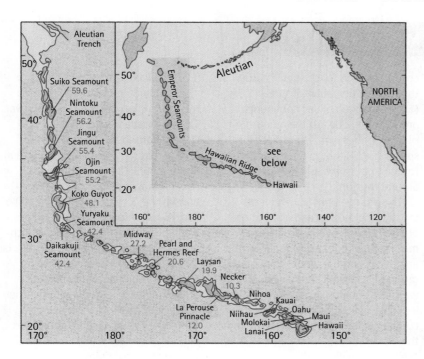

FIGURE 23.22
The Hawaiian Island/Emperor Seamount chain. The bend in the chain shows the change in direction of the Pacific Plate, which is likely a result of the collision of northern Mexico with the Pacific Ridge. The red numbers indicate the age (in millions of years) of the individual islands and seamounts.

Although this ice age continues today, there have been many alternations between glacial and interglacial conditions. During the glacial peak, as much as one-third of Earth's surface was covered by ice. On land, great thicknesses of ice formed continental glaciers. The great weight of ice depressed the land and altered the courses of many streams and rivers. The glaciers eroded and scratched the land in some places and deposited huge *moraines*, leaving behind abundant evidence of the extent of their former existence.

Cenozoic Life

After the mass extinctions at the end of the Mesozoic era, many environmental niches—habitats—were left vacant. These openings allowed the relatively rapid evolution of mammals. Bats, some large land mammals, and such marine animals as whales and dolphins evolved to occupy niches left vacant by the extinction of many of the Mesozoic reptiles. Later in the Cenozoic, as global temperatures cooled, the Ice Age began. The cool temperatures had a profound impact on life. Mammoths, rhinos, bison, reindeer, and musk oxen all evolved warm, woolly coats for protection from the frigid cold.

Humans evolved in the Quaternary period during the Pleistocene epoch. Extensive glaciation caused sea level to drop as water became bound up in glaciers and ice caps. Even though the distribution of landmasses was essentially the same as it is today, the lowered sea level resulted in "land bridge" connections between landmasses that are now separated by water. One of these land bridges existed across the present-day Bering Strait, and it provided the route for the human migration from Asia to North America.

The expansion of humans, not only into North America but also throughout the world, coincided with a period of extinctions that occurred during the Pleistocene. The extinctions involved primarily large terrestrial mammals, while marine animals were for the most part unaffected. In North America, many large mammals became extinct after humans arrived, and in Africa, mammalian extinctions can be related to the appearance of the Stone Age hunters.

> It would not have been possible for large mammals to evolve if dinosaurs had not gone extinct. Such mammals would have been tasty treats for large meat-eating dinosaurs!

LINK TO GLOBAL THERMODYNAMICS

Is It Cold Outside?

Yes it is, relatively speaking. For 90% of Earth's history, there were no glaciers of continental scale anywhere. In fact, because such glaciers exist today, mainly in the polar ice caps and in Greenland, we are technically now in an *ice age*. Because these continental-scale glaciers are restricted to polar regions, we are in what is known as an *interglacial* period of an ice age.

Ice ages have occurred five times over the course of Earth's history. The first one for which we have evidence occurred more than 2 billion years ago. Another began about 840 million years ago and lasted about 200 million years! There were two ice ages during the Paleozoic era, but none in the Mesozoic. For the first 50 million years or so of the Cenozoic era, there were also no ice ages. The present ice age actually began 8–10 million years ago, but the extensive glaciation that characterized the Pleistocene epoch began about 1 million years ago.

So what causes ice ages? There is, most likely, no single explanation, but most scientists agree that global-scale cooling leading to ice ages is caused by the right combination of three things: (1) the arrangement of continents around the globe, (2) the amount of sunlight being reflected back into space, and (3) the geometry of Earth's rotation on its axis and revolution around the Sun.

The arrangement of continents greatly influences ocean and atmospheric currents, which are the main mechanisms for redistributing thermal energy around the globe. Continents grouped in one location are easier to warm, as equatorial waters flow with less obstruction toward the poles. For continents that are spread out around the globe, as they are today, circulation "cells" are smaller and heat redistribution is more local and less efficient.

When sea level is lower, for whatever reason, more land area is exposed. The increased amount of land area tends to increase the amount of sunlight reflected back into space. This phenomenon results in cooler temperatures globally. Cloud cover and/or dust in the atmosphere also causes sunlight to be reflected back into space, reducing the absorption of solar radiation.

The *Milankovitch effect* refers to a combination of factors that affect the distribution of solar radiation over Earth's surface during the year: (1) variations in the angle at which Earth's rotational axis is tilted (currently about 23.5°), (2) the wobbling of Earth's rotational axis, and (3) variations in the eccentricity ("ovalness") of Earth's orbit around the Sun. Certain combinations of these factors, which recur periodically, lead to reduced solar radiation at high northern latitudes during the summer. If the reduction of that radiation is great enough, then not all the snow from the preceding winter melts and, if these conditions persist over many years, continental-scale glaciers eventually form.

The periodic nature of the Milankovitch effect may be the primary cause of glacial–interglacial cycles. According to this theory, the first two processes—the arrangement of continents and the amount of reflected radiation—make Earth cold enough, and the third—the Milankovitch effect—causes the climate to teeter-totter between glacial and interglacial periods.

The cause of the Pleistocene extinction is a much-debated issue. The extreme climatic variation during this time could have been partly responsible. However, though large-scale glaciation was occurring in some regions, the climate in many areas was relatively mild. This leads scientists to believe that harsh climate likely played only a small role in the Pleistocene extinctions and that humans essentially hunted and ate the large mammals to extinction. As with past extinctions, a single cause is unlikely. The Pleistocene extinction probably resulted from a combination of factors.

Following the Pleistocene is the Holocene epoch. To observe the Holocene environment, just look around. It is the most recent 10,000 years or so of Earth's history, including the present time. There has been some climatic variation. For example, the "Little Ice Age" occurred between about AD 1200 and 1700. However, in general, the Holocene has been a relatively warm interglacial period.

The Holocene is sometimes called the "human age." This is somewhat inaccurate because *Homo sapiens* had evolved and occupied many regions of the globe well before the start of the Holocene. Yet the Holocene has witnessed all of humanity's recorded history, including the rise and fall of all historic civilizations. Humanity has had a great impact on the Holocene environment. All organisms influence their environments to some degree, but few have ever

FYI The Anthropocene epoch is young and dates back to the time of the Industrial Revolution—the 1800s. The changes to our environment because of human population and economic development have been so drastic that post-industrialized Earth is very different from conditions in the Holocene.

changed Earth as much, or as fast, as we are doing in modern times. It is because of these changes that many scientists claim we have entered a new epoch—the Anthropocene.

23.8 Earth History in a Capsule

EXPLAIN THIS Is everything about Earth's history already known?

Our Earth has a long and exciting history—4.5 billion years' worth! To unravel this history we look at the rock record. Each event in Earth's long history is recorded in its rock layers. The order of geologic events is deciphered by relative dating, and absolute age is determined by radiometric dating. The geologic time scale provides a chronological listing of different time intervals according to the fossil record. When major changes in life occurred, a new era and period began.

Throughout geologic time, life has changed, landforms have changed, and climate has changed. Change in one area influences change in another area. Each change is recorded in the rocks. As mentioned earlier, the Earth scientist is like a crime scene investigator. The crime scene is Earth, and the rock record provides us with clues to unravel *what* happened and *when* it happened. Earth history is a like a big puzzle—we can understand the picture only by fitting in one piece at a time.

HUMAN GEOLOGIC FORCE

Although the "human age" amounts to only a brief 0.002% of geologic time, we are almost certainly the most clever and adaptable organisms to have evolved on the planet. All life forms alter their environments. Humans do it more, as we manipulate our environment to meet our needs. We have but to look at the irrigation systems of Mesopotamia, the cultivation of the Nile, the plowing of the prairies in the Great Plains, the invention of machines to further utilize the land, and the dams and locks on the Mississippi, Missouri, and Colorado rivers to illustrate the human role in geologic changes. Will the human activities that adversely affect our planetary life-support system—the large-scale burning of fossil fuels, the deterioration of the ozone layer, the destruction of the Amazon Rainforest, and of particular concern, global warming—ultimately lead to a new period of extinctions? Perhaps our own. Who knows? Because we have the capacity to affect geologic change, it is imperative that we take care of our terrestrial home. It's the only one we have!

For instructor-assigned homework, go to www.masteringphysics.com

---

SUMMARY OF TERMS (KNOWLEDGE)

Angular unconformity An unconformity in which older tilted rock layers are covered by younger, horizontal rock layers.

Cenozoic era The time of recent life, it began 65 million years ago and is ongoing.

Cross-cutting relationships Where an igneous intrusion or fault cuts through other rocks, the intrusion or fault is younger than the rock it cuts.

Faunal succession Fossil organisms succeed one another in a definite, irreversible, and determinable order.

Inclusions Any inclusion (pieces of one rock type contained within another) is older than the rock containing it.

Lateral continuity Sedimentary layers are deposited in all directions over large areas until some sort of obstruction, or barrier, limits their deposition.

Mesozoic era The time of middle life, from about 248 million years ago to 65 million years ago.

Original horizontality Layers of sediment are deposited evenly, with each new layer laid down nearly horizontally over the older sediment.

Paleozoic era The time of ancient life, from about 543 million years ago to 248 million years ago.

Pangaea The late-Paleozoic supercontinent made up of Gondwanaland (ancestral South America, Africa, Australia, Antarctica, and India) and Laurasia (ancestral North America, Europe, and Siberia/Asia).

Precambrian time The time of hidden life, which began about 4.5 billion years ago when Earth formed, lasted until about 543 million years ago (beginning of the Paleozoic), and makes up almost 90% of Earth's history.

Radiometric dating A method for calculating the age of geologic materials based on the nuclear decay of naturally occurring radioactive isotopes.

Relative dating The ordering of rocks in sequence by their comparative ages.

Superposition In an undeformed sequence of sedimentary rocks, each bed or layer is older than the one above and younger than the one below.

Unconformity A break or gap in the geologic record, caused by erosion of preexisting rock or by an interruption in the sequence of deposition.

READING CHECK QUESTIONS (COMPREHENSION)

23.1 The Rock Record—Relative Dating

1. What is relative dating?
2. What six principles are used in relative dating? Describe each one.
3. A granitic dike is found across a sandstone layer. What can be said about the relative ages of the dike and the sandstone? What principle applies here?
4. Why don't all rock formations show a continuous sequence from the beginning of time to the present?
5. How are fossils used in determining geologic time?
6. In a sequence of sedimentary rock layers, the oldest layer is on the bottom and the youngest layer is at the top. What relative dating principle applies here?
7. Explain how fossils of fishes and other marine animals are found at high elevations, such as the Himalayas.

23.2 Radiometric Dating

8. What is meant by radioactive half-life?
9. What are the half-lives of uranium-238, potassium-40, and carbon-14?
10. What isotope is preferred in dating very old rocks?
11. What isotope is commonly used for dating sediments or organic material from the Pleistocene?

23.3 Geologic Time

12. Which of the geologic time units spans the greatest length of time?
13. How old is Earth?

23.4 Precambrian Time

14. What key developments in life occurred during Precambrian time?
15. What evidence do we have of Precambrian life?

23.5 The Paleozoic Era

16. The Paleozoic era experienced several fluctuations in sea level. What effect did this have on life forms?
17. Name the periods of the Paleozoic era.
18. What life form emerged in the Silurian period?
19. What life forms are associated with the Devonian period?
20. Why are internal nostrils in the lobe-finned fishes an important step in the evolution of life on Earth?
21. Why do many geologists consider the lobe-finned fishes especially significant?
22. During what time period were most coal deposits laid down? Why was this period unique?
23. In what area of the United States do we find rich coal deposits?
24. What group evolved from the amphibians with the arrival of the amniote egg?

23.6 The Mesozoic Era

25. By what informal name is the Mesozoic era known?
26. What is the most likely cause of the Cretaceous extinction that wiped out the dinosaurs?
27. What effect did the breakup of Pangaea have on sea level?
28. How does the element iridium relate to the time of the extinction of the dinosaurs?

23.7 The Cenozoic Era

29. Which epochs make up the Tertiary period? The Quaternary period?
30. What important life forms evolved during the Cenozoic era?

ACTIVITIES (HANDS-ON APPLICATION)

31. Refer to the accompanying figure. Using the principles of relative dating, determine the relative ages of the rock bodies and other lettered features. Start with this question: What was there first?

Sequence of events

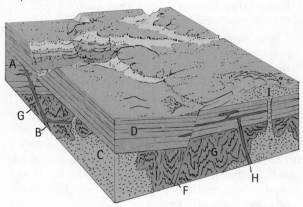

32. On a cross section, a dark wavy line is used to represent unconformities—missing layers of time. Match the following three types of unconformities—angular unconformity, nonconformity, and disconformity—to the diagrams.

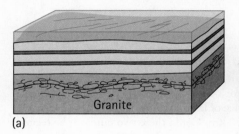

Granite

(a)

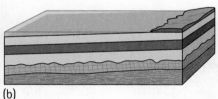

(b)

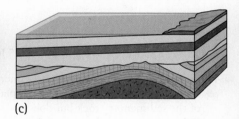

(c)

THINK AND SOLVE (MATHEMATICAL APPLICATION)

33. If fine muds were laid down at a rate of 1 cm/1000 yr, how long would it take to accumulate a sequence 1 km thick?

34. With the formation of Pangaea, disconnected continents merged into one huge landmass. This decreased the amount of shallow marine environments, which led to mass extinctions during the Permian. Let us compare the difference in coastlines for a single large landmass and a same-sized landmass that is broken up into several pieces.

(a) Find the perimeter for each shape shown here. Then add all the perimeters together.

(b) Like Pangaea, the shapes are joined into one landmass. Find the perimeter of the single landmass.

(c) Compare the the total perimeters for parts a and b. Do your results support the decrease in coastlines after the formation of Pangaea?

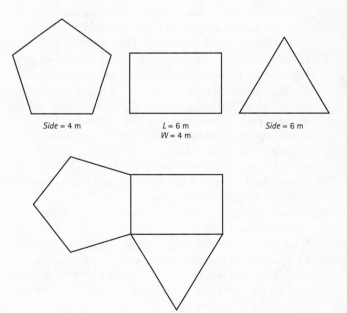

Side = 4 m

L = 6 m
W = 4 m

Side = 6 m

THINK AND RANK (ANALYSIS)

35. Going from oldest to youngest, rank these life forms: (a) rat bones, (b) fossil ferns, (c) trilobites, (d) dinosaur bones.

36. The geologic time scale is subdivided into eons, eras, periods, and epochs. Rank, from oldest to most recent, the following time periods: (a) Cretaceous, (b) Jurassic, (c) Triassic, (d) Tertiary, (e) Permian.

37. Earth has experienced several episodes of glaciation. Going from the most recent episode of glaciation to the oldest episode, rank the time periods: (a) late Precambrian, (b) Pleistocene, (c) early Permian, (d) Ordovician.

38. Throughout geologic time there have been several episodes of species extinction. Rank the following from most to least devastating: (a) Ordovician, (b) Permian, (c) Cretaceous, (d) Pleistocene.

39. Throughout geologic time there have been many changes to Earth's surface. From most recent to the distant past, rank these tectonic events: (a) formation of the Appalachian Mountains, (b) formation of Gondwanaland, (c) formation of Pangaea, (d) formation of the Gulf of Mexico.

40. The Cenozoic is known for many tectonic events. In sequential order, from first to last, list the following events: (a) formation of the Gulf of Mexico, (b) formation of the Himalayas, (c) carving of the Grand Canyon, (d) birth of the San Andreas fault.

41. Each period of the Paleozoic highlights changes in life forms. In sequential order, from first to last, rank these life forms: (a) reptiles, (b) age of fishes, (c) swamp lands, (d) terrestrial life, (e) vertebrates, (f) shelled organisms.

42. The Precambrian spans a huge amount of time. List the following Precambrian highlights from first to last: (a) stromatolites, (b) formation of the Moon, (c) primitive ozone layer, (d) formation of the oceans.

EXERCISES (SYNTHESIS)

43. Suppose you see a sequence of sedimentary rock layers covered by a basalt flow. A fault displaces the bedding of the sedimentary rock but does not cut into the basalt. Relate the fault to the ages of the two rock types in the formation.

44. If a sedimentary rock contains inclusions of metamorphic rock, which rock is older? Defend your answer.

45. How old are the oldest rocks on Earth? About how old are the oldest rocks on the Moon?

46. Which isotopes are most appropriate for dating rocks from the following ages: (a) early Precambrian time, (b) the Mesozoic era, (c) the late Pleistocene epoch?

47. Has the amount of uranium in Earth increased over geologic time? Has the amount of lead in Earth increased?

48. Granitic pebbles within a sedimentary rock have a radiometric age of 300 million years. What can you say about the age of the sedimentary rock? Nearby, a dike having a radiometric age of 200 million years intrudes into an outcrop of the same sedimentary rock. What can you say about the age of the sedimentary rock?

49. Before the discovery of radioactivity, how did geologists estimate the age of rock layers?

50. In dating a mineral, what is meant by "resetting the mineral's time clock"?

51. If we divide a number by 2, and then divide the result by 2, and so on indefinitely, the answer will never be zero. Why then is carbon dating useful only for materials that are no older than about 50,000 years? Hint: What is the half-life of carbon-14?

52. Geologists often refer to the early Paleozoic era as the *Cambrian explosion*. What do you think is meant by this term?

53. Two isolated rock outcrops that share similar characteristics—sequence of layers and fossil communities—are separated by 50 km. Could these two outcrops be related? What principle helps to confirm their relationship?

54. Two isolated rock outcrops share a few similar characteristics—sequence of layers, mineralogy, and certain fossil communities—but not all layers are present. The rocks are separated by 300 km. Could the two outcrops be related? How can you explain the missing layers?

55. Name the eons in Precambrian time. Describe each eon using no more than three words.

56. Suppose that in an undeformed sequence of rocks, you find a trilobite embedded in a black shale layer at the bottom of the formation and fossil leaves embedded in a green shale at the top of the formation. From your observation, what can you say about the ages of the formation?

57. In a sequence of sedimentary rock layers, the youngest layer is found at the bottom and the oldest layer at the top. What does this type of layering signify?

58. What is the difference between a nonconformity and an angular unconformity?

59. How did the Precambrian atmosphere become nitrogen enriched?

60. What factors are believed to have contributed to the generation of free oxygen during the late Precambrian? In what way did the increase in oxygen affect our planet?

61. Before entering the atmosphere, where did much of the newly released oxygen end up during the Precambrian?

62. Why are Paleozoic marine sedimentary rocks such as limestone and dolomite found widely distributed in the continental interiors?

63. Throughout the Paleozoic, sea level was variable; sometimes it was high and other times it was low. What was the primary cause of this variation in sea level?

64. What was the coldest period of the Paleozoic? What may have contributed to this cooling period?

65. During the Silurian, many continental landmasses were covered by shallow seas. Yet this time period marks the emergence of land plants. How can this be?

66. Coal beds form from the accumulation of plant material that has become trapped in swamp floors. Yet coal deposits are found on the continent of Antarctica, where no swamps or vegetation exists. What is your explanation?

67. A radiometric date is determined from mica that has been removed from a rock. What does the date signify if the mica is found in granite? What does the date signify if the mica is found in schist?

68. Most scientists think the iridium-rich sediments found straddling the Cretaceous-Tertiary boundary can be explained by a meteorite impact. Why are high concentrations of iridium significant?

69. During Earth's long history, life has emerged and life has perished. Briefly discuss the emergence of life and the extinction of life during each era.

70. In what ways can sea level be lowered? What effect might this have on existing life forms?

71. What can cause a rise in sea level? Is this likely to happen in the future? Why or why not?

72. What general assumption must be made to understand the processes that occurred throughout Earth's history?

73. What are some potential worldwide consequences that can occur when all landmasses are joined together as in the formation of Pangaea?

74. Was there a time when dinosaurs and humans coexisted?

75. Why does sea level rise when the rate of seafloor spreading increases?

76. If sea level were to rise today, what land areas would be most affected? What life forms would be in danger of extinction?

77. What circumstances are likely to lead to the formation of continental-scale glaciers?

78. How does basaltic lava in a rift zone separate two landmasses?

79. What is a likely cause of glacial–interglacial cycles? Are we currently in an ice age? Explain.

80. What impact did the breakup of Pangaea have on sea level?

81. What is the basis for the divisions of the geologic time scale?

82. What geologic event most likely resulted in the bending of the Hawaiian Island/Emperor Seamount chain?

83. How did Pleistocene glaciations affect the land surface?

84. How did the San Andreas fault form?

85. What was the impact on animal diversity as continents became separated from one another?

86. What event allowed the evolution of many mammals in the early Cenozoic era?

87. How was the Gulf of California formed?

DISCUSSION QUESTIONS (EVALUATION)

88. How did the loss of coastlines affect shallow marine life? Could this happen again?

89. How have modern humans affected geologic processes?

90. Throughout geologic time, species have become extinct. Many scientists believe that Earth is presently experiencing one of the fastest mass extinctions of all time. This extinction is mainly the result of human activity. What are some possible effects of this dramatic loss of species?

91. Species can go extinct when they are unable to adapt to changes in the environment or compete effectively with other organisms. The cause of extinction can be catastrophic or natural. What are some possible explanations for extinction?

READINESS ASSURANCE TEST (RAT)

If you have a good handle on this chapter, if you really do, then you should be able to score at least 7 out of 10 on this RAT. If you score less than 7, you need to study further before moving on.

Choose the BEST answer to each of the following.

1. The principle of superposition is that each new
   (a) sedimentary layer is older than the layer above.
   (b) sedimentary layer is younger than the layer above.
   (c) layer of sediment is laid down nearly horizontally.
   (d) layer of sediment is laid down accordingly.

2. Life forms throughout Earth's past have occurred in a definite order. This is called the principle of
   (a) fossil assemblage.
   (b) faunal succession.
   (c) conformable fossils.
   (d) fossil determination.

3. The time it takes for 50% of a radioactive substance to decay is known as
   (a) radiometric dating.
   (b) carbon-14.
   (c) the proportion of atoms remaining.
   (d) the half-life.

4. Development of Earth's oceans was likely due to
   (a) water-rich meteors bombarding Earth's surface.
   (b) volcanic outgassing in Precambrian time.
   (c) slow convection in the mantle.
   (d) volcanic outgassing in the early Paleozoic.

5. The buildup of free oxygen was crucial to the emergence of life on Earth because it led to the formation of
   (a) O_2 for plants.
   (b) ozone, O_3, which helped screen Earth from harmful incoming UV radiation.
   (c) ozone, O_3, which primitive organisms could breathe.
   (d) the oceans, where life emerged.

6. The Paleozoic experienced several fluctuations in sea level. When sea level rises,
   (a) shallow seas cover the continents.
   (b) more water is tied up in glaciers, thus making the climate colder.
   (c) the climate turns warmer and swamps form.
   (d) ocean basins become shallow.

7. The most important event during the Cambrian period was the
   (a) emergence of the fishes.
   (b) ability of organisms to form an outer skeleton.
   (c) emergence of the trilobite.
   (d) ability of organisms to develop lungs.

8. The formation of the supercontinent Pangaea
   (a) resulted from the collision of all major landmasses.
   (b) produced widespread mountain building in the Himalayas.
   (c) resulted in extensive volcanic activity and flood basalts.
   (d) all of these

9. Glaciation during the Cenozoic resulted in
   (a) lowering of sea level worldwide.
   (b) carving Earth's surface (for example, the Swiss Alps).
   (c) land-bridge connections between various continents.
   (d) all of these

10. The creation of the San Andreas fault corresponded to the
   (a) collision of the Pacific Plate and the Hawaiian Islands.
   (b) collision of the Pacific Ridge system and North America.
   (c) movement of the Pacific Plate over a turbulent zone.
   (d) collision of India and Eurasia.

Answers to RAT

1. a, 2. b, 3. d, 4. b, 5. b, 6. a, 7. b, 8. d, 9. d, 10. b

CHAPTER 24
The Oceans, Atmosphere, and Climatic Effects

THE VIEW of Earth from space shows that our planet is colored distinct shades of blue and silver. The blue is due to the water in the oceans and the silver to clouds in the atmosphere. How did our beautiful, swirling atmosphere and oceans come to be? What is the atmosphere made of? How do the atmosphere and oceans interact? So many questions come to mind.

We begin to answer these questions by learning about the evolution of Earth's atmosphere and oceans. We then explore important features of these two fluid shells as we investigate the transfer of heat between them—and how this heat transfer affects Earth's climate. We conclude with a look at the mechanisms that influence atmospheric and oceanic circulation patterns and the impact these patterns have on climate.

24.1   Earth's Atmosphere and Oceans

EXPLAIN THIS   Is the composition of Earth's atmosphere still changing?

Seventy-one percent of Earth's surface is covered by water (Figure 24.1). The remaining 29% is land, most of which is located in the Northern Hemisphere (Figure 24.2). Although the many oceans are named for their various locations, they are really one big continuous ocean.

FIGURE 24.1
Most of Earth's surface is covered by water. We can divide Earth into (a) an ocean-dominated hemisphere and (b) a land-dominated hemisphere.

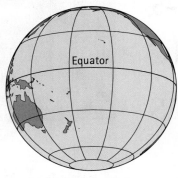

(a) Ocean hemisphere   (b) Land hemisphere

As we learned in Section 22.1, the oceans are the reservoir from which water evaporates into the atmosphere, only to precipitate later as rain and snow. The oceans play a major role in moderating Earth's temperature and climate. Recall from Chapter 6 that water has a high specific heat capacity—it is slow to heat up or cool down. As such, water transfers large amounts of thermal energy to its surroundings when it cools, and it absorbs large amounts of thermal energy from its surroundings when it warms. Because of water's high specific heat capacity, lands nearer to oceans have moderate temperatures. The moderating influence of the oceans can be seen when we look at seasonal temperature variations for two cities at the same latitude: coastal San Francisco, California, and continental Wichita, Kansas (Figure 24.3). Whereas temperatures in San Francisco tend to have small seasonal variations, temperatures in Wichita show strong seasonal fluctuations—cold winters and hot summers. The oceans do a great job of moderating climate, both by making summers cooler and by making winters warmer.

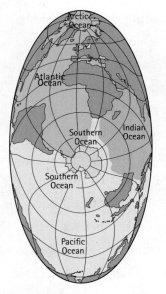

FIGURE 24.2
When a map is centered over Antarctica, the expanse of the world ocean can be seen. In terms of size and volume, the Pacific Ocean accounts for more than half of the world ocean and is thus the largest ocean. In fact, the Atlantic and Indian Oceans combined would easily fit into the space occupied by the Pacific.

Evolution of the Earth's Atmosphere and Oceans

Earth probably had an atmosphere before the Sun was fully formed. This primitive atmosphere was possibly composed of only hydrogen and helium, the two most abundant gases in the universe, along with trace amounts of ammonia and methane. There was no free oxygen in the early atmosphere. Then, when the temperature and pressure in the contracting center of the still-forming Sun became high enough to ignite thermonuclear reactions, our Sun was born. The release of energy from the Sun's formation likely produced a strong outflow of charged particles—an outflow strong enough to sweep Earth of its earliest atmosphere.

The next atmosphere formed as gases trapped in Earth's hot interior escaped through volcanoes and fissures at Earth's surface. The gases spewed out by these early eruptions were probably much like the gases found in the volcanic

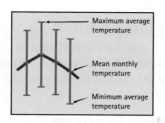

FIGURE 24.3
Comparison of seasonal temperature ranges for coastal San Francisco, California, and continental Wichita, Kansas.

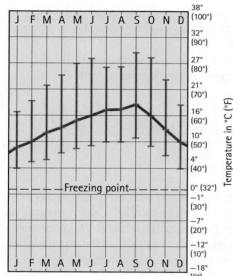

Station: San Francisco, California
Latitude/longtitude: 37°37' N, 122°23' W
Average annual temperature: 14°C (57.2°F)
Total annual precipitation: 47.5 cm (18.7 in.)
Elevation: 5 m (16.4 ft)
Population: 750,000
Annual temperature range: 9°C (16.2°F)

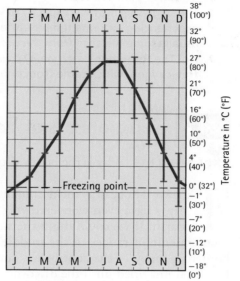

Station: Wichita, Kansas
Latitude/longtitude: 37°39' N, 97°25' W
Average annual temperature: 13.7°C (56.6°F)
Total annual precipitation: 72.2 cm (28.4 in.)
Elevation: 402.6 m (1321 ft)
Population: 350,000
Annual temperature range: 27°C (48.6°F)

eruptions of today—about 90%–95% water vapor, the rest mostly carbon dioxide. This early atmosphere still had no free oxygen and could not support the type of life we have today.

As Earth cooled, the rich supply of water vapor in the atmosphere condensed to form the oceans. Cometary debris from interplanetary space also contributed water to the oceans. These oceans, essential to the evolution of life and ultimately to the development of the present global environment, have remained for the rest of Earth's history.

With the oceans in place, and as blocks of land solidified, primitive bacteria—cyanobacteria—emerged to survive in the oxygen-poor environment. In time, these organisms, like the green plants that followed, developed a simple form of *photosynthesis* to convert carbon dioxide and water into carbohydrates and free oxygen:

$$CO_2 + H_2O + light \rightarrow CH_2O + O_2$$

As we learned in Chapter 23, much of this initial oxygen dissolved in the oceans, where it went into the production of limestone and iron oxide rocks. Then, as photosynthesizing plants flourished, free oxygen (O_2) finally began to accumulate in the air. With sufficient oxygen in the atmosphere, chemical reactions produced a primitive ozone (O_3) layer in the upper atmosphere. Because the ozone layer acts like a filter to reduce the amount of ultraviolet radiation reaching Earth's surface, the surface was then much more able to support life.

24.2 Components of Earth's Oceans

LEARNING OBJECTIVE
Describe the topography of the ocean floor and the factors that influence the ocean's salinity.

EXPLAIN THIS Why is the ocean salty?

The world ocean is the dominant feature of our planet; some philosophy types have even wondered why our planet isn't called Ocean instead of Earth! If we could drain the water from Earth's oceans, we'd see enormous mountain ranges in the middle of the ocean basins and deep trenches bordering many of the continents, not the flat, featureless ocean bottom that had been imagined until the discoveries of the mid-20th century.

The ocean is an immense body of salty water. Its salty nature contributes to its variable density, which contributes to the movement of ocean currents. Waves move through the ocean in response to atmospheric movements, but the ocean also responds to extraterrestrial influences—the Moon and the Sun, which create tides. With so much of Earth's surface covered by oceans, the oceans hold a wealth of scientific treasures. But because it is so vast, we still have much to discover—we know more about the surface of the Moon than we know about the ocean floor!

The Ocean Floor

The features of the ocean floor are very pronounced. In fact, land rises, on average, about 840 m above sea level, while the ocean bottom drops, on average, about 3800 m below sea level. If we compare the height of Mt. Everest in the Himalayas, at a majestic 8848 m above sea level, with the depth of the Marianas Trench in the Pacific Ocean, an astounding 11,035 m below sea level, we see that the oceans are much deeper than land mountains are high.

The ocean floor consists of continental margins, deep ocean basins, midocean ridges, and deep trenches. Away from the continental margins, the topography of the ocean floor varies greatly. The midocean ridges that encircle the globe are tall and variable, the sediment-covered ocean bottom is relatively flat, and seafloor trenches near continental margins can be very deep (Figure 24.4).

FYI The evidence of volcanic activity on the ocean bottom is impressive—more than 20,000 volcanic peaks have been found in the Pacific Ocean alone!

FIGURE 24.4
Map of the ocean floor showing variation in topography. (a) Atlantic profile. (b) Pacific profile.

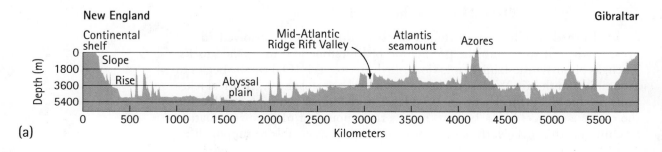

(a)

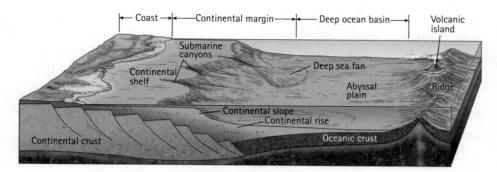

FIGURE 24.5
Profile of the continental margin going from land to the deep ocean bottom.

The **continental margin** marks the boundary between continent and ocean. As Figure 24.5 shows, the continental margin consists of a *continental shelf* (the submerged upper, landward portion of the margin), a *continental slope* (the breakpoint where the shelf steepens as it descends to depths of 2–3 km), and a *continental rise* (the area from the base of the slope seaward to the deep ocean floor).

Closest to the continent, the *continental shelf* is a gently sloping, underwater surface extending from the shoreline toward the ocean basin—it is an underwater extension of the continental crust. Due to fluctuations in sea level, the extent of the continental shelf changes over geologic time. When sea level is lower than at present, such as it was during the last glacial period, the continental shelf is narrower because less of the continental edge is submerged. On the other hand, when sea level is higher, as it would be if the ice caps on Antarctica and Greenland melted, the continental shelf would be wider because some of the continental land would be submerged. No matter what sea level is, the continental shelf varies widely from one geographic location to another. It is practically nonexistent in some locations, but extends seaward to a width of up to 1500 km in others.

The *continental slope* marks the boundary between continental crust and oceanic crust. The continental slope is steeper along mountainous coastlines that border convergent plate boundaries and more gently sloped along coastlines that lack plate boundaries (such as the eastern edge of North America and the Indian subcontinent).

The *continental rise* is the transition zone between the continental margin and the deep ocean floor. It is basically a wedge of continental sediment that has accumulated at the base of the continental slope. Not all continental margins have a continental rise. For example, along active convergent plate boundaries, the continental slope leads directly into a deep ocean trench.

The deep ocean basins account for about 30% of Earth's surface. Deep ocean basins are generally 3–5 km deep and are characterized by abyssal plains, ocean trenches, midocean ridges, and seamounts. Abyssal plains begin where the continental margin ends. They are some of the deepest and flattest regions

FYI Not all midocean ridges are truly "midocean." The East Pacific Rise, for example, is much closer to the eastern margin of the Pacific Ocean than it is to the middle. In any case, midocean ridges are where seafloor spreading occurs. Consequently, they mark divergent plate boundaries wherever they are located.

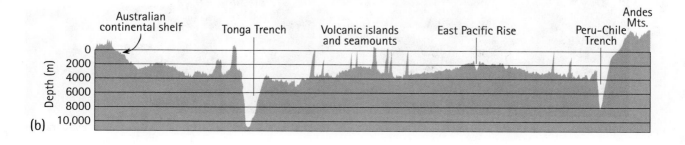

FYI Marine life influences the composition of seawater by removing salts, dissolved gases, and other solutes. Shellfish, tiny foraminifera (marine protozoa), and various crustaceans (such as crabs and shrimp) remove calcium salts to build their bodies; diatoms (microscopic marine algae) and some types of microscopic animals draw heavily on the ocean's dissolved silica to form their shells. Some animals concentrate elements present in seawater in minute, almost undetectable amounts. For example, lobsters extract copper and cobalt; certain seaweeds concentrate iodine; and sea cucumbers extract vanadium!

FIGURE 24.6
Salinity increases as the supply of fresh water decreases. Factors that increase salinity include formation of sea ice and evaporation. Salinity decreases as the supply of fresh water increases. Factors that decrease salinity include the runoff from streams and rivers, precipitation, and the melting of glacial ice, sea ice, and icebergs.

on Earth—because thick accumulations of fine-grained sediment bury the underlying uneven oceanic crust.

Sticking up through the accumulated sediment of the abyssal plain are a variety of volcanic peaks—seamounts and small abyssal hills. Abyssal plains end where the midocean ridge begins. Recall from Chapter 21 that midocean ridges form a continuous, underwater mountain chain that extends throughout all the oceans like the seams on a baseball. The entire ridge system is volcanic and composed of basalt. This is the zone of seafloor spreading where oceanic crust is made, and where tension from diverging plates causes normal faults to form and small earthquakes to rumble.

In ocean basins that border convergent plate boundaries, ocean trenches mark the deepest places on Earth. They occur at subduction zones, where lithosphere is forced down into the asthenosphere. This action causes earthquakes—sometimes quite large—and creates volcanoes that are capable of violent eruptions.

Seawater

Seawater is a complex solution of dissolved minerals, dissolved gases, and decomposing biological material. Interestingly, just about every natural compound can be found at some concentration in the ocean. Water, "the universal solvent," has a strong ability to dissolve salts (Chapter 18), so salts make up the majority of dissolved material in the sea. Overall, the composition of seawater is surprisingly simple, because only a few elements and compounds are present in abundance. Chloride, sodium, sulfate, magnesium, calcium, and potassium make up more than 99% of the salts in the sea (Table 24.1).

The amount of dissolved salts in seawater is measured as **salinity**—the mass of salts dissolved in 1000 g of seawater. On average, the salinity of seawater is about 35 g per 1000 g of seawater, or 35 parts per thousand. (Scientists note this with the symbol ‰.) So the salinity of the ocean is written 35‰.

Salinity varies from one part of the ocean to another, but the overall composition of seawater is the same from place to place—a mixture of 96.5% water and 3.5% salt. Salinity variation is, of course, influenced by factors that increase or decrease supplies of fresh water (Figure 24.6). Fresh water enters the ocean from three sources: runoff from streams and rivers, precipitation, and melting of ice.

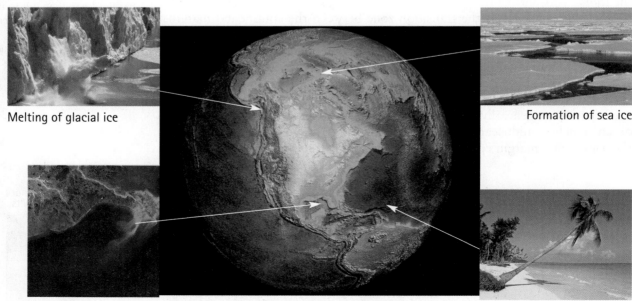

Melting of glacial ice

Formation of sea ice

Runoff

Evaporation

TABLE 24.1	PRINCIPAL ELEMENTS IN SEA SALTS	
Element	Chemical Symbol	Percentage by Weight
Chloride	Cl^-	55.07
Sodium	Na^+	30.62
Sulfate	SO_4^-	7.72
Magnesium	Mg^{2+}	3.68
Calcium	Ca^{2+}	1.17
Potassium	K^+	1.10
Total		99.36

Fresh water leaves the ocean by two methods: evaporation and the formation of sea ice. Evaporation increases salinity because only pure water vapor leaves the seawater solution—the salts are left behind. And when sea ice forms, only the water molecules freeze. Salts are once again left behind in solution. There are slight regional variations in salinity as well. The salinity of oceans in the dry subtropics where evaporation is high can reach 37%. And heavy precipitation dilutes the oceans in equatorial locations to as low as 33%. Overall balance is maintained when evaporation is offset by precipitation/runoff and ice formation is offset by ice melting.

CHECKPOINT

If you want to make up a batch of saltwater with the same salinity as seawater, how many grams of salt would it take to make 1 kg of seawater? How many grams of water?

Was this your answer?

The salinity of saltwater equals 35‰. So measure out 35 g of salt, then pour in 965 g of water to make 1 kg of seawater.

24.3 Ocean Waves, Tides, and Shorelines

EXPLAIN THIS In the moving ocean, why does a floating piece of wood appear to stay in one place?

Ocean waves come in a variety of sizes and shapes, from tiny ripples to the gigantic waves powered by hurricanes. Water waves, like all other waves, begin with some kind of disturbance. The most common disturbance that causes ocean waves is wind. Blow on a bowl filled with water and you'll see a succession of small ripples moving across the water's surface. The generation of waves in the ocean is similar. As wind speed increases, the ripples grow to full-sized waves. When stronger winds blow, larger waves are created. As waves travel away from their origin, they develop into regular patterns of smooth, rounded waves called *swell*—the mature undulations of the open ocean.

Recall from our study of waves in Chapter 10 that wave motion can be described in terms of a sine curve (Figure 24.7) and that it is the *disturbance* that is carried by a wave, not the material through which the wave is moving. The waveform travels across the ocean, while the water making up the wave remains, for the most part, in one place.

(a)

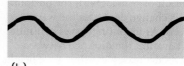
(b)

FIGURE 24.7
Ocean waves have characteristics of simple sine waves.

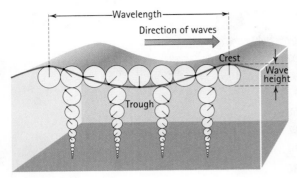

FIGURE 24.8
INTERACTIVE FIGURE (MP)

Movement of water particles with the passage of a wave. The particles move in a circular orbit. Circular motion is greatest at the surface and gradually decreases with depth. At depths greater than half a wavelength, circular motion is negligible.

When waves approach the beach head-on, so that the line of wave crests is parallel to the shore, rip currents form instead of longshore currents.

Waves on the ocean surface are *orbital* waves. As a water wave passes a given point, the water particles at that point move in a circular path. This circular motion can be seen by observing the behavior of a floating piece of wood on the ocean surface. The wood sways to and fro while bobbing up and down, actually tracing a circle during each wave cycle. This circular motion occurs near the water surface, and it decreases gradually with depth (Figure 24.8). At a depth of about one-half the wavelength, the circular component of the wave is negligible. For this reason, we can say, with reasonable accuracy, that wind-driven ocean waves occur mainly at the surface.

When a wave approaches shore, where the water depth decreases, the circular motion is interrupted by the ocean bottom. As the water gets shallower, the wave's circular path flattens and becomes elliptical. The frictional drag along the ocean bottom slows the wave down. As a result, incoming waves gain on slower leading waves and the distance between waves decreases. This bunching up of waves in a narrower zone produces higher, steeper waves. When wave height steepens to the point where the water can no longer support itself, the wave overturns and tumbles forward, breaking shoreward with a crash. The turbulent water created by the crash is called *surf*. The *surf zone* is the area of active water between the line of breakers and the shore (Figure 24.9).

Wave Refraction

As waves enter shallow water, their forward direction changes when they approach the shore at an angle. The portion of the wave closest to shore begins touching the ocean bottom at a depth of one-half the wave's wavelength. This near-shore part of the wave slows and lags behind portions of the wave still in the deeper water away from shore. As the next portion of the incoming wave touches bottom, it too slows. Thus, in a continuous fashion, the line of the wave crest bends as it moves into shallower water, in a sense pivoting around the slowest portion of the wave, to become more parallel to the coastline (Figure 24.10). This is wave refraction. Additionally, the oblique approach of waves causes a *longshore current* that flows parallel to shore.

FIGURE 24.9
INTERACTIVE FIGURE (MP)

Waves change form as they travel from deep water through shallow water to shore. In deep water, orbital motion is circular. In shallow water, orbital motion becomes elliptical as a result of contact with the bottom. This change decreases the wave speed. As incoming waves continue to advance, the distance between waves decreases, causing wave height to increase. When waves reach a critical height, they break and crash into the surf zone.

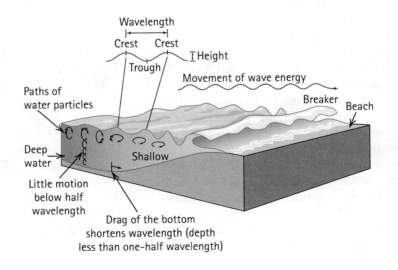

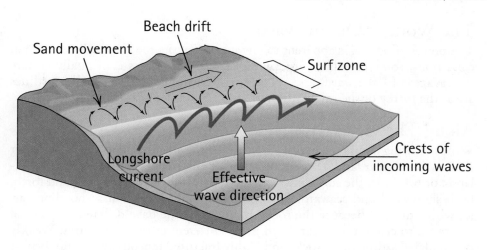

FIGURE 24.10
When waves approach a shoreline, they refract (bend) so that the crests of the approaching waves become more parallel to the shore as they move into shallower water. Because the overall direction of wave movement is oblique to the shore, a longshore current forms, which causes water and sand to move parallel to the shoreline.

Wave refraction has a significant impact on irregular shorelines, making them straighter and more regular. The effect is greatest at shorelines that have protruding headlands and small bays. Refraction causes the wave energy to be unevenly distributed (Figure 24.11). Wave energy is concentrated at headland areas where shorelines project into the water, because the wave encounters shallow water at such places first. The extra energy focused there increases the erosion rate. At adjacent bays, the shoreline is farther inland and wave energy is diluted. The gentler bay water facilitates sediment deposition. Headlands are worn back landward, while bays accumulate sediment seaward.

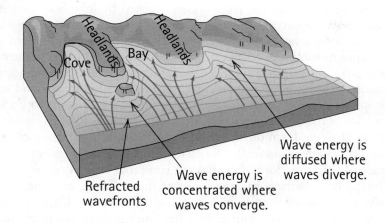

FIGURE 24.11
On irregular coastlines, wave energy is concentrated as it converges on headlands and diffused as it diverges in coves and bays.

CHECKPOINT

1. **Why does a resting surfer tend to stay at the same spot in an area of regular ocean swell?**
2. **What causes a wave to slow as it approaches the shoreline?**

Were these your answers?

1. Think of the bobbing piece of wood. The water under the resting surfer moves in a circular path, moving backward as much as it moves forward. So unless the surfer paddles to the shore or farther out to sea, the surfer basically bobs up and down in one place.
2. A wave slows when the bottommost part of its circular motion encounters the ocean bottom and flattens.

FYI About 280 barrier islands ring the Atlantic and Gulf coasts. They form a barricade between the coast and the open ocean. The lagoons that separate the narrow islands from the shore are relatively quiet waters. Small boats often use these lagoons as a "freeway" between Florida and New York, thus avoiding the potentially rough waters of the open Atlantic.

FYI Beach sand varies according to the area eroded. On tropical beaches, such as Florida or Hawaii, the beach is made up of organic material—sand-sized fragments from coral reefs and carbonate platforms. In contrast, the sand on many beaches of the continents is mostly inorganic—composed of silicate minerals eroded from continental rock.

The Work of Ocean Waves

The power of oceans is apparent to anyone who visits the coast. Under normal conditions, some coastlines are quite placid, while others continually endure the ravages of large waves and rough surf. During large storms, all coastlines are at the mercy of the mighty ocean.

Along The Coast

Winds blowing across the ocean surface generate waves, and as the waves enter shallow water near land, they become higher and steeper until they finally collapse, or break. In the surf zone, wave activity moves sediment back and forth, both shoreward and seaward. Because the amount of surf at a shoreline varies with time and because the rocks at any shoreline have different degrees of resistance to erosion, surf can form many different erosional features. Weakly consolidated sedimentary rocks and highly fractured igneous and metamorphic rocks erode fastest. Resilient, well-consolidated sedimentary rocks and unfractured igneous and metamorphic rocks erode more slowly.

Along shorelines consisting of hard rock, the pounding surf cuts into and notches the base of the land. As erosion proceeds, the notches deepen and the rocks above begin to jut out over the empty space at the base. As the overhanging rocks fall into the surf, the cliff progressively retreats. In time, waves cut into the cliff to form a relatively flat surface known as a *wave-cut platform* (Figure 24.12).

Rock particles eroded from the coast must sooner or later be deposited. Much of the material is deposited in the most common and well-known shoreline depositional environment—the *beach*. Sandy beaches are the result of turbulent motion in the surf zone. Beaches tend to be elongated by longshore currents that form when waves approach the shoreline at an oblique angle. For example, waves approaching a north–south coastline from the northwest would cause a longshore current southward down the beach (see Figure 24.10). These currents move sand down the length of the coast. Where the currents deposit sand, we have the formation of *spits*. Spits begin as submerged ridges of sand. As sand accumulates, the spit rises above the surface and projects from the coast into open water as a continuation of the beach, frequently as a fingerlike piece of land (Figure 24.13).

When sand ridges form offshore and parallel to the coast, they can eventually grow into *barrier islands*. Barrier islands form where ridges of sand break the surface of the water for a long enough time that vegetation begins to take hold. During large storms, surf washes over the lowlands, making inlets into the lagoon area between the barrier island and the shore (see Figure 24.13). The lagoon area is a quieter environment. It has finer-grained silts and muds that feature cross-bedding and small ripple marks caused by the oscillating motion of the lagoon water. On shore, smooth stones, rounded pebbles, and/or sand make up the beaches.

FIGURE 24.12
Characteristic coastal erosional landforms. Waves cut into cliffs to form platforms and terraces. Some notched cliffs erode to form sea caves. Sea arches can form if two sea caves, usually on opposite sides of a headland, become connected. When an arch collapses, a sea stack is left behind. In time, wave action also erodes away the sea stack.

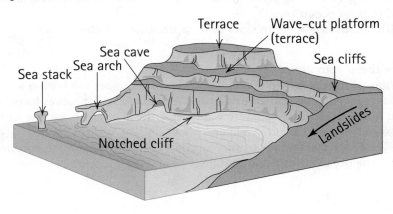

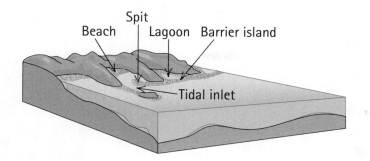

Beach   Spit   Lagoon   Barrier island

Tidal inlet

FIGURE 24.13
Characteristic coastal depositional landforms.

CHECKPOINT

1. **How are wave-cut platforms different from a beach?**
2. **Barrier islands provide a first line of defense against high-energy storm waves of the open ocean. Are barrier islands permanent features of the coast?**

FYI Ancient coral reefs are composed of alternating layers of porous material and impermeable muds. This layered quality gives ancient coral reefs the potential to act as traps for oil and gas, making their discovery economically important.

Were these your answers?

1. A wave-cut platform is formed by erosion; a beach is formed by deposition. Wave-cut platforms, terraces, sea arches, and sea stacks are erosional features. Beaches, spits, and barrier islands are depositional features.
2. Barrier islands are permanent features in the sense that they will always form. But they are impermanent in that they are vulnerable to the forces of nature. Barrier islands are made of transported sand—unconsolidated material susceptible to the erosive action of waves and high winds that constantly embattle the coastline.

LINK TO BIOLOGY

Coral Reefs, an Offshore Environment

Coral reefs are composed of actively growing, individual coral organisms that secrete calcium carbonate exoskeletons as they grow. When we look at a piece of coral or a coral reef, we see the calcium carbonate exoskeleton, not the soft coral organism. Most corals, but not all of them, form colonies, and the colony-forming corals are the major reef builders. Reefs grow outward with time, but they can also grow upward as new corals cement themselves to the exoskeletons of dead coral below.

Many types of coral reefs can survive only in shallow water because a major food source for these corals is photosynthetic algae, which need bright light to live. The coral and algae live in a *symbiotic* relationship. The coral provides protection for the algae, and the algae provide oxygen and nutrients for the coral. These types of corals require warm, clear, and relatively sediment-free water to flourish.

Carbonate platforms are much larger than coral reefs, but their existence is also due to organisms. Carbonate platforms are the graveyards of calcium-secreting organisms and are formed in shallow waters close to or attached to continents. They account for the largest portion of carbonate sediment produced in the ocean.

Both coral reefs and carbonate platforms can be partially destroyed by constant, pummeling wave action as the reefs grow and approach the water surface. This destruction is natural. But coral reefs are also destroyed by less natural processes; thus, they indicate ocean health. We know that warm climates favor carbonate deposition because carbonates dissolve more easily in cold water than in warm water. But coral organisms are living creatures and when the water gets too warm, the symbiotic algae leave the coral or the coral expels the algae. With nutrients cut off, the coral begins to lose its color—called *coral bleaching*—and slowly begins to die.* Scientists view coral bleaching as an indicator of global warming. Aside from warmer ocean temperatures, the symbiotic relationship between coral and algae may also be stressed by increased pollution; elevated ultraviolet (UV) radiation levels; changes in salinity, diseases, or predators; or a combination of these factors.

*The beautiful colors associated with coral reefs are the result of the relationship with the symbiotic algae. Corals by themselves are colorless.

Ocean Tides

Along any coastline, it is easy to notice that the level of the ocean rises and falls cyclically. These daily changes in ocean elevation are known as *tides*. Seafaring people have always known that there is a connection between ocean tides and the Moon. But until Isaac Newton, no one could offer a satisfactory explanation of why there are two high tides per day. Newton showed that ocean tides are caused by the *differences* in the gravitational pull exerted by the Moon on opposite sides of Earth. Gravitational force between the Moon and Earth is stronger on the side of Earth closer to the Moon, and it is weaker on the side of Earth farther from the Moon. This is simply because gravitational force is weaker with increased distance.

This difference in pulls across Earth slightly elongates both Earth and the Moon (Figure 24.14). So, rather than being spherical, they are pulled into a shape that slightly resembles a football. For Earth, the elongation is in the oceans. The oceans bulge out about 1 m on opposite sides of Earth. Because Earth spins once per day, a fixed point on Earth passes beneath both of these bulges each day. Any part of Earth that passes beneath one of the bulges has a high tide. Ideally, this produces two high tides per day, separated by almost 12.5 hours. But why aren't the high tides separated by 12 hours? It turns out that while Earth spins on its axis, the Moon moves in its orbit around Earth, such that the Moon appears at the same position in our sky every 24 hours and 50 minutes. So, the second high tide follows the first by 12 hours and 25 minutes, and so on. That is why tides do not occur at the same time every day.

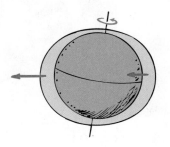

FIGURE 24.14
Two tidal bulges remain relatively fixed with respect to the Moon while Earth spins daily beneath them.

The Sun also contributes to ocean tides, although it is less than half as effective as the Moon in raising tides—even though its pull on Earth is 180 times greater than the pull of the Moon. So, why doesn't the Sun cause tides 180 times greater than lunar tides? The answer has to do with a key word: *difference*. The Sun is much farther away, so the difference in its gravitational pull on opposite sides of Earth is very small (Figure 24.15).

Nevertheless, the Sun still plays a role in ocean tides on Earth. When the Sun, Earth, and Moon are aligned, tides due to the Sun and Moon coincide. We then have higher-than-average high tides and lower-than-average low tides—**spring tides** (Figure 24.16). (Spring tides have nothing to do with the spring season.) Spring tides occur at the times of a new or full Moon.

When the Moon is halfway between a new Moon and a full Moon, the pulls of the Moon and Sun are perpendicular to each other. As a result, the solar and lunar tides do not overlap. Then high tides are not as high and low tides are not as low. These are called **neap tides** (Figure 24.17).

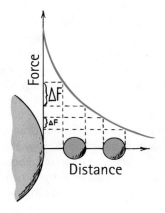

FIGURE 24.15
A plot of gravitational force versus distance (not to scale). The greater the distance from the Sun, the smaller the force *F*, which varies as $1/d^2$, and the smaller the difference in gravitational pulls on opposite sides of a planet, ΔF.

FIGURE 24.16
When the Sun, Earth, and Moon are aligned with one another, as during a full or new Moon, spring tides occur.

FIGURE 24.17
When the Sun and the Moon are about 90° apart (at the time of the half Moon), neap tides occur.

In the open ocean, the variation in water level due to tides is generally less than a meter. Along shorelines, the range varies. The most noteworthy tides are in the Bay of Fundy, between New Brunswick and Nova Scotia in eastern Canada. Here, the tidal differences sometimes exceed 15 m (Figure 24.18). This is largely due to the ocean floor, which funnels shoreward in a V shape. The tide often comes in faster than a person can run. Don't dig clams near the water's edge at low tide in the Bay of Fundy!*

FIGURE 24.18
Low and high tide at Hall's Harbour, Nova Scotia.

---

* Our treatment of tides is quite simplified here; tides are actually more complicated. Interfering land-masses and friction with the ocean floor, for example, complicate tidal motions. In many places, the tides break up into smaller "basins of circulation," where a tidal bulge travels like a circulating wave. These waves move around as if in a small basin of water that is tilted. For this reason the high tide may be hours away from an overhead Moon.

FYI Earth is not a rigid solid but is somewhat deformable (remember plastic flow in the mantle?). Because of this, we have Earth tides, though less pronounced than ocean tides. Twice each day the solid surface of Earth rises and falls as much as 25 cm (for comparison, the ocean rises and falls about 1 m each day). There are also atmospheric tides, which affect the intensity of cosmic rays that reach Earth's surface. These rays, regulated even more strongly by Earth's magnetic field, induce subtle changes in living things. Do any of your friends act strangely when there is a full Moon?

FYI Do you know why we see only one face of the Moon? The reason: Tides. The Moon has two tidal bulges for the same reason there are two tidal bulges on Earth—near and far sides of each body are pulled differently. But the Moon's tidal bulges do not rise and fall; they are "fixed," and this elongates the Moon's shape. The elongated shape causes the Moon's center of gravity to be slightly displaced from its center of mass. So whenever the Moon's long axis is not lined up toward Earth, Earth exerts a small torque on the Moon. This tends to twist the Moon to align with Earth's gravitational field, like the torque that aligns a compass needle with a magnetic field. So as the Moon takes 27.3 days to make a single revolution about its own axis (and also about the Earth–Moon axis), the same lunar hemisphere faces Earth all the time. That is why the Moon always shows us the same face! This will be further explored in Section 26.5.

CHECKPOINT

We know that both the Moon and the Sun produce our ocean tides. And we know the Moon plays the greater role because it is closer. Does closeness mean that the Moon pulls on Earth's oceans with more gravitational force than the Sun does?

Was this your answer?
No, the Sun's pull is much stronger. The difference in distance is the key to tidal forces. If the Moon were closer to Earth, the tides on both Earth and the Moon would increase. Too close could catastrophically tear the Moon into pieces—the likely cause of the planetary rings of Saturn and other planets.

LEARNING OBJECTIVE
Describe the vertical structure of Earth's atmosphere.

24.4 Components of Earth's Atmosphere

EXPLAIN THIS Why does an air balloon expand as it rises in elevation?

If gravity did not exist, gas molecules in Earth's atmosphere would fly off into outer space. Gases are compressible, which allows the invisible force of gravity to squeeze and hold a great number of gas molecules close to Earth's surface (where gravity is strongest). Thus the density of air molecules is greatest at Earth's surface and gradually decreases with height.

Because air has weight, it exerts pressure on Earth's surface. This pressure is known as *atmospheric pressure* or, simply, *air pressure*. The more weight, the more pressure. Like the atmosphere's density, air pressure also decreases with increasing height above Earth's surface. The higher you go, the lower the air pressure. Interestingly, the weight of the air on the ocean's surface keeps the ocean from boiling away. Recall from Chapter 7 that water boils at 0°C when no air pressure acts on it. So fish as well as birds appreciate the existence of the atmosphere.

Table 24.2 shows that Earth's atmosphere is a mixture of various gases—primarily nitrogen and oxygen, with small percentages of water vapor, argon, and carbon dioxide and trace amounts of other elements and compounds.

FYI Volcanic eruptions release a great deal of carbon dioxide, yet it is a minor constituent of Earth's atmosphere. That's because most carbon dioxide is gobbled up by the ocean, where it dissolves and ends up as calcium carbonate.

TABLE 24.2   COMPOSITION OF THE ATMOSPHERE

Permanent Gases			Variable Gases		
Gas	Symbol	Percentage by Volume	Gas	Symbol	Percentage by Volume
Nitrogen	N_2	78%	Water vapor	H_2O	0–4
Oxygen	O_2	21%	Carbon dioxide	CO_2	0.038**
Argon	Ar	0.9%	Ozone	O_3	0.000004*
Neon	Ne	0.0018%	Carbon monoxide	CO	0.00002*
Helium	He	0.0005%	Sulfur dioxide	SO_2	0.000001*
Methane	CH_4	0.0001%	Nitrogen dioxide	NO_2	0.000001*
Hydrogen	H_2	0.00005%	Particles (dust, pollen)		0.00001*

* Average value in polluted air.
** The amount of CO_2 in the atmosphere is increasing, as we will see later in this chapter and in Chapter 25.

Vertical Structure of the Atmosphere

If you have ever gone mountain climbing, you have probably noticed that the air grows cooler and thinner with increasing elevation. At sea level, the air is generally warmer and denser. The greater density near Earth's surface is due to gravity. The density of the air, like the density of a deep pile of feathers, is greatest at the bottom and least at the top. More than half of the atmosphere's mass lies below an altitude of 5.6 km, and about 99% lies below an altitude of 30 km. Unlike a pile of feathers, however, the atmosphere doesn't have a distinct top. It gradually thins to the near vacuum of outer space.

The atmosphere is divided into layers, each with distinct characteristics (Figure 24.19). The lowest layer, the **troposphere**, is where weather occurs—it contains 90% of the atmosphere's mass and almost all of its water vapor and clouds. The troposphere is thin and extends to a height of 16 km over the equatorial regions and 8 km over the polar regions. Commercial jets generally fly at the top of the troposphere to minimize the bumpiness caused by weather disturbances. Temperature in the troposphere decreases steadily (at about 6.5°C per kilometer) with increasing altitude. At the top of the troposphere, the temperature averages a freezing −50°C.

Above the troposphere is the **stratosphere**, which reaches a height of 50 km above the ground. Ozone molecules (O_3) form in the stratosphere and absorb UV radiation from the Sun. The absorption of UV radiation by the ozone layer causes temperature to rise from about −50°C at the bottom to about 0°C at the top.

Above the stratosphere, the **mesosphere** extends upward to about 80 km. The gases that make up the mesosphere absorb very little of the Sun's radiation. As a result, the temperature decreases again from about 0°C at the bottom of the layer to about −90°C at the top.

The situation is just the opposite in the layer above the mesosphere, the **thermosphere**. Here, temperature generally increases with altitude. This layer contains little air, but what air there is readily absorbs solar radiation. For this reason, temperatures are high, ranging from 500°C to 1500°C depending on solar activity. Because of low air density, however, this extreme temperature has

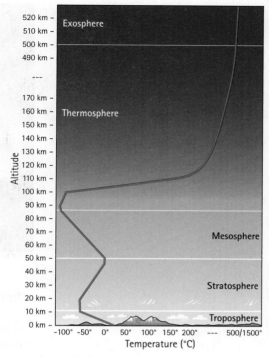

FIGURE 24.19
The atmospheric layers. The average temperature of the atmosphere varies in a zigzag pattern with altitude.

Mastering**PHYSICS**

TUTORIAL: Vertical Structure of the Atmosphere

FIGURING PHYSICAL SCIENCE

Dense as Air

Given the density of air (1.25 kg/m³), it's a straightforward calculation to find the mass of air for any given volume—simply multiply air's density by the volume. The volume of an average-sized room is assumed to be

4.00 m × 4.00 m × 3.00 m = 48.0

cubic meters. Thus the mass of the air in the room is

$$\frac{1.25 \text{ kg}}{m^3} \times 48.0 \text{ m}^3 = 60.0 \text{ kg}$$

If you're curious to know how many pounds this is, multiply by the conversion factor 2.20 lb/1 kg:

$$60.0 \text{ kg} \times \frac{2.20 \text{ lb}}{\text{kg}} = 132 \text{ lb}$$

Because we know the density of air (1.25 kg/m³), it's a straightforward calculation to find the mass of air for any given volume.

SAMPLE PROBLEM

What is the mass in kilograms of the air in a classroom that has a volume of 796 m³?

Solution:

Each cubic meter of air has a mass of 1.25 kg, so

$$796 \text{ m}^3 \times \frac{1.23 \text{ kg}}{m^3} = 995 \text{ kg}$$

which is as much as the combined mass of 17 students with a mass of about 60 kg each.

FIGURE 24.20
The *aurora borealis* over Alaska is created by solar-charged particles that strike the upper atmosphere and light up the sky (just as similar particles on a smaller scale light up a fluorescent lamp).

little significance. In fact, it would actually be quite chilly if you could visit the thermosphere.

The **ionosphere** is an ion-rich region within the thermosphere and uppermost mesosphere. The ions are produced by the interaction of high-frequency solar radiation with atoms of atmospheric gases. Incoming solar rays strip electrons from nitrogen and oxygen atoms, producing a large concentration of free electrons and positively charged ions in this layer. The degree of ionization depends on air density and on the amount of solar radiation. Ionization is greatest in the upper part of the ionosphere, where air density is low and solar radiation is high.

Ions in the ionosphere cast a faint glow that prevents moonless nights from becoming stark black. Near Earth's magnetic poles, fiery light displays called *auroras* occur as the solar wind (high-speed charged particles ejected by the Sun) strikes and excites molecules of atmospheric gases in the ionosphere (Figure 24.20). These auroral displays are particularly spectacular during times of solar flares—storms or eruptions of hot gases on the Sun.

Finally, above 500 km, in the **exosphere**, the thinning atmosphere gradually yields to the radiation belts and magnetic fields of interplanetary space.

LEARNING OBJECTIVE
Explain how the angle of the Sun affects the seasons, and describe the role of terrestrial radiation in warming the lower atmosphere.

24.5 Solar Energy

EXPLAIN THIS   Why are equatorial areas warm and polar areas cold?

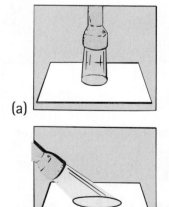

FIGURE 24.21
(a) When the flashlight is held directly above at a right angle to the surface, the beam of light produces a bright circle. (b) When the light shines at an angle, the beam is dispersed over a larger area and is therefore less intense.

Earth's equatorial regions are warmer than the polar regions. But why? Surface temperatures on Earth depend on the energy each part of Earth receives from the Sun each day. And how much energy is received depends on the angle at which the Sun's rays strike Earth's surface. The different angles at which sunlight strikes Earth can be mimicked by holding a flashlight vertically over a table and shining the light directly down on, and perpendicular to, the flat surface (Figure 24.21a). The light produces a bright circle. Now tip the flashlight at various angles and notice that the circle elongates into ellipses, spreading the same amount of energy over a greater area and, therefore, decreasing the intensity of the light. The same is true of sunlight on Earth's surface. High noon in equatorial regions is like the vertically held flashlight; high noon at higher latitudes is like the flashlight held at an angle.

The Seasons

The northern United States and Canada, both temperate regions, have distinct summer and winter seasons. These seasons change because the angle at which the Sun's rays strike these locations varies over the course of a year. Figure 24.22 shows how the tilt of Earth causes the variation in the rays' angle. Can you see that the rays are more perpendicular to the ground in the Northern Hemisphere when Earth's axis is tilted *toward* the Sun? When the Sun's rays are closest to perpendicular at any spot on Earth, that region experiences summer. Six months later, the rays fall upon the same region more obliquely, and we have winter. In between are fall and spring.

It is interesting to note that, because Earth follows an elliptical path around the Sun, Earth is farthest from the Sun when the Northern Hemisphere experiences summer. The angle of the Sun's rays, not the distance from the Sun, is most responsible for Earth's surface temperatures.

Another effect of the tilted rays is the length of daylight each day. Can you see in Figure 24.22 that a location in summer has more daylight per daily rotation of Earth than the same location when Earth is on the opposite side of the Sun in winter? If you have trouble visualizing this, look at the high latitudes near the poles. Consider the special latitude where daylight lasts nearly 24 hours during the summer solstice (around June 21) and night lasts about 24 hours at the winter solstice (around December 21). This latitude is called the Arctic Circle in the Northern Hemisphere and the Antarctic Circle in the Southern Hemisphere. During the summer solstice, the North Pole leans toward the Sun and the South Pole leans away from the Sun. (Summer and winter are reversed, of course, in the two hemispheres.)

Halfway between the peaks of the winter and summer solstice, around mid-September and mid-March, the hours of daylight and night are of equal length. These are called the *equinoxes* (Latin for "equal nights"). The equal hours of day and night during the equinoxes are not restricted to high latitudes but occur all over the world.

FYI As you travel north of the Arctic Circle or south of the Antarctic Circle, there are more summer days with the Sun always above the horizon and more winter days with the Sun always below the horizon. At the poles, there is a full six months of continuous sunlight followed by a full six months of continuous night! These 24-hour-long "days" in the polar regions are never very bright because the Sun is never very far above the horizon. Likewise, the 24-hour-long "nights" aren't all that dark because the Sun never sinks very far below the horizon.

Mastering**PHYSICS**
TUTORIAL: Seasons

CHECKPOINT

1. **Why are there fewer daylight hours in winter months?**
2. **A friend says, "The tilt of Earth's axis, not Earth's distance from the Sun, causes Earth's seasons." Do you agree or disagree? Explain.**

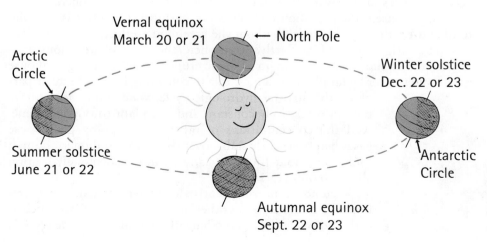

FIGURE 24.22
The tilt of Earth and the corresponding different spreading of solar radiation produce the yearly cycle of seasons.

Were these your answers?

1. Earth is tilted on its axis, like a top leaning in one direction all the time. As Earth revolves around the Sun, the Northern Hemisphere is tilted toward the Sun in the summer and away from the Sun during the winter. When the hemisphere is tilted away, the Sun is closer to the horizon. Hence, the Sun rises later and sets earlier, resulting in shorter days.
2. Agree. The tilt of Earth's axis affects the angle at which solar radiation strikes a given location. When a location experiences winter, the angle at which the Sun's rays strike Earth's surface is farthest from perpendicular. In summer, the rays strike more directly.

Terrestrial Radiation

Solar radiation covers a wide spectrum of wavelengths, mostly in the visible short-wavelength part of the spectrum. Earth absorbs some of this energy and, in turn, reradiates part of it back into space. Because this re-emitted energy is from Earth's surface, it is called *terrestrial radiation* (Figure 24.23). Terrestrial radiation is emitted in the infrared, long-wavelength part of the spectrum.

Interestingly, it is terrestrial radiation rather than solar radiation that directly warms the lower atmosphere. This explains why air close to the ground is so much warmer than air at higher elevations. The temperature of Earth's surface depends on the amount of solar radiation coming in compared with the amount of terrestrial radiation going out. In direct sunlight, the net effect is warming because Earth's surface absorbs more energy from the Sun than it emits. At night, the net effect is cooling because Earth's surface emits more energy than it absorbs. Cloud cover blocks both incoming solar radiation and outgoing terrestrial radiation. Hence, cloudy days are much cooler than sunny days, and cloudy nights are generally warmer than clear nights.

The Greenhouse Effect and Global Warming

Incoming short-wavelength solar radiation easily penetrates the atmosphere to reach and warm Earth's surface. The warmed surface then reradiates this energy back as long-wavelength terrestrial radiation. But not all of the outgoing long-wavelength terrestrial radiation can penetrate the atmosphere to escape into space. Much like the panes of glass in a greenhouse, atmospheric gases (mainly water vapor and carbon dioxide) trap long-wavelength terrestrial radiation, thereby warming the lower atmosphere. This warming of the lower atmosphere is called the **greenhouse effect**. The atmosphere acts like an insulating blanket, keeping Earth's surface warmer than it would be without an atmosphere.

In and of itself, the greenhouse effect is not harmful. In fact, it is essential to life on Earth as we know it. Without the greenhouse effect, Earth's surface would be a frigid −18°C! Like Earth's atmosphere, the origin of greenhouse gases is linked to the release of gases in volcanic eruptions. Of all the greenhouse gases, water vapor by far plays the largest role in confining Earth's heat. As part of Earth's natural hydrologic cycle, water vapor levels have remained relatively constant and abundant throughout time. And the concentrations of other heat-trapping greenhouse gases had been fairly stable over the course of modern human evolution (about 100,000 years) until the dawn of civilization about 10,000 years ago.

What now concerns scientists are increases in the concentration of greenhouse gases other than water vapor. Gases such as carbon dioxide (CO_2), methane (CH_4), nitrous oxide (N_2O),

FIGURE 24.23

The hot Sun emits visible short waves, and the cool Earth re-emits infrared long waves—*terrestrial radiation*. Not all of the re-emitted infrared long waves are bounced back to space. Some are absorbed and trapped in the atmosphere by greenhouse gases. In this way, the greenhouse gases act like a blanket that warms the lower atmosphere.

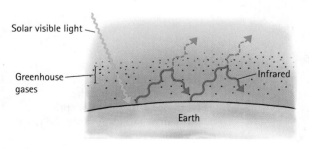

Solar visible light

Greenhouse gases

Infrared

Earth

ozone (O_3), and chlorofluorocarbons (CFCs) are on the rise because of human activities. These gases are potent contributors to the greenhouse effect. As their concentrations increase, the atmosphere's ability to absorb and trap reradiated terrestrial heat energy also increases.

Climate changes have happened over the course of geologic time, but the present environmental concern is that increased levels of atmospheric carbon dioxide and other greenhouse gases may make Earth too warm, causing rapid changes to Earth's natural systems that would negatively affect humans. Since the Industrial Revolution of the 1800s, for example, atmospheric CO_2 levels have been steadily increasing (Figure 24.24). The CO_2 increase has been linked to the warming of Earth's surface by about 0.7°C since the late 1800s. Many long-standing, famous alpine glaciers are quickly receding, and some have disappeared altogether. Scientists now believe that further warming will occur if CO_2 emissions are not held in check. The result is *climate change*.

The effects of climate change are not fully known. One consequence that is now occurring is the melting of the ice caps on Antarctica and Greenland. If this trend continues, sea level will rise and low-lying coastal lands will be flooded. Warming will also change rainfall patterns, and this will seriously affect agriculture. The grain-growing regions of North America and Asia might shift northward as local climates warm and growing seasons lengthen. On the other hand, deserts in continental interiors might spread to cover much larger areas. We don't know. We do know that Earth has experienced warmer and colder periods in times past and that global-scale climatic changes may have contributed to many of the extinctions discussed in Chapter 23. Ongoing research is becoming more focused on understanding the current impacts of this present-day climate change.

If planet Earth had *no* greenhouse effect, its average temperature would be −18°C. Brrrrr!

FYI Past global temperatures are determined by measuring the ratio between hydrogen and its heavier isotope, deuterium, in polar ice cores. When global temperatures are high, the ocean is warmer, and larger fractions of water containing deuterium evaporate from the ocean and fall as snow. A high deuterium-to-hydrogen ratio in preserved ice cores therefore indicates a warmer climate.

CHECKPOINT

1. **What does it mean to say the greenhouse effect is like a "one-way valve"?**
2. **Which gas in the atmosphere is the greatest contributor to the greenhouse effect?**
3. **What is the primary "natural" contributor to greenhouse gases in Earth's atmosphere?**

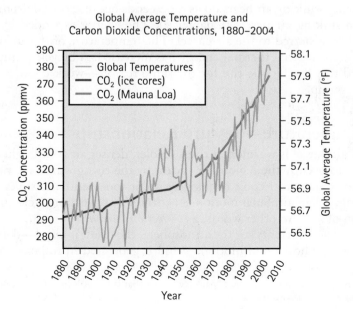

Global Average Temperature and Carbon Dioxide Concentrations, 1880–2004

Legend:
— Global Temperatures
— CO_2 (ice cores)
— CO_2 (Mauna Loa)

CO_2 Concentration (ppmv) / Global Average Temperature (°F) / Year

FIGURE 24.24
Since the Industrial Revolution of the 1880s, atmospheric CO_2 levels and global temperatures have increased. Do you think there is a correlation?

Were these your answers?
1. The transparent material—atmosphere for Earth and glass for the florist's greenhouse—allows only incoming short waves to pass through and blocks outgoing long waves. In other words, radiation travels only one way.
2. Water vapor.
3. Volcanic eruptions.

LEARNING OBJECTIVE
Describe how pressure differences produce winds and how winds are affected by the Coriolis force and surface frictional force.

24.6 Driving Forces of Air Motion

EXPLAIN THIS Why does air move from areas of high pressure to areas of low pressure?

The underlying driving force for air movement is the unequal heating of Earth's surface by the Sun. It is often said that "warm air rises." Why is this so? The fundamental answer is that warm air is less dense than cool air. As we learned in Chapter 5, a less-dense fluid rises buoyantly when surrounded by a denser fluid. This is why hot-air balloons rise—the air inside the balloon is heated, making it less dense compared to the surrounding air.

But why is warm air less dense than cool air? Air molecules within an air parcel* are constantly colliding and bouncing off each other. When temperatures are cool, the molecules move more slowly and the distance they travel after each bounce is small. When the air parcel is heated, the molecules move faster and bounce farther after each collision. Thus, warm air requires more space—a larger volume. Because the mass of the air parcel stays the same, density is lower when volume is increased.

To envision this, imagine that the air parcel is surrounded by a very thin, stretchable container. The air molecules need to do work to cause the parcel to expand—they expend thermal energy (Chapter 6). Because the air parcel loses thermal energy, it cools as it expands. According to Boyle's law (Chapter 5), our air parcel can achieve this expansion only if the pressure around it is reduced. Air pressure decreases with altitude, so air that rises always expands and cools. The air parcel continues to rise until it becomes the same temperature and density as the surrounding air.

Conversely, sinking air heats up as it descends because the environment does work on a sinking air parcel (Chapter 6). The energy expended by the environment is transferred to the air parcel. The temperature of the air parcel goes up because its internal thermal energy has increased. A bicycle pump heats up when used to fill a bicycle tire for the same reason—work is done on the air inside the pump.

The Temperature–Pressure Relationship

When an air parcel is warmed, it rises. Cooler, denser air from higher altitudes then sinks to occupy the region left vacant by the rising warm air. Because the rising, warm air parcel cools on its way up, it too eventually sinks to replace other rising, warm air. Such motion constitutes a convection cycle and thermal circulation of air—in other words, a *convection current*.

As convection currents stir the atmosphere, the result is *wind*—defined as air with an average horizontal motion. Wind is generated in response to pressure

*Consider an air parcel to be a distinct "package" of air that is free to expand and contract—but does not mix with the surrounding air.

differences in the atmosphere, which, as we will see, are largely the result of temperature differences. A difference in pressure between two different locations is called a *pressure gradient*, and the force that causes air to move is the **pressure-gradient force**. The pressure-gradient force drives air from areas of high pressure toward areas of lower pressure.

To see how temperature affects pressure, consider the two equivalent air columns shown in Figure 24.25a. The two air columns are equivalent because they contain the same number of air molecules, distributed uniformly. To aid visualization, we make three simplifying assumptions: (1) there is no change in air density with height within a column (the air density of an entire column can change, however), (2) air cannot enter or leave either column (mass is constant, so surface air pressure does not change), and (3) the width of each column remains constant. So each column has a fixed amount of air with uniform density, which can move only up or down.

When the cities are at the same temperature, the columns are the same height, as shown in Figure 24.25a. At any elevation—say, the elevation marked X in the drawing—the air pressure in each column is the same, because the number of air molecules above X is the same in both cases. This is true for any elevation. Thus, we have no pressure difference between the two cities at any elevation because we have no temperature difference.

Now suppose that the temperature in city 1 drops while the temperature in city 2 rises. In this case, we get the situation shown in Figure 24.25b. The cooling air over city 1 contracts and becomes denser, and the warming air over city 2 expands upward (remember, column width is fixed) and becomes less dense. We still have the same number of air molecules over each city, but there is a cool, short air column over city 1 and a warmer, much taller air column over city 2. The air pressure at ground level stays the same in each city, because equal

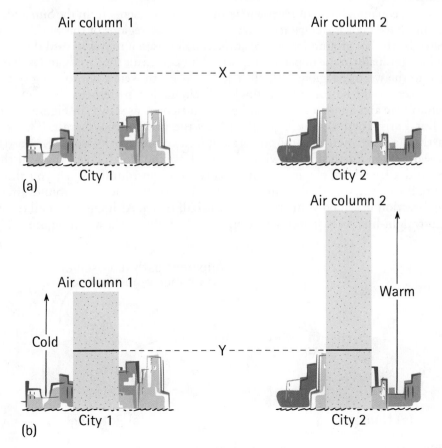

FIGURE 24.25
(a) Air columns of the same temperature over two cities. Note that the air pressure at any elevation, such as elevation X shown here, is the same for both cities. (b) When city 1 is cold and city 2 is warm, the air column over city 2 is taller because of expansion, which means that for any given elevation above the surface, pressure is greater in the taller column. This produces a pressure gradient between the two cities. So wind blows from city 2 to city 1.

FYI Of course, air density is not really constant in a tall vertical column of atmosphere. But this simplification greatly helps in understanding the connection between temperature and pressure and its effect on air movement.

numbers of air molecules are pressing down on both. But the warming and cooling of the respective air columns cause pressure changes in the air aloft.*

Now consider elevation Y, halfway up air column 1. Because Y is halfway up column 1, the top half of air column 1 contains half the total number of molecules (again, assuming constant density with height). Elevation Y is less than halfway up in air column 2, because column 2 is taller. Thus, the part of column 2 above elevation Y contains more than half of column 2's air molecules. Because more air molecules are bearing down at elevation Y in air column 2 than in air column 1, the air pressure at Y must be greater in column 2 than it is in column 1. In other words, a difference in temperature has led to a difference in pressure aloft. So the number of air molecules above a particular elevation (i.e., the mass of the air above) governs the atmospheric pressure at that elevation. We come to a very important concept: warm air aloft is associated with high atmospheric pressure aloft, and cold air aloft is associated with low atmospheric pressure aloft.

So what does this pressure difference at elevations of a kilometer or more have to do with wind at Earth's surface? Well, differences in pressure cause air to move and, hence, wind to blow. Let us now allow the air between city 1 and city 2 to mingle. Because air moves from areas of high pressure to areas of low pressure, the wind aloft blows from city 2 toward city 1. As air aloft moves from column 2, air density in city 2 decreases and the surface pressure in city 2 drops. Meanwhile, as air accumulates in cooler column 1, surface pressure in city 1 rises. So we find that cold days are associated with high surface pressure, and warm days are associated with low surface pressure. Now that the surface pressure is higher in city 1, surface winds blow from city 1 to city 2. This, in turn, changes the pressure distribution again, as air moves away from city 1. A dynamic system indeed!

Large-Scale Air Movement

Globally, equatorial regions receive maximum radiant energy from the Sun and, as a result, have higher average temperatures than other regions of the world. As air heated by the hot ground (or ocean) at the equator rises, it moves toward the poles, cooling gradually in the upper atmosphere. This cooled air then sinks and is drawn back to the warmer regions near the equator. If Earth were a nonrotating sphere, the effect would be one simple single-cell circulation pattern in the Northern Hemisphere and another in the Southern Hemisphere, as shown in Figure 24.26.

But Earth rotates, which greatly affects the path of moving air. Think of Earth as a large merry-go-round rotating in a counterclockwise direction (the same direction Earth spins, as viewed from above the North Pole). Imagine that you and a friend are playing catch on this merry-go-round. When you throw the ball to your friend, the circular movement of the merry-go-round affects the direction in which your friend sees the ball travel. Although the ball travels in a straight-line path, it appears to curve to the right, as shown in Figure 24.27.

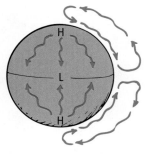

FIGURE 24.26
If Earth were simply a nonrotating sphere, air circulation would be in a single Northern Hemisphere cell and a single Southern Hemisphere cell. In each cell, heated air would rise at the equator and move toward the polar regions, where it would cool, sink, and be drawn back to the warmer regions of the equator.

Mastering**PHYSICS**
TUTORIAL: The Coriolis Effect

FIGURE 24.27
(a) On the nonrotating merry-go-round, a thrown ball travels in a straight line. (b) On the counter-clockwise-rotating merry-go-round, the ball also moves in a straight line. However, because the merry-go-round is rotating, the ball appears to deflect to the right of its intended path.

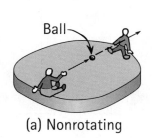
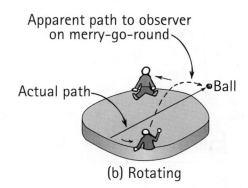

(a) Nonrotating   (b) Rotating

Air aloft is defined as air higher than 1 km above the ground.

(The ball travels straight but your friend never catches it, because the movement of the merry-go-round causes her position to change.) This apparent curving is similar to what happens on Earth. As Earth spins, all free-moving objects—air and water, aircraft and ballistic missiles, and even snowballs, to a small extent—appear to deviate from their straight-line paths as Earth rotates under them. This deflection due to the rotation of Earth is called the **Coriolis force**.*

A significant result of the Coriolis force is the deflection of winds toward the right in the Northern Hemisphere and toward the left in the Southern Hemisphere (Figure 24.28b). The impact of the Coriolis force varies according to the speed of the wind. The faster the wind, the greater the deflection. Latitude also influences the degree of deflection. Deflection is greatest at the poles and decreases to zero at the equator. As Figure 24.29 shows, the Coriolis force has a significant impact on atmospheric motion—and airplanes—in the midlatitudes.

Air moving close to Earth's surface also encounters a frictional force. The rougher the surface, the greater the friction, and so the greater the drag on the wind. Because surface friction reduces wind speed, it also reduces the effect of the Coriolis force. This causes winds in the Northern Hemisphere to spiral out clockwise from a high-pressure region and spiral counterclockwise into a low-pressure region (top part of Figure 24.28c). In the Southern Hemisphere, these circulation patterns are reversed (bottom part of Figure 24.28c).

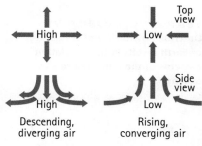

(a) Pressure-gradient force

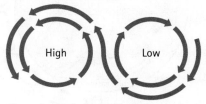

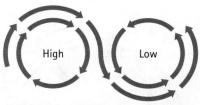

(b) Coriolis force

(c) Frictional force at the ground

FIGURE 24.28
The Coriolis force—the apparent deflection of winds from straight-line paths by Earth's rotation—is a principal force in the production of wind. It is, however, not the only force. (a) Air moves because of pressure differences—the pressure-gradient force. (b) Once the air is moving, it is affected by Earth's rotation—the Coriolis force. (c) As air moves close to the ground, it slows because of frictional force.

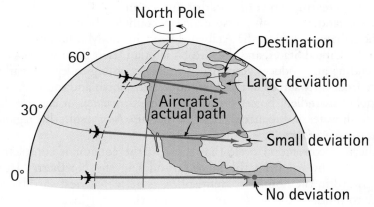

FIGURE 24.29
Latitude influences the apparent deflection resulting from the Coriolis force. A free-moving object heading east (or west) appears to deviate from its straight-line path as Earth rotates beneath it. Deflection is greatest at the poles and decreases to zero at the equator. The Coriolis force is evident only for large-scale motions such as atmospheric winds or oceanic currents, in which one part has a greater tangential speed than the other about Earth's axis.

*The Coriolis force is not a true force in the strictest sense and is sometimes called the *Coriolis effect*. On the merry-go-round, for example, the curved path is perceived only by someone on the merry-go-round. Someone in a tree directly over the merry-go-round would see the ball move in a straight line. But because we are all on Earth, and not suspended above it, the Coriolis effect truly causes wind to change direction, as if directed by a force. So because the wind is truly deflected, the cause is referred to as the Coriolis force.

24.7 Global Circulation Patterns

EXPLAIN THIS How is the movement of deeper waters different from the movement of water at the surface?

Cell-like circulation patterns are responsible for the redistribution of heat across Earth's surface and for global winds (Figure 24.30). At the equator, warmed air flows straight up with very little horizontal movement, resulting in a vast low-pressure zone at the surface. This rising motion creates a narrow, windless realm of air that is still, hot, and stagnant. Sailors of long ago cursed the equatorial seas as their ships floated listlessly for lack of wind, and they referred to the area as the *doldrums*. When the moist air from the doldrums rises, it cools and releases torrents of rain. Over land areas, such frequent rains give rise to the tropical rainforests that characterize the equatorial region.

The air of the sweltering doldrums rises to the boundary between the troposphere and stratosphere, where it divides and spreads out to the north and south. (Very little wind crosses the equator into the neighboring hemisphere.) Because of the Coriolis force, the wind is deflected and travels along longer paths poleward than if it traveled straight toward the poles. By the time the wind has reached about 30°N and 30°S latitude, the air has cooled enough to descend toward the surface. The descending air warms as it is compressed. A resulting high-pressure zone girdles Earth, creating a belt of hot, dry surface air.

On land, these high-pressure zones account for the world's great deserts—the Sahara in Africa, the Arabian Desert in the Middle East, the Mojave Desert in the United States, and the Great Victoria Desert in Australia. At sea, the hot, descending air produces very weak winds. According to legend, early sailing ships were frequently stalled at these latitudes, both north and south. As food and water supplies dwindled, horses on board were either eaten or cast overboard to conserve fresh water and reduce the load of the ship. As a result, this region is now known as the *horse latitudes*.

The thermal convection cycle that starts at the equator is completed when air flowing southward from the horse latitudes in the Northern Hemisphere and northward in the Southern Hemisphere is deflected westward by the Coriolis force to produce the *trade winds*. Air that flows northward from the horse latitudes in the Northern Hemisphere and southward in the Southern Hemisphere is deflected eastward to produce the prevailing *westerlies*.

Near the poles, frigid air continuously sinks, pushing the surface air outward. The Coriolis force is quite evident in the polar regions, as the wind deflects to the west to create the *polar easterlies* (see Figure 24.30). The cool, dry polar air meets the warm, moist air of the westerlies at latitudes 60°N and 60°S. This boundary, called the *polar front*, is a zone of low pressure where contrasting air masses converge, often resulting in storms.

The middle latitudes are noted for their unpredictable weather. Although the winds tend to be from the west, they are often quite changeable, as the temperature and pressure differences between the subtropical and polar air masses at the polar front produce powerful winds. As air moves from regions of high pressure, where air is denser, toward regions of low pressure, the result is a whirlwind effect (as we will learn in Chapter 25, a *midlatitude cyclone*).

Irregularities in Earth's surface also influence wind behavior. Mountains, valleys, deserts, forests, and great bodies of water all play a part in determining which way the wind blows.

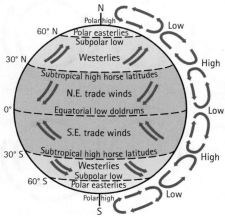

FIGURE 24.30
Global circulation of the atmosphere results from a combination of two main factors: unequal heating of Earth's surface (which sets up convection cells) and Earth's rotation. The atmosphere has six cell-like circulation patterns; prevailing winds blow in the directions indicated by the arrows. The major prevailing winds are the *westerlies*, the *easterlies*, and the *trade winds*.

FYI Air currents are named for the direction the wind is coming from. Thus, the westerlies blow from the west—but air moves to the east.

Upper Atmospheric Circulation

In the upper troposphere, "rivers" of rapidly moving air meander around Earth at altitudes of 9–14 km. These high-speed winds are the *jet streams*. With wind speeds averaging between 95 and 190 km/h, the jet streams play a critical role in the global transfer of heat from the equator to the poles.

The two most important jet streams, the *polar jet stream* and the *subtropical jet stream*, form in both the Northern and Southern Hemispheres. The formation of polar jet streams is due to a temperature gradient at the polar front—at about 60°N and 60°S latitude—where cool polar air meets warm tropical air. This temperature gradient causes a steep pressure gradient that increases the wind speed. During the winter, the polar jet stream is strong and migrates to lower latitudes, bringing strong winter storms and blizzards to the United States. In summer, the polar jet stream is weaker and migrates to higher latitudes.

The subtropical jet stream is generated by warm air carried poleward from the equator, which produces a sharp temperature gradient along the subtropical front—about 30°N and 30°S latitude. Once again, a pressure gradient caused by the temperature gradient generates strong winds.

The subtropical jet stream above Southeast Asia, India, and Africa merits special mention (Figure 24.31). The formation of this jet stream is related to the warming of the air above the Tibetan highlands. During the summer, the air above the continental highlands is warmer than the air above the ocean to the south. The warmer air rises, drawing in cooler, moist air from over the ocean. Thus, temperature and pressure gradients generate strong onshore winds that contribute to the region's *monsoon* climate.* During winter, the winds change direction to produce a dry season.

This cycle of winds characterizes the climates of much of Southeast Asia. The predictable rain-bearing summer wind from the sea that moves over the heated land is called the *summer monsoon*; the prevailing dry wind from land to sea in winter is called the *winter monsoon*.

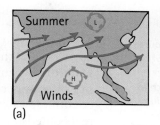
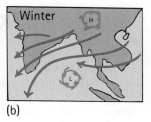

FIGURE 24.31

Winds over Southeast Asia. (a) During the summer months, air over the oceans is cooler than air over land. The summer monsoon brings heavy rains as the winds blow from sea to land. (b) During the winter months, air over continents is cooler than air over oceans. The winter monsoon generally has clear skies and winds that blow from land to sea.

CHECKPOINT

1. **What are the underlying causes of the trade winds, jet streams, and monsoons?**
2. **In the middle latitudes, airlines schedule shorter flight times for planes traveling west to east and longer flight times for planes traveling east to west. Why are eastbound planes faster?**

Were these your answers?

1. Simply enough, the unequal heating of Earth's surface coupled with Earth's rotation.
2. The upper-level westerly winds of the jet stream account for faster-moving eastbound aircraft. As the jet stream moves from west to east, it carries along everything in its path. To save time and fuel, airline pilots seek the jet stream when traveling west to east and avoid it when traveling east to west.

FYI Air circulation affects ocean surface currents, as was found during a severe storm in 1990. Five cargo containers of athletic shoes were washed overboard from freighters en route from South Korea to the Pacific Northwest. Thousands of sneakers, hiking boots, and other shoes were picked up along beaches from British Columbia to Oregon and as far into the mid-Pacific as Hawaii. Although months at sea left many shoes mismatched, most shoes were still wearable after washing. Beachcombers formed "swap meets" to search for mates of found shoes!

Monsoon is derived from the Arabic word *mausim*, which means "season." So monsoon is taken to mean "seasonal wind."

FYI Where do the ocean's salts originate? One source is the chemical weathering of continental rocks. As the rocks are weathered, elements such as sodium, calcium, and potassium dissolve into water. The water carrying these chemicals usually flows downstream and makes its way to the ocean. Another source of elements is Earth's interior. For eons, volcanic eruptions have delivered huge quantities of chlorine, as well as water vapor and other gases, to the ocean's waters. Chlorine plus sodium? Sodium chloride—common table salt. Hence, the ocean's salty taste.

Oceanic Circulation

The forces that drive the winds also affect the movement of seawater. In the open ocean, the movement of seawater results from two types of currents: wind-driven surface currents and density-driven deep-water currents. The density of seawater is controlled by two factors—temperature and salinity. Near coastlines, water movement is affected not only by surface and deep-water currents but also by tides and the presence of coastal boundaries.

Like the atmosphere, the ocean can be divided into different vertical layers—the *surface zone*, a *transition zone*, and the *deep zone*. Scuba divers notice an increase in water pressure when swimming to lower depths. The deeper the descent, the greater the water pressure. The pressure is simply the weight of the water above pushing down on you (Chapter 5). Another factor that generally changes as you descend is temperature. Deeper waters are colder. So in addition to variations in salinity, seawater varies in temperature and pressure. Because cold water is denser than warm water—for the same reasons that cold air is denser than warm air—cold seawater sinks below warm seawater. The salinity of the water also affects density: the greater the salinity, the greater the density. These variations are best illustrated when we look at the ocean's vertical structure (Figure 24.32).

FIGURE 24.32
The ocean's vertical structure. In the surface zone, water is well mixed as it moves vertically in response to temperature and density changes, and as it moves horizontally in response to wind. Water in the transition zone moves along density surfaces. Water in the deep zone is density driven, as it circulates from cold polar regions to warmer equatorial regions.

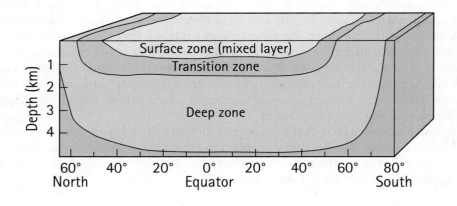

CHECKPOINT
Would you expect the pressure 100 m deep in high-salinity water to be the same as it is 100 m deep in lower-salinity water?

Was this your answer?
No. Higher salinity means higher density. A volume of denser water weighs more than an equal volume of less-dense water. Because pressure is weight per unit area, pressure at 100-m depth is greater if the water is more saline.

Surface Currents

As winds blow across the ocean, frictional forces set surface waters into motion. If distances are short, the surface waters move in the same direction as the wind. For longer distances, however, other factors come into play.

One such factor is the deflective Coriolis force, which causes surface water currents to deflect up to 45° to the right in the Northern Hemisphere and to the left in the Southern Hemisphere. So water in the surface zone, which can be

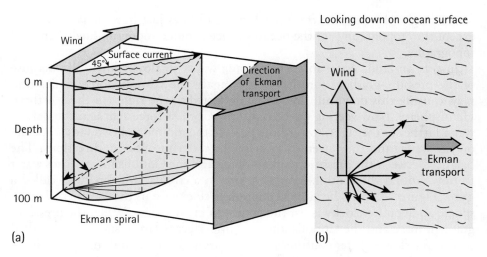

(a)

Looking down on ocean surface

(b)

FIGURE 24.33
Two views of the Ekman spiral and Ekman transport: (a) shows velocity decrease with depth, and (b) shows a bird's-eye view of the surface water direction and the overall direction of net water transport. With no outside force, water will move in the same direction as the wind. But the Coriolis force deflects surface water up to 45° to the right of the wind direction (Northern Hemisphere). As we move down in depth, deflection increases to the right but water moves at a slower speed, causing the net transport to be up to 90° from the wind direction.

thought of as a sequence of layers, moves in a different direction than the prevailing wind. Because of friction between these layers, water in the uppermost layer sets water layers below into motion. With increasing depth, each successive layer of water moves more slowly and is deflected more and more to the right. The result is a phenomenon known as the *Ekman spiral* (Figure 24.33a). Over the depth of the spiral (about 100 m), the average flow from each layer of water set in motion adds up to create a net flow that moves up to 90° to the right of the surface wind. This net flow, known as *Ekman transport,* represents the *overall* transport direction of surface water set in motion by the wind (Figure 24.33b).

Figure 24.34 shows the global air circulation pattern between 40°N and 40°S latitudes. Focusing for the moment on the Northern Hemisphere, note that south of 30°N, the trade winds blow from northeast to southwest, while north of 30°N, the westerlies blow from southwest to northeast. Because of Ekman transport, ocean surface currents south of 30°N flow to the northwest, and north of 30°N they flow to the southeast. Can you see in Figure 24.34 that Ekman-influenced currents flow toward each other, or converge, at about 30°N? Such convergence causes a mounding of water (Figure 24.35).

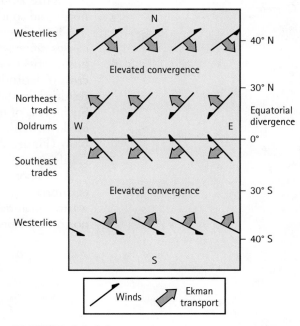

FIGURE 24.34
Ekman transport driven by prevailing winds causes currents to converge around 30°N and 30°S latitudes. The convergence causes the surface water in this area to mound.

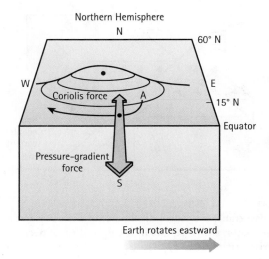

FIGURE 24.35
Mounding of water in the Northern Hemisphere due to Ekman transport. Water initially flows "downhill" but is deflected to the right, as shown by the arrow coming from point A. Now flowing around the mound, geostrophic flow develops as the Coriolis force and the pressure-gradient force balance each other.

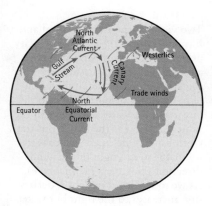

The gyre in the North Atlantic is composed of the Gulf Stream, the North Atlantic Current, the Canary Current, and the North Equatorial Current.

Interestingly, the ocean surface has highs and lows just like the land surface, except the "topography" of the ocean surface is much more subdued than land topography.

Just as on land, water in such mounds wants to flow "downhill" because of the pressure-gradient force. But once again the Coriolis force comes into play, causing this downhill flow to be deflected to the right (in the Northern Hemisphere). Now veering to the right, the Coriolis force acts again to deflect water farther to the right, toward the peak of the mound. So two forces operate on the water, and they attempt to drive the water in opposite directions. The balance between the pressure-gradient force and the Coriolis force causes water to flow around the central, mounded area in a circular whirl pattern called a **gyre** (Figure 24.36). The circular motion is clockwise in the Northern Hemisphere and counterclockwise in the Southern Hemisphere. When the pressure-gradient force and the Coriolis force are in perfect balance, such flow is called *geostrophic* flow—water continues flowing around the mound instead of down it. Gyres dominate the overall flow of surface currents in each ocean basin. As Figure 24.37 shows, gyres don't cross the equator. There are separate gyres north and south of the equator in each ocean basin.

The gyre-generating, convergent flow around 30°N and 30°S latitude also causes *downwelling* that drives some surface water to flow downward. The opposite effect occurs at the equator, where the southeast trade winds slightly cross 0° latitude. Here, Ekman transport causes water to diverge at the equator, resulting in the *upwelling* of deeper, colder, nutrient-rich water (Figure 24.38). Such areas of upwelling are full of life. Ekman transport also causes upwelling or downwelling along coasts, depending on which way the dominant winds blow (Figure 24.39). Two such areas of upwelling are the coasts of Peru and California.

An important consequence of these large gyres is the transport of heat from equatorial regions to higher latitudes. In the North Atlantic Ocean, for example, warm equatorial water flows westward into and around the Gulf of Mexico, then northward along the eastern coast of the United States. This warm-water current

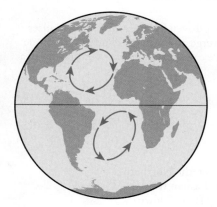

Separate gyres form north and south of the equator. Flow in the Northern Hemisphere is clockwise; flow in the Southern Hemisphere is counterclockwise.

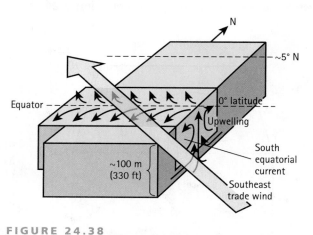

Ekman transport causes divergence of surface water at the equator, which results in the upwelling of deep, cold, nutrient-rich water.

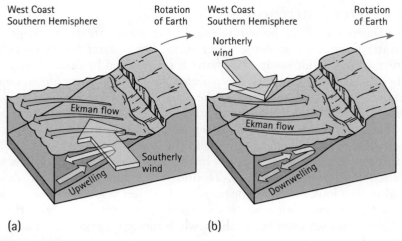

FIGURE 24.39

Ekman transport causes coastal upwelling (a) or downwelling (b), depending on the dominant wind direction.

is called the *Gulf Stream*. As the Gulf Stream flows northward along the North American coast, flow in the gyre steers the warm current eastward toward Europe (Figure 24.40). The warm water transfers heat to the westward-moving air, which influences weather and climate in Europe. Great Britain and Norway benefit from this heat, for lands at this northern latitude might well be colder were the heat not brought northward from equatorial regions and the Gulf of Mexico. As the warm current encounters Europe, it is turned southward toward the equator, where it continues to flow in the North Atlantic gyre. Eventually, this water moves westward toward the Gulf of Mexico and once again becomes part of the Gulf Stream.

Each ocean basin has a similar surface flow: a gyre flow pattern with a warm western boundary current. The Pacific counterpart of the Gulf Stream is the warm, northward-flowing current known as the *Kuroshio*. In the Southern Hemisphere, surface oceanic circulation (with the exception of the Antarctica Circumpolar Current) is similar, except that the gyres move counterclockwise. Warm western boundary currents occur in all three Southern Hemisphere ocean basins.

FYI Don't believe stories about swirling water in your sink being affected by the Coriolis force. Any effect due to the difference in speeds of one part of the sink compared with the other are minuscule, and they are masked by thermal motions in the water.

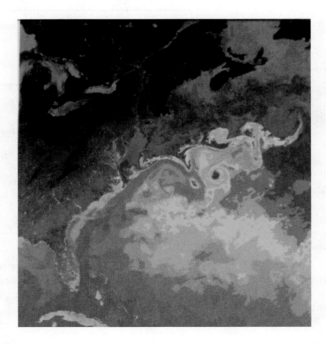

FIGURE 24.40

The reddish areas in this false-color satellite image indicate warm water. The Gulf Stream transports huge amounts of heat northward and then eastward toward Europe.

Deep-Water Currents

Surface waters are driven by winds, but deeper waters are driven by gravity—deep water flows because dense water sinks. Deep water moves more slowly than surface water, but vastly more water is transported by deep-water currents than by surface-water currents. Deep-water flow is like a huge global conveyor belt, moving water and heat all over the planet (Figure 24.41).

Across most of the ocean, warmer and less dense surface waters do not interact much with colder and denser deep waters. But in high latitudes, surface-zone seawater is much colder than it is in the lower to mid-latitudes. This cold surface water interacts with seawater in the deep zone, and a very slow worldwide circulation pattern develops. To understand how this pattern develops, we need to first look at what happens when seawater begins to freeze. Seawater does not freeze easily, but when it does, only the water freezes and the salt is left behind. Thus seawater below the newly forming ice experiences an increase in salinity, which in turn causes an increase in density. This cold, saltier, and denser seawater sinks, setting up vertical flow toward the ocean floor. Horizontal movement begins when this dense, sinking water reaches the ocean bottom and continues to flow along the ocean floor.

Conveyor-belt circulation begins in the North Atlantic as dense, cold, salty seawater around Greenland and Iceland sinks and flows along the ocean bottom toward the equator and then on to the Antarctic Ocean. Once near Antarctica, the water flows eastward around the continent, then northward into the Pacific and Indian Oceans. Thus deep-water currents flow in an overall north–south circulation pattern.

As the conveyor belt continues, some of the deep water makes it back to the ocean's surface. A combination of deep-water mixing by ocean-floor tidal stirring and upwelling due to favorable winds brings the deep waters slowly back to the surface. Figure 24.42 shows the cycle from cold, sinking water to warmer surface flow. Eventually, all of the deep water flows to the ocean surface, and vice versa, although the process may take hundreds or thousands of years.

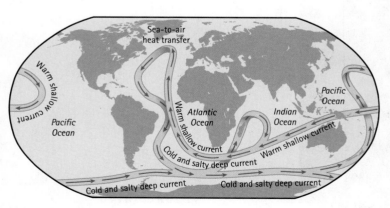

FIGURE 24.41

Deep-ocean currents act like a conveyor belt, transporting cold water from the North Atlantic past the equator and on to the Antarctic. From the Antarctic, water flows eastward and then northward into the Pacific Ocean and the Indian Ocean.

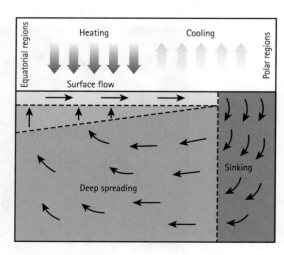

FIGURE 24.42

Cold, dense water sinks in the polar regions and then flows toward the equator. Eventually it is forced upward, where it warms, becomes part of the surface flow, and returns to the polar regions. Darker colors indicate colder water.

LINK TO WEATHER

The El Niño Condition

When weather is measured over time, an average weather pattern can be seen. Consistent behavior of weather over time is *climate*. During some periods, the *average* weather pattern departs from its norm. These variations from the average are expected and have recognizable short-term changes in the weather. A prime example of such a disruption is the El Niño condition.

Under normal conditions, weather patterns in the Pacific are controlled by warm, high-pressure systems located in both hemispheres near the equator. These high-pressure systems cause the trade winds to blow westward along the equator, dragging the warm equatorial surface waters along with them. As warm surface waters move westward, deeper, colder waters to the east rise upward to occupy the space left vacant by the warm surface water. The upwelling cold waters, rich in nutrients, attract a variety of sea life. Upwelling of these cold waters has

been especially important to the fishing industry along the coast of South America, where people earn their living catching anchovies that come to feed in the nutrient-rich waters.

Fishing is not always good, however. Each year in October the trade winds slacken, reversing the normal westward flow of warm tropical surface waters. As the warm surface waters drift eastward, upwelling decreases and so does the fishing industry. People along the South American coast refer to this occurrence as El Niño because it begins each year around the traditional December celebration of Christ's birth (Christ is *El Niño*, "The Child," in Spanish). Under normal conditions, the trade winds pick up again in early spring, the surface waters are again blown westward across the ocean, and everything returns to normal.

In some years, however, the trade winds fail to strengthen and the warm surface waters remain off the coast of South America for a year or longer. During these abnormal conditions,

upwelling of cold water ceases, and South American fishing industries fail. Although a small El Niño occurs each year, this extended El Niño is referred to as the *El Niño condition*.

The El Niño condition influences climate on both sides of the tropical Pacific Ocean. Under normal conditions, upwelling cold water on the eastern side of the Pacific coincides with dry cool air, high pressures, and clear skies. On the western side of the Pacific, surface waters—warmed and fueled by their long journey across the ocean—warm the surrounding air. As the warm moist air rises, low pressures and storms develop on the warm western side of the Pacific.

During an extended El Niño condition, the pattern is reversed. Warm water, rising warm moist air, low pressures, and storms are found on the eastern side of the Pacific rather than the western side. This exchange of pressure systems and weather patterns between east and west during an El Niño condition is sometimes referred to as the *southern oscillation*.

For instructor-assigned homework, go to www.masteringphysics.com

SUMMARY OF TERMS (KNOWLEDGE)

Continental margin The boundary between the continents and the ocean; it consists of a continental shelf, a continental slope, and a continental rise.

Coriolis force The apparent deflection from a straight-line path observed in any body moving near Earth's surface, caused by Earth's rotation.

Exosphere The fifth atmospheric layer above Earth's surface, extending from the thermosphere upward and out into interplanetary space.

Greenhouse effect Warming caused by short-wavelength radiant energy from the Sun that easily enters the atmosphere and is absorbed by Earth. This energy is then reradiated at longer wavelengths that cannot easily escape Earth's atmosphere.

Gyre A circular or spiral whirl pattern, usually referring to very large current systems in the open ocean.

Ionosphere An electrified region within the thermosphere and uppermost mesosphere where fairly large concentrations of ions and free electrons exist.

Mesosphere The third atmospheric layer above Earth's surface, extending from the top of the stratosphere to 80 km.

Neap tide A tide that occurs when the Moon is midway between new and full, in either direction. The pulls of the Moon and Sun are perpendicular, so the solar and lunar tides do not overlap. This makes high tides not as high and low tides not as low.

Pressure-gradient force The force that moves air from a region of high-pressure air to an adjacent region of low-pressure air.

Salinity The mass of salts dissolved in 1000 g of seawater.

Spring tide A high or low tide that occurs when the Sun, Earth, and Moon are aligned so that the tides due to the Sun and Moon coincide, making the tides higher or lower than average; occurs during the full Moon or new Moon.

Stratosphere The second atmospheric layer above Earth's surface, extending from the top of the troposphere up to 50 km. This is where stratospheric ozone forms.

Thermosphere The fourth atmospheric layer above Earth's surface, extending from the top of the mesosphere to 500 km.

Troposphere The atmospheric layer closest to Earth's surface, 16 km high over the equator and 8 km high over the poles, containing 90% of the atmosphere's mass and essentially all of its water vapor and clouds.

READING CHECK QUESTIONS (COMPREHENSION)

24.1 Earth's Atmosphere and Oceans

1. Why are temperatures moderate on lands bordering the oceans?

2. What were the main components of Earth's first atmosphere? What happened to this atmosphere?

3. Earth's present atmosphere likely developed from gases that escaped from its interior during volcanic eruptions. What three principal atmospheric gases did these eruptions produce?

4. Explain the importance of photosynthesis in the evolution of the atmosphere.

24.2 Components of Earth's Oceans

5. What are four features we may find between two continental margins (the ocean floor)?

6. The salinity of the ocean varies from one place to another. What two factors lead to an increase in salinity? What two factors lead to a decrease in salinity?

24.3 Ocean Waves, Tides, and Shorelines

7. Why do waves become taller as they enter shallow water?

8. What is wave refraction? Why does it occur in ocean waves?

9. Why is a barrier island's lagoon usually a quiet environment?

10. Why are all tides highest at the time of a full or new Moon?

11. When do the highest high tides occur: during a spring tide or during a neap tide?

24.4 Components of Earth's Atmosphere

12. What elements make up today's atmosphere?

13. Why doesn't gravity flatten the atmosphere against Earth's surface?

14. In which atmospheric layer does all our weather occur?

15. Does temperature increase or decrease as one moves upward in the troposphere? As one moves upward in the stratosphere?

24.5 Solar Energy

16. What does the angle at which sunlight strikes Earth have to do with the temperate and polar regions?

17. What does Earth's tilt have to do with the change of seasons?

18. Why are the hours of daylight equal all around the world on the two equinoxes?

19. How does radiation emitted from Earth differ from radiation emitted by the Sun?

20. How is the atmosphere near Earth's surface heated from below?

24.6 Driving Forces of Air Motion

21. What is the underlying cause of air motion?

22. What causes pressure differences to arise and hence causes the wind to blow?

23. In what direction does Earth spin: west to east or east to west?

24. What does the Coriolis force do to winds? To ocean currents?

25. How does the Coriolis force determine the general path of air circulation?

24.7 Global Circulation Patterns

26. Why are most of the world's deserts found in the area known as the horse latitudes?

27. What are the trade winds?

28. Why are eastbound aircraft flights usually faster than westbound flights?

29. What factors set surface ocean currents in motion?

30. Explain the circulation pattern of the Gulf Stream.

ACTIVITIES (HANDS-ON APPLICATION)

31. For an easy-to-make model of air pressure, fill a glass with water. Place a piece of cardboard over the top of the glass, and hold the cardboard in place with one hand as you turn the glass upside down. *Make sure the cardboard does not bend.* Now remove your hand from the cardboard. What happens? Does the cardboard fall off and the water flow out? No, the cardboard remains on the glass. Why? The air pressure pushing up on the cardboard is greater than the weight of the water pushing down. For fun, turn the cup sideways and then turn it all around. The cardboard continues to stick!

32. A wave forms when the water's surface is disturbed. The following terms are used to describe a wave.

> Crest—the highest elevation above the undisturbed sea surface
> Trough—the lowest depression below the undisturbed sea surface
> Wavelength—the distance between two successive crests or two successive troughs
> Wave height—the vertical distance from the top of the crest to the bottom of the trough

Amplitude— the distance from either the crest or the trough to the undisturbed water level, or one-half the wave height

Label the parts of a wave on the sketch.

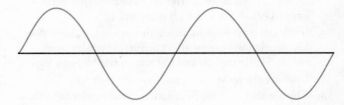

THINK AND SOLVE (MATHEMATICAL APPLICATION)

Refer to Figuring Physical Science: Dense as Air on page 661 to answer the following two questions.

33. What is the mass in kilograms of the air in an "empty" nonpressurized scuba tank that has an internal volume of 0.0100 m³?

34. What is the mass in kilograms of the air in a scuba tank that has an internal volume of 0.0100 m³ and is pressurized so that the density of the air in the tank is 240 kg/m³?

THINK AND RANK (ANALYSIS)

35. Going *from* continental land toward the deep ocean basin, place these parts of the continental margin in order: (a) slope, (b) shelf, (c) rise, (d) midocean ridge.

36. Rising through Earth's atmospheric layers, starting from ground level, place Earth's atmospheric layers in order: (a) exosphere, (b) stratosphere, (c) troposphere, (d) thermosphere.

37. Beginning with the underlying cause, rank the factors that contribute to global air circulation from most to least significant: (a) pressure differences, (b) temperature differences, (c) Coriolis force, (d) frictional force.

38. From the equator to the poles, place the following wind bands in order: (a) trade winds, (b) horse latitudes, (c) doldrums, (d) westerlies.

39. Deep-water ocean currents transport water and heat all over the planet. Starting from where the conveyor belt originates, place the following areas in the geographic sequence of the current's path: (a) Antarctica, (b) Pacific Ocean, (c) North Atlantic.

EXERCISES (SYNTHESIS)

40. If a gas fills all the space available to it, why doesn't the atmosphere go off into space?

41. If there were no water on Earth's surface, would weather occur? Defend your answer.

42. How do the wavelengths of radiant energy vary with the temperature of the radiating source? How does this affect solar and terrestrial radiation?

43. How is global warming affected by the relative transparencies of the atmosphere to long- and short-wavelength electromagnetic radiation?

44. Explain why your ears pop when you ascend to higher altitudes?

45. How does the density of air in a deep mine compare with the density of air at sea level? Defend your answer.

46. Earth is closest to the Sun in January, but January is cold in the Northern Hemisphere. Why?

47. How do the total number of hours of sunlight in a year compare for equatorial regions and polar regions of Earth? Why are polar regions so much colder?

48. If the composition of the upper atmosphere were changed so that it permitted a greater amount of terrestrial radiation to escape, what effect would this have on Earth's average temperature? How about if the atmosphere reduced the amount of terrestrial radiation that could escape?

49. As humans consume more energy, the average temperature of Earth's surface tends to rise. Yet no matter how much energy is consumed, the temperature doesn't rise without limit. What process prevents an indefinite rise? Explain.

50. As the world's population increases, the amount of carbon dioxide emissions from fossil fuel combustion also increases. Yet the amount of carbon dioxide emitted is greater than the amount found in the atmosphere. Where is the likely repository of excess atmospheric carbon dioxide?

51. Why is it important that mountain climbers wear sunglasses and apply sunblock even when the temperature is below freezing?

52. If Earth were not spinning, in what direction would the surface winds blow where you live? In what direction do surface winds actually blow on Earth at 15°S latitude, and why?

53. What is the relationship between global atmospheric circulation and ocean currents? Relate oceanic gyres to patterns of subtropical high pressure.

54. Relate the jet stream to upper-air circulation. How does this circulation pattern affect airline schedules from New York to San Francisco and the return trip to New York?

55. What are the jet streams, and how do they form?

56. Why are temperature fluctuations greater over land than over water? Explain.

57. Because seawater does not freeze easily, sea ice never gets very thick. So from where do large icebergs originate?

58. The Mediterranean Sea is highly salty. What can you say about the relative rates of evaporation and precipitation over the Mediterranean?

59. What role does the Sun play in the circulation of ocean currents?

60. How does the ocean influence weather on land?

61. What happens to the salinity of seawater when evaporation at the ocean surface exceeds precipitation? When precipitation exceeds evaporation? Defend your answers.

62. Why is there more concern about melting polar ice caps than about melting icebergs?

63. Water denser than surrounding water sinks. With respect to the densities of deeper water, how far does it sink?

64. As a volume of seawater freezes, the salinity of the surrounding water increases. Explain.

65. Why are headlands prime areas for erosion?

66. Suppose a breakwater is built offshore and perpendicular to the shore. How will this structure affect the longshore current and its transport of sand? Defend your answer.

67. Carbonate rocks are formed mainly in marine environments. Why do we find abundant carbonate deposits on continental land?

68. Why is the sand of some beaches composed of small pieces of seashells?

69. The oceans are composed of salt water, yet evaporation over the ocean surface produces clouds that precipitate fresh water. Why is there no salt?

70. As waves approach shallow water, those with longer wavelengths slow down before those with shorter wavelengths do. Why?

71. What effect does the formation of sea ice in polar regions have on the density of seawater? Explain.

72. Which receives more solar energy over the course of a year: tropical regions or temperate regions? How does this affect ocean salinity?

73. In tropical regions, solar energy exceeds terrestrial radiation. What effect does this have on the salinity of oceans in tropical regions?

74. Why do the temperate zones have unpredictable weather?

75. What happens to the water level in a glass of water when a floating ice cube in the glass melts? Similarly, what happens to the water level in the Great Lakes when floating chunks of ice melt?

76. Most people today know that the ocean tides are caused principally by the gravitational influence of the Moon. And most people therefore think that the gravitational pull of the Moon on Earth is greater than the gravitational pull of the Sun on Earth. What do you think?

77. Would ocean tides exist if the gravitational pull of the Moon (and Sun) were somehow equal on all parts of the world? Explain.

78. Why aren't high ocean tides exactly 12 hours apart?

79. With respect to spring and neap ocean tides, when are the lowest tides? That is, when is the tide best for digging clams?

80. When the ocean tide is unusually high, is the following low tide unusually low? Defend your answer in terms of "conservation of water." (If you slosh water in a tub so it is extra deep at one end, is the other end extra shallow?)

81. How does the Coriolis force influence the movement of surface waters?

82. What is the characteristic climate of the doldrums, and why does it occur?

83. Because our atmosphere developed as a result of volcanic eruptions, why aren't there higher traces of atmospheric carbon dioxide, one of the principal volcanic gases?

84. Why is the thermosphere so much hotter than the mesosphere?

85. What is the source of the ions that give the ionosphere its name?

86. If it is winter and January in Chicago, what are the corresponding season and month in Sydney, Australia?

87. What causes the fiery displays of light called the auroras?

88. Explain why most of the bottom water of the oceans forms in the North Atlantic and near Antarctica.

89. How does the density of seawater vary with changes in temperature? How does density change with salinity?

DISCUSSION QUESTIONS (EVALUATION)

90. During a heavy storm, which beach type will have a more protected shore: a gently sloping or a steep, narrow one?

91. Why do wind-generated waves change direction when they reach shallow water?

92. At the surface, does an Ekman spiral look like a whirlpool or gyre?

93. If a north wind blows across the sea surface, what direction does the surface current flow in the Northern Hemisphere? In the Southern Hemisphere?

94. How would air circulate in the Northern and Southern Hemispheres if there was no land and Earth did not rotate?

READINESS ASSURANCE TEST (RAT)

If you have a good handle on this chapter, if you really do, then you should be able to score at least 7 out of 10 on this RAT. If you score less than 7, you need to study further before moving on.

Choose the BEST answer to each of the following.

1. Earth's lower atmosphere is kept warm by
   (a) solar radiation.
   (b) terrestrial radiation.
   (c) short-wave radiation.
   (d) the ozone layer.

2. Compared to lands far from the oceans, lands that border the oceans tend to have
   (a) extreme seasonal variations.
   (b) small seasonal variations.
   (c) wet, cold winters.
   (d) dry, hot summers.

3. Which pulls with the greater *force* on Earth's oceans?
   (a) the Sun
   (b) the Moon
   (c) both the Sun and the Moon, with equal force
   (d) There is no force.

4. Air motion is greatly influenced by
   (a) pressure differences.
   (b) temperature differences.
   (c) the Coriolis force.
   (d) all of the above

5. Ocean tides are caused by differences in the
   (a) gravitational pull of the Sun on opposite sides of Earth.
   (b) force of the Moon.
   (c) gravitational pull of the Moon on opposite sides of Earth.
   (d) distance of the Sun from the Moon.

6. A factor that increases ocean salinity is
   (a) runoff from streams and rivers.
   (b) formation of sea ice.
   (c) precipitation.
   (d) glacial melting.

7. The wind blows in response to
   (a) frictional drag.
   (b) Earth's rotation.
   (c) pressure differences.
   (d) moisture differences.

8. Planet Earth experiences changes of the seasons because of
   (a) incoming solar radiation.
   (b) Earth's rotation around the Sun.
   (c) the Coriolis force.
   (d) the tilt of Earth's axis.

9. The most significant result of the Coriolis force is the
   (a) deflection of air and water currents.
   (b) creation of Ekman transport.
   (c) reflection of air currents and global air circulation.
   (d) creation of the jet streams.

10. The ultimate cause of ocean surface currents is
    (a) divergence in equatorial regions.
    (b) the gradient between the doldrums and the horse latitudes.
    (c) density contrasts.
    (d) frictional drag by prevailing winds.

Answers to RAT

1. b, 2. b, 3. a, 4. d, 5. c, 6. b, 7. c, 8. d, 9. a, 10. d

25

CHAPTER

Driving Forces of Weather

SURROUNDING OUR Earth is a life-giving shell of air—the atmosphere. Its mixture of gases provides us with air to breathe and protection from harmful ultraviolet radiation. In Chapter 24, we learned about the atmosphere's vertical structure and how most of the atmosphere's mass resides in the lowest layer—the troposphere. This is where Earth's weather occurs. And that brings us to our present topic—the weather. The factors that influence the weather are atmospheric moisture, temperature, air pressure, and the geographic arrangement of land and water features. We begin our discussion of weather by looking first at atmospheric moisture and at how the amount of moisture in the air influences atmospheric stability. Then we discuss the development of different air masses and some resulting weather patterns. We conclude with a look at the violent weather forces that greatly affect our planet's surface.

25.1 Atmospheric Moisture

EXPLAIN THIS Why is the grass wet first thing in the morning?

Water is certainly vital to life on Earth. But think for a moment about water's immense role in the physical processes of Earth—water shapes Earth's surface and governs its weather. And no matter how "dry" air may feel at times, it is never completely dry; some concentration of water vapor is always present.

But instead of saying "concentration of water vapor," we use the much shorter term **humidity**. Specifically, humidity is the mass of water vapor per volume of air, just as concentration is the mass of solute per volume of solution. It is important to note, however, that when you hear TV weather forecasters describing humidity, they are actually talking about **relative humidity**. Relative humidity depends on temperature—it describes the amount of water vapor currently in the air compared to the maximum amount of water vapor that could be in the air at that specific temperature. The maximum amount varies depending on the temperature:

$$\text{Relative humidity} = \frac{\text{water-vapor content}}{\text{water-vapor capacity}} \times 100\%$$

For example, a relative humidity of 50% means that the water-vapor content in the air is half the amount of the air's maximum capacity at that temperature.

Think of water-vapor capacity in the same way that you think of mineral solubility—the solubility of salt in water, for example. At a given temperature, the concentration of salt dissolved in water has an upper limit—the salt's *solubility* (Chapters 16 and 20). The solubility concentration cannot be exceeded: add more salt and it does not dissolve.

The same is true for the concentration of water vapor in air. At a given temperature, the humidity has an upper limit that cannot be exceeded, just as each mineral has a solubility limit. The humidity limit is called the **saturation vapor pressure**.* When air reaches its saturation vapor pressure, the humidity (or vapor pressure) cannot increase, and the relative humidity is 100%. The air is said to be *saturated*.

Warm air has a higher saturation vapor pressure than cold air (Table 25.1). For example, when saturated air at 10°C is warmed to 20°C—and no water vapor is added—the air is no longer saturated and the relative humidity drops below 100%—to about 52% in this case. Even though the *relative* humidity dropped, the *actual* humidity—the concentration of water vapor in the air—stays the same. It is just that the warm air has a higher saturation vapor pressure and can, in a sense, "hold" more water vapor.

The saturation vapor pressure varies with temperature because there are two competing processes—the evaporation rate and the condensation rate (Chapter 7). The evaporation rate depends on temperature, but the condensation rate does not—it depends only on the humidity (Figure 25.1). When the evaporation rate equals the condensation rate—at any temperature—air is saturated, and the relative humidity is 100%.

LEARNING OBJECTIVE
Distinguish between humidity and relative humidity, and describe the relationship between condensation and temperature change.

FYI Most raindrops are nearly perfect spheres, not the teardrop shapes often depicted by artists. The attraction of molecules in a liquid, called surface tension, tends to pull the droplet into the shape with the smallest surface area—a sphere. The exceptions are very large raindrops (more than 1 mm in diameter), which air resistance flattens to produce a hamburger-bun shape with a flat bottom and a slightly curved top.

FYI Relative humidity is a good indicator of comfort. For most people, conditions are ideal when the temperature is about 20°C and the relative humidity is between 50% and 60%. When the relative humidity is too high, moist air feels "muggy," as condensation counteracts the evaporation of perspiration. Cold air that has a high relative humidity feels colder than dry air of the same temperature because of increased conduction of heat from the body. When the relative humidity is high, hot weather feels hotter and cold weather feels colder.

---

* *Vapor pressure* measures the same thing as humidity—the amount of water vapor in air. Humidity and vapor pressure describe the same thing, but they have different units—grams per cubic meter for humidity and millibars (or millimeters of mercury) for vapor pressure.

TABLE 25.1	SATURATION VAPOR PRESSURE VS. TEMPERATURE	
Temperature (°C)	Temperature (°F)	Saturation Vapor Pressure (millibars, mb)
0	32	6.1
5	41	8.7
10	50	12.3
15	59	17.0
20	68	23.4
25	77	31.7
30	86	42.5
35	95	56.3

Saturation Vapor Pressure

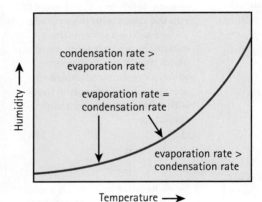

FIGURE 25.1
The curve in this figure is the saturation vapor pressure as a function of temperature and humidity (not relative humidity). On the curve, relative humidity is 100% and the evaporation rate equals the condensation rate. Below the curve, the relative humidity is less than 100% and the evaporation rate is greater than the condensation rate. Above the curve, the condensation rate is greater than the evaporation rate, and precipitation forms. This curve helps explain why hot summer days can feel very muggy. Your skin cannot evaporate water so easily. From the curve, can you see that warm air has a higher saturation vapor pressure than cold air?

Interestingly, evaporation and condensation are always happening—neither ever stops. When the evaporation rate is greater than the condensation rate, the relative humidity is always less than 100%. The greater the difference between the evaporation rate and the condensation rate, the lower the relative humidity and the drier the air feels.

On the other hand, if the condensation rate is greater than the evaporation rate, the air is saturated *and* there is an excess of water vapor. In such a case, the excess water vapor condenses to form liquid water, because the evaporation rate is not fast enough to turn the freshly created liquid water back to vapor before more water condenses. Because slower-moving molecules characterize lower air temperatures, saturation and condensation are more likely to occur in cool air than in warm air (Figure 25.2).

In the example given on page 683, at 10°C the evaporation rate was the same as the condensation rate, so the relative humidity was 100%. When the temperature increased from 10°C to 20°C, the condensation rate did not change, but the evaporation rate increased. As the temperature, evaporation rate, and saturation vapor pressure all increased, the air could accommodate more water vapor. The relative humidity also dropped, because no water vapor was added—the air was no longer saturated at the higher temperature.

What would happen if we started with 20°C air and a relative humidity of 60%, then lowered the temperature to 10°C? As the temperature decreased,

Fast-moving H₂O molecules rebound upon collision.

FIGURE 25.2
Condensation of water molecules.

Slow-moving H₂O molecules condense upon collision.

the evaporation rate would slow down, but water vapor would continue to condense at the same rate as it did when the air was 20°C. The saturation vapor pressure and the evaporation rate would both go down, so the relative humidity would go up. In fact, when the air temperature reaches 12°C, the evaporation rate and the condensation rate would be equal and the relative humidity would be 100%—the air would be saturated. When the temperature drops below 12°C, condensation rules and liquid water droplets form—rain.

Evaporation and condensation are essentially opposites of each other. They are phase-change processes between the liquid and gaseous states of water. And as we know, liquid water changes to a solid—it freezes—at 0°C. Water can also change phase from solid to gas—and from gas to solid—without becoming liquid in between. When ice changes phase directly to water vapor, the process is called *sublimation* (Chapter 7). The opposite process—water vapor changing directly to ice—is called *deposition.**

FYI In rainy weather, when your car windshield fogs up, make sure the air conditioner is on when in defrost mode, even when blowing heated air. What causes window fogging is the humidity in the car caused by rain, by wet clothes, and by the breath of the passengers. Because the air from the air conditioner is very dry, it clears a foggy windshield very nicely in a very short time.

Temperature Changes and Condensation

As an air parcel rises, it expands. The expansion occurs because the air moves to a region of lower air pressure. As we learned in Chapter 7, air cools when it expands. This seems like a contradiction, because warm air rises. But a rising air parcel must do work on the surrounding environment in order to expand. And doing that work uses energy, which is lost in the form of heat. As the air cools, water molecules move slower and condensation outpaces evaporation. Liquid water forms on microscopic particles of dust, smoke, and salt—*cloud condensation nuclei*—and this creates a cloud. As the sizes of the cloud droplets grow, they fall to Earth and we have rain. Rain is but one form of precipitation. Other familiar forms of precipitation are mist, hail, snow, and sleet.

Water vapor does not need to be high in a cloud to form precipitation—condensation can occur in air close to the ground as well. When condensation occurs at or near Earth's surface, we call it *dew, fog,* or *frost.* On cool, clear nights, objects near the ground cool down more rapidly than the surrounding air. As the air cools, the saturation vapor pressure drops and the air cannot accommodate as much water vapor as when the air was warmer. When the temperature drops below a certain threshold, called the *dew point temperature,* the air becomes saturated—relative humidity is 100%—and condensation dominates. In this case, cloud condensation nuclei are not needed. Water from the now-saturated air condenses on any available surface—a twig, a blade of grass, the windshield of a car, and so on. We often call this type of condensation *early-morning dew,* because it occurs when daily temperatures are the coldest, just before sunrise. When the dew point is at or below water's freezing point, we have frost.** When a large mass of air cools and reaches its dew point, we get a cloud near the ground—fog.

FIGURE 25.3
San Francisco is well known for its summer fog.

CHECKPOINT
What is the major difference between fog and a cloud?

Was this your answer?
Altitude.

---

* The term *deposition* in meteorology refers to the conversion of water vapor to a solid. This is different from the way we use this term in Chapters 20 and 22, where deposition refers to the laying down of sediment.
** Interestingly, frost is not frozen dew. Frost forms directly from water vapor as a result of *deposition.*

FIGURING PHYSICAL SCIENCE

Humidity

Humidity is the mass of water vapor per volume of air. Relative humidity is the ratio of the air's water-vapor content to the air's water-vapor capacity at a certain temperature.

For these three problems, consider a small, experimental air mass at 30°C that weighs 90 N.

SAMPLE PROBLEM 1

If the air density is 1.25 kg/m³, what is the volume of the air mass?

Solution:

Newton's second law tells us that

$$a = \frac{F}{m}$$

Rearrange the equation and get

$$\text{Mass} = \frac{\text{force}}{\text{acceleration due to gravity}} = \frac{90 \text{ N}}{10 \text{ m/s}^2} = 9 \text{ kg}$$

The volume of 9 kg can be found in this way:

$$9 \text{ kg} \times \frac{1 \text{ m}^3}{1.25 \text{ kg}} = 7.2 \text{ m}^3$$

SAMPLE PROBLEM 2

If there is 0.13 kg of water vapor in the air mass, what is the humidity of the air mass?

Solution:

$$\text{Humidity} = \frac{\text{mass of water}}{\text{volume of air}} = \frac{0.13 \text{ kg}}{7.2 \text{ m}^3} = 0.018 \text{ kg/m}^3$$

SAMPLE PROBLEM 3

At 30°C, the maximum amount of water vapor in the air mass is 30 g/m³ of water vapor. What is the relative humidity of the air mass?

Solution:

First, we must convert the units:

$$\frac{30 \text{ g}}{\text{m}^3} \times \frac{1 \text{ kg}}{1000 \text{ g}} = 0.03 \text{ kg/m}^3$$

Then

Relative humidity

$$= \frac{\text{amount of water vapor in air}}{\text{maximum amount of water vapor in air at 30°C}}$$

$$= \frac{0.018 \text{ kg/m}^3}{0.03 \text{ kg/m}^3} \times 100\% = 60\%$$

LEARNING OBJECTIVE
Describe the relationships among the three variables that control the weather.

Mastering**PHYSICS**®
TUTORIAL: Surface Temperature of Planets

25.2   Weather Variables

EXPLAIN THIS Why does warm air rise and cool air sink?

Air pressure, temperature, and density are three key variables that control how air behaves, and hence they control the weather. To understand and predict the weather, we must understand all three. First, consider air pressure. Air is a mixture of molecules that move randomly and collide with one another like billiard balls on a pool table. When a molecule bumps into something, it exerts a small push on whatever it hits. Such pushes by countless molecules produce *air pressure.*

The faster the air molecules move, the greater their kinetic energy. The greater the kinetic energy, the greater the impact of molecular collisions—and the greater the air pressure. All else being equal, air composed of fast-moving molecules—warm air—exerts more air pressure on its surroundings than cooler air.

Another factor that affects air pressure is density. The denser the air, the more molecules are present and hence the greater the number of molecular collisions. And more collisions means greater air pressure. Air becomes denser when it is compressed, and it becomes less dense when it expands. The changes in density occur because the volume of a given mass of air is made smaller by compression and larger by expansion.

Adiabatic Processes in Air

The concept of *heat exchange* shows us that air pressure, temperature, and density are interrelated. When heat is added to an air mass, the air temperature increases, the air pressure increases, or both increase. Heat can be added to air

Mastering**PHYSICS**®
VIDEO: Adiabatic Process

by solar radiation, by moisture condensation, or by contact with warm ground. When heat is subtracted from an air mass, the temperature or the pressure of the air falls. Heat can be subtracted from air by radiation to space, by the evaporation of rain falling through dry air,* or by contact with cold surfaces.

But air can change temperature without the loss or gain of heat. When heat transfer is zero, or nearly so, we have an **adiabatic** process. Adiabatic processes occur when air is expanded or compressed. For adiabatic temperature changes to occur, expansion or compression must be rapid enough that there is insufficient time for heat exchange to occur while the volume change is occurring. To illustrate how bodies of air behave, imagine a body of air enclosed in a very thin plastic garment bag—an *air parcel*. Like a free-floating balloon, the parcel can expand and contract freely without heat transfer to or from air outside the parcel.

Recall from Chapter 24 that air pressure decreases with increasing height. For example, as an air parcel flows up the side of a mountain, the pressure exerted on it decreases, allowing the air parcel to expand—and cool without any heat exchange. Such changes in temperature, caused by changes in pressure, are described by the *ideal gas law*:**

Pressure ~ density × temperature

Assuming relatively small changes in air density, which is generally the case in the lower atmosphere, the ideal gas law states that temperature goes down when pressure goes down. And we know that air pressure goes down when air expands. Conversely, temperature goes up when pressure goes up—air pressure goes up when air is compressed. Can you now see more clearly that the temperature of an air parcel decreases when lifted? And that air temperature increases when an air parcel descends? According to the ideal gas law, such temperature changes do not require the loss or gain of heat—they are adiabatic.

With adiabatic expansion, the temperature of a dry air parcel† decreases about 10°C for each kilometer it rises (Figure 25.4). This rate of cooling for dry air is called the *dry adiabatic lapse rate*. Air flowing up and over tall mountains or rising in thunderstorms may change elevation by several kilometers. Thus, when a dry air parcel at ground level at a comfortable 25°C rises 6 km, its temperature drops to a frigid −35°C! On the other hand, if air at 6 km elevation, at a typical temperature of −20°C, descends to the ground, its temperature rises to a whopping 40°C!

Adiabatic processes are not restricted to dry air. As rising air cools, its ability to accommodate water vapor decreases, increasing the relative humidity of the rising air. If the air cools to its dew point, the relative humidity climbs to 100%, water vapor condenses, and clouds form. Because condensation releases heat (Chapter 7), the air parcel is warmed. The heat counteracts to some extent the cooling due to expansion, making the air cool at a lesser rate—the *moist adiabatic lapse rate*. Although the moist adiabatic lapse rate varies according to the temperature and the moisture content of air, on average a saturated air parcel cools by about 6°C for every kilometer it rises.

A dramatic example of adiabatic warming is the *Chinook*—a dry wind that blows down from the Rocky Mountains across the Great Plains (Figure 25.5). Cold air moving down a mountain slope is compressed as it moves to lower elevations

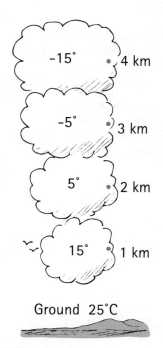

FIGURE 25.4
The temperature of a parcel of dry air that expands adiabatically changes by about 10°C for each kilometer of elevation.

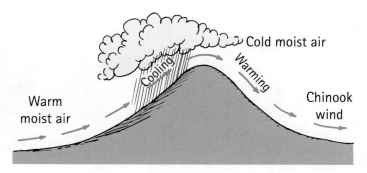

FIGURE 25.5
Chinooks—which are warm, dry winds—occur when high-altitude air descends and is adiabatically warmed.

* As we learned in Chapter 7, evaporation removes heat from the surrounding environment.
** The ideal gas law is related to Boyle's law, which was discussed in Chapter 5.
† A dry air parcel is an air parcel that is not saturated—its relative humidity is less than 100%.

(where air pressure is greater than at higher elevations), and it becomes much warmer. The effect of expansion or compression of gases is quite impressive.*

Atmospheric Stability

Let's consider a parcel of air descending from aloft. Air pressure at lower elevations is greater than air pressure aloft, so the air parcel is compressed as it descends, causing it to warm up. If the descending air parcel's temperature becomes warmer—and therefore less dense—than air at the lower elevation, the parcel will rise back to the elevation from which it came. Analogously, think of what happens when you try to submerge a flotation device, which is less dense than water, in a swimming pool. The flotation device pops right back to the surface. The flotation device and the descending air parcel in this example are both *stable*, with respect to their environments. They "want" to return to their original positions.

Now consider an air parcel that is forced upward. Such an air parcel expands as it rises, becoming cooler. If the temperature of the rising air becomes cooler than the surrounding air, it becomes denser and sinks back to the elevation where it started. This is also stable air.

On the other hand, if the rising air stays warmer than the surrounding air, it continues to rise instead of returning to its starting position. This is *unstable air*. Eventually, the air parcel expands and cools sufficiently to match the surrounding air. When the temperatures match, the air parcel rises no farther, but it does not sink back to its starting position. So unstable air rises, but stable air does not. Hot-air balloons use this strategy to rise or descend to a desired elevation. Heat is added to the air in the balloon to make it rise, and the air inside the balloon is allowed to cool to make it descend.

The surrounding air plays an important role in determining whether an air parcel is stable or unstable. Under normal conditions, air temperature decreases with altitude. The *environmental lapse rate* describes the manner in which these temperature changes occur with altitude. A lower environmental lapse rate indicates that temperatures decrease less for each meter of increased altitude than with a higher environmental lapse rate. The environmental lapse rate varies from place to place and from day to day—it can even vary over the course of a day! The average environmental lapse rate is a decrease of 6.5°C for each kilometer rise in elevation.

A rising parcel of air continues to rise as long as it is warmer and less dense than the surrounding air. Air that is cooler and denser than its surroundings does the opposite—it sinks. Under some conditions, large parcels of cold air sink and remain at low elevations. This results in air above that is warmer. When upper regions of the atmosphere are warmer than lower regions, which is the opposite of what normally occurs, we have a **temperature inversion**. In such cases, most rising air parcels can't pass through the upper layer of warmer air, because the rising air is cooler and denser. On a small scale, evidence of a temperature inversion is commonly seen over a cold lake when visible gases and small particles such as smoke spread out in a flat layer above the lake rather than rising and dissipating higher in the atmosphere (Figure 25.6).

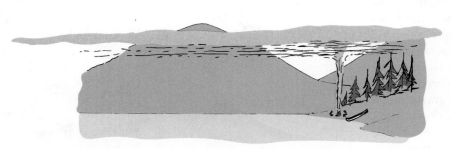

FIGURE 25.6
The layer of campfire smoke over the lake indicates a temperature inversion. The air above the smoke is warmer than the smoke, and the air below is cooler.

* When you're flying at high altitudes where the outside air temperature is typically −35°C, you're quite comfortable in the warm cabin—but not because of heaters. The process of compressing outside air to maintain a cabin pressure that is nearly the same as the air pressure at sea level normally heats the air to a roasting 55°C (131°F). So air conditioners must be used to extract heat from the pressurized air.

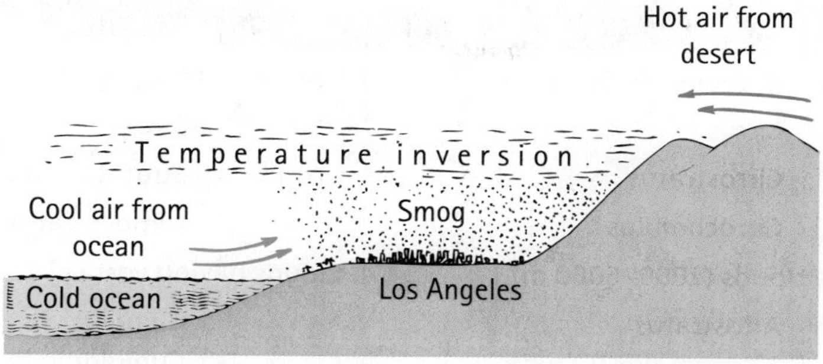

Hot air from desert

Temperature inversion

Cool air from ocean

Smog

Cold ocean

Los Angeles

FIGURE 25.7
Smog in Los Angeles is trapped by the mountains and a temperature inversion caused by warm air from the Mojave Desert overlying cool air from the Pacific Ocean.

The smog of Los Angeles is trapped by such an inversion, caused by cold air from the ocean being capped by a layer of hot air moving westward over the mountains from the hot Mojave Desert. The west-facing side of the mountains helps confine the trapped air (Figure 25.7). The Rocky Mountains on the western edge of Denver play a similar role in trapping smog beneath a temperature inversion.

FYI Stable air that is forced to rise spreads out horizontally. When clouds develop in stable air, they too spread out into thin horizontal layers having flat tops and bottoms. Unstable air favors upward movement. When unstable rising air is moist, billowy and towering clouds develop.

CHECKPOINT

1. If a parcel of dry air initially at 0°C expands adiabatically while flowing upward alongside a mountain, what is its temperature when it has risen 2 km? When it has risen 5 km?
2. What happens to the air temperature in a valley when dry, cold air blowing across the mountains descends into the valley?
3. When air is stable, it resists vertical movement. When air is unstable, it wants to rise to a stable level—where its air temperature equals the temperature of its surroundings. In the example of a temperature inversion, is the air stable or unstable?

Were these your answers?

1. The air cools at the dry adiabatic lapse rate of 10°C for each kilometer it rises. When the parcel rises to an elevation of 2 km, its temperature is −20°C. At an elevation of 5 km, its temperature is −50°C.
2. The air is adiabatically compressed, and so its temperature increases. Residents of some valley towns in the Rocky Mountains, such as Salida, Colorado, benefit from this adiabatic compression and enjoy "banana belt" weather in midwinter.
3. A temperature inversion occurs when warmer air overlies cooler air. Air in this situation is very stable—the denser, cooler air cannot rise above the less dense, warmer air, and so it resists vertical movement.

25.3 Cloud Development

EXPLAIN THIS Why do some clouds form as layers, while other clouds seem to form as towering castles?

LEARNING OBJECTIVE
Describe the different types of clouds, and explain how they form.

When an unstable air parcel rises, the relative humidity increases as the air cools at the dry adiabatic lapse rate. At a certain elevation, the relative humidity reaches 100% and the air parcel is saturated. This key elevation is the **lifting condensation level**.

As the air parcel continues to rise, air now cools at the moist adiabatic rate. But more important, the condensation rate exceeds the evaporation rate above the lifting condensation level—cloud droplets begin to form. The lifting

TABLE 25.2	**THE FOUR MAJOR CLOUD GROUPS**
1. High clouds (above 6000 m)	3. Low clouds (below 2000 m)
Cirrus	Stratus
Cirrostratus	Stratocumulus
Cirrocumulus	Nimbostratus
2. Middle clouds (2000–6000 m)	4. Clouds having vertical development
Altostratus	Cumulus
Altocumulus	Cumulonimbus

The altitude range of the major cloud groups varies somewhat with season and latitude. Also, some clouds extend vertically in more than one altitude range.

> Clouds don't float! They are buoyed up by an invisible conveyor belt of air. Clouds are always moving up.

condensation level marks the base of cloud formation. The air parcel continues to rise until it has cooled enough to match the temperature of the surrounding air. At this elevation, the *equilibrium level*, the air parcel stops rising and the condensation rate equals the evaporation rate—the air parcel is now stable. The equilibrium level marks the upper limit of cloud formation.

The elevation of the equilibrium level depends on the environmental lapse rate. Because the environmental lapse rate describes temperature changes of the surrounding air, the elevation at which a rising air parcel stabilizes can be quite variable.

To summarize, a rising air parcel cools at the dry adiabatic lapse rate until it reaches saturation. After saturation, the moist adiabatic lapse rate controls how thick the cloud becomes. Therefore, the height of the cloud base and how thick the cloud becomes depend on three variables: the environmental lapse rate, the dry adiabatic lapse rate, and the moist adiabatic lapse rate.

Clouds are classified according to their altitude and shape. There are ten principal cloud forms, each of which belongs to one of four major groups (Table 25.2).

CHECKPOINT
To what does the environmental lapse rate refer?

Was this your answer?
The environmental lapse rate refers to the change of air temperature with altitude. It describes the temperature changes of the air that a rising air parcel passes through. The environmental lapse rate varies with local conditions. On average, the rate is about 6.5°C for each kilometer rise in elevation.

High Clouds

High clouds are clouds that form at altitudes above 6000 m. High clouds (other than cirrus clouds) are denoted by the prefix *cirro-*. The air at this elevation is quite cold and dry, and so clouds this high are made up almost entirely of ice crystals.

The most common high clouds are thin, wispy *cirrus* clouds. Cirrus clouds are blown by high winds into their well-known wispy shapes, such as the classic "mare's tail" or "artist's brush." Cirrus clouds usually move across the sky from west to east (with the prevailing winds) and indicate fair weather.

Cirrocumulus clouds are the familiar rounded, white puffs. They are found in patches, and they seldom cover more than a small portion of the sky. Small

> **FYI** Montana is called "Big Sky Country" because of the high lifting condensation level. The dry climate means that cloud bases are usually high in the sky—hence the sky looks "bigger."

(a)

(c)

(b)

(d)

FIGURE 25.8
The four cloud groups. (a) High clouds: cirrus, cirrostratus, cirrocumulus. (b) Middle clouds: altostratus, altocumulus. (c) Low clouds: stratus, stratocumulus, nimbostratus. (d) Clouds with vertical development: cumulus, cumulonimbus.

ripples and a wavy appearance make the cirrocumulus clouds look like the markings on the body of a mackerel. Hence, cirrocumulus clouds are often said to make up a *mackerel sky*.

Cirrostratus clouds are thin and sheetlike, and they often cover the entire sky. The ice crystals in these clouds refract light and produce a halo around the Sun or the Moon. When cirrostratus clouds thicken, they give the sky a white, glary appearance—an indication of an advancing storm. In fact, rain or snow may come soon when cirrostratus clouds are followed by middle clouds.

Middle Clouds

Middle clouds form at altitudes between 2000 and 6000 m. Middle clouds are denoted by the prefix *alto-*. These clouds are composed of water droplets and, when temperature allows, ice crystals.

Altostratus clouds are gray to blue-gray, and they often cover the sky for hundreds of square kilometers. Altostratus clouds are often so thick that they diffuse incoming sunlight to the extent that objects on the ground don't produce shadows (they also don't produce halos). Altostratus clouds often form before a storm. Look at the ground the next time you're going on a picnic. If you don't see your shadow, cancel your plans!

Altocumulus clouds appear as gray, puffy masses in parallel waves or bands. The individual puffs are much larger than those in cirrocumulus clouds, and the color is also much darker. The appearance of altocumulus clouds on a warm, humid summer morning often indicates thunderstorms by late afternoon.

Low Clouds

Low clouds ranging from the surface up to 2000 m are called *stratus* clouds. They are almost always made up of water droplets, but, in cold weather, they may also contain ice crystals and snow.

The suffix *-nimbus* means "rain-producing."

Stratus clouds are uniformly gray, and they often cover the whole sky. They are very common in winter, and they account for the sky's "hazy shade of winter." They resemble a high fog that doesn't touch the ground. Although stratus clouds are not directly associated with falling precipitation, they sometimes generate a light drizzle or mist.

Stratocumulus clouds either form a low, lumpy layer that grows in horizontal rows or patches or, with weak rising motion, appear as rounded masses. Their color is generally light to dark gray. To tell the difference between altocumulus clouds and stratocumulus clouds, hold your hand at arm's length and point toward the cloud in question. An altocumulus cloud commonly appears to be the size of your thumbnail; a stratocumulus cloud appears to be about the size of your fist. Precipitation of rain or snow is not usually produced by stratocumulus clouds.

Nimbostratus clouds are dark and foreboding. The prefix "nimbo" means rain, so nimbostratus clouds are a wet-looking cloud layer associated with light to moderate rain or snow.

Clouds with Vertical Development

Cumulus clouds are the most familiar of the many cloud types. Cumulus clouds resemble pieces of floating cotton, with sharp outlines and flat bases. They are white to light gray, and they generally occur about 1000 m above the ground. The tops of cumulus clouds are often in the form of rising towers, showing the upward limit of the rising air. These are the clouds childhood daydreams are made of. Did you ever see castles or the shapes of animals in the clouds?

When cumulus clouds turn dark and are accompanied by precipitation, they are referred to as *cumulonimbus* clouds. In this case, they indicate a coming storm. As we shall see, cumulonimbus clouds often become *thunderheads*.

Precipitation Formation

Several things have to happen for precipitation to form. Each progressive step toward precipitation is part of the *collision–coalescence process*. The first requirement is the presence of dust—the condensation nuclei discussed earlier in this chapter.

Water vapor is less dense than air. But once cloud droplets form, the droplets are considerably denser than the air. The gravitational force pulling the droplets downward is enough to make them fall. So why don't all the water droplets in a cloud fall to the ground? The answer involves *updrafts*—the upward movement of air. A typical cumulus cloud has an updraft speed of at least 1 m/s, which is faster than the droplets can fall. So the droplets are "floated up" by the upward-rising air. Without updrafts, the droplets drift so slowly out of the bottom of the cloud and evaporate so quickly that they have no chance of reaching the ground. They are replaced by new droplets forming above.

In the collision–coalescence process, tiny droplets coalesce to form a range of droplet sizes. Early on, updrafts are stronger than the downward motion of the droplets, and all droplets are repeatedly blown upward—rain does not fall. As the droplets grow, they eventually fall at the same rate as the updraft, becoming more or less stationary. But the droplets are repeatedly bombarded from below by smaller droplets rising with the updraft. Such bombardment is like standing still on a beach when a strong wind blows and being constantly pelted by tiny sand grains blowing in the wind. This is when significant droplet growth occurs. Eventually, the stationary drops of water grow larger and become huge compared to typical cloud droplets—they become raindrops. Because raindrops fall faster than the updraft can push them upward, precipitation forms.

This process requires sufficient vertical development of the cloud; otherwise, there are not enough droplet collisions for individual droplets to grow big

FYI   Cumulus clouds are denser than the surrounding air. So why don't they fall? The answer is, they do fall! They fall at the same speed at which the air is rising, and therefore they remain fixed in elevation. Without updrafts, there would be no cumulus clouds.

FYI   Have you ever noticed the long white lines left in the sky after a jet plane passes? These are jet contrails—a *condensation trail*—that form as hot, humid jet exhaust mixes with the cold air surrounding the plane. Air at these high altitudes is very cold, and only a small amount of water vapor is needed for condensation to occur. Contrails can be used to help predict the weather—if a contrail does not form, or if it disappears quickly, fair weather will continue. If the contrail remains for a long time, a change in the weather may be expected.

TABLE 25.3	CLASSIFICATION OF AIR MASSES AND THEIR CHARACTERISTICS		
Typical Source Region	**Classification**	**Symbol**	**Characteristics**
Arctic	maritime arctic	mA	cool, moist, unstable
Greenland	continental arctic	cA	cold, dry, stable
North Atlantic and Pacific Oceans	maritime polar	mP	cool, moist, unstable
Canada, Siberia	continental polar	cP	cold, dry, stable
Caribbean Sea, Gulf of Mexico	maritime tropical	mT	warm, moist, usually unstable
Mexico, southwestern United States	continental tropical	cT	hot, dry, stable aloft, unstable at surface

enough to reach the stationary point. So the presence of thicker clouds means a higher chance of rain—and a higher chance that a hard rain will fall. Thicker clouds give droplets more time and space to coalesce into drops that are heavy enough to fall.

Raindrops form because the condensation rate exceeds the evaporation rate. But raindrops shrink as they fall because once they are out of the cloud, the evaporation rate exceeds the condensation rate. Sometimes, enough evaporation occurs that raindrops disappear before reaching the ground. Such precipitation is called *virga*.

25.4 Air Masses, Fronts, and Storms

EXPLAIN THIS In the Northern Hemisphere, why does a cyclone have a counterclockwise rotation?

LEARNING OBJECTIVE
Describe the three atmospheric lifting mechanisms, and explain the formation of a midlatitude cyclone.

An air mass is a volume of air much larger than the parcels of air we've discussed so far. Various distinct air masses cover large portions of Earth's surface. Each has its own characteristics. An air mass formed over water in the tropics is different from one formed over land in the polar regions. Air masses are divided into six general categories, according to the type of land or water they form over and the latitude at which their formation occurs (Table 25.3 and Figure 25.9). The type of surface over which air-mass formation occurs is designated by a lowercase letter (m for maritime, c for continental). The source region in which an air mass forms is designated by a capital letter (A for arctic, P for polar, T for tropical.)

Continental polar (cP) and continental arctic (cA) air masses generally produce very cold, dry weather in winter and cool, pleasant weather in summer. Maritime polar (mP) and maritime arctic (mA) air masses, picking up moisture as they travel across the oceans, generally bring cool, moist weather to a region. Continental tropical (cT) air masses are generally responsible for the hot, dry weather of summer, and warm, humid conditions are due to maritime tropical (mT) air masses.

So we see that different types of air masses have their own characteristics. When two different air masses meet, a variety of weather conditions can develop.

FIGURE 25.9
Source regions of air masses for North America.

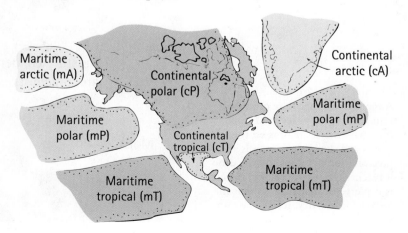

FYI   As any glider pilot will attest, there's no way all air can rise. Some has to come back down. Where air rises and water vapor condenses we see clouds; where it descends we see blue sky between the clouds.

FIGURE 25.10
Cumulus clouds are often found as individual towering white clouds separated from each other by expanses of blue sky.

Mastering**PHYSICS**
TUTORIAL: Rain Shadow Activity

FYI   Just as a "regular" shadow is a place where very little light falls because some obstacle blocks the light, a rain shadow is a place where little rain falls because some obstacle (such as a mountain) blocks precipitation.

Atmospheric Lifting Mechanisms

Clouds are great indicators of mechanisms weather. For clouds to form, air must be lifted. The three principal lifting mechanisms in the atmosphere are convectional lifting, orographic lifting, and frontal lifting.

Convectional Lifting

Earth's surface is heated unequally. Some areas are better absorbers of solar radiation than others, so they heat up more quickly. The air that touches these surface "hot spots" becomes warmer than the surrounding air, and so it rises, expands, and cools. This rising of air is accompanied by the sinking of cooler air from above. The circulatory motion produces **convectional lifting**.

If cooling occurs close to the air's saturation temperature, the condensing moisture can form a cumulus cloud. Air within the cumulus cloud moves in a cycle: warm air rises, cool air descends. Because descending cool air inhibits the expansion of warm air beneath it, small cumulus clouds usually have a great amount of blue sky between them (Figure 25.10).

Cumulus clouds often remain in the same place that they formed, dissipating and reforming many times. As they grow, they shade the ground beneath from the Sun. This slows surface heating and inhibits the upward convection of warm air. Without a continuous supply of rising air, a cumulus cloud begins to dissipate. Once the cloud is gone, the ground reheats, allowing the air above it to warm and rise. Thus, convectional lifting begins again, and another cumulus cloud begins to form at the same location.

Orographic Lifting

An air mass that is pushed upward over an obstacle, such as a mountain range, undergoes **orographic lifting**—the rising air cools. If the air is humid, clouds form. The types of clouds that form depend on the air's stability and moisture content. If the air is stable, a layer of stratus clouds may form. If the air is unstable, cumulus clouds may form. As the air mass moves down the other side of the mountain (the leeward slope), it warms adiabatically. This descending air is dry because most of its moisture was removed in the form of clouds and precipitation on the windward (upslope) side of the mountain. The descending air is warm because condensation releases heat. Because the dry leeward (downslope) sides of mountain ranges are sheltered from rain and moisture, they are often referred to as being in the *rain shadow* (Figure 25.11).

FIGURE 25.11
A mountain range may produce a rain shadow on its leeward slope. As warm, moist air rises on the windward slope, the air cools and precipitation develops. By the time it reaches the leeward slope, the air is depleted of moisture, so that the leeward side is dry. It lies in a rain shadow.

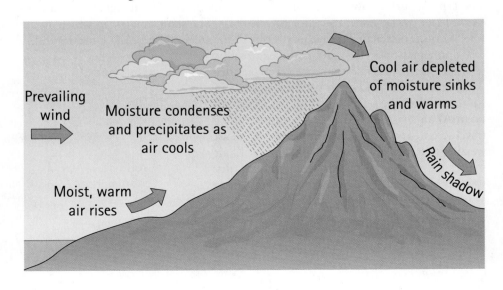

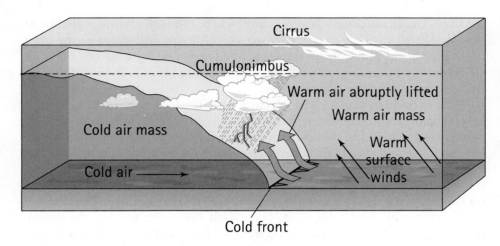

FIGURE 25.12
INTERACTIVE FIGURE (MP)

A cold front occurs when a cold air mass moves into a warm air mass. The cold air forces the warm air upward, where it condenses to form clouds. If the warmer air is moist and unstable, heavy rainfall and gusty winds develop.

Frontal Lifting

In weather reports, we often hear about fronts. A **front** is the contact zone between two different air masses. When two air masses make contact, differences in temperature, moisture, and pressure can cause one air mass to ride over the other. When this occurs, we have **frontal lifting**. If a cold air mass moves into an area occupied by a nonmoving warm air mass, the contact zone between them is called a *cold front*; and if warm air moves into an area occupied by a nonmoving mass of cold air, the zone of contact is called a *warm front*. Fronts are usually accompanied by wind, clouds, rain, and storms. An *occluded front* forms when a cold front overtakes a warm front, or vice versa. If neither air mass is moving, the contact zone is called a *stationary front*.

Meteorologists and other observers of the sky can often tell when a cold front is approaching by observing high cirrus clouds, a shift in wind direction, a drop in temperature, and a drop in air pressure. As cold air moves into a warm air mass, forming a cold front, the warm air is forced upward (Figure 25.12). As it rises, it cools, and water vapor condenses into a series of cumulonimbus or nimbostratus clouds. The advancing wall of clouds at the front develops into thunderstorms with heavy showers and gusty winds. After the front passes, the air cools and sinks, pressure rises, and rain ceases. Except for a few fair-weather cumulus clouds, the skies become clear, and we have the calm after the storm.

When warm air moves into a cold air mass, forming a warm front, the less-dense warmer air gradually rides up and over the colder, denser air (Figure 25.13). The approach of a warm front, although less obvious and more gradual than the approach of a cold front, is also indicated by cirrus clouds. Ahead of

Mastering**PHYSICS**
TUTORIAL: Cold Front Activity
TUTORIAL: Warm Front Activity

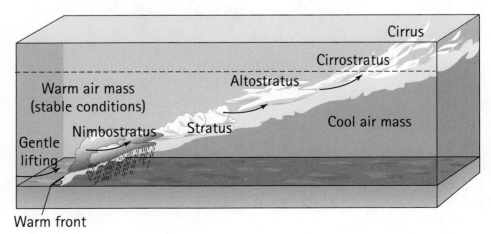

FIGURE 25.13

A warm front occurs when a warm air mass moves into a cold air mass. The less-dense warmer air rides up and over the colder, denser air, resulting in widespread cloudiness and light to moderate precipitation that can cover great areas.

the front, the cirrus clouds descend and thicken into altocumulus and altostratus clouds that turn the sky an overcast gray. Moving still closer to the front, light to moderate rain or snow develops, and winds become brisk. At the front, air gradually warms, and the rain or snow turns to drizzle. Behind the front, the air is warm and the clouds scatter.

When a cold front and a warm front merge, the result is an occluded front. Several steps lead to the formation of an occluded front. In the first step, the two fronts are joined at the center of a low-pressure area and they do not overlap. At this stage, warm air is in contact with the ground, between the two fronts (Figure 25.14a). When the cold front first overtakes the warm front, the two fronts meet at the ground, such that the warm air above each front no longer touches the ground (Figure 25.14b). As the cold front continues to invade the warm front, the warm front itself no longer touches the ground (Figure 25.14c). The advancing cold front wedges itself under the warm front so that the intersection between the two fronts is above ground. As you would imagine, a wide area of rainy weather accompanies occluded fronts.

When two different air masses are not strong enough to overtake each other, the boundary between them becomes a *stationary front*. A stationary front is

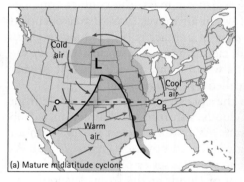

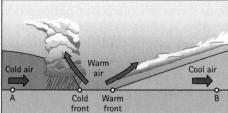

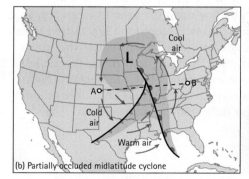

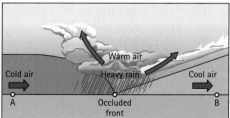

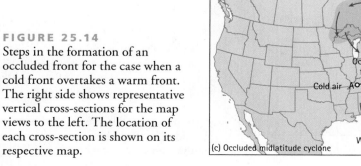

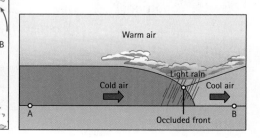

FIGURE 25.14

Steps in the formation of an occluded front for the case when a cold front overtakes a warm front. The right side shows representative vertical cross-sections for the map views to the left. The location of each cross-section is shown on its respective map.

like a stalemate between fronts and, as such, can remain over an area for several days. Eventually the stalemate ends and the front either dissipates or, depending on conditions aloft, changes into a cold or warm front.

Midlatitude Cyclones

Midlatitude cyclones are storm systems that typically sweep across the United States (and other places) from west to east. The formation of midlatitude cyclones is closely tied to the interactions between air masses at the polar front. The term *cyclone* refers to counterclockwise rotation, in the Northern Hemisphere, around a low-pressure center. *Anticyclonic* flow rotates clockwise around a high-pressure center.

The "typical" midlatitude cyclone progresses through six distinct steps, starting with *cyclogenesis*—the birth of the cyclone (Figure 25.15). This initial stage occurs at the polar front at a location with the characteristics of a stationary front, where both air masses are moving parallel to the front (Figure 25.15a).

> **FYI** So why do we call the contact zone between air masses a front? The word *front* is a military term used to describe the boundary between two different armies. During World War I, Norwegian meteorologists adopted *front* as a way to describe the boundary line between two "warring" air masses.

> **FYI** As opposing air masses slide past one another, the air starts to spin—a cyclone. Place a pencil in your hand and put your palms together. Push your right hand away from you (warm air flow) and draw your left hand toward you (cool air flow). How does the pencil rotate? The pencil should rotate in a counterclockwise direction—like a Northern Hemisphere midlatitude cyclone.

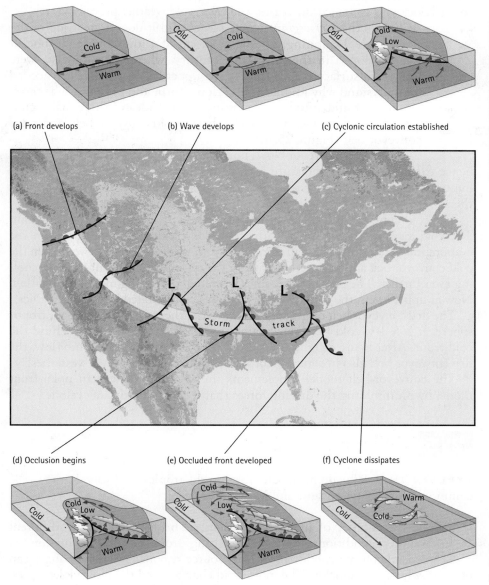

(a) Front develops   (b) Wave develops   (c) Cyclonic circulation established

(d) Occlusion begins   (e) Occluded front developed   (f) Cyclone dissipates

FIGURE 25.15
Steps in the life of a typical midlatitude cyclone.

Cyclogenesis happens when the linear front becomes disturbed and develops a wavelike curve, with warm air jutting into the cold air mass (Figure 25.15b). Usually, the wave tracks eastward with the westerlies.

If the wave doesn't die out, warm air continues to move poleward as cold air pushes toward the equator. At this stage two distinct fronts develop: a warm front on the leading edge of the system, trailed by a cold front. The fronts terminate at the center of what has become a roughly circular area of *low pressure* (Figure 25.15c). The pressure-gradient force drives air inward toward the center of the low, but the Coriolis force deflects the wind to the right. The deflected wind sets up a counterclockwise rotation around the central low. A pronounced temperature difference exists between the warmer east side of the system and the colder west side.

Under most conditions, the cold front moves eastward faster than the warm front. Cold air begins to lift the warm front, forming an occluded front (Figure 25.15d). As the occlusion continues to grow (Figure 25.15e), more and more warm air is displaced upward, until there is no significant temperature difference horizontally across the storm. But where did the warm air go? Figure 25.15f shows that all the warm air was forced up and over the steadily advancing cold front. Without the horizontal temperature difference, the storm is without an energy source. The counterclockwise cyclonic flow ceases, and clear, cold weather prevails.

This is the *Norwegian cyclone model*, and it does a great job of describing the beginning, middle, and end of a midlatitude cyclone. It was developed in the early 20th century, when meteorologists were limited to surface data. Data for the upper-level processes are needed to understand why midlatitude cyclones form and progress. The modern view of midlatitude cyclones invokes the idea of three major conveyor belts of air: a warm conveyor belt, a cold conveyor belt, and a dry conveyor belt (Figure 25.16).

The warm conveyor blows northward from the Gulf of Mexico, and hence carries significant moisture with it. Airflow originates near the surface but undergoes frontal lifting when it encounters the warm front. The warm air then cools adiabatically, leading to condensation and precipitation.

The cold conveyor belt blows in low from the east, north of the warm front. The cold air flows beneath the warm conveyor, picking up moisture from the evaporation of raindrops that fall through it from the warmer air above. When the cold conveyor belt approaches the center of the cyclone, the cold air is lifted, cooling even further. The rising air becomes saturated, producing more precipitation. Now at upper levels, the cold conveyor becomes incorporated into the westerlies.

The dry conveyor belt originates to the west of the storm, in the upper troposphere. It is a dry, cold flow that maintains the supply of cold air behind the cold front. After being caught and turned northward by the cyclonic flow, the dry conveyor belt also becomes incorporated into the upper-level westerlies.

The conveyor-belt model supplements the original Norwegian polar-front model by illuminating the driving forces that feed a midlatitude cyclone.

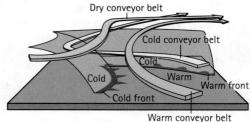

FIGURE 25.16
The conveyor-belt model of midlatitude cyclones, showing the relationship of the warm, cold, and dry conveyor belts to fronts observed at the surface.

25.5 Violent Weather

EXPLAIN THIS Why do hurricanes form in the tropics, and why do they commonly form between June and November?

The three types of lifting just discussed bring about many different weather conditions. Weather resulting from air masses in contact depends on the conditions of their source regions. Weather changes can occur slowly or very quickly. The most rapid changes, and the most violent ones, occur with three major types of storms: thunderstorms, tornadoes, and hurricanes.

Thunderstorms

A thunderstorm begins with humid, unstable air rising, cooling, and condensing into a single cumulus cloud. This cloud builds and grows upward as long as it is fed by an updraft of rising warm air from below. Cloud droplets grow larger and heavier within the cloud until they eventually begin to fall as rain. The falling rain drags some of the cool air along with it, creating a downdraft—the chilled air is colder and denser than the air around it. Together, the rising warm updraft and the sinking, chilled downdraft make the cloud into a storm cell. This is the mature stage, at which the thunderstorm cloud appears as a lonely giant—dark and brooding in the sky. It typically has a base several kilometers in diameter, and it can tower to altitudes up to 12 km. At such high altitudes, horizontal winds and lower temperatures flatten and stretch the thunderhead crown into a characteristic anvil shape (Figure 25.17). After the thunderstorm dissipates, it leaves behind the cirrus anvil as a reminder of its once-mighty presence.

At any given time, about 1800 thunderstorms are in progress in Earth's atmosphere. Wherever thunderstorms occur, there are lightning and thunder. As water droplets in the cloud bump into and rub against one another, the cloud becomes electrically charged. Rather than being uniformly distributed throughout the cloud, the electric charge separates—it is usually positively charged where there are ice crystals (at the colder cloud top) and negatively charged where the cloud is warmer (at the cloud bottom). As electric stresses between the oppositely charged regions build up, the charge becomes great enough that electric energy is released and passed to other points of opposite charge, which quite often means the ground. The electric energy flowing from cloud to ground is lightning (Figure 25.18). As lightning heats up the air, the air expands and we hear lightning's noisy companion—thunder. Lightning strikes Earth roughly 100 times every second, with some bolts having an electric potential of as much as 100 million volts. Lightning claims about 100 human victims per year in the United States alone.

Tornadoes

A revolving object, such as a whirling ball on a string, speeds up when pulled toward its axis of revolution, thus conserving its angular momentum. Similarly, winds slowly rotating over a large area speed up when the radius of rotation decreases.

FIGURE 25.17
The mature stage of a thunderstorm cloud appears as a towering cumulonimbus cloud that reaches up to about 12 km. Strong horizontal winds and icy temperatures flatten and distend the cloud's crown into a characteristic anvil shape.

FYI The spring of 2011 was an exceptionally destructive and deadly tornado season. From Minnesota to Louisiana and from Texas to Virginia, more than 1,383 tornadoes touched ground resulting in approximately 580 deaths! The deadliest strikes—from April 25 to 28—brought devastation to areas of the southern United States with Tuscaloosa, Alabama, being the hardest hit. From May 21 to 26, a series of tornadoes touched down in areas of Minnesota, Kansas, Oklahoma, and Missouri, with by far the most catastrophic damage in Joplin, Missouri. The Joplin tornado, with the highest intensity rating of EF-5 (winds in excess of 200 mph), is ranked as the 8th deadliest single tornado to strike the United States.

FIGURE 25.18
Time exposure of cloud-to-ground lightning during an intense thunderstorm. The blue, green, and red images of the flash were produced by a diffraction grating held over the camera lens when the photo was taken.

FIGURE 25.19
Like a gigantic vacuum cleaner, the strong wind of a tornado can pick up and obliterate everything in its path.

This increase in speed can produce a *tornado*, which is a funnel-shaped cloud that extends downward from a large cumulonimbus cloud. Produced as an extension of a powerful thunderstorm, the funnel cloud is called a tornado only after it touches the ground. The winds of a tornado travel in a counterclockwise direction (clockwise in the Southern Hemisphere) at wind speeds as low as 65 km/h but up to 450 km/h depending on the tornado's strength.

As a tornado moves across the land, at speeds from 45 to 95 km/h, it follows a path controlled by its parent thundercloud. The tornado can bounce and skip as it rises briefly from the ground and then touches back down again. A tornado acts like a gigantic vacuum cleaner, picking up everything in its path. It wreaks havoc not only by suction but also by the battering power of its whirling winds. In its wake it leaves a trail of flying dirt and debris (Figure 25.19).

Tornadoes occur in many parts of the world. In the flat central plains of the United States, a tornado zone extends from northern Texas through Oklahoma, Kansas, and Missouri. In this area, more than 300 tornadoes touch down each year. Hence the name for this area: Tornado Alley. Tornadoes are so frequent in this part of the country that many homes are built with underground storm shelters. The power of a tornado is terrifying and devastating.

Hurricanes

In the steamy tropics, where the Sun warms the oceans, heat transfer to the atmosphere by evaporation and conduction is so thorough that air and water temperatures are about equal. The high humidity in this part of the world favors the development of cumulus clouds and afternoon thunderstorms. Most of the individual storms are not severe, but thunderstorms sometimes become organized and begin to behave as a single system called a *tropical disturbance*. Warm, rising air creates a central low-pressure area that continuously draws more air toward the center of the disturbance. The inward-flowing air spirals because of the Coriolis force (Chapter 24). If this spiraling storm isn't broken apart by upper-level winds, it can develop into a *tropical depression* (wind speed less than 60 km/h), so called because of a central area of low pressure. If the storm intensifies, it progresses to a *tropical storm*, with increased wind speeds above 60 km/h. If favorable conditions continue, a more violent storm develops—a hurricane—with wind speeds above 120 km/h and up to nearly 300 km/h.

Hurricanes gain energy from the heat released by the condensation of water (Chapter 7). The condensation produces the vast amounts of rain that are typical of such storms. The heat warms the surrounding upper-level air, causing it to rise. As the upper-level air rises, surface air is sucked upward, intensifying the low-pressure center—the eye of the hurricane. Horizontal airflow spirals counterclockwise

FYI Hurricane Katrina formed on August 23, 2005, near the Bahamas. It struck the Gulf Coast of the United States six days later, wreaking havoc. New Orleans, Louisiana, and coastal Mississippi were hit particularly hard. Hurricane Katrina, however, was such a large storm that it devastated areas of the Gulf Coast up to 160 km from the eye of the storm. At least 1836 people were killed as a result of the hurricane and the floods that followed. It was the deadliest U.S. hurricane since the Okeechobee Hurricane of 1928. It was a costly storm, too. More than $100 billion was needed to repair storm damage and fund cleanup operations. Some areas still haven't recovered. It was the costliest natural disaster in U.S. history.

FIGURE 25.20
This satellite image of 2005's Hurricane Katrina shows the characteristic appearance of a hurricane. Bands of cumulonimbus clouds spiral around the low-pressure eye of the storm.

(in the Northern Hemisphere) around the eye. The spiral bands of cumulonimbus clouds give the hurricane its familiar appearance (Figure 25.20).

The airflow that sets up in the eye forms a positive feedback loop. Condensation releases heat, which draws moist air upward from the ocean surface. The moist air cools and more condensation occurs, which releases more heat, which draws more warm, moist air upward. This cycle—essentially a natural heat engine—continues unless strong, upper-level winds from outside the storm disrupt the upward flow pattern, or the hurricane moves over land. Once over land, the hurricane is deprived of its energy source.

> **FYI** We hear weather forecasters talk about short-, medium-, and long-range forecasting. What do these types of forecasting mean? In general, short-range forecasting predicts weather for the next two days with considerable detail about temperature, wind, and air. Medium-range forecasting predicts weather for the third to seventh days in less detail. Forecasting beyond seven days is considered long-range, and the predictions in such forecasts are in terms of conditions that are expected to be above or below normal.

25.6 The Weather—The Number One Topic of Conversation

> **LEARNING OBJECTIVE**
> List the information needed to predict the weather.

EXPLAIN THIS Can the weather be predicted with 100% accuracy?

Meteorologists have the important job of forecasting hurricanes and other storms. Weather forecasting is, in part, a matter of determining air-mass characteristics, predicting how and why the characteristics might change, and estimating the direction in which air masses might move. In the case of hurricanes and tornadoes, such predictions are lifesaving. Meteorologists have a long and remarkable record of saving human lives and reducing property loss.

There are several methods of weather forecasting. Some forecasts are based on the *continuity* of a weather pattern; such as rain today likely means rain tomorrow. Or, because surface weather systems tend to change direction or speed, a forecast may be based on the *trend* of the weather pattern. For example, if a cold front is moving eastward at an average speed of 20 km/h, it can be expected to affect the weather 80 km away in 4 hours. We also hear about

> **FYI** Swirling storms and disturbances caused by solar activity produce "space weather" between the Sun and Earth. Solar flares, coronal mass ejections, and magnetic storms affect not only Earth satellites but Earth's surface environment as well. Communications system failures, power blackouts, and brownouts are often attributed to space weather. As our use of space grows, so must our ability to predict its weather.

WEATHER MAPS

The weather forecaster's primary tool is the weather map, or chart. A weather map is essentially a representation of the frontal systems and the high-pressure and low-pressure systems that overlie the areas outlined in the map. Symbols on such a map are a shorthand notation to represent data gathered from various observation stations. These symbols are called *weather codes*.

This shorthand notation compiles 18 categories of data into a very small area called a *station model*. The circle at the center describes the overall appearance of the sky. Jutting from the circle is a wind arrow, its tail in the direction from which the wind comes and its feathers indicating wind speed. The other 15 weather elements are in standard position around the circle.

A weather map is covered with lines—*isobars*—that connect points of equal pressure. As air moves from a high-pressure region to a low-pressure region, it rises and cools, and the moisture in it condenses into clouds. In the vicinity of the low (L on a map), we see an extensive cloud cover. In the vicinity of the high (H on a map), we see clear skies. In a high-pressure region, air sinks and warms adiabatically. Because sinking air does not produce clouds, we find clear skies and fair weather. The heavy lines on a weather map represent fronts. Because fronts generally mean a change in the weather, they are of great importance on weather maps.

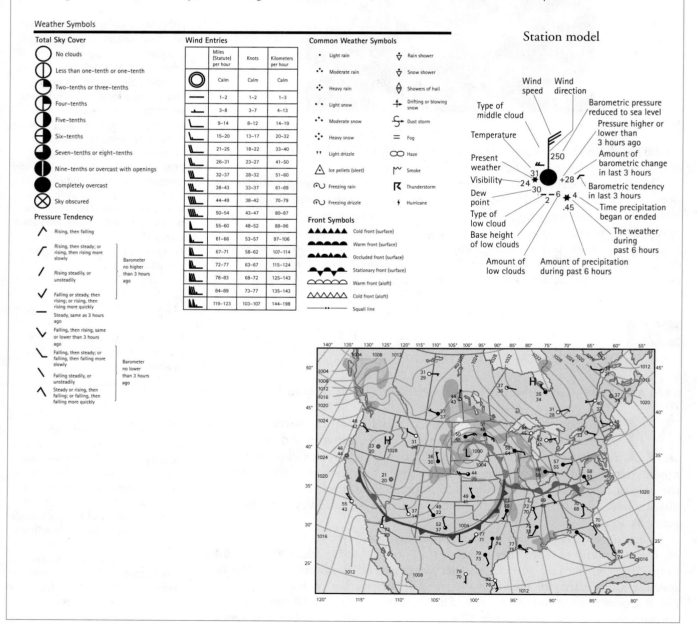

the *probability* of a weather condition—for example, the probability of rain is 70%. This is an expression of chance, meaning that there is a 70% chance that rain will fall somewhere in the forecast area. So you should probably carry an umbrella. Another forecast we often hear about is the *extended forecast*. This forecast is based on weather types that develop in certain areas. Recall the classification of air masses and their characteristics; if a continental polar air mass is approaching, we can expect cold, dry weather; whereas if a maritime polar air mass is approaching, we can expect cold, moist weather. All these methods of weather prediction are based on the statistical analysis of weather information.

Weather forecasting involves great quantities of data from all over the world. Meteorologists use numerical models and computers not only to plot and analyze data but also to help predict the weather. The computer draws maps of projected weather conditions, which the weather forecaster uses as a guide for predicting weather. Even so, the many variables involved often impede the making of accurate predictions; so don't count on an absence of rain on your parade!

For instructor-assigned homework, go to www.masteringphysics.com

SUMMARY OF TERMS (KNOWLEDGE)

Adiabatic A term that describes temperature change in the absence of heat transfer; expanding air cools and compressing air warms.

Convectional lifting An air-circulation pattern in which air warmed by the ground rises while cooler air aloft sinks.

Front The contact zone between two different air masses.

Frontal lifting The lifting of one air mass by another as two air masses converge.

Humidity A measure of the concentration or amount of water vapor in the air: the mass of water vapor per volume of air.

Lifting condensation level The height at which rising air cooling at the dry adiabatic rate becomes saturated and condensation begins.

Midlatitude cyclone A west-to-east-traveling storm with a central low-pressure area about which counterclockwise

flow develops (in the Northern Hemisphere) and from which usually extends a cold front and a warm front; generally forms at the polar front.

Orographic lifting The lifting of an air mass over a topographic barrier such as a mountain.

Relative humidity The amount of water vapor in the air at a given temperature, expressed as a percentage of the maximum amount of water vapor the air can accommodate at that temperature.

Saturation vapor pressure The maximum amount of moisture the air can accommodate at a given temperature; the upper limit for humidity.

Temperature inversion A condition in which the upper regions of the troposphere are warmer than the lower regions.

READING CHECK QUESTIONS (COMPREHENSION)

25.1 Atmospheric Moisture

1. What is the difference between humidity and relative humidity?

2. Why does relative humidity increase at night?

3. As air temperature rises, does relative humidity increase, decrease, or stay the same?

4. What does saturation point have to do with dew point?

5. What happens to the water vapor in saturated air as the air cools?

6. What factors are responsible for condensation?

7. Does condensation occur more readily at high temperatures or low temperatures? Explain.

8. When water vapor condenses to liquid water, is heat absorbed or released?

9. Distinguish between condensation and precipitation.

25.2 Weather Variables

10. Explain why warm air rises and cools as it expands.

11. When a parcel of air rises, does it become warmer, become cooler, or remain the same temperature?

12. What is an adiabatic process?

13. Name at least two ways that thermal energy in air can be increased.

14. Name at least two ways that thermal energy in air can be decreased.

15. What is a temperature inversion? Give one location where these inversions often occur.

16. What happens to the air pressure and temperature of an air parcel as it flows up the side of a mountain?

25.3 Cloud Development

17. How do clouds form?

18. Name the cloud form associated with (a) the hazy shade of winter, (b) a mackerel sky, (c) floating cotton, and (d) snowfall.

19. Name the cloud group to which each of the following cloud types belongs: (a) altocumulus, (b) cirrostratus, (c) nimbostratus, and (d) cumulus.

20. Rain or snow is most likely to be produced by which of the following cloud forms: (a) cirrostratus, (b) nimbostratus, (c) altocumulus, or (d) stratocumulus?

21. Are clouds that have vertical development characteristic of stable air, stationary air, unstable air, or dry air?

22. Which type of cloud can become a thunderhead?

25.4 Air Masses, Fronts, and Storms

23. Explain how convectional lifting plays a role in the formation of cumulus clouds.

24. Explain how a convection cycle is generated.

25. Does a rain shadow occur on the windward or leeward side of a mountain range? Explain.

26. Differentiate between a cold front and a warm front.

27. What are the three main atmospheric lifting mechanisms?

28. Under what conditions does orographic precipitation occur?

25.5 Violent Weather

29. How do downdrafts form in thunderstorms?

30. Briefly describe how thunder and lightning develop.

ACTIVITIES (HANDS-ON APPLICATION)

31. What happens when water vapor in a can suddenly condenses? Put a small amount of water in an aluminum soft-drink can, then heat the can on a stove until steam comes out of the opening. With a pair of tongs, invert the can into a pan of cool water. Air is driven from the can so that only water vapor molecules remain inside. When the can is inverted into the pan of water, the water vapor condenses. The condensation of molecules in the can leaves behind a vacuum. The pressure of the atmosphere, with nothing inside pushing back, crushes the can.

32. Search the sky for passing jets and look for the condensation trail—contrail—that the jet leaves behind. Because jet exhaust is mainly carbon dioxide and water vapor, when the water vapor mixes with the cold environment of the upper troposphere, it can condense into small water droplets to form a line of white "clouds."

Study the length and thickness of the contrails. You may even take a photo. After about 10 minutes, reexamine the contrails. Are they still there? Have they dissipated, or have they grown thicker? If they have disappeared, or if the jet did not leave a contrail, the humidity of the atmosphere is too low to create clouds and produce rain. So it is safe to say that tomorrow's weather will probably be very similar to today's weather. If the contrails are still present and have grown thicker, the humidity of the atmosphere is high enough for cloud formation. And with clouds can come precipitation. So tomorrow's weather may be very different from today's weather.

THINK AND SOLVE (MATHEMATICAL APPLICATION)

Refer to the Figuring Physical Science box on page 686 for the following two problems.

33. At 50°F the maximum amount of water vapor in air is 9 g/m^3. If the relative humidity is 40%, what is the mass of water vapor in 1 m^3 of air?

34. The relative humidity of an air parcel is 50%, and the pressure is 1000 millibars (mb). If the pressure increases to 1053 mb without changing the temperature or the water-vapor content of the air, what is the relative humidity? Hint: Use Boyle's law from Chapter 5.

THINK AND RANK (ANALYSIS)

35. The cooling rate of a rising air parcel depends on the type of air mass. From the greatest to the least temperature change, rank the following lapse rates for each kilometer of rising air: (a) environmental lapse rate, (b) moist adiabatic lapse rate, (c) dry adiabatic lapse rate.

36. Several steps need to happen for a cloud to fully develop. Going from first to last, rank these steps: (a) air rises to the lifting condensation level, (b) air rises and the condensation rate exceeds the evaporation rate, (c) air rises to a level where the condensation rate equals the evaporation rate.

37. Several steps need to happen before precipitation forms. Going from first to last, rank the steps in the collision–coalescence process: (a) cloud droplets form, (b) water vapor condenses on small particles, (c) cloud droplets are floated upward, (d) cloud droplets are bombarded by updraft droplets, (e) raindrops form.

38. Rank, from beginning to end, the stages in the formation of a warm front: (a) light to moderate rain and wind, (b) air warms and rain turns to drizzle, (c) cirrus cloud formation, (d) altocumulus and altostratus clouds darken the sky.

39. Rank, from beginning to end, the stages in the formation of a cold front: (a) warm air is forced upward where it cools; (b) air cools and sinks, pressure rises, rain stops; (c) formation of cumulonimbus or nimbostratus clouds; (d) thunderstorms with heavy showers and gusty winds.

40. Rank, from beginning to end, the stages in the formation of a hurricane: (a) central low pressure area intensifies, (b) sufficient warm air and high humidity, (c) inward flowing air spirals, (d) rising warm air creates a central low-pressure area, (e) increase in rotation and speed of wind.

EXERCISES (SYNTHESIS)

41. What is the difference between weather and climate?

42. Which produces precipitation: a rising moist air mass, a descending moist air mass, or both?

43. Why do clouds tend to form above mountain peaks?

44. Why does warm, moist air blowing over cold water result in fog?

45. Why does dew form on the ground during clear, calm summer nights?

46. Why does a July day in the Gulf of Mexico generally feel appreciably hotter than a July day in Arizona, even when temperatures are the same?

47. Would you expect a glass of water to evaporate more quickly on a windy, warm, dry, summer day or on a calm, cold, dry, winter day? Defend your answer.

48. Why does the surface temperature rise on a clear, calm night as low cloud cover moves overhead?

49. During a summer visit to Cancun, Mexico, you stay in an air-conditioned room. Getting ready to leave your room for the beach, you put on your sunglasses. The minute you step outside, your sunglasses fog up. Why?

50. After a day of skiing in the Rocky Mountains, you decide to go indoors to get a warm cup of cocoa. As you enter the ski lodge, your eyeglasses fog up. Why?

51. Can the temperature of an air mass change if no heat is added or subtracted? Explain.

52. Why must an air mass rise in order to produce precipitation?

53. As an air mass moves first upslope and then downslope over a mountain, what happens to the air's temperature and moisture content?

54. Why are saturation and condensation more likely to occur on a cold day than on a warm day?

55. The sky is overcast, and it is raining. What type of cloud is above you: nimbostratus or cumulonimbus?

56. Distinguish between dew and frost.

57. What accounts for the large spaces of blue sky between cumulus clouds?

58. Why don't cumulus clouds form over cool water?

59. Which cloud form is associated with a stable air mass?

60. What cloud form is associated with thunderstorms?

61. What is the relationship between saturation vapor pressure and humidity?

62. When the condensation rate is greater than the evaporation rate, what happens to the air?

63. Cite at least four types of information needed to predict the weather.

64. The accuracy of weather forecasts depends on great quantities of data and thousands of calculations. If the number of data points were decreased, would accuracy also decrease?

65. What is the difference between rainfall that accompanies the passage of a warm front and rainfall that accompanies the passage of a cold front?

66. When does an adiabatic process happen in the atmosphere?

67. How can a layer of altostratus clouds change into altocumulus clouds?

68. Antarctica is covered by glaciers and large ice sheets. Is the snowfall in Antarctica therefore heavy or light? Why?

69. How do fronts cause clouds and precipitation?

70. Explain why freezing rain is more commonly associated with warm fronts than with cold fronts.

71. What steps need to happen before precipitation occurs?

72. In simplest terms, what is an occluded front?

73. How does a rain-shadow desert form?

74. Sinking air warms, and yet the downdrafts in a thunderstorm are cold. Why?

75. Tornadoes form in the regions of a strong updraft, and yet they descend from the base of a cloud. Explain.

76. Why are hurricanes more likely to occur on the East Coast of the United States than on the West Coast?

77. Why are clouds that form over water more efficient in producing precipitation than clouds that form over land?

78. What is the source of the enormous amount of energy released by a hurricane?

79. What part of the United States has the highest frequency of tornadoes?

80. In which atmospheric layer does all our weather occur?

DISCUSSION QUESTIONS (EVALUATION)

81. Clouds can act like a blanket around Earth—they can trap terrestrial radiation to warm Earth's surface, and they can block incoming solar radiation to cool Earth's surface. Which cloud type is a better warmer and which is a better cooler: high, thin cirrus clouds or low, thick stratus clouds?

82. In old-growth redwood forests, some trees have survived for well over a thousand years. As such, they provide a record of climate change. By studying tree rings, scientists have discovered that redwoods have played a significant role in removing CO_2 from the atmosphere. They have also learned that the bigger the tree, the more it grows and hence the more carbon it sequesters. What other information can tree rings from these huge redwoods tell us about climate? By studying what has happened in the past, will we be better prepared for the future?

83. Our global climate is changing. There are many contributing factors to this change. One often ignored factor is soot: black and brown particles emitted by burning fossil fuel—diesel, coal, gasoline, and jet fuel—and from burning solid biofuels—wood, manure, and dung (used for heating homes and cooking in many locations). Just like wearing a black shirt on a sunny day, soot in the atmosphere absorbs solar radiation—heating the surrounding air. It also absorbs light reflecting off Earth's surface—a double whammy for further warming. Unlike CO_2, which lingers in the atmosphere for decades, soot lingers for only a few weeks before being washed out. So, reducing soot emissions can be very helpful in slowing the trend toward global warming. How can soot emission be reduced?

84. The year 2010 had several extreme weather events—the Moscow heat wave, disastrous flooding in Pakistan, Vietnam, and the South Pacific; drought in the Amazon; and several devastating winter storms throughout the Northern Hemisphere. Are these severe weather events related to global warming? What type of weather has been happening where you live?

READINESS ASSURANCE TEST (RAT)

If you have a good handle on this chapter, if you really do, then you should be able to score at least 7 out of 10 on this RAT. If you score less than 7, you need to study further before moving on.

Choose the BEST answer to each of the following.

1. Air that contains the maximum amount of water vapor for the temperature of the air mass is considered to
   (a) have a relative humidity of 100%.
   (b) be saturated.
   (c) have an evaporation rate equal to its condensation rate.
   (d) all of these

2. In most midlatitude cyclones, the warm front
   (a) is behind and to the west of the cold front.
   (b) is ahead and to the east of the cold front.
   (c) terminates in the high-pressure center.
   (d) convolutes back to a stationary front.

3. An adiabatic process has occurred when
   (a) air is warmed by solar radiation.
   (b) heat is subtracted by evaporation.
   (c) air expands and warms.
   (d) air expands and cools.

4. When air sinks, it
   (a) compresses and warms.
   (b) reaches its equilibrium level and then begins to sink.
   (c) expands and cools.
   (d) forms into clouds.

5. When upper regions of the atmosphere are warmer than lower regions, we have
   (a) convective lifting.
   (b) a temperature inversion.
   (c) absolute instability.
   (d) an adiabatic process.

6. A key factor needed for precipitation to occur is
   (a) updraft motion in relatively thick clouds.
   (b) a condensation rate that exceeds the evaporation rate.
   (c) condensation nuclei.
   (d) all of these

7. For clouds to form, air must be lifted. The principal lifting mechanisms are
   (a) convectional, orographic, and frontal lifting.
   (b) continental, orogeny, and occluded lifting.
   (c) conversational, orthodontic, and face lifting.
   (d) stationary, occluded, and contact lifting.

8. As air temperature decreases, relative humidity
   (a) increases.   (b) decreases.
   (c) stays the same.   (d) none of these

9. In the Northern Hemisphere, tornadoes and hurricanes rotate in a _____ direction because of the _____.
   (a) clockwise; Coriolis force
   (b) counterclockwise; Coriolis force
   (c) clockwise; cyclogenesis force
   (d) counterclockwise; cyclogenesis force

10. When air is saturated, the condensation rate
    (a) is greater than the evaporation rate.
    (b) is less than the evaporation rate.
    (c) equals the evaporation rate.
    (d) depends on temperature.

Answers to RAT

1. d, 2. b, 3. d, 4. a, 5. b, 6. d, 7. a, 8. a, 9. b, 10. c

PART FOUR

4

Astronomy

If I were standing on the Moon, I wonder what Earth would look like when the Moon passes into Earth's shadow—that is, during a lunar eclipse. I'm told Earth would look like a large dark orb surrounded by a ring of brilliant red as the light from the hidden Sun refracts through Earth's atmosphere—like a zillion sunsets all at once! Hmm. All those reddish sunsets would light up the Moon's surface. That must be why here on Earth the Moon appears red during a lunar eclipse!

26

CHAPTER 26

The Solar System

HOW DOES the Sun produce so much energy? How are the planets similar, and how are they different? How did our Moon form and how does it go through phases? Why do we see only one side of the Moon? What are solar and lunar eclipses and why are they rare? What are meteors, asteroids, and comets? How frequently do they collide with our planet, and why does a comet's tail always point away from the Sun?

For thousands of years, people have gazed into the night sky and pondered questions such as these. Countless generations have wondered about our place in the universe. Only recently have we begun to understand it. We begin with a tour of the universe with a focus on our solar system, which on a cosmic scale is Earth's own backyard.

26.1   The Solar System and Its Formation

EXPLAIN THIS   How is gravity responsible for solar energy?

Our solar system is the collection of objects gravitationally bound to the Sun. Along with the Sun itself, the solar system contains at least eight **planets**, which are large orbiting bodies massive enough for their gravity to make them spherical but small enough to avoid having nuclear fusion in their cores. The planets also have successfully cleared all debris from their orbital paths, which all lie roughly in the same plane. This plane, called the **ecliptic**, is defined as the plane of Earth's orbit. Within the solar system are also numerous moons (objects orbiting planets), asteroids (small, rocky bodies), comets (small, icy bodies), and a collection of miniature planets known as dwarf planets that orbit on the outer edges of the solar system. The most well-known dwarf planet is Pluto, which was downgraded from planet status in 2006. The planets and all these other objects are quite small compared to the Sun. Figure 26.1 shows the sizes of the planets relative to the Sun, which contains a whopping 99.86% of the solar system's mass.

The vast distances between the Sun and the objects orbiting it can be grasped by imagining the Sun reduced to the size of a large beach ball 1 m in diameter. The closest planet, Mercury, would be an apple seed located about 40 m (130 ft) away. The next closest planet, Venus, would be the size of a pea about 80 m (255 ft) distant. Earth, also about the size of a pea, would be about 110 m away, which is greater than the length of a football field. The next planet, Mars, which is only a bit larger than Mercury, would be an apple seed almost two football field lengths away from our solar beach ball.

These first four planets—Mercury, Venus, Earth, and Mars—are called the **inner planets** because of their relatively close proximity to the Sun. All the inner planets are solid rocky planets. The **outer planets** are larger gaseous planets located much farther away. The first outer planet is Jupiter, which on the scale mentioned would be the size of a softball more than half a kilometer away. The second outer planet, Saturn, famous for its extensive ring system, would be the size of a baseball more than a kilometer away. Planets Uranus and Neptune would both be about the size of Ping-Pong balls located 2 and 3 km away, respectively. We see that solar system objects are mere specks in the vastness of the space about the Sun.

Because of these vast interplanetary distances, astronomers use the *astronomical unit* to measure them. One **astronomical unit (AU)** is about 1.5×10^8 km (about 9.3×10^7 mi) or the distance from Earth to the Sun. Table 26.1 gives the distances of planets from the Sun in AU. The data in Table 26.1 also show the division of the planets into two groups with similar properties. The inner planets—Mercury,

Mastering**PHYSICS**

TUTORIAL: Formation of the Solar System
VIDEO: History of the Solar System
VIDEO: Orbits in the Solar System

The ancients could tell the difference between planets and stars because of the difference in their movements in the sky. The stars remain relatively fixed in their patterns in the sky, but the planets wander. The term *planet* is derived from the Greek for "wandering star."

FIGURE 26.1
This illustration shows the order and relative sizes of planets. Moving away from the Sun, we have in order: Mercury, Venus, Earth, Mars, Jupiter, Saturn, Uranus, and Neptune. The planets range greatly in size, but the Sun dwarfs them all—containing more than 99% of the mass in the solar system. Note: Distances are not to scale in this illustration.

TABLE 26.1 PLANETARY DATA

	Mean Distance from Sun (Earth distances, AU)	Orbital Period (years)	Diameter (km)	Diameter (Earth = 1)	Mass (kg)	Mass (Earth = 1)	Density (g/cm³)	Inclination to Ecliptic
Sun			1,392,000	109.1	1.99×10^{30}	3.3×10^5	1.41	
Terrestrial								
Mercury	0.39	0.24	4,880	0.38	3.3×10^{23}	0.06	5.4	7.0°
Venus	0.72	0.62	12,100	0.95	4.9×10^{24}	0.81	5.2	3.4°
Earth	1.00	1.00	12,760	1.00	6.0×10^{24}	1.00	5.5	0.0°
Mars	1.52	1.88	6,800	0.53	6.4×10^{23}	0.11	3.9	1.9°
Jovian								
Jupiter	5.20	11.86	142,800	11.19	1.90×10^{27}	317.73	1.3	1.3°
Saturn	9.54	29.46	120,700	9.44	5.7×10^{26}	95.15	0.7	2.5°
Uranus	19.18	84.0	50,800	3.98	8.7×10^{25}	14.65	1.3	0.8°
Neptune	30.06	164.79	49,600	3.81	1.0×10^{26}	17.23	1.7	1.8°
Dwarf Planets								
Pluto	39.44	247.70	2,300	0.18	1.3×10^{22}	0.002	1.9	17°
Eris	67.67	557	2,400	0.19	1.6×10^{22}	0.002	1.9	44°

FIGURING PHYSICAL SCIENCE

The Scale of the Solar System

Astronomical distances are mind-boggling. Try the following problems to better appreciate the sizes of bodies in the solar system and distances between them. (Use the distance formula: distance = speed × time and data from Table 26.1.)

SAMPLE PROBLEM 1
The distance between Earth and the Moon is 384,401 km. How many Earth diameters would fit between Earth and the Moon?

Solution:

$$\frac{38,401 \text{ km}}{12,760 \text{ km}} \approx 30$$

About 30 Earth-sized planets would fit in the distance between Earth and the Moon.

SAMPLE PROBLEM 2
How long would it take to drive from Earth to the Moon if you drive at 55 mi/h? State your answer in years.

Solution:

$$d = vt \rightarrow t = \frac{d}{v}$$

Convert all units to metric.
(55 mi/h) (1 km/0.62 mi) ≈ 89 km/h. Then t = 384,410 km/89 km/h = 4319 h (1 day/24 h) (1 yr/365 days) ≈ 0.5 yr. It would take about one-half year to drive to the Moon traveling at freeway speeds.

SAMPLE PROBLEM 3
If you could fly to the Sun on a jet that moves at 1000 km/h, how long would it take? State your answer in years.

Solution:

$$t = \frac{d}{v} = \frac{1.5 \times 10^8 \text{ km}}{1000 \text{ km/h}}$$

$$= 1.5 \times 10^5 \text{ h}$$

To convert to years, multiply by the following conversion factors:

$$(1.5 \times 10^5 \text{ h})\left(\frac{1 \text{ day}}{24 \text{ h}}\right)\left(\frac{1 \text{ yr}}{365 \text{ days}}\right)$$

$$= 17 \text{ yr}$$

So if you could fly to the Sun in a jet without becoming vaporized, it would take about 17 years.

SAMPLE PROBLEM 4
The diameter of the Sun is 1,390,000 km. What is its diameter in AU? How much greater is the mean distance between Earth and the Sun compared to the diameter of the Sun?

Solution:

$$(1.39 \times 10^6 \text{ km})$$

$$(1 \text{ AU}/1.5 \times 10^8 \text{ km}) \approx 0.01 \text{ AU}$$

$$\frac{1 \text{ AU}}{0.01 \text{ AU}} = 100$$

The diameter of the Sun is approximately equal to 0.01 AU, so the distance between Earth and the Sun is about 100 times the diameter of the Sun. Big as the Sun is, the solar system is mostly empty space.

Venus, Earth, and Mars—are solid and relatively small and dense. They are called the *terrestrial planets*. The outer planets are large, have many rings and satellites, and are composed primarily of hydrogen and helium gas. These are called the *jovian planets* because their large sizes and gaseous compositions resemble those of Jupiter.

Nebular Theory

Any theory of solar system formation must be able to explain two major regularities: (1) *the motions among large bodies of the solar system* and (2) *the division of planets into two main types*—terrestrial *and* Jovian. Further, a viable theory of solar system formation must explain other known features of the solar system, including the existence of asteroids, comets, and moons.

The modern scientific theory that satisfies these requirements is called the **nebular theory**, which holds that the Sun and planets formed together from a cloud of gas and dust, a *nebula* (Latin for "cloud"). According to the theory, the solar system began to condense from the cloud of gas and dust about 5 billion years ago. The cloud would have been very diffuse and large, with a diameter thousands of times larger than Pluto's orbit. A nebula currently visible within the constellation of Orion is shown in Figure 26.2.

Within the nebula from which our solar system formed, the gravitational pull on particles exceeded the tendency of gas to disperse and fill all available space. Once the gravitational collapse of a cloud begins, gravity ensures that it continues. The universal law of gravity (see Chapter 4) is an inverse-square law: the strength of the gravitational force increases dramatically as the particles become closer. The cloud maintained a constant mass as it shrank, gravitational forces grew ever stronger, and the cloud took on a spherical shape.

This nebula must also have had a slight net rotation, possibly due to the rotation of the galaxy itself. Then over the millions of years during which the solar nebula collapsed, it heated up, spun faster, and flattened into a disk shape. As a result, the nebula transformed from a large, diffuse, spherical cloud to a much smaller spinning disk with a hot center, as shown in Figure 26.3.

As the nebula shrank under the influence of gravity, much heat was released upon the collision of particles. Also, as the nebula shrank, it spun faster and faster (because of the *conservation of angular momentum* discussed in Chapter 3). As any spinning object contracts, the speed of its spin increases such that angular momentum is conserved. A familiar example is an ice skater whose spin rate increases when her extended arms are pulled inward. A nebula does the same.

What happens to the shape of a sphere as it spins faster and faster? The answer is, it flattens. A familiar example is the chef who turns a ball of pizza dough into a disk by spinning it on his hands. Even planet Earth is a slightly "flattened" sphere because of its daily spin. Saturn, with a greater spin, noticeably departs from a purely spherical shape. So the initially spherical nebula progressed to a spinning disk, the center of which became the *protosun*.

FIGURE 26.2
This photograph, provided by the Hubble Telescope, shows the Orion Nebula. The Orion Nebula, like the nebula from which our solar system formed, is an interstellar cloud of gas and dust and the birthplace of stars.

FIGURE 26.3
(a) The nebula from which the solar system formed was originally a large, diffuse cloud that rotated quite slowly. The cloud began to collapse under the influence of gravity.
(b) As the cloud collapsed, it heated up as gravitational potential energy converted to heat. It spun faster by the conservation of angular momentum. (c) The cloud flattened into a disk as a result of its fast rotation. A spinning, flattened disk was produced whose mass was concentrated at its hot center.

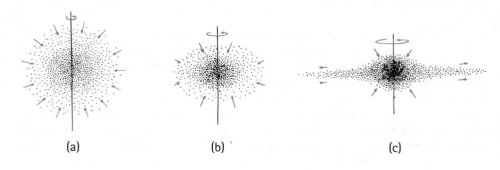

(a)   (b)   (c)

The formation of the spinning disk explains the motions of our solar system today. All planets orbit the Sun in nearly the same plane because they formed from that same flat nebular disk. The direction in which the disk was spinning became the direction of the Sun's rotation and the orbits of planets. It was also the preferred direction of rotation for planets—which is why most planets rotate in the same direction today.

CHECKPOINT

As a nebula contracts, its rate of spin increases. What rule of nature is at play here?

Was this your answer?
The rule that applies to all spinning bodies is the *conservation of angular momentum*. That rule and the conservation of energy are dominant players in the universe.

The hot, central portion of the solar nebula, the protosun, was a clump of gas and dust that became the Sun when thermonuclear fusion ignited within it. The surrounding disk was the source of material that became the planets. In the spinning disk, matter collected in some regions more densely than in others. Perhaps small particles of gas and dust stuck together via gravity or electrostatic attraction. Because of their extra mass, these clumps exerted a stronger gravitational force on one another than on neighboring regions of the disk, and so they pulled in even more material to them. This led to the accretion of the nebular disk into small objects called *planetesimals*, which ranged in size from boulders to objects several kilometers in diameter. Planetesimals grew larger through countless collisions until they gravitationally dominated surrounding matter and finally became full-grown planets.

The planetesimals grew into planets at about the same time that the protosun was commencing thermonuclear fusion. Once thermonuclear fusion occurred and the Sun began to radiate energy, the nebular disk warmed, with the inner portions reaching higher temperatures than the outer portions. As a result, the inner and outer planets developed differently. The inner planets formed from materials that remained solid at high temperatures; hence, the inner planets are rocky. The outer planets, by contrast, consist mainly of hydrogen and helium gas that coalesced in the cold regions of the solar system far from the Sun. Thus, we can see that the nebular theory accounts for the formation of the planets and the neat division of them into two groups.

26.2 The Sun

EXPLAIN THIS Why is the Sun's surface much cooler than its inner core?

LEARNING OBJECTIVE
Recognize the features of the Sun, including its interior, photosphere, sunspots, solar cycle, chromosphere, and corona.

The Sun produces energy from the thermonuclear fusion of hydrogen to helium. Each second, approximately 657 million tons of hydrogen is fused to 653 million tons of helium. The 4 million tons of mass lost is discharged as radiant energy. This conversion of hydrogen to helium in the Sun has been going on since it formed nearly 5 billion years ago, and it is expected to continue at this rate for another 5 billion years. A tiny fraction of the Sun's energy reaches Earth and is converted by photosynthesizing organisms to chemical energy stored in large molecules. These energy-rich molecules

MasteringPHYSICS
TUTORIAL: The Sun

are the primary energy source for almost all the organisms of this planet. The Sun, Earth's nearest star, is the solar system's power supply.

Solar energy is generated deep within the core of the Sun. The solar core constitutes about 10% of the Sun's total volume. It is very hot—more than 15,000,000 K. The core is also very dense, with more than 12 times the density of solid lead. Pressure in the core is 340 billion times Earth's atmospheric pressure! Because of these intense conditions, the hydrogen, helium,

FIGURE 26.4
Never directly look at the Sun! Instead, you can get a nice view of the Sun by focusing the image of the Sun from a pair of binoculars onto a white surface. If the Sun is eclipsed by the Moon, which is a rare event, the Sun is seen as a crescent. More commonly, the Sun's image may reveal sunspots.

and minute quantities of other elements exist in the plasma state. (Plasma, recall, is the phase of matter beyond gas, consisting of ions and electrons rather than atoms—electrons have been stripped from atoms by high energies.) The nuclei of this plasma move fast enough to undergo nuclear fusion, as discussed in Section 13.7. The energy released from this nuclear fusion rises to the surface, where it causes gases to emit a broad spectrum of electromagnetic radiation, centered in the visible region (see Figure 11.31).

The Sun's surface is a layer of glowing 5800 K plasma, which is much cooler than the Sun's core but hot enough to generate lots of light. This layer, called the *photosphere* (sphere of light), is about 500 km deep. Within the photosphere are relatively cool regions that appear as **sunspots** when viewed from Earth. Sunspots are cooler and darker than the rest of the photosphere and are caused by magnetic fields that impede hot gases from rising to the surface. As shown in Figure 26.4, sunspots can be seen by focusing the image of the Sun from a telescope or pair of binoculars onto a flat white surface. Sunspots are typically twice the size of Earth, move around because of the Sun's rotation, and last about a week or so. Often, they cluster in groups as shown in Figure 26.5.

The Sun spins slowly on its axis. Because the Sun is a fluid rather than a solid, different latitudes of the Sun spin at different rates. Equatorial regions spin once in 25 days, but higher latitudes take up to 36 days to make a complete rotation. This differential spin means the surface near the equator pulls ahead of the surface farther north or south. The Sun's differential spin wraps and distorts the solar magnetic field, which bursts out to form the sunspots mentioned earlier. A reversal of magnetic poles occurs every 11 years, and the number of sunspots also reaches a maximum every 11 years (currently). The complete cycle of solar activity is 22 years.

Above the Sun's photosphere is a transparent 10,000-km-thick shell called the *chromosphere* (sphere of color), seen during an eclipse as a pinkish glow surrounding the eclipsed Sun. The chromosphere is hotter than the photosphere, reaching temperatures of about 10,000 K. Its beautiful pink color, as shown in Figure 26.6, arises from the emission of light from hydrogen atoms.

Beyond the chromosphere are streamers and filaments of outward-moving, high-temperature plasmas curved by the Sun's magnetic field. This outermost region of the Sun's atmosphere is the *corona*, which

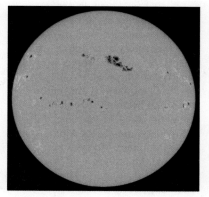

FIGURE 26.5
Sunspots on the solar surface are relatively cool regions. We say relatively cool because they are hotter than 4000 K. They look dark only in contrast with their 5800 K surroundings.

FIGURE 26.6
The pink chromosphere becomes visible when the Moon blocks most of the light from the photosphere during a solar eclipse.

FIGURE 26.7
The pearly white solar corona is visible only during a solar eclipse. Notice how this exceptional photo of the corona also captures some of the pink of the chromosphere as well as the face of the new Moon, which is faintly illuminated by light reflected from the full Earth. In other words, if you were standing on the Moon when this photo was taken, you would see your faint shadow cast by the fully lit Earth shining above you.

extends out several million kilometers (Figure 26.7). The temperature of the corona is amazingly high—on the order of 1 million K—and it is where most of the Sun's powerful X-rays are generated. Because the corona is not very dense, its brightness is not as intense as the Sun's surface, which makes the corona safe to observe during (and only during) a total solar eclipse. High-speed protons and electrons are cast outward from the corona to generate the *solar wind*, which powers the aurora on Earth and produces the tails of comets.

FYI We have seven days in a week because ancient Europeans decided to name days after the seven wandering celestial objects they could observe. The English day names were derived from the language of the Teutonic tribes who lived in the region that is now Germany. In Teutonic, the Sun is *Sun* (Sunday), the Moon is *Moon* (Monday), Mars is *Tiw* (Tuesday), Mercury is *Woden* (Wednesday), Jupiter is *Thor* (Thursday), Venus is *Fria* (Friday), and Saturn is *Saturn* (Saturday).

CHECKPOINT

1. **Was the Sun more massive 1000 years ago than it is today? Defend your answer.**
2. **Of the photosphere, chromosphere, and corona, which is thinnest? Which is hottest? Which is between the other two?**

Were these your answers?

1. Yes, although slightly compared with its great mass. The Sun loses mass as hydrogen nuclei combine to make helium nuclei.
2. The photosphere is the thinnest. The corona is the hottest. The chromosphere is the pinkish layer above the photosphere and below the vast corona.

LEARNING OBJECTIVE
Identify the major properties of the four inner planets: Mercury, Venus, Earth, and Mars.

26.3 The Inner Planets

EXPLAIN THIS Why does Venus, not Mercury, have the hottest surface of any planet in the solar system?

Compared with the outer planets, the four planets nearest the Sun are close together. These are Mercury, Venus, Earth, and Mars. These rocky planets each have a mineral-containing solid crust.

FIGURE 26.8
Mercury is heavily cratered from the impacts of many meteorites. Mercury is a small planet, with only about 6% of the volume and mass of Earth. This photo was taken by the Messenger spacecraft, which was launched from Earth in 2004 and reached orbit around planet Mercury in 2011.

Mercury

Mercury (Figure 26.8) is about 1.4 times larger than Earth's Moon and similar in appearance. It is the closest planet to the Sun. Because of this closeness it is the fastest planet, circling the Sun in only 88 Earth days—which thus equals one Mercury "year." Mercury spins about its axis only three times for each two revolutions about the Sun. This makes its daytime very long and very hot, with temperatures as high as 430°C.

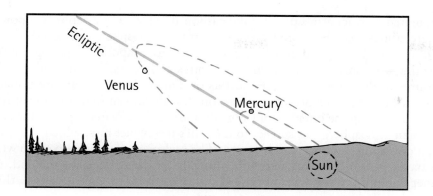

Because the orbits of Mercury and Venus lie inside the orbit of Earth, they are always near the Sun in our sky. Near sunset (or sunrise) they are visible as "evening stars" or "morning stars."

Because of Mercury's small size and weak gravitational field, it holds very little atmosphere. Mercury's atmosphere is only about a trillionth as dense as Earth's atmosphere—it's a better vacuum than laboratories on Earth can produce. So without a blanket of atmosphere, and because there are no winds to transfer heat from one region to another, nighttime on Mercury is very cold, about –170°C. Mercury is a fairly bright object in the nighttime sky and is best seen as an evening "star" during March and April or as a morning star during September and October. It is seen near the Sun at sunup or sunset.

Venus

Venus is the second planet from the Sun and, like Mercury, has no moon. Venus is frequently the first starlike object to appear after the Sun sets, so it is often called the evening "star," as illustrated in Figure 26.9. Compared with other planets, Venus most closely resembles Earth in size, density, and distance from the Sun. However, as shown in Figure 26.10, Venus has a very dense atmosphere and opaque cloud cover that generate high surface temperatures (470°C)—too hot for oceans. The atmosphere of Venus is about 96% CO_2. Remember from Chapter 24 that carbon dioxide is a "greenhouse gas." By this we mean that CO_2 blocks the escape of infrared radiation from Earth's surface and contributes to the warming of our planet. The thick blanket of CO_2 surrounding Venus effectively traps heat near the Venusian surface. This and Venus's proximity to the Sun make Venus the hottest planet in the solar system.

Another difference between Venus and Earth is in how the two planets spin about their axes. Venus takes 243 Earth days to make one full spin and only 225 Earth days to make one revolution around the Sun. So, on Venus, a day is longer than a year. Furthermore, Venus spins in a direction opposite to the direction of Earth's spin. So, on Venus, the Sun rises in the west and sets in the east. But because the cloud cover is so intense, a Venusian sunrise or sunset is never visible from its surface.

The slow spin of Venus means that the atmosphere is not disturbed by the Coriolis effect described in Chapter 24. As a result, there is very little wind and weather on the surface of Venus. Instead, the stifling hot dense air sits still through its long days and nights.

In recent years, 17 probes have landed on the surface of Venus. There have been 18 flyby spacecraft (notably *Pioneer Venus* in 1978 and *Magellan* in 1993). From spacecraft data, scientists have been able to account for why the atmospheres of Venus and Earth are so different. According to the generally accepted model, when the two planets first formed they had similar amounts of water. Venus, however, is a bit closer to the Sun than Earth is. It also rotates much more slowly. These two factors combined to make the Sun-facing side of Venus significantly warmer than Earth. With greater warmth, more of Venus's water

Venus is an Earth-sized planet barren of any oceans. The surface of Venus was first mapped in the early 1990s by the spacecraft Magellan, which used microwave radar ranging to "see through" the planet's thick atmosphere of carbon dioxide and sulfuric acid clouds.

FYI Ancient American cultures ran their lives using three calendars. Their secular calendar, which told them when to plant seeds and so forth, followed the 365-day orbit of Earth. Their religious calendar centered on the roughly 260-day orbit of Venus. Both of these calendars were cyclical and couldn't account for succeeding years. For that purpose they developed the "long count" calendar, which, interestingly enough, employed the concept of zero. They did this centuries before the concept of zero was recognized by accountants in India.

FYI Not all of the Venusian water was lost to space. Some instead reacted with volcano-generated sulfur dioxide to form sulfuric acid, which now laces the upper levels of the Venusian atmosphere. That which stayed behind to react with sulfur dioxide, however, was predominately water containing the heavier isotope of hydrogen, known as deuterium. This has been confirmed with direct measurements from space probes, which show an unusually high proportion of deuterium. This, in turn, supports the theory that Venus lost its water because of a runaway greenhouse effect—if such didn't happen, then the proportion of deuterium in the atmosphere would remain similar to that currently found on Earth.

evaporated into its atmosphere. Like carbon dioxide, water vapor is a powerful greenhouse gas. So more water vapor in the atmosphere caused even more warming, which caused more of the oceans to evaporate, causing further warming—runaway global warming! Venus's early oceans contained massive amounts of dissolved carbonates, just as on Earth. As the oceans evaporated, these carbonates transformed into carbon dioxide and moved into the atmosphere, increasing the intensity of the greenhouse effect. Water vapor in the early Venusian atmosphere was subject to the Sun's ultraviolet rays, which broke the water down into hydrogen and oxygen. The hydrogen escaped into space while the oxygen chemically reacted with minerals on the surface. In the end, the planet's supply of water was forever lost. All that remains we see today as a thick atmosphere of heat-trapping carbon dioxide.

CHECKPOINT

As the Sun gets older it also gets hotter. How will this affect the amount of water vapor in our atmosphere?

Was this your answer?
Initially, the additional heat from the Sun will increase the rate at which Earth's oceans evaporate. More water vapor in the atmosphere will then enhance the greenhouse effect, causing even warmer temperatures, which will provide for even more evaporation. As the oceans disappear, carbon dioxide levels will also rise dramatically, ensuring that the greenhouse effect keeps the water vaporized. Ultraviolet light from the Sun, however, will zap this water from the atmosphere. So in the end, as the Sun gets hotter, the amount of water vapor in the atmosphere will drop to near nothing. We will share the same fate as our sister planet Venus. You can breathe a bit easy because the time frame for this is about a billion years.

Earth

Our home planet Earth resides within the Sun's *habitable zone*, which is a region not too close and not too far from the Sun so that water can exist predominately in the liquid phase, as shown in Figure 26.11. Earth has an abundant supply of liquid water covering about 70% of Earth's surface, which makes our planet the blue planet (Figure 26.12).

Earth's oceans support the *carbon dioxide cycle*, which acts as a thermostat to keep global temperatures from reaching harsh extremes. For example, if Earth froze over completely, carbon dioxide released by volcanoes would no longer be absorbed by the oceans. Instead the carbon dioxide would build up in the atmosphere, which would warm the atmosphere and hence melt the frozen Earth. Conversely, if Earth became hotter and hotter, more water would evaporate. This would lead to more precipitation, which would remove carbon dioxide from the atmosphere. With less carbon dioxide in the atmosphere, the greenhouse effect would be minimized and Earth would cool. So not only are we a nice distance from the Sun, but our atmosphere contains just enough water vapor and carbon dioxide to keep temperatures favorable for life. Furthermore, our relatively high daily spin rate allows only a brief and small lowering of temperature on the nighttime side of Earth. So temperature extremes of day and night are also kept moderate.

FIGURE 26.11
Earth resides on the inner side of the Sun's habitable zone, which is where conditions are favorable for life as we know it.

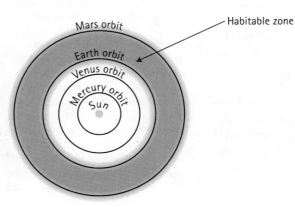

Solar System

Moreover, recall from Sections 9.5 and 21.2 that movements within Earth's molten core generate a strong magnetic field around our planet. This magnetic field, called the *magnetosphere*, extends thousands of kilometers into space and shields the Earth from *solar wind*, which is a flow of charged particles emanating from the Sun (see Figure 9.16). This form of cosmic radiation is harmful to living organisms. Furthermore, solar wind is capable of stripping a planet of its atmosphere. Planet Earth is indeed a sanctuary for life in an otherwise inhospitable universe.

For more information on the geology, weather, and history of planet Earth, review Chapters 20–25, which discuss Earth science.

FIGURE 26.12
Earth, the blue planet. This famous photo was taken by *Apollo 17* astronauts as they returned from the last manned mission to the Moon in 1972. It is the only existing photo of the full Earth from an appreciable distance. Can you see that it was taken in the summer months of the Southern Hemisphere?

CHECKPOINT

If Venus was also once protected by the carbon dioxide cycle, what went wrong?

Was this your answer?
The carbon dioxide cycle requires water. The Venusian carbon dioxide cycle, therefore, broke down as the Venusian water was split apart by the Sun's ultraviolet rays.

Mars

Mars captures our fancy as another world, perhaps even as a world with life because it resides on the outer fringes of the habitable zone. Mars is a little more than half Earth's size; its mass is about one-ninth that of Earth; and it has a core, a mantle, a crust, and a thin, nearly cloudless atmosphere. It has polar ice caps and seasons that are nearly twice as long as Earth's because Mars takes nearly two Earth years to orbit the Sun. Mars and Earth spin on their axes at about the same rate, which means the lengths of their days are about the same. When Mars is closest to Earth, a situation that occurs once every 15 to 17 years, its bright, ruddy color outshines the brightest stars.

The Martian atmosphere is about 95% carbon dioxide, with only about 0.15% oxygen. Yet because the Martian atmosphere is relatively thin, it doesn't trap heat via the greenhouse effect as much as Earth's and Venus's atmospheres do. So the temperatures on Mars are generally colder than on Earth, ranging from about 30°C in the day at the equator to a frigid −130°C at night. If you visit Mars, never mind your raincoat, for there is far too little water vapor in the atmosphere for rain. Even the ice at the planet's poles consists primarily of carbon dioxide. And don't give a second thought to waterproof footwear, for the very low atmospheric pressure won't permit the existence of any puddles or lakes.

(a)

(b)

FIGURE 26.13
(a) A model of NASA's Mars Exploration Rover, *Spirit*, with cameras mounted on the white mast. (b) *Spirit* took photographs in June 2004 for this composite, true-color image of the region named Columbia Hills on Mars. The vehicle later traveled to the hills to analyze their composition.

FYI Jupiter is near the point at which the addition of more matter would cause the size of this planet to contract. This is analogous to a stack of pillows. Start stacking pillows and the stack gets taller. Eventually, however, a point is reached at which the weight of upper pillows pushes down on lower pillows such that the column of pillows gets shorter. Interestingly, Jupiter is *larger* than the smallest stars, which, though smaller than Jupiter, are about 80 times as massive.

LEARNING OBJECTIVE
Identify the major properties of the four outer planets: Jupiter, Saturn, Uranus, and Neptune.

FIGURE 26.14
Jupiter, with its moons Io (orange dot over planet) and Europa (white dot to right of planet), as seen from the *Voyager 1* spacecraft in February 1979. The Great Red Spot (lower left), larger than Earth, is a cyclonic weather pattern of high winds and turbulence.

Surface features on Mars, such as channels, indicate that liquid water was once abundant on this planet. This implies a distant past that was much warmer—a situation likely made possible by the greenhouse effect of a thicker atmosphere.

What happened to this thicker atmosphere? A currently accepted model is that because Mars is such a small planet, its molten core cooled down and solidified relatively fast—within a billion years of formation. This would have had at least two major consequences. The first is a decrease in the activity of volcanoes, which are a prime source of atmospheric gases. The second is the loss of the planet's magnetosphere. Without a magnetosphere, the bulk of the early Martian atmosphere—no longer replaced by volcanoes—was carried away into space by solar winds.

Today, landings on Mars show it to be a very dry and windy place. Because the Martian atmosphere has a very low density, its winds are about 10 times as fast as winds on Earth.

In 2004, spacecraft orbiting Mars detected signs of the organic compound methane, CH_4, within the atmosphere. This is unusual because methane decomposes fairly rapidly, which tells us that this compound is currently being produced. The likely source is ongoing residual volcanic activity, which could potentially melt underground ice into liquid water. Indeed, scientists have found evidence of the leakage of underground liquid water onto the surface occurring since we started surveying Mars from space. Once on the surface, this water would evaporate, freeze, and sublime away. Underground pools of volcanically warmed liquid water, however, may harbor microscopic life forms.

Mars has two small moons—Phobos, the inner one, and Deimos, the outer. Both are potato-shaped and have cratered surfaces. They are likely captured asteroids. Phobos orbits in the same easterly direction in which Mars spins (like our Moon), at a distance of only 6000 km in a period of only 7.5 hours. From Mars it appears about half the size of our Moon. Deimos is about half the size of Phobos, and it orbits Mars in 30.3 hours at a distance of 20,000 km from the Martian surface.

26.4 The Outer Planets

EXPLAIN THIS The exteriors of the outer planets are gaseous, but their interiors are mostly liquid. Why?

The outer planets—Jupiter, Saturn, Uranus, and Neptune—are gigantic, gaseous, low-density worlds. They each formed from rocky and metallic cores that were much more massive than the terrestrial planets. The gravitational forces of these cores were strong enough to sweep up gases of the early planetary nebula, primarily hydrogen and helium. The cores continued to collect gases until the Sun ignited and the solar wind blew away all remaining interplanetary gases. The core of Jupiter was the first to develop, and hence it had the longest time to collect gas before solar ignition. This is why Jupiter is the largest of the outer planets. Another commonality is that they all have ring systems, Saturn's being the most prominent. We will explore the outer planets in the order of their distance from the Sun.

Jupiter
Jupiter is the largest of all the planets. Its yellow light in our night sky is brighter than that of any star. Jupiter spins rapidly about its axis in about 10 hours, a speed that flattens it so that its equatorial diameter is about 6% greater than its polar

diameter. As with the Sun, all parts do not rotate in unison. Equatorial regions complete a full revolution several minutes before nearby regions in higher and lower latitudes. The atmospheric pressure at Jupiter's rocky surface is more than a million times the atmospheric pressure of Earth. Jupiter's atmosphere is about 82% hydrogen, 17% helium, and 1% methane, ammonia, and other molecules.

Jupiter's average diameter is about 11 times Earth's, which means Jupiter's volume is more than 1000 times Earth's. Jupiter's mass is greater than the combined masses of all the other planets. Because of its low density, however—about one-fourth of Earth's—Jupiter's mass is barely more than 300 times Earth's. Investigations of Jupiter tell us that its core is a solid sphere about 15 times as massive as the entire Earth, and it is composed of iron, nickel, and other minerals.

More than half of Jupiter's volume is an ocean of liquid hydrogen. Beneath the hydrogen ocean lies an inner layer of hydrogen compressed into a liquid metallic state. In it are abundant conduction electrons that flow to produce Jupiter's enormous magnetic field.

Jupiter has more than 60 moons in addition to a faint ring. The four largest moons were discovered by Galileo in 1610; Io and Europa are about the size of our Moon, and Ganymede and Callisto are about as large as Mercury (Figure 26.16). Jupiter's moon Io has more volcanic activity than any other body in the solar system. Perhaps most intriguing of all, however, is Europa, whose surface is made of frozen water. As shown in Figure 26.17, deep beneath this ice is likely an ocean of water kept warm by the strong tidal forces from nearby Jupiter. If life were to be found anywhere in this solar system besides Earth, it would likely be on the floor of Europa's ocean adjacent to volcanic thermal vents. Such extraterrestrial life forms may be similar to the bizarre forms of life recently discovered adjacent to deep thermal vents on Earth's ocean floor. Alternatively, they may be single-cell organisms, such as bacteria. Then again, there may be nothing. So is life rare or common in this universe? The answer may be waiting for us in our own galactic backyard.

FIGURE 26.15
This artist's rendering shows aurorae (pink) in the upper atmosphere of Jupiter. Thunderclouds are seen below the aurorae, and the nearest major moon, Io, is seen at center left. Aurorae, like the northern lights on Earth, are caused by charged particles from the solar wind exciting gas molecules in the upper atmosphere. The gas molecules emit light as they return to an unexcited energy state.

FIGURE 26.16
Jupiter's four largest moons were discovered by Galileo, who was the first to point the recently invented telescope toward the heavens. He noted the changing positions of these moons and concluded that they were orbiting Jupiter, which was a violation of the then widely held belief that all heavenly objects orbited Earth. His discovery was revolutionary. In his honor, four moons are known as the Galilean moons—from left to right: Ganymede, Callisto, Io, and Europa.

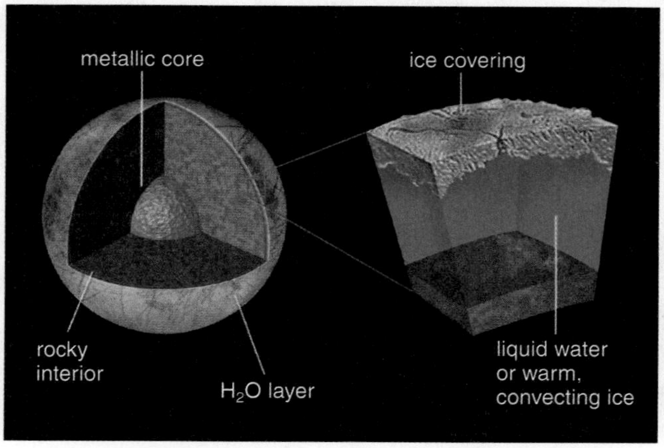

metallic core

ice covering

rocky interior

H_2O layer

liquid water or warm, convecting ice

FIGURE 26.17
A model of the interior of Europa with a zoomed-in view of its ice-capped ocean, which, according to magnetic measurements, likely covers the entire sphere.

FIGURE 26.18

Saturn surrounded by its famous rings, which are composed of rocks and ice.

Saturn

Saturn is one of the most remarkable objects in the sky, with its rings clearly visible through a small telescope. It is quite bright—brighter than all but two stars—and it is second only to Jupiter in mass and size. Saturn is twice as far from Earth as Jupiter is. Its diameter, not counting its ring system, is nearly 10 times that of Earth, and its mass is nearly 100 times Earth's. It is composed primarily of hydrogen and helium, and it has the lowest density of any planet, only 0.7 times the density of water. These characteristics mean that Saturn would easily float in a bathtub, if the bathtub were large enough. Its low density and its 10.2-hour rapid spin produce more polar flattening than can be seen in the other planets. Notice its oblong shape in Figure 26.18.

Saturn's rings, only a few kilometers thick, lie in a plane coincident with Saturn's equator. Four concentric rings have been known for many years, and spacecraft missions have detected many others. The rings are composed of chunks of frozen water and rocks, believed to be the material of a moon that never formed or the remnants of a moon torn apart by tidal forces. All the rocks and bits of matter that make up the rings pursue independent orbits about Saturn. The inner parts of the ring travel faster than the outer parts, just as any satellite near a planet travels faster than a more distant satellite.

Saturn has about 50 moons beyond its rings. The largest is Titan, 1.6 times as large as our Moon and even larger than Mercury. It spins once every 16 days and has a methane atmosphere with atmospheric pressure that is likely greater than Earth's. Its surface temperature is cold, roughly −170°C. A space probe built by NASA and the European Space Agency landed on Titan in 2005. Remarkably, photos revealed a landscape similar to Earth's despite the fact that the materials are completely different (Figure 26.19). Lakes and streams are filled with not water but liquid methane. Rocks are made of ice. Instead of lava, Titan has a flowing slush of ice and liquid ammonia. No life is expected to be found on this moon because of the intensely cold temperatures. Titan, however, holds an intriguing soup of organic molecules whose chemistry may provide a clue to what Earth was like during the time before life arose here.

FIGURE 26.19

Images from Saturn's largest moon, Titan, taken by the Cassini spacecraft and its space probe, the *Huygens*, which successfully descended to the surface.

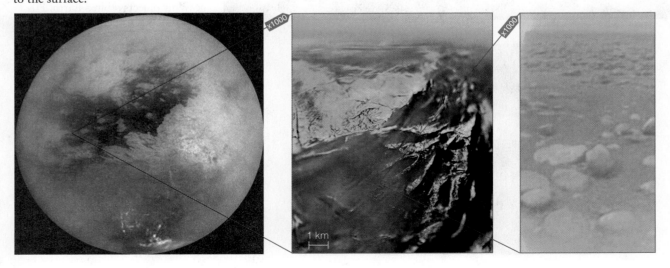

Uranus

Uranus (pronounced "YUR-uh-nus," accent on the first syllable) is twice as far from Earth as Saturn is, and it can barely be seen with the naked eye. Uranus was unknown to ancient astronomers and not discovered as a planet until 1781. The *Voyager 2* spacecraft first visited this planet in 1986. It has a diameter four times that of Earth and a density slightly greater than that of water. So if you could place Uranus in a giant bathtub, it would sink. The most unusual feature of Uranus is its tilt. Its axis is tilted 98° to the perpendicular of its orbital plane, so it lies on its side (Figure 26.20). Unlike Jupiter and Saturn, it appears to have no appreciable internal source of heat. Uranus is a cold place.

Uranus has at least 27 moons, in addition to a complicated faint ring system. Recall from Chapter 4 that perturbations in the planet Uranus led to the discovery in 1846 of a farther planet, Neptune.

Neptune

Neptune has a diameter about 3.9 times that of Earth, a mass 17 times as great, and a mean density about a third of Earth's. Its atmosphere is mainly hydrogen and helium, with some methane and ammonia, which makes Neptune bluer than Uranus (Figure 26.21).

The *Voyager 2* spacecraft flew by Neptune in 1989. It showed that Neptune has at least 13 moons in addition to a ring system. The largest moon, Triton, orbits Neptune in 5.9 days in a direction opposite to the planet's eastward spin. This suggests that Triton is a captured object. Triton's diameter is three-quarters of our Moon's diameter, and yet Triton is twice as massive as Earth's Moon. It has bright polar caps and geysers of liquid nitrogen.

FIGURE 26.20
Astronomers believe the rotational axis of Uranus is tilted as a result of a collision it had with a large body early in the solar system's history. Methane in the upper atmosphere absorbs red light, giving Uranus its blue-green color.

Recent studies of Galileo's notebooks show that Galileo saw Neptune in December 1612 and again in January 1613. He was interested in Jupiter at the time, and so he merely plotted Neptune as a background star.

FIGURE 26.21
Cyclonic disturbances on Neptune in 1989 produced a great dark spot, which was even larger than Earth and similar to Jupiter's Great Red Spot. The spot has now disappeared. The gray horizon in the foreground of this computer-generated montage is a close-up of Neptune's moon Triton, which has a composition and size similar to those of Pluto.

We see the aftermath of meteoroid bombardment on the Moon because it wears no makeup. Similar bombardment on Earth has been long erased by erosion.

FIGURE 26.22

Three steps to the formation of Earth's Moon. A Mercury-sized object collides with Earth, which turns molten. Debris collects in a ring that accretes into the Moon, which is quite close to the rapidly rotating Earth. Over the next billion years tidal forces slow the rate of Earth's rotation while also causing the Moon to move farther away.

26.5 Earth's Moon

EXPLAIN THIS When the Moon rises at sunset, its phase is always full. Why?

Earth's Moon is puzzling. It is close to the size of Mercury, which is a planet and *not* a moon. The composition of Earth's Moon is nearly the same as Earth's mantle. Furthermore, the Moon possesses a rather small iron core. To explain these and a multitude of other facts about the Moon, scientists have pieced together the following probable scenario for its origin.

During the early history of the solar system, the young Earth had a Mercury-sized companion form within an orbit close to that of Earth. Normally, if the companion's orbit were a bit closer to the Sun, then it would orbit faster than Earth and move ahead of Earth. Upon passing through a special point, known as the *Lagrangian point*, however, the object would find that the gravitational pull of Earth is strong enough to hold it back so that it orbited with Earth in unison. So early Earth may have had a twin that paraded with Earth around the Sun much like two horses running side by side on a circular track.

Eventually, a random event, such as the passing of an asteroid or comet, caused our companion to sway from the Lagrangian point and fall toward and collide with Earth. The collision would have been massively spectacular, spewing debris everywhere while turning Earth fully molten. Hitting askew, the impact sent Earth into a wild spin rotating once every five hours. The debris soon collected as a ring around Earth, and then, within about 1000 years, the ring coalesced into the Moon. This scenario is known as the *giant impact theory* of the origin of Earth's Moon. It explains why the Moon is so large (we started out as twin planets), why its composition is similar to Earth's (it formed from our mantle and our mantle formed from it), why it has such a small iron core (Earth's iron core had already differentiated and was

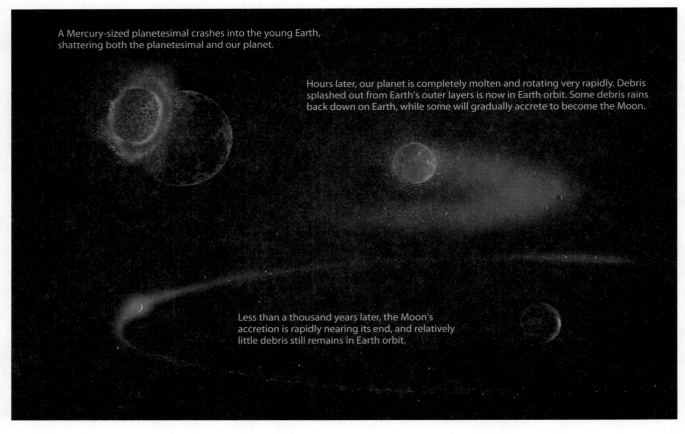

A Mercury-sized planetesimal crashes into the young Earth, shattering both the planetesimal and our planet.

Hours later, our planet is completely molten and rotating very rapidly. Debris splashed out from Earth's outer layers is now in Earth orbit. Some debris rains back down on Earth, while some will gradually accrete to become the Moon.

Less than a thousand years later, the Moon's accretion is rapidly nearing its end, and relatively little debris still remains in Earth orbit.

not sent up with the debris), and much more. This impact theory is still the subject of much research and is thus being continually refined. Although it was developed only within the past couple decades, scientists are excited by its explanatory powers.

From a distance, Earth and the Moon still resemble a twin planet system, as you can see in Figure 26.23. Compared to Earth, however, the Moon is relatively small, with a diameter of about the distance from San Francisco to New York City. It once had a molten surface, but it cooled too rapidly for the establishment of moving crustal plates, like those of Earth. In its early history, it was intensely bombarded by meteoroids (as was Earth). A little more than 3 billion years ago, meteoroid bombardment and volcanic activity filled basins with lava to produce its present surface. It has undergone very little change since then. Its igneous crust is thicker than Earth's. The Moon is too small with too little gravitational pull to have an atmosphere, and so, without weather, the only eroding agents have been meteoroid impacts.

FIGURE 26.23
Earth and the Moon as photographed in 1977 from the *Voyager 1* spacecraft on its way to Jupiter and Saturn.

The Phases of the Moon

Sunshine always illuminates half of the Moon's surface. The Moon shows us different amounts of its sunlit half as it circles Earth each month. These changes are the **Moon phases** (Figure 26.24). The Moon cycle begins with the **new Moon**. In this phase, its dark side faces us and we see darkness. This occurs when the Moon is between Earth and the Sun (position 1 in Figure 26.25).

FIGURE 26.24
The Moon in its various phases.

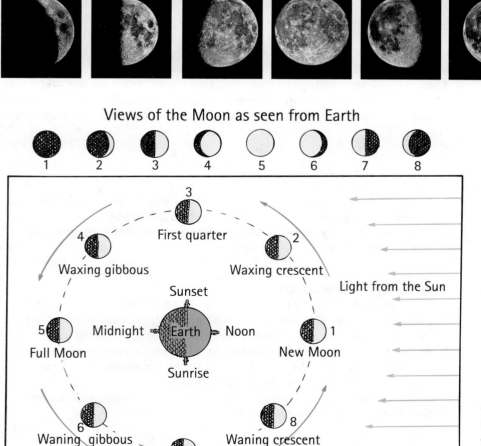

FIGURE 26.25
Sunlight always illuminates half of the Moon. As the Moon orbits Earth, we see varying amounts of its sunlit side. One lunar phase cycle takes 29.5 days.

FIGURE 26.26
Edwin E. Aldrin, Jr., one of the three *Apollo 11* astronauts, stands on the dusty lunar surface. To date, 12 people have stood on the Moon.

Most planets wobble significantly as they spin about their axes. The Moon, however, helps keep Earth's wobble to a minimum. As a result, our weather patterns are fairly consistent through the ages, which makes our planet even more favorable for the development of life. Thank you, Moon!

 FYI During the time of the dinosaurs, a day was only about 19 hours. Today our days are about 24 hours. Billions of years from now, as Earth continues to slow down, a day will last about 47 hours. At that time, Earth and the Moon will be gravity locked such that the Moon always appears in one location in the sky. To see the Moon, you will need to be on the Moon side of the planet, where perhaps real estate prices will be higher because of the view. One huge problem, however, is that by then, our Sun will have already gone through its dying phases, through which Earth will be subject to the fate of Venus, where it is forever cloudy. Nothing is permanent.

 If someone shone a flashlight on a ball in a dark room, you could tell where the flashlight was by looking at the illumination on the ball. The Moon is similarly lit by the Sun.

During the next seven days, we see more and more of the Moon's sunlit side (position 2 in Figure 26.25). The Moon is going though its waxing crescent phase (*waxing* means "increasing"). At the first quarter, the angle between the Sun, the Moon, and Earth is 90°. At this time, we see half the sunlit part of the Moon (position 3 in Figure 26.25).

During the next week, we see more and more of the sunlit part. The Moon is going through its waxing gibbous phase (position 4 in Figure 26.25). (*Gibbous* means "more than half.") We see a **full Moon** when the sunlit side of the Moon faces us squarely (position 5 in Figure 26.25). At this time, the Sun, Earth, and the Moon are lined up, with Earth in between. To view this full Moon you need to be on the nighttime side of Earth, at sunset when the full Moon rises from the east, or at sunrise when it sets in the west.

The cycle reverses during the following two weeks, as we see less and less of the sunlit side while the Moon continues in its orbit. This movement produces the waning gibbous, last quarter, and waning crescent phases. (*Waning* means "shrinking.") The time for one complete cycle is about 29.5 days.*

CHECKPOINT

1. **Can a full Moon be seen at noon? Can a new Moon be seen at midnight?**
2. **Astronomers prefer to view the stars when the Moon is absent from the night sky. When, and how often, is the Moon absent from the night sky?**

Were these your answers?

1. Figure 26.26 shows that at noontime, you would be on the wrong side of Earth to view the full Moon. Likewise, at midnight, the new Moon would be absent. The new Moon is in the sky in the daytime, not at night.
2. At the time of the new Moon and during the week on either side of the new Moon, the night sky does not show the Moon. Unless an astronomer wishes to study the Moon, these dark nights are the best time for viewing other objects. Astronomers usually view the night skies during two-week periods every two weeks.

* The Moon actually orbits Earth once every 27.3 days relative to the stars. The 29.5-day cycle is relative to the Sun and is due to the motion of the Earth–Moon system as it revolves about the Sun.

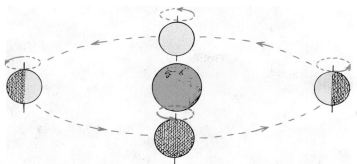

FIGURE 26.27
The Moon spins about its own polar axis just as often as it circles Earth. So as the Moon circles Earth, it spins so that the same side (shown in yellow) always faces Earth. In each of the four successive positions shown here, the Moon has spun one-quarter of a turn.

Why One Side Always Faces Us

The first images of the back side of the Moon were taken by the unmanned Russian spacecraft *Lunik 3* in 1959. The first human witnesses of the Moon's back were *Apollo 8* astronauts, who orbited the Moon in 1968. From Earth, we see only a single lunar side. The familiar facial features of the "man in the Moon" are always turned toward us on Earth. Does this mean that the Moon doesn't spin about its axis as Earth does daily? No, but, relative to the stars, the Moon in fact does spin, although quite slowly—about once every 27 days. This monthly rate of spin matches the rate at which the Moon revolves about Earth. This explains why the same side of the Moon always faces Earth (Figure 26.27). This matching of monthly spin rate and orbital revolution rate is not a coincidence. After you answer the following Checkpoint, we'll explore why.

> **CHECKPOINT**
>
> **A friend says that the Moon does not spin about its axis, and evidence for a nonspinning Moon is the fact that its same side always faces Earth. What do you say?**
>
> **Was this your answer?**
> Place a quarter and a penny on a table. Pretend the quarter is Earth and the penny the Moon. Keeping the quarter fixed, revolve the penny around it in such a way that Lincoln's head is always pointed to the center of the quarter. Ask your friend to count how many rotations the penny makes in one revolution (orbit) around the quarter. He or she will see that it rotates once with each revolution. The key concept is that the Moon takes the same amount of time to complete one rotation as it does to revolve around Earth.

Think of a compass needle that lines up with a magnetic field. This lineup is caused by a *torque*—a "turning force with leverage" (like that produced by the weight of a child at the end of a seesaw). The compass needle on the left in Figure 26.28 rotates because of a pair of torques. The needle rotates counter-clockwise until it aligns with the magnetic field. In a similar manner, the Moon aligns with Earth's gravitational field.

We know from the law of universal gravitation that gravity weakens with the inverse square of distance, so the side of the Moon nearer to Earth is gravitationally pulled more than the farther side. This stretches the Moon out slightly toward a football shape. (The Moon does the same to Earth and gives us tides.) If its long axis doesn't line up with Earth's gravitational field, a torque acts on it as shown in Figure 26.29. Like a compass in a magnetic field, it turns into alignment. So the Moon lines up with Earth in its monthly orbit. One hemisphere always faces us. It's interesting to note that for many moons orbiting other planets, a single hemisphere faces the planet. We say these moons are "tidally locked."

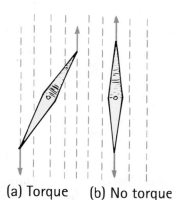

(a) Torque   (b) No torque
FIGURE 26.28
(a) When the compass needle is not aligned with the magnetic field (dashed lines), the forces represented by the blue arrows at either end produce a pair of torques that rotate the needle. (b) When the needle is aligned with the magnetic field, the forces no longer produce torques.

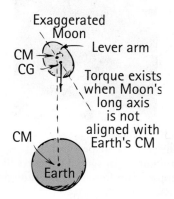

FIGURE 26.29
When the long axis of the Moon is not aligned with Earth's gravitational field, Earth exerts a torque that rotates the Moon into alignment. (CM: center of mass; CG: center of gravity)

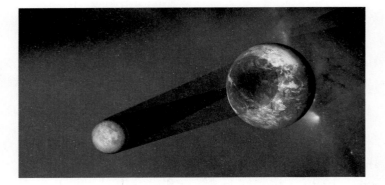

Eclipses

Although the Sun is 400 times as large in diameter as the Moon, it is also 400 times as far away. So from Earth, both the Sun and Moon measure the same angle (0.5°) and appear to be the same size in the sky. This coincidence allows us to see solar eclipses.

Both Earth and the Moon cast shadows when sunlight shines on them. When the path of either of these bodies crosses into the shadow cast by the other, an eclipse occurs. A **solar eclipse** occurs when the Moon's shadow falls on Earth. Because of the large size of the Sun, the rays taper to provide an umbra and a surrounding penumbra, as shown in Figures 26.30 and 26.31.

FIGURE 26.31
Geometry of a solar eclipse. During a solar eclipse, the Moon is directly between the Sun and Earth and the Moon's shadow is cast on Earth. Because of the small size of the Moon and tapering of the solar rays, a total solar eclipse occurs only on a small area of Earth.

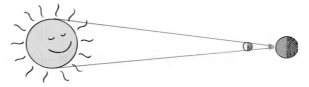

An observer in the umbra part of the shadow experiences darkness during the day—a total eclipse, *totality*. Totality begins when the Sun disappears behind the Moon, and ends when the Sun reappears on the other edge of the Moon. The average time of totality at any location is about 2 or 3 minutes, with a maximum no longer than 7.5 minutes. The eclipse time in any location is brief because of the Moon's motion. During totality, what appears in the sky is an eerie black disk surrounded by the pearly white streams of the corona, as was shown in Figure 26.7. It is an experience one can never forget. With binoculars, the features of the Moon can be seen because they are lit by the sunlight reflected from Earth. Pink flares from the chromosphere may also appear. But great caution is advised when viewing the totality, which must be a totality of 100%. The moment the first edges of the photosphere appear, which is the moment you now have 99.99% totality, is the very moment when you can seriously damage your eyes if you continue to look.* At that moment you have entered the penumbra, where the eclipse is partial. An ideal way to view the partial solar eclipse is to focus the light of the eclipse onto a white surface, as was shown in Figure 26.4. Alternatively, you can view the crescent Sun under the shade of a tree, which casts pinhole images of the Sun onto the ground, as shown in Figure 26.32. Check the map shown in Figure 26.33 to see if a solar eclipse is coming to your area soon. Many solar eclipse enthusiasts travel the world to view this inspiring natural phenomenon.

FIGURE 26.32
The crescent-shaped spots of sunlight are images of the partially eclipsed Sun formed as the sunlight passes through the leaves, which overlap to create image-forming pinholes. On a normal noneclipsed sunny day, these spots beneath a tree are round because the Sun is round. Look for round "Sun balls" on your next sunny day outside.

*People are cautioned not to look at the Sun at the time of a solar eclipse because the brightness and the ultraviolet light of direct sunlight damage the eyes. This good advice is often misunderstood by those who then think that sunlight is more damaging at this special time. However, staring at the Sun when it is high in the sky is harmful whether or not an eclipse occurs. In fact, staring at the bare Sun is more harmful than when part of the Moon blocks it. The reason for special caution at the time of an eclipse is simply that more people are interested in looking at the Sun during this time.

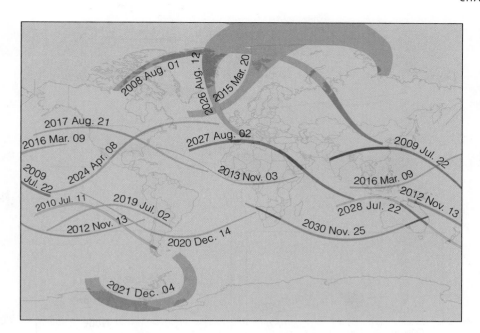

FIGURE 26.33
This map shows the paths of total solar eclipses from 2006 through 2030. More details about these and other future solar eclipses can be found on NASA's eclipse Web site, http://sunearth.gsfc.nasa.gov/eclipse/eclipse.html.

The alignment of Earth, the Moon, and the Sun also produces a **lunar eclipse** when the Moon passes into the shadow of Earth, as shown in Figure 26.34. Usually a lunar eclipse precedes or follows a solar eclipse by two weeks. Just as all solar eclipses involve a new Moon, all lunar eclipses involve a full Moon. They may be partial or total. All observers on the dark side of Earth see a lunar eclipse at the same time. Interestingly enough, when the Moon is fully eclipsed, it is still visible as is shown and discussed in Figure 26.35.

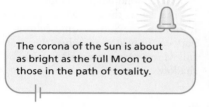

The corona of the Sun is about as bright as the full Moon to those in the path of totality.

FIGURE 26.34
A lunar eclipse occurs when Earth is directly between the Moon and the Sun and Earth's shadow is cast on the Moon.

Why are eclipses relatively rare events? This has to do with the different orbital planes of Earth and the Moon. Earth revolves around the Sun in a flat planar orbit. The Moon similarly revolves about Earth in a flat planar orbit. But the planes are slightly tipped with respect to each other—a 5.2° tilt, as shown in Figure 26.36. If the planes weren't tipped, eclipses would occur monthly. Because of the tip, eclipses occur only when the Moon intersects the Earth–Sun plane at the time of a three-body alignment (Figure 26.37). This occurs about two times per year, which is why there are at least two solar eclipses per year (visible from only certain locations on Earth). Sometimes there are as many as seven solar and lunar eclipses in a year.

CHECKPOINT
1. **Does a solar eclipse occur at the time of a full Moon or a new Moon?**
2. **Does a lunar eclipse occur at the time of a full Moon or a new Moon?**

Were these your answers?
1. A solar eclipse occurs at the time of a new Moon, when the Moon is directly in front of the Sun. Then the shadow of the Moon falls on part of Earth.
2. A lunar eclipse occurs at the time of a full Moon, when the Moon and Sun are on opposite sides of Earth. Then the shadow of Earth falls on the full Moon.

FIGURE 26.35
A fully eclipsed Moon is not completely dark in the shadow of Earth but is quite visible. This is because Earth's atmosphere acts as a lens and refracts light into the shadow region—sufficient light to faintly illuminate the Moon.

FIGURE 26.36
The Moon orbits Earth in a plane tipped 5.2° relative to the plane of Earth's orbit around the Sun. A solar or lunar eclipse occurs only when the Moon intersects the Earth–Sun plane (points A and B) at the precise time of a three-body alignment. Otherwise, the shadows are not aligned, as shown in the lower drawing.

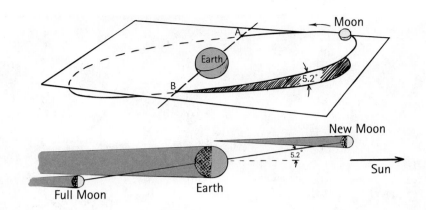

FIGURE 26.37
A total eclipse can occur only when the Moon's orbit intersects with the plane of Earth's orbit, which is the ecliptic. A solar eclipse occurs during the day as the new Moon passes in front of the Sun. A lunar eclipse happens only at night when the full Moon passes through Earth's shadow.

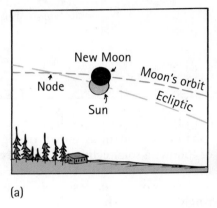

LEARNING OBJECTIVE
Compare and contrast asteroids, Kuiper belt objects, and the Oort cloud.

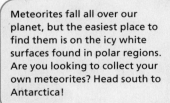

Meteorites fall all over our planet, but the easiest place to find them is on the icy white surfaces found in polar regions. Are you looking to collect your own meteorites? Head south to Antarctica!

26.6 Failed Planet Formation

EXPLAIN THIS If an asteroid and a comet of equal mass were on a collision course toward Earth, the asteroid would be easier to deflect. Why?

In three regions of our solar system, we find the remains of material that failed to collect into planets. These regions are the asteroid belt, the Kuiper belt, and the Oort cloud.

The Asteroid Belt and Meteors

The **asteroid belt** is a collection of rocks located between the orbits of Mars and Jupiter. More than 150,000 asteroids have been cataloged so far, but many more no doubt have yet to be discovered. They come in all shapes and sizes, but the largest asteroid, Ceres, is just under a thousand kilometers in diameter. Although Ceres is large enough to be fairly round, most asteroids are shaped more like a potato, as shown in Figure 26.38.

Evidence suggests that when the solar system was forming, the asteroid belt held much more mass than it does today. Massive Jupiter likely disrupted the orbits of this material, sending it off in many directions, including toward the inner planets and out of the solar system. The two moons of Mars, for example, are thought to be former asteroids. What remains of the asteroid belt is small. If all the presently remaining asteroids were scrunched together, they would make a sphere less than half the size of our Moon.

Jupiter also causes the collisions of asteroids with asteroids, which then break apart into smaller fragments. So rather than building into a planet, this material is slowly ground down and pushed off course. Asteroid fragments known as **meteoroids** frequently find their way to Earth, where they are heated white-hot by friction with the atmosphere. As they descend with a fiery glow, they are called **meteors** (Figure 26.39).

If the meteoroid is large enough, it may survive to reach the surface, where it is called a **meteorite**. Most meteoroids, meteors, and meteorites are from asteroids, but many are also from comets, as we discuss later. Fortunately, smaller meteorites hit us more frequently than larger ones do. About 200 tons of small meteorites strike Earth every day. Every 10,000 years or so we are hit with a meteorite big enough to create a large crater, such as the one shown in Figure 26.40. Every 100 million years or so we are hit with one big enough—about 10 km in diameter—to cause a mass extinction, as occurred 65 million years ago at the end of the Cretaceous period discussed in Chapter 23. Accordingly, one of NASA's goals is to detect up to 90% of all large, near-Earth objects. If we can detect a dangerous space fragment early enough, we can take actions to alter its orbital path sufficiently to avoid impending disaster.

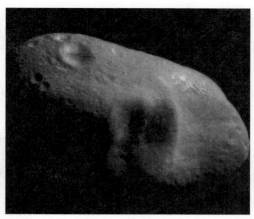

FIGURE 26.38
The asteroid Eros is about 40 km long, and like other small objects in the solar system, it is not spherical.

The Kuiper Belt and Dwarf Planets

Beyond Neptune at a distance from about 30 to 50 AU is a region known as the **Kuiper belt** (pronounced "KI-pur," "KI" as in *kite*, rhymes with *hyper*). The Kuiper belt is occupied by many rocky, ice-covered objects. The most well-known Kuiper belt object is Pluto, which until recently was classified as a planet. Since its discovery in 1930, however, astronomers knew that Pluto was quite different from all the other known planets. For example, Pluto orbits at an angle to the plane of Earth's orbit—the ecliptic. Also, Pluto is quite small, being only one-seventh as massive as our Moon. Then, starting in the 1990s, astronomers began discovering many more Kuiper belt objects, some as large as or larger than Pluto. So in 2006 these Pluto-sized Kuiper belt objects were officially classified as **dwarf planets**. The main reason they do not meet full planet status is that they have yet to accrete all the material in their orbital paths. In the outer edges of our solar system, however, matter is simply too sparse for that to happen. Interestingly, if the Kuiper belt were more dense with material, then these dwarf planets could have served as cores for additional jovian planets. But that never happened, and so the Kuiper belt is another zone of failed planet formation.

FIGURE 26.39
A meteor is produced when a meteoroid, usually about 80 km high, enters Earth's atmosphere. Most are sand-sized grains, which are seen as "falling" or "shooting" stars.

Space probes have yet to visit any of the dwarf planets of the Kuiper belt. Pluto and its moon, Charon, however, are due to be visited by the *New Horizons* spacecraft in 2015. We may have already had a preview, however, when the *Voyager 2* spacecraft took pictures of Neptune's moon Triton. Astronomers now suspect that Triton is a Kuiper belt dwarf planet pulled off course and captured into orbit around Neptune.

The larger Kuiper belt objects, such as Pluto, have a fair amount of inertia and so are not so easily thrown off course. Lighter Kuiper belt objects, however, are thrown off course quite frequently. Sometimes they are thrown toward the Sun, where the added heat and solar wind cause the ice and other volatile materials to be ejected, always in a direction away from the Sun. We see these objects as **comets**, which are characterized by their long and sometimes quite brilliant tails. The unusually bright comet McNaught passed close to the Sun in early 2007. Comets that come from the Kuiper belt tend to have orbital periods of less than 200 years. An example is Comet Halley, which returns to the inner solar system every 76 years—once in an average lifetime (Figure 26.43). Its next scheduled return is in 2061.

FIGURE 26.40
The Barringer Crater in Arizona, made 25,000 years ago by an iron meteorite with a diameter of about 50 m. The crater extends 1.2 km across and reaches 200 m deep.

FIGURE 26.41
This infrared image of Pluto being orbited by its moon, Charon, is fuzzy because of the small size of these bodies and their great distance from Earth.

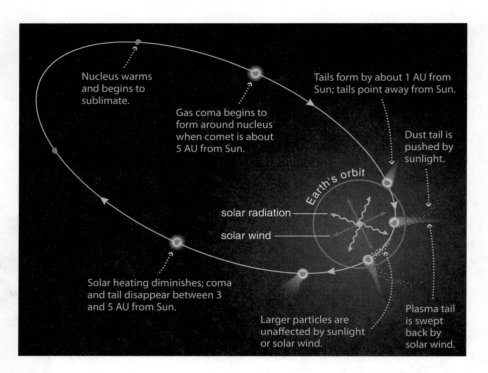

FIGURE 26.42
The comet warms as it gets closer to the Sun and initially develops a coma, which is a halo of gases surrounding the comet nucleus. From this coma arises the tail, which is blown outward by the solar wind. Note how the tail always extends away from the Sun. Most Kuiper belt objects never make this journey and instead remain perpetually frozen within the outer reaches of our solar system.

Comets apparently reside in at least two regions. The first is the Kuiper belt, which lies roughly within the same plane of the solar system. The second region lies much farther out and surrounds our entire solar system—like a cloud.

The Oort Cloud and Comets

As the jovian planets grew, their gravitational pulls became stronger, which made them even more effective at pulling in additional interplanetary debris. Not all debris, however, was pulled fully into the jovian planets. In many cases, a

FIGURE 26.43
Observations of Comet Halley have been recorded for thousands of years. Although it usually provides a brilliant display, its last visit in 1986 was not so spectacular when viewed from Earth. We were ready, though, with space probes that flew close enough to Halley to capture dramatic images of its nucleus.

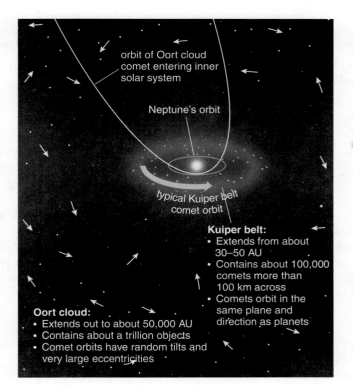

FIGURE 26.44
There are two major sources for comets: the Kuiper belt and the Oort cloud.

chunk of rock or ice just missing a planet was instead whipped around the planet and then flung violently outward in some direction. Over billions of years this created a sphere of far-out objects just barely held to our solar system. We refer to this collection of far-out objects as the **Oort cloud** (*Oort* rhymes with *court*). Evidence suggests that the Oort cloud consists of trillions of objects extending as far out as 50,000 AU, which brings the cloud about a quarter of the way to the nearest star. A few of these objects occasionally fall toward and then around the Sun, where they appear as comets. The orbital periods of comets originating from the Oort cloud are on the order of thousands or even millions of years. They come from nearly any angle.

Whether the comet comes from the Kuiper belt or the Oort cloud, it still has the potential for colliding with a planet. In 1994, Comet Shoemaker-Levy collided spectacularly with Jupiter, as shown in Figure 26.45. Also, the large meteorite that collided with Earth 65 million years ago causing the mass extinction of the dinosaurs may have been a comet.

Most comets usually last only a couple of orbits before they break up. But if the solar system is billions of years old, shouldn't they be depleted by now? This very question led to the idea of the Oort cloud, which provides a continual supply of new comets, replacing those that are destroyed.

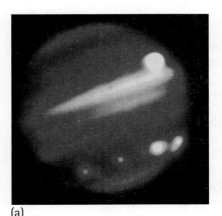

(a)

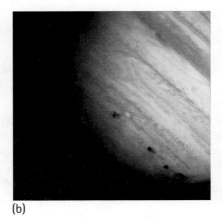

(b)

FIGURE 26.45
Comet Shoemaker-Levy was already broken up into a string of objects just before it collided with Jupiter in 1994. The image in (a) is an infrared view of the collision, which produced much heat as well as visible scars (the black dots), as shown in the photograph in (b).

FIGURE 26.46
When the orbiting Earth intercepts the debris from a comet, we see a meteor shower.

TABLE 26.2		METEOR SHOWER DATA		
Shower Name	Radiant*	Dates	Peak Dates	Meteors per Hour
Quadrantids	Pegasus	Jan. 1–6	Jan. 3	60
Eta Aquarids	Aquarius	May 1–10	May 6	35
Perseids	Perseus	Jul. 23–Aug. 20	Aug. 12	75
Orionids	Orion	Oct. 16–27	Oct. 22	25
Geminids	Gemini	Dec. 7–15	Dec. 13	75

* Meteors appear to radiate from a certain region of the sky, appropriately called a *radiant*. Radiants refer to constellations. See Chapter 27 for more on where the various constellations are located in the night sky.

The tail of the comet leaves behind a wide trail of particles. Each year Earth passes through the remnants of comet tails that create annual meteor showers, as indicated in Table 26.2. Meteor showers are beautiful to watch. Just go outside, look up at the sky, and with good eyesight every minute or so you will see a shooting star. Each streak is a tiny chip of a comet, once so very far away, that has fallen into Earth's neighborhood (Figure 26.46).

CHECKPOINT

Of the asteroid belt, the Kuiper belt, and the Oort cloud:
1. **Which is closest to the Sun?**
2. **Which generates comets?**
3. **Which gives us the most meteorites?**
4. **Which gives us the brightest meteor showers?**
5. **Which consists of fragments that never coalesced into planets?**

Were these your answers?
1. the asteroid belt
2. the Kuiper belt and the Oort cloud
3. the asteroid belt
4. the Kuiper belt and the Oort cloud
5. all of them

For instructor-assigned homework, go to www.masteringphysics.com

SUMMARY OF TERMS (KNOWLEDGE)

Asteroid belt A region between the orbits of Mars and Jupiter that contains small, rocky, planet-like fragments that orbit the Sun. These fragments are called *asteroids* ("small star" in Latin).

Astronomical unit (AU) The average distance between Earth and the Sun; about 1.5×10^8 km (about 9.3×10^7 mi).

Comet A body composed of ice and dust that orbits the Sun, usually in a very eccentric orbit, and that casts a luminous tail produced by solar radiation pressure when close to the Sun.

Dwarf planet A relatively large icy body, such as Pluto, that originated within the Kuiper belt.

Ecliptic The plane of Earth's orbit around the Sun. All major objects of the solar system orbit roughly within this same plane.

Full Moon The phase of the Moon when its sunlit side faces Earth.

Inner planets The four planets orbiting within 2 AU of the Sun, including Mercury, Venus, Earth, and Mars—all rocky and known as the *terrestrial* planets.

Kuiper belt (pronounced "KI-pur") The disk-shaped region of the sky beyond Neptune populated by many icy bodies and a source of short-period comets.

Lunar eclipse The phenomenon in which the shadow of Earth falls on the Moon, producing the relative darkness of the full Moon.

Meteor The streak of light produced by a meteoroid burning in Earth's atmosphere; a "shooting star."

Meteorite A meteoroid, or a part of a meteoroid, that has survived passage through Earth's atmosphere to reach the ground.

Meteoroid A small rock in interplanetary space, which can include a fragment of an asteroid or comet.

Moon phases The cycles of change of the "face" of the Moon, changing from *new*, to *waxing*, to *full*, to *waning*, and back to *new*.

Nebular theory The idea that the Sun and planets formed together from a cloud of gas and dust, a *nebula*.

New Moon The phase of the Moon when darkness covers the side facing Earth.

Oort cloud The region beyond the Kuiper belt populated by trillions of icy bodies and a source of long-period comets.

Outer planets The four planets orbiting beyond 2 AU of the Sun, including Jupiter, Saturn, Uranus, and Neptune—all gaseous and known as the *jovian* planets.

Planets The major bodies orbiting the Sun that are massive enough for their gravity to make them spherical and small enough to avoid having nuclear fusion in their cores. They also have successfully cleared all debris from their orbital paths.

Solar eclipse The phenomenon in which the shadow of the Moon falls on Earth, producing a region of darkness in the daytime.

Sunspots Temporary, relatively cool and dark regions on the Sun's surface.

READING CHECK QUESTIONS (COMPREHENSION)

26.1 The Solar System and Its Formation

1. How many known planets are in our solar system?
2. What dwarf planet was downgraded from planetary status in 2006?
3. How are the outer planets different from the inner planets aside from their location?
4. Why does a nebula spin faster as it contracts?
5. According to the nebular theory, did the planets start forming before or after the Sun ignited?

26.2 The Sun

6. What happens to the amount of the Sun's mass as it "burns"?
7. What are sunspots?
8. What is the solar wind?
9. How does the rotation of the Sun differ from the rotation of a solid body?
10. What is the age of the Sun?

26.3 The Inner Planets

11. Why are the days on Mercury very hot and the nights very cold?
12. What two planets are evening or morning "stars"?
13. Why is Earth called "the blue planet"?
14. What gas makes up most of the Martian atmosphere?
15. What evidence tells us that Mars was at one time wetter than it presently is?

26.4 The Outer Planets

16. What surface feature do Jupiter and the Sun have in common?
17. Which move faster: Saturn's inner rings or the outer rings?
18. How tilted is Uranus's axis?
19. Why is Neptune bluer than Uranus?

26.5 Earth's Moon

20. Why doesn't the Moon have an atmosphere?
21. Where is the Sun located when you view a full Moon?
22. Where are the Sun and the Moon located at the time of a new Moon?
23. Why don't eclipses occur monthly, or nearly monthly?
24. How does the Moon's rate of rotation about its own axis compare with its rate of revolution around Earth?

26.6 Failed Planet Formation

25. Between the orbits of what two planets is the asteroid belt located?
26. What is the difference between a meteor and a meteorite?
27. What is the Kuiper belt?
28. What is the Oort cloud, and what is it noted for?
29. What is a falling star?
30. What causes comet tails to point away from the Sun?

ACTIVITIES (HANDS-ON APPLICATION)

31. Find a Ping-Pong ball on the next clear day when the Moon is out. Hold the Ping-Pong ball with your arm stretched out toward where the Moon is so that the ball overlaps the Moon. Look carefully at how the ball is lit by the Sun. Notice that this is the same way the Moon is lit by the Sun! For an example, see the photograph accompanying Exercise 67. To see the different phases that the Moon would have if it were elsewhere in the sky, move your Ping-Pong ball around. Note as you bring the ball closer to the Sun, the crescent on the ball gets thinner. The same thing happens with the Moon. This activity is a great way of really experiencing the roundness of the Moon.

32. Simulate the lunar phases. Insert a pencil into a Styrofoam ball. This will be the Moon. Position a lamp (representing the Sun) in another room near the doorway. Hold the ball in front and slightly above yourself. Slowly turn yourself around, keeping the ball in front of you as you move. Observe the patterns of light and shadow on the ball. Relate this to the phases of the Moon.

33. When viewed from the North Pole, Earth spins counterclockwise, which is toward the east. This means that the stars appear to move in the opposite direction, which is toward the west. This is just like when you're sitting in a train that begins moving eastward. The only way you know that you're moving eastward is because things outside your window give the appearance of moving westward. Just as Earth spins counterclockwise, the Moon revolves around us counterclockwise, though not as fast as we spin. Look where the Moon is located one night at, say, 11:00. Look for the Moon the next night at the same time, and you'll see that it has moved eastward (a counterclockwise direction) from where it was on the previous night.

34. The crescent Moon always points toward the Sun. You can use this fact to estimate your latitude. Down by the equator (0° latitude), the setting crescent Moon lies flat on the horizon, while up close to the North Pole (90° latitude), the crescent Moon stands on its end. Deviations to this can arise because of Earth's 23° tilt and because the Moon's orbit lies 5° outside the ecliptic. Nonetheless, those who live in Alaska see a crescent Moon that is much more upright than those who live in Hawaii. With this in mind, next time you see the crescent Moon close to the horizon, look carefully at its angle and try correlating that angle to your local latitude.

35. The planets of our solar system orbit in roughly the same plane, which is the plane of our solar system. We can identify the plane of our solar system in the night sky by noting the positions of the planets, which, from our point of view on Earth, appear in a roughly straight line relative to one another. This straight line, which is a cross-section of our solar system, always intersects with the Sun and often with the position of our Moon. On a clear evening after sunset, several planets can often be seen in the western sky forming a line that points directly toward the Sun. Upon finding this alignment, you are looking directly at the plane of our solar system. Check it out! How might this line look different if viewed from the North Pole versus the equator?

36. The dates of meteor showers are rather predictable as listed in Table 26.2. How intense the shower might be, however, is still somewhat of a guess. So keep your eye to the nighttime sky for these meteor showers. While doing so, review in your mind the distinctions among meteoroids, meteors, meteorites, and comets. Also notable in our nighttime skies are the lunar eclipses, which can seen by anyone who happens to be on the nighttime side of Earth during the eclipse (assuming skies are clear). Here are the dates for upcoming lunar eclipses viewable in North America.

Date	Type
Dec. 10, 2011	Total
Jun. 4, 2012	Partial
Apr. 15, 2014	Total
Oct. 8, 2014	Total
Sep. 28, 2015	Total
Jan. 21, 2019	Total

THINK AND SOLVE (MATHEMATICAL APPLICATION)

37. Knowing that the speed of light is 300,000 km/s, show that it takes about 8 min for sunlight to reach Earth.

38. How many days does sunlight take to travel the 50,000 AU from the Sun to the outer reaches of the Oort cloud?

39. The light-year is a standard unit of distance used by astronomers. It is the distance light travels in one Earth year. In units of light-years, what is the approximate diameter of our solar system, including the outer reaches of the Oort cloud? (Assume that 1 light-year equals 63,000 AU.)

40. The nearest star to our Sun is Alpha Centauri, which is about 4.4 light-years away. Assume that it too has an Oort cloud about 1.6 light-years in diameter. Show that there is enough space between us and it to fit about 1.75 solar systems.

41. If the Sun were the size of a beach ball, Earth would be the size of a green pea 110 m away. Show that the nearest star, Alpha Centauri (4.4 light-years away), would be about 30,000 km distant. (Hint: Find the distance to Alpha Centauri in units of AU.)

THINK AND RANK (ANALYSIS)

42. Rank these planets in order from longest to shortest year: (a) Mercury, (b) Venus, and (c) Earth.

43. Rank these planets in order of increasing number of moons: (a) Mars, (b) Venus, and (c) Earth.

44. Rank in order of increasing average density: (a) Jupiter, (b) Saturn, and (c) Earth.

45. Rank in order of increasing pressure at the center of each planet: (a) Jupiter, (b) Saturn, and (c) Earth.

46. Rank in order of decreasing number of people who have seen a: (a) solar eclipse, (b) lunar eclipse, and (c) new Moon.

47. Rank in order of increasing average distance from the Sun: (a) Kuiper belt objects, (b) asteroids, and (c) Oort cloud objects.

EXERCISES (SYNTHESIS)

48. According to nebular theory, what happens to a nebula as it contracts under the force of gravity?

49. What happens to the shape of a nebula as it contracts and spins faster?

50. A TV screen is normally light gray when not illuminated. How is the darkness of sunspots similar to the black parts of an image on a TV screen?

51. When a contracting ball of hot gas spins into a disk shape, it cools. Why?

52. If Earth didn't spin on its axis but still revolved around the Sun, how long would an Earth day be?

53. If Earth didn't spin on its axis but still revolved around the Sun, would the Sun set on the eastern or western horizon or not at all?

54. The greenhouse effect is very pronounced on Venus but doesn't exist on Mercury. Why?

55. What is the cause of winds on Mars (and also on almost every other planet)?

56. Why is there so little wind on the surface of Venus?

57. If Venus were somehow transported into the habitable zone, would conditions once again become favorable for life?

58. Mercury and Venus are never seen at night straight up toward the top of the sky. Why not?

59. What is the major difference between the terrestrial and jovian planets?

60. What does Jupiter have in common with the Sun that the terrestrial planets don't? What differentiates Jupiter from a star?

61. When it comes to celestial bodies, such as planets and stars, why doesn't a larger size necessarily mean a larger mass?

62. Why are the seasons on Uranus different from the seasons on any other planet?

Uranus

63. Earth rotates much faster than Venus. How does the giant impact theory of the Moon account for this fact?

64. Why are many craters evident on the surface of the Moon but not on surface of Earth?

65. Why is there no atmosphere on the Moon? Defend your answer.

66. Is the fact that we see only one side of the Moon evidence that the Moon spins or that it doesn't spin? Defend your answer.

67. Photograph (a) shows the Moon partially lit by the Sun. Photograph (b) shows a Ping-Pong ball in sunlight. Compare the positions of the Sun in the sky when each photograph was taken. Do the photos support or refute the claim that they were taken on the same day? Defend your answer.

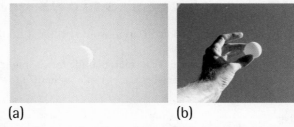

(a)   (b)

68. We always see the same face of the Moon because the rotation of the Moon on its axis matches the rate at which it revolves around Earth. Does it follow that an observer on the Moon always sees the same face of Earth?

69. If we never see the back side of the Moon, would an observer on the back side of the Moon ever see Earth?

70. In what alignment of the Sun, the Moon, and Earth does a solar eclipse occur?

71. In what alignment of the Sun, the Moon, and Earth does a lunar eclipse occur?

72. What does the Moon have in common with a compass needle?

73. If you were on the Moon and you looked up and saw a full Earth, would it be nighttime or daytime on the Moon?

74. If you were on the Moon and you looked up and saw a new Earth, would it be nighttime or daytime on the Moon?

75. Earth takes 365.25 days to revolve around the Sun. If Earth took this same amount of time to spin on its axis, what might we note about the Sun's position in the sky?

76. Do astronomers make stellar observations during the full Moon part of the month or during the new Moon part of the month? Does it make a difference?

77. Nearly everybody has witnessed a lunar eclipse, but relatively few people have seen a solar eclipse. Why?

78. Because of Earth's shadow, the partially eclipsed Moon looks like a cookie with a bite taken out of it. Explain with a sketch how the curvature of the bite indicates the size of Earth relative to the size of Moon. How does the tapering of the Sun's rays affect the curvature of the bite?

Use the following illustration for Exercises 79–82.

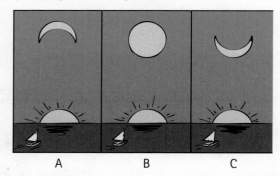

A   B   C

79. Which of the three orientations of the Moon at sunset is most correct?

80. Assuming the illustrations depict a sunset, within 24 hours of when this scene is depicted will the Moon appear to be farther from or closer to the Sun?

81. Is the sailboat sailing at a location closer to the North Pole or closer to the equator? How can you tell?

82. Where and how would the Moon be positioned if the scenes were close to the North Pole?

83. In what sense is Pluto a potential comet?

84. Smaller chunks of asteroids are sent hurling toward Earth much more frequently than larger chunks of asteroids. Why?

85. Why are meteorites so much more easily found in Antarctica than in other continents?

86. A meteor is visible only once, but a comet may be visible at regular intervals throughout its lifetime. Why?

87. What would be the consequence of a comet's tail sweeping across Earth?

88. Chances are about 50–50 that in any night sky there is at least one visible comet that has not been discovered. This keeps amateur astronomers busy looking night after night, for the discoverer of a comet gets the honor of having it named for him or her. With this high probability of comets in the sky, why aren't more of them found?

DISCUSSION QUESTIONS (EVALUATION)

89. Project what human civilization would be like if Earth had no Moon.

90. What are the chances that microbial life forms might one day be found elsewhere in our solar system? How much effort should we spend on searching for such life forms, and what precautions should we take upon such a discovery?

READINESS ASSURANCE TEST (RAT)

If you have a good handle on this chapter, if you really do, then you should be able to score 7 out of 10 on this RAT. If you score less than 7, you need to study further before moving on.

Choose the BEST answer to each of the following.

1. The Sun contains what percentage of the solar system's mass?
   (a) about 35%
   (b) 85%
   (c) the percentage varies over time
   (d) over 99%

2. The solar system is like an atom in that both
   (a) are governed principally through the electric force.
   (b) consist of a central body surrounded by objects moving in elliptical paths.
   (c) are composed of plasma.
   (d) are mainly empty space.

3. The nebular theory is based on the observation that the solar system
   (a) follows patterns indicating that it formed progressively from physical processes.
   (b) has a structure much like an atom.
   (c) is highly complex and appears to have been built by chaotic processes.
   (d) appears to be very old.

4. When a contracting hot ball of gas spins into a disk shape, it cools faster due to
   (a) increased radiation transfer.
   (b) increased surface area.
   (c) decreased insulation.
   (d) increased convection currents.
   (e) eddy currents.

5. Each second, the burning Sun's mass
   (a) increases.
   (b) remains unchanged.
   (c) decreases.

6. Compared to your weight on Earth, your weight on Jupiter would be about
   - (a) 3000 times as much.
   - (b) half as much.
   - (c) 3 times as much.
   - (d) 300 times as much.
   - (e) 100 times as much.

7. When the Moon assumes its characteristic thin crescent shape, the position of the Sun is
   - (a) almost directly in back of the Moon.
   - (b) almost directly behind Earth, so that Earth is between the Sun and the Moon.
   - (c) at right angles to the line between the Moon and Earth.

8. When the Sun passes between the Moon and Earth, we have
   - (a) a lunar eclipse.
   - (b) a solar eclipse.
   - (c) met our end.

9. Asteroids orbit
   - (a) the Moon.
   - (b) Earth.
   - (c) the Sun.
   - (d) all of the above
   - (e) none of the above

10. With each pass of a comet about the Sun, the comet's mass
   - (a) remains virtually unchanged.
   - (b) actually increases.
   - (c) is appreciably reduced.

Answers to RAT

1. d, 2. d, 3. a, 4. b, 5. c, 6. c, 7. a, 8. c, 9. c, 10. c

CHAPTER 27
Stars and Galaxies

ON A moonless night the unaided eye sees not more than 3000 stars, horizon to horizon. Many more stars become visible with a telescope, especially when the telescope is pointed toward a cloudlike band of light that stretches north to south. The ancient Greeks called this diffuse band of light the Milky Way.

Today we know the Milky Way to be a vast collection of more than 100 billion stars. When viewed from afar, all of these stars—along with our own star, the Sun—appear as a great swirl of stars known as a *galaxy*.

In this chapter we will explore the nature of stars—how they form, how they die, and how they are organized within galaxies. We will explore how there are many different types of stars, just as there are many different types of galaxies. We begin by taking a closer look at the stars visible to our unaided eyes. Compared to the size of the entire Milky Way galaxy, these thousands of stars form our immediate neighborhood.

27.1 Observing the Night Sky

EXPLAIN THIS When can winter constellations be seen in the summer?

Early astronomers divided the night sky into groups of stars, called *constellations*, such as the group of seven stars we now call the Big Dipper. The names of the constellations today carry over mainly from the names assigned to them by early Greek, Babylonian, and Egyptian astronomers. The Greeks, for example, included the stars of the Big Dipper in a larger group of stars that outlined a bear. The large constellation Ursa Major (the Great Bear) is illustrated in Figure 27.1. The groupings of stars and the significance given to them have varied from culture to culture. To some cultures, the constellations stimulated storytelling and the making of great myths; to other cultures, the constellations honored great heroes, such as Hercules and Orion; to yet others, they served as navigational aids for travelers and sailors. To many cultures, including the African Bushmen and Masai, the constellations provided a guide for planting and harvesting crops because they were seen to move in the sky in concert with the seasons. Charts of this periodic movement became some of the first calendars. We can see in Figure 27.2 why the background of stars varies throughout the year.

The stars are at different distances from Earth. However, because all the stars are so far away, they appear equally remote. This illusion led the ancient Greeks and others to conceive of the stars as being attached to a gigantic sphere surrounding Earth, called the **celestial sphere**. Though we know it is imaginary, the celestial sphere is still a useful construction for visualizing the motions of the stars (Figure 27.3).

The stars appear to turn around an imaginary north–south axis once every 24 hours. This is the *diurnal motion* of the stars. Diurnal motion is easy to visualize as a rotation of the celestial sphere from east to west. This motion is a consequence of the daily counterclockwise rotation of Earth on its axis. When we speak of the diurnal motion of the stars, we are referring to the motions of celestial objects as a whole; this motion does not change the relative positions of objects. Figure 27.4 shows the diurnal motion of the stars making up the Big Dipper. Time-exposure photographs show that the Big Dipper appears to move in circles around the North Star (Figure 27.5). The North Star appears stationary as the celestial sphere rotates because it lies very close to the projection of Earth's rotational axis.

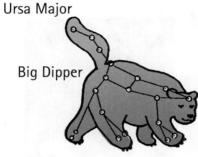

Ursa Major

Big Dipper

FIGURE 27.1
The constellation Ursa Major, the Great Bear. The seven stars in the tail and back of Ursa Major form the Big Dipper.

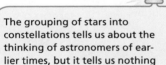

The grouping of stars into constellations tells us about the thinking of astronomers of earlier times, but it tells us nothing about the stars themselves.

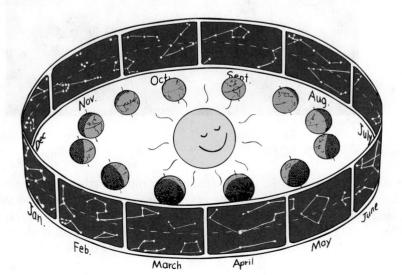

FIGURE 27.2
The night side of Earth always faces away from the Sun. As Earth circles the Sun, different parts of the universe are seen in the nighttime sky. Here the circle, representing 1 year, is divided into 12 parts—the monthly constellations. The stars in the nighttime sky change in a yearly cycle.

FIGURE 27.3
The celestial sphere is an imaginary sphere to which the stars are attached. We see no more than half of the celestial sphere at any given time. The point directly over our heads at any time is called the *zenith*.

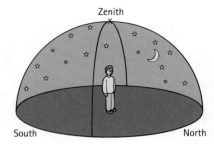

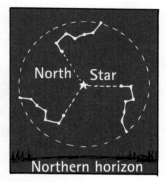

FIGURE 27.4
The pair of stars in the end of the Big Dipper's bowl point to the North Star. Earth rotates about its axis and therefore about the North Star, so over a 24-hour period the Big Dipper (and other surrounding star groups) makes a complete revolution.

In addition to the diurnal motion of the sky, there is *intrinsic motion* of certain bodies that change their positions with respect to the stars. The Sun, the Moon, and planets, called "wanderers" by ancient astronomers, appear to migrate across the fixed backdrop of the celestial sphere. Interestingly, the stars themselves have intrinsic motion. They are so far away, however, that this motion is not apparent on the time scale of a human life. As shown in Figure 27.6, over thousands of years, the intrinsic movement of stars results in new patterns of stars. In other words, the constellations we see today are quite different from the ones that appeared to our earliest ancestors.

CHECKPOINT
1. **Which celestial bodies appear fixed relative to one another, and which celestial bodies appear to move relative to the others?**
2. **What are two types of observed motions of the stars in the sky?**

Were these your answers?
1. The stars appear fixed as they move across the sky. The Sun, the Moon, and planets move relative to one another as they move across the backdrop of the stars.
2. One type of motion of the stars is their nightly rotation as if they were painted on a rotating celestial sphere; this is due to Earth's rotation on its own axis. Stars also appear to undergo a yearly cycle around the Sun because of Earth's revolution about the Sun.

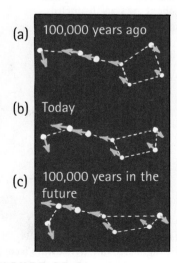

FIGURE 27.6
The present pattern of the Big Dipper is temporary. Here we can see its pattern (a) 100,000 years ago; (b) as it appears at present; and (c) as it will appear in the future, about 100,000 years from now.

FIGURE 27.5
A time exposure of the northern night sky.

Some stars on the celestial sphere are actually much farther away than others from Earth. Astronomers measure the vast distances between Earth and the stars using *light-years*. One **light-year** is the distance that light travels in 1 year, nearly 10 trillion km. For perspective, the diameter of Neptune's orbit is about 0.001 light-year. The distance from the Sun to the outer edges of the Oort cloud (the full radius of our solar system) is about 0.8 light-year. The star closest to our Sun,

Proxima Centauri, is about 4.2 light-years away. The diameter of the Milky Way galaxy is about 100,000 light-years. The next closest major galaxy, the Andromeda galaxy, is about 2.3 million light-years distant. Figure 27.7 shows the distances to the seven stars making up the Big Dipper in light-years.

The speed of light (as we know from Chapter 11) is 3×10^8 m/s. Although this is very fast, it nevertheless takes light appreciable time to travel large distances. And so when you see the light emitted by a very distant object, you are actually seeing the light it emitted long ago—you are looking back in time. Consider the example of Supernova 1987a (a supernova is the explosion of a star, as you will learn more about in Section 27.4). This supernova occurred in a galaxy 190,000 light-years from Earth. Although we witnessed the supernova in 1987, the light from this explosion took 190,000 years to reach our planet, so the explosion actually occurred 190,000 years earlier. "News" of the supernova took 190,000 years to reach Earth!

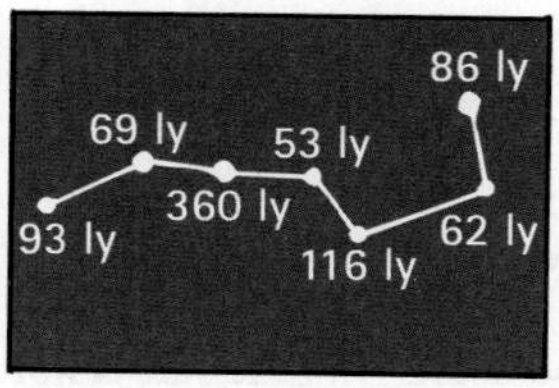

FIGURE 27.7

Interestingly, the seven stars of the Big Dipper are at varying distances from Earth. Note their varying distances in light-years (ly).

27.2 The Brightness and Color of Stars

EXPLAIN THIS How do astronomers gauge the temperature of a star?

Stars are born from clouds of interstellar dust with roughly the same chemical composition as the Sun (see Chapter 26). About three-fourths of the interstellar material from which a star forms is hydrogen; one-fourth is helium; and no more than 2% of the material from which a star forms consists of heavier chemical elements. Stars shine brilliantly for millions or billions of years because of the nuclear fusion reactions that occur in their cores. And all stars, the Sun included, ultimately exhaust their nuclear fuel and die. Yet not all stars are the same. If you look into the night sky, you will see that stars differ in two very visible ways: brightness and color.

Brightness relates to how much energy a star produces. However, although a star's brightness is related to its energy output, its brightness also depends on how far away it is from Earth. Recall from earlier chapters the inverse-square law: the intensity of light diminishes as the reciprocal of the square of the distance from the source. For example, the stars Betelgeuse and Procyon appear equally bright even though Betelgeuse emits about 5000 times as much light as Procyon. The reason? Procyon is much closer to Earth than Betelgeuse is.

To avoid confusing brightness with energy output, astronomers clearly distinguish between apparent brightness and the more important property, *luminosity*. Apparent brightness is the brightness of a star as it appears to our eyes. Luminosity, on the other hand, is the total amount of light energy that a star emits into space. Luminosity is usually expressed relative to the Sun's luminosity, which is noted L_{Sun}. For example, the luminosity of Betelgeuse is 38,000 L_{Sun}. This indicates that Betelgeuse is a very luminous star emitting about 38,000 times as much energy each second into space as the Sun. On the other hand, Proxima Centauri is quite dim, with a luminosity of 0.00006 L_{Sun}. Astronomers have measured the luminosity of many stars and found that stars vary greatly in this respect. The Sun is somewhere in the middle of the luminosity range. The most luminous stars are about a million times as luminous as the Sun, while the dimmest stars produce about 1/10,000 as much energy per second as the Sun.

Besides apparent brightness, a star's color is another property that varies widely among stars. Figure 27.8, a photograph of stars taken with the Hubble Telescope, shows this—stars come in every color of the rainbow. A star's color directly tells you about its surface temperature—for example, a blue star is hotter than a yellow star, and a yellow star is hotter than a red star. In fact,

LEARNING OBJECTIVE
Distinguish between a star's apparent brightness and luminosity, and identify its temperature by its color.

FIGURE 27.8
Most of the stars in this photograph are approximately the same distance—2000 light-years—from the center of the Milky Way galaxy. A star's color indicates its surface temperature—a blue star is hotter than a yellow star, and a yellow star is hotter than a red star. This photo was taken by the Hubble Telescope.

astronomers use color to measure the temperatures of stars. Why is it that a star's color corresponds to its temperature?

Radiation Curves of Stars

As you learned in Chapters 7 and 11, all objects with a temperature emit energy in the form of electromagnetic radiation. The peak frequency $\overline{f}$ of the radiation is directly proportional to the absolute temperature T of the emitter:

$$\overline{f} \sim T$$

Stars have different colors because they emit different frequencies of electromagnetic waves in the visible range. Our eyes sense different frequencies of visible radiation as different colors. Figure 27.9 shows the radiation curves, which are graphs of the intensity of emitted radiation versus wavelength for two stars of the same size with different temperatures. The radiation curves show that the hotter a star is, the shorter the wavelength of its peak frequency and the bluer it looks. So the blue stars in the night sky have higher temperatures than the red ones. The Sun, for example, with its approximately 5800 K surface temperature, emits most strongly in the middle of the visible spectrum and so appears yellow. Betelgeuse, on the other hand, appears red because of its cooler surface temperature (about 3400 K). Betelgeuse emits more red light than blue light.

FYI Interestingly, the Earth's atmosphere is transparent to a narrow band of light centered upon the Sun's peak frequency. Creatures here on Earth's surface evolved to be sensitive to these most abundant frequencies, which we now perceive as visible light. Within the spectrum of visible light we are most sensitive to a greenish-yellow, which is why many emergency vehicles are commonly painted greenish yellow.

FIGURE 27.9
These radiation curves for stars of the same size and different surface temperatures show two important facts: (1) hotter stars emit radiation with higher average frequency than cooler stars, and (2) hotter stars emit more radiation per unit surface area at every frequency than cooler stars.

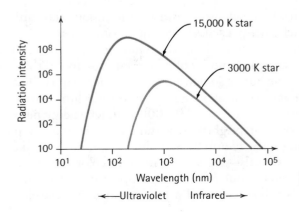

Notice also from Figure 27.9 that the hotter a star is, the more radiant energy it emits. Thus we see that hot blue stars are more luminous than cooler red stars of the same size.

CHECKPOINT

The temperature of Sirius is about 9400 K. What color is this star—and why?

Was this your answer?
Sirius has a slightly blue color. It emits more blue light than red light because of its high surface temperature.

27.3 The Hertzsprung–Russell Diagram

EXPLAIN THIS When is a cool star larger than a hot star?

LEARNING OBJECTIVE
Describe the relationship between stellar luminosity and surface temperature as portrayed in the Hertzsprung–Russell diagram.

When you compare the luminosity of stars to their temperature, interesting patterns emerge. Early in the 20th century, Danish astronomer Ejnar Hertzsprung and American astronomer Henry Norris Russell did just this. They produced a diagram known as the **Hertzsprung–Russell diagram**, or **H–R diagram**, which is of key importance in astronomy (Figure 27.10). The H–R diagram is a plot of the luminosity versus surface temperature of stars. Luminous stars are near the top of the diagram, and dim stars are toward the bottom. Hot bluish stars are toward the left side of the diagram and cool reddish stars are toward the right side.

The H–R diagram shows several distinct regions of stars. Most stars are plotted on the band that stretches diagonally across the diagram. This band is called the **main sequence**. Stars on the main sequence, including our Sun, generate energy by fusing hydrogen to helium. As we would expect, the hottest main-sequence stars are the brightest and bluest stars and the coolest main-sequence stars are the most dim and red stars. Take a moment to locate the Sun on the H–R diagram. Can you see that the Sun is a roughly average main-sequence star in terms of its luminosity and temperature?

Toward the upper right of the diagram is a distinct group of stars—the **giant stars**. These stars clearly do not follow the pattern of the hydrogen-burning main-sequence stars. Because these stars are red, we know they must have low surface temperatures. If they were main-sequence stars, the giants would be dim. Yet notice how high the giants are on the luminosity scale—they are very bright. The fact that the giants are both much cooler and much brighter than the Sun tells us that these stars must also be much larger than the Sun. (Hence the name *giant*.) Above the giants on the H–R diagram are a few rare stars, the *supergiants*. The supergiants are even larger and brighter than the giants. As you will see in the next section, the giants and supergiants are stars nearing the end of their lives because the hydrogen fuel in their cores is running out.

Toward the lower left are some stars that are so dim they cannot be seen with the unaided eye. The surfaces of these stars can be hotter than the Sun, which makes them blue or white. Yet their luminosities are quite low—on the order of 0.1 L_{Sun} to 0.0001 L_{Sun}. To be so hot and radiate so little light, these stars must be very small—they are called the **white dwarfs**. White dwarfs are

Because the giants and supergiants are so luminous, they are easy to see in the night sky even if they are not close to Earth. You can often identify them by their reddish color.

FYI The H–R diagram is to astrophysicists what the periodic table is to chemists—an extremely important tool. A star's position on the H–R diagram can reveal its age. The age of our galaxy can be estimated by looking at the positions of our oldest stars and their white-dwarf remnants.

FIGURE 27.10
The H–R diagram shows a star's surface temperature on the horizontal axis and its luminosity on the vertical axis. The giant and supergiant stars shown here as circles are not drawn to scale. The red supergiant, Antares, for example, is so large that, if drawn to scale, it would reach the ceiling of your classroom. Interestingly, although the radius of Antares is 700 times that of our Sun, its mass is only about 15 times greater. So, although Antares is much larger, it is also much less dense.

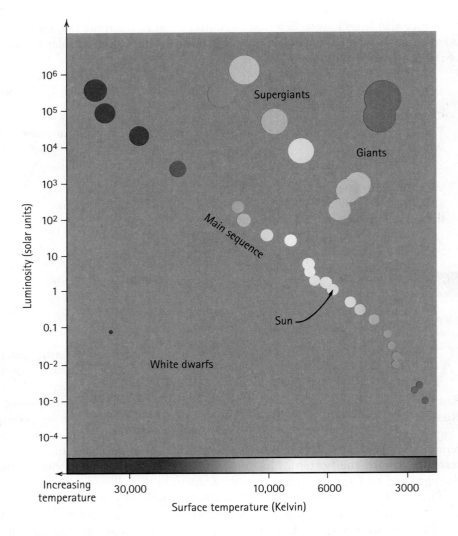

typically the size of Earth or even smaller, yet they have mass comparable to the Sun. The density (or mass per volume) of a white dwarf is thus extremely high—about a million g/cm^3. For comparison, gold has a density of about 19 g/cm^3, while the average density of Earth is about 5.4 g/cm^3. As you will learn in the next section, white dwarfs are dead stars, the remnants of stars that have exhausted their nuclear fuel.

CHECKPOINT
1. What characteristic do all main-sequence stars share?
2. Giants have cool surface temperatures yet are highly luminous. Does this mean that the frequency of light emitted by a giant does not depend on its surface temperature as described by Figure 27.9?

Were these your answers?
1. All main-sequence stars generate energy by the nuclear fusion of hydrogen to helium.
2. No, radiation curves hold for a giant star as for any other radiating body. Giants do have a relatively low energy output per unit surface area; they are highly luminous only because they are very large.

ASTROLOGY

There is more than one way to view the cosmos and its processes—astronomy is one and astrology is another. Astrology is a belief system that began more than 2000 years ago in Babylonia. Astrology has survived nearly unchanged since the second century AD, when some revisions were made by Egyptians and Greeks who believed that their gods moved heavenly bodies to influence the lives of people on Earth. Astrology today holds that the position of Earth in its orbit around the Sun at the time of birth, combined with the relative positions of the planets, has some influence over one's personal life. The stars and planets are said to affect such personal things as one's character, marriage, friendships, wealth, and death.

Could the force of gravity exerted by these celestial bodies be a legitimate factor in human affairs? After all, the ocean tides are the result of the Moon's and Sun's positions, and the gravitational pulls between the planets perturb one another's orbits. Because slight variations in gravity produce these effects, might not slight variations in the planetary positions at the time of birth affect a newborn? If the influence of stars and planets is gravitational, then credit must also be given to the effect of the gravitational pull between the newborn and Earth itself. This pull is enormously greater than the combined pull of all the planets, even when lined up in a row (as occasionally happens). The gravitational influence of the hospital building

on the newborn far exceeds that of the distant planets. So planetary gravitation cannot be an underlying agent for astrology.

Astrology is not a science, because it doesn't change with new information as science does, nor are its predictions borne out by fact. Rather, its predictions depend on coincidence and also on the tendency of many people to seek external explanations for their fates or personal behaviors. Astrological beliefs are built on anecdotal evidence that is neither reproducible nor testable. Astrology means different things to different people, but in any case, it is far outside the realm of science. It is a pseudoscience lying within the realm of superstition.

27.4 The Life Cycles of Stars

EXPLAIN THIS Why doesn't a neutron star emit beta particles?

In Chapter 26, we discussed the nebular theory, which explains how the Sun formed from an expansive, low-density cloud of gas and dust called a *nebula* (Figure 27.11). Other stars form in the same way. That is, over time, a nebula flattens, heats, and spins more rapidly as it gravitationally contracts. The center of the nebula becomes dense enough to trap infrared radiation so that this energy is no longer radiated away. The hot central bulge of a nebula is called a *protostar*.

Mutual gravitation between the gaseous particles in a protostar results in an overall contraction of this huge ball of gas, and its density increases still further as matter is crunched together, with an accompanying rise in pressure and temperature. When the central temperature reaches about 10 million K, hydrogen nuclei begin fusing to form helium nuclei. This thermonuclear reaction, converting hydrogen to helium, releases an enormous amount of radiant and thermal energy, as discussed in Chapter 26. The ignition of nuclear fuel marks the change from protostar to star. Outward-moving radiant energy and the gas accompanying it exert an outward pressure called *thermal pressure* on the contracting matter. When nuclear fusion occurs fast enough, thermal pressure becomes strong enough to halt the gravitational contraction. At this point, outward thermal pressure balances inward gravitational pressure, and the star's size stabilizes.

CHECKPOINT

What do the processes of thermonuclear fusion and gravitational contraction have to do with the physical size of a star?

Was this your answer?
The size of a star is the result of these two continually competing processes. Energy from thermonuclear fusion tends to blow the star outward like hydrogen bomb explosion, and gravitation tends to contract its matter in an implosion. The outward thermonuclear expansion and inward gravitational contraction produce an equilibrium that accounts for the star's size.

Though all stars are born in the same way from contracting nebulae, they do not all progress through their lives in the same way. A star's mass determines the stages a star will go through from birth to death. There are limits on the mass that a star can attain. A protostar with a mass less than 0.08 times the mass of the Sun ($0.08 M_{Sun}$) never reaches the 10 million K threshold needed for sustained fusion of hydrogen. On the other hand, stars with masses above $100 M_{Sun}$ would undergo fusion at such a furious rate that gravity could not resist thermal pressure and the star would explode. So stars exist within the limits of about a tenth of the mass of the Sun and 100 times the solar mass.*

Most stars have masses not very different from that of the Sun. Such stars inhabit a central place on the main sequence of the H–R diagram. If you plot the life-cycle stages of average stars on an H–R diagram, they trace a curve similar to the one for our Sun, which is shown in Figure 27.12. The Sun was born about 4.5 billion years ago at position 1, when the fusion of hydrogen ignited. The Sun will spend most of its lifetime—some 10 billion years—on the main sequence, with thermal pressure keeping gravity at bay. Speaking more generally, a star's hydrogen-burning lifetime lasts for a period of a few million to 50 billion years, depending on its mass.

More-massive stars have shorter lives than less-massive stars. This may sound counterintuitive, because if they have more mass, they have more fuel to burn longer, right? High-mass stars, however, are more luminous than low-mass stars, meaning that they burn their hydrogen fusion fuel at a faster rate.

FIGURE 27.11
This image of the Trifid Nebula was obtained by the Spitzer Space Telescope. This nebula is located 5400 light-years from Earth in the constellation Sagittarius. Within each of the four red dust clouds are developing stars.

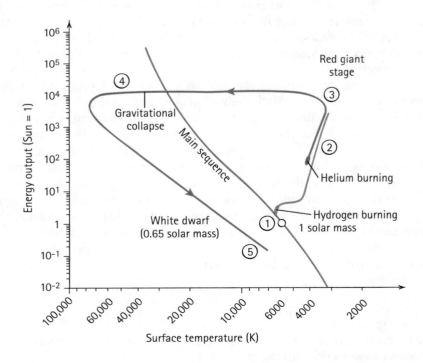

FIGURE 27.12
The stages of the Sun's life cycle are plotted on this H–R diagram. The short segment labeled Hydrogen burning lasts about 10 billion years. The later segments are much shorter.

* One solar mass, $1 M_{Sun}$, is a unit of mass equivalent to that of the Sun: 2×10^{33} kg.

Massive stars *must* be more luminous than small-mass stars so that the outward pressure of their nuclear fusion can offset the greater gravitational force of their contraction. Massive stars start out with more hydrogen fuel than small-mass stars, but they consume their fuel so much faster that they die billions of years younger than smaller stars.

No star lasts forever. In the old age of an average-mass star like our Sun, as the supply of hydrogen fuel diminishes, gravity overwhelms thermal pressure and the star pulls inward. Temperature rises as the burned-out hydrogen core contracts because of gravity. At a certain point, the temperature becomes high enough in the core to launch *helium burning*—the fusion of helium to carbon. The star then has a structure consisting of concentric shells. Helium fuses to carbon at the star's center while hydrogen fuses to helium in a surrounding shell. Energy output soars, moving the star off the main sequence.

With such intensified nuclear fusion within a star, the outward force of thermal pressure wins out over the inward force of gravity. The star balloons to become a giant (position 2). When our Sun reaches the giant stage about 5 billion years from now, its swelling and increased energy output will escalate Earth's temperatures. Earth will be stripped of its atmosphere and the oceans boiled dry. Ouch!

As fusion continues, carbon will continue to accumulate in the Sun's core, but temperatures will never become hot enough to allow the carbon to undergo fusion. Instead, carbon "ash" accumulates inside the star and fusion gradually tapers off. Then gravity predominates and the star contracts, which boosts its temperature. With higher temperatures, the color of the shrinking Sun will shift from red to blue, and its position will shift to the left in the H–R diagram.

A star's life cycle depends on its mass. The lowest-mass stars are brown dwarfs, dim but long-lived stars. Medium-mass stars progress from main-sequence stars to red giants or super-giants, then to white dwarfs. Very massive stars have short lives and die in massive explosions called *supernovae*.

When our Sun turns into a red giant billions of years from now, its diameter will encompass the orbit of Venus.

CHECKPOINT
Why does a star shrink when its core runs out of nuclear fuel?

Was this your answer?
Outward thermal expansion and inward gravitational contraction produce an equilibrium that accounts for the star's size. As the heat from the inner thermonuclear reactions begins to die down, gravity predominates and the star shrinks. Upon shrinking, matter becomes compressed, which is an additional source of heat to ignite further nuclear fusion. For a star the size of our Sun, compression raises the temperature enough to fuse elements to carbon. The fusion of heavier elements in our Sun, however, is not possible.

Our fuel-exhausted Sun will continue to shrink until the electrons within the Sun are so squeezed that they resist any further compression. Interestingly, the reason they resist further compression has to do with the ideas of the quantum hypothesis introduced in Chapter 12. Briefly, each subatomic particle has its own *quantum state* and no two subatomic particles can share the same quantum state. The Sun will shrink, but only up to the point at which electrons resist trespassing into the quantum states of their neighboring electrons. Having spent all of its nuclear fuel, our dead Sun, now quite small, will no longer be producing energy.

As our Sun goes through this final collapse, the layers of plasma and gas surrounding the core will be ejected in a brilliant display, forming what is called a **planetary nebula** (Figure 27.13). Despite its name, a planetary nebula has

Astronomers have found evidence suggesting that the carbon within the center of many white dwarfs crystallizes into diamond. They expect that when our Sun transforms into a white dwarf 5 billion years from now, its ember core will crystallize as well, leaving a planet-sized diamond at the center of our solar system.

FIGURE 27.13
The Cat's Eye planetary nebula, seen here with the Hubble Space Telescope, measures about 1.2 light-years across, which is about a thousand times the diameter of Neptune's orbit. This planetary nebula is about 3000 light-years away, which places it within our galaxy. Clearly visible are the hot gases exploding away from the central Sun-sized star, which is in the process of transforming into a white dwarf.

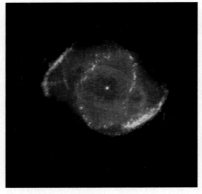

nothing to do with planets. The name is derived from the fact that the planetary nebula looks like a nebula from which planets could form. The planetary nebula, however, will disperse within a million years, leaving the Sun's cooling carbon core behind as a white dwarf. White dwarfs have the mass of a star but the volume of a planet, and are thus far more dense than anything on Earth. Because the nuclear fires of a white dwarf have burned out, it is not actually a star anymore, but is more accurately called a stellar remnant. In any case, a white dwarf cools for eons in space until it becomes too cold to radiate visible light (Figure 27.14).

Novae and Supernovae

There is another possible fate for a white dwarf, if it is part of a *binary star*. A binary star is a double star—a system of two stars that revolve about a common center, just as Earth and the Moon revolve about each other. If a white dwarf is a binary and close enough to its partner, the white dwarf may gravitationally pull hydrogen from its companion star. It then deposits this material on its own surface as a very dense hydrogen layer. Continued compacting increases the temperature of this layer, which ignites to embroil the white dwarf's surface in a thermonuclear blast that we see as a **nova**, which appears in the nighttime sky as a new star (*nova* is Latin for "new"). A nova is an event, not a stellar object. After a while, a nova subsides until enough matter accumulates to repeat the event. A given nova flares up at irregular intervals that may range from decades to hundreds of thousands of years.

Although low- and medium-mass stars become white dwarfs, the fate of stars more than about $10M_{Sun}$ is quite different. When such a massive star contracts after its giant or supergiant phase, more heat is generated than in the contraction of a small star. Such a star does not shrink to become a white dwarf. Instead, carbon nuclei in its core fuse and liberate energy while synthesizing heavier elements, such as neon and magnesium. Thermal pressure halts further gravitational contraction until all the carbon is fused. Then the core of the star contracts again to produce even greater temperatures, and a new fusion series produces even heavier elements. The fusion cycles repeat until the element iron is formed.

Fusion of elements with atomic numbers greater than those of iron consumes energy rather than liberating energy. (The reason for this, as you may recall from Chapter 13, is that the average mass per nucleon is lower for iron than for any other element.) Once nuclei transform to iron, the fusion process stops. Thermal expansion that pushes against gravity, therefore, also stops. Gravity thus predominates and the entire star begins its final contraction.

Recall that with a dying medium-sized star, such as our Sun, contraction continues until gravity is counteracted by the resistance of electrons. With a supermassive supergiant, however, the gravitational forces are strong enough to overcome this resistance. The electrons, however, do not merge into one another. Instead, they combine with protons to form neutrons. What happens next is an astounding event called a **supernova**. Within minutes, the supergiant's iron core, about the size of Earth, collapses into a ball of neutrons only several kilometers in diameter. Massive amounts of energy are released—enough to outshine an entire galaxy. During this brief time of abundant energy, the heavy

FIGURE 27.14
A white dwarf, shown here in an artist's sketch, is the final stage in the evolution of low- and medium-mass stars. After a star has used all its nuclear fuel, its outer layers escape into space, leaving the dense core behind as a white dwarf. The strong gravitational field of a white dwarf causes it to attract matter from surrounding space to form an accretion disk. The disk is heated by friction where it meets the star, causing it to glow brightly.

elements beyond iron are synthesized, as protons and neutrons outside the core mash into other nuclei to produce such elements as silver, gold, and uranium. These heavy elements are less abundant than the lighter elements because of the brief time available for synthesizing them.

Most of the energy during the collapse of the iron core is released in the form of *neutrinos*—nearly massless subatomic particles that rarely interact with matter. Concentrations of neutrinos released from the collapse of the iron core are great enough to blow the outer shells of the star outward at speeds in excess of 10,000 km/s, which is fast enough to travel 1 AU in about four hours. Over time, this supernova wind of heavy elements spreads to far reaches of the galaxy where the elements are taken in by nebulae destined to become new stars. The gold and platinum we wear for jewelry here on Earth, as well as the bulk of Earth itself, are dust from supernovae that exploded many years before our solar system came to be.

> **FYI** Do planets also orbit other stars? The answer is a resounding yes. These "exoplanets" reveal themselves by causing slight but detectable wobbles in the star they orbit. In some cases, the exoplanet transits in front of the star, which causes the star to become slightly dimmer, again at detectable levels. Over a thousand exoplanets have so far been discovered. Most are Jupiter-sized planets, but some near-Earth-sized planets have also been detected, such Gliese 581g, which is about 20 light-years distant.

CHECKPOINT

A star can undergo a nova more than once. Can a star also go through multiple supernovae? Why or why not?

Was this your answer?

A nova is a thermonuclear explosion that occurs when a white dwarf collects sufficient mass from a very close neighboring star. As long as the neighboring star provides mass, this explosion can be repeated multiple times. A supernova is such an energetic release of energy that it is an end-all event occurring never more than once for a particular supergiant star.

A supernova flares up to millions of times its former brightness. In AD 1054, Chinese astronomers recorded their observation of a star so bright that it could be seen by day as well as by night. This was a supernova (a "super new star"), its glowing plasma remnants now making up the spectacular Crab Nebula, shown in Figure 27.15. A less spectacular but more recent supernova

FIGURE 27.15
The Crab Nebula is the remnant of a supernova explosion first observed on Earth in AD 1054. The explosion took place within our galaxy at a distance of about 7000 light-years from Earth. Had the blast occurred within 50 light-years, most life on Earth would have likely gone extinct. Is there any nemesis star right now within this limit ready to supernova? Good question. Check the Internet for information about Betelgeuse.

FIGURE 27.16
This image of the 1987A super-nova was captured by the Hubble Telescope about 20 years after the initial explosion was sighted. Note the development of the ring systems, which continue to expand outward. This supernova occurred safely outside the Milky Way galaxy some 160,000 light-years away within a nearby smaller galaxy called the Large Magellanic Cloud. Some of its neutrinos were detected on Earth.

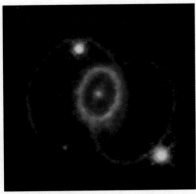

A neutron star is a kilometer-sized atomic nucleus!

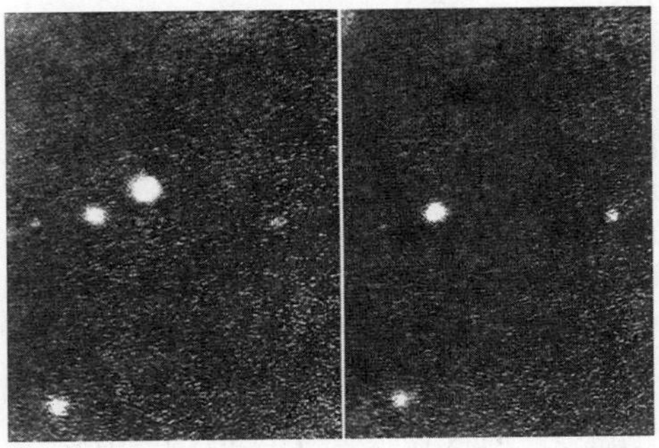

FIGURE 27.17
The pulsar in the Crab Nebula rotates like a searchlight, beaming visible light and X-rays toward Earth about 30 times per second, blinking on and off: (a) pulsar on, (b) pulsar off.

(a)   (b)

was witnessed in 1987. The progress of this supernova, shown in Figure 27.16, has been very carefully monitored by modern scientific equipment.

The superdense neutron core that remains after the supernova is called a **neutron star**. In accord with the law of conservation of angular momentum, these tiny bodies, with densities hundreds of millions times greater than those of white dwarfs, can spin at fantastic speeds. Neutron stars provide an explanation for the existence of *pulsars*. **Pulsars**, which are neutron stars, are rapidly varying sources of low-frequency radio emissions. As a pulsar spins, the beam of radiation it emits sweeps across the sky. If the beam sweeps over Earth, we detect its pulses. Of the approximately 300 known pulsars, only a few have been found emitting X-rays or visible light. One is in the center of the Crab Nebula (Figure 27.17). It has one of the highest rotational speeds of any pulsar studied, rotating more than 30 times per second. This thousand-year-old pulsar is relatively young. It is theorized that X-rays and visible light are emitted only during a pulsar's early history.

We saw earlier that a medium-sized star, such as our Sun, can collapse no further than a white dwarf because the force of gravity is not strong enough to overcome the resistance of electrons, which refuse to trespass into the quantum states of neighboring electrons. Similarly, a neutron star stops collapsing because neutrons, like electrons, resist trespassing into their neighboring neutrons. For a dying star, however, the bigger they are, the harder they fall. When the collapsing star is the biggest of the big, gravitational forces can be strong enough to overcome even the resistance of neutrons. The collapse continues beyond the stage of a neutron star, and the star disappears altogether from the observable universe. What is left is a *black hole*.

27.5 Black Holes

LEARNING OBJECTIVE
Identify the major attributes of black holes, such as the photon sphere, event horizon, and singularity.

Mastering**PHYSICS**
TUTORIAL: Black Holes

EXPLAIN THIS What happens to a light beam bouncing between two upright and perfectly parallel mirrors here on Earth?

A **black hole** is the remains of a supergiant star that has collapsed into itself. Upon this collapse, the force of gravity at the surface increases dramatically. Consider this from the perspective of Newton's law of gravity. According to this law, as discussed in Section 4.1, the force of gravity depends on the inverse square of the distance. If a star collapses to a tenth of its original size, the distance between the surface and the center of the star is one-tenth as much. The inverse square of one-tenth ($1/0.1^2$) equals 100. The weight at the surface, therefore, is 100 times as much, as suggested in Figure 27.18. So the gravitational force at the surface of a collapsing star increases because the star is getting smaller.

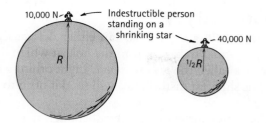

10,000 N — Indestructible person standing on a shrinking star

40,000 N

R

½R

FIGURE 27.18
If a star collapses to half its radius with no mass change, gravitation at its surface increases fourfold (in accordance with the inverse-square law). If the star collapses to one-tenth its radius, gravitation at its surface increases a hundredfold.

As the force of gravity increases, so does the *escape speed*. Recall from Section 4.9 that escape speed is the speed a moving object needs to fly away without ever falling back. For planet Earth, the escape speed is 11.2 km/s. This means that an object shot outward at 11.2 km/s (about 25,000 mi/h) will never fall back to Earth. The escape speed from the surface of our Sun is 618 km/s. For a supergiant star that has collapsed past the neutron star stage, the escape speed increases to the speed of light, which is 300,000 km/s.

In the early 20th century, Einstein proposed that light, like matter, is affected by gravity. We don't normally witness light being affected by gravity because light moves so fast, but with careful observations it is quite measurable. Starlight grazing the eclipsed Sun, for example, is seen to bend inward as the light passes through the Sun's strong gravitational field. So light is pulled downward by gravity. Sunlight can leave our Sun because the speed of light is much greater than the escape velocity. If a star such as our Sun, however, were to collapse to a radius of 3 km, the escape velocity from its surface would exceed the speed of light, and nothing—not even light—could escape. The Sun would be invisible. It would be a black hole.

The Sun, in fact, has too little mass to experience such a collapse, but when some stars with core masses over 40 times greater than the mass of the Sun reach the end of their nuclear resources, they undergo collapse; their collapse continues until the stars reach infinite densities. Gravitation near the surfaces of these shrunken stars is so enormous that light cannot escape from them. They have crushed themselves out of visible existence.

A black hole has the same amount of mass after its collapse as before its collapse, so the gravitational field in regions at and beyond the star's original radius is no different in either case. An orbiting planet would keep on orbiting as though nothing happened. But closer distances near the vicinity of a black hole, beneath the star's original radius, are nothing less than the collapse of space itself, with a surrounding warp into which anything that passes too close—light, dust, or a spaceship—is drawn (Figure 27.19). Astronauts in a powerful spaceship could enter the fringes of this warp and still escape. Below a certain distance, however, they could not, and they would disappear from the observable universe.

FYI Contrary to stories about black holes, they're nonaggressive and don't reach out and swallow objects at a distance. Their gravitational fields are no stronger than the original fields about the stars before their collapse—except at distances less than the radius of the original star. Except when they are too close, black holes shouldn't worry future astronauts.

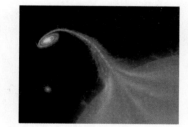

FIGURE 27.19
A rendering of a black hole stealing matter from a companion star.

CHECKPOINT

If the Sun somehow suddenly collapsed to a black hole, what change would occur in the orbital speed of Earth?

Was this your answer?
None. This is best understood classically; nothing in Newton's law of gravitation, $F = G\frac{mM}{d^2}$, changes. The fact that the Sun is compressed doesn't change its mass, M, or its distance, d, from Earth. Because Earth's mass, m, and G don't change either, the force, F, holding Earth in its orbit does not change.

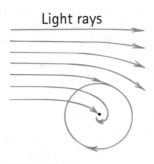

FIGURE 27.20
Light rays deflected by the gravitational field around a black hole. Light aimed far from the black hole is slightly deflected. Light aimed close to the black hole is drawn into the hole. In between, there is a particular radius at which photons can orbit the black hole.

Black Hole Geometry

As shown in Figure 27.20, a black hole can either deflect light or capture it. Also, there is one particular distance from the hole at which light can orbit in a circle. This distance is called the *photon sphere*. Light orbiting at that distance is highly unstable. The slightest disturbance will send it off into spare or spiraling into the black hole.*

An indestructible astronaut with a powerful enough spaceship could venture into the photon sphere of a black hole and come out again. While inside the photon sphere, she could still send beams of light back into the outside universe as shown in Figure 27.21. If she directed her flashlight sideways and toward the black hole, the light would quickly spiral into the black hole, but light directed vertically and at angles close to the vertical would still escape. As she drew closer and closer to the black hole, however, she would need to shine the light beams closer and closer to the vertical for escape. Moving closer still, our astronaut would find a particular distance where *no* light can escape. No matter in what direction the flashlight pointed, all the beams would be deflected into the black hole. Our unfortunate astronaut would have passed within the **event horizon**, the boundary where no light within can escape. Once inside the event horizon, she could no longer communicate with the outside universe; neither light waves, radio waves, nor any matter could escape from inside the event horizon. Our astronaut would have performed her last experiment in the universe as we conceive it.

The event horizon surrounding a black hole is often called the *surface* of the black hole, the diameter of which depends on the mass of the hole. For example, a black hole resulting from the collapse of a star 10 times as massive as the Sun has an event-horizon diameter of about 30 km. Calculated radii of event horizons for black holes of various masses are shown in Table 27.1. The event horizon is not a physical surface. Falling objects pass right through it. The event horizon is simply the boundary of no return.

When a collapsing star contracts within its own event horizon, the star still has substantial size. No known forces, however, can stop the continued contraction, and the star quickly shrinks in size until finally it is crushed, presumably to the size of a pinhead, then to the size of a microbe, and finally to a realm of size smaller than ever measured by humans. At this point, according to theory, what remains has infinite density. This point is the **black-hole singularity**.

Locating black holes is very difficult. One way to find them is to look for a binary system in which a single luminous star appears to orbit about an invisible companion, as was illustrated in Figure 27.19. If they are closely situated,

FIGURE 27.21
Just beneath the photon sphere, an astronaut can still shine light to the outside. But as she gets closer to the black hole, only light directed nearer to the vertical gets out, until finally even vertically directed light is trapped. This occurs at a distance called the event horizon.

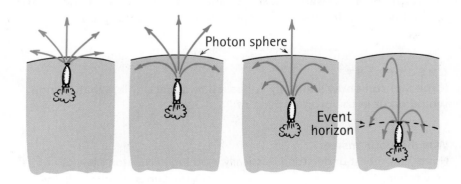

---

* This discussion applies to a nonrotating black hole. The situation for a rotating one is more complicated.

TABLE 27.1	CALCULATED RADII OF EVENT HORIZONS FOR NONROTATING BLACK HOLES OF VARIOUS MASSES
Mass of Black Hole	**Radius of Event Horizon**
1 Earth mass	0.8 cm
1 Jupiter mass	2.8 m
1 solar mass	3 km
2 solar masses	6 km
3 solar masses	9 km
5 solar masses	15 km
10 solar masses	30 km
50 solar masses	148 km
100 solar masses	296 km
1000 solar masses	2961 km

matter ejected by the normal companion and accelerating into the neighboring black hole should emit X-rays. The first convincing candidate for a black hole, the X-ray star Cygnus X-1, was discovered in 1971. Many additional black hole candidates have since been found, which suggests that black holes are common. Studies of the center of our galaxy strongly suggest the presence of a black hole some

FALLING INTO A BLACK HOLE

Imagine yourself exploring a black hole on some futuristic scientific mission. Your spacecraft is cruising within a safe orbit around the black hole. Your first experiment is to launch a clock-bearing probe toward the black hole. The clock consists of a large LED display of blue lights. Through a telescope you watch as the probe descends. Remarkably, the closer the probe gets to the black hole, the slower the clock appears to run. Furthermore, the light coming from the clock shifts from blue to a lower-frequency red. As the probe gets closer still, the clock runs even slower. Soon you can't make out the clock at all because it has shifted to the infrared. So you switch to your infrared telescope and see that as the probe gets closer to the black hole its clock slows to a creep. Furthermore, the probe seems to be taking an unusually long time to descend. Eventually, the light from the clock is visible only with your microwave telescope, followed by your radio telescope as the frequency of the light from the clock gets lower and lower. Ultimately, just as the clock disappears completely, emitting no light whatsoever, you note that the

clock has frozen in time. To get to this point, however, would take, from your point of view, forever.

But you don't have forever, and you soon grow tired of watching the ultra-slow clock as it creeps ever so slowly toward the black hole. So you decide to move on to the second experiment, for which you have volunteered to place yourself within a second probe equipped with a blue clock and an array of telescopes. As you descend toward the black hole you note that your clock runs perfectly normally without changing color. The clock on the mother ship, however, is running rather fast. Furthermore, its color is shifting toward ultraviolet and beyond. Your old shipmates are moving rather fast as well. As you descend even deeper, their peculiar speed grows even faster. Soon they grow impatient waiting for you and they leave you behind for dead. Before you know it, the remaining visible stars quickly pass through their life spans, their light coming to you in a flash of ultrahigh frequencies but through a narrowing field of view. And then there is nothing. At that moment you pass through

the event horizon, which is a mathematical boundary, not a physical one. Singularity is still kilometers beneath you, but you are caught within its unrelenting grip. The universe you left behind has run through infinite time and exists no longer.

Unfortunately, such a fall through the event horizon of a regular-sized black hole would not be survivable. As you come closer, the gravitational pull on your feet would far exceed that on your head. As a result, your body would be stretched. You would be "ripped" of the opportunity to experience what it would be like inside the event horizon. Furthermore, as you continued to fall toward the black-hole singularity, your atoms would be compressed to an infinitely small size, which you would not survive. What would happen next is only conjecture. Perhaps your mass would explode like a Big Bang into another universe. Perhaps what happens to the mass that falls into the singularity is even stranger than we're capable of imagining. Maybe one day we crafty humans will come to understand such processes. If our species lasts that long!

6 billion km in diameter, which is as large as our solar system! The origin of this mega black hole is likely related to the formation of the galaxy itself. It is currently thought that most, if not all, large galaxies contain central mega-sized black holes.

As described in the box on page 753, tidal forces would rip you apart before you fall into a regular-sized black hole. For a mega-sized black hole, like the one at the center of our galaxy, the tidal forces would be negligible—your spaceship would survive passage through the event horizon.

CHECKPOINT

What determines whether a star becomes a white dwarf, a neutron star, or a black hole?

Was this your answer?
The mass of a star is the principal factor that determines its fate. Stars that are about as massive as the Sun, and those that are less massive, evolve to become white dwarfs; stars with masses of $10M_{Sun}$ or greater evolve to become neutron stars; the most massive stars of about $40M_{Sun}$ or greater ultimately become black holes.

LEARNING OBJECTIVE
Describe the discovery of galaxies; their classification as elliptical, spiral, or irregular; and how they are organized into superclusters.

27.6 Galaxies

EXPLAIN THIS All the celestial objects discussed so far in this and the preceding chapter are located in what galaxy?

Look up into the clear nighttime sky away from the city lights and you will see plenty of stars. In between the stars you'll also see plenty of black. Before the early 20th century, the abundance of black in the night sky led many people to conclude that the universe consisted of an island of millions of stars nestled within a vast sea of emptiness. In addition to stars, however, are the cloudlike nebulae, some of them with a distinct spiral-shaped structure. As early as the 1750s, the German philosopher Immanuel Kant proposed that these spiral clouds were other islands of stars called *galaxies*. But without powerful telescopes, there was no way to tell whether that was true.

The debate about whether the universe consisted of one or many islands of stars was settled by the American astronomer Edwin Hubble. In 1927, working with the newly built largest telescope in the world at Mt. Wilson in California, Hubble made out individual stars within the Andromeda spiral nebula (Figure 27.22). Some of these stars he noticed to be *Cepheids*, which are stars that change their luminosity over short periods of time. By measuring the rate at which they changed luminosity he estimated their distance, which he found to be much farther away than any star within our own galaxy. Spiral nebulae were not simply clouds—they were neighboring islands of stars within a vast emptiness that potentially extended forever.

But Hubble took his research a step further and discovered something even more amazing. He knew that the color of light emitted by a star or galaxy receding away from us shifts to the red because of the Doppler effect (see Section 10.8). The degree of redshift could be measured quantitatively by focusing on the line spectrum of hydrogen (see Section 12.6). The greater the shift in the lines of hydrogen's spectrum, the faster the receding speed. His research team measured both the distances and redshifts of numerous galaxies and discovered that the farther the galaxy, the greater the redshift. This meant that the galaxies were not static islands. Rather, they were receding from us in every direction, which meant that the universe itself was expanding.

If distant galaxies were all moving away from one another, that could only mean that they were once much closer together. Running the cosmic movie backward

FYI Galaxies are cataloged by two systems. The first catalog is based on the work of Charles Messier, who in 1781 published a list of heavenly structures, such as galaxies, relatively easy to observe with small telescopes. The Andromeda galaxy, for example, is the 31st entry of this catalog and is thus listed as M31. A "New General Catalog" was begun in 1888 that was subsequently used to identify all structures, including the many more that became visible with the advent of more powerful telescopes. Under this system, the Andromeda galaxy is cataloged as NGC 224. You can use these catalog numbers in your Internet search engine to learn more about these objects, including their location in the nighttime sky.

FIGURE 27.22
Hubble showed that the great spiral nebula within the Andromeda constellation was not just a swirling cloud of gas, but a neighboring galaxy of stars, which is now called the Andromeda galaxy and catalogued as M31. You can see the Andromeda galaxy for yourself by looking between the constellations of Cassiopeia and Pegasus in the late fall nighttime sky. The galaxy appears huge, covering an area six times that of the full Moon. It is, of course, much dimmer than the Moon. Best viewing comes with a good pair of binoculars far away from city lights.

would inevitably lead to a moment when all the galaxies were gathered together, perhaps within a single point. The universe as we know it, therefore, had a beginning. This moment has come to be known as the **Big Bang**, which we will discuss in more detail in Chapter 28. For the remainder of this chapter, however, we will simply describe the different kinds of galaxies and how they are organized within the observable universe.

A **galaxy** consists of a large assemblage of stars, interstellar gas, and dust. Galaxies are the breeding grounds of stars. Our own star, the Sun, is an ordinary star among more than 100 billion others in an ordinary galaxy known as the Milky Way galaxy (Figure 27.23). With unaided eyes, we see the Milky Way as

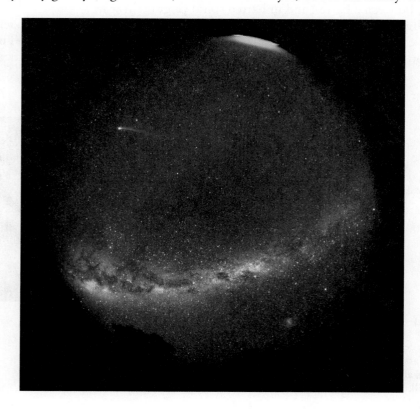

FIGURE 27.23
A wide-angle photograph of the Milky Way, which appears as a north–south cloudlike band of light. The dark lanes and blotches are interstellar gas and dust obscuring the light from the galactic center. If it weren't for this dust, the Milky Way would be a much more spectacular nighttime display. This photograph also shows Comet Hyakutake, which appeared in 1996.

a faint band of light that stretches across the sky. The early Greeks called it the "milky circle" and the Romans called it the "milky road" or "milky way." The latter name has stuck.

The masses of galaxies range from about a millionth the mass of our galaxy to some 50 times more. Galaxies are calculated to have much more mass than can be seen with the telescope. A small proportion of the invisible mass is simply matter that has grown so cold that it doesn't emit enough light for us to see. The bulk of the invisible mass, however, is likely an unknown form of matter, called *dark matter*, that does not absorb or emit light. It does, however, possess mass and so its gravitational effects are quite measurable. In the next chapter we describe how dark matter likely played a key role in the formation and distribution of galaxies.

FIGURE 27.24
This small elliptical galaxy, Leo I, found within the constellation Leo, is only about 2500 light-years in diameter. For comparison, the diameter of our Milky Way galaxy is about 100,000 light-years.

FYI The Andromeda galaxy is our closest spiral neighbor, being only some 2.5 million light-years away. It contains many more stars than the Milky Way, which makes it more luminescent. Also, its diameter is about 220,000 light-years, compared to the Milky Way's 100,000 light-years. Thus, our view of the Andromeda is likely more spectacular than the Andromeda's view of us.

Elliptical, Spiral, and Irregular Galaxies

The millions of galaxies visible in photographs can be separated into three main classes—elliptical, spiral, and irregular. *Elliptical galaxies* are the most common galaxies in the universe. They are spherical, with the stars more crowded toward the center. Most contain little gas and dust, which makes them easy to see through. They also tend to be yellow, which tells us that they consist primarily of older stars—older stars are yellow, while hot young stars tend to be blue. Most ellipticals are small, consisting of fewer than a billion stars (Figure 27.24). An exception is the giant elliptical galaxy M87 (Figure 27.25). The largest ellipticals are about 5 times as large as our galaxy, and the smallest are 1/100 as large.

Spiral galaxies, such as the Andromeda galaxy, shown in Figure 27.22, are perhaps the most beautiful arrangements of stars. Some spirals, such as the Sombrero galaxy of Figure 27.26, have a spheroid central hub. Others, like the one shown in Figure 27.27, have a hub shaped like a bar. The Milky Way galaxy is thought to look much like the NGC 6744 spiral galaxy, which is an intermediate between a barred and unbarred spiral (Figure 27.28).

Elliptical and/or spiral galaxies sometimes cross paths or even collide. In such cases, gravity causes the shape of the galaxy to become distorted. These distorted looking galaxies are called *irregular galaxies*. Most irregular galaxies

FIGURE 27.25
The giant elliptical galaxy M87, one of the most luminous galaxies in the sky, is located near the center of the Virgo cluster, some 50 million light-years from Earth. It is about 120,000 light-years across and about 40 times as massive as our own galaxy, the Milky Way.

FIGURE 27.26
The Sombrero galaxy, cataloged as M104, is about 80,000 light-years in diameter and about 32 million light-years from Earth. At its center is one of the most supermassive black holes measured in any nearby galaxy.

are small and faint and are difficult to detect. They tend to contain large clouds of gas and dust mixed with both young (blue) and old (yellow) stars. The irregular galaxy first described by the navigator on Magellan's voyage around the world in 1521 is our nearest neighboring galaxy—the Magellanic Clouds. This galaxy consists of two "clouds," called the Large Magellanic Cloud (LMC) and the Small Magellanic Cloud (SMC), both of which are slowly being pulled into the Milky Way. The LMC is dotted with hot young stars with a combined mass of some 20 billion solar masses, and the SMC contains stars with a combined mass of about 2 billion solar masses (Figure 27.29). Some irregular galaxies, such as NGC 4038 shown in Figure 27.30, are the aftermaths of galactic collisions.

FIGURE 27.27
The beautiful barred spiral galaxy NGC 1300 is about 100,000 light-years across and some 70 million light-years away.

CHECKPOINT
Is it possible for one type of galaxy to turn into another?

Was this your answer?
Yes, and this occurs as two symmetrically shaped galaxies collide to form an asymmetrically shaped irregular galaxy.

Active Galaxies

Galaxies differ greatly in the activity going on inside them. This activity may include star formation, supernova, or energy-releasing processes at the galactic core. Those galaxies with notable activity are sometimes called *active galaxies*.

When it comes to star formation, our Milky Way is a relatively calm place, producing on average about one new star per year. By comparison, one type of active galaxy, known as a **starburst galaxy**, can produce more than 100 new

FIGURE 27.28
The NGC 6744 galaxy is an intermediate between a barred and unbarred spiral galaxy. Studies of the Milky Way suggest that it too is an intermediate spiral. In other words, this is what we may look like from afar.

(a)   (b)

FIGURE 27.29
(a) The Large Magellanic Cloud and (b) the neighboring Small Magellanic Cloud are a pair of irregular galaxies. The Magellanic Clouds are our closest galactic neighbors, about 150,000 light-years distant. They likely orbit the Milky Way.

Shown in black and white is the ground-telescope view of an irregular galaxy resulting from the collision of two galaxies. Note the remnant arms that suggest two former spiral galaxies. The inset shows a close-up color view taken by the Hubble Telescope. Evident is the rapid formation of new stars (blue) occurring as the two galaxies combined.

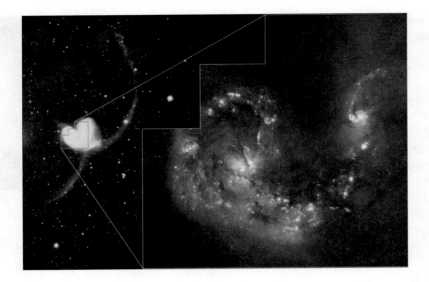

stars per year. A starburst's high rate of star formation is often the result of some violent disturbance, such as a collision between two galaxies. The irregular galaxy shown in Figure 27.30 is an example of a starburst galaxy. Another example is the Cigar galaxy, M82, which is being deformed by the tidal forces from its much larger neighbor, M81 (Figure 27.31). A starburst tends to die down once the disturbance is removed or after the starburst galaxy consumes all its interstellar fuel. Many elliptical galaxies are thought to be former starburst galaxies because of their low abundance of interstellar dust and gases.

Other active galaxies are active by virtue of their galactic core, which hosts a black hole more massive than millions or even billions of Suns. The event horizons of these black holes are about as large as our solar system! Most large galaxies, including the Milky Way, have such black holes in their centers, and these massive black holes can be a source of much activity.

For example, in 2010, using NASA's Fermi gamma ray space telescope, astronomers discovered two massive gamma-ray–emitting bubbles extending north and south from the center of our galactic disk, as shown in Figure 27.32. Together, these bubbles, known as *Fermi bubbles*, extend about 50,000 light-years, which is about half the diameter of our galaxy. Their edges are well defined (not diffuse), which suggests they are a fairly recent occurrence, perhaps

The Cigar galaxy, M82, is a spiral galaxy tilted away from us so that we see it from an edge-on view. Tidal forces from the nearby M81 galaxy disturb the distribution of matter within M82, which clumps, allowing for the formation of many new stars, as evidenced by M82's remarkable blue color. The red gases above and below the galactic plane are primarily hydrogen being pushed out by abundant stellar wind.

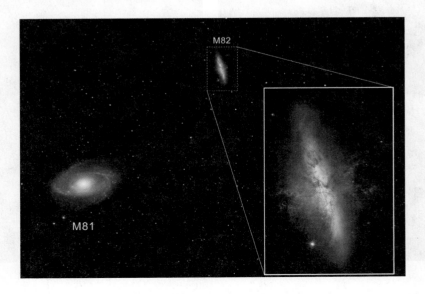

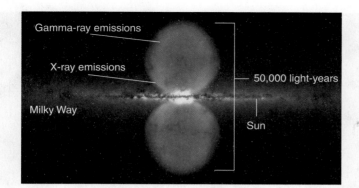

FIGURE 27.32
Vast gamma-ray–emitting bubbles, known as Fermi bubbles, extend north and south from the center of our galaxy. Though invisible to the naked eye, these bubbles span over half the sky when viewed from Earth through gamma ray detectors.

arising only several million years ago. Their nature and origin are not yet fully understood. One model suggests they arose from large amounts of material falling into our galaxy's mega-size central black hole. Alternatively, they may have resulted from the outgassing of a brief but intense period of starburst activity occurring at the galactic core.

The Milky Way's Fermi bubbles, however, are minor compared to the activity we see arising from the centers of other large galaxies. In such cases, supermassive amounts of matter are likely falling into central supermassive black holes. Before falling into the black hole, the doomed mass forms a rapidly spinning disk, called an *accretion disk*, around the equator of the black hole. Charged particles in this hyperspinning disk create a narrow yet ultrastrong magnetic field that rises from the black hole's poles. Rather than falling into the black hole, some of the charged particles, such as electrons, are accelerated outward through these magnetic fields to nearly the speed of light. This results in two extremely long streams of particles, called *jets*, extending over 100,000 light-years away from the galactic center, which is called an **active galactic nucleus (AGN)**.

A relatively close AGN is found within the large elliptical galaxy M87, which was shown in Figure 27.25. High-resolution images of this galaxy, as shown in Figure 27.33, reveal a jet of material streaming away from the center of this galaxy. Interestingly, the jet is angled toward us. This plus the great speed of the jet (99.5% of the speed of light) helps make the jet appear more luminous. The opposite "counterjet" receding away from us at such great speeds is barely visible.

Nearby active galactic nuclei, such as that of M87, give us a clue as to the possible nature of the most energetic galaxies of all—the *quasars*. Starting in the 1960s, astronomers began discovering extremely energetic bodies hundreds of times more luminous than our own galaxy, yet farther away than any observed object. Because they looked like radio-emitting stars, they were dubbed "quasi-stellar radio sources," which was shortened to **quasars**. Because all quasars are so very far away, they occurred a very long time ago—up to 13 billion years ago, which was close to the beginning of the universe. As we look to quasars, therefore, we are peering into the early lives of galaxies (Figure 27.34). During this galactic youth, much material was still falling into the supermassive black holes found within their galactic cores. The dynamics of this process allow for an efficient conversion of mass into energy, which would provide for colossal jets of highly energetic particles and light. When one of these ancient jets faces our direction, the result is an unusually brilliant display of energy we call a **blazar**.

FIGURE 27.33
Material falling into the supermassive black hole in the center of M87 generates powerful jets that shoot out at near light speed.

FIGURE 27.34
FIGURE 27.34
Each disk in this deep-space image taken by the Hubble Telescope is a galaxy. The quasar shown in the center is billions of light-years behind this cluster of galaxies. Interestingly, the cluster's gravity has bent the light from the quasar like a lens so that multiple images of the quasar are also seen.

CHECKPOINT
Are there any quasars found within the Milky Way galaxy?

Was this your answer?
No. A quasar is the active galactic nucleus of a galaxy as it appeared toward the beginning of the universe. All quasars are billions of light-years away from our galaxy.

Clusters and Superclusters

Galaxies are not the largest structures in the universe; they tend to cluster into distinguishable groups. Our Milky Way galaxy, for example, is part of a cluster of local galaxies that include two other major spiral galaxies—namely, the Andromeda galaxy and the Triangulum galaxy. Also included are more than a dozen smaller elliptical galaxies, including the Leo I galaxy shown in Figure 27.24, and a few irregular galaxies, such as the Large Magellanic Cloud. Altogether this cluster of galaxies is called the **Local Group**. Their approximate distributions are shown in Figure 27.35. If drawn correctly to scale, Andromeda is only about 20 Milky Way diameters away from the Milky Way. The Triangulum galaxy, so named because it completes a triangle between the spirals, is even closer to Andromeda, but farther away from us.

Our Local Group of galaxies is also under the gravitational influence of neighboring galactic clusters. Our cluster plus all these other clusters makes for what is called a *supercluster*, which is a cluster of galactic clusters. Our Local Group is actually a rather minor component of our **Local Supercluster**, as is illustrated in Figure 27.36.

Our Local Supercluster is tied in with an elaborate network of many other superclusters, as shown in Figure 27.37. Together, these superclusters appear as though they reside on the surface of a foam inside which are large voids of empty space. Zooming out farther we find that the network of superclusters extends to the edges of the *observable universe*, as illustrated in Figure 27.38. By "observable universe" we mean all that we are able to see given the fact that the

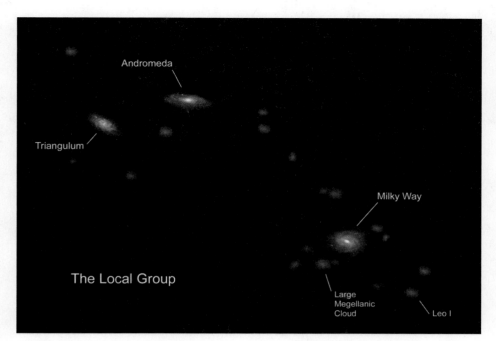

FIGURE 27.35
This two-dimensional composition shows the approximate relative distances between the members of our Local Group of galaxies. These galaxies are all moving toward each other and will one day collide into a larger supergalaxy.

universe is only about 14 billion years old. The light coming from any object farther than 14 billion light-years has not had sufficient time to reach us.

So our observable universe is huge. Utterly huge. How large, then might be the entire universe—the whole shebang? We don't know, and maybe never will. But that doesn't stop cosmologists from developing models proposing possible answers. One such model suggests that if the observable universe were the size of a proton, the entire universe would be about as large as planet Earth. Consider how many protons fit within the dimensions of Earth. This would be the number of observable universes within the entire universe. The number is so enormous that if you could travel $10^{10^{118}}$ m in any direction, you would have a large probability of coming across another observable universe that looks very much like the one you left. Continuing this speculation, you'd look for somebody like yourself who is reading a book exactly like the one you are reading now.

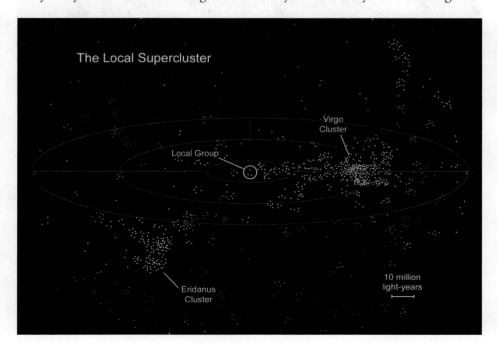

FIGURE 27.36
A supercluster is a cluster of galactic clusters. Each dot represents a galaxy. Note that our Local Group is found midway between two much larger clusters, the Virgo and Eridanus clusters.

FIGURE 27.37
Each cloud represents a supercluster. Note that the superclusters are strung together as though on the surface of a foam.

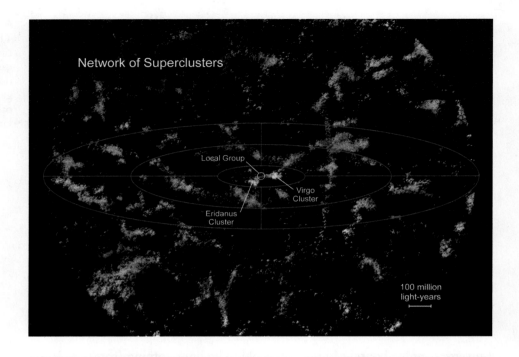

FIGURE 27.38
The network of superclusters extends to the edges of the observable universe, which is no farther than 14 billion light-years away. This illustration, however, shows a hypothetical bird's-eye view of this observable universe fully matured to the present moment, which, because of cosmic expansion, would place those most distant objects now some 42 billion light-years away.

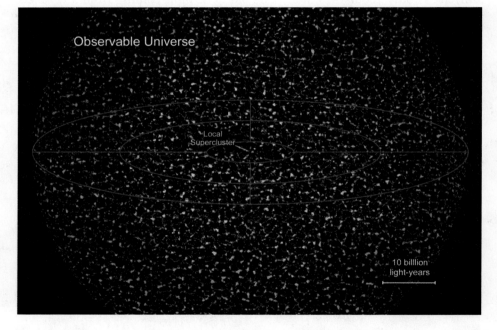

What if this person is actually a future you? Besides living very, very far away, the only measurable difference is that he or she has already finished reading this paragraph. Scientists call this the "multiverse" model, in which each observable universe is static, representing one possible arrangement of matter. We don't move through time. Rather, we jump from one observable universe to the next, which gives the appearance of moving through time. Welcome. You have just jumped to a new observable universe. The old one you left six sentences ago is now ever so distant.

Polls show that about half of American adults do not know that it takes one year for Earth to go around the Sun. Many of us, therefore, are still struggling with scientific ideas of 400 years ago. Your knowledge is much greater, and being aware of the amazing possibilities that science continues to reveal puts you in a very privileged minority. Celebrate that!

FIGURING PHYSICAL SCIENCE

The Scale of the Galaxy

SAMPLE PROBLEM 1
Earth is about 0.000016 light-year from the Sun and about 4.2 light-years from the next nearest star, which is Proxima Centauri. How far away is Proxima Centauri in Earth–Sun distances?

Solution:
Divide our distance from Proxima Centauri by our distance to the Sun:

$$\frac{4.2 \text{ light-years}}{0.000016 \text{ light-year}} = 262{,}500$$

So the nearest star is about 260,000 times as far from us as the Sun is.

SAMPLE PROBLEM 2
Our distance from the center of the Milky Way galaxy is about 26,000 light-years. How many Earth–Proxima Centauri distances is that?

Solution:
Divide the distance to the center of the galaxy by the distance to Proxima Centauri:

$$\frac{26{,}000 \text{ light-years}}{4.2 \text{ light-years}} = 6190$$

So the center of our galaxy is about 6200 times as far away as the nearest star.

SAMPLE PROBLEM 3
The Milky Way and Andromeda galaxies are about 2,300,000 light-years apart. The diameter of the Milky Way galaxy is about 100,000 light-years. How many Milky Way diameters distant is the Andromeda galaxy?

Solution:
Divide the distance to the Andromeda galaxy by the diameter of the Milky Way galaxy:

$$\frac{2{,}300{,}00 \text{ light-years}}{100{,}000 \text{ light-years}} = 23$$

So the Andromeda galaxy is about 23 Milky Way galaxy diameters away.

SAMPLE PROBLEM 4
The Andromeda galaxy is moving toward the Milky Way galaxy at a rate of about 300,000 mi/h. In how many years will these two galaxies collide?

Solution:
Convert 300,000 mi/h to light-years per year. First convert the 300,000 mi to light-years:

(300,000 mi)(1.61 km/1 mi) ×
(1 light-year/10,000,000,000,000 km)
$= 4.83 \times 10^{-8}$ light-year

Second, convert hours to years:

(1 h)(1 day/24 h)(1 yr/365.25 days)
$= 1.14 \times 10^{-4}$ year

Put the two converted values together:
300,000 mi/h
$= 4.83 \times 10^{-8}$ light-year/1.14×10^{-10} yr
$= 4.23 \times 10^{-10}$ light-year/yr

Use the equation for speed from Chapter 1:

Speed = distance/time

4.23×10^{-4} light-year/yr
$= 2{,}300{,}000$ light-years/x yr

Solve for x:

$x = 5{,}400{,}000{,}000$ yr

So in roughly 5.4 billion years, the Andromeda and Milky Way galaxies will collide. By this time our Sun will have exhausted almost all of its nuclear fuel, so this is not something we Earthlings will be around to witness. Collisions between galaxies, however, are fairly common, and astronomers have photographed many such occurrences now in progress.

CHECKPOINT
Which is greater: the number of stars in our galaxy or the number of galaxies in the observable universe?

Was this your answer?
There are far more galaxies within the entire universe than there are stars in our galaxy. Recall from the beginning of this chapter that astronomers estimate that there are about 100 billion stars in our galaxy and about 100 billion galaxies in our observable universe. If true, that means there are about 10^{22} stars in our observable universe, which is about the same number of water molecules in a drop of water. As large as the observable universe is large is as small as the fundamental building blocks of our body are small. As humans we are nicely situated between these two extremes.

For instructor-assigned homework, go to www.masteringphysics.com (MP)

SUMMARY OF TERMS (KNOWLEDGE)

Active galactic nucleus The central region of a galaxy in which matter is falling into a supermassive black hole and emitting huge amounts of energy.

Big Bang The primordial explosion of space at the beginning of time.

Black hole The remains of a giant star that has collapsed upon itself. It is so dense, and has a gravitational field so intense, that light itself cannot escape from it.

Black-hole singularity The object of zero radius into which the matter of a black hole is compressed.

Celestial sphere An imaginary sphere surrounding Earth to which the stars are attached.

Event horizon The boundary region of a black hole from which no radiation may escape. Any events within the event horizon are invisible to distant observers.

Galaxy A large assemblage of stars, interstellar gas, and dust, usually catagorized by its shape: elliptical, spiral, or irregular.

Giant stars Cool giant stars above main-sequence stars on the H–R diagram.

H–R diagram (Hertzsprung–Russell diagram) A plot of luminosity versus surface temperature for stars. When so plotted, stars' positions take the form of a main sequence for average stars, with exotic stars above or below the main sequence.

Light-year The distance light travels in one year.

Local Group Our immediate cluster of galaxies, including the Milky Way, Andromeda, and Triangulum spiral galaxies plus a few dozen smaller elliptical and irregular galaxies.

Local Supercluster A cluster of galactic clusters in which our Local Group resides.

Main sequence The diagonal band of stars on an H–R diagram; such stars generate energy by fusing hydrogen to helium.

Neutron star A small, extremely dense star composed of tightly packed neutrons formed by the welding of protons and electrons.

Nova An event in which a white dwarf suddenly brightens and appears as a "new" star.

Planetary nebula An expanding shell of gas ejected from a low-mass star during the latter stages of its evolution.

Pulsar A celestial object (most likely a neutron star) that spins rapidly, sending out short, precisely timed bursts of electromagnetic radiation.

Quasar The core of a distant galaxy early in its lifespan when its central black hole has not yet swept much matter from its vicinity, leading to a rate of radiation greater than that from entire older galaxies.

Starburst galaxy A galaxy in which stars are forming at an unusually fast rate.

Supernova The explosion of a massive star caused by gravitational collapse with the emission of enormous quantities of matter and radiation.

White dwarf A dying star that has collapsed to the size of Earth and is slowly cooling off; located at the lower left on the H–R diagram.

READING CHECK QUESTIONS (COMPREHENSION)

27.1 Observing the Night Sky

1. What are constellations?

2. Why does an observer at a given location see one set of constellations in the winter and a different set of constellations in the summer?

3. Why do the stars appear to turn on an imaginary north–south axis once every 24 hours?

4. Is the light-year a measurement of time or distance?

27.2 The Brightness and Color of Stars

5. Which is hotter: a red star or a blue star?

6. What is the difference between apparent brightness and luminosity?

27.3 The Hertzsprung–Russell Diagram

7. What is an H–R diagram?

8. Where are the great majority of stars plotted on an H–R diagram?

9. Where does our Sun reside on an H–R diagram?

10. Among stars originating from the main sequence, which are larger: red stars or yellow stars?

27.4 The Life Cycles of Stars

11. What process changes a protostar into a full-fledged star?

12. What are the outward forces that act on a star?

13. What are the inward forces that act on a star?

14. When will our Sun reach the red-giant stage?

15. Is the lifetime of a high-mass star longer or shorter than that of a low-mass star?

16. What is the relationship between the heavy elements that we find on Earth today and supernovae?

17. What is the relationship between a supernova and a neutron star?

18. What is the relationship between a neutron star and a pulsar?

27.5 Black Holes

19. What is the relationship between a supergiant star and a black hole?

20. Why don't we think the Sun will eventually become a black hole?

21. How does the mass of a star before its collapse compare with the mass of the black hole that it becomes?

22. If black holes are invisible, what is the evidence for their existence?

23. Is a black hole's event horizon a physical or mathematical boundary?

27.6 Galaxies

24. What type of galaxy is the Milky Way?

25. What are the consequences of galaxies colliding?

26. What is a starburst galaxy?

27. How does the brightness of a quasar compare with that of a large galaxy?

28. How many spiral galaxies are in the Local Group?

29. Is the Local Group a relatively small or large cluster of galaxies?

30. Name three galactic clusters found in our Local Supercluster.

ACTIVITIES (HANDS-ON APPLICATION)

31. To observe the daily diurnal motion of the stars, go star watching tonight. Pick a star or constellation that lines up with a stationary landmark such as a tree or house. Then come back in an hour or so and you will see that the star has moved away from the landmark but remains in place relative to the other stars. In what direction has that star moved? To the east? West? Southwest? Northwest? Where will it be when the Sun rises? Will it be in the same location in 24 hours?

32. To observe the revolutionary motion of Earth around the Sun, go star watching and make note of the stars directly above you. Sketch their pattern on a piece of paper and write down the date and time at which you make this observation. If these stars are not already a well-known constellation, make up a new constellation and give it a creative name. After a month has passed, look for this same constellation at the same time of night. Why isn't it still directly above you? In what direction has your constellation shifted? Why?

33. The nighttime sky is full of more than just stars and planets. If you are so lucky as to be in a location where the clear night sky is free from city lights, you must try to find the Andromeda galaxy. The best viewing times in the Northern Hemisphere are during the late fall, early winter months. First, look for a large rectangular box, which is the body of the flying horse, Pegasus. Viewing Andromeda is best when the Pegasus box is directly overhead. Second, look to the north (close to where the Big Dipper is often seen) to find the letter "W," which is the Cassiopeia constellation. From Pegasus you will see a "V" of stars that extends from the box to the left of Cassiopeia. These are Pegasus's rear legs, which overlap with the Andromeda constellation. The Andromeda galaxy lies just above the upper leg, directly between the Cassiopeia "W" and the Pegasus box. Look carefully at Figure 27.22 to help you pinpoint the exact location. With the unaided eye, it looks like small dim fuzz. You'll see it best by not looking at it directly. The fuzz takes on more of an oval shape when viewed through a good pair of binoculars. What you see with the binoculars is mostly the central core. The full galaxy, visible only with more powerful telescopes (and time exposure), is about six times the diameter of the Moon!

THINK AND SOLVE (MATHEMATICAL APPLICATION)

34. Suppose Star A is four times as luminous as Star B. If these stars are both 500 light-years away from Earth, how will their apparent brightness compare? How will the apparent brightness of these stars compare if Star A is twice as far away as Star B?

35. The brightest star in the sky, Sirius, is about 8 light-years from Earth. Show that if you could somehow travel there at jet-plane speed, 2000 km/h, the trip would take about 4.3 million years. (Note: 1 light-year equals 9.46×10^{12} km.)

36. If you were to travel straight up from the core of our galaxy and then look back, you would have a grand view of the Milky Way's spiral shape. If the distance from the core to the outer edges was 50,000 light-years, how much surface area are you looking at? Assume the galaxy is a circle whose area can be found by the equation area $= \pi r^2$.

37. Assume the Milky Way contains 100 billion stars evenly distributed with none concentrated toward the center. What would be the surface area density of stars? Use the equation surface area density $=$ number of stars/surface area.

38. Use the information in the preceding problem to figure out the average amount of space around a single star in units of AU (Note: 1 light-year $= 63,000$ AU.)

39. From your answer to the preceding problem, would it be possible for two galaxies with stars evenly distributed to pass right through each other?

THINK AND RANK (ANALYSIS)

40. Rank the appearance of the North Star in order of increasing height from the horizon as seen from (a) Alaska, (b) Florida, and (c) Vermont.

41. Rank the objects in order of increasing intrinsic motion as viewed from Earth: (a) the Moon, (b) Venus, and (c) the North Star.

42. Rank the following stars in order of increasing radius:

	Star A	Star B	Star C
Surface temperature (K):	6,000	4,000	30,000
Luminosity (solar units):	1	100	0.01

43. Rank these stages of stellar development from earliest to latest: (a) white dwarf, (b) nova, and (c) red giant.

44. Rank the nuclear fuels in order of being consumed, from first to last: (a) carbon, (b) helium, and (c) hydrogen.

45. Rank the following features of a black hole in order of increasing radius: (a) photon sphere, (b) singularity, and (c) event horizon.

46. Rank in order of increasing size: (a) solar system, (b) Local Group, and (c) galaxy.

EXERCISES (SYNTHESIS)

47. The 19th-century author and social commentator Thomas Carlyle wrote, "Why did not somebody teach me the constellations and make me at home in the starry heavens, which are always overhead and which I don't half know to this day?" What besides the names of the constellations didn't Thomas Carlyle know?

48. Is any star bright enough for us to see on a sunny day?

49. On the Moon, stars other than the Sun can be seen during the daytime. Why?

50. Which figure in this chapter best shows that a constellation seen in the background of a solar eclipse is one that will be seen six months later in the night sky?

51. We see the constellations as distinct groups of stars. Discuss why they would look entirely different from some other location in the universe, far distant from Earth.

52. Distinguish between the diurnal and intrinsic motions of celestial objects.

53. Which moves faster from horizon to horizon: the Sun or the Moon? Explain.

54. The Big Dipper is sometimes right-side up (can hold water) and at other times upside down (cannot hold water). What length of time is required for the Big Dipper to change from one position to the other?

55. Why does the Big Dipper change its position in the night sky over the course of the evening but Polaris, the North Star, remains relatively fixed in its position?

56. What is the relationship between a planetary nebula and a white dwarf?

57. What do the outward and inward forces acting on a star have to do with its size?

58. What is the relationship between a white dwarf and a nova?

59. What event marks the birth of a star? When does a star die?

60. What does the color of a star tell you about the star?

61. What is expected to happen to the Sun in its old age?

62. When can a burnt-out collapsing star rekindle itself?

63. In what sense are we all made of star dust?

64. How is the gold in your mother's ring evidence of ancient stars that ran through their life cycles long before the solar system came into being?

65. Would you expect metals to be more abundant in old stars or in new stars? Defend your answer.

66. What is the evidence for believing our Sun is a relatively young star in the universe?

67. Some stars contain fewer heavy elements than our Sun contains. What does this indicate about the age of such stars relative to the age of our Sun?

68. Why is there a lower limit on the mass of a star? (What can't happen in a low-mass accumulation of hydrogen atoms and other interstellar material?)

69. What keeps a main-sequence star from collapsing?

70. How does the energy of a protostar differ from the energy that powers a star?

71. Why don't nuclear fusion reactions occur on the outer layers of stars?

72. Why are massive stars generally shorter lived than low-mass stars?

73. Why are supermassive stars relatively rare?

74. What does the spin rate of a star have to do with whether or not it has a system of planets?

75. With respect to stellar evolution, what is meant by the statement, "The bigger they are, the harder they fall"?

76. Why isn't the Sun able to fuse carbon nuclei in its core?

77. Which has the highest surface temperature: a red star, white star, or blue star?

78. In terms of the life cycle of the Sun, explain why life on Earth cannot last forever.

79. Elements heavier than iron are created in stars. Are they formed in the same way as elements lighter than iron? Explain.

80. A black hole is no more massive than the star from which it collapsed. Why, then, is gravitation so intense near a black hole?

81. What happens to the radial distance of the event horizon as more and more mass falls into the black hole? Explain.

82. What is the difference between a black hole's photon sphere and its event horizon.

83. Will the Sun become a supernova? A black hole? Defend your answer.

84. Are there galaxies other than the Milky Way that can be seen with the unaided eye? Discuss.

85. From where does a quasar release its energy?

86. Does the Milky Way galaxy contain an active galactic nucleus?

DISCUSSION QUESTIONS (EVALUATION)

87. Compare and contrast astronomy and astrology.

88. Project what human civilization would be like if our Sun were hidden in a dusty part of the galaxy such that no stars were ever visible to us at night.

89. It takes an infinite amount of time to watch an object fall through a black hole's event horizon. In what sense, therefore, can it be said that black-hole singularity does not exist?

90. Why is it important to have a science-based understanding of the structure of our universe? Try condensing your answer into a single philosophical sentence.

READINESS ASSURANCE TEST (RAT)

If you have a good handle on this chapter, if you really do, then you should be able to score 7 out of 10 on this RAT. If you score less than 7, you need to study further before moving on.

Choose the BEST answer to each of the following.

1. Summer and winter constellations are different because
   (a) of the spin of Earth about its polar axis.
   (b) the night sky faces in opposite directions in summer and winter.
   (c) of the tilt of Earth's polar axis.
   (d) the universe is symmetrical and harmonious.

2. Polaris is always directly over
   (a) the North Pole.
   (b) any location north of the equator.
   (c) the equator.
   (d) the South Pole.

3. The star nearest Earth is
   (a) Proxima Centauri.   (c) Mercury.
   (b) Polaris.   (d) the Sun.

4. The property of a star that relates to the amount of energy per unit time it is producing is its
   (a) luminosity.   (d) volume.
   (b) apparent brightness. (e) mass.
   (c) color.

5. The longest-lived stars are those of
   (a) low mass.   (c) intermediate mass.
   (b) high mass.   (d) infinite mass.

6. We do not see stars in the daytime because
   (a) the Sun blocks them.
   (b) they simply don't exist in the daytime part of the sky.
   (c) skylight overwhelms starlight.
   (d) of the lack of contrast with moonlight.
   (e) the solar wind obscures them from view.

7. After our Sun burns its supply of hydrogen, it will become a
   (a) white dwarf.   (d) red giant.
   (b) black dwarf.   (e) blue giant.
   (c) black hole.

8. A black hole is
   (a) an empty region of space with a huge gravitational field.
   (b) a small region that has the mass of many galaxies.
   (c) the remains of a giant collapsed star.
   (d) as large as its photon sphere.

9. The shape of an active starburst galaxy tends to be
   (a) elliptical.   (c) irregular.
   (b) spiral.   (d) all of the above

10. Scientists estimate the age of our universe to be about
   (a) 5000 years old.
   (b) 1 billion years old.
   (c) 14 billion years old.
   (d) 42 billion years old.

Answers to RAT

1. c, 2. a, 3. d, 4. a, 5. a, 6. c, 7. d, 8. c, 9. c, 10. c

CHAPTER 28

The Structure of Space and Time

IN THE preceding two astronomy chapters we discussed our galactic backyard, the solar system; our neighborhood of stars, the Milky Way galaxy; and our neighborhood of galaxies, the Local Group. We left off with a discussion on how superclusters of galaxies extend throughout the universe. Note in the photograph shown above that each spot of light is not a star, but an entire galaxy! Consider further that this photo reveals only one tiny portion of our observable universe.

Now we conclude with discussions of two very broad-reaching and most fascinating topics: *cosmology* and *relativity*. Through cosmology we attempt to answer such questions as How did the universe come into being? and What might be its ultimate fate? Relativity, first postulated by Einstein, is the study of how space, energy, and mass are related to time.

28.1 Looking Back in Time

EXPLAIN THIS What three major lines of evidence strongly support the Big Bang theory?

LEARNING OBJECTIVE
Describe the Big Bang and three lines of evidence that support it.

Imagine some fantastic optical instrument through which we can see the actual history of human civilization. Tune into some 40,000 years ago and you are able to witness the migration of humans into the Australian subcontinent. Re-focus to some 4500 years ago and see the building of the Egyptian pyramids. With such a device we would have an amazing window into our past. The accuracy of history books would be assured.

When it comes to the history of the universe, we have exactly such a device. It is called the *telescope*. The speed of light is fast, but the universe is exceedingly large. The light from our nearest star, Proxima Centauri, takes 4.2 years to reach us. Thus, as we view this star, we are literally looking at the star as it appeared 4.2 years ago. Similarly, the Andromeda galaxy we see today is the Andromeda galaxy as it appeared 2.5 million years ago.

Want to see the history of our universe? All we need to do is look through our telescopes. The farther away the object, the older it is. Start cataloging the past as we see it today, and soon we come to an appreciable understanding of how the universe itself formed. The study of the overall structure and evolution of the universe is called **cosmology**. With the advent of modern technology, especially with the development of space telescopes, the field of cosmology is currently in a golden era of discovery.

In this chapter we aim to present some of the more well-established findings of cosmology, which necessarily include the nature of time and space as spelled out by Einstein (Section 28.3). To provide a sense of the current excitement, however, we also dip into some speculations, such as the possible final fate of the universe. At minimum, this chapter will serve as your launching pad for exploring the latest news as it happens through online resources such as those listed in Table 28.1.

We begin with how it all began with a bang—the *Big Bang*.

MasteringPHYSICS
TUTORIAL: Hubble's Law
VIDEO: From the Big Bang to Galaxies

TABLE 28.1   WEB RESOURCES FOR COSMOLOGY

- The Harvard-Smithsonian Center for Astrophysics
  http://www.cfa.harvard.edu

- The Hubble Space Telescope
  http://Hubblesite.org

- The Wilkinson Microwave Anisotropy Probe
  http://map.gsfc.nasa.gov

The Big Bang

Not so long ago it was commonly thought that our Milky Way galaxy comprised the whole universe. Then in the early 1920s, the astronomer Edwin Hubble, using the new Mt. Wilson telescope, discovered that the Andromeda "nebula" was, in fact, a separate galaxy farther away than the outermost stars of the Milky Way. This was a monumental discovery, but Hubble didn't stop there. As discussed in the previous chapter, Hubble identified and measured the distances to numerous other galaxies. What he discovered next was most astonishing—the galaxies are all receding from one another. Furthermore, the farther away the galaxy, the greater the velocity with which it is receding.

FIGURE 28.1

INTERACTIVE FIGURE

Every ant on the expanding balloon sees all other ants moving farther away. Each ant may therefore think that it is at the center of the expansion. Not so! There is no center to the surface of the balloon, just as there are no edges.

FIGURE 28.2
Edwin Hubble (1889–1953), astronomer and cosmologist, is shown here in 1923 using the 100-inch telescope at the Mt. Wilson Observatory, where he worked for most of his life. In 1929, he announced Hubble's law, which states that the farther apart galaxies are, the faster they move.

Hubble's observations had two major implications. The first was that if you could run the movie backward, you would find a time when the whole universe was compressed to a very small point. Perhaps the universe as we know it has not always existed. Perhaps there was a moment at which the universe itself was born. This beginning point from which all matter and energy within our universe arose is referred to as the **Big Bang**.

The second major implication of Hubble's observations is that the universe is not contained within a region of space. Rather, space is *in* the universe and this space is rapidly expanding. This is peculiar because you may first think that the Big Bang occurred within an already existing infinite space and that matter and energy flew outward from this Big Bang to occupy this space. If this were the case, however, the distribution of galaxies we observe today and their relative motions would be quite different. To reemphasize this often confused concept: when we talk about the expansion of the universe, we are referring to an expansion of the very structure of space itself. A useful analogy is a group of ants on a balloon that is expanding, as shown in Figure 28.1. As the balloon is inflated, every ant sees every other ant moving farther away. Likewise, in an expanding universe, any observer sees all other galaxies moving away. So the Big Bang marked not only the beginning of time, but the beginning of space.

So how was Hubble able to measure the distances to incredibly distant galaxies? He needed the power of the newly constructed Mt. Wilson telescope, which allowed him to distinguish individual stars within neighboring galaxies (Figure 28.2). As Hubble studied these distant stars he discovered that some of them were of a certain type, called *Cepheids*, that regularly change how bright they are over a period of a few days.

At the time of Hubble's observations, astronomers could calculate the luminosity of a Cepheid from its periodic changes in brightness. So by measuring the periods of the distant Cepheids, Hubble was able to calculate their luminosities. Astronomers use the term *luminosity* to mean how much energy a star puts out per second. The more luminous a star, the brighter it appears. Note that the apparent *brightness* of a star is not its luminosity. Why? Because brightness diminishes with distance. The farther away you are from a luminous star, the dimmer it appears. Like gravity, light grows weaker via the inverse-square law—if you double your distance from the light source, you see that the light is $\left(\frac{1}{2}\right)^2 = \frac{1}{4}$ as bright. So by comparing a Cepheid's luminosity with its brightness and plugging this into the inverse-square law, Hubble was able to calculate the distance to any galaxy containing Cepheids.

CHECKPOINT

Star A is four times as bright as star B, yet the two stars have the same luminosity.
1. **How can this be?**
2. **If star A is 100 light-years away, how far away is star B?**

Were these your answers?
1. Light from a light source becomes dimmer with distance. Star A, therefore, is brighter because it is closer to us.

2. The inverse-square law tells us that the brightness of a star is related to the inverse distance to the star squared:

$$\text{Apparent brightness} \sim \left(\frac{1}{d}\right)^2$$

So star A is four times as bright if star B is twice as far away:

$$\text{Apparent brightness of star A} \sim \left(\frac{1}{1}\right)^2 = 1$$

$$\text{Apparent brightness of star B} \sim \left(\frac{1}{2}\right)^2 = 0.25$$

Notice that a brightness of 1 is four times as great as a brightness of 0.25. From these values, we see that star A has a distance factor of 1, while Star B has a distance factor of 2. Star B is twice as far, which is 200 light-years.

> **FYI** The North Star, also known as Polaris, is a Cepheid variable with a period of about four days. Its change in apparent brightness is not discernable to the naked eye. Interestingly, the North Star is in orbit around two companion stars, which means that the North Star is actually a triple star. These three stars are too close together and too far away (about 430 light-years) to be distinguished with the naked eye.

How was Hubble able to calculate the velocity with which these distant galaxies were receding? Recall the Doppler effect, discussed in Chapter 10. Recall that sound waves stretch out when the sound source recedes and compress when the sound source approaches. Light waves do the same. As a galaxy recedes, the wavelength of light reaching us is stretched out. This longer wavelength means a lower frequency, which for visible light means a shift toward the red end of the color spectrum. Hubble studied the spectra of light coming from distant galaxies and measured the degree to which that light was shifted toward the red. The greater the *redshift*, the greater the velocity of the receding galaxy. Hubble's great accomplishment was collecting both the distance and redshift data for many galaxies and then correlating these data on a graph. His graph showed a clear relationship—the farther away the galaxy, the greater the velocity of the galaxies from us, and the greater the redshift (Figure 28.3).

Galaxies recede from one another because the space between them is expanding. Astronomers thus follow an alternate explanation for why the spectra of galaxies show redshifts. As light waves travel through expanding space, the light waves themselves stretch out. This elongation of light waves due to the expansion of space is called the **cosmological redshift**. The farther away the galaxy, the longer its light has been traveling through expanding space, and hence the greater the cosmological redshift. The following Checkpoint illustrates this idea with an analogous rubber band.

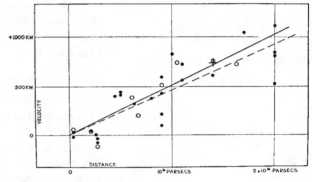

FIGURE 28.3
Hubble's original graph of velocity versus distance showing that more distant galaxies recede with greater velocities.

> Light from a distant galaxy is the light from glowing elements, which emit spectra of particular frequencies as discussed in Section 12.6. An examination of the full spectrum of a galaxy's light shows a pattern of peaks that are the sum of the spectra of all of the many glowing elements, primarily hydrogen and helium. If these peaks are shifted toward the red, then we know the galaxy must be receding away from us. The amount of redshift tells us how fast that galaxy is receding.

CHECKPOINT
You draw three dots equally spaced along the length of a rubber band. Assume the distance between these dots is 5 mm. So the distance between the first and second dots is 5 mm, while the distance between the first and third dots is 10 mm. In 1 s you stretch the rubber band to 10 times its original length.
1. **Show that the second dot recedes from the first dot with a speed of 45 mm/s. (Hint: speed = distance/time.)**

2. Show that the third dot recedes from the first dot with a speed of 90 mm/s.

3. Why do farther galaxies recede from us at faster rates?

Were these your answers?

1. The original distance was 5 mm, which stretched to 50 mm. The change in distance, therefore, was 45 mm over a time of 1 s. So the speed equals 45 mm/s.

2. The difference between 100 mm and 10 mm is 90 mm over a time of 1 s, which equals 90 mm/s, or twice the speed of 45 mm/s.

3. For an expanding rubber band, we see that a dot twice as far away travels twice as fast. The same goes with the universe: a galaxy twice as far away from us is moving twice as fast. The reason for these receding velocities is the expansion of space itself. (A few nearby galaxies, including Andromeda, buck the trend and move toward us.)

Hubble suggested a simple relationship between the distance of an object from Earth and its velocity of recession. This simple relationship, which has been confirmed by numerous measurements over many decades, is known as **Hubble's law**:

$$\text{Hubble's law: } v = H \times d$$

where v is the velocity of a galaxy as deduced from its cosmological redshift, H is a constant known as Hubble's constant, and d is the distance of the galaxy from Earth. This law tells us, for example, that if one galaxy is twice as far away as another, the farther galaxy recedes twice as fast from us. Furthermore, if a galaxy were to move from where we are now to its present location with a velocity v, then the time of this trip would be the distance it travels divided by its velocity:

$$t = \frac{d}{v}$$

We use Hubble's law to substitute for v:

$$t = \frac{d}{H \times d} = \frac{1}{H}$$

When we enter the value of H into this equation, we have an estimate of the interval of time for the expansion. Said another way, we have determined the age of the universe. Plugging the currently accepted value of H into the equation indicates that the universe is nearly 14 billion years old. Wow!

Cosmic Background Radiation

In addition to the expansion of the universe, a second line of evidence supporting the Big Bang theory is *cosmic background radiation*. In 1964, scientists Arno Penzias and Robert W. Wilson, working at Bell Labs in New Jersey, used a simple radio receiver to survey the heavens for radio signals (Figure 28.4). No matter which way they directed their receiver, they detected microwaves with a wavelength of 7.35 cm coming toward Earth. Penzias and Wilson were puzzled. With no specific source of the radiation, where were the microwaves coming from and why?

FIGURE 28.4
Arno Penzias and Robert Wilson in front of the microwave receiver they used to detect the afterglow of the Big Bang.

Remember that any object above absolute zero emits energy in the form of electromagnetic radiation. The frequency of this radiation is proportional to the absolute temperature of the emitter. Theorists at Princeton, working around the same time as Penzias and Wilson, showed that if the universe began in a primordial explosion as described by the Big Bang, it would still be cooling off. Further, they showed that the temperature of the early universe would have cooled by today to an average temperature of about 3 K. A universe of this temperature would be expected to emit microwave radiation of just the frequency observed by Penzias and Wilson. Thus the influx of microwave radiation that initially puzzled Penzias and Wilson was found to be emitted by the cooling universe itself. This faint microwave radiation is now referred to as **cosmic background radiation** and is taken as strong evidence of the Big Bang (Figure 28.5).

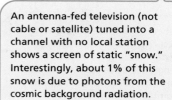

An antenna-fed television (not cable or satellite) tuned into a channel with no local station shows a screen of static "snow." Interestingly, about 1% of this snow is due to photons from the cosmic background radiation.

FIGURE 28.5
This all-sky map of the cosmic background radiation taken by the Wilkinson Microwave Anisotropy Probe (WMAP) satellite reveals an average temperature of about 2.73 K everywhere. This is the cooled-off remnants of the Big Bang. The color shows minor temperature variation on the order of $\pm$ 0.0001 K.

BIG BANG HELIUM

As the universe expands, it cools. The cosmic background microwave radiation tells us that it has cooled to an average of about 3 K. Working backward from the present, scientists estimate a temperature in excess of 100 billion K within a few seconds after the Big Bang. At this extreme temperature, protons would be turning into neutrons and neutrons would be turning into protons. The rates of these transformations would be the about the same, meaning the ratio of protons to neutrons in this super early universe would be about 1 : 1.

Over the next three minutes, the temperature dropped to less than 100 billion K, which favored the formation of protons.* Soon protons outnumbered neutrons. By the time the ratio of protons to neutrons was 7 : 1, the universe would have *cooled enough* to allow for nuclear fusion. We say "cooled enough" because the universe was still quite hot, but not as hot as before. At this point, the protons and neutrons would have begun fusing into deuterium nuclei (consisting of one proton and one neutron). Deuterium nuclei would have then fused to form helium. They would continue doing so until the deuterium

grew sparse. (Incidentally, this explains why deuterium is such a rare isotope today.) Continued cooling would prevent the further fusion of helium into heavier elements, such as carbon. By the time the three minutes of helium synthesis were over, the universe would be left with about 75% hydrogen and 25% helium, which is what we observe in the universe today. The Milky Way is actually about 28% helium. The 3% excess helium is likely from the fusion of hydrogen in the stars. No galaxy detected so far, however, has a helium abundance of less than 25%, just as predicted by the Big Bang theory.

* Neutrons are slightly more massive than protons. The transformation of a proton into a neutron, therefore, requires a slightly greater input of energy in accordance with $E = mc^2$. At temperatures below 100 billion K, there is insufficient energy to allow this transformation. The conversion of neutrons into protons, however, releases energy at lower temperatures in accordance with the second law of thermodynamics.

Where exactly did the Big Bang occur? Was it at some now far distant point from which we have long since traveled? The answer is an astounding NO! Rather, every point in the universe was present at the Big Bang. It's just that all of these points are now quite far away from one another. So if you want to point to the location of the Big Bang, just point your finger to the tip of your nose, or anywhere for that matter. You can't miss.

The Abundance of Hydrogen and Helium

The Big Bang answers another cosmic mystery involving the element helium. Measurements show that matter in the universe is about 75% hydrogen and 25% helium. (Heavier elements such as those found on Earth are very minor contributors to the total amount of matter in the universe.) Hydrogen is the simplest of all elements, consisting of a single proton nucleus. It makes sense that hydrogen was the original element. Helium, however, is a more complex element containing a nucleus of two protons and two neutrons. We know helium is produced from the fusion of hydrogen in stars. But the number of stars is insufficient to account for all the observed helium—not more than 10% of the observed helium could have originated in stars. Most of the helium observed in the universe must have been created elsewhere. As described in the Big Bang Helium box, the Big Bang model predicts that the early universe would have been favorable to the formation of helium, but not the formation of other elements. A more detailed analysis shows that the amount of helium created just after the Big Bang should be about that which we observe in the universe today.

In summary, three major lines of evidence strongly support the Big Bang theory. The first is the current expansion of space, which causes galaxies to recede from one another. The second is the discovery of cosmic background radiation, which is the Big Bang's afterglow. The third is the Big Bang's ability to explain the observed proportions of elements. Because of these and other similar lines of evidence, the Big Bang has come to be widely accepted by the scientific community as the most viable explanation for the beginning of our universe.

28.2 Cosmic Inflation

LEARNING OBJECTIVE
Identify three successes of the
theory of cosmic inflation.

EXPLAIN THIS Does the universe expand into space or does the space of the universe expand?

FIGURE 28.6
For their efforts in developing the theory of cosmic inflation, Alan Guth (left) and Andrei Linde (right) in 2004 together shared the prestigious Cosmology Prize of the Peter Gruber Foundation.

A *theory* is a comprehensive idea that can be used to explain a broad range of phenomena. As was discussed in the prologue of this textbook, theories are not carved in stone. Rather, they pass through stages of refinement. With each successive refinement, the theory grows stronger. In the early 1980s, the Big Bang theory gained such a refinement from the insights of physicists Alan Guth and Andrei Linde (Figure 28.6). The Big Bang theory had predicted that particles known as *magnetic monopoles* should be abundant in our current universe. Such monopoles, however, have never been detected despite countless attempts to find them. Guth realized that monopoles would become exceedingly rare if within an instant after the Big Bang, space expanded in a dramatic burst. (The monopoles would have been diluted beyond our ability to detect them.) This burst would have begun 10^{-38} second after the start of the Big Bang and ended a mere 10^{-36} second later. During this ultrabrief speck of time, the space within the universe inflated by a factor of 10^{30}. This moment of dramatic expansion is known as **cosmic inflation**. Soon after Guth first recognized the significance of cosmic inflation, Linde followed through with many important refinements on the theory. To the delight of Guth, Linde, and many others, the idea of cosmic inflation solved many other mysteries that had been plaguing the Big Bang theory.

One outstanding question was why the cosmic background radiation is so uniform in its temperature. Normally, two regions need to be in contact with each other in order to come to the same temperature. A cup of hot water and a cup of cold water, for example, won't blend to make a cup of warm water unless they are able to mix. In contrast, a constant expansion of space after the Big Bang with different regions so far apart would result in regions unable to mix their thermal energy with that of other regions, even after the lifetime of the universe. We would therefore expect some regions to be warmer than others. But studies of the cosmic background radiation clearly show a nearly uniform temperature. According to cosmic inflation, this uniform temperature was achieved in the moments before cosmic inflation. During these moments, all regions of the universe were still bound together. Being so close, they were able to intermingle and thus come to a homogeneous temperature. So the Big Bang didn't explode right away. Instead, it held tight, expanding "slowly" for as long as it could before bursting like a bubble, which then calmed down almost immediately, but not before a dramatic inflation of space.

As indicated by the cosmic background radiation, the distribution of energy within the expanded universe is remarkably uniform, but minor fluctuations still appear. According to cosmic inflation, the reason for these fluctuations can be traced to the realm of quantum mechanics, which tells us that physical attributes such as position and momentum become fuzzy on the scale of the very, very small—such as the size of subatomic particles. Indeed, the universe before inflation was very, very small—at one point even *smaller* than the size of subatomic particles. Assuming today's laws of quantum mechanics held true at the time of

This image from NASA depicts the early history of the universe starting with the Big Bang. Quantum fluctuations were amplified by a sudden burst in size during a period known as inflation. Some 380,000 years later, particles settled such that the universe became transparent. It's from this moment that we see the cosmic background radiation afterglow. The first stars didn't form until about 400 million years later.

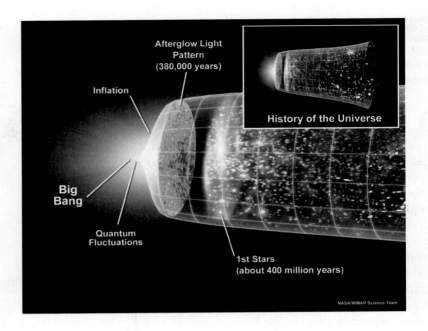

the Big Bang, then quantum fuzziness had to be a feature of the universe's birth (Figure 28.7). Upon inflation, ultrasmall quantum variations in position and momenta would have been magnified. The distribution of matter and energy in the post-inflation era would not have been perfectly uniform. The result would be a bit of clumpiness. Gravitational forces would then ensure that these bits of clumpiness would be the seeds for further clumpiness, which would eventually give rise to such things as galaxies and superclusters of galaxies. Thus, when we look at the distribution of galaxies and superclusters of galaxies in the universe today, we are looking at the world of quantum mechanics magnified to a universal scale. So cosmic inflation explains not only the uniformity of matter and energy in the universe, but how minor fluctuations gave rise to the cosmic structure we see through our telescopes today. (See Figure 27.38 from the preceding chapter.)

A third success of cosmic inflation has to do with the very shape of the universe. According to Einstein's *theory of general relativity*, which we discuss next, we live in three dimensions of space plus one dimension of time. Together these add up to a four-dimensional universe of **spacetime**. Mass has the effect of *curving* spacetime. With such curvature, parallel lines may eventually meet or diverge. By analogy, consider the surface of planet Earth. Draw two parallel lines from the equator pointing exactly north–south. Because of the curvature of Earth, these parallel lines eventually merge at the poles. You'll see this happen with the longitudinal lines on any globe. Similarly, mass causes spacetime to curve in such a way that parallel lines could eventually merge. Calculations of the total mass of the universe showed that the universe itself should be curved in very detectable ways. But this is not what astronomers observe. Instead, light travels in rather straight paths (unless the light passes through intense gravitational fields surrounding a star or black hole). We find that the observable universe is flat. Parallel lines remain parallel.

An explanation involves inflation. Consider an ant on a balloon. It may find that by walking straight it also walks in circles—it comes back to its starting point. Looking up, an unusually smart ant may see the curved horizon and conclude that its environment is highly curved. Take this balloon and inflate it to the size of the Sun. At this point the ant would walk on and on to new territories on what looked to the ant to be an infinite flat plane. Parallel lines would appear to remain parallel. Similarly, we humans on the surface of Earth drive our cars around as though the world is flat. But from afar, we know that Earth's surface is not flat—it is curved. Likewise for the universe: what we observe within our own corner of the universe appears to be flat, in a four-dimensional sort of way.

So is the universe as a whole curved? We do not know the answer to that question, but interestingly enough, most astrophysical data to date are consistent with a flat universe. This has major implications regarding the ultimate fate of the universe. But before delving into these implications, we need to explore the notion of spacetime, as well as the now firmly established idea that gravity is best viewed not as a force, but as a curvature of spacetime.

28.3 General Relativity

EXPLAIN THIS What is the downward acceleration of a light beam at Earth's surface?

LEARNING OBJECTIVE
Summarize general relativity and the three successful predictions made by this theory.

In 1915, Einstein published his now famous **general theory of relativity**, which was a huge reworking of Newton's well-tested laws of gravity. According to Einstein, gravity need not be viewed as a force exerted by one object on another through distance. Instead, gravity is the effect we witness when a large mass—such as a planet, star, or galaxy—causes a curvature in the shape of the spacetime in which it resides. The curvature of four-dimensional spacetime (three of distance and one of time) can be expressed mathematically, but is rather impossible for us to visualize. We can get a glimpse of this curvature by considering a simplified analogy in two dimensions: a heavy ball resting in the middle of a large rubber sheet. The more massive the ball, the more it warps the two-dimensional surface. A marble rolling past this warped surface would trace a bent path, as shown in Figure 28.8. If the marble were to roll closer to the ball, it might trace an elliptical curve around the ball. Rolled at the proper angle and speed (assuming no friction to slow it down), the marble could enter a perpetual orbit around the ball. In this analogy, we see here there is no "force" holding the marble to the ball. Rather, the marble is simply following the natural curvature of the rubber sheet.

In a similar manner, the Sun's mass curves the spacetime around it. Our planet is moving sideways along a path that follows this curvature. We are moving at just the proper speed so that the orbit is sustained. If we were to slow down, the steepness of the curvature would cause us to fall into the Sun. If we were to speed up sufficiently, we would escape this curvature and leave the Sun for good.

FIGURE 28.8
A two-dimensional analogy of four-dimensional warped spacetime. Spacetime near a star is curved in a way similar to the surface of a rubber sheet when a heavy ball rests on it.

CHECKPOINT

If the Sun suddenly became less massive, what would happen to our orbit?

Was this your answer?
With less mass, the Sun would have less of a warping effect on spacetime. If Earth maintained the same speed, it would be compelled to break away from the Sun. By Newton's laws, we would say that we escaped the Sun's now weaker gravitational pull. According to Einstein, however, we would say that spacetime became more flat, which allowed us to move along our merry way.

Einstein was led to this new theory of gravity by thinking about observers in accelerated motion. He imagined himself in a spaceship far away from gravitational influences (Figure 28.9). In such a spaceship at rest or in uniform motion relative to the distant stars, he and everything within the ship would float freely; there would be no "up" and no "down." But if rocket thrusters were activated to accelerate the ship, things would be different; phenomena similar to gravity would be observed. The wall adjacent to the rocket thrusters would push up against any occupants and become the floor, while the opposite wall would become the ceiling. Occupants in the ship would be able to stand on the floor and even jump up and down. If the acceleration of the spaceship were equal to *g*, the occupants could well be convinced the ship was not accelerating but was simply at rest on the surface of Earth.

Einstein concluded that gravity and acceleration are related, a conclusion now called the **principle of equivalence**:

Local observations made in an accelerated frame of reference cannot be distinguished from observations made in a gravitational field.

To examine this new "gravity" due to acceleration (as opposed to acceleration due to gravity, discussed in Chapter 4), Einstein considered the consequence of releasing two balls—say, one of wood and the other of lead. When released in a uniformly moving spaceship, the balls would continue to move with the spaceship at the moment of release. If the spaceship were moving at a *constant velocity* (zero acceleration), the balls would remain suspended in the same place, because both the spaceship and the balls would be moving by the same amount. But if the spaceship is accelerating, the floor moves upward faster than the balls, which are soon intercepted by the floor (Figure 28.10). Both balls, regardless of their masses, would meet the floor at the same time. Recall the legend of Galileo's demonstration at the Leaning Tower of Pisa. Occupants of the spaceship might be prone to attribute their observations to the force of gravity.

The two interpretations of the falling balls are equally valid, and Einstein incorporated this equivalence, or the impossibility of distinguishing between gravitation and acceleration, into the foundation of his general theory of relativity. The principle of equivalence states that observations made in an accelerated reference frame are indistinguishable from observations made in a Newtonian gravitational field. This equivalence would be interesting but not revolutionary if it applied only to mechanical phenomena, but Einstein went further and stated that the principle holds for *all* natural phenomena; it holds for optical and all electromagnetic phenomena as well.

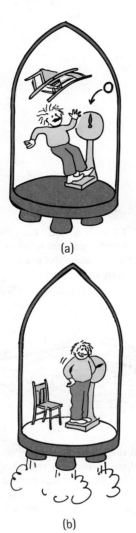

FIGURE 28.9
(a) Everything is weightless inside a nonaccelerating spaceship far away from gravitational influences. (b) When the spaceship accelerates, an occupant inside feels "gravity."

FIGURE 28.10
To an observer inside the accelerating ship, a lead ball and a wood ball appear to fall together when released.

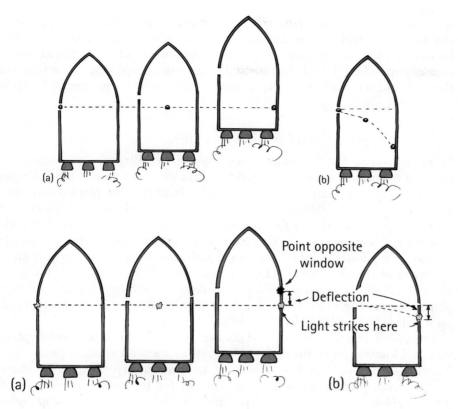

Point opposite
window

Deflection

Light strikes here

FIGURE 28.11
(a) An outside observer sees a horizontally thrown ball travel in a straight line, and, because the spaceship is moving upward while the ball travels horizontally, the ball strikes the wall below a point opposite the window. (b) To an inside observer, the ball bends as if in a gravitational field.

FIGURE 28.12
(a) An outside observer sees light travel horizontally in a straight line, but, like the ball in Figure 28.11, it strikes the wall slightly below a point opposite the window. (b) To an inside observer, the light also bends as if responding to a gravitational field.

Consider a ball thrown sideways in a stationary spaceship in the absence of gravity. The ball follows a straight-line path relative to both an observer inside the spaceship and a stationary observer outside the spaceship. But if the spaceship is accelerating, the floor overtakes the ball and it hits the wall below a point opposite the window (Figure 28.11). An observer outside the spaceship still sees a straight-line path, but to an observer in the accelerating spaceship the path is curved; it is a parabola. The same result holds true for a beam of light (Figure 28.12). The only difference is the curvatures of both. If the ball were somehow thrown at the speed of light, both curvatures would be the same.

According to Newton, tossed balls curve because of a force of gravity. According to Einstein, both tossed balls and light curve when close to a planet or star because the spacetime in which they travel is curved (Figure 28.13).

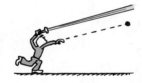

FIGURE 28.13
The trajectory of a flashlight beam is identical to the trajectory of a baseball "thrown" theoretically at the speed of light. Both paths curve equally in a uniform gravitational field.

CHECKPOINT
Whoa! We learned previously that the pull of gravity is an interaction between masses. And we learned that light has no mass. Now we say that light can be bent by gravity. Isn't this a contradiction?

Was this your answer?
There is no contradiction when the mass–energy equivalence (see Chapter 13) is understood. It's true that light has no mass, but it is not "energyless." The fact that gravity pulls downward on light provides evidence that gravity pulls on the energy of light. Energy indeed is equivalent to mass!

General relativity, then, calls for a new geometry: a geometry not only of curved space but of curved time as well—a geometry of curved four-dimensional spacetime.* General relativity tells us that the presence of mass produces the curvature or warping of spacetime. Instead of visualizing gravitational forces between masses, we abandon altogether the notion of force and instead think of masses responding in their motion to the curvature or warping of the spacetime they inhabit. It is the bumps, depressions, and warpings of geometric spacetime that *are* the phenomena of gravity.

Tests of General Relativity

Using equations for four-dimensional spacetime, Einstein recalculated the orbits of the planets about the Sun. Beyond the planets, space is almost flat, and objects travel along nearly straight-line paths. Near the Sun, planets and comets travel along curved paths because of the curvature of space. With only one minor exception, his theory gave almost exactly the same results as Newton's law of gravity. The exception was that Einstein's theory predicted that the elliptical orbits of the planets should slip forward with each revolution, a process known as *precession* (Figure 28.14). This precession would be very slight for distant planets and more pronounced for planets close to the Sun. Mercury is the only planet close enough to the Sun for the curvature of space to produce an effect on it that is not predicted by Newton's law.

Precession in the orbits of planets caused by the gravity of other planets was well known. Since the early 1800s, astronomers measured a precession of Mercury's orbit—about 574 seconds of arc per century. The gravity exerted by the other planets was found to account for the precession—except for 43 seconds of arc per century. Even after all known corrections due to possible effects by other planets had been applied, the calculations of physicists and astronomers failed to account for the extra 43 seconds of arc. Either Venus was extra massive or some other, previously undiscovered planet was pulling on Mercury. And then came the explanation of Einstein, whose general-relativity equations, when applied to Mercury's orbit, predict the extra 43 seconds of arc per century!

As a second test of his theory, Einstein predicted that measurements of starlight passing close to the Sun would be slightly deflected but large enough to be measured. This deflection of starlight can be observed during an eclipse of the Sun. (Measuring this deflection has become a standard practice at every total eclipse since the first measurements were made during the total eclipse of 1919.) A photograph taken of the darkened sky around the eclipsed Sun reveals the presence of the nearby bright stars. The positions of the stars then are compared with those in other photographs of the same area taken at other times during the night. In every instance, the deflection of starlight has supported Einstein's prediction (Figure 28.15).

FIGURE 28.14
A precessing elliptical orbit when viewed from directly above is seen to advance forward in its orbit.

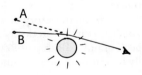

FIGURE 28.15
Starlight bends as it grazes the Sun. Point A shows the apparent position; point B shows the true position.

CHECKPOINT
Why don't we notice the bending of light by gravity in our everyday environment?

Was this your answer?
Only because the deflection by Earth is so tiny and too small to measure.

---

* Don't be discouraged if you cannot visualize four-dimensional spacetime, let alone the curving of spacetime. Einstein himself often told his friends, "Don't try. I can't do it either." Perhaps we are not too different from the great thinkers around Galileo who couldn't think of a moving Earth!

Einstein made a third prediction—that gravity causes a slowing down of time. He predicted, for example, that a minute of time on the surface of a massive planet should last longer compared to a minute of time on the surface of a less massive planet. Thus, a clock sitting on the more massive planet would be seen to be running slower than an identical clock sitting on the less massive planet. Gravity alters time! The stronger the gravity, the greater the effect.

Einstein suggested a way to measure this effect when he formulated the principle of equivalence. He knew that all atoms emit light at specific frequencies characteristic of the vibrational rate of electrons within the atom. Every atom is therefore a "clock," and a slowing down of atomic vibrations indicates the slowing down of such atomic clocks. An atom on the Sun, therefore, should emit light of a lower frequency (slower vibration) than light emitted by the same atom on Earth. Astronomers have found that the slowing down of time for atoms on our Sun is obscured by their thermal motion. The time-slowing effect of gravity, however, has since been observed and accurately measured in white dwarf stars, which are dimmer and have gravitational fields much stronger than that of our Sun. Furthermore, ultraprecise experiments here on Earth have shown that time runs slower at the bottom of a tall tower, than at the top of the tower. The difference is exceedingly small, but the results of the experiment are reproducible and in strict accordance with Einstein's prediction (Figure 28.16).

Einstein's general theory of relativity was an elaboration of his **special theory of relativity**, which he had proposed ten years earlier in 1905. Through special relativity, Einstein showed how matter and energy are really two forms of the same thing as related by his now famous equation $E = mc^2$, introduced in Chapter 13. In the *Conceptual Physical Science Practice Book*, we present a fascinating aspect of special relativity that tells us how time changes with motion. For example, an astronaut taking a two-year round trip into space at high speeds could come back to find that Earth has aged 2000 years! For now, however, we turn to something even more mysterious, which is the potential presence of a form of matter completely invisible to our sight and sense of touch.

FIGURE 28.16
If you move from a distant point down to the surface of Earth, you move in the direction in which the gravitational force acts—toward a location where clocks run more slowly. A clock at Earth's surface runs slower than a clock located farther away.

28.4 Dark Matter

EXPLAIN THIS How is it possible to detect something in space that we can't see through our telescopes?

LEARNING OBJECTIVE
Describe the evidence for dark matter and its role in the formation of galaxies.

Evidence now suggests that the Big Bang generated matter in at least two different forms—one we can see and another that we can't. The visible form of matter is the "ordinary matter" made of subatomic particles such as protons, neutrons, and electrons. As you learned in earlier chapters, these particles combine to make the atoms of the periodic table. Atoms then combine to make molecules, such as those that make our bodies. This matter we can touch. We interact with it directly. You, the planets, and the stars are made of this form of matter, which we will from here on refer to as **ordinary matter**.

The second form of matter generated from the Big Bang is quite unlike ordinary matter. This form of matter does not recognize the strong nuclear force, which means it cannot clump to form atomic nuclei. Neither does this second form of matter recognize the electromagnetic force, which makes it invisible to light as well as our sense of touch. As was explained in Chapter 12, the electromagnetic force is responsible for the repulsion between electrons. The reason you can't walk through a wall is because of the repulsions between the electrons

Dark matter

Luminous matter

FIGURE 28.17

As large as a galaxy is, its diffuse halo of dark matter is much larger. This halo may measure up to 10 times the diameter of the luminous galaxy and be about six times as massive.

FYI An alternative theory to dark matter is Modified Newtonian Dynamics (MOND), proposed by the physicist Mordehai Milgrom in the early 1980s. According to MOND, Newton's equation $a = F/m$ fails when the force is exceedingly weak, perhaps because of quantum effects. A modified version of this equation was thus created to account for the observed orbital velocities of stars within galaxies. The theory is controversial. To date, most astronomers find the dark matter theory to be more acceptable.

in your body and the electrons in the wall. If the wall were made of this invisible matter you would be able to walk right through it. Of course, you wouldn't be able to see the wall either. This invisible form of matter that we cannot see or touch is known as **dark matter**.

So if dark matter is invisible to us, how do we know it's there? The answer is that this ghostly form of matter gives itself away by its gravitational effects. One of the first clues came to us as we were mapping the speeds at which stars orbit our galactic center. According to the laws of gravity, orbital speed is a function of the force of gravity between the orbiting object and the object being orbited—the greater the force of gravity, the greater the orbital speed. The inner planets of our solar system, for example, orbit the Sun much faster than the outer planets because they are closer to the Sun and experience greater gravitational forces. Relative to our galaxy, we might expect the same trend—stars closest to the galactic center should have faster orbital speeds than stars farther out. Interestingly, that's not what we observe! Instead, stars closer to the galactic center and those farther out orbit with about the same speed. How can this be?

For a solar system, planets orbit as they do because most of the solar system's mass is concentrated within the central sun. For a galaxy such as the Milky Way or Andromeda, it sure looks as though most of the mass is concentrated within the central bulge. The measured orbital speeds of stars, however, tell us that the bulk of the galaxy's mass lies outside the galaxy itself within a diffuse yet massive invisible halo many times the diameter of the visible galaxy, as shown in Figure 28.17. We know it's invisible because all of our telescopes see right through it! But something must be there affecting stellar orbital speeds.

Another bit of evidence for dark matter comes from measuring the speeds of galaxies as they orbit one another within clusters. The measured speeds tell us that the masses of these galaxies are many times greater than the total mass of all their stars.

Lastly, we know that the path of light is bent by gravity much as it is bent by an optical lens. A cluster of galaxies, therefore, can bend the light from an even farther cluster lying directly behind it—we say the foreground cluster behaves as a *gravitational lens*. Such a gravitational lensing effect was shown in Figure 27.33. The degree to which the light from the distant cluster bends is a function of the mass of the foreground cluster. Once again, the degree of light bending tells us that the mass of the closer cluster far exceeds that which we would expect based solely on the cluster's luminosity. So by carefully studying the bending of light from distant galaxies, we can build a map of the dark matter's distribution. Such a study using the Hubble Space Telescope is illustrated in Figure 28.18.

So the evidence for dark matter is strong. Our current problem, however, is trying to figure out exactly what dark matter is made of. It's clear that dark matter is not simply ordinary matter, such as expired stars, that have gotten so cold that they emit no light. Although numerous theories abound about the fundamental nature of dark matter, no dark matter particles have been detected. Until that happens, we are left with yet another fascinating mystery of our universe.

Galaxy Formation

We can speculate that when the universe formed, ordinary matter plus an even greater amount of dark matter was produced. Held together by gravity, the ordinary matter and dark matter would have been strewn outward in a clumpy

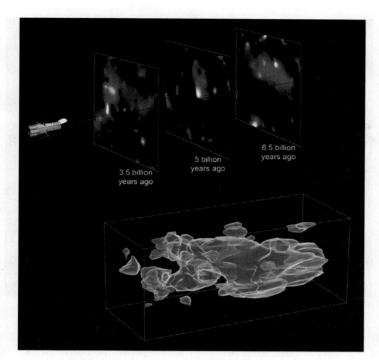

FIGURE 28.18
Images of dark matter, shown in blue, were created through the Hubble Space Telescope's Cosmic Evolution Survey. Of course, the dark matter is not blue. This graphic, however, shows dark matter's distribution over a narrow region of the sky back to about 6.5 billion years ago.

fashion. Within a clump, ordinary and dark matter may initially have been uniformly mixed together. These two forms of matter differ significantly in that when ordinary matter collides with ordinary matter, energy is released as heat. With this loss of energy, the ordinary matter loses orbital speed and thus falls closer to the center of the clump. Over time, while all dark matter stayed distributed throughout the clump, the ordinary matter became concentrated at the center (Figure 28.19). This concentration of ordinary matter at the center of the clump allowed for the formation of stars. Also, as ordinary matter congregated toward the center, the rate of rotation would increase—angular momentum would have been conserved. If the original clump of ordinary and dark matter was just barely spinning, then the stars forming at the center would take on the form of an elliptical galaxy. If the original clump was spinning a bit faster, then the new stars would be spinning fast enough to flatten the galaxy, much like a rapidly spinning ball of pizza dough. The resulting disk would take on the form of a spiral galaxy. So from a clump of ordinary and dark matter, ordinary matter condensed to form a central galaxy. The dark matter remained diffuse, forming an invisible halo surrounding the newly formed galaxy.

FIGURE 28.19
Ordinary matter condensed out of a mixture of dark and ordinary matter.

28.5 Dark Energy

EXPLAIN THIS What began to accelerate around 7.5 billion years ago?

LEARNING OBJECTIVE
Discuss the significance of the presence of dark energy.

In the years just before Hubble's discovery of the expansion of the universe, Einstein was struggling to understand why gravity wasn't causing the universe to collapse in a Big Crunch. He was thinking of the universe as static, neither collapsing nor expanding. But in order for the universe to remain static against the inward curvature of gravity, there would need to be another fundamental outward force counteracting gravity. In other words, if gravity is the "pull inward," there should be a phenomenon that creates a "push outward."

FIGURE 28.20
The expansion of space started to accelerate about 7.5 billion years ago, which is shown on this diagram as a gradual widening just after the development of galaxies. The cause of this accelerated expansion has been given the name *dark energy*. But just because we can name something doesn't mean we understand it. Such is the case with dark energy.

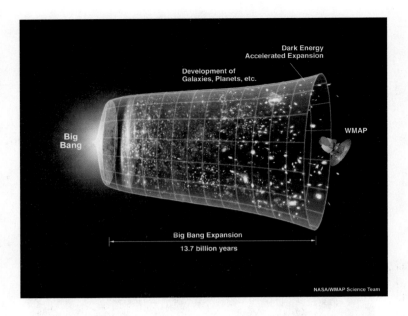

To allow for such a balance, he introduced into his equations the idea of what he called the *cosmological constant*. He had no proof for the existence of such a phenomenon. Rather, he just postulated it to account for the apparent stability of the universe.

About 10 years later, Hubble announced that the universe was *not* stable, but very dynamic and expanding. Einstein later remarked that his own failure to predict a dynamic universe was the "greatest blunder of his life." Subsequent workings of Einstein's equations showed that a static universe would not be stable. Just the slightest push this way or that would cause it to either collapse or expand. Einstein abandoned his notion of a cosmological constant, which for the next 75 years remained a historical curiosity.

Then in the 1990s, some 40 years after Einstein's death, two teams of astronomers made a startling discovery. High-resolution data from very distant galaxies showed that space, beginning about 7.5 billion years ago, started to accelerate in its expansion. Galaxies are not simply coasting away from each other and slowing down. Rather, some unknown form of energy is causing an increase in the rate at which galaxies are receding. Here was a phenomenon that acted as the opposite of gravity, possible evidence of Einstein's once-proposed cosmological constant.

There remains much speculation about how these findings affect the fate of the universe. This form of unknown energy is generally described as **dark energy** (Figure 28.20). Theorists have various ideas about the nature of dark energy, including the possibility that it may be Einstein's famed cosmological constant.

Some current models suggest that dark energy finds its source within the emptiness of spacetime. Matter has the effect of pulling spacetime together so that it contracts. The classic example is what happens upon the formation of a black hole—spacetime contracts to a point of zero volume (and infinite density). In the absence of matter—within a perfect vacuum—empty spacetime seethes with an energy that creates an opposite curvature, which allows spacetime to expand. As more empty spacetime is created, the *vacuum energy* becomes more predominant, which accelerates the formation of even more empty spacetime. Distant galaxies are thus seen to be accelerating from each other. It is as though gravity and dark energy are diametrically opposed. When gravity gains full rein, the result is infinite density—a black hole. When dark energy gains full rein, the result may be an infinite vacuum—an empty hole.

FYI The space through which our planets orbit contains about 100 hydrogen atoms per liter. The space between the stars of our galaxy contains about 2 hydrogen atoms per liter. If you want really empty space, you'll need to travel to the vast voids that separate the superclusters as was shown in Figure 27.36. It's only from these regions that dark energy appears to be taking hold. The space within and between our local galaxies is too dense!

Many other intriguing models attempt to explain the nature of dark energy. Which model is best can be determined only after the collection of more evidence. Stay tuned for science news reports. In particular, the European Space Agency's *Planck Surveyor* may answer many of our current questions. However, this powerful space telescope and its many successors will, no doubt, also raise more questions than they answer. One thing is for sure: the universe holds no shortage of mysteries.

28.6 The Fate of the Universe

EXPLAIN THIS   When should speculations be discounted?

The universe is expanding. Matter, however, has the effect of reversing this expansion. Might enough matter exist within the universe to halt or even reverse this expansion? Before the discovery of dark energy and dark matter, astronomers had calculated that the ordinary matter in the universe was only 4% of the mass needed to halt the expansion. Dark matter was then discovered to be about six times as abundant as ordinary matter, making up about 23% of the mass needed to halt the expansion. If the remaining $100\% - (4\% + 23\%) = 73\%$ of matter could be accounted for, then the mass of the universe would be sufficient to one day halt the expansion. The geometric shape of such a universe would be flat. In other words, parallel lines would never touch. Because we observe the universe to be rather flat, astronomers puzzled over where this remaining 73% of matter might be.

Then came the discovery of dark energy. Recall that matter and energy are related by Einstein's equation $E = mc^2$. Both dark matter and dark energy, therefore, need to be included in tabulations of the total composition of the universe. Both would have an effect on the curvature of the observable universe. The abundance of dark energy then fits the bill for making up the remaining 73% of the composition of the observable universe, as shown in Figure 28.21. So it makes sense that our observable universe is flat, or quite close to it. There's a twist, however, in that dark energy causes the *expansion* of space, not contraction. The current thinking of most cosmologists, therefore, is that our flat universe is destined for an eternal expansion. This gives rise to a select number of possible scenarios for the fate of our universe.

In one scenario, called **heat death**, the universe will continue to expand, approaching absolute zero and a state of maximum entropy. After 10^{14} years, all stars will have exhausted all possible fuel. The universe will be fully dark. After

LEARNING OBJECTIVE
Compare and contrast three scientifically possible fates of the universe.

Although *dark energy* and *dark matter* both begin with the word *dark*, they are uniquely different. One is a form of matter; the other is a form of energy. Both are still mysterious and may yet be disproven by new evidence or alternate explanations. By the next edition of this textbook, the picture may look quite different!

Mastering**PHYSICS**
TUTORIAL: Fate of the Universe

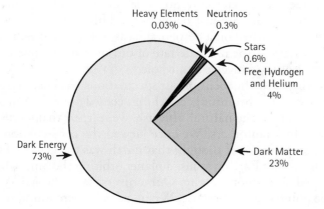

FIGURE 28.21
Ordinary matter, the stuff from which we and the galaxies we live in are made, makes up not more than 4% of the composition of the universe. The remainder is primarily dark matter (23%) and dark energy (73%), both of which we know very little about.

FIGURE 28.22
If the strength of dark energy
remains constant, we can expect
our universe to suffer heat death. If
dark energy gains strength, then the
ultimate demise of our universe may
be the Big Rip.

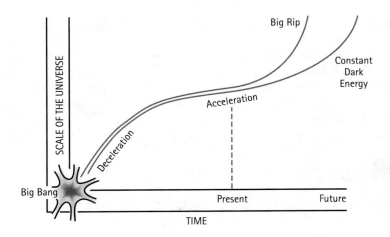

FIGURE 28.22
If the strength of dark energy remains constant, we can expect our universe to suffer heat death. If dark energy gains strength, then the ultimate demise of our universe may be the Big Rip.

10^{16} years, planets and stars will be flung from their orbits because of random collisions, most then falling into supermassive black holes and the rest forming scattered stellar debris. After 10^{40} years, all protons and neutrons will have decayed, leaving behind gamma radiation and *leptons*, of which the electron is an example. From this time to about 10^{100} years in the future, supermassive black holes will likely be the dominant form of mass in the universe. But by the passing of 10^{150} years, they too will depart as they evaporate into photons and leptons. From then to perhaps 10^{1000} years in the future, the wavelengths of photons as well as all other remaining particles will be stretched to the lowest energy states possible. Entropy will have won supreme victory.

A second scenario, known as the **Big Rip**, recognizes that the influence of dark energy may grow stronger over time (Figure 28.22). As the universe expands exponentially, clusters of galaxies will be pulled farther apart, past the point of being visible to each other. Neighboring galaxies will then be pulled out of each other's sight. About 60 million years before the end, stars within galaxies will fly off in every direction. Some three months before the end, solar systems will disperse. In the last few minutes, stars and planets will become unbound. Then in the last instant, all atoms will be ripped apart, followed by their subatomic particles. The time frame for the Big Rip is estimated to be about 35 billion years after the Big Bang, which would be about 21 billion years from now. And where will all the matter fly off to? Good question. Perhaps to subsequent Big Bangs?

Alan Guth, one of the early developers of cosmic inflation, supports the idea that the universe will not end everywhere at once. Some 14 billion years ago, a tiny patch of primordial material inflated to form our own observable universe. But inflation will keep on going for other patches of this primordial material and will continue to do so eternally. So while our region of the universe expands to infinity, other regions are just being born. This scenario, in which observable universes are spawned on a perpetual basis, is known as **eternal inflation**. With eternal inflation, the ultimate fate of our own observable universe is not the same as that of the universe as a whole.

It is interesting how the focus of our speculations has been narrowing down over our history. In the beginning, anything seemed possible. But as we opened our eyes and minds to the natural universe, we learned that some speculations were more worthy than others. We once viewed the constellations as heavenly gods. Of course, it was once thought that Earth was the center of the universe. Then we realized that Earth was just a planet orbiting the Sun, which itself was a medium-sized star among many. Our universe was the Milky Way galaxy until Hubble pointed out otherwise. When galaxies were found to be receding, many hypothesized that gravity might be strong enough to pull them back together in a Big Crunch. This possibility has since been eliminated. Now we are at the point of wondering what might happen after eternal expansion. Our current speculations are just that—speculations. But they are highly refined speculations based on a great deal of collected evidence. As we continue to look at the natural universe with ever more powerful telescopes, we can expect that our speculations will become ever more refined. This is the mind-opening art of science, which seeks to learn the nature of the universe for what it is—not for what we might wish it to be. Stay tuned.

For instructor-assigned homework, go to www.masteringphysics.com

SUMMARY OF TERMS (KNOWLEDGE)

Big Bang The primordial creation and expansion of space at the beginning of time.

Big Rip A model for the end of the universe in which dark energy grows stronger over time and causes all matter to rip apart.

Cosmic background radiation The faint microwave radiation emanating from all directions that is the remnant heat of the Big Bang.

Cosmic inflation The moment of the sudden and brief burst in the size of the universe immediately after the Big Bang.

Cosmological redshift The elongation of light waves due to the expansion of space.

Cosmology The study of the overall structure and evolution of the universe.

Dark energy An unknown form of energy that appears to be causing an acceleration of the expansion of space; thought to be associated with the energy exuded by a perfect vacuum.

Dark matter Invisible matter that has made its presence known so far only through its gravitational effects.

Eternal inflation A model of the universe in which cosmic inflation is not a one-time event but rather progresses to continuously spawn an infinite number of observable universes in its wake.

General theory of relativity The theory first proposed by Einstein discussing the effects of gravity on spacetime.

Heat death A model for the end of the universe in which all matter and energy disperse to the point of maximum entropy.

Hubble's law The farther away a galaxy is from Earth, the more rapidly it is moving away from us: $v = H \times d$.

Ordinary matter Matter that responds to the strong nuclear, weak nuclear, electromagnetic, and gravitational forces. This is matter made of protons, neutrons, and electrons, which includes the atoms and molecules that make us and our immediate environment.

Principle of equivalence Local observations made in an accelerated frame of reference cannot be distinguished from observations made in a Newtonian gravitational field.

Spacetime The continuum in which we live, consisting of three dimensions of space plus the fourth dimension of time.

Special theory of relativity The theory first proposed by Einstein discussing the effects of uniform motion on space, time, energy, and mass.

READING CHECK QUESTIONS (COMPREHENSION)

28.1 Looking Back in Time

1. Is the universe in space or is space in the universe?
2. What is a Cepheid?
3. Which depends on distance: a star's brightness or its luminosity?
4. What is the approximate age of the universe?
5. According to the cosmic background radiation, what is the average temperature of the universe today?

28.2 Cosmic Inflation

6. According to cosmic inflation theory, how long did it take for the universe to increase its size by a factor of 10^{30}?
7. At what point did the universe's temperature even out?
8. What did inflation do to the quantum fluctuations found within the early universe?
9. What would happen to the curvature of a balloon if it could be inflated to the size of the Sun?
10. How many dimensions are there in spacetime?

28.3 General Relativity

11. In what year did Einstein publish his general theory of relativity?

12. Can an accelerated frame of reference be distinguished from a gravitational field?

13. You release a ball while standing against the floor of an accelerating spaceship. What happens to the ball?

14. What did general relativity predict about the orbit of Mercury?

15. What happens to starlight as it passes close to the Sun?

28.4 Dark Matter

16. What type of matter is visible?

17. If we can't see dark matter, how do we know it is there?

18. Is dark matter found mostly within or just outside a galaxy?

19. The closer a planet is to the Sun, the faster it orbits. Is it also true that the closer a star is to the center of the galaxy, the faster it orbits?

20. A huge cloud of ordinary matter and dark matter are uniformly mixed together. Over time, the ordinary matter becomes concentrated toward the center of this cloud. Why?

28.5 Dark Energy

21. Did Einstein first believe that the universe was static or dynamic?

22. What was Einstein's cosmological constant?

23. What was Einstein's "greatest blunder of his life"?

24. According to recent evidence, how long ago did the expansion of the universe start accelerating?

25. What does WMAP stand for?

28.6 The Fate of the Universe

26. What is likely the major constituent of our universe?

27. Which is more abundant: dark matter or ordinary matter?

28. According to the heat death scenario, about how long will it take for the black holes of the universe to evaporate?

29. What does the Big Rip scenario assume about dark energy?

30. What scenario for the fate of the universe proposes that cosmic inflation is not a one-time event?

ACTIVITIES (HANDS-ON APPLICATION)

31. Find a quiet and comfortable place to sit where you won't be disturbed. Place your hands on your legs and straighten your back. Imagine a string attached to the top of your head lifting you upward. Gaze at the floor a couple feet ahead of you. Relax your jaw so that your lips remain slightly open. Once you are comfortable, begin focusing on your out-breath. Simply notice your out-breath as it leaves your body and disperses into your surroundings. When you find yourself carried away by a thought, label the thought as "thinking" and let the thought drift away. New thoughts will keep popping into your mind, but return your attention to your out-breath. Try doing this for at least 10 minutes. Using this technique, you may come to the point of being able to distinguish your bare presence from your discursive thoughts.

32. If you tried the preceding activity, you will know it is not easy to find yourself amid your many thoughts. Nonetheless, consider it a worthy endeavor and healthy habit to get to know yourself in such a fashion. Once you feel comfortable with staying with your breath, you can practice welcoming in a particular thought, which becomes your focus while your breathing recedes to the background. A good thought to start with is the idea that your bare presence and the bare presence of every sentient being shine with a wisdom we might call "basic goodness." Underneath all the veneer, no matter what the flavor, there is basic goodness in everyone. Try doing this for at least 10 minutes.

33. After you have given the contemplation of basic goodness a fair shake, there are a zillion other thoughts also worthy of your contemplation. Within the context of this chapter, try contemplating the sheer size of the universe. Envision yourself in your chair, then mentally zoom out to see yourself in the room, then in the building, then on the planet. Zoom out stepwise so that you can experience each stage. Zoom out to the solar system circling within a galaxy that orbits neighboring galaxies within a supercluster that is one supercluster of billions upon billions of superclusters. Once you have zoomed fully out, focus on that sensation of bigness. When you lose the sensation, start over moving as slowly or as fast as is comfortable.

THINK AND RANK (ANALYSIS)

34. Rank these elements in order of increasing abundance in the universe: (a) helium, (b) hydrogen, (c) carbon.

35. Rank the following in order of increasing abundance in the universe: (a) dark matter, (b) ordinary matter, (c) dark energy.

36. Rank the events in order of oldest to most recent: (a) star formation, (b) inflation, (c) Big Bang.

37. Rank the following in order of increasing duration: one minute on the surface of (a) Earth, (b) Mercury, and (c) the Moon.

EXERCISES (SYNTHESIS)

38. When was most of the helium in the universe created?

39. What does the expansion of space do to light passing through it?

40. What is the relationship between dark energy and Einstein's cosmological constant?

41. A police officer pulls you over for speeding. He tells you that his radar tracked you moving at a rate of 45 mph away from his parked police car. Were you really speeding away from him, or was the space between the two of you simply expanding?

42. Is space just the absence of matter?

43. The average temperature of the universe right now is about 2.73 K. Will this temperature likely go up or down over the next billion years?

44. Mass can transform into energy and energy can transform into mass in accordance with Einstein's equation $E = mc^2$. So if the amount of mass-energy in the universe remains constant, what happens to the amount of space within the universe? What then is happening to the density of the universe?

45. If the initial universe remained hotter for a longer period of time, would there likely be more or less helium?

46. If not from the Big Bang, then where do elements heavier than helium come from?

47. No galaxy found so far is made of less than 25% helium. If not from the stars, where did this helium come from?

48. Are astronomers able to point their telescopes in the direction of where the Big Bang occurred?

49. True or False: A helium balloon here on Earth pops, releasing direct remnants of the Big Bang. Explain.

50. If we are made of stardust, what are stars made of?

51. Astronomers tell us that the average temperature of the universe is a rather homogeneous 2.73 K ($+/- 0.0001$ K). But how can this be when we know the temperatures of stars are ultrahot?

52. What are three lines of evidence supporting cosmic inflation?

53. What if there was a symmetry to cosmic background radiation such that the pattern of temperature fluctuations in one direction was exactly the same pattern seen in the exact opposite direction? What might you conclude about the curvature of the universe?

54. Explain how weight can be caused by both gravity and acceleration.

55. If gravity is not a force, then what is it?

56. You toss a tennis ball up and down in front of you as you sit in a jet airplane accelerating down the runway for takeoff. Why is the tennis ball difficult to catch?

57. You toss a tennis ball up and down in front of you as you sit in a jet airplane cruising at a constant speed of 500 mph. Why is the tennis ball easy to catch despite the fact that the jet is moving so fast?

58. You are free-floating in a spaceship at uniform motion deep in outer space. A ball is hovering in front of you. Suddenly the ball starts moving to the floor. What is happening to the spaceship? What happens to you?

59. Where does a clock run slower: at the front end or back end of an accelerating spaceship?

60. Several billion years in the future our Sun will grow in size to become a red giant, whose surface will extend to the present orbit of Earth. After this happens, will the slowing of time on the Sun's surface be more or less pronounced than it is today?

61. An astronaut is provided a "gravity" when the ship's engines are activated to accelerate the ship. This requires the use of fuel. Is there a way to accelerate and provide "gravity" without the sustained use of fuel? Explain. (Hint: Consider Appendix A.)

62. Should a person who wants a long life live at the top or at the bottom of a tall apartment building?

63. You shine a beam of colored light to a friend up in a high tower. Will the color of light your friend receives be the same color you send? Explain.

64. An identical twin leaves his identical twin brother behind to live on another planet that is half as massive. After living apart for many years, the two brothers reunite. According to general relativity, which of the two brothers is now younger than the other? Explain.

65. Why does the gravitational attraction between the Sun and Mercury vary?

66. When do clocks move slowest on Mercury?

67. What might we assume about the distribution of dark matter if the planets in our solar system all orbited the Sun at about the same speed?

68. Early astronomers such as Kepler and Newton developed the laws of gravity based on the motion of the planets around the Sun. How might these laws have been different if our solar system was surrounded by a thick halo of dark matter?

69. What force allows dark matter to clump?

70. Why doesn't dark matter clump together as effectively as ordinary matter?

71. If dark matter is affected by gravity, might there be lots of it surrounding us here on the surface of the Earth?

72. What is one important difference between dark matter and dark energy?

73. Why isn't dark energy called the dark force?

74. If the universe were unchanging and there were an infinite number of stars, what effect might this have on the darkness of a clear night sky?

75. If there are so many stars and galaxies, why do we see so much darkness in the clear night sky?

76. If we can't even predict the weather, how can we ever expect to predict the fate of the universe?

DISCUSSION QUESTIONS (EVALUATION)

77. Compare and contrast the Big Bang with a black hole.

78. Discuss your understanding of and thoughts on the potential fates of the universe as described in Section 28.6. For more background, review the references in Table 28.1.

READINESS ASSURANCE TEST (RAT)

If you have a good handle on this chapter, if you really do, then you should be able to score 7 out of 10 on this RAT. If you score less than 7, you need to study further.

Choose the BEST answer to each of the following.

1. Which of the following is not accepted evidence for the Big Bang?
   (a) cosmic background radiation
   (b) homogeneity of the temperature of the universe
   (c) the abundance of helium
   (d) dark energy

2. If the universe stopped expanding at this very moment, the cosmological redshift of distant galaxies would
   (a) promptly disappear.
   (b) gradually turn into a cosmological blueshift.
   (c) stop getting more redshifted than it already is.
   (d) not matter because such a sudden halt to expansion would knock all galaxies and planets off their orbits and we would no longer be here to talk about it.

3. What percentage of galaxies were created during cosmic inflation?
   (a) 100%
   (b) about 70%
   (c) about 24%
   (d) 0%

4. What do cosmic inflation and dark energy have in common?
   (a) Both are responsible for an expansion of the universe.
   (b) Both are still occurring.
   (c) They both arise from the vast voids of space between superclusters.
   (d) nothing

5. Light bends in a gravitational field. Why isn't this bending taken into consideration by surveyors who use laser beams as straight lines?
   (a) The bending of light occurs only in outer space.
   (b) Red light doesn't bend so much, which is why lasers are red.
   (c) Laser light is an exception to this rule.
   (d) Light travels so fast that the curvature is not noticeable.

6. Time slows in a gravitational field. Would time slow in the artificial gravity produced in a rotating space habitat?
   (a) no, because there is no acceleration
   (b) yes, because of the principle of equivalence
   (c) no, because the space station is spinning at a constant rate
   (d) yes, even though the gravity is artificial

7. Should it be possible in principle for a photon to orbit a star?
   (a) yes, if the photon is moving at the speed of light
   (b) yes, if the star's mass-density is huge enough to make it a black hole
   (c) no, because photons, by definition, always move in straight paths
   (d) no, because stars already produce countless photons of their own

8. Dark matter is
   (a) ordinary matter that is no longer emitting light.
   (b) altering the orbits of our planets.
   (c) attracted to ordinary matter.
   (d) repelled by ordinary matter.

9. Space in our local universe is
   (a) not empty.
   (b) seething with energy.
   (c) not well understood.
   (d) all of the above

10. Which theory for the fate of the universe assumes that dark energy will grow stronger?
    (a) heat death
    (b) Big Rip
    (c) eternal inflation
    (d) all of the above

Answers to RAT

1. d, 2. c, 3. d, 4. a, 5. d, 6. b, 7. b, 8. c, 9. d, 10. b

Linear and Rotational Motion

WHEN WE DESCRIBE THE MOTION of something, we say how it moves relative to something else (Chapter 1). In other words, motion requires a reference frame (an observer, origin, and axes). We are free to choose this frame's location and to have it moving relative to another frame. When our frame of motion has zero acceleration, it is called an *inertial frame*. In an inertial frame, force causes an object to accelerate in accord with Newton's laws. When our frame of reference is accelerated, we observe fictitious forces and motions. Observations from a carousel, for example, are different when it is rotating and when it is at rest. Our description of motion and force depends on our "point of view."

We distinguish between *speed* and *velocity* (Chapter 1). Speed is how fast something moves, or the time rate of change of position (excluding direction): a *scalar* quantity. Velocity includes direction of motion: a *vector* quantity whose magnitude is speed. Objects moving at constant velocity move the same distance in the same time in the same direction.

Another distinction between speed and velocity has to do with the difference between distance and net distance, or *displacement*. Speed is *distance per duration* while velocity is *displacement per duration*. Displacement differs from distance. For example, a commuter who travels 10 kilometers to work and back travels 20 kilometers, but has "gone" nowhere. The distance traveled is 20 kilometers and the displacement is zero. Although the instantaneous speed and instantaneous velocity have the same value at the same instant, the average speed and average velocity can be very different. The average speed of this commuter's round-trip is 20 kilometers divided by the total commute time—a value greater than zero. But the average velocity is zero. In science, displacement is often more important than distance. (To avoid information overload, we have not treated this distinction in the text.)

Acceleration is the rate at which velocity changes. This can be a change in speed only, a change in direction only, or both. Negative acceleration is often called *deceleration*.

In Newtonian space and time, space has three dimensions—length, width, and height—each with two directions. We can go, stop, and return in any of them. Time has one dimension, with two directions—past and future. We cannot stop or return, only go. In Einsteinian space-time, time is treated as a fourth dimension.

Computing Velocity and Distance Traveled on an Inclined Plane

Recall from Chapter 1 Galileo's experiments with inclined planes. We considered a plane tilted such that the speed of a rolling ball increases at the rate of 2 meters per second each second—an acceleration of 2 m/s^2.

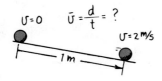

FIGURE A.1

The ball rolls 1 m down the incline in 1 s and reaches a speed of 2 m/s. Its average speed, however, is 1 m/s. Do you see why?

So at the instant it starts moving its velocity is zero, and 1 second later it is rolling at 2 m/s, at the end of the next second 4 m/s, the end of the next second 6 m/s, and so on. The velocity of the ball at any instant is simply Velocity = acceleration × time. Or, in shorthand notation $v = at$. (It is customary to omit the multiplication sign, ×, when expressing relationships in mathematical form. When two symbols are written together, such as the at in this case, it is understood that they are multiplied.)

How fast the ball rolls is one thing; how *far* it rolls is another. To understand the relationship between acceleration and distance traveled, we must first investigate the relationship between instantaneous velocity and *average velocity*. If the ball shown in Figure A.1 starts from rest, it will roll a distance of 1 meter in the first second. What will be its average speed? The answer is 1 m/s (it covered 1 meter in the interval of 1 second). But we have seen that the *instantaneous velocity* at the end of the first second is 2 m/s. Since the acceleration is uniform, the average in any time interval is found the same way we usually find the average of any two numbers: add them and divide by 2. (Be careful not to do this when acceleration is not uniform!) So if we add the initial speed (zero in this case) and the final speed of 2 m/s and then divide by 2, we get 1 m/s for the average velocity.

In each succeeding second we see the ball roll a longer distance down the same slope in Figure A.2. Note the distance covered in the second time interval is 3 meters. This is because the average speed of the ball in this interval is 3 m/s. In the next 1-second interval the average speed is 5 m/s, so the distance covered is 5 meters. It is interesting to see that successive increments of distance increase as a *sequence of odd numbers*. Nature clearly follows mathematical rules!

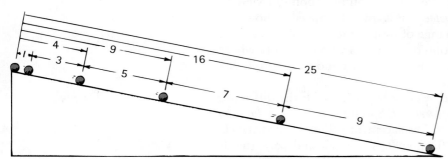

FIGURE A.2

If the ball covers 1 m during its first second, then in each successive second it will cover the odd-numbered sequence of 3, 5, 7, 9 m, and so on. Note that the total distance covered increases as the square of the total time.

Investigate Figure A.2 carefully and note the *total* distance covered as the ball accelerates down the plane. The distances go from zero to 1 meter in 1 second, zero to 4 meters in 2 seconds, zero to 9 meters in 3 seconds, zero to 16 meters in 4 seconds, and so on in succeeding seconds. The sequence for *total distances* covered is of the *squares of the time*. We'll investigate the relationship between distance traveled and the square of the time for constant acceleration more closely in the case of free fall.

CHECKPOINT

During the span of the second time interval, the ball begins at 2 m/s and ends at 4 m/s. What is the *average speed* of the ball during this 1-s interval? What is its *acceleration*?

Were these your answers?

$$\text{Average speed} = \frac{\text{beginning + final speed}}{2} = \frac{2 \text{ m/s} + 4 \text{ m/s}}{2} = 3 \text{ m/s}$$

$$\text{Acceleration} = \frac{\text{change in velocity}}{\text{time interval}} = \frac{4 \text{ m/s} - 2 \text{ m/s}}{1 \text{ s}} = \frac{2 \text{ m/s}}{1 \text{ s}} = 2 \text{ m/s}^2$$

Computing Distance When Acceleration Is Constant

How far will an object released from rest fall in a given time? To answer this question, let us consider the case in which it falls freely for 3 seconds, starting at rest. Neglecting air resistance, the object will have a constant acceleration of about 10 meters per second each second (actually more like 9.8 m/s^2, but we want to make the numbers easier to follow).

$$\text{Velocity at the } \textit{beginning} = 0 \text{ m/s}$$

$$\text{Velocity at the } \textit{end} \text{ of 3 seconds} = (10 \times 3) \text{ m/s}$$

$$\textit{Average} \text{ velocity} = \tfrac{1}{2} \text{ the sum of these two speeds}$$

$$= \tfrac{1}{2} \times (0 + 10 \times 3) \text{ m/s}$$

$$= \tfrac{1}{2} \times 10 \times 3 = 15 \text{ m/s}$$

$$\text{Distance traveled} = \text{average velocity} \times \text{time}$$

$$= \left(\tfrac{1}{2} \times 10 \times 3\right) \times 3$$

$$= \tfrac{1}{2} \times 10 \times 3^2 = 45 \text{ m}$$

We can see from the meanings of these numbers that

$$\text{Distance traveled} = \tfrac{1}{2} \times \text{acceleration} \times \text{square of time}$$

This equation is true for an object falling not only for 3 seconds but for any length of time, as long as the acceleration is constant. If we let d stand for the distance traveled, a for the acceleration, and t for the time, the rule may be written, in shorthand notation,

$$d = \tfrac{1}{2}at^2$$

This relationship was first deduced by Galileo. He reasoned that if an object falls for, say, twice the time, it will fall with *twice the average speed*. Since it falls for *twice* the time at *twice* the average speed, it will fall *four* times as far. Similarly, if an object falls for *three* times the time, it will have an average speed *three* times as great and will fall *nine* times as far. Galileo reasoned that the total distance fallen should be proportional to the *square* of the time.

In the case of objects in free fall, it is customary to use the letter g to represent the acceleration instead of the letter a (g because acceleration is due to *gravity*). While the value of g varies slightly in different parts of the world, it is approximately equal to 9.8 m/s^2 (32 ft/s^2). If we use g for the acceleration of a freely falling object (negligible air resistance), the equations for falling objects starting from a rest position become

$$v = gt$$

$$d = \tfrac{1}{2}gt^2$$

Much of the difficulty in learning physics, like learning any discipline, has to do with learning the language—the many terms and definitions. Speed is somewhat different from velocity, and acceleration is vastly different from speed or velocity. Please be patient with yourself as you find learning the similarities and the differences among physics concepts is not an easy task.

FIGURE A.3

When Chelcie Liu releases both balls simultaneously, he asks, "Which will reach the end of the equal-length tracks first?" (Hint: On which track is the average speed of the ball greater? Then, double hint: Which wins, the fast ball or the slow ball?)

CHECKPOINT

1. **An auto starting from rest has a constant acceleration of 4 m/s². How far will it go in 5 s?**
2. **How far will an object released from rest fall in 1 s? In this case the acceleration is $g = 9.8$ m/s².**
3. **If it takes 4 s for an object to freely fall to the water when released from the Golden Gate Bridge, how high is the bridge?**

Were these your answers?

1. Distance $= \frac{1}{2} \times 4 \times 5^2 = 50$ m
2. Distance $= \frac{1}{2} \times 9.8 \times 1^2 = 4.9$ m
3. Distance $= \frac{1}{2} \times 9.8 \times 4^2 = 78.4$ m

Notice that the units of measurement when multiplied give the proper units of meters for distance:

$$d = \tfrac{1}{2} \times 9.8 \text{ m} \times 16 = 78.4 \text{ m}$$

FIGURE A.4

When a phonograph record turns, a ladybug farther from the center travels a longer path in the same time and has a greater tangential speed.

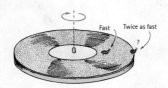

FIGURE A.5

The entire disk rotates at the same rotational speed, but ladybugs at different distances from the center travel at different tangential speeds. A ladybug twice as far from the center moves twice as fast.

Circular Motion

Linear speed is what we have been calling simply *speed*—the distance traveled in meters or kilometers per unit of time. A point on the perimeter of a merry-go-round or turntable moves a greater distance in one complete rotation than a point nearer the center. Moving a greater distance in the same time means a greater speed. The speed of something moving along a circular path is **tangential speed,** because the direction of motion is tangent to the circle.

Rotational speed (sometimes called angular speed) refers to the number of rotations or revolutions per unit of time. All parts of the rigid merry-go-round turn about the axis of rotation *in the same amount of time.* All parts share the same rate of rotation, or *number of rotations or revolutions per unit of time.* It is common to express rotational rates in revolutions per minute (rpm).* Phonograph records that were common a few years ago rotate at 33 1/3 rpm. A ladybug sitting anywhere on the surface of the record revolves at 33 1/3 rpm.

Tangential speed is *directly proportional* to rotational speed (at a fixed radial distance). Unlike rotational speed, tangential speed depends on the distance from the axis (Figure A.5). Something at the center of a rotating platform has no tangential speed at all, and merely rotates. But, approaching the edge of the platform, tangential speed increases. Tangential speed is directly proportional to the distance from the axis (for a given rotational speed). Twice as far from the rotational axis, the speed is twice as great. Three times as far from the rotational axis, there is three times as much tangential speed. When a row of people locked arm in arm at the skating rink makes a turn, the motion

---

* Physics types usually describe rotational speed in terms of the number of "radians" turned in a unit of time, for which they use the symbol ω (the Greek letter *omega*). There's a little more than 6 radians in a full rotation (2π radians, to be exact).

of "tail-end Charlie" is evidence of this greater speed. So tangential speed is directly proportional both to rotational speed and to radial distance.*

CHECK POINT

On a rotating platform similar to the disk shown in Figure A.5, if you sit halfway between the rotating axis and the outer edge and have a rotational speed of 20 rpm and a tangential speed of 2 m/s, what will be the rotational and tangential speeds of your friend who sits at the outer edge?

Was this your answer?

Since the rotating platform is rigid, all parts have the same rotational speed, so your friend also rotates at 20 rpm. Tangential speed is a different story; since she is twice as far from the axis of rotation, she moves twice as fast—4 m/s.

Torque

Whereas force causes changes in velocity, *torque* causes changes in rotation. To understand torque (rhymes with *dork*), hold the end of a meterstick horizontally with your hand. If you dangle a weight from the meterstick near your hand, you can feel the meterstick twist. Now if you slide the weight farther from your hand, the twist you feel is greater, although the weight is the same. The force acting on your hand is the same. What's different is the torque.

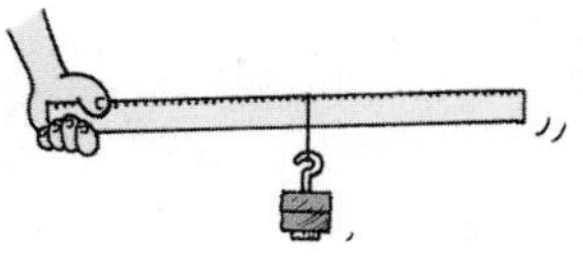

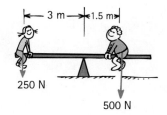

FIGURE A.6

If you move the weight away from your hand, you will feel the difference between force and torque.

$$\text{Torque} = \text{lever arm} \times \text{force}$$

Lever arm is the distance between the point of application of the force and the axis of rotation. It is the shortest distance between the applied force and the rotational axis. Torques are intuitively familiar to youngsters playing on a seesaw. Kids can balance a seesaw even when their weights are unequal. Weight alone doesn't produce rotation. Torque does, and children soon learn that the distance they sit from the pivot point is every bit as important as weight (Figure A.7). When the torques are equal, making the net torque zero, no rotation is produced.

Recall the equilibrium rule in Chapter 1—that the sum of the forces acting on a body or any system must equal zero for mechanical equilibrium. That is, $\Sigma F = 0$. We now see an additional condition. The *net torque* on a body or on a system must also be zero for mechanical equilibrium. Anything in mechanical equilibrium doesn't accelerate—neither linearly nor rotationally.

Suppose that the seesaw is arranged so that the half-as-heavy girl is suspended from a 4-meter rope hanging from her end of the seesaw (Figure A.8). She is now 5 meters from the fulcrum, and the seesaw is still balanced. We see that the lever-arm distance is 3 meters, not 5 meters. The lever arm about any axis of rotation is the perpendicular distance from the axis to the line along

FIGURE A.7

No rotation is produced when the torques balance each other.

FIGURE A.8

The lever arm is still 3 m.

---

* When customary units are used for tangential speed v, rotational speed ω, and radial distance r, the direct proportion of v to both r and ω becomes the exact equation $v = r\omega$. So the tangential speed will be directly proportional to r when all parts of a system simultaneously have the same ω, as for a wheel, disk, or rigid wand. (The direct proportionality of v to r is not valid for the planets because planets don't all have the same ω.)

FIGURE A.9

Although the magnitudes of the force are the same in each case, the torques are different.

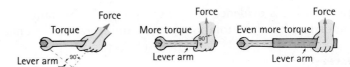

which the force acts. This will always be the shortest distance between the axis of rotation and the line along which the force acts.

This is why the stubborn bolt shown in Figure A.9 is turned more easily when the applied force is perpendicular to the handle, rather than at an oblique angle, as shown in the first figure. In the first figure, the lever arm is shown by the dashed line and is less than the length of the wrench handle. In the second figure, the lever arm is equal to the length of the wrench handle. In the third figure, the lever arm is extended with a pipe to provide more leverage and a greater torque.

> **CHECKPOINT**
> 1. If a pipe effectively extends a wrench handle to three times its length, by how much will the torque increase for the same applied force?
> 2. Consider the balanced seesaw in Figure A.7. Suppose the girl on the left suddenly gains 50 N, such as by being handed a bag of apples. Where should she sit in order to balance, assuming the heavier boy remains in place?
>
> **Were these your answers?**
> 1. Three times more leverage for the same force gives three times more torque. (This method of increasing torque sometimes results in shearing off the bolt!)
> 2. She should sit half-a-meter closer to the center. Then her lever arm is 2.5 m. This checks: 300 N × 2.5 m = 500 N × 1.5 m.

Angular Momentum

Things that rotate, whether a cylinder rolling down an incline or an acrobat doing a somersault, keep on rotating until something stops them. A rotating object has an "inertia of rotation." Recall, from Chapter 3, that all moving objects have "inertia of motion" or *momentum*—the product of mass and velocity. This kind of momentum is **linear momentum.** Similarly, the "inertia of rotation" of rotating objects is called **angular momentum.**

For the case of an object that is small compared with the radial distance to its axis of rotation, like a tetherball swinging from a long string or a planet orbiting around the sun, the angular momentum can be expressed as the magnitude of its linear momentum, mv, multiplied by the radial distance, r, (Figure A.10).* In shorthand notation, angular momentum = mvr. Like linear momentum, angular momentum is a vector quantity and has direction as

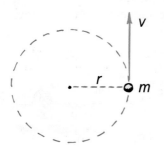

FIGURE A.10

A small object of mass m whirling in a circular path of radius r with a speed v has angular momentum mvr.

* For rotating bodies that are large compared with radial distance—for example, a planet rotating about its own axis—the concept of *rotational inertia* must be introduced. Then angular momentum is rotational inertia × rotational speed. See any of Hewitt's *Conceptual Physics* textbooks for more information.

well as magnitude. In this appendix, we won't treat the vector nature of angular momentum (or even of torque, which also is a vector).

Just as an external net force is required to change the linear momentum of an object, an external net torque is required to change the angular momentum of an object. We can state a rotational version of Newton's first law (the law of inertia):

An object or system of objects will maintain its angular momentum unless acted upon by an unbalanced external torque.

We see application of this rule when we look at a spinning top. If friction is low and torque also low, the top tends to remain spinning. The Earth and planets spin in torque-free regions, and once they are spinning, they remain so.

Conservation of Angular Momentum

Just as the linear momentum of any system is conserved if no net forces are acting on the system, angular momentum is conserved if no net torque acts on the system. In the absence of an unbalanced external torque, the angular momentum of that system is constant. This means that its angular momentum at any one time will be the same as at any other time.

Conservation of angular momentum is shown in Figure A.11. The man stands on a low-friction turntable with weights extended. To simplify, consider only the weights in his hands. When he is slowly turning with his arms extended, much of the angular momentum is due to the distance between the weights and the rotational axis. When he pulls the weights inward, the distance is considerably reduced. What is the result? His rotational speed increases!* This example is best appreciated by the turning person, who feels changes in rotational speed that seem to be mysterious. But it's straight physics! This procedure is used by a figure skater who starts to whirl with her arms and perhaps a leg extended and then draws her arms and leg in to obtain a greater rotational speed. Whenever a rotating body contracts, its rotational speed increases.

The law of angular momentum conservation is illustrated by the motions of the planets and the shape of the galaxies. When a slowly rotating ball of gas in space gravitationally contracts, the result is an increase in its rate of rotation. The conservation of angular momentum is far-reaching.

FIGURE A.11

Conservation of angular momentum. When the man pulls his arms and the whirling weights inward, he decreases the radial distance between the weights and the axis of rotation, and the rotational speed increases correspondingly.

---

* When a direction is assigned to rotational speed, we call it *rotational velocity* (often called *angular velocity*). By convention, the rotational velocity vector and the angular momentum vector have the same direction and lie along the axis of rotation.

APPENDIX B

Vectors

Vectors and Scalars

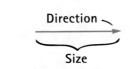

A *vector* quantity is a directed quantity—one that must be specified not only by magnitude (size) but by direction as well. Recall from Chapter 1 that velocity is a vector quantity. Other examples are force, acceleration, and momentum. In contrast, a *scalar* quantity can be specified by magnitude alone. Some examples of scalar quantities are speed, time, temperature, and energy.

Vector quantities may be represented by arrows. The length of the arrow tells you the magnitude of the vector quantity, and the arrowhead tells you the direction of the vector quantity. Such an arrow drawn to scale and pointing appropriately is called a *vector*.

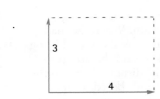

FIGURE B.2

Adding Vectors

Vectors that add together are called *component vectors*. The sum of component vectors is called a *resultant*.

To add two vectors, make a parallelogram with two component vectors acting as two of the adjacent sides (Figure B.2). (Here our parallelogram is a rectangle.) Then draw a diagonal from the origin of the vector pair; this is the resultant (Figure B.3).

Caution: Do not try to mix vectors! We cannot add apples and oranges, so velocity vectors combine only with velocity vectors, force vectors combine only with force vectors, and acceleration vectors combine only with acceleration vectors—each on its own vector diagram. If you ever show different kinds of vectors on the same diagram, use different colors or some other method of distinguishing the different kinds of vectors.

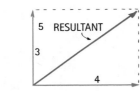

FIGURE B.3

Finding Components of Vectors

Mastering**PHYSICS**
TUTORIAL: Vectors

Recall from Chapter 2 that to find a pair of perpendicular components for a vector, first draw a dashed line through the tail of the vector (in the direction of one of the desired components). Second, draw another dashed line through the tail end of the vector at right angles to the first dashed line. Third, make a rectangle whose diagonal is the given vector. Draw in the two components. Here we let **F** stand for "total force," **U** stand for "upward force," and **S** stand for "sideways force."

FIGURE B.4

FIGURE B.5

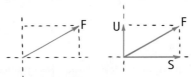

FIGURE B.6

EXAMPLES

1. Ernie Brown pushes a lawnmower and applies a force that pushes it forward and also against the ground. In Figure B.7, **F** represents the force applied by the man. We can separate this force into two components.

   The vector **D** represents the downward component, and **S** is the sideways component, the force that moves the lawnmower forward. If we know the magnitude and direction of the vector **F**, we can estimate the magnitude of the components from the vector diagram.

   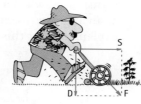

   FIGURE B.7

2. Would it be easier to push or pull a wheelbarrow over a step? Figure B.8 shows the force at the wheel's center. When you push a wheelbarrow, part of the force is directed downward, which makes it harder to get over the step. When you pull, however, part of the pulling force is directed upward, which helps to lift the wheel over the step. Note that the vector diagram suggests that pushing the wheelbarrow may not get it over the step at all. Do you see that the height of the step, the radius of the wheel, and the angle of the applied force determine whether the wheelbarrow can be pushed over the step? We see how vectors help us analyze a situation so that we can see just what the problem is!

   FIGURE B.8

3. If we consider the components of the weight of an object rolling down an incline, we can see why its speed depends on the angle. Note that the steeper the incline, the greater the component **S** becomes and the faster the object rolls. When the incline is vertical, **S** becomes equal to the weight, and the object attains maximum acceleration, 9.8 m/s^2. There are two more force vectors that are not shown: the normal force **N**, which is equal and oppositely directed to **D**, and the friction force **f**, acting at the barrel-plane contact.

   FIGURE B.9

4. When moving air strikes the underside of an airplane wing, the force of air impact against the wing may be represented by a single vector perpendicular to the plane of the wing (Figure B.10). We represent the force vector as acting midway along the lower wing surface, where the dot is, and pointing above the wing to show the direction of the resulting wind impact force. This force can be broken up into two components, one sideways and the other up. The upward component, **U**, is called *lift*. The sideways component, **S**, is called *drag*. If the aircraft is to fly at constant velocity at constant altitude, then lift must equal the weight of the aircraft and the thrust of the plane's engines must equal drag. The magnitude of lift (and drag) can be altered by changing the speed of the airplane or by changing the angle (called *angle of attack*) between the wing and the horizontal.

   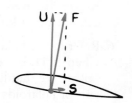

   FIGURE B.10

5. Consider the satellite moving clockwise in Figure B.11. Everywhere in its orbital path, gravitational force **F** pulls it toward the center of the host planet. At position A we see **F** separated into two components: **f**, which is tangent to the path of the projectile, and **f′**, which is perpendicular to the path. The relative magnitudes of these components in comparison to the magnitude of **F** can be seen in the imaginary rectangle they compose: **f** and **f′** are the sides, and **F** is the diagonal. We see that component **f** is along the orbital path but against the direction of motion of the satellite. This force component reduces the speed of the satellite. The other component, **f′**, changes the direction of the satellite's motion and pulls it away from its tendency to go in a straight line. So the path of the satellite curves. The satellite loses speed until it reaches position B. At this farthest point from the planet (apogee), the gravitational force is somewhat weaker but perpendicular to the satellite's motion, and component **f** has reduced to zero. Component **f′**, on the other hand, has increased and is now fully merged to become **F**. Speed at this point is not enough for circular orbit, and the satellite begins to fall toward the planet. It picks up speed because the component **f** reappears and is in the direction of motion as shown in position C. The satellite picks up speed until it

   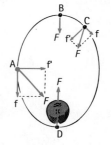

   FIGURE B.11

whips around to position D (perigee), where once again the direction of motion is perpendicular to the gravitational force, $\mathbf{f}'$ blends to full $\mathbf{F}$, and $\mathbf{f}$ is nonexistent. The speed is in excess of that needed for circular orbit at this distance, and it overshoots to repeat the cycle. Its loss in speed in going from D to B equals its gain in speed from B to D. Kepler discovered that planetary paths are elliptical, but never knew why. Do you?

6. Refer to the Polaroids held by Ludmila back in Chapter 11, in Figure 11.58. In the first picture (a), we see that light is transmitted through the pair of Polaroids because their axes are aligned. The emerging light can be represented as a vector aligned with the polarization axes of the Polaroids. When the Polaroids are crossed (b), no light emerges because light passing through the first Polaroid is perpendicular to the polarization axes of the second Polaroid, with no components along its axis. In the third picture (c), we see that light is transmitted when a third Polaroid is sandwiched at an angle between the crossed Polaroids. The explanation for this is shown in Figure B.12.

FIGURE B.12

Sailboats

Sailors have always known that a sailboat can sail downwind, in the direction of the wind. Sailors have not always known, however, that a sailboat can sail upwind, against the wind. One reason for this has to do with a feature that is common only to recent sailboats—a fin-like keel that extends deep beneath the bottom of the boat to ensure that the boat will knife through the water only in a forward (or backward) direction. Without a keel, a sailboat could be blown sideways.

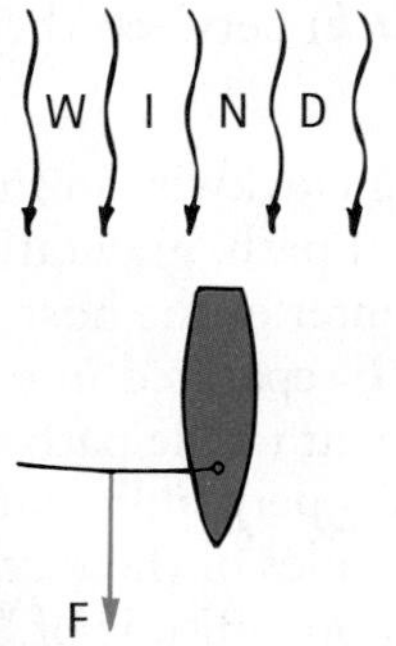

Figure B.13 shows a sailboat sailing directly downwind. The force of wind impact against the sail accelerates the boat. Even if the drag of the water and all other resistance forces are negligible, the maximum speed of the boat is the wind speed. This is because the wind will not make impact against the sail if the boat is moving as fast as the wind. The wind would have no speed relative to the boat and the sail would simply sag. With no force, there is no acceleration. The force vector in Figure B.13 *decreases* as the boat travels faster. The force vector is maximum when the boat is at rest and the full impact of the wind fills the sail, and is minimum when the boat travels as fast as the wind. If the boat is somehow propelled to a speed faster than the wind (by way of a motor, for example), then air resistance against the front side of the sail will produce an oppositely directed force vector. This will slow the boat down. Hence the boat when driven only by the wind cannot exceed wind speed.

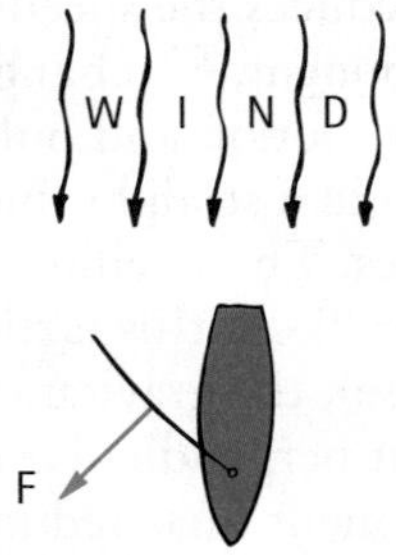

If the sail is oriented at an angle, as shown in Figure B.14, the boat will move forward, but with less acceleration. There are two reasons for this:

1. The force on the sail is less because the sail does not intercept as much wind in this angular position.

2. The direction of the wind impact force on the sail is not in the direction of the boat's motion, but is perpendicular to the surface of the sail. Generally speaking, whenever any fluid (liquid or gas) interacts with a smooth surface, the force of interaction is perpendicular to the smooth surface.* The boat does not move in the same direction as the perpendicular force on the sail, but is constrained to move in a forward (or backward) direction by its keel.

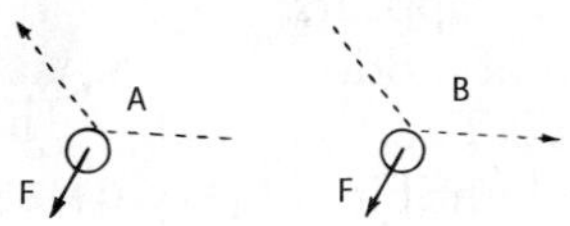

FIGURE B.13

FIGURE B.14

* You can do a simple exercise to see that this is so. Try bouncing a coin off another on a smooth surface, as shown. Note that the struck coin moves at right angles (perpendicular) to the contact edge. Note also that it makes no difference whether the projected coin moves along path A or path B. See your instructor for a more rigorous explanation, which involves momentum conservation.

We can better understand the motion of the boat by resolving the force of wind impact, **F**, into perpendicular components. The important component is that which is parallel to the keel, which we label **K**, and the other component is perpendicular to the keel, which we label **T**. It is the component **K**, as shown in Figure B.15, that is responsible for the forward motion of the boat. Component **T** is a useless force that tends to tip the boat over and move it sideways. This component force is offset by the deep keel. Again, maximum speed of the boat can be no greater than wind speed.

Many sailboats sailing in directions other than exactly downwind (Figure B.16) with their sails properly oriented can exceed wind speed. In the case of a sailboat cutting across the wind, the wind may continue to make impact with the sail even after the boat exceeds wind speed. A surfer, in a similar way, exceeds the velocity of the propelling wave by angling his surfboard across the wave. Greater angles to the propelling medium (wind for the boat, water wave for the surfboard) result in greater speeds. A sailcraft can sail faster cutting across the wind than it can sailing downwind.

As strange as it may seem, maximum speed for most sailcraft is attained by cutting into (against) the wind, that is, by angling the sailcraft in a direction upwind! Although a sailboat cannot sail directly upwind, it can reach a destination upwind by angling back and forth in a zigzag fashion. This is called *tacking*. Suppose the boat and sail are as shown in Figure B.17. Component **K** will push the boat along in a forward direction, angling into the wind. In the position shown, the boat can sail faster than the speed of the wind. This is because as the boat travels faster, the impact of wind is increased. This is similar to running in a rain that comes down at an angle. When you run into the direction of the downpour, the drops strike you harder and more frequently, but when you run away from the direction of the downpour, the drops don't strike you as hard or as frequently. In the same way, a boat sailing upwind experiences greater wind impact force, while a boat sailing downwind experiences a decreased wind impact force. In any case the boat reaches its terminal speed when opposing forces cancel the force of wind impact. The opposing forces consist mainly of water resistance against the hull of the boat. The hulls of racing boats are shaped to minimize this resistive force, which is the principal deterrent to high speeds.

Iceboats (sailcraft equipped with runners for traveling on ice) encounter no water resistance and can travel at several times the speed of the wind when they tack upwind. Although ice friction is nearly absent, an iceboat does not accelerate without limits. The terminal velocity of a sailcraft is determined not only by opposing friction forces but also by the change in relative wind direction. When the boat's orientation and speed are such that the wind seems to shift in direction, so the wind moves parallel to the sail rather than into it, forward acceleration ceases—at least in the case of a flat sail. In practice, sails are curved and produce an airfoil that is as important to sailcraft as it is to aircraft, as discussed in Chapter 5.

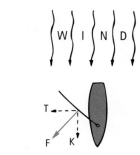

FIGURE B.15

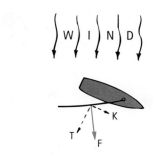

FIGURE B.16

FIGURE B.17

Exponential Growth and Doubling Time*

One of the most important things we seem unable to perceive is the process of exponential growth. We think we understand how compound interest works, but we can't get it through our heads that a fine piece of tissue paper folded upon itself 50 times (if that were possible) would be more than 20 million kilometers thick. If we could, we could "see" why our income buys only half of what it did several years ago, why the price of everything has doubled in the same time, why populations and pollution proliferate out of control.**

When a quantity such as money in the bank, population, or the rate of consumption of a resource steadily grows at a fixed percent per year, we say the growth is exponential. Money in the bank may grow at 4 percent per year; electric power generating capacity in the United States grew at about 7 percent per year for the first three-quarters of the 20th century. The important thing about exponential growth is that the time required for the growing quantity to double in size (increase by 100 percent) is also constant. For example, if the population of a growing city takes 12 years to double from 10,000 to 20,000 inhabitants and its growth remains steady, in the next 12 years the population will double to 40,000, and in the next 10 years to 80,000, and so on.

There is an important relationship between the percent growth rate and its *doubling time,* the time it takes to double a quantity:†

$$\text{Doubling time} = \frac{69.3}{\text{percent growth per unit time}} \approx \frac{70}{\%}$$

So to estimate the doubling time for a steadily growing quantity, we simply divide the number 70 by the percentage growth rate. For example, the 7 percent growth rate of electric power generating capacity in the United States means that in the past the capacity had doubled every 10 years [70%/(7%/year) = 10 years]. A 2 percent growth rate for world population means the population of the world doubles every 35 years [70%/(2%/year) = 35 years]. A city planning commission that accepts what seems like a modest 3.5 percent growth rate may not realize that this means that doubling will occur in 70/3.5 or 20 years; that requires doubling the capacity for such things as water supply, sewage-treatment plants, and other municipal services every 20 years.

What happens when you put steady growth in a finite environment? Consider the growth of bacteria that grow by division, so that one bacterium

---

* This appendix is drawn from material by University of Colorado physics professor Albert A. Bartlett, who strongly asserts, "The greatest shortcoming of the human race is man's inability to understand the exponential function." See Professor Bartlett's still-timely article, "Forgotten Fundamentals in the Energy Crisis" (*American Journal of Physics,* September 1978) or any of his revised articles on the web.

** K. C. Cole, *Sympathetic Vibrations* (New York: Morrow, 1984).

† For exponential decay we speak about half-life, the time required for a quantity to reduce to half its value. This case is treated in Chapter 13.

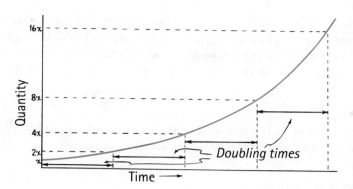

FIGURE C.1

An exponential curve. Notice that each of the successive equal time intervals noted on the horizontal scale corresponds to a doubling of the quantity indicated on the vertical scale. Such an interval is called the doubling time.

becomes two, the two divide to become four, the four divide to become eight, and so on. Suppose the division time for a certain strain of bacteria is 1 minute. This is then steady growth—the number of bacteria grows exponentially with a doubling time of 1 minute. Further, suppose that one bacterium is put in a bottle at 11:00 A.M. and that growth continues steadily until the bottle becomes full of bacteria at 12 noon. Consider seriously the following question.

CHECKPOINT
When was the bottle half-full?

Was this your answer?

11:59 A.M.; the bacteria will double in number every minute!

It is startling to note that at 2 minutes before noon the bottle was only 1/4 full. Table C.1 summarizes the amount of space left in the bottle in the last few minutes before noon. If you were an average bacterium in the bottle, at which time would you first realize that you were running out of space? For example, would you sense there was a serious problem at 11:55 A.M., when the bottle was only 3% filled, (1/32), and had 97% of open space (just yearning for development)? The point here is that there isn't much time between the moment that the effects of growth become noticeable and the time when they become overwhelming.

Suppose that at 11:58 A.M. some farsighted bacteria see that they are running out of space and launch a full-scale search for new bottles. Luckily, at 11:59 A.M. they discover three new empty bottles, three times as much space as they had ever known. This quadruples the total resource space ever known to

FIGURE C.2

TABLE C.1	THE LAST MINUTES IN THE BOTTLE	
Time	Part Full (%)	Part Empty
11:54 A.M.	1/64 (1.5%)	63/64
11:55 A.M.	1/32 (3%)	31/32
11:56 A.M.	1/16 (6%)	15/16
11:57 A.M.	1/8 (12%)	7/8
11:58 A.M.	1/4 (25%)	3/4
11:59 A.M.	1/2 (50%)	1/2
12:00 noon	full (100%)	none

FIGURE C.3

A single grain of wheat placed on the first square of the chessboard is doubled on the second square, this number is doubled on the third, and so on, presumably for all 64 squares. Note that each square contains one more grain than all the preceding squares combined. Does enough wheat exist in the world to fill all 64 squares in this manner?

the bacteria, for they now have a total of four bottles, whereas before the discovery they had only one. Further suppose that, thanks to their technological proficiency, they are able to migrate to their new habitats without difficulty. Surely, it seems to most of the bacteria that their problem is solved—and just in time.

CHECKPOINT
If the bacteria growth continues at the unchanged rate, what time will it be when the three new bottles are filled to capacity?

Was this your answer?
12:02 P.M.!

We see from Table C.2 that quadrupling the resource extends the life of the resource by only two doubling times. In our example the resource is space—but it could as well be coal, oil, uranium, or any nonrenewable resource.

Continued growth and continued doubling lead to enormous numbers. In two doubling times, a quantity will double twice ($2^2 = 4$;

TABLE C.2	EFFECTS OF THE DISCOVERY OF THREE NEW BOTTLES
Time	Effect
11:58 A.M.	Bottle 1 is 1/4 full
11:59 A.M.	Bottle 1 is 1/2 full
12:00 noon	Bottle 1 is full
12:01 P.M.	Bottles 1 and 2 are both full
12:02 P.M.	Bottles 1, 2, 3, and 4 are all full

quadruple) in size; in three doubling times, its size will increase eightfold ($2^3 = 8$); in four doubling times, it will increase sixteenfold ($2^4 = 16$); and so on.

This is best illustrated by the story of the court mathematician in India who years ago invented the game of chess for his king. The king was so pleased with the game that he offered to repay the mathematician, whose request seemed modest enough. The mathematician requested a single grain of wheat on the first square of the chessboard, two grains on the second square, four on the third square, and so on, doubling the number of grains on each succeeding square until all squares had been used. At this rate there would be 2^{63} grains of wheat on the 64th square. The king soon saw that he could not fill this "modest" request, which amounted to more wheat than had been harvested in the entire history of the Earth!

It is interesting and important to note that the number or grains on any square is one grain more than the total of all grains on the preceding squares. This is true anywhere on the board. Note from Table C.3 that when eight grains are placed on the fourth square, the eight is one more than the total of seven grains that were already on the board. Or the 32 grains placed on the sixth square is one more than the total of 31 grains that were already on the board. We see that in one doubling time we use more than all that had been used in all the preceding growth!

So if we speak of doubling energy consumption in the next however many years, bear in mind that this means in these years we will consume more energy than has heretofore been consumed during the entire preceding period of steady growth. And if power generation continues to use predominantly fossil

TABLE C.3	FILLING THE SQUARES ON THE CHESSBOARD	
Square Number	Grains on Square	Total Grains Thus Far
1	1	1
2	2	3
3	4	7
4	8	15
5	16	31
6	32	63
7	64	127
.	.	.
.	.	.
.	.	.
64	2^{63}	$2^{64} - 1$

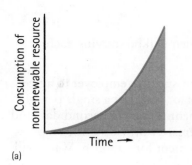

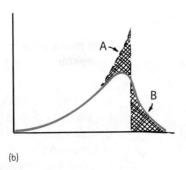

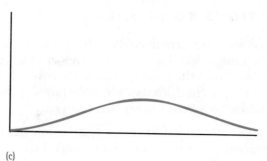

(a)   (b)   (c)

FIGURE C.4

(a) If the exponential rate of consumption for a nonrenewable resource continues until it is depleted, consumption falls abruptly to zero. The shaded area under this curve represents the total supply of the resource. (b) In practice, the rate of consumption levels off and then falls less abruptly to zero. Note that the crosshatched area A is equal to the crosshatched area B. Why? (c) At lower consumption rates, the same resource lasts a longer time.

fuels, then except for some improvements in efficiency, we would burn up in the next doubling time a greater amount of coal, oil, and natural gas than has already been consumed by previous power generation, and except for improvements in pollution control, we can expect to discharge even more toxic wastes into the environment than the millions upon millions of tons already discharged over all the previous years of industrial civilization. We would also expect more human-made calories of heat to be absorbed by Earth's ecosystem than have been absorbed in the entire past! At the previous 7 percent annual growth rate in energy production, all this would occur in one doubling time of a single decade. If over the coming years the annual growth rate remains at half this value, 3.5 percent, then all this would take place in a doubling time of two decades. Clearly this cannot continue!

The consumption of a nonrenewable resource cannot grow exponentially for an indefinite period, because the resource is finite and its supply finally expires. The most drastic way this could happen is shown in Figure C.4 (a), where the rate of consumption, such as barrels of oil per year, is plotted against time, say in years. In such a graph the area under the curve represents the supply of the resource. We see that when the supply is exhausted, the consumption ceases altogether. This sudden change is rarely the case, for the rate of extracting the supply falls as it becomes more scarce. This is shown in Figure C.4(b). Note that the area under the curve is equal to the area under the curve in (a). Why? Because the total supply is the same in both cases. The principal difference is the time taken to finally extinguish the supply. History shows that the rate of production of a nonrenewable resource rises and falls in a nearly symmetric manner, as shown in (c). The time during which production rates rise is approximately equal to the time during which these rates fall to zero or near zero.

Production rates for all nonrenewable resources decrease sooner or later. Only production rates for renewable resources, such as agriculture or forest products, can be maintained at steady levels for long periods of time (Figure C.5), provided such production does not depend on waning nonrenewable resources such as petroleum. Much of today's agriculture is so petroleum-dependent that it can be said that modern agriculture is simply the process whereby land is used to convert petroleum into food. The implications of petroleum scarcity go far beyond rationing of gasoline for cars or fuel oil for home heating.

The consequences of unchecked exponential growth are staggering. It is important to ask: Is growth really good? In answering this question, bear in mind that human growth is an early phase of life that continues normally through adolescence. Physical growth stops when physical maturity is reached. What do we say of growth that continues in the period of physical maturity? We say that such growth is obesity—or worse, cancer.

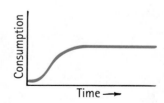

FIGURE C.5

A curve showing the rate of consumption of a renewable resource such as agricultural or forest products, where a steady rate of production and consumption can be maintained for a long period, provided this production is not dependent upon the use of a nonrenewable resource that is waning in supply.

QUESTIONS TO PONDER

1. According to a French riddle, a lily pond starts with a single leaf. Each day the number of leaves doubles, until the pond is completely covered by leaves on the 30th day. On what day was the pond half covered? One-quarter covered?

2. In an economy that has a steady inflation rate of 7 percent per year, in how many years does a dollar lose half its value?

3. At a steady inflation rate of 7 percent, what will be the price every 10 years for the next 50 years for a theater ticket that now costs $30? For a coat that now costs $300? For a car that now costs $30,000? For a home that now costs $200,000?

4. If the sewage treatment plant of a city is just adequate for the city's current population, how many sewage treatment plants will be necessary 42 years later if the city grows steadily at 5 percent annually?

5. If world population doubles in 40 years and world food production also doubles in 40 years, how many people then will be starving each year compared to now?

6. Suppose you get a prospective employer to agree to hire your services for wages of a single penny for the first day, 2 pennies for the second day, and double each day thereafter providing the employer keeps to the agreement for a month. What will be your total wages for the month?

7. In the preceding exercise, how will your wages for only the 30th day compare to your total wages for the previous 29 days?

8. If fusion power were harnessed today, the abundant energy resulting would probably sustain and even further encourage our present appetite for continued growth and in a relatively few doubling times produce an appreciable fraction of the solar power input to the earth. Make an argument that the current delay in harnessing fusion is a blessing for the human race.

Odd-Numbered Solutions

PROLOGUE

1. *Science* is the product of human curiosity about how the world works—an organized body of knowledge that describes order and causes within nature and an ongoing human activity dedicated to gathering and organizing knowledge about the world. **3.** The equations are guides to thinking that show the connections between concepts in nature. **5.** Competent scientists must be experts at changing their minds. **7.** A discredited scientist does not get a second chance in the community of scientists. The penalty for fraud is professional excommunication. **9.** *Pseudoscience* is "fake science" that doesn't meet scientific standards. **11.** The concern of science is the physical universe; the concern of religion is spiritual matters, such as belief and faith. Whereas scientific truth is a matter of public scrutiny, religion is a deeply personal matter. **13.** One benefit is an open and exploring mind. **15.** The other sciences build on physics, not the other way around. **17.** (a) is a scientific hypothesis because there is a test for wrongness. For example, you can extract chlorophyll from grass and note its color. (b) is without a means of proving it wrong and is therefore not a scientific hypothesis. (c) is a scientific hypothesis and can be proved wrong—for example, by showing that tides don't correspond to the positions of the Moon in the sky. **19.** For Russell to publicly change his mind about his ideas is a sign of strength rather than weakness. It takes more courage to change your ideas when confronted with counter evidence than to hold fast to your ideas. If a person's ideas and view of the world are no different after a lifetime of varied experience, then that person either was miraculously blessed with unusual wisdom at an early age or learned nothing. The latter is more likely. Education is learning about what you don't yet know. It is arrogant to think you know it all in the later stages of your education, and stupid to think so at the beginning of your education.
21. Examples of pros and cons are open ended.

CHAPTER 1

1. Aristotle believed that light things rose like smoke and heavy things fell like boulders. **3.** Galileo found that moving things, once moving, continued in motion *without* the application of forces. **5.** The name is *inertia*. **7.** Your weight is greater on Earth than on the Moon. Your mass is the same in both locations or in any location. **9.** The object would weigh less on the Moon. **11.** The two necessary quantities are magnitude and direction. **13.** The tension is 20 N. **15.** Because the support force is perpendicular to a surface, it is called a normal force (*normal* meaning perpendicular). **17.** Both are in equilibrium because both are not accelerating. **19.** The two are in opposite directions; friction opposes motion. **21.** Yes, friction acts to counter your push so the net force on the furniture is zero. **23.** Velocity involves direction as well as magnitude; speed involves only magnitude. **25.** Relative to the couch you're sitting on, you're at rest; relative to the Sun, you're moving at 100,000 km/h. **27.** The acceleration is zero for an object that moves at constant velocity. The net force is zero in this case. **29.** Open ended. **31.** This illustrates Newton's first law—the law of inertia. **33.** Open ended.

35. Average speed $= \dfrac{\Delta d}{\Delta t} = \dfrac{1.0 \text{ m}}{0.5 \text{ s}} = 2$ m/s.

37. Acceleration $= \dfrac{\text{change of velocity}}{\text{time}} = \dfrac{\Delta v}{\Delta t} = \dfrac{10 \text{ m/s}}{2 \text{ s}} = 5$ m/s^2.

39. Distance of fall $= \frac{1}{2}gt^2 = \frac{1}{2}(10 \text{ m/s}^2)(3 \text{ s})^2 = 45$ m. Similarly, in 10 s it falls 500 m. **41.** Since each scale reads 350 N, Lucy's total weight is 700 N. **43.** From the equilibrium rule, $\Sigma F = 0$, the upward forces are 800 N, and the downward forces are 500 N + the weight of the scaffold. The scaffold must weigh 300 N (800 N − 500 N = 300 N). **45.** (a) The net force is zero (because the velocity is constant). (b) $\Sigma F = 0$, so *friction* = −120 N, which is opposite in direction to the push. (c) Zero, since no skidding or attempted skidding occurs. **47.** (a) The landing speed is 60 km/h − 40 km/h = 20 km/h. (b) The landing speed is 60 km/h + 40 km/h = 100 km/h (now you see why planes land in a

headwind and not a tail wind). (c) In a headwind that matches the air speed, the plane would "hover" above the ground as it touches down! So its landing speed is zero.

49. (a) Speed $= \dfrac{\Delta d}{\Delta t} = \dfrac{5 \text{ km}}{0.5 \text{ h}} = 10$ km/h.

(b) $\dfrac{10 \text{ km}}{\text{h}} \times \dfrac{1000 \text{ m}}{1 \text{ km}} \times \dfrac{1 \text{ h}}{3600 \text{ s}} = 2.8$ m/s

51.

Time (s)	Velocity Acquired (m/s)	Distance Fallen (m)
6	60	180
7	70	245
8	80	320
9	90	405
10	100	500

53. (a) $\bar{v} = \dfrac{d}{t} = \dfrac{2\pi r}{t}$. (b) $\bar{v} = \dfrac{2r}{t} = \dfrac{2r(100 \text{ m})}{14 \text{ s}} = 45$ m/s.

55. (a) The velocity of the ball at the highest point of its vertical trajectory is instantaneously zero. (b) One second before reaching its highest point, its velocity is 10 m/s. (c) The change in velocity is 10 m/s during this 1-s interval (or any other 1-s interval). (d) One second after it reaches its highest point, its velocity is −10 m/s—equal in magnitude but oppositely directed to its value 1 s before reaching the top. (e) The change in velocity during this (or any) 1-s interval is 10 m/s. (f) In 2 s, the change in velocity, from 10 m/s up to 10 m/s down, is 20 m/s (not zero). (g) The acceleration of the ball is 10 m/s^2 before it reaches its highest point, when it reaches the highest point, and after it reaches the highest point. In all cases acceleration is downward, toward Earth. **57.** (a) From $v = v_0 + at$, and with v_0 being 0, $v = at = (4.0 \text{ m/s}^2)(15 \text{ s}) = 60$ m/s. (b) With v_0 being 0, $d = \frac{1}{2}at^2 = \frac{1}{2}(4.0 \text{ m/s}^2)(15 \text{ s})^2 = 450$ m. **59.** C, A, B, D. **61.** (a) A = B = C (no force). (b) C, B, A. **63.** The Leaning Tower experiment discredited the idea that heavy things fall proportionally faster. The inclined-plane experiments discredited the idea that a force is needed for motion. **65.** You say that nothing, no force, keeps the probe moving when the rocket disengages. The probe moves of its own inertia. **67.** Mass. **69.** A person on a diet loses mass. To lose weight, the person could go to the top of a mountain where the force of gravity is less. But the amount of matter would be the same. **71.** Divide your weight in pounds by 2.2, and you'll have your mass in kilograms. Multiply this number by 10 (or more precisely 9.8), and you'll have your weight in newtons. **73.** The upward force is the tension in the vine. The downward force is that due to gravity. Both are equal when the monkey hangs in equilibrium. **75.** No, the upward force is the normal force, which is the same whether the book is on slippery ice or sandpaper. Friction plays no role unless the book slides or tends to slide along the table surface. **77.** Yes, because it doesn't change its state of motion (accelerate). Strictly speaking, some friction does act, so it is close to being in equilibrium. **79.** Constant speed implies that the net force on the cabinet is zero. So the force of friction is 550 N in the opposite direction. **81.** The support force on the jug is W. When water of weight w is added, the support force is $W + w$. **83.** The impact speed is 2 km/h (100 km/h − 98 km/h), a small bump. Head-on, the impact speed is 198 km/h (100 km/h + 98 km/h), lethal! **85.** Emily will not be successful because her speed will be zero relative to the land. **87.** The equation $d = \frac{1}{2}gt^2$ is most appropriate. Distance increases as the square of time, so each successive distance covered is greater than the preceding distance covered. **89.** Alex is correct because Gracie is describing *speed*. Alex is stating the rate at which speed changes (how fast you get fast), which is *acceleration*. **91.** Acceleration is 10 m/s^2, constant, all the way down. (Velocity, however, is 50 m/s at 5 s and 100 m/s at 10 s.) **93.** Whatever initial forces set the asteroids in motion are irrelevant now. No force pushes them; they move of their own inertia. **95.** The tendency of the ball is to remain at rest. From a point of view outside the wagon, the ball stays in place as the back of the wagon moves toward it. (Because of friction, the ball may roll along the wagon surface; without friction, the surface would slide beneath the ball.) **97.** Your body tends to remain at rest, in accord with Newton's first law. In a car, the back of the seat pushes you forward. If your car is hit from the rear and the back of your head is not supported by a headrest, your head is not pushed forward with the rest of your body. This may result in whiplash.   **99.**   The car has *no* tendency to resume its original twice-as-fast speed. Instead, in accord with Newton's first law, it tends to continue at the speed it had when the engine stopped running, half speed, decreasing further over time due to air resistance and road friction.   **101.** If the upward force were the only force acting, the book indeed would rise. But another force, gravity, results in the net force being zero.   **103.** At the top of its path (and everywhere else along its path), the force of gravity acts to change the ball's motion. Even though it momentarily stops at the top, the net force on the ball is not zero and it therefore is not in equilibrium.   **105.** The coin is moving along with you when you toss it. While in the air, it maintains this forward motion, so the coin lands in your hand. If the train slows while the coin is in the air, the coin lands in front of you. If the train rounds a curve while the coin is in the air, the coin lands off to your side. The coin continues in its horizontal motion, in accord with the law of inertia.   **107.** (a) Yes, ball B rolls faster along the lower part of the track. (b) Yes, the speed ball B gained going down the extra dip equals its loss in speed going up the dip, so both balls end up with the same speed. (c) Yes, the average speed of ball B dipping down and up is greater than the average speed of ball A. (d) The average speed of ball B is greater than that of ball A. That means ball B wins the race. But the final instantaneous speeds (not the times) are the same because the speed ball B gained on the down-ramp is equal to the speed it lost on the up-ramp. (Many people give the wrong answer to the preceding question because they assume that because the balls end up with the same speed, they roll for the same time. Not so.)

CHAPTER 2

1. Every object continues in a state of rest or at uniform speed in a straight line unless acted on by a nonzero force.   **3.** They missed the concept of inertia.   **5.** In the absence of a force, the planets would follow straight-line paths.   **7.** Acceleration is directly proportional to force. Push on a brick and it accelerates (see Figure 2.7).   **9.** The acceleration is the same as before.   **11.** A heavy object has proportionally more mass. The heavy weight is offset by the greater mass, so the ratio of force to mass remains the same.   **13.** Both speed and frontal area affect the force of air resistance on a falling object.   **15.** The faster one encounters greater air resistance.   **17.** Both. In the simplest sense, a force is a push or pull. In a broader sense, a force makes up an interaction between one thing and another.   **19.** The force is the wall pushing back on your fingers.   **21.** Whenever one object exerts a force on a second object, the second object exerts an equal and opposite force on the first.   **23.** The forces are different because the masses are different, in accord with Newton's second law.   **25.** No, internal forces don't affect the acceleration of the system. Acceleration needs an interaction with an external force.   **27.** The force that propels a rocket is the reaction to the force the rocket exerts on its exhaust gases.   **29.** Newton's third law; every action has a reaction.

31. Open ended.   **33.** As the hand deflects air downward, the air deflects the hand upward, in accord with Newton's third law.

35. $a = \dfrac{F_{net}}{m} = \dfrac{15\ \text{N}}{3.0\ \text{kg}} = \dfrac{5.0\ \text{N}}{\text{kg}} = 5.0\ \text{m/s}^2.$

37. $F_{net} = ma = (12\ \text{kg})(7.0\ \text{m/s}^2) = 84\ \text{kg} \cdot \text{m/s}^2 = 84\ \text{N}.$ **39.** The given pair of forces produce a net force of $220\ \text{N} - 180\ \text{N} = 40\ \text{N}$ forward, which accelerates the car. To make the net force zero, a force of 40 N backward must be exerted on the car.

41. Acceleration $a = \dfrac{F_{net}}{m} = \dfrac{160\ \text{N} - 80\ \text{N}}{20\ \text{kg}} = \dfrac{80\ \text{N}}{20\ \text{kg}} = 4.0\ \text{m/s}^2.$

43. For the jet: $a = \dfrac{F_{net}}{m} = \dfrac{2(30,000\ \text{N})}{30,000\ \text{kg}} = 2\ \text{m/s}^2.$

45. $F_{net} = ma = (1\ \text{kg})(10\ \text{m/s}^2) = 10\ \text{kg} \cdot \text{m/s}^2 = 10\ \text{N}.$

47. $a = \dfrac{f}{m} = \dfrac{50\ \text{N}}{20\ \text{kg}} = 2.5\ \text{m/s}^2;\ d = \dfrac{1}{2}at^2 = \dfrac{1}{2}(2.5\ \text{m/s}^2)(2\text{s})^2 = 5\text{m}.$

49. (a) The force of air resistance will be equal to her weight, mg, or 500 N. (b) She'll reach the same air resistance, 500 N, but at a lower speed. (c) The answers are the same, but for different speeds. In each case she attains equilibrium (no acceleration).

51. (a) From $a = \dfrac{F_{net}}{m}$; $a = \dfrac{F - f}{m}$ (one step!).

(b) $a = \dfrac{F - f}{m} = \dfrac{12.0\ \text{N} - 6.0\ \text{N}}{4.0\ \text{kg}} = \dfrac{6.0\ \text{N}}{4.0\ \text{kg}} = \dfrac{1.5\ \text{N}}{\text{kg}} = 1.5\ \text{m/s}^2.$

53. The force on the bus is Ma.

New acceleration $= \dfrac{\text{same force}}{\text{new mass}} = \dfrac{Ma}{M + M/6} = \dfrac{6Ma}{6M + M} = \dfrac{6Ma}{7M} = \dfrac{6}{7}a.$

(b) New acceleration $= \dfrac{6}{7}a = \dfrac{6}{7}(1.2\ \text{m/s}^2) = 1.0\ \text{m/s}^2.$

55. C, B, A.   **57.** (a) C, A, B. (b) B, A, C.   **59.** Agree, acceleration (slowing the car) is opposite to velocity (the direction the car is moving).   **61.** The cleaver tends to keep moving when it encounters the vegetables, thus cutting them more effectively.   **63.** There are two horizontal forces on the car: road friction acting on the tires and air resistance in the opposite direction. At constant velocity there is no acceleration. The net force is therefore zero—meaning the force on the tires is equal and opposite to the force of air resistance.   **65.** You run your engine to provide a force large enough to overcome friction. A net force of zero requires that you provide this force.   **67.** The force that you exert on the scale is greater than your weight because you momentarily accelerate upward. Your weight reading is momentarily greater.   **69.** With air resistance, water drops fall at a tolerable terminal speed. Without air resistance, water drops would be in free fall and, depending on the height of the clouds, speeds of impact would be hazardous.   **71.** The net force is $10\ \text{N} - 2\ \text{N} = 8\ \text{N}$ downward (or more precisely $9.8\ \text{N} - 2\ \text{N} = 7.8\ \text{N}$ downward).   **73.** (a) Two force pairs act: Earth's pull on the apple (action) and the apple's pull on Earth (reaction). Your hand pushes the apple upward (action), and the apple pushes your hand downward (reaction). (b) If air resistance can be neglected, one force pair acts: Earth's pull on the apple, and the apple's pull on Earth. If air resis-tance counts, then the air pushes upward on the apple (action), and the apple pushes downward on the air (reaction).   **75.** The acceleration of any object is $a = F_{net}/m$, and F_{net} in free fall $= mg$. So $a = mg/m = g$. The greater the weight, the greater the mass.   **77.** (a) A skydiver encountering *no* air resis-tance is in free fall. But a diver falling in air at terminal velocity encounters air resistance and is not in free fall. (b) The only force acting on a satellite is gravity, so a satellite is in free fall (more about this in Chapter 4).   **79.** Your friend is wrong; the skydiver is in fact speeding up as acceleration decreases. Eventually the acceleration will become zero, in which case the diver has reached terminal velocity.   **81.** No, each hand pushes equally on the other in accord with Newton's third law.   **83.** When the barbell is accelerated upward, the force exerted by the athlete is greater than the weight of the barbell (the barbell simultaneously pushes with greater force against the athlete). When acceleration is downward, the force supplied by the athlete is less.   **85.** When you pull up on the handlebars, the handlebars in turn pull down on you. This downward force is transmitted to the pedals.   **87.** As in the preceding exercise, the force on each cart is the same. But, since the masses are different, the accelerations differ. The twice-as-massive cart undergoes only half the acceleration of the less massive cart and gains only half the speed.   **89.** The forces on each person have the same magnitude and their masses are the same, so their accelerations are the same. They slide equal distances of 6 m to meet at the midpoint.   **91.** The winning team pushes harder against the ground. The ground then pushes harder on them, producing a net force in their favor.   **93.** Neither a stick of dynamite nor anything else, a fist or a hammer, "contains" force. We will see later that a stick of dynamite contains *energy*, which like a fist or hammer is capable of producing forces when an interaction of some kind occurs.   **95.** Newton's first law applies again. When you jump, you tighten the disks similar to tightening a hammerhead (see Figure 2.2). So you're shorter at the end of the day. At night, while you are lying prone, relaxation undoes the compression and you get taller!   **97.** Let Newton's second law guide the answer to this: $a = F/m$. As m decreases (much is the mass of the fuel), acceleration a increases while force F remains constant.   **99.** Once the ball leaves your hand, only the force of gravity acts, ignoring air resistance. So the ball's acceleration will be 10 m/s² (even though its speed will be greater).   **101.** The stone's acceleration is g. This is consistent with Newton's second law: $a = F/m = mg/m = g$. It is common to confuse velocity with acceleration and get this question wrong; velocity (which is zero at the top) is not acceleration (which is a rate of change of velocity). If air resistance isn't a factor, then the acceleration of the tossed stone everywhere is g while in flight—even at the top—and tossed at any angle.   **103.** There are usually two terminal speeds: one before the parachute opens, which is faster, and one after opening, which is slower. The difference involves the different frontal areas while falling. The large area presented by the open chute results in a lower terminal speed, slow enough for a safe landing.   **105.** Air resistance is not really negligible for such a high drop, so the heavier ball does strike the ground first (see Figure 2.10). Although a twice-as-heavy ball strikes the ground first, it falls only a little faster, not twice as fast, which is what followers of Aristotle believed. Galileo recognized that the small difference is due to friction with the air (air resistance) and this difference wouldn't be present if there were no friction.   **107.** When you push the car, you exert a force on the car. When the car simultaneously pushes back on you, that force is on *you*—not the car. You don't cancel a force on the car with a force exerted on you. For cancellation, the forces have to be equal and opposite and act on the *same* object. Two equal and opposite forces would have to act *on the car* for cancellation.   **109.** Ken's pull on the rope is transmitted to Joanne, causing her to accelerate toward him. By Newton's third law, the rope pulls back on Ken, causing him to accelerate toward Joanne. So both will accelerate and move toward each other.

CHAPTER 3

1. A moving skateboard because it has speed, or velocity.   **3.** Giving the ball more time to change momentum means less force on your hand.   **5.** By swift execution, the time of contact is very brief, which makes the force of impact huge.   **7.** (c) because it is (a) and (b) combined.   **9.** When momentum, or any quantity in physics, does not change, we say it is *conserved*.   **11.** The speeds are the same because momentum is simply transferred from one car to the other.   **13.** Energy is most evident when it is changing.   **15.** The work required is the same in both cases: (50 kg)(2 m) = (25 kg)(4 m).   **17.** The more massive car has twice the potential energy.   **19.** The brakes must supply 16 times more work, and the stopping distance is 16 times longer ($4^2 = 16$).   **21.** Its gain in KE will equal its decrease in PE, 10 kJ.   **23.** Power is the amount of work done per time it takes to do it.   **25.** As force is increased, distance is decreased by the same factor.   **27.** The energy ordinarily becomes thermal energy.   **29.** Radioactivity is the ultimate source.   **31.** Predictions should be consistent with the principle of energy conservation.   **33.** The temperature after shaking should be higher because the energy of shaking is transformed into thermal energy.   **35.** Momentum = mv = (2 kg) (8 m/s) = 16 kg·m/s.   **37.** Ft = (100 N)(0.5 s) = 50 N·s.   **39.** Ft = (25 N)(2 s) = 50 Ns = 50 kg·m/s.   **41.** W = (20 N)(3.5 m) =70 J.   **43.** PE = (10 kg)(10 N/kg)(5 m) = 500 J.   **45.** KE = $\frac{1}{2}mv^2$ = $\frac{1}{2}$(84 kg) (2 m/s)^2 = 168 J.

47. ΔKE = Fd = (50 N)(20 m) = 1000 J.

49. P = W/t = (100 J)/(2 s) = 50 W.

51. From $Ft = \Delta(mv)$, $F = \Delta(mv)/t$ = (10 kg)(3 m/s)/2 s = 15 N.

53. (a) $Ft = \Delta(mv)$; $F = \Delta(mv)/t$ = (6 kg)(3 m/s)/0.5 s = 36 N. (b) 36 N, according to Newton's third law.   **55.** From the conservation of momentum, momentum$_{Atti}$ = momentum$_{(Judy+Atti)}$
(15 kg)(3.0 m/s) = (40 kg + 15 kg)v
$$45 \text{ kg·m/s} = (55 \text{ kg})v$$
$$v = 0.8 \text{ m/s}$$

57. Momentum$_{before}$ = momentum$_{after}$
(5 kg)(1 m/s) + (1 kg)v = 0
$v = \dfrac{-5 \text{ kg·m/s}}{1 \text{ kg}}$ = −5 m/s.   So if the little fish approaches the big fish at 5 m/s, the momentum after lunch will be zero.

59. $(F \times d)_{in} = (F \times d)_{out}$
60 N × 1.2 m = W × 0.2 m
W = [(60 N)(1.2 m)]/0.2 m = 360 N.

61. $(F \times d)_{in} = (F \times d)_{out}$
F × 2 m = 6000 N × 0.2 m
$F = \dfrac{(6000 \text{ N})(0.2 \text{ m})}{2 \text{ m}}$ = 600 N.

63. Power = Fd/t = (2 J)/(1 s) = 2 W.   **65.** (a) From Fd = ΔKE = $\Delta\frac{1}{2}mv^2$; $d = \dfrac{\frac{1}{2}mv^2}{F} = \dfrac{mv^2}{2F}$. (b) If both the distance and the force are doubled, four times as much work is done, which produces four times as much change in kinetic energy.   **67.** From PE = ΔKE, $mgh = \frac{1}{2}mv^2$, and so $v = \sqrt{(2gh)}$ = $\sqrt{[2(10 \text{ m/s}^2)(4.0 \text{ m})]}$ = $\sqrt{80}$ m/s (= 8.9 m/s).

69. (a) B = D, A = C. (b) D, C, A = B.   **71.** (a) C, B = D, A. (b) C, B = D, A. (c) A, B = D, C.   **73.** A, B = C.   **75.** Airbags lengthen the time of impact, thereby reducing the force of impact.   **77.** Although the impulses may be the same for the two cases, the times of impact are not. When the egg strikes the wall, the impact time is short and the impact force is correspondingly large; the egg breaks. When the egg strikes the sagging sheet, the impact time is long and the impact force is correspondingly small.   **79.** The total (net) momentum is zero. That's true whatever their masses because each will have the same amount of momentum but in the opposite direction from the other.   **81.** By Newton's third law, the force on the bug is equal in magnitude and opposite in direction to the force on the car windshield. The rest is logic: Since the time of impact is the same for both, the amount of impulse is the same for both, which means they both undergo the same change in momentum. The change in momentum of the bug is evident because of its large change in speed. The same change in momentum of the considerably more massive car is not evident because the change in speed is correspondingly very small. Nevertheless, $m\Delta V$ for the bug is equal to $M\Delta v$ for the car!   **83.** When you leap, you impart the same momentum to both yourself and the canoe. You leap from a canoe that is moving away from the dock, reducing your speed relative to the dock, so your leaping distance is less than expected.   **85.** Regarding Exercise 82, if you don't exert a net force on the ball, it won't exert a force on you. No net force means no change in motion. Regarding Exercise 83, you exert a force on the boat (action), and the boat exerts a force on you (reaction). If your jump lacks sufficient force, you end up in the water. Regarding Exercise 84, if one throws clothing, the force that accelerates the clothes is paired with an equal and opposite force on the thrower. This force can provide recoil toward shore.   **87.** Yes, you exert an impulse to throw a ball. You also exert an impulse to catch a ball. Since you change the ball's momentum by the same amount in both cases, the impulse you exert in both cases is the same. To catch the ball and then throw it back again at the same speed requires twice as much impulse. On a skateboard, you'd recoil and gain momentum when you throw the ball, you'd gain the same momentum when you catch the ball, and you'd gain twice the momentum if you did both—catch and then throw the ball at its initial speed in the opposite direction.   **89.** This is similar to the preceding exercise. In terms of force, Freddy's feet are brought up to speed when they make contact with the moving board. The friction force that brings him up to speed is countered by the same amount of force on the board in the opposite direction, slowing the board. In terms of momentum conservation, since no external forces act in the horizontal direction, the momentum after the board catches Freddy equals the momentum before. Since Freddy's mass is added, velocity must decrease.   **91.** Your friend does twice as much work ($4d$ × half the force > d × the same force).   **93.** Both people do the same amount of work because they reach the same height. The one who climbs in 30 s uses more power because the work is done in a shorter time.   **95.** The KE of a pendulum bob is maximum where it moves fastest, at the lowest point; the PE is maximum at the uppermost points. When the pendulum bob swings by the point that marks half its maximum height, it has half its maximum KE, and its PE is halfway between its minimum and maximum values. If we define PE = 0 at the bottom of the swing, the place where KE is half its maximum value is also the place where PE is half its maximum value, and KE = PE at this point. (From energy conservation: Total energy = KE + PE.)   **97.** If the ball is given an initial KE, it will return to its starting position with the same KE moving toward and hitting the instructor. (The usual classroom procedure is to release the ball from the nose at rest. Then when it returns, it has no KE and stops short of bumping the instructor's nose.)   **99.** Both will have the same speed. This is easier to see here because both balls convert the same PE to KE. Which gets to the end first, however, is a different question! (Ball B wins due to its greater *average* speed.)   **101.** When air resistance is a factor, the snowball returns with less speed (air resistance never increases speed!). It therefore has less KE. You can see this directly from the fact that the snowball loses mechanical energy to the air molecules it encounters, so when it returns to its starting point and to its original PE, it has less KE. This does not contradict the law of energy conservation because energy is dissipated from the moving-snowball system, not destroyed.   **103.** In a conventional car, braking converts KE to heat. In a hybrid car, braking charges up the batteries. In this way, braking energy can later be transformed into KE.   **105.** When the velocity is doubled, the momentum is doubled and the KE is increased by a factor of 4. Momentum is proportional to speed; KE is proportional to speed squared.   **107.** Zero KE means zero speed, so momentum is also zero.   **109.** The net momentum before the lumps collide is zero, and it is still zero after the collision. Momentum is indeed conserved. KE is zero after the collision, but it was greater than zero before the collision. The lumps are warmer after colliding because the initial KE of the lumps is transformed into thermal energy. Momentum has only one form. There is no way to "transform" momentum from one form into another, so it is conserved. But energy comes in various forms and can easily be transformed. No single form of energy such as KE need be conserved. **111.** Without this slack, a locomotive might simply sit still and spin its wheels. The loose coupling gives the entire train a longer time to gain momentum, requiring less force of the locomotive wheels against the track. In this way, the overall required impulse is divided into a series of smaller impulses. (This loose coupling can be very important for braking as well.) **113.** The craft will move to the right because two impulses act on the craft: the wind against the sail and the fan recoiling from the wind it produces. These impulses are oppositely directed, but are they equal in magnitude? No, because of bouncing. The wind bounces from the sail and produces a greater impulse than if it merely stopped. This greater impulse on the sail produces a net impulse in the forward direction, toward the right. There are two force pairs to consider: the fan–air force pair and the air–sail force pair. Because of bouncing, the air–sail force is greater. So the net force on the craft is forward, to the right. (The principle described here is applied in thrust reversers used to slow jet planes after they land.) **115.** Removing the sail and turning the fan around is the best means of propelling the craft! Then maximum impulse is exerted on the craft. (Such propeller-driven boats are used in very shallow water, as in the Florida Everglades.) If the fan is not turned around, the boat is propelled backward, to the left. **117.** Yes, an object can have energy without momentum if we consider PE (if an object has KE, then it must have momentum because it is moving). An object can have PE without being in motion and therefore without having momentum. And every object has "energy of being"—stated in the celebrated equation $E = mc^2$. So, whether an object moves or not, it has some form of energy. If it has KE, then with respect to the frame of reference in which its KE is measured, it also has momentum. **119.** The physics is the conservation of energy applied to machines. Sufficient work occurs because with each pump of the jack handle, the force she exerts acts over a much greater distance than the car moves. A small force acting over a long distance can do significant work. **121.** The tension in the string supporting the 10-kg block is 100 N (which is the same all along the string). Block B is supported by two strands of string, each 100 N, which means the mass of block B is twice that of block A. So block B has a mass of 20 kg. **123.** You need to know the distance the rock penetrates into the ground, or the time it takes to do so. The work that the rock does on the ground is equal to its PE before being dropped, $mgh = 100$ J. The force of impact, however, depends on the distance the rock penetrates into the ground. If we do not know this distance, we cannot calculate the force. If we knew the time during which the impulse occurs, we could calculate the force from the impulse–momentum relationship; however, not knowing the distance or time of the rock's penetration into the ground, we cannot calculate the force. **125.** The work–energy theorem applies here. Your friend is correct: Changing the KE requires work, which means more fuel consumption and decreased air quality. **127.** In accord with energy conservation, a person who takes in more energy than is expended stores what's left over as added chemical energy in the body, which in practice means more fat. A person who expends more energy than is taken in gets extra energy by "burning" body fat. An undernourished person who performs extra work does so by consuming stored chemical energy in the body—something that cannot occur very long without health, and life. **129.** (a) In accord with Newton's second law, the component of gravitational force that is parallel to the incline produces an acceleration parallel to the incline. (b) In accord with the work–energy theorem, that force component multiplied by the distance the ball travels is equal to the change in the ball's KE. **131.** This is similar to the preceding two questions. In a circular orbit, the force of gravity is everywhere perpendicular to the satellite's motion. With no component of force parallel to its motion, no work is done and the satellite's KE remains constant.

CHAPTER 4

1. Newton discovered that gravity is universal. **3.** Every body in the universe attracts every other body with a force that, for two bodies, is directly proportional to the product of their masses and inversely proportional to the square of the distance separating them:

$$F = G\frac{m_1 m_2}{d^2}$$

5. The gravitational force is 9.8 N. **7.** You're closer to Earth's center at sea level, so you weigh more there than on any mountain peak. **9.** The

compression does not change at constant velocity. **11.** Your weight is mg when you are firmly supported and in equilibrium. **13.** The perturbations of Uranus led to the discovery of Neptune. **15.** Dark energy makes up some 73% of the universe; dark matter makes up about 23%. **17.** Neglecting air resistance, the vertical component of velocity decreases as the stone rises and increases as it descends, the same as with any freely falling object. **19.** The projectile returns at the same speed of 100 m/s, as indicated in Figure 4.22. **21.** A satellite must remain above Earth's atmosphere because air resistance would not only slow it down but also incinerate it at its high speed. A satellite must not contend with either of these. **23.** With no gravity, the path of a moving object would be a straight line, in accord with the law of inertia. **25.** As in the preceding question, speed doesn't change when there is no component of gravitational force in the direction of its motion. **27.** In an elliptical orbit there *is* a component of force in the direction of motion. **29.** Such a speed is called the *escape speed* because the satellite escapes the influence of Earth. **31.** Open ended. **33.** Open ended. **35.** Discovering this connection between falling water and falling satellites was an "Aha" moment for Paul Hewitt while whirling a water-filled bucket during a rotational-motion classroom demonstration—on a day when a much-publicized satellite launch was being discussed. How exhilarating to discover connections in nature!

37. $F = G\dfrac{m_1 m_2}{d^2} = 6.67 \times 10^{-11} \text{ N} \cdot \text{m}^2/\text{kg}^2 \times \dfrac{(1 \text{ kg})(6 \times 10^{24} \text{ kg})}{[2(6.4 \times 10^6 \text{ m})]^2}$
   $= 2.5 \text{ N}.$

39. $F = G\dfrac{m_1 m_2}{d^2}$

   $= 6.67 \times 10^{-11} \text{ N} \cdot \text{m}^2/\text{kg}^2 \times \dfrac{(6.0 \times 10^{24} \text{ kg})(2.0 \times 10^{30} \text{ kg})}{(1.5 \times 10^{11} \text{ m})^2}$

   $= 3.5 \times 10^{22} \text{ N}$

41. $F = G\dfrac{m_1 m_2}{d^2} = 6.67 \times 10^{-11} \text{ N} \cdot \text{m}^2/\text{kg}^2 \times \dfrac{(3.0 \text{ kg})(100 \text{ kg})}{(0.5 \text{ m})^2}$

$= 8.0 \times 10^{-8}$ N. The obstetrician exerts twice the gravitational force on the baby! **43.** From $F = G\dfrac{mM}{d^2}$, $F = G\dfrac{2m2M}{d^2} = G\dfrac{4mM}{d^2}$, which means the force of gravity between them is 4 times greater. **45.** From $F = G\dfrac{mM}{d^2}$, if d is made 10 times smaller, $1/d^2$ is made 100 times larger, which means the force is 100 times greater.

47. $\dfrac{\text{Force in orbit}}{\text{Force on ground}} = \dfrac{GmM/(d + 200 \text{ km})^2}{GmM/d^2} = \dfrac{d^2}{(d + 200 \text{ km})^2}$
   $= \dfrac{(6380 \text{ km})^2}{(6380 \text{ km} + 200 \text{ km})^2} = 0.94$

49. One second after being thrown, the ball's horizontal component of velocity is 10 m/s, and its vertical component is also 10 m/s. By the Pythagorean theorem, $v = \sqrt{(10^2 + 10^2)} = 14.1$ m/s. (It is moving at a 45° angle.)

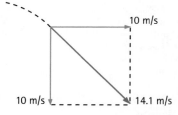

51. Total energy = PE + KE = 5000 MJ + 4500 MJ = 9500 MJ. So KE = total energy − PE = 9500 MJ − 6000 MJ = 3500 MJ.
53. (a) This is very much like the preceding problem. The height the ball reaches depends only on the vertical component of motion, and the time to reach that height is $t/2$.

So from $y = \frac{1}{2}gt^2$, $h = \frac{1}{2}g\left(\dfrac{t}{2}\right)^2 = \frac{1}{2}g\left(\dfrac{t^2}{4}\right) = \dfrac{gt^2}{8}$. (b) $h = gt^2/8 = g(4 \text{ s})^2/8 = 2g = 19.6$ m. (c) Yes, the time would be the same for the same height, whatever the launching angle. **55.** (a) $d = v_x t = vt$. The time t is the same time for the ball to fall $(y - 0.1y) = 0.9y$ (due to the height of the coffee can). From $0.9 = \frac{1}{2}gt^2$, $t^2 = \sqrt{\frac{2(0.9y)}{g}}$, and $t = \sqrt{\frac{2(0.9y)}{g}}$. So

$d = v\sqrt{\frac{2(0.9y)}{g}}$. (b) $v\sqrt{\frac{2(0.9y)}{g}} = 4.0$ m/s $\sqrt{\frac{2(0.9)(1.5\text{ m})}{9.8\text{ m/s}^2}} = 2.1$ m.
57. C, B, A. **59.** (a) A = B = C. (b) A = B = C. (c) A = B = C.
(d) B, A, C. **61.** (a) A, B, C, D. (b) A, B, C, D. (c) A, B, C, D. (d). A, B,
C, D. (e) D, C, B, A. (f) A = B = C = D. (g) A, B, C, D. **63.** The force
of gravity is the same on each because the masses are the same, as Newton's
equation for gravitational force verifies. **65.** Use the equation for gravitation to guide your thinking: Half the diameter is half the radius, which corresponds to 4 times as much weight as on Earth. **67.** Your weight would
decrease if Earth expanded with no change in its mass and would increase
if Earth contracted with no change in its mass. Your mass and Earth's mass
don't change, but the distance between you and Earth's center does change.
Force is proportional to the inverse square of this distance. **69.** You are
weightless because there is no support force. The force of gravity is nevertheless acting on you and will be the only force on you until air resistance builds up. **71.** The two forces are the normal force and *mg*, which
are equal when the elevator doesn't accelerate and unequal when the elevator accelerates. **73.** In accord with the principle of horizontal and vertical projectile motion, the time to hit the floor is independent of the ball's
speed. **75.** There are no forces acting horizontally (neglecting air resistance), so there is no horizontal acceleration; hence the horizontal component of velocity doesn't change. Gravitation acts vertically, which is why
the vertical component of velocity changes. **77.** Both balls have the same
range (see Figure 4.19). The ball with the initial projection angle of 30°,
however, is in the air for a shorter time and hits the ground first. **79.** The
hang time is the same, in accord with the answer to the preceding exercise.
Hang time is related to the vertical height attained in a jump, not to the
horizontal distance moved across a level floor.
81. Neither the speed of a falling object (without air resistance) nor the
speed of a satellite in orbit depends on its mass. In both cases, a greater
mass (greater inertia) is balanced by a correspondingly greater gravitational
force, so the acceleration remains the same ($a = F/m$, Newton's second
law). **83.** Gravity changes the speed of a cannonball when the cannonball moves in the direction of Earth's gravity. At low speeds, the cannonball curves downward and gains speed because there is a component of the
force of gravity along its direction of motion. If the cannonball is fired fast
enough, however, the curvature matches the curvature of Earth, so the cannonball moves at right angles to the force of gravity. With no component of
force along its direction of motion, its speed remains constant. **85.** Hawaii
is closer to the equator and therefore has a greater tangential speed about
the polar axis. This speed is added to the launch speed of a satellite and
thereby saves fuel. **87.** A satellite travels faster when closer to the body
it orbits. Therefore Earth travels faster around the Sun in January than in
July. **89.** If a box of anything is "dropped" from an orbiting space vehicle, it has the same tangential speed as the vehicle and remains in orbit.
If a box is dropped from a high-flying jumbo jet, it too has the tangential
speed of the jet. But this speed is insufficient for the box to fall around and
around Earth. Instead it soon falls to Earth's surface. **91.** Communication satellites have tangential velocities that ensure they fall around Earth.
They appear motionless only because their orbital period coincides with the
daily rotation of Earth. **93.** There is nothing to be concerned about on
this consumer label. It simply states the universal law of gravitation, which
applies to *all* products. It looks like the manufacturer knows some physics
and has a sense of humor. **95.** The force of gravity is the same on each
because the masses are the same, as Newton's equation for gravitational
force verifies. When dropped, the crumpled foil falls faster only because it
encounters less air resistance than the uncrumpled foil. **97.** The force of
gravity on moon rocks at the Moon's surface is considerably stronger than
the force of gravity on the distant Earth. Rocks dropped on the Moon fall
onto the Moon's surface. (The force of the Moon's gravity is about $\frac{1}{6}$ of the
weight the rock would have on Earth; the force of Earth's gravity at that
distance is only about $\frac{1}{3600}$ of the rock's Earth-weight.) **99.** You disagree
because the force of gravity on orbiting astronauts is almost as strong as at
Earth's surface. They feel weightless because of the absence of a support
force. **101.** The forces between the apple and Earth are the same in magnitude. Force is the same either way, but the corresponding accelerations
of each are different. **103.** For the planet half as far from the Sun, light

would be 4 times as intense. For the planet five times as far, light would be $\frac{1}{25}$
as intense. **105.** At noon you're closer to the Sun, so gravitation by the Sun
on you is greater. Tomorrow at midnight you're one Earth diameter farther
from the Sun and gravitation is weaker. **107.** More fuel is required for a
rocket that leaves Earth to go to the Moon than the other way around. This
is because a rocket must move against the greater gravitational field of Earth
most of the way. (If launched from the Moon to Earth, the rocket would be
traveling with Earth's field most of the way.) **109.** For very slow-moving
bullets, the dropping distance is comparable to the horizontal range, and the
resulting parabola is easily noticed (the curved path of a bullet tossed sideways
by hand, for example). For high-speed bullets, the same drop occurs in the
same time, but the horizontal distance traveled is so far that the trajectory is
"stretched out" and hardly seems to curve at all. But it does curve. All bullets
drop equal distances in equal times, whatever their speed. (It is interesting to
note that air resistance plays only a small role, since the air resistance acting
downward is practically the same for both a slow-moving and a fast-moving
bullet.) **111.** The Moon has no atmosphere (because the escape velocity at
the Moon's surface is less than the speeds of any atmospheric gases). A satellite 5 km above Earth's surface is still in considerable atmosphere as well as in
the range of some mountain peaks. Atmospheric resistance is the factor that
most determines orbiting altitude. **113.** Consider "Newton's cannon" fired
from a tall mountain on Jupiter. To match the wider curvature of much larger
Jupiter, and to contend with Jupiter's greater gravitational pull, the cannonball would have to be fired significantly faster. (Orbital speed about Jupiter is
about 5 times that for Earth.) **115.** The half brought to rest fall vertically to
Earth. The other half, in accord with conservation of linear momentum, have
twice the initial velocity, overshoot the circular orbit, and enter an elliptical
orbit whose apogee (highest point) is farther from Earth's center.

CHAPTER 5

1. Liquid and gas are both fluids because they flow. **3.** Mass density is
mass per volume; weight density is weight per volume. **5.** According to
the formula liquid pressure = density × depth, pressure increases with
an increase in depth and pressure increases with an increase in density.
7. Pressure at the same depth is the same in both the pond and the
lake. **9.** The buoyant force acts upward because pressure is greater against
the bottom where the water is deeper. **11.** An immersed body is buoyed
up by a force equal to the weight of the fluid displaced. **13.** The buoyant force
is equal to the weight of the water displaced. **15.** The volume of the water displaced is 0.5 L, and the buoyant force is 4.9 N. **17.** Since 100 tons of water
are displaced by a 100-ton floating ship, the buoyant force on the ship is 100
tons. **19.** The pressure is doubled when the balloon is squeezed to half its volume. **21.** Both pressures are the same. **23.** Because the density of water is
1/13.6 that of mercury, a water column needs to be 13.6 times taller than a mercury column to have the same weight and produce the same pressure. **25.** An
increase in pressure in one part of a confined fluid is transmitted throughout the
fluid to all parts. **27.** When the balloon is in equilibrium, its buoyant force
equals its weight, 1 N. If the buoyant force decreases, it will move downward;
if it increases, the balloon will rise. **29.** Bernoulli's principle refers only to the
internal pressure changes in a fluid. **31.** The egg floats in salt water, evidence
that salt water is more dense than fresh water. **33.** When the can comes to an
abrupt stop, the water forces the submerged Ping-Pong ball to shoot into the air.
Quite impressive! **35.** The pressures should be approximately the same. The
rigid walls of the tire prevent the pressure calculations from being closer. The
calculated value should therefore be somewhat greater. **37.** You have a barometer of sorts, but since the medium is water, the glass would have to be 10.3 m
tall to give the same pressure as a column of mercury 76 cm tall. **39.** The
gurgling is due to air entering the jar. No gurgling would occur if you tried
this experiment on the Moon where there is no atmosphere. **41.** This is more
dramatic than the larger can of the preceding activity. The aluminum cans
implode dramatically. What occurs is described in Chapter 7 (p. 176). This is a
must-do activity! **43.** When your finger closes off the top of the water-filled
straw, atmospheric pressure no longer acts on the top part of the water, which is
easily lifted. When you lift your finger, the water spills out the bottom. This is a
nice procedure for transferring liquids from one test tube to another. **45.** The
blown air that spreads between the spool and card is of low pressure, low
enough that the greater atmospheric pressure on the outside part of the card

presses the card to the spool. **47.** Pressure = weight density × depth = 9800 N/m³ × 1 m = 9800 N/m² = 9.8 kPa. **49.** Pressure = weight density × depth = 9800 N/m³ × 220 m = 2,160,000 N/m² = 2160 kPa. **51.** A 2-kg ball weighs 20 N, so the pressure is 20 N/cm² = 200 kPa. **53.** Pressure = weight density × depth = 9800 N/m³ × (5 + 1) m = 9800 N/m³ × 6 m = 58,800 N/m² = 58.8 kPa. **55.** If each crate will push the barge 4 cm deeper, the question becomes: How many 4-cm increments will make 15 cm? 15 cm/4 cm = 3.75, so three crates can be carried without sinking. Four crates will sink the barge. **57.** The displaced water, with a volume of 90% of the vacationer's volume, weighs the same as the vacationer (to provide a buoyant force equal to her weight). Therefore her density is 90% of the water's density. Vacationer's density = (0.90)(1025 kg/m³) = 923 kg/m³. **59.** The relative areas are as the squares of the diameters: 6²/2² = 36/4 = 9. The larger piston can lift 9 times the input force applied to the smaller piston. **61.** According to Boyle's law, the pressure increases to three times its original pressure.

63. From $P = \dfrac{F}{A} = \dfrac{\text{density} \times g \times \text{vol}}{A}$

$= \dfrac{\text{density} \times g \times A \times h}{A} = \text{density} \times g \times h;$

$h = \dfrac{P}{\text{density} \times g} = \dfrac{100{,}000 \text{ N/m}^2}{1.2 \text{ kg/m}^3 \times 10 \text{ N/kg}} = 8300 \text{ m} = 8.3 \text{ km}.$ The

top of the atmosphere would be about 8.3/40 or 0.2 of the almost-top of our atmosphere. **65.** c, a, b. **67.** a, b, c. **69.** a, c, b. **71.** Water. **73.** The pressure is appreciably greater in the high heels of the woman, which would hurt you more. **75.** A person lying on a waterbed experiences less body weight pressure because more of the body is in contact with the supporting surface. The greater area reduces the support pressure. **77.** The force needed is the weight of 1 L of water, which is 9.8 N. If the weight of the carton is not negligible, then the force needed is 9.8 N minus the carton's weight because the carton is "helping" to push itself down. **79.** Mercury is more dense (13.6 g/cm³) than iron. A block of iron will displace its weight and still be partially above the mercury's surface. Hence it floats in mercury. In water it sinks because it cannot displace its weight. **81.** When the ball is held beneath the surface, it displaces a greater weight of water. **83.** In the case of the wooden block, the weight on the scale doesn't change. The weight of the brim-full beaker is the same whether or not a block floats in it, because the block's weight equals the weight of water that overflows (see Figure 5.16). But not so with an iron block that sinks in the water and is heavier than the water displaced. The scale reading increases when the iron block is in the beaker. **85.** The low energy required to rotate the Falkirk Wheel is due to balanced caissons, which weigh the same as long as the water in them has the same depth. If nonfloating material is carried, then balance and the associated low energy input would be problematic. This is because only in floating does an object displace a weight of water that matches the weight of the floating object. **87.** The buoyant force does not change. The buoyant force on a floating object is always equal to that object's weight, no matter what the fluid. **89.** Unlike water, air is easily compressed—and also foam rubber. In fact, air's density is proportional to its pressure. So, near Earth's surface, where the pressure is greater, the air's density is greater (and a foam brick is more squashed), and at high altitude, where the pressure is less, the air's density is less. **91.** A vacuum cleaner wouldn't work on the Moon. A vacuum cleaner operates on Earth because the atmospheric pressure pushes dust into the machine's region of reduced pressure. On the Moon there is no atmospheric pressure to push the dust anywhere. **93.** Drinking through a straw is slightly more difficult on top of a high mountain. This is because the reduced atmospheric pressure is less effective in pushing soda up into the straw. **95.** One's lungs, like an inflated balloon, are compressed when submerged in water, and the air within them is compressed. Air will not of itself flow from a region of low pressure into a region of high pressure. The diaphragm in one's body reduces lung pressure to permit breathing, but this limit is strained when nearly 1 m below the water surface. The limit is exceeded at more than 1-m deep. Breathing through a hose that extends to the surface can't occur because air will not move from low pressure at the surface to higher pressure below in your lungs. **97.** An object rises in air only when the buoyant force exceeds its weight. A steel tank of anything weighs more than the air it displaces, so it doesn't rise. A helium-filled balloon weighs less than the air it displaces and rises. **99.** Some of the molecules in Earth's atmosphere *do* go off into outer space—those like helium with speeds greater than the escape speed.

But the average speeds of most molecules in the atmosphere are well below the escape speed, so the atmosphere is held to Earth by Earth's gravity. **101.** The force of the atmosphere is on both sides of the window; the net force is zero, so windows don't normally shatter under the weight of the atmosphere. In a strong wind, however, pressure is reduced on the windward side (Bernoulli's principle) and the forces no longer cancel to zero. Many windows are blown *outward* in strong winds. **103.** In accord with Bernoulli's principle, the sheets of paper will move inward together because the air pressure between them is reduced and is lower than the air pressure on the outside surfaces. **105.** Part of whatever pressure you add to the water is transmitted to the hungry crocodiles, via Pascal's principle. If the water were confined—that is, not open to the atmosphere—the crocs would receive every bit of the pressure you exert. But even if you were able to slip into the pool to quietly float without exerting pressure via swimming strokes, your displacement of water raises the water level in the pool. This ever-so-slight rise, and the accompanying ever-so-slight increase in pressure at the bottom of the pool, is an ever-so-welcome signal to the hungry crocodiles. **107.** An airplane flies upside down by tilting its fuselage so that there is an angle of attack of the wing with oncoming air. (It does the same when flying right side up, but then, because the wings are designed for right-side-up flight, the tilt of the fuselage may not need to be as great.) **109.** Your upper arm is at the same level as your heart, so the blood pressure in your upper arm is the same as the blood pressure in your heart. **111.** The diet drink is less dense than water, whereas the regular drink is denser than water. (Water with dissolved sugar is denser than pure water.) Also, the weight of the can is less than the buoyant force that would act on it if it were totally submerged. So it floats, where the buoyant force equals the weight of the can. **113.** The use of a water-filled garden hose as an elevation indicator is a practical example of water seeking its own level. The water surface at one end of the hose will be at the same elevation above sea level as the water surface at the other end of the hose. **115.** When a ship is empty, its weight is lowest and it displaces the least amount of water and floats higher. Carrying a load of anything increases the ship's weight and makes it float lower. It will float as low when carrying a few tons of Styrofoam as when carrying the same number of tons of iron ore. So the ship floats lower in the water when loaded with Styrofoam than when empty. If the Styrofoam were outside the ship, attached below the water line, then the ship would float higher, just as a person would with a life preserver. **117.** While the submarine is floating, its buoyant force equals its weight. When the submarine is submerged, the buoyant force equals the submarine's weight *plus* the weight of water taken into its ballast tanks. Looked at another way, the submerged submarine displaces a greater weight of water than the same submarine floating. So we can't assume the buoyant force is greater on floating things! **119.** When an ice cube melts, the water level at the side of the glass is unchanged (neglecting temperature effects). To see this, suppose the ice cube is a 5-g cube; then while floating it displaces 5 g of water. But when melted, it becomes the same 5 g of water. Hence the water level is unchanged. The same occurs when the ice cube with the air bubbles melts. Whether the ice cube is hollow or solid, it will displace as much water floating as it will when melted. If the ice cube contains grains of heavy sand, however, when it melts, the water level at the edge of the glass will drop. This is similar to the case of the scrap iron of Question 116. **121.** The Magdeburg hemispheres demonstration used two teams of horses for showmanship and effect: a single team and a strong tree would have provided the same force on the hemispheres. So if two teams of nine horses each could pull the hemispheres apart, then a single team of nine horses could do the same if a tree or some other strong object held the other end of the rope. **123.** If the bag is airtight, then whatever air is inside the bag when it is on the ground expands against the decreased cabin pressure when the plane is aloft. The air pressure in the bag is greater than the surrounding atmospheric pressure. In the case of the crumpled drum in the preceding question, the surrounding air has more pressure than whatever air remains inside the barrel. Opposite results for opposite situations. **125.** The strong man will be unsuccessful. He will have to push with 50 times the weight of the 10 kg. That's 5000 N, more than his weight. The hydraulic arrangement works to his disadvantage. Ordinarily, the input force is applied to the smaller piston and the output force to the large piston. This arrangement is just the opposite. **127.** Your partner isn't distinguishing between the internal pressure in a fast-moving fluid and the external pressure that fluid can exert on something in its way. Bernoulli's principle refers to the internal pressure changes in a fluid, not to pressures the fluid may exert when it interacts with objects in the fluid. (This distinction eludes many people!)

CHAPTER 6

1. Water freezes at 0°C and 32°F, and boils at 100°C and 212°F. **3.** The necessary condition is thermal equilibrium because only then do the thermometer and thing being measured have the same temperature. **5.** The pressure would be zero at −273°C. **7.** No energy can be removed from a system at 0 K. **9.** Hot objects contain thermal energy, not heat. **11.** The direction of thermal energy flow is from objects at higher temperatures to objects at lower temperatures. **13.** The energy needed is 4.19 J. **15.** All are units of energy. A calorie is the energy needed to change the temperature of 1 g of water by 1°C; a Calorie is 1000 cal; a joule is the modern unit of energy, where 1 cal = 4.19 J. **17.** Added heat that doesn't raise the temperature increases the thermal energy of the system and/or does external work if it leaves the system. **19.** Thermal energy can move from lower to higher temperatures only when external work is done on the system. **21.** Entropy is a measure of disorder. The greater the disorder, the greater the entropy. So entropy increases with disorder in a system. **23.** Silver heats more quickly and has the higher specific heat capacity. **25.** Thermal energy is carried in the Gulf Stream from tropical waters to the North Atlantic, where it warms the otherwise cold climate. **27.** Liquids generally expand more for an equal increase in temperature. **29.** Ice is less dense than water due to its ice crystals that have open structures. **31.** See Think and Solve 36. **33.** $Q = cm\Delta T =$ (1 cal/g·°C)(30 g)(30°C − 20°C) = 300 cal. **35.** 300 cal(4.19 J/1 cal) = 1257 J. **37.** The work the hammer does on the nail is $F \times d$, and the temperature change of the nail can be found from $Q = cm\Delta T$. First, we get everything into convenient units for calculating: 6.0 g = 0.006 kg; 8.0 cm = 0.08 m. Then $F \times d = 600$ N × 0.08 m = 48 J, and 48 J = (0.006 kg)(450 J/kg·°C)(ΔT), which we can solve to get $\Delta T = 48$J/(0.006 kg × 450 J/kg·°C) = 17.8°C. (You will notice a similar effect when you remove a nail from a piece of wood. The nail that you pull out is noticeably warm.) **39.** Raising the temperature of 10 kg of steel by 1°C takes 10 kg(450 J/kg·°C) = 4500 J. Raising the temperature by 100°C takes 100 times as much, or 450,000 J. By formula, $Q = cm\Delta T$ = (450 J/kg·°C)(10 kg)(100°C) = 450,000 J. Heating 10 kg of water through the same temperature difference takes 1000 calories, which is [1000 cal(4.18 J/cal)] = 41,800 J, more than ten times for the piece of steel—another reminder that water has a high specific heat capacity. **41.** If a 1-m-long bar expands 0.6 cm when heated, a bar of the same material that is 100 times as long will expand 100 times as much, 0.6 cm for each meter, or 60 cm. (The heated bar will be 100.6 m long.) **43.** If a snugly fitting steel pipe that girdled the world were heated by 1°C, it would stand about 70 m off the ground! The most straightforward way to see this is to consider the radius of the 40,000-km-long pipe, which is the radius of Earth, 6370 km. Steel will expand 11 parts in a million for each 1°C increase in temperature; the radius as well as the circumference will expand by this fraction. So 11 millionths of 6370 km = 70 m. Isn't this astounding? **45.** c, a, b. **47.** c, a, b. **49.** The average speed of molecules in both containers is the same. There is more thermal energy in the full glass (twice the matter at the same temperature). More heat will be required to increase the temperature of the full glass by 1°C—twice as much, in fact. **51.** The hot coffee has a higher temperature but not a greater amount of thermal energy. Although the iceberg has less thermal energy per mass, its enormously greater mass gives it a greater total amount of energy than that in the small cup of coffee. **53.** Work is done in compressing the air, which, in accord with the first law of thermodynamics, increases its thermal energy. This is evident by its higher temperature. **55.** Friction occurs between the road and the tires, which warms up the tires and warms the air within. The molecules in the warmed air move faster, which increases the air pressure in the tires. **57.** A greater change in temperature occurs for iron because it has a lower specific heat capacity. **59.** Sand has a low specific heat capacity, as evidenced by the relatively large temperature changes for small changes in thermal energy. A substance with a high specific heat capacity, on the other hand, must absorb or give off large amounts of thermal energy for comparable temperature changes. **61.** The water in the melon has more "thermal inertia"—a higher specific heat capacity—than sandwich ingredients. Be glad water has a high specific heat capacity the next time you're enjoying cool watermelon on a hot day! **63.** No, the different rates of expansion are what bends the strip or coil. Yes, it is important: Without the different rates of expansion, a bimetallic strip would not bend when heated (or when cooled). **65.** Thin glass is used because temperature changes can be sudden. If the glass were thicker, unequal expansions and contractions would break the glass when sudden temperature changes occur. **67.** Cool the inner glass and heat the outer glass. If it's done the other way around, the glasses will stick even tighter (if not break). **69.** The gap in the ring becomes wider when the ring is heated. Try this: Draw a couple of lines on a ring where you imagine a gap to be. When you heat the ring, the lines are farther apart—the same amount as if a real gap were there. Every part of the ring expands proportionally when heated uniformly—thickness, length, gap, and all. **71.** The atoms and molecules of most substances are more closely packed in solids than in liquids. So most substances, such as iron, aluminum, and most metals, are denser in the solid phase than in the liquid phase. Water is different. In its solid phase, the structure is open spaced. Ice is less dense than water, and hence ice floats in water. **73.** There is no temperature difference between your hand and forehead. If the temperature of your forehead is a couple degrees higher than normal, then the temperature of your hand is also a couple degrees higher. **75.** Only the second law is a probabilistic statement and has exceptions. **77.** The iron brick will cool off quickly, and you'll be cold in the middle of the night. Bring a jug of hot water with its higher specific heat capacity to bed, and you'll stay warm throughout the night. **79.** As the ocean off the coast of San Francisco cools in the winter, the lost heat warms the atmosphere it comes in contact with. This warmed air blows over the California coastline to produce a relatively warm climate. If the winds were easterly instead of westerly, the climate of San Francisco would be chilled by winter winds from dry, cold Nevada. The climate would be reversed also in Washington, DC, because air warmed by the cooling of the Atlantic Ocean would blow over Washington, DC, and produce a warmer climate in winter in that region. **81.** The natural state of the perfume molecules is to spread out, to reach a state of increased disorder and increased entropy. Perfume molecules and their smell soon drift from the corner to all parts of the room. **83.** Temperature differences cause differences in expansion and contraction, which produce sounds as structures expand and contract. **85.** When the rivets cool, they contract, and this tightens the fit of the plates. **87.** Water has its maximum density at 4°C; therefore, either cooling or heating at this temperature results in an expansion of the water. A small rise in the water level would be ambiguous and make a water thermometer impractical in this temperature region.

CHAPTER 7

1. Conduction, convection, and radiation. **3.** Wood is a good insulator even when it's red hot. Therefore very little thermal energy is transferred to the feet. **5.** Heat is transferred by the movement of fluids. **7.** Her hand is not in steam, but in a jet of condensed vapor that has expanded and cooled. **9.** Peak frequency and absolute temperature are directly proportional: $f \sim T$. **11.** Temperatures don't continuously decrease because all objects are also absorbing radiant energy. **13.** By Newton's law of cooling, the hot poker in the cold room radiates more because of the greater temperature difference between the poker and the room. **15.** Without the greenhouse effect, Earth would be a very cold place, with an average temperature about −18°C. **17.** The four commons phases of matter are solids, liquids, gases, and plasmas. **19.** Evaporation is the change of phase from liquid to gas. As fast-moving molecules in a liquid escape into the air, slower ones on average are left behind, thereby cooling the water. **21.** Condensation is the opposite of evaporation. Gas molecules near the surface of a liquid are attracted to the liquid, strike it with increased KE, and thereby warm the liquid. **23.** Evaporation is a phase change at the surface of a liquid; boiling is a phase change that occurs throughout a liquid, producing bubbles. **25.** High temperature cooks food, not the bubbles associated with boiling. **27.** Molecular motion slows, thus allowing molecules to bind together. **29.** A liquid absorbs energy when it changes into a gas; when a liquid changes into a solid, it gives off energy. **31.** Open ended. **33.** This activity nicely shows that water is not a good conductor of heat. **35.** A geyser and a coffee percolator work on the same principle. **37.** The rainfall seen resembles actual rain in that condensation of vapor leads to drops of water. It differs in that natural rain is the result of cooling in clouds of vapor, rather than by condensation on a chilled surface. **39.** Water at 100°C will boil when the heat needed for a phase change is added. Since the temperatures of water in the pan and in the inner container are the same, no heat passes from the water in the pan to the water in the container. So the water in the inner container doesn't boil. This technique is common for cooks in the kitchen. **41.** $Q = cm\Delta T$ = (1 cal/g·°C)(20 g)(90°C − 30°C) = 1200 cal.

43. $Q = mL_f = (200 \text{ g})(80 \text{ cal/g}) = 16{,}000$ cal.
45. $Q = mL_v = (200 \text{ g})(540 \text{ cal/g}) = 108{,}000$ cal.
47. $Q = cm\Delta T + mL_v = (1 \text{ cal/g} \cdot {}^\circ\text{C})(15 \text{ g})(100{}^\circ\text{C} - 20{}^\circ\text{C}) + (15 \text{ g})(540 \text{ cal/g}) = 9300$ cal.
49. (a) From $-273{}^\circ$C "ice" to $0{}^\circ$C ice requires $(1 \text{ g})(0.5 \text{ cal/g} \cdot {}^\circ\text{C})$ $(273{}^\circ\text{C}) = 140$ cal. From $0{}^\circ$C ice to $0{}^\circ$C water, mL, requires 80 cal. From $0{}^\circ$C water to $100{}^\circ$C water, $cm\Delta T$, requires 100 cal. The total, 140 cal + 80 cal + 100 cal = 320 cal. (b) Boiling this water at $100{}^\circ$C takes 540 cal, considerably more energy than it took to bring the water all the way from absolute zero to the boiling point! (In fact, at very low temperature, the specific heat capacity of ice is less than 0.5 cal/g · °C, so the true difference is even greater than calculated here.) **51.** (a) Half the PE warms the ball: $0.5mgh = cm\Delta T$. $\Delta T = 0.5mgh/cm = 0.5gh/c = (0.5)$ $(9.8 \text{ N/kg})(100 \text{ m})/450 \text{ J/kg} = 1.1{}^\circ$C. (b) Note that the mass cancels, so the same temperature holds for a ball of any mass, assuming half the heat generated goes into warming the ball. Note also that the units check, since $1 \text{ J/kg} = 1 \text{ m}^2/\text{s}^2$ $(1 \text{ J/kg} = 1 \text{ N} \cdot \text{m/kg} = 1 \text{ kg} \cdot \text{m/s}^2 \cdot \text{m/kg} = 1 \text{ m}^2/\text{s}^2)$. **53.** The final temperature of the water will be the same as that of the ice, $0{}^\circ$C. The quantity of heat given to the ice by the water is $Q = cm\Delta T =$ $(1 \text{ cal/g} \cdot {}^\circ\text{C})(50 \text{ g})(80{}^\circ\text{C}) = 4000$ cal. This heat melts the ice. How much? From $Q = mL_f$, $m = Q/L_f = (4000 \text{ cal})/(80 \text{ cal/g}) = 50$ g. So water at $80{}^\circ$C will melt an equal mass of ice at $0{}^\circ$C. **55.** Note that the heat of vaporization of ethyl alcohol L_v (200 cal/g) is 2.5 times greater than the heat of fusion of water (80 cal/g), so in a change of phase for both, 2.5 times as much ice will change phase; $2.5 \times 4 \text{ kg} = 10$ kg. Or via formula, the refrigerant would draw away $Q = mL_v = (4000 \text{ g})(200 \text{ cal/g}) = 800{,}000$ cal. The mass of ice formed is $(800{,}000 \text{ cal})/(80 \text{ cal/g}) = 10{,}000$ g, or 10 kg. **57.** b, c, a (the Sun is a yellow-green-hot star). **59.** c, b, a.
61. Wrapping the potatoes in aluminum foil retains the heat after the potatoes are removed from the oven. Also, heat transfer by radiation is minimized because radiation from the potatoes is internally reflected; heat transfer by convection is minimized because circulating air cannot make contact with the shielded potatoes. The foil also retains moisture. **63.** Air is an excellent insulator. Fiberglass is a good insulator principally because of the vast amount of air space trapped in it. **65.** The conductivity of wood is relatively low whatever the temperature—even at the stage of red-hot coals. You can safely walk barefoot across red-hot wooden coals if you step quickly because very little heat is conducted to your feet. Because of the poor conductivity of the coals, energy from within the coals does not readily replace the energy that transfers to your feet. This is evident in the diminished redness of the coal after your foot has left it. Stepping on red-hot *iron* coals, however, is a different story. Because of the excellent conductivity of iron, large amounts of heat would injure your feet. Ouch! **67.** Agree; at thermal equilibrium the gases have the same temperature, which is to say, the same average KE. **69.** Because of the high specific heat capacity of water, sunshine warms water much less than it warms land. As a result, air is warmed over the land and rises. Cooler air from above the cool water takes its place and convection currents are formed. If the land and water were heated equally by the Sun, such convection currents (and the winds they produce) wouldn't occur. **71.** The energy given off by rock at Earth's surface transfers to the surroundings practically as fast as it is generated. Hence there is no buildup of energy at Earth's surface. **73.** When it is desirable to reduce the radiant energy coming into a greenhouse, as in the summer, whitewash is applied to the glass simply to reflect much of the incoming sunlight. Energy reflected by the greenhouse is energy not absorbed. **75.** Alcohol produces more cooling because of its higher rate of evaporation. **77.** Hot coffee poured into a saucer cools because (1) the greater surface area of the coffee permits more evaporation to take place, and (2) by the conservation of energy, the thermal energy that heats up the saucer comes from the coffee, cooling it. **79.** Evaporation would not cool the remaining liquid because the energy of exiting molecules would be no different from the energy of molecules left behind. Although the thermal energy of the liquid would decrease with evaporation, the energy per molecule would not change. The temperature of the liquid would not change. (The surrounding air, on the other hand, would be cooled. Molecules flying away from the liquid surface would be slowed by the attractive force of the liquid acting on them.) **81.** A bottle wrapped in a wet cloth will cool by the evaporation of liquid from the cloth. As evaporation progresses, the average temperature of the liquid left behind in the cloth can easily drop below the temperature of the cool water that wet it in the first place. So, to cool a bottled beverage at a picnic, wet a piece of cloth in a bucket of cool water and then wrap it around the bottle.

As evaporation progresses, the temperature of the water in the cloth drops and cools the bottle to a temperature below that of the water in the bucket. **83.** You could not cook food in low-temperature water that is boiling by virtue of reduced pressure. Food is cooked by the high temperature it is subjected to, not by the bubbling of the surrounding water. For example, put room-temperature water in a vacuum and it will boil. But this doesn't mean the water will transfer more thermal energy to an egg than before boiling—an egg in this boiling water won't cook at all! **85.** The cooking time will be no different for vigorously boiling water and gently boiling water because both have the same temperature. The reason spaghetti is cooked in vigorously boiling water is simply to ensure that the spaghetti doesn't stick to itself and the pot. For fuel economy, stir your spaghetti in gently boiling water. **87.** You can add heat without raising the temperature when the substance is undergoing a change of phase. Small amounts of heat added to $0{}^\circ$C ice or to $100{}^\circ$C water, for example, don't increase the temperature. **89.** No water leaks from inside the unit because water vapor in warm air condenses on its low-temperature metal surface. That's the water that drips. **91.** Dogs have no sweat glands (except between the toes for most dogs) and therefore they cool by the evaporation of moisture from the mouth and the respiratory tract. So dogs literally cool from the inside out when they pant. **93.** A feather quilt is an excellent insulator (poor conductor), which slows the transfer of heat from your body to the surroundings. **95.** If the hydrogen and oxygen have the same temperature, then by definition they have the same average kinetic energies per molecule. But the hydrogen, with less mass, has the higher average speed $(\frac{1}{2}mV^2 = \frac{1}{2}Mv^2$, where $V > v)$. **97.** The heat you felt was caused by radiation, which travels at the speed of light. Heating the glass, however, takes longer. **99.** Every student radiates about the same amount of heat as a 100-W incandescent bulb. So the heat radiated by you and your classmates increases the temperature of the room. **101.** A good reflector is a poor radiator of heat, and a poor reflector is a good radiator of heat. **103.** Both remove thermal energy from one place and put it in another by the phase change of a refrigerant. **105.** Turn the air conditioner off altogether to keep ΔT small, as in the preceding answer. Heat leaks at a faster rate into a cold house than into a not-so-cold house. The greater the rate at which heat leaks into the house, the greater the amount of fuel consumed by the air conditioner. **107.** The wood melts more ice because its higher specific heat capacity means it releases more energy in cooling. **109.** Every gram of water that freezes releases 80 cal of thermal energy to the cellar. This continual release of energy by the freezing water keeps the temperature of the cellar from going below $0{}^\circ$C. Sugar and salts in the canned goods prevent them from freezing at $0{}^\circ$C. Only when all the water in the tub freezes will the temperature of the cellar go below $0{}^\circ$C and then freeze the canned goods. The farmer must, therefore, replace the tub before or just as soon as all the water in it has frozen. **111.** The black mailboxes, better absorbers, are somewhat warmer in sunlight or even on an overcast day than the light-colored boxes, so snow is more likely to melt on the black surfaces than on the light ones. Hence the accumulation of snow on the lighter mailboxes.

CHAPTER 8

1. The nucleus and its protons are positively charged; electrons are negatively charged. **3.** The masses of electrons are much less than the masses of protons (1:2000). **5.** When electrons are stripped from an object, it is left with a positive charge. **7.** Both laws are inverse-square laws. How they differ is mainly that gravitation is only attractive, whereas electrical forces can repel. **9.** By the inverse-square law, particles twice as far apart have 1/4 the force; particles three times as far apart have 1/9 the force. **11.** Two force fields are gravitational and electric. (Magnetic fields also, which we'll learn about in the next chapter.) **13.** Electric potential energy is measured in joules; electric potential in volts. **15.** A sustained flow needs a sustained difference in potential across a conducting medium, such as a battery or generator. **17.** Electric charge flows through a circuit. Voltage doesn't flow at all but is impressed across a circuit. **19.** A battery produces dc. A generator normally produces ac. **21.** The unit of electrical resistance is the ohm, symbol Ω. **23.** Dry skin has considerably more electrical resistance than wet skin. **25.** The source of electrons producing a shock is your own body. **27.** The voltage across the second lamp would be 4 V, so the sum of the two is 6 V. **29.** The more branches in both cases, the less the overall resistance. **31.** Open ended. **33.** The fact that the stream is deflected in the same way by both a negatively and a positively charged object indicates that the deflection is due to charge polarization. Water molecules in the water are themselves polarized.

35. $F = k\dfrac{q_1 q_2}{d^2} = (9 \times 10^9 \text{ N} \cdot \text{m}^2/\text{C}^2)\dfrac{(0.1 \text{ C})(0.1 \text{ C})}{(0.1 \text{ m})^2} = 9 \times 10^9 \text{ N}.$

37. $I = V/R = (6 \text{ V})/(1000 \ \Omega) = 0.006 \text{ A}.$

39. $P = I \times V = (0.5 \text{ A})(120 \text{ V}) = 60 \text{ W}.$

41. From Coulomb's law, $F = k\dfrac{q_1 q_2}{d^2} = (9 \times 10^9 \text{ N} \cdot \text{m}^2/\text{C}^2)$

$\dfrac{(1.0 \times 10^{-6} \text{ C})^2}{(0.03 \text{ m})^2} = 10 \text{ N}.$ This is the same as the weight of a 1-kg mass.

43. From Coulomb's law, the force is given by $F = \dfrac{kq^2}{d^2}$, so the square of the charge is $q^2 = \dfrac{Fd^2}{k} = \dfrac{(20 \text{ N})(0.06 \text{ m})^2}{9 \times 10^9 \text{ Nm}^2/\text{C}^2} = 8.0 \times 10^{-12} \text{ C}^2.$ Taking the square root of this gives $q = 2.8 \times 10^{-6}$ C, or 2.8 microcoulombs.

45. From voltage = energy/charge, we get energy = voltage $\times$ charge = $(12 \text{ V})(4 \text{ C}) = 48 \text{ J}.$ **47.** Energy is charge $\times$ potential: $PE = qV = (2 \text{ C})(100 \times 10^6 \text{ V}) = 2 \times 10^8 \text{ J}.$ **49.** From $I_{total} = I_1 + I_2 + I_3 + \dots + I_n$, a substitution of $I = \dfrac{V}{R}$ for each current gives

$\dfrac{V}{R_{eq}} = \dfrac{V}{R_1} + \dfrac{V}{R_2} + \dfrac{V}{R_3} + \dots + \dfrac{V}{R_n}.$ Dividing each term by V gives

$\dfrac{1}{R_{eq}} = \dfrac{1}{R_1} + \dfrac{1}{R_2} + \dfrac{1}{R_3} + \dots + \dfrac{1}{R_n}.$ **51.** From current = voltage/resistance, we get resistance = voltage/current = $(120 \text{ V})/(20 \text{ A}) = 6 \ \Omega.$

53. Ohm's law can be stated $V = IR$. Then $P = IV = I(IR) = I^2R.$

55. From $P = \dfrac{V^2}{R}$, $R = \dfrac{V^2}{P} = \dfrac{(120 \text{ V})^2}{1320 \text{ W}} = 10.9 \ \Omega$, or about 11 Ω.

57. First, 25 W = 0.025 kW. Energy = $Pt = (0.025 \text{ kW})(24 \text{ h}) = 0.6 \text{ kW} \cdot \text{h}.$ Cost = $(0.6 \text{ kWh})(\$0.080/\text{kW} \cdot \text{h}) = \$0.048$, or about 5 cents. **59.** The dryer's power is $P = IV = (120 \text{ V})(8.4 \text{ A}) = 1008 \text{ W} = 1008 \text{ J/s}.$ The heat energy generated in 1 min is $E = \text{power} \times \text{time} = (1008 \text{ J/s})(60 \text{ s}) = 60{,}400 \text{ J}$, or about 60 kJ. **61.** The bulb is designed for use in a 120-V circuit. With an applied voltage of 120 V, the current in the bulb is $I = V/R = (120 \text{ V})/(95 \text{ W}) = 1.26 \text{ A}.$ The power dissipated by the bulb is then $P = IV = (1.26 \text{ A})(120 \text{ V}) = 151 \text{ W}$, close to the rated value. If this bulb is connected to 220 V, it would carry twice as much current and would dissipate four times as much power (twice the current $\times$ twice the voltage), more than 600 W. It would likely burn out. (This question can also be answered by first performing some algebraic manipulation. Since current = voltage/resistance, we can write the formula for power as $P = IV = \left(\dfrac{V}{R}\right)V = \dfrac{V^2}{R}.$ Solving for V gives $V = \sqrt{PR}.$ Substituting for the power and the resistance gives $V = \sqrt{[(151 \text{ W})(95 \ \Omega)]} = 119 \text{ V}$, close to 120 V.) **63.** A, C, B. **65.** A = B = C (all the same).
67. A, B, C. **69.** Something is electrically charged when it has an excess or deficiency of electrons compared with the number of protons in the atomic nuclei of the material. **71.** The objects aren't charged because of their equal number of protons. **73.** The law would be written the same.
75. Doubling one charge doubles the force. The magnitude of the force does not depend on the sign of the charge. **77.** The mechanism of sticking is charge induction. If it's a metal door, the charged balloon induces an opposite charge on the door. It accomplishes this by attracting opposite charges to it and repelling like charges to parts of the door farther away. The balloon and the oppositely charged part of the door are attracted and the balloon sticks. If the door is an insulator, the balloon induces polarization of the molecules in the door material. Oppositely charged sides of the molecules in the surface of the door face the balloon and attraction results. So whether you consider the door to be an insulator or a conductor, the balloon sticks by induction. **79.** The bits of thread become polarized in the electric field, one end positive and the other negative, and become the electric counterparts of the north and south poles of the magnetic compass. Opposite forces on the ends of the fibers (or compass needle) produce torques that orient the fibers along the field direction (look ahead to Figure 9.4 in Chapter 9). **81.** For both electricity and heat, the conduction is via electrons, which in a metal are loosely bound, easy flowing, and easy to get moving. (Many fewer electrons in metals, however, take part in heat conduction than in electric conduction.) **83.** Six liters per minute (10 L − 4 L = 6 L). **85.** The cooling system of an automobile is a better analogy to an electric circuit because it is a closed system and it contains a pump, analogous to the battery or

other voltage source in a circuit. The water hose does not recirculate the water as the auto cooling system does. **87.** Most, more than 90%, of the electric energy fed into an incandescent lamp goes directly to heat. Thermal energy is the graveyard of electric energy. **89.** Less damage is done plugging a 220-V appliance into a 110-V circuit. Damage generally occurs by excess heating when too much current is driven through an appliance. For an appliance that converts electric energy directly to thermal energy, overheating occurs when excess voltage is applied. So don't connect a 110-V iron, toaster, or electric stove into a 220-V circuit. Interestingly, if the appliance is an electric motor, then applying too *little* voltage can result in overheating and burn up the motor windings. (This is because the motor will spin at a low speed and the reverse "generator effect" will be small and allow too great a net current in the motor.) So don't connect a 220-V power saw or any 220-V motor-driven appliance into a 110-V circuit. To be safe use the recommended voltages for all appliances. **91.** Electric power in your home is likely supplied at 60 Hz and 110–120 V through electrical outlets. This is ac (and delivered to your home via transformers between the power source and your home—we will see in Chapter 10 that transformers require ac power for operation). Electric power in your car needs to be supplied by the battery. Since the + and − terminals of the battery do not alternate, the current produced by the battery does not alternate either. It flows in one direction and is dc. **93.** More branches reduce resistance to motion in both circuits and toll booths. **95.** The sign is a joke. High voltage may be dangerous, but high resistance is a property of all nonconductors. **97.** From $I = V/R$, if both voltage and resistance are doubled, current remains unchanged. Likewise if both voltage and resistance are halved. **99.** Cosmic rays produce ions in air, which offer a conducting path for the discharge of charged objects. Cosmic-ray particles streaming downward through the atmosphere are attenuated by radioactive decay and by absorption, so the radiation and the ionization are stronger at high altitudes than at low altitudes. Charged objects lose their charge more quickly at higher altitudes. **101.** Yes; after the initial charge of the battery is spent, recharging occurs as the motor runs, and this energy is from the fuel. **103.** Most of the electric energy in a lamp filament is transformed into heat. For low currents in the bulb, the heat that is produced may be enough to feel but not enough to make the filament glow red or white-hot. **105.** There is less resistance in the higher-wattage lamp. Since power = current $\times$ voltage, more power for the same voltage means more current. And by Ohm's law, more current for the same voltage means less resistance. (Algebraic manipulation of the formulas $P = IV$ and $I = \dfrac{V}{R}$ leads to $P = \dfrac{V^2}{R}.$) **107.** Agree with your friend. The hairs act like leaves in an electroscope. If your arms were as light, they'd stand out too. **109.** Only circuit 5 is complete and will light the bulb. (Circuits 1 and 2 are "short circuits" and will quickly drain the cell of its energy. In circuit 3 both ends of the lamp filament are connected to the same terminal and are therefore at the same potential. Only one end of the lamp filament is connected to the cell in circuit 4.) **111.** An electric device does not "use up" electricity but rather *energy*. And strictly speaking, it doesn't "use up" energy but transforms it from one form into another. It is common to say that energy is used up when it is transformed to less concentrated forms—when it is degraded. Electric energy ultimately becomes heat energy. In this sense it is used up. **113.** The current is greater in the bulb connected to the 220-V source. Twice the voltage would produce twice the current if the resistance of the filament remained the same. (In practice, the greater current produces a higher temperature and greater resistance in the lamp filament, so the current is greater than that produced by 110 V but appreciably less than twice as much for 220 V.) A bulb rated for 110 V has a very short life when operated at 220 V.) **115.** The equivalent resistance of resistors in parallel is less than the lower resistance of the two. So connect a pair of resistors in parallel for less resistance. **117.** Agree; in series, more resistances add to the circuit resistance. But in parallel, the multiple paths provide less resistance (just as more lines at a checkout counter lessen resistance to flow). **119.** Bulb C is the brightest because the voltage across it equals that of the battery. Bulbs A and B share the voltage of the parallel branch of the circuit and have half the current of bulb C (assuming resistances are independent of voltages). If bulb A is unscrewed, the top branch is no longer part of the circuit and current ceases in both bulbs A and B. They no longer give light, while bulb C glows as before. If bulb C is instead unscrewed, it goes out and bulbs A and B glow as before. **121.** Yes, because of the increased current that flows through the battery. The internal voltage drop increases with current in the battery, which means reduced voltage supplied at its terminals to the circuit it powers. (If the parallel circuit is powered by a stronger source, such as the power provided via common wall sockets, no dimming of bulbs will be seen as more and more parallel paths are added.) **123.** Current divides in a branch, with more passing in the branch of lower resistance. But current in a branch never reduces to zero unless the resistance of the branch becomes infinite. A voltage across a non-infinite resistor will produce current in accord with Ohm's law.

CHAPTER 9

1. Refrigerator magnets are shorter range than common bar magnets. **3.** Magnetic poles cannot be isolated. **5.** The two kinds of motion are electron spin and electron revolution about the nucleus. **7.** Iron has magnetic domains; wood does not. **9.** The magnetic field also reverses direction. **11.** A piece of iron strengthens the electromagnet because the alignment of domains in the piece of iron adds to the overall field. **13.** Earth's magnetic field deflects many of the incoming cosmic rays, reducing the intensity of rays striking Earth's surface. **15.** When the current is reversed, the magnetic force acts in the opposite direction. **17.** Yes; the difference is that the rotation in a motor continues, whereas in a galvanometer the deflection is momentary. **19.** The induced voltage in a coil is proportional to the number of loops multiplied by the rate at which the magnetic field changes within those loops. **21.** Both frequencies are the same. **23.** The current is ac because the induced voltage is ac. **25.** No; a generator transforms energy from one form into another. **27.** A transformer changes voltage and current, but not energy and power. **29.** An alternating magnetic field is induced. **31.** Open ended. **33.** Cans contain iron. Domains in the can tend to line up with Earth's magnetic field. When the cans are left stationary for several days, they become magnetized by induction, aligning with Earth's magnetic field.

35. $\dfrac{120 \text{ V}}{10 \text{ turns}} = \dfrac{x \text{ V}}{100 \text{ turns}}$, where $x = 100 \text{ turns} \times \dfrac{120 \text{ V}}{10 \text{ turns}} = 1200 \text{ V}$.

37. $\dfrac{120 \text{ V}}{500 \text{ turns}} = \dfrac{6 \text{ V}}{x \text{ turns}}$, so $x \text{ turns} = 500 \text{ turns} \times \dfrac{6 \text{ V}}{120 \text{ V}} = 25 \text{ turns}$.

39. From the transformer relationship,

$\dfrac{\text{primary voltage}}{\text{number of primary turns}} = \dfrac{\text{secondary voltage}}{\text{number of secondary turns}}$. So $\dfrac{120 \text{ V}}{24 \text{ V}} = \dfrac{5}{1}$. So there are 5 times as many primary turns as secondary turns.

41. The transformer steps up voltage by a factor $36/6 = 6$. Therefore a 12-V input will be stepped up to $6 \times 12 \text{ V} = 72 \text{ V}$. **43.** The voltage step-up is $(12,000 \text{ V})/(120 \text{ V}) = 100$, so there should be 100 times as many turns on the secondary as on the primary. **45.** B, C, A. **47.** The tiny flakes of iron can be removed with a simple magnet. Try it and be amazed! **49.** Striking the nail shakes up the domains, allowing them to realign themselves with Earth's magnetic field. The result is a net alignment of domains along the length of the nail. (Note that if you hit an already magnetized piece of iron that is not aligned with Earth's field, the result can be to weaken, not strengthen, the magnet.) **51.** Refrigerator magnets have narrow strips of alternating north and south poles. They are strong enough to hold sheets of paper on a refrigerator door, but they have a very short range because the north and south poles cancel a short distance from the magnetic surface. Ordinary magnets have longer ranges. **53.** Domains in the paper clip are induced into alignment in a manner similar to the electric charge polarization in an insulator when a charged object is brought nearby. Either pole of a magnet will induce alignment of domains in the paper clip: Attraction results because the pole of the aligned domains closest to the magnet's pole is always the opposite pole. **55.** Apply a small magnet to the door. If it sticks, your friend is wrong because aluminum is not magnetic. If it doesn't stick, your friend might be right (but not necessarily—there are lots of nonmagnetic materials). **57.** Yes, because the compass aligns with Earth's magnetic field, which extends from the magnetic pole in the Southern Hemisphere to the magnetic pole in the Northern Hemisphere. **59.** Yes. Since the magnet exerts a force on the wire, the wire, according to Newton's third law, must exert a force on the magnet. **61.** Use Newton's third law again: Yes, the paper clip, as part of the interaction, certainly does exert a force on the magnet—just as much as the magnet pulls on it. The magnet and the paper clip pull equally on each other to make up the single interaction between them. **63.** An electron has to be moving across lines of magnetic field in order to feel a magnetic force. So an electron at rest in a stationary magnetic field feels no force to set it in motion. In an electric field, however, an electron is accelerated whether or not it is already moving. (A combination of magnetic and electric fields is used in particle accelerators such as cyclotrons. The electric field accelerates the charged particle in its direction, and the magnetic field accelerates it perpendicular to its direction, causing it to follow a nearly circular path.) **65.** Charged particles moving through a magnetic field are deflected most when they move at right angles to the field lines, and they are deflected least when they move parallel to the field lines. If we consider cosmic rays heading toward Earth from all directions and from great distances, those descending toward northern Canada are moving nearly parallel to the magnetic field lines of Earth. They are not deflected very much, and the secondary particles they create high in the atmosphere also stream downward with little deflection. Over regions closer to the equator, like Mexico, the incoming cosmic rays move more nearly at right angles to Earth's magnetic field, and many of them are deflected back out into space before they reach the atmosphere. The secondary particles they create are less intense at Earth's surface. (This "latitude effect" provided the first evidence that cosmic rays from outer space consist of charged particles—mostly protons, as we now know.) **67.** Cosmic-ray intensity at Earth's surface would be greater when Earth's magnetic field passed through a zero phase because magnetic shielding is minimal. Fossil evidence suggests that the periods of no protective magnetic field may have been as important in changing life forms as X-rays have been in the famous heredity studies of fruit flies. **69.** Singly charged ions traveling with the same speed through the same magnetic field will experience the same magnetic force. The extent of their deflections will then depend on their accelerations, which in turn depend on their respective masses. The least massive ions will be deflected the most, and the most massive ions will be deflected the least. **71.** Magnetic induction will not occur in nylon, since it has no magnetic domains. That's why electric guitars use steel strings. **73.** Work must be done to move a current-carrying conductor in a magnetic field. This is true whether the current is externally produced or produced as a result of the induction that accompanies the motion of the wire in the field. It's also a matter of energy conservation. There has to be more energy input if there is more energy output. **75.** Part of Earth's magnetic field is enclosed in the wide loop of wire embedded in the road surface. If this enclosed field is somehow changed, then in accord with the law of electromagnetic induction, a pulse of current is produced in the loop. Such a change is produced when the iron parts of a car pass over it, momentarily increasing the strength of the field. A practical application is triggering highway traffic lights. (When small ac voltages are used in such loops, small "eddy currents" are induced in metal of any kind that passes over the loop. The magnetic fields so induced are then detected by the circuit.) **77.** The changing magnetic field of the moving tape induces a voltage in the coil. A practical application is a message or tape recorder. **79.** Agree; any coil of wire spinning in a magnetic field that cuts through magnetic field lines is a generator. **81.** If the lightbulb is connected to a wire loop that intercepts changing magnetic field lines from an electromagnet, voltage will be induced, which can illuminate the bulb. Change is the key, so the electromagnet should be powered with ac. **83.** The iron core increases the magnetic field of the primary coil. The greater field means a greater magnetic field change in the primary and a higher voltage induced in the secondary. The iron core in the secondary further increases the changing magnetic field through the secondary and further increases the secondary voltage. Furthermore, the core guides more magnetic field lines from the primary to the secondary. The effect of an iron core in the coils is the induction of appreciably more voltage in the secondary. **85.** A step-up transformer multiplies voltage in the secondary; a step-down transformer does the opposite—decreases voltage in the secondary. **87.** No, no, no, a thousand times no! No device can step up energy. This principle is at the heart of physics. Energy cannot be created or destroyed. **89.** Electromagnetic waves depend on mutual field regeneration. If the induced electric fields did not in turn induce magnetic fields and transfer energy to them, the energy would be localized rather than "waved" into space. Electromagnetic waves would not exist. **91.** The slowness of the fall is due to the interaction of the magnetic field of the falling magnet with the field induced in the conducting tube. No conduction, as in a cardboard tube, means no induced field and no slowness of fall. **93.** Copper wires were not insulated in Henry's time. A coil of non-insulated wires touching one another would be a short circuit. Silk was used to insulate the wires so that current would flow along the wires in the coil rather than across the loops touching one another. **95.** The needle is not pulled toward the north side of the bowl because the south pole of the magnet is equally attracted southward. The net force on the needle is zero. (The net torque, on the other hand, is zero only when the needle is aligned with Earth's magnetic field.) **97.** Associated with every moving charged particle—electrons, protons, or whatever—is a magnetic field. Since a magnetic field is not unique to moving electrons, there is a magnetic field about moving protons as well. However, it differs in direction. The field lines about the proton beam circle in one direction, whereas the field lines about an electron beam circle in the opposite direction. (Physicists use the "right-hand rule." If the right thumb points in the direction of motion of a positive particle, then the curved fingers of that hand show the direction of the magnetic field. For negative particles, the left hand can be used.) **99.** The hum heard when a transformer is operating on a 60-Hz ac line is a 60-Hz forced vibration of the iron slabs in the transformer core as their magnetic polarities alternate. The hum is louder if any other mechanical parts are set into vibration. **101.** Yes; each experiences a force because each is in the magnetic field generated by the other. Interestingly, currents in the same direction attract, and currents in opposite directions repel. **103.** The scheme violates both the first and second laws of thermodynamics. Because of inherent inefficiencies, the generator will produce less electricity than is used by the adjoining motor to power the generator. A transformer will step up voltage at the expense of current, or current at the expense of voltage, but it will not step up both simultaneously—that is, a transformer cannot step up energy or power. Like all practical systems, more energy is put in than is supplied for useful purposes.

CHAPTER 10

1. The source of all waves is a vibrating object. **3.** Frequency and period are inverses of each other. **5.** No; the propagation of energy travels, not the medium itself. **7.** Vibrations in a transverse wave are perpendicular to wave travel, and parallel to wave travel in longitudinal waves. **9.** Sound travels faster in warm air because the air molecules in warm air themselves travel faster and therefore don't take as long before they bump into one another. A shorter time for the molecules to bump against one another results in a faster speed of sound. **11.** The angle of incidence is equal to the angle of reflection. **13.** Refraction occurs when parts of the wave fronts of sound travel at different speeds and produce bending in the direction of the waves. **15.** Dolphins emit ultrasound and then time its echoes. **17.** Forced vibrations occur when a surface is forced to vibrate. When the forcing frequency matches the natural frequency of the surface, the result is resonance. **19.** Troops "break step" in order to avoid marching in rhythm with the natural frequency of the bridge—to prevent resonance. **21.** Wave amplitude is increased in constructive interference, reduced in destructive interference. **23.** Interference is the phenomenon underlying beats. **25.** Only frequency and wavelength change in the Doppler effect, not wave speed. **27.** The speed of the waves and the speed of the source are the same when a wave barrier is produced. Wave speed is greater when a bow wave is produced. **29.** False, because a sonic boom occurs for aircraft moving faster than the speed of sound. **31.** The greater the frequency of shaking, the greater the number of nodes. **33.** This is one of Paul Hewitt's favorite classroom demonstrations! **35.** The sound intensity varies with different glasses and metal bowls. **37.** This is a good group activity! **39.** Frequency = 1/period = 1/(3 s) = $\frac{1}{3}$ Hz. **41.** Wave speed = $f\lambda$ = (1.5/s)(3 m) = 4.5 m/s. **43.** $v = f\lambda$ = (256 Hz)(1.33 m) = 340 m/s. **45.** (a) Frequency = 2 bobs/s = 2 Hz. (b) Period = $1/f$ = 1/2 s. (c) The amplitude is the distance from the equilibrium position to maximum displacement, one-half the 20-cm peak-to-peak distance, or 10 cm. **47.** (a) The skipper notes that 15 m of wave pass each 5 s or, equivalently, 3 m pass each 1 s, so speed = distance/time = (15 m)/(5 s) = 3 m/s. (b) In wave terminology, speed = frequency × wavelength = ($\frac{1}{5}$ Hz)(15 m) = 3 m/s. **49.** For the highest-frequency sound, $\lambda = v/f$ = (340 m/s)/(20,000 Hz) = 0.017, or 1.7 cm. For the lowest-frequency sound, $\lambda = v/f$ = (340 m/s)/(20 Hz) = 17 m. **51.** $v = f\lambda$, so $\lambda = v/f$ = (1530 m/s)/(7 Hz) = 219 m. **53.** Assume the speed of sound is 340 m/s. The time to reach the wall is half of 0.1 s because of the round trip (0.05 s to reach the wall and 0.05 s to return). So distance = speed × time = 340 m/s × 0.05 s = 17 m. **55.** Sound goes from the hermit to the mountain in 4 h and back in another 4 h to wake him. The distance from the hermit to the mountain = speed of sound × time = 340 m/s × 3600 s/h × 4 h = 4.9 × 10⁶ m = 4900 km (about the distance from New York to San Francisco)! (Very far and, due to the inverse-square law, also very weak!) **57.** (a) Wavelength = speed/frequency = (1500 m/s)/(256 Hz) = 5.86 m. (b) By the time the vibration completes one cycle, the wave travels farther in water than in air, so the wavelength—which is the distance the wave travels in one period—is longer in water. **59.** Speed of plane = 1.41 × speed of sound (Mach 1.41). In the time it takes sound to go from A to C, the plane goes from A to B. Since the triangle ABC is a 45-45-90 triangle, the distance AB is $\sqrt{2}$ = 1.41 times as long as the distance AC. **61.** A, B, D, C. **63.** A, C, B. **65.** b, a, c, d. **67.** As you dip your fingers more frequently into the puddle, the waves you produce are of a higher frequency (we see the relationship between "how frequently" and "frequency"). The crests of the higher-frequency waves are closer together; their wavelengths are shorter. **69.** Violet light has the higher frequency. **71.** To produce a transverse wave with a Slinky, shake it to and fro in a direction that is perpendicular to the length of the Slinky itself (as with the garden hose in the preceding exercise). To produce a longitudinal wave, shake a Slinky to and fro along the direction of its length to create a series of compressions and rarefactions. **73.** The echoed chirps have a higher frequency due to the Doppler effect. **75.** Light travels about a million times faster than sound—hence the delay between what you see and what you hear. **77.** Between us and other planets is a vacuum, and sound does not travel in a vacuum. **79.** The pitch of sound emitted by bats is higher than humans can hear, well beyond 20,000 Hz. **81.** Amplitude is likely increasing. **83.** The pitch of the sound gets lower as the glass is filled. As the mass of the system (glass plus water) increases, its natural frequency decreases. For systems of a given size, more mass usually means lower frequency. This can be seen on a guitar, where the most massive string has the lowest natural pitch. (If you've completed this exercise without actually trying it, shame on you!) **85.** Marchers at the end of the parade are out of step with marchers near the band because time is required for the sound of the band to reach the marchers at the back. They march to the delayed beat they hear. **87.** A harp produces softer sounds because its sounding board is smaller and lighter. **89.** The glass shatters because of resonance, when the buildup of vibrations in the glass exceeds the breaking point of the glass. **91.** The piano tuner should loosen the piano string. When he first hears three beats per second, the tuner knows he is 3 Hz off the correct frequency. But this could be either 3 Hz above or 3 Hz below. When he tightens the string and increases its frequency, a lower beat frequency would have told him he was on the right track. But the higher beat frequency tells him he should have loosened the string. When there is no beat frequency, the frequencies match. **93.** No; the effects of shortened waves and stretched waves would cancel one another. **95.** As in the preceding exercise, radar waves are reflected from moving balls. From the shift in the returned frequencies, the speed of the balls is determined and displayed. **97.** Oops, be careful. The Doppler effect is about changes in *frequency*, not speed. **99.** The fact that you hear an airplane in a direction that differs from where you see it means the airplane is moving, but not necessarily faster than sound (a sonic boom is evidence of supersonic flight). If the speed of sound and the speed of light were the same, then you'd hear a plane where it appears in the sky. But because the two speeds are so different, the plane you see appears ahead of the plane you hear. **101.** The carrier frequency of electromagnetic waves emitted by the radio station is 101.1 MHz. **103.** The fact that we can see a ringing bell but can't hear it indicates that light is a distinctly different phenomenon than sound. When we see the vibrations of the "ringing" bell in a vacuum, we know that light can pass through a vacuum. The fact that we can't hear the bell indicates that sound does not pass through a vacuum. Sound needs a material medium for its transmission; light does not. **105.** If the speed of sound were different for different frequencies—say, faster for higher frequencies—then the farther a listener is from the music source, the more jumbled the sound would be. In that case, higher-frequency notes would reach the ear of the listener first. The fact that this jumbling doesn't occur is evidence that sounds of all frequencies travel at the same speed. (Be glad this is so, particularly if you sit far from the stage or if you like outdoor concerts.) **107.** An echo is weaker because sound spreads and is therefore less intense with distance. If you are at the source, the echo sounds as if it originated on the other side of the wall from which it reflects (just as your image in a mirror appears to come from behind the glass). Also contributing to its weakness is the wall, which likely is not a perfect reflector. The farther away the wall, the weaker the echo. **109.** If a single disturbance sends longitudinal waves at one known speed and transverse waves at a lesser-known speed, and you measure the difference in the arrival times of the waves, you can calculate the distance. The wider the gap in time, the greater the distance—which could be in any direction. If you use this distance as the radius of a circle on a map, you know the disturbance occurred somewhere on that circle. If you phone two friends who have made similar measurements of the same event from different locations, you can transfer their circles to your map, and the point where the three circles intersect is the location of the disturbance. **111.** Waves of the same frequency can interfere destructively or constructively, depending on their relative phase, but to *alternate* between constructive and destructive interference, two waves must have different frequencies. Beats arise from such alternation between constructive and destructive interference.

CHAPTER 11

1. The principal difference among radio, light, and X-rays is frequency. **3.** The resonant frequency of electrons in glass is in the ultraviolet region. **5.** The energy in infrared light becomes thermal energy. **7.** The incident and emerging speeds of light are the same. **9.** Image distance and object distance are the same. **11.** The law of reflection holds locally at each tiny part of the irregular surface, but not for the diffuse surface as a whole. **13.** The speed of light slows when the light is refracted in a medium. **15.** A mirage is a distorted view in which refracted light appears as if it is reflected light. **17.** The peak frequency of sunlight is yellow-green, which is also the color that our eyes are most sensitive to. **19.** Two colors that add to produce white are called complementary colors. **21.** High-frequency blue light is scattered all along the path of sunlight, so the long path at sunrise or sunset finds much blue missing. What remains is light of lower frequencies, which accounts for the reddish color of the Sun at these times. **23.** Infrared and red light are absorbed by water. Absorption of red produces the complementary color, cyan. **25.** The ground gets in the way, cutting off the view of a whole circular rainbow. **27.** Polarization is a property of transverse waves, never longitudinal waves. **29.** Light will pass through the filters when their axes are aligned, but not when their axes are at right angles to each other. **31.** Activities that tell you more about yourself are valuable ones. **33.** This illustrates the scattering of blue to the side and the transmission of light of lower frequencies through longer regions. **35.** Amazing, indeed! **37.** Open ended. **39.** The Earth–Moon distance is 3.84×10^8 m, so the round-trip distance is 7.68×10^8 m. Rearranging $d = vt$, we have $t = \dfrac{d}{v} = \dfrac{7.86 \times 10^8 \text{ m}}{3.00 \times 10^8 \text{ m/s}} = 2.56$ s.

41. Speed $= \dfrac{\text{distance}}{\text{time}} = \dfrac{300{,}000{,}000 \text{ km}}{1300 \text{ s}} = 231{,}000$ km/s. This value is 77% of the modern value.

43. From $c = f\lambda$, $\lambda = \dfrac{c}{f} = \dfrac{3.00 \times 10^8 \text{ m/s}}{6 \times 10^{14} \text{ Hz}} = 5 \times 10^{-7}$ m, or 500 nm. This is 5000 times larger than the size of an atom, which is about 0.1 nm. (The nanometer is a common unit of length in atomic and optical physics.) **45.** The spider's image is 30 cm in back of the mirror, so the distance from the image to your eye is 30 cm + 30 cm + 65 cm = 125 cm = 1.25 m. **47.** If 96% is transmitted through the first face, and 96% of 96% is transmitted through the second face, we have [(0.96)(0.96) = 0.92], so 92% is transmitted through both faces of the glass. **49.** A = B = C (all the same). **51.** B, C, A. **53.** The fundamental source is vibrating electric charges, which emit vibrating electric and magnetic fields. **55.** The wavelengths of radio waves are longer than those of light waves, which are longer than the wavelengths of X-rays. **57.** The speed of X-rays is c, the speed of light. **59.** Radio waves travel at the same speed as every other electromagnetic wave—the speed of light. Don't confuse radio waves with the sound waves they can make. **61.** The instantaneous speed of the bullet after it passes through the board is slower than its incident speed, but not so with light. The instantaneous speed of light before it meets the glass, while passing through it, and when emerging is a constant, c. The fundamental difference between a bullet fired through a board and light passing through glass is that in the case of the bullet, the *same* bullet strikes and later emerges. Not so for light. The "bullet of light" (photon) that is incident upon glass is absorbed by its interaction with an atom or molecule. The atom or molecule in turn then emits, with some time delay, a new "bullet of light" in the same direction. This process cascades through the glass, with the result being that the "bullet of light" emerging is not the same "bullet" that was first incident. In the space between the atoms in matter, the instantaneous

speed of light is c. Because of the time delay of the interactions, only its *average speed* is less than c. The light that emerges has speed c. **63.** The greater number of interactions per distance tends to slow the light, and the result is a slower average speed. **65.** Clouds are transparent to ultraviolet light, which is why clouds offer no protection from sunburn. Glass, however, is opaque to ultraviolet light and therefore shields you from sunburn. The UV light needed to activate transition eyeglasses doesn't get through the window glass. **67.** Only light from card 2 reaches her eye. **69.** Such lettering is seen in readable form in the rearview mirrors of cars ahead. **71.** The minimum length of a vertical mirror must be half your height in order for you to see a full-length view of yourself. The part of the mirror above and below your line of sight to your image isn't needed, as the sketch shows.

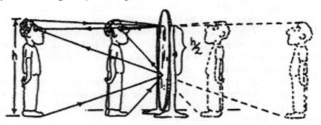

73. The amount of your face you can see is twice the size of the mirror—whether you hold the mirror close or at arm's length. (You can win bets on this question!)

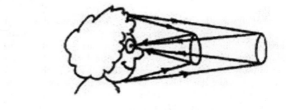

75.

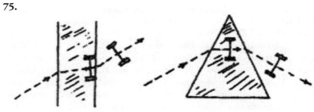

77. The customer is being reasonable. Under fluorescent lighting, with its predominant higher frequencies, the bluer colors rather than the redder colors are accented. Colors appear quite different in sunlight. **79.** If the yellow clothes are illuminated with a complementary blue light, they will appear black. **81.** The red shirt in the photo is seen as cyan in the negative, and the green shirt appears magenta—the complementary colors. When white light shines through the negative, red is transmitted where cyan is absorbed. Likewise, green is transmitted where magenta is absorbed. **83.** Blue illumination (the complementary color of yellow) will produce black. **85.** Particles in the smoke scatter predominantly blue light, so against a dark background you see the smoke as blue. But against the bright sky what you see is predominantly the sky minus the light scattered by the smoke. You see yellow. **87.** The net absorption typical of large drops in rain clouds produces the darkness of the rain clouds. **89.** The water is broken up into a multitude of droplets of different sizes when the wave breaks, and like the droplets in the clouds overhead, light of many visible frequencies is scattered to produce the white color. **91.** The fact that two observers standing apart do not see the same rainbow can be understood by exaggerating the situation: Suppose the two observers are several kilometers apart. Obviously they are looking at different drops in the sky. Although they may both see a rainbow, they are looking at different rainbows. Likewise if they are closer together. Only if their eyes are at the very same location will they see exactly the same rainbow. **93.** With their polarization axes aligned, a pair of filters transmit all the components of light along the axes. Half of the light gets through the first filter, and all of that light gets through the second. So that's half, or 50%. With axes at right angles, no light is transmitted. **95.** Light doesn't get through two crossed Polaroid filters—say, 1 and 2—because their axes are at 90° to each other. But with filter 3 sandwiched at an angle between them, some light gets through 1 and 3 because their axes are at 45°, not crossed. This light continues on to 2 and, again, because the light is at 45° to the axis of 2, light gets through! **97.** The terms are misleading because they imply that ultraviolet and infrared are forms of visible light. More correctly, they are forms of electromagnetic radiation. So, in the sense used, *light* means "electromagnetic radiation." This usage may originate from the fact that the ultraviolet and infrared regions of the spectrum are adjacent to the visible light region. The terms "radio light" and "X-ray light" are rarely if ever used, likely because the radio part and X-ray part of the spectrum are far removed from the visible part. **99.** Rays do not converge as with a glass lens, so a pinhole image is sharp in all positions.

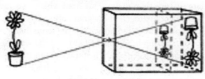

101. With polarization axes aligned, a pair of Polaroid filters will transmit all components of light along the axes. That's 50%. Half of the light gets through the first Polaroid, and all of that gets through the second. With axes at right angles, no light will be transmitted.

CHAPTER 12

1. Hydrogen is the oldest element. **3.** The atomic nucleus is at the center of the atom. **5.** An atom is the smallest particle of an element that has all the chemical properties of the element. An element is made up of only one type of atom. **7.** Elements are listed in the periodic table in order of increasing atomic number. **9.** Mass number is the total number of nucleons in an isotope. Atomic mass is a measure of the total mass of an atom. **11.** The periodic table has 7 periods and 18 groups. **13.** The atoms in the baseball would be the size of Ping-Pong balls. **15.** A physical model attempts to replicate an object on a different scale. A conceptual model describes a system. **17.** Atoms give off light as they are subjected to energy. **19.** A beam of light consists of zillions of small, discrete packets of energy. Each packet is called a quantum. **21.** No; Bohr's model merely illustrated the different energy levels of an electron in an atom. **23.** An electron travels around the nucleus at around 2 million m/s. **25.** The shell model is used to explain the organization of elements in the periodic table. With the shell model, for example, we can understand how the atoms of elements get larger as we move down a group or how they get smaller as we move from left to right across a period. **27.** The number of electrons that each shell can hold corresponds to the number of elements in the period. **29.** Open ended. **31.** When every student scores 80%, the class average is 80%. If one student scored higher, it would raise this average to 81%. Similarly, the atomic mass of an element is the average mass of all the various isotopes of that element. The heavier isotopes have the effect of slightly raising the average. Carbon, for example, has an atomic mass of 12.011, which is slightly greater than 12.000 because of the few heavier carbon-13 isotopes found in naturally occurring carbon. **33.** (c). A 50:50 mix of Br-80 and Br-81 would result in an atomic mass of about 80.5, and a 50:50 mix of Br-79 and Br-80 would result in an atomic mass of about 79.5. Neither of these is as close to the value reported in the periodic table as is a 50:50 mix of Br-79 and Br-81, which would result in an atomic mass of about 80.0. **35.** Helium, chlorine, argon. **37.** (a) Helium, argon, aluminum. (b) Helium, aluminum, argon. (c) Helium, aluminum, argon. **39.** The cat leaves a trail of molecules across the yard. These molecules leave the ground, mix with the air, and then enter the dog's nose and activate its sense of smell. **41.** The atoms that make up a newborn baby or anything else in this world originated in the fires of ancient stars. This process is explored in Chapter 13. **43.** The carbon atoms that make up Leslie's hair or anything else in this world originated in the fires of ancient stars. This process is explored in Chapter 13. **45.** Yes, as is everyone else's and everything around you. We interact with our environment, like bumping our head against a cabinet, because of the repulsive electric fields that prevent atoms from overlapping one another. **47.** The atomic mass is 99 amu, and the element is technetium, Tc, with atomic number 43. **49.** The neutron was elusive because of its lack of electric charge. **51.** The atomic masses are average numbers that reflect the variety of isotopes that exist for an

element. **53.** From the periodic table we see that an oxygen atom has a mass of about 16 amu. Two oxygen atoms have a mass of about 32 amu, as does a single oxygen molecule, O_2. **55.** A water molecule, H_2O, has a mass of about 18 amu, while a carbon dioxide molecule, CO_2, has a mass of about 44 amu. So a carbon dioxide molecule is more than twice as heavy as a water molecule. **57.** The tree gains weight. In fact, the tree gets mass (carbon, oxygen, and hydrogen) directly from the air, not from the water and nutrients it absorbs through its roots. **59.** The neutrons carry no electric charge and thus have a greater likelihood of passing through the tissue. **61.** Helium is on the far right-hand side of the periodic table in group 18 because its physical and chemical properties are most similar to those of the other group 18 elements. **63.** Calcium is readily absorbed by the body for the building of bones. Since calcium and strontium are in the same group, they have similar physical and chemical properties. The body, therefore, has a hard time distinguishing between the two and absorbs strontium as though it were calcium. **65.** No; the colors are merely an artificial computer rendering. Atoms do not have color in the same sense that a macroscopic object may have color. Atoms, however, can be a source of light, which is what we observe as we look at an element's atomic spectrum. **67.** A scanning probe microscope shows only the relative sizes and positions of atoms. It does this by detecting the electrical forces between the tip of the microscope's needle and the outer electrons of the atom. Recall from Section 12.1 that the atom itself is made of mostly empty space, so the best "image" of the inside of an atom would be a picture of nothing. It doesn't make sense to talk about taking an "image" of the inside of an atom. Instead, we develop models that provide a visual handle on the behavior of the components of atoms. **69.** Many objects or systems may be described just as well by a physical model as by a conceptual model. In general, a physical model is used to replicate an object or system of objects on a different scale. A conceptual model, by contrast, is used to represent abstract ideas or to demonstrate the behavior of a system. A gold coin, a car engine, and a virus would be described using a physical model. A dollar bill (which represents wealth but is really only a piece of paper), air pollution, and the spread of sexually transmitted disease would be described using a conceptual model. **71.** The higher the frequency of the light, the greater the energy of the electrons. **73.** As the electron vibrates in the atom, it generates electromagnetic waves. The electron vibrates back and forth between different energy levels. **75.** Blue light has a higher frequency than red light and therefore corresponds to a greater energy transition. **77.** An electron not restricted to particular energy levels would release light continuously as it spiraled closer into the nucleus. We would see a broad spectrum of colors rather than distinct lines. **79.** A shell is just a region of space in which electrons may reside. This region exists with or without the electrons. The space defined by the shell exists whether or not an electron is to be found there. **81.** Both a potassium atom and a sodium atom are in group 1 of the periodic table. The potassium atom, however, is larger because it contains an additional shell of electrons. **83.** Both a lithium atom and a beryllium atom are in the second period of the periodic table, which means that both contain two shells of electrons. The nuclear charge of the lithium atom, however, is not as strong as that of the beryllium atom. So the electrons of the lithium atom are not drawn in as close to the nucleus. **85.** If you think scientists know all there is to know about the universe, think again. Although we certainly know much more than we used to know, much still remains unknown. One unknown, as of the writing of this textbook, is the nature of dark matter. Astronomers find evidence of massive amounts of this stuff surrounding each galaxy. This form of matter, however, interacts only with the gravitational force. It does not recognize the strong nuclear force (see Chapter 13), which means it cannot clump to form atomic nuclei. Nor does it recognize the electromagnetic force (see Chapter 8), which is responsible for light and electric charge. Thus dark matter is invisible to light as well as to our sense of touch. The reason you can't walk through a wall is because of the repulsions between the electrons in your body and the electrons in the wall (see Section 12.1). If the wall were made of this invisible matter, you would be able to walk right through it. Of course, you wouldn't be able to see this wall either. This invisible form of matter that we cannot see or touch is known as *dark matter*. Stay tuned for current developments.

CHAPTER 13

1. Alpha radiation decreases the atomic number of the emitting element by 2 and the atomic mass number by 4. Beta radiation increases the atomic number of an element by 1 and does not affect the atomic mass number.

Gamma radiation does not affect the atomic number or the atomic mass number. So alpha radiation results in the greatest change in atomic number, and hence charge, and mass number as well.   **3.** Coal-fired power plants produce more atmospheric radiation than nuclear power plants. Global combustion of coal produces about 13,000 tons of radioactive substances; nuclear plants produce about 10,000 tons.   **5.** The attractive strong nuclear force exerted by all nucleons is able to overcome the repulsive electric force of protons.   **7.** Neutrons act like "nuclear cement" to hold the nucleus together.   **9.** Emission of an alpha particle decreases the atomic number by 2. Emission of a beta particle increases the atomic number by 1.   **11.** The radioactive half-life is the amount of time it takes for half of the original quantity of a radioactive element to decay.   **13.** The nitrogen atom becomes radioactive carbon-14.   **15.** All uranium isotopes eventually decay to lead. So, any deposit of uranium ore will contain some lead that has been converted from uranium.   **17.** A chain reaction is more likely to occur in two pieces of uranium-235 stuck together.   **19.** Albert Einstein made this discovery.   **21.** The mass per nucleon is greater in uranium than in the fission fragments of uranium.   **23.** Magnetic containers hold high-temperature plasmas.   **25.** This activity demonstrates how radioactive decay is a statistical phenomenon. Notice that eventually you get down to the last coin, which may or may not decay on the next throw. Also notice that some coins last for a long time while others decay right away. This is similar to radioactive atoms. Although the half-life might be 10 years, any individual atom might last for only a few days, while others last for well beyond 10 years. For a coin with only two sides, you'll find that, on average, half the coins will have decayed after one toss. This is true whether you have 25 coins, 50 coins, or even a million. So the half-life is not a function of how many atoms you have. Rather, half-life is a function of the stability of the nucleus. Atoms with stable nuclei persist much longer than atoms with unstable nuclei.   **27.** Open ended.   **29.** Intensity decreases with distance according to the inverse-square law. Twice the distance is 1/4 the intensity, 25 counts per min. Three times the distance is 1/9 the intensity, 11 counts per min.   **31.** The count decreases to 40 counts per second (one half-life), 20 (2 half-lives), 10 (3 half-lives), and 5 (4 half-lives). If 4 half-lives equals 8 h, then a single half-life equals 8/4 = 2 h.   **33.** c, b, a.   **35.** c, a, b. (The longer the half-life, the lower the radioactivity.)   **37.** Yes. The strong nuclear force is effective only over an extremely short distance. Once the alpha particle leaves the nucleus, its attraction to the nucleus by the strong nuclear force is no longer significant. The electrical repulsion between the alpha particle and the protons of the nucleus, however, is still significant. This repulsive force causes the alpha particle to accelerate to high velocities as it moves away from the nucleus. Might the surrounding negatively charged electrons cause the alpha particle to slow down? Not significantly. The alpha particle is about 8000 times more massive than an electron. Once it starts moving away from the nucleus, its great mass allows it to plow right through the surrounding electrons.   **39.** Chemical properties have to do with electron structure, which is determined by the number of protons in the nucleus, not the number of neutrons.   **41.** All uranium isotopes eventually decay to lead. So, any deposit of uranium ore contains some lead that has been converted from uranium.   **43.** After beta emission from polonium, the atomic number increases by 1 and becomes 85, and the atomic mass is unchanged at 218. If the polonium emits an alpha particle, the atomic number decreases by 2 and becomes 82, and the atomic mass decreases by 4 and becomes 214.   **45.** Film badges monitor gamma radiation, which is very high frequency X-rays. As in photographic film, the greater the exposure, the darker the film upon processing.   **47.** Radium-226 is created by the radioactive decay of uranium, which persists for billions of years.   **49.** Radioactive decay rates are statistical averages from large numbers of decaying atoms. Because of the relatively short half-life of carbon-14, only trace amounts would be left after 50,000 years—too little to be statistically accurate.   **51.** For radioactive atoms, the chance of "dying" (undergoing decay) is always the same, regardless of the age of the atom. A young atom and an old atom of the same type have exactly the same chance to decay in the next equal interval of time. For adults, the chance of dying increases with age.   **53.** If strontium-90 (atomic number 38) emits betas, it should become the element yttrium (atomic number 39); hence the physicist can test a sample of strontium for traces of yttrium by spectrographic means or other techniques. To verify that it is a "pure" beta emitter, the physicist can check to make sure that the sample is emitting no alphas or gammas.   **55.** When a neutron bounces from a carbon nucleus, the nucleus rebounds, taking some energy away from the

neutron and slowing it down so it will be more effective in stimulating fission events. A lead nucleus is so massive that it scarcely rebounds at all. The neutron bounces with practically no loss of energy and practically no change of speed (like a marble bouncing from a bowling ball).   **57.** Iron is at the bottom of the "energy hill" of the curves shown in Figures 13.30, 13.31, and 13.32, so whether it undergoes fusion or fission, the product nuclei are more massive than the iron. This process absorbs energy rather than releasing it.   **59.** Energy would be released from the fissioning of gold and from the fusion of carbon, but by neither fission nor fusion for iron. Neither fission nor fusion will result in a decrease of mass for iron nucleons.   **61.** The mass of an atomic nucleus is less than the sum of the masses of the nucleons that compose it. One way to see why is to think about the work that must be done to separate a nucleus into its component nucleons. This work, according to $E = mc^2$, adds mass to the system, so the separated nucleons are more massive than the nucleus from which they came. Notice the large mass per nucleon of hydrogen in the graph of Figure 13.32. The hydrogen nucleus, a single proton, is already "outside" in the sense that it is not bound to other nucleons.   **63.** Fusing heavy nuclei (which is the way the heavy transuranic elements are made) costs energy. The total mass of the products is greater than the total mass of the fusing nuclei.   **65.** A hydrogen bomb produces fission energy as well as fusion energy. Some of the fission in the fission bomb "trigger" used to ignite the thermonuclear reaction, and some is in fissionable material that surrounds the thermonuclear fuel. Neutrons produced in fusion cause more fission in this blanket. Fallout results mainly from the fission.   **67.** No fusion-based nuclear power plants are in operation today. They are all of the fission type.   **69.** The fusion product would have atomic number 10 and atomic mass number 8, which is neon-8.   **71.** You don't get something for nothing. There is great misunderstanding about hydrogen. To release it from water or other chemicals costs more energy than you get back when you burn it. Hydrogen represents stored energy, like a battery. It's made in one place and used in another. It burns without pollution, a big advantage, but it should be regarded as a storage and transport medium for energy, not as a fuel.   **73.** Perhaps it's a fundamental human characteristic to seek control. People may fear loss of control more than anything else.   **75.** Such speculation could fill volumes. The energy and material abundance that is the expected outcome of a fusion age will likely prompt fundamental changes. Obvious changes would occur in the fields of economics and commerce, which would be geared to relative abundance rather than scarcity. Already our present price system, which is geared to and in many ways dependent on scarcity, often malfunctions in an environment of abundance. Hence we see instances where scarcity is created to keep the economic system functioning. Gem diamonds, for example, are abundant, but the gem diamond industry works hard to maintain the sense of scarcity. Changes at the international level will likely be worldwide economic reform, and at the personal level a reevaluation of the idea that scarcity is the basis of value. A fusion age will likely see changes touching every facet of our way of life.

CHAPTER 14

1. Chemistry is often called a central science because it touches all of the sciences.   **3.** Members have pledged to manufacture their products without causing environmental damage.   **5.** A biological cell is microscopic, which means it is best viewed through a microscope.   **7.** The particles in the solid vibrate around fixed positions, while those in a liquid tumble loosely around one another.   **9.** One gram of water vapor, the gaseous phase of water, occupies the greatest volume.   **11.** A physical property is the description of the physical attributes of a substance, such as color, hardness, density, texture, and phase.   **13.** Chemical properties characterize the tendency of a substance to react with or transform into other substances.   **15.** During a chemical reaction there is a change in the way atoms are bonded together.   **17.** Both physical and chemical changes involve changes in appearance.   **19.** If, by returning to the original set of conditions, you return to the original physical appearance, then the change was physical. If the substance that has undergone the change has new physical properties, then the change was chemical.   **21.** The elemental formula S_8 indicates that there are eight atoms in the sulfur molecule.   **23.** There are eight atoms in H_3PO_4: three H atoms, one P atom, and four O atoms.   **25.** Compounds have physical and chemical properties that are different from the properties of their elemental components.   **27.** KF is potassium fluoride.   **29.** Common names are more convenient to use than systematic names.   **31.** In the top-down approach, nanostructures are carved out of larger materials. In the bottom-up approach, nanostructures are pieced together atom by atom.   **33.** Open ended.   **35.** To represent the liquid phase, shake the balloon very gently so that the beads remain together but tumble over one another. To represent the solid phase, gently vibrate the balloon. Notice that in both cases you are adding energy to the balloon. Holding the balloon still would represent absolute zero! **37.** Hydrogen peroxide, H_2O_2, is a relatively unstable compound. In solution with water, it slowly decomposes to produce oxygen gas. In describing oxygen's physical properties, you should have noted that it is an invisible gas with no odor detectable over that of the yeast. Oxygen is light enough to rise out of the glass once it is released from the bubbles. A chemical property of oxygen is that it intensifies burning. **39.** c, b, a. **41.** c, a, b. **43.** It is easy to hide a lack of understanding by using big words with which others are unfamiliar. If you have truly mastered a concept, you should be able to explain it using language that everyone can understand. **45.** Biology is based on the principles of chemistry as applied to living organisms, while chemistry is based on the principles of physics as applied to atoms and molecules. Physics is the study of the fundamental rules of nature, which more often than not are rather simple in their design and readily described by mathematical formulas. Because biology sits at the top of these three sciences, it can be considered to be the most complex. **47.** The 50 mL plus 50 mL do not add up to 100 mL because, within the mix, many of the smaller BBs are able to fit inside the pockets of space that were empty in the 50 mL of large BBs. **49.** At 25°C a certain amount of thermal energy is available to all the submicroscopic particles of a material. If the attractions between the particles are not strong enough, the particles may separate to form a gaseous phase. If the attractions are strong, however, the particles may be held together in the solid phase. We can assume, therefore, that the attractions among the submicroscopic particles in the solid phase at 25°C are stronger than they are in the gas phase. **51.** In the middle box, you should have drawn all the particles aligned as shown in the left side of the first box. This indicates the solid phase. In the box on the right, you should have drawn all the particles in random places as in the right side of the first box. This represents the liquid phase. The first box illustrates ice melting, which occurs at 0°C. **53.** As each kernel is heated, the water within each kernel is also heated to the point that it would turn into water vapor. The shell of the kernel, however, is airtight and this keeps the water as a superheated liquid. Eventually the pressure exerted by the superheated water exceeds the holding power of the kernel and the water bursts out as a vapor, which causes the kernel to pop. These are physical changes. The starches within the kernel are also cooked by the high temperatures, and this is an example of a chemical change. **55.** The popcorn in the skillet can represent the alcohol molecules lying on the surface of the tabletop. As the kernels are heated, they eventually leave the skillet. Likewise, as the alcohol molecules absorb heat from the tabletop, they evaporate away from the tabletop. Do you see how easy it is to explain the disappearance of the alcohol by supposing the existence of alcohol molecules? The concept of molecules can be used to readily explain so many different phenomena. This is why many people believed in their existence years before the discovery of direct evidence. **57.** Even though the water appears to be still, the water molecules are bustling with kinetic energy. The red dye of the Kool-Aid gets knocked around by these molecules until the dye is eventually dispersed throughout the water. This is another case where the existance of molecules helps to explain the observed phenomenon. **59.** Boiling down the maple syrup involves the evaporation of water. As the syrup hits the snow, it warms the snow, causing it to melt while the syrup becomes more viscous. These are physical changes. As the maple syrup is boiled, the sugar within the syrup begins to caramelize, which is a chemical change. **61.** The reversibility of this process suggests it is a physical change. As you sleep in a reclined position, pressure is taken off of the discs in your spinal column, which allows them to expand so that you are significantly taller in the morning. Astronauts returning from extended space visits may be up to 2 inches taller upon their return. **63.** (a) Physical, (b) chemical, (c) chemical. **65.** (a) Chemical, (b) chemical, (c) physical, (d) physical. **67.** The changes that occur as we age involve the chemical reformation of our biomolecules, so they are chemical changes. **69.** All the oxygen in water is bound to hydrogen atoms to make water molecules. Water is uniquely different from the elements oxygen, O_2, and hydrogen, H_2, from which it can be made. The oxygen our bodies are designed to breathe is gaseous molecular oxygen, O_2. A person can drown when breathing in water because it contains so little O_2. Although both contain oxygen, gaseous oxygen, O_2, and water, H_2O, are vastly different. **71.** Water used to be classified as an element, but that was before people recognized that the basic building blocks of matter are the tiny particles

called atoms. Today an element is identified as a material consisting of only one kind of atom. **73.** Box A contains a mixture, box B a compound, and box C an element. Three different types of molecules are shown: one with two small blue circles joined, one with a larger red circle and smaller blue circle joined, and one with two larger red circles joined. **75.** Tribarium dinitride or, more simply, barium nitride because no ratio other than three bariums to two nitrogens is possible. **77.** Oxygen, O_2. **79.** The idea that matter is made up of molecules came about only after many questions were asked about the behavior of matter. If questions were not asked, then the conclusion that matter is made up of molecules probably would not have been reached. People began to accept the idea of molecules because they saw how well it answered questions and explained observations. For example, is cinnamon-scented air a single material or a mixture of two materials? If it is two materials, shouldn't it be heavier than the same volume of fresh air? (It is.) Can this air be made fresh by passing it through a filter of activated charcoal? (It can.) Does the charcoal now smell like cinnamon? (It does.) Does the cinnamon smell of the charcoal increase or decrease as it is warmed? (It increases but eventually tapers off.) Does the charcoal lose or gain weight as the cinnamon smell tapers off? (It loses weight.) Might the charcoal be losing weight as tiny particles (molecules) of cinnamon evaporate from its surface? (That would make sense.) Notice that throughout this process you never need to say that molecules exist because someone told you they exist. Instead, you conclude that molecules exist because they offer the best explanation for what you observe. **81.** During Feynman's time, scientists did not have the technology to manipulate single atoms and molecules. Today this technology is growing rapidly. Chemist Rick Smalley (1943–2005), however, was quick to point out that atoms and molecules tend to be sticky little entities. Putting atoms together one by one is much like trying to arrange peanut butter–coated Ping-Pong balls: You can't move the ball without it sticking to your fingers. Smalley emphasized that chemists already know how to combine atoms (in bulk) to create novel molecules (also in bulk). Great things will no doubt be achieved through nanotechnology, but as a complement to the great things achieved through novel chemical reactions. **83.** In the beginning of the 20th century, many were opposed to the introduction of electricity because of its inherent dangers. Fear, however, usually arises from a lack of understanding. As people learned more about electricity—understanding both its dangers and benefits—they came to accept this new technology. The same may hold true for nanotechnology. It is the obligation of the vendors of nanotechnology to keep us informed. But who will watch over these nanotech companies to make sure that greed does not take priority over safety? That is the role of an attentive government and a well-informed consumer. **85.** Demographers know that birth rates are lowest in countries where life expectancies are highest. If a cure for aging were found, the rate of population growth would not be so high as one might expect. The people who live into their 100s or 200s would not be frail and helpless. If they were, then medicine would not be successful at keeping them alive. Instead, these people would continue to be productive, vibrant, and perhaps highly respected members of society. They would continue to work and not draw on their retirement plan until years later. Of course, if people really do start to live that much longer, then it's a safe bet that there would be a shift toward privatized retirement accounts. So what would it be like if there were a sizable population of 200-year-olds in society? What effect might their longevity have on social reforms, such as civil rights? Would healthy 200-year-olds be set in their ways or still open to change?

CHAPTER 15

1. Two electrons fit into the first shell. Eight electrons fit into the second shell. **3.** The electron-dot structures of elements in the same group have the same number of valence electrons. **5.** A gain of electrons creates a negatively charged ion. **7.** Elements on opposite sides of the periodic table tend to form ionic bonds. **9.** An ionic crystal is composed of a multitude of ions grouped together in a highly ordered three-dimensional array. **11.** An alloy is a mixture composed of two or more metallic elements. **13.** Elements that tend to form covalent bonds are primarily nonmetallic elements. **15.** Oxygen can form two covalent bonds. **17.** A dipole is an uneven distribution of electrons in a bond caused by a difference in electronegativity between two atoms. **19.** A carbon–oxygen bond is more polar. **21.** A nonpolar substance tends to have weak attractions to itself, which causes a low boiling point. **23.** Oil and water do not mix because water molecules are more attracted to themselves than to oil molecules. **25.** The ion–dipole attraction is stronger. **27.** No; induced dipoles are only temporary. They occur (are induced) only when they are in close proximity to a water molecule or another dipole. **29.** Your model of CH_2F_2 should look like the methane in Figure 15.19b, except that two hydrogen atoms are replaced by two fluorine atoms. The structure of ethane is made by joining the carbons of two methane molecules (minus two hydrogens). The structure of hydrogen peroxide is $H—O—O—H$, where each $H—O—O$ bond is bent to an angle (similar to water's structure). For hydrogen peroxide, the $O—O$ bond should be rotated so that the two hydrogen atoms are as far apart as possible. The structure of acetylene is linear like a pencil with a triple bond between the two carbons. **31.** According to Table 15.1, the carbonate ion carries a $2-$ charge, which means it has picked up two electrons. These two electrons must have come from the single manganese ion to which it is bound. **33.** The formula is $MgCl_2$ (two single negatively charged chlorine ions are needed to balance the one doubly positively charged magnesium ion). A shortcut way of solving this sort of problem is to take the charge of one ion and make it the subscript of the opposite ion. For example, take the $2+$ charge of the magnesium and make it the subscript of the chlorine. Then take the $1-$ charge of the chlorine and make it the subscript of the magnesium. Because subscripts of 1 are implied when not written, we have not Mg_1Cl_2 but $MgCl_2$. **35.** a, c, b (the greater the difference in electronegativity between bonded atoms, the greater the polarity of the bond). **37.** c, b, a. **39.** Electron-dot structures of elements in the same group have the same number of valence electrons. **41.** Only one additional electron can fit in the valence shell of a hydrogen atom. **43.** Two inner shells of electrons shield the valence electron from the nucleus. Consider that the charge of sodium's nucleus is $11+$. Ten inner-shell electrons $(10-)$ shield the electron in the outermost shell from this positive charge. This means that the outer-shell electron experiences the nucleus as though it were $1+$, which is significantly less than the actual nuclear charge of $11+$. **45.** The number of unpaired valence electrons is the same as the number of bonds the atom can form. **47.** The charge of neon's nucleus is $10+$. There are only two inner-shell electrons $(2-)$ that shield the outer-shell electrons from this positive charge. This means that the outer-shell electrons experience the nucleus as though it were $8+$, which is enough to hold the electrons tightly. Also, neon doesn't gain additional electrons because there is no more room available in its outermost shell. **49.** Sulfuric acid loses two protons to form the sulfate ion, SO_4^{2-}. A water molecule loses a single proton to form the hydroxide ion, OH^-. **51.** The electrical force of attraction weakens with increasing distance. So the force of attraction between a smaller sodium ion and a chloride ion is stronger than the force of attraction between a larger potassium ion and a chloride ion. This explains, in part, why potassium chloride crystals are weaker (softer) than sodium chloride crystals. **53.** The mold should be bigger than 6 inches because the metal will shrink (contract) as is cools. Unfortunately, as the molten metal cools and shrinks, it also pulls away from the inside of the mold while it is still soft, which means that it doesn't retain the form of the mold very well. Some exceptions are low-melting-point metals, such as lead. In general, metals are best stamped or machined to desired dimensions. **55.** The valence electrons of a potassium atom are weakly held by the nucleus. The potassium atom has a hard enough time holding onto its one valence electron, let alone a second one, which is what would happen if the potassium joined in a covalent bond. Potassium atoms are joined together by the metallic bond. **57.** When bonded to an atom with low electronegativity, such as any group 1 element, the nonmetal atom pulls the bonding electrons so closely to itself that it forms an ion. **59.** The atoms that are closer to the lower left corner of the periodic table bear the positive charge: hydrogen, bromine, carbon, neither! **61.** Sometimes true and sometimes false! Within any one period of the periodic table, it is true that electronegativity increases with increasing nuclear charge. The nuclear charge of a fourth-period potassium atom, however, is much greater than the nuclear charge of a second-period fluorine atom, but its electronegativity is much less. **63.** Water is a polar molecule because in its structure the dipoles do not cancel. Polar molecules tend to stick to one another, which gives rise to relatively high boiling points. Methane, on the other hand, is nonpolar because of its symmetrical structure, which results in no net dipole and a relatively low boiling point. The boiling points of water and methane are less a consequence of the masses of their molecules and more a consequence of the attractions among their molecules. **65.** The single greedy kid ends up being slightly negative, while the two more generous kids are slightly positive (deficient of electrons). The greedy negative kid is twice as

negative as one of the positive kids is positive. In other words, if the greedy kid had a charge of $2-$, each positive kid would have a charge of $1+$. This is a polar situation where the electrons are not distributed evenly. If all three kids were equally greedy, then the situation would be more balanced—that is, nonpolar. **67.** The charges in sodium chloride are balanced, but they are not neutralized. As a water molecule gets close to the sodium chloride, it can distinguish the various ions and it is thus attracted to an individual ion by ion–dipole forces. This works because sodium and chloride ions and water molecules are the same scale. **69.** Bromine atoms are larger, and this makes the formation of induced dipole–induced dipole attractions more favorable. **71.** If you haven't done so already, hold a charged balloon up to a thin stream of water from a faucet. First, charge the balloon by rubbing it on your hair. Then hold the balloon up to the thin steam and look for the effect. The dipoles of the water molecules all turn toward the charged balloon to which they are attracted. A similar effect can be seen by holding a charged balloon close to (but not touching) the corner of an ice cube. With some persistence you can get the ice cube to rotate back and forth. Narrow rectangular ice cubes (the kind that often come out of automatic ice makers) work best. **73.** We understand that muscles require exercise in order to stay in shape. What many people don't understand is that the brain works the same way: It too requires exercise in order to stay in shape. Furthermore, just as your muscles become stronger with extra exercise, so do your mental capacities become stronger with extra mental exercise. This is one of the main reasons we humans go to school for so many years—we understand the value and benefits of a well-exercised and in-shape mind. And as we hope you will discover, learning continues throughout one's life. But learning about the molecular nature of our environment is valuable for more than just the mental exercise. By understanding nature at this level, we gain a deeper appreciation, and with deeper appreciation comes greater respect. More than ever, humans are having a great impact on the environment. Should we do so mindlessly or mindfully? By studying chemistry you have decided for the latter. We thank you! **75.** There are many obstacles to recycling. There's confusion about what can and can't be recycled. There's concern about how clean a container must be before it can be recycled. Some don't know or understand the value of recycling. Others may feel a certain degree of laziness. To help overcome obstacles such as these, there can be campaigns to educate the general public. An easy-to-read pamphlet describing the dos and don'ts about recycling can be effective. These are typically produced by local governments in cooperation with local recyling companies. State and federal governments can also play a role in educating the general public and in setting packaging standards for companies that use recyclable materials to package their goods. Many states offer refunds for empty containers. Some communities even impose fines on individuals who do not sort plastics correctly. Other communities ask their citizens not to sort at all, reasoning that the task of sorting inhibits people's tendency to recycle. Ultimately, this is a global issue: Peoples of all nations should be encouraged not to waste material resources.

CHAPTER 16

1. A material is a mixture if it contains more than one substance. **3.** During distillation, one of the components of the mixture is boiled and the vapor is collected in another container. **5.** In a solution all the components are in the same phase, whereas the components in a suspension are in different phases. **7.** The volume of a sugar solution gradually increases as more sugar is dissolved in it. **9.** The more solute in a given volume of solvent, the more concentrated the solution is. **11.** The solubility of a gas decreases with increasing temperature because the gas molecules have more kinetic energy and are more likely to escape from solution. **13.** Sugar is very polar, as evidenced by its solubility in water. **15.** A detergent is a synthetic soap that has stronger grease penetration. Detergents are cheaper than soaps. **17.** Soap molecules are attracted to calcium and magnesium ions because both have a $2+$ charge. Soap is more attracted to these ions than to its own sodium ions (with a $1+$ charge). **19.** People disinfect their water by boiling it or adding disinfecting iodine tablets. **21.** Hawaii can have less stringent requirements because it is surrounded by a very deep ocean. **23.** A third level of wastewater treatment is expensive, and not all communities are in locations where this level of treatment is necessary. **25.** It would be humorous to scrape the residue from your boiled-down drinking water into sealable containers labeled as drinking water from your particular region, such as "Rocky Mountain Drinking Water." Think of the potential market. You could ship these containers to customers around the world and, because the containers are not weighted down with water, shipping costs would be low. Of course, each bottle would have to come with the instruction "Just add distilled water." Would you want to push it by adding the word *Pure* to your label? With your classmates, discuss the science and ethics of such a venture. **27.** The water level rises just as it would if you were adding sand. It does not matter that what you add also dissolves. **29.** Interesting crystals can also be made from supersaturated solutions of Epsom salts ($MgSO_4 \cdot 7\ H_2O$) and alum ($KAl(SO_4)_2 \cdot 12\ H_2O$), which is used for pickling and is available in the spice section of some grocery stores. Compare the crystal shapes of Epsom salts, alum, and sugar. **31.** Divide the number of water molecules by the total number of molecules and multiply by 100 to get the percentage:

999,999 million trillion/1,000,000 million trillion $\times$ 100 = 99.9999%

33. Mass = concentration $\times$ volume = (3.0 g/L)(15 L) = 45 g. **35.** The total volume of solution should be 20.0 g/10.0 g/L = 2.00 L, but this is the volume of the solution, *not* the volume of the solvent. Because the volume of the solution is equal to the combined volume of the solute and the solvent, the volume of water (solvent) is equal to the volume of the solution minus the volume of the sodium chloride (solute). If we ignore the rules for significant figures (see Appendix B) and assume that the 20.0 g of sodium chloride occupies 7.50 mL (0.00750 L), this volume of water is:

Volume of solution	2.00000 L
− Volume of solute	0.00750 L
Volume of water	1.99250 L

Don't waste time measuring out 1.99250 L of water. A far better approach is to make the solution by first adding the sodium chloride (solute) to an empty container calibrated for 10 L and then adding water (solvent) as needed to make 10.0 L of solution. **37.** Hexanol, butanol, ethanol. The carbon–hydrogen structures are nonpolar. The structure for hexanol is therefore the most nonpolar of these three molecules; hence, it has the lowest solubility in water. Another way to look at this is: The OH bond is a polar bond and this is what is needed to allow for good solubility in water. Only a small percentage of the hexanol molecule is made up of the OH bond. A greater percentage of the ethanol molecule (about 33%) is made up of the OH bond. Water, therefore, has an easier time being attracted to an ethanol molecule than to a hexanol molecule. **39.** Add the mixture of sand and salt to some water. Stir, and then filter the sand. Rinse the sand several times with fresh water to make sure all the salt has been removed. Collect all the salty water and evaporate away the water. The residue that remains will be the salt. After the sand dries, you've got just the sand. For a mixture of iron and sand, take advantage of the fact that only iron is attracted to a magnet. **41.** The transformation of elements into a compound is necessarily a chemical change. To go backward—from the compound back into the elements—would also be a chemical change. So the only way to separate an element from a compound is by chemical means. **43.** (a) Table salt is generally a heterogeneous mixture of the compound sodium chloride plus desiccants that absorb moisture and prevent the salt from clumping. (b) Blood is a suspension, an example of a homogeneous mixture. (c) Steel is a solid solution, a homogeneous mixture consisting of mostly iron plus smaller amounts of carbon and nickel. (d) Planet Earth is a heterogeneous mixture. **45.** Salt, sodium chloride, is a compound. Stainless steel is a mixture of iron and carbon. Tap water is a mixture of dihydrogen oxide plus impurities. Sugar, chemical name sucrose, is a compound. Vanilla extract, butter, and maple syrup, all natural products, are mixtures. Aluminum, a metal, is an element in pure form (sold commercially as a mixture of mostly aluminum with trace metals, such as magnesium). Ice, dihydrogen oxide, is a compound in pure form and a mixture when made from impure tap water. Milk, a natural product, is a mixture. Cherry-flavored cough drops, a pharmaceutical, are a mixture. **47.** Box A shows oval molecules dissolved within smaller circle molecules. **49.** The boiling points go up because of an increase in the number of molecular interactions between molecules. Remember that when we talk about the "boiling point" of a substance, we are referring to a pure sample of that substance. We see that the boiling point of 1-pentanol (the molecule at the bottom) is relatively high because 1-pentanol molecules are so attracted to one another (by induced dipole–induced dipole as well as by dipole–dipole and dipole–induced dipole attractions). When we refer to the "solubility" of a substance, we are referring to how well that substance interacts with a

second substance—in this case, water. Note that water is much less attracted to 1-pentanol because most of 1-pentanol is nonpolar (its only polar portion is the OH group). For this reason 1-pentanol is not very soluble in water. Put yourself in the place of a water molecule and ask yourself how attracted you might be to the methanol molecule (the one at the top) compared to the pentanol molecule (the one at the bottom). **51.** Nitrogen atoms are bigger, so nitrogen molecules should be more soluble in water due to stronger dipole–induced dipole attractions. **53.** The helium is less soluble in the bodily fluids, so less dissolves for a given pressure. Upon decompression, there is less helium to "bubble out" and cause potential harm. **55.** A saturated solution of sodium nitrate, $NaNO_3$, is more concentrated than a saturated solution of sodium chloride, NaCl. **57.** Assuming concentration is given in units of mass (or moles) of solute per volume of solution, the concentration necessarily decreases with increasing temperature. **59.** After a bottle of seltzer water is resealed, the carbon dioxide continues to come out of solution to fill the head space inside the bottle. The larger the head space, the greater the amount of carbon dioxide that comes out. The bottle that is three-fourths empty has more head space to fill, but the reserve of dissolved carbon dioxide is less because it contains less seltzer water. The near-empty bottle of seltzer water thus becomes quickly depleted of carbon dioxide, which leads to little fizz. Contrast this with the bottle that is three-fourths full, where there is only a small amount of head space to fill, but plenty of reserve. **61.** The fresh water can absorb more carbon dioxide because it contains more water (solvent) than the sugar water. **63.** The nonpolar molecules have a hard time passing the ionic heads of the fatty acid molecules, which are surrounded by water molecules. Polar molecules may be attracted to the ionic heads of the fatty acid, but they have a hard time passing by the nonpolar tails. **65.** Water and soap are attracted to each other primarily by strong ion–dipole attractions between the polar heads of each soap molecule and each water molecule. Polar water molecules, however, are also able to induce dipoles within the nonpolar tails of the soap molecules. So there is an attraction between water and the nonpolar tails of soap. This attraction is similar to that which occurs between water and oil molecules. Note, however, that this attraction is weaker than the attraction water molecules have for themselves, which is why oil and water are immiscible and appear to repel. **67.** Sodium carbonate (Na_2CO_3) has a 2− charge in the carbonate ion $(CO_3)^{2-}$ to which calcium and magnesium are more attracted than to the 1− charge in a molecule of soap. The hard water ions, calcium and magnesium, bind to the carbonate ions, which "softens" the water. **69.** Osmosis is a process whereby water is drawn from a region of low solute concentration (high water concentration) to a region of high solute concentration (low water concentration). Because the cells at the top of the tree have a higher concentration of sugars than the cells at the bottom, water is forced to travel up the tree against gravity by osmosis. **71.** The net flow of water is reversed. Osmosis transfers fresh water into salt water, but reverse osmosis transfers salt water to fresh water. **73.** Distilled water is pure only before you drink it. Once in your stomach, it mixes with everything else to make up a nutrient-filled solution. Tap water may contribute a few more milligrams of hard-water ions, such as calcium, which your body actually uses as a mineral. But there's nothing wrong with drinking water that has been distilled. In fact, it is about as pure as any water you'll ever be able to drink. **75.** The decomposition of food by bacteria in our digestive system is primarily anaerobic because little oxygen makes it from our mouths to our intestines where food decomposition takes place. As a consequence, gases that come out our other ends are frequently of the odoriferous sort. Composting toilets add air to our waste products, which moves the mode of decomposition from anaerobic to less smelly aerobic microorganisms. This is the reason fresh dog poop always smell more foul than week-old dog poop. **77.** Open ended. **79.** Open ended.

CHAPTER 17

1. Coefficients show the ratios in which reactants combine and products form in a chemical equation. **3.** A chemical equation must be balanced because the law of conservation of mass says that mass can be neither created nor destroyed. There must be the same number of each atom on both sides of the equation. **5.** The mass of golf balls is greater than the mass of Ping-Pong balls, so a given mass contains more Ping-Pong balls than golf balls. **7.** The formula mass of NO is 30.01 amu. **9.** For water, 18 g is 1 mole. **11.** One mole of water contains 6.02×10^{23} water molecules. **13.** In order to form a product, reactants must collide in a certain orientation with sufficient energy. **15.** The fastest-moving reactant molecules are the first to pass over the energy barrier. **17.** Atomic chlorine is a catalyst for the destruction of ozone. **19.** A catalyst is unchanged by a chemical reaction. **21.** The amount of energy released when the bond is formed equals the amount of energy needed to break the bond, which is 436 kJ. **23.** Energy is consumed by an endothermic reaction. **25.** Entropy is always increasing. **27.** The energy of the lightning or electrostatic sparks passing through the air converts oxygen molecules into ozone molecules. This reaction is endothermic because it requires an input of energy. **29.** The balanced equation is $2 H_2O_2 \rightarrow O_2 + 2 H_2O$. Bubbles of oxygen gas form when hydrogen peroxide and baker's yeast are mixed. That oxygen gas is formed can be demonstrated by inserting a glowing wood splint into the bubbles. The splint will flame up as soon as it makes contact with the oxygen. Students should be directed to do this only under careful supervision. **31.** Dissolving the salt in the water causes a decrease in temperature, which tells us that this is an endothermic (energy-absorbing) process. At the molecular level, a number of things are going on; for example, energy is used to help separate the bonds between the sodium and chloride ions. **33.** (5.00 g gold)(1 mole gold/197 g gold)(6.02 × 10²³ atoms/1 mole) = 1.53 × 10²² gold atoms. **35.** The coefficients in this balanced equation tell us the ratio in which reactants react to form products. We see that 3 moles of oxygen gas are produced for the reaction of every 2 moles of $KClO_3$ solid. Do you also see that only 1.5 moles of oxygen gas would be produced from the reaction of 1 mole of $KClO_3$ solid? **37.** Use the periodic table to find the masses of all the atoms in each molecule. Add these masses to get the given formula masses. Use these masses to help answer the next question. **39.** Is there enough oxygen to react with all of the methane? Is there enough methane to react with all of the oxygen? We must work out how much of one reactant is needed in order for all of the other reactant to be consumed. According to the following calculation, 16 g of CH_4 requires 64 g of O_2:

$$(16 \text{ g } CH_4)(1 \text{ mole } CH_4/16 \text{ g } CH_4)(2 \text{ moles } O_2/1 \text{ mole } CH_4)$$
$$(32 \text{ g } O_2/1 \text{ mole } O_2) = 64 \text{ g } O_2$$

But we have only 16 g of O_2, which means that not all of the CH_4 is going to be able to react. How much of the CH_4 will react? That can be calculated as follows:

$$(16 \text{ g } O_2)(1 \text{ mole } O_2/32 \text{ g } O_2)(1 \text{ mole } CH_4/2 \text{ moles } O_2)$$
$$(16 \text{ g } CH_4/1 \text{ mole } CH_4) = 4 \text{ g } CH_4$$

The maximum amount of CO_2 that can be formed is calculated as follows:

$$(16 \text{ g } O_2)(1 \text{ mole } O_2/32 \text{ g } O_2)(1 \text{ mole } CO_2/2 \text{ moles } O_2)$$
$$(44 \text{ g } CO_2/1 \text{ mole } CO_2) = 11 \text{ g } CO_2$$

41.
(a)

Energy to break bonds	Energy released from bond formation
N—N = 159 kJ	H—H = 436 kJ
N—H = 389 kJ	H—H = 436 kJ
N—H = 389 kJ	H—H = 436 kJ
N—H = 389 kJ	H—H = 436 kJ
N—H = 389 kJ	N≡N = 946 kJ
Total = 1715 kJ absorbed	Total = 2690 kJ released

Net = 1715 kJ absorbed − 2690 kJ released = −975 kJ released (exothermic)

(b)

Energy to break bonds	Energy released from bond formation
O—O = 138 kJ	
H—O = 464 kJ	O=O = 498 kJ
H—O = 464 kJ	H—O = 464 kJ
O—O = 138 kJ	H—O = 464 kJ
H—O = 464 kJ	O—H = 464 kJ
H—O = 464 kJ	O—H = 464 kJ
Total = 2132 kJ absorbed	Total = 2354 kJ released

Net = 2132 kJ absorbed − 2354 kJ released = −222 kJ released (exothermic)

43. c, a, b. The endothermic reaction, c, will likely take place slower than the exothermic reaction, a, because it requires a decrease in entropy. The fastest reaction will be b, which has no energy of activation. **45.** c, a, b. Contrary to popular opinion, the entropy of a deck of playing cards has nothing to do with whether or not it has been shuffled. Historically, there has been a misunderstood association between entropy and disorder, but now you know better. Entropy has nothing to do with what our minds perceive as being orderly or disorderly. From the point of view of the

molecules in the cards, it makes no difference whether the deck is shuffled. Entropy is merely a measure of the tendency of energy to disperse. The greater the difference between the temperature of the cards and the room in which the cards reside, the greater the amount of energy that gets dispersed. Of course, a burning deck of cards would produce the most entropy of all. **47.** (a) 2, 3, 1. (b) 1, 6, 4. (c) 2, 1, 2. (d) 1, 2, 1, 2. **49.** Only two diatomic molecules are represented (not three!). These are the two shown in the left box, one of which is also shown in the right box. Remember, the atoms before and after the arrow in a balanced equation are the same atoms only in different arrangements. **51.** Equation d best describes the reacting chemicals. **53.** $Fe_2O_3 + 3 CO \rightarrow 2 Fe + 3 CO_2$ **55.** (a) 1 mole. (b) 1 mole. (c) 2 moles. (d) 1 mole. **57.** A single water molecule has a very small mass, 18 amu. **59.** No, because 14 amu is less than the mass of a single oxygen atom. **61.** There are 69.7 g of gallium, Ga (atomic mass 69.7 amu). Note that 145 g is the formula mass for this compound. **63.** As the carbon-based fuel combusts, it gains mass as it combines with the oxygen from the atmosphere to form carbon dioxide, CO_2, which comes out in the exhaust. **65.** Pure oxygen has a greater concentration of one of the reactants (oxygen) in the chemical reaction (combustion). As discussed in this chapter, the greater the concentration of reactants, the faster the rate of reaction. **67.** Photosynthesis produces oxygen, O_2, which migrates from Earth's surface to high up in the stratosphere, where it is converted by the energy of ultraviolet light into ozone, O_3. Plants and all other organisms that live on Earth's surface benefit from this ozone because of its ability to shade the planet's surface from ultraviolet light. **69.** Hydrogen chloride, HCl, does not stay in the atmosphere very long because it is quite soluble in water, as can be deduced from its polarity (see Chapter 15). Thus, atmospheric hydrogen chloride mixes with atmospheric moisture and precipitates with rain. **71.** The chemical reactions in a disposable battery are exothermic, as evidenced by the electrical energy they release. Recharging a rechargeable battery requires the input of electrical energy; hence, the reactions that occur during the recharging process are endothermic. **73.** As an exothermic reaction proceeds from reactants to products, the result is a release (dispersion) of thermal energy, which favors the formation of products. Typically, the amount of energy dispersion is significantly larger than the difference in chemical entropies of the products and reactants. **75.** The solar energy is more readily dispersed by the water molecules in the gaseous phase. **77.** The reaction shows the formation of a single molecule from three molecules plus two atoms. This is a "coming together" of matter, which suggests a decrease in entropy. The reaction, however, also involves the breaking of three H—H bonds (1308 kJ/mol) and the forming of one C—C bond and six C—H bonds (2831 kJ/mol), which provides a net release of 1523 kJ/mol of energy. (To understand this, you need to be able to deduce the chemical structures H—H for H_2 and H_3C—CH_3 for C_2H_6.) The large exothermic nature of this reaction, therefore, provides a large enough increase in the entropy of the surroundings that this reaction is favored. **79.** Endothermic reactions require the input of energy, which can include the input of thermal energy. This gives the molecules greater kinetic energy, which can help their collisions to be more effective. The elevated temperature also helps to minimize the unfavorable decrease in entropy due to the heat absorbed by the reaction. Some exothermic reactions are so exothermic that they explode if not run at cold temperatures. The cold temperatures slow the reactive molecules down, which gives the chemist greater control. Also, the heat generated by the reaction is more efficiently dispersed under the colder conditions. This allows for a greater increase in entropy, which helps with the formation of products. **81.** Putting more ozone into the atmosphere to replace what has been destroyed is a bit like throwing more fish into a pool of sharks to replace those fish that have been eaten. The solution is to remove the CFCs that destroy the ozone. Unfortunately, CFCs degrade slowly, and the ones that are there now will remain there for many years to come. Our best bet is to stop producing CFCs and hope that we haven't already caused too much damage.

CHAPTER 18

1. The Brønsted–Lowry definition of an acid and base says that an acid is any chemical that donates a hydrogen ion and a base is any chemical that accepts a hydrogen ion. **3.** A chemical that loses a hydrogen ion is behaving as an acid. **5.** A solution of a strong acid has more ions in solution and can conduct electricity better. **7.** Water is a weak acid. **9.** The pH indicates the acidity of a solution as judged by the concentration of hydronium ions. **11.** CO_2 and H_2O react to form carbonic acid, H_2CO_3.   **13.** The burning of fossil fuels contributes to SO_2 in the air.   **15.** The elements in the upper right of the periodic table (except for the noble gases) have the greatest tendency to behave as oxidizing agents.   **17.** $K \rightarrow K^+ + 1e^-$ **19.** Electrochemistry is the study of electrical energy and chemical changes.   **21.** Reduction occurs at the cathode (remember the "red cat"). Oxidation occurs at the anode (remember "an ox" pulls the "red cat" in a cart).   **23.** Fuel cells also produce the chemical products of electricity-generating chemical reactions. A hydrogen fuel cell, for example, produces clean water suitable for drinking.   **25.** Aluminum is produced primarily by electrolysis.   **27.** Zinc coats galvanized nails.   **29.** Iron is forced to accept electrons from either zinc or magnesium atoms.   **31.** The cabbage is purple when you buy it at the store. This color indicates that it is only slightly acidic, so the juices in the cabbage are less acidic than vinegar. The baking soda will react with the vinegar to form gaseous carbon dioxide. The two should neutralize each other so that the color moves toward purple.   **33.** The carbon dioxide of your breath reacts with the water to form carbonic acid, which reacts with the washing soda to lower the pH. As described in this chapter, the carbon dioxide in the atmosphere reacts with rainwater to produce carbonic acid, which makes rainwater acidic. The more washing soda you add, the more carbonic acid you'll need to neutralize it. While a couple breaths can neutralize a pinch of washing soda, it would take quite a while to neutralize a tablespoon full.   **35.** As silver tarnishes, it is oxidized ($Ag \rightarrow Ag^+ + 1\ e^-$). Baking soda is an ionic compound, and it is needed in this activity to help the electrons flow through the hot solution. The aluminum behaves as a reducing agent as it gives electrons back to the silver ($Al \rightarrow Al^{3+} + 3\ e^-$).   **37.** Try this activity with tap water instead of salt water to see the difference dissolved ions make—the ions are needed to conduct electricity between the two electrodes. The primary reaction occurs at the negative electrode (anode), where water molecules accept electrons to form hydrogen gas and hydroxide ions. Recall from Section 18.3 that an increase in hydroxide-ion concentration causes the pH of the solution to rise. You can track the production of hydroxide ions by adding a pH indicator to the solution. The indicator of choice is phenolphthalein, which you might obtain from your instructor. Alternatively, you can use red cabbage extract. Whichever indicator you use, note the swirls of color forming at the anode as hydroxide ions are generated. The battery is quickly ruined because placing it in the conducting liquid short-circuits the terminals, which results in a large drain on the battery. You may wonder why oxygen gas is not generated along with the hydrogen gas. For reasons beyond the scope of this text, oxygen gas is generated only when the positive electrode (cathode) is made of certain metals, such as gold or platinum. The steel electrode of the 9-V battery does not suffice.   **39.** The pH of this solution is 10, which is basic. The pH of the second solution is 4, which is acidic.   **41.** The concentration of hydronium ions in the pH = 1 solution is 0.1 M. Doubling the volume of the solution with pure water means that its concentration is cut in half. The new concentration of hydronium ions after the addition of 500 mL of water, therefore, is 0.05 M. To calculate for pH:

$$pH = -\log[H_3O^+] = -\log(0.05) = -(-1.3) = 1.3$$

43. Set this up as a unit conversion:

$$(1.6 \times 10^7\ mt)(1000\ kg/1\ mt)(1000\ g/1\ kg) = 1.6 \times 10^{13}\ g\ of$$
$$aluminum\ (16\ quadrillion)$$

45. a, b, c. All these solutions have the same concentration. They differ in their acid strength. As discussed in the text, hydrogen chloride is a strong acid, which means nearly all of the hydrogen chloride molecules donate hydrogen ions to form hydronium ions. Acetic acid is a weak acid, which means only a few of the acetic acid molecules in solution donate hydrogen ions to form hydronium ions. The ammonia behaves better as a base than as an acid, which means that it contributes very, very few hydrogen ions. The ammonia solution, therefore, has the lowest concentration of hydronium ions. The acetic acid solution has more hydronium ions, but not as many as the hydrogen chloride solution.   **47.** c, b, a. The pH of the rain decreases (becomes more acidic) with increasing atmospheric concentrations of carbon dioxide, as in Figure 18.17.   **49.** a, b, c. A reducing agent causes other materials to gain electrons because of its tendency to lose electrons. Atoms with low electronegativity tend to lose electrons easily and, therefore, also behave as strong reducing agents. Sodium, Na, has the weakest electronegativity, which means it is the strongest reducing agent of these three elements. (It is, however, the weakest oxidizing agent of these three.) Sulfur, S, is in between, and chlorine, Cl, with the greatest electronegativity, is the weakest reducing agent.   **51.** The base accepted the hydrogen ion, H^+, and

thus gained a positive charge. So the base forms the positively charged ion. Conversely, the acid donated a hydrogen ion and thus lost a positive charge. The acid forms the negatively charged ion.   **53.** Squeeze lemon juice on the fish. The citric acid in lemon juice reacts with the smelly alkaline compounds to form less smelly salts. The smell and taste of the lemon also help to mask any additional fishy odors.   **55.** The positive sodium ion of sodium hypochlorite combines with the negative chloride ion of hydrochloric acid. Meanwhile, the negative hypochlorite ion combines with the positive hydrogen ion to form hypochlorous acid, HOCl. Interestingly, the hypochlorous acid continues to react with hydrogen chloride to form water and poisonous chlorine gas, Cl_2, which is why bleach and toilet bowl cleaner should *never* be mixed together.   **57.** Ammonia, NH_3, is the stronger base. To understand why, ask yourself which is more willing to lend its lone pair to the positive charge on another molecule. The fluorines in nitrogen trifluoride, NF_3, have the effect of pulling the lone-pair electrons closer to the nitrogen. They do so because of their high electronegativities (electron pulling power). This makes the lone pair in NF_3 less available to attack the positive charge on another molecule.   **59.** (a) pH is a measure of the hydronium-ion concentration: The higher the hydronium-ion concentration, the lower the pH. As water warms, the hydronium-ion concentration increases, albeit only slightly. Thus, pure water that is hot has a slightly lower pH than pure water that is cold. (b) Yes. As water warms up, the hydronium-ion concentration increases and so does the hydroxide-ion concentration—and by the same amount. Thus, the pH decreases and yet the solution remains neutral because the hydronium- and hydroxide-ion concentrations are still equal. At 40°C, for example, the hydronium- and hydroxide-ion concentrations of pure water are both 1.71×10^{-7} moles/L (the square root of K_w). The pH of this solution is the negative log of this number, or 6.77. This is why most pH meters need to be adjusted for the temperature of the solution being measured. Except for this exercise, which probes your powers of analytical thinking, this textbook ignores the slight role that temperature plays in pH. Unless noted otherwise, continue to assume that K_w is a constant 1.0×10^{-14}; in other words, assume that the solution being measured is at 24°C   **61.** This solution has a hydronium-ion concentration of 10^3 M, or 1000 moles/L. The solution would be impossible to prepare because only a certain amount of acid can dissolve in water before the solution is saturated and no more will dissolve. The highest concentration possible for hydrochloric acid, for example, is 12 M. Beyond this concentration, any additional HCl, which is a gas, added to the water simply bubbles back out into the atmosphere.   **63.** No. Consider this analogy: You are at a bottle cap convention where you hope to sell your bottle caps. You soon realize, however, that you're a lousy salesperson and all the other vendors at the convention are excellent salespeople. Not only do they steal away all your potential customers, but they're so good that you end up buying bottle caps from them. In the end, you have given away no bottle caps. Likewise, carbonic acid is unable to give away any hydrogen ions in a concentrated solution of such a strong acid. If it did, an HCl molecule would come along and quickly give one right back. Adding carbon dioxide gas will not lower the pH even though the carbon dioxide readily transforms into carbonic acid.   **65.** As the soda water loses carbon dioxide molecules, it is losing the carbonic acid that these carbon dioxide molecules form when in solution. Thus, the pH of flat soda is typically higher than the pH of the same soda when carbonated.   **67.** The warmer the ocean, the lower the solubility of any dissolved gases such as carbon dioxide, CO_2. Less CO_2 would be absorbed and more of it would remain to perpetuate global warming.   **69.** The tin ion, Sn^{2+}, is the oxidizing agent because it causes the silver, Ag, to lose electrons. The silver atoms, Ag, are the reducing agents because they cause the tin ion to gain electrons.   **71.** The unsaturated fatty acids are gaining hydrogen atoms, so they are being reduced.   **73.** The copper-coated zinc penny is not an example of a voltaic cell, a device that allows the flow of electrons by permitting a reverse flow of positive charge. An immediate buildup of charge in either the zinc or the copper prevents continued oxidation–reduction from occurring.   **75.** As the carbon of propane, C_3H_8, forms carbon dioxide, CO_2, it is losing hydrogen and gaining oxygen, so the carbon is being oxidized, which is the opposite of what happens to carbon during photosynthesis.   **77.** The first step is to balance the atoms by showing two iodide ions, I^-:

$$Fe^{3+} + 2\ I^- \rightarrow Fe^{2+} + I_2$$

The charges can then be balanced by showing two Fe^{2+} ions and two Fe^{3+} ions:

$$2\ Fe^{3+} + 2\ I^- \rightarrow 2\ Fe^{2+} + I_2$$

79. One of the products of combustion is water vapor. **81.** If copper has a greater tendency to become reduced than iron, then electrons will preferentially flow from the iron to the copper (and then to oxygen as indicated in the answer to Exercise 80). Corrosion is thus accelerated at the interface of these two metals. This was one of the main reasons the Statue of Liberty required a full restoration in 1976. **83.** There is a lower ratio of hydrogen atoms in the acetaldehyde product, so the grain alcohol is oxidized. **85.** The digestion and subsequent metabolism of foods and drugs tend to make the molecules of foods and drugs more polar. Oxidation is one way the body does this. **87.** Adding iron to the ocean to enhance its ability to absorb carbon dioxide may help to decrease the amount of carbon dioxide in the atmosphere. At the same time, however, it might drastically alter the ocean's ecology. For example, the excess carbon dioxide would lower the pH of the ocean. Shelled marine organisms would die, which could affect the whole food chain. We would end up trading one problem for another. **89.** Is it fair that developed nations mandate through international treaties that developing nations never produce as much carbon dioxide as the developed nations have produced? If the developed nations were allowed to emit all that pollution, why can't the developing nations do the same? Imagine this same mentality applied to corporations whose primary goal is to increase their profits. Yes, countries and industries have the capability to self-regulate, but forces being what they are, this self-regulation is not always ideal for everyone. This is one of the main responsibilities of government, which, ideally, is there to represent the collective will. Consumers speak with two voices: their pocket book and the voting booth.

CHAPTER 19

1. Structural isomers have different arrangements of their carbon atoms, but the number of carbon atoms in each is the same. **3.** The boiling points of hydrocarbons are used for fractional distillation. **5.** A carbon atom in a saturated hydrocarbon is bonded to four atoms. **7.** A hydrocarbon must have at least one multiple (double or triple) bond to be unsaturated. **9.** A heteroatom is any atom other than carbon or hydrogen in an organic molecule. **11.** Small alcohols are soluble in water because they have polar oxygen–hydrogen bonds similar to those in water. Like dissolves like. **13.** An alcohol has a hydroxyl (oxygen bonded to hydrogen) group; an ether contains an oxygen atom bonded to two carbon atoms. **15.** Amines tend to be basic. The lone electron pair on nitrogen atoms makes them basic because this pair is able to accept a hydrogen ion. **17.** Morphine and caffeine are two alkaloids. **19.** Both contain carbonyl groups. Ketones have the carbonyl carbon bonded to two adjacent carbon atoms, whereas aldehydes have the carbonyl carbon bonded either to one hydrogen and one carbon or to two hydrogens. **21.** Aspirin is made from salicylic acid. **23.** In the formation of a condensation polymer, a small molecule (i.e., water or hydrochloric acid) is released from each monomer. **25.** A copolymer is a polymer composed of two or more different monomers. **27.** PETE has a T_g of around 69°C, which is why a PETE 2-L bottle deforms so easily when placed in boiling water. Ziplock sandwich bags are also commonly made of PETE. People who wash their ziplock bags in hot water (rather than throwing them away) will note that the ziplock bags are much softer and more flexible in hot water than at room temperature. The T_g of polystyrene is around 100°C. This is good news for polystyrene (Styrofoam) coffee cups, which would otherwise not be able to hold hot water. **29.** a, c, b. **31.** a, b, c. **33.** a, b, c. Molecule a is an ether, which has limited solubility in water. Molecule b is an amine. The hydrogen on the nitrogen is able to participate in hydrogen bonding with water, which makes this compound somewhat soluble in water. Molecule c is a salt consisting of ions, which are nicely soluble in water. **35.** The melting point increases because of a greater number of induced dipole–induced dipole molecular attractions between adjacent hydrocarbon molecules. **37.** Only two structural isomers are drawn. The one in the middle and the one on the right are actually two conformations of the same isomer. **39.** To make it to the top of the fractionating column, a substance must remain in the gaseous phase. Only substances with very low boiling points, such as methane (bp −160°C), are able to make it to the top. According to Figure 19.3, gasoline travels higher than kerosene, so it must have a lower boiling point. Kerosene, therefore, has the higher boiling point. **41.** The percentage of carbon increases as the hydrocarbon gets bigger: methane's is 20%; ethane, 25%; propane, 27%; butane, 29%. **43.** (a) C_7H_{16}. (b) C_7H_{12}. (c) $C_4H_{10}O$. (d) C_4H_8O. **45.** The second and the fourth structures are the same. In all, three different structures are shown. **47.** The 80-proof vodka is 40% ethanol by volume and 60% water.

49. Like many natural oils derived from fats, the carbons in cetyl alcohol are arranged in sequence with no branching. **51.** Three ethanols surround a central nitrogen atom.
53.

55. No! This label indicates that it contains the hydrogen chloride salt of phenylephrine, not acidic hydrogen chloride. This organic salt is as different from hydrogen chloride as is sodium chloride (table salt), which may also go by the name "the hydrogen chloride salt of sodium." Think of it this way: Assume you have a cousin named George. Now, you are George's cousin, not George. In a similar fashion, the hydrogen chloride salt of phenylephrine is made using hydrogen chloride, but it is not hydrogen chloride. A chemical substance is uniquely different from the elements or compounds from which it is made.
57. 1. ether 2. amide 3. ester 4. amide 5. alcohol 6. aldehyde 7. amine 8. ether 9. ketone

59. The transformation of benzaldehyde to benzoic acid is an oxidation. **61.** EDTA has a strong affinity for lead ions, Pb^{2+}, because the +2 charge of the lead ion is attracted to the two −1 charges of the EDTA molecule. EDTA is used to help remove the lead ions from people, usually children, suffering from lead poisoning.

63. Polypropylene consists of a polyethylene backbone with methyl groups attached to every other carbon atom. This side group interferes with the close packing that could otherwise occur among the molecules. As a consequence, polypropylene is actually less dense than polyethylene—even low-density polyethylene. **65.** Note the similarities between the structures of SBR and polyethylene and polystyrene, all of which possess no heteroatoms. SBR is an addition polymer made from the monomers 1,3-butadiene and styrene mixed together in a 3:1 ratio. Notably, SBR is the key ingredient that allows bubbles to form in bubble gum. **67.** There are eight isoprene units needed to make a single beta-carotene molecule. A quick way to figure this out is to count up the number of carbon atoms in beta-carotene, which is forty, and divide by five. Initially, the beta-carotene structure looks complex.

Upon careful examination, however, we see that this molecule is simply the result of smaller units joining together. Similarly, many of the molecules that you have studied in this chapter may have initially looked intimidating. With a basic understanding of the concepts of chemistry, however, you already have much insight into their properties. **69.** The HCl would react with the free base to form the water-soluble, but diethyl ether–insoluble, hydrochloric acid salt of caffeine. With no water available to dissolve this material, it precipitates out of the diethyl ether as a solid that may be collected by filtration.

CHAPTER 20

1. As heavy materials moved toward Earth's center and lighter materials moved toward the surface, Earth's interior became layered according to density. **3.** Earth is 33% iron. **5.** A mineral is a naturally formed, inorganic, crystalline solid composed of an ordered array of atoms and having a specific chemical composition. **7.** Minerals are identified by crystal form, hardness, cleavage, color, density, luster, and streak. **9.** A polymorph is two or more minerals that contain the same elements in the same proportions but with different arrangements of atoms. So, their crystalline structure and properties are different. Examples are graphite and diamond. **11.** Ferromagnesian silicates contain iron (Fe) and magnesium (Mg); they tend to be dark and dense. Nonferromagnesian silicates do not contain significant amounts of iron or magnesium, so they have lower densities and a lighter color. **13.** Crystallization is the growth of a solid from a liquid or gas whose atoms come together with a specific chemical composition and crystalline arrangement. Beginning with the formation of a single microscopic crystal, the crystal grows as more atoms bond to the crystal faces. **15.** Minerals with lower amounts of silica crystallize first. **17.** Igneous rocks formed from the cooling and crystallization of magma. Sedimentary rocks originated from weathered material carried by water, wind, or ice. Metamorphic rocks formed from preexisting rocks transformed by high temperature, high pressure, or both—without melting. **19.** Partial melting is the incomplete melting of rock. Each mineral type has its own melting point, which results in magmas of different compositions. **21.** Silica content determines the melting point. Rocks with low silica have a high melting point, and rocks with high silica have a low melting point. **23.** Weathering breaks down and decomposes surface rock. Erosion is the process by which weathered rock particles are removed and transported away by water, wind, or ice. **25.** Shale, sandstone, and conglomerate. **27.** Most carbonate rocks originate biologically as a result of shell growth. When shell-bearing organisms die, their shells accumulate on the seafloor, where they dissolve to form a noncrystalline ooze of calcium carbonate. This eventually crystallizes into calcite, which then forms limestone. **29.** Foliated metamorphic rocks have a layered appearance, like the parallel flakes, or plates, of mica. Nonfoliated rocks do not have the aligned, platy mineral configuration. **31.** After shaking, there is a pattern in the settling. The mixture of materials settles into layers. Denser materials settle to the bottom, and less-dense materials migrate to the top. Earth's elements separated according to density. Heavy, iron-rich elements sank to early Earth's center; lighter, silicate elements migrated toward the surface. **33.** Halite has an isometric crystal structure (three axes of equal length that make right angles) and a cubic form. It has perfect cleavage—it breaks on all sides, so it always breaks into cubes. **35.** (a) Ice melts first. The cheese and butter do not melt. The melt is different from the original "rock" because it is richer in H_2O (the water plus ice). (b) High-silica minerals melt; low-silica minerals do not melt. The partial melt differs from the original rock because it is richer in silica. High-silica minerals are the first to melt. (c) Low-silica minerals (rocks) are the first to form. The remaining melt becomes depleted in the constituents of minerals that have crystallized and enriched in the constituents of minerals that are yet to crystallize. **37.** Looking at the periodic table, we calculate that the formula masses for MgO, FeO, and SiO_2 are 40 amu, 72 amu, and 60 amu, respectively. Therefore the formula mass of pyroxene is

$$1\ MgO + 1\ FeO + 2\ SiO_2 = 232\ amu$$

and the mass percentages are

$$\frac{40\ amu}{232\ amu} \times 100\% = 17\%\ MgO$$

$$\frac{72\ amu}{232\ amu} \times 100\% = 31\%\ FeO$$

$$2 \times \frac{60\ amu}{232\ amu} \times 100\% = 52\%\ SiO_2$$

(a) $0.52 \times 225\ kg = 117\ kg$ of silica. (b) The total silica removed from the magma:

114 kg from olivine (from Figuring Physical Science box) + 117 from pyroxene = 231 kg silica

500 kg silica in original magma − 231 kg silica removed = 269 silica in remaining liquid

The mass of the remaining liquid is 1000 kg − 325 kg − 225 kg = 450 kg. So the mass percentage of silica is

$$\frac{269\ kg\ silica}{450\ kg\ magma} \times 100\% = 60\%$$

39. c, a, d, b. **41.** (c) Slate, (b) schist, (a) gneiss. **43.** b, d, c, a. **45.** c, d, b, a. **47.** c, a, d, b. **49.** In a direct sense, no. In an indirect sense, yes. The tendency of silicon to bond with oxygen is so strong that silicon is never found in nature as a pure element; it is always combined with oxygen. Quartz, composed of only oxygen and silicon, is a primary source of silicon for making microchips. **51.** The oxides and sulfides make up the majority of ore minerals. **53.** Mineral A, because high-silica minerals are the first to melt. **55.** The factors are strength of the ionic charges (high charge = great attraction = strong bond = hard mineral), size of the atoms (small atoms pack closer, large atoms have more space between them), and packing of the atoms (closely packed atoms have more attractive forces, loosely packed atoms have less attractive forces). **57.** Mineral B, because low-silica minerals are the last to melt. **59.** "Last to crystallize" means "crystallizes at the lowest temperature," which means the last crystallizers have the lowest melting and freezing points. Having a low melting/freezing point means that such minerals tend to melt easily, so they melt first. **61.** Yes; limestone is formed predominantly from the shells of dead marine organisms. **63.** Chemical impurities affect color, and many minerals come in multiple colors. **65.** Small, closely packed atoms have a smaller distance between them and thus form stronger bonds than do minerals in which the atoms are not so closely packed. Gold, because of its large atomic size, is soft (hardness < 3), and diamond, with its small tightly packed carbon atoms, is hard (hardness = 10). **67.** No. The planar surfaces in cleavage are where a mineral breaks due to a weakness in the crystal structure or bond strength. The planar surfaces in a crystal form are the external shape from the crystal's internal arrangement of atoms. **69.** Poorly sorted, angular particles of various shapes imply a short transportation distance. Well-sorted, well-rounded particles imply a long distance. Glacial deposits tend to be poorly sorted and angular, whereas wind-blown deposits tend to be well sorted with small particles. **71.** (a). **73.** The Hawaiian Islands are made up of predominantly volcanic igneous rock. **75.** Characteristic patterns are an increase in crystal size due to recrystallization and changes in the water content of the metamorphosed rock. Crystals are largest at the contact, and get smaller with distance from that point. The water content of the rock also changes with distance. At the contact, where the temperature is high, the water content is low. So we find dry, high-temperature minerals such as garnet and pyroxene at the contact. Farther away are water-rich, low-temperature minerals such as muscovite and chlorite. **77.** Yes. High-silica minerals have lower melting points and do not require very high temperatures to melt. They are "easier" to melt than minerals with low silica content and a higher melting point. **79.** Smoothness and roundness of rock particles indicate travel time and hence distance. If particles are angular, then a short travel time is indicated. Small, rounded particles indicate a longer travel time and a longer distance. Size is also an indication of distance traveled: Particles become smaller the farther they travel. **81.** Halite weathers first because it has a high solubility (precipitates last in an evaporating body of water) and so dissolves easily in a humid environment. **83.** Granite, composed of predominantly quartz and feldspar minerals, is resistant to chemical weathering. Marble is metamorphosed limestone that succumbs more easily to chemical weathering. With time, marble may dissolve from the conglomerate. **85.** Fossils are the remains of ancient life and are used to interpret Earth's geologic past. They play an important role as time indicators and in matching rocks from different places of similar age. **87.** Metamorphic and sedimentary rocks are made from previously existing rock. Sedimentary rock does not require high temperature or pressure. **89.** Muscovite and biotite. **91.** Regional metamorphism is associated with compressive stress and mountain building. Compressive stresses push rocks together. So, as Earth's crust is compressed, the rock layers become deformed; they become folded and fractured. **93.** Foliation is the distinguishing feature. Sheet-structured minerals such as the micas orient themselves perpendicular to the direction of maximum pressure. These parallel flakes give schist and gneiss a layered look—foliation. Foliation does not develop if the rock does not have the right chemical composition for micas to form. The chemical compositions of marbles and quartzites do not favor the formation of micas. **95.** Schist rocks are shiny with obvious foliation. Because schist rocks have large crystals, the minerals can be easy to identify. **97.** As magma cools, high-melting-point and low-silica minerals crystallize first. The remaining liquid magma becomes depleted in the constituents of minerals that have crystallized and enriched in the constituents of minerals that have yet to crystallize. Because high-silica minerals are the last to crystallize, the magma becomes enriched with silica. **99.** Obsidian does not have a crystalline structure, so it is not a mineral. Glassy volcanic rocks are classified by composition as a whole (not by individual minerals). The rock simply cooled too quickly for individual minerals to grow, but it is still made up of the elements that would have formed minerals. **101.** Prospectors look for sedimentary rock. Petroleum formation begins with the accumulation of sediment from areas rich in plant and animal remains. As buried organic-rich sediment is heated over time, chemical changes take place that create oil. Under the pressure of overlying sediments, oil droplets are squeezed out into overlying porous sedimentary rocks that become reservoirs.

CHAPTER 21

1. P-waves are longitudinal; they compress and expand rock as they move through it, they are fast (register first on a seismograph), and they can travel through any type of material—solid, liquid, or gas. S-waves are transverse; they vibrate the particles of their medium up and down and side to side, they are not as fast (register second on a seismograph), and they can travel through solids only. **3.** Rayleigh waves move in an up-and-down, rolling motion. Love waves move in a side-to-side, whiplike motion. Both travel at lower speeds than P- and S-waves and so register last on a seismograph. **5.** Both P- and S-waves are strongly refracted at the core–mantle boundary, causing a wave shadow. The reflection and refraction of P-waves through the core identified the solid inner core. **7.** S-waves cannot travel through liquids. The presence of the S-wave shadow—no S-waves are found in this zone—indicates that the outer core is liquid. **9.** They differ in density, composition, and thickness. Oceanic crust is thin and compact (about 10 km thick) and composed of dense basaltic rocks. Continental crust is between 20 and 60 km thick and composed of less dense granitic rocks. **11.** Wegener's evidence was the "jigsaw puzzle" fit of continents at their margins, similar rock types on separated continents, and data from paleoclimatology and paleontology. **13.** Paleomagnetism revealed that, over the past 500 million years, the position of the magnetic north pole had *apparently* wandered throughout the world. So, either the magnetic poles had migrated through time or the continents had drifted. Because the apparent path of polar movement varied from continent to continent, it was more reasonable that the continents had moved. Later on, seafloor spreading was confirmed by the magnetic reversal recorded in the seafloor. **15.** Hess discovered seafloor spreading. The seafloor is constantly being renewed at midocean ridges located above upwelling convection cells in the mantle. As rising material from the mantle oozes upward, new lithosphere is formed. Old lithosphere is simultaneously destroyed in deep ocean trenches near the edges of continents. **17.** Seafloor spreading initiates the movement of the continents. The ocean floor moves in a conveyor-belt fashion, with new lithosphere formed at a spreading center and older lithosphere pushed from the ridge crest to be recycled back into the mantle at a deep ocean trench. **19.** Gravity and heat cause convection currents. When rock is heated, it rises and expands—making it less dense. Then, closer to the surface, the rock cools and contracts—making it more dense. Convection in the hot mantle occurs because gravity pulls the denser rock downward relative to the less dense rock, which continues to rise. As the less dense rock rises, it takes the place of the sinking dense rock pulled downward by gravity: convection! **21.** A divergent boundary, or midocean spreading ridge, separates the two plates. **23.** A transform boundary is where two plates are neither colliding nor pulling apart, but rather sliding horizontally passed each other. A strike-slip fault "transforms" motion from one ridge segment to another. Because forces are neither tensional nor compressional, there is no creation or destruction of lithosphere. **25.** An anticline folds upward like an arch. Rocks in the core of an anticline are the oldest, and away from the axis they get younger. A syncline folds downward—it sags. Rocks in the core of a syncline are the youngest, and away from the axis they

get older. **27.** Tensional forces create normal faults. Compressional forces create reverse faults. **29.** Most volcanoes are formed near plate boundaries. About 80% of the world's volcanoes are in the region known as the "Ring of Fire." **31.** The pull-push force creates vibrations (waves) in the Slinky. The waves move from one person to the other along the length of the Slinky. Part of the Slinky compresses and part of it stretches as the waves travel back and forth in the direction of propagation. This is like P-waves that compress and expand as they move through Earth's interior. The side-to-side force creates vibrations that move perpendicular to the direction of propagation. This is like an S-wave that moves from side to side as it travels. **33.** The four-straw frame is weak. Adding the fifth straw across the middle gives the frame more support. Placing the fifth straw diagonally from corner to corner reinforces the frame even more. **35.** From distance = speed × time, we have time = distance/speed. To get a numerical value we express km in cm; there are 1000×100 cm in 1 km, so

$$\text{Time} = \frac{5000 \text{ km} \times 1000 \text{ m/km} \times 100 \text{ cm/m}}{2.5 \text{ cm/yr}} = 200 \text{ million years}$$

37. To find the solution (distance), we need to know the time period and the rate of movement. The time period is $2025 - 2005 = 20$ years. We know the time and speed, so we can rearrange the basic equation as distance = rate × time. Thus, the two parts of the fence will be separated by 3.5 cm/yr × 20 yr = 70 cm. **39.** b, a, d, c. **41.** d, b, a, c. **43.** a, b, c. **45.** P-waves travel through both solids and liquids, whereas S-waves travel through solids only. So when S-waves fail to traverse part of Earth's interior, a liquid phase is indicated. By studying the passage of both P- and S-waves through Earth, scientists can identify the solid and liquid layers. **47.** Differences in P- and S-wave propagation through the core are evidence. At the core–mantle boundary, a very pronounced wave shadow develops. P-waves are reflected and refracted at the boundary, but S-waves are only reflected. S-waves cannot travel through liquids, which implies a liquid outer core. The P-waves continue to propagate through the outer core, and at a certain depth they refract and increase in speed as they enter a new medium. The faster-traveling wave indicates a solid inner core. **49.** Although the inner core is very hot, intense pressure from the weight of Earth above prevents the material of the inner core from melting. Because less weight is exerted on the outer core, the pressure is less and so the iron and nickel are liquid. **51.** Part of Earth's mantle is rigid, and part is hot enough to flow as a plastic solid. The crust is embedded in the rigid lithosphere, which floats on the plastic asthenosphere. **53.** The continental crust stands higher because it is composed of buoyant granitic material. Like an iceberg, the thicker it is, the farther it extends into the supporting medium. Because the oceanic crust is thinner, it doesn't extend as deep into the mantle. **55.** The apparent path of polar wandering as determined from North American rocks is different from the path determined from European rocks. If North America and Europe had been stationary, the paths would be the same. **57.** The driving force is compression force due to oceanic–continental convergence. **59.** Mountain ranges are the result of plate convergence. Since plate boundaries are typically long, the mountain ranges that form near them are long. The ranges are relatively narrow because most of the deformation related to plate interaction does not propagate far from the plate boundary. **61.** Metamorphic rock can be found at all three plate boundaries. At convergent boundaries, regional metamorphism involves mechanical deformation and elevated temperatures and pressures. Divergent boundaries have thermally metamorphosed rocks. Transform boundaries have mechanically deformed rocks. The majority of metamorphic rocks are associated with convergent boundaries. **63.** One possible answer is the granitic Sierra Nevada range, which are the batholiths left over from subduction-derived partial melting and magma crystallization. A second is the occurrence of metamorphic rocks both in the Sierra Nevada and near the trench deposits. **65.** The oceans have been around since very early in Earth's history. The present-day ocean basins, however, are not a permanent feature. For example, the present-day Atlantic Ocean began as a tiny rift area between continental lands. **67.** Ocean floors are subducted at convergent plate boundaries, whereas continental crust is not subducted—it remains at Earth's surface. **69.** Apparent polar wandering and magnetic surveys of the ocean floor. Paleomagnetic studies during the 1950s revealed that the position of the magnetic poles had moved around the globe. This apparent polar wandering suggested that it was the continents that moved and not the poles. Support for this idea came from magnetic surveys of the ocean floors that showed alternating stripes of normal and reversed polarity, paralleling either side of the spreading rift areas. The seafloor had been growing when magnetic pole reversals occurred. **71.** The north magnetic pole periodically reverses and becomes the south magnetic pole, and vice versa. Pole reversals and paleomagnetism provide strong evidence for seafloor spreading. When rock forms, magnetic minerals in the rock align with the magnetic field. So when new basalt is extruded at the oceanic ridge it is magnetized according to the existing magnetic field. Magnetic surveys of the ocean floor show alternating normal and reversed polarity, paralleling both sides of the rift area. So Earth's magnetic history is recorded in the spreading ocean floors. Because the dates of pole reversals can be determined by dating ocean-floor rock, the rate of seafloor spreading can also be determined. **73.** The focus is the point where the rock actually breaks. Directly above the focus is the epicenter, the point on Earth's surface where most of an earthquake's energy does its damage. **75.** The Richter scale is more precise. It uses the amplitude of seismic waves recorded on a seismograph. Richter magnitude is determined from the logarithim of the amplitude, which tells the energy released by the earthquake. **77.** Lithosphere is created at spreading centers (divergent boundaries) and destroyed at subduction zones (convergent boundaries). They are in equilibrium if Earth is neither growing nor shrinking, which is believed to be the case. **79.** The ground shakes 10 times more. **81.** The Himalayas are the result of continent–continent collision between the Indian Plate and the Eurasian Plate. The Andes are the result of volcanic eruptions and uplift related to the subduction of the Nazca Plate beneath the South American Plate. **83.** The lithosphere is pushed and pulled as it moves. This creates stress. The upper part of the lithosphere—the crust—responds to stress in two ways: by breaking (faults) and by flowing (folds). **85.** The entire state of Nevada, eastern California, southern Oregon, southern Idaho, and western Utah are greatly affected by normal faulting. **87.** You would consider the ages of the different rock layers. If the rocks in the fold axis are younger than those away from the axis, the fold is a syncline. If the rocks in the fold axis are older than those away from the axis, the fold is an anticline. **89.** At divergent boundaries, basaltic magma is generated by the partial melting of rising mantle rock. As the rock rises, pressure on the rock decreases, lowering the melting point enough for melting to occur. At convergent boundaries, andesitic magma dominates. Water migrating upward from the descending plate lowers the melting point of the mantle rock above the sinking slab, causing partial melting. As the magma rises and/or is impeded by overlying lithosphere, crystallization occurs. Assimilation of surrounding rock and crystallization increase the silica content of the magma, producing andesitic magma and, given enough time, granitic magma. **91.** It has taken about 190 million years for a mere fracture in an ancient continent to turn into the Atlantic Ocean. **93.** No; smaller quakes—foreshocks—often precede the main quake. Likewise, smaller quakes—aftershocks—often follow the main quake. **95.** Earthquakes occur at plate boundaries; about 80% of them are found in the area of the Pacific Rim. People who live in areas with frequent earthquakes are generally better prepared than those in areas where earthquakes are rare. Developed countries, with more resources, tend to have better building construction, which adds to their preparedness and a speedier recovery. Countries with less money (for example, Haiti) do not necessarily have the resources to prepare and are dependent on help from other countries. Politics also affects recovery after an earthquake, and this can lead to heated debate on many levels. No matter where a large earthquake occurs, loss of life and devastation result. **97.** If Earth spins faster, day length shortens. If Earth spins slower, day length increases. Ice, water, and rocks have mass, and as they move around, they exert a change on Earth's mass balance. The movement of mass on and beneath Earth's surface affects Earth's rotational inertia, and by the conservation of angular momentum, Earth's rotational velocity must also change (consider the ice skater analogy). **99.** Mountain ranges, volcanoes, plutonic rocks, metamorphic rocks, and folded and faulted rocks are all explained by plate tectonics. Virtually all geologic processes can be traced back to plate tectonics, although sometimes the link is indirect.

CHAPTER 22

1. Earth's surface area is mostly ocean, so most evaporation and precipitation occur over the oceans. **3.** Porosity is the volume of open space in a soil or rock sample compared to the total volume of solids plus voids. Hydraulic conductivity is the ability of a material to transmit water. **5.** The unsaturated zone is where pore spaces are partly filled with water and partly filled with air. The saturated zone is where pore spaces are filled (saturated) with water. **7.** Artesian systems form where aquifers become sandwiched between low-permeability rocks, forming a confined aquifer. Recharge to confined aquifers occurs only in places where the aquifer has access to percolating rainwater—where the aquifer is unconfined, in the recharge area. Natural recharge areas are always higher than confined portions of the aquifer, so some water is under pressure and can flow out of any opening in the aquifer. **9.** Rainwater chemically reacts with carbon dioxide in the air and soil to produce carbonic acid. This carbonic acid partially dissolves limestone rock and creates unusual erosional features. **11.** Caverns, caves, and sinkholes. **13.** The *gradient* of a stream is the ratio of the vertical drop to the horizontal distance of that drop. Because of gravity, stream velocity tends to be greater where the gradient is steep. **15.** The cross-sectional shape of a stream channel determines the amount of water that is in contact with the channel. The larger the contact area, the greater the friction. The greater the friction, the slower the flow. If the stream channel is rounded and deep, as opposed to flat-bottomed and relatively shallow, the stream will flow faster because there is less contact with the channel. **17.** In chemical weathering, the stream water contains dissolved substances that react with rock material. Hydraulic action is the sheer force of running water, and abrasion is the powerful scouring of rocks and sediments against the channel. **19.** As a stream flows downhill and its gradient becomes gentler, the focus of its energy changes from eroding downward to eroding laterally in a side-to-side motion. A result of this lateral action is the stream's sinuous form. **21.** Faster currents can carry larger particles. A fast-moving, turbulent flow can move and carry a range of particle sizes—from the smallest particles of clay to large pebbles and cobbles. **23.** Glaciers are formed from the accumulation of recrystallized snow. As more snow falls, the pressure compacts and recrystallizes the bottom layers of icy snow into glacial ice. When the ice mass is 50 m thick, it becomes a glacier as it moves under its own weight. **25.** When accumulation exceeds ablation, the glacier advances—it grows. When ablation exceeds accumulation, the glacier retreats—it shrinks. **27.** Cirque, tarn, arête, horn, moraines (lateral, medial, and terminal), hanging valleys, and a characteristic U-shape valley. **29.** A sand mound starts moving downwind as sand grains on the windward slope move up and over the crest of the dune to fall on the leeward slope. Over time, this continuous process moves the entire dune. **31.** The surface of the soap is scratched in the direction it is rubbed. Glacial striations found on land can be used to tell the direction of ice movement.

33. $2.14 + 0.61 + 0.009 + 0.005 = 2.76$

Ice caps and glaciers	$2.14/2.76 = 0.77$ or 77%
Groundwater	$0.61/2.76 = 0.22$ or 22%
Streams, lakes, and rivers	$0.009/2.76 = 3.25 \times 10^{-3}$ or 0.32%
Soil moisture	$0.005/2.76 = 1.81 \times 10^{-3}$ or 0.18%

35. Calculations are not necessary. Flow rate = hydraulic conductivity × cross-sectional area × hydraulic gradient. If the gradient increases 10 times and all else stays the same, the flow rate increases 10 times as well.

37. The first step is to calculate the hydraulic gradient:

$$\text{Hydraulic gradient} = \frac{\text{head change}}{\text{distance}} = \frac{210\text{ m} - 209\text{ m}}{300\text{ m}} = 0.0033333$$

(keep the extra significant figures for an accurate solution). Then

$$\text{Specific discharge} = \text{hydraulic conductivity} \times \text{hydraulic gradient}$$
$$= 150\text{ m/day} \times 0.0033333 = 0.5\text{ m/day}$$

39. b, a, c, d. **41.** b, a, c. **43.** a, b, c, d. **45.** c, a, b, d. **47.** Most rainfall ends up in the oceans. **49.** Water can rise to the level of the water table, until it is pumped. **51.** No; to be groundwater, all open pore spaces must be completely filled with water. In the unsaturated zone, open pore spaces are filled with water *and* air, and the water is called soil moisture. **53.** More water means more discharge, so discharge increases. If the stream channel is unchanged, then discharge and stream speed increase. If the stream channel changes, discharge increases and stream speed may or may not increase. (The increase in speed depends on how big the increase in area is compared to the increased discharge.) **55.** The stream speed also doubles. **57.** Stream gradient, stream discharge, and channel geometry. **59.** Groundwater flow into a stream channel can provide a base flow. At the surface, precipitation and surface runoff contribute to stream flow during and after storms. **61.** As the main stream channel becomes choked with sediment, it takes more energy for water to push through the accumulated sediment than to go around it. So new, smaller channels form off the main channel like branches on a tree. **63.** Ice has the greatest ability to transport sediment particles and can carry the largest loads. Glaciers moving across a landscape loosen and lift up blocks of rock and incorporate them into the ice. They literally pick up everything in their path. As the ice melts, rock debris is left behind and is deposited. **65.** At times in Earth's history, shallow seas covered the continental land, allowing carbonate rocks to be deposited. The shallow seas are now gone, so the carbonate rocks are exposed. **67.** Yes; stream speed can increase without an increase in water volume. Stream speed can increase with an increase in gradient or a decrease in channel width or depth. If stream speed increases, a laminar flow can become turbulent. **69.** Surface water erodes rocks and sediments and transports them from their original locations. Surface water also deposits sediments as a stream's ability to carry sediments declines with speed. **71.** Roche moutonnées are an erosional feature in which the steep side points in the direction of glacial advance. Drumlins are depositional features in which the steep side points in the direction of glacial retreat. **73.** A glacier abrades and picks up everything in its path, so glacial deposits are characteristically composed of unsorted rock fragments in a variety of shapes and sizes. River-deposited rocks tend to be more sorted and more rounded. **75.** No; a stream's sediment load can increase even if the stream speed decreases. The key factor is discharge. For example, if a stream's speed stays the same while the discharge increases, the stream carries more water and so can carry more sediment. Don't confuse sediment *size* with sediment *load*. **77.** The uppermost portion of the glacier, carried along both by basal sliding and by internal plastic deformation, behaves like a rigid, brittle mass that may fracture. Huge, gaping cracks called *crevasses* may develop in this surface ice. **79.** Large, out-of-place boulders provide proof of a glacier's ability to transport heavy loads for great distances. If a bedrock outcrop is found that matches the rock type of an out-of-place boulder, then the distance and direction of glacial transport can be estimated. **81.** Gravity by itself causes mass movement. All forms of moving water are assisted by gravity—groundwater, rivers and streams, and ice. **83.** Polar ice is one possible answer. Desalinization of seawater is another. In many countries (the Caribbean, North Africa, and the Middle East), desalinized water makes up most of the municipal water supply. The combined capacity of these salinization plants is about 16 billion liters of water a day. **85.** It can cause flooding. Increased urbanization means areas of vegetation are replaced by impervious surfaces. Because water cannot soak into the ground, storm water runoff increases. This runoff must be collected by extensive drainage systems that combine curbs, storm sewers, and ditches to carry storm water runoff directly to stream channels. When the volume of storm water runoff exceeds the stream channel's capacity, the result is flooding.

CHAPTER 23

1. Relative dating is the ordering of rocks in sequence by their comparative ages. It doesn't reveal the date when an event occurred but rather its timing relative to other episodes in Earth's past. **3.** The sandstone must be older than the granite because the sandstone had to be there in order for the granite to cut through it. This illustrates the principle of cross-cutting relationships. **5.** Fossils record the evolution of life. Fossil organisms succeed one another in a definite and irreversible order, so they are used to help correlate rocks of similar age in different regions because any time period can be recognized by its unique fossil content. **7.** Areas at high elevations were once much lower and covered by a shallow sea; they have now been uplifted. **9.** From Table 23.1, we find uranium-238, 4.5 billion years; potassium-40, 1.3 billion years; and carbon-14, 5760 years. **11.** Carbon-14. **13.** Earth is approximately 4.5 billion years old. **15.** Stromatolites and other fossils found in rocks are known to be Precambrian in age. **17.** The periods are the Cambrian, Ordovician, Silurian, Devonian, Carboniferous (Mississippian and Pennsylvanian), and Permian. **19.** In the Devonian period (known as the "age of fishes"), fish diversified into many new groups (the sharks and bony fishes are still present today). The lobe-finned fishes are an important link to amphibians, which made their appearance during the late Devonian. **21.** The lobe-finned fishes evolved internal nostrils, which enabled some species to breathe air. Also, the fins were lobed and muscular with jointed appendages that enabled the animals to walk. **23.** Coal deposits are found in some eastern states, such as Pennsylvania and West Virginia. **25.** The Mesozoic era is known as the "age of reptiles." **27.** A worldwide rise in sea level occurred during the Cretaceous period, probably due to the breakup of Pangaea. **29.** The Tertiary period includes the Paleocene, Eocene, Oligocene, Miocene, and Pliocene epochs. The

Quaternary period includes the Pleistocene and Holocene. **31.** From oldest to youngest, the sequence is G, A, B, C, D, I, H, F. **33.** It would take 100 million years. **35.** c, b, d, a. **37.** b, c, d, a. **39.** d, c, a, b. **41.** f, e, d, b, c, a. **43.** The fault is older than the basalt and younger than the sedimentary rock. The sedimentary rock had to be there before the fault in order for the fault to displace it. The reverse argument holds for the basalt. **45.** The oldest rocks on Earth are the Acasta gneisses, dated at 4.03 billion years old. Moon rocks can be no older than the time of the Moon's formation, about 4.5 to 4.3 billion years ago. **47.** The amount of uranium has decreased, and the amount of lead has increased (via radioactive decay). **49.** Geologists used (and still use) the principles of original horizontality, superposition, cross-cutting relationships, inclusions, lateral continuity, and faunal succession. **51.** The half-life of carbon-14 is 5760 years. When a material is older than 50,000 years, the amount of carbon-14 that is left in it is too small to measure, so all we can tell is that the material is older than about 50,000 years. **53.** Yes; the principle of lateral continuity states that sedimentary layers are continuously deposited over large areas until some sort of obstruction, or barrier, limits their deposition. **55.** The Hadean eon—old, hot, hell, oceanless. The Archean eon—rain, seas, banded iron formations, photosynthesis, stromatolites. The Proterozoic eon—Rodinia, glaciation, single-celled life, O_2 and O_3. **57.** This sequence must have been overturned by some structural deformation event, such as mountain building. According to the principles of original horizontality and superposition, the layers could not have been deposited in this order. **59.** During the Archean, Earth cooled to the point where water vapor in the atmosphere condensed to form rain clouds. Rain washed the atmosphere of carbon dioxide, which dissolved in the newly formed oceans. Nitrogen was essentially left behind as the dominant gas in Earth's atmosphere. **61.** It ended up in rocks! Much of the oxygen released by photosynthesizing plants dissolved in the oceans, where it reacted with iron to produce layers of iron oxide minerals on the seafloor. This became the banded iron formations. This process lasted for 700 million years and generated the great quantities of iron ores that are crucial to our industrial age. **63.** Glaciation was the cause. During glaciated periods, water was tied up in glacial ice. When the climate warmed, the glaciers melted and sea level rose. **65.** These land plants occupied low wetland areas. Like wetlands today, Silurian wetlands were productive habitats with considerable biodiversity. **67.** In granite, the date signifies the age of the granite (when the mineral crystallized from magma). In schist, it signifies the age of the metamorphic event.

69. Precambrian—first life; stromatolites, bacteria, algae; soft-bodied animals. Paleozoic—trilobites, shelled animals, first life on land, first fishes, first amphibians, first reptiles; major extinctions in the Ordovician and Permian. Mesozoic—age of reptiles, dominance and diversification of dinosaurs, first mammals, first birds, first flowering plants; major extinction at the end of the Cretaceous (bye-bye dinosaurs!!).

Cenozoic—age of mammals, diversification of mammals, expansion of flora, emergence of humans; extinction of many large mammals.

71. Melting of the polar ice caps can cause a rise in sea level. The breakup of a supercontinent and an increase in heat flow to rift zone areas, resulting in accelerated seafloor spreading, can cause sea level to rise. It is likely that sea level will rise in the future, as it has in the past. **73.** Redistribution of land and water affects the world's climate—the oceanic and atmospheric circulation patterns. Inland areas are more arid and coastal areas monsoonal. With the landmasses all joined together, there is a global reduction of continental area relative to global oceanic area—less land area, less water displacement. A reduction in the amount of coastline lowers sea level and exposes more of the continental shelf, thus endangering shallow marine organism habitats. In the case of Pangaea, sea level was low for a long time: 20–25 million years! This undoubtedly placed a great deal of stress on marine organisms. **75.** Faster seafloor spreading means warmer and hence less ocean crust. The newer, warmer ocean crust "rides" higher, forcing seawater onto the continents. **77.** Global-scale cooling leading to continental-scale glaciers is most likely caused by the right combination of three things: the arrangement of continents around the globe, the amount of sunlight reflected back into space, and the geometry of Earth's rotation on its axis and revolution around the Sun. **79.** The cause is the Milankovitch effect. By definition, we are currently in an ice age because continental-scale glaciers are present on Earth. **81.** The divisions are based on changes in life forms. Eons (hidden life and visible life) and eras (ancient, middle, and recent life) mark very large time spans. Periods mark changes within eras and are considered the most useful time intervals to document change. **83.** During the height of glaciation, as much as one-third of Earth's surface was covered by ice. The heavy ice gouged the land surface. Because these gouges (glacial striations) still scar many rock surfaces, scientists are able to trace the extent of glaciation. **85.** Isolated animal species had to adapt to new environments; this resulted in new species and increased animal diversity. **87.** About 30 million years ago, when the westward-moving North American Plate collided with the Pacific Ridge system (when the San Andreas fault formed), Baja California tore away from the Mexican mainland, forming the Gulf of California. **89.** Among other things, humans have dammed rivers, built irrigation systems, and caused pollution, extinctions, and global warming. **91.** Catastrophic causes are meteorite impacts and comet showers. Natural causes are volcanism, glaciation, variations in sea level (also changes in oxygen or salinity), and global climate change.

CHAPTER 24

1. Water, with its high specific heat capacity, retains heat longer. **3.** The atmosphere was probably much like the gases found in volcanic eruptions of today—about 85% water vapor, 10% carbon dioxide, and 5% nitrogen, by mass. **5.** The continental margins (shelf, slope, and rise), deep ocean basins (abyssal plains and hills), midocean ridges, and deep trenches. **7.** The waves slow down. Incoming waves gain on leading slower waves, and the distance between waves decreases. This concentration of water in a narrower zone produces taller, steeper waves. **9.** Barrier islands form where ridges of sand break the surface of the water and allow vegetation to take hold. Lagoons form between the ocean-side coast and the shore. They are quiet environments because they are sheltered from the battered coast. **11.** The highest tides occur during a spring tide. **13.** Energy from the Sun energizes atoms and molecules and keeps them in motion. **15.** Temperature decreases as one moves up in the troposphere, increases as one moves up in the stratosphere. **17.** The tilt of the Earth is responsible for the seasons. **19.** Earth radiates long-wavelength radiation; the Sun emits short-wavelength radiation. **21.** The unequal heating of Earth's surface causes air motion. **23.** Earth rotates in a counterclockwise direction—west to east. **25.** The Coriolis force deflects winds toward the right in the Northern Hemisphere and toward the left in the Southern Hemisphere. The magnitude of the Coriolis force varies according to the speed of the wind (the faster the speed, the greater the deflection) and the latitude (deflection is greatest at the poles and zero at the equator). **27.** From the horse latitudes, surface air moves back toward the equator. As it moves, it is deflected to the northeast in the Northern Hemisphere and to the southeast in the Southern Hemisphere. The winds are steady, which benefited sailors of yesteryear—hence the name *trade winds*. **29.** Surface friction between water and wind, driven by temperature and pressure differences, set the currents in motion. **31.** Open ended. **33.** The density of the air in the tank is 1.25 kg/m³. The mass of this air is found by multiplying the density by the tank volume: $(1.25 \text{ kg/m}^3)(0.0100 \text{ m}^3) = 0.0125 \text{ kg}$. **35.** b, a, c, d. **37.** b, a, c, d. **39.** c, a, b. **41.** Yes. Unequal heating of Earth's surface is responsible for weather; this is greatly affected by the presence of oceans but is not completely dependent on oceans. Winds and other weather conditions occur on other planets, all without oceans. And on Earth, far inland away from bodies of water, weather conditions such as Chinook winds and tornadoes occur. **43.** Earth absorbs short-wavelength radiation from the Sun and reradiates it as long-wavelength terrestrial radiation. Incoming short-wavelength solar radiation easily penetrates the atmosphere to reach and warm Earth's surface, but all the outgoing long-wavelength terrestrial radiation cannot penetrate the atmosphere to escape into space. Instead, atmospheric gases (mainly water vapor and carbon dioxide) absorb the long-wavelength terrestrial radiation, keeping Earth's surface warmer than it would be if the atmosphere were not present. **45.** The air in a deep mine is denser because there is a greater mass of air over a deep mine than at sea level. This greater mass causes the air pressure to be higher, which in turn creates denser air. (Pressure is directly proportional to density.) **47.** The number of hours of sunlight (and solar energy) depends on the incidence of the Sun's rays on Earth's surface. In equatorial regions, the Sun's rays are concentrated as they strike perpendicular to Earth's surface. These regions receive twice as much solar energy as polar regions. In polar regions, the Sun's rays strike at an angle, so solar energy is spread out and dispersed and these regions are cool. Earth's tilt causes polar regions to have nearly 24 hours of sunlight (albeit dispersed sunlight) during half the year and nearly 24 hours of darkness during the other half. **49.** Cooling by radiation prevents Earth's temperature

from rising indefinitely. **51.** The relevant factor is not air temperature but solar radiation. At high elevations there is less atmosphere above to filter out ultraviolet rays, so climbers are exposed to more high-energy radiation. **53.** Cell-like circulation patterns set up by atmospheric temperature and pressure differences are responsible for the redistribution of heat across Earth's surface and global winds. Because winds set surface waters in motion, atmospheric circulation and oceanic circulation are interrelated. What affects one affects the other. Ocean currents do not follow the wind pattern exactly, however; they spiral in a circular whirl pattern—a gyre. In the Northern Hemisphere, as prevailing winds blow clockwise and outward from a subtropical high, the ocean currents move in a more or less circular, but clockwise, pattern. The Gulf Stream, a warm-water current in the North Atlantic Ocean, is actually part of a huge gyre. **55.** The jet streams are high-speed winds in the upper troposphere. They play an essential role in the global transfer of thermal energy from the equator to the poles. The *polar jet* and the *subtropical jet* form in response to temperature and pressure gradients. The polar jet results from a temperature gradient at the polar front, where cool polar air meets warm subtropical air. The subtropical jet is generated as warm air is carried from the equator toward the poles, producing a sharp temperature gradient along the subtropical front. **57.** Large icebergs come from, or calve off of, glaciers on land. **59.** Like the circulation of atmospheric currents, oceanic currents are driven by the heat of the Sun. **61.** When evaporation exceeds precipitation, salinity increases. In ocean water it is the water that evaporates; the salt is left behind. When precipitation exceeds evaporation, salinity decreases as a new influx of fresh water dilutes the salt solution. **63.** It sinks until it reaches a point of equilibrium—the point where it encounters either water of the same density or the seafloor, whichever comes first. **65.** Headlands stick out from the rest of the shoreline, so they receive the full impact of waves. **67.** At times in Earth's history, shallow seas covered the continental land, allowing for the deposition of carbonate rocks. Now that the shallow seas are gone, the carbonate rocks are exposed. **69.** As evaporation occurs over the ocean surface, only the water evaporates; the salts are left behind, thus making the seawater saltier. Although most of the salt is left behind, minute salt particles in the ocean spray can act as condensation nuclei, which aid in the formation of water vapor droplets. The amount of salt particles, however, is so minute that precipitation is essentially pure fresh water. **71.** When seawater in polar regions freezes, only the water freezes and the salt is left behind. The seawater that does not freeze experiences an increase in salinity, which in turn brings about an increase in density. **73.** Although the salinity varies from one part of the ocean to another, the overall composition of seawater is fairly uniform—about 96.5% water and 3.5% salt. With greater amounts of solar energy at the tropics, one would expect evaporation to exceed precipitation and cause an increase in salinity. Although this is a reasonable assumption, in reality evaporation and precipitation tend to pretty much balance. **75.** The water level remains the same when the ice melts. Water expands when it turns to ice, which is why part of it sticks above the surface. When it melts, it shrinks back down to its original size, which is why the water level doesn't change. So, when floating chunks of ice in the Great Lakes melt, the water level of the lakes doesn't change. **77.** No; tides are caused by differences in gravitational pulls. **79.** Spring tides are the lowest tides as well as the highest tides. So the spring tide cycle consists of higher-than-average high tides followed by lower-than-average low tides (the best time for digging clams). **81.** Friction is a primary force that sets surface waters in motion. If distances are short, the surface waters move in the same direction as the wind. For longer distances, the deflective Coriolis force is influenced by Ekman transport, which causes water to spiral in a gyre. The circular motion is clockwise in the Northern Hemisphere and counterclockwise in the Southern Hemisphere. **83.** Carbon dioxide dissolves in the ocean. **85.** The source is nitrogen and oxygen ions (atoms stripped of electrons). **87.** High-speed ions ejected from the Sun stir up the ionosphere. **89.** Cold water is denser than warm water. The greater the salinity, the greater the density. **91.** When approaching the shoreline at an angle, waves are bent as they touch bottom, which slows part of the wave. The nearshore end of the wave feels the bottom and slows, while the speed of the other end of the wave remains unchanged because it is still in deeper water. The difference in wave speed causes the wave to turn. **93.** Moving water is deflected by the Coriolis force—to the right in the Northern Hemisphere and to the left in the Southern Hemisphere. So a north wind, which blows north to south in the Northern Hemisphere, creates a surface current that moves to the west-southwest. In the Southern Hemisphere, the surface current moves to the east-southeast.

CHAPTER 25

1. Humidity is the mass of water vapor per volume of air. Relative humidity is the ratio of the amount of water vapor currently in the air to the maximum amount that could be in the air at a given temperature. **3.** The relative humidity increases because air is approaching its saturation point. **5.** The water vapor condenses. **7.** Condensation occurs more readily at low temperatures because slower-moving molecules can condense. **9.** Condensation is the change of phase from vapor to water. Precipitation occurs when the size of condensed drops grows to fall as rain, sleet, or snow. **11.** The air gets cooler due to adiabatic expansion. **13.** When heat is added to an air mass, its temperature, its pressure, or both increase. Heat can be added to air by solar radiation, by moisture condensation, or by contact with warm ground. **15.** In a temperature inversion, upper regions of air are warmer than lower regions. Inversions occur in the Los Angeles basin and the Denver basin. **17.** As warm, moist air rises, it cools and becomes less able to accommodate water vapor. Clouds form as the water vapor condenses into tiny droplets. **19.** (a) Middle clouds, (b) high clouds, (c) low clouds, (d) low clouds. **21.** These clouds are characteristic of unstable air. **23.** Earth's surface is heated unequally, with some areas heating up more quickly than others. Air that is in contact with these surface "hot spots" rises, expands, and cools. This rising of air is accompanied by the sinking of cooler air aloft—convectional lifting. When the rising air cools close to the air's saturation temperature, the condensing moisture forms a cumulus cloud. Air within the cumulus cloud moves in a cycle: Warm air rises and cool air descends. **25.** A rain shadow occurs on the leeward side. As an air mass moves down the leeward slope, it warms. The descending air is dry because most of its moisture was removed in the form of clouds and precipitation on the windward (upslope) side of the mountain. **27.** Convectional lifting, orographic lifting, and frontal lifting. **29.** As particles of precipitation grow larger and heavier within a cloud, they begin to fall as rain, which drags some of the cool air along with it to create a downdraft. **31.** Open ended.

33. $\dfrac{\text{Amount of water in air}}{\text{Maximum amount of water vapor in air at } 50°F}$

$= \dfrac{\text{amount of water in air}}{9 \text{ g/m}^3} = 0.4$

Amount of water in air $= (9 \text{ g/m}^3)(0.4) = 3.6 \text{ g/m}^3$. So the mass of water vapor in 1 m^3 of air is 3.6 g.

35. c, a, b. **37.** b, a, c, d, e. **39.** a, c, d, b. **41.** Weather is the state of the atmosphere with respect to temperature, moisture content, and atmospheric stability or instability at any given place and time. Climate is the consistent behavior of weather over time. **43.** As moist air is lifted upslope against a mountain, it cools adiabatically. As rising air cools, its capacity for holding water vapor decreases, which increases the relative humidity of the rising air. If the air cools to its dew point, the water vapor condenses and a cloud forms. Stable air forced upward forms stratus-type clouds; unstable air tends to form cumulus-type clouds. **45.** At night the ground is often cooler than the surrounding air. As air makes contact with these cold surfaces, it cools and its ability to hold water vapor decreases. Water vapor condenses onto the nearest available surface. **47.** Warm, dry air holds more water vapor than cold, dry air. Wind keeps the air above the glass dry by blowing away moist air formed from evaporation. Hence, a glass of water evaporates more quickly on a windy, warm, dry summer day. **49.** Your glasses fog up because of the change from cold to warm. As you leave your air-conditioned room, the warm air outside comes in contact with the cold surface of the sunglasses. The cold surface cools the air by conduction, and the warm air's ability to hold water vapor decreases. As the air cools to its dew point, water vapor condenses on the sunglasses. **51.** Yes! This is adiabatic expansion or compression. **53.** The rising air cools, and if the air is humid, clouds form and precipitation occurs. As the air mass moves down the leeward slope, it warms. This descending air is dry because most of its moisture was removed in the form of clouds and precipitation on the windward (upslope) side of the mountain. **55.** The clouds are nimbostratus. Although cumulonimbus clouds are associated with precipitation, they do not produce an overcast sky. **57.** The large spaces are caused by warm air rising and cool air sinking. As cool air sinks, the expansion of warm air beneath it is inhibited, so we see single cumulus clouds with a great deal of blue sky between them. **59.** Stable air forced to rise spreads out horizontally, so

we see stratus-type clouds: cirrostratus, altostratus, nimbostratus, or stratus. **61.** Saturation vapor pressure is the upper limit of humidity. At this level humidity cannot increase because the air is saturated. **63.** Temperature, air pressure, humidity, types of clouds, level of precipitation, wind direction, and wind speed. **65.** In a warm front, warm air slides upward over a wedge of cooler air near the ground, producing stratus and nimbostratus clouds and drizzly rain showers. In a cold front, warm, moist air is forced upward more quickly by the advancing cold air. This abrupt lifting produces cumulonimbus clouds, which are often accompanied by heavy showers, lightning, thunder, and hail. **67.** A change in atmospheric stability can cause this change. Altostratus clouds are a stable, layered type of cloud that often covers the sky for hundreds of kilometers. If the top of the cloud cools as the bottom warms, the cloud becomes unstable to the point that small convection currents develop within the cloud. The up-and-down motions make the cloud develop a puffy appearance—the transformation into an altocumulus cloud. **69.** When two air masses meet, differences in temperature, moisture, and pressure can cause one air mass to ride over the other, forming clouds and precipitation. **71.** The collision–coalescence process needs to occur: (1) condensation nuclei, (2) water vapor to cloud droplets, (3) updrafts, (4) droplet bombardment, and (5) droplet growth. A cloud must have sufficient vertical development; otherwise, there are not enough droplet collisions for individual droplets to grow. Thicker clouds have a higher chance of producing rain because the droplets have more time and space to coalesce into drops that are heavy enough to fall. Raindrops form when the condensation rate exceeds the evaporation rate. **73.** As an air mass is pushed upward over a mountain range, the rising air expands and cools; if it is humid, clouds form. As the air mass moves down the leeward slope, it warms as it is compressed. This descending air is dry because most of its moisture was removed in the form of clouds and precipitation on the windward side of the mountain. Because the dry leeward sides of mountain ranges are sheltered from rain and moisture, rain-shadow deserts often form. **75.** Tornadoes evolve from thunderstorms that form in regions of strong vertical wind shear. Rapidly increasing wind speed and changing wind direction with height cause the updraft within the storm to rotate. Rotation begins in the middle of the thunderstorm and then works its way downward. As air rushes into the low-pressure vortex, it expands, cools, and condenses into a funnel cloud. As the air beneath the funnel is drawn into the core, the funnel cloud descends toward the surface. When the funnel cloud reaches the ground surface, it is called a tornado. **77.** They contain more moisture. **79.** The highest frequency is in Texas through Oklahoma, Kansas, and Missouri, a zone known as Tornado Alley. **81.** High, thin cirrus clouds are normally warming clouds: They let sunlight through but are good at trapping terrestrial heat. Low, thick stratus clouds, on the other hand, are typically cooling clouds: They tend to be more efficient at blocking and reflecting sunlight than they are at trapping radiated heat. **83.** Curb the source of soot emissions. The most immediate, effective, and low-cost way to reduce soot emissions is to put particle traps on vehicles, diesel trucks, buses, and construction equipment. Particle traps filter out soot particles from exhaust fumes. New green technology is another fix—converting vehicles to run on clean, renewable electric power. Providing electricity (for home heating and cooking) to rural developing areas is effective in reducing the burning of biofuels.

CHAPTER 26

1. There are eight known planets. **3.** The outer planets are gaseous and much larger than the inner planets, which are solid. **5.** Accretion of the planets began before the Sun ignited. **7.** Sunspots are relatively cool regions on the solar surface that move with the Sun's rotation and are created by strong magnetic fields. **9.** Equatorial regions spin faster than regions at higher latitudes. **11.** Mercury is very close to the Sun and rotates slowly on its axis. **13.** Blue oceans dominate Earth's surface. **15.** The dry ocean beds and channels cut from water on Mars indicate a wetter past. **17.** The inner rings move faster, like the greater speed of inner planets or any close-orbiting satellites. **19.** Neptune contains greater amounts of methane and ammonia. **21.** The Sun is directly in back of you. **23.** Eclipses occur only when the plane of the Moon's orbit intersects the plane of Earth's orbit about the Sun, which seldom happens. **25.** The asteroid belt is between the orbits of Mars and Jupiter. **27.** The Kuiper belt is a disk-shaped region beyond Neptune that is populated by many icy bodies and considered the source of short-period comets. **29.** A falling star is a meteor visible in the sky as it burns in the atmosphere. **31.** Open ended. **33.** Open ended. **35.** Toward the equator the plane of our solar system appears perpendicular to the horizon. When viewed from the North Pole, it appears more parallel to the horizon.   **37.** Use the formula speed = distance/time and solve for time; then convert the units from seconds to minutes:

$$300,000 \text{ km/s} = \frac{150,000,000 \text{ km}}{\text{time}}$$

$$\text{Time} = \frac{150,000,000 \text{ km}}{300,000 \text{ km/s}} = 500 \text{ s}$$

$$(500 \text{ s})(1 \text{ min}/60 \text{ s}) = 8.3 \text{ min}$$

39. The diameter of our solar system including the Oort cloud is twice 50,000 AU, or about 100,000 AU. Use the given conversion factor to convert to units of light-years:

$$(100,000 \text{ AU}) \frac{1 \text{ light-year}}{63,000 \text{ AU}} = 1.6 \text{ light-years}$$

41. From Problem 39 you are given that 63,000 AU equal about 1.0 light-year. Use this relationship to calculate the distance to Alpha Centauri in units of AU:

$$(4.4 \text{ light-years})\left(\frac{63,000 \text{ AU}}{1.0 \text{ light-year}}\right) = 277,000 \text{ AU}$$

If, according to the analogy, 1 AU corresponds to 110 m, then 277,000 AU corresponds to (277,000 AU)(110 m/1 AU) = 30.5 million m. Divide by 1000 to get about 30,000 km. For perspective, this is about a tenth of the way to the Moon, which is about 300,000 km distant.
43. b, c, a. Mars has two small moons, Earth has one moon, and Venus has no moon, which is not specifically mentioned in Section 26.3.   **45.** c, b, a. The more massive a planet, the greater the weight of material pushing downward, which increases the internal pressure.   **47.** b, a, c.   **49.** As a nebula contracts and spins faster, it flattens out into a disk.   **51.** The gas has more surface area in the disk shape, which allows it to radiate more energy.   **53.** The Sun would set in the east and rise in the west. To understand why, you need to complete the preceding exercise. As you read this answer and act out your movement around the solar beach ball, envision writing "New York City" on your left arm and "Los Angeles" on your right arm. You will see that after 90° the Sun is setting over New York City, which is east of Los Angeles. So the revolution of Earth causes the Sun to appear to move backward across the sky. Our rate of rotation is much faster, so the Sun appears to move toward the west. The eastward-moving effect of revolution, however, is enough to push the Sun eastward by 3 min 57 s every day. One Earth rotation actually takes 24 h 3 min 57 s. If we subtract the eastward movement of the Sun, the solar day we experience becomes 24 h. So it takes Earth longer than a day to spin around once!   **55.** Winds are caused by the unequal heating of the surface and therefore the atmosphere.   **57.** No, because the water from Venus has already been lost.   **59.** The jovian planets are large, gaseous, low-density worlds with rings. The terrestrial planets are rocky and have no rings.   **61.** Adding more mass means stronger gravitational forces that compress the volume. A more massive object, therefore, can have a smaller volume. In Chapter 27 we'll explore black holes, which are very, very massive yet quite small.   **63.** The impacting object must have hit the young Earth not dead-on but askew, which gave the Earth a fast rotation.   **65.** The gravitational pull at the Moon's surface is too weak; the escape velocity at the Moon's surface is less than the speeds that molecules of gas would have at regular Moon temperatures, so any gases on the Moon escape.   **67.** In both photos the Sun is off to the right by just a little bit such that the Sun is almost behind the two spheres—one sphere being the Moon and the other a Ping-Pong ball. If the photos were taken on the same day (or one month apart), then the Ping-Pong ball must have been held up to the sky aimed right at the Moon so that it overlapped with the Moon. If the photos were taken on different days, say a week apart, then the Ping-Pong ball must have been held up at the same position relative to the Sun, but away from the Moon, which would have moved toward the east where it would be a half moon. Do yourself a favor and hold a Ping-Pong ball up to the sunlight to see how the position of the sphere determines its phase.   **69.** An observer on the back side of the Moon would never see Earth. To see Earth, the observer would need to travel to the front side of the Moon. As she did so, she would finally reach a point where she would see Earth coming up from the horizon. She is midway between the back and front sides. If she stopped at this point for a picnic, would Earth continue to rise? No, it wouldn't. In fact, on the Moon, as long as you don't move about, Earth remains in the

same place in the sky for the same reason we see only one face of the Moon. When astronauts walked on the Moon, they could count on Earth being in the same place in the sky for their entire stay. What does change, however, is Earth's phase, which takes a month to cycle from full Earth to new Earth and back to full Earth.   **71.** A lunar eclipse occurs when all three are aligned, with Earth between the Sun and the Moon.   **73.** It would be nighttime because you would be directly between Earth and the Sun. Earth would be so bright, however, that you could clearly see your shadow. From Earth people would look up and see a dark new Moon. But it would be daytime on Earth, and the sky would be so bright and the new Moon so dim that people wouldn't see a thing. Unless, of course, it was during a spectacular solar eclipse.   **75.** The Sun would always appear in the same location in the sky. One side of Earth would have constant light, and the other side would have constant dark.   **77.** A lunar eclipse can be seen from the whole hemisphere of Earth that faces the Moon. But a solar eclipse can be seen from only a small part of the hemisphere that faces the Sun at that time.   **79.** C; when close to the Sun, the Moon is necessarily in a crescent phase.   **81.** The sailboat is sailing close to the equator, as evidenced by the perpendicular orientation of the crescent moon.   **83.** Comets are icy bodies that orbit the Sun. When they get close to the Sun, they lose volatile compounds, such as water, which escape and appear as the comet's tail. Pluto is an ice body. If it were somehow knocked off its orbit (unlikely because it's so massive), then it too would show a tail as it came closer to the Sun. Because of its great mass, it would be a most spectacular and potentially dangerous tail for those viewing from Earth.   **85.** Many meteorites are embedded in ice, which indicates that they came from above. On regular ground, they are not so obvious.   **87.** There would be spectacular meteor showers high in the atmosphere.   **89.** If Earth had no moon, then Earth's rotation would be much slower and our planet might not be amenable to life. Assuming the same habitability, however, we can suggest how human civilization might be different by considering how the Moon has influenced our history and worldview. For starters, the year would likely not be divided into 12 months. Consider also the role our Moon has played in the timing of the seasons, navigation, nighttime activities, religions, and our perspective of our place in the universe.

CHAPTER 27

1. Constellations are visible groups of stars in the nighttime sky.   **3.** This motion of the stars is due to the rotation of Earth.   **5.** A blue star is hotter than a red star.   **7.** An H–R diagram is a plot of intrinsic brightness versus surface temperature for stars. When plotted, stars' positions take the form of a main sequence for average stars, with exotic stars above or below the main sequence.   **9.** Our Sun is along the middle of the main sequence.   **11.** The process is the ignition of nuclear fuel and subsequent thermonuclear fusion.   **13.** Gravitational forces act on a star.   **15.** High-mass stars have shorter lifetimes than low-mass stars.   **17.** A neutron star, enormously dense, is what remains after a supernova.   **19.** A black hole is a collapsed supergiant.   **21.** The masses are the same.   **23.** Anything can pass into a black hole through its event horizon. Once it passes beyond this mathematical boundary, however, there is no possible return.   **25.** Some collisions form new stars.   **27.** The quasar is brighter!   **29.** The Local Group is a relatively small cluster situated between two larger clusters, the Virgo and Eridanus clusters.   **31.** The star has moved westward. When the Sun rises, the star will have already set and is obscured by the ground. After 24 hours it will be slightly more to the west because of the revolution of Earth around the Sun.   **33.** Open ended.

35. $\text{Time} = \dfrac{\text{distance}}{\text{speed}} = \dfrac{8 \text{ light-years}}{2000 \text{ km/h}}$

$= \dfrac{(8)(9.46 \times 10^{12} \text{ km})}{2 \times 10^3 \text{ km/h}} = 3.8 \times 10^{10} \text{ h}$

Convert the units from hours to years:

$$(3.8 \times 10^{10} \text{ h})\left(\frac{1 \text{ day}}{24 \text{ h}}\right)\left(\frac{1 \text{ yr}}{365 \text{ days}}\right) = 4.3 \times 10^6 \text{ yr}$$

37. Surface area density = $(1.00 \times 10^{11} \text{ stars})/(7.85 \times 10^9 \text{ light-years})^2$ = 12.7 stars/light-years^2 **39.** Yes. The stars within a galaxy are spaced so far apart that the collision of two galaxies is analogous to the collision of two swarms of mosquitoes—it is possible for the swarms to pass right through each other with no mosquito–mosquito collisions. The stars within a galaxy, however, are more concentrated toward the center, so if the galactic collision were dead-center, then more stellar collisions would occur. The combination of the cores of the two galaxies would most likely result in an active galactic nucleus. Generally, when two galaxies collide, they sideswipe each other. They pass through each other, but then gravity tugs them back into a single irregular galaxy, which eventually transforms into an elliptical galaxy. As the two galaxies merge, the interstellar gases and dust mix. This allows for the creation of many new stars, which is why most starburst galaxies are galaxies in the process of colliding. **41.** c, b, a. The closer the celestial object, the faster the appearance of the intrinsic motion. By analogy, imagine you're standing next to a railroad track watching a train coming toward you at a constant 100 mph. While still far away, the train may not appear to be moving very fast. The closer it gets to you, however, the more apparent its intrinsic motion. **43.** c, a, b. A star inflates to a red giant before it collapses into a white dwarf, which has the potential to form a nova when accompanied by a binary star. **45.** b, c, a. **47.** Carlyle didn't know that the stars above his head were only a small portion of a much larger conglomeration of stars called a galaxy. Nor did he know that our galaxy is just one of billions and billions. What would he have given to have access to the science textbooks of the 21st century? Don't take our modern-day understanding of the universe for granted. Likewise, have respect for all that we will surely be learning in the future. **49.** There is no atmosphere on the Moon, so the sky appears black and even dim stars are visible. **51.** Both near and faraway stars appear as if on the inner surface of one great sphere, with us at the center. Two stars that appear very close together are on the same line of sight, but they may actually be an enormous distance apart and would not appear close together when viewed from another location. **53.** Celestial objects generally appear to move westward across the sky. This motion, called diurnal motion, is due to Earth's spin, which is eastward. Because of diurnal motion, the Moon appears to travel westward. The Moon, however, orbits Earth in an eastward direction, which explains why the Moon always appears east of where it was 24 hours earlier. So, because of the Moon's eastward orbit, the Moon takes a bit longer than the Sun to travel from horizon to horizon. **55.** Within the celestial sphere, Polaris is closely aligned with the North Pole, which marks the axis of Earth's spin. Over the course of a night the other stars of the Big Dipper are seen to rotate around this axis, which is diurnal motion. Interestingly, a thousand years from now Polaris will have drifted from this unique position due to its intrinsic motion, which is its motion relative to other stars as it orbits around the Milky Way galaxy. **57.** When the outward and inward forces are equal, they determine the size of the star. **59.** A star is born upon the ignition of thermonuclear fusion within the core of a large mass of gasses compressed by gravity. A star "dies" when it is no longer able to burn thermonuclear fuel. **61.** The Sun will expand into a red giant before ultimately becoming a white dwarf. **63.** The nuclei of atoms that make up our bodies were once parts of stars. All nuclei beyond iron in atomic number were in fact manufactured in supernovae. **65.** Since all the heavy elements are manufactured in supernovae, the newer the star, the greater percentage of heavy elements available for its construction. Very old stars were made when heavy elements were less abundant. **67.** Stars with fewer heavier elements formed at an earlier time than the Sun. **69.** Thermonuclear fusion reactions produce an outward pressure that counteracts the inward pressure that would lead to collapse due to gravity. **71.** Thermonuclear fusion is caused by gravitational pressure, which squashes hydrogen nuclei together. Gravitational pressures in the outer layers are insufficient to produce fusion. **73.** One reason supermassive stars are rare is that the more massive a star, the quicker it burns. Supermassive stars, therefore, burn out in short periods of time, such as a few million years. If there were once many supermassive stars, they have long since died out. A second reason is that the matter within a typical galaxy just isn't "clumpy" enough to allow for the frequent formation of supermassive stars. So supermassive stars are born in smaller numbers to begin with. **75.** Bigger stars live shorter lives and collapse more energetically when they burn out. **77.** Blue stars are hottest, then white stars, and red stars are coolest. **79.** As discussed in Chapter 13, elements heavier than iron require the net input of energy in order to be created. Elements heavier than iron, therefore, are created within the rare moments of a supernova blast when there is abundant energy to allow the endothermic (energy requiring) fusion of iron into heavier elements. All the elements lighter than iron are formed through the more common stellar process of exothermic thermonuclear fusion. **81.** The radius decreases as the mass of the hole increases because the increased mass, and hence gravity, makes it harder for light to escape. **83.** The Sun does not have sufficient mass to supernova or to become a black hole. **85.** A quasar releases its energy from its active galactic nucleus. **87.** *Astronomy* is a science dedicated to the study of celestial objects. The mission of astronomy is to learn about the nature and origins of these objects so that we may better understand the natural universe in which we live. *Astrology* is an ancient pseudoscience founded in the belief that the positions of celestial objects influence our physical, mental, and social well-being. Astronomy relies on discoveries made using advanced technologies, such as space telescopes. As a field, it has matured greatly over the past hundred years, but it is still in a golden age of making new and astounding discoveries. Astrology, by contrast, relies on non-confirmable anecdotal evidence and, though very popular, has not changed significantly over the past hundred years. **89.** From our point of view outside of the black hole, the singularity does not exist at present, only in the infinite future and therefore outside the realm of our observable universe.

CHAPTER 28

1. Space is in the universe. The Big Bang marked not only the beginning of time but also the beginning of space. **3.** The luminosity of a star is the amount of energy it puts out per second. The brightness of the star diminishes with distance. **5.** The average temperature is about 2.73 K. **7.** Just prior to cosmic inflation, the material of the universe was compact enough so that differing temperatures could equilibrate. **9.** Its apparent curvature would become negligible. This analogy is used to explain how inflation diminished the apparent curvature of the observable universe. **11.** 1915. **13.** The ball appears to fall to the floor. **15.** The path of the starlight is bent by the Sun's distortion of spacetime. **17.** We can detect dark matter by its gravitational effects. **19.** A star should orbit faster, but it doesn't, which is evidence that our galaxy is surrounded by a massive halo we cannot detect with our telescopes. **21.** Einstein first believed, as most people did during his time, that the universe was static. **23.** Einstein failed to predict the dynamic nature of the universe. **25.** The Wilkinson Microwave Anisotropy Probe, which provided a high resolution of cosmic background radiation. **27.** Dark matter is about six times more abundant than ordinary matter. **29.** The Big Rip scenario assumes that dark energy will grow stronger over time. **31.** Open ended. **33.** Open ended. **35.** b, a, c. **37.** c, b, a. Time slows down within a greater gravitational field. So clocks run slower on the surface of Earth than they do on the surface of Mercury. And clocks run slower on Mercury than they do on the Moon, which has the least mass of these three celestial bodies. **39.** The light waves are stretched, which means the frequency of the light is lowered. **41.** Space does not expand locally within our galaxy or between neighboring galaxies. Rather the expansion of space is seen only on the widest of scales, which is between superclusters of galaxies. Here on Earth gravity keeps space nice and tight. So, apologize to the police officer and pay for the speeding ticket. **43.** As the universe continues to expand, its average temperature should decrease. **45.** If the initial universe remained hotter for a longer time, then more hydrogen would have been able to fuse into helium and the proportion of helium in the universe would be greater. **47.** Helium was produced in the few minutes after the Big Bang. **49.** False. Most of the helium on Earth arises from the radioactive decay of heavy isotopes, as was discussed in Chapter 13. These heavy isotopes, however, were forged within past supernovae. The primary fuel for these stars that underwent the supernova was hydrogen, which originated with the Big Bang. So the helium that pops out of a balloon is better described as an *indirect* remnant of the Big Bang. **51.** The cosmic background radiation is radiation that came from the universe when it was only about 380,000 years old. At that time the average temperature of the universe was rather homogeneous. It should be emphasized that 2.73 K is the average *background* temperature of the universe. **53.** Seeing the same thing in two opposite directions might imply that the curvature of the universe was "closed," much like Earth's surface—if you could see around the world, you might be able to see the back of your head way off in the distance! Astronomers have been looking for such repeating patterns but none have been found, which supports the notion that the universe is "open," expanding in infinite directions. **55.** Mass has the effect of curving spacetime. We detect this curvature as gravity. **57.** It doesn't matter how fast the jet is moving. As long as you and the jet are moving uniformly, catching the ball will be as easy as when the jet is parked at the gate. **59.** A clock runs slower at the bottom of a skyscraper than at the top. According to the principle of equivalence, the same should apply in an accelerating spaceship. The clock at the back end of the ship by the engines, which corresponds to the bottom floor of the skyscraper, will run slower than the clock at the front end, which corresponds to the top of the skyscraper. **61.** Yes, by centripetal acceleration as occurs in a rotating system. In a uniformly rotating giant wheel, inhabitants could feel normal *g* on the inner rim, provided the rotational speed was correct for the radial distance of the wheel. **63.** The photons of light are climbing against the gravitational field and losing energy. Less energy means lower frequency. Your friend sees the light red-shifted. The frequency she receives is lower than the frequency you sent. **65.** Mercury follows an elliptical path in its orbit about the Sun. On one side of the ellipse, called the *perihelion*, the planet is closer to the Sun than on the other, called the *aphelion*. During the perihelion, the gravitational attraction between the Sun and Mercury is strongest because that is when the distance between the Sun and Mercury is the smallest. **67.** If the planets all orbited the Sun at about the same speed, we might conclude that most of the mass of the solar system is not concentrated in the center of the solar system. Some unknown form of matter might lurk just outside our solar system, perhaps within the Oort cloud. **69.** Dark matter has mass and therefore has the effect of curving spacetime, which we perceive as gravity. Dark matter stays clumped together by way of this "force" of gravity. (In this answer we avoid using the term *force* to describe gravity. Nonetheless, gravity does in fact behave as though it is a force, so even scientists who are comfortable with general relativity use this term in their discussions of gravity.) **71.** Dark matter is invisible to the electromagnetic force, which is what holds us up off the ground. If any dark matter did come to Earth, it would fall right through the surface and probably accumulate in Earth's core. **73.** To account for the flatness of the universe, there needs to be large amounts of mass-energy to make up for the deficiency of dark and ordinary matter. An *energy* that causes an expanion of the universe fits the bill. If dark energy turns out to be the "antigravity," remember that gravity itself is best described not as a "force" but as a warping of spacetime. So, we can speculate that while mass warps spacetime in one direction, dark energy warps it in the other. **75.** The darkness of the clear night sky was a mystery to astronomers. Many early astronomers believed the universe consisted of a finite number of stars enclosed within some sort of huge black wall. As late as the beginning of the 20th century, many astronomers believed the universe consisted of a finite number of stars bunched together within an infinite space. Today, Big Bang theory tells us that the universe is changing. There are zillions of galaxies we don't see because the universe is not old enough for their light to have reached us. **77.** The Big Bang is an event; a black hole is an object. Matter and energy are greatly compacted within both. The Big Bang, however, contained the entire universe. To talk about what existed outside the Big Bang is meaningless. The Big Bang was not like a firecracker that someone would have been able to watch explode. There was no time or space outside of this primordial event. A black hole, though infinitely dense, is an object found in our universe. We can observe its gravitational effects on objects in the space surrounding it.

Glossary

Ablation The amount of ice lost, and the process of losing ice, from a glacier.

Absolute zero The temperature at which no further energy can be taken from a system.

Acceleration The rate at which velocity changes with time; the change in velocity may be in magnitude or direction or both, usually measured in m/s^2.

Accumulation The amount of snow added, and the process of adding snow, to a glacier.

Acid A substance that donates hydrogen ions.

Acidic Description of a solution in which the hydronium-ion concentration is higher than the hydroxide-ion concentration.

Activation energy The minimum energy required in order for a chemical reaction to proceed.

Active galactic nucleus The central region of a galaxy in which matter is falling into a supermassive black hole and emitting huge amounts of energy.

Addition polymer A polymer formed by the joining together of monomer units with no atoms being lost as the polymer forms.

Additive primary colors The three colors—red, green, and blue—that, when mixed in certain proportions, can produce any color in the spectrum.

Adiabatic A term that describes temperature change in the absence of heat transfer; expanding air cools and compressing air warms.

Air resistance The force of friction acting on an object due to its motion in air.

Alcohol An organic molecule that contains a hydroxyl group bonded to a saturated carbon.

Aldehyde An organic molecule containing a carbonyl group, the carbon of which is bonded either to one carbon atom and one hydrogen atom or to two hydrogen atoms.

Alloy A mixture of two or more metallic elements.

Alpha particle A subatomic particle consisting of the combination of two protons and two neutrons ejected by a radioactive nucleus. The composition of an alpha particle is the same as that of the nucleus of a helium atom.

Alternating current (ac) An electric current that repeatedly reverses its direction; the electric charges vibrate about relatively fixed points. In the United States, the vibrational rate is 60 Hz.

Amide An organic molecule containing a carbonyl group, the carbon of which is bonded to a nitrogen atom.

Amine An organic molecule containing a nitrogen atom bonded to one or more saturated carbon atoms.

Ampere The unit of electric current; the rate of flow of 1 coulomb of charge per second.

Amphoteric Description of a substance that can behave as either an acid or a base.

Amplitude For a wave or vibration, the maximum displacement on either side of the equilibrium (midpoint) position.

Angular unconformity An unconformity in which older tilted rock layers are covered by younger, horizontal rock layers.

Anode The electrode where chemicals are oxidized.

Anticline An up-fold in rock with relatively old rocks at the fold core; rock age decreases with horizontal distance from the fold core.

Applied research Research that focuses on developing applications of knowledge gained through basic research.

Archimedes' principle An immersed body is buoyed up by a force equal to the weight of the fluid it displaces (for both liquids and gases).

Aromatic compound Any organic molecule containing a benzene ring.

Artesian system A system in which confined groundwater under pressure can rise above the upper boundary of an aquifer.

Asteroid belt A region between the orbits of Mars and Jupiter that contains small, rocky, planet-like fragments that orbit the Sun. These fragments are called asteroids ("small star" in Latin).

Asthenosphere A subdivision of the upper mantle situated below the lithosphere, a zone of plastic, easily deformed rock.

Astronomical unit (AU) The average distance between Earth and the Sun; about 1.5×10^8 km (about 9.3×10^7 mi).

Atmospheric pressure The pressure exerted against bodies immersed in the atmosphere resulting from the weight of air pressing down from above. At sea level, atmospheric pressure is about 101 kPa.

Atomic mass The mass of an element's atoms listed in the periodic table as an average value based on the relative abundance of the element's isotopes.

Atomic nucleus The dense, positively charged center of every atom.

Atomic number A count of the number of protons in the atomic nucleus.

Atomic spectrum The pattern of frequencies of electromagnetic radiation emitted by the atoms of an element, considered to be an element's "fingerprint."

Atomic symbol The abbreviation for an element or atom.

Avogadro's number The number of particles—6.02×10^{23}—contained in 1 mole of anything.

Barometer Any device that measures atmospheric pressure.

Base A substance that accepts hydrogen ions.

Basic research Research that leads us to a greater understanding of how the natural world operates.

Basic Description of a solution in which the hydroxide-ion concentration is higher than the hydronium-ion concentration; also sometimes called alkaline.

Beats A series of alternate reinforcements and cancellations produced by the interference of two waves of slightly different frequency, heard as a throbbing effect in sound waves.

Bernoulli's principle The pressure in a fluid moving steadily without friction or external energy input decreases when the fluid velocity increases.

Beta particle An electron emitted during the radioactive decay of a radioactive nucleus.

Big Bang The primordial creation and expansion of space at the beginning of time.

Big Rip A model for the end of the universe in which dark energy grows stronger over time and causes all matter to rip apart.

Black hole The remains of a giant star that has collapsed upon itself. It is so dense, and has a gravitational field so intense, that light itself cannot escape from it.

Black-hole singularity The object of zero radius into which the matter of a black hole is compressed.

Body wave A type of seismic wave that travels through Earth's interior.

Boiling A rapid state of evaporation that takes place within the liquid as well as at its surface. As with evaporation, cooling of the liquid results.

Bond energy The amount of energy required to pull two bonded atoms apart, which is the same as the amount of energy released when the two atoms are brought together into a bond.

Bow wave The V-shaped wave made by an object moving across a liquid surface at a speed greater than the wave speed.

Boyle's law The product of pressure and volume is a constant for a given mass of confined gas regardless of changes in either pressure or volume individually, so long as the temperature remains unchanged: $P_1V_2 = P_2V_2$.

Buoyant force The net upward force that a fluid exerts on an immersed object.

Carbon-14 dating The process of estimating the age of once-living material by measuring the amount of radioactive carbon-14 present in the material.

Carbonyl group A carbon atom double-bonded to an oxygen atom; found in ketones, aldehydes, amides, carboxylic acids, and esters.

Carboxylic acid An organic molecule containing a carbonyl group, the carbon of which is bonded to a hydroxyl group.

Catalyst Any substance that increases the rate of a chemical reaction without itself being consumed by the reaction.

Cathode The electrode where chemicals are reduced.

Celestial sphere An imaginary sphere surrounding Earth to which the stars are attached.

Cenozoic era The time of recent life, it began 65 million years ago and is ongoing.

Chain reaction A self-sustaining reaction in which the products of one reaction event initiate further reaction events.

Channel geometry The shape of a stream channel; the cross sectional area.

Chemical bond The force of attraction between two atoms that holds them together.

Chemical change The formation of new substance(s) by rearranging the atoms of the original material(s).

Chemical equation A representation of a chemical reaction in which reactants are shown to the left of an arrow that points to the products.

Chemical formula A notation that indicates the composition of a compound, consisting of the atomic symbols for the different elements of the compound and numerical subscripts indicating the ratio in which the atoms combine.

Chemical property Any property that characterizes the ability of a substance to change into a different substance under specific conditions.

Chemical reaction A term synonymous with chemical change.

Chemical sediments Sediments that form by the precipitation of minerals from water on Earth's surface.

Chemistry The study of matter and the transformations it can undergo.

Cleavage The tendency of a mineral to break along planes of weakness.

Combustion An exothermic oxidation–reduction reaction between a nonmetallic material and molecular oxygen.

Comet A body composed of ice and dust that orbits the Sun, usually in a very eccentric orbit, and that casts a luminous tail produced by solar radiation pressure when close to the Sun.

Complementary colors Any two colors that, when mixed, produce white light.

Compound A material in which atoms of different elements are bonded to one another.

Compression A condensed region of the medium through which a longitudinal wave travels.

Concentration A quantitative measure of the amount of solute dissolved in a solution.

Conceptual model A representation of a system that helps us predict how the system behaves.

Condensation polymer A polymer formed by the joining together of monomer units accompanied by the loss of small molecules, such as water.

Condensation The change of phase from gas to liquid; the opposite of evaporation. Warming of the liquid results.

Conduction The transfer of thermal energy by molecular and electron collisions within a substance.

Conductor Any material having free charged particles that easily flow through it when an electrical force acts on them.

Configuration A description of how the atoms within a molecule are connected. For example, two structural isomers consist of the same number and same kinds of atoms, but in different configurations.

Conformation One of a wide range of possible spatial orientations of a particular configuration.

Conservation of energy for machines The work output of any machine cannot exceed the work input. In an ideal machine, where no energy is transformed into thermal energy,

$$\text{work}_{\text{input}} = \text{work}_{\text{output}} \text{ and } (Fd)_{\text{input}} = (Fd)_{\text{output}}.$$

Continental drift A hypothesis by Alfred Wegener that the world's continents are mobile and have moved to their present positions as the ancient supercontinent Pangaea broke apart.

Continental margin The boundary between the continents and the ocean; it consists of a continental shelf, a continental slope, and a continental rise.

Convection The transfer of thermal energy in a gas or liquid by means of currents in the heated fluid.

Convectional lifting An air-circulation pattern in which air warmed by the ground rises while cooler air aloft sinks.

Convergent plate boundary A plate boundary where tectonic plates move toward one another; an area of compressive stress where lithosphere is recycled into the mantle or shortened by folding and faulting.

Core The central layer of Earth's interior, divided into an outer liquid core and an inner solid core.

Coriolis force The apparent deflection from a straight-line path observed in any body moving near Earth's surface, caused by Earth's rotation.

Corrosion The deterioration of a metal, typically caused by atmospheric oxygen.

Cosmic background radiation The faint microwave radiation emanating from all directions that is the remnant heat of the Big Bang.

Cosmic inflation The moment of the sudden and brief burst in the size of the universe immediately after the Big Bang.

Cosmological redshift The elongation of light waves due to the expansion of space.

Cosmology The study of the overall structure and evolution of the universe.

Coulomb The SI unit of electric charge. One coulomb (symbol C) is equal in magnitude to the total charge 6.25×10^{18} of electrons.

Coulomb's law The relationship among electrical force, charge, and distance: If the charges are alike in sign, the force is repelling; if the charges are unlike, the force is attractive.

Covalent bond A chemical bond in which atoms are held together by their mutual attraction for two or more electrons they share.

Covalent compound A substance, such as an element or chemical compound, in which atoms are held together by covalent bonds.

Critical mass The minimum mass of fissionable material needed for a sustainable chain reaction.

Cross-cutting relationships Where an igneous intrusion or fault cuts through other rocks, the intrusion or fault is younger than the rock it cuts.

Crust Earth's outermost layer.

Crystal form The outward expression of the orderly internal arrangement of atoms in a crystal.

Crystallization The growth of a solid from a liquid or gas whose atoms come together in specific chemical proportions and crystalline arrangements.

Cut bank A steep bank on the outside bend of a river's channel; an area of erosion.

Dark energy An unknown form of energy that appears to be causing an acceleration of the expansion of space; thought to be associated with the energy exuded by a perfect vacuum.

Dark matter Invisible matter that has made its presence known so far only through its gravitational effects.

Delta An accumulation of sediments, commonly forming a triangular or fan-shaped plain, deposited where a stream flows into a standing body of water.

Density The amount of matter per unit volume:

$$\text{Density} = \frac{\text{mass}}{\text{volume}}$$

Weight density is expressed as weight per unit volume.

Deposition The stage of sedimentary rock formation in which eroded particles come to rest.

Diffuse reflection Reflection in irregular directions from an irregular surface.

Dipole A separation of charge that occurs in a chemical bond because of differences in the electronegativities of the bonded atoms.

Direct current (dc) An electric current flowing in one direction only.

Discharge The volume of water that passes a given location in a stream channel in a certain amount of time.

Dispersion The separation of light into colors arranged by frequency.

Dissolving The process of mixing a solute in a solvent to produce a homogeneous mixture.

Distillation A purifying process in which a vaporized substance is collected by exposing it to cooler temperatures over a receiving flask, which collects the condensed purified liquid.

Divergent plate boundary A plate boundary where lithospheric plates move away from one another—a spreading center; an area of tensional stress where new lithospheric crust is formed.

Doppler effect The change in frequency of wave motion resulting from motion of the sender or the receiver.

Drift A general term for all glacial deposits.

Dwarf planet A relatively large icy body, such as Pluto, that originated within the Kuiper belt.

Earthquake The shaking or trembling of the ground that happens when rock under Earth's surface moves or breaks.

Ecliptic The plane of Earth's orbit around the Sun. All major objects of the solar system orbit roughly within this same plane.

Efficiency The percentage of the work put into a machine that is converted into useful work output:

$$\text{Efficiency} = \frac{\text{useful energy output}}{\text{total energy input}}$$

(More generally, efficiency is useful energy output divided by total energy input.)

Elastic collision A collision in which colliding objects rebound without lasting deformation or the generation of heat.

Electric current The flow of electric charge that transports energy from one place to another.

Electric field Defined as force per unit charge, it can be considered an energetic aura surrounding charged objects. About a charged point, the field decreases with distance according to the inverse-square law, like a gravitational field. Between oppositely charged parallel plates, the electric field is uniform.

Electric potential energy The energy a charge possesses by virtue of its location in an electric field.

Electric potential The electric potential energy per amount of charge, measured in volts and often called voltage.

Electric power The rate of energy transfer, or the rate of doing work; the amount of energy per unit time, which can be measured by the product of current and voltage:

$$\text{Power} = \text{current} \times \text{voltage}$$

It is measured in watts (or kilowatts), where $1\text{ A} \times 1\text{ V} = 1\text{ W}$.

Electrical resistance The property of a material that resists the flow of an electric current through it, measured in ohms (Ω).

Electrically polarized Term applied to an atom or molecule in which the charges are aligned so that one side has a slight excess of positive charge and the other side a slight excess of negative charge.

Electrochemistry The study of the relationship between electric energy and chemical change.

Electrode Any material that conducts electrons into or out of a medium in which electrochemical reactions are occurring.

Electrolysis The use of electric energy to produce chemical change.

Electromagnet A magnet whose field is produced by an electric current. It is usually in the form of a wire coil with a piece of iron inside the coil.

Electromagnetic induction The induction of voltage when a magnetic field changes with time.

Electromagnetic spectrum The range of electromagnetic waves that extends in frequency from radio waves to gamma rays.

Electromagnetic wave An energy-carrying wave emitted by vibrating electric charges (often electrons) and composed of oscillating electric and magnetic fields that regenerate each other.

Electron An extremely small, negatively charged subatomic particle found outside the atomic nucleus.

Electron-dot structure A shorthand notation of the shell model of the atom, in which valence electrons are shown around an atomic symbol. The electron-dot structure for an atom or ion is sometimes called a Lewis dot symbol, while the electron-dot structure of a molecule or poly-atomic ion is sometimes called a Lewis structure.

Electronegativity The ability of an atom to attract a bonding pair of electrons to itself when bonded to another atom.

Element Any material that is made up of only one type of atom.

Elemental formula A notation that uses the atomic symbol and (sometimes) a numerical subscript to denote how many atoms are bonded in one unit of an element.

Ellipse The oval path followed by a satellite. The sum of the distances from any point on the path to two points called foci is a constant. When the foci are together at one point, the ellipse is a circle. As the foci get farther apart, the ellipse becomes more eccentric.

Endothermic Description of a chemical reaction in which there is a net absorption of energy.

Energy The property of a system that enables it to do work.

Entropy The measure of energy dispersal of a system. Whenever energy freely transforms from one form to another, the direction of transformation is toward a state of greater disorder and, therefore, toward one of greater entropy.

Equilibrium rule The vector sum of forces acting on a non-accelerating object equals zero:

$$\Sigma F = 0.$$

Erosion The wearing away of rocks, and the processes by which rock particles are transported by water, wind, or ice.

Escape speed The speed that a projectile, space probe, or similar object must reach to escape the gravitational influence of Earth or of another celestial body to which it is attracted.

Ester An organic molecule containing a carbonyl group, the carbon of which is bonded to one carbon atom and one oxygen atom bonded to another carbon atom.

Eternal inflation A model of the universe in which cosmic inflation is not a one-time event but rather progresses to continuously spawn an infinite number of observable universes in its wake.

Ether An organic molecule containing an oxygen atom bonded to two carbon atoms.

Evaporation The change of phase at the surface of a liquid as it passes to the gaseous phase.

Event horizon The boundary region of a black hole from which no radiation may escape. Any events within the event horizon are invisible to distant observers.

Exosphere The fifth atmospheric layer above Earth's surface, extending from the thermosphere upward and out into interplanetary space.

Exothermic Description of a chemical reaction in which there is a net release of energy.

Fact A phenomenon about which competent observers who have made a series of observations are in agreement.

Faraday's law The law of electromagnetic induction, in which the induced voltage in a coil is proportional to the number of loops multiplied by the rate at which the magnetic field changes within those loops. (The induction of voltage is actually the result of a more fundamental phenomenon: the induction of an electric field.)

$$\text{Voltage induced} \sim \text{number of loops} \times \frac{\text{change in magnetic field}}{\text{time}}$$

Fault A fracture along which movement of rock on one side relative to rock on the other side has occurred.

Faunal succession Fossil organisms succeed one another in a definite, irreversible, and determinable order.

First law of thermodynamics A restatement of the law of energy conservation, usually as it applies to systems involving changes in temperature: The heat added to a system is equal to the system's gain in thermal energy plus the work that it does on its surroundings.

Floodplain A wide plain of almost flat land on either side of a stream channel. Submerged during flood stage, the plain is built up by sediments discharged during floods.

Force pair The action and reaction pair of forces that occur in an interaction.

Force Simply stated, a push or a pull.

Forced vibration The setting up of vibrations in an object by a vibrating force.

Formula mass The sum of the atomic masses of the elements in a chemical formula.

Fracture A break that does not occur along a plane of weakness.

Free fall Falling only under the influence of gravity—falling without air resistance.

Freezing The process of changing phase from liquid to solid, as from water to ice.

Frequency For a vibrating body or medium, the number of vibrations per unit time. For a wave, the number of crests that pass a particular point per unit time.

Friction The resistive force that opposes the motion or attempted motion of an object past another with which it is in contact, or through a fluid.

Front The contact zone between two different air masses.

Frontal lifting The lifting of one air mass by another as two air masses converge.

Full Moon The phase of the Moon when its sunlit side faces Earth.

Functional group A specific combination of atoms that behaves as a unit in an organic molecule.

Fundamental frequency The lowest frequency of vibration, or the first harmonic. In a string, the vibration makes a single segment.

Galaxy A large assemblage of stars, interstellar gas, and dust, usually categorized by its shape: elliptical, spiral, or irregular.

Gamma ray High-frequency electromagnetic radiation emitted by radioactive nuclei.

General theory of relativity The theory first proposed by Einstein discussing the effects of gravity on spacetime.

Generator An electromagnetic induction device that produces electric current by rotating a coil within a stationary magnetic field.

Giant stars Cool giant stars above main-sequence stars on the H–R diagram.

Glacier A large mass of ice formed by the compaction and recrystallization of snow, moving downslope under its own weight.

Gradient The vertical drop in the elevation of a stream channel divided by the horizontal distance for that drop; the steepness of the slope.

Greenhouse effect Warming caused by short-wavelength radiant energy from the Sun that easily enters the atmosphere and is absorbed by Earth. This energy is then reradiated at longer wavelengths that cannot easily escape Earth's atmosphere.

Groundwater Underground water in the saturated zone.

Group A vertical column in the periodic table, also known as a family of elements.

Gyre A circular or spiral whirl pattern, usually referring to very large current systems in the open ocean.

Half reaction Half of an oxidation–reduction reaction, represented by an equation showing electrons as either reactants or products.

Half-life The time required for half the atoms in a sample of a radioactive isotope to decay.

Hang time The time that one's feet are off the ground during a vertical jump.

Hard water Water containing large amounts of calcium and magnesium ions.

Harmonic A partial tone that is an integer multiple of the fundamental frequency. The vibration that begins with the fundamental vibrating frequency is the first harmonic, twice the fundamental is the second harmonic, and so on in sequence.

Heat death A model for the end of the universe in which all matter and energy disperse to the point of maximum entropy.

Heat of fusion The amount of energy needed to change a unit mass of any substance from solid to liquid (and vice versa). For water, this is 334 J/g (or 80 cal/g).

Heat of vaporization The amount of energy needed to change a unit mass of any substance from liquid to gas (and vice versa). For water, this is 2256 J/g (or 540 cal/g).

Heat The thermal energy that flows from a substance of higher temperature to a substance of lower temperature, commonly measured in calories or joules.

Hertz The SI unit of frequency; one hertz (symbol Hz) equals one vibration per second.

Heteroatom Any atom other than carbon or hydrogen in an organic molecule.

Heterogeneous mixture A mixture in which the different components can be seen as individual substances.

Homogeneous mixture A mixture in which the components are so finely mixed that any one region of the mixture contains the same ratio of substances as any other region.

H–R diagram (Hertzsprung–Russell diagram) A plot of luminosity versus surface temperature for stars. When so plotted, stars' positions take the form of a main sequence for average stars, with exotic stars above or below the main sequence.

Hubble's law The farther away a galaxy is from Earth, the more rapidly it is moving away from us: $v = H \times d$.

Humidity A measure of the concentration or amount of water vapor in the air: the mass of water vapor per volume of air.

Hydraulic conductivity A measure of the ability of a porous rock or sediment to transmit fluid.

Hydrocarbon A chemical compound containing only carbon and hydrogen atoms.

Hydrogen bond An unusually strong dipole–dipole attraction occurring between molecules that have a hydrogen atom covalently bonded to a small, highly electronegative atom, usually nitrogen, oxygen, or fluorine.

Hydrologic cycle The natural circulation of all states of water from ocean to atmosphere to land and then back to ocean.

Hydronium ion A polyatomic ion made by adding a proton (hydrogen ion) to a water molecule.

Hydroxide ion A polyatomic ion made by removing a proton (hydrogen ion) from a water molecule.

Hypothesis An educated guess; a reasonable explanation of an observation or experimental result that is not fully accepted as factual until tested over and over again by experiment.

Igneous rocks Rocks formed by the cooling and crystallization of hot, molten rock material called magma (or lava).

Impulse The product of the force acting on an object and the time during which it acts.

Impulse–momentum relationship Impulse is equal to the change in the momentum of an object that the impulse acts upon. In symbol notation:

$$Ft = \Delta(mv)$$

Impure The state of a material that is a mixture of more than one element or compound.

Inclusions Any inclusion (pieces of one rock type contained within another) is older than the rock containing it.

Induced dipole A temporarily uneven distribution of electrons in an otherwise nonpolar atom or molecule.

Inelastic collision A collision in which the colliding objects become distorted, generate heat, and possibly stick together.

Inertia The property by which objects resist changes in motion.

Inner planets The four planets orbiting within 2 AU of the Sun, including Mercury, Venus, Earth, and Mars—all rocky and known as the terrestrial planets.

Insoluble Not capable of dissolving to any appreciable extent in a given solvent.

Interaction Mutual action between objects during which each object exerts an equal and opposite force on the other.

Interference A property of all types of waves; a result of superimposing different waves, often of the same wavelength. Constructive interference results from crest-to-crest reinforcement; destructive interference results from crest-to-trough cancellation.

Inverse-square law The intensity of an effect from a localized source spreads uniformly throughout the surrounding space and weakens with the inverse square of the distance:

$$\text{Intensity} = \frac{1}{\text{distance}^2}$$

Gravity follows an inverse-square law, as do the effects of electric, light, sound, and radiation phenomena.

Ion An atom having a net electric charge because of either a loss or gain of electrons.

Ionic bond A chemical bond in which there is an electric force of attraction between two oppositely charged ions.

Ionic compound A chemical compound containing ions.

Ionosphere An electrified region within the thermosphere and uppermost mesosphere where fairly large concentrations of ions and free electrons exist.

Isostasy The process by which oceanic crust and continental crust come into vertical equilibrium, with respect to the mantle; the dense oceanic crust sits lower in the mantle than the less-dense continental crust.

Isotopes Members of a set of atoms of the same element whose nuclei contain the same number of protons but different numbers of neutrons.

Ketone An organic molecule containing a carbonyl group, the carbon of which is bonded to two carbon atoms.

Kilogram The unit of mass; one kilogram (symbol kg) is the mass of 1 liter (L) of water at 4°C.

Kinetic energy Energy of motion, quantified by the relationship:

$$\text{Kinetic energy} = \frac{1}{2}mv^2$$

Kuiper belt (pronounced "KI-pur") The disk-shaped region of the sky beyond Neptune populated by many icy bodies and a source of short-period comets.

Laminar flow Water flowing smoothly and fairly slowly in straight lines with no mixing of sediment.

Lateral continuity Sedimentary layers are deposited in all directions over large areas until some sort of obstruction, or barrier, limits their deposition.

Lava Molten magma that moves upward from inside Earth and flows onto the surface. The term lava refers both to the molten rock itself and to the solid rocks that form from it.

Law A general hypothesis or statement about the relationship of natural quantities that has been tested over and over again and has not been contradicted; also known as a principle.

Law of conservation of energy Energy cannot be created or destroyed; it may be transformed from one form into another, but the total amount of energy never changes.

Law of conservation of momentum In the absence of an external force, the momentum of a system remains unchanged. Hence, the momentum before an event involving only internal forces is equal to the momentum after the event:

$$mv_{\text{before event}} = mv_{\text{after event}}$$

Law of mass conservation Matter is neither created nor destroyed during a chemical reaction; atoms merely rearrange, without any apparent loss or gain of mass, to form new molecules.

Law of reflection The angle of incidence equals the angle of reflection. The incident and reflected rays lie in a plane that is normal to the reflecting surface.

Law of universal gravitation Every body in the universe attracts every other body with a force that, for two bodies, is directly proportional to the product of their masses and inversely proportional to the square of the distance separating them:

$$F = G\frac{m_1 m_2}{d^2}$$

Lever A simple machine consisting of a rigid rod pivoted at a fixed point called the fulcrum.

Lifting condensation level The height at which rising air cooling at the dry adiabatic rate becomes saturated and condensation begins.

Light-year The distance light travels in one year.

Lithosphere The entire crust plus the rigid portion of the mantle that is above the asthenosphere.

Local Group Our immediate cluster of galaxies, including the Milky Way, Andromeda, and Trangulum spiral galaxies plus a few dozen smaller elliptical and irregular galaxies.

Local Supercluster A cluster of galactic clusters in which our Local Group resides.

Longitudinal wave A wave in which the medium vibrates in a direction parallel (longitudinal) to the direction in which the wave travels. Sound consists of longitudinal waves.

Lunar eclipse The phenomenon in which the shadow of Earth falls on the Moon, producing the relative darkness of the full Moon.

Machine A device, such as a lever or pulley, that increases (or decreases) a force or simply changes the direction of a force.

Magma Molten rock in Earth's interior.

Magnetic domains Clustered regions of aligned magnetic atoms. When these regions themselves are aligned with one another, the substance containing them is a magnet.

Magnetic field The region of magnetic influence around a magnetic pole or a moving charged particle.

Magnetic force (1) Between magnets, it is the attraction of unlike magnetic poles for each other and the repulsion between like magnetic poles. (2) Between a magnetic field and a moving charge, it is a deflecting force due to the motion of the charge: The deflecting force is perpendicular to the velocity of the charge and perpendicular to the magnetic field lines. This force is greatest when the charge moves perpendicular to the field lines and is smallest (zero) when it moves parallel to the field lines.

Main sequence The diagonal band of stars on an H–R diagram; such stars generate energy by fusing hydrogen to helium.

Mantle The middle layer in Earth's interior, between the crust and the core.

Mass The quantity of matter in an object. More specifically, the measure of the inertia or sluggishness that an object exhibits in response to any effort made to start it, stop it, deflect it, or change in any way its state of motion.

Maxwell's counterpart to Faraday's law A magnetic field is induced in any region of space in which an electric field is changing with time. Correspondingly, an electric field is induced in any region of space in which a magnetic field is changing with time.

Mechanical deformation Metamorphism caused by stress, such as increased pressure.

Melting The process of changing phase from solid to liquid, as from ice to water.

Mesosphere The third atmospheric layer above Earth's surface, extending from the top of the stratosphere to 80 km.

Mesozoic era The time of middle life, from about 248 million years ago to 65 million years ago.

Metallic bond A chemical bond in which positively charged metal ions are held together within a "fluid" of loosely held electrons.

Metamorphic rocks Rocks formed from preexisting rocks that have been changed or transformed by high temperature, high pressure, or both.

Metamorphism The changes in rock that happen as physical and chemical conditions change.

Meteor The streak of light produced by a meteoroid burning in Earth's atmosphere; a "shooting star."

Meteorite A meteoroid, or a part of a meteoroid, that has survived passage through Earth's atmosphere to reach the ground.

Meteoroid A small rock in interplanetary space, which can include a fragment of an asteroid or comet.

Midlatitude cyclone A west-to-east-traveling storm with a central low-pressure area about which counterclockwise flow develops (in the Northern Hemisphere) and from which usually extends a cold front and a warm front; generally forms at the polar front.

Mineral A naturally formed, inorganic, crystalline solid composed of an ordered arrangement of atoms with a specific chemical composition.

Mixture A combination of two or more substances in which each substance retains its chemical properties.

Mohorovičić discontinuity (Moho) The crust–mantle boundary; marks one of the depths where the speed of P-waves traveling through Earth increases.

Mohs scale of hardness A ranking of a mineral's hardness, which is its resistance to scratching.

Molar mass The mass of 1 mole of a substance.

Molarity A common unit of concentration equal to the number of moles of a solute per liter of solution.

Mole The very large number 6.02×10^{23}; usually used in reference to the number of atoms, ions, or molecules in a macroscopic amount of a material.

Molecule The fundamental unit of a chemical compound, which is a group of atoms held tightly together by covalent bonds.

Momentum Inertia in motion, given by the product of the mass of an object and its velocity.

Monomers The small molecular units from which a polymer is formed.

Moon phases The cycles of change of the "face" of the Moon, changing from new, to waxing, to full, to waning, and back to new.

Nanotechnology The manipulation of individual atoms or molecules.

Natural frequency The frequency at which an elastic object naturally tends to vibrate, so that minimum energy is required to produce a forced vibration or to continue vibration at that frequency.

Neap tide A tide that occurs when the Moon is midway between new and full, in either direction. The pulls of the Moon and Sun are perpendicular, so the solar and lunar tides do not overlap. This makes high tides not as high and low tides not as low.

Nebular theory The idea that the Sun and planets formed together from a cloud of gas and dust, a nebula.

Net force The combination of all forces that act on an object.

Neutral Description of a solution in which the hydronium-ion concentration is equal to the hydroxide-ion concentration.

Neutralization A reaction between an acid and a base.

Neutron star A small, extremely dense star composed of tightly packed neutrons formed by the welding of protons and electrons.

Neutron An electrically neutral subatomic particle of the atomic nucleus.

New Moon The phase of the Moon when darkness covers the side facing Earth.

Newton The scientific unit of force.

Newton's first law of motion Every object continues in a state of rest, or in a state of motion in a straight line at constant speed, unless acted on by a net force.

Newton's law of cooling The rate of loss of transfer of thermal energy from a warm object is proportional to the temperature difference between the object and its surroundings:

$$\text{Rate of cooling} \sim \Delta T$$

Newton's second law of motion The acceleration produced by a net force on an object is directly proportional to the net force, is in the same direction as the net force, and is inversely proportional to the mass of the object.

Newton's third law of motion Whenever one object exerts a force on a second object, the second object exerts an equal and opposite force on the first object.

Nonbonding pairs Two paired valence electrons that are not participating in a chemical bond.

Nonpolar Description of a chemical bond or molecule that has no dipole. In a nonpolar bond or molecule, the electrons are distributed evenly.

Nonsilicate A mineral that does not contain silica (silicon + oxygen).

Nova An event in which a white dwarf suddenly brightens and appears as a "new" star.

Nuclear fission The splitting of the atomic nucleus into two smaller halves.

Nuclear fusion The combining of nuclei of light atoms to form heavier nuclei.

Nucleon Any subatomic particle found in the atomic nucleus; another name for either a proton or a neutron.

Ohm's law The current in a circuit varies in direct proportion to the potential difference or voltage and inversely with the resistance:

$$\text{Current} = \frac{\text{voltage}}{\text{resistance}}$$

A current of 1 A is produced by a potential difference of 1 V across a resistance of 1 Ω.

Oort cloud The region beyond the Kuiper belt populated by trillions of icy bodies and a source of long-period comets.

Opaque The property of absorbing light without re-emission (opposite of transparent).

Ordinary matter Matter that responds to the strong nuclear, weak nuclear, electromagnetic, and gravitational forces. This is matter made of protons, neutrons, and electrons, which includes the atoms and molecules that make us and our immediate environment.

Organic chemistry The study of carbon-containing compounds.

Original horizontality Layers of sediment are deposited evenly, with each new layer laid down nearly horizontally over the older sediment.

Orographic lifting The lifting of an air mass over a topographic barrier such as a mountain.

Osmosis The net flow (diffusion) of water across a semipermeable membrane from a region of low solute concentration to a region of high solute concentration.

Outer planets The four planets orbiting beyond 2 AU of the Sun, including Jupiter, Saturn, Uranus, and Neptune—all gaseous and known as the jovian planets.

Oxidation The process whereby a reactant loses one or more electrons.

Paleomagnetism The natural, ancient magnetization in a rock that can be used to determine the polarity of Earth's magnetic field and the rock's location of formation.

Paleozoic era The time of ancient life, from about 543 million years ago to 248 million years ago.

Pangaea The late-Paleozoic supercontinent made up of Gondwanaland (ancestral South America, Africa, Australia, Antarctica, and India) and Laurasia (ancestral North America, Europe, Siberia, Asia).

Parabola The curved path followed by a projectile under the influence of constant gravity only.

Parallel circuit An electric circuit with two or more devices connected in such a way that the same voltage acts across each one, and any single one completes the circuit independently of all the others.

Partial melting The incomplete melting of rocks, resulting in magmas of various compositions.

Partial tone One of the frequencies present in a complex tone. When a partial tone is an integer multiple of the lowest frequency, it is a harmonic.

Pascal's principle A change in pressure at any point in an enclosed fluid at rest is transmitted undiminished to all points in the fluid.

Period The time required for a vibration or a wave to make a complete cycle; equal to 1/frequency.

Period A horizontal row in the periodic table.

Periodic table A chart in which all the known elements are listed in order of atomic number.

pH A measure of the acidity of a solution, equal to the negative logarithm of the hydronium-ion concentration.

Phase The molecular state of a substance: solid, liquid, gas, or plasma.

Phenol An organic molecule in which a hydroxyl group is bonded to a benzene ring.

Physical change A change in a substance's physical properties but with no change in its chemical identity.

Physical model A representation of an object on some convenient scale.

Physical property Any physical attribute of a substance, such as color, density, or hardness.

Pitch The subjective impression of the frequency of sound.

Planetary nebula An expanding shell of gas ejected from a low-mass star during the latter stages of its evolution.

Planets The major bodies orbiting the Sun that are massive enough for their gravity to make them spherical and small enough to avoid having nuclear fusion in their cores. They also have successfully cleared all debris from their orbital paths.

Plate tectonics The theory that Earth's lithosphere is broken into pieces (plates) that move over the asthenosphere; boundaries between plates are where most earthquakes and volcanoes occur and where lithosphere is created and recycled.

Plutonic rock Intrusive igneous rock formed from magma that cools beneath Earth's surface. Granite is a plutonic rock.

Point bar A sandy, gentle bank on the inside bend of a river's channel; an area of deposition.

Polar Description of a chemical bond or molecule that has a dipole. In a polar bond or molecule, electrons are congregated to one side. This makes that side slightly negative, while the opposite side (lacking electrons) becomes slightly positive.

Polarization The alignment of the transverse electric vectors that make up electromagnetic radiation. Such waves of aligned vibrations are said to be polarized.

Polyatomic ion An ionically charged molecule.

Polymer A long organic molecule made of many repeating units.

Polymorphs Two or more minerals that contain the same elements in the same proportions but have different crystal structures.

Porosity The volume of open space in rock or sediment compared to the total volume of solids plus open space.

Potential difference The difference in potential between two points, measured in volts and often called voltage difference.

Potential energy The energy that matter possesses due to its position:

$$\text{Gravitational PE} = mgh$$

Power The rate of doing work (or the rate at which energy is expended):

$$\text{Power} = \frac{\text{work}}{\text{time}}$$

Precambrian time The time of hidden life, which began about 4.5 billion years ago when Earth formed and lasted until about 543 million years ago (beginning of the Paleozoic), and makes up almost 90% of Earth's history.

Precipitate A solute that has come out of solution.

Pressure The ratio of force to the area over which that force is distributed:

$$\text{Pressure} = \frac{\text{force}}{\text{area}}$$

$$\text{Liquid pressure} = \text{weight density} \times \text{depth}$$

Pressure-gradient force The force that moves air from a region of high-pressure air to an adjacent region of low-pressure air.

Primary wave (P-wave) A longitudinal body wave that compresses and expands the material through which it moves; it travels through solids, liquids, and gases and is the fastest seismic wave.

Principal quantum number n An integer that specifies the quantized energy level of an atomic orbital.

Principle of equivalence Local observations made in an accelerated frame of reference cannot be distinguished from observations made in a Newtonian gravitational field.

Principle of flotation A floating object displaces a weight of fluid equal to its own weight.

Products The new materials formed in a chemical reaction.

Projectile Any object that is projected by some means and continues its motion by its own inertia.

Proton A positively charged subatomic particle of the atomic nucleus.

Pseudoscience Fake science that pretends to be real science.

Pulsar A celestial object (most likely a neutron star) that spins rapidly, sending out short, precisely timed bursts of electromagnetic radiation.

Pure The state of a material that consists solely of a single element or compound.

Quality The characteristic timbre of a musical sound, which is governed by the number and relative intensities of partial tones.

Quantum hypothesis The idea that light energy is contained in discrete packets called quanta.

Quantum A small, discrete packet of light energy.

Quasar The core of a distant galaxy early in its lifespan when its central black hole has not yet swept much matter from its vicinity, leading to a rate of radiation greater than that from entire older galaxies.

Rad A quantity of radiant energy equal to 0.01 J absorbed per kilogram of tissue.

Radiation The transfer of energy by means of electromagnetic waves.

Radioactivity The high-energy particles and electromagnetic radiation emitted by a radioactive substance.

Radiometric dating A method for calculating the age of geologic materials based on the nuclear decay of naturally occurring radioactive isotopes.

Rarefaction A rarefied region, or a region of lessened pressure, of the medium through which a longitudinal wave travels.

Reactants The reacting substances in a chemical reaction.

Reaction rate A measure of how quickly the concentration of products in a chemical reaction increases or the concentration of reactants decreases.

Recrystallization A process that occurs when rocks are subjected to high temperatures and pressures and go through a change in minerals; often accompanied by the loss of H_2O or CO_2.

Reduction The process whereby a reactant gains one or more electrons.

Reflection, light The return of light rays from a surface in such a way that the angle at which a given ray is returned is equal to the angle at which it strikes the surface (also called specular reflection).

Reflection, sound The return of a sound wave; an echo.

Refraction, light The bending of an oblique ray of light when it passes from one transparent medium to another. This is caused by a difference in the speed of light in the transparent media. When the change in medium is abrupt (say, from air to water), the bending is abrupt; when the change in medium is gradual (say, from cool air to warm air), the bending is gradual, which accounts for mirages.

Refraction, sound The bending of a wave, either through a non-uniform medium or from one medium to another, caused by differences in wave speed.

Relative dating The ordering of rocks in sequence by their comparative ages.

Relative humidity The amount of water vapor in the air at a given temperature, expressed as a percentage of the maximum amount of water vapor the air can accommodate at that temperature.

Rem A unit for measuring the ability of radiation to harm living tissue.

Resonance The response of a body when a forcing frequency matches its natural frequency.

Reverberation Re-echoed sound.

Reverse osmosis A technique for purifying water by forcing it through a semipermeable membrane into a region of lower solute concentration.

Rift (rift valley) A long, narrow gap that forms as a result of two plates diverging.

Rock cycle A sequence of events involving the formation, destruction, alteration, and reformation of rocks as a result of the generation and movement of magma; the weathering, erosion, transportation, and deposition of sediment; and the metamorphism of preexisting rocks.

Rock An aggregate of minerals. Some rocks are aggregates of fossil shell fragments, solid organic matter, or any combination of these components.

Salinity The mass of salts dissolved in 1000 g of seawater.

Salt An ionic compound commonly formed from the reaction between an acid and a base.

Sand dune A landform created when airflow is blocked by an obstacle, slowing air speed and therefore promoting the deposition of airborne sand.

Satellite A projectile or small celestial body that orbits a larger celestial body.

Saturated hydrocarbon A hydrocarbon containing no multiple covalent bonds, with each carbon atom bonded to four other atoms.

Saturated solution A solution containing the maximum amount of solute that will dissolve in its solvent.

Saturation vapor pressure The maximum amount of moisture the air can accommodate at a given temperature; the upper limit for humidity.

Scanning probe microscope A tool of nanotechnology that detects and characterizes the surface atoms of materials by way of an ultrathin probe tip, which is detected by laser light as it is mechanically dragged over the surface.

Science The collective findings of humans about nature, and a process of gathering and organizing knowledge about nature.

Scientific method Principles and procedures for the systematic pursuit of knowledge involving the recognition and formulation of a problem, the collection of data through observation and experiment, and the formulation and testing of hypotheses.

Seafloor spreading The moving apart of two oceanic plates at a rift in the seafloor.

Second law of thermodynamics Heat never spontaneously flows from a cold substance to a hot substance. Also, in natural processes, high-quality energy tends to transform into lower-quality energy—order tends to disorder.

Secondary wave (S-wave) A transverse body wave that vibrates the material through which it moves side to side or up and down; it cannot travel through liquids and so does not travel through Earth's outer core.

Sedimentary rocks Rocks formed from the accumulation of weathered material (sediments) that has been eroded by water, wind, or ice.

Sedimentation The stage of sedimentary rock formation in which deposited sediments accumulate and change (lithify) into sedimentary rock through the processes of compaction and, usually, cementation.

Semipermeable membrane A membrane containing submicroscopic pores that allow the passage of water molecules but not of larger solute ions or solute molecules.

Series circuit An electric circuit with devices connected in such a way that the current is the same in each device.

Shell A region of space around the atomic nucleus within which electrons may reside.

Shock wave The cone-shaped wave made by an object moving at supersonic speed through a fluid.

Silicate A mineral that contains both silicon and oxygen and (usually) other elements in its chemical composition; silicates are the largest and most common rock-forming mineral group.

Solar eclipse The phenomenon in which the shadow of the Moon falls on Earth, producing a region of darkness in the daytime.

Solubility The ability of a solute to dissolve in a given solvent.

Soluble Capable of dissolving to an appreciable extent in a given solvent.

Solute Any component in a solution that is not the solvent.

Solution A homogeneous mixture in which all components are dissolved in the same phase.

Solvent The component in a solution that is present in the largest amount.

Sonic boom The loud sound resulting from a shock wave.

Spacetime The continuum in which we live, consisting of three dimensions of space plus the fourth dimension of time.

Special theory of relativity The theory first proposed by Einstein discussing the effects of uniform motion on space, time, energy, and mass.

Specific heat capacity The quantity of heat required to raise the temperature of a unit mass of a substance by 1°C.

Spectroscope A device that uses a prism or diffraction grating to separate light into its color components.

Speed The distance traveled per time.

Spring tide A high or low tide that occurs when the Sun, Earth, and Moon are aligned so that the tides due to the Sun and Moon coincide, making the tides higher or lower than average; occurs during the full Moon or new Moon.

Standing wave A stationary wave pattern formed in a medium when two sets of identical waves pass through the medium in opposite directions.

Starburst galaxy A galaxy in which stars are forming at an unusually fast rate.

Stratosphere The second atmospheric layer above Earth's surface, extending from the top of the troposphere up to 50 km. This is where stratospheric ozone forms.

Streak The name given to the color of a mineral in its powdered form.

Strong nuclear force The attractive force between all nucleons, effective at only very short distances.

Structural isomers Molecules that have the same molecular formula but different chemical structures.

Subduction The process in which one tectonic plate bends and descends beneath another plate at a convergent boundary.

Sublimation The change of phase directly from solid to gas.

Submicroscopic The realm of atoms and molecules, where objects are too small to be detected by optical microscopes.

Subtractive primary colors The three colors of absorbing pigments—magenta, yellow, and cyan—that, when mixed in certain proportions, can reflect any color in the spectrum.

Sunspots Temporary, relatively cool and dark regions on the Sun's surface.

Superconductor Any material with zero electrical resistance, in which electrons flow without losing energy and without generating heat.

Supernova The explosion of a massive star caused by gravitational collapse with the emission of enormous quantities of matter and radiation.

Superposition In an undeformed sequence of sedimentary rocks, each bed or layer is older than the one above and younger than the one below.

Support force The force that supports an object against gravity, often called the normal force.

Surface wave A type of seismic wave that travels along Earth's surface.

Suspension A homogeneous mixture in which the various components are finely mixed, but not dissolved.

Syncline A down–old in rock with relatively young rocks at the fold core; rock age increases with horizontal distance from the fold core.

Tectonic plates Sections into which Earth's crust is broken up; they move in response to heat flow and convection in Earth's interior.

Temperature A measure of the hotness of substances, related to the average translational kinetic energy per molecule in a substance, measured in degrees Celsius, degrees Fahrenheit, or kelvins.

Temperature inversion A condition in which the upper regions of the troposphere are warmer than the lower regions.

Terminal speed The speed at which the acceleration of a falling object terminates when air resistance balances its weight.

Terminal velocity Terminal speed in a given direction (often downward).

Terrestrial radiation The radiant energy emitted by Earth.

Theory A synthesis of a large body of information that encompasses well-tested and verified hypotheses about certain aspects of the natural world.

Thermal energy The total energy (kinetic plus potential) of the submicroscopic particles that make up a substance.

Thermodynamics The study of thermal energy and its relationship to heat and work.

Thermonuclear fusion Nuclear fusion brought about by high temperatures.

Thermosphere The fourth atmospheric layer above Earth's surface, extending from the top of the mesosphere to 500 km.

Third law of thermodynamics No system can reach absolute zero.

Transform plate boundary A plate boundary where two plates are sliding horizontally past each other, without appreciable vertical movement.

Transformer A device for transferring electric power from one coil of wire to another by means of electromagnetic induction.

Transmutation The changing of an atomic nucleus of one element into an atomic nucleus of another element through a decrease or increase in the number of protons.

Transparent The term applied to materials through which light can pass without absorption, usually in straight lines.

Transverse wave A wave in which the medium vibrates in a direction perpendicular (transverse) to the direction in which the wave travels. Light consists of transverse waves.

Troposphere The atmospheric layer closest to Earth's surface, 16 km high over the equator and 8 km high over the poles, containing 90% of the atmosphere's mass and essentially all of its water vapor and clouds.

Turbulent flow Water flowing rapidly and erratically in a jumbled manner, stirring up everything it touches.

Unconformity A break or gap in the geologic record, caused by erosion of preexisting rock or by an interruption in the sequence of deposition.

Unsaturated hydrocarbon A hydrocarbon containing at least one multiple covalent bond.

Unsaturated solution A solution that is capable of dissolving additional solute.

Valence electrons The electrons in the outermost occupied shell of an atom.

Valence shell The outermost occupied shell of an atom.

Vector quantity A quantity whose description requires both magnitude and direction.

Vector An arrow that represents the magnitude and direction of a quantity.

Velocity The speed of an object and specification of its direction of motion.

Vibration A wiggle in time.

Volcanic rocks Extrusive igneous rocks formed by the eruption of molten rock at Earth's surface. Basalt is a volcanic rock.

Volume The quantity of space an object occupies.

Water table The upper boundary of the saturated zone, below which every pore space is completely filled with water.

Wave speed The speed with which waves pass a particular point:

$$\text{Wave speed} = \text{frequency} \times \text{wavelength}$$

Wave A wiggle in both space and time.

Wavelength The distance between successive crests, troughs, or identical parts of a wave.

Weathering Disintegration or decomposition of rock at or near Earth's surface.

Weight The force that an object exerts on a supporting surface (or, if suspended, on a supporting string), which is often, but not always, due to the force of gravity.

Weightless Being without a support force, as in free fall.

White dwarf A dying star that has collapsed to the size of Earth and is slowly cooling off; located at the lower left on the H–R diagram.

Work The product of the force and the distance moved by the force:

$$W = Fd$$

(More generally, work is the component of force in the direction of motion multiplied by the distance moved.)

Work–energy theorem The net work done on an object equals the change in kinetic energy of the object:

$$\text{Work} = \Delta \text{KE}$$

Photo Credits

Index

Eon	Era	Period	Subperiod	Epoch	Ma
Phanerozoic	Cenozoic	Quaternary		Holocene	0.01
				Pleistocene	1.8
		Tertiary		Pliocene	5.3
				Miocene	23.8
				Oligocene	33.7
				Eocene	54.8
				Paleocene	65
	Mesozoic	Cretaceous			144
		Jurassic (first bird)			206
		Triassic			248
	Paleozoic	Permian (first reptiles)			290
		Carboniferous	Pennsylvanian		323
			Mississippian		354
		Devonian (first amphibians)			417
		Silurian (first insect fossils)			443
		Ordovician (first vertebrate fossils)			490
		Cambrian (first plant fossils)			543
Precambrian Time		Proterozoic			2500
		Archean			3800
		Hadean			4500